High Energy Physics-1980
(XX International Conference, Madison, Wisconsin)

AIP Conference Proceedings
Series Editor: Hugh C. Wolfe
Number 68
Particles and Fields Subseries No. 22

High Energy Physics-1980

(XX International Conference, Madison, Wisconsin)

Part 1

Editors
Loyal Durand and Lee G. Pondrom
University of Wisconsin

American Institute of Physics
New York 1981

L.C. Catalog Card No. 81-65032
ISBN 0-88318-167-3
DOE CONF- 800724

FOREWORD

The XX International Conference on High Energy Physics was held at the University of Wisconsin-Madison from July 17-23, 1980. The conference was sponsored by the International Union of Pure and Applied Physics, and was supported in part by IUPAP, the U.S. National Science Foundation, the U.S. Department of Energy, and the University of Wisconsin-Madison.

The 1980 conference was the twentieth in the series started by Robert Marshak at the University of Rochester in 1950, and was the first of the series to be run as an open conference. It was attended by 1229 delegates from 45 countries, (perhaps one-eighth of the particle physicists active world-wide) with 550 (45%) from outside the United States, and was larger by about 40% than its recent predecessors.

The conference spanned seven days, with three days of parallel sessions on the University of Wisconsin campus, followed by a day of "rest" which featured, among other excursions, a vigorous canoe trip down a stretch of the Wisconsin River, and three days of plenary sessions at the Madison Civic Center. The weather was excellent, prime Wisconsin summer, and most delegates seemed to find the facilities and the amenities of Madison and the University - the lakes, the social activities, the easy access to the downtown area, and the miniscule cost of the food and dormitories - much to their liking. In the words of one, "it was like a visit to an underdeveloped country - hot and cheap".

The conference ran very smoothly even with the larger-than-usual number of delegates. To quote the closing remarks of Robert Marshak, "I do not know how it has been achieved, but I have had the same feeling of excitement, lively debate, and intimacy in Madison with 1200 delegates as I had thirty years ago in Rochester with 50". The excitement and lively debate we attribute to the presence of the large number of younger physicists and the physicists from smaller institutions and countries who could attend the open conference. The intimacy and smoothness of operation resulted from the excellent facilities and support provided by the University and the Civic Center.

The main work of organizing the conference was carried out by our local high energy physics staff under the direction of James Kolonko, the Conference Secretary, and Debra Goff. They made the multitudinous arrangements for housing, meals, transportation, excursions, and facilities; arranged for the reception at the State Capitol, the picnic, and the concerts by the Fine Arts Quartet and the Karlos Moser group; and, not least important, obtained the conference T-shirts.

The overall scientific program was organized by the Wisconsin high energy physics group, with generous advice from the national and international advisory committees, a number of whose members also contributed directly at the conference as session chairmen or organizers. Most of our high energy faculty, postdocs, and graduate students were also directly involved in organizational tasks, or in the running of the sessions. However, the credit for the scientific success of the conference must go to the physicists who organized or spoke in the parallel sessions, and to those who presented the excellent summary talks contained in these Proceedings.

We also included in the program a special session on the future of high energy physics which is not reported in the Proceedings. This session, organized by Professor Bernice Durand, was intended to give the participants, Professors Leon van Hove, Leon Lederman, Murray Gell-Mann, and Abdus Salam, a chance to discuss informally their ideas on, and concerns about, the future of our subject. Professor Yuval Ne'eman gave a brief historical introduction, and acted as moderator for the discussion, which was videotaped, but not reduced to writing. The discussion ranged over a variety of subjects, including speculations on future directions in particle theory, and the existence or non-existence of a "desert" above present energies (Gell-Mann and Salam); concerns that the field be able to attract - and provide opportunities for - outstanding young physicists (Lederman and van Hove); constraints imposed by present accelerator technology, and the critical necessity for more research on machine physics (Salam, Lederman, van Hove); and the opportunities and challenges for particle physics in the developing countries (Salam). The debate was lively, and stimulated, we hope, further discussion of the future of high energy physics among the other delegates. We hope that sessions of this type will be continued at future conferences.

The format of the Proceedings follows that of the conference. The first volume and the beginning of the second volume contain the papers given in the parallel sessions, grouped by subject area. The remainder of the second volume contains, with three exceptions, the plenary session talks in the order in which they were presented, and lists of the delegates, the countries represented, and the titles of papers contributed to the conference. We unfortunately did not receive a written version of the excellent review of the characteristics of inclusive and exclusive hadronic interactions at low p_T presented by Professor H. Miettinen; nor do we have Professor L. Susskind's stimulating survey of new theoretical ideas and speculations. Finally, Professor L.B. Okun', who had been scheduled to give the conference summary, was regrettably unable at the last minute to attend the conference. Both his presence and his views on the status of high energy physics were missed. We were unable to obtain a written version of the talk he had prepared for inclusion in these Proceedings.

We feel that the present experiment with an open conference was notably successful - that the excitement and sense of international community engendered by the broadened attendance more than compensated for any extra organizational problems - and trust that future conferences in this series will also be open. Should the conference return to Madison at some future time, we would propose only to find enough canoes to take the entire delegation paddling.

Loyal Durand
Lee G. Pondrom
Co-chairmen
December, 1980

SPONSORED AND SUPPORTED IN PART BY:

The International Union of Pure and Applied Physics

The National Science Foundation

The United States Department of Energy

The University of Wisconsin-Madison

TABLE OF CONTENTS

NOTE: The symbol # in place of a page number means that no manuscript was available.

NEW PARTICLE PRODUCTION

e^+e^- PHYSICS AND ELECTROMAGNETIC PROPERTIES, EXPERIMENT AND THEORY

COSMIC RAYS

NEW ACCELERATORS AND NEW EXPERIMENTAL TECHNIQUES

GAUGE THEORIES AND MATHEMATICAL PHYSICS

Chapter 1

Hadronic Interactions, Theory and Experiment

HADRON SPECTROSCOPY, EXPERIMENT

B. French, CERN, Organizer

EVIDENCE FOR NEUTRAL A_1 AND H RESONANCES IN CHARGE EXCHANGE*

J.A. Dankowych, J.F. Martin, A.J. Pawlicki,
J.D. Prentice and T.-S. Yoon
University of Toronto, Toronto, Ontario M5S 1A7, Canada

R.S. Longacre
Brookhaven National Laboratory, Upton, N.Y. 11973, USA

K.W. Edwards and D. Legacey
Carleton University, Ottawa, Ontario K1S 5B6, Canada

N.R. Stanton
The Ohio State University, Columbus, Ohio 43210, USA

P. Brockman, J. Gandsman, P.M. Patel, E. Shabazian and C. Zanfino
McGill University, Montreal, Quebec H3A 2T8, Canada

Presented by N.R. Stanton

ABSTRACT

We present results from an isobar-model partial wave analysis of a high statistics sample of the neutral 3π system produced in forward charge exchange. Large forward phase motions relative to an exotic reference wave are seen for partial waves with the quantum numbers of the A_1 and of the previously unobserved H mesons.

We have performed an isobar-model partial wave analysis (PWA) of the $\pi^+\pi^-\pi^0$ system produced in the reaction

$$\pi^- p \rightarrow \pi^+\pi^-\underset{\hookrightarrow\gamma\gamma}{\pi^0}\, n$$

at 8.45 GeV/c, and observe resonant signals for both the neutral A_1 ($J^{PC} = 1^{++}$) and the previously-unseen H($J^{PC} = 1^{+-}$) mesons.

The data were taken at the Argonne ZGS with the Charged and Neutral Spectrometer[1] which detected both $\pi^{\pm}$ and both γ's. The data sample, discussed elsewhere[1], is roughly ten times larger than previously available in charge exchange. Stringent cuts on $n\pi$ mass[1] have greatly reduced the 3π background from ρN^* reflections. The events used in the PWA (Fig. 1a) survive these cuts and also the momentum transfer restriction $0.0 \leq t' \leq 0.45(\mathrm{GeV/c})^2$. The PWA was performed with a modified version of the Berkeley-SLAC amplitude program[2]. Partial waves occur in four incoherent groups having either natural or unnatural exchange parity (NPE, UPE), and either helicity flip or nonflip at the nucleon vertex, determined from the

*Work supported in part by DOE (USA) and NSERC/IPP (Canada).

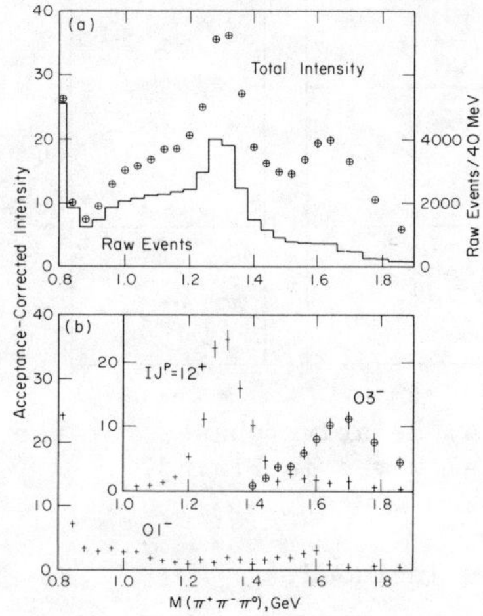

Fig. 1a. Raw events and total acceptance-corrected intensity.
Fig. 1b. Intensities for IJP = 01-, 12+ and 03-.

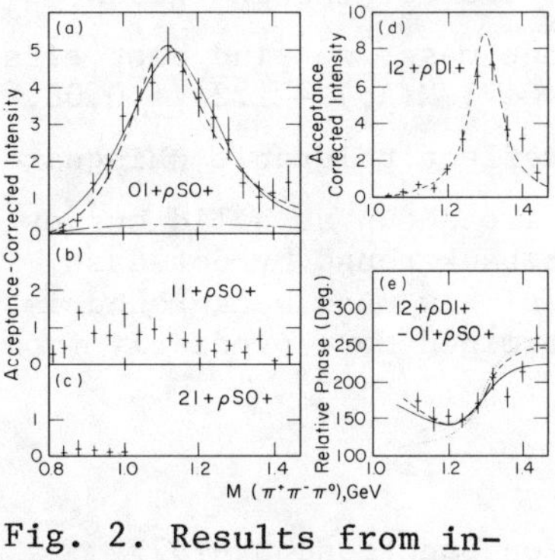

Fig. 2. Results from incoherent group with NPE and nucleon helicity nonflip.

forward t'-dependence.

The circled points in Fig. 1a are the total acceptance-corrected intensity from the PWA. Total intensities from the partial waves with $IJ^P = 01^-$, 12^+ and 03^- (Fig. 1b) show peaks for the known ω, A_2 and $\omega_g(1670)$. The absence of feedthrough from the large A_2 peak into the small ω tail is good evidence that the analysis can distinguish small partial waves despite finite experimental acceptance.

Fig. 2 presents intensities and relative phases from the incoherent group having NPE and nucleon helicity nonflip. We use the notation IJP(isobar)LMη with IJP = isospin, spin, parity; L = orbital angular momentum of bachelor pion; M = z-component of J(t-channel); η = exchange parity. The partial wave with quantum numbers of the H, 01+ρS0+, has a large intensity peak (Fig. 2a) described quite well by a Breit-Wigner form (dashed curve) with M = 1.13, Γ = 0.28 GeV. In addition, the 12+ρD1+ phase relative to this wave (Fig. 2c) does not show the full phase motion expected for the A_2 (dotted curve), but rather the limited relative motion expected if the H reference were also resonant (dashed curve).

Results from the incoherent group with NPE and nucleon helicity flip are shown in Figs. 3 and 4. In this group the 01+ρS0+ (H) wave (Fig. 3c) is again quite large and is roughly described by the above resonance parameters (dashed curve). The M = 1 A_1 wave 11+ρS1+ (Fig. 3a) is also sizable and can be approximately described by the same resonance parameters as the H. The smaller exotic partial wave 21+ρS0+ (Fig. 3d) is still large enough to serve as a phase reference.

The phase of 11+ρS1+ relative to 01+ρS0+ (Fig. 4a) is nearly flat, implying either that both partial waves resonate similarly or not at all. If the exotic wave 21+ρS0+ is used as a reference, however, both 11+ and 01+ phases show large forward motions ~150° (Figs. 4b,c), establishing resonant behavior for both A_1 and H. The partial wave 12+ρD1+ (A_2) also shows very large forward phase motion relative to this reference which may fit very well (solid curve in Fig. 4d)

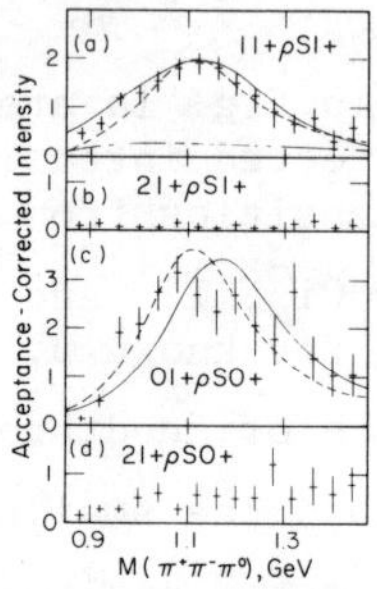

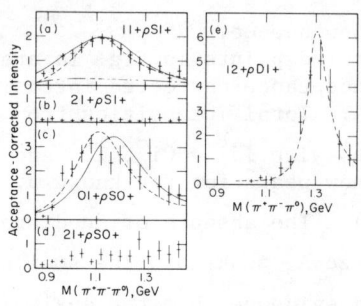

Fig. 3. Intensities from incoherent group with NPE and nucleon helicity flip.

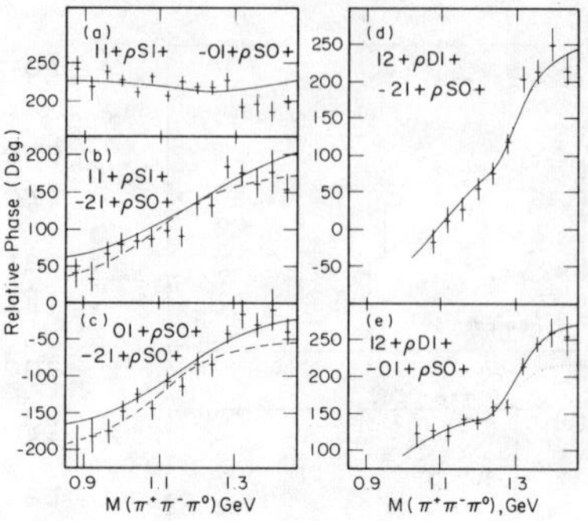

Fig. 4. Relative phases for the partial waves in Fig. 3.

if a 3% coherent background under the A_2 is included.

If the A_1 and H signals we see include no coherent (Deck-like) background, the neutral A_1 and H masses are both ~1.13 GeV, lower than those of the B(1235) and of the charged A_1 (1.28 ± .04 GeV) seen[3] in diffractive production. If however we add a coherent background of the Bowler type as was done in Ref. 3 for the charged A_1 and plot chisquare as a function of resonance mass, we find best fits for M(H) = 1.19 ± 0.06, Γ(H) = 0.32 ± 0.05 GeV; M(A_1) = 1.24 ± 0.08, Γ(A_1) = 0.38 ± 0.10 GeV, where the errors reflect tolerable chisquare range around the minimum. These best fits are shown as solid curves in Figs. 2, 3, 4, and the level of coherent background by dot-dash curves. If it were possible to know a priori how much background is present, the resonance masses could be determined more precisely from our data.

REFERENCES

1. K.W. Edwards et al., in Experimental Meson Spectroscopy-1977, edited by E. von Goeler and R. Weinstein (Northeastern University Press, Boston, 1977).
2. See, e.g., D.J. Herndon et al., Phys. Rev. D11 3183 (1975).
3. C. Daum et al., Phys. Lett. 89B, 281 (1980).

GENERAL FEATURES OF THE REACTION $\pi^+p \rightarrow \Delta^{++}\pi^0\pi^0$ AT 8 GeV/c

N.M. Cason, M.U. Ahmed, A.E. Baumbaugh, P.E. Cannata**, N.N. Biswas, J.M. Bishop. V.P. Kenney, M.J. Lawson***, C.A. Rey††, R.C. Ruchti, W.D. Shephard, and J.M. Watson
University of Notre Dame*, Notre Dame, Indiana 46556
Argonne National Laboratory†, Argonne, Illinois 60439

ABSTRACT

Characteristics of the reaction $\pi^+p \rightarrow \Delta^{++}\pi^0\pi^0$ at 8 GeV/c are presented. The sample consists of some 27,000 events for which mass spectra, t-distributions, and $\pi\pi$ angular distributions are presented. The $\pi^0\pi^0$ mass spectrum rules out the "narrow ϵ(700)" solution for the $\pi\pi$ S_0 phase shift. Data in the 1300 MeV region show substantial S - D wave interference.

In this paper we report some general features of the reaction $\pi^+p \rightarrow \Delta^{++}\pi^0\pi^0$ produced at 8 GeV/c. Our analysis is continuing, and detailed amplitude analyses will be reported elsewhere.

The experiment was performed at the Argonne National Laboratory using the 1.5 m streamer chamber in combination with a 68-element lead-glass hodoscope. The experiment was designed to have good acceptance in the $\pi^0\pi^0$ decay angular distribution up to about 1700 MeV/c^2 in $M(\pi^0\pi^0)$. The data consist of 27,112 events consistent with the $p\pi^+\pi^0\pi^0$ final state after χ^2 and production vertex cuts. The $\pi^0\pi^0$ effective mass spectrum for these events is shown in Fig. 1, both weighted and unweighted. Weights have been calculated on an event-by-event basis. Monte Carlo events were generated at a random azimuthal angle about the beam axis and then tested to see if they would trigger our system. For each real event, Monte Carlo events were generated until the detection probability for the real event was known to better than about 5%. The average detection probability for our data was 0.48. As can be seen from Fig. 1, this detection probability does not vary strongly with $\pi\pi$ mass. The most prominent feature of Fig. 1 is the f^0 peak at ~1260 MeV/c^2. In addition a broad enhancement in the 700

* Work supported in part by the National Science Foundation
† Work supported by the U.S. Department of Energy
** Present address: Bell Laboratories, Murray Hill, N.J.
*** Present address: Bell Laboratories, Warrenville, Illinois
†† Present address: Inter Sonics Corp., Chicago, Illinois

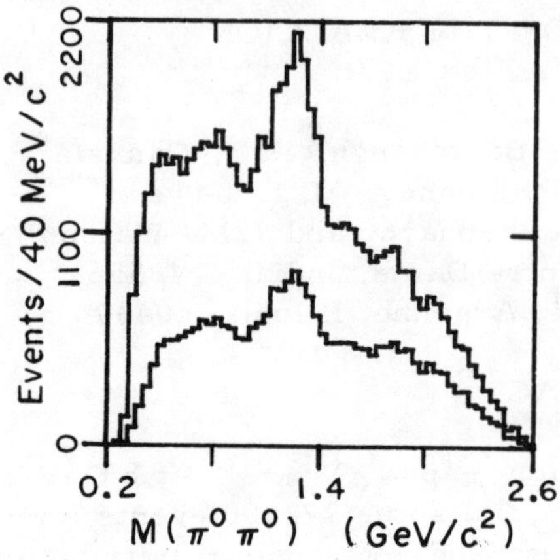

Fig. 1. The $\pi^0\pi^0$ effective mass distribution (upper histogram weighted and lower histogram unweighted) for all events kinematically consistent with the reaction $\pi^+p \rightarrow \pi^+p\pi^0\pi^0$.

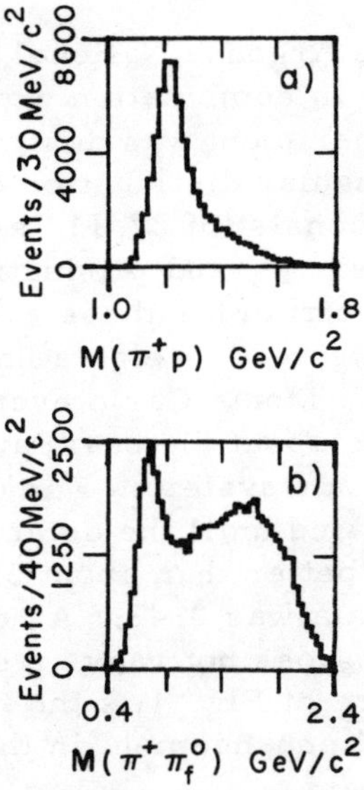

Fig. 2. The weighted a) π^+p and b) $\pi^+\pi^0_f$ effective mass distribution for all events.

MeV/c^2 region can be attributed to the broad ϵ(700). The data clearly do not favor a "narrow ϵ" solution.

Fig. 2 shows the weighted π^+p and $\pi^+\pi^0_f$ effective mass distribution for the data. (The fast π^0 in the final state is signified by π^0_f.) Strong Δ^{++} production is obvious, and a ρ^+ signal is also clear. For the remainder of the paper, we require a Δ^{++} to be present ($M_{\pi^+p} < 1.36$ GeV/c^2) and exclude the ρ^+ signal ($0.66 < M_{\pi^+\pi^0_f} < 0.86$ GeV/c^2).

The t-dependence of the data can be seen in the Chew-Low plot of Fig. 3. The highly peripheral nature of the data,

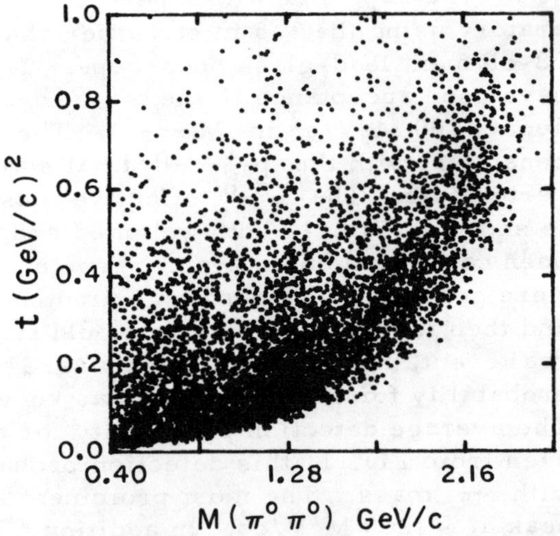

Fig. 3. Scatterplot of the $\pi^0\pi^0$ effective mass versus -t, the momentum transfer from the target to the π^+p system.

expected for one pion exchange, is obvious, and in Fig. 4 is shown the $\pi^0\pi^0$ mass spectrum with $t' < 0.2\ (GeV/c)^2$ and with Δ^{++} and ρ^+ cuts. The final sample with these cuts shows features similar to the total data sample except that below 600 MeV/c^2 a greater fraction of events is cut out, and at 1270 MeV/c^2 the f^0 signal is enhanced.

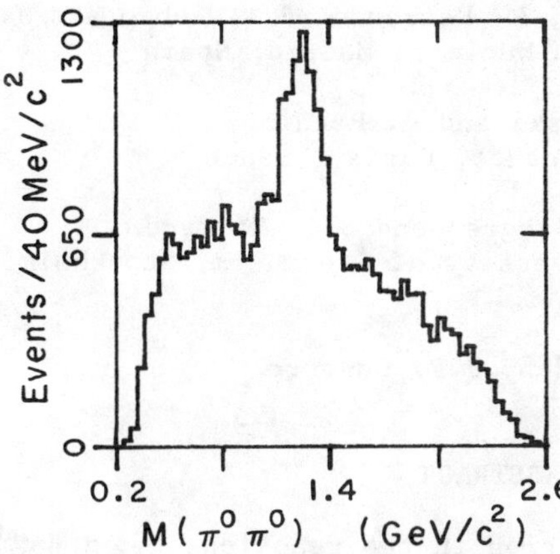

Fig. 4. The weighted $\pi^0\pi^0$ effective mass distributions for the final data sample.

Finally, we show in Fig. 5 the weighted cos θ_J decay angular distributions for four $\pi^0\pi^0$ mass bins. The data shown have been satisfactorily fitted using S and D waves only, and an amplitude analysis of the moments of the spherical harmonics as a function of $\pi\pi$ mass will be published soon.

In conclusion, although the amplitude analysis of our $\pi^0\pi^0$ data is still to be completed, it appears that the "narrow ϵ(700)" solution can be ruled out; in addition, substantial S-wave is present in the 1300 MeV region.

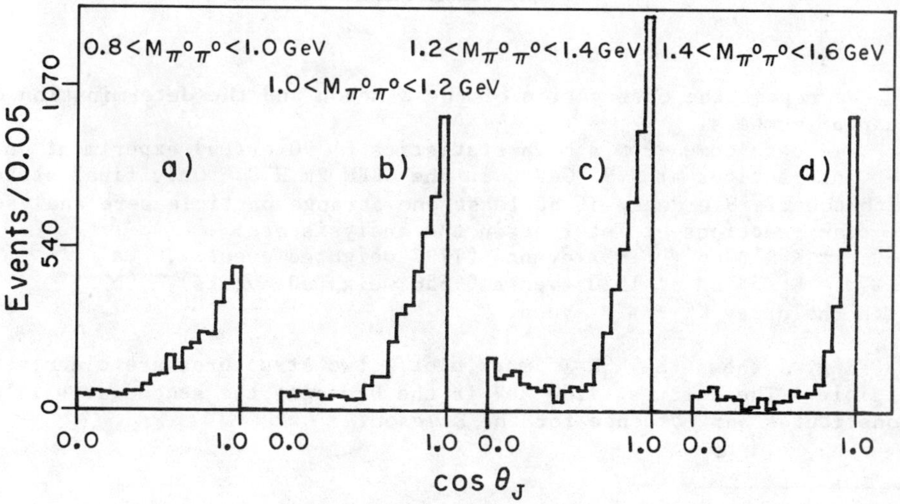

Fig. 5. Weighted decay angular distribution of the $\pi^0\pi^0$ system (t-channel) for four $\pi^0\pi^0$ mass regions.

E MESON OBSERVATION IN $\pi^- p$ INTERACTIONS AT 3.95 GeV/c

CERN-Collège de France-Madrid-Stockholm Collaboration

C. Dionisi, Ph. Gavillet, R. Armenteros, M.J. Losty(*), P.F. Loverre, M. Mazzucato and L. Montanet
CERN, European Organization for Nuclear Research, Geneva, Switzerland

M. Aguilar Benitez, C. Fernandez, A. Ferrando, J.A. Rubio and J. Salicio
Junta de Energia Nuclear, Madrid, Spain

L. Dobrzynski and D. Pennino
Collège de France, Paris, France

S. Rodebäck, I. Sjögren and S.O. Holmgren
Department of Physics(**), University of Stockholm, Stockholm, Sweden

Presented by P.F. Loverre

ABSTRACT

Evidence for the E meson production in the reactions $\pi^- p \to K^0 K^{\pm} \pi\ n$ (σ = (8.2 ± 1.0)μb) is presented. The mass and width of the meson are M = (1426 ± 6)MeV and Γ = 40 ± 15 MeV. Our analysis shows that the quantum numbers are, most likely, $I^G J^P = 0^+ 1^+$. The favoured decay mode is $E \to K^* \bar{K}$ + c.c.

We report the observation of the E meson and the determination of its quantum numbers.

The data come from a high statistics (∿ 90 ev/μb) experiment on $\pi^- p$ interactions at 3.95 GeV/c in the CERN 2m HBC. Only final states with the visible decay of at least one strange particle were analysed.

The reactions of interest in the analysis are:
- $\pi^- p \to K_S^0 K^+ \pi^- n$ 1652 events (1797 weighted events), (a)
- $\pi^- p \to K_S^0 K^- \pi^+ n$ 1520 events (1650 weighted events), (b)

with the decay $K_S^0 \to \pi^+ \pi^-$ seen.

Fig. 1 shows the $K_S^0 K^{\pm} \pi$ mass plot: two structures are clearly visible. The first at 1280 MeV is the D meson; the second at ∿ 1420 MeV constitutes our evidence for the E meson.

(*) Present address: National Research Council, Ottawa, Canada.
(**) Work supported by the Swedish Natural Science Research Council.

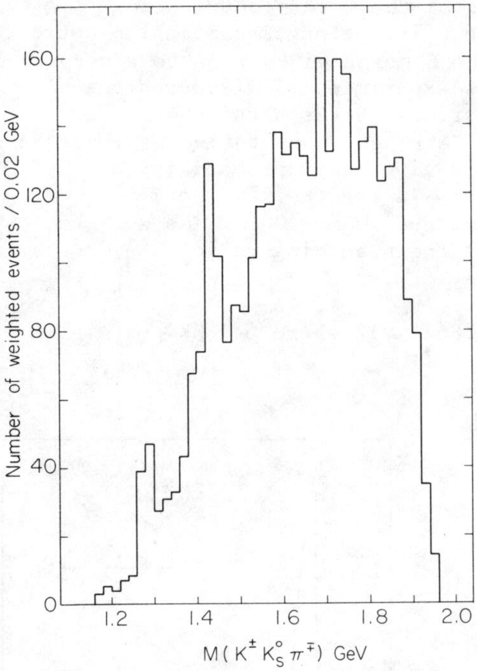

Fig. 1 - $K^\pm K^0_S\pi^\mp$ mass distribution of the reactions $\pi^- p \to K^0_S K^\pm \pi^\mp n$

We will now discuss the quantum numbers assignment.

Isospin. No signal at $m(\bar{K}K\pi) = 1420$ MeV is observed in the reactions $\pi^- p \to K^0_S K^- \pi^0 p$ and $\pi^- p \to K^0_S (K^0)\pi^- p$; thus we agree with previous experiments[1-5] in assigning I = 0 to the E meson.

J^{PC} and branching ratios. Following ref. 6), we have performed a mass independent analysis of the $K\bar{K}\pi$ system using the Dalitz plot variables (the integration over the other variables is required by the limited statistics). For the maximum likelihood fits of the Dalitz plot of the different mass regions (four mass bins from $m(K\bar{K}\pi) = 1.31$ GeV to 1.65 GeV) we have used quasi-two-body amplitudes where the angular part was constructed using the Zemach representation. The isobars considered were the $K^*(890)$ and the $\delta(980)$. The E being an eigenstate of G, its KK^* amplitudes have the form $A = (K^*\bar{K} + G\cdot\bar{K}^*K)$ where G = +1(-1) gives rise to constructive (distructive) interference. Together with a set of coherent amplitudes (COH) allowing for interference between waves of the same J^P, we have included as background an incoherent contribution (INC) of $K^*\bar{K}$ and $\bar{K}^*K$ waves. Two separate sets of fits, for the hypotheses G = +1 and -1 respectively, have been performed. The fits with G = -1 do not reproduce the data so ruling out the C = -1 assignment. The result of the fit with G = +1 is shown in fig. 2.

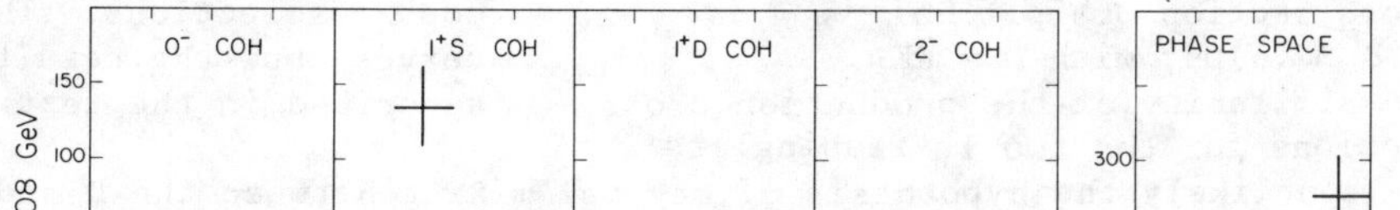

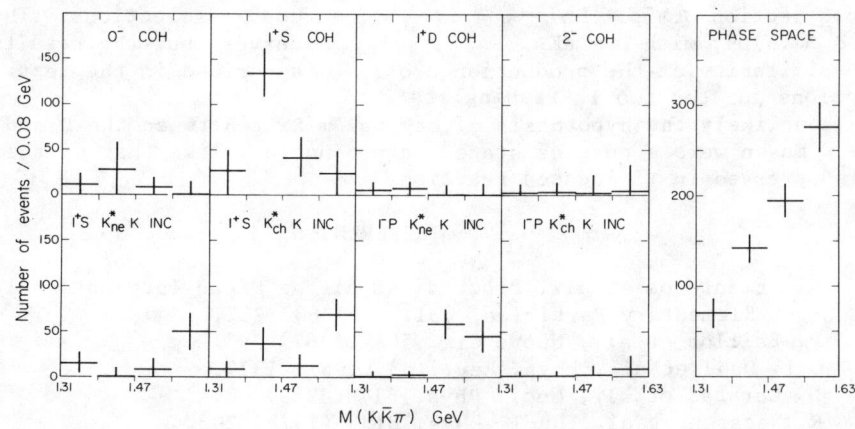

Fig. 2 - Results of the partial wave analysis for the fit described in the text.

There we have plotted the intensities of the waves contributing significantly [the notation is J^PL where L is the angular momentum between the isobar and the remaining meson]. The E meson gives rise to a very clear signal in the 1^+ wave. Since a previous experiment [4] favoured the $J^P = 0^-$ assignment, we have performed various checks showing that our partial wave decomposition is reliable. A rough quantitative estimate based on the theorem of the likelihood ratios[7] gives a χ^2 probability of $\sim$ 85% for the $J^P = 1^+$ assignment against $\sim$ 1% for the 0^-. We conclude that the E meson quantum numbers are $I^GJ^P = 0^+1^+$. As a byproduct of the J^P analysis we also get the branching ratio

$$\frac{E \to [K^*\bar{K} + c.c.]}{E \to [(K^*\bar{K} + c.c.) + \delta\pi]} = .86 \pm .12 \text{ where } \delta \to K\bar{K}.$$

Mass and width. The Dalitz plot analysis shows that the E decays mainly to $K^*\bar{K}$ + c.c. populating the K^*'s interference region. We have therefore selected the K^* in the $K\pi$ mass combinations and cut on the K^* helicity decay polar angle $\theta_{K^*}(\cos\theta_{K^*} > .5)$. The resulting $K\bar{K}\pi$ mass spectrum is shown in fig. 3. A clear gain in the signal/background ratio is achieved and a fit to this spectrum with an S-wave relativistic Breit Wigner plus polynomial background gives $M_E = (1426 \pm 6)$MeV and $\Gamma_E = (40 \pm 15)$MeV.

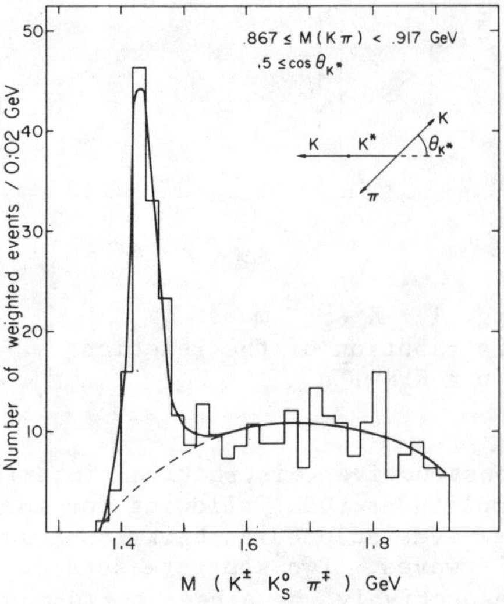

Fig. 3 - $K^{\pm}K^0_S\pi$ mass spectrum after the K^* and $\cos\theta_{K^*}$ selections. The curves show the result of the fit described in the text.

Production properties. The production cross section has been determined to be: $\sigma(\pi^- p \to En)$ = $(12.3 \pm 1.5)\mu b$ with $E \to \bar{K}K\pi$. The ratio between forward and total cross section has been estimated as $.7 \pm .2$.

We have also measured the D meson cross section $\sigma(\pi^- p \to Dn)$ = $(5.8 \pm .6)\mu b$, with $D \to \bar{K}K\pi$. The similarity of the production cross sections for the two 1^+ isosinglets makes unlikely the hypothesis of magical mixing between the D and the E: if the E meson were a pure $s\bar{s}$ state its production, like that of the φ, would be suppressed in π^- induced reactions.

REFERENCES

1) R. Armenteros et al., Proceedings of the Siena International Conference of Elementary Particles, vol. 1 (1963) 287.
2) P. Baillon et al., Nuovo Cim. 50A (1967) 393.
3) O.I. Dahl et al., Phys. Rev. 163 (1967) 1377.
4) B. Lorstad et al., Nucl. Phys. B14 (1969) 63.
5) R. Nacasch et al., Nucl. Phys. B135 (1978) 203.
6) Ph. Gavillet, thesis, Orsay (1979) 2147.
7) W.T. Eadie et al., Statistical Methods in Experimental Physics (North-Holland, Amsterdam, 1971) 230.

SEARCH FOR NARROW $B\bar{B}(\pi)$ STATES IN 16 GeV/c π^-p AND 5 GeV/c $\bar{p}p$ INTERACTIONS

S. U. Chung*
Physics Department, Brookhaven National Laboratory
Upton, New York 11973

INTRODUCTION

We report here on the searches for narrow baryon-antibaryon ($B\bar{B}$) bound states from two recent experiments carried out at the BNL Multiparticle Spectrometer. First, we present the results of a search for narrow states in the $\bar{p}p$, $\Lambda p\pi^{\pm}$ and $\bar{p}p\pi^+$ systems from π^-p interactions at 16 GeV/c. The data come from an experiment carried out by physicists from BNL, Brandeis, CCNY, Southeastern Massachusetts, University of Massachusetts.[1] Next, we show results of a search for narrow $\bar{p}p$ states from $\bar{p}p$ interactions at 5 GeV/c, an experiment with collaborators from BNL, Brandeis, University of Cincinnati, Florida State University, and Southeastern Massachusetts University.[2]

π^-p DATA AT 16 GeV/c

This experiment was conceived in part to further study the two $\bar{p}p$ states with widths less than 25 MeV and masses 2020 and 2200 MeV seen in a CERN Ω-spectrometer experiment by Benkheiri, et al.[3] Although a number of experiments looked for these states in formation as well as production processes, ours is the first to have searched for them in the same reaction as that of the CERN experiment with similar trigger techniques and acceptance.

The reaction studied is

$$\pi^-p \to (p_f\pi^-)\ (\bar{p}p_s) \tag{1}$$

where $p_f(p_s)$ refers to a fast (slow) proton in the laboratory. The trigger for this study required detection of a fast forward proton with momentum between 8 and 12 GeV/c, implemented via two three-dimensional coincidence-matrix (RAM) logic systems and two gas Čerenkov counters.[4] A total of 3.4×10^6 proton triggers were recorded, and ~80% of them have been analyzed to date, corresponding to a raw sensitivity of 62 ev/nb.

The events have been processed through our chain of data-reduction programs, yielding finally a total of ~7 K events with acceptable four-constraint kinematic fits to reaction (1). We estimate that the contamination from non-4C background in this final

*Research supported by the U. S. Department of Energy under contract DE-AC02-76CH00016.

sample is less than 3%. In Fig. 1a we present the $M(p_f\pi^-)$ spectrum showing clear $\Delta^0(1238)$ and $N^0(1520)$ peaks in our data. Fig. 1b shows the spectrum $M(\bar{p}p_s)$ of the recoil system. There is no evidence for the production of 2020 and 2200 MeV states in our data. We have gone to considerable lengths in an attempt to enhance the baryon-exchanged production of the $\bar{p}p_s$ system, but no significant peaks have been observed in the entire $\bar{p}p$ mass range.

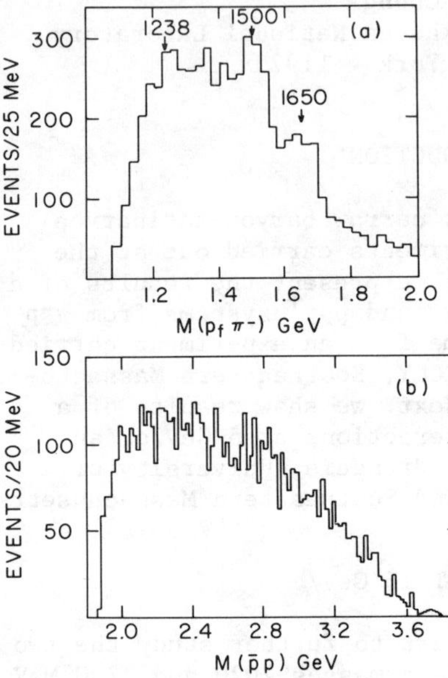

Fig. 1(a,b). $M(p_f\pi^-)$ and $M(\bar{p}p)$ spectra for reaction (1).

Using the calculated acceptance and the experimental and data-reduction losses, we estimate that the overall visible sensitivity of our present data ranges from 8 ev/nb at $M(\bar{p}p_s) = 2.02$ GeV with $\Delta^0(1238)$ to 5 ev/nb at $M(\bar{p}p_s) = 2.20$ GeV with $N^0(1520)$. From these we conclude that the 2 σ upper limit cross sections are 3.0 nb for the 2020-MeV state (obtained from the $\bar{p}p$ spectrum with Δ^0 and N^0 selections) and 2.0 nb for the 2200-MeV state (from Δ^0 events alone). These values are to be compared to the corresponding cross sections quoted in Benkheiri, et al.[3]: 36 ± 9 nb for 2200-MeV state and 21 ± 5 nb for 2200 MeV state at 12 GeV/c. The rms mass resolution of our data is better than 7(11) MeV at 2020 (2200) MeV, sufficient for observation of narrow $\bar{p}p$ states, if produced in our data.

We have in addition searched for narrow states in the meson-exchange reactions

$$\pi^- p \rightarrow (\Lambda\bar{p}\pi^+)X^0 \ (2); \quad (\Lambda\bar{p}\pi^-)X^{++} \ (3); \quad (\bar{p}p\pi^+)X^- \ (4);$$

where X denotes the recoiling system off that being studied. Note that the $\Lambda\bar{p}\pi^-$ system is explicitly exotic, whereas the $\Lambda\bar{p}\pi^+$ and $\bar{p}p\pi^+$ systems require I = 3/2 and Q = 2 exotic meson exchanges. The protons in (4) and those from the Λ decay in (2) and (3) are the triggered particles, identified by two Cerenkov counters, and the $\bar{p}$'s in (2), (3) and (4) were required off-line to go through one Cerenkov counter without yielding light. The latter requirement eliminated π^-'s, but some K^-'s do remain in our sample, contributing to the general background in the mass spectra being studied.

No significant narrow peak is observed in any of the spectra (not shown). At 2.5 GeV the 2 σ upper limits for $\Lambda\bar{p}\pi^+$ ($\Lambda\bar{p}\pi^-$) states with width less than 40 MeV are ~25 nb (30 nb), and for $\bar{p}p\pi^+$ with width less than 20 MeV it is ~25 nb. The rms mass resolution at 2.5 GeV is estimated to be less than 20 MeV for all three mass spectra.

$\bar{p}p$ DATA AT 5 GeV/c

We have investigated the $\bar{p}p$ system produced in the baryon-exchange process

$$\bar{p}p \to p_f \bar{p} \pi^0 (\text{or } \rho^0) \qquad (5)$$

at 5 GeV/c. The p_f is the fast forward proton ($p \gtrsim 1.2$ GeV/c), triggered with the aid of the RAM logic system and a high-pressure Cerenkov counter.[6] Both p_f and $\bar{p}$ were required to have momenta greater than 1.8 GeV/c for the maximum efficiency of the Cerenkov counter.

The advantages of this experiment over previous experiments for narrow $B\bar{B}$ searches are two-fold: first, our beam momentum at 5 GeV/c favors baryon-exchange process compared to a meson beam at 12 GeV/c, where Benkheiri, et al.[3] saw two narrow $\bar{p}p$ states at 2.0 and 2.20 GeV. Second, the four-momentum transfer squared from the $\bar{p}$ beam to the $\bar{p}p$ final state can become positive, thus allowing for closer approach to the baryon poles. We estimate that the combined effects amount to an enhancement factor of $\sim$30 compared to the experiment of Benkheiri, et al.

The sample of data presented here corresponds to a raw sensitivity of 6.5 nb^{-1}. The missing mass spectrum recoiling off $\bar{p}p$ (not shown) shows clear peaks at π^0 and ρ^0, respectively, demonstrating our substantial acceptance for the reaction (5). We estimate that non-π^0 (or non-ρ^0) background in our data is less than 50%. Figs. 2a and b show the $\bar{p}p$ mass spectra for the π^0 and ρ^0 events. No significant peak with width $\lesssim$ 20 MeV are seen in our data. The superposed curves are our estimate of the overall acceptance, obtained via MC events.[5] The rms mass resolution at 2.20 GeV is estimated to be less than 15 MeV. We obtain a 2 σ upper-limit cross section of 130 nb (75 nb) for a $\bar{p}p$ state at 2.0 GeV (2.2 GeV) with width less than 20 MeV for π^0 events. For ρ^0 events the corresponding upper limits are $\sim$50% higher than those of the π^0 events.

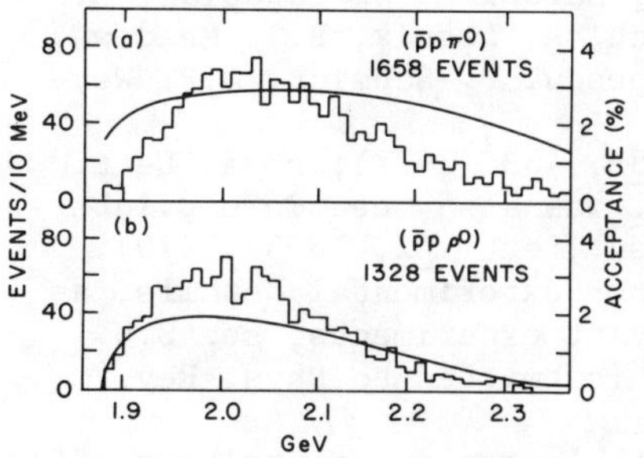

Fig. 2(a,b). M($\bar{p}p$) spectra for reaction (5).

SUMMARY AND CONCLUSIONS

We have presented here the latest results of a systematic search for narrow $B\bar{B}$ states in two BNL MPS experiments requiring fast forward protons in the trigger. We have found no significant narrow peaks with widths less than 40 MeV in all the spectra we have examined so far.

In the experiment with $\pi^- p$ interactions at 16 GeV/c, we obtain 2σ upper limits of less than 3 nb for $\bar{p}p$ states at 2.0 to 2.2 GeV with widths less than 20 MeV. Based on the cross sections for 2.02 and 2.20 $\bar{p}p$ states quoted by Benkheiri, et al.[3] and assuming baryon-exchange processes for production of these states, we should have seen better than 5 σ signals at our energy. We have in addition searched for narrow states in $\Lambda\bar{p}\pi^{\pm}$ and $\bar{p}p\pi^{+}$ systems; 2 σ upper limits are $\lesssim$ 30 nb for the states with widths less than 40 MeV.

In the experiment with $\bar{p}p$ interactions at 5 GeV/c, we have looked for narrow $\bar{p}p$ states produced by baryon-exchange with either π^{o} or ρ^{o} as recoiling particles. Given an enhancement factor of $\sim$30 for our experiment and the cross section of $\sim$30 nb quoted in the paper by Benkheiri, et al.,[3] we should have seen the $\bar{p}p$ states with cross sections of the order of $\sim$1 μb. Instead, we find that 2 σ upper limits are $\lesssim$ 130 nb for a $\bar{p}p$ state with less than 20 MeV and mass below 2.2 GeV.

REFERENCES

1. Z. Bar-Yam, J. Bensinger, J. Button-Shafer, S. Dhar (Present address: Raytheon Corp., Wayland, MA 01420), J. Dowd, A. Etkin, R. Fernow, K. Foley, J.H. Goldman (Present address: Florida State University, Tallahassee, FL 32306), W. Kern, H. Kirk, J. Kopp, M.A. Kramer, A. Lesnik (Present address: Fairchild Republic Co., Farmingdale, NY 11735), R. Lichti (Present address: Texas Tech University, Lubbock, TX 79409), S.J. Lindenbaum, W. Love, U. Mallik, T. Morris, W. Morris, S. Ozaki, E. Platner, S.D. Protopopescu, A. Saulys, D.P. Weygand, C.D. Wheeler, E. Willen, M. Winik.
2. J. Albright, Z. Bar-Yam, J. Bensinger, R. Diamond, J. Dowd, H. Fenker, R. Fernow, J.H. Goldman, V. Hagopian, S. Hagopian, M. Jenkins, W. Kern, H. Kirk, L. Kirsch, A. Lesnik, B.T. Meadows, W. Morris, R. Poster, S.D. Protopopescu, P. Schmidt, D.P. Weygand, M. Winik.
3. P. Benkheiri et al., Phys. Lett. 68B, 483 (1977); Phys. Lett. 81B, 380 (1979). A marginal signal was also seen in $e^{-}p$ interactions by Gibbard et al., Phys. Rev. Lett. 42, 1593 (1979).
4. For the experimental layout and other experimental details, as well as for the references to relevant experiments, see S.U. Chung, et al., BNL preprint OG 548 (submitted to Phys. Rev. Lett.).
5. The curves do not include additional losses due to trigger efficiency, impurities in the beam, absorption in LH_2, etc.
6. For the experimental layout and other experimental details, as well as for the references to relevant experiments, see S.U. Chung, et al., BNL preprint OG 545 (submitted to Phys. Rev. Lett.)

SEARCH FOR NARROW BARYONIUM STATES IN K^+p REACTIONS

D. Frame, I. Hughes, J. Lynch, D. McFadzean, P. Minto,
L. Scott, D. Stewart, A. Thompson, R. Turnbull, I. Wilkie
Department of Natural Philosophy, The University, Glasgow G12 8QQ

D. Colley, P. Harper, M. Jobes, G. Jones, S. O'Neale
Department of Physics, University of Birmingham B15 2TT

T. Armstrong, CERN, Geneva

ABSTRACT

Data are presented from a high statistics K^+p experiment designed to trigger on final states including a $\bar{\Lambda}$ or $\bar{p}$. The $\bar{\Lambda}p$, $\bar{\Lambda}n$, $\bar{\Lambda}p\pi^+$ and $\bar{p}p$ mass spectra observed in the reactions

$$K^+p \to \bar{\Lambda}pp$$
$$K^+p \to \bar{\Lambda}pp\pi^0$$
$$K^+p \to \bar{\Lambda}pn\pi^+$$
$$K^+p \to K^+\bar{p}pp$$

are presented. No narrow states are observed in these spectra.

INTRODUCTION

We report the results of a study of K^+p interactions relevant to the search for narrow four quark (baryonium) states decaying to $\bar{\Lambda}$-nucleon, $\bar{\Lambda}p\pi^+$ and $\bar{p}p$. There is evidence from an earlier experiment of a narrow resonance in the $\bar{\Lambda}p\pi^+$ system at a mass of 2.46 GeV/c^2[1]. A resonance in this system requires a minimum of 4 quarks to account for its quantum numbers. Narrow resonances have also been reported in the $\bar{p}p$ combination[2] while a large number of theoretical papers have discussed the existence and properties of baryonium states. (See references quoted in ref.1).

EXPERIMENT

The experiment was performed using the OMEGA spectrometer at the CERN SPS with a trigger designed primarily to select $\bar{\Lambda}$ production events by detection of fast antiprotons. An R.F. separated beam[3] of about 10^5 K^+/burst (50% of the total beam flux) at 13 GeV/c was incident on a 60 cm liquid hydrogen target. The incident positive kaons were identified by a differential Cerenkov counter in the beam line.

The layout of the OMEGA spectrometer and the trigger elements is shown in figure 1. The trigger required that a track giving correlated hits in the hodoscopes H1 and H2 gave no light in the corresponding cell of the Cerenkov counter thus selecting events with a $\bar{p}$ or K^- of momentum > 3 GeV/c originating from the target

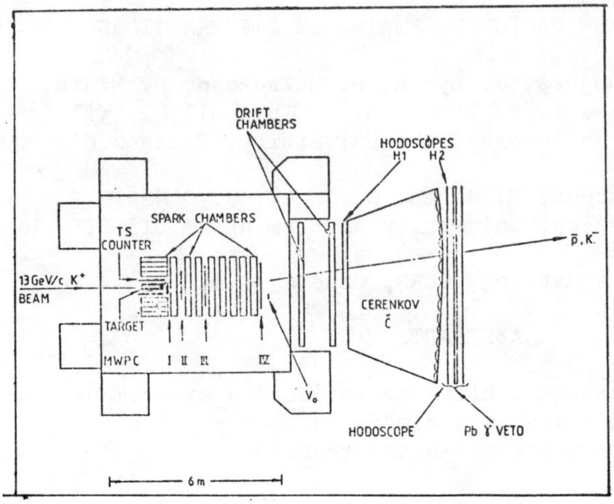

Fig. 1 Layout of the spectrometer.

region. Additional trigger requirements on the hits in the cylindrical scintillation counter TS, MWPC1 and the beam veto V_0 ensured a target interaction.

The data were taken in two runs. In the first run with the above trigger conditions the beam flux sensitivity was 3 events/nb (0.7×10^6 triggers). The second run included a lead scintillator γ detection system to veto π^0s and in this run the beam flux sensitivity was 10 events/nb (1.7×10^6 triggers).

DATA AND RESULTS

All the data were processed through the pattern recognition and geometrical reconstruction programme ROMEO. An additional programme specifically designed to find $\bar{\Lambda}$ decays (particularly within the target) proved to be essential. Four constraint channels were selected by cuts on a function related to energy and momentum balance. One constraint events were selected by cuts on missing mass squared. The numbers of events obtained in each channel were

$K^+p \rightarrow \bar{\Lambda}pp$ 8949 events (1)
$K^+p \rightarrow \bar{\Lambda}pp\pi^0$ 3826 events (run 1 only) (2)
$K^+p \rightarrow \bar{\Lambda}pn\pi^+$ 12109 events (3)
$K^+p \rightarrow \bar{p}ppK^+$ 2149 events (4 constraint unambiguous only) (4)

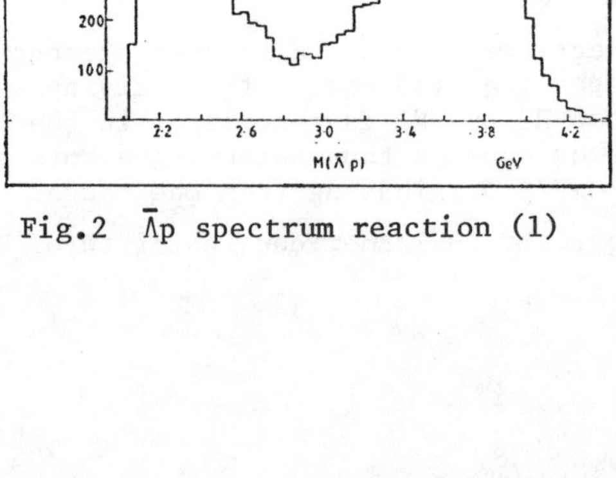

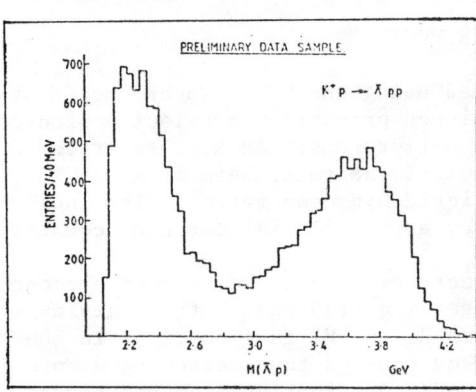

Fig.2 $\bar{\Lambda}p$ spectrum reaction (1)

The $(\bar{\Lambda}p)$ effective mass spectrum from reaction (1) is shown in fig.2. No narrow peaks are evident. At 2.2 GeV/c^2 σ . (B.R.) < 0.17 μb for a 3σ peak. The t slope is constant at $\sim -4(\text{GeV}/c)^2$ throughout the low mass ($\bar{\Lambda}$-fast proton) peak. No narrow peaks are seen in the various $\bar{\Lambda}$-nucleon combinations of reactions (2) and (3). The combined $\bar{\Lambda}$-nucleon spectrum from reactions (1), (2) and (3) also shows no narrow effects.

The $(p\pi^+)$ and $(\bar{\Lambda}\pi^+)$ effective mass distributions from reaction (3) show production of the $\Delta^{++}(1232)$ and $\bar{\Sigma}(1385)$. The $(\bar{\Lambda}p\pi^+)$ spectrum is shown in fig.3 and the corresponding spectrum with Δ^{++} selected in fig.4. No evidence is seen for narrow peaks at 2.46 GeV/c^2 or elsewhere. No narrow peaks are seen if $\bar{\Sigma}(1385)$ is selected nor in the corresponding combinations from the final state $\bar{\Lambda}pp\pi^+\pi^-$.

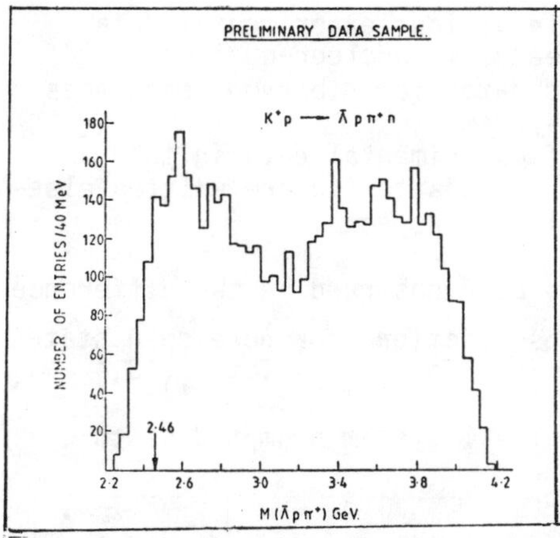

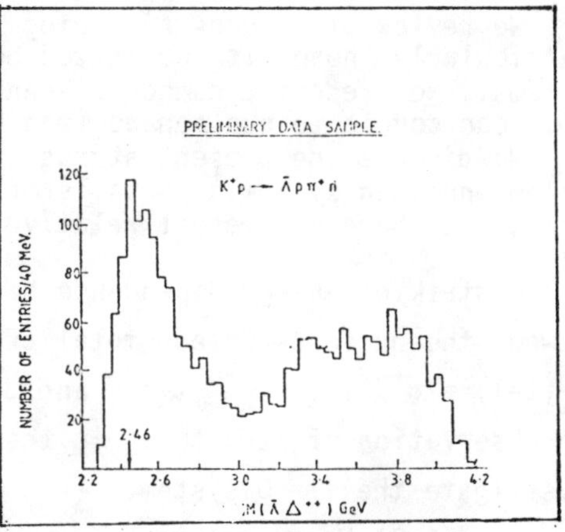

Fig.3 $M(\bar{\Lambda}p\pi^+)$ from reaction (3). Fig.4 $M(\bar{\Lambda}p\pi^+)$ with $\Delta^{++}(1232)$ selected.

Reaction (4) shows strong $\bar{\Lambda}(1520)$ ($\rightarrow K^+\bar{p}$) production and the final state $\bar{\Lambda}(1520)pp$ closely resembles $\bar{\Lambda}(1115)pp$. No narrow structures are seen in $M(\bar{p}p)$ or in $M(\bar{\Lambda}(1520)p)$. For a 3σ effect in $M(\bar{p}p)$ at a mass $\sim$ 1940 MeV/c^2 σ . B.R. $\lesssim$ 30 nb.

REFERENCES

1. T.A. Armstrong et al., Phys. Lett. 77B, 447 (1978).
2. See review talk by L. Montanet at this conference.
3. A. Citron et al., N.I.M. 155, 73 (1978).

REVIEW OF N-N SCATTERING

A. Yokosawa High Energy Physics Division Argonne National Laboratory
9700 South Cass Avenue Argonne, Illinois 60439

Abstract

We review structures appearing in various experimental data (particularly those with polarized beams) in nucleon-nucleon systems. We present a number of candidates for dibaryon resonances which can couple to nucleon-nucleon systems.

We discuss the present status of experimental data in the nucleon-nucleon system.[1] Details of N-N scattering are written elsewhere,[2] and here we present relatively new aspects.

A striking energy dependence has been observed in the difference between the nucleon-nucleon total cross sections for pure spin states, $\Delta\sigma_L(I=1) = \sigma^{Tot}(\underset{\leftarrow}{\rightarrow}) - \sigma^{Tot}(\underset{\rightarrow}{\rightarrow})$, and $\Delta\sigma_T = \sigma^{Tot}(\uparrow\downarrow) - \sigma^{Tot}(\uparrow\uparrow)$.[3,4] The observation of structures in the I = 1 system prompted us to investigate the I = 0 system.

1) $\Delta\sigma_L$ (I=0) Measurements

The Argonne PPT group has recently measured the difference between isoscalar nucleon-nucleon total cross sections for pure longitudinal initial spin states, $\Delta\sigma_L$(pd), using a polarized proton beam and a polarized deuteron target.[5] In the simplest approximation, $\Delta\sigma_L(pd) \approx \Delta\sigma_L(pp) + \Delta\sigma_L(pn)$. One can extract $\Delta\sigma_L$(I=0) data using both $\Delta\sigma_L$(pd) and $\Delta\sigma_L$(pp) as shown in Fig. 1; a significant structure is observed around 1.5 GeV/c. This seems to suggest the existence of a new isoscalar spin-singlet dinucleon resonance. We note here that there exists a clear shoulder in the np total cross-section data[6] in the vicinity of 1.5 GeV/c.

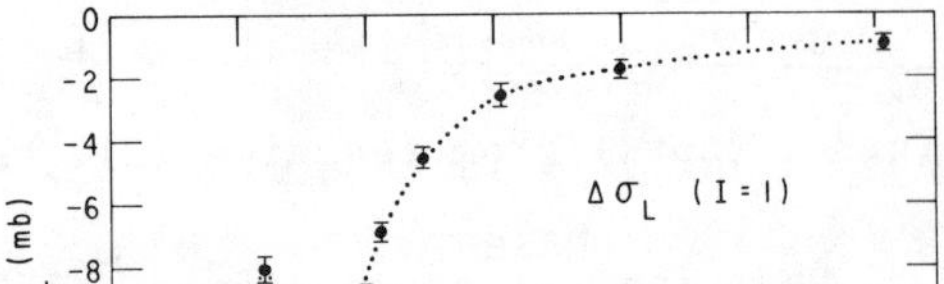

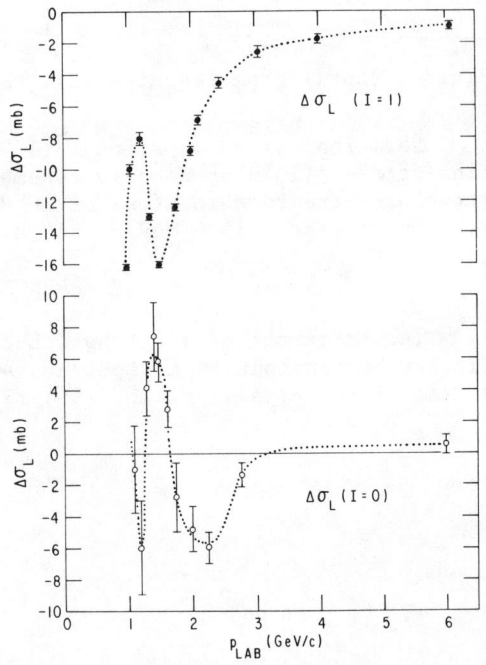

$\Delta\sigma_L$(I=0) together with $\Delta\sigma_L$(I=1) .

A recent phase-shift analysis using these preliminary data by Hoshizaki et al.[7] suggests that there exists a partial wave whose behavior is consistent with the Breit-Wigner resonance formula, namely, the spin singlet $^1F_3$ wave. From the dispersion analysis of a forward I = 0 scattering amplitude using the data on $\Delta\sigma_L(I = 0)$, Grein and Kroll[8] showed that the Argand plot of the amplitude has a resonancelike behavior around 1.5 GeV/c, and that suggests the existence of a spin-singlet dibaryon resonance.

2) C_{LL}(pp) and C_{SL}(pp) Measurements

Our group has collected a very large amount of data in proton-proton scattering, using a longitudinally (L) polarized target and a longitudinally (L) or transversely (S) polarized proton beam at the following seven different incident proton momenta: p_{lab} = 1.17, 1.35, 1.48, 1.70, 2.00, 2.25, 2.50 GeV/c, and data were obtained in the channels pp → pp, π^+d covering essentially the whole kinematic range of angles in these two reactions, namely $\theta_{c.m.} \sim 10^0 - 90^0$ for pp-events. The data analysis is in progress. These data are compared with predictions from phase-shift analyses by Hoshizaki and independently Arndt which give solutions consistent with the existence of two diproton resonances, namely the $^1D_2$ (2140) and $^3F_3$ (2220).

3) A Close Investigation of $\Delta\sigma_L(I = 1)$ and $\Delta\sigma_T(I = 1)$ Data

The existence of $^1D_2$ and $^3F_3$ dinucleon resonance in the I = 1 system has been discussed by many authors.[9] Here we discuss additional structure obtained with I = 1 system. $\Delta\sigma_T$ data[4] contain only singlet structures. Then we expect to see only the triplet structure in $(\Delta\sigma_T - \Delta\sigma_L)$.[1] We observe a new triplet structure at 2.0 GeV/c in addition to the one at 1.5 GeV/c. We can deduce that the quantum number of the triplet peak at 2.0 GeV/c is as follows:

$$\Delta\sigma_T - \Delta\sigma_L \sim (2J + 1)\mathrm{Im}\, R_{JJ} - (J + 2)\mathrm{Im}\, R_{J+1,J} - (J - 1)\mathrm{Im}\, R_{J-1,J}$$

only the R_{JJ} term has positive sign, therefore the triplet peak at 2.0 GeV/c is due to a partial wave R_{JJ}.

4) A Close Investigation of Polarization Data

We note that there is no $^3F_3$ partial-wave contribution to the polarization data at $\theta_{c.m.} = 63°$. We see an interesting structure in a plot of $k^2P\ (d\sigma/d\Omega)/\sin 2\theta_{c.m.}$ vs. p_{lab}. This quantity is proportional to $(2\,\mathrm{Im}^3P_0 + 3\,\mathrm{Im}^3P_1)(\mathrm{Re}^3P_2) - (2\,\mathrm{Re}^3P_0 + 3\,\mathrm{Re}^3P_1)(\mathrm{Im}^3P_2)$

by neglecting higher partial waves. Since the $^3P_2$ partial wave has very little energy dependence,[10,11] there is a good chance that either $^3P_0$ or $^3P_1$ partial wave is resonating at ~ 1.3 GeV/c.

5) Conclusion on I = 0 and I = 1 Resonances

Candidates for dibaryon resonances that can couple to nucleon-nucleon systems are summarized in the table below.

Candidates for Dinucleon Resonances

i) I = 1 Isospin State

	B_1^2 (2.14)	B_1^2 (2.18)	B_1^2 (2.22)	B_1^2 (2.43)	B_1^2 (2.43)
Mass, GeV	2.14 - 2.17	2.18 - 2.20	2.20 - 2.25	2.43 - 2.50	2.43 - 2.50
Width, MeV	50 - 100	100 - 200	100 - 200	~150	~150
Quantum State	$^1D_2$	Triplet P ?	$^3F_3$	probably $^1G_4$	Triplet R_{JJ} ?

ii) I = 0 Isospin State

	B_0^2 (2.14)	B_0^2(2.22)	B_0^2 (2.43)
Mass, GeV	2.14 - 2.17	2.20 - 2.26	2.40 - 2.50
Width, MeV	50 - 100	100 - 200	
Quantum State	Triplet ?	$^1F_3$	Triplet ?

References

1. I. P. Auer et al., paper sumbitted to this conference (paper number 581).
2. A. Yokosawa to be published in Phys. Report (1980).
3. I. P. Auer et al., Phys. Rev. Lett. 41, 354 (1978).
4. W. deBoer et al., Phys. Rev. Lett. 34, 558 (1975).
   E. K. Biegert et al., Phys. Lett. 73B, 235 (1978).
5. D.Underwood et al., Bull. of Am. Phys. Soc. 24, 636 (1979),to be published.
6. T. J. Devlin et al., Phys. Rev. D8, 136 (1973).
7. N. Hoshizaki et al., Proceedings of the 1979 INS Symposium, Tokyo, Japan (November 1979).
8. Private communication.
9. K. Hidaka and A. Yokosawa, Surveys in High Energy Physics, London, England (1979), and references contained therein.
10. N. Hoshizaki, Prog. Theor. Phys. 60, 1796 (1978); 61, 129 (1979).
11. R. A. Arndt, Talk given during LAMPF Nucleon-Nucleon Workshop, July, 1978; also private communication (1979).

HADRON SPECTROSCOPY, THEORY

R. JAFFE, MIT ORGANIZER

MULTIQUARK STATES AND EXOTICS

Anthony J. G. Hey
Physics Department
University of Southampton
Southampton, England SO9 5NH

ABSTRACT

The P-matrix approach to low-mass multiquark spectroscopy is reviewed, along with a critique of "colour chemistry" models for high-mass states. These questions are discussed in most detail for the $Q^2\bar{Q}^2$ sector but the $Q^4\bar{Q}$ and Q^6 sectors are briefly mentioned. The possibility of a Q^3 Glue exotic is also discussed.

1 LOW MASS $Q^2\bar{Q}^2$ STATES AND THE P-MATRIX

The first detailed predictions for low mass $Q^2\bar{Q}^2$ states were made in 1977 by Jaffe[1] in the context of the MIT Bag. In this model, the mass of a state has three types of contribution

$$M = \sum_{Q,\bar{Q}} E_Q + U + H_{gluon} \tag{1}$$

corresponding to kinetic energy, gluon confining energy, U, and a perturbative correction from one gluon exchange between quarks, H_{gluon}. Detailed calculations show that the dominant contribution to H_{gluon} is a short-range 'magnetic' colour-spin interaction of the form

$$H_{gluon} \sim \alpha_s \sum_{i>j} \underline{\lambda}_i \cdot \underline{\lambda}_j \; \underline{S}_i \cdot \underline{S}_j \tag{2}$$

where α_s is the QCD fine structure constant, and the factors $\underline{\lambda}_i \cdot \underline{\lambda}_j$ and $\underline{S}_i \cdot \underline{S}_j$ describe the colour and spin couplings, respectively, of the quarks i and j. For multiquark states, this colour-spin factor can produce large splittings. The predicted masses for the ground state Q^2Q^2 mesons range from 650 MeV to 2350 MeV. The interaction also has the intriguing property that the lightest states have non-exotic SU(3) flavour quantum numbers - Jaffe's "crypto-exotic" states. For decay widths, the Bag model can only make qualitative statements. In general, one expects these states to be much broader than "normal" $Q\bar{Q}$ or Q^3 states since they can "fall apart" by quark rearrangement into two $Q\bar{Q}$ mesons - a so-called "Zweig superallowed" transition (Figure 1).

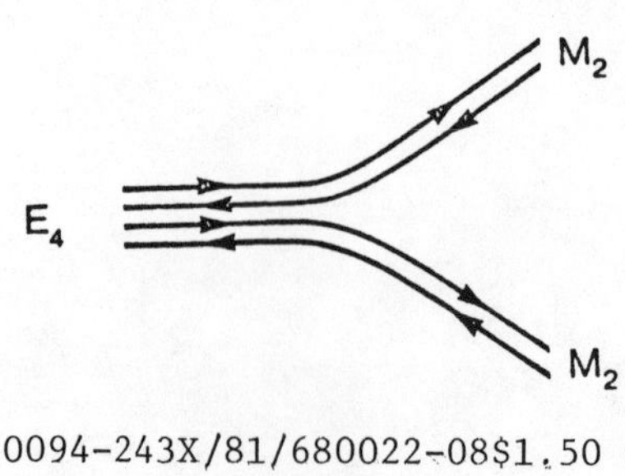

Figure 1: Quarkline diagram for Zweig "superallowed" decay of $Q^2\bar{Q}^2$ state.

On this basis, Jaffe identified the observed 0^{++} mesons around 1 GeV, previously supposed to be $Q\bar{Q}$ states with orbital angular momentum L=1, with his $Q^2\bar{Q}^2$ states:

$$\frac{1}{\sqrt{2}}(u\bar{u} + d\bar{d})\, s\bar{s} \quad \sim S^*\ (993)$$

$$u\bar{d}s\bar{s},\ \text{etc} \quad \sim \delta\ (976) \qquad (3)$$

$$u\bar{u}d\bar{d} \quad \sim \varepsilon\ (700)$$

The ε is expected to be broad, because of its fall-apart decay to $\pi\pi$, while the S* and δ states must couple to $K\bar{K}$ and are narrow only because of the closeness of $K\bar{K}$ threshold. Despite this encouraging start, the Bag also predicts an $I = \frac{1}{2}$, 0^{++} κ around 900 MeV and a genuine $I = 2$, 0^{++} flavour exotic at about 1100 MeV. The relevant phase-shifts are certainly non-resonant at these energies. What are we to make of this conflict?

Jaffe and Low[2] have re-examined the question of multiquark states and conclude that "masses" calculated for $Q^2\bar{Q}^2$ configurations in the Bag will not in general, correspond to resonant phase shifts. Their starting point is the observation that $Q^2\bar{Q}^2$ colour singlet states contain projections on to two types of $(Q\bar{Q})$ $(Q\bar{Q})$ subsystems

$$[Q^2\bar{Q}^2]_{1_c} \begin{cases} \nearrow [(Q\bar{Q})_{8_c} - (Q\bar{Q})_{8_c}]_{1_c} \\ \searrow [(Q\bar{Q})_{1_c} - (Q\bar{Q})_{1_c}]_{1_c} \end{cases} \qquad (4)$$

The component with $(Q\bar{Q})$ coupled to an octet of colour is expected to be confined; but the component with colour singlet $(Q\bar{Q})$'s should not experience strong binding forces. In this sense, a piece of the $Q^2\bar{Q}^2$ wavefunction has been "artificially" confined in the Bag calculation, instead of being treated as an open decay channel. In the absence of a detailed decay model, Jaffe and Low introduce a quantity they call the P-matrix, whose poles are expected to correspond to the energies calculated in the Bag.

The basic physical idea is that meson-meson scattering in the centre of mass can be divided into two regions - an "inside", where quarks and gluons are the relevant dynamical variables, and an "outside", where interactions between colour singlet hadrons are comparatively negligible. The radius of the boundary between these two regions may be estimated by matching the effective 2-body density of the $(Q\bar{Q})$ - $(Q\bar{Q})$ subsystems in the $Q^2\bar{Q}^2$ Bag state to that of the free meson-meson wavefunction. The effective 2-body density vanishes as the separation approaches twice the radius of the Bag, 2R: this vanishing is identified with the first zero of the meson-meson wavefunction at some distance b, of the order of R. Explicitly, Jaffe and Low find the relation $b \simeq 1.4R$ for $Q^2\bar{Q}^2$ systems. Now, the Bag eigenstate for the $Q^2\bar{Q}^2$ system, - what Jaffe and Low call the "primitive" - is obtained by insisting that all bag wavefunctions vanish for r = R. By the correspondence above, even in the case when the $Q^2\bar{Q}^2$ system has a large coupling to unconfined channels, the Bag calculations will describe meson-meson scattering at an energy where the meson-meson wavefunction has a zero at r = b(R). The connection between meson-meson scattering and Bag "primitives" is therefore formulated by means of a <u>boundary condition</u>. The P-matrix is the quantity whose poles correspond to the vanishing of the meson-meson wavefunction at r = b, and is calculable in terms of the experimentally determined phase shifts. The energies at which these poles occur are then to be identified with the calculated Bag "primitives". The magnitude of the residue of the P-matrix pole is related to the coupling of the $Q^2\bar{Q}^2$ system to open channels. A small residue corresponds to the usual resonant phase shift behaviour and thus would be classified as a resonance by the Particle Data Group. Whether or not other P-matrix poles, corresponding to non-resonant phase shifts and large coupling to open channels, are included in the Data Tables becomes, from this point of view, a matter of taste!

The explicit connection between the primitives and scattering phase shifts is best illustrated by an example from non-relativistic scattering theory. An S-wave interaction between colour singlet systems is modelled by a weak square-well potential with no bound states or resonances (Figure 2(a)). On the other hand, the Bag approximation,

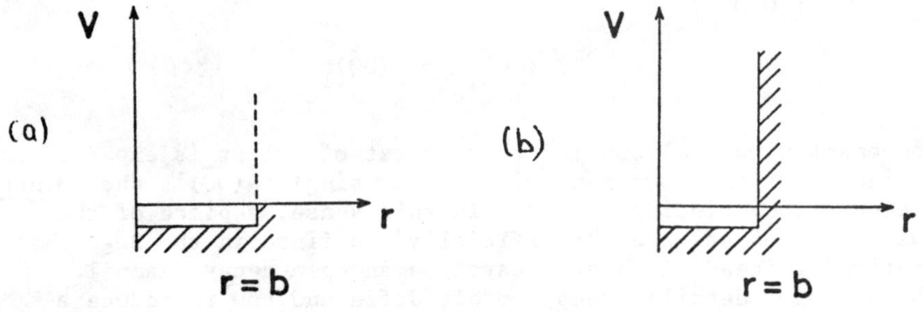

<u>Figure 2</u>: $(Q\bar{Q})$, - $(Q\bar{Q})$, potentials: (a) Realistic potential and (b) Bag "approximation"

which has confined the colour singlet $(Q\bar{Q})$'s, is analogous to the infinite square-well potential of Figure 2(b). What is the connection between these two potential problems? In the case of the weak potential, the scattering phase shift is found by solving the Schroedinger equation for r < b and for r > b, and matching $\psi'(r)/\psi(r)$ at r = b. One obtains the condition

$$q \cot qb = k \cot (kb + \delta(k)) \tag{5}$$

where $q^2 = k^2 - 2mV$. For the infinite square-well, on the other hand, we find an infinite sequence of bound states - the "primitives" - whose positions are determined by the imposition of the boundary condition $\psi(b) = 0$. This corresponds to poles in the logarithmic derivative

$\psi'(r)/\psi(r)$ at $q_n b = n\pi$. However, from the definition of the phase shift given above, we see that these energies q_n, of the infinite well, can also be found by constructing the quantity

$$P(k) = k \cot (kb + \delta(k)) \tag{6}$$

with the phase shift of the weak square-well, and determining the positions of the poles.

This is, in essence, the P-matrix of Jaffe and Low - although things are considerably more complicated when generalized to the multichannel situation, including both open and closed channels. The upshot of this is that multiquark "primitives" calculated in the Bag can, in certain cases, be associated with non-resonant phase shifts. In particular, this resolves the problem of the $\kappa(900)$. Furthermore, if one accepts this framework, there is evidence for a flavour exotic - the $I = 2$ 0^{++} primitive - corresponding to a repulsive phase shift, but a P-matrix pole in about the right place! (Figure 3). Jaffe and Low

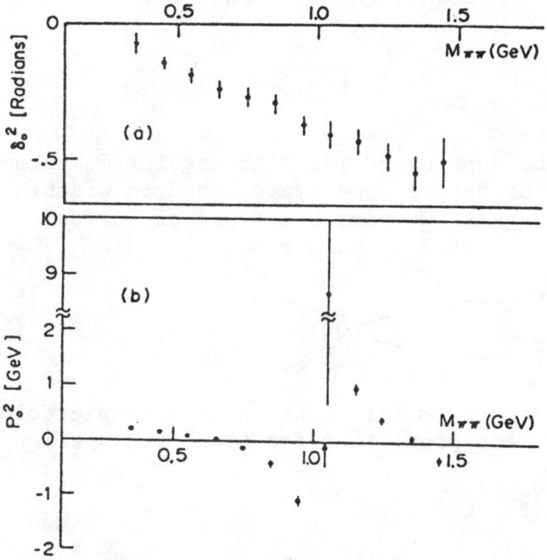

Figure 3: $\pi\pi$ $I = 2$ s-wave: (a) Phase shift and (b) P-matrix (the arrow marks the position of the pole)

also claim that the relative positions of these P-matrix poles are evidence for the colour-spin interaction from one-gluon-exchange.

It is clear that the P-matrix framework requires a disturbing rethinking of what is meant by a resonance: the absence of a resonant phase shift becomes evidence for a flavour exotic primitive. Nevertheless, disturbing or not, I believe that the basic physics is sound and that Jaffe and Low have identified an important and interesting interpretation of quark model calculations.

Two papers submitted to this conference are concerned with other ways of checking such $Q^2\bar{Q}^2$ assignments. Bramou and Masso[3] construct a pole model for $\eta' \to \eta\pi\pi$ and conclude that their results favour $\delta \sim (Q\bar{Q})$: Achasov et al.[4] perform a coupled channel analyses of the S* and δ region and conclude that $S^* \sim (Q^2\bar{Q}^2)$! Neither analysis seems compelling.

Another such attempt is by Aaron and Longacre[5] who look at the 1^+ $\rho\pi$ channel: they find two P-matrix poles but only one S-matrix pole, and suggest this is evidence for a 1^+ $(Q^2\bar{Q}^2)$ primitive. I await a copy of their paper with interest.

2 HIGH MASS $Q^2\bar{Q}^2$ STATES AND THE DEMISE OF COLOUR CHEMISTRY

For high mass, high angular momentum states, Johnson and Thorn[6] showed that round Bags were inappropriate, and a long, thin configuration was preferred. For baryons, this leads to a $Q - Q^2$, quark-diquark structure; for $Q^2\bar{Q}^2$ mesons, it suggests a diquark-antidiquark "molecule" as shown in Figure 4. Recombination into $(Q\bar{Q})$ colour

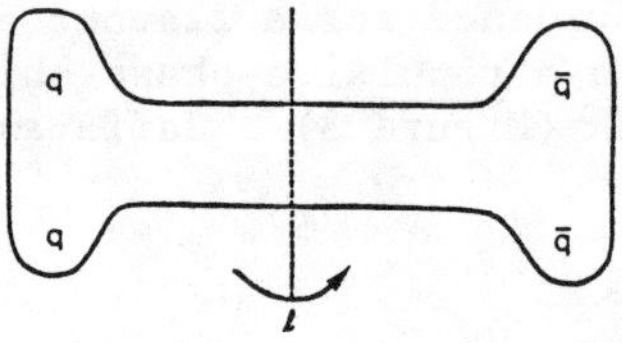

Figure 4: Diquark picture of $Q^2\bar{Q}^2$ meson: the MIT Bone

singlet pairs is expected to be suppressed for high angular momentum ℓ, so these states have a chance of having some typical hadron width. In fact, colour considerations suggest two possible types of baryonium, corresponding to the diquark in 6 or $\bar{3}$ of colour

$$(QQ) \begin{array}{c} 6 \;—\; \bar{6} \\ \bar{3} \;—\; 3 \end{array} (\bar{Q}\bar{Q}) \qquad (7)$$

Furthermore, a $(Q\bar{Q})$ creation mechanism for decay, plus a simple colour argument, suggested very different properties for these two species of "baryonium".

T-Baryonium: $(Q^2)_3 - (Q^2)_{\bar{3}}$

- Unusually large coupling to $B\bar{B}$ channels
- Typical hadronic widths $\Gamma \sim 100$ MeV

M-Baryonium: $(Q^2)_6 - (\bar{Q}^2)_{\bar{6}}$

- Couplings to $B\bar{B}$ suppressed by colour
- small widths $\Gamma \sim 10$ MeV.

The promising appearance of narrow $B\bar{B}$ status in production led advocates of this "colour chemistry" approach to make detailed phenomenological predictions and comparisons[7]. It is now well-known that experiments designed to confirm and find new narrow states, in fact found nothing[8]. On the other hand, the detailed phase shift analyses of $pp \to \pi\pi$ in the T, U and V regions find many possible resonant states not inconsistent with T-Baryonium[9]. What can we conclude? Alas, only that colour chemistry advocates were unreasonably optimistic, both in theory and phenomenology. For example[9], attempts have been made to construct detailed potential models of baryonium using non-relativistic potentials from baryon spectroscopy. Such models typicaaly find no marked diquark clustering for M-baryonium, even at rather large values of ℓ, so that M-baryonium is unstable against fall-apart decays. In any event, the naive first guesses about the dynamics of colour molecules are obviously wrong.

3 $Q^4\bar{Q}$ STATES

Strottman[10] has investigated low-mass $Q^4\bar{Q}$ states in the MIT Bag. From the previous discussion, these predictions must be interpreted as "primitives": Roisnel[11] has carried out a P-matrix analysis of $\frac{1}{2}^-$ states in πN, KN and $\bar{K}$N. He finds P-matrix poles in general agreement with Bag primitives.

For high mass states, it seems irresponsible to continue to canvas predictions of the colour molecule approach, after their unmitigated disaster for the $Q^2\bar{Q}^2$ sector. I shall therefore not review papers that do just that. However, there is evidence for a narrow Y* at a mass of 3.17 GeV with a width of less than 20 MeV[12]. But one cannot base a whole spectroscopy on just one state.

4 Q^6 STATES

I have time only for some rapid remarks:

(i) For the low mass Q^6 states, Jaffe and Shatz[13] have made a P-matrix investigation. In particular, they observe that weakly bound states, such as the deuteron, correspond to zeroes of the P-matrix.

(ii) For high ℓ states, one must understand the failure in the $Q^2\bar{Q}^2$ sector, before one can have any confidence in colour molecule predictions for Q^6.

(iii) A recent paper by Harvey[16] adds some controversy to the question of the Q^6 contribution to the NN-potential, in that he finds no repulsive core, unlike previous calculations[15]. An interesting feature of his calculation is that in his harmonic oscillator approximation, the ground state does not correspond to a configuration with all quarks in the lowest S-state $(S)^6$. Since the colour-spin interactions can cause a splitting greater than the S-P state splitting, his preferred ground state configuration is (S^4P^2).

5 (Q^3 GLUE) 'EXOTIC'

In 1976 Cutkosky[16] suggested that the ΔD35 (1925) resonance belonged to a $[\underline{56}, 1^-]$ multiplet which was due to the excitation of (Q^3) relative to a gluonic "core" in the baryon. In both SU(6) mass fits and explicit string models of baryons[18] it proved difficult to produce a genuine Q^3 [56, 1^-] so low in mass. This conclusion has recently been re-examined[19] in the framework of Isgur and Karl's non-relativistic quark model, which has had spectacular success in resolving many old problems in baryon spectroscopy[20]. In fact, although their papers are concerned only with the harmonic oscillator N=0, 1 and 2 bands, the same parameters also predict the mean masses of three of the N=3 band multiplets, the [56, 1^-] included! The numerical predictions are shown in Figure 5. The disappointing conclusion is that the ΔD35 (1925) is easily compatible with an ordinary Q^3 state.

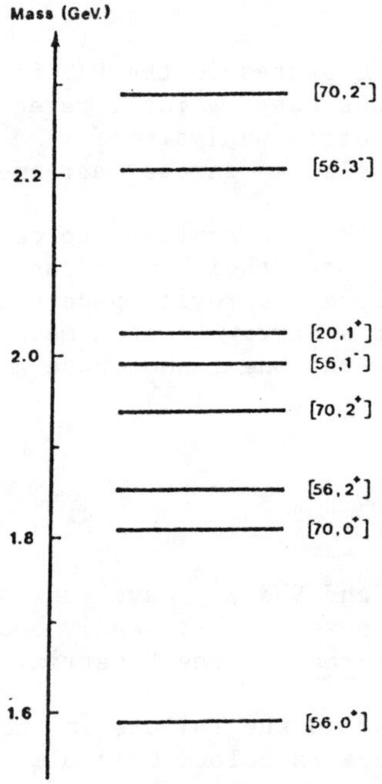

Figure 5: Mass spectrum of the N=2 and some N=3 multiplets

6 CONCLUSIONS

(i) We have seen that the naive calculations of multiquark states either need care in interpretation, as in the P-matrix approach, or make too simplistic guesses for molecular structure.

(ii) A final word of warning. In the spherical, static approximation to the Bag model all quarks are treated as independent particles: Predictions can very simply be generated for $Q\bar{Q}$, Q^3, $Q^2\bar{Q}^2$, Q^6 and so on. In QCD, however, with the conventional string picture of confinement, it is not at all obvious that there need be a simple connection between the forces in these difference sectors (Figure 6).

Figure 6: String picture of hadrons

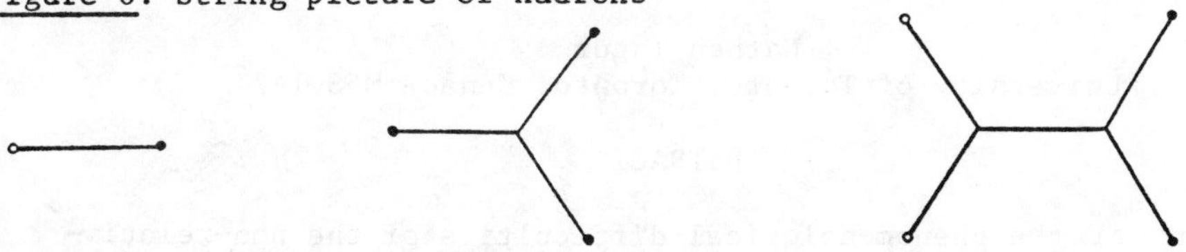

(a) $Q\bar{Q}$ (b) Q^3 (c) $Q^2\bar{Q}^2$

REFERENCES AND FOOTNOTES

1. R L Jaffe, Phys Rev D15, 267 (1977).
2. R L Jaffe and F E Low, Phys Rev D19, 2105 (1979).
3. A Bramon and E Masso, paper submitted to this conference.
4. N N Achasov, S A Devyanin and G N Shestakov, paper submitted to this conference.
5. R Aaron and R Longacre, private communication.
6. K Johnson and C Thorn, Phys Rev D13, 1934 (1976).
7. It would be uncharitable to list all these papers.
8. See the experimental reviews of the 1980 International Conference on Experimental Meson Spectroscopy, to be published by AIP.
9. For more details, see the review by this author in the conference proceedings cited in reference 8.
10. D Strottman, Phys Rev D20, 748 (1979).
11. C Roisnel, Phys Rev D20, 1646 (1979).
12. J Amirzadeh et al., Phys Lett 89B, 125 (1979).
13. R L Jaffe and M P Shatz, Caltech preprint CALT-68-775 (1980).
14. M Harvey, Chalk River preprint (1980).
15. D Liberman, Phys Rev D16, 1542 (1977); C DeTar, Phys Rev D17, 323 (1978); M Oka, PhD Thesis, Tokyo University 1979; C S Warke and R Shanker, Phys Rev C21, 2643 (1980).
16. R E Cutkosky, contribution to the 1976 Oxford Conference on Baryon Spectroscopy, edited by R T Ross and D H Saxon.
17. R H Dalitz, R R Horgan and L J Reinders, J Phys G, L195 (1977).
18. R E Cutkosky and R E Hendrick, Phys Rev D16, 793 (1977).
19. K C Bowler, P J Corvi, A J G Hey and P D Jarvis, Phys Rev Lett 45, 97 (1980).
20. For reviews of this model, see the proceedings of the 1980 Toronto Conference on Baryon Spectroscopy, to be published.

BARYONS WITH CHROMODYNAMICS

Nathan Isgur
University of Toronto, Toronto, Canada M5S 1A7

ABSTRACT

Many of the phenomenological difficulties of the non-relativistic quark model for baryons are overcome when some current prejudices from chromodynamics about quark forces are imposed. The effects of flavour independent confinement, symmetry breaking through quark masses, and colour hyperfine interactions are most prominent, leading to a satisfactory understanding of both the spectroscopy of low-lying baryons and of the signs and magnitudes of baryon couplings. The previously worrisome absence in partial wave analyses of a large number of the states expected in the non-relativistic quark model is explained in terms of decouplings of the resonances from their elastic channels.

INTRODUCTION

I have just come from Baryon 1980 in Toronto where we agreed that baryons are much too little appreciated. One does of course pay a price theoretically in having to deal with a three body system, but in return one gains immensely from the fact that baryons are known experimentally much better than mesons. Unlike the old mesons where the first orbital excitations are only now beginning to be sorted out, the data tables are filled with hundreds of baryons; because they can (usually) be produced as s-channel resonances their properties are also quite well known with each resonance typically having three or four signed decay amplitudes measured. There is a great deal of information on quark dynamics contained in all of this data.

Though I have (mercifully) been commissioned only to speak about the recent work in which I have been directly involved --- mostly in collaboration with Gabriel Karl and Roman Koniuk --- I must begin by recalling the origins of modern quark models of baryons in the work of Dalitz, Greenberg, and their contemporaries who, in the absence of a detailed dynamical scheme, analyzed the baryon spectrum in terms of a general SU(6) tensor structure. These models provided powerful evidence for the quark model. Successful as these models were, however, they had several faults:

1. without a concrete dynamical basis, many parameters had to be introduced to describe each SU(6) multiplet
2. the decay amplitudes of the states were sometimes in conflict with the data
3. (perhaps most serious) a great number of the states expected by the model were not seen

The work I am going to describe takes this old framework and applies to it prejudices about quark dynamics from QCD. These prejudices are that

1. the quarks are confined in a flavour and spin independent potential

2. at short distances the quarks are perturbed by one gluon exchange

3. all flavour symmetry breaking arises exclusively from (constituent) quark mass differences

In concluding this section I would like to anticipate possible misgivings about our use of a non-relativistic framework. I especially feel the need to make such comments in the presence of my relativistic chairman. Our approach, as will be seen in the following, has been to rely as little as possible on the numerical consequences of assuming non-relativistic dynamics. We have tried instead to use this framework to extract the dominant physical effects in a simple context free of the difficulties of the relativistic bound state problem. We believe that most of these effects will survive an eventual 'relativisation'. (Compare, e.g., the physics here and in the bag for the ground states.)

THE HAMILTONIAN AND ITS SOLUTION

The actual Hamiltonian we use is extremely simple, namely,

$$H = \sum_i (m_i + \frac{p_i^2}{2m_i}) + \sum_{i<j} (V^{ij} + H^{ij}_{hyp}) \qquad (1)$$

where V^{ij} is a flavour and spin independent potential which one may visualise as containing both the confining and Coulomb potentials and where H^{ij}_{hyp} is the colour hyperfine interaction depicted in Figure 1. We solve this system approximately by writing $V^{ij} = 1/2kr^2_{ij} + U(r_{ij})$; we then treat the anharmonic terms U and H^{ij}_{hyp} perturbatively in the harmonic basis. It should be noted that one gluon exchange spin orbit terms have been discarded from (1) based on a) their apparent absence in the data (e.g. $\Delta 7/2^+ \simeq \Delta 5/2^+ \simeq \Delta 3/2^+ \simeq \Delta 1/2^+$), b) the recognition that they tend to cancel with the effects of Thomas precession in the confinement potential, and c) the success of this latter cancellation in the charmonium spectrum.

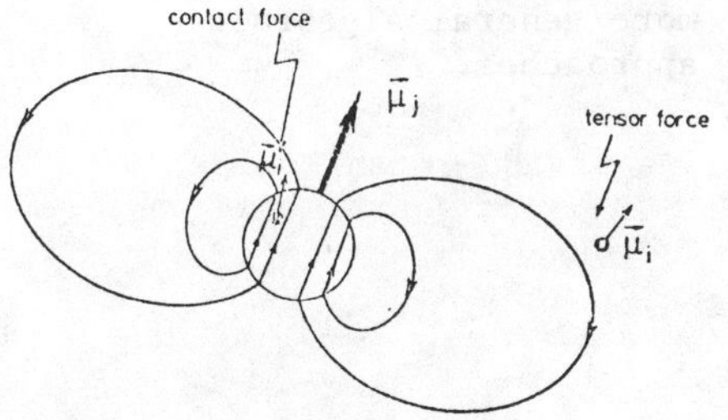

Fig. 1. A representation of the origin of the tensor and contact parts of the hyperfine interaction.

For equal mass quarks in the harmonic limit the spectrum has at N = 0 a $[56,0^+]$, at N = 1 a $[70,1^-]$, and at N = 2 degenerate $[56',0^+]$, $[70,0^+]$, $[56,2^+]$, $[70,2^+]$, and $[20,1^+]$ SU(6) supermultiplets, with equal spacings between these three N-values. Surprisingly, it is possible to show that in the presence of an arbitrary anharmonic perturbation U the energies of these seven SU(6) supermultiplets always follow the pattern shown in Figure 2. This allows us to avoid commitment to a specific potential V^{ij} as well as a heavy reliance on non-relativistic dynamics by simply choosing E_0, Ω, and Δ of Figure 2 as our "potential parameters".

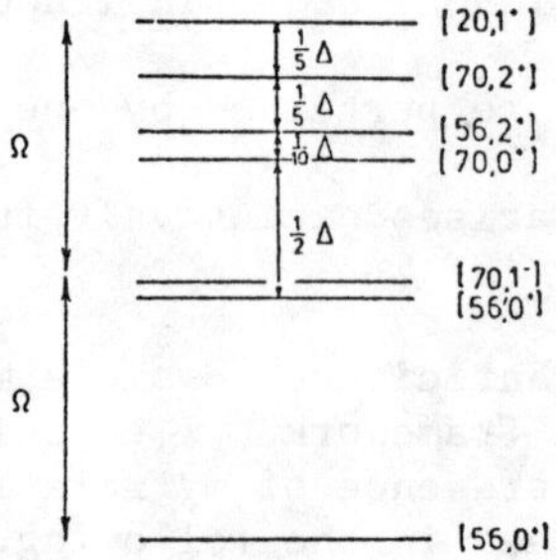

Fig. 2. The pattern of low-lying supermultiplets from first order perturbation theory.

When the quarks do not have equal masses, as in the strangeness −1 sector, there is a peculiar and vital new feature of the solutions to the confinement problem. In this case the two previously degenerate modes of the three body oscillator (corresponding to excitations of the variables $\vec{\rho} \equiv \frac{1}{\sqrt{2}}(\vec{r}_1 - \vec{r}_2)$ and $\vec{\lambda} \equiv \frac{1}{\sqrt{2}}(\vec{r}_1 + \vec{r}_2 - 2\vec{r}_3)$) are split; the result is that SU(3) is "maximally" violated, somewhat in analogy to the $\omega - \phi$ system.

The final step in the solution of (1) is then to turn on the hyperfine interactions and diagonalize the resulting matrices of fixed isospin and J^P.

THE MODEL FOR BARYON COUPLINGS

At this point the model can provide predictions for the masses and quantum numbers of the low-lying baryons, but has little to say about their properties until supplemented by a model for their couplings to photons and mesons. We adopted the simplest such model, depicted in Figure 3, and applied it in as unprejudicial a way as we could; in this form it has much in common with the more general algebraic approaches.

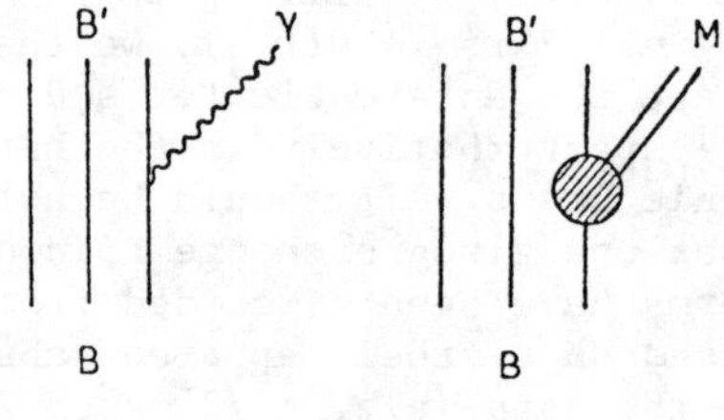

Fig. 3. a) The $B \to B'\gamma$ coupling b) The $B \to B'$+ meson coupling.

THE RESULTS AND DISCUSSION

The results of this full program are partly summarised in Figures 4 and 5 which compare the predicted spectrum of low-lying

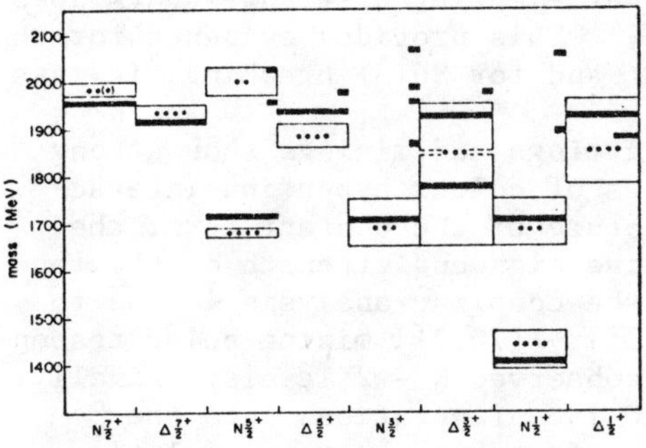

Fig. 4. The S = 0 positive parity excited baryons.

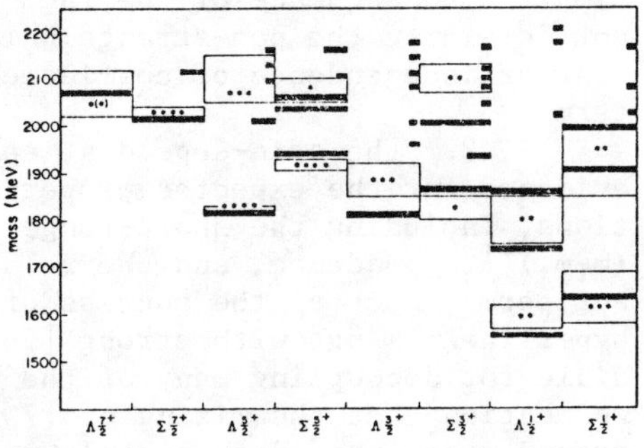

Fig. 5. The S = -1 positive parity excited baryons.

positive parity states to experiment; relatively weakly coupled resonances are denoted by "stubs" instead of full bars, showing a remarkable correspondence between the observed states and those which ought to be observable. In addition, detailed examination of the couplings of observed states are in reasonable agreement with experiment. Table I shows two examples from the hundreds in

Table I. The Decay Amplitudes of S11(1535) and S11(1650)

	S11(1535)		S11(1650)	
	theory	experiment	theory	experiment
Nπ	5	8±3	9	9±2
Nη	+ 5	+ 9±2	− 2	− 2±1
ΣK	no	no	− 2	± 2±1
ΛK	no	no	− 3	− 4±1
Δπ	− 2	(−) 1±1	− 8	− 4±2
γp	+145	+ 80±20	+90	+50±15
γn	−120	−110±35	−35	−45±25

our tables. The agreement is consistent with the very rough nature of the model, but seems to indicate that the dominant features of baryon dynamics have been taken into account.

Finally, we would like to try to highlight those aspects of the results which we consider to be evidence for the QCD-motivated components of the model:

1. the observed value of the "potential parameter" Δ indicates the presence of an attractive short range anharmonicity, con-

sistent with the expected Coulomb-like term,

2. the non-degeneracy of the ρ and λ modes is not only responsible for observed spectroscopic effects (like $\Lambda 5/2^- > \Sigma 5/2^-$) but also for many of the $\bar{K}N$ decouplings in the S = -1 sector (six out of seven $\Lambda 3/2^+$ states, for example). (These decouplings are simply a consequence of the fact that ejecting a strange quark does not de-excite the non-strange pair.) This provides evidence for the flavour independence of confinement and for SU(3) breaking via mass terms,

3. the spin-dependent splittings and mixings show strong evidence for the expected properties of colour hyperfine interactions, including the short-range nature of the contact term, the $(m_i m_j)^{-1}$ dependence, and the relative sign and strength of the tensor term. Much of the success of the coupling analysis is due to hyperfine mixing, with strong $[56,2^+] - [70,2^+]$ mixing being responsible for decoupling many of the unobserved N = 2 levels. Finally, we mention that the mixing of $[70,0^+]$ configurations into the ground state provides an explanation for many observed violations of SU(6) symmetry in baryon couplings.

ACKNOWLEDGEMENTS AND APOLOGIES

I would like to acknowledge once again the collaboration of Gabriel Karl and Roman Koniuk in the work on which I have reported. Les Copley, Kuang-Ta Chao, and Kim Maltman have also been involved in related papers. I have learned a great deal about this subject from discussions with and papers by many people whose work I have been unable to mention adequately here, including R.H. Dalitz, O.W. Greenberg, G. Morpurgo, L.J. Reinders, D. Gromes, P. Fage, H.J. Schnitzer, H.J. Lipkin, D. Faiman, A.W. Hendry, and J. Rosner.

REFERENCES

This subject is too old and too broad for it to be possible for me to present here an adequate set of references. However, references to the work summarized here as well as to related work can be found in the recent baryon coupling paper by Koniuk and myself published in Phys. Rev. D21, 1868 (1980).

DIFFICULT STATES IN THE QUARK MODEL: GLUEBALL AND THE PION

John F. Donoghue
Massachusetts Institute of Technology, Cambridge, MA 02139

ABSTRACT

Work on the spectroscopy of glueballs and on the pion is reviewed.

INTRODUCTION

Quark models are now mature enough to start confronting the really hard questions. In this talk, I will give mini-reviews of some efforts on two such topics-glueballs and the pion. Due to space limitations my comments will tend to be compact; hopefully the references cited will satisfy further curiosity.

GLUEBALLS

It is widely believed that the spectrum of QCD includes bound states of gluons-glueballs. However, there is less unanimity as to their properties. There are some common expectations for glueball spectroscopy,[1-8] and I will review these first. Then I go through a series of comments, elaborations and other points of view in order to illuminate the difficult aspects of glueballs. The hope is to convince some theorists that there are interesting and worthwhile areas for further study in this system. I have elsewhere reviewed the more phenomenological aspects of glueballs.[7]

There are two common ways to treat gluons in bound states. One is to consider color electric and magnetic fields in a region of space,[1, 3] essentially as a boundary condition problem in E and M. In this method the various modes can be classified as transverse electric (TE) or transverse magnetic (TM). With the confining boundary conditions of the bag model, the lowest mode is TE, ℓ=1. The other method envisions what can be called "lumps of glue", i.e., the gluon as a massive spin one particle in a spin independent potential.[2, 4] The problem is then one of combining together what could be viewed as colored ω or ϕ mesons in various orbital and spin states.

Both methods agree on the two gluon sector. All states here will have positive charge conjugation. The ground states will consist of 0^{++} and 2^{++}. In the first excited state one can form 0^{-+}, 1^{-+}, 2^{-+}. (Note, however, that later on I will argue that the 1^{-+} does not belong here.) Many more excited states can be formed through radial and orbital excitations, although, until the low lying states are found, these are of only academic interest.

The three gluon sector is more subtle, and much of the literature is confused on this subject.[8] Color singlets can be formed with the f_{ABC} or d_{ABC} coefficients. Because of the intrinsic odd charge conjugation, states formed with f_{ABC} will have C=+1 while those with d_{ABC} will have C=-1. In the bag model the lowest mode is TE, which is dominantly a magnetic field, a 1^+ mode. The ground state of these gluons are 0^{++}, 1^{+-} and 3^{+-}. In contrast, the potential models combine up three 1^- particles in an S wave. Their spectrum has the opposite parity: 0^{-+}, 1^{--} and 3^{--}. The reason for this difference is that for a massless gluon the $\ell=0$ mode (which would be 1^-) is the Coulomb mode which does not exist in the absence of sources. The first transverse mode is then either the $\ell=1$ TE mode (1^+) or $\ell=1$ TM (1^-). The requirement that no flux leave the surface ($\eta_\mu F^{\mu\nu}=0$) favors the TE mode. It is exciting that experiment can then provide a clear distinction between the bag model picture and the potential models.

A rough guide to the mass spectrum can be found by using the bag model without intergluon interactions.[1] Not surprisingly the spectrum then starts at 1 GeV. In particular the predictions are:

$$
\begin{aligned}
M\left((TE)^2 = 0^{++},\ 2^{++}\right) &= 960 \text{ MeV} \\
M\left((TE)(TM) = 0^{-+},\ 1^{-+},\ 2^{-+}\right) &= 1290 \text{ MeV} \qquad (1) \\
M\left((TE)^3 = 0^{++},\ 1^{+-},\ 3^{+-}\right) &= 1460 \text{ MeV}
\end{aligned}
$$

One interesting feature of the standard picture is the existence of a light exotic 1^{-+} state in the first excited multiplet of two gluons. Since it can not be formed by $q\bar{q}$ its presence would be good evidence for glueballs. However, I feel that this state is spurious, for the reasons to be discussed below.

There are good reasons to be cautious about naive models for glueballs. Potential models describe a massive field with 3 spin degrees of freedom instead of two for massless gluons. Bag models deal with a fixed bag which acts as an extra body in the problem. These limitations can lead to spurious states in the spectrum.

An alternative trial definition of a two gluon glueball is as a state that is a strong resonance in gluon-gluon scattering. Years ago Yang analyzed this situation for photons,[10] and found that the families (0^{++}, 2^{++}, 4^{++}...), (0^{-+}, 2^{-+}, 4^{-+}...) and (3^{++}, 5^{++}, 7^{++}...) were the only ones allowed by gauge and Lorentz invariance. In particular 1^{-+} cannot be formed by two real gluons, while its partners 0^{-+} and 2^{-+} can.

Yang's results require real gluons; however, there are other ways to obtain these results even for bound gluons. A useful technique in quark model physics to describe a quark state with a given set of quantum numbers is to form quark bilinears and then project out the appropriate state. For example, the fields

$$\bar{\psi}\gamma_5\psi \text{ or } p_\mu\bar{\psi}\gamma^\mu\gamma_5\psi$$
$$\epsilon^\mu\bar{\psi}\gamma_\mu\psi \text{ or } V_{\mu\nu}\bar{\psi}\sigma^{\mu\nu}\psi \tag{2}$$

with $V_{\mu\nu} = (p_\mu\epsilon_\mu - p_\nu\epsilon_\nu)$ project out the ground state pseudoscalar and vector states of two quarks. Similar methods can be applied to glueballs. Gauge invariance requires that we work with the field tensors $F^A_{\mu\nu}$: The only nonvanishing combinations with no derivatives are

$$\begin{aligned} 0^{++} &\sim F^A_{\mu\nu}F^{A\mu\nu} \\ 2^{++} &\sim \epsilon^{\mu\nu}F^A_{\mu\lambda}F^{A\lambda}{}_\nu \\ 0^{-+} &\sim F^A_{\mu\nu}\tilde{F}^{A\mu\nu} \\ 2^{-+} &\sim \epsilon^{\mu\nu}F^A_{\mu\lambda}\tilde{F}^{A\lambda}{}_\nu \end{aligned} \tag{3}$$

where $\tilde{F}^{\mu\nu} = \epsilon^{\mu\nu\alpha\beta}F_{\alpha\beta}$. The 1^{-+} candidate $(V_{\mu\nu}F^{A\mu\lambda}F^{A\nu}_\lambda)$ vanished by the symmetry of the Lorentz indices. Even with one derivative the 1^{-+} candidate $(\epsilon_\mu F^A_{\nu\lambda}\partial^\nu F^{A\lambda\mu})$ can be shown to vanish. Since my talk, Johnson has informed me of a third method of removing the 1^{-+} state in the bag model by use of the transversality of the gluon field and a bag model method for removing spurious states introduced by the fixed cavity.[11] All three methods use gauge and Lorentz invariance to say that the 1^{-+} does not belong with the 0^{-+} and 2^{-+} in the first excited state. It is possible to form 1^{-+} states with three gluons (or perhaps at high excitation with two gluons). In the bag model this state is in the first excited state of three gluons, and lies near 1.8 GeV. While such a state should be looked for, it will probably be much more difficult to find than a light two gluon state.

In $q\bar{q}$ states the spin-spin interaction, mediated by one gluon exchange, are responsible for splitting the ρ and π. Similar forces for gluons will split the multiplets of the above models. Because the gluon-gluon force is 9/4 of the $q\bar{q}$ force, we expect the splittings to be considerably larger in glueballs. Common expectation is that low spin states are pushed down in mass while high spin states move up. No systematic investigation has been done, but Thorn has studied the two gluon ground state.[12] He finds that the 2^{++} goes up to 1290 MeV, while the 0^{++} drops to 110 MeV! (With a center of mass correction[18] this would have $m^2<0$). This would appear to be disastrous; however, there is a reasonable way out. Another scalar state exists: the vacuum. When one finds such a low mass state one is suspicious that one has found a component of the vacuum. The two 0^{++} states must be orthogonalized, pushing the nonvacuum state up in mass. This program has not yet been carried out.

Giles has found a potential instability in the bag model glueballs.[13] The bag shape is usually assumed to be spherical. However, Giles found that a single gluon wavefunction in a cavity can also satisfy the bag boundary condition for a family of nonspherical surfaces, which could have lower total energy. This result does not always hold for two superimposed gluons so its consequences for glueballs is not yet understood.

Alternative descriptions of glueballs exist. For example Suura[6] uses a gauge invariant wavefunction of E and B fields at different space time points with links connecting them. He then derives a wave equation with a potential, similar to the Breit equation, for the spin zero sector. He finds 0^{++} and 0^{-+} degenerate, and when using a linear potential from light quark states bounds the mass to be less than 2 GeV. Adding a coulomb piece lowers this considerably and no lower bound can be obtained.

Another viewpoint is to study glueballs on a lattice.[5] A glueball can be formed by creating a closed path of flux linking lattice sites. The lightest configuration involves four sites-"a boxiton". Depending on orientations for the boxes, 0^{++}, 2^{++} and 1^{++} states may be formed. Kogut, Sinclair, and Susskind have calculated mass ratios to be

$$\frac{M(2^{++})}{M(0^{++})} = 1.003 \qquad \frac{M(1^{++})}{M(0^{++})} = 1.575 \tag{4}$$

Six link states (including 0^{-+} and 1^{-+}) are heavier. In principle the absolute glueball mass can be computed in lattice theories (by comparison to the string tension, for example), but this apparently is difficult and has not been done.

Finally, it seems appropriate to comment on the widths of glueballs. Some authors have favored widths which are the mean between standard quark widths and Zweig suppressed (a "square root of Zweig" suppressed) on the assumption that the OZI rule comes half from converting quarks into glue and half from turning the glue back into quarks. However I have never seen this argument made convincing, and it is particularly suspect for light glueballs. It is quite plausible that glueball widths are more or less typical of hadronic widths. In the limit of a large number of colors, N_C, glueballs are only narrower than quark states by a factor $1/N_C$. In the limit of a large number of light flavors, N_F, glueballs are broader by a factor of N_F. I have recently done a calculation which obtains the remaining space time factors by doing a bag model calculation using the P-matrix,[14] averaging over 16 levels of glueballs and 15 levels of quark states.[15] The resulting factor is of order two:

$$\frac{\Gamma(gg)}{\Gamma(q\bar{q})} = \frac{N_C}{N_C^2-1} N_F \quad .48 \tag{5}$$

Since the widths of quark states fluctuate more than a factor of three, all these numbers fall in a "typical" hadronic range.

To sum up, glueballs should exist and we think that we know some of their features. However, the subject needs much more theoretical work. We need more models for the spectrum, and a better understanding of the dynamics of mixing with $q\bar{q}$, decays, gauge invariance, and good phenomenology. In addition the role of glueballs in the known meson spectrum is an interesting subject which I don't have time to treat. Most importantly, experimenters need to find glueballs!

THE PION

It is well known that the pion is difficult to account for in the quark model. In the first approximation the mass of $q\bar{q}$ states ($\rho+\pi$) is roughly 2/3 of the nucleon's mass. Spin splittings from transverse gluon exchange make the pion the lightest state, but generally not light enough. In particular there is no natural limit where $M_\pi=0$. Worse than simply the problem of the light mass is the connection with chiral symmetry. QCD is nearly invariant under the chiral transformation (exact in the limit $M_u=M_d=0$), but the vacuum presumably is not. In this limit, a massless pion is required, with well defined chiral couplings. For M_q small we get a small pion mass

$$M_\pi^2 = (M_u+M_d)\langle\pi|\bar{q}q|\pi\rangle$$

Standard quark models do not contain the chiral symmetry and generally are not compatibles with it.

Perhaps the best paper on what a theory of a pion is all about is the 1961 article by Nambu and Jona Lasinio.[16] They considered a massless quark with a local four fermion interaction (with a cutoff). By summing bubble graphs they found that it became the theory of a <u>massive</u> quark and a <u>massless</u> bound state. They calculated the chiral couplings, and found the vacuum as a condensate of $q\bar{q}$ pairs. What is needed is a similar model in QCD which also ties in with other quark model calculations.

Pagels and Stokar[17] have attempted to do something like this using a Bethe-Salpeter formalism. They sum ladder graphs and use a dynamical quark mass $M(p^2) = 4M_D^3/p^2$. This allows them to calculate f_π and $F_\pi(Q^2)$ in terms of M_D

$$f_\pi = \left[\frac{2^{1/3}}{2\pi\sqrt{3}}\right]^{1/2} M_D \sim 83 \text{ MeV (vs 93 MeV)} \qquad (6)$$

$$Q^2F_\pi(Q^2) \underset{Q^2\to\infty}{\to} \frac{4\ell n^2}{3\pi(2)^{1/3}} \frac{M_D^4}{f_\pi^2} \sim .17 \text{ GeV}^2 \text{ (vs .38 GeV}^2\text{)}$$

The absolute numbers come from an estimate of M_D=244 MeV by Hagiwara and Sanda. The agreement is reasonably good. The drawbacks of this framework are that it doesn't explain the pion's relation to other states, and it disagrees with the perturbative QCD results on $F_\pi(Q^2)$. However its advantage is that it does incorporate the chiral properties of the pion.

From the other side of the fence the pion has been reconsidered in the bag model by Johnson and myself.[18] We argue that while the bag model does not contain the chiral symmetry, it is consistent with it. The point is that the bag model is really a guess at the vacuum structure of QCD, and the vacuum should not be chirally invariant. The bag has two forms of vacuum; where fields are strong (inside hadrons) one has the perturbative vacuum while outside of hadrons is the true vacuum. The order parameter is $\bar{q}q$, as in chiral theories, leading to the Lagrangian

$$L_{Bag} = (L_{QCD}-B)\theta(\bar{q}q) \qquad (7)$$

where B is the Energy/Volume difference between the two vacuums. Since the bag builds the vacuum into L_{Bag} it isn't chirally invariant but then again it shouldn't be. A massless pion will not be automatic; however if this is the vacuum structure of QCD, a massless pion should be possible. With old techniques[19] one had M_π=280 MeV. However standard quark model techniques do not apply to light states. We developed the appropriate techniques for handling light static states, through use of wave packets. Using these, one can easily obtain M_π=0. We feel that this should be imposed in the chiral limit to be consistent with QCD. Expanding about this limit, we can obtain the chiral perturbation formula

$$M_\pi^2 = (M_u+M_d)\langle\pi|\bar{q}q|\pi\rangle \qquad (8)$$

and evaluate it to obtain $1/2(M_u+M_d)$=33 MeV, M_s=330 MeV. Most chiral estimates[20] calculate a different mass $M_q^*=M_qZ$ where $Z=\langle H|\bar{q}q|H\rangle$. In the bag model $Z\simeq1/2$, so that $1/2(M_u^*+M_d^*)$ = 17 MeV and M_s^*=160 MeV, to be compared with the chiral SU(3) estimate of 7 MeV and 150 MeV. Note that even though we obtain the expansion Eq. 8 , we don't obtain the chiral SU(3) mass ratios. This is because the expansion fails for the kaon; M_K is larger than other scales in the problem and the expression for its mass is not the simple linear expansion. This may be correct and chiral SU(3) could fail. There is some indication of this from the sigma term.[15] We also calculate $f_\pi = .5/R_\pi \simeq .15$ GeV, which is closer than previous bag estimates and, most importantly, finite as $M_\pi \to 0$. The conclusion is that some hints of chiral symmetry seem to emerge from the bag model, although much more lies uncovered.

In a separate talk at this conference, R. Haymaker has described an approach which is somewhere between the two above viewpoints, done in collaboration with T. Goldman.

In summary, the pion is being brought slowly into line. In particular the mass turns out not to be a major problem, but the connection between chiral symmetry and quark models is not yet understood.

REFERENCES

1. R.L. Jaffe and K. Johnson, Phys. Lett. 34, 1645 (1976).
2. D. Robson, Nucl. Phys. B130, 328 (1977).
3. J.D. Bjorken, SLAC Summer Institute on Particle Physics (1979).
4. J.J. Coyne, P.M. Fishbane and S. Meshkov, Phys. Lett. 91B, 259 (1980).
5. J. Kogut, D.K. Sinclair and L. Susskind, Nucl. Phys. B114, 199 (1976).
6. H. Suura, Phys. Rev. Lett. 44, 1319 (1980). See also K. Ishikawa, Phys. Rev. D20, 731 (1979).
7. J.F. Donoghue, Invited talk at the VI International Conf. on Experimental Meson Spectroscopy, April 1980, Brookhaven National Lab., MIT preprint CTP #854, to be published in the proceedings.
8. Work on other aspects of glueballs is contained in H. Fritszch and P. Minkowski, Nuovo Cimento 30A, 393 (1975); P.G.O. Freund and Y. Nambu, Phys. Rev. Lett. 34, 1645 (1975); J. Willemsen, Phys. Rev. D13, 1327 (1976); P. Roy and T. Walsh, Phys. Lett. 78B, 62 (1978); V. Novikov, M. Shifman, A. Vainshtein and V. Zakharov, Nucl. Phys. B165, 55 (1980); ibid B165, 67 (1980); R.H. Capps, Purdue preprint (1980); H. Goldberg, Northeastern preprint (1980); C. Carlson, J. Coyne, P. Fishbane, F. Gross and S. Meshkov, contributions to this conference.
9. The confusion about three gluon states that I know of are: Jaffe and Johnson (Ref. 1) have the charge conjugation reversed, and list a spurious spin two $(TE)^3$ state, but have the other spins and parities correct. Robson (Ref. 2) is incorrect in criticizing Jaffe and Johnson's parity assignments. Bjorken (Ref. 3) lists states as if E fields have lower energy than B fields. With confining boundary conditions the reverse is true, and his tables should be read from the bottom up. In my previous talk (Ref. 7) I had incorrectly assumed that the TE mode was 1^- (I thank K. Johnson for correcting this misconception.).
10. C.N. Yang, Phys. Rev. 77, 242 (1950).
11. These methods will be discussed more fully in a paper in preparation by J.F. Donoghue, K. Johnson and B. Li.

12. C. Thorn, unpublished.
13. R. Giles, to be published.
14. R.L. Jaffe and F.E. Low, Phys. Rev. D19, 2105 (1979).
15. J.F. Donoghue, to be published.
16. Y. Nambu and Jona Lasinio, Phys. Rev. 122, 345 (1961); 124, 246 (1961).
17. H. Pagels and S. Stokar, Phys. Rev. D20, 2947 (1979).
18. J.F. Donoghue and K. Johnson, Phys. Rev. D21, 1975 (1980).
19. T. De Grand, R.L. Jaffe, K. Johnson and J. Kiskis, Phys. Rev. D12, 2060 (1975).
20. S. Weinberg in Festschrift for I.I. Rabi, edited by Lloyd Motz (New York Academy of Sciences, New York 1977).

HIGH ENERGY HADRONIC INTERACTIONS, EXCLUSIVE PROCESSES, EXPERIMENT AND THEORY

L.W. Jones, Michigan, Organizer

SMALL ANGLE HADRON PROTON ELASTIC SCATTERING AT FERMILAB ENERGIES

J. Lach
Fermilab, Batavia, Illinois

ABSTRACT

This paper summarizes contribution 476[1] and 477[2]; two high statistics experiments on hadron proton elastic scattering. The first is a high statistics study of $\pi^{\pm}p$ and pp elastic scattering at 200 GeV/c in a range of

$$0.60 < -t < 0.02 (GeV/c)^2.$$

The logarithmic slope, b, is found to increase substantially at small values of -t in agreement with a simple additive quark model (AQM). In the second contribution elastic scattering cross sections are measured on hydrogen in the Coulonb-nuclear interference region for $\pi^{\pm}p$, $K^{\pm}p$ and $p^{\pm}p$ from 70 to 200 GeV/c.

The measurement of the slope of the hadron proton forward elastic scattering peak is in itself a fundamental measurement, but it is also a prerequisite for extracting ρ, the ratio of the real to imaginary part of the forward scattering amplitude by the Coulonb-nuclear interference method. Both experiments described in these papers utilize a high resolution forward spectrometer to examine different t ranges of the elastic peak. The statistical accuracy of the data is ($\sim 10^6$ events in the pp case) sufficiently precise so that local values for b can be extracted and compared to other experiments. Figure 1 shows that, in general, the agreement is good. We see that b is dependent on t in this region. The data of figure 1 are fit well by a simple AQM which attributes the major part of the small t elastic cross section variation to the hadronic form factors of the target and projectile. The form factors describe the spatial distribution of the "clothed" quarks; in the small t region the scattering is dominated by single quark-quark scattering. The only parameters in this model are the hadron sizes which enter into the form factors and a size associated with the "clothed" quark. Fits are obtained which give hadronic sizes comparable to the electromagnetic form factors and a "clothed" quark radius of 0.35-0.45 fm.

In Fajardo et al[2] the same apparatus is used, but the t acceptance is optimized for the Coulonb-nuclear interference region. Values of ρ were presented for $\pi^{\pm}p$, $K^{\pm}p$, and $p^{\pm}p$ scattering from 70-200 GeV/c. These are the first high energy measurements of ρ for the pp and Kp system. In the determination of ρ the nuclear amplitude is parameterized in terms of the AQM. Agreement with the predictions of dispersion relations is good except for the pp case.

Our pp results are higher than dispersion relation predictions and the experimental results of the internal target group working at Fermilab[3]. We believe this is due to the steeper slope we have measured in the forward direction. In order to verify this we have refit the data of reference 3 with our AQM parameterization. In figure 2 we have plotted the refitted points of reference 3 as well as our data and that of other experiments. It is seen that our data and that of reference 3 are in good agreement, but both are now somewhat higher than the dispersion relation predictions.

The high precision elastic scattering data has also been used to look for possible violation of unitarity. None are seen.

REFERENCES

1. A. Schiz et al, A High Statistic Study of $\pi^{\pm}p$, π-p and pp Elastic Scattering at 200 GeV/c. Fermilab Pub. 79/81-Exp. (Submitted to Phys. Rev.)

2. L. A. Fajardo et al, The Real Part of the Forward Elastic Nuclear Amplitude of pp, $\bar{p}p$, $\pi^{\pm}p$, $\pi^{-}p$, $K^{\pm}p$, and $K^{-}p$ Scattering Between 70 and 200 GeV/c. Fermilab Pub. 80/27-Exp. (Submitted to Phys. Rev.)

3. E. Jenkins et al, Fermilab Pub. 78/35-Exp. (1978): Submitted to Sov. J. Nucl. Phys.

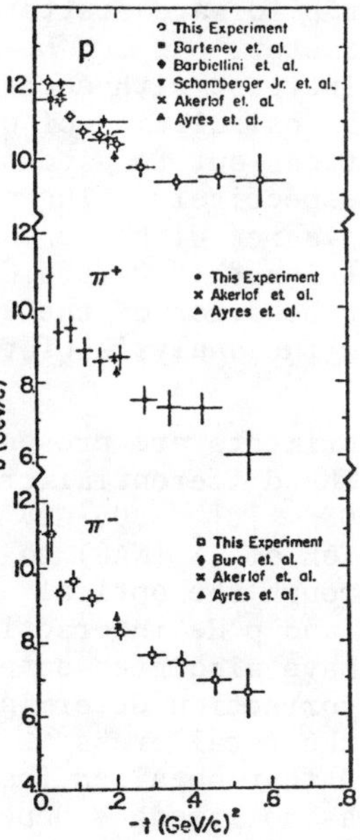

Figure 1

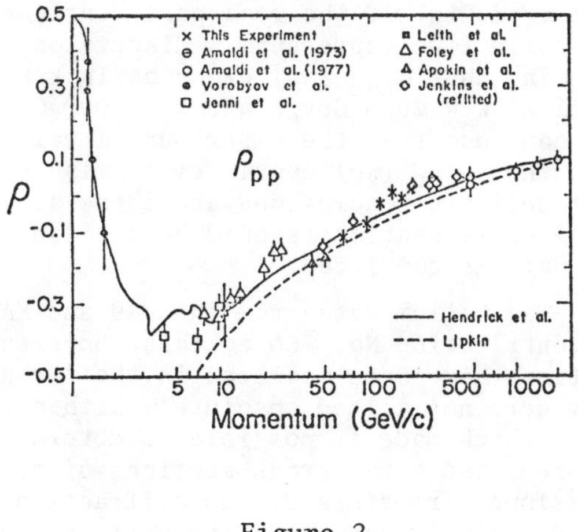

Figure 2

FORWARD π^-p, $\pi^\pm$ He, AND pHe ELASTIC SCATTERING AT THE SPS ENERGIES

A. A. Vorobyov
Leningrad Nuclear Physics Institute, Gatchina, U.S.S.R.

ABSTRACT

Differential cross sections for small-angle elastic scattering of pions and protons on hydrogen and helium have been measured at incident momenta ranging from 50 to 345 GeV/c. From the analysis of the helium data, diffraction slope parameters and total cross sections have been obtained. Preliminary results on π^-p scattering at 345 GeV/c are presented. It is shown that the real part of the π^-p forward scattering amplitude continues to increase with energy.

The Clermont-Ferrand-Leningrad-Uppsala collaboration has recently undertaken a study of πp, pp, π^4He, and p^4He small angle elastic scattering in the WA9 and NA8 experiments at the CERN SPS. The results from the WA9 experiment on the π^-p real part in the energy range 30-140 GeV have already been published.[1] In the NA8 experiment these measurements were extended to 345 GeV. Data have been taken in the energy range 100 to 345 GeV (8 energy points) for π^-p scattering; at 100 and 150 GeV for π^+p scattering; and at 100, 150, 250, and 300 GeV for pp scattering. More than 10^5 events have been detected at each energy in the t-range $.002 < |t| < .05$ GeV^2. Preliminary results from the analysis of 50,000 events at 345 GeV (1/4 of the full statistics at this energy) are presented to this Conference (No. 757): $\rho = .08 \pm .02$ and $b = (11.0 \pm .3)$ GeV^{-2} where ρ is the real to imaginary part ratio of the π^-p forward scattering amplitude and b is the differential slope parameter. As can be seen from Fig. 1, the real part continues to increase with energy. The solid line represents a dispersion relation calculation of ρ assuming that σ_{tot} (πp) grows as $\ln^2 k$ but flattens out to a constant level at k = 2000 GeV/c and k = 1000 GeV/c, respectively. The authors conclude that the experimental values of ρ agree with the assumption that σ_{tot} (πp) continues to rise up to at least 1000-2000 GeV/c. More definite conclusions about the asymptotic behavior of the πp total cross sections should be available after the analysis of the NA8 data is completed.

The helium data from the WA9 and NA8 experiments are presented in contribution No. 756 to this Conference. The differential cross sections have been measured in the t-range $.008 < |t| < .05$ GeV^2. They were normalized absolutely with a precision of 1% (NA8) to 2% (WA9) which made it possible to determine, through the optical theorem, the total cross sections of the π^4He and p^4He interactions. The slope parameters of the diffraction cone have also been determined. Fig. 2 shows the inelastic shadowing correction determined as the difference $\Delta\sigma_{in} = \sigma_{tot}^{Gl} - \sigma_{tot}$ between the total cross section calculated using the Glauber model and that obtained from the experiment. It is relatively small and has no strong s-dependence. Note that $\Delta\sigma_{in}$ (pHe) proved to be considerably less than those obtained in the jet-target experiment.[2] The hadron-helium slope parameters were found to exhibit some shrinkage effect: b_1 (pHe) = (0.7 ± 0.3) GeV^{-2} and b_1 (πHe) = (0.3 ± 0.2) GeV^{-2} where $b(s) = b_0 + b_1 \ln s$. However, contrary to the conclusions of ref. 2, the present data show no evidence for the additional shrinkage of the hadron-helium diffraction cone if compared with the shrinkage of the hadron-proton cone. The measured differential cross section proved to be well described by the Glauber formula with the double scattering term increased by 10% (pHe) and by 25% (πHe) to take into account the inelastic shadowing effect.

REFERENCES

1. J.P. Burg, et al, Phys. Lett. 77B, 438 (1979).
2. E. Jenkins et al, Phys. Rev. to be published.
V.A. Nikitin, Proc. Int. Conf. on High Energy Physics, Geneva, 547 (1979)

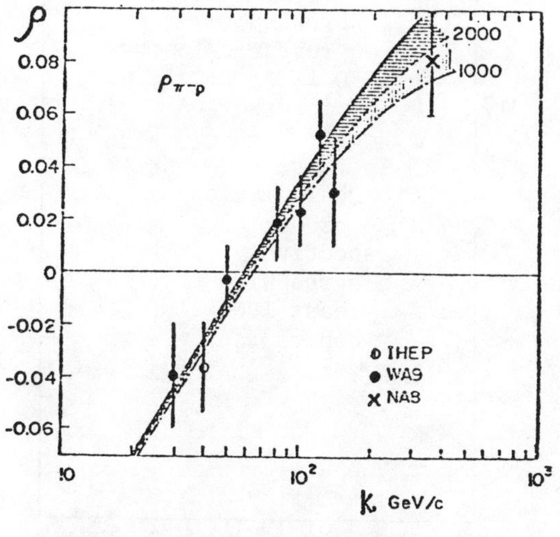

Fig. 1: Real part of the $\pi^- p$ forward scattering amplitude as a function of the incident particle momentum.

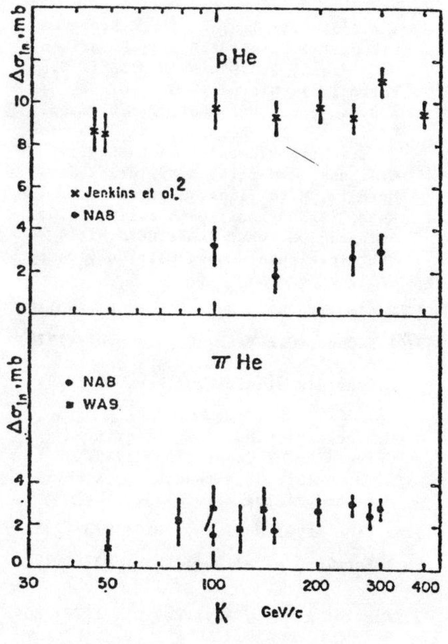

Fig. 2: Inelastic shadowing correction in helium.

FIRST OBSERVATION OF A DIP AT $t = -1.4$ $(GeV/c)^2$ IN $\bar{p}p$ DIFFERENTIAL ELASTIC CROSS SECTION AT 50 GeV/c

M. Poulet
LAPP, BP 909, 74019 Annecy-le-Vieux - France

Preliminary results of the CERN WA7 collaboration[1)] concerning the $\bar{p}p$ elastic differential cross-section measurements at 50 GeV/c are presented. The t-range covered extends from -.7 to -5. $(GeV/c)^2$

The set-up is an extended version of the arrangement used at very large t in $\pi^{\pm}p$[2)]. The results are shown in Fig. 1. The overall normalization is not yet finished. A clean dip-bump structure is observed with a pronounced minimum at $t = -1.4$ $(GeV/c)^2$. This dip could possibly be of the same origin as the one seen in pp scattering at a similar t-value and for the same value of the total cross-section. Different models based on the geometrical pictures[3)] or multiple quark[4)] mechanism could describe these data.

FIG.1 $\bar{p}p$ elastic scattering
preliminary July 1980

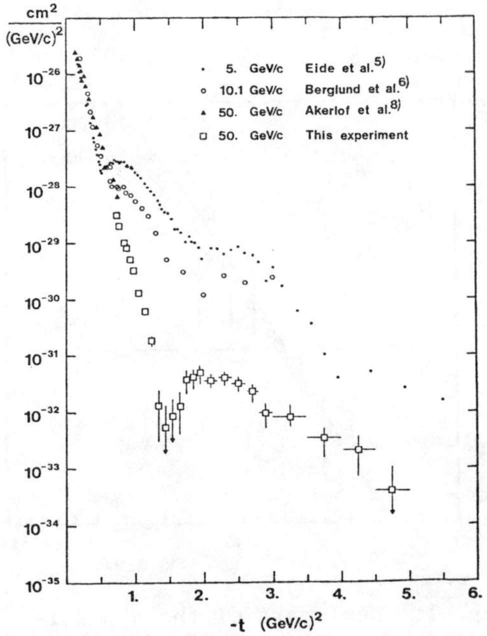

REFERENCE

1. Z.Asa'd, C.Baglin, R.Böck, K.Brobakken, L.Bugge, T.Buran, A.Buzzo, P.Carlson, M.Coupland, D.G.Davis, B.G.Duff, S. Ferroni, I. Gjerpe, V.Gracco, J.D.Hansen, P.Helgaker, F.F.Heymann, D.C.Imrie, T. Jacobsen, K.E.Johansson K.Kirsebom, R.Lowndes, A.Lundby, G.J.Lush, M.Marcri, R.Møllerud, J. Myrheim, M.Phillips, M. Poulet, L. Rossi, A.Santroni, G.Skjevling, S.O. Sorensen, M.Yvert (LAPP, CERN, Niels Bohr Institute, Genova, Oslo, University College London)
2. R.Almas et al., Phys.Lett.93B(1980)199
3. T.T.Chou, C.N.Yang, Phys.Rev.D19(1979) 3268.
   C.Bourrely et al. Phys.Rev.D19(1979) 3249.
4. C.W.Heines, M.M.Islam, Preprint University of Connecticut 1979.
   A. Donnachie, P.W.Landshoff, Z Phys. Part.and Fields 2)1979)55.
5. A.Eide et al.Nucl.Phys.B60(1973)173
6. A.Berglund et al.Nucl.Phys.B137(1979) 276.
7. Akerlof et al., Phys.Rev.D15(1977)2864

HADRON ELASTIC SCATTERING AND DIFFRACTION DISSOCIATION AT HIGH ENERGIES AND SMALL MOMENTUM TRANSFER

R.L. Cool, K. Goulianos, S.L. Segler, H. Sticker and S.N. White
The Rockefeller University, New York, N.Y., 10021

ABSTRACT

We have measured the small $|t|$ elastic and diffractive cross sections of hadrons on protons at 100 and 200 GeV/c. Both the elastic and inelastic data display a simple exponential t dependence. The diffraction cross sections vary predominantly as $1/M_x^2$, decrease slowly with s, and agree well with factorization rules.

— — — — — — — — —

We report measurements of the differential cross sections for elastic scattering and diffraction dissociation of hadrons at 100 and 200 GeV/c in the range $0.025 < |t| < 0.10$ (GeV/c)^2. The experimental equipment, shown in Fig. 1, was situated in the M6W beamline at Fermilab. This apparatus measured the kinetic energy (T_p) and the angle of the recoil proton, from which one obtains the momentum transferred squared $|t| = 2M_pT_p$ and the missing mass squared, M_x^2, or equivalently, the scaling variable $1-x = (M_x^2 - M_h^2)/s$. The t dependence of the elastic and inelastic cross sections is exponential. The slopes are consistent with other measurements[1,2] in our $|t|$ range or with extrapolations from higher $|t|$. A recent measurement[3] of the π^-p elastic slope at 100 GeV/c for $0.002 < |t| < 0.04$ (GeV/c)^2 reported $b = 11.3 \pm .3$ (GeV/c)$^{-2}$ which is to be compared with our value of 8.95 ± 0.31 for $|t| > 0.025$. This difference indicates that the elastic slope must change rapidly within a small $|t|$ range.

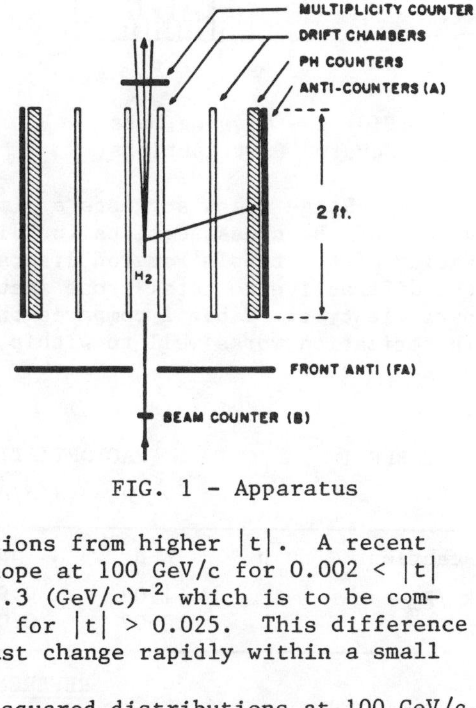

FIG. 1 - Apparatus

Fig. 2 shows the missing mass squared distributions at 100 GeV/c. For each reaction, there is an elastic peak at $(M_x^2 - M_h^2) \sim 0$ GeV^2, some resonance structure at $(M_x^2 - M_h^2) \sim 1\text{-}3$ GeV^2, and a smooth falloff at large M_x^2. The cross sections at large M_x^2 can be fitted by the function

$$\frac{d^2\sigma}{dtdx} = \frac{A}{(1-x)} + B(1-x) \tag{1}$$

where A arises from the triple Pomeron term and B from lower lying trajectories in the Regge model. The data and the fits, shown in Fig. 3, illustrate the dominant $1/(1-x)$ behavior.

0094-243X/81/680049-02$1.50

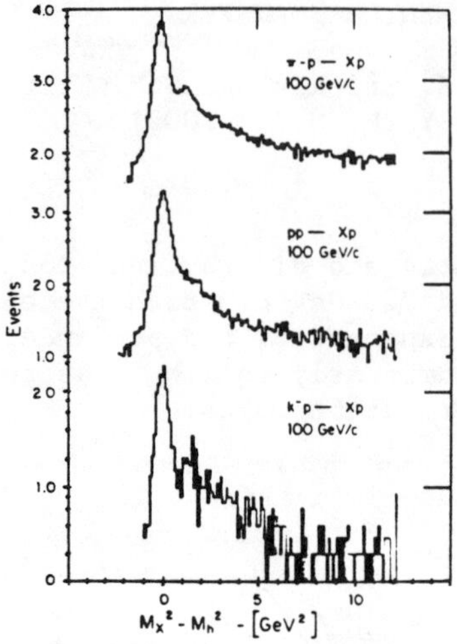

FIG. 2 - Missing Mass Squared Distributions

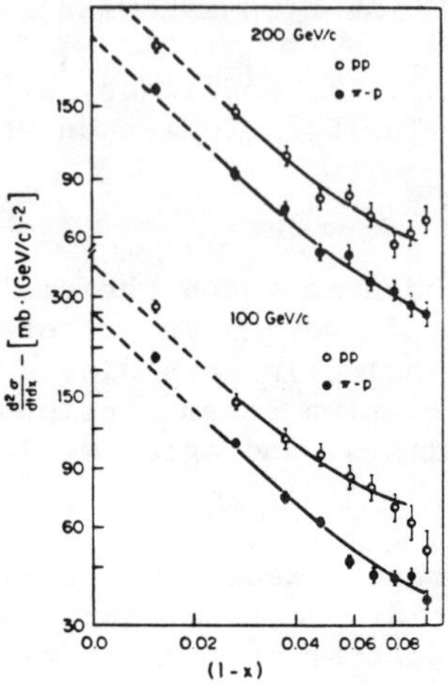

FIG. 3 - Diffractive Cross Sections at $|t| = 0.05$ $(GeV/c)^2$

The Regge model suggests a simple way to compare the absolute values of the cross sections for different hadrons. The factorization of the triple Pomeron diagram predicts that the ratio of the diffractive to total cross sections be independent of incident particle type. Table I compares the ratios and shows that this factorization works well to within the experimental uncertainty of a few percent.

TABLE I FACTORIZATION TEST

$R = (d^2\sigma/dtdx)/\sigma_T$ $p_{lab} = 100$ GeV/c

Reaction	π^-p	K^-p	$\bar{p}p$	π^+p	K^+p	pp
R_h/R_{π^-}	1	1.08 ±0.07	0.94 ±0.06	1.02 ±0.04	0.85 ±0.12	0.97 ±0.03

REFERENCES

1) Elastic: V. Bartenev et al., Phys. Rev. Lett. 31, 1088 (1973); D. Ayres et al., Phys. Rev. D15, 3105 (1977); A. Schiz et al., Fermilab Report 79/81 (1980).
2) Diffractive: Y. Akimov et al., Phys. Rev. Lett. 35, 766 (1975); R.L. Anderson et al., Phys. Rev. Lett. 38, 880 (1977).
3) J.P. Burg et al., Phys. Lett. 77B, 438 (1978).

MEASUREMENT OF THE TOTAL CROSS SECTIONS OF Σ^- AND Ξ^- ON PROTONS AND DEUTERONS BETWEEN 75 AND 135 GeV/c

Bristol, Cambridge, Geneva, Heidelberg, Lausanne,
Queen Mary College, Rutherford Collaboration.

Presented by A.A. Carter

SUMMARY

The Σ^-p and Σ^-d total cross sections have been measured to a statistical accuracy of 1.0% and 0.5% respectively at 5 momenta between 74.5 and 136.9 GeV/c, and the Ξ^-p and Ξ^-d cross sections were measured to the same accuracy at 101.5 and 133.8 GeV/c. The systematic uncertainty at each momentum is estimated to be of order 0.5%. The results are shown in the figure, together with the straight line fits to the Σ^- data. Each total cross section is seen to rise with energy, and the data serve to discriminate between existing total cross section models. The most successful predictions come from the two-component Pomeron of Lipkin[1].

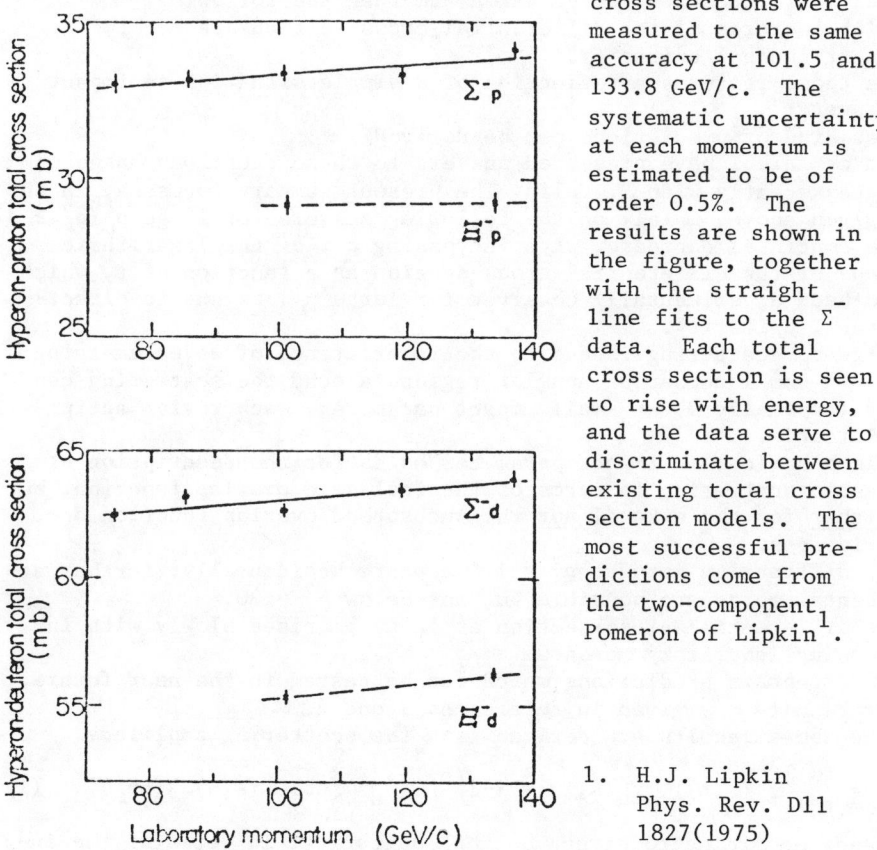

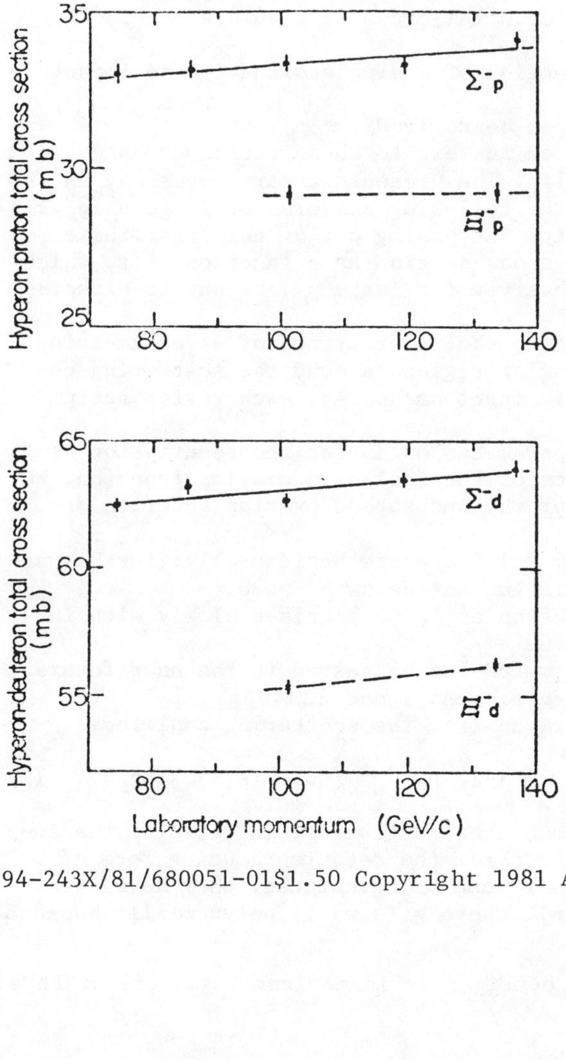

1. H.J. Lipkin Phys. Rev. D11 1827(1975)

ELASTIC SCATTERING THEORY

R. Henzi
McGill University, Montreal, Que., Canada H3A 2T8

Unitarity in the direct channel is a fundamental constraint on elastic scattering leading naturally to the question whether or not it has directly observable consequences. At present, the vast amount of data on proton-proton elastic scattering and total cross sections which have been collected in the past decade provide the most promising starting point for the search of such effects. A particularly useful data sample is provided by the CERN-ISR and the FNAL high laboratory momentum data (q_L = 200-2000 GeV/c) covering transverse momenta in the center of mass up to $p_\perp^2 = |t| = 14$ (GeV/c)^2. For these values of the kinematic parameters real part and spin are expected to give rise only to secondary corrections; hence, direct channel unitarity effects are likely to be more pronounced.

The basic questions to be asked are then the following:

1. Which features of the differential cross section are due to direct channel unitarity?
2. Are these features a reflection of a simple situation in impact parameter?
3. What kind of predictions can be derived?

In my talk I have presented answers to these questions based on work done recently with P. Valin. The present summary focusses, in the order given above, mainly on the following answers for large $p_\perp$ ($p_\perp^2 \geq 3$):

1. The continual decrease, with increasing $p_\perp$, of the logarithmic slope of the differential cross section as a function of t, which has been experimentally observed for large $p_\perp$, is due to direct channel unitarity.

2a. Large $p_\perp$ scattering is due to the interference of waves emerging from a small number of annular regions around the scattering center (typically 3) at small impact parameter, each region acting coherently.

2b. This behaviour in impact parameter holds for the description of elastic scattering in terms of the inelastic overlap function, but <u>neither</u> for the eikonal <u>nor</u> the unabsorbed overlap function descriptions.

3a. The diffraction zero at $p_\perp^2 \simeq 1.4$ appears accidentally; further accidental zeros are possible but not below $p_\perp^2 \simeq 30$.

3b. The differential cross section at large $p_\perp$ rises slowly with increasing laboratory momentum.

3a and 3b contain predictions which can be tested in the near future. Further results are given in references 1 and 2.

The above results are derived from the scattering amplitude

$$f(s,t) = \sum_{n=1}^{\infty} f_n(s,t); \quad f_n(s,t) = (-1)^{n+1} \binom{1/2}{n} \int_0^\infty db\, b\, G^n(s,b) J_0(bp_\perp) \qquad (1)$$

expressed, according to direct channel unitarity, in terms of the inelastic overlap function G(s,b). From the data reproducing form of G determined in ref. 1 the terms f_n can be calculated; they have the form $f_n(s,t) = g_n(s,t)\exp(at/n)$, where $g_n(s,t)$ is polynomially bounded in t.

In figure 1 $f_n(s,t)$ at $q_L = 1486$ is plotted in the vertical direction as function of n and $p_\perp$ (successive lines for $p_\perp$ = const. are multiplied by $10^4\exp(2.56p_\perp)$): At fixed $p_\perp$ the terms f_n group into five lobes $L^0 \ldots L^4$ (from right to left) with alternating signs. The contribution of L^i to the series (1) is $\Delta n^i f_{n^i}$ where $n^i(p_\perp)$ and $\Delta n^i(p_\perp)$ are the position of the lobe maximum and the lobe width (properly defined), respectively. $n^i \propto p_\perp$ for sufficiently large $p_\perp$ leading to an $\exp(-\text{const.}p_\perp)$ contribution due to L^i which is modulated by a slowly varying function of $p_\perp$. For $2 \le p_\perp \le 5$ only L^0, L^1 and L^2 are important and their sum can be lumped into a single effective $\exp(-\text{const.}p_\perp)$ law accounting for the slope decrease described above (Answer 1).

To each lobe L^i corresponds a characteristic profile $bG^{n^i}(s,b)$. For L^0, L^1 and L^2 these profiles are shown in figure 2 (profile area normalized to lobe contribution) and compared with $bG(s,b)$ (arbitrarily normalized) which defines the size ($\simeq$ 1 fm) of the interaction volume. These profiles give rise to the annular, coherently acting regions described above (Answer 2a). A similar analysis shows that no coherence effects exist in the eikonal and unabsorbed overlap function descriptions (Answer 2b). L^1 cancels L^0 near $p_\perp = 1.4$ producing an accidental diffraction zero, and dominates for $2 \le p_\perp \le 5$; further accidental zeros, due to several lobes cancelling, might be possible but not below $p_\perp \simeq 30$ (Answer 3a). The energy dependence prediction (Answer 3b) has been derived and illustrated in detail in ref. 1.

The work I have summarized leads to the conclusion that direct channel unitarity produces directly observable effects which are properly described in terms of the inelastic overlap function.

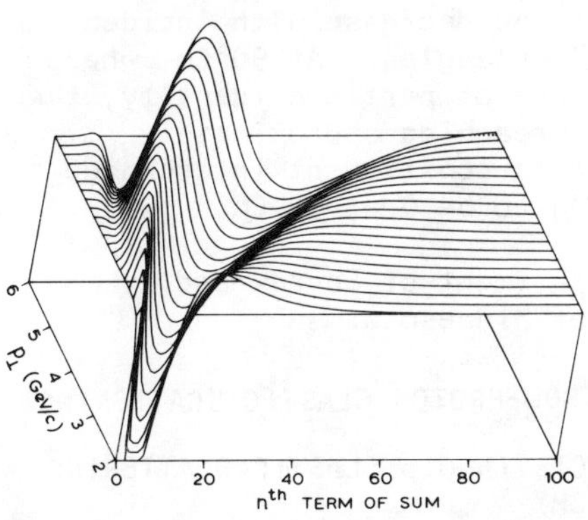

Fig. 1. $f_n(s,t)$ as a function of n and $p_\perp$ (q_L = 1486).

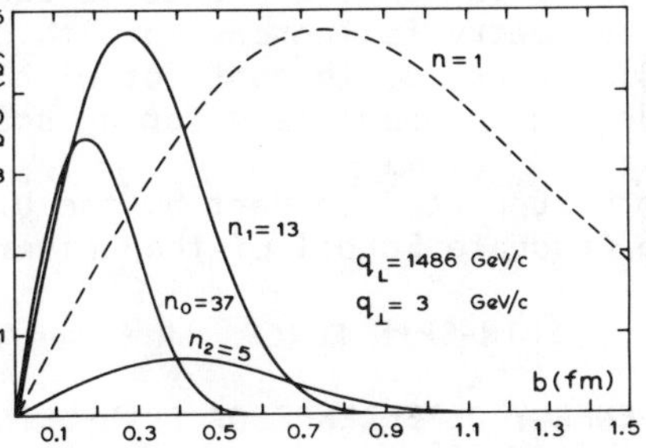

Fig. 2. Profiles $bG^n(s,b)$.

In my talk I have also presented a new model[3] in which the proton-proton total cross section rise is due to glue-glue annihilation into heavy resonances. Details are given by B. Margolis in these Proceedings.

REFERENCES

1. R. Henzi and P. Valin, Nucl. Phys. B148, 513 (1979). Contr. paper #444.
2. R. Henzi and P. Valin, to be published. Contr. paper #728.
3. Y. Afek, C. Leroy, B. Margolis and P. Valin, Phys. Rev. Letters 45, 85 (1980). Contr. paper #557.

SPIN EFFECTS IN NUCLEON-NUCLEON ELASTIC SCATTERING

Kent M. Terwilliger
The University of Michigan, Ann Arbor, Michigan 48109

Three papers involving spin effects in high momentum transfer nucleon-nucleon elastic scattering are briefly summarized.

POLARIZATION IN LARGE ANGLE PROTON-NEUTRON ELASTIC SCATTERING*

Y. Makdisi, M.L. Marshak, B. Mossberg, E.A. Peterson, K. Ruddick
School of Physics and Astronomy, University of Minnesota
Minneapolis, Minnesota 55455

J.B. Roberts
Physics Department and T.W. Bonner Nuclear Laboratories
Rice University, Houston, Texas 77001

R.D. Klem
Argonne National Laboratory, Argonne, IL 60439

The authors have measured the large angle polarization asymmetry A in the proton-neutron elastic scattering at 2,3, and 6 GeV/c using the polarized proton beam at the Argonne ZGS and a liquid deuterium target. These measurements, the first at high energy, show that A is large (20-40%) and negative at the larger angles, larger and opposite sign to pp scattering, and with no decrease with incident energy, unlike the earlier data at smaller angles. At 90°CM, where A for pp is constrained to be zero because of particle identity, the np asymmetry is increasing with energy, reaching approximately -.3 at 6 GeV/c, in conflict with the basic constituent interchange model which predicts A for np scattering to be 0 at 90°CM.

*Work supported in part by the U.S. Department of Energy and by the Graduate School of the University of Minnesota.

SPIN-SPIN FORCES IN 6 GeV/c NEUTRON-PROTON ELASTIC SCATTERING

ENERGY DEPENDENCE OF SPIN-SPIN EFFECTS IN p-p ELASTIC SCATTERING AT 90°CM

D.G. Crabb, R.C. Fernow, P.H. Hansen, A.D. Krisch, B. Sandler, T. Shima and K.M. Terwilliger
Randall Laboratory of Physics, The University of Michigan
Ann Arbor, Michigan 48109

J.R. O'Fallon
Argonne Universities Association, Argonne, Illinois 60439

0094-243X/81/680054-02$1.50

E.A. Crosbie, L.G. Ratner, P.F. Schultz, G.H. Thomas
Argonne National Laboratory, Argonne, Illinois 60439

N.L. Karmarkar
University of Kiel, Kiel, Germany

S.L. Linn and A. Perlmutter
Department of Physics and Center for Theoretical Studies
The University of Miami, Coral Gables, Florida 33124

A. Lin
Abadan Institute of Technology, Abadan, Iran

A.J. Salthouse
Bell Laboratories, Murray Hill, New Jersey 07974

P. Kyberd
Nuclear Physics Laboratory
Oxford University, Oxford, England

In the two-spin experiment $n\uparrow+p\uparrow\rightarrow n+p$ we measured $d\sigma/dt$ at $P_\perp^2 = 0.8$ and $1.0\ (GeV/c)^2$ at 6 GeV/c. We used the 6 GeV/c 53% polarized neutrons from the 12 GeV/c polarized deuteron beam at the Argonne ZGS, and scattered them from our 75% polarized proton target. Both spins were oriented perpendicular to the scattering plane. We found interesting spin-spin effects in n-p elastic scattering: $A_{nn}=-.17\pm.05$ at $P_\perp^2 = 0.8$, and $A_{nn}=-.19\pm.05$ at $P_\perp^2 = 1.0\ (GeV/c)^2$. These values are larger in magnitude and opposite in sign from A_{nn} in pp elastic scattering at 6 GeV/c at the same $P_\perp^2$. The basic consitituent interchange model predicts the np A_{nn} to be -.44.

In the two-spin experiment $p\uparrow+p\uparrow\rightarrow p+p$ the energy dependence of the spin-parallel and spin-antiparallel cross-sections at 90^0CM, with spins normal to the scattering plane, was measured for beam momenta between 6 GeV/c and 12.75 GeV/c. The ratio $(d\sigma/dt)_{parallel}$: $(d\sigma/dt)_{antiparallel}$ is about 1.2 up to 8 GeV/c and then increases rapidly to a value of almost 4 near 11 GeV/c (A_{nn} goes from approximately .1 to .6). The highest momenta points suggest that the ratio may reach a limiting value of about 4. When plotted against $P_\perp^2$ this rapid increase in cross section ratio closely matches that observed earlier at the fixed laboratory momentum of 11.75 GeV/c, where the scattering angle was varied. This close correspondence suggests that the pure spin cross sections may be mainly dependent on $P_\perp^2$ in the hard scattering region. The data are in strong disagreement with the basic constituent interchange model, which predicts a cross-section ratio of 2.

This work was supported by the U.S. Department of Energy

pp SPIN CORRELATIONS AT HIGH p_T

I. P. Auer, E. Colton, W. R. Ditzler, D. Hill, H. Spinka, N. Tamura, J.-J. Tavernier, G. Theodosiou, K. Toshioka, D. Underwood, R. Wagner and A. Yokosawa
High Energy Physics Division Argonne National Laboratory
9700 South Cass Avenue Argonne, Illinois 60439

An experiment was performed at the ZGS to test some of the theoretical predictions of spin effects in pp reactions using quark-parton models. The beam momentum was 11.75 GeV/c and the beam and target were both longitudinally polarized. Both elastic scattering and inclusive π^+ and π^- spin asymmetries were measured. The forward going particles were detected in a magnetic spectrometer, and the recoil protons from elastic scattering were detected in coincidence. Preliminary results are shown below. The measured inclusive asymmetries are consistent with zero. Asymmetries predicted by models based on QCD[1,2] are also shown for comparison. Combining the elastic C_{LL} data with the Michigan C_{NN} results[3], it appears that the assumption of quark helicity conservation (used in many quark-parton models) is verified within our statistical errors at large cm-angles.

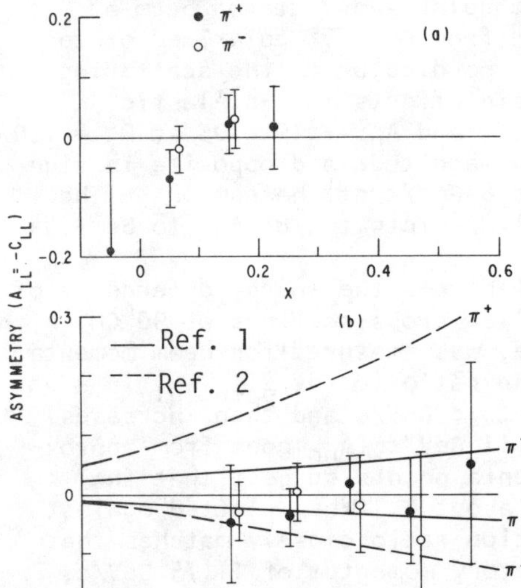

Fig. 1 Inclusive asymmetries

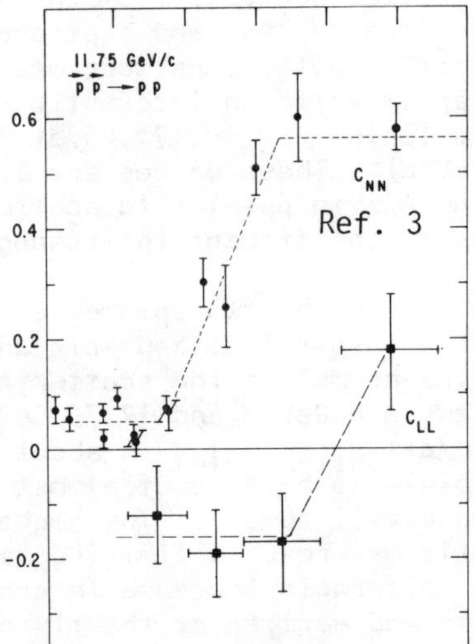

Fig. 2 Elastic scattering asymmetries

References

1. J. Babcock et al., Phys. Rev. D19, 1483 (1979).
2. H. Cheng and E. Fischbach, Phys. Rev. D19, 860 (1979).
3. J. R. O'Fallon et al., Phys. Rev. Lett. 39, 733 (1977).

MEASUREMENT OF THE POLARIZATION PARAMETER IN pp ELASTIC SCATTERING AT 200 GeV/c

W. Bartl[4], R. Birsa[3], F. Bradamante[3], G. Fidecaro[1], M. Fidecaro[1], R. Frühwirth[4], F. Gasparini[2], M. Giorgi[3], Ch. Gottfried[4], L. Lanceri[1], G. Leder[4], W. Majerotto[4], A. Meneguzzo[2], G. Neuhofer[4], S. Nurushev[1], A. Penzo[3], M. Pernicka[4], L. Piemontese[3], M. Posocco[2], Ch. Poyer[1], M. Regler[4], P. Sartori[2], P. Schiavon[3], V. Solovianov[1], M. Steuer[4], H. Stradner[4], A. Vascotto[1], A. Villari[3] and C. Voci[2]

CERN[1]-PADOVA[2]-TRIESTE[3]-VIENNA[4] Collaboration

The differential cross section and the polarization parameter in pp elastic scattering at 200 GeV/c, and in the momentum transfer interval $0.6 < -t < 4.0$ GeV^2, have been measured in the CERN SPS WA6 experiment. The results are preliminary and obtained on ∿ 40% of the overall statistics.

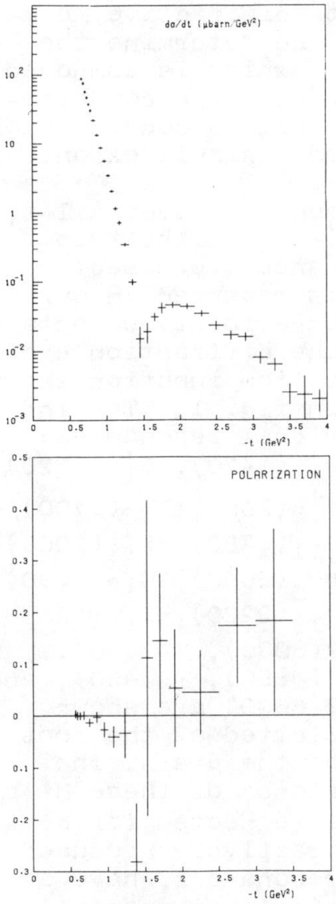

The apparatus is based on scintillation counters and hodoscopes, and MWPC telescopes. Both forward and backward proton momenta are measured by means of magnetic spectrometers. Fast Decision Logics and Matrix Correlations on track geometry and kinematics of the elastic events are used at the trigger level. Two targets have been placed inside the cryostat, one pentanol polarized target, 10 cm long, and one CH_2 target, 4 cm long, so that data have been taken at the same time both on polarized and unpolarized protons. For background evaluation, data have also been collected in separate runs with a carbon target, with a comparable content of unpolarized nucleons, but hydrogen free.

The differential cross section shows a well developed dip around $-t = 1.5$ GeV^2 thus confirming previous measurements.

The polarization behaviour is consistent with that already observed at 150 GeV.[1)] Negative values around $-t \simeq 1.0$ GeV^2 seem to be confined to positive values in correspondence to the interference region.

(1) G. Fidecaro et al., CERN-Padova-Trieste-Vienna Collaboration, Measurement of the polarization parameter in pp elastic scattering at 150 GeV/c, submitted to Nuclear Physics.

PROTON DIFFRACTION EXCITATION FUNCTION*

J. Lee-Franzini
SUNY at Stony Brook, Stony Brook, New York 11794

In a high precision $p+p \to p+X$ experiment at Fermilab's ITA we have collected over 2×10^7 events covering the s range 18 GeV^2 to 754 GeV^2, t range $-.014\ (GeV/c)^2$ to $-.563\ (GeV/c)^2$, and x ($\simeq 1-M^2/s$) range 1 to 0.75. Typical M^2 resolution in this experiment is 0.1 $GeV^2 x(E_{beam}/100\ GeV)$ at $t=-0.15\ (GeV/c)^2$. A simultaneous fit to the data at all energies taking into account the resolution function (vs. s,t,M^2) allowed us to separate the inclusive spectra into their energy independent (diffractive) component and Feynman scaling component and determine the proton diffraction excitation function which is composed of fifteen natural parity N^* resonances. We also determined the x and t dependences of the scaling contribution; in particular the latter deviates from a simple exponential t dependence in our t region.

By integrating all the fitted resonances over all t, (~88% of the resonance cross-section is measured in our experiment) we obtain the diffraction excitation function shown in fig. 1. The individual resonances: $P'_{11}(1470)$, $D'_{13}(1520)$, $F'_{15}(1688)+D''_{13}(1700)$, $P''_{11}(1780)$, $F''_{15}(2000)+D'''_{13}(2040)$, $G_{17}(2190)+H_{19}(2220)$, $I_{111}(2650)$, ?(3030), ?(3750), ?(4850), ?(6450), and ?(8000) are shown plotted at the foot of the graph. The slopes of these N*'s, as expected for diffractively produced resonances, have a simple exponential, exp(-bt) dependence. The total σ (inelastic diffractive) is $4.26\pm.09$ mb.

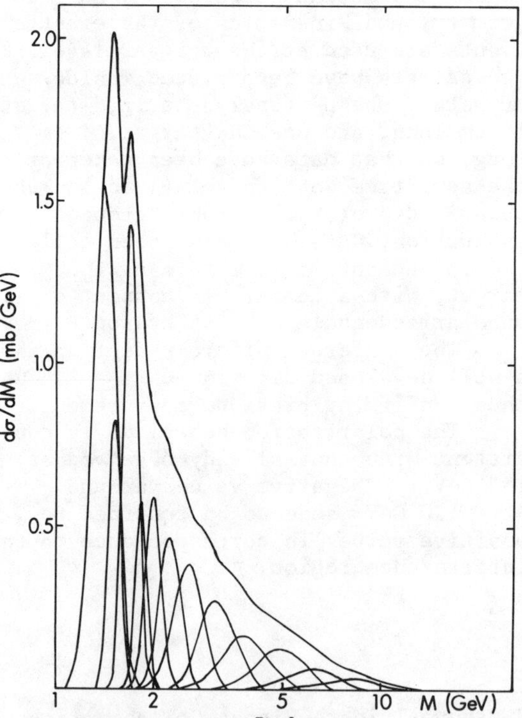

Fig 1

*Supported in part by the NSF.

0094-243X/81/680058-01$1.50

DIP AND KINK STRUCTURES IN HIGH ENERGY DIFFRACTION DISSOCIATION PROCESSES*

T. T. Chou
University of Georgia, Athens, Ga. 30602

ABSTRACT

The dip and shoulder structures observed in high energy diffraction dissociation experiments are discussed in the context of the geometrical picture.

Recently there has been some experimental data concerning the angular distribution of hadron-nucleus and hadron-hadron diffraction dissociation. One of the conspicuous features of all these experiments is the existence of dip or kink structures similar to that observed in pp elastic scattering. We suggest that the geometrical model can offer a natural explanation of these dip structures. (See Ref. 1.)

Consider the passage of an incoming hadron through an extended target. At an impact parameter b the dissociation can take place at any point along its path during its traversal. The probability for the process to occur is approximately proportional to the thickness of the material traversed, or $\Omega(b)$. There is also absorption of the incoming wave before dissociation, and of the outgoing wave after dissociation. Assuming equal mean free path for incoming and outgoing waves, the total absorption factor can be written as $\exp[-\Omega(b)]$. Thus the source distribution for the outgoing hadron in diffraction dissociation may be approximated by $\Omega\exp(-\Omega)$, in contrast to $1 - \exp(-\Omega)$ for the elastic case. This approximation was first used in charge exchange scatterings and was given the name "coherent droplet model." Using the proposed approximate source functions we can show qualitatively that the existence of a dip in diffraction dissociation occurs in general at a smaller -t value than the dip position in elastic scattering as experiments have indicated.

With Ω determined from electron scattering experiments together with hadron-hadron total cross sections, numerical computations for differential cross section in hadron-hadron and hadron-nucleus diffraction dissociation processes have been made. The computation contains no adjustable parameters. The calculated dip positions for both elastic scattering and diffraction dissociation process are in very good agreement with experimental values, justifying the geometrical origin of the dips and kinks in the data.

REFERENCES

1. T. T. Chou and C. N. Yang, Phys. Rev. D22, 610 (1980).

*Work supported in party by U.S. Department of Energy under Contract No. DE-AS09-76ER00946.

MEASUREMENT OF EXCLUSIVE HYPERCHARGE-EXCHANGE REACTIONS AT 35 TO 140 GeV/c[†]

E. N. May, M. W. Arenton, D. S. Ayres, D. Cohen,[#] R. Diebold, J. R. Sauer, A. B. Wicklund, and P. D. Zemany[##]
Argonne National Laboratory, Argonne, Illinois 60439

J. E. Elias
Fermi National Accelerator Laboratory, Batavia, Illinois 60510

S. Michalowski, K. Rich, and D. M. Ritson
Stanford Linear Accelerator Center, Stanford, California 94305

We have obtained data on the reactions $\pi^+p \to K^+\Sigma^+$ and $\pi^+p \to K^+\Sigma^+(1385)$ using the Fermilab Single Arm Spectrometer Facility at momenta of 35, 70 and 140 GeV/c, for momentum transfers $|t| \leq 1$ GeV^2. The line-reversed reactions $K^-p \to \pi^-\Sigma^+$ and $K^-p \to \pi^-\Sigma^+(1385)$ were also studied at 70 GeV/c in the same apparatus. Both the incident and forward scattered particles were identified and momentum analyzed. Particular reactions of interest were isolated by the missing-mass technique, and cross sections and Σ^+ polarizations were determined.

To summarize our preliminary findings, the s dependence of the cross sections for the reactions $\pi^+p \to K^+\Sigma^+$ and $\pi^+p \to K^+\Sigma^+(1385)$ is well described by a simple Regge pole parametrization using the $K^*(892)$ and $K^{**}(1430)$ trajectories. The shape of the differential cross section for the reaction $\pi^+p \to K^+\Sigma^+(1385)$ is not a simple exponential and indicates a substantial helicity-flip contribution. The differential cross section for the reaction $\pi^+p \to K^+\Sigma^+$ is well fitted by the simple form Ae^{Bt} for the range $0.0 \leq -t \leq 0.4$ GeV^2, but at $-t = 0.5$ GeV^2 a distinct break to a shallower slope is observed; in this reaction we find Σ^+ polarization values of ~40% at $|t| \sim 0.3$ GeV^2, independent of incident momentum. We obtain a cross-section ratio ($\pi^+p \to K^+\Sigma^+/K^-p \to \pi^-\Sigma^+$) of 0.66 ± 0.05 over the range $0.0 \leq -t \leq 0.4$ GeV^2 at 70 GeV/c. Thus, the Σ^+ polarization and the line-reversal cross-section comparison indicate breaking of both strong and weak exchange degeneracy.

[†]Work done under the auspices of the U. S. Department of Energy.

[#]Current address: 10640 S. W. 77th Ave., Miami, Florida 33156.

[##]Current address: Diano Corp., 8 Commonwealth Ave., Woburn, Massachusetts 01801.

HYPERON POLARIZATION AT FERMILAB

K. Heller
University of Minnesota, Minneapolis, Minnesota
55455

The polarization of Λs produced by 300 GeV protons discovered at Fermilab[1] has since been observed at energies from 24 to 1500 GeV.[2] This is a summary of new results from the group which made the original discovery. The Michigan-Minnesota-Rutgers-Wisconsin collaboration consists of: Beretvas, Bunce, Cox, Deck, Devlin, Dukes, Dworkin, Grobel, Handler, Heller, Luk, Lundberg, Overseth, Pondrom, Rameika, Sheaff, Skubic, and Wilkinson.

The reaction $p+p \rightarrow \Lambda + X$ (400 GeV protons) was studied at 10 production angles from 1 to 10 mrad. This allows, for the first time, the separation of polarization behavior depending on transverse (p_T) and longitudinal variables (x). Figure 1 shows the Λ polarization is both a function of x and p_T. Roughly speaking, it depends primarily on p_T for small p_T as required by symmetry. However for $p_T > .8$ GeV/c the polarization depends mainly on x. This data is compared with that taken using a Be target[3] in Figure 2. No target dependence was observed.[4]

Figure 2 shows the hyperon polarization from $p+Be \rightarrow H+X$ where H is Λ, Ξ^0, Ξ^-, or Σ^+. The agreement between the polarization of Λ, Ξ^0 and Ξ^- is striking. The Σ^+ polarization has the opposite sign[5] however its magnitude agrees with that of other hyperons. This agreement between Λ, Ξ^0 and Ξ^- polarizations as well as the reversed sign of the Σ^+ polarization follows from a quark model in which the produced valence quarks are polarized. The baryon polarization is then calculated using the standard quark wavefunctions.[6]

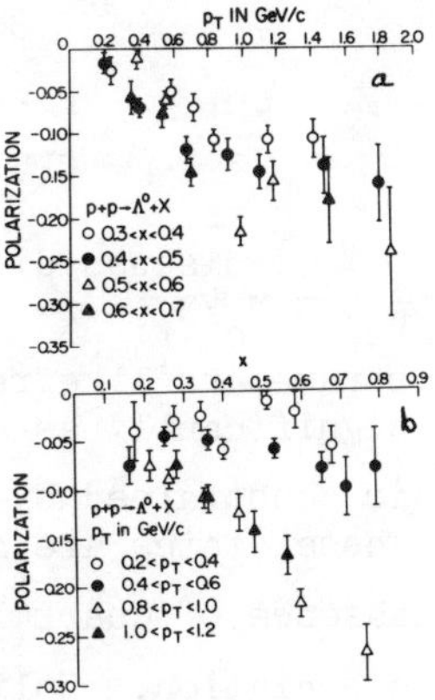

Figure 1

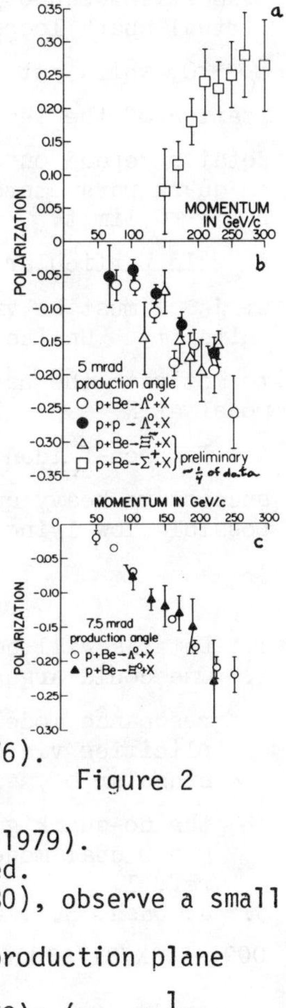

Figure 2

REFERENCES

1. G. Bunce et al., Phys. Rev. Lett. 36, 1113 (1976).
2. K. Heller et al., Phys. Lett. 68B, 480 (1977); F. Lomanno et al., Phys. Rev. Lett. 43, 1905 (1979); S. Erhan et al., Phys. Lett. 82B, 301 (1979).
3. L. Schachinger, Ph.D. Thesis (1978); unpublished.
4. Raychandhuri, et al., Phys. Lett. 90B, 319 (1980), observe a small target dependence at 28 GeV.
5. Positive polarization is perpendicular to the production plane ($\hat{k}_{in} \times \hat{k}_{out}$).
6. K. Heller et al., Phys. Rev. Lett. 41, 607 (1978); ($p_{\Sigma^+} = \frac{1}{3} p_\Lambda$ should read $p_{\Sigma^+} = -\frac{1}{3} p_\Lambda$).

COLOR, FLAVOR, AND THE POMERON [1]

Jan W. Dash
CPT 2, CNRS, Marseille, France F13288

I discuss diffractive hadron-hadron scattering at $\rho_{cf} \equiv N_c/N_f = \infty$ and at $N_c = 3$. Unlike some soft hadronic parameters like resonance widths, diffraction can change significantly as a function of N_c. The reason is that diffraction is constrained by s-channel unitarity to inelastic hadron states. These states are not realistically described at $\rho_{cf} = \infty$ where the absence of quark loops is tied to the $\alpha_G \approx 1$ glueball-Pomeron association. Unlike deep inelastic processes, the $g \to q\bar{q}$ transitions cannot be innocuously bypassed here.[2] Specifically, virtual quark loops can damp the $\rho_{cf} = \infty$ glueball intercept to $\alpha_G \approx 0$, while cut quark loops provide a large positive renormalization of the leading planar trajectory $\alpha_{P\ell}$ to $j \approx 1$. The details depend on ordinary light and heavy hadron masses dependent on quark parameters and hence not describable at the no-quark $\rho_{cf} = \infty$ limit.

In particular, at $N_c = 3$ the origin of the bare Pomeron pole at $j \approx 1$ must be viewed through the quark content of s-channel unitarity. Unlike the situation at $\rho_{cf} = \infty$, it seems difficult to motivate the usual "Pomeron plus nearly-EXD-ideally mixed f" model at $N_c = 3$. In contrast, it seems more natural to motivate the Pomeron-f identity including trajectory renormalization due to successive heavy quark excitation ("flavoring")[3] along with possible low lying glueball effects.

REFERENCES

1. Details will be presented separately (Zeit. Phys. C, to appear).
2. One could argue that the $\rho_{cf} = \infty$ limit is tied to a $g = 0$ dual resonance model which does contain finite particle multiplicities via cascade decays. However this also implies a singular $\rho_{cf} = \infty$ limit in which quark loops remain, obscuring the no-quark glueball-Pomeron identification. Moreover, the $g = 0$ dual model contains various unrealistic features. See ref. 1.
3. J. Dash, S.T. Jones, and E. Manesis, Phys. Rev. D18, 303 (1978).

HIGH ENERGY HADRONIC INTERACTIONS

R.C. HWA, OREGON, ORGANIZER

DESCRIPTION OF INCLUSIVE π^+/π^- PRODUCTION RATIOS, $\bar{p}p$ ANNIHILATION AND PROTON DIFFRACTION DISSOCIATION AT MEDIUM ENERGIES WITH THE QUARK MODEL

M. Markytan

Institut für Hochenergiephysik der Österreichischen Akademie der Wissenschaften, Vienna, Austria

COMPARISON OF INCLUSIVE π^+/π^- PRODUCTION RATIOS FOR $\pi^\pm p$, K^-p, pp AND $\bar{p}p$ INTERACTIONS

The x distributions of $R_H = \int(dp_T^2 E^* d\sigma(\pi^+)/dp_T^2 dx)/\int(dp_T^2 E^* d\sigma(\pi^-)/dp_T^2 dx)$ for hadron-proton interactions in the proton fragmentation region at medium energies (fig.1) show a strong dependence on the quantum numbers of the beam particle, but a weak energy dependence for pp [1]. This result, reminiscent of Regge behaviour, is at variance with the conjecture by most quark models that target proton fragmentation is independent of the projectile valence quarks.

To that purpose the quark recombination model of Das and Hwa [2] was extended in ref. 3 to include the annihilation into gluons of a valence quark and antiquark, both assumed to be wee [4,1], of the target proton and the projectile respectively. Thus

$$(x/\sigma_H).d\,\sigma/dx = \int dx_1 \int dx_2\, G_{q_1q_2}(x_1,x_2).R(x_1,x_2,x) + $$
$$+\, C_H.\int_0^{x_0} dx_j \iint dx_1 dx_2\, G_{q_jq_1q_2}(x_j,x_1,x_2).R(x_1,x_2,x), \qquad (1)$$

where the three quarks structure function G_{qjq1q2} of the proton is required to have one valence quark q_j with momentum less than $x_o = \text{const}/\sqrt{s}$ in the non-scaling term. C_H contains the wee valence antiquark probability of the projectile. $G_{q1q2} = G^{VS} + G^{SS}$ and $G_{qjq1q2} = G^{VVS} + G^{VSS}$ (V=valence, S=sea). These structure functions are calculated [3] using the Kuti-Weisskopf model [5] in which Regge behaviour is translated into the x distribution of valence quarks. The sea quark and gluon input structure functions parametrized as $g_a^2(1-x)^{k_a}$ were determined from deep inelastic ep and en scattering and from R_p in the 1oo - 4oo GeV energy range to have the following numerical values: $g^2_{u,\bar{u},d,\bar{d}}$ =o.1375, $g^2_{s,\bar{s}}$ = o.oo95, g^2_G=3, $k_{u,\bar{u},d,\bar{d},s,\bar{s}}$ =1.2 and k_G = 2.7.

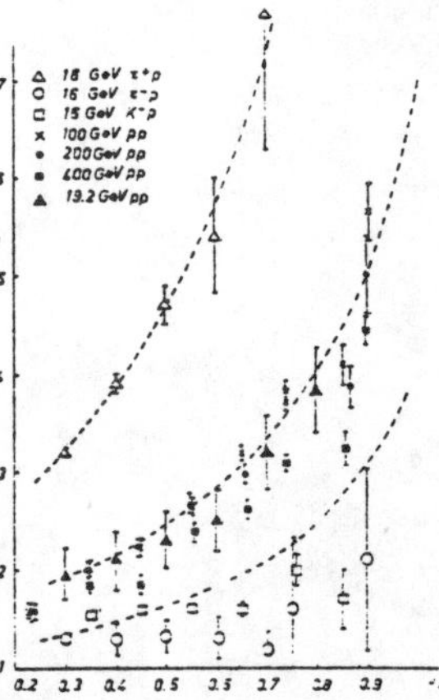

Fig. 1. Comparison of the proton fragmentation ratio R_H of π^+/π^- production in $H p$ interactions for $H = \pi^\pm$, K^- and p.

Fitting equ.(1) to the x distributions of R_{π^+} at 16,1oo and 147 GeV/c, which show the largest variation with energy, determines C_H so that R_{π^-}, R_{K^-}

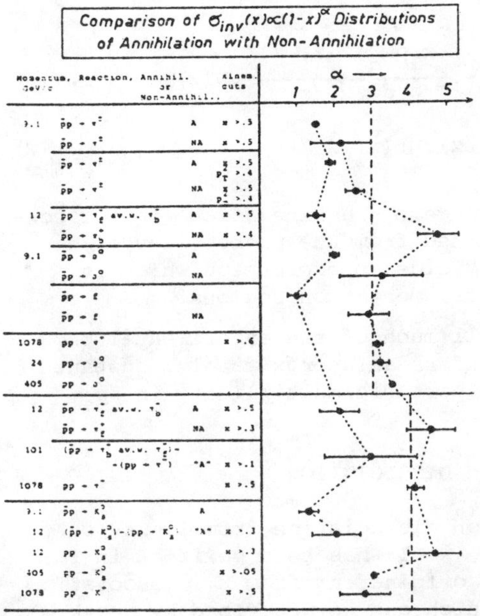

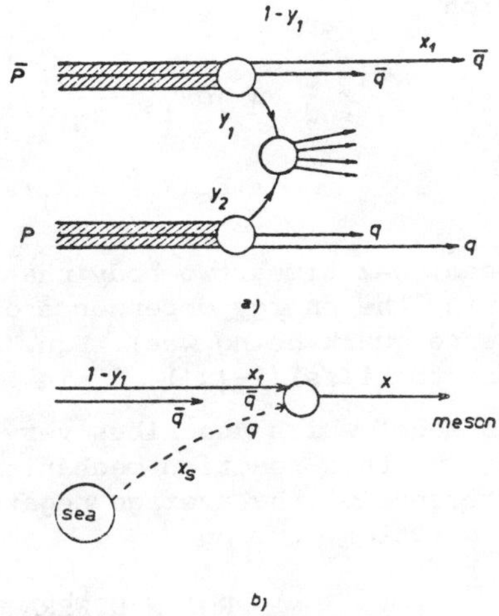

Fig. 3. Quark model picture of the reaction mechanism of $\bar{p}p$ annihilation.

Fig. 2. x distributions of mesons produced by $\bar{p}p$ annihilation in comparison to pp and $\bar{p}p$ non-annihilation.

and $R_{\bar{p}}$ could be predicted for $|x| > 0.2$ and arbitrary energies without free parameters. Some of the fits and predictions are given by the curves in fig. 1. It can be concluded that the extended quark recombination model of Buschbeck et al. yields a very good description of the x and energy dependence of R_H for $H = \pi\pm$, K^-, p and $\bar{p}$.

The dominance of valence quark-antiquark annihilation over fusion is argued from the results of $\pi^\pm p \rightarrow \rho^0$ and $\rho^\pm X$ at 16 GeV/c [6] which infer that if $\rho^\pm$ production is described by fusion alone a much greater cross section of ρ^0 production would be the consequence.

$\bar{p}p$ ANNIHILATION

The energy dependence of the $\bar{p}p$ annihilation cross section has been found proportional to $p_{lab}^{-0.61}$. It is shown in fig.2 that mesons sharing a $u(\bar{u})$ quark with the incident (anti)proton are distributed in x as $(1-|x|)^{1.5\div 2.1}$ in annihilation in contrast to $(1-|x|)^3$ in pp and $\bar{p}p$ non-annihilation. This difference is confirmed by the SLAC experiment BC-64 [7].
The energy dependence allows, according to fig.3a, to interpret the pp annihilation mechanism as being initialized by the annihilation of a wee valence quark of the proton and a wee valence antiquark of the antiproton [4]. The formation of valence diquark clusters [8] distributed as $(1-|x|)^1$ ensues. In contrast to their immediate recombination to baryons they are assumed to disintegrate as the two-body phase space before recombining with sea quarks to mesons

in annihilation (fig.3b). The quark recombination model of Das and Hwa[2] has been extended [9] to account for this picture of $\bar{p}p$ annihilation:

$$x\,\frac{d\sigma}{dx} \sim \int dx_1 \int \frac{dx_S}{x_S} \left(\int_{x_1}^{1} \frac{dz}{z} . \frac{x_1 dN_{qq}(x_1/z)}{z\, d\,(x_1/z)} . D(z) \right) . $$

$$. F_S(x_S) . R(x_1/x, x_S/x) . \rho\,(x_1, x_S) \qquad (2)$$

(S=sea, D=z times two body phase space, ρ = a kinematical factor containing the energy dependence obtained from the probability of a valence quark being wee). Equ.(2) yields an expression which is close to $(1/\sqrt{s})(1-|x|)^{1.5}$ in the fragmentation regions.

This model which describes very well much of the experimental knowledge of this reaction mechanism infers an approximately constant difference of the average negative particle multiplicity in $\bar{p}p$ annihilation from pp.

PROTON DIFFRACTION DISSOCIATION

The extraction of proton diffraction dissociation from 11oGeV/c K^-p interactions is shown in fig.4 [1o]. It has been verified in 16 GeV/c K^-p that negative hadrons h^- originating from K^- dissociation behave similar to the background which can be described by quark spectator counting rules.

This method was applied to events with visible strange particle decays (fig.5). The cross sections obtained are summarized in table I. Assuming equal probabilities among the K+pions and $NK\bar{K}$ +pions

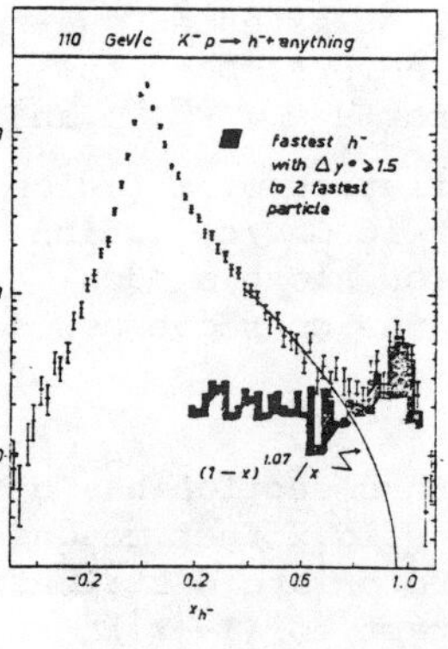

Fig. 4. Proton diffraction dissociation in $K^-p \rightarrow$ negative hadrons at 110 GeV/c above a background determined by quark spectator counting rules.

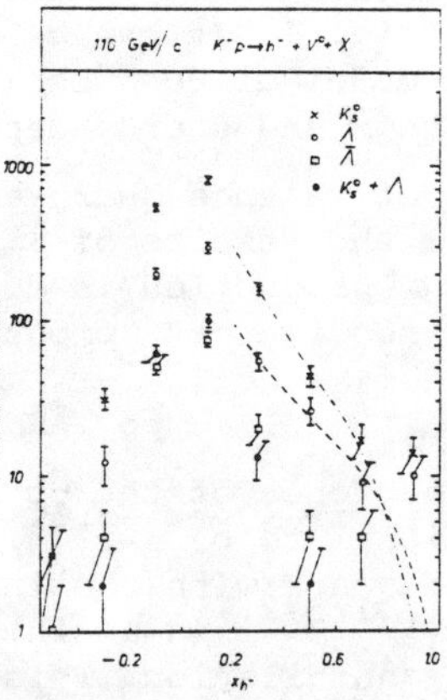

Fig. 5. Estimation of proton diffraction dissociation into K^o and Λ in 110 GeV/c K^-p interactions.

p_{lab}, GeV/c	V^o Particle	Proton Diffraction Dissociation for Given V^o, μb	x Interval Used for Background Fit	Power n of Background
16	Λ	31 ± 5	o.4 ÷ o.6	1.25 ± o.3
	K^o	-----	o.4 ÷ o.6	1.64 ± o.31
32	Λ	3o ± 1o } predicted for $s\bar{s}$: 7o ± 1o	o.4 ÷ o.7	o.77 ± o.o7
	K^o	6o ± 2o	o.4 ÷ o.7	2.o6 ± o.27
11o	Λ	72 ± 28	o.2 ÷ o.8	1.o6 ± o.o4
	K^o	153 ± 55	o.2 ÷ o.8	1.85 ± o.6o
		Kaon Diffraction Dissociation for Given V^o, b		
11o	K^o	634 ± 111		
	Λ	1o ± 1o		
	$s\bar{s}$ System	Proton Diffraction Cross Section for given $s\bar{s}$ system, estimated		
11o	ΛK	72 ± 28 } predicted for $s\bar{s}$: 112 ± 2o		
	$K\bar{K}$	142 ± 54		

Table I Cross sections of proton diffraction dissociation into strange particles from K^-p interactions

states, an estimate for inclusive proton diffraction into $s\bar{s}$ was obtained. Using the model of Gustafson and Peterson [11] with a factor $\left(\int_{2m_Q/\sqrt{s}}^{1} dx\,(1-x)^5/x \right)^2$, the total inclusive proton diffraction dissociation cross section yields an $s\bar{s}$ value in reasonable agreement with experiment. This model infers cross section of 1.18 mb and 21.6 μb for diffractive $s\bar{s}$ and $c\bar{c}$ production respectively in s=63 GeV pp interactions. Assuming Pomeron exchange to be double gluon exchange to one of which a heavy $Q\bar{Q}$ pair is coupling, the following expression for diffraction dissociation follows:

$$\sigma_D(Q\bar{Q}) \propto \int_0^{1-\varepsilon} dx/(1-x) \int_{2m_Q/\sqrt{s}}^{1-x} dx_1(1-x_1)^{n_1}/x_1 \cdot \frac{1}{M_Q^2} \cdot \int_{2m_Q/\sqrt{s}}^{1} dx_2(1-x_2)^5/x_2. \quad (3)$$

n_1 depends on the nature of the beam particle.

REFERENCES

1. B. Buschbeck et al., Zeitschr. f. Physik C3, 97 (1979).
2. K.P. Das and R.C. Hwa, Phys. Lett. 68B, 459 (1977).
3. B. Buschbeck et al., paper 3o9 submitted to this conference.
4. K. Kinoshita and S. Pokorski, Lett. al Nuovo Cimento 16, 498 (1976).
   M. Markytan, Proc. II Int. Symp. on Hadron Structure and Multiparticle Production, Kazimierz, Poland, 2o-26 May, 1979.
5. J. Kuti and V.F. Weisskopf, Phys.Rev. D4, 3418 (1971).
   R. Gerhold, HEPHY Vienna Preprint 198o.
6. R. Kinnunen, Univ. of Helsinki preprint HU-P-178 (198o).
7. D.R. Ward, Cavendish Lab., Cambridge, England. Paper submitted to the 5th Eur. Symp. on Nucleon Anti-Nucleon Interactions, Bressanone, Italy, June 198o.
8. J. Benecke, Proc. 18th Int. Conf. on High Energy Physics, Tbilisi, July 1976, p. A3-12.
9. H. Muirhead et al., paper 293 submitted to this conference.
1o. J. MacNaughton et al., paper 294 submitted to this conference.
11. G. Gustafson and C. Peterson, Phys. Lett. 67B, 81 (1977).

SMALL p_T HADRON REACTIONS BY MONTE CARLO SIMULATION

Y. ISHIHARA and C. ISO (Presented by C. I.)
Tokyo Institute of Technology, Meguro-ku, Tokyo, 152-Japan

Small p_T hadron jets are generated by Monte Carlo method using our quark fragmentation model[1] with recombination mechanism. We analyse inclusive spectra, particle spectra under the other particle trigger and correlation functions.

The hadron jet generation in the proton fragmentation region by our recombination model proceeds as is illustrated in Fig. 1. At first the proton separates into two u quarks and one d quark, whose Feynman variables are assigned x_0', x_0'' and x_0''', respectively. We parametrize the x distribution of the initial quarks as follows;

$$6/(3-p)\,dx_0'dx_0''dx_0'''(1-px_0''')\,\delta(x_0'+x_0''+x_0'''-1), \quad \text{(p:parameter)},$$

where p=0 corresponds to the equipartition distribution. We chose p=1 in order to fit our prediction to the observed π^+/π^- ratio. Secondly, the successive type meson fragmentation occurs according to the meson emission function $f(x_0-x,x_0)=(1+\alpha)/x_0(x_0-x/x_0)^{\alpha}$ and broken SU(3) type meson emission probabilities parametrized by 4 parameters. We took α=1.5 and set the ratio ps/v=1. Finally, the remaining three quarks recombine to a baryon. The recombination parameter λ', the probability ratio of recombination to meson emission, is chosen to be 0.5 in case of hadron reactions.

We got the following results;[2,3]

(a) Considering the decay of the vector mesons, we successfully reproduce the inclusive spectra of the reactions $p\to pX$, $\Delta^{++}X$, $\pi^{\pm}X$, $K^{\pm}X$, $\pi^+\to\pi^{\pm}X$ etc. at small p_T including the particle ratio R(p $\to$ π^+/π^-). We also succeed to describe $e^+e^-\to$meson$^{\pm}$X in an unified way by taking the c-quark effect into account, and by taking λ'=0.01.

(b) Charged pion spectra (given in x) and charged pion trigger particle (parametrized in x_{trig}) scale in the proton fragmentation region with the variable $\tilde{x}=x/(1-x_{trig})$ in our recombination model. On the contrary, when we specialize charges for those pions, small deviation from $\tilde{x}$ scaling is seen owing to charge conservation. The $\pi^{\pm}$ meson spectra under π^+ meson trigger deviate from scaling in an opposite way, which causes a remarkable $\tilde{x}$ scaling violation to the π^+/π^- ratio under π^+ meson trigger; our prediction nicely fits to de Wolf et. al's data (Fig. 2). The predicted $\pi^{\pm}$ spectra, and π^+/π^- ratio under π^- meson and Λ particle trigger also deviate from $\tilde{x}$ scaling and fit to the observed data. The $\tilde{x}$ scaling

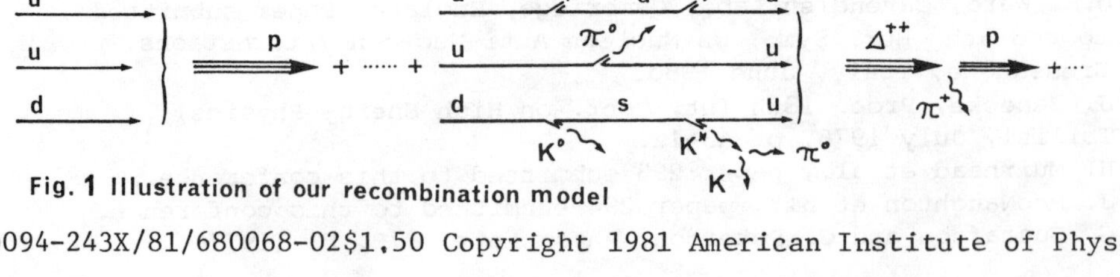

Fig. 1 Illustration of our recombination model

violation in our recombination model contradicts the scaling behaviour predicted using Das and Hwa's recombination model. The experimental check on the existence of such a scaling violation may be a good test of these models.

(c) In order to fit the quantities stated above with the experimental data, the asymmetrical initial condition (p=1) is essential.

(d) The two meson correlation functions for the reactions $u \to \pi^{ch}(x_1) + \pi^{ch}(x) + X$, $\pi^+ \to \pi^{ch}(x_1) + \pi^{ch}(x) + X$, and $p \to \pi^{ch}(x_1) + \pi^{ch}(x) + X$ are plotted in Fig. 3. The correlation functions depend much on the recombination mechanism and, as we see in Fig. 3, they depend on the value of λ' in the $x \sim 0.1$ region. According to our model the correlation function for the reaction $\pi^+ \to \pi^{ch} + \pi^{ch} + X$ near $x \sim 0.8$ is remarkably dirrerent from that for the reaction $p(\text{or } u) \to \pi^{ch} + \pi^{ch} + X$. This is the effect of the recombined meson spectrum in the pion's fragmentation region. The measurement of the correlation function will be useful to check the recombination mechanism.

REFERENCES

1. H. Fukuda and C. Iso, Prog. Theor. Phys. 57, 483 (1977).
2. H. Fukuda, Y. Ishihara and C. Iso, Contributed paper 529.
3. Y. Ishihara and C. Iso, "π^+/π^- Ratio and Two Particle Production in Proton's Fragmentation Region by Monte Carlo Simulation".

For the other references see the above papers.

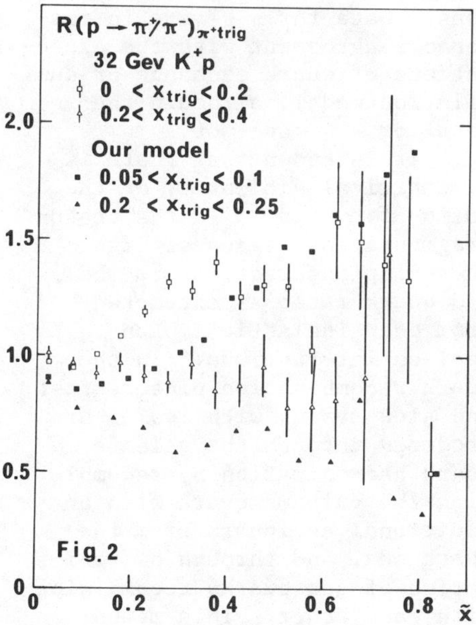

Fig. 2 π^+/π^- ratio under π^+ meson trigger vs. x.

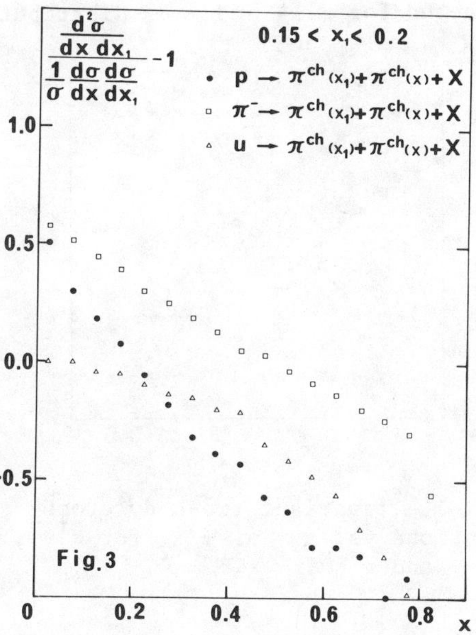

Fig. 3 Correlation function vs. x for fixed x_1=0.15.

"Massive Meson Production, Correlations, Primordial p_T etc."

K. Takahashi
KEK-OCU-KINKI Collaboration, KEK, Japan

We present results on inclusive single- and double-pion distributions in 405 GeV/c pp interactions from a Fermilab experiment using the 30-in. hydrogen bubble chamber[1)2)]. The Feynman x distributions and particle production ratios of double-pion systems are analysed. The x distributions are compared with the prediction of quark-counting rule. Pion production ratios are analysed on the basis of the quark recombination model[3)]. Primordial motion of quarks studied from the analysis of the average values of transverse momenta p_T of leading $\pi^{\pm}$ in the fragmentation region is also presented.

There have been only very scarce data of inclusive two-pion systems with various charge combinations in the same fragmentation direction. In Fig. 1 we show the x distributions of $\pi^+\pi^+$, $\pi^+\pi^-$ and $\pi^-\pi^-$ combinations produced in the backward hemisphere of the c.m.s., where $x = x_1 + x_2$, and 1 or 2 stands for each one of two pions. Data with $|x| \geq 0.2$ are fitted to the functional behavior $(1 - |x|)^n$. Exponents are obtained to be 3.7 ± 0.2 for $\pi^+\pi^+$, 3.1 ± 0.2 for $\pi^+\pi^-$ and 4.7 ± 0.2 for $\pi^-\pi^-$ respectively. The quark exchange or annihilation model predicts $(1 - |x|)^3$ for $\pi^+\pi^{\pm}$ and $(1 - |x|)^7$ for $\pi^-\pi^-$. According to gluon exchange, $(1 - |x|)^5$ and $(1 - |x|)^9$ behaviors are expected for $\pi^+\pi^{\pm}$ and $\pi^-\pi^-$ distributions. Data for $\pi^+\pi^{\pm}$ are in reasonable agreement with the expectations of quark exchange or annihilation model, although the data for $\pi^-\pi^-$ are not.

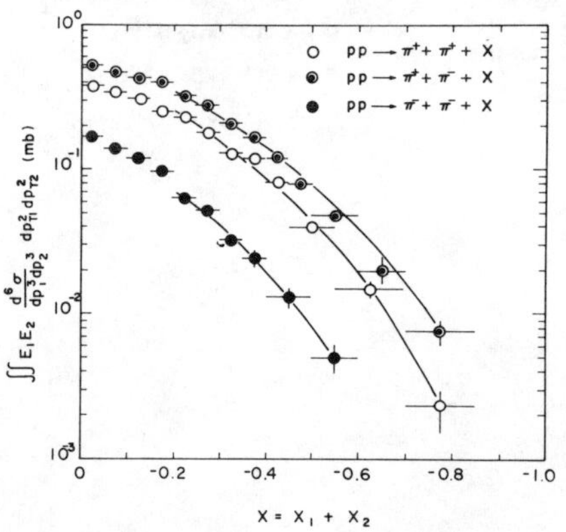

Fig. 1. Invariant two-body cross-sections vs. $x = x_1 + x_2$ for $\pi^+\pi^+$, $\pi^+\pi^-$ and $\pi^-\pi^-$.

It is known that there is an empirical similarity of the ratio of π^+ and π^- in the proton fragmentation region of high energy pp interactions to the u/d quark ratio as extracted from deep inelastic lepton-nucleon interactions. In the quark recombination picture fast and slow mesons with low p_T are produced through the valence quark hadronization by recombining the valance quark with an additional antiquark from the quark sea, and through hadronization of sea quarks recombining with each other. This means that a particle ratio of produced hadrons in the interactions is a good measure to the quark distribution and its hadronization function. Here we report the x dependence of double particle production ratios which we define as

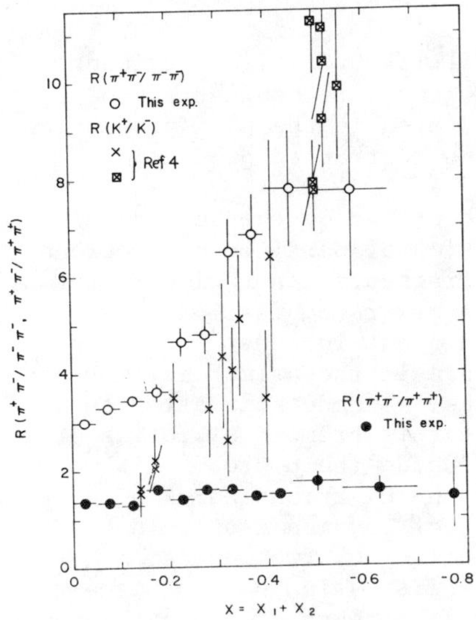

Fig. 2. Double pion production ratios vs. $x = x_1 + x_2$. Ratios of $\pi^+\pi^-$ to $\pi^-\pi^-$ are compared with those of K^+ to K^- in Ref. 4.

$$R(h_1h_2/h_3h_4) =$$

$$\iint E_1E_2\frac{d^6\sigma}{dp_1^3dp_2^3}(h_1h_2)dp_{T1}^2dp_{T2}^2/$$

$$\iint E_3E_4\frac{d^6\sigma}{dp_3^3dp_4^3}(h_3h_4)dp_{T3}^2dp_{T4}^2.$$

Fig. 2 gives the results. A rapid rise with increasing x in $R(\pi^+\pi^-/\pi^-\pi^-)$ is observed. This fact is understandable by the following situation that in the fragmentation region of the proton, a fast $(\pi^+\pi^-)$ pair can contain valence u_V and d_V quarks, and in a fast $(\pi^-\pi^+)$ pair, only one π^- can take the valence d_V quark and the other π^- consists of two sea quarks. Such recombination picture suggests an interesting approximate identity between $R(\pi^+\pi^-/\pi^-\pi^-)$ and $R(K^+/K^-)$. Both ratios are similar in the shapes of x dependence and in the absolute magnitudes in the region of $|x| \gtrsim 0.3$.

The ratio of $\pi^+\pi^-$ to $\pi^+\pi^+$ is also shown in Fig. 2. One can find almost constant x dependence in the whole x region.

The ratio of π^+ to π^- in association with a backward $\pi_b^\pm$ or Λ_b trigger in the target fragmentation region is also investigated by using selected events which contain leading backward $\pi_b^\pm$ or Λ_b production. Here we define a leading backward particle in the target fragments as a particle, having not only the largest $|x|$ in the event, but also the value of $|x|$ greater than 0.2. In the quark recombination picture, π^+ and π^- produced in association with a leading backward π_b^- or Λ_b trigger in the target fragments contain $(u_V \bar{d}_S)$ for π^+ and $(d_S \bar{u}_S)$ for π^-. It is therefore expected that the ratio $R(\pi^+/\pi^-)$ for the $\pi_b^\pm$ trigger, exceeds the value for the Λ_b trigger by a factor of two. Moreover in the ratio of two-pion systems, the ratio, $R(\pi^+\pi^-/\pi^-\pi^-)$ is considered to have approximately the same x dependence as $R(\pi^+/\pi^-)$ in the leading backward π_b^- triggered events by the quark recombination model.

Fig. 3(a) shows the $\tilde{x}$ dependence of $R(\pi^+/\pi^-)$ in association with $\pi_b^\pm$ and Λ_b triggers, where $\tilde{x}$ is equal to $x(\pi^\pm)/[1 - x(\text{trigger particle})]$. Fig. 3(b) shows the comparison of the $\tilde{x}$ dependence of three different pion production ratios, $R(\pi^+/\pi^-)$ for the leading π_b^-, $2 \times R(\pi^+/\pi^-)$ for the leading Λ_b trigger, and $R(\pi^+\pi^-/\pi^-\pi^-)$ in the two-pion system, where $\tilde{x} = x_1 + x_2$ for $R(\pi^+\pi^-/\pi^-\pi^-)$. The agreement of three distributions looks pretty good, particularly in larger $|\tilde{x}|$

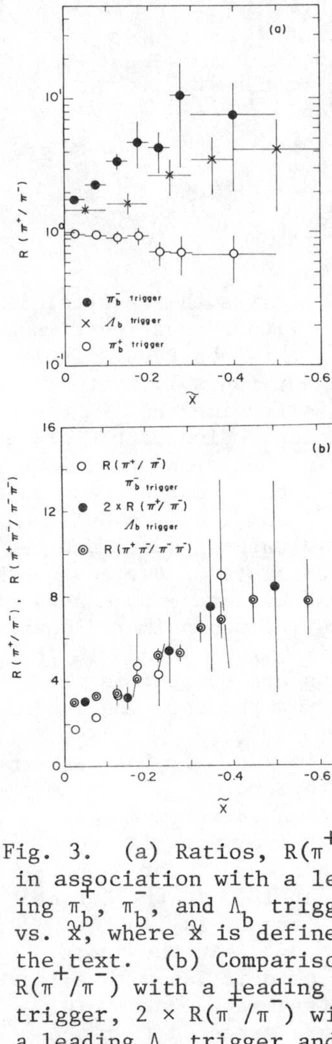

Fig. 3. (a) Ratios, $R(\pi^+/\pi^-)$ in association with a leading π_b^+, π_b^-, and Λ_b trigger vs. $\tilde{x}$, where $\tilde{x}$ is defined in the text. (b) Comparison of $R(\pi^+/\pi^-)$ with a leading π_b^- trigger, $2 \times R(\pi^+/\pi^-)$ with a leading Λ_b trigger and $R(\pi^+\pi^-/\pi^-\pi^-)$ in the two-pion system.

region, $|\tilde{x}| > 0.2$. It is evident that the quark recombination model is valid even in the pion production under the restricted production processes.

Encouraged by the successful description of particle productions in the fragmentation region in terms of the quark recombination model, and by a great interest in recent literature in the primordial motion of quarks, we made some attempts to estimate this primordial motion of quarks inside the proton[5]. We studied the distribution of p_T as a function of Feynman x only in the backward ($y \leq 0$) fragmentation.

We first selected the highest p_T track from each event in the rapidity region $y < 0.0$ and plotted the average of these maximum p_T tracks as a function of x (Fig. 4(a)). $\langle p_T \max\rangle$ reaches a constant value of 0.6 - 0.7 GeV/c which is consistent with the result of $\mu^+\mu^-$ pair production. This behavior of $\langle p_T \max\rangle$ suggests to us that the constant value reflects the maximum primordial p_T-value of quarks inside the target proton. In terms of the recombination model, we may expect at large $|x|$-values;

$$x(\pi^+) = x(u_V) + x(\bar{d}_S) \gtrsim x(u_V) \text{ and}$$
$$x(\pi^-) = x(d_V) + x(\bar{u}_S) \gtrsim x(d_V).$$

This makes us able to measure the x-dependence of the primordial p_T of the quarks from the x-dependence of the outgoing pions we select.

In Fig. 4(b) we have changed the rapidity cut to be $y \leq -2.0$. The behavior is still the same and even more distinct.

Fig. 4(c) shows $\langle p_T \min\rangle$ versus x with the restriction that $\langle p_T \max\rangle \neq \langle p_T \min\rangle$. We expect to see a different behavior in the x-distribution of $\langle p_T \max\rangle$ and $\langle p_T \min\rangle$. The values of $\langle p_T \min\rangle$ reaches again a "constant" for $|x| \gtrsim 0.2$, but it is lower, $\sim$0.3 GeV/c, than in the previous case. This lower constant value which $\langle p_T \min\rangle$ reaches reflects that either we select quarks with primordial p_T-values lower than the maximum values the valence quarks can obtain, or our selected outgoing pions now contain more sea-quarks than they

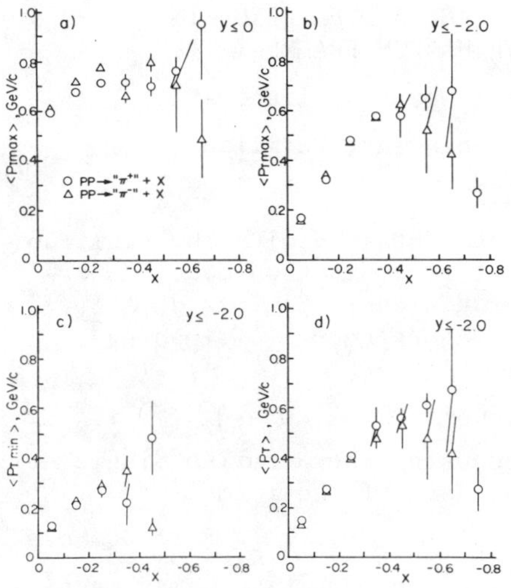

Fig. 4. (a) $<p_T>$ of tracks with maximum p_T, $<p_T \max>$, as a function of x, in $y \leq 0$. (b) The same as (a) except the region of rapidity y which is now taken as $y \leq -2.0$. (c) $<p_T>$ of tracks with minimum p_T, $<p_T \min>$, as a function of x in $y \leq -2.0$. In this case we fixed $<p_T \max> \neq <p_T \min>$. (d) Average p_T of maximum y tracks of π^+, and π^- vs. x in $y \leq -2.0$.

otherwise would have done. Therefore the x-distribution for $<p_T \min>$ will give us information about the $<p_T>$-values of the sea-quark at least for low $|x|$. Fig. 4(d) gives the most significant description of the phenomena. We have plotted $<p_T>$ versus x with the y-cut, $y \lesssim -2.0$, but restricting rapidity to be maximal. The $<p_T>$ values are clearly rising up to a constant value 0.6 - 0.7 GeV/c for $|x| \gtrsim 0.2$. This again suggests that the constant value for $<p_T \max> \sim 0.6 - 0.7$ GeV/c, would be associated with the maximum primordial momentum of quarks inside the proton.

References

1. A. Suzuki et al., Contributed papers no's 497, 563, 564 and 565.
2. For massive meson production; see A. Suzuki et al., Nuclear Phys. to be published 1980.
3. K. P. Das and R. Hwa, Phys. Lett. 68B, 459(1977), and ibid. 73B,504(1978).
4. P. Capiluppi et al., Nucl. Phys. B70, 1(1974); and ibid. B79, 189(1974); B. Alper et al., ibid. B100, 237(1975); J. Singh et al., Nucl. Phys. B140, 189(1978).
5. G. Altarelli, G. Parisi and R. Petronzio, Phys. Lett. 76B, 351(1978).

MESON STRUCTURE FUNCTIONS AND A-DEPENDENCE IN SINGLE PARTICLE INCLUSIVE HADRON FRAGMENTATION

Bari-Brown-CERN-Fermilab-MIT-Warsaw Collaboration

Presented by Peter H. Garbincius, Fermilab

ABSTRACT

Low p_t hadronic fragmentation data obtained with the Fermilab Single Arm Spectrometer were used in a recombination model determination of pion and kaon structure functions. Similar data for nuclear targets was used to study the A-dependence of leading particle and fragmentation reactions.

INTRODUCTION

Hadron fragmentation was studied using the Fermilab Single Arm Spectrometer. Data on inclusive processes of the form

$$a + b \rightarrow c + x$$

where a and c may be any combination of $\pi^{\pm}$, $K^{\pm}$, p, or $\bar{p}$ and b may be p, C, Al, Cu, Ag, or Pb were taken over the range $0.3 \leq$ x-Feynman ≤ 0.9 and $0.18 \leq p_t \leq 0.75$ GeV/c, for ± 100 and ± 175 GeV/c beam momenta. The beam line contained one threshold, one DISC, and two differential Cerenkov counters, while the spectrometer contained one differential and three threshold counters, allowing simultaneous and unambiguous identification of all reaction types.

MESON STRUCTURE FUNCTIONS[1]

The recombination model[2] describes the invariant cross section for the fragmentation of beam meson a into meson $b(q_1,\bar{q}_2)$ as

$$\frac{x}{\sigma^{abs}} \frac{d\sigma^{a\rightarrow b}}{d\,x}(x) = \int\int f^a_{q_1}(x_1)\, f^a_{\bar{q}_2}(x_2)$$

$$* K^b \frac{x_1}{x}\frac{x_2}{x}\,\delta\left(\frac{x_1}{x} + \frac{x_2}{x} - 1\right)\frac{dx_1}{x_1}\frac{dx_2}{x_2}$$

where $f^a_q(x)$ is the structure function for finding quark q, at x, in particle a, and $K^b\,(x_1/x)\,(x_2/x)\,\delta(x_1/x + x_2/x - 1)$ represents the probability of the quarks, q_1 and $\bar{q}_2$ recombining to form meson b. Data for all charged mesons fragmenting into different charged mesons have been simultaneously fit using valence structure functions of the form

$$V_q(x) = a\sqrt{x}(1-x)^n \text{ with normalization } \int_0^1 V_q(x)\frac{dx}{x} = 1,$$

and sea structure functions of the form

$$S_q(x) = a(1-x)^n \,.$$

The results for the powers n are:

	V_u^π	S_{n-s}^π	S_s^π	V_u^K	S_{n-s}^K	S_s^K
n	1.0	3.5	2.2	2.5	3.6	7.5
	± 0.1	± 0.2	± 0.2	± 0.6	± 0.3	± 2.6

The strange and the non-strange valence quark distributions may be different for the kaon. However, we have little sensitivity for the strange valence distribution. The sum of the quark momenta is 1.2 ± 0.1 for the pion and 1.0 ± 0.1 for the kaon. The pion valence structure function, Figure 1a, agrees well with those obtained for other reactions and different phenomenologies,[3,4,5] and with some theories[6]. There is some disagreement, however, between fragmentation,[1] large p_t π^0 production,[5] and Drell-Yan[7] determinations of the kaon valence structure functions. This is shown as the ratio of kaon to pion valence structure functions in Figure 1b. The non-strange sea distributions have identical powers for both pions and kaons.

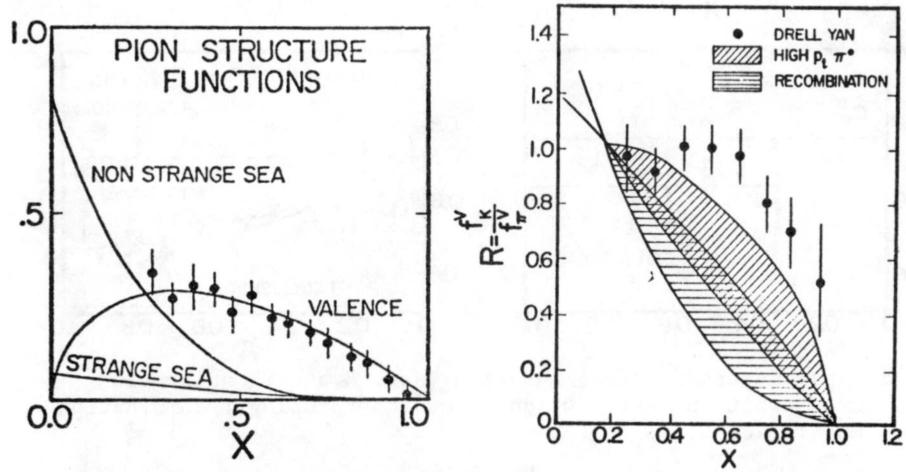

Fig. 1 a) Pion structure functions from recombination model fit. Data points are from Ref. 3.

b) Ratio of kaon to pion valence structure functions using recombination model (Ref. 1), high p_t π^0 production (Ref. 5) and Drell-Yan di-muon production (Ref. 7).

A-DEPENDENCE OF LEADING PARTICLE AND FRAGMENTATION REACTIONS[8]

The A-dependence of the invariant cross section per nucleus for each channel at each x and p_t has been fitted using the form

$$E \frac{d\sigma^{a\to b}}{d^3p}(x,p_t,A) = f_1^{a\to b}(x,p_t)\; A^{\alpha(x,p_t,a,b)}$$

over the range $12 \le A \le 207$. Values of α for typical channels are depicted in Figure 2. Also included are α values for 24 GeV/c pA data,[9] absorption cross sections,[10] a single collision model where the nucleus is transparent to all secondary particles, and the model of Dar and Takagi.[11] It is observed that the effectiveness of nucleons at producing particles decreases with A for both leading and non-leading inclusive reaction types. There is no discernible change of α over the range $24 \leq E \leq 100$ GeV and $0.18 \leq p_t \leq 0.5$ GeV/c. The quark stripping recombination model of Dar and Takagi represents the A-dependence of reaction types with leading quarks quite well.

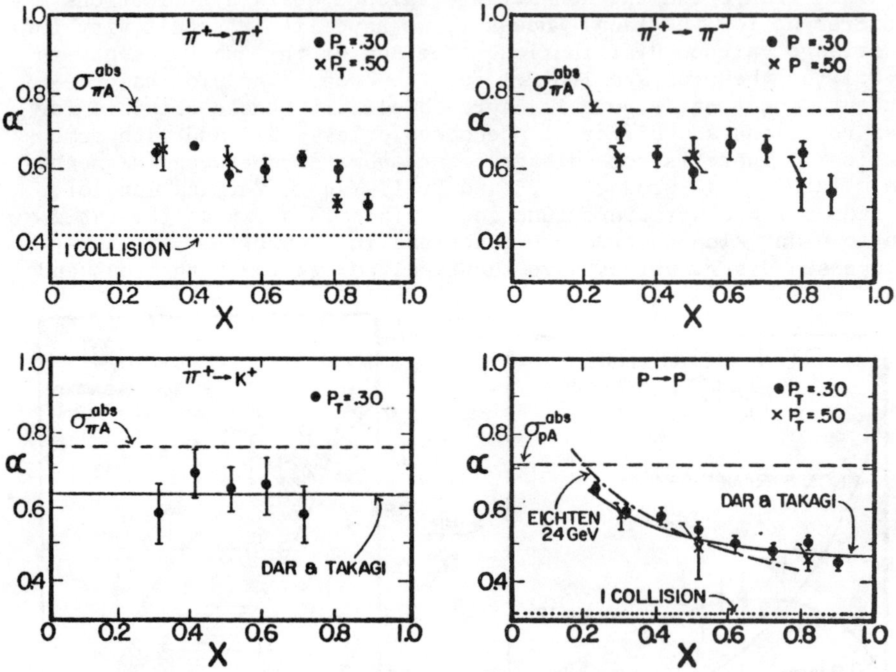

Fig. 2 α parameters for fragmentation (Ref. 8 and 9), absorption cross section (Ref. 10), and a quark stripping/recombination model (Ref. 11).

REFERENCES

1. W. Aitkenhead, et al, Phys. Rev. Lett. 45, 157 (1980).
2. See Ref. 1 for a fairly complete list of references.
3. C. B. Newman, et al, Phys. Rev. Lett. 42, 951 (1979).
4. J. Badier, et al, Phys. Lett. 89B, 145 (1979).
5. K.-W. Lai and R.L. Thews, Phys. Rev. Lett. 44, 1729 (1980).
6. R. D. Field and R. P. Feynman, Phys. Rev. D15, 2590 (1977), S. J. Brodsky and J. F. Gunion, Phys. Rev. D17, 848 (1978), and E. L. Berger and S. J. Brodsky, Phys. Rev. Lett. 42, 940 (1979).
7. J. Badier et al, Phys. Lett. 93B, 354 (1980).
8. A. E. Brenner, et al, Preprint Fermilab - Conf-80/47-Exp(1980).
9. T. Eichten, et al, Nucl. Phys. B44, 333 (1972).
10. A. S. Carroll, et al, Phys. Lett. 80B, 319 (1979).
11. A. Dar and F. Takagi, Phys. Rev. Lett. 44, 768 (1980).

π^--NEON COLLISIONS AT HIGH ENERGY

W. D. Walker
Duke University, Durham, N. C. 27706

ABSTRACT

Data on π-Neon collisions is summarized. Data from π^+ and π^--Ne collisions at 30 GeV confirm the presence of near relativistic protons among the collision products. The rapidity distributions from π^--Ne collisions at 200 GeV are shifted to lower average rapidity than the corresponding π^--p collisions. The plateau height for these rapidity distributions is higher by a factor of nearly 1.3 than the π^--p counterpart. A short discussion of the comparison with various models is given.

INTRODUCTION

We have known for some while that the multiplicity of π mesons produced in hadron-nucleon and in hadron-nucleus collisions do not differ very much. We report here data obtained in a bubble chamber study of collisions between π mesons and neon nuclei with π^+ and π^- incident at energies of 10.5[1] and 30 GeV[2] as well as π^- at 200 GeV[3]. The following table summarizes the results which have been obtained from the measurements with the help of charge symmetry.[1]

Table I. Multiplicities — π-Neon

	10.5 GeV/c	30 GeV/c	200 GeV/c
(Min. Tracks)	3.91 ± .03	5.8 ± .03	9.41 ± .21
(Slow Protons)	1.50 ± .08		1.77 ± .1
(Fast Protons)	.46 ± .12	.68 ± .05	.44 ± .14
(π^+, π^-)	3.45 ± .05	5.18 ± .04	8.97 ± .25
$R = \frac{(\pi^+ + \pi^-)_{Ne}}{(\pi^+ + \pi^-)_p}$	1.11 ± .03	1.11 ± .05	1.21 ± .04
(π°)	1.77 ± .05		4.55 ± .44

We note the following. 1. The multiplicity factor R increases with increasing energy and is quite close to one. 2. The number of relativistic nucleons produced per interaction is approximately one. The number of slow nucleons is above 3 per interaction. We also compare the multiplicity distribution at 10.5

and 200 GeV/c with those from π^--p collisions. From these distributions we have concluded that KNO scaling seems to describe the multiplicity distribution for $n_\pi/(\bar{n}_\pi) < 2.5$. In Figure 1 we show the rapidity distributions for π^+ and π^- produced in π^--Ne collisions and π^--p collisions. In Figure 2 we show the ratio of the multiplicity of particles in π^--Ne to those in π^--p collisions as a function of the rapidity. π^--Ne and π^--p collisions are extremely similar so far as these distributions are concerned. The increase in multiplicity in the neon as compared to hydrogen collisions is mainly due to the higher plateau region as shown in Figure 2. The differential distribution is higher by a factor of $1.28 \pm .05$ in the plateau region than for π^--p collisions. The average rapidity for π^+, π^-, π° is less by $.32 \pm .03$ units of rapidity as compared to π^--p. This difference in rapidity is consistent with an increase in target mass to two nucleons from the usual one nucleon in π^--p collisions. The width of the rapidity distribution of the secondary pions is essentially the same in π^--Ne and π^--p collisions.

We show in Figure 3 the limiting fragmentation limit. As bombarding energy is increased, the number of slow pions (target fragmentation region pions) decreases. This behavior is very similar to that found for π^--p collisions. In the region of interest ($P_{long} \leq .4$ GeV/c) the multiplicity ratio, R, of Neon target to nucleon ((n + p)/2) is $2.56 \pm .25$.

CONCLUSIONS

π-Ne collisions are very similar to π-Nucleon collisions, and consequently there must be only a small amount of cascading occurring inside of the struck nucleus. The production process for energetic pions has a characteristic time which is longer than the collision time. Various models explain some features of the data, but none are completely satisfactory. The Energy Flux[4] model elucidates the small increase in multiplicity with increasing A, the Coherent Tube[5] model is consistent with the more massive target found kinematically in π-Ne collisions, and the Independent Parton[6] model can explain the higher plateau region in the rapidity distribution which is observed at 200 GeV. The E. F. model fails with respect to the observed increase of R with energy; the C. T. model predicts an increase in the width of the rapidity distribution which is not observed. The I. P. model allows for the composite structure of the pion, and consequently that the height of the rapidity distribution is greater than in the case of π^--p collisions without an increase in width.

REFERENCES

1. W. M. Yeager et al, Phys. Rev. D 16, 1294 (1977).
2. T. H. Burnett et al, Univ. of Washington, Strasbourg, Warsaw collaboration -- paper submitted to the 20th ICHEP.
3. H. R. Band et al, Duke Univ. and SUNY-Albany -- paper submitted to the 20th ICHEP.
4. K. Gottfried, Phys. Rev. Lett. 32 , 957 (1974).

5a. G. Berlad et al, Phys. Rev. D 13, 161 (1976).

b. F. Takagi -- Proc. of the 19th ICHEP -- Tokyo.

6. S. Brodsky et al, Phys. Rev. Lett. 39, 1120 (1977).

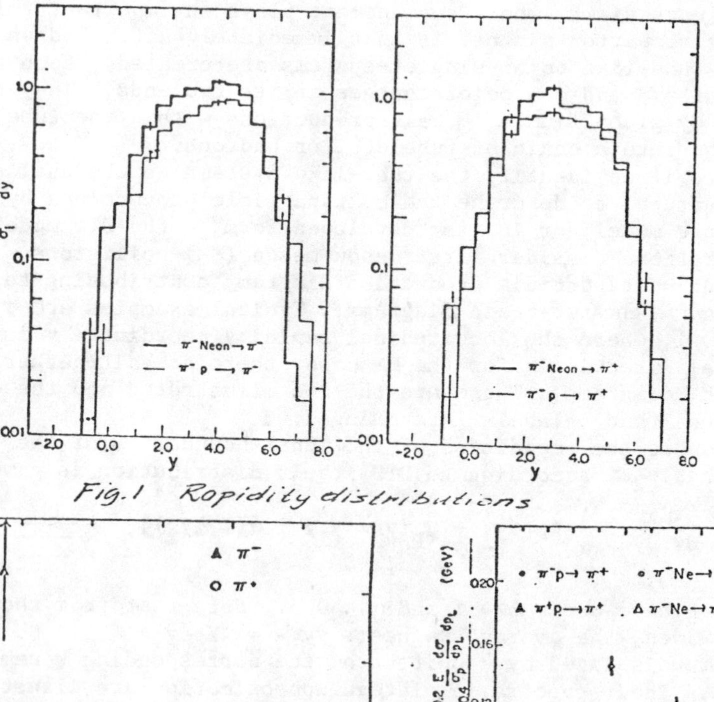

Fig. 1 Rapidity distributions

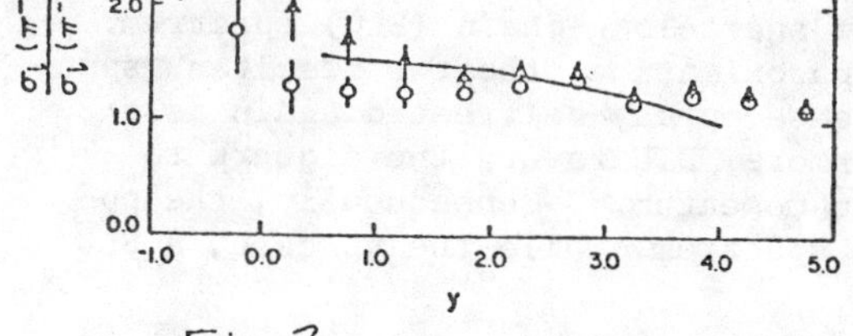

Fig. 2

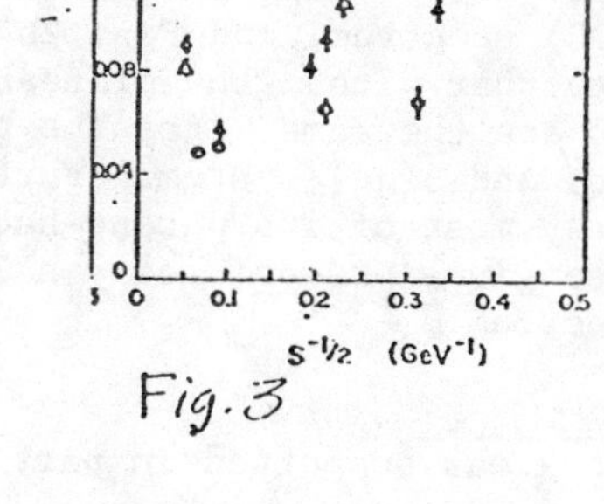

Fig. 3

A REVIEW ON DTU-PARTON MODEL FOR hh AND hA COLLISIONS*

Charles B. Chiu
Center for Particle Theory, Univeristy of Texas, Austin, Texas

Recently several groups[1-5] have considered small-p_T models, which combine features from both the parton model and the DTU model. We shall refer to them loosely as the DTU-parton model. In this talk, we take a definite point of view to motivate this model, and based on this framework we briefly survey its phenomenological applications to hadron-hadron and hadron-nucleus collisions.

Data indicate that small p_T-multiparticle productions constitute a large fraction of final particles in hadron collisions. Within QCD, these productions are presumably dominated by nonperturbative mechanisms, where confinement plays an important role. A suggestive parton picture is that immediately after hadron collisions, tube-like color-singlet-systems are created. Each tube consists of a 3 and a $\overline{3}$ color charges at its two ends. They are connected by gluon flux.[6,7] Pair productions within the tube break up the tube into a chain of tubelets, or hadrons.

Now if we identify the tube-like systems as the dual strings, it is natural to describe the multiparticle productions by the dual resonance model, or in its "developed form": the DTU model.[8] For illustration, consider first meson-meson (MM) collisions. Here the dominant multiparticle production diagrams contributing to the Pomeron are the two-chain diagrams. Typical examples are illustrated in Fig. 1, where the longitudinal rapidity coordinate y is along the vertical direction. For the MM case, there are altogether four "y-ordered" diagrams. These are the two illustrated and the other two with the 3 and $\overline{3}$ labels interchanged.

Denote the rapidities of the inner boundaries of the two chains by y_1 and y_2. According to DTU, their distribution is given by:

$$\frac{d^2n}{dy_1dy_2} \sim \exp[\alpha y_1 + \alpha_P(y_2 - y_1) + \alpha(Y - y_2)] \sim \frac{1}{\sqrt{-x_1x_2}} , \qquad (1)$$

where we have used $\alpha_P \approx 1$ and $\alpha \approx 0.5$. Note that from the distributions given, the average values: $\langle y_1\rangle = \langle Y - y_2\rangle \approx 2$. So the inner ends are displaced by 2 units from the corresponding kinematic boundaries. Their spectra in plateau approximation are illustrated in Figs. 2a,b. Fig. 2a will be referred to as the 2-displaced-chain (2DC) spectrum, and Fig. 2b the short-long-chain (SLC) spectrum. Note that within the plateau approximation, the two resultant spectra are the same. For the pp case, p may be treated again as a 3(q) and $\overline{3}$(qq) system. Furthermore, DTU favors the diquark to carry most of its parent-hadron momentum.[3] Consequently, the pp case gives predominantly a 2DC spectrum, while the $\overline{p}$p case, a SLC spectrum.

---

*Work was supported in part by the U.S. Department of Energy.

The corresponding more realistic DTU spectra are sketched in 2c and 2d, where the spectrum near each end approaches zero like $(1-x)^n$, with x being the magnitude of the Feynman variable defined in the rest frame of the chain. The power n varies for different types of endings. Also n should in general depend on t, where for $a \to cX$, $t = (p_a - p_c)^2$. This very t-dependent power behavior has been confirmed: e.g.[9] in $\pi^\pm p \to \pi^0 X$. More specifically, the triple Regge prediction of DTU gives $n = 1 - 2\alpha(t)$, with $\alpha(t)$ being the ρ trajectory. The power behavior extracted from their data compared quite favorably with the ρ trajectory. (See Fig. 3.)

Analyses on inclusive distributions have been reported by Capella et al.[1] and by Cohen-Tannoudji et al.[3] on the following reactions:

$$x{=}0 \to 1 \begin{cases} pp \to \pi_{ch} X, \\ \pi^\pm p \to (\text{charge poles})X, \\ \bar{p}p \to (\text{charge poles})X,\ \pi^+ X. \end{cases} \quad ;0.1 \leq x \leq 0.9 \begin{cases} \pi^\mp p \to \pi^\pm X \\ \pi^+ p \to \pi^+ X \ (\text{nondiffr.}) \\ K^+ p \to \pi^- X \end{cases}$$

Since only t-integrated data have been considered, the power n at different ends have been treated as constant parameters. In the future, it is worthwhile to determine n as a function t.

For central plateau heights at nonasymptotic energies there is the following regularity[1]: $h_{pp} < h_{\pi p} < h_{\bar{p}p}$. This stems from the fact that: at present energies h_{pp} is a sum of heights of two shoulders (see Fig. 2c), while $h_{\bar{p}p}$ is a sum of two central maxima (see Fig. 2d). The πp case is contributed by both the pp type and the $\bar{p}p$ type. So it falls in between. The energy dependence of these central plateau heights is illustrated in Fig. 4. The pp data is also included for comparison. Notice that the rise of the central plateau is naturally explained here.

In the context of the DTU-parton model, $\bar{p}p$ annihilation process has recently been considered by Sukhatme.[2] DTU predicts that the 3-chain annihilation processes should dominate all annihilation diagrams with its cross section goes like $1/\sqrt{s}$. The energy dependence of $h_{\bar{p}p}$ of annihilation is also illustrated in Fig. 4. Asymptotically the ratio of annihilation to non-annihilation heights approaches 3/2, a reflection of 3-chain versus 2-chain topologies.

Now we turn to hA collisions. Earlier we have assumed that it is the pair production within the tube, which leads to the breakup of the tube. It is plausible that the pair production time τ_0, defined in the rest frame of the pair should be a constant. In fact, in the flux tube model[6] τ_0 can be estimated qualitatively. In particular, the probability of the pair production per unit time per volume has been estimated to be[6]: $P \approx 10^{-3}\ \text{GeV}^4$. So $1/\tau_0 \approx PV$, where the available volume V is approximately $\pi r^2 \tau_0$. Assume $r \sim 1/m_\pi$, we get $1/\tau_0 \sim 0.4$ GeV. Now the time taken for each pair with rapidity y to travel a distance d is $t = d/\tanh y$. This implies that within the spatial interval d, there is a critical rapidity $y_c = \sinh^{-1}(d/\tau_0)$, such that only for $y \leq y_c$, $q\bar{q}$ pairs will have time to be produced. Taking $d \sim 1/m_\pi$ we get $y_c \approx 1.8$. So the upshot is that, consider two consecutive collisions, immediately before the second collision, only pairs up to $y \approx 2$ will be produced.

For a proton projectile, this next collision is expected to be dominated between the energetic tube system N* (the unbroken tube II in Fig. 1a) and the target nucleon. We assume that the spectrum of N*N collision is similar to that of pp collision. This then leads to the spectrum of Fig. 5, where there are two chain-I spectrum and one chain-II spectrum. In general for $\bar{\nu}$ collisions, there would be $\bar{\nu}$ chain-I's and one chain-II. So the ratio

$$R = \langle n \rangle_{hA} / \langle n \rangle_{hh} = (\bar{\nu}+1)/2 \ . \qquad (2)$$

For the pA case, Chao et al.[4] showed that the comparison between their model and the available dn/dy pA data are satisfactory. Their comparison: R vs $\bar{\nu}$ for 200 GeV pA data is shown in Fig. 6, where "a" corresponds to having only N*N collisions as the successive collisions. The agreement is satisfactory. The dotted line: "b", is the prediction including also the remaining interactions between the less energetic tube and target nucleons, which have not interacted with N*, the energetic projectile system. The agreement here is even better.

Next we turn to the $\bar{p}$A case (see Fig. 7). Using similar argument as that for the pp case and the approximate relation indicated above: $y_c \sim y_1 \sim 2$, we get e.g. the plateau spectrum of Fig. 8. A comparison with Fig. 5 leads to the statement of $\bar{\nu}$-universality: i.e. for $\bar{\nu}_{pA} = \bar{\nu}_{pA'} = \bar{\nu}_{MA''}$

$$dn/dy|_{pA} \approx dn/dy|_{pA'} \approx dn/dy|_{MA''} \ . \qquad (3)$$

In ref. 5, it has also been shown that the above $\bar{\nu}$ universality condition persists, so long as the inequality $y_c \geqq y_1$ is satisfied. For instance, denote $\Delta = y_c - y_1$, within the same plateau approximation, both pA and $\bar{p}$A' cases lead to

$$R = (\bar{\nu}+1)/2 + \Delta(\bar{\nu}-1)/[2(Y - y_1)] \ . \qquad (4)$$

We recall that y_c is related to the pair production rate within the confining flux tube. So the above discussion illustrates how nuclear targets provide a unique arena for confinement related information, not available in hh collisions.

Finally we mention that Capella and Tran[10] have also considered a hA model, assuming that projectile sea quarks also form additional tubes in small-p_T productions. This assumption leads to the asymptotic prediction: $R \sim \bar{\nu}$. However, at present energies, phenomenological predictions of the models of ref. 4 and ref. 10 have been shown to be similar for the pA case.

We thank Rudy Hwa, Uday Sukhatme and Chung-I Tan for stimulating discussions.

REFERENCES

1. Capella et al., Phys. Lett. 81B, 68 (1979); Z. Phys. C3, 329

(1980).
2. Sukhatme, Phys. Rev. Lett. 44, 518 (1980).
3. Cohen-Tannoudji et al., Saclay preprint Dph-T/79-92.
4. Chao et al., Phys. Rev. Lett. 44, 518 (1980).
5. Chiu and Tow, CPT preprint, 1980 (DOE-ER-397), revised.
6. Casher et al., Phys. Rev. D20, 179 (1979); refs. therein.
7. Donoghue, MIT preprint, 1980 (CTP-829).
8. Chan et al, NP B86, 479 (1975); Chew et al, Phys. Rep. 41C, 263 (1978).
9. Caltech-LBL collaboration, Nucl. Phys. B145, 45 (1978).
10. Capella and Tran, Orsay preprint, 1980.

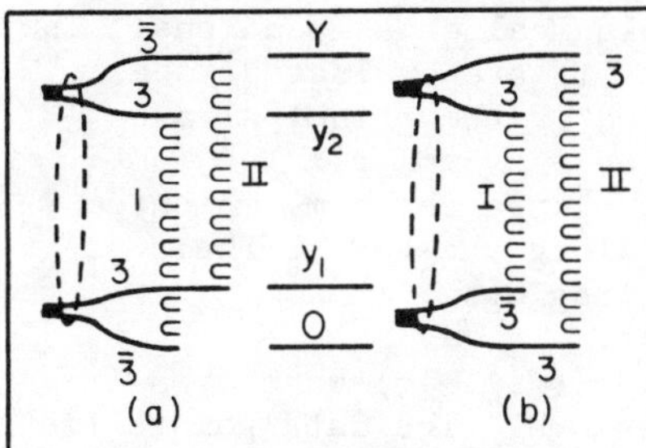

Fig. 1. 2-chain diagrams:
a. 2-displaced chains (2DC)
b. short-long chains (SLC)

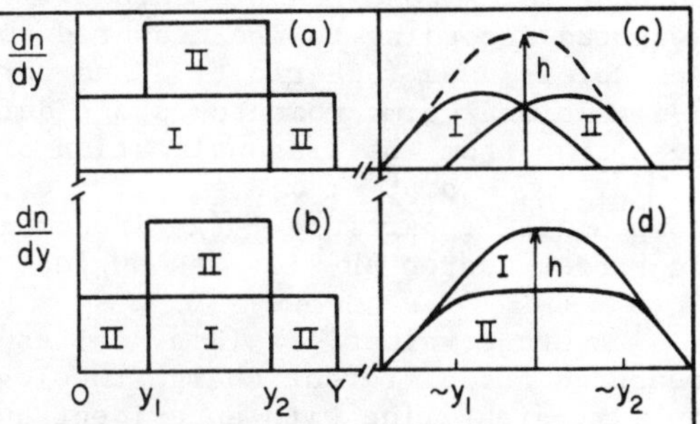

Fig. 2. y-spectra of 2-chain diagrams:
a,c. 2DC spectra
b,d. SLC spectra

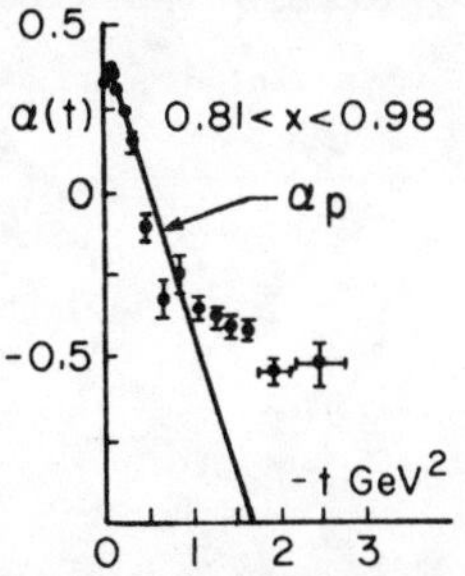

Fig. 3. α extracted from $(1-x)^{1-2\alpha(t)}$

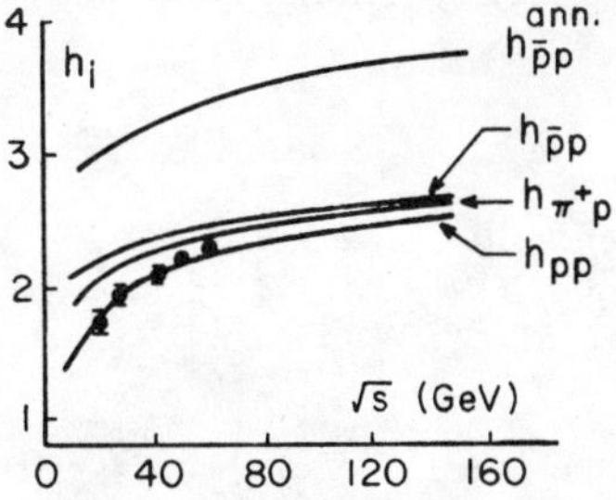

Fig. 4. Energy dependence of central plateaus.

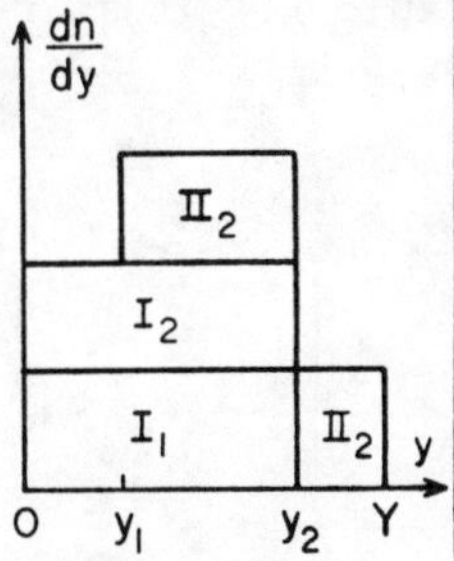

Fig. 5. pA spectra for $\bar{\nu} = 2$.

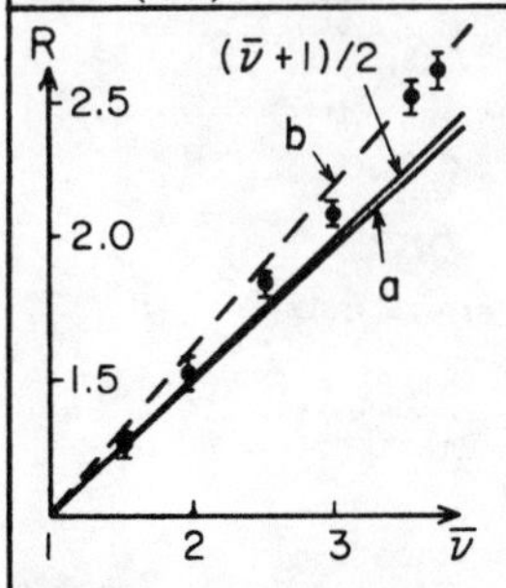

Fig. 6. $R = \langle n \rangle_{hA} / \langle n \rangle_{hh}$. See text for lines a,b.

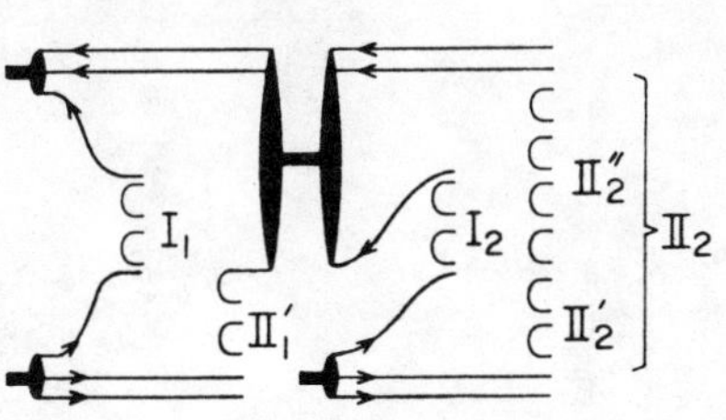

Fig. 7. pA collision for $\bar{\nu} = 2$.

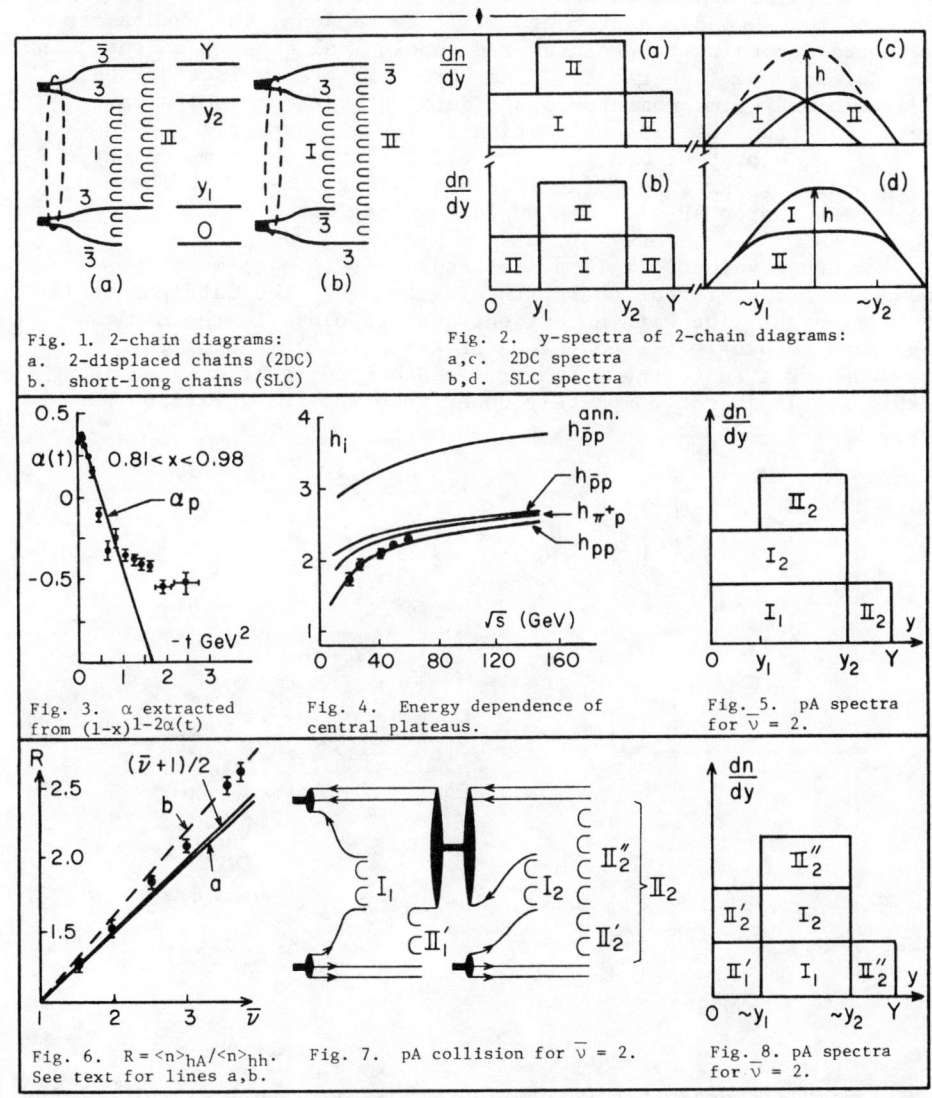

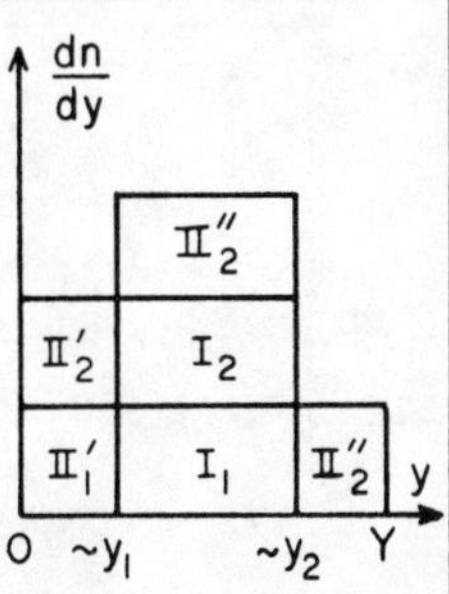

Fig. 8. pA spectra for $\bar{\nu} = 2$.

MULTIPARTICLE JETS FROM $\pi^+/K^+/p$ p COLLISIONS AT 147 GeV/C COMPARED TO e^+e^- ANNIHILATIONS

Sergio P. Ratti
Istituto di Fisica Nucleare and Sezione INFN - I27100 Pavia (Italy)

The data presented here have been collected by the International Hybrid Spectrometer Consortium in the 30" B.C. hybrid spectrometer exposed to a 147 GeV/C positive tagged beam made of 47% π^+; 45% p; 8% K^+ ($\sqrt{s}$ = 16.7 GeV) at Fermilab. The sample consists of 15802 events with at least 4 prongs (11822 events with at least 6 prongs) fully measured and reconstructed with an accuracy $\Delta p/p < 0.5\%$. To account for second-measurements still to be done, the events are weighted according to the measured topological cross sections. The variables[1] used (Sphericity S and thrust T) are evaluated in a 3-dimensional c.m. momentum space defined by the principal axes resulting from the diagonalization of the 3 x 3 matrix $T^{\alpha\beta} = \sum_i \left[p_i^2\, \delta_{\alpha\beta} - p_i^\alpha p_i^\beta\right]$; $\alpha,\beta = x,y,z$; p_i^x, p_i^y, p_i^z are the c.m. momentum components of particle i (x-direction along the beam). Their use in hadron hadron physics was suggested long ago.[2]

Average values <T> (Fig. 1a) and <S> (Fig. 1b) have been measured for different charge topologies N_{CH}. The data points lie on a straight line with an evident cut-off close to the maximum observed topologies. The values of the parameters of a form f = a + b N_{CH}, fitted to the data for $4 \leq N_{CH} \leq 14$, are reported in Table I. It is not completely understood why the variation has to be strictly linear.

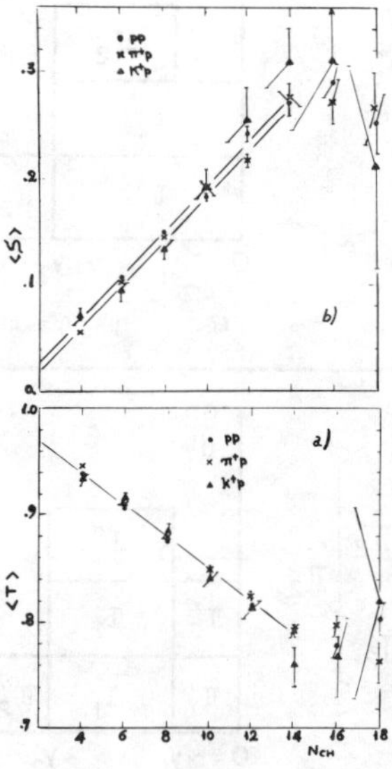

Fig.1

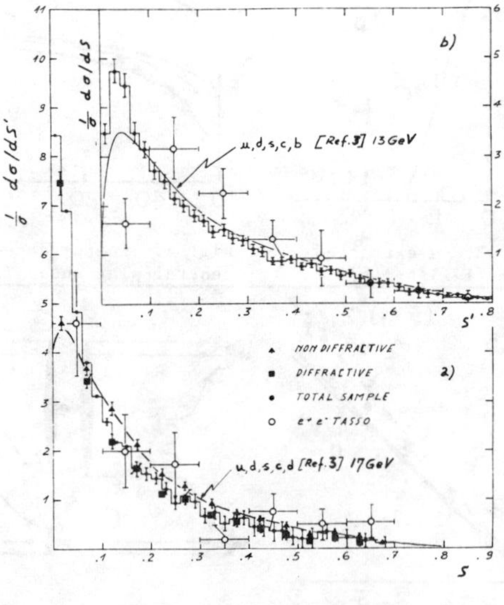

Fig.2

The "shape" of the events, as given by the tensor $T^{\alpha\beta}$, has been compared to e^+e^- data at the same c.m. energy.[3]

The different baryon number involved, the different role of the gluons, the fragmentation of a diquark instead of a quark, should introduce a visible difference between the mechanisms initiating e^+e^- annihilations and hadron hadron collisions. In an attempt to separate the main mechanisms contributing to jet production in hadron hadron interactions, "diffractive-like" events (which correspond to completely different diagrams in the dual topological model) have been selected in a standard way.[4] The sphericity distributions obtained are compared to e^+e^- data at 17 GeV[3a] in Fig. 2a. The solid curve is a quark model prediction for u,d,s,c,b contributions at 17 GeV (from Ref. 3a). In an attempt to account for the different baryon number and the different average (qq) c.m. system, both the fastest and the slowest particle have been neglected and new T' and S' evaluated in the c.m. system of the remaining hadrons having an energy spectrum with mean value $\langle E_{HADR}\rangle = 12.12$ GeV and spread $\sigma_E = 2.97$ GeV. The sphericity distribution obtained is compared to e^+e^- data at 13 GeV[3a] in Fig. 2b. The solid curve is the same quark model prediction at 13 GeV from Ref. 3a. The similarity between our data and e^+e^- data is striking. Our "non-diffractive data" in Figs. 2a are in excellent agreement with the quark model predictions. Our data in Fig. 2b are in better agreement with the quark model than e^+e^- data at 13 GeV for $S' \gtrsim 0.1$. The departure of our data from the curve for $S' \lesssim 0.1$ can be due to an incomplete removal of the leading particles in the charge exchange reactions.

TABLE I. Straight line fits for $4 < N_{CH} < 14$

	a_S	b_S	x^2/NDF	a_T	b_T	x^2/NDF
pp	$-.014\pm.008$	$.020\pm.001$	2.5/4	$1.00\pm.010$	$-.015\pm.001$	1.4/4
π^+p	$-.026\pm.010$	$.021\pm.001$	2.54	$1.00\pm.007$	$-.015\pm.001$	5.2/4
K^+p	$-.018\pm.032$	$.020\pm.005$	3.9/4	$1.00\pm.016$	$-.015\pm.002$	4.0/4

REFERENCES

1. J. D. Bjorken et al. P.R. D1, 1416 (1970);
2. A. Giovannini et al. Riv. N. Cim. Vol. 2, n. 10 (1979), p. 1.
3. a - TASSO coll. P.L. 83B, 261; 86B, 243 (1979); DESY 79/61,73.
   b - PLUTO coll. P.L. 78B, 176; 81B, 410; 86B, 418 (1979); DESY 79/57.
4. W. Burdett et al. N. Phys. B48, 13 (1972); J. Benecke et al. N. Phys. B86, 29 (1974); U. Idschok et al. N. Phys. B140, 365 (1978).

STUDY OF THE BASIC PROPERTIES OF LOW-p_T MULTIPARTICLE SYSTEMS PRODUCED IN pp COLLISIONS REMOVING LEADING PROTONS

M. Basile[1], G. Cara Romeo[2], L. Cifarelli[2], A. Contin[3], G. D'Ali[2], P. Di Cesare[2], B. Esposito[4], P. Giusti[2], T. Massam[2], R. Nania[5], F. Palmonari[2], G. Sartorelli[2], G. Valenti[2] and A. Zichichi[3]

(Presented by B. Esposito)

ABSTRACT

The results of a study of the basic properties of multiparticle systems produced in pp collisions are presented. The experiment was performed at the CERN Intersecting Storage Rings at $\sqrt{s}$ = 62 GeV, using the Split-Field Magnet facility. The key point of the analysis is to remove the leading proton and thus redefine the correct available energy for particle production. Our results show that, taking into account the correct hadronic energy, the multiparticle system produced in pp collisions, at low p_T, have similar properties with respect to the multihadron jets observed in e^+e^- annihilations.

1. INTRODUCTION

The study of jet-like properties of multiparticle systems produced in e^+e^- annihilations is an important means to investigate the properties and the dynamics of elementary constituents (quarks and gluons)[1]. The properties of the multiparticle systems produced in hadron reactions at low p_T appear to be different if analysed in the standard way. A striking difference between hadron-hadron processes and e^+e^- annihilation is the leading-particle effect.

To study multiparticle production in pp reactions we have used a new approach. The key point of our method is to remove the leading proton and thus redefine the correct hadronic energy, E_{had}, available for particle production. With this method a remarkable similarity between pp and e^+e^- processes in the fractional momentum distributions was observed[2]. This result opened up the field to the study of other relevant properties of the multiparticle systems, taking into account the correct energy E_{had}. For details of the analysis we refer the reader to our previous publications[2-5].

---

1 Istituto di Fisica dell'Università, Bologna, Italy.

2 Istituto Nazionale di Fisica Nucleare, Bologna, Italy.

3 CERN, Geneva, Switzerland.

4 Istituto Nazionale di Fisica Nucleare, Laboratori Nazionali di Frascati, Italy.

5 Istituto Nazionale di Fisica Nucleare, Bari, Italy.

2. RESULTS

The results reported are based on the analysis of about 40,000 minimum bias events, collected at the CERN ISR at $\sqrt{s} = 62$ GeV, using the Split-Field Magnet facility.

Owing to lack of time, only some of the most relevant results we have obtained will be reported here:

i) the fractional momentum distributions

$$\frac{1}{N_{ev}}\frac{dN}{dx_R^*}, \qquad \left(x_R^* = \frac{p}{E_{had}}\right)$$

for three ranges of $2E_{had}$;

ii) the average charged multiplicity $\langle n_{ch}\rangle$ versus $2E_{had}$;

iii) the inclusive p_T^2 distribution

$$\frac{1}{N_{ev}}\frac{dN}{dp_T^2}$$

for "low" and "high" $2E_{had}$ ranges;

iv) the average $\langle p_T^2\rangle$ distribution of tracks in the plane and out of the plane of the event:

$$\frac{1}{N_{ev}}\frac{dN}{d\langle p_T^2\rangle_{in}}, \qquad \frac{1}{N_{ev}}\frac{dN}{d\langle p_T^2\rangle_{out}}$$

for "low" and "high" $2E_{had}$ ranges.

These results are shown in Figs. 1 to 4, together with the data from PETRA at the corresponding c.m. energies: $(\sqrt{s})_{e^+e^-} = 2E_{had}$.

3. CONCLUSIONS

Our method of removing the leading proton and redefining the hadronic energy has allowed the observation of striking similarities in some of the main basic properties of multiparticle systems produced in pp and e^+e^- reactions.

Wherever leading particle effects are present, this method should provide a clue to a deeper understanding of multiparticle hadronic systems produced in different reactions.

REFERENCES

1. B. Wiik, Proceedings of this Conference.
2. M. Basile et al., preprint CERN-EP/80-26 and Phys. Lett. 92B, 367 (1980).
3. M. Basile et al., preprint CERN-EP/80-111 (1980).
4. M. Basile et al., preprint CERN-EP/80-112 (1980).
5. M. Basile et al., preprint CERN-EP/80-113 (1980).

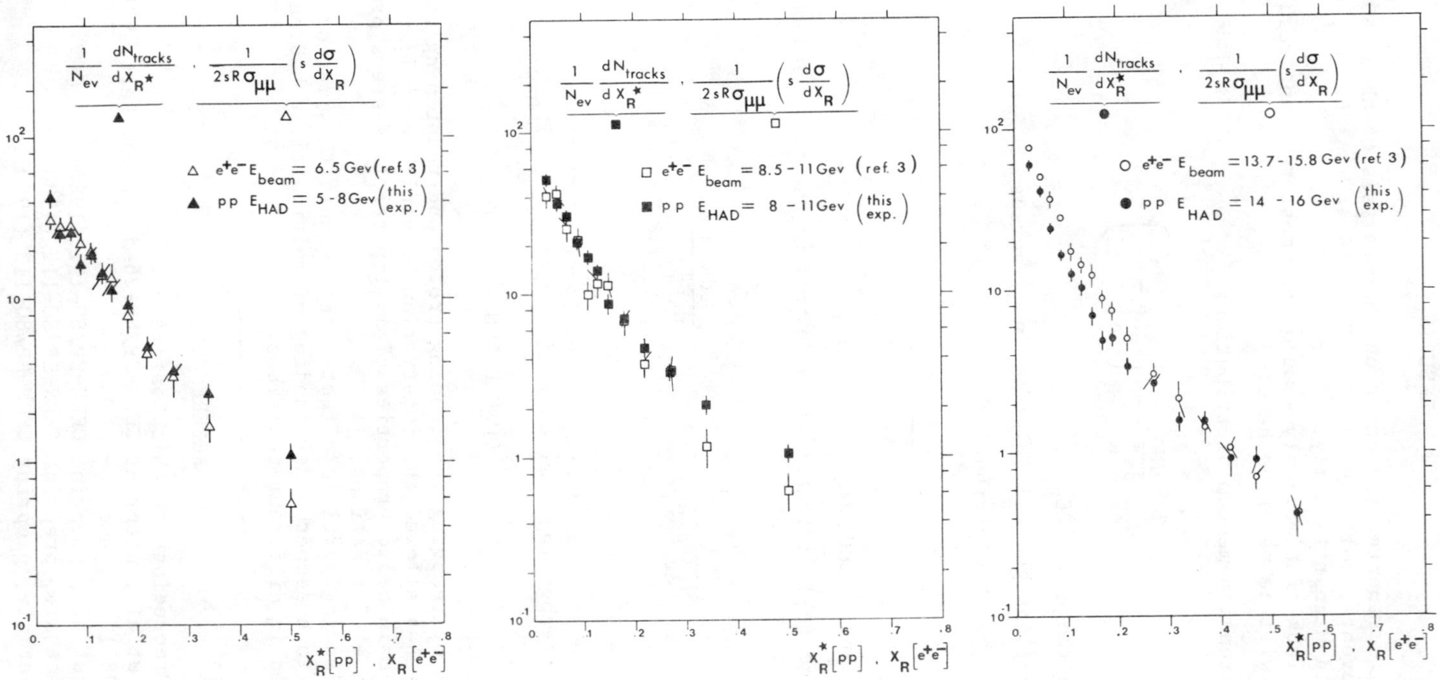

Fig. 1 Fractional momentum distribution for three ranges of $2E_{had}$

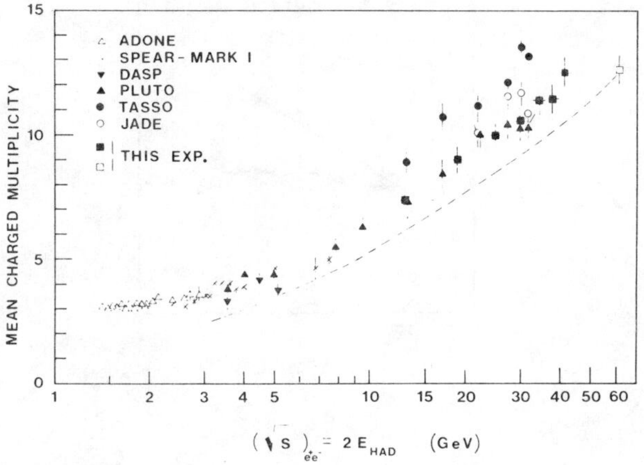

Fig. 2 Average charged multiplicity versus $2E_{had}$

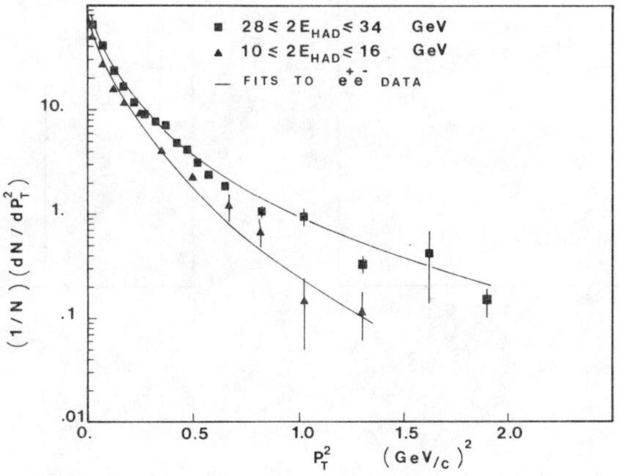

Fig. 3 Inclusive p_T^2 distribution for "low" and "high" $2E_{had}$ range

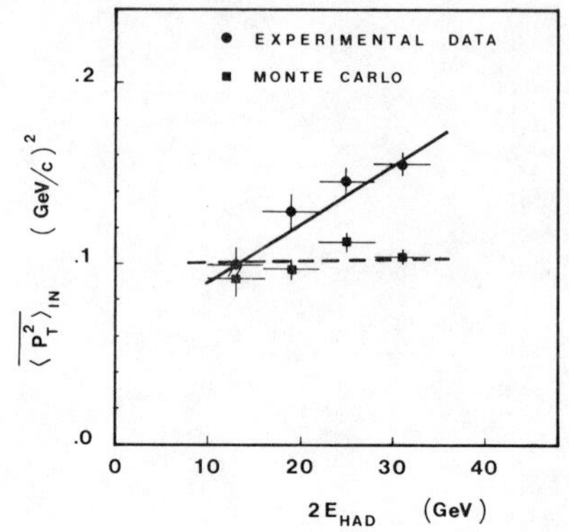

a)

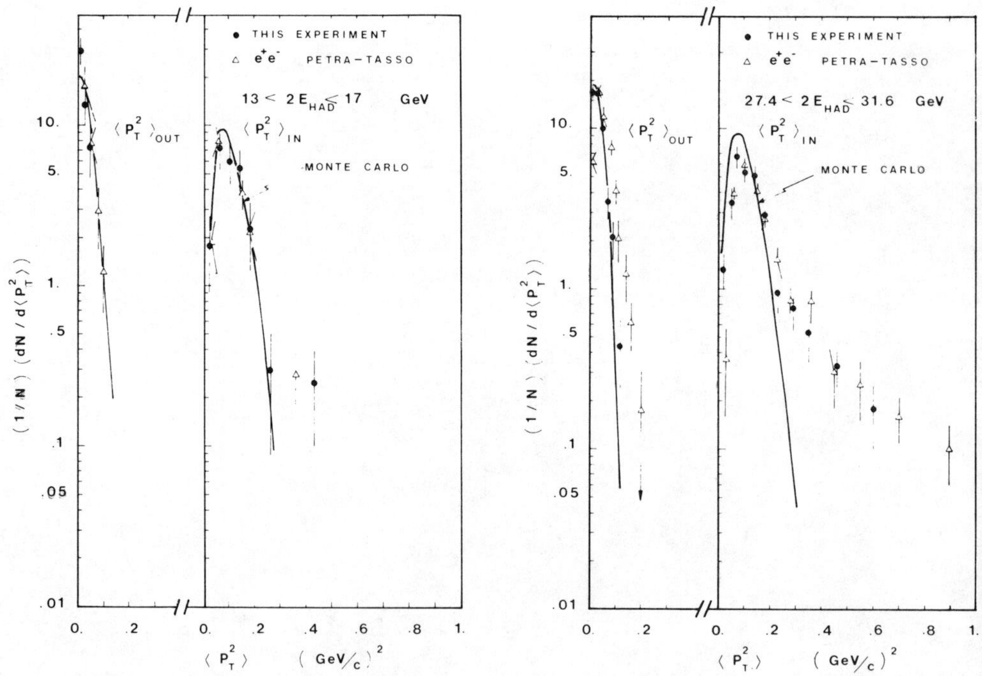

b)

Fig. 4 a) Average p_T^2 distribution versus $2E_{had}$, and b) inclusive $\langle p_T^2 \rangle$ distributions for tracks in the plane and out of the plane of the event (preliminary results)

HIGH P_T HADRONIC INTERACTIONS, EXPERIMENT

R. Cool, Rockefeller, Organizer

PRELIMINARY LARGE TRANSVERSE ENERGY CROSS SECTIONS MEASURED WITH A 2π CALORIMETER TRIGGER

Presented by K. Pretzl

C. Favuzzi, G. Germinario, L. Guerriero, P. Lavopa,
G. Maggi, C. de Marzo, M. de Palma, F. Posa,
A. Ranieri, G. Selvaggi, P. Spinelli and F. Waldner
University of Bari, Bari, Italy

A. Białas, W. Czyz, T. Coghen, A. Eskreys,
K. Eskreys, D. Kisielewska, P. Malecki,
K. Olkiewicz, K. Sliwa and P. Stopa
University of Krakow, Krakow, Poland

W.H. Evans, J.R. Fry, C. Grant, M. Houlden,
A. Morton, H. Muirhead and J. Shiers
University of Liverpool, Liverpool, U.K.

M. Antič, W. Baker, H. Bechteler, I. Derado,
W. Eckardt, J. Fent, P. Freund, H.J. Gebauer,
T. Kahl, R. Kalbach, A. Manz, P. Polakos,
K.P. Pretzl, N. Schmitz, P. Seyboth,
J. Seyerlein, D. Vranic and G. Wolf
Max-Planck-Institut für Physik und Astrophysik
Munchen, Germany

F. Crijns, W. Metzger, C. Pols
T. Schouten and T. Spuijbroek
University of Nijmegen, Nijmegen-NIKHEF, Netherland

N. Sarma
Visitor at CERN from Bhabha Institute, Bombay, India

ABSTRACT

Preliminary large transverse energy cross sections of 150, 300 GeV pions and protons on hydrogen measured with a large acceptance, segmented calorimeter are presented. Processes other than the scattering of two constituents appear to dominate this deep inelastic hadron scattering process.

INTRODUCTION

The main motivation for this experiment was to study deep inelastic hadron hadron collisions using a large acceptance calorimeter trigger. With this trigger we selected events in which a large transverse energy $\Sigma|p_T|$ was deposited in the calorimeter. We examined these events for constituent scattering processes leading to two jets at large p_T and two forward/backward spectator jets. Large p_T jets should manifest themselves as energy clusters observed in the segmented calorimeter showing a planar event structure.

APPARATUS

The layout of the experiment is shown in fig. 1. The data presented here were obtained with 150 GeV/c π^- and 300 GeV/c p, π^-

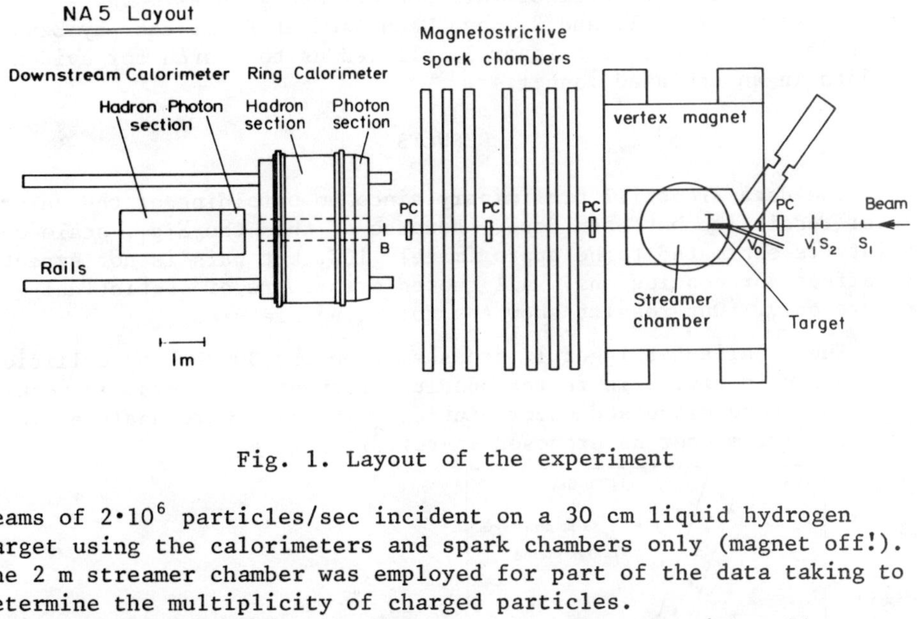

Fig. 1. Layout of the experiment

beams of $2 \cdot 10^6$ particles/sec incident on a 30 cm liquid hydrogen target using the calorimeters and spark chambers only (magnet off!). The 2 m streamer chamber was employed for part of the data taking to determine the multiplicity of charged particles.

The ring calorimeter, a barrel of 3 m diameter with a 56 cm central hole, has a lead-scintillator sandwich photon section (16 × 0.55 cm Pb-sheets) followed by an iron-scintillator sandwich hadronic section (20 × 5 cm Fe-sheets). Both sections are subdivided into 240 independent cells, each subtending about 9° in the c.m.s. polar angle and 15° in azimuthal angle. Combined wave-length shifting acrylic rods (doped with Yellow 323 and BBQ) were used to draw separated signals from the photon and hadron part of the calorimeter onto 240 pairs of photomultiplier tubes[1]. The obtained energy resolutions were $\sigma/E = 0.23/\sqrt{E}$ for electrons and $\sigma/E = 0.71/\sqrt{E} + 0.06$ for hadrons. The downstream calorimeter, which covered the central hole of the ring calorimeter, measured the energy flow at small angles. The combined information from both calorimeters was used to reject possible background. Scaling was measured by varying the distance of the calorimeters to the target with the incident beam energy such that the ring calorimeter always covered 50° to 130° in the c.m.s. polar angle. The $\Sigma|p_T|$ trigger was derived from the sum of the analog signals of all or part of the ring calorimeter cells weighted by their radial distance from the beam axis. The shape of the trigger pulses was recorded. Occasional background triggers due to Cerenkov light produced in the acrylic rods was eliminated off-line by pulse shape analysis.

Data were taken with 3 types of trigger:

1. one arm trigger: sector of $\pi/2$ in azimuth;
2. two arm trigger: two opposite sectors each of $\pi/2$ in azimuth;
3. full calorimeter trigger: all sectors of 2π in azimuth.

Triggers similar to 1. and 2. have been used at Fermi Lab. by E260 [2] and E395 [3]. The use of trigger 3. allowed us to search for evidence of jets in an unbiased manner.

RESULTS

The data are still preliminary since no unfolding of the calorimeter resolution has been done. Because of this the $\Sigma|p_T|$ scale uncertainty is estimated to be approximately 10%, but this is not expected to affect our conclusions. All quoted errors are of statistical nature only. Our results show the following features:

1. The invariant trigger 1. cross section for 300 GeV pp collisions shown in fig. 2 is in reasonable agreement with refs.[2,3] within the above discussed uncertainties. The data were analysed in a similar manner as proposed in ref.[3].

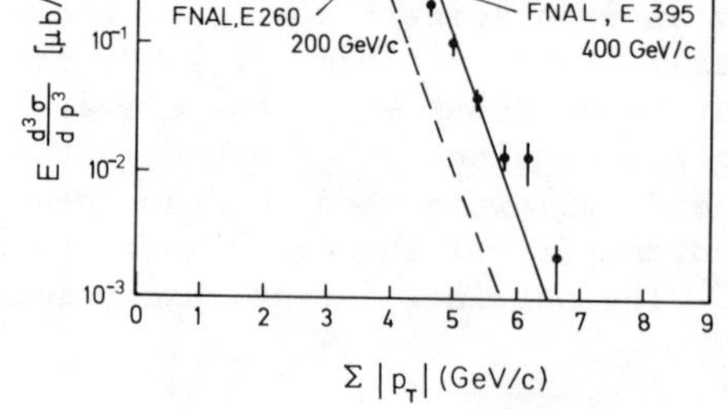

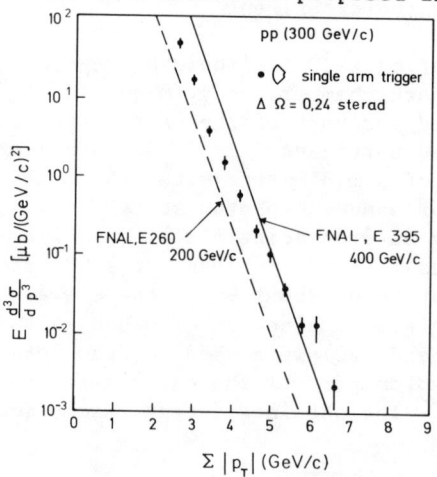

Fig. 2. Invariant cross sections measured with a sector of $\pi/2$ of the calorimeter. Results from FNAL experiments E260, E395 are shown for comparison.

2. We observe that the cross sections measured with trigger 3. are 10-100 times larger than with trigger 2. (see fig. 3). A QCD model calculation [4] comes close to explain the trigger 2. cross sections but remains a factor 5 below that for trigger 3.

3. For all triggers the cross sections for pp decrease faster with rising $x_\perp = 2\ p_T/\sqrt{s}$ than for $\pi^- p$ (see fig. 4). A similar dependence was observed for jet cross sections as reported by refs.[2,3], where it has been interpreted as support for constituent scattering, since the partons in the pion contribute on average more energy to the hard scattering process. A constituent scattering model leads one to expect a planar event structure.

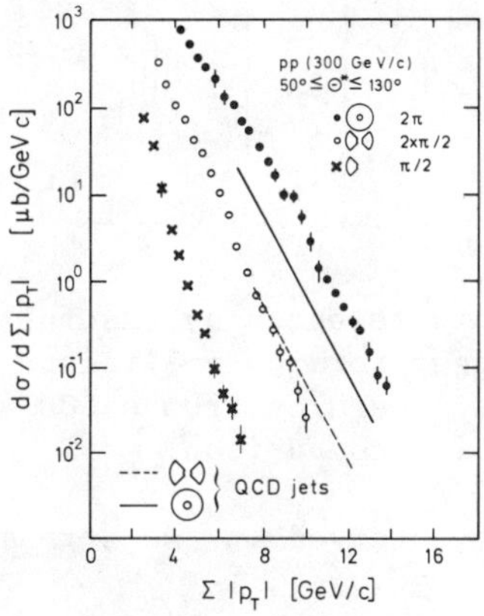

Fig. 3. Cross sections versus transverse energy $\Sigma|p_T|$ measured by the full calorimeter or some sectors of the calorimeter.

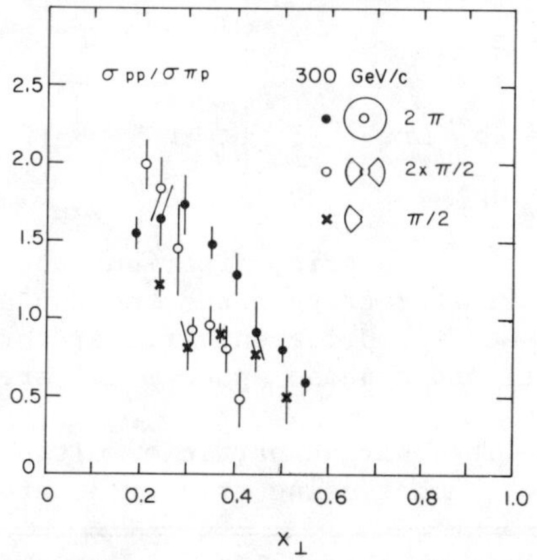

Fig. 4. Ratio of pp to $\pi^- p$ cross sections as a function of $x_\perp$.

4. Planar events do not seem to dominate the trigger 3. data. However the number of events with a planarity close to 1 increases with an increasing $\Sigma|p_T|$-trigger threshold (see fig. 5). The planarity P of the events has been calculated by performing a principal axis analysis of the transverse momentum distribution for each event as measured by the calorimeter. The planarity was defined as $P = (a-b)/(a+b)$ with a(b) being the sum of the squares of the projected transverse momenta to the maximum (minimum) principal p_T-axis. For comparison fig. 5 shows the planarity of Monte-Carlo simulated events for the trigger 3. using a phase space [5] and a QCD jet model [4]. For both models the detector resolution and acceptance were simulated.

Planarity

$$P = \frac{a-b}{a+b}$$

$$a = \Sigma p_T^2 \text{ max}$$

$$b = \Sigma p_T^2 \text{ min}$$

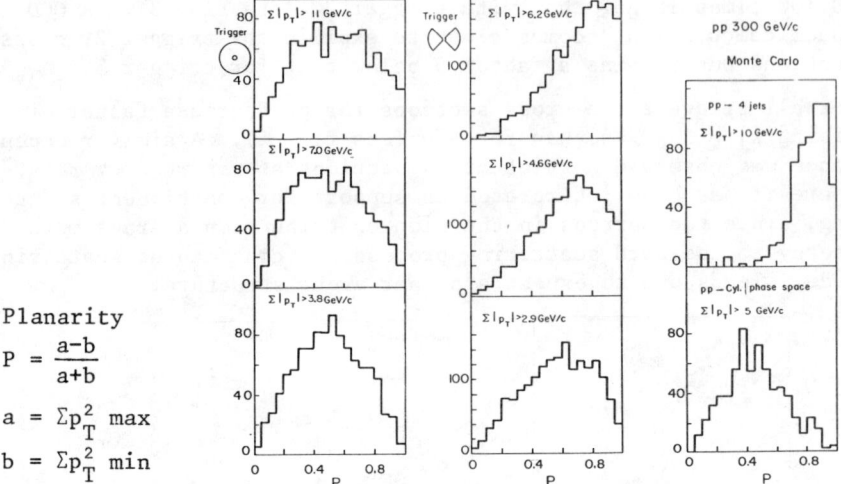

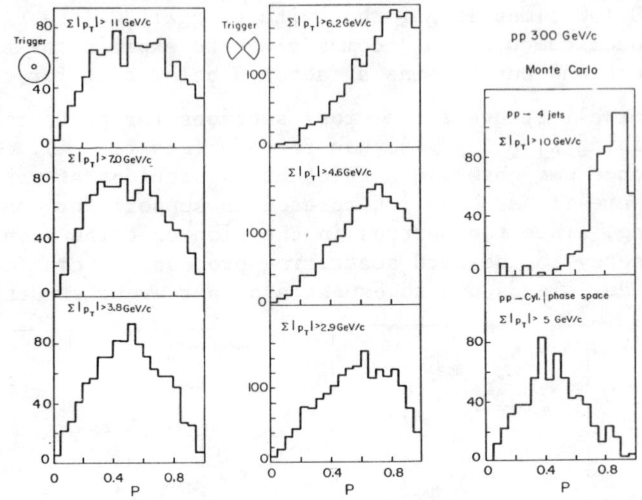

Fig. 5. Planarity distributions of events selected by the full calorimeter trigger and the two arm trigger form pp collisions at 300 GeV for different trigger thresholds. Results from a QCD 4-jet model and a phase space model are shown for comparison.

5. The charged particle multiplicities observed in the streamer chamber using trigger 3. are very large.

Trigger threshold: $\Sigma\|p_T\|$ (GeV/c)	$\langle n_{CH} \rangle$
10	26.5 ± 1 (12 into calorimeter)
7.5	23.5 ± 2
5	21.0 ± 1
∿ 0	8

6. The scaling parameter n derived from the energy dependence of the trigger 1. and 3. cross sections parameterized by

$$E \frac{d^3\sigma}{dp^3} \sim p_T^{-n} f(x_\perp)$$

rises with increasing $x_\perp$ (see fig. 6). Constituent scattering models predict a constant n = 4 at large $x_\perp$.

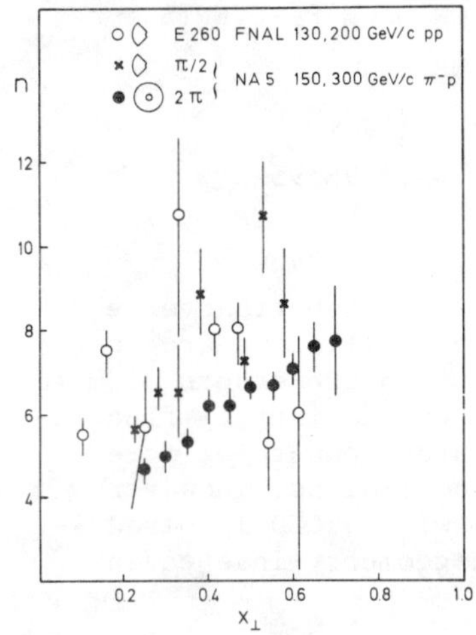

Fig. 6. The scaling behaviour of the full calorimeter and one arm trigger cross sections for π^-p collisions at 300 and 150 GeV/c. Results from FNAL experiment E260 are shown for comparison.

CONCLUSION

The events selected by the full calorimeter trigger show no dominant jet structure. They appear to originate from processes other than two constituent scattering.

REFERENCES

1. V. Eckardt et al., Nucl. Instr. & Meth. 155, 389 (1978).
2. C. Bromberg et al., Preprint CALT-68-738 (1979); Phys. Rev. Lett. 43, 565 (1979).
3. W. Selove, High p_T jet studies at Fermilab, Proceedings of the 14th Rencontre de Moriond 1979, Vol. I, 401.
4. In the QCD-model calculation of the cross sections the relevent processes of ref. 6, the constituent distributions of ref. 7 and the coupling constant $\alpha_s = 12\pi/25 \log(Q^2/\Lambda^2)$ with $\Lambda = 0.5$ GeV were used. The model had 2 large p_T jets with a multiplicity derived from e^+e^- data and 2 forward/backward spectator jets.
5. The phase space model has a KNO multiplicity distribution, an exponential p_T cut off with $\langle p_T{}^2\rangle^{\frac{1}{2}} = 0.3$ GeV/c, a leading particle effect and was normalised to $\sigma_{tot\ inel.}$ (pp).
6. B.L. Combridge et al., Phys. Lett. 70B, 234 (1977).
7. V. Barger, R.J.N. Phillips, Nucl. Phys. B73, 269 (1974).

RESULTS FROM FERMILAB EXPERIMENT E236 ON HIGH-P_t EVENTS FROM PROTON-PROTON COLLISIONS.*†

Robert W. Williams
University of Washington, Seattle, WA. 98195

ABSTRACT

Large-aperture calorimeter measurements of high transverse momentum events from 100, 200, and 340 GeV pp collisions are reported. The cross sections are higher by one to two orders of magnitude than the yield expected from uncorrelated statistical fluctuations in multiparticle events. The events are shown to be more diffuse than pure jets as produced in e^+e^- collisions. However, these events are in most details consistent with a QCD inspired 4-jet model. The possibility of extracting a geometry-independent jet cross section is briefly discussed.

INTRODUCTION

This paper presents measurements of the properties of high-energy p-p collisions in which a group of particles carries off transverse momentum which is an appreciable fraction, typically $x_t \sim 0.3$ to 0.5 of the maximum possible ($x_t = 2P_t/\sqrt{s}$).

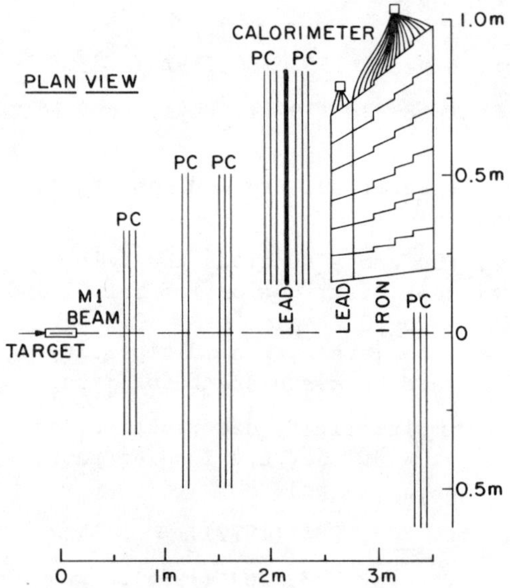

Fig. 1. Schematic plan view of the experiment.

The experiment was motivated by the possibility that total event measurements may be relatively unbiased reflections of elementary hard-scattering processes.[1] To carry out such measurements, we used a segmented calorimeter of very large solid angle(2.2 to 2.4 steradians in the center-of-mass)(Fig. 1) and we have obtained the yield and profiles of inclusive multiparticle high-P_t events.

JET TRIGGER EVENTS

The characteristics of the events we recorded were clearly not what was expected from jets of particles similar to those previously

*Work supported in part by the National Science Foundation.
†University of Washington-Fermilab-Tufts University collaboration.

observed in e^+e^- collisions.[2] Our Monte Carlo model reproduces the features of e^+e^- jets described in Reference 2. The angular width of the energy distribution of our observed events is a measure of the transverse momentum of the particles with respect to their central axis. This distribution is obtained from the horizontal counters in the calorimeter. It was found to be ∿1.5 times as wide as predicted by the Monte Carlo model of pure e^+e^--like jets. In addition it exhibited large tails to the edge of the calorimeter. Figure 2 illustrates the discrepancy between our measured azimuthal distribution of event axes and that expected for e^+e^--like jets.

Both the four-jet model and a no-azimuthal correlation exponentially cut-off phase space model resemble the data - i.e., the energy of these events is widely distributed.

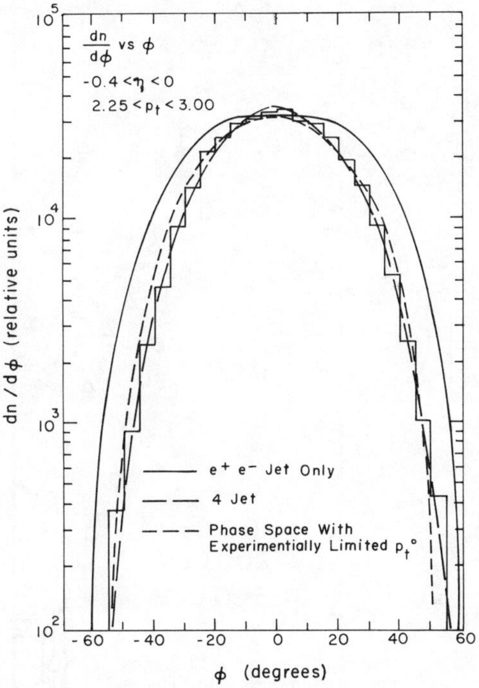

Fig. 2. Azimuthal distribution of the jet axis for 340 GeV. The histogram gives the data while the curves give the predictions for the indicated models.

EVENT ANALYSIS

We have measured the level of P_t contribution from particles far from the central axis of such events. This contribution, and other features of our data, force us to introduce a non-jet component in the analysis, representing the beam and target "debris" in the manner of four-jet models.[3] We try to fix the parameters of the debriw by matching the features of the data; the azimuth (ϕ) and polar angle (pseudorapidity η) of the momentum vector of the event, and the angular distribution of the energy.

CROSS SECTIONS

In these circumstances a cross section, derived by selecting events with momentum axis in a fiducial area near the center of the calorimeter, has a limited, geometry-dependent meaning. Fig. 3 shows such cross sections, corrected for calorimeter response.

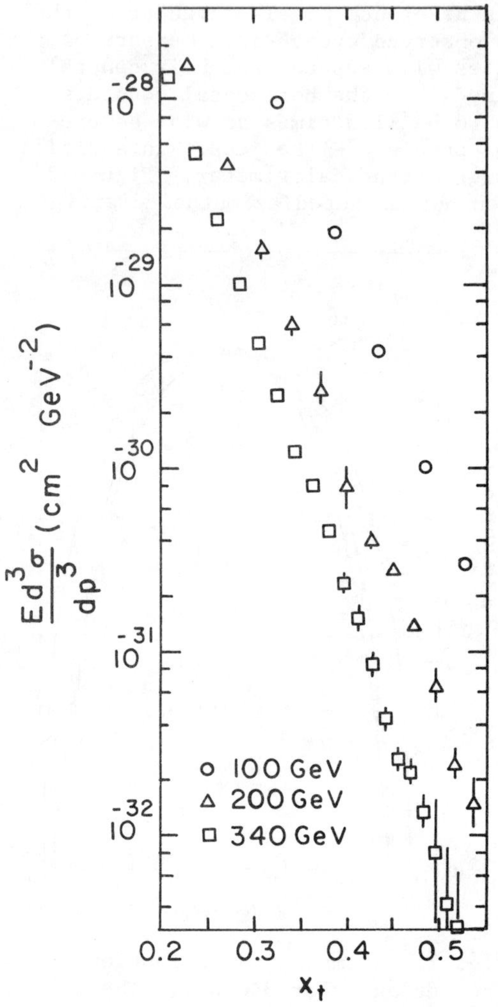

Fig. 3. Proton+proton→jet+anything invariant cross sections for the three energies.

CONCLUSIONS

We have determined that the observed rate of these events is far greater than expected from random fluctuations of particles obeying single-particle distributions. To extract physics from these results one can compare the yields in detail with a comprehensive theory, or one can attempt to extract a "pure jet" cross section, using empirical studies of the data. When we use the latter approach we find cross sections which scale reasonably, which agree at 200 GeV with the extracted jet cross section of Ref. 3, and which suggest (with considerable systematic uncertainty at this stage of the analysis) an n-value, for

$$E\frac{d\sigma}{d^3p} = \sqrt{s}^{\,-n} f(x_t),$$

in the region of 4 or 5.

REFERENCES

1. S. D. Ellis and M. B. Kislinger, Phys. Rev. D 9, 2027 (1974); R. D. Field and R. P. Feynman, Phys. Rev. D 15, 2590 (1976). For a review, and other references, see M. Jacob and P. V. Landshoff, Physics Reports 48, 285 (1978).
2. G. Hanson, Proceedings of the XIII Rencontre de Moriond (Savoie, France, 1968) Vol. II, p. 16; J. Bell et al., Phys. Rev. D 19, 1 (1979); W. A. Loomis et al., Fermilab-PUB-78/94 (unpublished).
3. C. Bromberg et al., Phys. Rev. Lett. 38, 1447(1977) and FERMILAB-Conf-77/62-EXP (unpublished)(1977 Kaysersberg Conference).

ANGULAR DEPENDENCE OF HIGH-p_T π° PRODUCTION

D. Lloyd Owen, G. W. Abshire, G. Finocchiaro,
P. D. Grannis, H. Jostlein, R. D. Kephart,
and R. Thun
State University - New York at Stony Brook

G. Bellettini, P. L. Braccini, R. Castaldi,
V. Cavasinni, T. Del Prete, P. Laurelli,
G. Sanguinetti, and M. Valdata-Nappi
INFN, University of Pisa and
Scuola Normale Superiore, Pisa, Italy

We have measured inclusive π° production cross-sections at the CERN ISR for $p_T < 5$ GeV/c at c.m. angles of $\theta = 90°$ and $5 < \theta < 22°$ at c.m. energies $\sqrt{s} = 23$ and 53 GeV.

The data are in agreement with previous measurements of π° production at FNAL and the ISR. We have compared these data with the hypothesis of radial scaling, which asserts that the inclusive cross section should be a function of x_R and p_T alone. Here x_R is the fraction of available energy carried by the produced particle. We find that radial scaling is not satisfactory over the energy and angle range of this experiment. Comparing our data with radial scaling fits to mainly FNAL data, we find that our cross sections approach ten times the prediction at large p_T and x_R. As a more direct test of the hypothesis, we compare our data for which x_R variation occurs as θ varies at fixed s, with that from ISR experiments at fixed $\theta = 90°$ and variable s. The trend of the two data sets is in marked disagreement.

We have also compared our results with predictions from a QCD-based model. The angular distribution at fixed p_T is well described by the model.

The results described here are more fully discussed in a recent publication[1].

1. D. Lloyd Owen et al., Phys. Rev. Letters 45, 89 (1980).

A MEASUREMENT OF THE TRANSVERSE MOMENTA OF PARTONS AND OF JET FRAGMENTATION AS A FUNCTION OF $\sqrt{s}$ IN p-p COLLISIONS
(CCOR Collaboration)

H.-J. Besch, L. Camilleri, C. del Papa, L. Di Lella,
C.B. Newman, B.G. Pope, S.H. Pordes, A.M. Smith and K.K. Young
CERN, Geneva, Switzerland

B.J. Blumenfeld, R.J. Hollebeek,
L.M. Lederman, D.A. Levinthal, R.W. Rusack, R.A. Vidal
Columbia University, New York, NY, USA

A.L.S. Angelis, N. Phinney,
A.M. Segar, J.S. Wallace-Hadrill and J.M. Yelton
Oxford University, Oxford, UK

T.J. Chapin, R.L. Cool, Z. Dimcovski, J.T. Linnemann,
A.F. Rothenberg and M.J. Tannenbaum
The Rockefeller University, New York, NY, USA

ABSTRACT

A large solid-angle apparatus consisting of a superconducting solenoid magnet, cylindrical drift chambers and two arrays of lead-glass counters was used to examine particles associated with a high transverse momentum trigger in p-p collisions with three $\sqrt{s}$ values at the CERN ISR.

The trigger was given by energy deposition in lead-glass arrays centered at 90°. The trigger transverse momentum range covered was $3 < p_{Ttrig} < 11$ GeV/c. Results are given for p_{out} for both individual charged particles, and also for the sum of charged particle momenta in the hemisphere opposite to the trigger. Mean values are then deduced for the parton transverse momentum k_T, and for the jet fragmentation momentum j_T. Results of a jet analysis are also presented.

- - - - - - - - - - - - - - - - - - -

The CCOR Collaboration is engaged in a general study of high transverse momentum processes arising from p-p collisions at the CERN Intersecting Storage Rings (ISR)[1-4]. This contribution reports a study of the particles on the opposite side to that of a high-momentum trigger particle. This is the first experiment capable of momentum-analysis of charged particles over the full azimuth. Thus the measurement of particle momenta out of the trigger plane p_{out} does not need acceptance corrections and jet analysis is simplified.

The apparatus consists of two arrays of lead-glass Cerenkov counters to provide a high-energy trigger[5], and an inner detector to momentum-analyse charged particles[6]. The lead glass gives a transverse momentum measurement with an r.m.s. resolution of $\Delta p/p = \Delta E/E = 0.004 + 0.043/\sqrt{E}$ (E in GeV) for each counter, a 5% r.m.s. counter-to-counter variation, and a systematic scale uncertainty of 5%. The energy deposited is located in azimuth with an r.m.s. uncertainty of

$\Delta\phi$ = 20 mrad. The track-finding efficiency is estimated to be 75 ± 10%, and the spatial resolution is 350 μm. This together with the number of measurements and multiple Coulomb scattering gives an r.m.s. momentum resolution $\Delta p_T/p_T = \sqrt{(0.07\ p^2) + 0.02^2}$ (p_T in GeV/c). The r.m.s. azimuthal resolution for tracks is $\Delta\phi$ = 10 mrad.

Data have been taken with the ISR operating at $\sqrt{s}$ = 62, 45, and 31 GeV, and at luminosities of up to 4×10^{31} cm^{-2} s^{-1}. The data were divided into nine different sets according to their $p_{T\ trig}$ and $\sqrt{s}$ values, where $p_{T\ trig}$ is the transverse momentum of the trigger particle calculated in the p-p centre-of-mass system.

Tracks were reconstructed and vertices fitted for all events. The two lead-glass arrays were denoted 'inside' and 'outside'. Each charged particle was required to have p_T > 300 MeV/c. Those in the hemisphere opposite to the outside array were required to have $|\eta| < 0.7$, and those in the hemisphere opposite to the inside array were required to have $|\eta| < 0.9$. These cuts ensured a uniform ϕ acceptance for the charged particles within each hemisphere. An estimate of some of the possible systematic errors involved was made by a comparison of the results of the two sides. The results agreed within the errors quoted.

For each charged track, p_{out} is the component of momentum out of the plane formed by the two beams and the trigger particle, and $x_E = -\vec{p}_{T\ track} \cdot \vec{p}_{T\ trig}/|p_{T\ trig}|^2$. Corrections were applied to $<|p_{out}|>^2$ for charged particle momentum resolution and ϕ_{trig} resolution. These corrections, which are bigger at higher x_E^2, always decrease the value of $<|p_{out}|>^2$, but are always less than 20%.

In parton models, finite values of p_{out} are believed to be produced by two effects: the transverse momentum (k_T) of the partons that enter the hard scattering process, and the transverse momentum relative to the jet axis (j_T) given to a particle during the fragmentation of its parent parton after scattering. In this picture the parameters of the model may be obtained from the approximate relationship[7]:

$$<|p_{out}|>^2 = 2<|k_{Ty}|>^2\ x_E^2 + <|j_{Ty}|>^2\ (1 + x_E^2)\ , \qquad (1)$$

where $<|k_{Ty}|>$ and $<|j_{Ty}|>$ are the average values of the components of k_T and j_T out of the scattering plane. The data do not satisfy Eq. (1) for the complete range of x_E^2. At low x_E^2 a departure from linearity is expected because j_{Ty} is kinematically constrained to be small when track momenta are small. If only those points are used which correspond to $p_{T\ track}$ > 1.4 GeV/c, for which the kinematic constraint is small, reasonable χ^2/d.o.f. may be obtained for straight-line fits. Even in this region, the model's assumption that x_E is equivalent to the jet fragmentation variable z can influence the numerical results of $<|k_{Ty}|>$. However, the trends in the data should not be sensitive to any errors introduced by this assumption.

Figure 1 shows $<|j_{Ty}|>$ as a function of $p_{T\ trig}$ (Fig. 1a) and $\sqrt{s}$ (Fig. 1b). For Fig. 1a, $<|j_{Ty}|>$ was constrained to be the same value for all $\sqrt{s}$ values, but was allowed to vary with $p_{T\ trig}$. There is no apparent variation of $<|j_{Ty}|>$ with $p_{T\,trig}$. Similarly, for Fig. 1b $<|j_{Ty}|>$ was constrained to be the same for all $p_{T\,trig}$ values but was allowed to vary with $\sqrt{s}$. There is no indication that $<|j_{Ty}|>$ is a function of $\sqrt{s}$. Thus it is reasonable to constrain $<|j_{Ty}|>$ to one value for all nine data samples, and fit for $<|k_{Ty}|>$ and $<|j_{Ty}|>$. This yields a χ^2 of 62 for 50 degrees of freedom, and the $<|j_{Ty}|>$ found is 0.393 ± 0.007 GeV/c. The values of $<|k_{Ty}|>$ (Fig. 2a) show an increase with both $p_{T\,trig}$ and $\sqrt{s}$, rising to $<|k_{Ty}|> \sim 0.8$ GeV/c at the highest $\sqrt{s}$ and $p_{T\,trig}$.

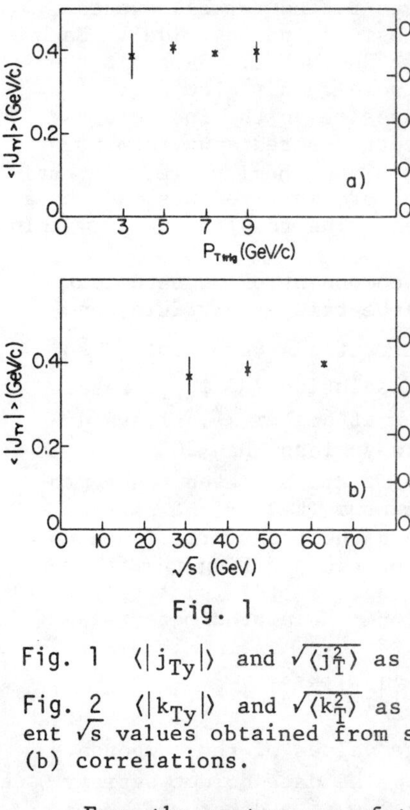

Fig. 1

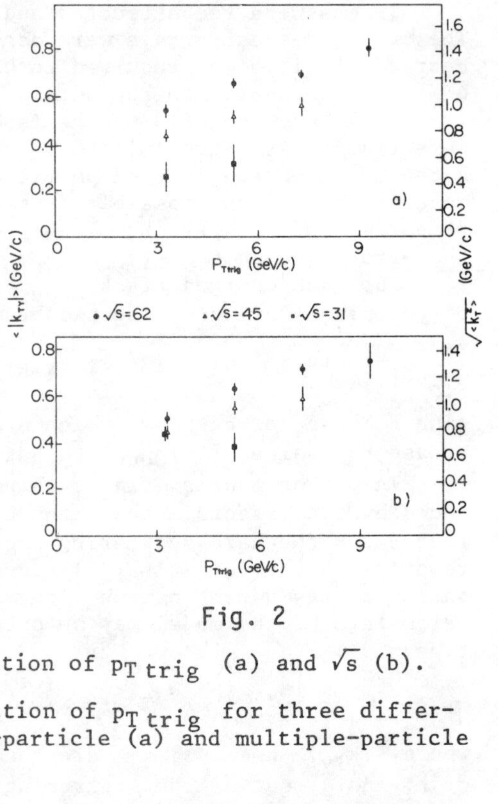

Fig. 2

Fig. 1 $\langle|j_{Ty}|\rangle$ and $\sqrt{\langle j_T^2\rangle}$ as a function of $p_{T\,trig}$ (a) and $\sqrt{s}$ (b).

Fig. 2 $\langle|k_{Ty}|\rangle$ and $\sqrt{\langle k_T^2\rangle}$ as a function of $p_{T\,trig}$ for three different $\sqrt{s}$ values obtained from single-particle (a) and multiple-particle (b) correlations.

From the vector sum of the away-side charged particles another measurement of $<|k_{Ty}|>$ may be obtained, which has different systematic uncertainties but requires no extra physics assumptions. If one selects events where the sum of the charged particle transverse momenta in the hemisphere opposite to the trigger balances the trigger transverse momentum, i.e., $\Sigma x_E \sim 1$, then the same model predicts

$$\langle|\Sigma p_{out}|\rangle = \sqrt{2\,\langle|k_{Ty}|\rangle^2 + \langle|j_{Ty}|\rangle^2}\ . \qquad (2)$$

Using the value of $<|j_{Ty}|>$ found above, $<|k_{Ty}|>$ may be determined. The model does not require corrections for 'jet' containment or track efficiency. In addition, corrections to $<|\Sigma p_{out}|>$ due to momentum resolution and ϕ_{trig} resolution were found to be negligible. Events were selected with $0.9 < \Sigma x_E < 1.1$, and a small correction was applied to allow for the fact that Σx_E in this region is not exactly 1. The results for $<|k_{Ty}|>$ thus obtained are shown in Fig. 2b.

These values are equal, within statistical errors, to the values obtained from single particles, with the exception of the $\sqrt{s}$ = 31 GeV, $p_{T\,trig}$ = 3 GeV/c point, which is higher. It is thought that formula (1) is least reliable when $k_{Ty} \sim j_{Ty}$, and thus the low threshold values in Fig. 2a have large systematic errors.

It should be noted that $<|k_{Ty}|>$ is the average value of the component out of the scattering plane of a parton's transverse momentum. If one assumes a Gaussian distribution in k_{Ty}, then $\sqrt{<k_{Ty}^2>} = <|k_{Ty}|> \times \sqrt{\pi/2}$. If one assumes, in addition, that the two components of k_T are equal, $\sqrt{<k_T^2>} = \sqrt{<k_{Ty}^2>} \times \sqrt{2}$, and these values can be read using the right-hand scale of Fig. 2. Similarly high values of k_T have previously been reported at the ISR[2,8,9]. A calorimeter experiment[10] has measured k_T in the scattering plane as a function of $p_{T\,trig}$ and $\sqrt{s}$, and found the same trends as this experiment.

The value of $<|j_{Ty}|>$, of 0.393 ± 0.007 GeV/c, gives $\sqrt{<j_{Ty}^2>}$ of 0.493 ± 0.009 GeV/c assuming a Gaussian, and $\sqrt{<j_T^2>}$ of 0.697 ± 0.013 GeV/c taking the two components to be equal[2]. This value refers only to fragments with $p_T > 1.4$ GeV/c and should not be directly compared with results obtained by integrating over all jet fragments. It is in reasonable agreement with the value obtained by another ISR group (0.62 ± 0.06 GeV/c)[9], but somewhat higher than results for e^+e^- data ($\sim$0.55 GeV/c)[11] obtained including only high-momentum hadrons. It should be stressed that the error quoted is statistical only.

A study has been made of the charged and neutral particles produced in events taken at $\sqrt{s}$ = 62.4 GeV with $P_{TRIG} > 7.0$ GeV/c. For each event, the vector sum $\vec{P}_s = \Sigma\, j_i$ is formed for all observed charged and neutral particles in the hemicylinder opposite to the π° trigger and $\vec{P}_J = \Sigma\, \vec{j}_i$ for those on the trigger side. The subsample of events selected for further analysis are those for which $\vec{P}_s$ and $\vec{P}_J$ fall within a fiducial range $|y| < 0.4$ and $|\Delta\phi| < 0.4$ rad, where $\Delta\phi$ is the deviation from the horizontal axis of the detector. Since charged particles are observed for $|y| < 0.7$ for all ϕ and neutrals within $|\overline{y}| < 0.6$ and $|\overline{\Delta\phi}| < 0.5$, approximately 93% of the momentum of the away-side and 97% of the trigger-side 'jets' is contained within the apparatus for the selected events. A correction for this loss and for charged-track inefficiency are made as appropriate.

If it is assumed that the vectors $\vec{P}_J$ and $\vec{P}_s$ represent the outgoing momenta of two quarks (gluons) which have scattered elastically, we may transform to the c.o.m. system of the colliding constituents. In that c.o.m. system (see Fig. 3), the magnitude of the equal and opposite momenta are designated as Q, at angles θ^* and ϕ^*. From this analysis, two results are given in this report.

First, for the particles which form the away-side 'jet' described by the vector $\vec{P}_s$, we have studied the distribution of the component of their momentum, $j_\perp$, transverse to the axis defined by $\vec{P}_s$ as a function of Q, the c.o.m. momentum of the scattered constituent. The observed values of $j_\perp$ at each Q are fitted by Gaussian distributions characterized by an r.m.s. width. However, it should be noted that approximately 5% of the observed particles fall in a "tail" at large $j_\perp$ which is not well described by the Gaussian fit. These particles could be due to experimental backgrounds or, possibly, to fragmentation of a radiated gluon. The results for $\langle j_\perp^2 \rangle^{\frac{1}{2}}$ are shown in Fig. 4 together with data from SLAC and PETRA . Since our data is an average value over particle momenta above 0.4 GeV, while those from PETRA include particles down to lower momenta, it is to be expected that the PETRA values should be somewhat lower due to the well-known "sea-gull" effect[14].

Second, we have studied the distribution of the momentum component $k_\perp$ which is required to transform from the p-p system to the c.o.m. of the colliding constituents. This component is perpendicular to the plane formed by the beam axis and the scattered constituents (see Fig. 3). In QCD models, $k_\perp$ can arise from "intrinsic" momenta of the constituents from non-perturbative binding effects and from radiation of gluons whose fragmenting particles are not observed in our apparatus. At each Q, the values of $k_\perp$ are adequately fitted by Gaussian distributions characterized by an r.m.s. value $\sigma_\perp$. They are displayed in Fig. 5. The observed values of $\sigma_\perp$ as a function of Q are adequately fit by a linear form $\sigma_\perp = (0.075 \pm 0.013)\ Q + (0.55 \pm 0.12)$. The large and Q dependent values, which are not expected from "intrinsic" effects, imply that the QCD radiative corrections due to gluon emission must be very important at these large momentum transfers.

The values of $\sigma_\perp$ deduced from the jet-jet analysis may be compared to those derived from the single particle correlation data of Fig. 2. To do so, we must make use of the approximate relation, derived from our data, that $Q \cong 0.79\ p_{TRIG}$. The results are shown in Fig. 6. The agreement between the two methods is entirely adequate when systematic uncertainties are taken into account.

REFERENCES

1. A.L.S. Angelis et al., Phys. Lett. 79B, 505 (1978).
2. A.L.S. Angelis et al., Phys. Scripta 19, 116 (1979).
3. A.L.S. Angelis et al., Phys. Lett. 87B, 398 (1979).
4. A.L.S. Angelis et al., preprint CERN-EP/80-68, to appear in Physics Letters.
5. J.S. Beale et al., Nucl. Instrum. Methods 117, 50 (1974).
6. L. Camilleri et al., Nucl. Instrum. Methods 156, 275 (1978).
7. E.M. Levin et al., Sov. Phys. JETP 69, 1537 (1975); R.P. Feynman, R.D. Field and G.C. Fox, Nucl. Phys. B128, 1 (1977).
8. C. Kourkoumelis et al., Nucl. Phys. B158, 39 (1979).
9. A.G. Clark et al., Nucl. Phys. B160, 397 (1979).
10. M.D. Corcoran et al., Phys. Rev. D 21, 641 (1980).
11. P. Söding, Proc. EPS Int. Conf. on High-Energy Physics, Geneva, 1979 (CERN, Geneva, 1980), Vol. 1, p. 271.
12. G. Hanson, Proc. XIIIth Rencontre de Moriond, Vol. II, p. 33 (1978).
13. TASSO COLLABORATION Preprint, DESY 79/50 (1979).
14. For a discussion of this effect see, for example, Ref. 2.

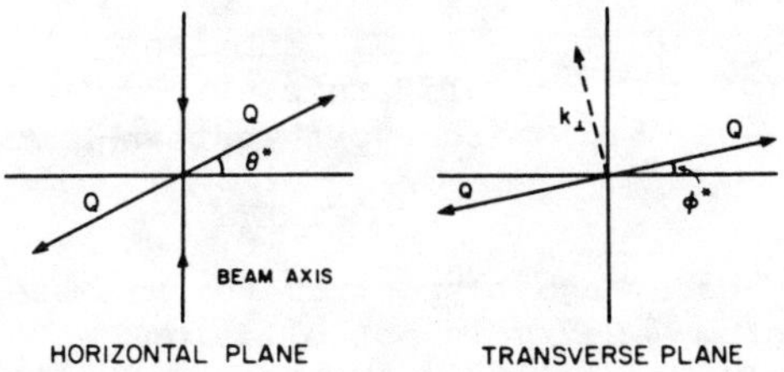

FIG. 3

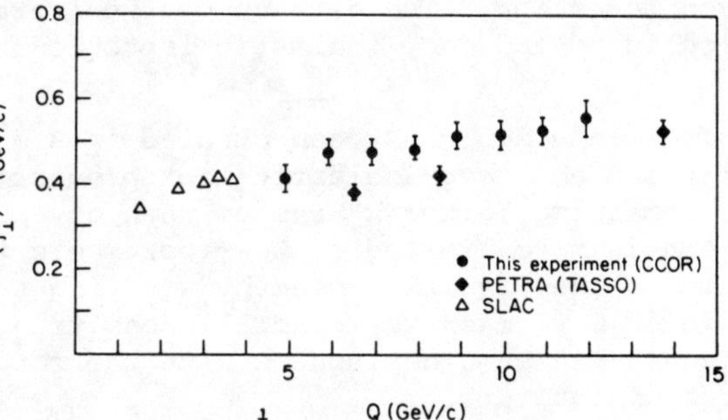

FIG. 4 - $\langle j_{\perp}^2 \rangle^{\frac{1}{2}}$ of jet fragmentation vs. Q.

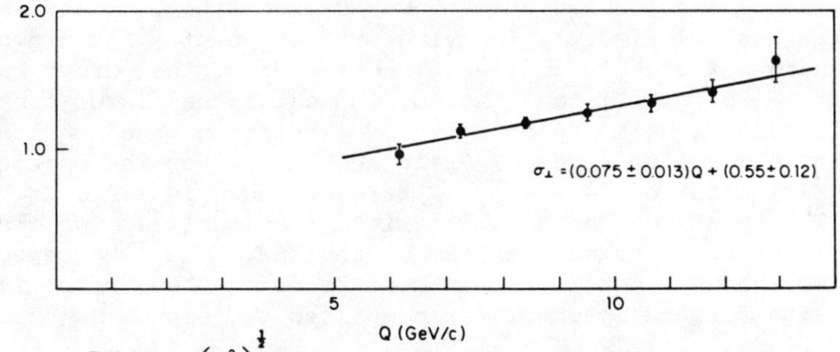

FIG. 5 - $\langle k_{\perp}^2 \rangle^{\frac{1}{2}}$ of "apparent" constituent momentum vs. Q.

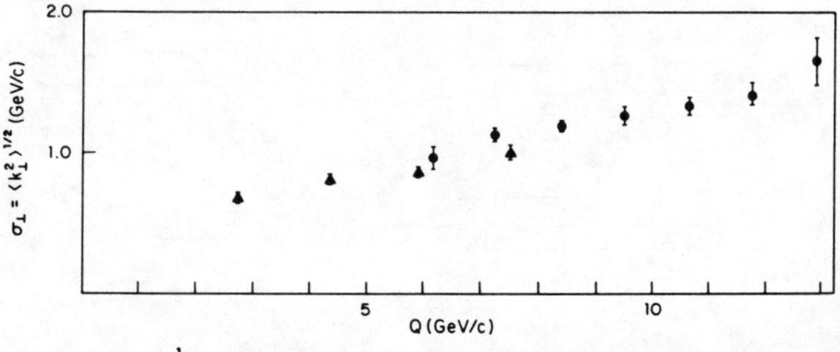

FIG. 6 - $\langle k_{\perp}^2 \rangle^{\frac{1}{2}}$ vs. Q; from dijet analysis, from $\langle p_{out} \rangle$ measurement.

DIQUARK FRAGMENTATION STUDIES AT THE CERN ISR

G.J. VanDalen
University of California, Riverside, Ca. 92521
(Aachen-CERN-Harvard-Munich-Northwestern-Riverside Collaboration)

ABSTRACT

The Lamp Shade Magnet spectrometer was used to trigger on 30° pions of momentum >5 GeV/c in pp collisions at the CERN ISR. Forward pion and proton production in the trigger hemisphere indicate that hadronization of diquarks is primarily recombination resulting in high x baryon formation. The data support universality of diquark fragmentation in pp and deep inelastic leptoproduction reactions.

Quark fragmentation has been studied in e^+e^- annihilation, hadron collisions and deep inelastic leptoproduction experiments. Data on the remaining "diquark" system when a valence quark is taken from a nucleon is more limited[1]. We report here the results of an investigation of the diquark system by studying small angle hadron production in 63 Gev cm energy pp collisions triggered by a high momentum pion in the same hemisphere with the forward hadrons identified by Cerenkov counters.

In the naive quark-parton model, the trigger pion is the decay product of a quark ejected in a hard scattering process, as shown in Fig. 1a. Pions of high p_T(>2.5 GeV/c) are likely to have a parent valence quark (x_{bj}>0.2), implying production of a forward diquark. The charge of the pion reflects the flavor of the parent quark, (u for π^+, d for π^-), and thereby indicates the flavor content of the forward diquark (ud or uu). A calculation based on the more realistic model of Feynman, Field and Fox[2] gives the parton composition for a 3 GeV/c p_T $\pi^\pm$ trigger shown in Table I[3].

The apparatus has been described previously[4]. Two coaxial spectrometers surround one beam of the ISR. The outer spectrometer or Lamp Shade Magnet (LSM) was used to define the triggering pion at 30°. The forward spectrometer comprised two septum magnets covering

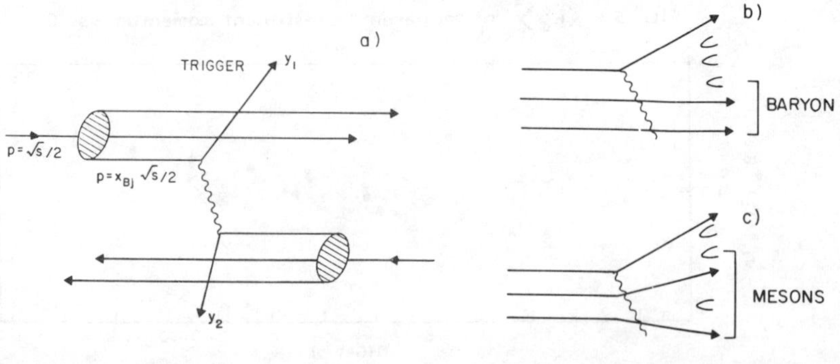

Fig. 1. Parton diagrams for: (a) quark-quark elastic scattering, diquark hadronization by (b) baryon formation, (c) meson formation.

Table I Fractional composition of parent partons for $\pi^{\pm}$ produced at 30° with p_T of 3 GeV/c in pp collisions at 63 GeV cm energy.

Parent Parton	Fraction of π^+ Trigger	Fraction of π^- Trigger
u quark	.57 ± .03	.17 ± .03
d quark	.06 ± .03	.37 ± .03
gluon	.32 ± .03	.40 ± .03
anti-quark	.05 ± .03	.05 ± .03

1° to 6° from the beam direction. Segmented Cerenkov counters in the magnets provided particle identification with π, K and p thresholds of 2.6, 9.3 and 17.7 GeV/c respectively. We distinguish between mesons and protons using the notation π for the mesons, which are mostly pions. Data was taken for an integrated luminosity of 1.4 pb^{-1}. After track reconstruction a total of 40000 π^+ and 32000 π^- triggers with $p_T > 2.5$ GeV/c remained.

The kinematics of the forward particles are summarized in the invariant: $F(z) = (z/N)\ (dn^h/dz)$, where n^h is the number of type h hadrons seen with fraction z of the diquark momentum and N is the number of triggers. The diquark momentum is $p=(1-x_{bj})\sqrt{s}/2$, where x_{bj} is the incident proton momentum fraction carried by the trigger quark. From the kinematics of the trigger pion, and assuming that the triggering pion carries on the average 90% of the jet momentum[5], we get $x_{bj} \simeq 0.08p_T$, where p_T refers to the trigger pion.

Figure 2 shows the F(z) versus (1-z) for forward, p, π^+ and π^- for π^+ and π^- triggers in the p_T range 2.5-4.0 GeV/c. Phase space considerations suggest a (1-z) power behavior for F(z)[6]. The forward mesons are consistent with a single power for both trigger charges over the z range 0.2-0.7, but the proton data requires two powers, 0.5 for π^- trigger, and 1.0 for π^+ trigger. The general features of the distributions are: (1) protons are produced more frequently and with a harder z distribution than mesons; (2) positive correlation between proton production and negative trigger; (3) meson distribution independent of trigger sign.

There are two classes of models for diquark fragmentation[1,7]: (a) The diquark behaves coherently, eventually recombining to form a leading baryon (Fig. 1b). Mesons are produced from newly created quark pairs and are not correlated with the sign of the trigger.

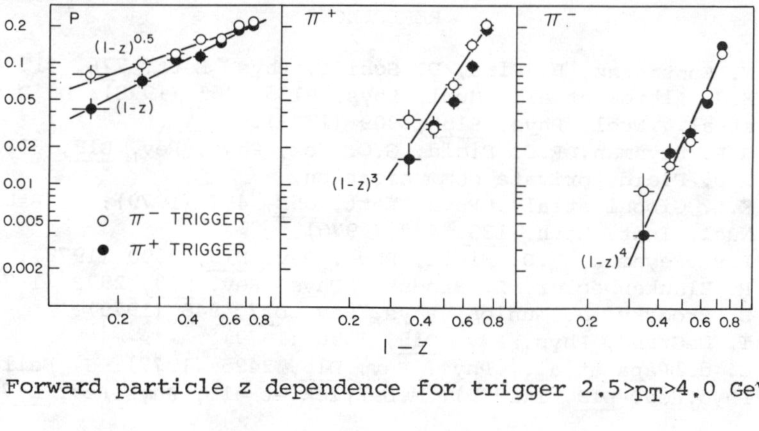

Fig. 2. Forward particle z dependence for trigger $2.5>p_T>4.0$ GeV/c.

(b) The diquark behaves incoherently (Fig. 1c), with the two valence quarks hadronizing into mesons. Fast mesons will contain valence quarks and be charge correlated with the trigger. Note that if the trigger comes from a gluon (30-40% of the events), no charge correlation between trigger and forward particle is expected.

The results of this experiment suggest model (a). The absence of a trigger charge dependence for meson production is the strongest evidence against direct diquark fragmentation into mesons.

We can investigate universality for diquarks by comparing our results with target fragmentation data in deep inelastic lepton scattering, in which the gauge boson (photon, W or Z) ejects a quark from the target nucleon leaving behind a diquark system. Figure 3 shows the comparison between π^- distributions in our data and in three lepton experiments. Valence u-quarks are removed in $\bar{\nu}p$ and in μp scattering (8 times more often than valence d's) leaving a ud diquark as in our π^+ trigger data. Similarly we compare the π^- trigger data with νp results. Agreement in overlap regions is good.

In summary, we conclude that hadronization of diquarks occurs predominantly via recombination of the diquark with a quark to form a baryon, and that universality of diquark fragmentation is supported.

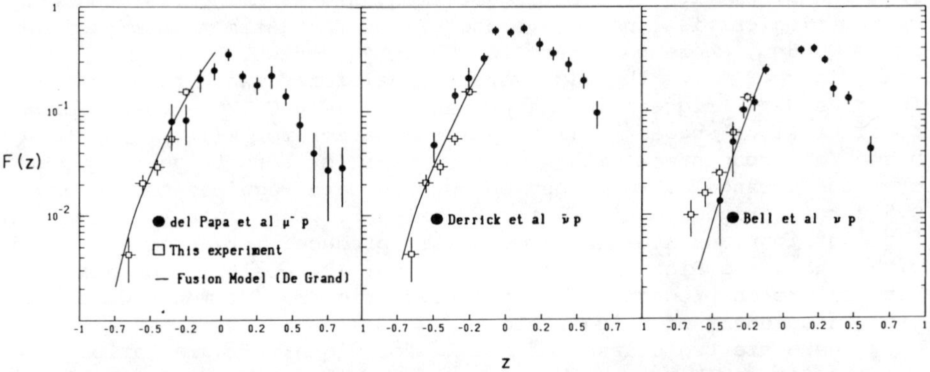

Fig. 3. Comparison of π^- production in target fragmentation in this experiment with (a) μp, (b) $\bar{\nu}p$ and (c) νp interactions.

REFERENCES

1. M. Fontannaz, B. Pire, D. Schiff, Phys. Lett. 77B, 315 (1978); M.G. Albrow et al., Nucl. Phys. B135, 461 (1978); D. Drijard et al., Nucl. Phys. B156, 309 (1979).
2. R.P. Feynman, R.D. Field, G.C. Fox, Phys. Rev. D18, 3320 (1978).
3. R.D. Field, private communication.
4. K.L. Giboni et al., Phys. Lett. 85B, 437 (1979); L. Baksay et al., Nucl. Inst. Meth. 133, 437 (1976).
5. R.P. Feynman, R.D. Field, Phys. Rev. D15, 2590 (1978).
6. R. Blankenbecler, S. Brodsky, Phys. Rev. D10, 2973 (1974); S. Brodsky, J. Gunion, Phys. Rev. D17, 848 (1979).
7. T. DeGrand, Phys. Rev. D15, 1398 (1979).
8. C. delPapa et al., Phys. Rev. D15, 2425 (1977); J. Bell et al., Phys.Rev. D19, 1(1979); M.Derrick et al., Phys.Rev. D17,1(1978).

PRODUCTION OF HIGH P_T π^o's AT THE CERN ISR

Athens - Brookhaven - CERN Collaboration

Presented by T. Fields

INTRODUCTION

We have now completed an extended program of measurements of inclusive π^o production at 90° c.m. in pp collisions. In this very brief report, we emphasize our recent results in which the two photons from π^o decay are each resolved for p_T up to 10 GeV/c. A full description of the entire π^o experiment with final results covering the range $3 < p_T < 16$ GeV/c and $31 < \sqrt{s} < 63$ GeV is being published[1].

DETECTOR

Liquid-argon/lead calorimeters of high spatial resolution and large solid angle acceptance were used for the detection of photons from π^o decay. For the data given here, the calorimeters were in their farthest position, 2.1 m from the interaction region, in order to resolve the decay $\pi^o \to 2\gamma$ up to $p_T \gtrsim 10$ GeV/c.

RESULTS

Table 1 gives the π^o cross-section results at $\sqrt{s} = 44.8$ and 62.8 GeV. The errors given do not include a 27% scale factor uncertainty which arises from a 2.5% uncertainty in the energy (p_T) scale.

DISCUSSION OF RESULTS

This is the first experiment to allow unambiguous π^o identification in this p_T range. Earlier π^o measurements without this feature have not distinguished direct photon events from π^o events. In fact, the cross-section results in Table 1 for $p_T \gtrsim 6$ GeV/c are somewhat lower than those we obtained earlier without π^o identification as would be expected since the present results do not include direct photon events[2,3,4].

Much interest has been focused upon the experimental determination of the power n in the scaling formula

$$[E \frac{d^3\sigma}{dp^3}]_{\pi^o} = Ap_T^{-n} f(x_T)$$

However, existing evidence for a decrease of n from its medium-p_T value of $\sim$8 has been obtained entirely from data on unresolved photon detection with $p_T \gtrsim 10$ GeV/c. This situation is illustrated by our own data in Fig. 1, where it can be seen that only data for unresolved photon detection (solid points) show evidence for a decrease in n. Thus, correct determination of the behaviour of n will require the use of data on single photon production, especially for $p_T \gtrsim 10$ GeV/c.

REFERENCES

1. C. Kourkoumelis et al., Zeitschrift für Physik C5, 95 (1980).
2. M. Diakonou et al., Phys. Lett. 87B, 292 (1979).
3. M. Diakonou et al., Phys. Lett. 91B, 296 (1980).
4. A.L.S. Angelis et al., Phys. Lett. 94B, 106 (1980).

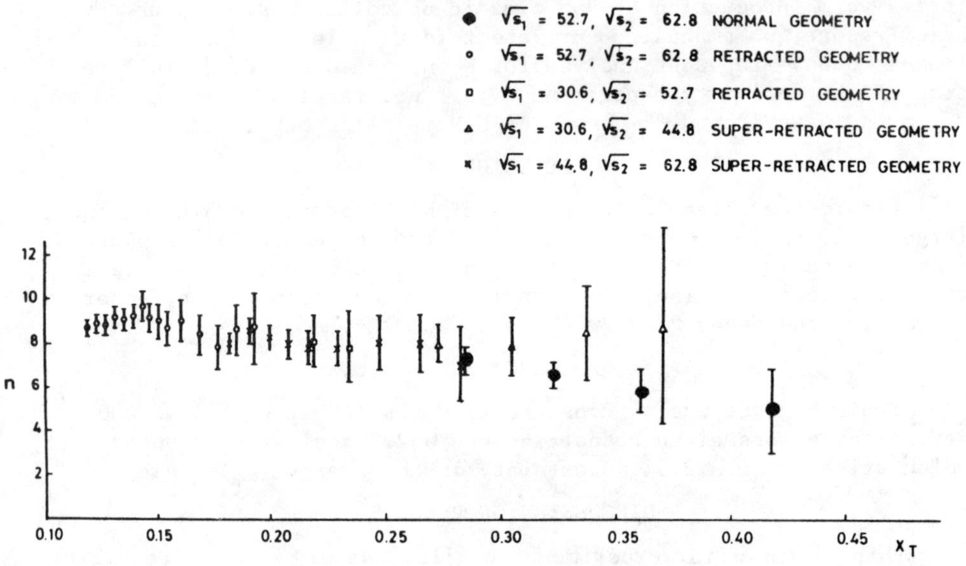

Figure 1: Variation of n with x_T

Table 1:
Resolved π^o
Cross-sections

p_T range (GeV/c)	$\langle p_T \rangle$ (GeV/c)	$E(d^3\sigma/dp^3)$ $(cm^2/GeV^2\ c^{-3})$ $\sqrt{s}$ = 44.8 GeV	$\sqrt{s}$ = 62.8 GeV
4.0- 4.2	4.10	$(1.03 \pm 0.11) \times 10^{-32}$	$(2.29 \pm 0.23) \times 10^{-32}$
4.2- 4.4	4.30	$(6.44 \pm 0.66) \times 10^{-33}$	$(1.43 \pm 0.15) \times 10^{-32}$
4.4- 4.6	4.50	$(3.92 \pm 0.40) \times 10^{-33}$	$(8.96 \pm 0.92) \times 10^{-33}$
4.6- 4.8	4.70	$(2.46 \pm 0.25) \times 10^{-33}$	$(5.86 \pm 0.60) \times 10^{-33}$
4.8- 5.0	4.89	$(1.56 \pm 0.16) \times 10^{-33}$	$(3.84 \pm 0.40) \times 10^{-33}$
5.0- 5.4	5.18	$(8.00 \pm 0.83) \times 10^{-34}$	$(2.15 \pm 0.22) \times 10^{-33}$
5.4- 5.8	5.58	$(3.49 \pm 0.37) \times 10^{-34}$	$(9.52 \pm 0.99) \times 10^{-34}$
5.8- 6.2	5.97	$(1.53 \pm 0.17) \times 10^{-34}$	$(4.17 \pm 0.45) \times 10^{-34}$
6.2- 6.6	6.37	$(6.83 \pm 0.84) \times 10^{-35}$	$(2.33 \pm 0.26) \times 10^{-34}$
6.6- 7.0	6.80	$(3.67 \pm 0.49) \times 10^{-35}$	$(1.25 \pm 0.15) \times 10^{-34}$
7.0- 7.5	7.24	$(1.49 \pm 0.23) \times 10^{-35}$	$(5.78 \pm 0.73) \times 10^{-35}$
7.5- 8.0	7.70	$(5.99 \pm 1.25) \times 10^{-36}$	$(2.80 \pm 0.40) \times 10^{-35}$
8.0- 9.0	8.32	$(2.56 \pm 0.55) \times 10^{-36}$	$(1.14 \pm 0.17) \times 10^{-35}$
9.0-10.0	9.40	$(2.84 \pm 1.23) \times 10^{-37}$	$(2.28 \pm 0.52) \times 10^{-36}$

HIGH p_T HADRONIC INTERACTIONS, THEORY

J. Owens, Florida State, Organizer

QCD CORRECTIONS TO HADRONIC LARGE p_T SCATTERING*

Ian Hinchliffe
Lawrence Berkeley Laboratory, Berkeley, California 94720

ABSTRACT

QCD corrections to large p_T scattering are calculated. The status of the perturbation expansion is discussed.

This talk is an account of the work by R.K. Ellis, H. Haber, M. Furman, and myself,[1] however, the contents of this talk reflect my views and not necessarily those of my collaborators.

The aim of this work is to decide whether QCD perturbation theory makes reliable predictions for the production of particles at large transverse momentum in hadron-hadron collisions. I will briefly recall the ingredients of the QCD prediction. The process $h_1 + h_2 \to h_3 + X$ is indicated in Fig. 1. Hadrons h_1 and h_2 of momenta P_1 and P_2 emit constituents of momenta x_1P_1 and x_2P_2 which then scatter at wide angle resulting in parton of momentum p_3 which then fragments into the observed hadron h_3 with momentum P_3. (Throughout upper case letters refer to hadrons and lower case refers to partons.)

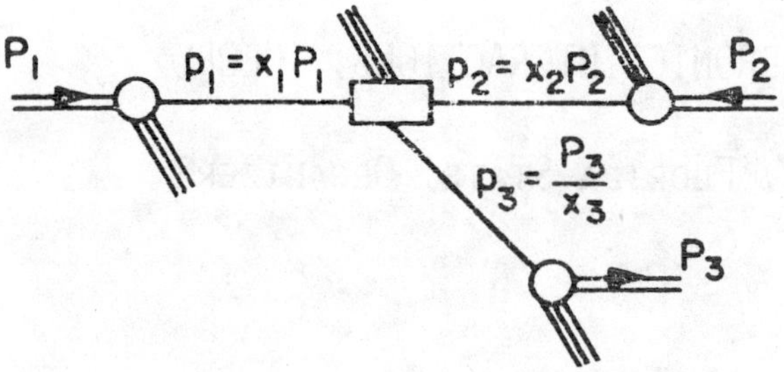

Fig. 1

Defining $t = (p_1 - p_3)^2$, $s = (p_1 + p_2)^2$, $u = (p_2 - p_3)^2$ and $v = 1 + t/s$, $w = -u/(s + t)$, the cross-section for production of h_3 is

$$\frac{1}{SV}\frac{d\sigma}{dVdW} = \sum_{ijk}\int_0^1 dx_1 dx_2 \frac{dx_3}{sx_3^2} \times \left[F_i(x_1,M^2)\, F_j(x_2,M^2)\, D_K(x_3,M^2)\right] \cdot$$

$$\times \left(\frac{1}{v}\frac{d\sigma_{ij}^K}{dv}(s,v)\,\delta(1-w) + \frac{\alpha_s^3}{2\pi} K_{ij}^K(s,v,w)\right) \quad (1)$$

---

* This work was supported by the High Energy Physics Division of the U.S. Department of Energy under contract No. W-7405-ENG-48.

where $F_i(x,M^2)\left[D_i(x,M^2)\right]$ is the probability of extracting a parton (hadron) of momentum fraction x from a hadron (parton). $\frac{d\sigma}{dv}$ is the lowest order parton parton scattering cross-section. $\frac{d\sigma}{dv}$ is proportional to α_s^2 and has support only at the elastic limit. The aim of the work I describe is to calculate the quantity K(s,v,w). If it is small then the phenomenology based on the lower order[2] is believable, if it is large then some doubt must be cast on the validity of the phemomenology.

Since only the size of K is relevant we chose[1] the simplest process (≡ simplest gauge invarient set of diagrams). Consider $q_1 + q_2 \rightarrow q_3 + X$ where q_1 and q_2 are quarks of distinct flavour, and we will ignore (initially) incoming gluons. A physical process can be found where this is the dominant process (e.g. $K^+ + \pi^- \rightarrow K^+ + X$, at large p_T where valence quarks dominate). In $\frac{d\sigma}{dv}$ there is only one graph, that involving t channel gluon exchange. The calculation of K is tedious and I will merely outline it here (details are given in ref. 1).

1.) Calculate the graphs which contribute through order α_s^3.

2.) Renormalize the ultraviolet divergence.

3.) Remove the mass singularities. The graph shown in Fig. 2, has a mass divergence when the propagator A goes on mass shell. These divergences must be reabsorbed into F and D. We are guaranteed that this proceedure will work by the factorization proof.[3] In order to avoid confusion, one merely has to remember where the F's and D come from. F is extracted from deep inelastic lepton scattering. The structure function

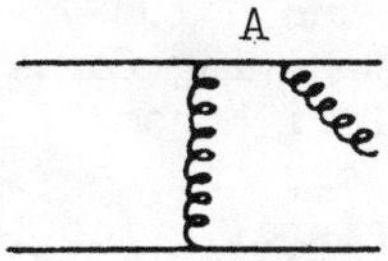

Fig. 2

$$F_2(x,M^2) = \sum_i e_i^2 F_i(x,M^2) \qquad (2)$$

where the sum i runs over quarks of charge e_i and M^2 is the scale at which the structure function is measured. If F_2 is calculated in perturbation theory the same mass divergences are encountered so that one uses F_i defined by equation (2) all the mass singularities cancel. The singularities in D are removed in a similar way by defining it from one particle inclusive $e\bar{e}$ annihilation.

4.) There is a complication associated with gluons of Fig. 3. A singularity arises when the propagator A goes on shell. This process corresponds to the incoming quark decaying into a gluon which then undergoes a hard scattering against the other quark. Now the gluon distribution is not measured in any other process,

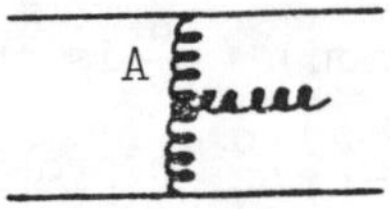

Fig. 3

so we are forced to define it. This leads to a (finite) ambiguity in K. However, the gluon distribution is not totally arbitrary. The singular part is defined and there is a constraint on the finite part from energy momentum conservation.

Having calculated K we now substitute back into equation (1), using some parton distributions and look at the effect of K on $\frac{d\sigma}{dVdW}$. It is important to discuss the ambiguities in K.

1.) The scale M^2 is characteristic of the hard scattering process; a change in M is equivalent to a change in K. In lowest order M is arbitrary (a popular choice is $\frac{2stu}{s^2 + t^2 + u^2}$)[4].

2.) Part of K is the expansion of $\alpha_s^2(\mu)$ which appears in lowest order. Again μ is same scale characteristic of the hard scattering ($\mu = M$ is usually used in lowest order but there is no reason why they should be the same), and a change in μ results in a change in K. There is an additional ambiguity in the prescription used to define α_s. We used the so called $\overline{MS}$ prescription,[5] but one could equally use a momentum space scheme.[6]

3.) There is an ambiguity associated with the gluon distribution $G(x,M^2)$ as discussed earlier.

It is "obvious" that for some values of μ, M^2, and $G(x,M^2)$, K can be made small, and it is also probably true that having found those values somebody will come up with an a posteriori "proof" that these are the natural values. I will try to guess the scales before doing the calculation. $\mu = t$ seems natural since the coupling constant renormalization comes from dressing the lowest order (t channel) graph. We do not know what M is. An obvious upper value is s, and a lower one p_T^2 of the outgoing parton ($= tu/s$ at 90° in the center mass frame). Figure 4 shows the function

$$R = \left.\frac{1}{SV}\frac{d\sigma}{dWdV}\right|_{\text{including K}} \Big/ \left.\frac{1}{SV}\frac{d\sigma}{dVdW}\right|_{K=0}$$

at 90° in the center of mass as a function of $x_\perp = 2p_T/\sqrt{S}$. If $R \approx 1$ the correction is small and the perturbation expansion is reliable. The dashed line at $M^2 = tu/s$ illustrates the size of the ambiguity associated with $G(x,M^2)$. It disappears at large p_T where gluons are irrelevant. $M^2 = t$ and $M^2 = \frac{2stu}{s^2 + t^2 + u^2}$ produce curves

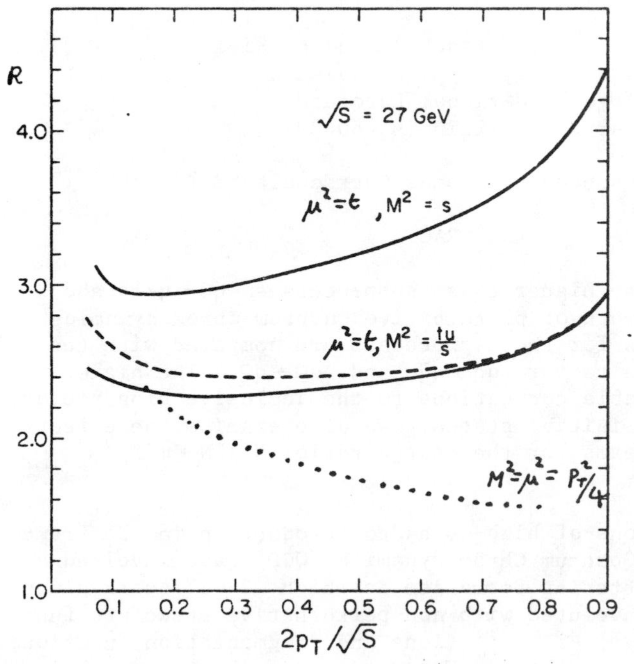

Fig. 4

which lie between the solid lines. I have shown an unnatural value of μ and M in the dotted line on Fig. 4 $\mu^2 = M^2 = p_T^2/4$. The correction is still large. The correction can be reduced by using a momentum space subtracted coupling constant. It will reduce the correction by about 40%. However, even for $p_T^2/4 = \mu^2 = M^2$ the corrections are still substantial.

In conclusion it appears that QCD perturbation theory does not make sound predictions for large p_T hadron scattering at current energies. As the energy increases the corrections reduce but only logarithmically, not until $\sqrt{S} \approx 1000$ GeV do they become reasonably small.

REFERENCES

1. R.K. Ellis, M. Furman, H. Haber, I. Hinchliffe, LBL-10304, to be published in Nucl. Phys. B.
2. B.L. Combridge, J. Kripfganz and J. Ranft, Phys. Lett. 70B, 234 (1977); R. Cutler and D. Sivers, Phys. Rev. D17, 196 (1978); J.F. Owens, E. Reya, and M. Gluck, Phys. Rev. D18, 1501 (1978); R.P. Feynman, R.D. Field, and G.C. Fox, Phys. Rev. D18, 3320 (1978); R. Baier, J. Engles, and B. Petersson, Z. Physik C2, 265 (1979).
3. R.K. Ellis, H. Georgi, M. Machacek, H.D. Politzer, and G.C. Ross, Nucl. Phys. B152, 285 (1979); D. Amati, R. Petronzio and G. Veneziano, Nucl. Phys. B146, 29 (1978); A.H. Mueller, Phys. Rev. D18, 3705 (1978); S.B. Libby and G. Sterman, Phys. Rev. D18, 3252 (1978).
4. R.P. Feynman, et al Ref. 2.
5. W. A. Bardeen and A. Buras, Phys. Rev. D20, 166 (1979).
6. W. Celmaster and R. Gonsalves, Phys. Rev. D20, 1420 (1979).

HIGHER TWIST QCD TERMS IN HIGH-P_T PION PRODUCTION

E. L. Berger, T. Gottschalk and D. Sivers
High Energy Physics Division
Argonne National Laboratory
Argonne, Illinois 60439

Presented by Thomas Gottschalk

ABSTRACT

We investigate the higher twist subprocesses $qG \to q\pi^{\pm}$ and $q\bar{q} \to G\pi^{\pm}$ in the framework of perturbative Quantum Chromodynamics (QCD). Cross sections for these processes are compared with the minimum twist QCD results for $q\bar{q} \to q\bar{q}$ and $qG \to qG$. The higher twist terms give sizeable corrections to the inclusive pion yields, particularly for $q\bar{q}$ initial states. We also examine the effects of higher twist QCD terms for the charge ratio, $N(\pi^- N \to \pi^- X)/N(\pi^- N \to \pi^+ X)$.

Most investigations of high-p_T hadron production in the framework of perturbative Quantum Chromodynamics (QCD) have involved a conventional hard-scattering formalism in which $2 \to 2$ quark/gluon cross sections are convoluted with non-perturbative structure functions and fragmentation functions. However, we now have reason to believe that this minimum-twist scenario is not the only important hadron production mechanism expected in QCD. It has been emphasized for some time that, in a hard-scattering expansion involving hadrons, a very significant role may be played by subprocesses involving "constituents" other than isolated quarks and gluons. Pairs of quarks or gluons from a given hadron may participate in a coherent fashion in a hard scattering process. Such higher-twist QCD mechanisms have been shown to be important in the Drell-Yan process and in semi-inclusive deep inelastic scattering. In this note, we summarize an extension of higher-twist physics to inclusive, high-p_T pion production in hadron-hadron collisions.[1]

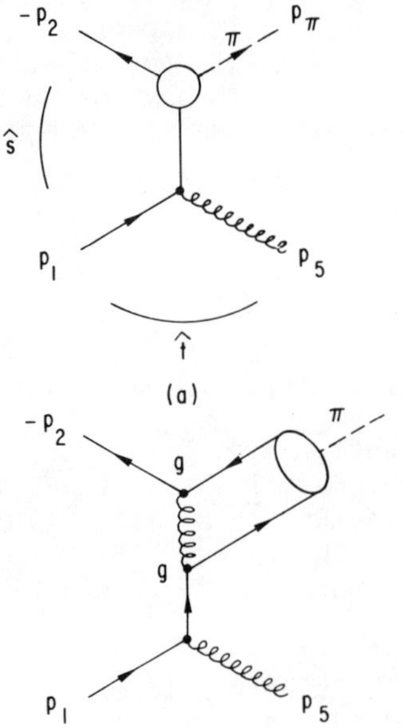

Fig.(1) (a) Quark-exchange amplitude for $q\bar{q} \to \pi G$.
(b) Hard-gluon approximation.

Our basic approach is illustrated in Fig.(1). In the single quark-exchange process sketched in Fig.(1.a), the pion emerges from an a priori complicated $q\bar{q}\pi$

vertex. However, if we consider only high p_T^2 scattering, and if we view the pion as a $q\bar{q}$ bound state, then the large-p_T process in Fig.(1.a) probes the pion wave function with one of the constituents far off shell. We represent this large off-shell momentum dependence by the exchange of a single hard gluon, as shown in Fig.(1.b). The unshaded oval encompasses all soft binding effects, and can be described in terms of the pion weak decay constant f_π.

The results presented below are only the beginning of a more comprehensive program. Just as there are <u>many</u> $2\to 2$ elementary processes involved in the conventional, minimum twist hard scattering expansion ($qq\to qq$, $qG\to qG$, etc.), there are many potentially important higher twist processes. We will examine the specific $2\to 2$ processes

$$q_A\bar{q}_B \to \pi^\pm G \tag{1}$$

$$q_A G \to q_B\pi^\pm \tag{2}$$

where q_A, q_B are distinct quark flavors. In quark-quark or gluon-gluon scattering, $2\to 3$ processes are required: $qq\to\pi qq$, $GG\to\pi q\bar{q}$. Our results thus provide a <u>lower bound</u> on the estimated size of higher-twist effects.

To construct amplitudes for $q\bar{q}\to\pi G$ and $qG\to q\pi$, we follow methods developed by Farrar and Jackson and extended by Brodsky and Lepage for exclusive hadronic reactions and electromagnetic form factors. The absolute normalization is determined in terms of the pion weak decay constant f_π. Our approach is consistent with that used by Berger and Brodsky for $\pi^- N\to\gamma^* X$, by Berger for $\nu N\to\mu^-\pi X$, and by Farrar and Fox in their study of $\pi q\to\pi q$. We write the total amplitude in the symbolic form

$$\langle q_A\bar{q}_B|\pi G\rangle = \langle q_A\bar{q}_B|q_A\bar{q}_B G\rangle\langle q_A\bar{q}_B|\pi\rangle \tag{3}$$

The first factor in Eq.(3) is the full $2\to 3$ QCD amplitude (Fig.(1.b) and its four gauge invariance partners) with the following restrictions: (i) The final $q\bar{q}$ pair is constrained to the parallel configuration, $p(q_A) = z_1 p_\pi$, $p(\bar{q}_B) = z_2 p_\pi$. (ii) The final state $q\bar{q}$ pair is projected onto a pseudo-scalar, color singlet configuration. For the pion wave function in Eq.(3) we use the asymptotic form

$$\langle q_A\bar{q}_B|\pi\rangle = \sqrt{3}\, f_\pi z_1 z_2\,\delta(1-z_1-z_2)\,dz_1 dz_2 \tag{4}$$

Evaluation of the resulting higher-twist cross sections is now quite straightforward; details are given in Ref.(1).

We write the elementary higher-twist cross sections in the form

$$\frac{d\sigma}{d\hat{t}}\,(ab\to c\pi) = \left(\frac{\pi\alpha_s^2}{\hat{s}^2}\right)\left(\frac{\alpha_s}{4\pi}\right)\left(\frac{s_0}{\hat{s}}\right) F(\cos\theta) \tag{5}$$

The fixed higher-twist scale is $s_0 = 16\pi^2 f_\pi^2$. The angular functions are

$$F(z;q\bar{q} \to \pi G) = \left(\frac{256}{81}\right)\left[\frac{1+z^2}{(1-z)^2(1+z)^2}\right] \qquad (6)$$

where $z = \cos(q,\pi)$.

$$F(z;qG \to q\pi) = \left(\frac{2}{27}\right)\left[1+\frac{4}{(1+z)^2}\right](1-z) \qquad (7)$$

with $z = \cos\theta(q,q)$.

It is worth noting that Eqs.(5)-(7) involve no arbitrary normalization parameters. The higher twist scale $s_0 = 16\pi^2 f_\pi^2$ is specified in a corresponding treatment of the pion form factor. Moreover, results numerically similar to Eqs.(6),(7) can also be obtained using a different, probabilistic approach in which a $2 \to 3$ QCD cross section is convoluted with a $q\bar{q} \to \pi$ recombination function. This suggests that our results are not strongly dependent on specific assumptions about the transition from $q\bar{q}$ to π.

Since we have not evaluated all important higher twist cross sections, it would be premature to use Eqs.(6),(7) to estimate the total high-p_T pion yield in actual hadronic processes. Instead,

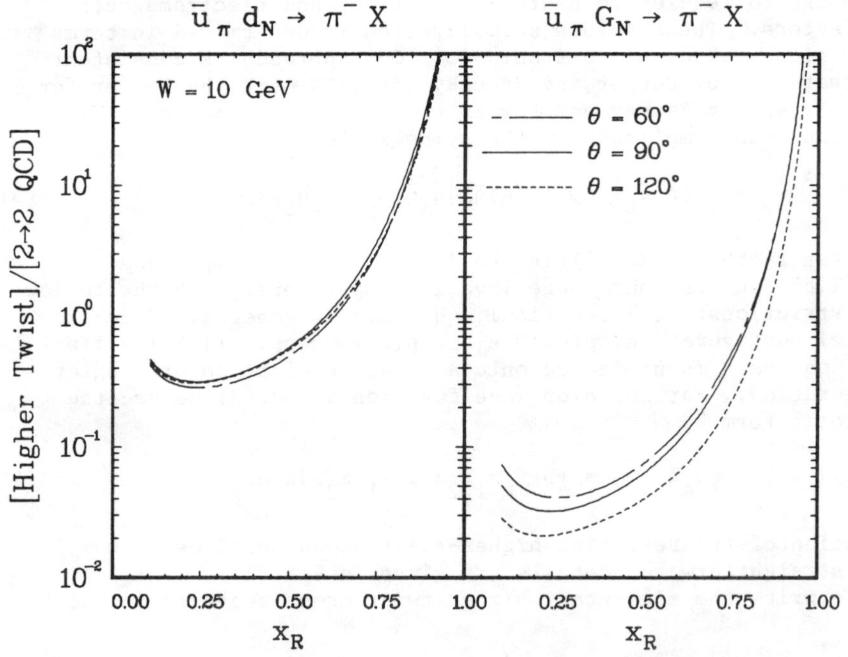

Fig.(2) Higher twist corrections to π^- production by $\bar{u}d$, $\bar{u}G$ initial states in $\pi^- N \to \pi^- X$.

we will view these results as corrections to minimum twist QCD for the specific initial states $\bar{u}_\pi d_N$ and $\bar{u}_\pi G_N$ in the hard scattering expansion for $\pi^- N \to \pi^- X$. In Fig.(2) we show results of such a comparison, using simple scaling parton densities and fragmentation functions. θ is the CM scattering angle between the initial and final pions and $x_R \equiv 2E_\pi/\sqrt{s}$. The curves in Fig.(2) scale as $s^{-1} \equiv W^{-2}$. We note that (i) the higher twist process $q_A \bar{q}_B \to \pi G$ provides a substantial correction to $q_A \bar{q}_B \to \pi X$, (ii) the higher twist corrections to $qG \to \pi X$ are about a factor of 5 less important than those for $q\bar{q} \to \pi X$, (iii) for both processes, the corrections are most important for large x_R (i.e., large p_T at fixed $\sqrt{s}$). Using simple, power-law parton densities it is easy to estimate

$$(\text{Higher Twist})/(\text{Minimum Twist}) \underset{x_R \to 1}{\longrightarrow} (1 - x_R)^{-3} \qquad (8)$$

Unfortunately, the large-x_R region is hard to reach experimentally.

It is also useful to examine the charge ratio

$$R(x_R) \equiv N(\pi^- N \to \pi^- X)/N(\pi^- N \to \pi^+ X) \ . \qquad (9)$$

The higher twist processes in Eqs.(1),(2) are maximally "charge-retaining" in that the final pions carry the quantum numbers of the incident quarks. Only $G_\pi u_N \to \pi^+ \bar{d}$ contributes to $\pi^- N \to \pi^+ X$ (ignoring sea quarks). The higher twist charge ratio is thus enormous. However, these higher twist processes are corrections to minimum twist QCD, not alternatives. To the order we are working,

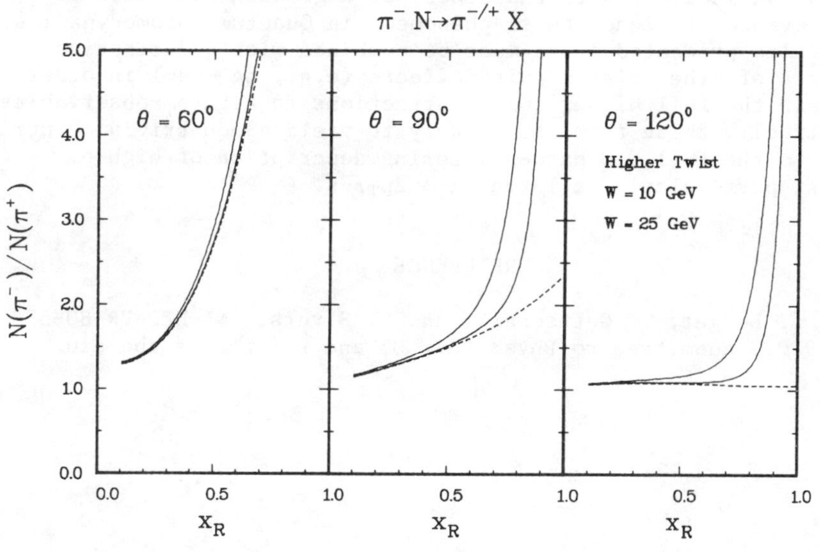

Fig.(3) $R(x_R)$ predictions, Eq.(10), for $\pi^- N \to \pi X$.

the total charge ratio can be written as

$$R(x_R) = \frac{A^-(x_R) + B^-(x_R)/s}{A^+(x_R) + B^+(x_R)/s} \tag{10}$$

where A,B describe minimum twist and higher twist respectively. Results for $R(x_R)$ are shown in Fig.(3); the dashed curves are the minimum twist results, A^-/A^+. We note that (i) for fixed s, the higher twist effects are unimportant for small x_R, but dominate for $x_R \to 1$, (ii) for fixed x_R, higher twist effects decrease with increasing s, roughly as s^{-1}. Presently available data do not extend to high enough x_R to see expected higher twist effects.

The 60° panel in Fig.(3) is interesting, primarily because preliminary data for this forward angle charge ratio are consistent with $R \simeq 1$, in marked contrast to the curves. The difficulties at 60° are not related to higher twist corrections to πN scattering but rather, reflect the strong charge retention properties of conventional, minimum twist fragmentation. Should the forward angle charge ratio data continue to differ substantially from the expectations of Fig.(3), a likely explanation might be that the empirical minimum twist fragmentation functions D(z) contain too much higher twist "contamination". The functions D(z) are deduced from minimum twist phenomenological fits to relatively low energy data (e.g., $\nu N \to \mu\pi^{\pm}X$ for $W \sim 4$ GeV). Higher twist effects in such data could well be sizeable. Removing expected higher twist effects from the data before extracting D(z) will soften the charge retention properties of the minimum twist fragmentation functions, thus lowering the dashed curves in Fig.(2).

To conclude, we note that there is a growing awareness of the relevance of higher twist phenomena in Quantum Chromodynamics. The results presented above must be combined with quantitative estimates of other higher twist effects (e.g., $qq \to qq\pi$) in order to obtain the full higher twist corrections to high-p_T observables. Collectively, these terms are likely to yield a non-trivial contribution to the full QCD hard-scattering description of high-p_T physics, particularly at large $x_T = 2p_T/\sqrt{s}$.

REFERENCE

1. E. L. Berger, T. Gottschalk, and D. Sivers, ANL-HEP-PR-80-58 (1980), submitted to Phys. Rev. D, and references therein.

DIRECT EVIDENCE FOR THE THREE-GLUON COUPLING

E. Reya
Institut für Physik, Universität Dortmund
46oo Dortmund, West-Germany

ABSTRACT

It is shown that the Yang-Mills structure of QCD can be directly tested using presently available high-p_T data. This is achieved by demonstrating that the gluon self-coupling, specifically the three-gluon vertex, dominates in the quark-gluon scattering subprocess whose importance has been experimentally established in high-p_T charge correlation measurements.

INTRODUCTION

Most of the quantitative tests of QCD done or suggested so far are mainly sensitive to the structure of the quark-gluon coupling, but not to the non-Abelian self-couplings of colored vector gluons. These tests include the well known scaling violations in electroproduction and neutrinoproduction, jets in e^+e^- annihilation, Drell-Yan process, and many others.[1,2] Although deep-inelastic scattering data provided us with the first strong evidence that gluons carry spin J=1 <u>and</u> color[1-3] which via gauge invariance[4] leaves QCD unrivalled, present high-statistics experiments are rather insensitive to the gluon self-interactions; the latter being so very typical for the Yang-Mills structure of QCD and are so very essential for asymptotic freedom. Needless to say that even qualitatively QCD remains unchallenged: Approximate chiral symmetry for massless quarks, i.e. $\partial^\mu(\bar{q}\gamma_\mu\gamma_5\vec{\tau}q)=0$, requires J=1 interactions (or trival scalar-gluon theories with <u>no</u> interactions) which together with the classical necessity for colored quarks[2,4] ($R_{e^+e^-}$, $\pi^0\to 2\gamma$, Pauli principle, spectroscopic considerations, etc.) implies QCD via gauge invariance.[4]

Nevertheless, it is intriguing to ask whether it is possible to observe the gluon self-couplings <u>directly.</u> In the past there have been numerous suggestions and attempts for delineating the effects of the triple-gluon vertex which, if experimentally feasible at all, turned out to be rather small. For example, the T-odd forward-backward asymmetry[5] of the normal to the heavy quarkonia decay plane to be measured in longitudinally polarized electron and positron annihilations turns out to be exceedingly small ($\sim 10^{-3}$), and therefore may be totally masked by non-perturbative effects due to final-state interactions. Similarly, it has been proposed[6] to test the 3g-vertex by looking for the Q^2 evolution of gluon jets in heavy quarkonium decays, $e^+e^-\to Q\bar{Q}\to 3g\to\pi$ + anything, where the gluon decay function $D_g^\pi(z,Q^2)$ could be measured experimentally; the Q^2 evolution of this latter quantity depends critically on the gluon$\to$gluon decay probability P_{gg} or, equivalently, on the purely gluonic anomalons dimension γ_{VV}^V. This requires, however, detailed

0094-243X/81/680123-05$1.50

measurements of $D_g^{\pi}(z,Q^2)$ for $0.3 \lesssim z \lesssim 1$ and $100\ GeV^2 \lesssim Q^2 \lesssim 1000\ GeV^2$, where the lower value of Q^2 refers to the Υ-family, where three gluon jets appear already to dominate, and the upper value of Q^2 refers to a possible yet undiscovered heavier quarkonium resonance, the toponium $t\bar{t}$, say. Similar effects of the three-gluon coupling can be obtained from the Q^2 evolution of the gluon distribution $G(x,Q^2)$ in the nucleon which can be measured via σ_L/σ_T. Again measurements of this latter quantity (or $F_L \equiv F_2 - 2xF_1$) are exceedingly difficult and not yet available.

Other suggestions[8-9] to test the triple-gluon vertex are either experimentally not feasible because of limited statistics, or involve[10] additional model assumptions. For example, the large charm multiplicities in e^+e^- annihilation on resonance expected in QCD[8] appear to be observable only at heavy yet unobserved $t\bar{t}$ resonances, since at the Υ the phase space is too small to allow for multiple charm production. Furthermore, it is commonly believed that the fact that in QCD gluon jets are broader than quark jets[11,12] provides a definite signature for the three-gluon coupling. Such a conclusion, however, is premature since a similar relative broadening of gluon jets can take place in other (conventional) quark-gluon field theories as well:[13] The ratio of quark and gluon jet opening angles are, to leading order, in QCD given by $\ln\delta_q/\ln\delta_g = C_2(G)/C_2(R) = 3/(4/3) = 2.25$ and for an Abelian vector gluon theory $\ln\delta_q/\ln\delta_g = 4N_f/(-4\ln 2\varepsilon - 3) = 4.65$ for an energy fraction $\varepsilon = 0.1$. A unique signature for QCD in e^+e^- annihilation would be indicated if <u>all</u> observed jet opening angles $\delta_q(\varepsilon)$ <u>and</u> $\delta_g(\varepsilon)$ <u>decrease</u> with ε. Again, because of limited statistics, such measurements are not feasible for the time being.

DIRECT OBSERVATION OF THE THREE-GLUON COUPLING

We therefore face the question whether it is possible to observe the effects of the gluon self-couplings with <u>presently</u> available data. The answer to this question is strikingly simple and affirmative. The three-gluon vertex is directly seen in high-p_T inclusive hadron collision via the observation[14,15] of a non-negligible gluon-quark scattering subprocess $gq \to gq$ as expected[16] in QCD. This is inferred from the charge ratios in the away-side jet[14] or between the target and beam <u>spectator</u> jets[15] accompanying a high-p_T ($p_T \simeq 2.5$ GeV/c) $\pi^{\pm}$ trigger in the forward direction ($\Theta \simeq 20^o$) for pp collisions at $\sqrt{s} = 52$ GeV. This clearly demonstrates that the gluon-quark subprocess is as important as the quark-quark subprocess in the total single particle inclusive production cross section for $pp \to \pi X$ which in a straightforward notation[16] is given by

$$E\frac{d\sigma}{d^3p} = \frac{1}{\pi}\sum_{\substack{a,b\\c,d}} \int dx_a \int dx_b\ P_a(x_a)\ P_b(x_b)\ \frac{d\hat{\sigma}^{ab\to cd}}{dt}\ \frac{1}{z}\ D_c^{\pi}(z) \quad (1)$$

where the sum over partons includes gluons as well as quarks and the parton distributions are $P_u \equiv u$, $P_g \equiv G$, etc.

The lowest order Feynman diagrams for the gq→gq subprocess are shown in Fig.1 and yield[16]

$$\frac{d\hat{\sigma}^{gq\to gq}}{dt} = \frac{\pi\alpha_s^2}{s^2}\left[-\frac{4}{9}\left(\frac{s}{u}+\frac{u}{s}\right)+\frac{s^2+u^2}{t^2}\right]. \qquad (2)$$

The first two terms correspond to diagrams (a) and (b) of Fig.1, and the last term is due to the genuine QCD diagram (c) involving

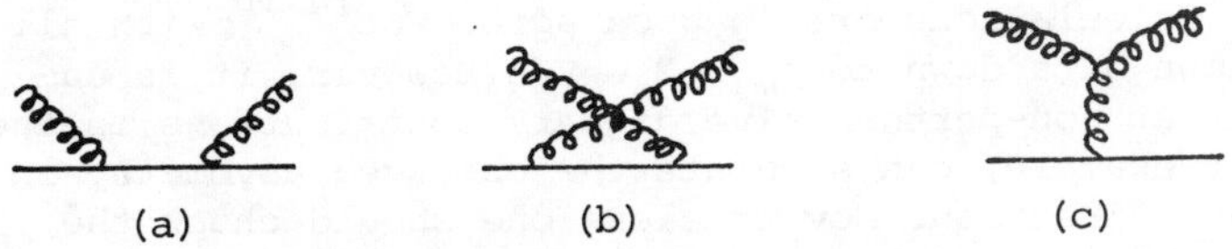

Fig.1. The gq→gq subprocess contribution to Eq.(1). The dominant contribution is due to the triple-gluon vertex induced diagram (c) which is responsible for the $1/t^2$ term in Eq.(2).

the gluon self-coupling. The analogous contribution of the quark-quark scattering one-gluon exchange diagram is given by[16]

$$\frac{d\hat{\sigma}^{q_a q_b\to q_a q_b}}{dt} = \frac{4}{9}\frac{\pi\alpha_s^2}{s^2}\frac{s^2+u^2}{t^2} \quad . \qquad (3)$$

Obviously, the contributions of the gluon exchange diagrams ($\sim 1/t^2$) for both processes are comparable and dominant for all realistic situations in which the average values of s and $|u|$ exceed the average value of $|t|$ by about one order of magnitude.[17] Without the last term in Eq.(2), the contribution of the gluon-quark scattering subprocess would be suppressed by more than an order of magnitude and could not account for the observed[14,15] equality of the qq and gq contributions at $p_T^\pi \simeq 3$ GeV/c!

To demonstrate this more quantitatively we have calculated[17] the ratio $R^\pi(gq/qq; p_T, \Theta)$ of the gq→gq and qq→qq contributions to the pion production cross section in Eq.(1) with and without the last term in Eq.(2). This ratio turns out to be rather insensitive to drastically different choices of gluon distributions and fragmentation functions, to scaling violations, etc. At $\sqrt{s}$ = 52 GeV, p_T = 2.5 GeV/c and $\Theta = 20^\circ$ we thus predict[17]

$$R^\pi \simeq 3, \qquad (4)$$

a result which is entirely dominated by the triple-gluon vertex diagram in Fig.1(c), i.e. by the last term in Eq.(2), and which is in full agreement with the experimental observation[14,15] $R^\pi \gtrsim 1$. On the other hand, the contributions of diagrams (a) and (b) in Fig.1 alone, i.e. the first two terms in Eq.(2), yield

$$R^{\pi} \simeq 0.1 \quad \text{(no 3g vertex)} \qquad (5)$$

which by itself would be in striking disagreement with experiment by more than an order of magnitude. Therefore, we conclude that the experimental observation that[14,15] $R^{\pi} \gtrsim 1$ proves unambiguously and directly the Yang-Mills structure of QCD!

One might, however, question the applicability and reliability of leading order QCD at $p_T \simeq 3$ GeV/c, although conventional perturbative calculations are in good agreement[14,16] with all presently known data down to $p_T \simeq 2$ GeV/c. However, it is unconceivable that non-perturbative diquark contributions in the low p_T region, for example, can simulate the observed asymmetry in Eq.(4) in the forward direction. Nevertheless one should check the p_T dependence of R^{π} in Eq.(4) at larger values of p_T where non-perturbative effects are obviously negligible, in order to make sure that R^{π} is entirely dominated by the perturbative triple-gluon vertex. Such measurements are presently under way at CERN - ISR and, for $\sqrt{s} = 52$ GeV and $\Theta = 20^{\circ}$, the p_T dependence of Eq.(4) is predicted to be[17] $R^{\pi} = 1.7$ at $p_T = 5$ GeV/c and $R^{\pi} = 0.9$ at $p_T = 7$ GeV/c. These predictions would be reduced by more than an order of magnitude if one contemplates a strong interaction theory without gluon self-couplings.

REFERENCES

1. For recent reviews see, for example, J.Ellis, Proceedings of NEUTRINO 79 (Bergen), vol.1, p.451; Proceedings of the Int. Symposium on Lepton and Photon Interactions (Fermilab), p.412; H.L. Anderson, Los Alamos report LA-8310-MS(1980).
2. For a recent review see, for example, E.Reya, Perturbative QCD, to appear in Physics Reports.
3. M. Glück and E.Reya, Phys.Lett. 69B, 77 (1977); Phys.Rev.D16, 3242 (1977); Nucl. Phys. B156, 456 (1979).
4. For a recent review see, for example, C.H.Llewellyn Smith, 1979 Boulder lectures, Oxford Univ. report 54/79.
5. A.de Rújula, R. Petronzio and B.Lautrup, Nucl. Phys. B146, 50 (1978).
6. K.Koller, T.F. Walsh and P.M. Zerwas, Phys. Lett. 82B, 263(1979).
7. E. Reya, Phys.Rev. Lett.43, 8 (1979).
8. W. Furmanski, R. Petronzio and S. Pokorski, Nucl. Phys. B155, 253 (1979); I.I.Y. Bigi, Phys. Lett. 86B, 355 (1979).
9. G. Eilam and M. Glück, Phys. Rev. Lett. 43, 185 (1979); A. Devoto, J.Pumplin, W. Repko and G.L. Kane, ibid. 43, 1062, 1540 (E) (1979); R.W. Brown, E.M. Haacke and J.D. Stronghair, Case Western Reserve Univ. report CWRUTH-80-2 (1980).
10. A. Krzywicki, J. Engels, B. Petersson and U. Sukhatme, Phys. Lett. 85B, 407 (1979).
11. S.J. Brodsky and J.F. Gunion, Phys. Rev. Lett.37, 402 (1976).

12. K.Shizuya and S.-H.H. Tye, Phys. Rev. Lett. 41, 787, 1195 (E) (1978); Phys. Rev. D2o, 11o1 (1979); M.B.Einhorn and B.G.Weeks, Nucl. Phys. B146, 445 (1978); R.K. Ellis and R.Petronzio, Phys. Lett. 8oB, 249 (1979); G.Grunberg, Y.J. Ng and S.-H.H. Tye, Phys. Lett. 93B, 281 (198o).
13. M. Glück and E. Reya, Univ. of Dortmund report DO-TH 8o/4,to appear in Phys. Lett. B.
14. R.P. Feynman, R.D. Field and G.C. Fox, Phys. Rev. D18, 332o (1978).
15. D. Drijard et al., ACCDHW collab., XI Int. Symposium on Multiparticle Dynamics, Bruges, 198o.
16. B.L. Combridge, J. Kripfganz and J. Ranft, Phys. Lett. 7oB,234 (1977); R. Cutler and D. Sivers, Phys. Rev. D17, 196 (1978); J.F. Owens, E. Reya and M. Glück, Phys. Rev. D18, 15o1 (1978).
17. M. Glück and E. Reya, Univ. of Dortmund report DO-TH 8o/11 (198o).

PROPERTIES OF MESON STRUCTURE FUNCTIONS

R. L. Thews
Department of Physics and Division of Theoretical Physics
University of Arizona, Tucson, Az. 85721

ABSTRACT

Meson structure functions have been extracted from high transverse momentum pion production by various hadron beams. The agreement with structure functions obtained by leptonic probes is amazingly good. The scaling violations in the pion structure function are extracted, and are consistent with those expected from asymptotic freedom. No indication of higher twist power law corrections is found.

INTRODUCTION

In a recent paper[1], the ratio of proton to pion quark structure functions has been extracted directly from the beam ratio of high-p_t $\pi°$ production cross sections. Essentially, the method consists of expressing the cross sections in terms of the average x-values of a parton in the beam or target which could have produced the observed $\pi°$ via two-body elastic scattering and subsequent fragmentation. Its success is apparent in the single-valued functions which result. A numberical test with sample distribution functions reveals that this procedure can accurately extract the $x \to 1$ behavior, but not the overall magnitude. The proton/pion ratio decreases rapidly as x approaches unity, indicating a harder quark distribution for quarks in pions than in protons. A fit to a power of 1 - x is shown in Figure 1a. More revealing is the comparison of this behavior with the ratio of proton to pion structure functions extracted by direct factorization from the Drell-Yan production of dileptons in π - p interactions.[2,3] Aside from an overall normalization, these two ratios are identical, strongly indicating that hard quark scattering is responsible for high-p_T particle production in hadronic interactions.

SCALING VIOLATIONS

We now turn to the scaling violations, or Q^2-dependence, contained in the distributions. For protons, this behavior can be parameterized by a varying power law

$$xf^p(x) \propto x^{\eta_1(\bar{s})}(1 - x)^{\eta_2(s)} \tag{1}$$

with the η_i linear functions of $\bar{s} = \ell n \frac{\ell n^{Q^2}/\Lambda^2}{\ell n^{Q_0^2}/\Lambda^2}$, where Q_0^2 is a reference point and Λ a scale of the strong interactions. The data comes from deep inelastic lepton scattering, and the parameters obtained are consistent with the moment predictions of asymptotic freedom in QCD.[4] An alternative scheme attributes the Q^2-dependence to so-called higher-twist effects, which give $1/Q^2$ terms in the distributions.[5] Similar effects occur in the pion structure function. A comparison of x-dependence at low-Q^2 (dilepton mass) and high-Q^2 shows a significant increase in the effective power of $(1 - x)$.[6-7] An alternative approach fixes the scaling part with the helicity counting rules at $(1 - x)^2$ for $Q^2 \rightarrow \infty$, and adds a higher twist term C/Q^2 from quark binding effects in the pion.[8] Physically, one expects $C = \frac{2}{9} \langle K_t^2 \rangle$, where K_t incorporates both intrinsic transverse momentum and quark mass effects. The high-Q^2 structure function from dilepton production can be fit by this form, but the range of Q^2 is not large enough to establish the Q^2-dependence of the x-independent term.

The high-p_t π° production data contains a large range[9] $1 \leq Q \leq 56$ GeV^2 of Q^2 values, so that we may hope to extract some details of scaling violation. We use the Buras-Gaemers parameterization of Reference 4 for the u-quark in the proton, which should be the dominant factor for x near unity. The resulting data points for the pion structure function $xf^\pi(x)$ are shown in Figure 1b. There is a substantial amount of scatter in the points, much more than in the ratio. This indicates a similar scaling violation (Q^2-dependence) in the individual structure functions which cancelled in the ratio (see Ref. 1). A trial fit shown by the solid line has the form[10]

$$xf^\pi(x) = A\sqrt{x}(1 - x)^B \tag{2}$$

We find $A = .19 \pm .01$, $B = .26 \pm .04$, and a relatively poor X^2 of 384 for 222 degrees of freedom. To accommodate the scaling violations, we perform another fit to

$$xf^\pi(x) = A\sqrt{x}\,(1 - x)^{\alpha(Q^2)} \tag{3}$$

with $\alpha(Q^2) = B + C\bar{s}$. We use the $Q_\circ^2 = 1.8$ GeV^2 and $\Lambda = 0.3$ GeV of Ref. 4. The results are $A = .31 \pm .02$, $B = .21 \pm .04$, and $C = .95 \pm .07$, with a much better X^2 of 255 for 221 degrees of freedom. Figure 1c shows the same data points, adjusted to their $Q^2 = 36$ GeV^2 values. The reduced scatter indicates the improvement due to including scale-violating terms in the pion structure function. Alternatively, we divide the data into various Q^2 bins, and perform separate fits of the form (3). In Figure 2a, we show the Q^2-variation of the powers $\alpha(Q^2)$,[11] along with the linear variation for the overall fit of Figure 1c. Also shown by open circles are the effective powers for fitting the dilepton data[6,7] at average $Q^2 = 3.17$ GeV^2 and $Q^2 = 36$ GeV^2. Although there is substantial scatter due to binning effects, the trend is clearly to favor increasing powers with higher Q^2. In addition, the absolute magnitudes of these powers are in good agreement with those extracted from the dilepton data.

We now examine the possibility of fitting this Q^2 dependence with higher twist terms alone. An overall fit to the data in the form

$$xf^{\pi}(x) = A\sqrt{x}\,[(1-x)^2 + C/Q^2] \qquad (4)$$

yields A = .80 ± .02 and C = .53 ± .03 GeV^2, with a X^2 of 435 for 222 degrees of freedom. This fit is shown in Figure 2b, with the data points adjusted to $Q^2 \to \infty$. This fit is certainly no better than the Q^2-independent fit of Figure 1b, indicating that the C/Q^2 dependence cannot adequately explain the scaling violations in the pion structure function. Separate fits to the various Q^2-bins have also been performed. The constant parts of the structure functions are shown in Figure 2c, along with the C/Q^2 term of the overall fit (solid line). The wide discrepancy just emphasizes the inadequancy of the higher twist fit. Even if one eliminates the low-Q^2 terms by appeal to some minimum Q^2 for such an expansion to be valid, the effective constant in a C/Q^2 fit to the high-Q^2 points is of the order 10 GeV^2, much too large to be interpreted as a quark binding effect as in Reference 8.

CONCLUSIONS

It is obvious that a much more definite statement about scaling violation must await high-x data at large Q^2. This means very high-p_t events at moderate c.m. rapidities are needed. Such events are now becoming available for protons at the ISR, but we must await new data for pion beams at the highest possible energies. What we can say now is that the logarithmic scaling violations are much larger in this kinematic region than any possible power-law terms.

REFERENCES

1. Kwan-Wu Lai and R. L. Thews, Phys. Rev. Lett. 44, 1729 (1980).
2. F. T. Dao et. al., Phys. Rev. Lett. 39, 1388 (1977).
3. G. E. Hogan et. al., Phys. Rev. Lett. 42, 951 (1979).
4. A. J. Buras and K. J. F. Gaemers, Nuclear Physics B132, 249 (1978).
5. L. F. Abbott and M. Barnett, SLAC-Pub-2325 (1979), and L. F. Abbott, W. B. Atwood, and M. Barnett, SLAC-Pub-2400 (1979). See however M. Glück and E. Reya, Universität Dortmund preprint TH 80/7, March 1980.
6. D. McCal et. al., Phys. Lett. 85B, 432 (1979).
7. C. B. Newman et. al., Phys. Rev. Lett. 42, 951 (1979).
8. E. L. Berger and S. J. Brodsky, Phys. Rev. Lett. 42, 940 (1979).

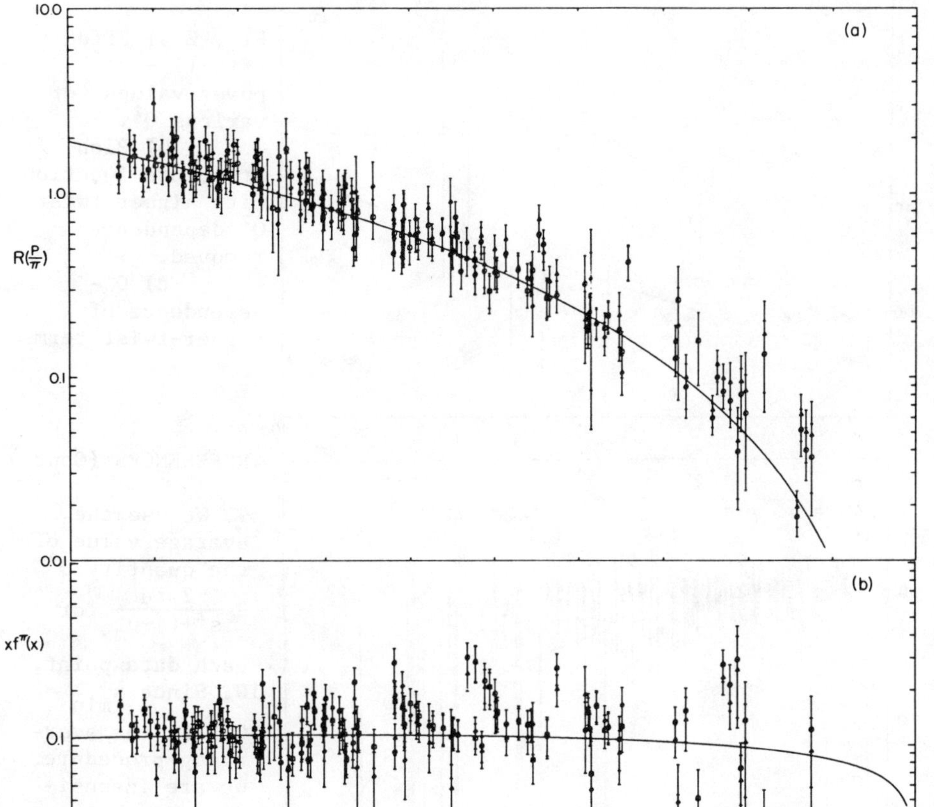

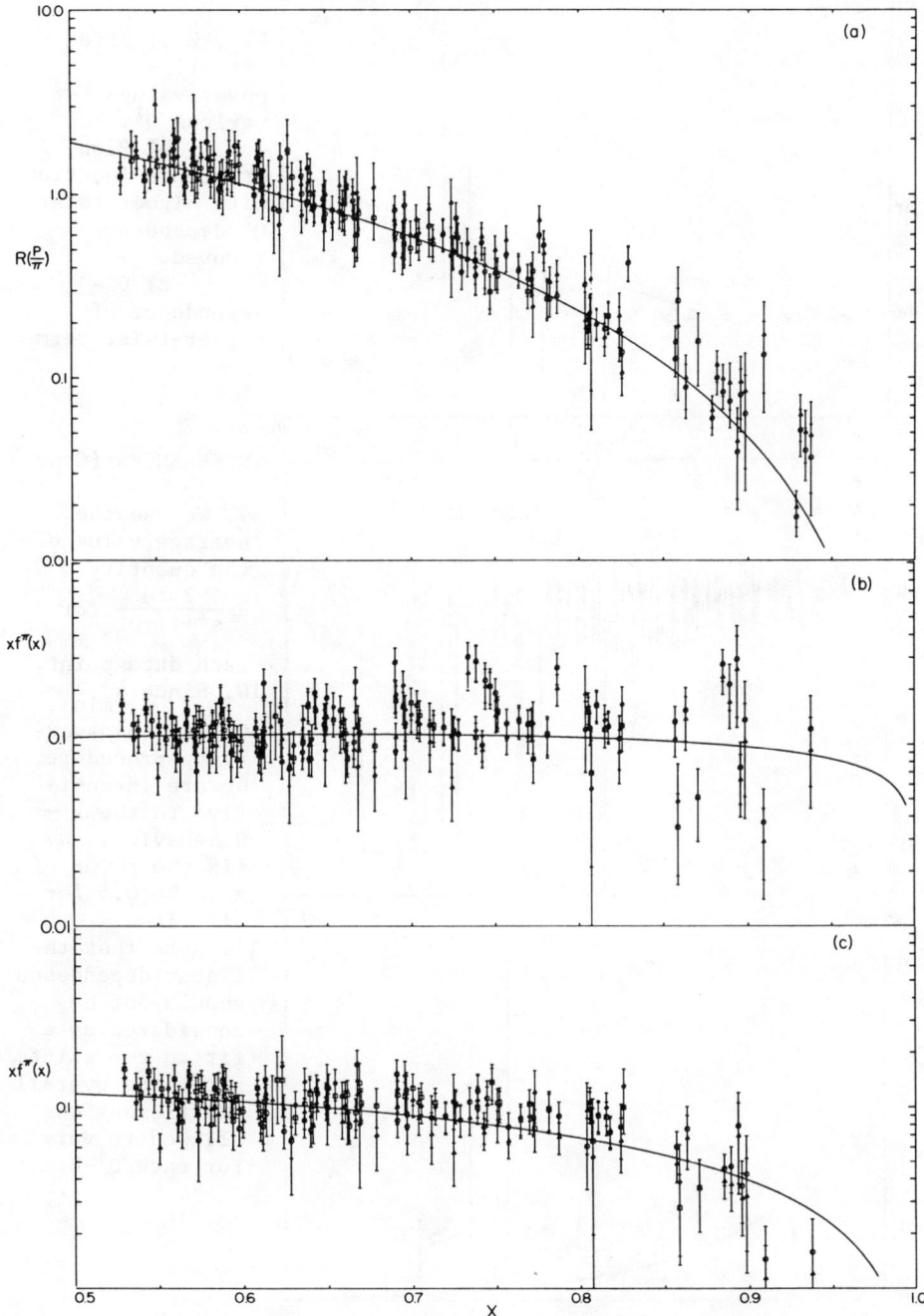

Fig. 1 a) Ratio of proton to pion structure functions.
b) Pion structure function with all Q^2 included.
c) Q^2-dependence removed as discussed in text.

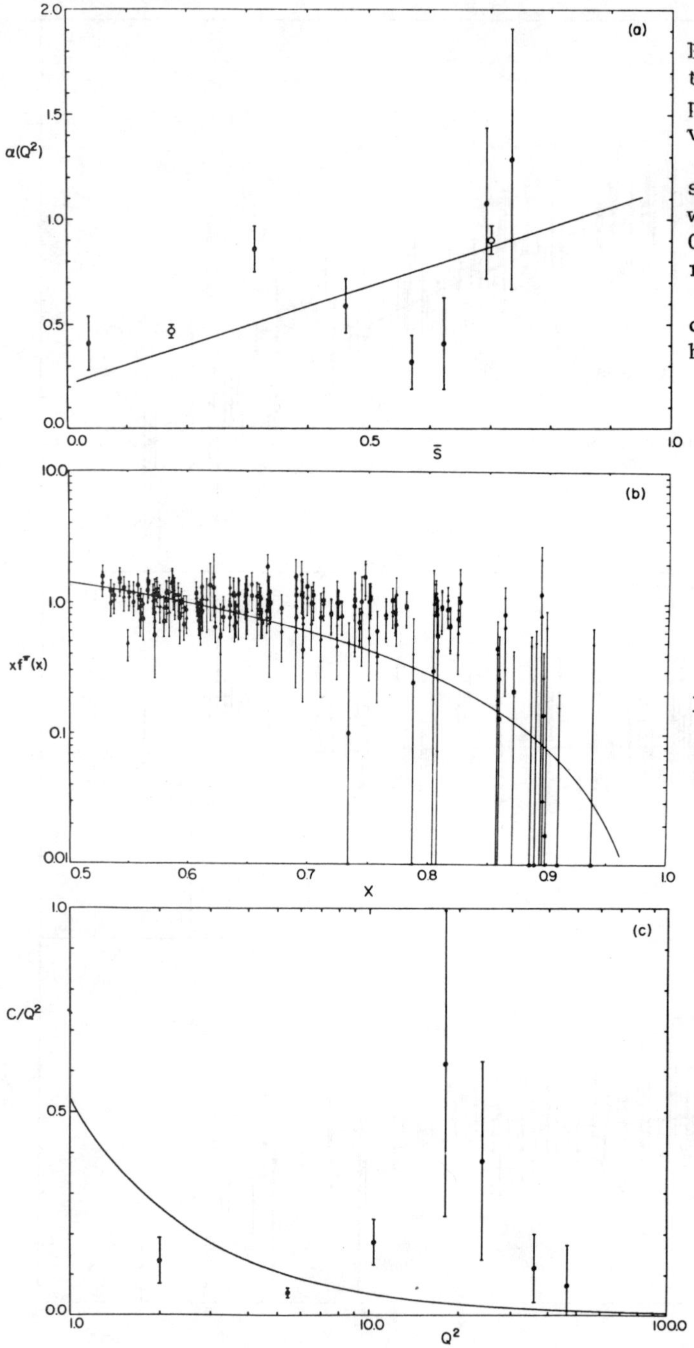

Fig. 2 a) Effective $(1 - x)$ power values for various Q^2.
b) Pion structure function with higher twist Q^2 dependence removed.
c) Q^2-dependence of higher-twist terms.

REFERENCES (Cont.)

9. We use the average value of the quantity $Q^2 \equiv \frac{2stu}{s^2+t^2+u^2}$ for each data point.
10. Since $x_{min} = 0.5$ in the averaging procedure, we are insensitive to the $x \to 0$ behavior. We fix the power of x to be 0.5 for all fits.
11. Note that the linear dependence should not be considered as a fit to the points, since the overall coefficient was allowed to vary for each Q^2-bin.

DIRECT LEPTONS AND PHOTONS IN HADRONIC INTERACTIONS, EXPERIMENT

L. Resvanis, Athens, Organizer

A STUDY OF DIRECT SINGLE PHOTONS AND CORRELATED PARTICLES IN PROTON-PROTON COLLISIONS AT $\sqrt{s}$ = 62.4 GeV

(CCOR Collaboration)

H.-J. Besch. L. Camilleri, C. del Papa, L. Di Lella,
C.B. Newman, B.G. Pope, S.H. Pordes, A.M. Smith and K.K. Young
CERN, Geneva, Switzerland
B.J. Blumenfeld, R.J. Hollebeek,
L.M. Lederman, D.A. Levinthal, R.W. Rusack, R.A. Vidal
Columbia University, New York, NY, USA
A.L.S. Angelis, N. Phinney,
A.M. Segar, J.S. Wallace-Hadrill and J.M. Yelton
Oxford University, Oxford, UK
T.J. Chapin, R.L. Cool, Z. Dimcovski, J.T. Linnemann,
A.F. Rothenberg and M.J. Tannenbaum
The Rockefeller University, New York, NY, USA

- -

Recent interest in the production of single photons in hadron-hadron collisions has been stimulated because significant experimental data became available[1] and theoretical studies showed[2], that this process provides a good determination of the gluon distribution in the proton, mainly via the QCD-Compton-Diagram.

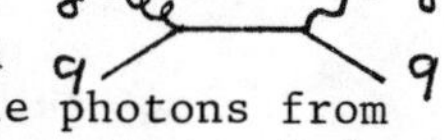

The main experimental problem for all single photon experiments is the large number of background single photons from π^0, η, η'... decays. That means, it is much easier to find a "signal" than to exclude one. Data taken at $\sqrt{s}$ = 62 GeV and with thresholds of 3, 5, 7 and 9 GeV/c in p_T are presented here. The apparatus is shown in Fig. 1. Essential for this experiment is the lead glass shower detector, the scintillation counters "B", and the coil and cryostat of the solenoid. Because the spatial resolution of the lead glass did not allow to separate high p_T γ, π^0, η... an unconventional method has been developed to extract single photon data with a statistical procedure. If an incident high energy photon hits an absorber t radiation length thick, the probability that this photon passes through the absorber without converting to an e^+e^- pair is given by $\nu_1 = \exp(-7t/9)$. So, the probability that out of the two photons of a π^0 both do not convert is given by $\nu_0 = (\nu_1)^2$. In this experiment the coil and cryostat (1 radiation length thick), were used as a converter so $\nu_1 = .46$ and $\nu_0 = .21$. Because of acceptance corrections ν_0 is slightly p_T-dependent and not the same for both of the lead glass walls. (ISR CM-motion). As a detector for converted photons the counters "B" just outside the cryostat have been used. To avoid confusion, events with a track pointing to the relevant B-counter have been excluded from the analysis. So, the observed non-conversion probability ν_{obs} is obviously related to the fraction $f_\gamma = \gamma$/all of single photon events in the data sample by $\nu_{obs} = \nu_E\,(1-f_\gamma) + \nu_1\, f_\gamma$ where ν_E is the expected non-conversion probability for all processes contributing to the trigger other than single γ. The results for f_γ are shown in Fig. 2a. Multiplying the measured invariant cross-section for the whole sample with f_γ gives the invariant cross section for direct photon production (Fig. 2b).

Further analysis of this production of single γ's gives the

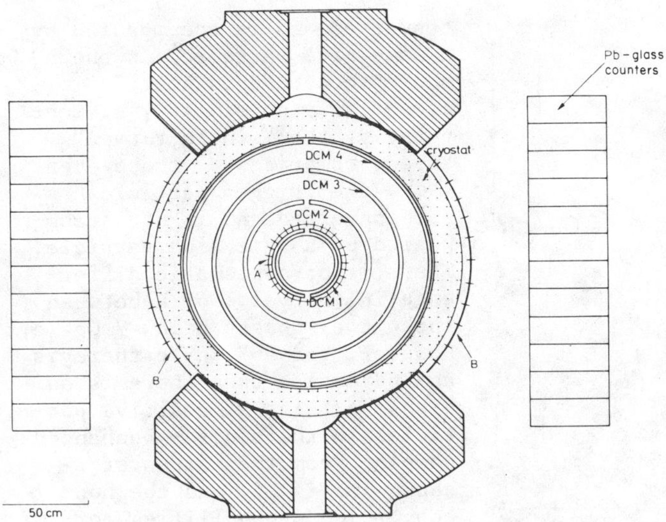

Fig. 1. View of the apparatus

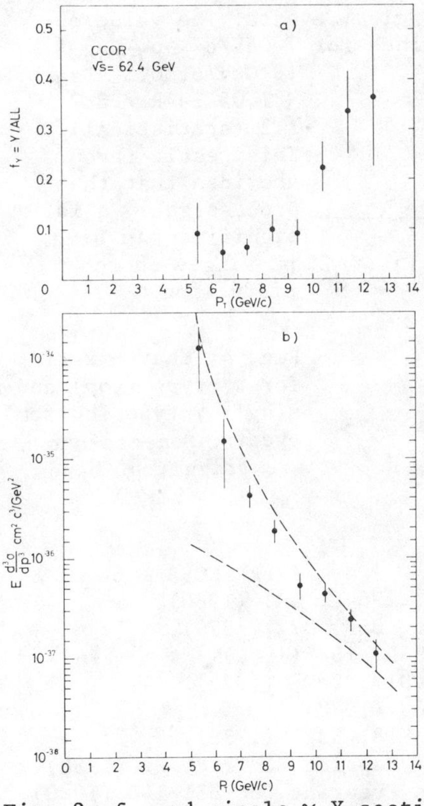

Fig. 2. f_{γ} and single γ X-section

following preliminary results. In some models, directly produced photons are expected to be unaccompanied by particles on the same side. Thus the value of $f_{\gamma} = \gamma/\text{all}$ should be enhanced for such events. By a method similar to the above, f_{γ} has been obtained separately for two classes of events. Those in Fig. 3a have at least one charged or neutral particle other than the trigger in the trigger hemisphere, while those in Fig. 3b do not. The f_{γ} for the accompanied trigger particles are consistent statistically with zero for $p_{T\,trig} < 11$ GeV/c. However, for the unaccompanied particles, f_{γ} clearly rises with $p_{T\,trig}$, and for $7 < p_{T\,trig} < 11$ GeV/c, 20% ± 2% (statistical) of this sample may be attributed to direct single γ's.

If the process yielding direct single photons is $gq \to \gamma q$ then theory predicts that because of the quark charges and abundances in the proton the quark involved will be a u quark eight times more often than a d quark. This large excess of positive to negative in the parent quark should be reflected in the structure of the 'jet' opposite to the single γ[3]. This has been investigated by the measurement of the charge ratio R = positive particles/negative particles in the hemisphere opposite the trigger, as a function of $x_E = -\vec{P}_{T\,track} \cdot \vec{P}_{T\,trig} / |P_{T\,trig}|^2$ Fig. 4 shows R as a function of x_E for four trigger bands and two subsets of the data. The "π^0-type" subset consists of those events where the trigger particle has

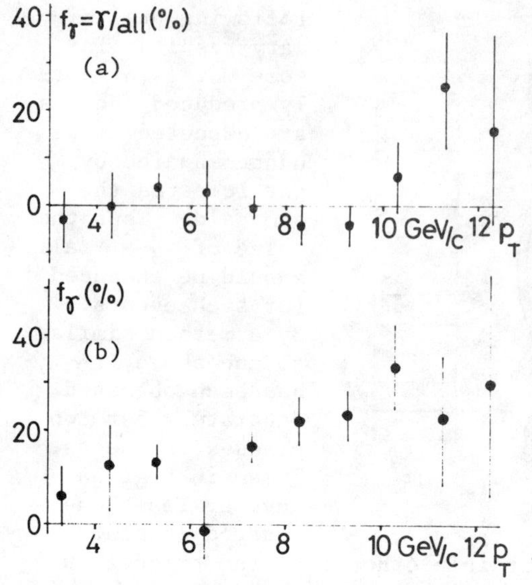

Fig. 3. f_γ for events with accompanied (top) and unaccompanied (bottom) trigger particles

converted and is accompanied by a same side particle. As shown above these events contain essentially no single γ's. Conversely, the "single γ-type" subset comprises those events where the trigger particle has not converted and is not accompanied by a same side particle. There is no systematic difference in the value of R between these two subsets below 7 GeV/c. However, above 7 GeV/c there is an indication that there is an excess of high x_E positive particles in the single-γ enhanced sample. From the values of f_γ shown in Fig. 4, and the non-conversion probabilities for γ's and for other particles, it is possible to extract R for single-γ events. The value obtained for 7 GeV/c < $p_{T\,trig}$ < 13 GeV/c, 0.3 < x_E < 1.05 is R = 3.7 ± 1.2 (statistical). This result favours the idea that the process $gq \to \gamma q$ is significant in high p_T single photon production.

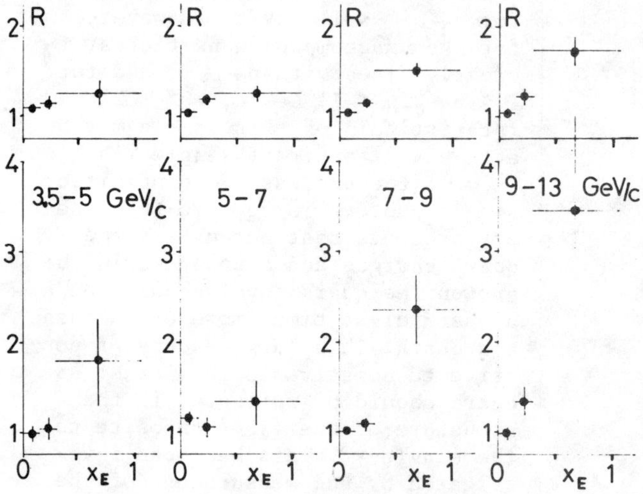

Fig. 4. Charge ratio for π^0-type (top) and single γ-type (bottom) events. Momenta shown are trigger p_T bands.

REFERENCES :

1. M. Diakonou et al., Phys. Lett. 91B, 296 et 301 (1980).
   R.M. Baltrusaitis et al., Phys. Lett. 88B, 372 (1979).
   A.L.S. Angelis et al., Phys. Lett. 94B, 106 (1980).
2. A.P. Contogouris et al., Phys. Rev. D19, 2607 (1979).
   R. Ruckl et al., Phys. Rev. D18, 2469 (1978).
   F. Halzen and D.M. Scott, Phys. Rev D18, 3378 (1978).
3. R. Baier et al., Univ. of Bielefeld Preprint BI-TP 80/07 (1980).
   F. Halzen et al., Univ. of Wisconsin Preprint Coo-881-140 (1980).

DIRECT SINGLE PHOTON PRODUCTION AND A SEARCH FOR DIPHOTON RESONANCE OR CONTINUUM PRODUCTION AT FERMILAB ENERGIES

B. Cox
Fermi National Accelerator Laboratory, Batavia, Ill.

ABSTRACT

Direct single photon production at Fermilab energies is reviewed and the high mass diphoton spectrum produced in pp collisions at 400 GeV/c is examined for any evidence of diphoton continuum or resonance production.

SUMMARY

Evidence for direct single photon production in proton-nucleon has been found both at Fermilab and CERN.[1-7] The level of the cross sections are shown in figure 1 along with the QCD prediction as presented by Cormell and Owen.[8] The direct photon production as measured by the Fermilab-John Hopkins collaboration with their lead glass spectrometer in the Proton West branch of the Proton Laboratory over a range of center of mass angles rises with both $P_\perp$ and X_F with an average $\gamma/\pi^0 \sim .07 \pm .025$ in the region $1.5 < P_\perp < 4.0$ GeV/c and $90^0 < \Theta_{CMS} < 160^0$ and 200 and 300 GeV/c.

A search[9] has been made with this same double arm spectrometer for high mass diphoton resonance or continuum production in 400 GeV/c proton-beryllium interactions. The observed diphoton mass spectrum is shown in figure 2. Curve a is the level of false diphotons arising from the misidentification of a coalescing π^0 as a single photon. No diphoton resonance production is observed. An upper limit $\sigma.\beta(\eta_c(3.0)\to\gamma\gamma)$ of $.9 \times 10^{-32} cm^2$/nucleon has been set by an analysis of this spectrum using the expected diphoton mass resolution of $\Delta m/m \sim .06$ FWHM.

We have attempted to ascertain the origin of the observed diphoton continuum production. In figure 3 we show the $\pi^0\pi^0$ mass spectrum as measured in this experiment at $y \sim -.55$. Taking this spectrum and a $(1-|X_F|)^5$ dependence of the $\pi^0\pi^0$ cross section which is indicated by the comparison of the $\pi^0\pi^0$ cross section with the $y = 0$ $\pi^+\pi^-$ cross sections measured in other experiments[10,11] we have, by Monte Carlo techniques, predicted the $\gamma\gamma$ mass spectrum which would result from events in which only a diphoton combination was seen in the spectrometer. This spectrum is shown in figure 2 as curve b. In order to estimate the contributions from final states such as $\eta\pi^0$ or $\eta\eta$, the cross section relationship $\sigma(\eta\eta) = \frac{1}{2}(\eta\pi^0) = \frac{1}{4}\sigma(\pi^0\pi^0)$ suggested by our measurement[2] of the ratio of inclusive η to inclusive π^0 production has been used. The resulting $\gamma\gamma$ spectrum when these contributions are added to the contributions from the $\pi^0\pi^0$ final state is given by curve c in figure 2. The observed $\gamma\gamma$ mass spectrum can be explained as arising from the decay photons from final states containing π^0 or η combinations.

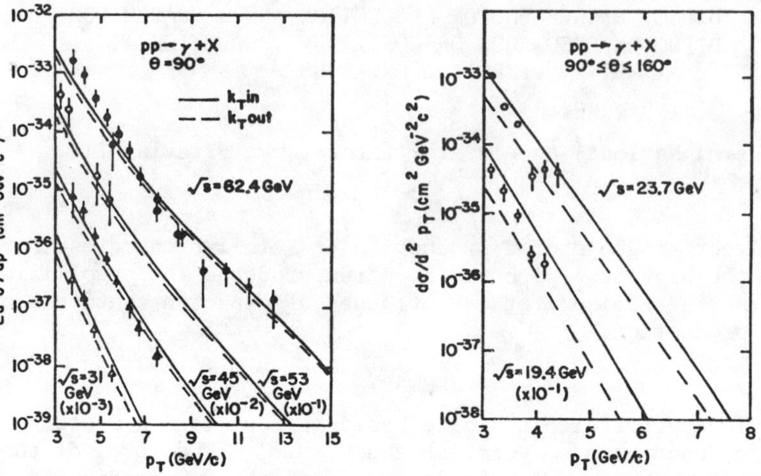

Fig. 1: Data for direct photon production as are now available as presented in reference 8. The 90^0 data of the ISR experiments are given in the left most plot. The data of the Fermilab-JHU collaboration which are an average over the angular range $90^0 < \Theta < 160^0$ are given in the right hand plot. The curves are the QCD calculation of reference 8 with and without intrinsic parton transverse momentum.

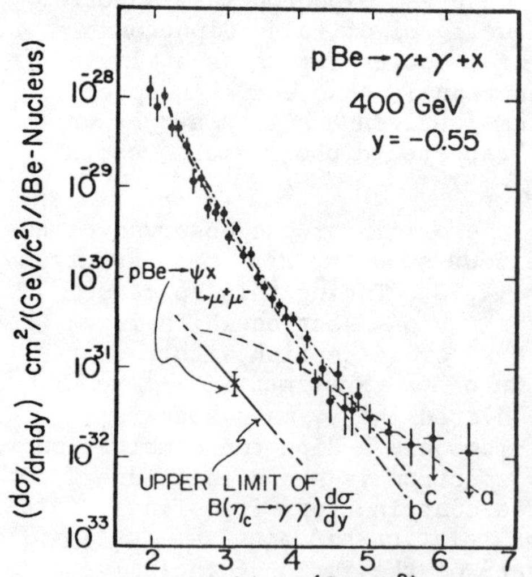

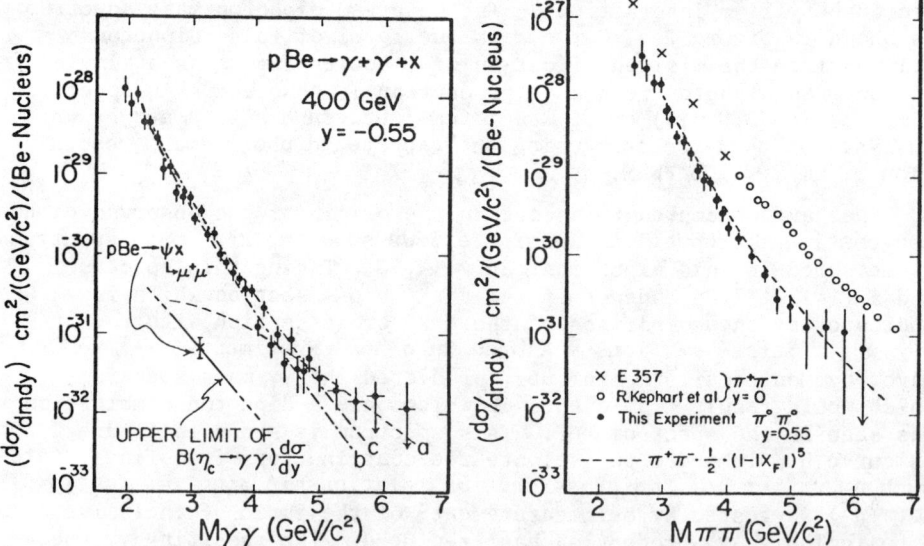

Fig. 2: Diphoton mass spectrum from 400 GeV/c p Be interactions

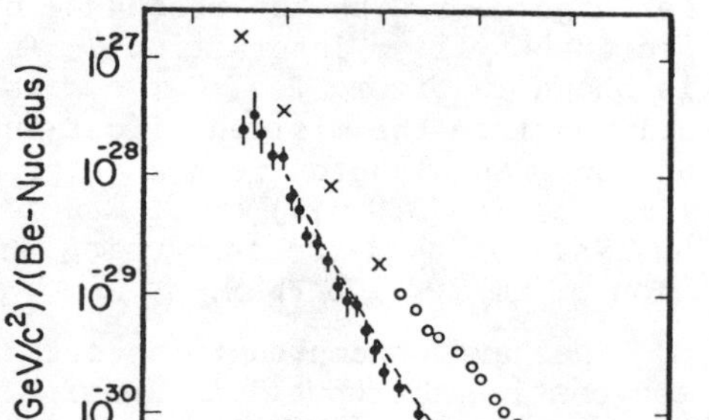

Fig. 3: Di pizero mass spectrum from 400 GeV/c p Be interactions.

REFERENCES

1. J. Cronin, Proceedings of the 1979 International Symposium on Lepton and Photon Interactions, 597 (1979).
2. R. M. Baltrusaitis et al., Phys. Lett. 88B, 372 (1979).
3. E. Amaldi et al., Phys. Lett. 84B, 360 (1979).
4. M. Diakonou et al., Phys. Lett. 87B, 292 (1979).
5. B. Cox, Proceedings of the 1979 International Symposium on Lepton and Photon Interactions at High Energies, 602 (1979).
6. H. J. Besch, contributed paper, this conference.
7. D. Rahm, contributed paper, this conference.
8. L. Cormell and J. F. Owens, contributed paper 0245, this conference.
9. R. M. Baltrusaitis et al., Fermilab Pub. 79/39 - EXP 7160.095, (submitted to Phys. Rev. Lett.)
10. R. D. Kephart et al., Phys. Rev. Lett. 39, 1440 (1977).
11. D. A. Finley, Ph.D. Thesis, Purdue Univ., unpublished, (1978).

OBSERVED DIFFERENCES IN EVENT STRUCTURES OF HIGH-p_T π^0 AND SINGLE PHOTON EVENTS PRODUCED IN pp COLLISIONS IN THE CERN ISR

Athens-BNL-CERN- Copenhagen-Lund-Rutherford-Tel Aviv Collaboration *)
CERN, CH-1211 Geneva 23, Switzerland
presented by D.C. Rahm

ABSTRACT

The direct photon production in pp collisions for c.m. energies $31 \leq \sqrt{s} \leq 63$ GeV and transverse momenta of up to 9 GeV/c were measured at the ISR using a segmented lead/liquid argon calorimeter. The observed γ/π^0 ratio is found to be significantly larger than zero at 4 GeV/c in p_T increasing to 0.4 at 9 GeV/c. The average multiplicity on the trigger side for the single-photon events was found to be significantly lower than for π^0 events. The correlations in Δy and $\Delta\phi$ between the trigger particle and an additional particle are found to differ mainly at small Δy and $\Delta\phi$.

The possibility of using direct photon production to probe the constituent structure of hadrons has been extensively investigated [1-12]. According to present theoretical views, direct photons are produced in proton-proton interactions predominantly via the process $qg \rightarrow \gamma q$ with important contributions also coming from the $qq \rightarrow qq\gamma$ process, where the photon is radiated in a bremsstrahlung process from the scattered quark. The production of high p_T mesons, on the other hand, is thought to proceed via the hard scattering of the proton constituents (quarks, gluons and antiquarks) and their subsequent fragmentation. The different production mechanisms for high p_T mesons and single photons are thus likely to be reflected in the event structure associated with each particle.

We have previously reported the observation of direct photons and differences in the event structures associated with π^0 or photon-triggered events [11-13]. Here we report on recent results on the same subjects.

The apparatus used for the direct photon measurement consists of two lead/liquid-argon calorimeters, which are described in more detail elsewhere [11-14]. These were subdivided both longitudinally (in the direction of shower development) for effective hadron-photon discrimination, and laterally. We were able to reconstruct each of the showers from the π^0 decay and thus separate π^0 from single photon events on an event-by-event basis.

Figure 1 shows the apparent γ/π^0 ratio and the background contribution expected from all the known meson decays. This background is mainly due to decays where one photon falls in the calorimeter and the other outside. Effects due to cosmic rays, beam gas interactions, and hadrons simulating electromagnetic showers have all been included.

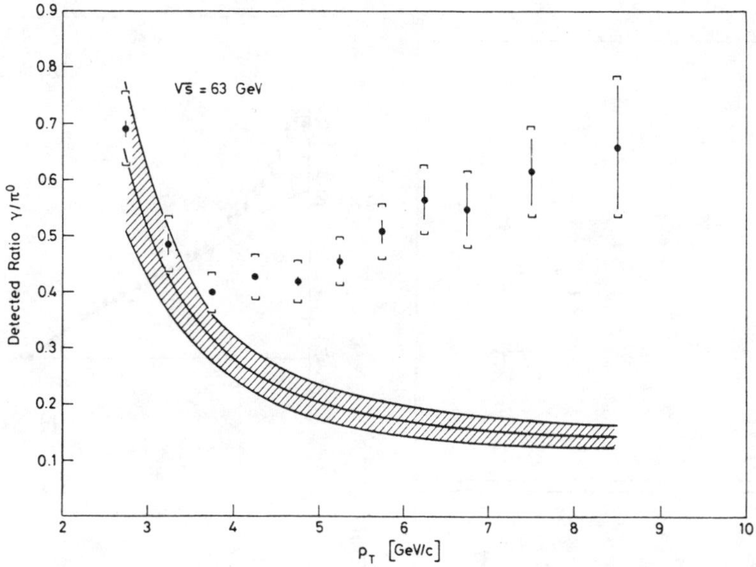

Fig. 1

In addition to the liquid-argon calorimeters, there was an array of 44 scintillation counters surrounding the interaction region at a distance of 18 cm. Each counter covered 8° in azimuth and about 2 units in rapidity. In Fig. 2 we show the multiplicities in these counters for π^0 and single-photon triggered events. Figure 2a gives the total multiplicities and 2b the same side multiplicities in the seven counters centred on the trigger direction. On the away side, the observed difference for the two different triggers was small. The trigger-side multiplicity for the π^0 events is consistent with being constant for all p_T bins, while the single-photon sample shows a decrease with increasing p_T. If one corrects for the meson decay background in the single-photon sample, assuming it has a multiplicity structure similar to that of the events with identified π^0's, one obtains a value of the multiplicity for the real direct single-photon events which is consistent with being constant for all p_T bins, but which is lower than that of the π^0 events.

In another configuration of the apparatus where proportional[13] chambers were in front of the calorimeters and the whole apparatus was placed much nearer the intersection region, we were able to obtain more detailed information on charged particles correlations. This set-up had the advantage that we were able to go to higher p_T, but had the disadvantage that the π^0 single photon separation was not as clean. In Fig. 3 we show the relative number of events versus the pseudo rapidity differences (Δy) between the trigger particle and charged tracks in the region $20^{o} < \Delta\phi < 70^{o}$ ($\Delta\phi$ is azimuthal difference) and for $p_T > 6.5$ GeV/c. The solid lines represent the expected distributions for uncorrelated tracks calculated assuming a random distribution and our experimental acceptance.

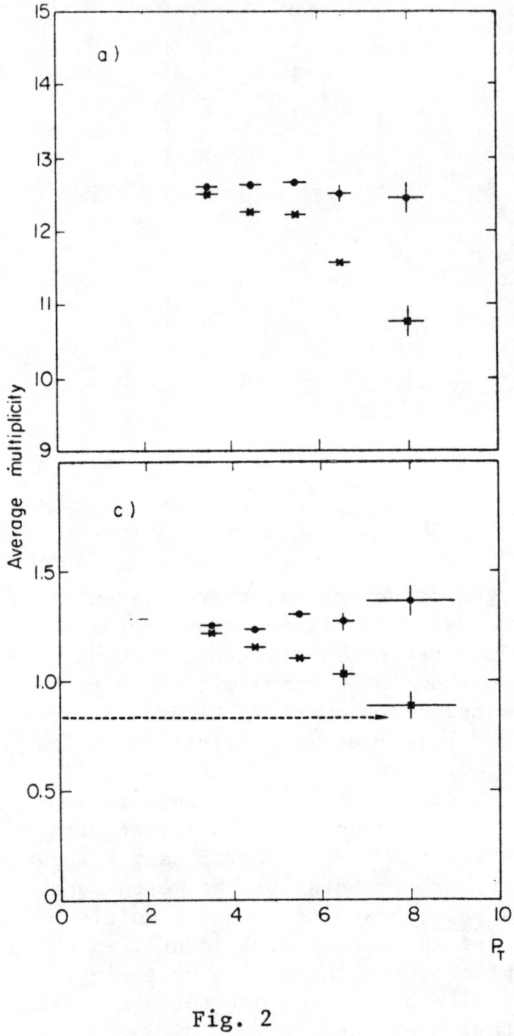

Fig. 2

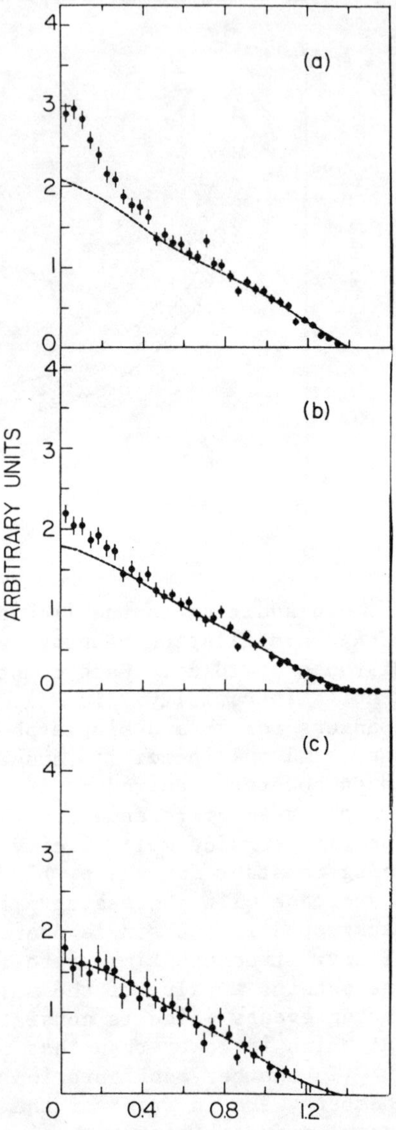

Fig. 3

The curves are normalized to the number of events at large Δy where most of the tracks are expected to be uncorrelated. In Fig. 3a this distribution is shown for π^0 events and a clear excess is observed for small Δy. Figure 3b shows the same distribution for the single photon candidate sample. If we subtract the calculated background due to misidentified π^0's, we obtain the distribution in Fig. 3c which shows no excess above that of the uncorrelated particles expected to come from the fragmentation of the other proton constituents.

In conclusion, we observe that the single photon events have considerably lower numbers of associated charged tracks as well as π^0 particles on the trigger side. Most of the difference is concentrated in the small $\Delta\phi$ and Δy region, where the accompanying particles of the same "jet" are expected to be found. The relative magnitude of the effect increases with increasing transverse momentum of the second particle. These observations are consistent with the hypothesis that high p_T single photons produced in pp collisions are unaccompanied, while high p_T π^0's are part of a "jet" of particles.

REFERENCES

*) Members of the collaboration are: C. Kourkoumelis, L.K. Resvanis, T.A. Filippas, E. Fokitis, S. Karpathopoulos, C. Trakkas (Athens), A.M. Cnops, T. Killian, T. Ludlam, R.B. Palmer, D.C. Rahm, P. Rehak, I. Stumer (BNL), D. Cockerill, C.W. Fabjan, A. Hallgren, I. Mannelli, W. Molzon, P. Mouzourakis, B.S. Nielsen, Y. Oren, L. Rosselet, W.J. Willis (CERN), O. Botner, H. Bøggild, E. Dahl-Jensen, I. Dahl-Jensen, G. Damgaard, K.H. Hansen, J. Hooper, R. Møller, S.Ø. Nielsen, B. Schistad (Copenhagen), T. Akesson, S. Almehed, G. von Dardel, S. Henning, G. Jarlskog, B. Lörstad, A. Melin, U. Mjörnmark, A. Nilsson, L. Svensson (Lund), M.G. Albrow, N.A. McCubbin (Rutherford), O. Benary, S. Dagan, D. Lissauer (Tel Aviv).

1. G.R. Farrar and S.C. Frautschi, Phys. Rev. Lett. 36 (1976) 1017.
2. C.O. Escobar, Nucl. Phys. B98 (1975) 173.
3. E.L. Feinberg, Nuovo Cimento 34A (1976) 391.
4. H. Fritzsch and P. Minkowski, Phys. Lett. 69B (1977) 316.
5. R. Rückl, S.J. Brodsky, J.F. Gunion, Phys. Rev. D18 (1979) 2469.
6. R.D. Field, Proc. 19th Int. Conf. on High-Energy Physics, Tokyo 1978 (Phys. Soc. Japan, Tokyo, 1979), p. 743.
7. A.P. Contogouris, S. Papadopoulos, M. Hongoh, Phys. Rev. D19 (1979) 2607.
   A.P. Contogouris, J. Kripfganz, McGill University preprint (1979).
8. P. Aurenche, J. Lindfors, preprint CERN TH.2768 (1979).
9. U. Amaldi et al., Nucl. Phys. B150 (1979) 326.
10. R.M. Baltrusaitis et al., Fermilab Pub-79/38-Exp (1979).
11. M. Diakonou et al., Phys. Lett. 87B (1979) 292.
    M. Diakonou et al., Phys. Lett. 91B (1980) 296.
12. M. Diakonou et al., Phys. Lett. 91B (1980) 301.
13. C. Kourkoumelis et al., Observed differences in the structure of events with a high p_T π^0 or single photon. Submitted to Nucl. Phys. B.
14. J.H. Cobb et al., Nucl. Instr. and Meth. 140 (1977) 413.
    J.H. Cobb et al., Nucl. Instr. and Meth. 159 (1977) 93.

MASSIVE ELECTRON PAIR PRODUCTION AT THE ISR

Athens-Athens-Brookhaven-CERN-Syracuse-Yale collaboration(1)
C.Kourkoumelis
University of Athens,Greece

A study of the production of high mass electron pairs from pp collisions at the CERN ISR has been performed. Results include an extension of the scaling function down to $\sqrt{\tau}=0.07$ as well as measurements of the angular distribution and average transverse momentum distribution of the pairs.

We have studied the production of the electron pairs from pp collisions at the CERN Intersecting Storage Rings (ISR) at the centre -of-mass energies from $\sqrt{s}=30$ GeV to $\sqrt{s}=63$ GeV.

The apparatus consisted of four modules each composed of proportional chambers, scintillators, lithium foil transition radiators and a lead/liquid-argon calorimeter. The corresponding luminosities for the running at each centre-of-mass energy were 0.095, 1.214 and 8.748×10^{37} $cm^{2}sec^{-1}$ at $\sqrt{s}=30$,53 and 63 GeV.

Details about the apparatus, trigger and analysis methods can be found elsewhere(2).

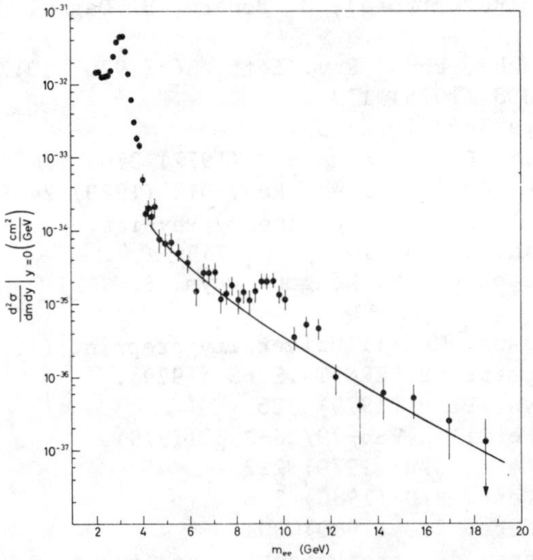

Fig. 1.The cross section $d^2\sigma/dmdy\big|_{y=0}$ as a function of m.

The cross section for the production of electron pairs as a function of their mass for the combined data at $\sqrt{s}=53$ and 63 GeV is shown in fig. 1.

The above data were divided into centre-of-mass energies, mass bins, and after resonance subtraction, were used for the investigation of scaling in $\sqrt{\tau}=m/\sqrt{s}$. Fig. 2 shows our data on scaling, together with other experimental results. Scaling seems approximately successful in the range $\sqrt{s}=28$-63 GeV down to values of $\sqrt{\tau}=0.07$-which

0094-243X/81/680144-03$1.50

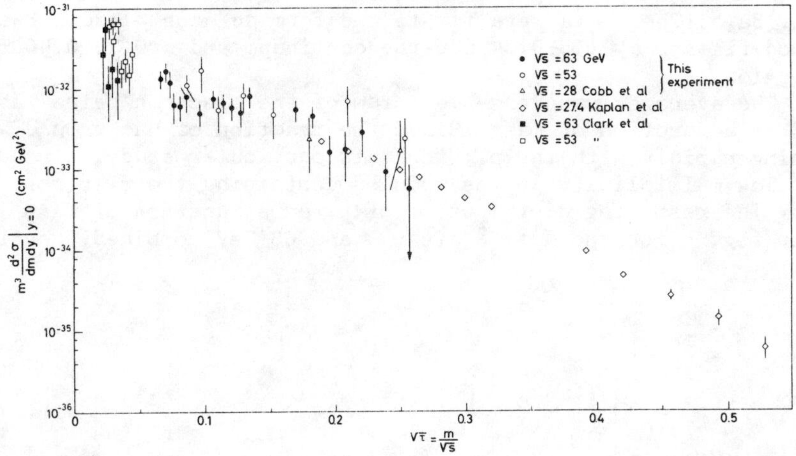

Fig. 2. Plot of the scaling cross section as a function of $\sqrt{\tau}$.

is the relevant region for the production of intermediate vector bosons at the new accelerators. Our data can be parametrized according to the form:

$$m^3 \frac{d^2\sigma}{dmdy} = (2.32\pm0.12) \times 10^{-32} \times \exp\left[(-11.6\pm0.5)\frac{m}{\sqrt{s}}\right] \text{cm}^2\text{GeV}^2$$

(χ^2=25 for 30 d.o.f)

All the data of fig. 2 fit well the form:

$$m^3 \frac{d^2\sigma}{dmdy} = (2.60\pm0.13) \times 10^{-32} \times \exp\left[(-2.0\pm0.7)\frac{m}{\sqrt{s}}\right]\left(1-\frac{m}{\sqrt{s}}\right)^{9.7\pm0.4} \text{cm}^2\text{GeV}^2$$

(χ^2=72 for 57 d.o.f)

This last function is shown superimposed on fig. 1.

The study of the angular distribution of the electron pairs was done for two different mass intervals. For the continuum

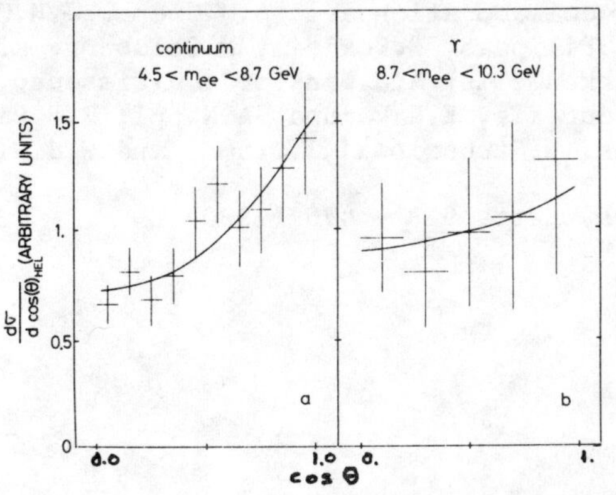

Fig. 3. Angular distribution of the electron pairs.

$4.5 < m < 8.7$ GeV and for the Υ region $8.7 < m < 10.3$ GeV and is shown in figs. 3a,b. The data were fit to the form $d\sigma/d\cos\vartheta = 1 + \alpha\cos^2\vartheta$ and gave good fit for $\alpha = 1.15 \pm 0.34$ for the continuum and $\alpha = 0.79 \pm 0.40$ for the Υ region.

The average transverse momentum of the electron pairs was studied as a function of mass. Since the fraction of background was increasing rapidly with the $\langle p_T \rangle$ for this particular study, a requirement on low multiplicity in the modules containing the electrons was imposed. The resulting distribution of $\langle p_T \rangle$ as a function of mass is given in fig. 4 for the data at $\sqrt{s} = 53$ and 63 GeV combined, together

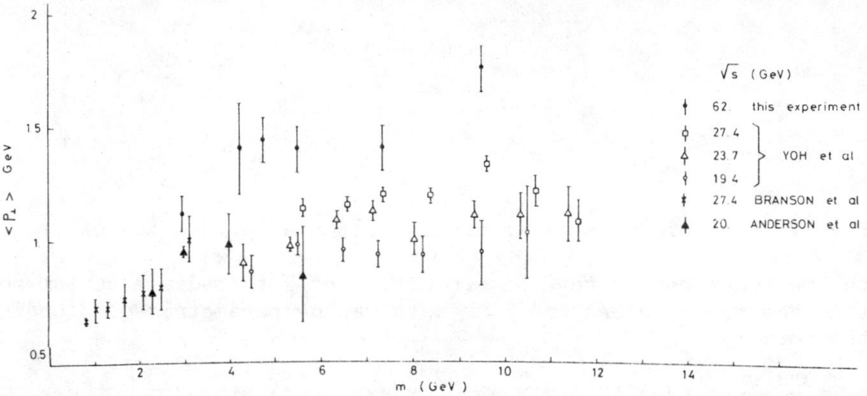

Fig. 4. $\langle p_T \rangle$ versus m for different $\sqrt{s}$ energies.

with other experimental results for comparison. The above figure shows that $\langle p_T \rangle$ is independent of mass, $\langle p_T \rangle = 1.43 \pm 0.07$ GeV, but rises with rising $\sqrt{s}$.

In conclusion scaling seems to be holding down to $\sqrt{\tau} = 0.07$ and the angular distribution is consistent with the virtual photon picture of the simple Drell-Yan mechanism. The only observed departure from the simple Drell-Yan mechanism is the rise of the $\langle p_T \rangle$ with $\sqrt{s}$ for fixed τ.

REFERENCES

(1) The members of the AABCSY collaboration are: A.M.Cnops, J.H.Cobb, C.W.Fabjan, T.Fields, T.A.Filippas, E.Fokitis, M.Goldberg, R.Hogue, N.Horwitz, S.Iwata, C.Kourkoumelis, A.J.Lankford, D.Lissauer, I.Mannelli, G.M.Moneti, P.Mouzourakis, K.Nakamura, A.Nappi, R.B.Palmer D.C.Rahm, P.Rehak, L.K.Resvanis, W.Struczinski, I.Stumer and W.J.Willis

(2) C.Kourkoumelis et al, Phys. Lett. 91B , 475 (1980).

MUON PAIR PRODUCTION ($M > 4.1$ GeV/c^2) BY π^- AT 150, 200 AND 280 GeV/c IN THE CERN - NA 3 EXPERIMENT

CEA, Saclay - CERN, Geneva - COLLEGE DE FRANCE, Paris
ECOLE POLYTECHNIQUE, Palaiseau, LAL, Orsay

Presented by Ph. MINE

ABSTRACT

We have performed a high statistics analysis of muon pair production. Our sample of events of mass between 4.1 and 8.5 GeV/c^2 consists of 2.2×10^4 events at 150 GeV, 5×10^3 events at 200 GeV and 5×10^3 events at 280 GeV incident pion energies. We present results on scaling, on analysis of our hydrogen data, and on the P_T distribution.

SCALING

All data presented on fig.1 are on platinum target, and in the range $M >$ 4.1 GeV/c^2 and $X_F \geqslant 0$. The scaling behaviour of $M^3 d\sigma/dM$, as a function of τ, is well verified within the experimental accuracy estimated to be ±15%. Scaling violation observed in deep inelastic scattering (2) and predicted by QCD first order leading log calculations is too small an effect to be observed.

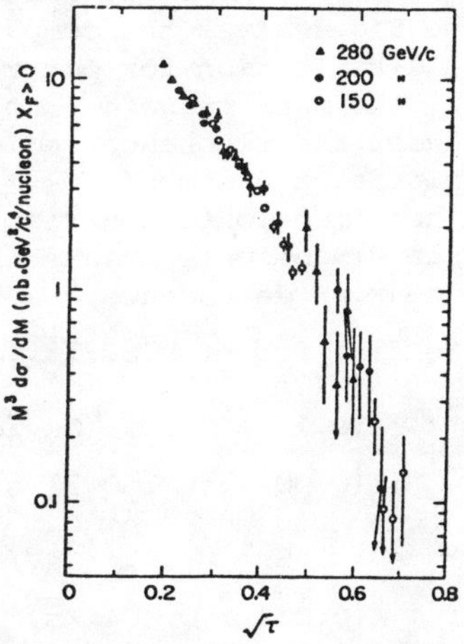

Fig.1 $M^3 d\sigma/dM$ function of $\sqrt{\tau} = M/\sqrt{S}$, upsilon region not plotted.

PRODUCTION OF DIMUONS ON HYDROGEN

Using our 541 events produced by 150 GeV/c π^- on hydrogen, we obtain a measurement of the absolute Drell-Yan cross-section on free proton and a check of the corresponding value of the K factor ($K = (d\sigma/dM)$ exp/$(d\sigma/dM)$ D.Y. model). We find $K = 2.4 \pm 0.4$, in agreement with our previous measurement at 200 GeV/c (1).

Possible nuclear effects in dimuon production are tested by direct comparison of dimuon events produced simultaneously by the same beam traversing two targets : one platinum and one hydrogen. Such a test is free of absolute normalization errors. Neglecting the meson sea contribution, the ratio of the x_2 distributions per nucleon, assuming linear A dependance, can be expressed as :

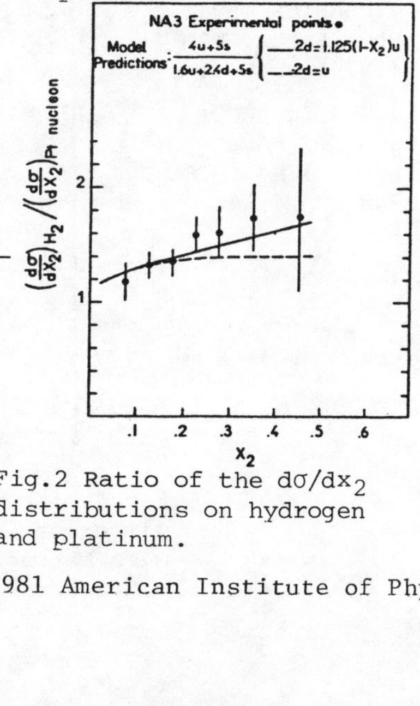

Fig.2 Ratio of the $d\sigma/dx_2$ distributions on hydrogen and platinum.

$$\frac{(d\sigma/dx_2)H_2}{(d\sigma/dx_2)pt} \simeq \frac{K_{H_2}}{K_{Pt}} \frac{4\ u^p(x_2) + 5\ s^p(x_2)}{1.6\ u^p(x_2) + 2.4\ d^p(x_2) + 5\ s^p\ (x_2)}$$

The behaviour of the data (Fig.2) agrees with the prediction of the Drell-Yan model (solid curve), using the CDHS structure function for the proton (2). Our data does not prefere a parametrization $u(x) = 2d(x)$, althrough it cannot be completely excluded. We find $K_{H2}/K_{Pt} = 1.03 \pm 0.09$. If we use an A dependance parametrization A^{α} and the assumption $K_{H2} = K_{Pt}$, we obtain $\alpha = 0.994 \pm 0.015$.

TRANSVERSE MOMENTUM DISTRIBUTIONS

Fig.3 shows the differential cross section of the 150 GeV/c data, for different mass bins. Fig.4 gives the comparison of the same distribution for our three beam energies. The mass range 4.1 to 8.5 GeV/c^2 excludes the resonance region. In Fig.5 is plotted the moment $\langle p_T^2 \rangle$ as a function of the dimensionless variable τ. On simple dimensional arguments, QDC predicts a linear s dependence.

We find for $\sqrt{\tau} = 0.275$:

$$\langle p_T^2 \rangle = (0.74 \pm 0.10)\ (\text{GeV/c})^2 + (0.0029 \pm 0.0003)\ s$$

$$\langle p_T \rangle = (0.49 \pm 0.08)\ (\text{GeV/c}) + (0.034 \pm 0.004)\sqrt{s}$$

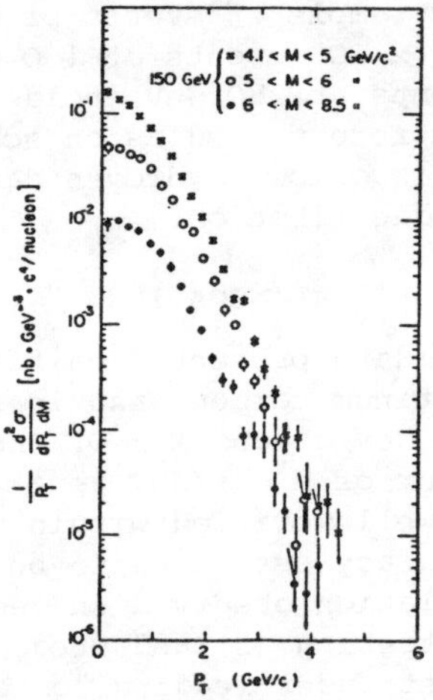

Fig.3. $1/p_T d^2\sigma/dp_T dM$ of 150 GeV/c pions for different mass bins.

REFERENCES

1. J. BADIER et al., Phys.Lett., 86B (1979) 98.
2. J.G.H de GROOT et al., Phys. Lett. 82B (1978) 948.

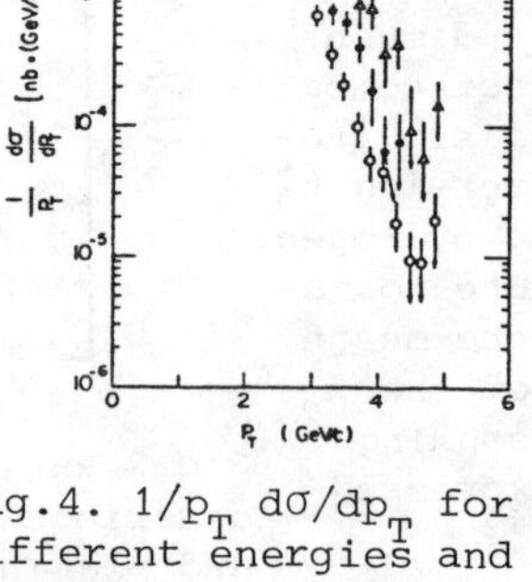

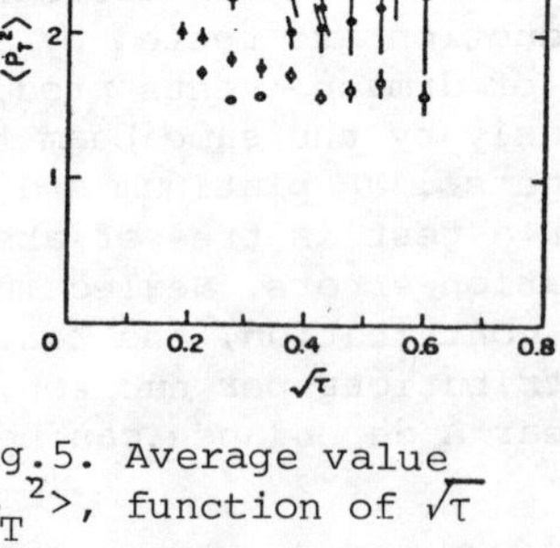

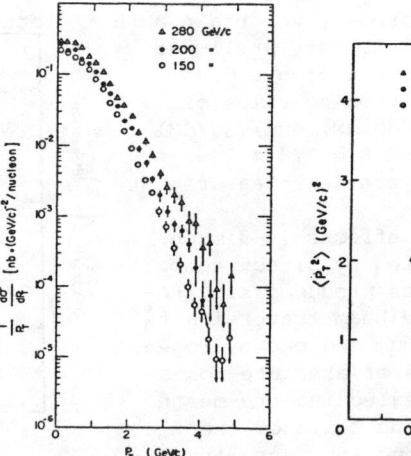

Fig.4. $1/p_T\ d\sigma/dp_T$ for different energies and mass between 4.1 and 8.5 GeV/c^2.

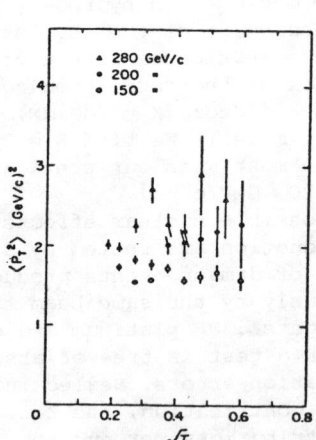

Fig.5. Average value $\langle p_T^2 \rangle$, function of $\sqrt{\tau}$

EXPERIMENTAL DETERMINATION OF NUCLEON, π AND K STRUCTURE FUNCTIONS FROM MASSIVE DIMUONS PRODUCED AT 150 AND 200 GeV/c.

NA3 Collaboration
CEN Saclay-CERN,Geneva-Collège de France,Paris-Ecole Polytechnique, Palaiseau-Lab. de l'Accélérateur Linéaire,Orsay.

Presented by D. DECAMP
Lab. de l'Accélérateur Linéaire, Orsay, F-91405

ABSTRACT

We present results on the measurement of the shape and absolute normalization of the nucleon, pion and kaon structure functions by the analysis of massive dimuon events produced by antiprotons, protons, π and K at 150 and 200 GeV/c. In the case of the nucleon, agreement is found with the results of the neutrino DIS experiments. In all channels, the experimental cross section turns out to be higher than the predicted value by the D.Y. model, by a factor of 2.3±0.4.

EXPERIMENTAL DATA

The data are obtained in the CERN-NA3 experiment using an high intensity hadron beam (up to 5×10^7 particles/burst) instrumented with high resolution Cerenkov counters for K and $\bar{p}$ identification and threshold Cerenkov counters for π^+-p separation. Numbers of dimuon events, collected at 150 and 200 GeV on 2 targets (6 cm Pt and 30 cm H_2) in the mass interval 4.1 to 8.5 GeV, are given in table I.

Table I

	Energy	π^-	K^-	$\bar{p}$	π^+	p	π^- (H_2)
Pt	150	21200	700	275	-	35	540
	200	4996	80	30	1770	1080	138

THE QUARK ANNIHILATION MODEL

The experiment measures the invariant mass M and the longitudinal momentum P_L^* of the muon pair. Then through the relations $M^2=x_1x_2s$ and $x=2P_L^*/\sqrt{s}=x_1-x_2$ (transverse momentum neglected), we determine the kinematic variables x_1 and x_2 which are the fractional momenta of the quarks in the projectile and target particle respectively. The cross section of the D.Y.[1] process is given by

$$\frac{d^2\sigma}{dx_1dx_2} = \frac{4\pi\alpha^2}{3s}\,\frac{1}{3}\sum_i \frac{Q_i^2}{x_1^2x_2^2}\left[f_i^h(x_1)\,f_{\bar{i}}^{h'}(x_2)+f_{\bar{i}}^h(x_1)\,f_i^{h'}(x_2)\right] \quad (1)$$

where Q_i are the quark charges and $f_i(x_1)$, $f_i(x_2)$ are the structure functions of the projectile and target particle respectively.

We define a scale factor K as the ratio of the measured lepton pair cross section to the one expected from the D.Y. model

$$(d^2\sigma/dx_1dx_2)_{exp} = K(d^2\sigma/dx_1dx_2)_{DY\ model}$$

The structure functions are extracted by fitting the (x_1,x_2) data sample to expression (1) with a Buras Gaemers type of parametrization of the structure function :

Parametrization	Nucleon	Pion
valence	$u^p = A^u_{\alpha\beta}\ x^{\alpha}(1-x)^{\beta}$; $\int u^p dx/x=2$ $d^p = A^d_{\alpha\beta}\ x^{\alpha}(1-x)^{\beta+1}$; $\int d^p dx/x=1$ $(u^n = d^p,\ d^n = u^p \ldots)$	$u^{\pi} = A_{\pi} x^{\alpha_{\pi}}(1-x)^{\beta_{\pi}}$; $\int u^{\pi} dx/x = 1$
sea	$S^p = \bar{u} = \bar{d} = A_s(1-x)^{\beta}$ $\bar{s} = 1/4(\bar{u} + \bar{d})$	$S^{\pi} = A_s{}^{\pi}(1-x)^{\beta_s{}^{\pi}}$

ANTIPROTON-NUCLEON DATA

From the expression of the Drell-Yan cross section, we can deduce the difference between the antiproton and proton differential cross section on platinum target nucleon N(Z/A=0.4). In this case, all contributions from the sea are cancelled in the subtraction and, up to an accuracy of 2%, the difference of the cross sections factorizes as:

$$\frac{d^2\sigma}{dx_1 dx_2}\Big|_{\bar{p}N-pN} = \frac{4\pi\alpha^2}{3s}\,\frac{1}{3x_1^2 x_2^2}\,\frac{1}{9}\,f(x_1)g(x_2)\ \left\{\begin{array}{l} f(x_1)=4u(x_1)+d(x_1) \\ g(x_2)=0.4u(x_2)+0.6d(x_2) \end{array}\right. \quad (2)$$

Each of the two functions $f(x_1)$ and $g(x_2)$ can be determined in turn from expression (2) after integration over the variable x_2 and x_1 respectively. For $f(x_1)$ we have :

$$f(x\ )=\left(\frac{L}{L_{\bar{p}}}\frac{dN}{dx_1}\Big|_{\bar{p}N} - \frac{1}{L_p}\frac{dN}{dx_1}\Big|_{pN}\right)\Big/\left(\frac{4\pi\alpha^2}{9sx_1{}^2}\int\frac{g(x_2)}{9}\,\frac{A(x_1,x_2)}{x_2{}^2}\,dx_2\right)$$

where dN/dx_1 is the measured event distribution in x_1, L is the integrated luminosity and $A(x_1,x_2)$ is the spectrometer acceptance. We evaluate $g(x_2)$ in expression (2) using the parametrization with $\alpha=0.51\pm 0.02$, $\beta=2.8\pm0.1$ as obtained from CDHS results at $Q^2=20\ GeV^2$ (3).

The two sructure functions $f(x_1)$ and $g(x_2)$ thus obtained are shown in fig.1 and 2, where the experimental points (full points) are compared to the Drell-Yan model based on the CDHS(3) determination of the valence nucleon structure function. We note a very good agreement of the shape of $f(x_1)$ and $g(x_2)$ data points to the model in the interval : $0.25\leqslant x_1\leqslant 1.0$ and $0.15\leqslant x_2\leqslant 0.45$. The measured yield requires a nor-

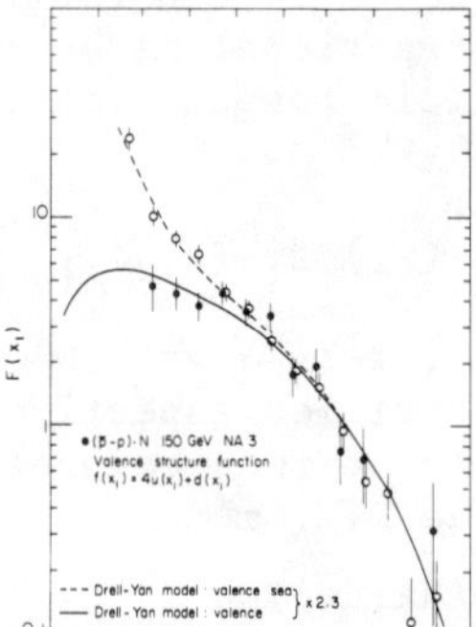

Fig. 1

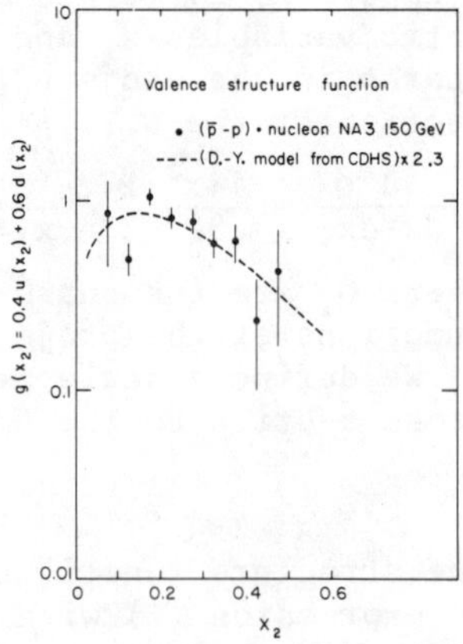

Fig. 2

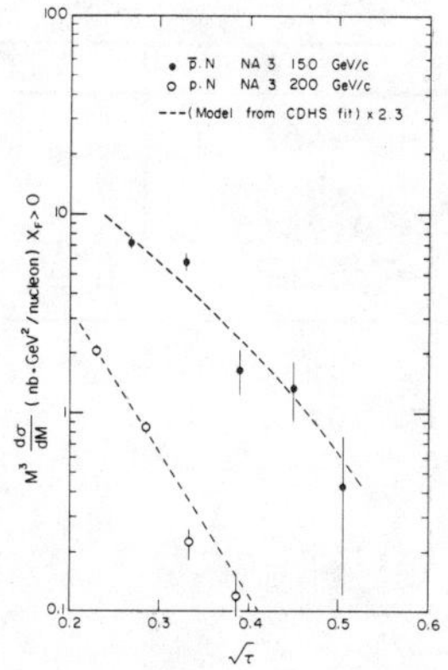

Fig. 3

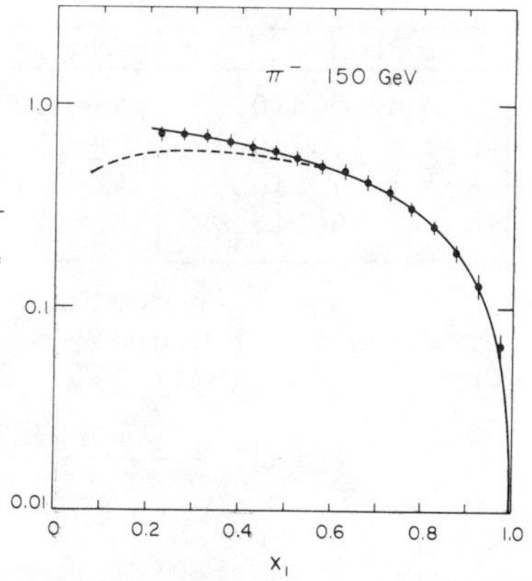

Fig. 4

malization factor K=2.3±0.4, also in good agreement with our previous experimental determination(4). Fig.3 shows the behaviour of $M^3 d\sigma/dM$ plotted versus $\sqrt{\tau}$ for our antiproton data of 150 GeV and our proton data of 200 GeV. We can compare the valence structure function of antiprotons with the overall valence+sea structure function previously obtained by our experiment with 200 GeV/c protons, both being determined using the Drell-Yan mechanism. This is shown in fig.1. As it can be seen the two structure functions superimpose for $x_1>0.4$. At lower values of x_1 they separate, clearly showing the sea contribution which is present in the proton data (open circles).

PION-NUCLEON DATA

Using the cross section of the Drell-Yan process (1) and the parametrization defined above, we can express the cross section for pions as follows :

$$\frac{d^2\sigma}{dx_1 dx_2} = \frac{4\pi\alpha^2}{3s}\,\frac{1}{3}\,\frac{1}{x_1{}^2 x_2{}^2}\left[u^\pi(x_1)\,G(x_2) + s^\pi(x_1)\,H(x_2)\right] \quad (3)$$

where $G(x_2)$ and $H(x_2)$, for platinum target (Z/A=0.4), take the following forms :

$$G(x_2) = 1/9(1.6u^p + 2.4d^p + 5S^p) \quad \text{for } \pi^-$$
$$G(x_2) = 1/9(0.6u^p + 0.4d^p + 5S^p) \quad \text{for } \pi^+$$
$$H(x_2) = 1/9(2.2u^p + 2.8d^p + 11S^p) \quad \text{for } \pi^+, \pi^-$$

a) Parametrization method : a global fit of the 200 GeV π^+ and π^- data allows a determination of both the valence and sea structure functions (results in table II). As in $\bar{p}N$ and pN data, the nucleon structure function obtained are in good agreement with the results of the ν DIS experiment(3). To improve the accuracy of the pion structure function we fix the nucleon parameters α=0.5 and β=2.8 (CDHS fit), we obtain the results given in Table IIIa). Table IIIb) shows the results of the fit of the 150 GeV π^- data where we impose the π sea from our 200 GeV π^+-π^- fit and we use the nucleon valence and sea from the CDHS fit.

Table II

Pion	Nucleon
$\alpha^{\pi} = 0.4\pm0.15$	$\alpha = 0.49\pm0.2$
$\beta^{\pi} = 1.07\pm0.12$	$\beta = 3.3\pm0.5$
$A_s^{\pi} = 0.32\pm0.2$	$A_s = 0.25\pm0.1$
$\beta_s^{\pi} = 6.9\pm2.9$	$\beta_s = 7.7\pm1.4$

Table III

a) 200 GeV	b) 150 GeV
$\alpha^{\pi} = 0.45\pm0.1$	$\alpha^{\pi} = 0.4\pm0.1$
$\beta^{\pi} = 1.04\pm0.1$	$\beta^{\pi} = 0.9\pm0.1$
$A_s^{\pi} = 0.25\pm0.15$	A_s^{π}, β_s^{π} } fixed from 200 GeV $\pi^+\pi^-$ fit
$\beta_s^{\pi} = 5.4\pm2.0$	

b) Projection method : by projecting the content of the x_1, x_2 array on the two axes, we get the distribution dN/dx_1 and dN/dx_2. We obtain from equation (3) an expression depending only on x_1 :

$$F_{\pi}(x_1) = \frac{dN/dx_1}{\frac{\sigma_0}{3}\frac{L}{x_1^2}\, I(x_1)} = K\left[u^{\pi}(x_1) + \frac{J(x_1)}{I(x_1)}\, s^{\pi}(x_1)\right] \qquad (4)$$

where : L is the integrated luminosity; $\sigma_0 = 4\pi\alpha^2/3s$; $J(x_1)$ and $I(x_1)$ are integrals involving the nucleon structure function $G(x_2)$ and $H(x_2)$ and the calculated acceptance of the apparatus $A(x_1,x_2)$

$$I(x_1) = \int G(x_2)A(x_1,x_2)dx_2/x_2^2 \qquad J(x_1) = \int H(x_2)A(x_1,x_2)dx_2/x_2^2$$

We use for u and d the results of ν DIS parametrization. In fig.4 we present the values of $F_{\pi}(x_1)$ obtained at 150 GeV with the curves calculated from our fit of the pion structure function.

Following the suggestion of Berger et al.[5] we have attempted to fit our π structure function at $x>0.5$ with the form $\sqrt{x}((1-x)^2+(2/9)\cdot\langle K_T^2\rangle/M^2)$ with $\langle K_T^2\rangle=1$. The result of this parametrization (corrected for experimental errors) is shown in fig.5. The experimental points are not compatible with a $(1-x)^2$ form of the pion structure function at $x_1<0.7$.

By assuming the K factor to be independent of x_1, we can derive its value after integration of $F_{\pi}(x_1)$ over x_1, from equation (4). At 200 GeV we obtain $K=2.05\pm0.4$ and at 150 GeV we obtain $K=2.4\pm0.4$. Thus the experimental cross section for production of massive μ pairs is larger than the prediction of the simple Drell-Yan model computation. We recall that in the case of pN and $\bar{p}$N collisions we had found $K_{pN}=2.3\pm0.4$ and $K_{\bar{p}N}=2.3\pm0.4$.

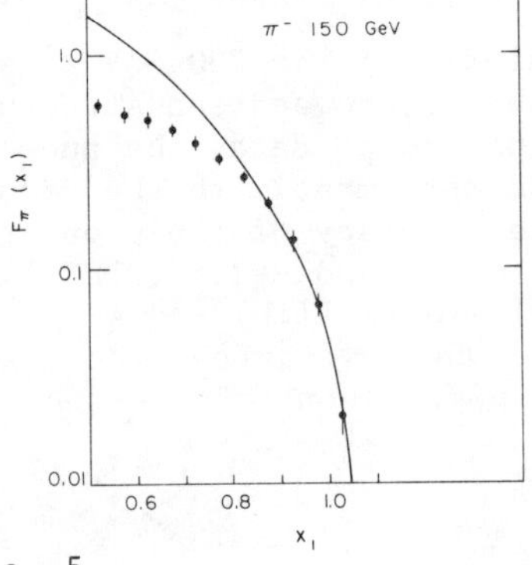

Fig. 5

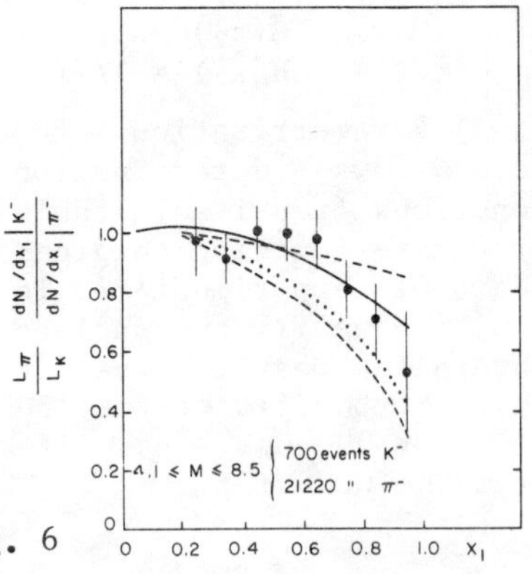

Fig. 6

We also collected μ pairs produced by pions on a hydrogen target[6]. The analysis of these data gives $K=2.4\pm0.4$. Thus the increase in cross section cannot be due to coherent nuclear effects on platinum.

It is therefore most probable that the K factor implies a fundamental correction to the leading log Drell-Yan computation. Such a correction, non leading log first order in QCD, has been calculated to give $K\sim1.8$[7], in the $\tau=M_{\mu\mu}/\sqrt{s}$ interval ~0.2 to ~0.4 where our experimental data are available.

K-NUCLEON DATA[8]

Using a formalism similar to the projection method described above we can obtain a $F_K(x_1)$ and $F_\pi(x_1)$. Since the K^- data are at values of $x_1>0.2$ and their statistical accuracy is limited, we neglected contributions of the meson sea. Up to an accuracy of $\sim10\%$ in the ratio $\bar{u}^K(x_1)/\bar{u}_\pi(x_1)$, we can also neglect terms corresponding to $d^\pi\bar{d}^p$ and $s^K\bar{s}^p$ annihilation. The only terms left are from the annihilation of the valence meson antiquark $\bar{u}$ with the valence quark of the nucleon. Assuming the ratios of the measured cross section to the prediced D.Y. value are equal for πN and KN data, we obtain (fig.6)

$$\bar{u}_K(x_1)/\bar{u}_\pi(x_1) = L_\pi(dN/dx_1)_K/L_K(dN/dx_1)_\pi$$

this ratio is independent of experimental acceptance and trigger efficiency due to simultaneous K and π data collection in our experiment. From fig.6 it appears that for values of $x_1>0.7$ the momentum spectrum of the $\bar{u}$ quark in the kaon decreases faster than the corresponding one for the $\bar{u}$ in the pion. This can also be expressed by parametrizing the data with the analytic form $R(1-x)^A$, giving $A=0.18\pm0.07$. In fig.6, we compare our data with theoretical models based on Regge considerations. These models limit the value of A in the range 1/8 to 1/2 (dashed curves). Recently a non relativistic calculation[9] of the π and K structure function in the framework of QCD has been performed. This is shown in fig.6 by the solid curve which agree satisfactorily with the data.

REFERENCES

(1) S.D. Drell and T.M. Yan, Phys. Rev. Lett. 25 (1970) 316.
(2) A.J. Buras and K.J.F. Gaemers, Nucl. Phys. B132 (1978) 249.
(3) A. Para, Proc. of the Lepton-Photon Symposium, Fermilab 1979, p343.
(4) J. Badier et al., Phys. Lett. 89B (1979) 145.
(5) E.L. Berger and S.J. Brodsky, Phys. Rev. Lett. 42 (1979) 940.
(6) J. Badier et al., Massive dimuon transverse momentum, scaling and A-dependence, submitted to Int. Conf. on High Energy Physics, Madison (1980).
(7) see for example : G. Altarelli, EPS Int. Conf. on High Energy Physics , page 727 (Geneva, 1979) and references therein.
(8) J. Badier et al., Phys. Lett. 93B (1980) 354.
(9) F. Martin, Preprint CERN TH 2845 and Proc. XV Rencontre de Moriond, Les Arcs, 1980.

DRELL-YAN MUON PAIRS AT 40GeV/c FROM TUNGSTEN USING $\pi^{\pm}$, $K^{\pm}$, $\bar{P}$ AND P BEAMS

M. J. Corden et al.
University of Birmingham, UK

P. Sonderegger
CERN, Geneva Switzerland

B. Chaurand et al.
Ecole Polytechnique, France

ABSTRACT

At the Ω spectrometer the inclusive hadro-production of muon pairs with masses >2 GeV/c^2 has been measured. Comparisons are made with the simple Drell-Yan model and quark structure function fits have been made to the data.

RESULTS AND CONCLUSIONS

For the mass bins presented here the contributions from ψ and ψ' have been removed using fits made to the mass spectra. Also like sign muon pairs have been subtracted (∿10% at 2 GeV/c^2 and negligible at the ψ mass).

The cross section ratios are shown (x_F >0) as a function of mass in fig. 1, where the two important features to note at large masses are (1) that π^+/π^- approaches the expected asymptotic value of ¼* (isoscalar target) and (2) the small values of $\bar{p}/\pi^-$ and K^+/π^- (the P and K^+ having no valence $\bar{u}$ or $\bar{d}$).

The curves on this figure are predictions calculated using the structure functions determined at 200 GeV/c [1].

In fig. 2 the differential cross sections $\frac{d\sigma}{dm}$ are given for π^-, π^+ and ($\pi^- - \pi^+$) together with the simple Drell-Yan cross sections (again using the same structure functions), multiplied by the factor K. This correction factor is believed to be necessary from Q.C.D. arguments [2].

The scaling plot $M^3 \frac{d\sigma}{dm}$ versus $\frac{M}{\sqrt{S}}$ is given in fig. 3 for π^- results at 40, 200 and 280 GeV/c [3]. Scaling violations would predict approximately a 20% decrease going from 40 to 280 GeV/c at $\frac{M}{\sqrt{S}}$ = 0.5. Such an effect is present but this difference is comparable with the normalisation uncertainties of the measurements.

*Nearer 0.3 for tungsten.

A pion structure function fit has been made to our π^- and π^+ data, using nucleon structure functions from CDHS and NA3 [1]. For the pion valence quark we find

$$U(x) \sim x^{0.5 \pm 0.1} (1-x)^{0.9 \pm 0.1}$$

REFERENCES

1. J. Badier et al., Phys. Lett. 89B 145 (1979).
2. See plenary talks by J. Lefrancois and C. Llewellyn-Smith.
3. J. Badier et al., CERN/EP 79-68.

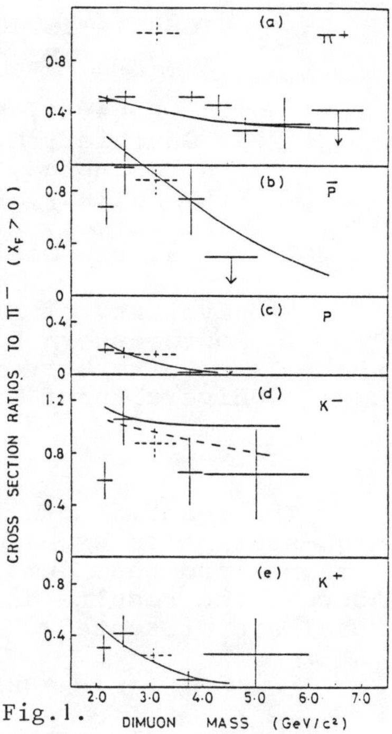

Fig.1.

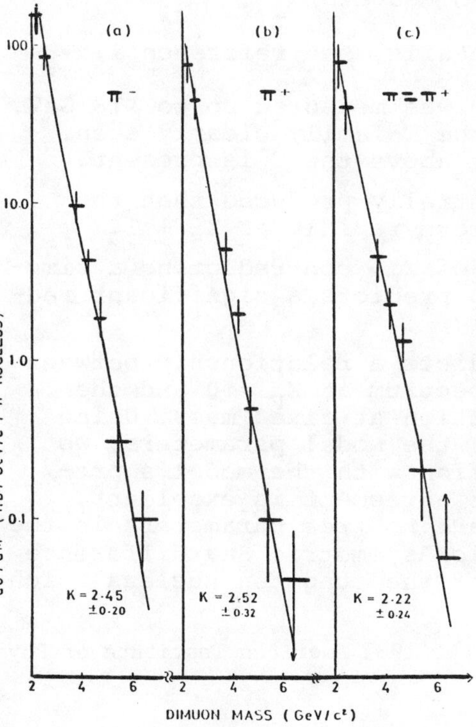

Fig. 2

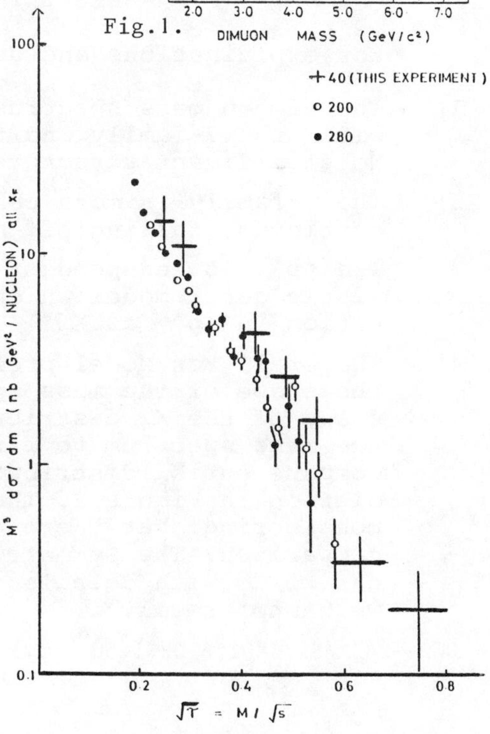

Fig. 3

A HIGH-STATISTICS STUDY OF DIMUON PRODUCTION BY 400 GeV/c PROTONS

T.J. Roberts, S. Childress, D.A. Garelick,
P.S. Gauthier, M.J. Glaubman, H.R. Gustafson,
H. Johnstad, L.W. Jones, M.J. Longo,
M.L. Mallary, P.M. Mockett, J. Moromisado,
W.P. Oliver, J.P. Rutherfoord, S.R. Smith,
E. von Goeler, and M.R. Whalley.

University of Michigan, Ann Arbor, MI 48109
Northeastern University, Boston, MA 02155
Tufts University, Medford, MA 02155
University of Washington, Seattle, WA 98195

ABSTRACT

The reaction $p+W \rightarrow \mu^+\mu^- + X$ has been measured in a high-statistics experiment using a 400 GeV/c proton beam, a magnetized beam dump, and a wide-acceptance detector. Abbreviated results and a comparison with the Drell-Yan Model are presented.

ABBREVIATED RESULTS

For explainations and details, see reference 1.

1) The dimuon mass spectrum was measured up to ∿18 GeV, with the Ψ-family and the ϒ-family clearly seen. No significant structure above the ϒ is present.

2) The ϒ-family is more centrally produced than the continuum, falling off more rapidly as $X_F \rightarrow 1$.

3) The $\langle p_t \rangle$ is independent of X_F, contradicting a simplistic quark model which predicts a significant reduction in $\langle p_t \rangle$ as $X_F \rightarrow 1$.

4) The Drell-Yan Model predicts a relationship between the shape of the mass spectrum at $X_F = 0$ and the shape of the X_F distribution at fixed mass. Using the mass spectrum to fix the model parameters, we compare our X_F distribution with the model's prediction in Figure 1. The agreement is excellent, considering that there are no free parameters in the comparison. The Symmetric/Asymmetric Sea difference has to do with details of the tungsten nucleus which we cannot resolve.

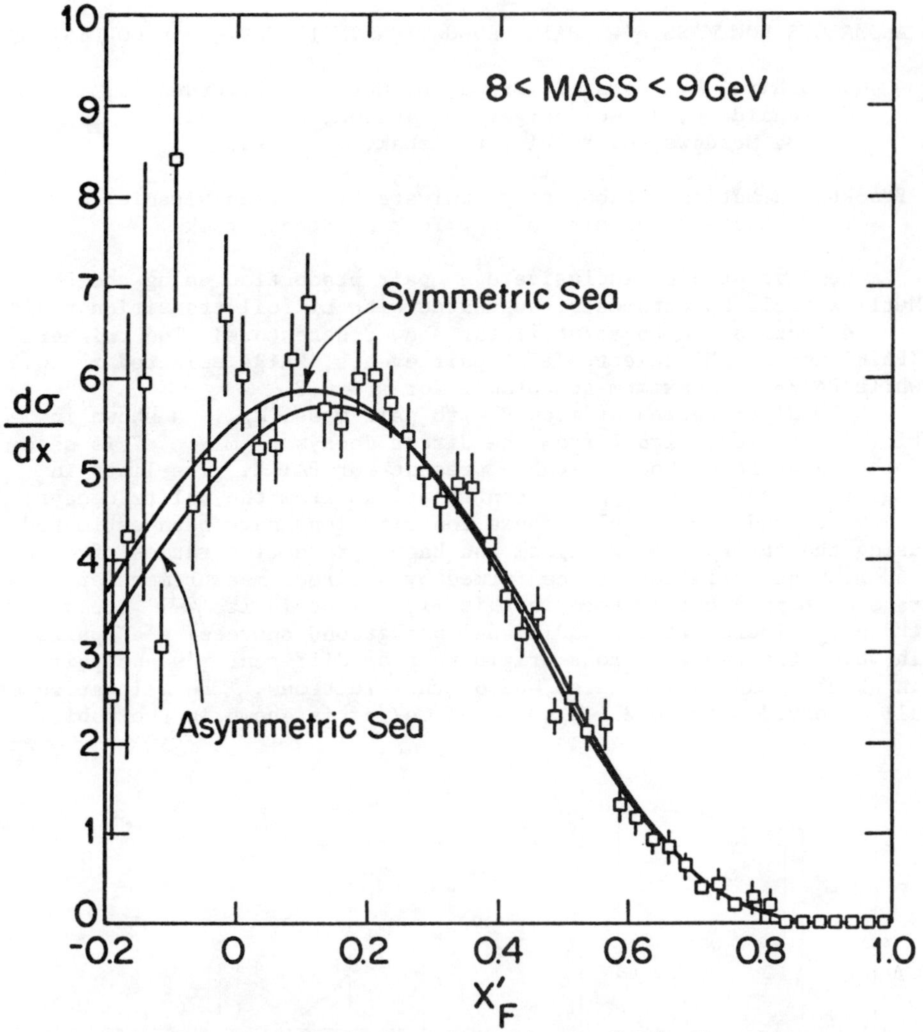

Fig. 1. Cross-section as a function of X_F' for dimuon masses between 8 and 9 GeV. The curves are Drell-Yan Model predictions with model parameters determined by the mass spectrum.

REFERENCES

1. W.P. Oliver, et al, Proc. 3rd Intl. Conf. on New Results in High Energy Physics, Vanderbilt, p 93. AIP Conf. Proc. No. 45 (1978).
P.M. Mockett, et al, Proc. 19th Intl. Conf. on High Energy Physics, Tokyo, p 187 (1978).
H.R. Gustafson, et al, paper submitted to this conference. Univ. of Michigan preprint UM HE 80-23.

ANOMALOUS LOW MASS e^+e^- PAIR PRODUCTION IN 17 GeV/c π^-p COLLISIONS

G. Abshire, M. Adams, C. Brown, L. Cormell, E. Crandall,
G. Donaldson, J. Goldberger, H. Gordon, P. Grannis,
B. Meadows, G. Morris, P. Rehak, J. Stekas

Brookhaven National Laboratory, University of Pennsylvania, and
State University of New York at Stony Brook

We have studied inclusive e^+e^- pair production using the BNL Mutliparticle Spectrometer, augmented with Li foil transition radiation detectors and Pb-scintillator shower detectors. Two triggers (PAIRA and PAIRB) selected e^+e^- pair events; PAIRA selected $x > 0.5$ while PAIRB had maximum acceptance for $x \sim 0.3$.

The distribution of events with pair mass, M_{ee}, is shown in Fig. 1. A clear signal from the direct decays $\rho, \omega \to e^+e^-$ is seen for trigger PAIRA but is weakly present for PAIRB. The lines in Fig. 1 indicate the expected contributions from the Dalitz decays, $\eta \to \gamma e^+e^-$ and $\omega \to \pi^\circ e^+e^-$. These contributions have been estimated using the observed ρ, ω signal and known production ratios of ρ, ω, and η. The estimates are confirmed by a direct measurement of the rate of η production, seen in this experiment in its γe^+e^- decay channel. There are two additional background sources: e^+e^- pairs in which the two electrons originate from different γ's and pairs in which a hadron simulates one of the electrons. The allocation of all observed pairs $0.2 < M_{ee} < 0.55$ GeV/c^2 is shown in the table.

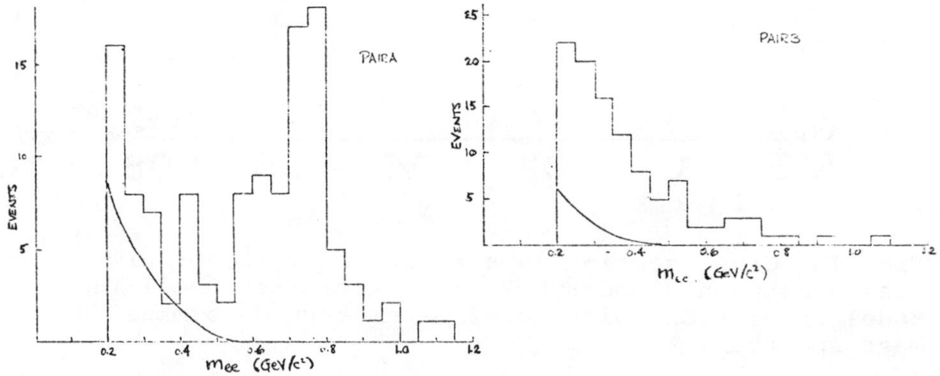

Fig. 1 Number of events versus pair mass.

	PAIRA	PAIRB
Total Signal	46 ± 6.8	90 ± 9.5
η,ω Dalitz	21.3 ± 10.3	12.0 ± 4.7
Different γ pairs	7.3 ± 3	1.5 ± 1
Hadron background	1 ± 1	23 ± 10
Unexplained excess	16.4 ± 12.7	53.5 ± 14.6

We find that our observed pairs are consistent with known sources for $x > 0.5$ (PAIRA) but leave a significant, unexplained excess for $x < 0.5$ (PAIRB). This is shown more clearly in Fig. 2, where the x-distribution of pairs ($0.2 < M_{ee} < 0.6$ GeV/c^2 and after subtraction of 2γ and hadron backgrounds) is given. The shaded band indicates the level of contribution due to η, ω Dalitz decays, we see that the excess is present only for $x < 0.5$.

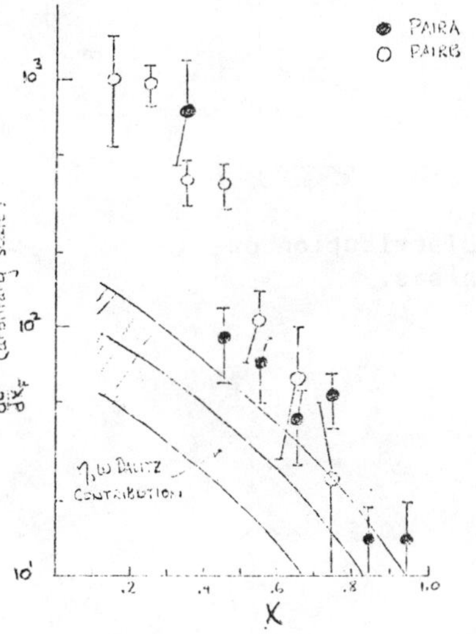

Fig. 2 Feynman x distribution of e^+e^- pairs, 0.2 M 0.6 GeV/c .

In this experiment, we have good efficiency for detecting all charged particles and γ's produced in association with a direct e^+e^-. We have examined several effective mass combinations for events in the anomalous region ($0.2 < M_{ee} < 0.6$ GeV/c^2; $x < 0.4$) in order to search for possible decay modes of known states which could explain our excess. Figures 3 and 4 give the (γe^+e^-) and ($\pi^\circ e^+e^-$) spectra; apart from possible signals due to η or ω Dalitz decays consistent with our estimates above, there is no significant structure observed. Figure 5 shows the distribution of ($\pi^+\pi^-e^+e^-$) masses; this channel is appropriate to a search for $f^\circ \to \rho^\circ\gamma_V$ (γ_V is a virtual gamma). No evidence of the f° is seen. Similarly, no evidence is found for $A_2^\pm \to \rho^\pm\gamma_V$.

We have made crude estimates of the single $e^\pm/\pi$ ratios which might be expected from our observed excess e^+e^- pairs. We find that this ratio might be expected to be in the range (0.5 - 1.) x 10^{-4} for production at p_T = 0.5 GeV/c and x = 0. We can also estimate the γ/π ratio, assuming that the distribution in mass of virtual γ's is similar to that for η Dalitz decays. We obtain a γ/π ratio of order 10^{-1} at x = 0.3.

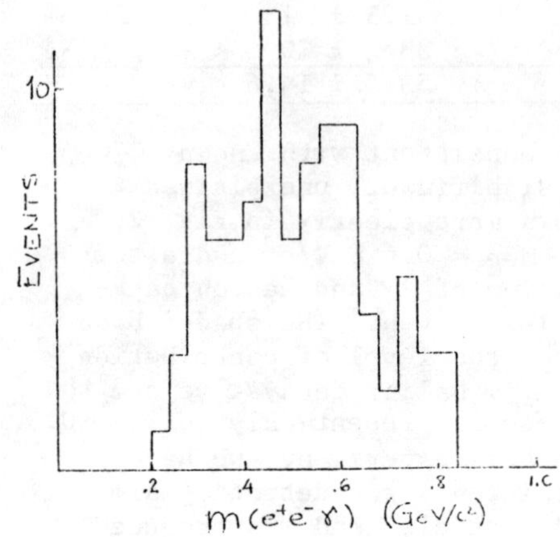

Fig. 3 Distribution of $e^+e^-\gamma$ masses.

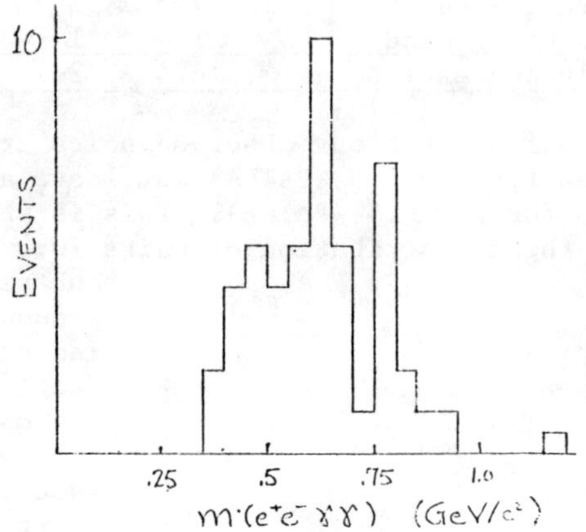

Fig. 4 Distribution of $e^+e^-\gamma\gamma$ masses.

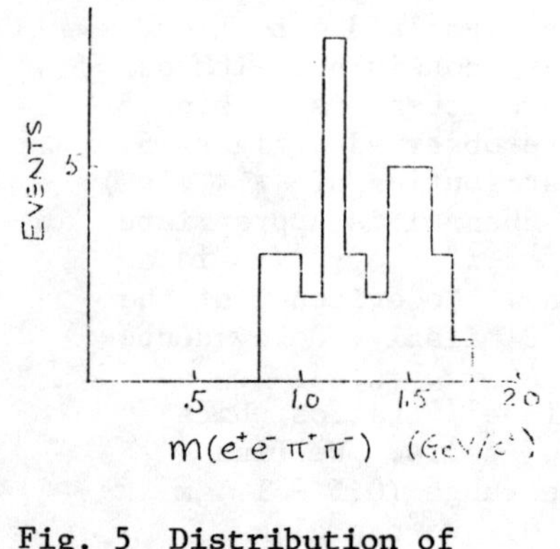

Fig. 5 Distribution of $e^+e^-\pi^+\pi^-$ masses.

OBSERVATION OF A DIRECT LOW-MASS e^+e^- CONTINUUM IN π^-p INTERACTIONS AT 16 GeV/c*

Presented by R. Stroynowski

SLAC — Johns Hopkins University — Caltech Collaboration †

The LASS spectrometer at SLAC has been used to measure the production of direct e^+e^- pairs in π^-p interactions at 16 GeV/c. Fully inclusive single $e^\pm$ trigger allowed for the unbiased study of background processes. The final data sample (after removal of γ conversions) consisted of 291 events in the kinematical region $0.2 \le M(e^+e^-) \le 1.2$ GeV/c^2, $0.1 \le x \le 0.45$ and $0 \le p_T \le 0.8$ GeV/c. About a third of these contained same sign electron pair combinations representing an irreducible background consisting of pairs formed by electrons originating from different γ conversions and pairs formed by an electron and misidentified hadron or by two misidentified hadrons. After subtraction of same sign background the remaining spectrum corresponds to a cross section of (955 ± 170) nb for $0.2 \le M(e^+e^-) \le 0.7$ GeV/c^2. The contributions from direct and Dalitz decays of known resonances have been estimated and subtracted using a Monte Carlo method. The observed excess signal of (700 ± 180) nb decreases slowly with increasing mass but exhibits very steep x and p_T dependence.

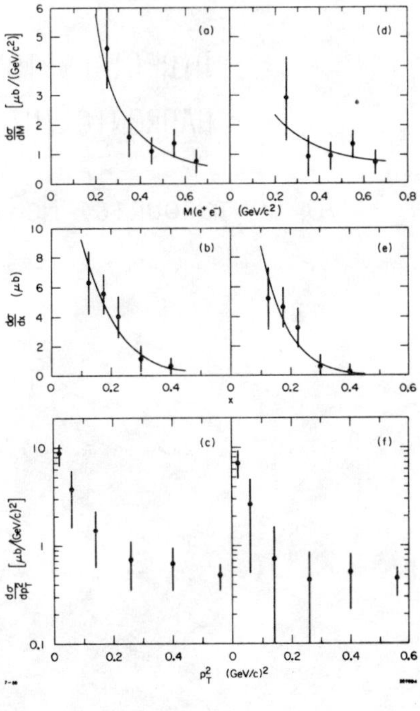

Fig. 1. The acceptance corrected mass, x and p_T^2 distributions for the continuum: (a)-(c) before, (d)-(f) after, subtraction of the η and ω Dalitz decay contributions.

* This work was supported by the Department of Energy under contract DE-AC03-SF00515.

† R. Stroynowski, D. Blockus, W. Dunwoodie, D. W. G. S. Leith, M. Marshall, C. L. Woody, B. Barnett, C. Y. Chien, T. Fieguth, M. Gilchriese, D. Hutchinson, W. B. Johnson, P. Kunz, T. Lasinski, L. Madansky, W. T. Meyer, A. Pevsner, B. Ratcliff, P. Schacht, J. Scheid, S. Shapiro, S. Williams.

DIRECT LEPTONS AND PHOTONS IN HADRONIC INTERACTIONS, THEORY

A. Contogouris, McGill University, Organizer

LARGE-p_T DIRECT PHOTON PRODUCTION AND PHOTON-HADRON CORRELATIONS IN QCD

A.P. Contogouris

Department of Physics, McGill University, Montreal, Canada

BSTRACT: The present status of large-p_T direct photon production in per-urbative QCD is reviewed. Recent calculations of the QCD correction due to hoton Bremsstrahlung (quark-quark subprocess) are presented. Theoretical redictions on $p+p \to \gamma+X$ are discussed and compared with data. It is shown hat opposite-side photon-hadron correlations provide a good determination of he gluon fragmentation to a hadron.

Direct photons as a test of QCD.

We consider inclusive production of photons at 90° in the c.m. of the olliding hadrons A and B and denote the ratio:

$$\frac{Ed\sigma(A+B \to \gamma+X)/d^3p}{Ed\sigma(A+B \to \pi^o+X)/d^3p} \equiv \frac{\gamma}{\pi^o} \qquad (1)$$

Some important mechanisms for production of large-p_T photons are the llowing:

Vector Dominance Model

The photon is produced by means of a large-p_T vector meson V^o via:

$$A+B \to V^o+X, \quad V^o \to \gamma \qquad (2)$$

is easy to estimate the photon yield in this mechanism. Clearly:

$$\left(\frac{\gamma}{\pi^o}\right)_{VDM} \sim \alpha \frac{V^o}{\pi^o} \qquad \alpha = \frac{1}{137}$$

erimentally (at $p_T \simeq 4$ GeV):

$$\frac{\rho^o}{\pi^o} \simeq \frac{\omega}{\pi^o} \simeq 10\frac{\phi}{\pi^o} \simeq 1$$

Therefore

$$\frac{\gamma}{\pi^\circ} = O(\alpha) = \text{a few percent} \tag{3}$$

Notice that this prediction is independent of the mechanism by which V° is produced at large p_T.

II. Quantum Chromodynamics

To the extent that QCD is a theory of interacting quarks and gluons photon production at large p_T proceeds via the perturbation subprocesses (Born terms, Fig. 1 (a), (b)):

$$\text{(a)} \quad q+g \rightarrow q+\gamma \qquad \text{(b)} \quad q+\bar{q} \rightarrow g+\gamma \tag{4}$$

(a) gives the dominant contribution in $p+p \rightarrow \gamma+X$ and (b) in $\pi+p \rightarrow \gamma+X$ and in $\bar{p}+p \rightarrow \gamma+X$.

III. Constituent Interchange Model (CIM) - Higher Twists

Photon production proceeds e.g. via[2]

$$q+M \rightarrow q+\gamma \qquad q+\bar{q} \rightarrow M+\gamma \tag{5}$$

This may lead to significant photon yields. Unfortunately, the absolute magnitude (normalization) of these contributions is controversial at present.

We concentrate on the mechanism II. It has been argued[2][3] that at the highest ISR energies and $p_T \sim 10$ GeV or Fermilab energies and $p_T \sim 8$ QCD predicts $\gamma/\pi^\circ \sim 1$. The reason is that $A+B \rightarrow \gamma+X$ involves a fragmentation function $D_{\pi^\circ/q}$ or $D_{\pi^\circ/g}$ (for $q \rightarrow \pi^\circ$ or $g \rightarrow \pi^\circ$), which suppresses the cross-section as p_T increases; in contrast, in the subprocesses of Fig. 1 the photon has a pointlike coupling to the quark ($\sim e$ = constant).

However, scale violations (Q^2-dependence) in the quark and particularly in the gluon distribution are important.[4] For large-p_T processes roughly $Q^2 \sim 2p_T^2$. Thus as p_T increases the distributions $F_{q/A}(x,Q^2)$, $F_{g/B}(x,Q^2)$ decrease (for $x \gtrsim 0.1$) and this somewhat diminishes the rate of $A+B \rightarrow \gamma+X$. Thus very roughly

$$\left(\frac{\gamma}{\pi^\circ}\right)_{QCD} \simeq 0.2 \sim 0.3 \tag{6}$$

(see below for details). Still this prediction is much larger than of VDM.

2. The photon Bremsstrahlung correction

Recently the contribution of the QCD subprocess (Fig. 1(c))

$$q+q \rightarrow q+q+\gamma \tag{7}$$

has also been calculated [5][6]; this provides a correction to (4).

There are two important kinematic configurations: (i) The intermediate gluon collinear with one quark (first graph of Fig. 1(c)) and (ii) The emitted photon collinear with one quark (second graph of Fig. 1(c)). The leading part of (i) ($\sim \log Q^2$) implements the Q^2-dependence in the gluon distribution $F_{g/A}(x,Q^2)$ and it is absorbed by standard procedures (absorption of mass singularities); however it does leave a finite correction.[7] The result of the perturbation calculation is of the form[6]:

$$\frac{d\sigma}{d^4Q}\,\text{Brems} \sim \left(\frac{\alpha_s}{\pi}\right)^2 \int \frac{d\alpha}{\alpha} \{P_{gq}(\alpha)\left[\log \frac{1+\alpha-x_T}{x_T^2\alpha} + 2\log\alpha + 2 - \frac{1}{\alpha}\right]|M_{qg}|^2 + \tag{8}$$
$$+ P_{\gamma q}(x_T)\log \frac{s(1-x_T)}{\mu^2}\,|M_{qq}|^2\}$$

where $x_T = 2p_T/\sqrt{s}$, M_{qg} and M_{qq} proportional to the amplitudes for $gq \rightarrow q\gamma$ and $qq \rightarrow qq$, P_{gq} and $P_{\gamma q}$ Altarelli-Parisi probability functions and μ a regularization mass. In (8) the first part comes from the configuration (i) and the second from the configuration (ii). The quantity $P_{\gamma q}\log(s/\mu^2)$ is related with the fragmentation of a quark to a photon. The contribution of (8) to $p+p \rightarrow \gamma+X$ is obtained by convolution with two quark distributions $F_{q/p}(x,Q^2)$.

The result is shown in Fig. 2 (dashed lines) and compared with the Born contributions (subprocesses 4(a) and (b); 4(a) is calculated with a gluon distribution $F_{g/p}(x,Q_o^2) \sim (1-x)^5$). Clearly, as p_T increases at fixed s (as $x_T \rightarrow 1$) Bremsstrahlung becomes significant. The reason is that it involves valence quark distributions ($F_{q/p}(x,Q_o^2) \sim (1-x)^3$), which, as $x \rightarrow 1$, are larger than $F_{g/p}$. Of course, the relative importance of Bremsstrahlung depends on the input $F_{g/p}(x,Q_o^2)$.

3. Comparison with experiment.

We present calculations of $p+p \to \gamma+X$ with two sets of input parton distributions[4]: One with a "weak glue":

$$F_{g/p}(x,Q_o^2) \sim (1-x)^5 \qquad (9)$$

and one with a "strong glue":

$$F_{g/p}(x,Q_o^2) \sim (1+9x)(1-x)^4 \qquad (10)$$

Notice the factor 1+9x in (10), which enhances the predictions at intermediate x ($\sim x_T$), as well.[8] In both cases we introduce QCD scale violations and partons' primordial transverse momenta with a modest $\langle k_T \rangle = 0.5$ GeV. When we compare with data on γ/π^o we divide our predictions with the $p+p \to \pi^o+X$ cross section of the same collaboration.

Fig. 3 shows comparison with data on γ/π^o of the A^2BC collaboration.[9] Our predictions are somewhat below, not only for a weak glue (dashed lines), but also for a strong glue (solid lines). There are several ways one can improve agreement: (i) Use of an even stronger glue, which appears to be consistent with CDHS analyses;[10] this is done in Refs. 11. (ii) Use of larger $\langle k_T \rangle$; this is done in Ref. 12 that uses $\langle k_T^2 \rangle = 1\ GeV^2$. (iii) Addition of higher twists; this is done by Horgan and Scharbach.[13] In fact, with their normalization ($a_M = 1.3\ GeV^2$) these authors achieve values of γ/π^o somewhat above the A^2BC data.[14]

We would rather await before any further action, for the following reason: In Fig. 4, together with our predictions and the data of A^2BC (all given in terms of $Ed\sigma(pp \to \gamma X)/d^3p$) we also give data of the CCOR collaboration[15] (squares, $\sqrt{s} = 63$). We see that they agree with our predictions and, in fact, support a weak glue at not too large p_T.

Nevertheless we may conclude the following:

(A) On the whole, the data are in fair agreement with the QCD predictions and far above those of VDM.[16]

(B) Data on $p+p \to \gamma+X$ provide a good determination of the gluon distribution in the proton. A^2BC data support a strong glue,[17] but we feel that a weak glue cannot yet be excluded (in view of CCOR).

4. Direct photon production with pion and antiproton beams.

Significant yields of large-p_T photons arise also from π^- beams on protons. Some ambiguity in the valence quark distribution in the pion is now settled by the CERN NA3 experiment that gives [18]

$$F_{u/\pi^+}(x,Q^2) \sim \sqrt{x}(1-x) \qquad Q^2 \simeq 20\ \text{GeV}^2 \tag{11}$$

This gives e.g. at $p_T = 6$ GeV: $(\gamma/\pi^o)_{QCD} \simeq 0.2 \sim 0.3$ at $\sqrt{s} \simeq 20$ and $(\gamma/\pi^o)_{QCD} \simeq 0.25 \sim 0.35$ at $\sqrt{s} \simeq 27$ GeV.

In $\bar{p}+p \to \gamma+X$ we predict even larger photon yields (Fig. 5).[6] The reason is that both the Born terms Eq. (4) (a) and (b) give large contributions. Here the Bremsstrahlung correction is negligible.

5. Opposite-side γ-hadron correlations and the gluon fragmentation function

Large-p_T hadron (h) production opposite-side a photon trigger may offer the best determination of the gluon fragmentation function $D_{h/g}$.[6][12] Consider

$$A+B \to \gamma(p_{T1}) + h(p_{T2}) + X \tag{12}$$

with γ and h at 90° and in opposite directions ($\phi \simeq 180°$). We consider distributions in transverse momentum sharing (x_e).[1] In the present case $x_e \simeq p_{T2}/p_{T1}$. Let $d\sigma_{ab}/d\hat{t}$ be the cross-section[(2)-(4)] for the subprocess $a+b \to c+\gamma$ (i.e. either of the two subprocesses (4) (a) and (b); then we find that the x_e-distribution for (12) is proportional to the quantity:[6]

$$S_A(h) \equiv \frac{2}{x_e} \sum_{a,b,c} F_{a/A}(x_{T1},Q^2) F_{b/B}(x_{T1},Q^2) \frac{d\sigma_{ab}}{d\hat{t}} D_{h/c}(x_e,Q^2) \tag{13}$$

In general this receives contributions from both subprocesses (4) (a) and (b), so that both $D_{h/q}$ and $D_{h/g}$ are involved. However, we can isolate $D_{h/g}$ by taking differences of appropriate x_e-distributions. Take B = proton (target) and consider the case $h=\bar{h}$ (e.g. $h=\pi^o$ or $h=\pi^++\pi^-$). Then

$$S_A(h)-S_{\bar{A}}(h) = \frac{2}{x_e} D_{h/g} \sum_q \frac{d\sigma_{q\bar{q}}}{d\hat{t}} (F_{q/A}-F_{\bar{q}/A})(F_{\bar{q}/p}-F_{q/p}) \tag{14}$$

Notice that the r.h. side involves only valence quark distributions, so that (14) is not affected by uncertainties in sea. This allows a good determination of $D_{h/g}$ in terms of data on $S_A(h)-S_{\bar{A}}(h)$.

For $h=\pi^++\pi^-$ and $A=\pi^-$ or A = proton, reasonable choices of $D_{h/g}$ give an estimate of $S_A(h)-S_{\bar{A}}(h)$ in the order of $10 - 10^{-3}$ nb, depending on p_{T1}, x_e, beam A and on the type of geometry in the experiment.

It is a pleasure to thank J. Rosner for arousing my interest in large-p_T photon physics.

REFERENCES AND FOOTNOTES

1. P. Darriulat, Ann. Rev. of Nucl. and Particle Science (to appear). - M. Jacob and P. Landshoff, Phys. Reports 48, 285 (1978).
2. R. Rückl, S. Brodsky, and J. Gunion, Phys. Rev. D18, 2469 (1978).
3. F. Halzen and D. Scott, Phys. Rev. Lett. 40, 1117 (1978); Phys. Rev. D18, 3378 (1978) and D21, 1320 (1980).
4. A.P. Contogouris, S. Papadopoulos, and M. Hongoh, Phys. Rev. D19, 2607 (1979).
5. P. Aurenche and J. Lindfors, Nucl. Phys. B168, 296 (1980).
6. A.P. Contogouris, S. Papadopoulos, and C. Papavassiliou, contributed paper #400; see also A.P. Contogouris, J. Kripfganz et al #402.
7. There is much confusion concerning the magnitude of the Bremsstrahlung. This is partly due to the fact that the results of Ref. 5, although in agreement with ours, are presented in a way that is misleading and gives the impression that the whole contribution from (8) is always completely negligible. Ref. 5 does not count as correction the second part of (8) (configuration (ii)). However, this part also produces observable photons in addition to those of the subprocesses (4) (a) and (b).
8. For a derivation of a glue very similar to Eq. (10) see M. Glück, E. Hoffmann, and E. Reya, Dortmund DO-TH 80/13.
9. M. Diakonou et al, Phys. Lett. B91, 296 (1980) and C. Kourkoumeli, L. Resvanis et al, contributions to this Conference.
10. J. de Groot et al, Phys. Lett. B82, 45 (1979); Z. Phys. C (Particles and Fields) 1, 143 (1979). See also Ref. 8.
11. K. Kato and H. Yamamoto, contributed paper #469. - F. Halzen et al, #135. - R. Baier, J. Engels and B. Peterson #302.
12. L. Cormell and J. Owens, #245. This work also uses as large variable $Q^2 = 2\hat{s}\hat{t}\hat{u}/(\hat{s}^2+\hat{t}^2+\hat{u}^2)$, which is known to somewhat enhance the predictions over e.g. our choice $Q^2 = 2p_T^2$.
13. R. Horgan and P. Scharbach, CERN TH.2836 (1980).
14. G. Farrar and G. Fox, Nucl. Phys. B167, 205 (1980) argue that $a_M \sim 10^{-2}\,GeV^2$
15. A. Angelis et al, CERN-EP/80-68 and contribution to this Conference.
16. Notice also that since the subprocesses (4) (a) and (b) involve gluons verification of the QCD predictions is an important test of the existence of the gluon.
17. Recent data on the associated charged particle multiplicity in direct γ events show rather few hadrons alongside the γ trigger (M. Diakonou et al Phys. Lett. B91, 301 (1980). This is certainly an argument against a large Bremsstrahlung contribution but not against a modest one.
18. J. Lefrançois, Rapporteur review (these Proceedings, Session T3).

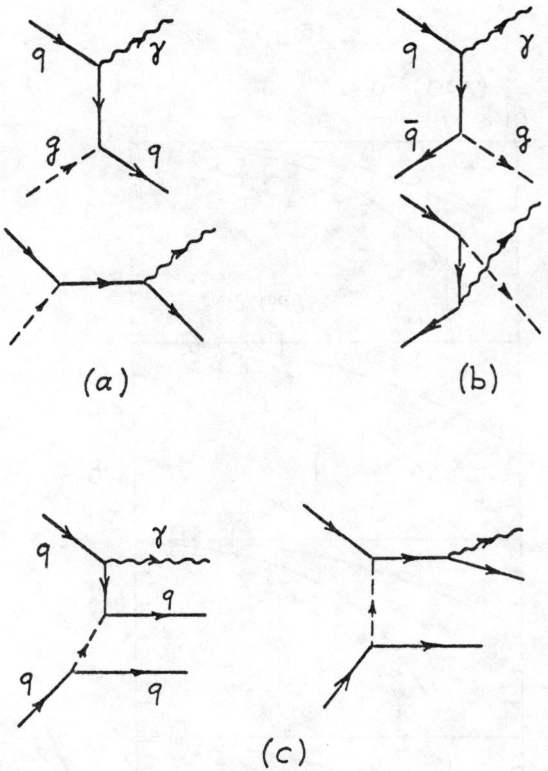
q
γ
g
q
(a)
q
γ
q̄
g
(b)
q
γ
q
q
q
(c)

Fig. 1

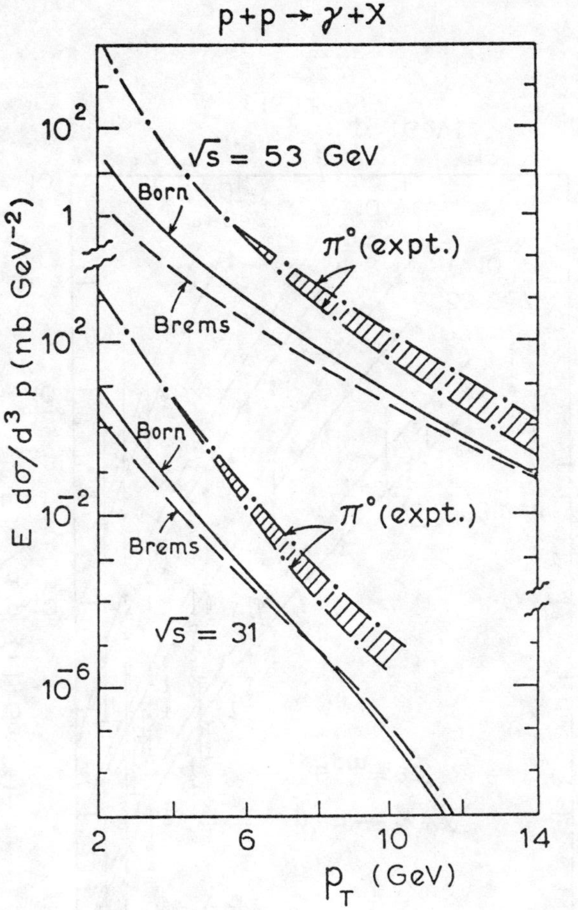
p+p → γ+X
$\sqrt{s}$ = 53 GeV
Born
Brems
π° (expt.)
Born
Brems
π° (expt.)
$\sqrt{s}$ = 31
E dσ/d³ p (nb GeV⁻²)
10²
1
10²
10⁻²
10⁻⁶
2
6
10
14
p_T (GeV)

Fig. 2

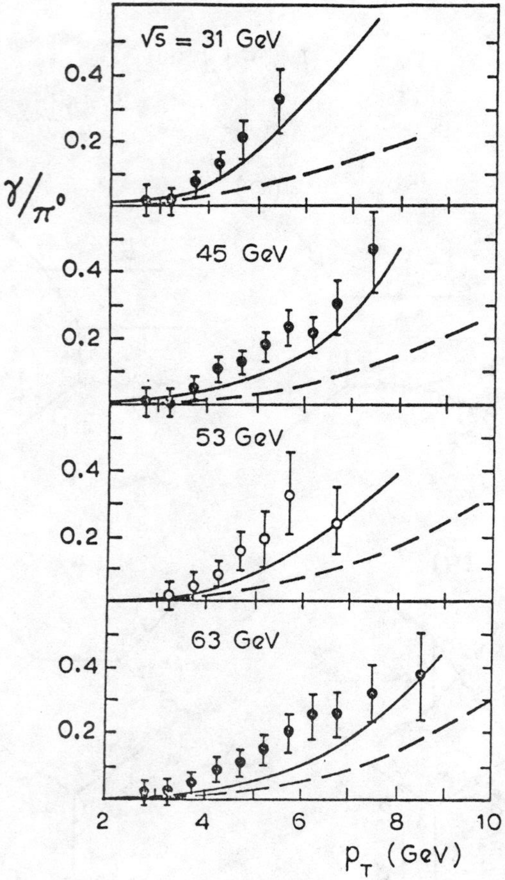

√s = 31 GeV
45 GeV
53 GeV
63 GeV
γ/π°
0.4
0.2
2
4
6
8
10
p_T (GeV)

Fig. 3

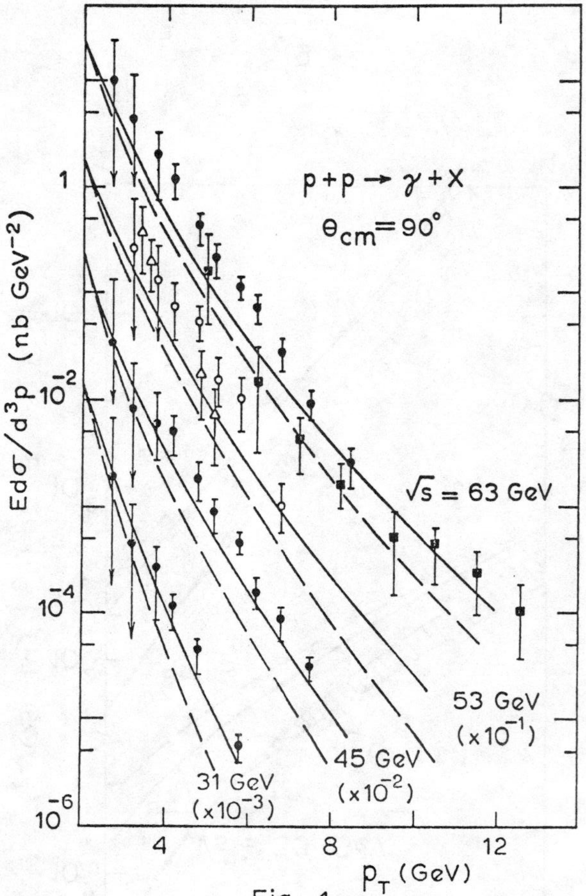

p+p → γ+X
θcm=90°
Edσ/d³p (nb GeV⁻²)
1
10⁻²
10⁻⁴
10⁻⁶
√s = 63 GeV
53 GeV
(x10⁻¹)
45 GeV
(x10⁻²)
31 GeV
(x10⁻³)
4
8
12
p_T (GeV)

Fig. 4

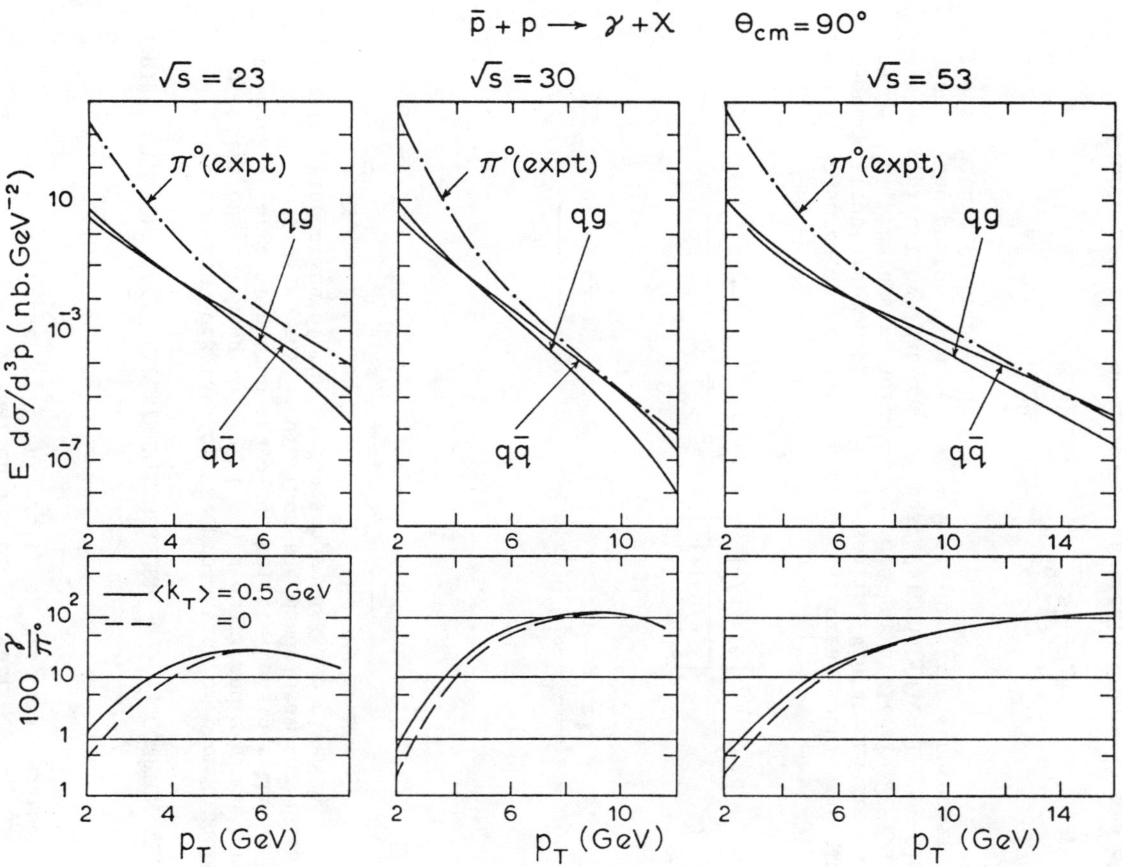
p̄ + p ⟶ γ + X θcm = 90°
√s = 23
√s = 30
√s = 53
π°(expt)
qg
qq̄
E dσ/d³p (nb.GeV⁻²)
10
10⁻³
10⁻⁷
100 γ/π°
10²
1
⟨kT⟩ = 0.5 GeV
= 0
pT (GeV)

Fig. 5

DIRECT PHOTONS

F. Halzen and D. M. Scott
Physics Department, University of Wisconsin-Madison

DIRECT PHOTON[1,2,3,4] EXPERIMENTS MEASURE $gq \to \gamma q$ COMPTON SCATTERING

The dominant diagram for the production of prompt photons in pp collisions is shown in Fig. 1. A u-quark and a gluon form back-to-back a large p_T photon and u-quark jet. Consider the case where their rapidities $y_\gamma = y_u = 0$. Then the Mandelstam variables defining the kinematics of the parton subprocess are given by

$$\hat{s} = x_T^2 s, \quad x_T = \frac{2p_T}{\sqrt{s}} \tag{1}$$

$$-\frac{\hat{t}}{\hat{s}} = -\frac{\hat{u}}{\hat{s}} = \frac{1}{2} \tag{2}$$

and

$$\frac{d\sigma}{dy_\gamma dy_u dx_T} = g(x_T, Q^2)\,[x_T u(x_T, Q^2)]\,\frac{d\hat{\sigma}}{d(-\frac{\hat{t}}{\hat{s}})} \tag{3}$$

$$\frac{d\hat{\sigma}}{d(\frac{\hat{t}}{\hat{s}})} = \frac{\pi\alpha\alpha_s}{3} e_u^2 \frac{1}{\hat{s}} \left[\frac{1+(-\hat{u}/\hat{s})^2}{(\hat{u}/\hat{s})}\right] . \tag{4}$$

$d\sigma/dx_T$ measures quark-gluon Compton scattering modulo the gluon and u-quark structure functions, their scaling violations are controlled by $Q^2 \simeq p_T^2$, the exact choice being somewhat arbitrary. Figs. 2 and 3 show explicit evaluations[4] of Eqs. (3) and (4). The p_T-distribution of the photon with $y_\gamma = y_u = 0$ measures the subprocess (given the structure functions) at 90°

$$\frac{d\hat{\sigma}}{d\hat{t}} = \frac{5\pi}{6} \alpha\alpha_s\, e_u^2 \frac{1}{\hat{s}^2} . \tag{5}$$

Single-particle inclusive cross sections $pp \to \gamma X$ are obtained by integrating over the opposite side quark rapidity. The result agrees with the data[5] shown in Fig. 4.

All the calculations shown in Figs. 2-4 include small contributions from d- and sea quarks. u dominates d by a factor 2 to 1 in proton valence content and by a charge factor 4 to 1 because of e_q^2 appearing in Eq. (4). Sea quarks can also contribute via the supplementary process $q\bar{q} \to \gamma g$. One a priori expects sizeable contributions from this process in $\bar{p}p$, $\pi^{\pm}p$ collisions. Except for $\pi^- p$, this is not the case because of the abundant gluon content of hadrons which we will discuss further on. This process is, however, responsible for large enhancements in γ yields with $\bar{p}$ and π beams when x_T becomes large (see Fig. 5).

DIRECT PHOTONS "SEE" THE GLUONS

It probably has not escaped your attention that there is an alternative way to explore Eq. (3). Given $\hat{\sigma}$ and the well known quark distributions, direct photon cross sections measure the gluon content of the proton $g(x,Q^2)$. Analysis of the data teaches us that effectively $g(x) \sim (1-x)^{n_g}$ with $n_g = 5$, or if one properly includes the scaling violations $g(x,Q_o^2) \sim (1-x)^3$, which yields an approximate $(1-x)^5$ distribution in the kinematic range of the data. Old-fashioned counting rules $g(x,Q_0^2) \sim (1-x)^5$ are ruled out. The gluon content of the proton is large, in fact, for $x \lesssim 0.4$ there are more gluons than valence quarks in the proton. Fig. 6 quantitatively illustrates these points: the variation in direct photon cross sections by changing n_g from 3 to 5 is shown for calculations including and omitting scaling violations in the structure functions.

THIS IS "THE" WAY TO DO HIGH-p_T PHYSICS WITH HADRON BEAMS

We list below our reasons of (probably too unqualified) support of this (probably too strong) statement:

(i) The structure of the underlying subprocesses is simple, basically one diagram.

(ii) No trigger bias: the trigger particle is not associated with the basic subprocess via a decay function, which is not computable in the theory. The trigger photon has a point coupling to the quark.

(iii) Do jet physics -- get one jet free (the photon): the ultimate purpose of any jet experiment is to directly measure parton subprocesses. This can be achieved here by just measuring the direction of the opposite-side jet (see Fig. 1). The momentum of the jet does not have to be experimentally determined, which avoids cloudy issues regarding its invariant mass.

(iv) Intrinsic or form-factor p_T effects are significantly reduced: at e.g., $\sqrt{s} = 31$ GeV, $p_T = 6$ GeV; smearing with a Gaussian transverse form factor is a factor 2 effect in the inclusive π^0 cross sections. It affects the γ yield only at the 60% level and is less than a 20% effect in γ/π.

(v) Direct photon cross section measurements allow us to perform clean factorization tests of perturbative QCD: a) large p_T Drell-Yan pairs: the relation is illustrated in Fig. 7a. This is, in fact, how we originally obtained our large direct photon prediction: from a QCD calculation, successfully describing the production of lepton pairs of mass m^2 and transverse momentum p_T^2, one removes the $\gamma^* \to \mu^+\mu^-$ conversion and takes the limit $m^2 \to 0$. We thus anticipated the large gluon content of the proton by insisting on simultaneously describing the Fermilab data on large p_T leptons pairs. Note that this is a corner of Drell-Yan physics which has not been discredited (yet?) by the "constant term crisis", which comes about only after integrating over the dilepton p_T^2. One can still test color factors here. Also, the non-leading term for direct photon production, $q\bar{q} \to \gamma g$, has a simple factorization property with the integrated Drell-Yan cross section (see Fig. 7b). Both are of the form

$$\sum_q e_q^2 \; q(x) \; \bar{q}(x) \quad \text{[computable cross sections]}$$

b) It would, of course, be interesting to study the decay functions of the opposite side quark jet (see Fig. 7c). If z is the fraction of quark momentum (which is equal to the γ momentum p_T for $\theta = 90^o$) carried by the hadron, then

$$\frac{dN}{dz} = \frac{\sum e_q^2 \; q(x) \; D_q^h(z)}{\sum e_q^2 \; q(x)} . \tag{6}$$

This is exactly the same combination of structure functions one measures in the inclusive leptoproduction of hadrons, as can be seen by comparing Figs. 7c, d. The abundance of u-quarks will result in an enhancement of positive over negative particles as shown in Fig. 8. Predictions are for $pp \to \gamma h^\pm X$, data from $\mu p \to \mu h^\pm X$.

COMMENT: BREMSSTRAHLUNG PHOTONS

In $O(\alpha_s^2)$ diagrams exist[4,6] where quarks radiate photons after being scattered to large angles. The bremsstrahlung diagram where a quark scatters off a gluon and subsequently radiates a photon is the only sizeable contribution with an opposite-side gluon. It can be further isolated by measuring the charge flow opposite the trigger photon (see Fig. 9). It does not look like the easiest experiment, but the price is worth the effort: the three-gluon coupling.

COMMENT: HIGHER TWISTS

Because of lack of any alternatives, let us assume, for the sake of argument, that the old-fashioned CIM model provides us with a parametrization of higher twists. It provides us with a way to include hadronic wave function effects in a specific calculational framework. Direct photons are predominantly produced via the process $Mq \to \gamma q$. Here M is a meson state contained in the colliding hadrons. This diagram can in principle explain the observed levels of γ's. One just has to require that the effective coupling of the meson M to the proton be $\alpha_M = 1 \sim 2$ (see Fig. 10). This unpleasant higher twist interference can fortunately be diffused on the basis of recent data[7] measuring $\pi p \rightarrow \pi^{\pm} X$. In CIM $\pi^- \rightarrow \pi^-$ is preferred over the identity change $\pi^- \rightarrow \pi^+$ by about a factor 10 if $\alpha_M \simeq 1 \sim 2$. This effect is not observed implying $\alpha_M < 1$ (see Fig. 11). We, therefore, believe that higher twists (at least á la CIM) will not spoil QCD phenomenology with direct photons.

COMMENT: DIRECT LEPTONS

We can now reverse the old speculations connecting direct leptons and direct photons.[3] Abundant large p_T photons will convert into abundant prompt leptons with $(\ell/\pi) \gg 10^{-4}$ (see Fig. 12).

REFERENCES

1. E. L. Feinberg, Izv. Akad. Nauk SSSR, Ser. Fiz. 26, 622 (1962).
2. F. Halzen and D. M. Scott, Phys. Rev. Lett. 40, 1117 (1978); Phys. Rev. D18, 3378 (1978); R. Ruckl, S. Brodsky, and J.Gunion, Phys. Rev. D18, 2469 (1978); A. P. Contogouris, S.Papadopoulos, and M. Hongoh, Phys. Rev. D19, 2607 (1979); R. Field, in Proceedings of the 19th International Conference on High Energy Physics, Tokyo, 1978, edited by S. Homma, M. Kawaguchi, and H. Miyazawa (Phys. Soc. of Japan, Tokyo, 1979); lectures at La Jolla Workshop, 1978, Caltech Report No. CALT-68-696 (unpublished); D. Jones and R. Ruckl, Phys. Rev. D20, 232 (1979); S. Petravca and F. Rapuano, Phys. Lett. B88, 167 (1979).
3. Previous related speculations include P. V. Landshoff (private communication); C. O. Escobar, Nucl. Phys. B98, 173 (1975); Phys. Rev. D15, 355 (1977); G. R. Farrar and S. C. Frautschi, Phys. Rev. Lett. 36, 1017 (1976); E. L. Feinberg, Nuovo Cimento 34A, 391 (1976); G. R. Farrar, Phys. Lett. 67B, 337 (1977); F. Halzen, Rutherford Report No. RS-77-049/A (unpublished); H. Fritzsch and P. Minkowski, Phys. Lett. 69B, 316 (1977).
4. This talk reviews a series of "second generation" papers on direct photon physics: F. Halzen and D. M. Scott, Phys. Rev. D21, 1320 (1980) and Univ. of Wisconsin Preprint COO-881-140; K. Kato and H. Yamamoto, Univ. of Tokyo Preprint UT-335; L. Cormell and J. F. Owens, Florida State Preprint; R. Baier, J. Engels and B. Petterson, Univ. of Bielefeld Preprint BI-TP 80/07; A. P. Contogouris, S. Papadopoulos and C. Papauassiliou, McGill Univ. Preprint, R. R. Horgan and P. N. Scharbach, CERN-TH-2836; A. Kotanski and J. Kubar, Univ. of Nice Preprint N TH 80/5.
5. For an extensive review of the data, see preceding session organized by L. Resvanis, these proceedings.

6. P. Aurenche and J. Lindfors, CERN-TH 2768.
7. H. J. Frisch et al., Phys. Rev. Lett. 44, 511 (1980).

FIGURE CAPTIONS

Fig. 1 Dominant diagram for the production of direct photons in hadron collisions.

Fig. 2 The cross section for $pp \to \gamma$ + jet + X at 90° in the center-of-mass for $\sqrt{s}$ = 27.4 and 63 GeV (Halzen et al., Ref. 4].

Fig. 3 The normalized distribution of the awayside jet in rapidity y_J in $pp \to \gamma$ + jet + X at $\sqrt{s}$ = 63 GeV for p_T= 6, 10 GeV [Halzen et al., Ref. 4].

Fig. 4 Comparison of direct photon cross sections, computed from lowest order QCD, and experimental π meson yields at large transverse momenta [Halzen et al., Ref. 2].

Fig. 5 Ratio of direct photon cross section for different beam particle types [Halzen et al., Ref. 2]

Fig. 6 Ratio of direct photon cross sections for two choices of n_g, defined as $g(x) \sim (1-x)^{n_g}$. The calculations are done including (solid line) and omitting (dashed line) scaling violations [Kato et al., Ref. 4].

Fig. 7 (a) How to calculate direct photon cross sections from lepton pairs with large p_T.
(b) Factorization of pair annihilation $q\bar{q} \to \gamma g$ with the Drell-Yan process.
(c) Inclusive production of hadrons opposite a direct photon trigger.
(d) Inclusive leptoproduction of hadrons.

Fig. 8 Radio of positive to negative hadrons opposite a direct photon ($pp \to \gamma h^{\pm} X$) as a function of z for p_T = 4, 8 GeV. The solid curves include photon bremsstrahlung. Data from leptoproduction [Baier et al., Ref. 4].

Fig. 9 Mean charge (units of 1/3) recoiling against a direct photon as a function of the away-side angle. Solid line shows the prediction with the three-gluon coupling removed [Horgan et al., Ref. 4].

Fig.10 CIM calculation compared to data on γ/π [Halzen et al., Ref. 4].

Fig.11 Comparison of $\pi p \to \pi^{\pm} X$ data with CIM and QCD prediction [H. J. Frisch et al., Ref. 13].

Fig.12 Rise of e/μ at large p_T due to conversion of direct photons [Halzen et al., Ref. 4].

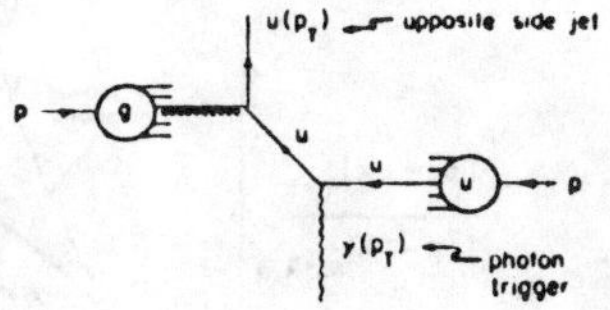

Fig. 1

Fig. 2

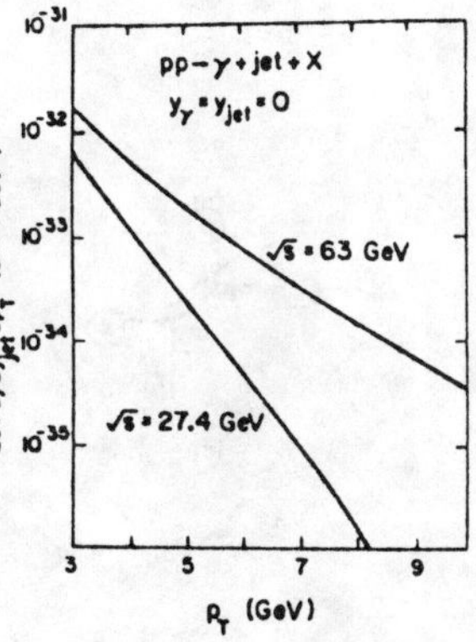

Fig. 3

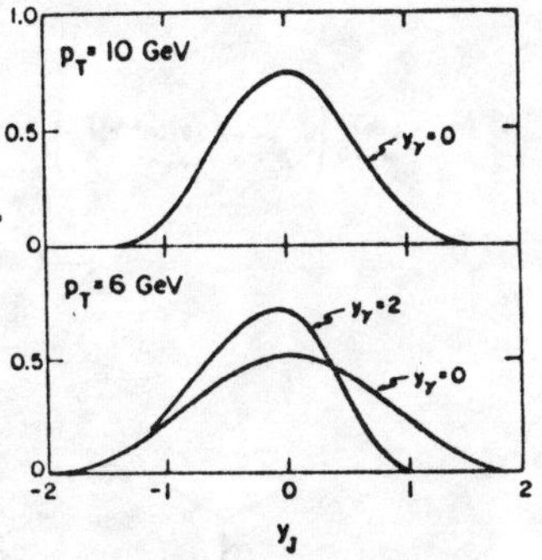

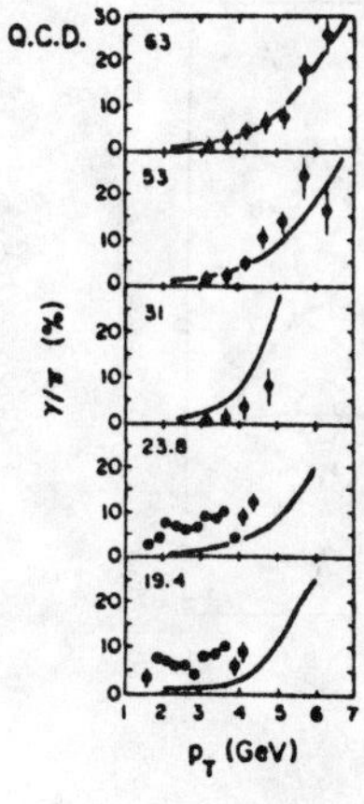

Fig. 4

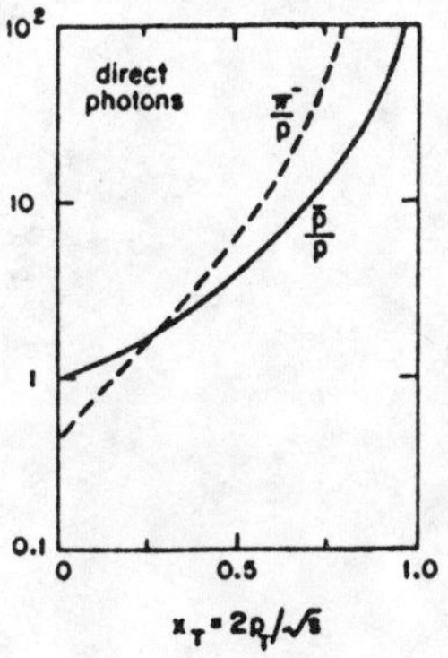

Fig. 5

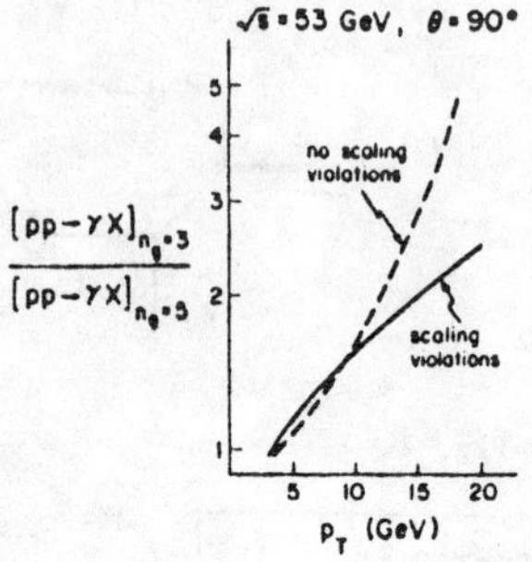

Fig. 6

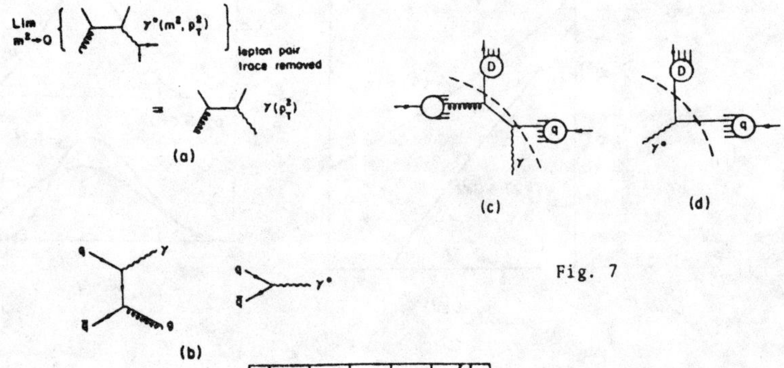

Fig. 7

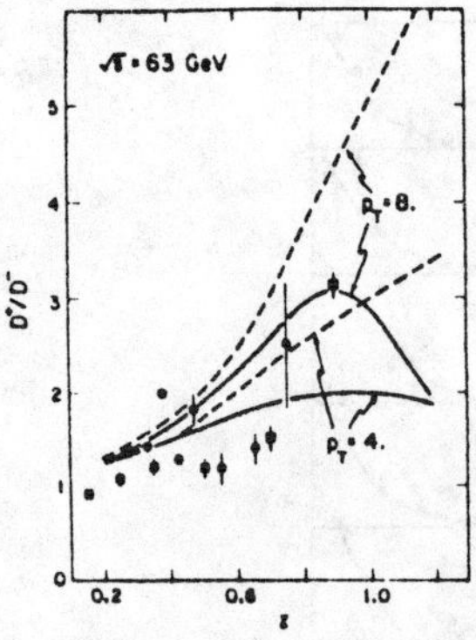

Fig. 8

Fig. 9

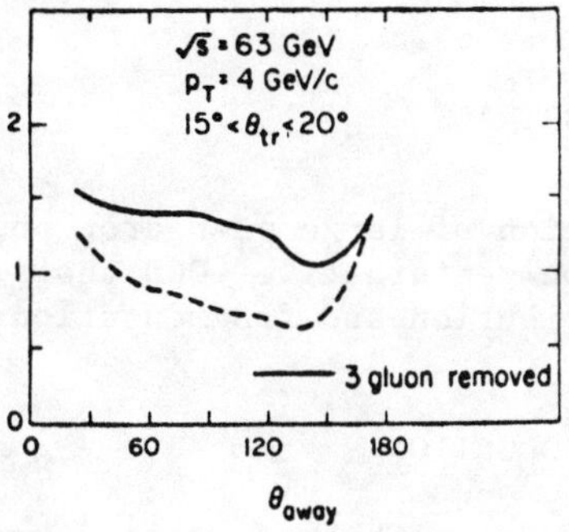

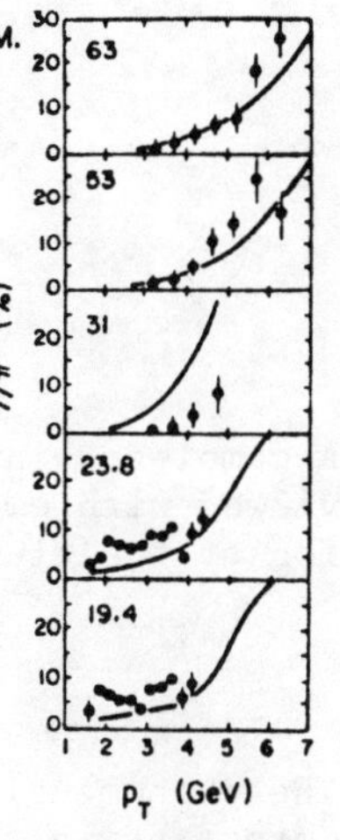

Fig. 10

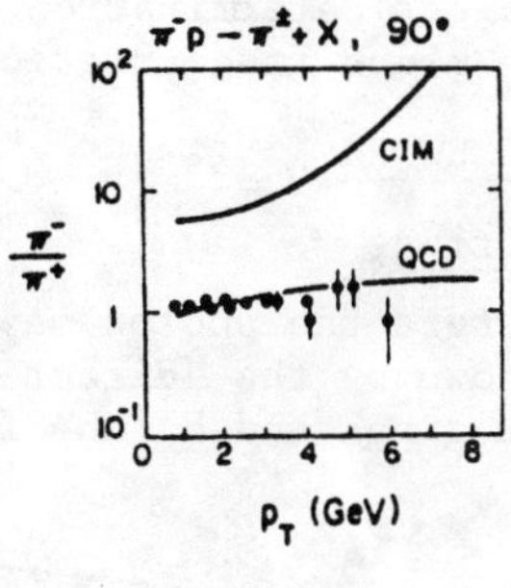

Fig. 11

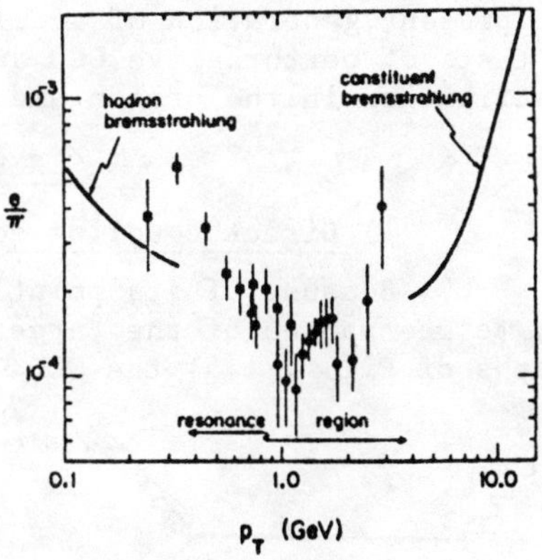

Fig. 12

PERTURBATIVE QCD AND LARGE P_T PHOTOPRODUCTION

M. Fontannaz*, A. Mantrach*, B. Pire**, D. Schiff*

*LPTHE, Université Paris-Sud, 91405 Orsay Cedex, France

**CPT, Ecole Polytechnique, 91128 Palaiseau, France

Presented by D. Schiff

ABSTRACT

A complete investigation of large p_T hadron photoproduction is reviewed with emphasis on perturbative QCD tests and on the determination of the gluon distribution and fragmentation functions.

INTRODUCTION

The phenomenology of Quantum Chromodynamics[1] has achieved tremendous progress in the understanding of hard scattering processes The case of large p_T photoproduction[2] is particularly interesting since gluons appear already at the Born term level. We show here that in the present generation of accelerator experiments it will yield precise tests of perturbative QCD and a direct determination of the gluon distribution in the proton and of the gluon fragmentation function.

GENERAL FRAMEWORK

1) Direct coupling contribution

Because of its pointlike nature the photon may couple directly to one parton of the target as shown by the "direct coupling" graphs of Fig. 1 : a) the QCD Compton graph and b) the fusion graph.

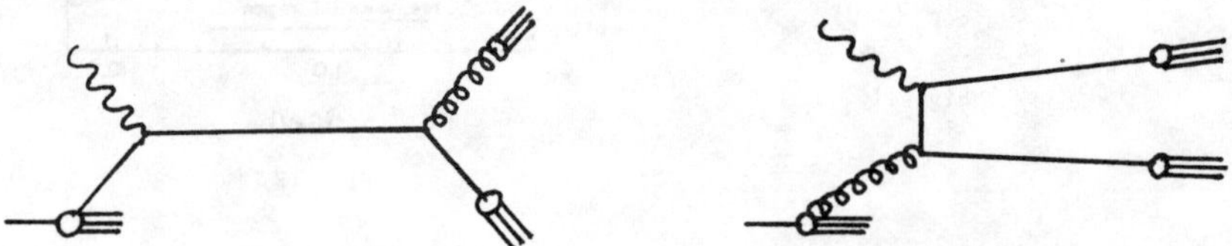

(a) (b)

fig. 1

These simple subprocesses have some very nice features :

a) differences of cross-sections $d\sigma(\gamma p \to h^+ X) - d\sigma(\gamma p \to h^- X)$

isolate the Compton graph of fig.1a while remaining of the same order of magnitude as individual cross-sections because of the electric charge factor in the coupling of the photon with the parton.

b) the internal Fermi motion of the constituents in the hadron has negligible effects on the cross-section, which is a very favorable situation as compared to hadron-hadron collisions.

c) we shall see below that, as they present a specific 3-jet topology, they are characterized by kinematical constraints in correlation distributions ; this property may be used to isolate these terms.

2) Anomalous photon distribution

The photon may convert into an almost colinear quark-antiquark pair ; this mechanism yields a structure function proportional to log Q^2 which is renormalized by the QCD emission of colinear gluons ; this is the anomalous photon structure function[3]. The interest of this contribution is that it is exactly calculable in the framework of perturbative QCD.

3) The hadronized photon contribution

The photon may hadronize in a quasi bound state of quarks and gluons. This yields a contribution to the structure function which is not determined by perturbative QCD calculations. It may be estimated in the framework of the Vector Dominance Model. We shall look for situations where this is suppressed compared to contributions 1) and 2) and may be viewed as a small background.

TESTS OF Q.C.D.

We are able to separate these different contributions and thus devise clear tests of QCD as shown in ref.4. First, the QCD Compton effect (fig.1a) may be isolated by studying the quantity $\Delta(\vec{p})$ defined as

$$\Delta^{\pi}(\vec{p}) = \Delta^{\pi}(\vec{p}_T,y) = E\frac{d\sigma}{d\vec{p}}(\gamma p \to \pi^+ X) - E\frac{d\sigma}{d\vec{p}}(\gamma p \to \pi^- X) \quad (1)$$

We find indeed that subtracting cross-sections as in eq.(1) allows to get rid of most of the anomalous and the hadronized photon contributions, the latter being not well under control. Moreover, the theoretical predictions for $\Delta^{\pi}(\vec{p})$ are not obscured by the uncertainties encountered in large p_T hadronic collisions, namely those due to the treatment of the internal Fermi motion of partons inside hadrons and those implied by the current ignorance of the gluon structure and fragmentation functions.

On fig.2 we compare the y integrated cross-section to already available data taken at energies between 45 and 70 GeV (see ref.6 for the detailed calculation). The direct coupling term dominates so that it is meaningful to compare $\Delta^{\pi}(\vec{p})$with the data. The difference between theory and experiment in the region $p_T > 2$ GeV/c is consistent with what is expected for the hadron-like background. Future experiments planned at the CERN-SPS and at FNAL with p_T up to 5 GeV/c will thus provide meaningful measurements of the QCD Compton process. Predictions for the rapidity distribution at $p_T = 3$ GeV/c and $s = 200$ GeV^2 are shown on fig.3.

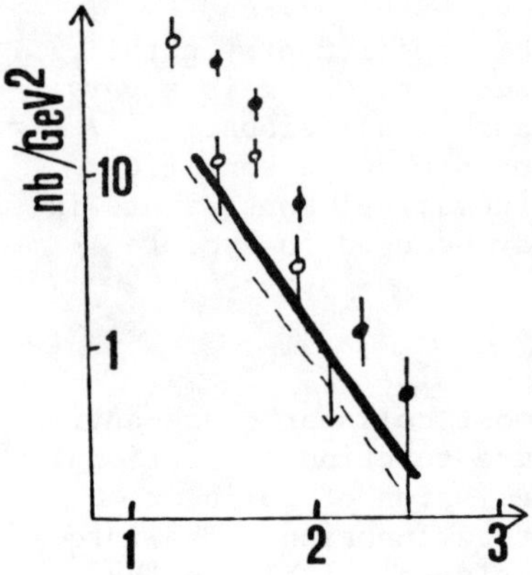

Fig. 2 : $d\sigma^{\pi^+ - \pi^-}/dp_T^2$ *for* $\theta_{lab} < 15°$ *(full curve and full dots) and* $y > .5$ *(dashed curve and open dots) for* $45 < E^\gamma < 70$ GeV. *Data are from ref. 5.*

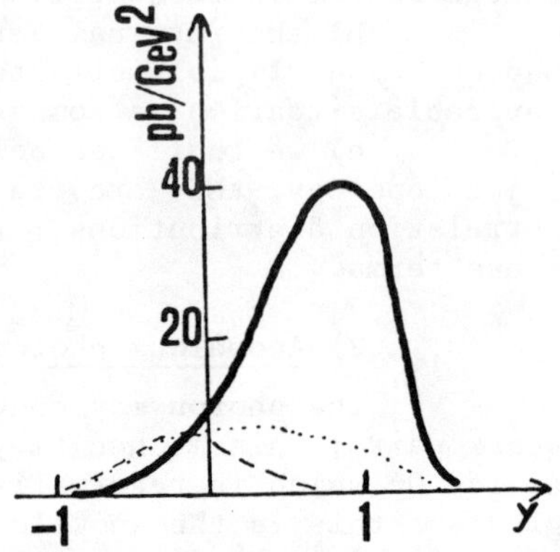

Fig. 3 : $d\sigma^{\pi^+ - \pi^-}/d\vec{p}_T\, dy$ *at* $p_T = 3$GeV/c *and* $s = 200$ GeV^2. *Compton term (full curve), anomalous (dashed curve), and VDM (dotted curve) contributions.*

GLUON FRAGMENTATION AND DISTRIBUTION FUNCTION DETERMINATIONS

A beautiful feature of large p_T photoproduction experiments is that they provide information about the gluon through the study of two particle (toward-away) inclusive distributions. Since the direct coupling term is the dominant contributions to cross-section differences, one may concentrate on it when discussing the following quantities.

$$\Delta^{\pm}(\vec{p}_1\vec{p}_2) = \frac{d\sigma(\gamma p \to \pi^+\pi^{\pm} X}{d\vec{p}_{T_1}\, dy_1\, dp_{2x} dy_2} - \frac{d\sigma(\gamma p \to \pi^-\pi^{\pm} X}{d\vec{p}_{T_1}\, dy_1 dp_{2x} dy_2} \,. \qquad (2)$$

The study of $\Delta^+(\vec{p}_1\vec{p}_2) + \Delta^-(\vec{p}_1\vec{p}_2)$ (resp. $\Delta^+(\vec{p}_1\vec{p}_2) - \Delta^-(\vec{p}_1\vec{p}_2)$ turns out to give a direct access to the gluon fragmentation (resp. distribution in the proton) function, since only the Compton (resp. the fusion) graph contributes.

1) Gluon fragmentation function

It is useful to define (p_T^{jet} is the outgoing jet transverse momentum) :

$$\frac{d\sigma^c}{dz_2\, dp_T^{jet}} = \int_{p_{T_1}^{min}} d\vec{p}_{T_1} \int dp_{2x}\, \delta\Big(z_2 - \frac{|p_{2x}|}{p_T^{jet}}\Big) \int_{y_1^{min}}^{y_1^{max}} dy_1\, dy_2$$

$$\delta\left(p_T^{jet} - \frac{\sqrt{s}}{e^{y_1} + e^{y_2}}\right) \cdot \left[\Delta^+(\vec{p}_1,\vec{p}_2) + \Delta^-(\vec{p}_1,\vec{p}_2)\right] \qquad (3)$$

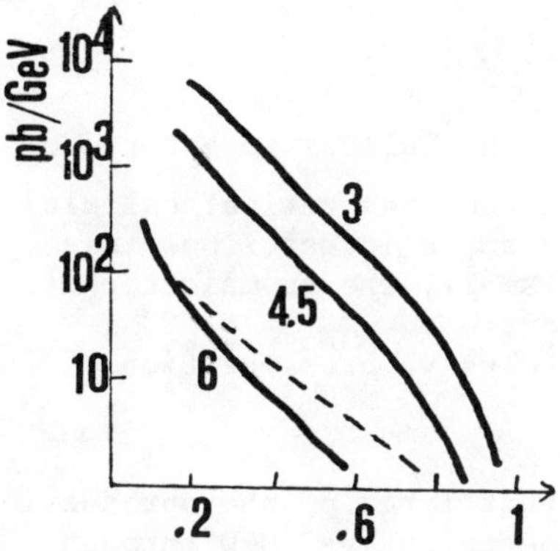

Fig.4 : $d\sigma^c/dz_2\, dp_T^{jet}$. *Full (dashed) curves correspond to the gluon fragmentation function of ref.9)(8)). Curves are labelled with the value of* p_T^{jet}. ($s = 200$ GeV^2).

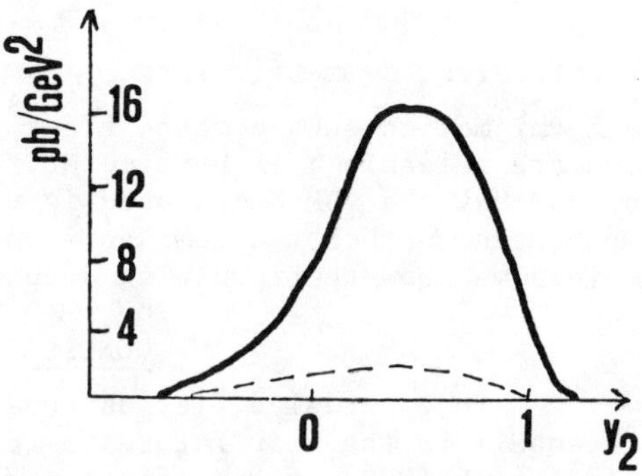

Fig.5 : $d\sigma/dp_{T1}\, dy_1\, dp_{2x}\, dy_2$ for $y_1 = 0$, $p_{T1} = p_{2x} = 3$ GeV/c and $s = 200$ GeV^2. *Full (dashed) curve is the anomalous (hadronized) contribution.*

which can be simply written as

$$\frac{d\sigma^c}{dz_2\, dp_T^{jet}} = D_g^{h_2}(z_2,\overline{Q}^2)\; d(p_{T_1}^{min}, p_T^{jet}, s, y_1^{min}, y_1^{max}) \tag{4}$$

with $\overline{Q}^2 \simeq (p_T^{jet})^2$ and d is straightforwardly calculated[7] from valence quark distribution and fragmentation functions. As shown in fig.4, $d\sigma^c/dz_2 dp_T^{jet}$ is fairly large and thus should allow a precise measurement of $D_g(z,Q^2)$. The scaling violation may also be investigated[7].

2) Gluon distribution function in the proton

A similar study may be performed using $\Delta^-(\vec{p}_1\vec{p}_2)-\Delta^+(\vec{p}_1\vec{p}_2)$.

We may define

$$\frac{d\sigma^F}{dx} = \int dy_1\, d\vec{p}_{T1}\, dp_{2x} \quad \frac{d\sigma^{\Delta^- - \Delta^+}}{dy\, d\vec{p}_{T1}\, dp_{2x}\, dx} \tag{5}$$

which is proportional to $G_g(x,\overline{Q}^2)/x$ where $\overline{Q}^2$ is the mean value of Q^2 in the integration range. Estimates of this quantity indicate the feasibility of the experimental measurement[7].

OBSERVATION OF THE ANOMALOUS PHOTON STRUCTURE FUNCTION

It is of great interest to measure the anomalous photon structure function. The strategy we propose is

- first, keep only 4-jet events, taking advantage of the kinematical constraint satisfied by direct coupling 3-jet events at the level of double inclusive jet cross-sections

$$d\sigma/dp_T^{jet}\ dy_1\ dy_2 \propto \delta(2 - x_T(e^{y_1} + e^{y_2}))$$

- then, select symmetric large p_T pairs. This allows to get rid of the Fermi motion enhancement effect[8] and thus theoretical calculations are reliable ; in particular, we discuss in ref.7 how this implies that the VDM background is very small, the anomalous photon component being then the dominant contribution.

In fig.5 we show the resulting double inclusive cross-section.

CONCLUSION

In conclusion, let us repeat that large p_T photoproduction will enable in the near future to test perturbative QCD through isolating the QCD Compton effect and extracting the anomalous photon structure function. Moreover, it will yield very important informations on some of the gluon properties upon which theorists have been speculating for a long time.

REFERENCES

1. For a review, see C.H. Llewellyn-Smith's talk at this conference.
2. H. Fritzsch and P. Minkowski, Phys. Lett. 69B (1977), 316. For a complete list of references, see ref.6.
3. E. Witten, Nucl. Phys. B120 (1977), 189 ; C.H. Llewellyn-Smith, Phys. Lett. 79B (1978), 83.
4. M. Fontannaz et al., Phys. Lett. 89B (1980), 263.
5. WA4 collaboration preliminary results.
6. M. Fontannaz et al., to be published in Z. Physic C.
7. M. Fontannaz et al., to be published in Phys. Lett.B. and Orsay preprint LPTHE 80/21.
8. R. Baier, J. Engels and B. Petersson, Z.Physik C2 (1979), 265.
9. J. Owens and E. Reya, Phys. Rev. D17 (1978), 3003.

*

* *

*

TRANSVERSE MOMENTUM DISTRIBUTION OF DIMUONS IN THE DRELL-YAN PROCESS

Davison E. Soper
University of Oregon, Eugene, OR 97403

John C. Collins
Illinois Institute of Technology, Chicago, IL 60616

ABSTRACT

We describe recent progress in the theory of the Q_T distribution in the Drell-Yan process for $0 \lesssim Q_T^2 \lesssim Q^2$. The basic theoretical result is in transverse position ($\underset{\sim}{b}$) space. In this talk we discuss in some detail how to get from $\underset{\sim}{b}$-space to $\underset{\sim}{Q}_T$-space.

INTRODUCTION

Drell and Yan[1] proposed in 1971 that the hadronic inclusive μ-pair production process $A+B \to \mu^+ + \mu^- + X$ would proceed predominantly at high energy through the annihilation of a quark from hadron A and an antiquark from hadron B. Recent theoretical and experimental work has tended to confirm this proposal. Thus the momentum Q^μ of the dimuon is the sum of the momenta of the two annihilating partons. The distribution of dimuons in Q_T measures the distribution of the partons in transverse momentum.

In the original parton model days it was thought that the transverse momentum distribution would be steeply falling for Q_T bigger than 300 MeV or so. Later it was recognized[2] that in a renormalizable field theory like QCD the Q_T distribution would spread out and fall off slowly at very high energy, like $1/Q_T^2$ times logarithms of Q_T for $1\ \mathrm{GeV}^2 << Q_T^2 << Q^2$. More recently Parisi and Petronzio[3] argued on the basis of a soft gluon emission picture that

$$\frac{d\sigma}{d^2Q} \propto \frac{1}{Q_T^2}\,\frac{\ln(Q^2/Q_T^2)}{\ln(Q_T^2/\Lambda^2)}\,\exp\left[-\frac{4}{3}\,\frac{\alpha_s(Q^2)}{2\pi}\,\ln^2\left(\frac{Q_T^2}{Q^2}\right)\right]$$

in an approximation in which possible corrections of order $\alpha_s(Q^2)\,\ln(Q_T^2/Q^2)$ are neglected.

We are presently completing a more detailed investigation of this process in which corrections of the form $\alpha_s(Q^2)^N\,\ln^M(Q_T^2/Q^2)$ are not neglected. Following Parisi and Petronzio, we work in transverse coordinate space. Let $H_F(b^2)$ be the fourier transform of the cross section and define $w = \ln(Q^2/\Lambda^2)$ and $z = \ln(4/b^2\Lambda^2)/\ln(Q^2/\Lambda^2)$. Then one is concerned with large w; $z \approx 0$ is the "confinement region" about which we cannot say anything, $z \approx 1$ is the region in which ordinary renormalization group improved perturbation theory is useful, and $0 < z < 1$ is the more difficult two scale region that we are interested in. We find that

$H_F(\underset{\sim}{b}^2)$ has the form

$$H_F(b^2) = \frac{1}{4\pi} e^{-wI(z)} R(w,z) \tag{1}$$

where

$$I(z) = \frac{16}{25} \int_z^1 d\bar{z} \frac{1-\bar{z}}{\bar{z}} \tag{2}$$

and R(w,z) can be computed (except for its normalization, which is proportional to the quark x-distributed) in a perturbation scheme in which the error after N terms is of order w^{-N} (all terms of order $1/Q^2$ are neglected from the start). We find that ℓn R(w,z) grows only like ℓn w for large w.

There are certain unproved assumptions made in this calculation that are related to the cancellation of infrared divergences in QCD processes with colored particles in the initial state, which is a difficult unsolved problem. (See the talk of Frenkel at this conference.) The crossed version of the Drell-Yan process, $e^+ + e^- \to A + B + X$ with nearly back to back hadrons A and B, should be free of these possible difficulties.

In this talk we assume the validity of the form (1) for the cross section in transverse position space and show how to extract the leading form for the cross section in momentum space. The essential physical point was first made by Parisi and Petronzio: as $Q^2 \to \infty$, the function $H_F(b)$ falls off so fast for large $\underset{\sim}{b}$ that the small $\underset{\sim}{b}$ region, in which eq. (1) for $H_F(\underset{\sim}{b})$ is reliable, dominates the cross section even at small $\underset{\sim}{Q}_T$.

CALCULATION

To calculate $H(\underset{\sim}{Q}_T)$ we use an intermediate step. Define the mellin transform

$$H_M(s) = \int_0^\infty \frac{db^2}{b^2} \left(\frac{\Lambda^2 b^2}{4}\right)^s H_F(b^2) \tag{3}$$

then

$$H(\underset{\sim}{Q}_T) = \frac{4\pi}{\underset{\sim}{Q}^2} \frac{1}{2\pi i} \int_{\frac{1}{2}-i\infty}^{\frac{1}{2}+i\infty} ds \frac{\Gamma(1-s)}{\Gamma(s)} \left(\frac{Q^2}{\Lambda^2}\right) H_M(s). \tag{4}$$

There are two steps, $H_F \to H_M$ then $H_M \to H$.

In the first step we insert the form (1) for H_F into eq. (3):

$$H_M(s) = \frac{w}{4\pi} \int_{-\infty}^{\infty} dz\, e^{-w(zs+I(z))} R(w,z).$$

For $w \to \infty$ with fixed s, we can use the saddle point approximation. (Here it is essential that R(w,z) grows only like ℓn w as $w \to \infty$). The saddle point is at $z_0(s) = A/(A+s)$, A = 16/25 in four flavor QCD. The result of the saddle point approximation is

$$H_M(s) \sim \sqrt{\frac{Aw}{8\pi}} \frac{1}{A+s} e^{-w(z_0 s+I(z_0))} R(w,z_0) \tag{5}$$

with computable 1/w corrections.

In the second step, $H_M \to H$, we insert (5) into (4). Using the variable $u = \ell n(Q_T^2/\Lambda^2)/\ell n(Q^2/\Lambda^2)$ one has

$$H(u) \sim \frac{\sqrt{2\pi A w}}{Q_T^2} \frac{1}{2\pi i} \int_{\frac{1}{2}+i\infty}^{\frac{1}{2}+i\infty} ds \, \frac{\Gamma(1-s)}{\Gamma(s)} \frac{1}{A+s} R(w, z_o(s)) \tag{6}$$

$$\times \exp\left(-w\left[(z_o(s)-u)s + I(z_o(s))\right]\right)$$

There are three cases to consider. First, for $w \to \infty$ with $u > 16/41$ one can again use the saddle point approximation. The saddle point is at $s_o(u) = (16/25)(1-u)/u$; there $z_o(s_o(u)) = u$. The result is

$$H(u) \sim \frac{1}{Q_T^2} \frac{\Gamma\left(1 - \frac{16}{25}\frac{1-u}{u}\right)}{\Gamma\left(\frac{16}{25}\frac{1-u}{u}\right)} e^{-wI(u)} R(w,u) \tag{7}$$

In the second case, $u < 16/41$, the saddle point $s_o(u)$ lies to the right of the pole of $\Gamma(1-s)$ at $s = 1$. When the integration contour is moved past the pole so as to run through the saddle point, $H(u)$ picks up a contribution from the integral around the pole. This pole contribution dominates the saddle point contribution for large enough w.

Neither the $u > 16/41$ approximation, nor the $u < 16/41$ approximation is accurate when $u \approx 16/41$. (For practical values of w, $u \approx 16/41$ means $0 < u < 1$.) In this interpolating region one treats the pole of $\Gamma(1-s)$ exactly and obtains an error function. There is no one "best" interpolating approximation. One fairly simple approximation is

$$H(Q_T^2) \sim \frac{1}{Q_T^2} \sqrt{\frac{\pi A w}{2}} \frac{\Gamma(2-\hat{s})}{\Gamma(\hat{s})(A+\hat{s})} R(w,\hat{u}) \tag{8}$$

$$\times \exp\left(-w\left[I(\hat{u})+\hat{u}-u + \frac{u^2}{2A}(\hat{s}-s_o)^2\right)\right. e^{\phi^2} \mathrm{erfc}(\phi)$$

where

$$\phi = \frac{A+1}{\sqrt{2A}} \sqrt{w} \left(u - \frac{16}{41}\right)$$

$$\hat{u} = \max\left(u, \frac{16}{41}\right), \quad \hat{s} = \min(s_o, 1), \quad s_o = \frac{16}{25}\frac{1-u}{u}$$

$$A = 16/25; \quad w = \ell n(Q^2/\Lambda^2); \quad wu = \ell n(Q_T^2/\Lambda^2)$$

DISCUSSION

Several features of the result (7) and (8) as u varies from 1 to $-\infty$ deserve discussion. For $u \to 1$, the factor $\Gamma[(16/25)(1-u)/u]$ in the denominator produces a factor $(1-u) \propto \ell n(Q^2/Q_T^2)$, in agreement with the first order perturbative result. For somewhat smaller u one can see the exponential of a double log: $\exp(-wI(u)) \sim \exp[-\frac{8}{25} w(1-u)^2]$. As Q_T decreases toward 0, the effective inverse impact parameter $4/b_{eff}^2 = \Lambda^2(Q^2/\Lambda^2)^{\hat{u}}$ does not decrease past

$$(4/b_{eff}^2)_{min} = \Lambda^2(Q^2/\Lambda^2)^{16/41} .$$

Thus the cross section is determined by short distance effects even near $Q_T^2 = 0$. Finally both the shape and normalization of the cross section for small Q_T are very sensitive to the QCD scale parameter Λ^2. For instance, $H(Q_T^2)/H(0) \sim \frac{1}{2}$ at $Q_T^2 = (Q^2)^{16/41}(\Lambda^2)^{25/41}$.

REFERENCES

1. S. D. Drell and T.-M. Yan, Phys. Rev. Lett. 25, 316 (1970).
2. D. E. Soper, Phys. Rev. Lett. 38, 461 (1977).
3. G. Parisi and R. Petronzio, Nucl. Phys. B154, 427 (1979); see also Y. Dokshitser, D. D'Yakanov, and S. Troyan, Phys. Rep. 58, 270 (1980).

Chapter 2

New Particle Production

NEW PARTICLE PRODUCTION I

C. BROWN, FERMILAB, ORGANIZER

DIFFRACTIVE HADROPRODUCTION OF CHARMED D MESONS*

L. J. Koester, G. Alverson, G. Ascoli, D. Bender, J. Cooper,
L. Holloway, U. Kruse, W. MacKay, R. Sard, M. Shupe, E. Smith
University of Illinois at Urbana-Champaign, 61801

T. B. W. Kirk, R. Raja
Fermi National Accelerator Laboratory, Batavia, Illinois

A. Loomis, A. Sessoms, R. Wilson, C. Tao
Harvard University, Cambridge, Massachusetts 02138

J. Davies, T. Quirk
University of Oxford, Oxford, England OXI 3RH

R. Thornton, R. Milburn
Tufts University, Medford, Massachusetts 02155

ABSTRACT

We have observed charmed D meson production from a hydrogen target in a 217 GeV/c π^-beam. Hadronic decays in the $K^{\mp}\pi^{\pm}\pi^{\pm}$ channels were reconstructed in the Chicago Cyclotron Spectrometer at Fermilab. The proton recoil angle and momentum transfer were programmed so that objects of mass near that of two D mesons would be produced in the forward direction. A mass peak in the 40-MeV bin at 1875 MeV was found in each channel. The cross section for $D^{\pm}$ production is 6-10 μb.

METHOD

This experiment measured hadroproduction of charmed D mesons by reconstructing their hadronic decays in the Chicago Cyclotron Spectrometer. A 217 GeV/cπ^-beam impinged on a liquid hydrogen target. The trigger required a recoil proton between 60° and 75°, with momentum transfer in the interval $-0.1 \geq t \geq -0.4$ (GeV/c)^2, to be in coincidence with a muon from the semileptonic decay of one of the two D's produced. The events analyzed for this report were of the form

$$\pi^- p \rightarrow p\ D^+ D^- X^-;\ \text{with}\ D_h^{\pm} \rightarrow K^{\mp}\pi^{\pm}\pi^{\pm};\ D_{sl}^{\pm} \rightarrow \mu^{\pm}\ X^o.$$

The restrictions on the proton insured that the blob produced would have an invariant mass bracketing the mass of two D's and would go forward through the spectrometer. The particles were observed in multiwire proportional chambers, wire spark chambers, scintillation hodoscopes, and an 18-cell gas Cherenkov detector which discriminated between K and π mesons of momentum 8 to 26 GeV/c. The muon momentum had to exceed 15 GeV/c to penetrate the steel

*Research supported in part by the Department of Energy contract DE-AC02-76ER01195

absorber.

RESULTS

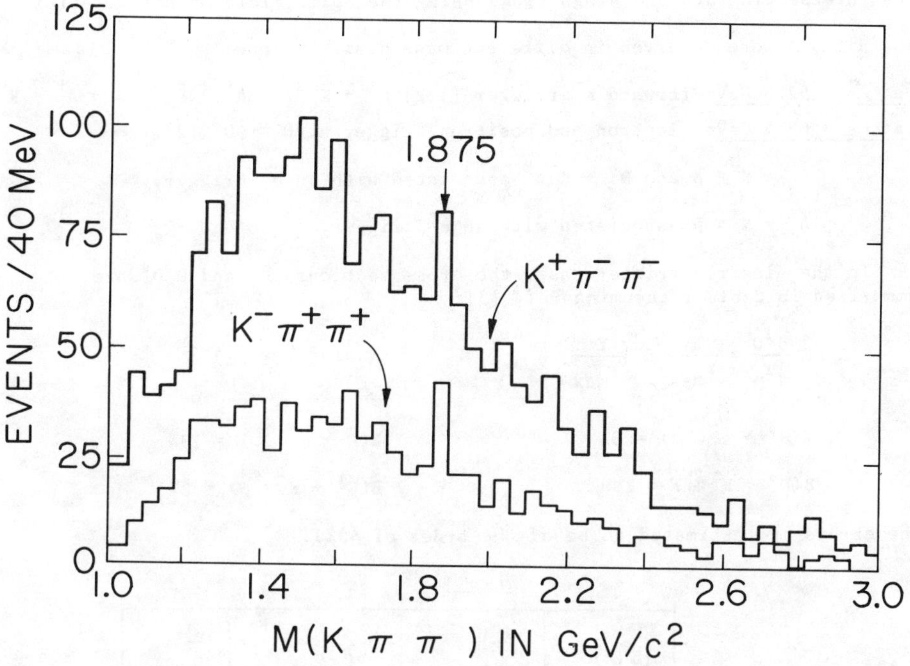

Fig. 1. Invariant mass distributions of reconstructed $D^+ \to K^-\pi^+\pi^+$ decays (lower curve) and, to the same scale, $D^- \to K^+\pi^-\pi^-$ decays (upper curve).

For both D^+ and D^- decays, a peak of about 25 events above background appears in a single 40-MeV bin at 1875 MeV corresponding to the known D mass (1868 MeV). No such peak appears in the Cabibbo forbidden D decay modes $K^-\pi^+\pi^-$ or $K^+\pi^+\pi^-$. The Feynman x of the observed D's is peaked at 0.4.

A Monte Carlo acceptance program based on a peripheral production model yielded, for the combined 50 ± 12 $D^\pm$ events, a production cross section σ = 6 to 10 μb.

PRODUCTION OF CHARM IN THE SFM AT THE ISR

CCHKK and ACCDHW Collaborations
Presented by: G. Sajot, Collège de France

Charm production has been observed in two data samples obtained at the CERN Intersecting Storage Rings (ISR) using the Split Field Magnet detector.

Signals were observed in different mass distributions:

- at $\sqrt{s}$ = 52.5 GeV: forward K^- trigger [1,2] $D^+ \to \bar{K}^{*0}\pi^+$, $\Lambda_c^+ \to \bar{K}^{*0}p$ and $\Lambda_c^+ \to K^-\Delta^{++}$.
- at $\sqrt{s}$ = 62.5 GeV: electron and positron trigger at $\theta \simeq 90°$ [2].

$\Lambda_c^+ \to K^-\pi^+p$ and $D^0 \to K^-\pi^+$ associated with an e^- trigger.

$\bar{\Lambda}_c^- \to K^+\pi^-\bar{p}$ associated with an e^+ trigger.

In the electron trigger case, the cross sections $\bar{C}\Lambda_c^+$ and $\bar{D}^0D^0$ are summarized in table 1 assuming ref. [3]:

$$E\frac{d^3\sigma}{d^3p} \propto \frac{1}{Y_{max}} f\left(\frac{Y}{Y_{max}}\right) e^{-Ap_T} \qquad A = 2.\ (\mathrm{GeV}/c)^{-1}$$

$B(\bar{D}^0 \to e^- \ldots) = 8\%$ $\qquad$ $B(\bar{C} \to e^- \ldots) = 10\%$

$B(D^0 \to K^-\pi^+) = 2.6\%$ $\qquad$ $B(G_c^+ \to K^-\pi^+p) = 2\%$

(the errors are estimated to be of the order of 40%).

TABLE 1

	$\sigma(\bar{D}^0D^0)$ μb	$\left.\frac{e^-}{\pi^-}\right)_{x=0}$ 10^{-4}	$\sigma(\bar{C}\,\Lambda_c^+)$ μb	$\left.\frac{e^-}{\pi^-}\right)_{x=0}$ 10^{-4}
$\frac{d\sigma}{dy}$ = cte	487	3.5	$\bar{C} = \bar{\Lambda}_c$ 637 $\bar{C} = \bar{D}$ 379	4.9 3.5
$\frac{d\sigma}{dx}$ = cte	1350	3.2	$\bar{C} = \bar{\Lambda}_c$ 1330 $\bar{C} = \bar{D}$ 1285	3.3 3.4
$\frac{d\sigma}{dx} \propto (1-x)^3$	400	4.7		
$\bar{D}$ $\frac{d\sigma}{dx} \propto (1-x)^3$ Λ_c^+ $\frac{d\sigma}{dx}$ = cte			254	3.7

REFERENCES

[1] D. Drijard et al., Phys. Lett. 81B (1979) 250.
[2] D. Drijard et al., Phys. Lett. 85B (1979) 452.
[3] G. Sajot, Ph.D.Thesis Paris 6 University, 21 April 1980.

PRODUCTION OF CHARMED PARTICLES AT THE CERN ISR

(Aachen-CERN-Harvard-Munich-Northwestern-U.C. Riverside)

presented by F. Muller (CERN)

The experiment used the LSM detector. A high-multiplicity diffractive trigger allowed the first observation[1] of hadronic production of the Λ_c and of its $K^-p\pi^+$ decay mode at a mass[2] $m = 2262 \pm 10$ MeV. New results are reported in two contributions at this Conference:

1) Cross-sections for diffractive charm production (J. Eickmeyer et al.)

The work described in Ref. 1 has been updated, yielding at $\sqrt{s} = 62.5$ GeV, in the 10-28 GeV range of the excited protonic mass M, $\Delta\sigma/\Delta x$ (mb) $= 0.24 \pm 0.12$ for the Λ_c in the 0.5-0.8 x-range, using $B(\Lambda_c \rightarrow K^-p\pi^+) = 0.022$. The 95% C.L. limits for D^0 ($\bar{D}^0$) are $\Delta\sigma/\Delta x < 0.16$ (0.32) mb, in the 0.2-0.65 and 10-28 GeV x and M ranges.

2) Production of charmed particles with an electron trigger (J. Irion et al.)

The $K^-p\pi^+$ mass distribution for events with an e^- candidate ($p_T > 0.4$ GeV/c at 30°) shows a sharp 44 events peak at $m \simeq 2.26$ GeV. A similar, though smaller (17 ev.), peak is seen in the $K^+\bar{p}\pi^-$ mass distribution obtained from e^+ events. No such peaks occur with the wrong-sign electron, nor with a pion trigger.

Attributing these peaks to Λ_c and $\bar{\Lambda}_c$ production, with $\sim 3\sigma$ and $\sim 2\sigma$ statistical significance, we consider that (44-17) Λ_c's come from $pp \rightarrow \Lambda_c\bar{D}X$ and 17 from $pp \rightarrow \Lambda_c\bar{\Lambda}_cX$. Cross-sections are calculated assuming that charmed particles follow $E\,d_3\sigma/\sigma dp^3 \sim e^{-4p_T^2}f$ (x or y). We use two models: 1) $f(x_D) = (1-x)^3$ and $f(x_\Lambda) = E$, 2) all f(y)'s = c^t, together with $B(\bar{D} \rightarrow e^-) = 0.08$ and $B(\bar{\Lambda} \rightarrow e^-) = 0.1$ (guess). Preliminary results are, in mb, $\sigma(\Lambda_c\bar{D}) = 0.66 \pm 0.52$ (model 1) or 1.5 ± 1.2 (model 2), $\sigma(\Lambda_c\bar{\Lambda}_c) = 1.9 \pm 1.4$ mb (model 2).

The statistical and systematic (±50%) errors are large, but the total Λ_c cross-section is significantly high [if $\sigma(\bar{\Lambda}_c)$ is assumed to be 0, $\sigma(\Lambda_c) \simeq 1.0 \pm 0.4$ mb with model 1]. Correspondingly the resulting e/π ratio, $\sim (14 \pm 6) \times 10^{-4}$, is large compared to the once measured[3] value of $\sim 2 \times 10^{-4}$ for $0.4 < p_T < 1$ GeV/c. However, the results for the Λ_c are compatible with those of another ISR experiment[4] and a $\bar{\Lambda}_c/\Lambda_c$ ratio of ~ 0.5 would be comparable to the $\bar{\Lambda}^0/\Lambda^0$ ratio found at the ISR[5].

REFERENCES

1) K.L. Giboni et al., Phys. Lett. 85B, 437 (1979).
2) K.L. Giboni, Thesis, Aachen University.
3) M. Barone et al., Nucl. Phys. B132, 29 (1978).
4) G. Sajot, communication at this Conference.
5) S. Ehran et al., Phys. Lett. 85B, 447 (1979).

Λ_c^+ PRODUCTION IN ν-D INTERACTIONS

T. Kitagaki

IIT-Maryland-Stony Brook-Tohoku-Tufts (E545)

The Λ_c^+ peaks are first observed in the $\Lambda\pi^+$ and $K^0 p$ inclusive mass distributions from ν-D charged current events.

The data sample used for this analysis is based on measurements of 90% of the Vee events in a 328,000 frame exposure of the Fermilab 15-foot deuterium-filled bubble chamber to a wide-band single-horn focused neutrino beam produced by 350 GeV/c protons. The neutrino energies range from 10 to 250 GeV, with an average energy, $\langle E_\nu \rangle \approx 50$ GeV.

The charged current events are selected by applying the kinematic method with the cuts, $\Sigma P_L^{vis} > 5$ GeV/c and $P_{TR} > 0.7$ GeV/c. With the restriction of the charged multiplicity, < 8 prongs, the number of charged current events is estimated to be 13,100, of which 11,000 events have hadronic energy $W > 2.2$ GeV. The mass resolutions are ± 5.1 MeV for $K_s^0 \rightarrow \pi^+\pi^-$ and 1.7 MeV for $\Lambda \rightarrow p\pi^-$.

Fig. 1(a), (b) and (c) show, respectively, the invariant mass distributions for the $\Lambda\pi^+$, $K^0 p$ and sum of these two channels, which are among the expected Cabibbo favored weak decays of the Λ_c^+. We observe peaks at the Λ_c^+ mass region in these three distributions. The fits to a polynominal background plus a Gaussian give the excesses of events 9.5 ± 4.7, 9.8 ± 5.7 and 19.3 ± 7.3 above the background curves for $\Lambda\pi^+$, $K^0 p$ and their sum in the mass interval of 2.24 to 2.32 GeV.

To reduce non-resonant backgrounds, the cut on the decay angle of the two body system in its rest frame, $\cos\theta > -0.75$ for $\Lambda\pi^+$ and $\cos\theta > -0.9$ for $K^0 p$, is applied. (shaded area in Fig. 1) The cut keeps the peaks at the Λ_c^+ mass region. Since the sum of the two channels gives a peak of 3.3σ and 2.6σ for the shaded and unshaded events above the background, we interpret this peak as a production of the charmed baryon, Λ_c^+. The peak at 2.0 GeV corresponds to the reported Y*(2030).

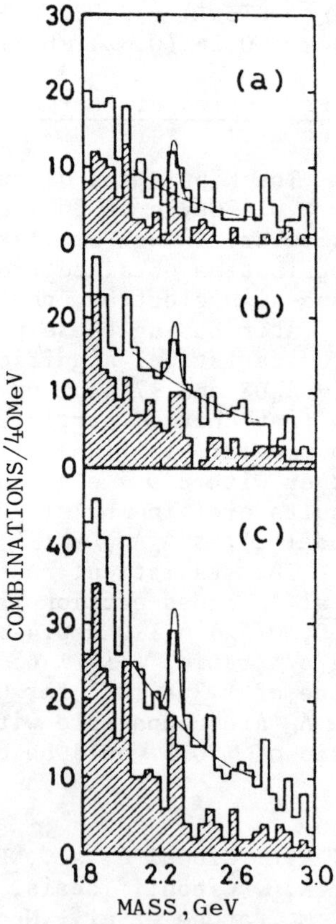

Fig.1. Inclusive mass distributions: (a) $\Lambda\pi^+$, (b) $K^0 p$ and (c) $\Lambda\pi^+ + K^0 p$. Shaded area with the decay angle cut of two body system.

Fig. 2(a) shows the ideogram for the sum of the $\Lambda\pi^+$ and $K^o p$ channels. The mass value of Λ_c^+ is obtained from the fit, to the ideogram to be

$$m = 2.275 \pm 0.010 \text{ GeV}.$$

We also search for charmed baryon production in exclusive events from the Vee event sample. Requiring the χ^2-probability to be greater than 5%, we find 48 Λ events which fit 3-constraint hypotheses with the $\Delta S = -\Delta Q$ signature of charm production. Of course, the 3-constraint fit does not always give a correct hypothesis, but certainly enhances the signal of charmed events.

Fig. 2(b) shows the ideogram of the $\Lambda\pi^+$ mass from the exclusive events. In spite of the low statistics data, the ideogram shows a clear peak at the Λ_c^+ mass. The peak position in Fig. 2(b), 2.272 $\pm$ 0.010 GeV, agrees well with the inclusive value.

Table I shows the production, rates,

$$\sigma_{\nu D}(\Lambda_c^+)\cdot Br/\sigma_{\nu D}(CC, W > 2.2 \text{ GeV}).$$

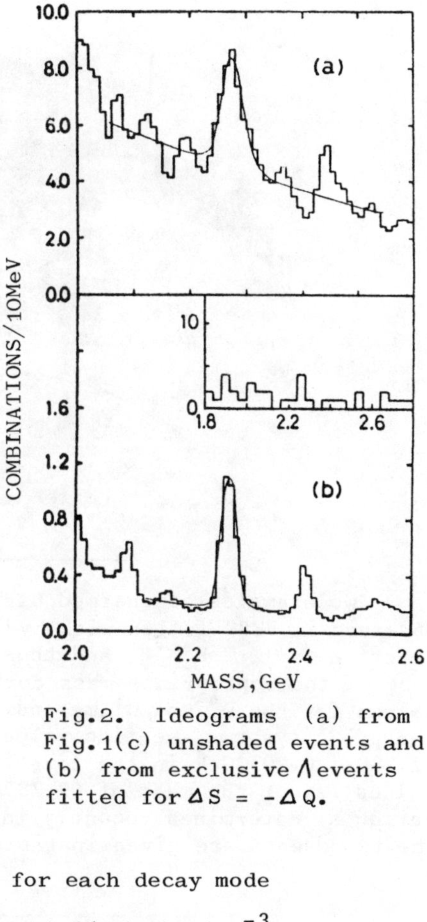

Fig.2. Ideograms (a) from Fig.1(c) unshaded events and (b) from exclusive Λ events fitted for $\Delta S = -\Delta Q$.

Table I Production rates of Λ_c^+ for each decay mode

Decay mode	$\sigma_{\nu D}(\Lambda_c^+)\cdot Br/\sigma_{\nu D}(CC)$ (10^{-3})
$\Lambda_c^+ \to \Lambda\pi^+$	1.8 ± 0.9
$\to K^o p$	3.5 ± 2.0
$\to \Lambda\, 2\pi^+\pi^-$	< 1.6
$\to K^o p\pi^+\pi^-$	< 4.8

$$Br(\Lambda\pi^+)/Br(K^o p) = 0.51 {}^{+0.62}_{-0.27}$$

OBSERVATION OF CHARMED BARYON PRODUCTION IN νp INTERACTIONS

Aachen-Bonn-CERN-Munich (MPI)-Oxford Collaboration

Presented by P.C. Bosetti
CERN, Geneva, Switzerland

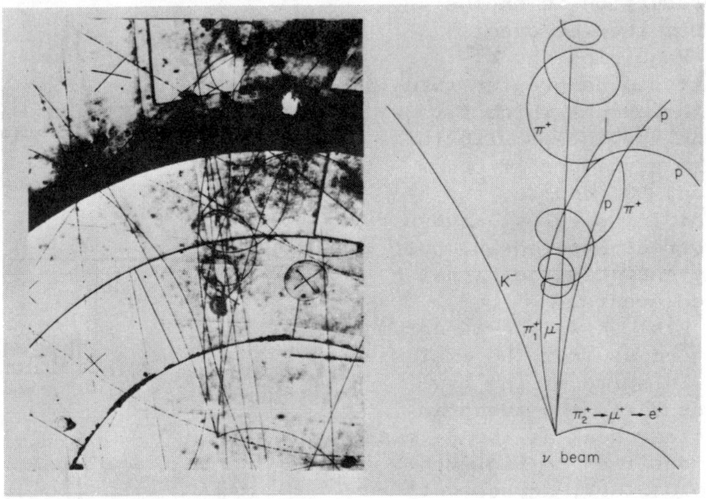

Two examples of charmed baryon production by neutrinos have been observed in BEBC filled with hydrogen. Both events fit uniquely the reaction $\nu p \rightarrow \mu^- p K^- \pi^+ \pi^+$ and thus apparently violate the $\Delta S = \Delta Q$ rule. None of the appropriate mass combinations is consistent with the mass values for the D^0 and D^+ mesons. However, for the mass combinations of the $pK^-\pi^+$ systems, we find values of (2.285 ± 0.005)GeV and (2.280 ± 0.003)GeV in the first and second event, respectively. These values favour the mass of (2.285 ± 0.006)GeV for the Λ_c^+ charmed baryon as determined recently in e^+e^- collisions at SLAC. Details of the two events are given in table 1. The figure shows event 1.

TABLE 1 - Description of the two charmed baryon events

Particle	Event 1		Event 2	
	Momentum (GeV/c)	Identification	Momentum (GeV/c)	Identification
μ^-	48.654 ± 1.613	EMI	10.348 ± 0.028	EMI
K^-	2.574 ± 0.011	FIT	2.514 ± 0.005	FIT
P	1.973 ± 0.015	Interaction	0.935 ± 0.005	IONIZATION
π_1^+	1.255 ± 0.006	Interaction	1.081 ± 0.003	FIT
π_2^+	0.225 ± 0.001	Decay	1.483 ± 0.005	FIT
$P(\chi^2)$	62%		40%	
E_ν, (GeV)	54.051 ± 1.615		15.879 ± 0.029	
W, (GeV)	2.808 ± 0.005		2.796 ± 0.005	
$M(pK^-\pi^+)$,(GeV)	2.285 ± 0.005		2.280 ± 0.003	

PRODUCTION OF J/ψ AND OTHER PARTICLES IN π^-N COLLISIONS AT π^- MOMENTA 140-197 GeV/c

R. Barate*, P. Bareyre, P. Bonamy, P. Borgeaud, M. David,
F. X. Gentit, G. Laurens, Y. Lemoigne, G. Villet and S. Zaninotti.
Centre d'études nucléaires, Saclay, Gif-sur-Yvette, France.

P. Astbury, A. Duane, G. J. King, B. C. Nandi, R. Namjoshi,
D. Pittuck, D. M. Websdale and J. Wiejak.
Imperial College, London, England.

J. G. McEwen.
Southampton University, England.

B. Pietrzyk and R. D. Tripp**.
CERN, Geneva, Switzerland.

B. Brabson, R. Crittenden, R. Heinz, J. Krider and T. Marshall.
Indiana University, Bloomington, Indiana, USA.

ABSTRACT

π^-N interactions at 185 GeV/c have been observed in a large acceptance spectrometer triggered by $J/\psi \to \mu\mu$. From a sample of 40 000 J/ψ with associated hadrons a search for naked beauty states decaying into $J/\psi K^o$ or $J/\psi K^o \pi^\pm$ leadsto limits on cross section x branching ratio of 0.08 and 0.51 nb/nucleon, respectively. The kinematic relation between the observed longitudinal momentum distribution of the J/ψ and the x-distributions of the fusing partons leads to the conclusion that J/ψ production at these energies is mainly through gluon fusion and gives the momentum distribution of gluons in the pion and the nucleon. The photons detected in a calorimeter in association with the J/ψ have shown that 36 ± 5 % of J/ψ are produced from decay of an intermediate χ state. The analysis of e^+e^- pairs in the spectrometer shows peaks in the invariant mass of the J/ψγ system at the masses of $\chi(1^{++})$ and $\chi(2^{++})$.

INTRODUCTION

The WA-11 experiment[1] at CERN SPS has used the large multi-particle spectrometer in the GOLIATH magnet (Fig. 1) to study π^-Be interactions triggered by $J/\psi \to \mu\mu$. We report here on the search for bottom mesons and present results on the logitudinal momentum J/ψ distribution, the contribution of χ decay from calorimeter measurements and the identification of more than one χ state; the latter three measurements are important in elucidating the mechanisms of J/ψ production.

* Present address: CERN, Geneva, Switzerland.
** On leave from Berkeley, University California, USA.

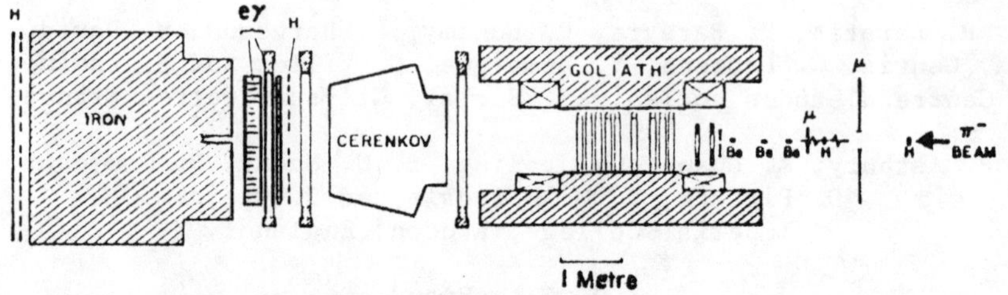

Fig. 1. Multiparticle spectrometer in GOLIATH with Be targets, eγ calorimeter and muon filter.

EXPERIMENTAL RESULTS

Limits on Bottom (/Beauty) meson production. This collaboration reported earlier[2, 3] a possible observation of a B meson at a mass of 5.3 GeV/c^2 based on a sample of 9,000 J/ψ which appeared as a 4σ effect in the sum of two channels: $J/\psi K^o \pi^\pm$, $J/\psi K^- \pi^+$. To test the reality of the peak we have taken 31,000 additional J/ψ (average $\bar{\pi}$ momentum 190 GeV/c). So far a search in the total sample of 40,000 J/ψ has been made in the channels $J/\psi K^o \pi^\pm$ and $J/\psi K^o$, and shows no significant peak (see for example $J/\psi K^o \pi^\pm$ in Fig. 2). The sensitivity is estimated to be 2.5 pb/J/ψ event. The mass resolution at 5.3 GeV/c is calculated to be σ = ± 20 MeV/c^2. (An experimental check for $J/\psi\, \pi^+\pi^-$ confirmed expectations at ψ′ mass). Based on an average of $\bar{q}q$ and gg production models for $\bar{B}B$, the limits with 90% confidence are:

$$\sigma_{\bar{B}B} \cdot Br_{B \to J/\psi K^0 \pi^\pm} < 0.51 \text{ nb/nucleon}$$

$$\sigma_{\bar{B}B} \cdot Br_{B \to J/\psi K^0} < 0.08 \text{ nb/nucleon}$$

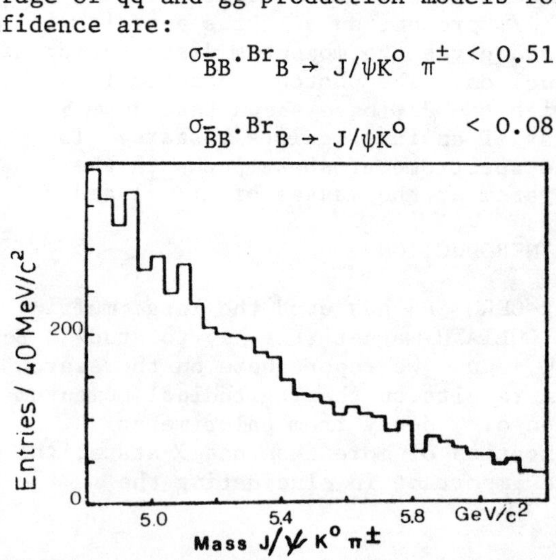

Fig. 2. The $J/\psi K^o \pi^\pm$ effective mass spectrum for the total sample of 40,000 J/ψ shows that the additional data has removed the peak seen with the original 9,000 J/ψ.

The Longitudinal momentum distribution of J/ψ. The x_F distribution of J/ψ ($x_F = 2p_L^*/\sqrt{s}$) is narrow and is centred about a positive, non-zero value of x_F which decreases slightly with increasing π^- momentum (Fig. 3 and 4). The natural explanation of this is that J/ψ production is the result of parton fusion. If quark-antiquark annihilation is assumed to form a J/ψ directly, the resulting x_F distribution can be calculated from the known x-distributions of quarks[4]. Fig. 3 shows that this is a bad fit.

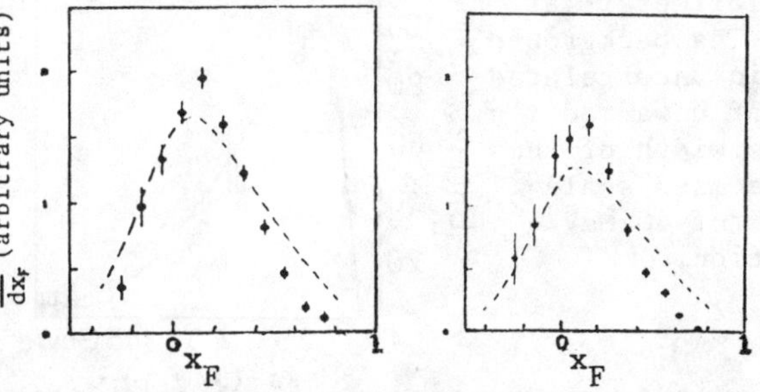

Fig. 3. Samples of the data (150 and 175 GeV/c) plotted with curves calculated for direct production of J/ψ by $q\bar{q}$ annihilation using the structure functions of Badier et al[4].

Good fits can be obtained (Fig. 4) at all energies with minimal $q\bar{q}$ and the assumption of J/ψ production from gluon fusion with gluon momentum distributions of the form $xg_i(x) \sim (1 - x)^a$. The fits give the values of the exponent a = 2.3 ± 0.3 for gluons in the pion and a = 5.1 (+1.5, −0.6) for gluons in the nucleon. The values of a found are altered by less than the present experimental errors if it is assumed that gluon fusion produces χ states which then decay to give J/ψ; further they are insensitive to the fraction of $\bar{q}q$.

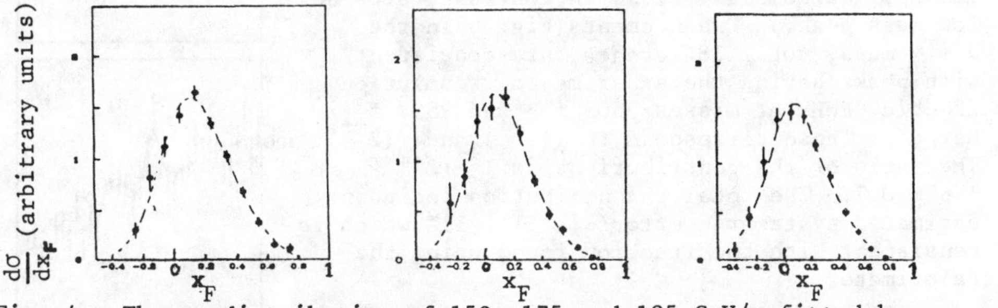

Fig. 4. The x_F distribution of 150, 175 and 185 GeV/c fitted by curves corresponding to direct production by partons with momentum distributions of the form $(1-x)^a$.

<u>The contribution of χ decay</u>. Photons produced in association with a J/ψ were identified in a Pb/scintillator sandwich calorimeter for a sample of 2493 J/ψ events at 175 GeV/c which contains an estimated 7% background events. The J/ψ γ mass spectrum, after exclusion of photons which gave a π^0 mass when combined with another photon, shows a peak above the background (Fig. 5). The number of events in the peak was estimated in two ways which in fact give similar results: (a) by fitting a Gaussian plus an exponential to the mass range 3.3 to 4.3 GeV/c^2 and (b) by making an independent model of the J/ψ γ background by combining J/ψ's with photons from independent events and making a correction for the unwanted χ photons. The efficiency of shower-finding was evaluated to be 60% by superimposing 10 GeV/c π^0 events (representing photons

from χ decay) on real data and reprocessing. The estimated photon acceptance of 22% then leads to a value of 36 ± 5% (statistical errors only) for the fraction of J/ψ resulting from χ decay.

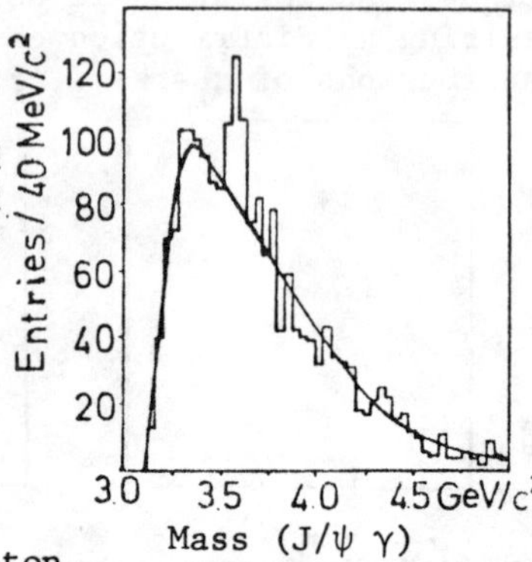

Fig. 5. The J/ψ γ mass spectrum for photons detected in the calorimeter (at 175 GeV/c; after π^0 cut). The background curve is derived from γ's in uncorrelated events with a correction for unwanted γ's from χ's. The peak has the width of the resolution, 90 MeV/c^2. The mass scale may have a systematic error of 40 MeV/c^2 due to calorimeter calibration.

<u>The identification of χ states</u>. Photon conversion to electron pairs in the targets and in the material of the spectrometer enables measurement of J/ψ γ masses with high resolution but low efficiency. The e^+e^- vertex can appear separate from the main vertex (Fig. 6a) or associated with it (Fig. 6b). Masses of e^+e^- pairs below 50 MeV/c^2 have been taken as gammas. (Like sign pairs, in contrast, slow no low mass peak.) Enhancements(Fig. 6)in the J/ψ γ mass, total 184 events, are consistent with peaks having the experimental resolution 22MeV/c^2 FWHM at masses 3505 ± 3 and 3557 ± 4 MeV/c^2. These correspond to $\chi(1^{++})$ and $\chi(2^{++})$. The ratio of the contributions is $1^{++}/2^{++}$ = 1.6 ± 0.7. The total χ contribution including estimated systematic errors is 31 ± 11% which is consistent with the fraction found using the calorimeter.

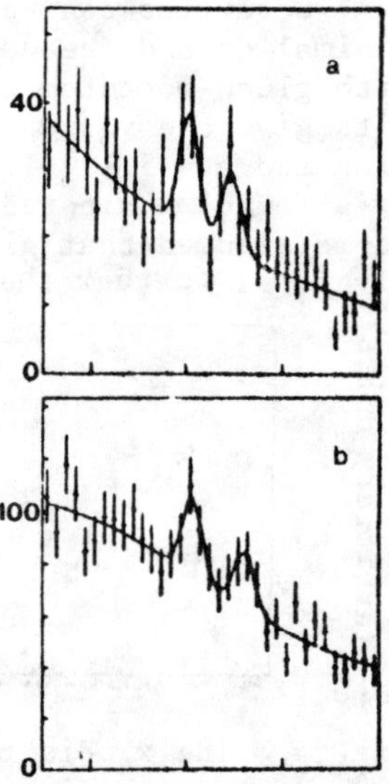

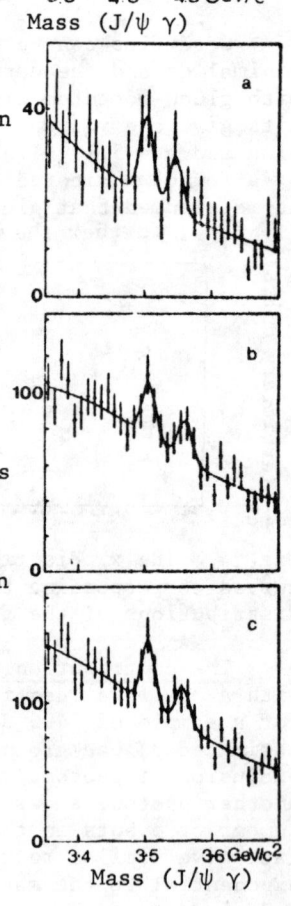

Fig. 6. J/ψ γ mass spectra with a 10 MeV/c^2 bin width in which the γ has been detected as an e^+e^- pair measured in the spectrometer.
a) events in which the e^+e^- vertex is distinct.
b) events in which the e^+ and e^- are associated with hadron tracks at the main vertex.
c) the sum of (a) and (b).

1. M. A. Abolins et al., Phys. Lett. <u>82B</u> , 145 (1979).
2. D. Treille, Rapporteur talk, Geneva Intl. Conf. June 1979.
3. R. Barate et al., Saclay/DPhPE 79-17 and Lepton, Photon Symp. 524 (1979).
4. J. Badier et al., CERN/EP 79/67 (1979).

HIGH STATISTICS STUDY (~10^6 EVENTS) OF J/ψ PRODUCTION AND ϒ PRODUCTION IN THE ENERGY RANGE 150 TO 280 GeV BY $\pi^{\pm}$, $p^{\pm}$ INCIDENT PARTICLE.

CEN, Saclay[1]-CERN, Geneva[2]-Collège de France, Paris[3]-
Ecole Polytechnique, Palaiseau[4]-Laboratoire de l'Accélérateur Linéaire, Orsay[5]

J. Badier[4], J. Boucrot[5], J. Bourotte[4], G. Burgun[1], O. Callot[5], Ph. Charpentier[1], M. Crozon[3], D. Decamp[2], P. Delpierre[3], A. Diop[3], R. Dubé[5], P. Espigat[3], B. Gandois[1], R. Hagelberg[2], M. Hansroul[2], J. Karyotakis[5], W. Kienzle[2], A. Lafontaine[1], P. Le Dû[1], J. Lefrançois[5], Th. Leray[3], J. Maillard[3], G. Matthiae[2], A. Michelini[2], Ph. Miné[4], H. Nguyen Ngoc[5], G. Rahal[1], O. Runolfsson[2], P. Siegrist[1], A. Tilquin[3], J. Timmermans[2], J. Valentin[3], R. Vanderhaghen[4], S. Weisz[2].

Presented by P. Delpierre

ABSTRACT

We have performed in the NA3 experiment the study of high mass dimuon production by a hadronic unseparated beam on hydrogen and platinum targets. The comparison of the production cross-section for proton and antiproton together with the differential cross-section dσ/dx allows us to compare the data with a production mechanism involving quark-antiquark and gluon-gluon interactions. The $\cos\theta^*$ distribution of the same J/ψ data have also been analysed and results will be presented. Finally we have observed ϒ production from 150 GeV/c incident pions.

INTRODUCTION

New measurements of dimuon production have been obtained in the NA3 experiment at CERN. They provide results on J/ψ and ϒ production from hadron-hadron interaction at 150 GeV/c. Statistics on the J/ψ region for hydrogen and platinum targets are given in Table I.

Table I Statistics of J/ψ events ($x_F > 0$)

Energy GeV/c	Target	π^-	K^-	$\bar{p}$	π^+	K^+	p
150	Pt	605600	19276	6619	7936	409	3642
	H_2	16330	474	194	202	15	118
200	Pt	145000	2800	1000	108000	16000	101000
	H_2	3000	56	17	2200	340	2300

1. A-DEPENDENCE AND J/ψ CROSS SECTION

a) J/ψ production by positive and negative pions.

On the mass spectrum, the events called J/ψ are those corresponding to 2.7 < M < 3.5 GeV from which we have subtracted a continuum background determined by events at higher and lower dimuon masses.

Using our $\pi^{\pm}$ data on hydrogen and on platinum we can compute :

$$R^+ = \frac{N(\pi^+H_2 \to J/\psi)}{N(\pi^+Pt \to J/\psi)} , \quad R^- = \frac{N(\pi^-H_2 \to J/\psi)}{N(\pi^-Pt \to J/\psi)} \quad \text{and } R = R^+/R^-$$

which are independant of luminosity and acceptance. Let us call
$\sigma^+ = \sigma(\pi^+p \to J/\psi) = \sigma(\pi^-n \to J/\psi)$ and $\sigma^- = \sigma(\pi^-p \to J/\psi) = \sigma(\pi^+n \to J/\psi)$

Assuming a power law for the A-dependence

$$R = (\sigma^+/A^{\alpha}(0.4\sigma^+ + 0.6\sigma^-))/(\sigma^-/A^{\alpha}(0.4\sigma^- + 0.6\sigma^+))$$

The A^{α} cancels and we deduce the ratio σ^+/σ^- directly from the measured values of R. The results are given in Table II, where a systematic error of ±0.05 is included. If the valence quark annihilation dominated we would find about 0.5. From these values we deduce

$$R_{Pt} = \sigma(\pi^+Pt \to J/\psi)/\sigma(\pi^-Pt \to J/\psi) = (0.4(\sigma^+/\sigma^-) + 0.6)/(0.4+0.6(\sigma^+/\sigma^-))$$

The ratios R_{Pt} are given in Table II. These results allow us to calibrate the π^+/π^- relative luminosity with an accuracy of 2%.

Table II

Energy GeV/c	σ^+/σ^-	R_{Pt}
150 200	0.94 ± 0.08 0.99 ± 0.06	1.01 ± 0.02 1.00 ± 0.01

b) A-dependence

Using the π^- data on hydrogen and platinum targets to compute

$$\sigma^- = \sigma(\pi^-p \to J/\psi) \text{ and } \sigma^-_{Pt} = \sigma(\pi^-Pt \to J/\psi)$$

we obtain $\qquad A^{\alpha} = \sigma^-_{Pt}/\sigma^-(0.4 + 0.6(\sigma^+/\sigma^-))$

With our values of σ^+/σ^- we find for α the values given in Table III.

Table III A- dependence for incident pions

P_{inc}(GeV/c)	α
150 200 280	0.935 ± 0.025 0.95 ± 0.03 0.935 ± 0.025

The mean value of our measurement is $\alpha = 0.94 \pm 0.02$

c) Proton over antiproton J/ψ production

The relative luminosity for protons and antiprotons in the positive and negative beam have been determined by threshold and differential (CEDAR) Cerenkov counters respectively with an accuracy of 10%. The $p/\bar{p}$ cross section ratios are reported on Table IV. The errors are mainly due to the protons and antiprotons identification. The proton to antiproton ratio decreases with energy as it is expected since the light quark fusion mechanism contribution increases when the energy decreases. This result is confirmed by the 40 GeV/c beam dump experiment at CERN[1].

d) J/ψ cross section

The J/ψ cross section per platinum nucleus, for $x_F > 0$ is reported on Table IV.

Table IV J/ψ cross section ratio for $x_F > 0$

p_{inc} GeV/c	Target	$\sigma(\pi^+\to\psi)/\sigma(\pi^-\to\psi)$	$\sigma(p\to\psi)/\sigma(\bar{p}\to\psi)$	$B\sigma^{\pi^-}$ ($x_F > 0$) (μb/nucleus)
150	Pt H_2	1.01 ± 0.02 0.94 ± 0.08	0.50 ± 0.08 0.51 ± 0.09	0.77 ± 0.09
200	Pt H_2	1.00 ± 0.01 0.99 ± 0.06	0.71 ± 0.16	0.95 ± 0.12
280	Pt			1.05 ± 0.24

2. $d\sigma/dx$ DIFFERENTIAL CROSS SECTION

Assuming that the J/ψ is produced only by light quark-antiquark annihilation and gluon-gluon interaction (in fact charmed quark-production seems to be negligible[2]) we can write the antiproton over proton differential cross section ratio as

$$R(x) = \frac{(d\sigma/dx)\bar{p}}{(d\sigma/dx)p} = \frac{\sigma_g f_g(x) + \sigma_q f_q^{\bar{p}}(x)}{\sigma_g f_g(x) + \sigma_q f_q^{p}(x)} \quad (1) \quad \text{where} \quad x = 2P_L^*/\sqrt{s}$$

The function f_g and $f_q^{p,\bar{p}}$, corresponding to the gluon-gluon and quark-antiquark interaction respectively can be calculated using the nucleon structure functions determined by the CDHS experiment[3].

The proportion of J/ψ events produced by gluon-gluon interaction for incident protons and antiprotons can be written as :

$$R_g^{p}(x) = 1 - \frac{R(x) - 1}{A(x) - 1} \quad \text{and} \quad R_g^{\bar{p}}(x) = 1 - \frac{A(x)}{A(x) - 1}\,\frac{R(x) - 1}{R(x)}, \text{ where}$$

$$A(x) = f_q^{\bar{p}}(x)/f_q^{p}(x) = ((\text{valence}\cdot\text{valence})+(\text{valence}\cdot\text{sea}))/(\text{valence}\cdot\text{sea})$$

which can be calculated using the quark and antiquark structure functions from CDHS experiment[3]. For example, at $x = 0$, using our measured value

$$R(0) = 2.0 \pm 0.4 \text{ (systematic error included)}$$

we have computed the proportion of J/ψ events produced by gluon-gluon interaction in pp reactions ($R_g^{p}(0)$) and in $p\bar{p}$ reaction ($R_g^{\bar{p}}(0)$)

$$R_g^{p}(0) = 0.69 \pm 0.15 \qquad R_g^{\bar{p}}(0) = 0.35 \pm 0.15$$

notice that these results are independent of the gluon structure function. Using the values and the gluon structure function in nucleon from CDHS, one can compute $\sigma_q/\sigma_g = 5.5 \pm 2.2$.

Equation (1) allows to compute R(x) from the value of σ_q/σ_g. In Fig.1, one can see that the shape of the computed value R(x) is rather flat and compatible with the experimental points within the statistical errors.

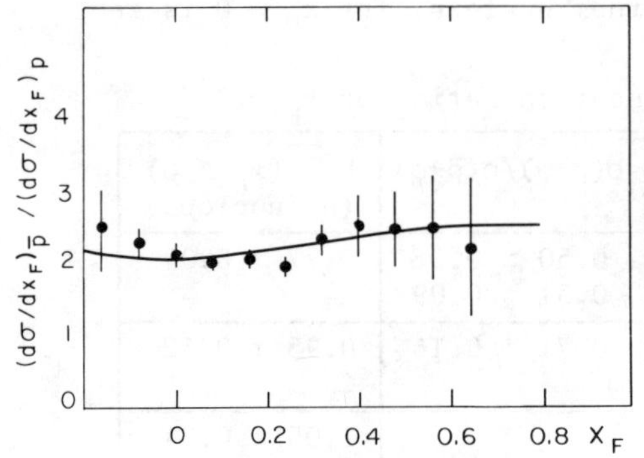

Fig.1 $(d\sigma/dx)\bar{p}/(d\sigma/dx)p$
solid line is R(x) calculated with CDHS structure functions.

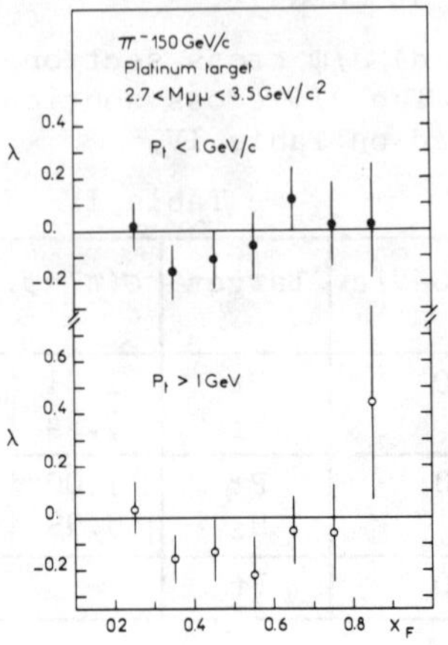

Fig.2 λ versus x_F

3. ANGULAR DISTRIBUTION

Using the π^- data we have fitted the J/ψ angular distribution, corrected by acceptance, in the Collins-Soper frame, for different slices in x_F and p_T by the term :$d\sigma/d(\cos\theta)\propto 1+\lambda\cos^2\theta$. Errors on the subtracted Drell-Yan type background are also included. In Fig.2 we show the values of λ vs. x_F.

The λ parameter is always close to zero, independent of x_F and p_T. At 200 GeV/c incident pions we found $\langle\lambda\rangle = 0.05 \pm 0.07$(4). This result on λ seems to indicate that, at our energies, the direct J/ψ production by light quark annihilation occurs mainly via an intermediate state.

4. UPSILON PRODUCTION

From a fit of the mass spectrum we have calculated the cross section for the upsilon production by 150 GeV/c pions. It is indicated in Table V together with the previously published values(5).

Table V ϒ production cross section by pions

Particle	π^-	π^-	π^-	π^+
Energy (GeV/c)	150	200	280	200
Number of ϒ events	63 ± 15	55 ± 15	66 ± 20	53 ± 12
Bσ(ϒ,ϒ',ϒ'')(pb/nucleus)	70 ± 30	292 ± 100	468 ± 216	370 ± 117

REFERENCES

(1) CERN WA39 experiment. Talk given at CERN April 1980.
(2) M. Binkley et al., Phys. Rev. Lett. 37 (1976) 571.
(3) A. Para, CERN EP 79/154, Proc. Lepton-Photon Symp. (FNAL 1979)
(4) J. Badier et al., CERN EP Preprint 79-61.
(5) J. Badier et al., Phys. Lett. 86B (1979) 98.

MULTIMUON PRODUCTION IN 280 GeV μ^+ IRON INTERACTIONS

The European Muon Collaboration
presented by R.P. Mount
University of Oxford

ABSTRACT

Results are presented on dimuon and trimuon final states in 280 GeV μ^+ iron interactions. Both dimuon and trimuon data show clear evidence for open charm production and suggest strongly that the dominant production process is photon-gluon fusion. Similar amounts of elastic and inelastic (shower energy > 5 GeV) J/ψ production are measured in the trimuon sample. Elastic J/ψ production is consistent with photon-gluon fusion plus naive assumptions. Inelastic J/ψ production is inconsistent with this simple model.

THE EXPERIMENT

The dimuon and trimuon events reported here were obtained using the EMC apparatus [1]. Positive muons at 280 GeV were incident on a 3.75 m. long iron/scintillator target which could measure the produced shower energy with a resolution of $0.56/E^{0.4}$. Only muons and neutrinos emerged from the target. Muons were measured by an air cored spectrometer using drift and proportional chambers with a 4Tm magnet followed by an iron absorber and further drift chambers to provide certain muon identification. The trigger required >1 cluster in hodoscopes beyond the absorber, and at least one muon at >13 mrad from the median plane to suppress QED tridents. To enable apparatus to withstand high rates most detectors were deadened in the central region so that fast muons emerging with angles below 7 mrad had only their direction measured.

DIMUON EVENTS

The events with a scattered muon and one extra muon could be:

a) QED tridents with one muon missed
b) Events with a π or K decay in flight
c) Events with a semileptonic decay of a charmed particle.

Since a missed muon is almost always the fast scattered muon, QED tridents contaminate only the $\mu^+\mu^-$ sample, whereas π, K decays or charm production should produce comparable $\mu^+\mu^-$ and $\mu^+\mu^+$ signals. Missing energy ($E_{beam} - E_{target} - E_{\mu_1} - E_{\mu_2}$) should be large for misidentified tridents, on average 9 GeV for π, K decays, and somewhat larger for charm production.

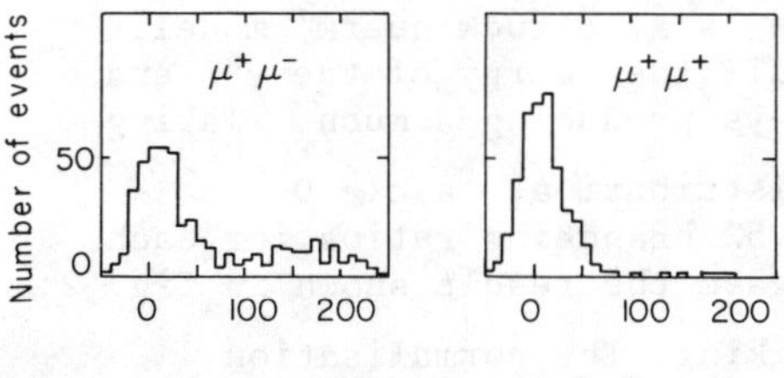

Fig. 1. Missing energy (GeV) for dimuons.

Fig. 1 shows the missing energy for the $\mu^+\mu^-$ and $\mu^+\mu^+$ samples where events with a signal from a third muon have been removed. Both samples are dominated by a peak at an average missing energy of 20 GeV. After a cut at 90 GeV missing energy, reducing the leakage from QED tridents, distributions in other variables were similar so that the $\mu^+\mu^-$ and $\mu^+\mu^+$ samples were merged for further analysis.

The events were weighted by the inverse of their acceptance probability, and for the $\mu^+\mu^+$ sample the faster μ^+ was chosen as the scattered muon. The region of adequate acceptance was defined by the following cuts:

$$E(\mu\text{-scattered}) > 20 \text{ GeV}$$
$$\theta(\mu\text{-scattered}) > 7 \text{ mrad}$$
$$Q^2 > 1 \text{ GeV}^2$$
$$E_{\mu 2} > 16 \text{ GeV}$$

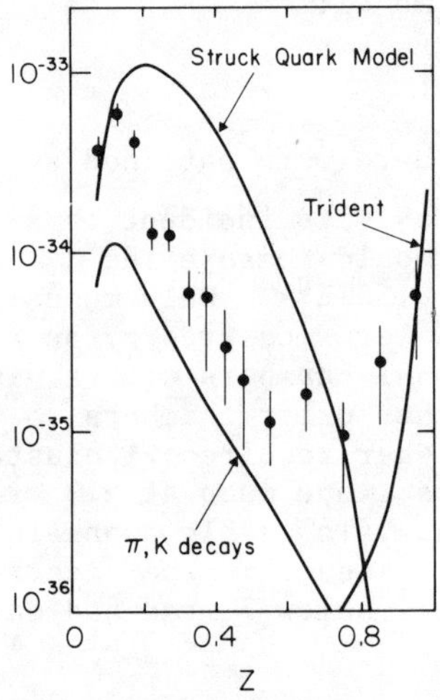

Fig. 2. Dimuons and some possible sources.

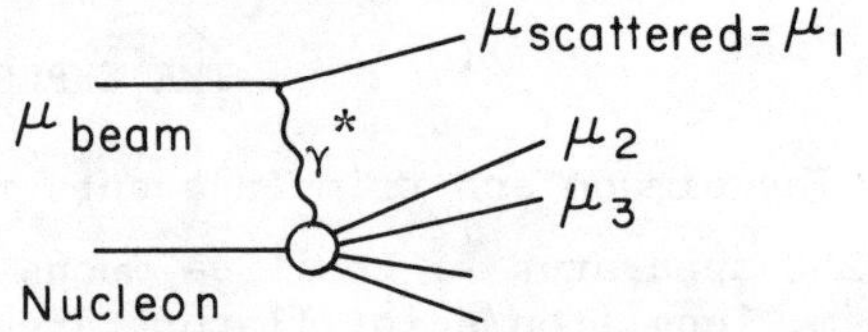

$\nu = \gamma^*$ energy

$-Q^2 = \gamma^*$ mass^2

$W = (\gamma^*$, nucleon) mass

$z = E_{\mu 2}/\nu$, $(E_{\mu 2} + E_{\mu 3})/\nu$

$p_T = p_T$ of μ_2 (or $\mu_2\mu_3$ pair) with respect to γ^*

$x =$ fraction of nucleon momentum carried by a constituent

Fig. 3. Variable definitions.

Fig. 2 shows the cross section, within these cuts, for dimuon production as a function of z (fig. 3 defines variables) in comparison with the expected background from π, K decays and QED tridents. There is a clear signal above these backgrounds, and other processes, apart from charm, are estimated to be negligible. The most simple model for charm production is a 'struck quark' model, where a single c or $\bar{c}$ quark absorbs all the energy of the γ^* and fragments to a D or $\bar{D}$ meson which decays producing a muon. Taking the Buras and Gaemers [2] charmed sea distribution, a $c \rightarrow D$ fragmentation function flat in z, and 5% branching ratios for each of the decay modes $D \rightarrow K\mu\nu$, $D \rightarrow K^*\mu\nu$ gave the result shown in fig. 2.

The disagreement with the data is striking. The normalisation difference probably reflects more on the Buras and Gaemers charmed sea distribution than on the struck quark model, but even disregarding normalisation, the model predicts harder muons than are seen.

The failure of the struck quark model may be due to the improbability of resolving the c or $\bar{c}$ independently; these quarks are too heavy to pursue an independent existence for long. A more realistic approach might be to use a photon-gluon fusion model [3] (γgF) which treats as one process, the formation of a $c\bar{c}$ pair from a gluon and their scattering by the γ^*. Model cross sections were calculated using the same assumptions about fragmentation and decay as above, plus, in accordance with conventional wisdom: $\Lambda_{QCD} = 0.5$ GeV, $m_c = 1.5$ GeV, $xG(x) = 3(1-x)^5$, $\alpha_s = 12\pi/(33-2n_f)\ln(Q^2/\Lambda^2)$. The data were cleaned up by rejecting events with $z > 0.6$ (trident contamination) and $p_T^2 < 0.25$ GeV^2 (large π, K contamination) after which the remaining 10% π, K contribution was subtracted.

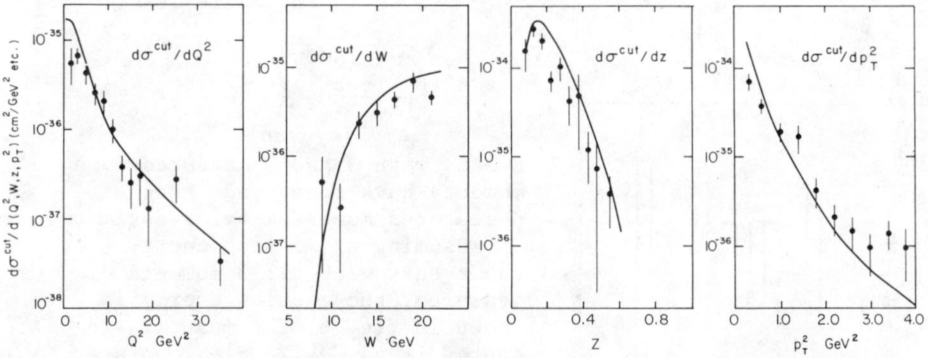

Fig. 4. Dimuons compared with the photon-gluon fusion model.

Fig. 4 shows cross sections for dimuon production in comparison with the γgF calculation. The agreement in all four variables is surprisingly good, especially considering that no adjustments, not even normalisation, have been made to the model.

TRIMUON EVENTS

The dimuon signal is consistent with γgF. When both c and $\bar{c}$ quarks produce muons the model predicts that both muons could be energetic enough to enter the acceptance of the spectrometer. After the exclusion of elastic QED tridents, trimuon signals may also be expected from the following processes:

a) Radiative tail from elastic QED tridents
b) Inelastic QED tridents
c) Two fold π, K decay to μ
d) Vector meson production and subsequent decay to $\mu^+\mu^-$
e) Drell-Yan ($\gamma \rightarrow q\bar{q}$, $q_\gamma + q_{target} \rightarrow \mu^+\mu^-$)

Fig. 5 shows tha mass spectra for the $\mu_2\mu_3$ combination (μ_1 is the scattered muon) for all 3μ events, and for 'inelastic' events.

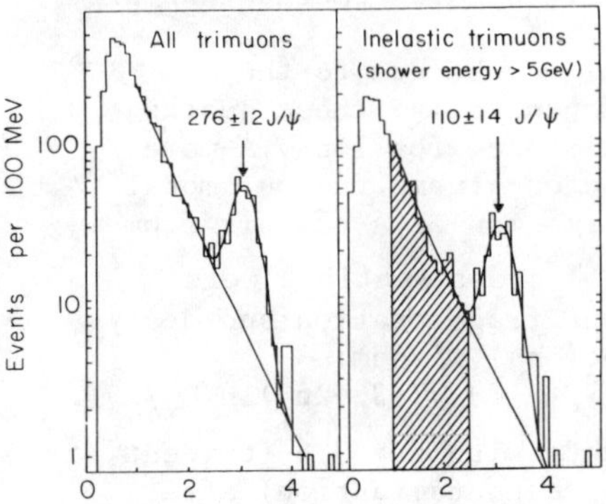

Fig. 5. Trimuons, mass of the ($\mu_2\mu_3$) pair (GeV)

Events were designated 'inelastic' if more than 5 GeV was deposited in the target calorimeter. Discussion of the obvious J/ψ peak is postponed until later. The charm/QED ratio was enhanced by confining attention to the inelastic events, and the region $1.0 < M_{\mu_2\mu_3} < 2.5$ GeV had adequate acceptance and no J/ψ contamination. Further suppression of the QED background was obtained by using information from shower profiles in the target calorimeter. Other cuts which improved the signal to background ratio and removed regions of low acceptance are:

$$p_{\mu i} > 10 \text{ GeV } (i=1,3)$$

$$E_{shower} > 20 \text{ GeV}$$

$$z = (E_{\mu_2} + E_{\mu_3})/\nu < 0.6$$

$$x_{feynman} > 0.0$$

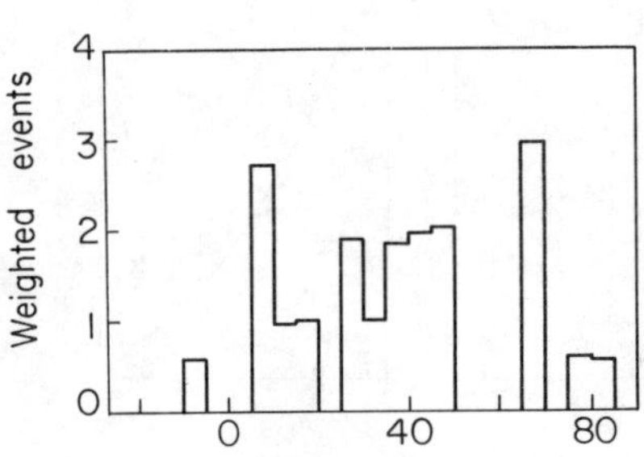

Fig. 6. Trimuon missing energy (GeV).

Events with 3 and 2 measured muon momenta have been used, the unmeasured momentum being estimated by assuming no missing energy. For the events with all 3 momenta measured, the missing energy is shown in fig. 6. The mean missing energy is 44 GeV, a clear pointer to the dominance of charm production. The acceptance corrected cross sections for all events are shown in fig. 7. The solid curves are the predictions of the γgF model, the

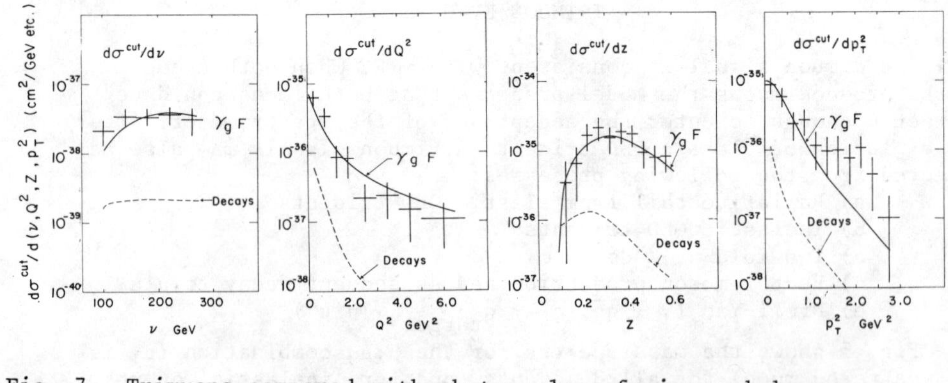

Fig. 7. Trimuons compared with photon-gluon fusion and decays.

background from π, K decays being shown as a dotted curve. Other backgrounds are estimated to be small. Clearly these data agree well with γgF, and their very existence rules out the simple struck quark model as a description of the dominant process.

ELASTIC J/ψ PRODUCTION

Since these data have already been published [4], only a comparison with γgF is presented here. To explain charmonium production γgF needs additional input. It is commonly assumed [5] that $c\bar{c}$ masses in the range $2m_c$ to $2m_D$ result in charmonia, and that a fixed fraction (say 1/6) goes directly to J/ψ accompanied by very little hadronic energy.

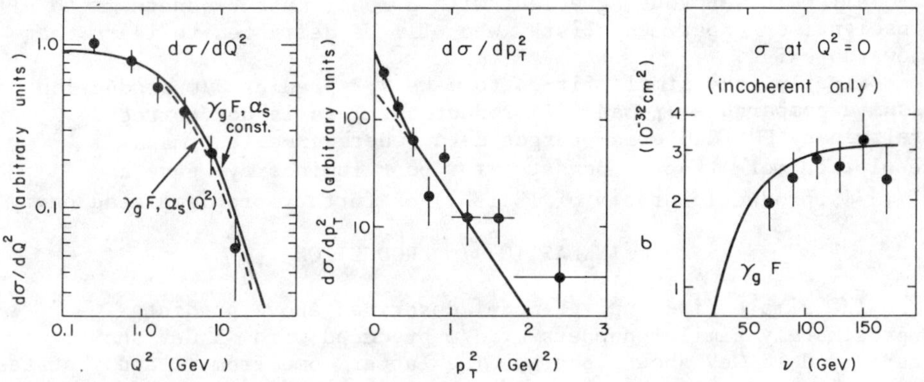

Fig. 8. Elastic J/ψ production compared with photon-gluon fusion.

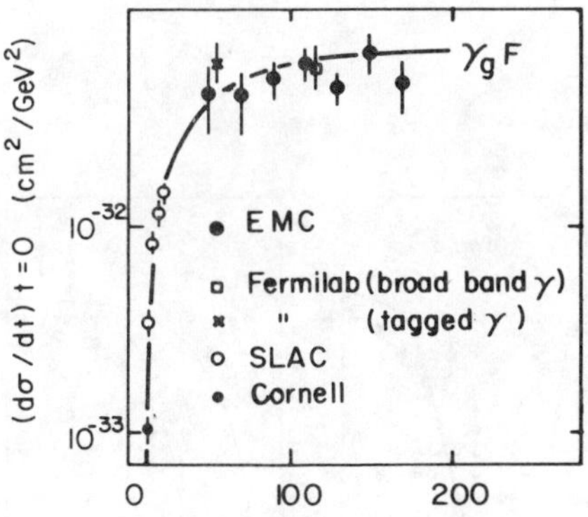

Fig. 9. Photon energy (GeV) dependence of J/ψ photoproduction

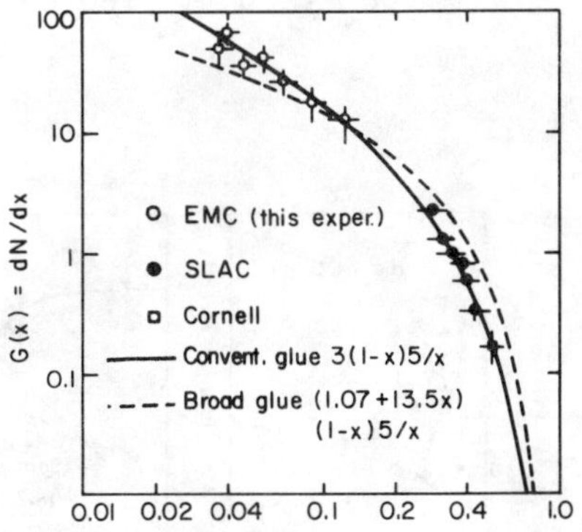

Fig. 10. Fractional momentum carried by the gluon.

Fig. 8 shows acceptance corrected elastic (<5 GeV shower) J/ψ cross sections. The $c\bar{c}$ p_T^2 is not predicted by γgF (it is 0 in the simple model). The data show a strong coherent peak and an incoherent background which have been fitted with a simple parameterisation. The Q^2 and ν dependences agree well with γgF, and from the Q^2 and p_T^2 dependence it is possible to calculate $d\sigma/dt$ at $t = 0$, $Q^2 = 0$ for comparison both with γgF and photoproduction experiments as shown in fig. 9. The agreement is good.

Furthermore, the small range of $c\bar{c}$ masses involved in the γgF calculation allows the transformation of the ν dependence into the x (momentum fraction) dependence of the gluon distribution, which had up to now been assumed. Fig. 10 shows that the conventional $3(1-x)^5/x$ distribution is indeed favoured by the data.

All this wondrous agreement with a model must be anathema to all conscientious experimentalists, who will be delighted to learn that flaws do exist.

Firstly, the simple fix-up to make γgF predict J/ψ production assumed comparable J/ψ and ψ' production. This is not correct. Preliminary EMC deuterium target data (where excellent mass resolution more than compensates for poor luminosity) gave a $\psi'/(J/\psi)$ production ratio of 13±13%. For further problems read on.

INELASTIC J/ψ PRODUCTION

The simple fixed-up γgF model described above predicts approximately similar numbers of J/ψ produced with <5 GeV shower energy and >5 GeV shower energy. The latter come from ψ' and χ states decaying to J/ψ. An inelastic J/ψ sample has been selected and cross sections calculated within the following cuts:

Energy deposited in target > 5 GeV
$z(J/\psi) > 0.3$
$60 < \nu < 180$ GeV

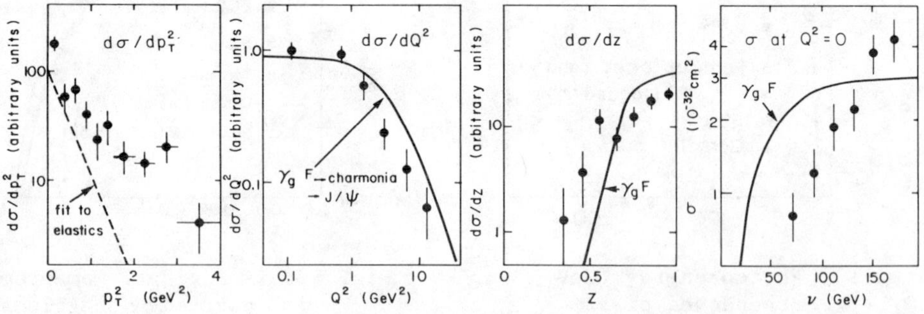

Fig. 11. Inelastic J/ψ production compared with $\gamma gF \to$ charmonia $\to J/\psi$.

These cross sections are compared with $\gamma_g F \rightarrow$ charmonia $\rightarrow J/\psi$ in fig. 11. Again the p_T^2 dependence is not predicted, but should be little broader than for 'elastic' J/ψ production. The fit to the elastic p_T^2 is shown as a dotted line; the inelastic p_T^2 distribution is clearly much broader, although a suggestion of a coherent peak remains. The Q^2 dependence is not markedly different from that predicted by $\gamma_g F$, but the z and ν dependences are in gratifying disagreement. Indeed the data strongly suggest that <40% of the observed cross section arises through the simple $\gamma_g F$ process.

Is the model mortally wounded? Possibly not. Some additional sources of 'inelastic' J/ψ events are:

a) Drell-Yan ($\gamma \rightarrow q\bar{q}$, $q_\gamma + q_{target} \rightarrow \gamma \rightarrow c\bar{c}$)

b) Deep inelastic scatter with gluon bremsstrahlung $\rightarrow c\bar{c}$

c) Higher order processes modifying γgF (gluon brems. from c or $\bar{c}$, etc.)

The first two have been estimated to be small[6], but some calculations of higher order processes in γgF have been performed recently[7] and are able to explain the relative magnitude and shapes of the p_T^2 dependence of the elastic and inelastic J/ψ cross sections. It remains to be seen whether these calculations will arrive at the measured z and ν dependences, and whether, when applied to open charm production, they spoil the presently excellent agreement.

REFERENCES

1. EMC, J.J. Aubert et al., CERN-EP/80-134, submitted to Nucl. Instr. and Methods.
2. A.J. Buras and K.J.F. Gaemers, Nucl. Phys. B132, 249 (1978).
3. J.P. Leveille and T. Weiler, Nucl. Phys. B147, 147 (1979); M. Glück and E. Reya, Phys. Lett. 83B, 98 (1979).
4. EMC, J.J. Aubert et al., Phys. Lett. 89B, 267 (1980).
5. H. Fritzch and K.H. Streng, Phys. Lett. 72B, 385 (1978).
6. J.P. Leveille and T. Weiler, Phys. Lett. 86B, 377 (1979).
7. D.W. Duke and J.F. Owens, Florida State University preprint, FSU-HEP-800709, (1980).

OPEN AND HIDDEN CHARM MUOPRODUCTION

A.R. Clark, K.J. Johnson, L.T. Kerth, S.C. Loken, T.W. Markiewicz, P.D. Meyers, W.H. Smith, M. Strovink, and W.A. Wenzel
Physics Department and Lawrence Berkeley Laboratory,
University of California, Berkeley, California 94720

R.P. Johnson, C. Moore, M. Mugge, and R.E. Shafer
Fermi National Accelerator Laboratory,
Batavia, Illinois 60510

G.D. Gollin[a], F.C. Shoemaker, and P. Surko[b]
Joseph Henry Laboratories, Princeton University
Princeton, New Jersey 08544

ABSTRACT

New results are presented on open and hidden charm and bottom production by 209-GeV muons interacting in a magnetized steel calorimeter. The upper limit on the production of T states by muons is $\sigma(\mu N\rightarrow\mu TX)B(T\rightarrow\mu\mu)<22\times10^{-39}$ cm^2 (90% confidence level). The distributions of elastically produced ψ's are consistent with s-channel helicity conservation (SCHC) and disagree with ψ dominance. From analysis of dimuon final states the cross section for diffractive charm muoproduction is $6.9^{+1.9}_{-1.4}$ nb. The structure function $F_2(c\bar{c})$ for diffractive charmed-quark pair production is presented.

INTRODUCTION

New results are presented from 209-GeV muon interactions in the Berkeley-Fermilab-Princeton Multimuon Spectrometer at Fermilab[1]. Because of space limits on this text, we have omitted many figures and made the discussions brief. The reader should consult the references for complete details of each analysis.

LIMIT ON T CROSS SECTION

Our data have yielded 102 678 trimuon final state events. In every event, all three outgoing muons are fully momentum analyzed and are subject to an energy-conserving 1-C fit using the calorimetric measurement of the shower energy.

A detailed analysis of the dimuon mass spectrum results in a limit on the T muoproduction cross section of $\sigma(\mu N\rightarrow\mu TX)B(T\rightarrow\mu\mu)<22\times10^{-39}$ cm^2 (90% confidence level). The reader is referred to the recently published report of this analysis for a detailed discussion[2].

---

[a] Now at Enrico Fermi Institute, Chicago, Illinois 60637.

[b] Now at Bell Laboratories, Murray Hill, NJ 07974.

ψ DISTRIBUTIONS

Our first results on ψ final states have been published[1]. Here we present the angular distributions of the full elastic data set and their effect on the measurement of the Q^2 distribution[3].

If we assume s-channel helicity conservation (SCHC), natural parity exchange (NPE), and no single spin flip contributions, then the angular distribution of $\psi\to\mu^+\mu^-$ is[4]:

$$W(\theta,\phi,\eta,R)=\frac{1}{1+\varepsilon R}\cdot\frac{3}{16\pi}\{(1+\cos^2\theta)+2\varepsilon R\sin^2\theta-\eta\varepsilon\sin^2\theta\cos2\phi\}, \qquad (1)$$

where $R=\sigma_L/\sigma_T$ is the ratio of psi production cross sections by, and $\varepsilon=\Gamma_L/\Gamma_T$ is the flux ratio of, longitudinally and transversely polarized virtual photons. We have inserted a factor η to monitor the size of the polarization angle asymmetry term; η=1 if SCHC and NPE are exactly obeyed.

The vector meson dominance (VMD) model of lepton scattering suggests that $R=\xi^2Q^2/m_\psi{}^2$. Any Q^2-dependence in the angular distribution, together with a non-uniform spectrometer acceptance in cosθ, can bias the interpretation of the overall Q^2 distribution. To study these effects the data were binned in a 4x5x3 Q^2, |cosθ|, and φ space. An individual mass-continuum subtraction was performed for each of the 60 bins; the resulting data were fit with the product of the angular function W(η,R) and a propagator $P(\Lambda)=(1+Q^2/\Lambda^2)^{-2}$, under various assumptions for η, R and Λ. An additional complication is the possibility of a Q^2 dependence in the amount of nuclear matter seen by the incident virtual photon. We have fit recently summarized data[5] measuring this effect for A∿200, scaled for use in Fe:

$$(A_{eff}/A)_{Fe}\equiv S(x')=(1-0.328e^{-28.3x'})^{0.760} \quad ; \quad x'\equiv\frac{Q^2}{2m_p\nu+m_p{}^2} \qquad (2)$$

All fits were made with S(x′) both included and ignored.

The results of the fits are presented in Table I; the angular data and the results of fits 1-4 are shown in Figure 1. For plotting purposes only, the data and fits have been summed over φ or |cosθ|. While there is little difference between fits of the general SCHC form, it is clear that the polarization angle data rule out a flat angular distribution (fit 3).

The Q^2 distribution for $\sigma_{eff}(\gamma_V N\to\psi X)$ is present in Fig. 2. Our insensitivity to the exact form of R and to the possible nuclear effects results in a propagator mass Λ between 1.9 and 2.6 GeV/c^2. The VMD prediction ($\Lambda=m_\psi$, fit 5) is ruled out. A photon-gluon-fusion (γGF) prediction has also been fit to the data (fit 7); the data fall faster than the γGF prediction. A complete discussion appears in Ref. 3.

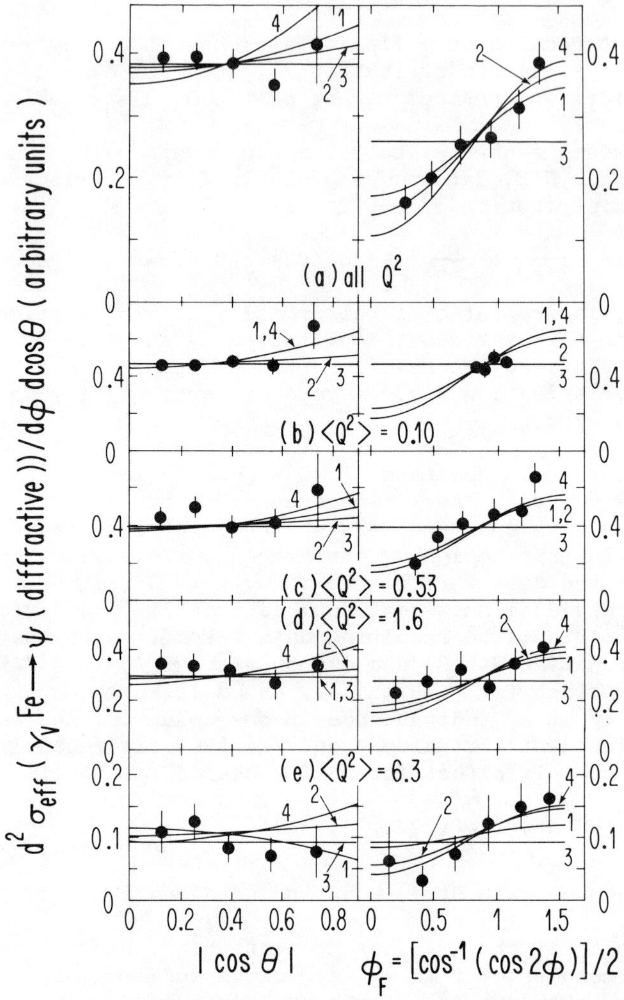

FIG. 1. Angular dependence of the effective cross section for the reaction $\gamma_V Fe \rightarrow \psi X$ (energy(X)<4.5 GeV). Data and statistical errors are presented vs. $|\cos\theta|$ (left column) and ϕ (right column), with ϕ folded into one quadrant. All data ($\langle Q^2 \rangle$=0.71) are shown in (a); (b)-(e) divide the data into four Q^2 regions. Numbered solid lines exhibit the results of fits 1-4 in Table I. Fits 1, 2, and 4 are to the SCHC formula with $\sigma_L/\sigma_T=\xi^2 Q^2/m_\psi{}^2$, constant, and zero, respectively; fit 3 corresponds to the production of unpolarized ψ's. Each fit is made to all the data with one adjustable normalization constant.

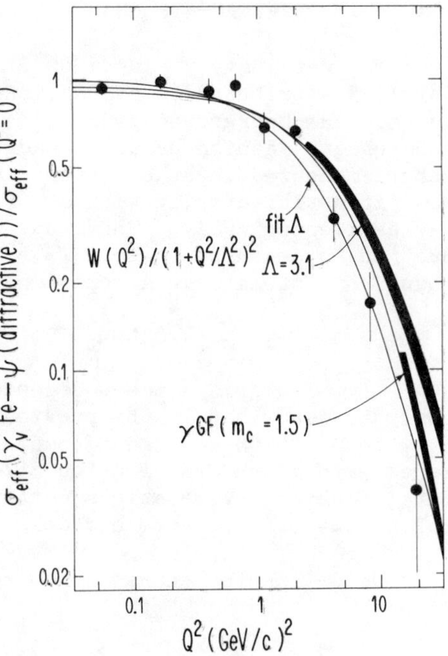

FIG. 2. Q^2-dependence of the effective cross section for the reaction $\gamma_V Fe \to \psi X$ ($E_X < 4.5$ GeV). Data and fits have been summed over $|\cos\theta|$ and ϕ. Statistical errors are shown. The data are fit to $(1+Q^2/\Lambda^2)^{-2}$ multiplied by the function $W(\eta,R)$ shown in Eqn. 1. The weak Q^2-dependence of W results from the Q^2-dependence of R and the particular average values of the angular factors $\cos\theta$ and $\cos 2\phi$. The best fits with free Λ (Table I, fit 1) and fixed $\Lambda=3.1$ (fit 5) are shown. The data are normalized so that fit 1 is unity at $Q^2=0$. Also exhibited is the γGF prediction (fit 7). At high Q^2, the two latter fits are displayed as a solid band, with the upper (lower) edge including (omitting) the screening factor $S(x')$.

Table I. Fits to the Q^2, ϕ, and θ-dependence of the effective cross section σ_{eff} for the reaction $\gamma_V Fe \to \psi X$ ($E_X < 4.5$ GeV). Errors on the fit parameters are statistical. Fit 6 is the same as fit 1 except that W is multiplied by $(1+\varepsilon R)$; Λ then parameterizes the Q^2-dependence of σ_T rather than σ_{eff}. Fit 7 compares the data integrated over ϕ and $\cos\theta$ with the Q^2-dependence predicted by γGF.

Fit No.	Function	$S(x')$	χ^2/DF	Λ(GeV/c^2)	η	ξ^2 or R
1	$W(\eta,R)\times P(\Lambda)$, $R=(\xi Q/m_\psi)^2$	in	45.4/56	$2.03^{+0.18}_{-0.12}$	$1.02^{+0.28}_{-0.23}$	$3.3^{+4.9}_{-3.0}$
		out	45.5/56	$2.18^{+0.18}_{-0.13}$	$1.04^{+0.28}_{-0.23}$	$4.0^{+4.8}_{-3.4}$
2	$W(\eta,R)\times P(\Lambda)$, R=constant	in	42.0/56	2.24±0.13	$1.09^{+0.31}_{-0.24}$	$.35^{+.26}_{-.18}$
		out	42.4/56	2.43±0.15	$1.10^{+0.31}_{-0.24}$	$.37^{+.27}_{-.22}$
3	$1\times P(\Lambda)$	in	73.3/58	2.06±0.11		
		out	73.3/58	2.22±0.13		
4	$W(1,0)\times P(\Lambda)$	in	48.6/58	2.21±0.12	≡1	≡0
		out	49.3/58	2.40±0.14		
5	$W(\eta,0)\times P(m_\psi)$	in	89.1/58	≡3.1	0.96±0.13	≡0
		out	68.5/58		0.93±0.14	
6	$(1+\varepsilon R)\times$Fit 1	in	47.0/56	2.08±0.24	0.86±0.17	$.24^{+.61}_{-.39}$
		out	47.6/56	2.20±0.29	0.87±0.17	$.34^{+.75}_{-.43}$
7	γGF -- Q^2 projection	in	32.1/8	m_c≡1.5 GeV/c^2		
		out	14.6/8			

DIFFRACTIVE CHARM MUOPRODUCTION CROSS SECTION

The data have yielded 20072 dimuon final state events, with (81±10)% attributed to production of charmed states decaying to muons. The background from π, $K\to\mu$ decay was simulated in a model-independent fashion by using hadron muoproduction and decay parameters measured in other experiments. The background-subtracted data was fit satisfactorily with a γGF model producing D mesons, which decay semileptonically. The cross section for diffractive charm production is measured to be $\sigma_{diff}(\mu N\to\mu c\bar{c}X)=6.9^{+1.9}_{-1.4}$ nb, where the errors are systematic. A report of this analysis has been published[6].

CHARM STRUCTURE FUNCTIONS

Fig. 3 displays the ν-dependence of $\sigma_{eff}(\gamma_V N\to c\bar{c}X)$ from the analysis described in the previous section. The insensitivity of σ_{eff} to Q^2 in this range decouples its Q^2 and ν-dependence. The γGF model with gluon distribution $3(1-x)^5/x$ successfully describes the data; however, systematic uncertainties prevent the analysis from ruling out other possible models (see Ref. 7).

We define the charm structure function $F_2(c\bar{c})$ through the relation

$$Q^4\nu d^2\sigma(c\bar{c})/dQ^2d\nu=4\pi\alpha^2(1-y+y^2/2)F_2(c\bar{c}) \quad ; \quad y=\nu/\nu_{max}. \qquad (3)$$

The Q^2 dependence of $F_2(c\bar{c})$ is shown in Fig. 4 for two values of ν. At its peak $F_2(c\bar{c})$ is $\sim$4% of F_2. The predictions of the γGF model resemble the data, but none of the models adequately fit the data.

In the energy range of the data in Fig. 5, $F_2(c\bar{c})$ is clearly scale-noninvariant for $Q^2<10$ $(GeV/c)^2$, or $x_B\lesssim 0.07$. To model the charm contribution to F_2 for smaller photon energies, we normalize the γGF model to the data and damp it at high Q^2 by the factor $(1+Q^2/(10\ GeV/c)^2)^{-2}$. The resulting family of dashed curves in Fig. 5 adequately matches the data.

Table II compares the fit[9] inclusive $\partial F_2/\partial \ln Q^2$ at fixed x_B to $\partial F_2(c\bar{c})/\partial \ln Q^2$ augmented for charmonium production, calculated with the γGF model that has been matched to the muoproduction data. Where the charm scale-noninvariance is most important, the calculation is reliable to $\sim\pm 40$%.

We conclude that diffractive charm production is responsible for $\sim$1/3 of the total inclusive scale-noninvariance observed in F_2 in a region bounded by $2<Q^2<13$ $(GeV/c)^2$ and $50<\nu<200$ GeV and centered at $x_B\sim 0.025$. A more complete discussion can be found in Ref. 7.

This work was supported by the High Energy Physics Division of the U.S. Department of Energy under Contract Nos. W-7405-Eng-48, DE-AC02-76ER03072, and EY-76-C-02-3000.

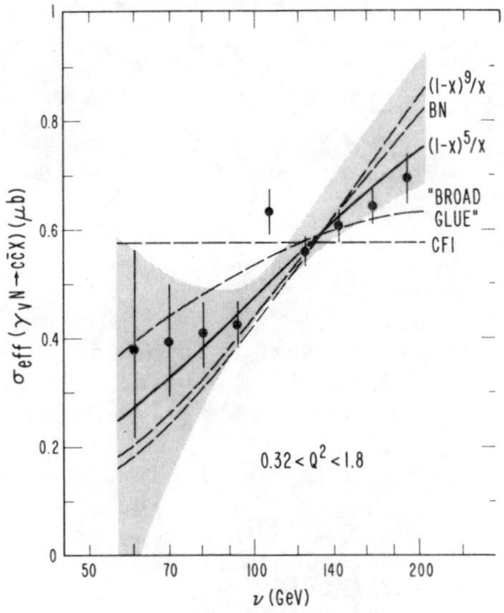

FIG. 3. Energy-dependence of the effective cross section σ_{eff} for diffractive charm production by virtual photons. For $0.32<Q^2<1.8$ (GeV/c), σ_{eff} varies with Q^2 by $\lesssim 20\%$. Errors are statistical. The solid curve exhibits the ν-dependence of the photon-gluon-fusion model with the "counting rule" gluon x distribution $3(1-x)^5/x$, and represents the data with 13% confidence. Other possible models indicated by dashed lines are described in Ref. 7. Curves are normalized to the data. The shaded band exhibits the range of changes in shape allowed by systematic error. For clarity it is drawn relative to the solid curve.

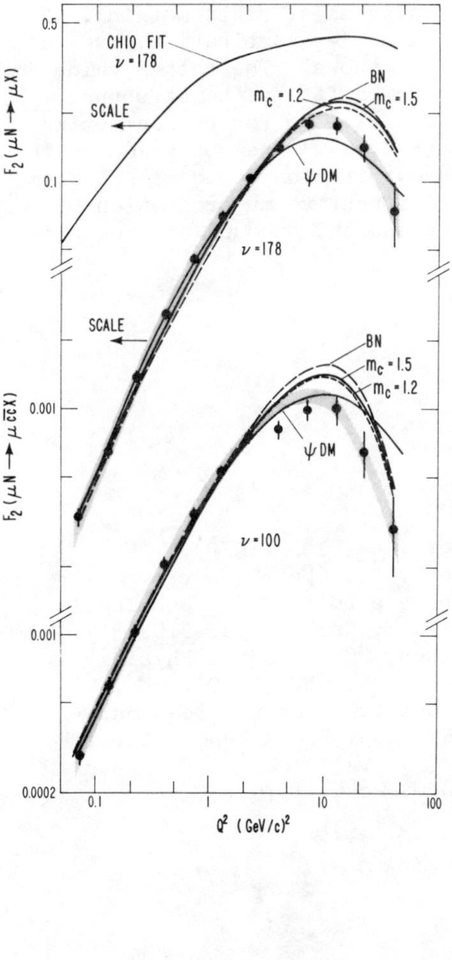

FIG. 4. Q^2-dependence of the structure function $F_2(c\bar{c})$ for diffractive charm muoproduction. At each of the two average photon energies, each curve is normalized to the data. Errors are statistical. The solid (short dashed) curves labelled m_c=1.5 (1.2) exhibit the photon-gluon-fusion prediction with a charmed quark mass of 1.5 (1.2) GeV/c^2. Solid curves labelled ψDM correspond to a ψ-dominance propagator, and long-dashed curves labelled BN are the model of Ref. 8. Shown at the top is a fit adapted from Ref. 9 to the inclusive structure function F_2 for isospin-0 μN scattering. The shape variations allowed by systematic errors are represented by the shaded bands.

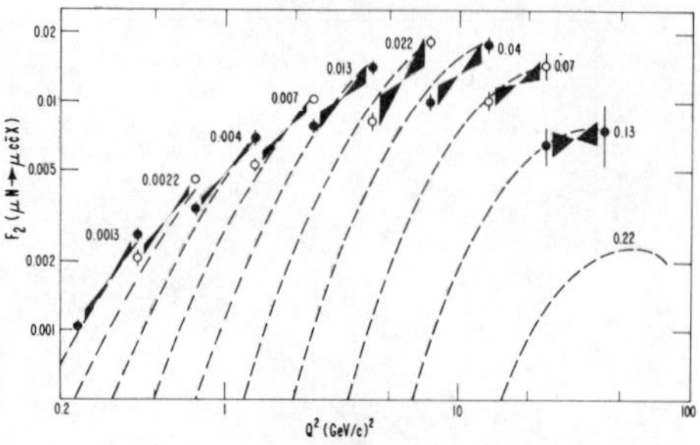

FIG. 5. Scale-noninvariance of $F_2(c\bar{c})$. Data points are arranged in pairs, alternately closed and open. The points in each pair are connected by a solid band and labelled by their common average value of $x_B = Q^2/2m_p\nu$. Errors are statistical. The dashed lines are the prediction of the photon-gluon-fusion model with $m_c = 1.5$ GeV/c^2 except that the model is renormalized and damped at high Q^2 as described in the text. The solid bands represent the slope variations allowed by systematic errors.

ν(GeV)	27	42	67	106	168	
Q^2 (GeV/c)^2	$10^4\partial F_2(c\bar{c})/\partial \ln Q^2$ / $10^4\partial F_2(\mu N)/\partial \ln Q^2$					x_B
0.63	17 / 1070	30 / 1090	43 / 1110	54 / 1120	58 / 1130	
1.0	23 / 980	43 / 1010	63 / 1040	77 / 1050	84 / 1060	0.002
1.6	30 / 650	59 / 680	87 / 700	107 / 720	116 / 730	0.003
2.5	36 / 310	73 / 340	110 / 350	139 / 360	146 / 360	0.005
4.0	36 / 320	80 / 390	128 / 430	162 / 460	163 / 480	0.008
6.3	29 / 210	75 / 330	128 / 410	165 / 460	154 / 490	0.013
10	15 / 50	54 / 220	104 / 340	138 / 430	112 / 480	0.020
16	4 / -130	27 / 50	64 / 230	90 / 360	52 / 440	0.032
25	-2 / -189	7 / -126	26 / 50	40 / 230	0 / 370	0.050
40	0 / -31	-1 / -171	6 / -122	10 / 50	-22 / 240	0.080
63		0 / -23	1 / -154	1 / -119	-16 / 50	0.130

Table II. Calculated $10^4\partial F_2/\partial \ln Q^2$ at fixed x_B vs ν (top), Q^2 (left margin), and x_B (diagonals, right margin). For each Q^2-ν combination, two values are shown. The bottom value is from a fit to the structure function F_2 for μN scattering (Ref. 9). The top value is the contribution $F_2(c\bar{c})$ to F_2 from diffractive muoproduction of bound and unbound charmed quarks.

REFERENCES

1. A.R. Clark et al., Phys. Rev. Lett. 43, 183 (1979).
2. A.R. Clark et al., Phys. Rev. Lett. 45, 686 (1980).
3. A.R. Clark et al., LBL-11562 (to be submitted for publication).
4. K. Schilling, P. Seyboth and G. Wolf, Nucl. Phys. B15, 397 (1970).
5. See these proceedings, H. Miettinen, "Soft Hadron Physics".
6. A.R. Clark et al., Phys. Rev. Lett. 45, 682 (1980).
7. A.R. Clark, et al., LBL-10879, submitted for publication.
8. F. Bletzacker and H.T. Nieh, SUNY-Stony Brook Report No. ITP-SB-77-44 (unpublished).
9. B.A. Gordon et al., Phys. Rev. D20, 2645 (1979).

PHOTOPRODUCTION, EXPERIMENT

T. NASH, FERMILAB, ORGANIZER

Measurement of the J/ψ Photoproduction Cross Section

J. P. Cumalat
Fermilab, Batavia, IL 60510

ABSTRACT

This paper reports on the preliminary measurement of the energy dependence of the J/ψ photoproduction cross section over the energy range 60 to 225 GeV. The data was acquired using the Broadband Photon Beam and spectrometer in the Proton Area at Fermilab by a Fermilab-University of Illinois collaboration, E-401.

The J/ψ was photoproduced in a liquid deuterium target and was detected in both the $\mu^+\text{-}\mu^-$ and $e^+\text{-}e^-$ decay channels. An event is considered a candidate if exactly two tracks are observed in the proportional chambers and no unassociated energy is found in the shower detectors. A mass plot of the event samples used for this paper is presented in Fig. 1. The total J/ψ photoproduction cross sections per nucleon are shown in Fig. 2. A significant rise in the cross section is observed over the energy interval 60-225 GeV for both decay channels.

Analysis is currently continuing on independent checks of the rising J/ψ cross section. The high mass Bethe-Heitler cross section and the ρ photoproduction cross section are being determined. These processes should not rise appreciably with increasing energy. The photon beam spectrum is being determined by comparing the energy spectrum of low mass Bethe-Heitler pass to the shape observed by dumping the photon beam into a lead glass block. Finally, more data exists which will increase our energy range to 300 GeV.

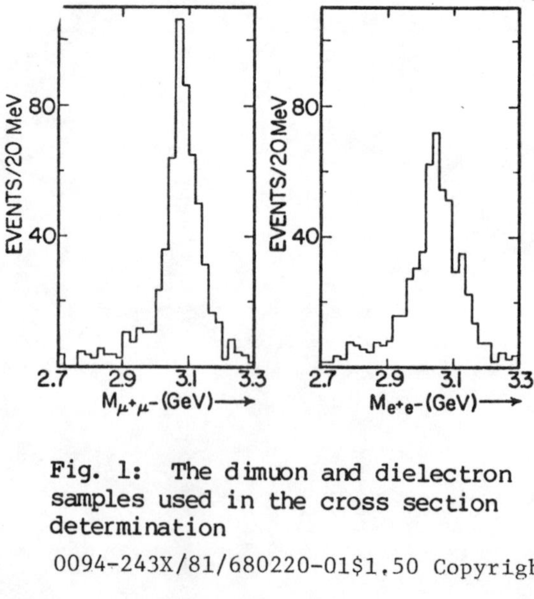

Fig. 1: The dimuon and dielectron samples used in the cross section determination

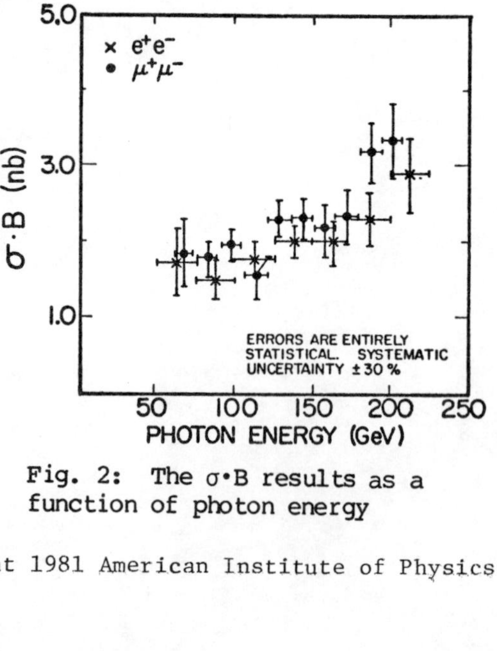

Fig. 2: The σ•B results as a function of photon energy

RESULTS OF THE WA4* PHOTOPRODUCTION EXPERIMENT AT THE CERN SPS

B. D'Almagne
Laboratoire de l'Accélérateur Linéaire, 91405 ORSAY, FRANCE

ABSTRACT

The results of an experiment using a 20-70 GeV tagged photon beam in the CERN Omega spectrometer are presented. Evidence is given for associated $C\bar{D}$ production and observation of $F^{\pm}$ production. The study of exclusive channels contributes to the spectroscopy of vector mesons.

INTRODUCTION

The experimental set up is shown on fig. 1. The total bending power was 3 T.m from the hydrogen target to the drift chambers. The 32 cell Cerenkov counter was filled with CO2 at atmospheric pressure. The main trigger required a multiplicity between 4 and 9 in MWPC 3. Special triggers selected two prong events and events giving a photon of transverse momentum larger than .8 GeV/c. An overall sensitivity of 60 events/nb has been achieved in the main trigger data.

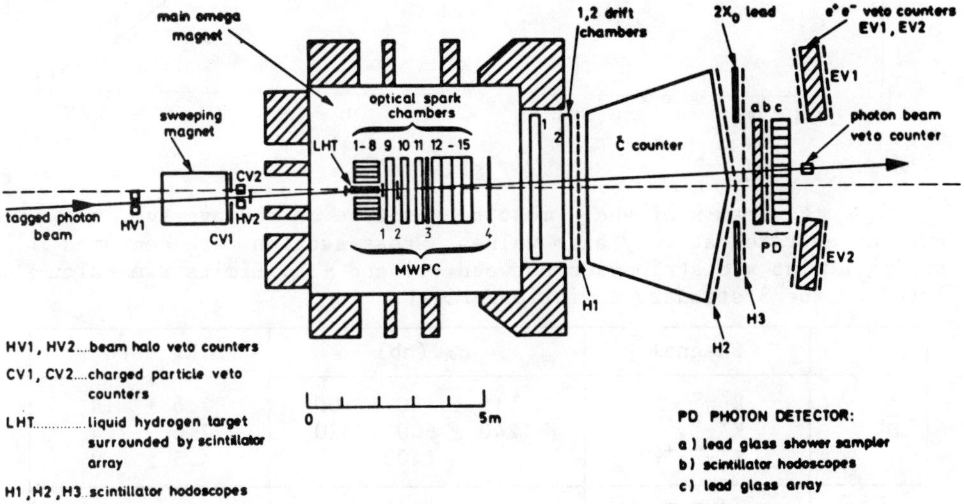

LAYOUT OF THE OMEGA SPECTROMETER FOR PHOTOPRODUCTION EXPERIMENTS

Fig. 1

* WA4 is a collaboration between the following laboratories : Bonn, CERN, Ecole Polytechnique, Glasgow, Lancaster, Manchester, Orsay, Paris VI, Paris VII, Rutherford, Sheffield.

I. CHARM PHOTOPRODUCTION

1) Inclusive spectra

Just requiring the beam energy to be greater than 40 GeV, a clear $\bar{D}^\circ$ signal is seen in the $K^+\pi^-$ and $K^+\pi^-\pi^\circ$ decay modes, while no indication appears for corresponding D° production (fig. 2).

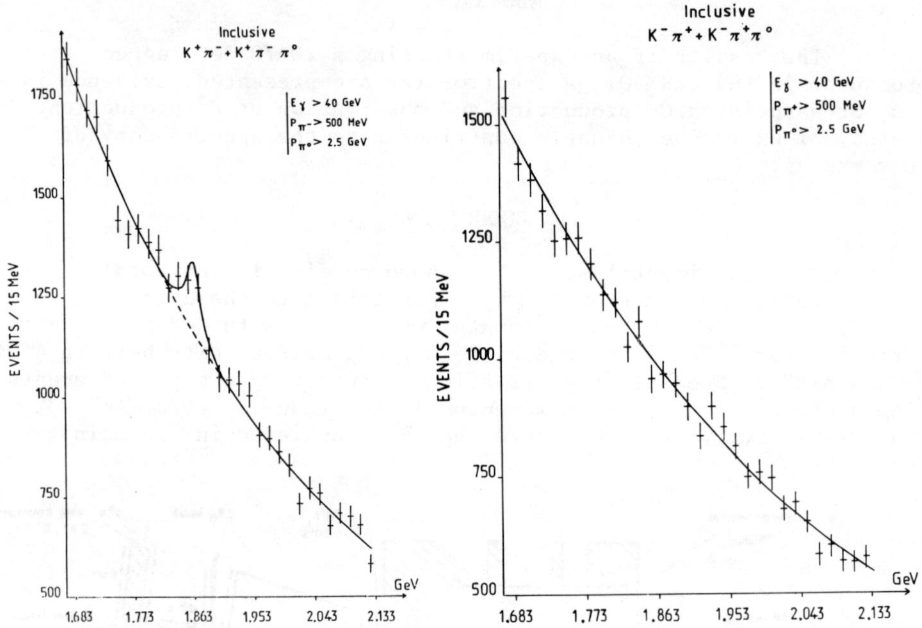

Fig. 2

Feynman x of the $\bar{D}$ must be positive on the average, but are not concentrated at very high values. Cross sections are computed taking a flat x distribution between -.1 and +.5. Limits are calculated at the 3 standard deviations level.

	Channel	σ_D (nb)	B.R. used %
$\bar{D}^\circ$	$K^+\pi^-$ $K^+\pi^-\pi^\circ$ $K^+\pi^-\pi^+\pi^-$	515 ± 160 ± 100 1240 ± 600 ± 320 < 1400	2.6 ± .4 7.6 ± 2.4 4.5 ± .9
D^-	$K^+\pi^-\pi^-$	< 650	4.6 ± .7
D°	$K^-\pi^+$ $K^-\pi^+\pi^\circ$ $K^-\pi^+\pi^-\pi^+$	< 450 < 1500 < 1150	
D^+	$K^-\pi^+\pi^+$	< 650	

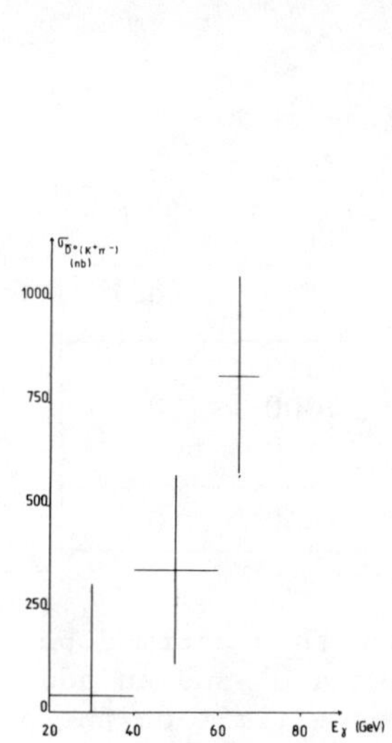

Fig. 3

The energy dependance of the $\bar{D}°$ cross section, shown on fig. 3, exhibits a rise between 40 and 70 GeV. The cascade $D^{*-} \rightarrow \bar{D}°\pi^-$ has been searched for, making use of the excellent resolution on the $(K\pi\pi)$ - $(K\pi)$ mass difference. A weak signal is seen giving the

$$\text{ratio} : \frac{N(D^{*-})}{N(\bar{D}°)} = .07 \pm .04.$$

2) Evidence for a $C\bar{D}$ associated production mechanism

The proton resulting from the decay of a charmed baryon can be seen as a "non π" particle in the Cerenkov counter. Requiring an additional K^+/p in the event improves the $\bar{D}°$ to background ratio by a factor $\sim$ 2.5 (fig. 4)

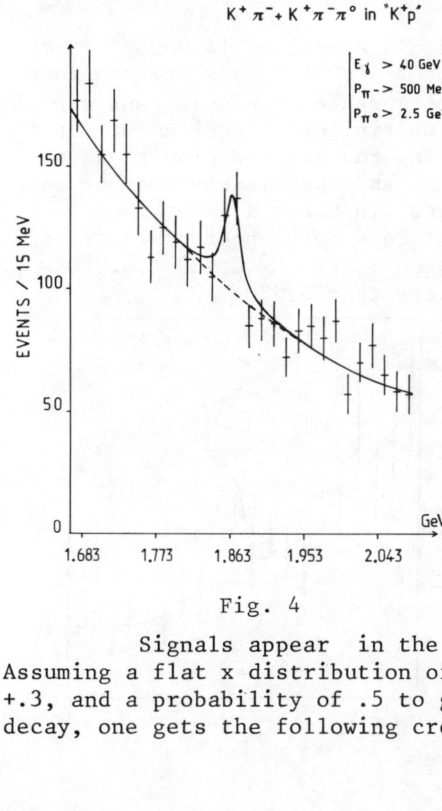

Fig. 4

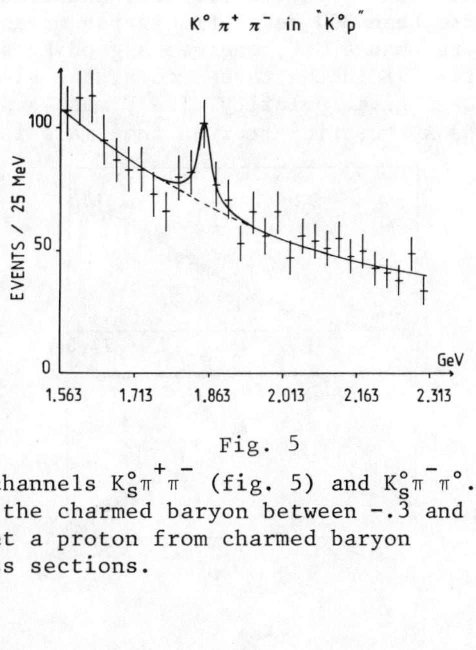

Fig. 5

Signals appear in the channels $K^°_S\pi^+\pi^-$ (fig. 5) and $K^°_S\pi^-\pi^°$. Assuming a flat x distribution of the charmed baryon between -.3 and +.3, and a probability of .5 to get a proton from charmed baryon decay, one gets the following cross sections.

	Channel	$\sigma_{\gamma p \to C\bar{D}X}$ (nb)
$\bar{D}^\circ$	$K^+\pi^-$	510 ± 220
	$K^+\pi^-\pi^\circ$	1300 ± 700
	$K^\circ\pi^+\pi^-$	1150 ± 600
D^-	$K^\circ\pi^-\pi^\circ$	450 ± 310

Hence, most $\bar{D}^\circ$ are produced in association with a charmed baryon. On the contrary, no correlation appears between a $\bar{D}^\circ$ and an additional K^-, or a D° and an additional K^+. This gives a limit on the pair production cross section :

$$\sigma(\gamma p \to D^\circ\bar{D}^\circ X) < 410 \text{ nb}$$

Attempts to see the charmed baryon have failed. Inclusive spectra of possible decay modes are structureless. This is not in contradiction with the previous result but just means that none of these decay channels has a large branching ratio. The upper limits obtained on branching ratios do not contradict the SPEAR results.

3) F meson production

The main trigger data provided a sample of 14 000 $\eta^\circ \to \gamma\gamma$. After fit of the η°, the $\eta^\circ\pi^\pm\pi^+\pi^-$ and $\eta^\circ\pi^\pm\pi^+\pi^-\pi^+\pi^-$ mass spectra show evidence for F production (fig. 6). The absence of significant signal in the $\eta^\circ\pi^\pm$ channel may be due to the multiplicity requirement. Using the high P_T photon trigger, and requiring the charged pion to have more than 300 MeV/c transverse momentum, and the visible mass to be more than 4 GeV, one get a good $F^\pm$ signal in the $\eta^\circ\pi$ decay mode (fig. 7). In the three modes, the signal does not appear concentrated in a given polarity. The F mass averaged is $m_F = 2.020 \pm 0.010$ GeV. The systematic error on this mass is less than 0.020 GeV.

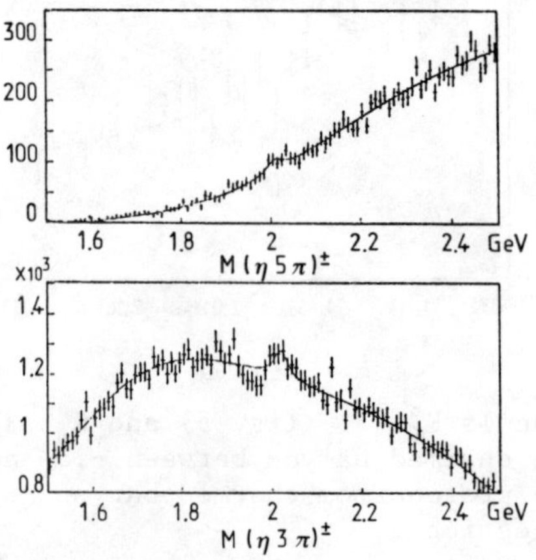

Fig. 6

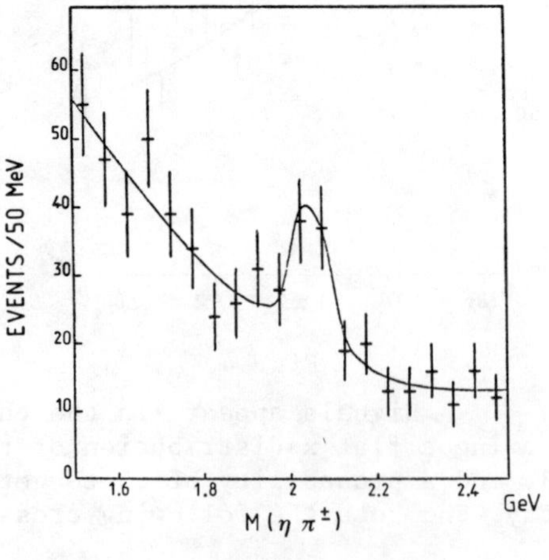

Fig. 7

II. OLD FLAVOUR PHOTOPRODUCTION

1) Inclusive production of $n\pi$, n even

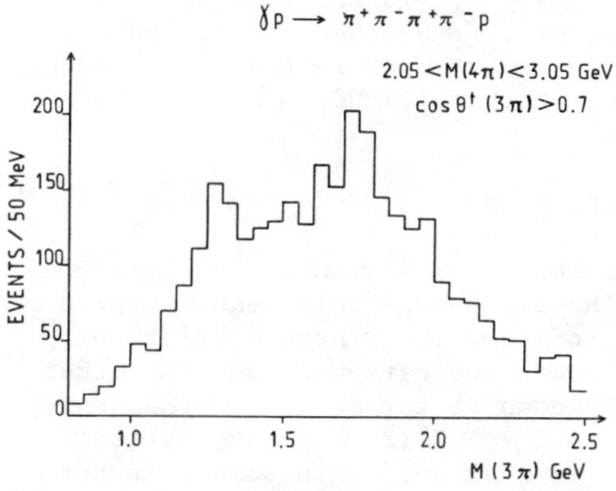

Fig. 8

At 4π masses below 2 GeV, the channel $\gamma p \rightarrow 2\pi^+ 2\pi^- p$ is dominated by the diffractive production of the $J^P = 1^- \rho'(1600)$, which is seen with a width of about 0.4 GeV and decays into $\rho^\circ \pi^+ \pi^-$. At higher 4π masses, there are indications of Deck-like production of the A2 meson, and of another 3π resonance at a mass of 1.76 GeV (Fig. 8). In the channel $\gamma p \rightarrow \pi^+ \pi^- p$, the $\rho'(1600)$ is seen with a width of 0.23 ± 0.08 GeV, significantly smaller than in 4π, and with a production cross section of 0.13 ± 0.02 μb.

The channel $\gamma p \rightarrow \omega^\circ \pi^\circ p$ gives evidence for an $\omega^\circ \pi^\circ$ bump of mass 1.25 GeV and width ~ 0.3 GeV, produced with a cross section of 1.6 ± 0.3 μb. The spin parity is dominantly 1^-, ruling out the interpretation of this bump as being the B meson.

2) Exclusive production of $n\pi$, n odd

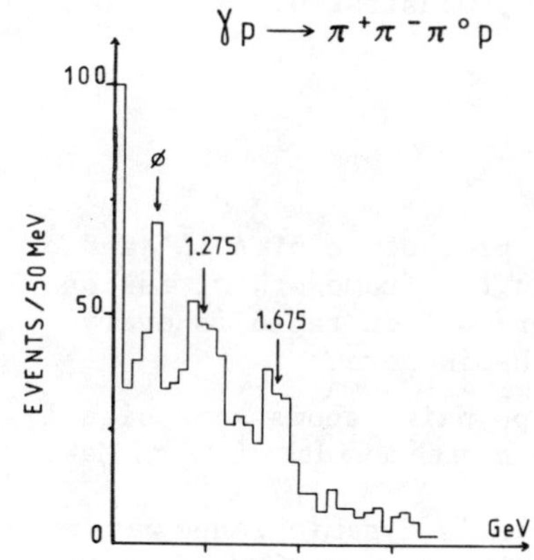

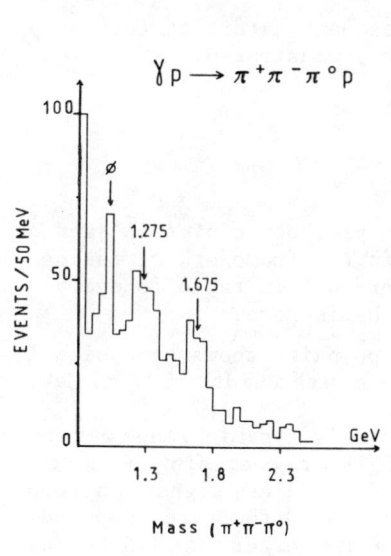

Fig. 9

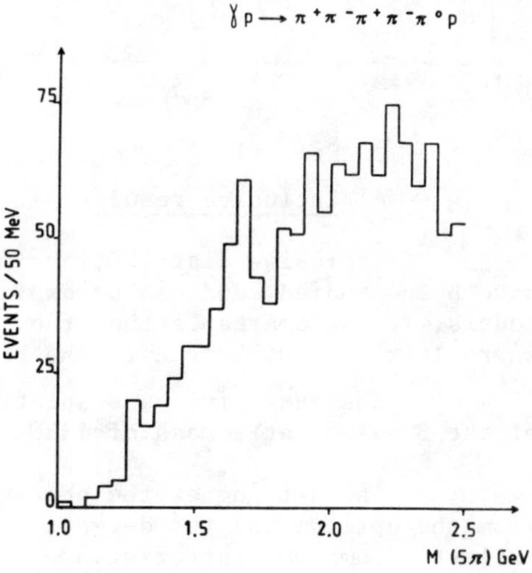

Fig. 10

Beyond ω° and ϕ, two structures appear in the 3π mass distribution of the reaction $\gamma p \rightarrow \pi^+\pi^-\pi^\circ p$, with widths of about 0.1 GeV and masses 1.27 and 1.67 GeV respectively, (Fig.9). Fine binning of the 5π mass spectrum in the $\gamma p \rightarrow 2\pi^+2\pi^-\pi^\circ p$ also indicates the second of these rather narrow structures (Fig. 10). In the same channel, a broad threshold enhancement ($M = 1.7$ GeV, $\Gamma \sim 0.5$ GeV) is produced in the $\omega^\circ\pi^+\pi^-$ mode with a cross section of ~ 0.1 μb. In the $\eta^\circ\pi^+\pi^-$ mode, a bump is seen which, if attributed to the $\rho'(1600)$, gives the ratio : $B(\rho' \rightarrow \eta^\circ\pi^+\pi^-/\rho' \rightarrow 2\pi^+2\pi^-) \sim 0.1$.

3) Exclusive channels with kaons

The channel $\gamma p \rightarrow K^+K^-p$ exhibits a K^+K^- mass peak at 1.75 GeV with a width of 80 MeV (Fig.11). The dip before this peak is typical of some interference effect between the narrow ϕ' and the other known vector mesons. In the channel $\gamma p \rightarrow K^*(890)Kp$, the statistics is weak but again some structure is present in the same mass region.

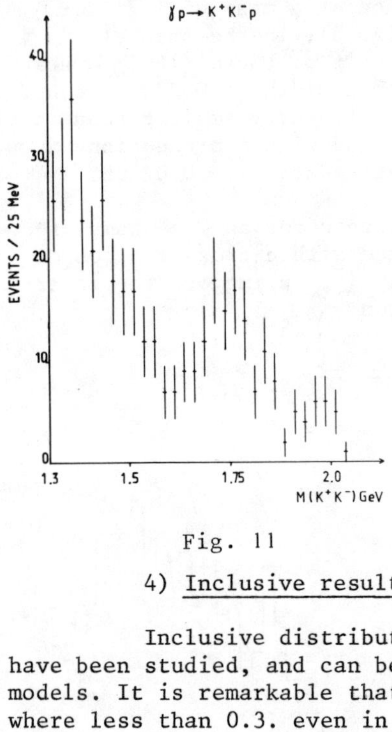

Fig. 11

The reaction $\gamma p \rightarrow K^+K^-\pi^+\pi^-p$ is dominated by the production of $K^*(890)K\pi$ systems for which a broad bump of mass 1.9 GeV and width 0.4 GeV is produced. The production cross-section of the $K^*K\pi \rightarrow K^+K^-\pi^+\pi^-$ channel is 60^{+30}_{-20} nb. The resonant nature of this bump is not demonstrated.

4) Inclusive results

Inclusive distributions for the production of $\Lambda^\circ, \bar{\Lambda}^\circ$ and ϕ have been studied, and can be explained in the framework of fusion models. It is remarkable that the $\bar{\Lambda}^\circ/\Lambda^\circ$ production ratio is everywhere less than 0.3. even in the forward hemisphere.

The inclusive mass spectrum of $p\bar{p}$ pairs shows some signal of the S meson, at a mass of 1930 ± 2 MeV and with a width of 12 ± 7 MeV.

The setting of the photon detector's dynamic range was far from the optimum for the detection of single γ rays arising from deep inelastic Compton scattering. Extraction of a single γ signal has been attempted however, and the data are consistent with the rate computed using the Bjorken-Paschos formula, at P_T values larger than 1.5 GeV/c.

EVIDENCE FOR $D\bar{D}$ DIFFRACTIVE PHOTOPRODUCTION OFF SILICON

E.Albini[2], S.R.Amendolia[3], R.Baldini Celio[1], G.Batignani[6], F.Bedeschi[3], G.Bellini[2], E.Bertolucci[3], G.Bologna[5], L.Bosisio[3], C.Bradaschia[3], M.Budinich[3], F.Celani[1], A.Codino[1], B.D'Ettorre Piazzoli[5], M.DeVincenzi[4], F.L.Fabbri[1], F.Fidecaro[3], L.Foà[6], E.Focardi[3], A.Giazotto[3], M.A.Giorgi[3], P.Laurelli[1], M.Leopold[2], F.Liello[6], P.F.Manfredi[2], G.Mannocchi[5], P.S. Marrocchesi[3], A.Menzione[3], E.Meroni[2], L.Moroni[2], C.Palazzi Cerrina[2], L.Petrillo[4], P.Picchi[5], M.Quaglia[3], F.Ragusa[2], P.G.Rancoita[2], L. Ristori[3], G.Rivellini[1], L.Rolandi[6], S.Sala[2], L.Satta[1], A.Scribano[3], M.Severi[4], P.Spillantini[1], A.Stefanini[3], R.Stanga[3], M.L.Vincelli[3], A.Zallo[1].

Laboratori Nazionali di Frascati[1], Istituto di Fisica e Sezione INFN di Milano[2], Pisa[3], Roma[4], Torino[5] e Trieste[6].

presented by E. Bertolucci

As shown in Fig. 1 an electron beam (5.10^6, 150 GeV electrons per pulse) radiates photons in a 0.1 rad.length converter and is deflected by a magnet into a pair of hodoscopes. These are used to tag the energy of the radiated photons with a resolution of 3-5% and to select photons with energy above 40 GeV. The photon beam crosses a collimator and interacts with the silicon target, 15% of a rad.length thick. A set of 4 magnets interspaced with MWPC and drift chambers measures the momentum of all particles with momentum larger than 0.5 GeV/c and produced within a cone of 90 mrad half aperture; this system provides a rather uniform resolution between 1 and 150 GeV/c, generally better than 1%. A set of counters positioned around the target vetoes all events in which particles are produced at large angles. However, not to loose those events in which the π from the D* decay is produced at large angle, a single particle is allowed to be detected even if up to 30°, outside the magnetic spectrometer acceptance. Two multicell Cerenkov counters are installed inside the first two magnets and select kaons from pions in the momentum range 4-20 GeV/c.

Five shower detectors subdivided with a very fine granularity measure all photons produced below 30°. They provide standard energy and a very good space (±1.5 mm) resolutions. Longitudinally each of them is subdivided into two blocks, the first one 4 rad.lengths thick, the second one ranging between 10 and 20 rad.lengths. The target consists of 40 layers of silicon detectors, 300 μm thick, spaced by 100 μm. Each detector provides a signal proportional to the energy deposited in it and therefore proportional to the multiplicity of crossing minimum ionizing particles (m. i. p.).

If the reaction is incoherent, the fragments release a large quantity of energy in one or, in most of the cases, several layers. If the event is coherent, the signal due to the nucleus recoil is negligible for low mass meson production, but it is equivalent to 2-8 m.i.p. if the mass of the meson is of the order of 4 GeV. Therefore the presence of an isolated signal of such height provides a precise indication of the layer in which the production takes place; in this case the distance form this layer to the following steps gives the paths travelled by the D-mesons.

Fig. 2 shows a few examples of the pulse height patterns shown by the silicon target for typical events identified and reconstrued in

0094-243X/81/680227-03$1.50

the spectrometer as: a) Coherent e^+e^- pair; b) Coherent production of a state of 6 charged particles with a mass of 3.5 GeV; c) Incoherent production of $\pi^+\pi^-$; d) D^{*0}-$\bar{D}^0$ production followed by the decay of the $\bar{D}^0$ into 2 and of the D^{*0} into 4 charged particles; e) D^{*0}-$\bar{D}^0$ production followed by the decay of the D^{*0} into 4 particles, while the decay of the $\bar{D}^0$ takes place outside the target; f) D^+D^- production followed by the decay of D^+ into 3 prongs.

A sample of data has been selected by means of a fit to the target pattern of the events searching for steps in multiplicity and "coherence signals". The selection program was very rough because it disregarded all short path events (gaps of 1-2 and often 3 layers) and it was very inefficient in the identification of D^+D^- events. It is clear thus, that this sample missed the majority of our events and that the bias introduced in this selection does not enable us to evaluate the D's lifetime. Thus we show here only the procedure adopted and the evidence for diffractive photoproduction of $D\bar{D}$ pairs.

Each event was analyzed in the following way:
a) It was checked that the total charge was 0, that the energy was compatible with the photon energy as measured in the tagging system, that the overall transverse momentum was smaller than 0.5 GeV/c;
b) π^0's were reconstructed;
c) All particles were subdivided into two groups, the charge of each group being fixed on the basis of the target pattern;
d) Charged particles with momentum between 4 and 20 GeV/c and seen by the Cerenkov conters were identified as pions. More energetic particles or particles not seen by the Cerenkov conters were labelled "π or K".
e) For each group the request was made that one charged particle was a K or that a pair of pions had the mass of the K^0. If more than one particle was a K-candidate, all configurations were accepted.
f) All groups of particles compatible with the previous requests were finally plotted as points on a m_1 versus m_2 diagram.

This plot is shown in Fig.3 for a 10% of the data analysed. A clustering of points in the region $m_1 + m_2 = 4$ GeV, where D-$\bar{D}$ pairs are expected to fall, is evident. A projection of this plot along the diagonal $m_1 + m_2$ within a cut of $|m_1 - m_2| \leq 0.5$ GeV is shown in Fig. 4 for the full sample of data. The combinatorial spectrum shows again a clear peak around 4 GeV. If we select m_1 to fall in the interval (1.95 - 2.2) GeV (D^* hypothesis), the m_2 spectrum, plotted in Fig. 5 is peaked at 1.86 GeV suggesting that the majority of the analysed events are D-D^*. A D^*-$\bar{D}^*$ component may also be suggested by the data. However this category of events is surely depressed in this preliminary analysis by the request for the total charge of the event to be 0 in the spectrometer whereas at trigger level we allow a single π to escape detection. Altogether we have identified in this way $\sim$ 30 D^0-$\bar{D}^0$ pairs and 10 D^+D^- pairs. On the basis of these results we are now convinced that our data contain a substantial sample of charmed meson pairs, that the apparatus is able to identify them and that for most of them the target provides a value for the decay path of at least one meson. The unbiased analysis of the complete sample of data should therefore allow a precise measurement of the D's lifetime.

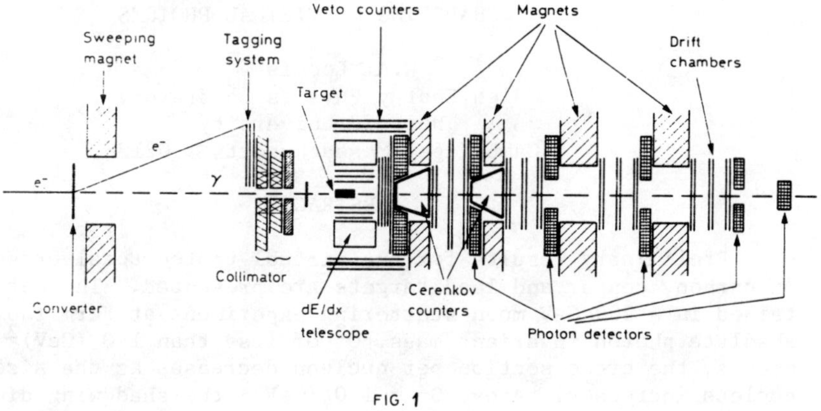

FIG. 1

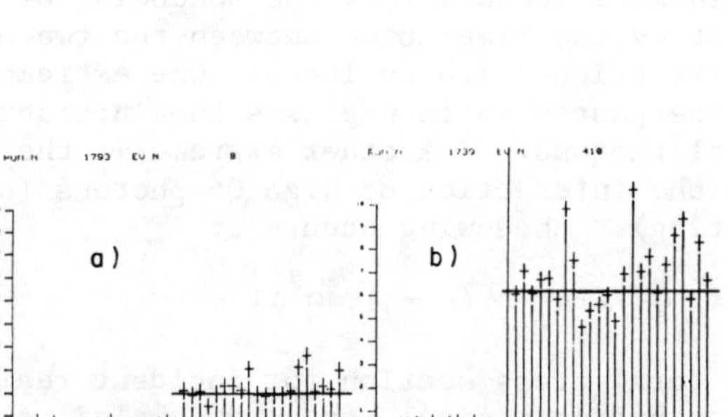

FIG. 2

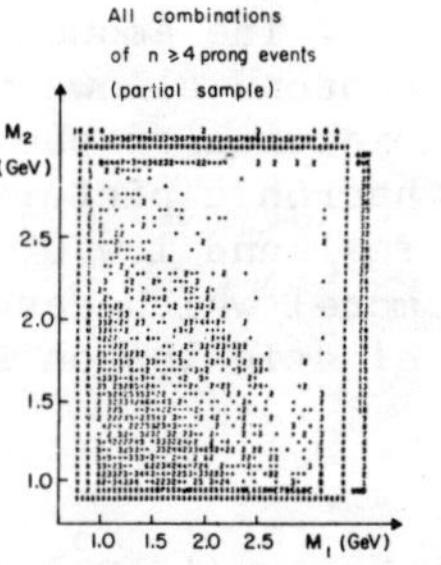

FIG. 3

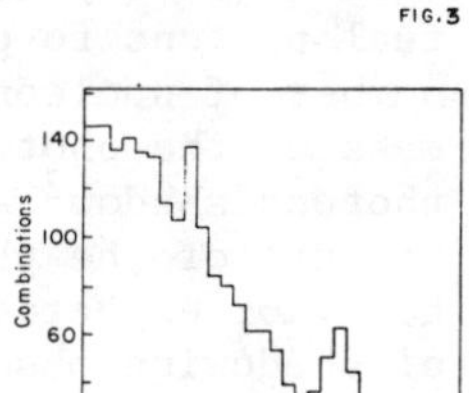

FIG. 4

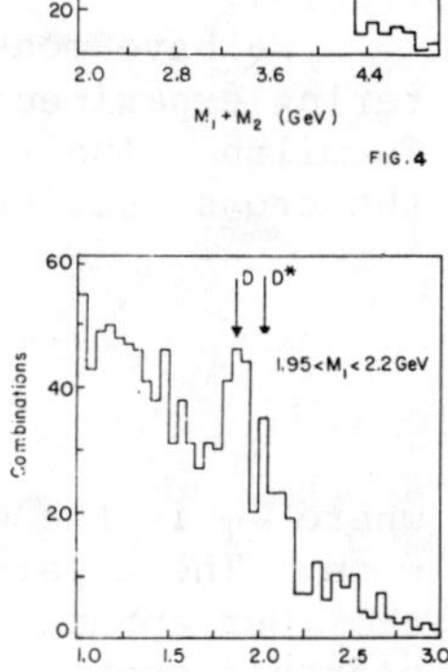

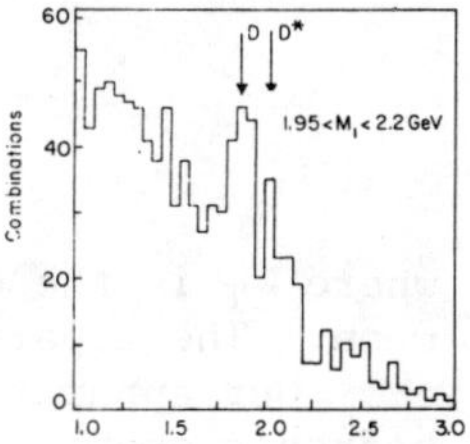

FIG. 5

SHADOWING OF VIRTUAL PHOTONS

W.A. Loomis*
High Energy Physics Laboratory
Harvard University
Cambridge, Massachusetts 02138

ABSTRACT

Preliminary results for the virtual photon total cross section on carbon, copper and lead targets are presented. The data were obtained in a 209 GeV muon scattering experiment at Fermilab. For absolute photon invariant mass, Q^2 of less than 1.0 $(GeV)^2$ shadowing occurs; the cross section per nucleon decreases as the size of the nucleus increases. Above $Q^2 = 1.0$ $(GeV)^2$ the shadowing diminishes.

The amount and kinematic behavior of the shadowing of virtual photons allows us to study the transition between the two extreme pictures of photon interactions with nucleons. One extreme is the hadronic picture[1] of the photon which explains the interactions of real and low Q^2 virtual photons. The other extreme is the quark model which describes the interaction of high Q^2 photons (deep inelastic lepton scattering). Shadowing occurs if

$$\sigma(A,\ Q^2)/[A \cdot \sigma(A = 1,\ Q^2)] < 1 \qquad (1)$$

where $\sigma(A, Q^2)$ is the total cross section for incident real or virtual photons to produce hadronic final states on nuclei, A is the number of nucleons in the nucleus and Q^2 is the absolute invariant mass of the photon and is zero for real photons. Real and virtual photons shadow[2] at small Q^2 as would be expected in the hadronic picture of the photon. However the transition to non-shadowing behavior expected at high Q^2 is not well mapped out and the amount of shadowing observed for real and virtual photons has been controversial.

We have measured $\sigma(Q^2, A)$ for virtual photons in a muon scattering experiment with the Chicago cyclotron magnet spectrometer at Fermilab. The total cross section for virtual photons is taken from the cross section for muon scattering:

$$\frac{1}{\Gamma_T}\left(\frac{d^2\sigma_\mu}{dQ^2\ d\nu}\right) = \sigma(Q^2,\ A) \qquad (2)$$

where Γ_T is the effective flux of virtual photons created by the muon. The apparatus is shown in Figure 1.

Our apparatus has allowed us, for the most part, to identify hadronic events explicitly. The basis for the hadron identification

---

*Representing M.S. Goodman, M. Hall, R. Wilson, F.M. Pipkin, R.K. Thornton, R. Hicks, T. Kirk, J. McAllister, T. Quirk, S.C. Wright, G. Brandenburg, C. Young, W.R. Francis and I. Kostoulas.

0094-243X/81/680230-03$1.50

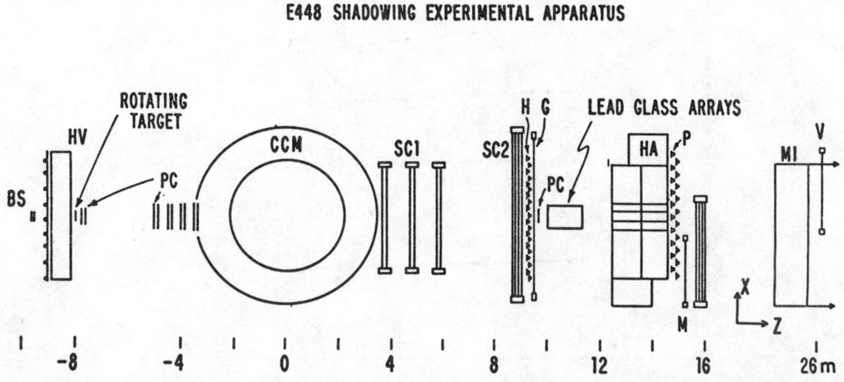

Figure 1. The CCM Spectrometer: BS are beam defining counters, HV is a halo veto wall, PC are multiwire proportional chambers, SC are spark chambers, H, G, M, and P are counter hodoscopes, HA is a 2M steel muon identifier and V is the beam veto.

Our apparatus has allowed us, for the most part, to identify hadronic events explicitly. The basis for the hadron identification is the typical transverse momentum squared of hadron secondaries, approximately 0.18 $(GeV/c)^2$, with respect to the virtual photon direction. This is much larger than the transverse momentum of the photons and/or electrons in the much more numerous wide angle bremsstrahlung (WAB) events. Most of the latter events are clearly identified in the lead glass shower counter array.

Figure 2 shows the total photon cross section per nucleon for various Q^2 ranges as a function of A. The cross sections are averaged over the Q^2 intervals shown. For the lowest Q^2 range, the photon energy (energy lost by the muon) is between 120 and 200 GeV. This range increases to 40 to 200 GeV above Q^2 of 3.0 $(GeV)^2$. The data are corrected for loss of hadronic events, muon acceptance, reconstruction efficiency, and empty target background. We have ignored the difference between neutron and proton in this analysis because the effect should be small at these energies. We have also not made the iterative correction for radiative smearing of Q^2 and photon energy because this depends on the shadowing effect. Hydrogen data[3] also appear in Figure 2. These come from a previous experiment but are treated similarly with respect to the radiative corrections. We summarize the shadowing effect by fitting the cross section per nucleon with:

$$\sigma(Q^2, A)/A = C \cdot A^{\rho - 1} \quad (3)$$

If $\rho - 1$ is less than one then there is shadowing. The results of these fits with and without the hydrogen data are shown in Figure 3. We observe shadowing out to Q^2 of 1 $(GeV)^2$. Beyond that the shadowing appears to diminish with increasing Q^2. We expect this data to have important theoretical consequences.

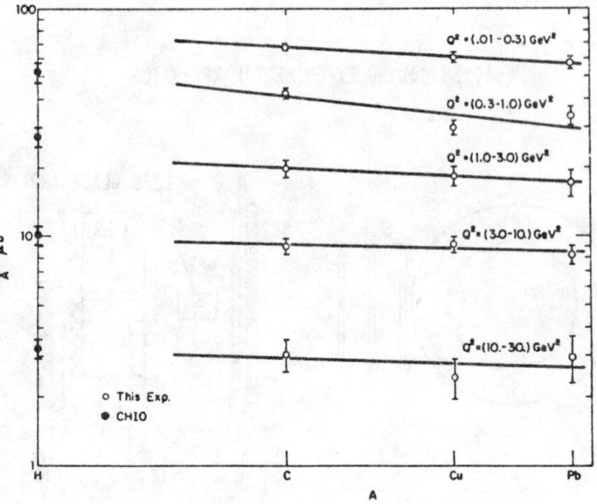

Figure 2. Average virtual photon cross section for various Q ranges as a function of A.

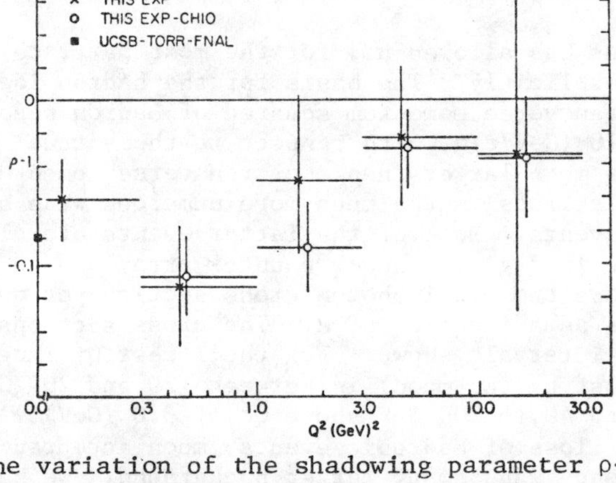

Figure 3. The variation of the shadowing parameter ρ-1 as a function of Q^2. The crosses are fits to the C, Cu, and Pb data only. The open circles include hydrogen data from previous experiments on this apparatus (CHIO collaboration). The square point is from a high energy real photon experiment (Santa Barbara-Toronto-Fermilab).

REFERENCES

1. T.H. Bauer, et al., Rev. Mod. Phys., 50, 261 (1978).
2. D.O. Caldwell, et al., Phys. Rev. Lett., 42, 553 (1979).
   J. Eickmeyer, et al., Phys. Rev. Lett., 36, 289 (1976).
   S. Michalowski, et al., Phys. Rev. Lett., 39, 737 (1977).
3. B.A. Gordon, et al., Phys. Rev. D20, 2645 (1980).

COMPTON SCATTERING AND QUASI-ELASTIC OMEGA PHOTOPRODUCTION AT 50-130 GeV

A.M. Breakstone, D.C. Cheng, D.E. Dorfan, A.A. Grillo, C.A. Heusch, V. Palladino, T. Schalk, A. Seiden, D.B. Smith
Presented by D.B. Smith

University of California, Santa Cruz, California 95064

ABSTRACT

We report a study of elastic Compton scattering ($\gamma p \to \gamma p$) and quasi-elastic ω photoproduction ($\gamma p \to \omega p$) where ω was detected in its π^0 decay mode. These reactions were measured for incident photon beam energies of between 50 and 130 GeV, for the following range of four-momentum transfer squared: $.1\ \text{GeV}^2 < |t| < 1.2\ \text{GeV}^2$.

We measured these reactions in the Fermilab Tagged Photon Laboratory. Using a 5.5% radiation length copper radiator, we received during each spill approximately 2×10^5 tagged photons of between 40% and 95% of the secondary electron beam energy (135 GeV). Our target was a 3 cm diameter by 75 cm long vessel of liquid hydrogen.

We detected the final state of these reactions with a recoil spectrometer and a forward electromagnetic calorimeter. The recoil spectrometer consisted of: a recoil azimuthal angle hodoscope (a barrel of sixteen 150 cm long scintillation counters at one meter around the beam line) and drift chambers (four quadrants of eight drift chamber planes each). The angle of the scattered proton was typically measured with a precision of .01 radian. In addition to the hodoscope for recoiling charged particles, counters not included in the trigger were used to detect charged particles going into the forward hemisphere of the reaction center-of-mass: a 20 cm x 20 cm scintillation counter with a 6 cm diameter hole centered on the beam line (just downstream of the target) to measure charged particles going into the forward hemisphere but not hitting our forward detector; and a bank of six charged particle counters (twenty meters downstream) which shadowed our forward detector. Our forward electromagnetic calorimeter sat 28 meters downstream of the target. A lead-plexiglass sandwich counter with sampling every 1.1 radiation lengths, it had dimensions of 1.58 m x 1.58 m transverse to the beam with a 8 cm by 8 cm square hole for the beam. At a depth of 9 radiation lengths, two fine-grained hodoscopes with 3 cm wide scintillator strips measured the vertical and horizontal profiles of the showers. A further 11 radiation lengths of absorber followed. Our typical azimuthal angular resolution was 0.1 radians.

We had 2.3 million triggers, each a coincidence between the tag, itself a coincidence between the tagging lead glass and hodoscope counters in the absence of signals from associated veto counters, and a signal coming from our forward detector. (The forward detector pulse had to exceed an energy deposition of at least 35 GeV with at least 1.75 GeV in the upstream absorption counters.) To select ("filter out") events of the two reactions of interest, we made

0094-243X/81/680233-03$1.50

essentially the following requirements: Consistency in the measurement of the tagged electron energy by the hodoscope pattern and the lead glass pulse heights. Showers present in the forward detector consistent with the reaction hypothesis: one shower (consistent with single shower widths seen in calibration runs) for the Comptons; and two or three showers for the quasi-elastic ω. Energy in the forward shower counter (after position corrections) within 17% of the nominal photon beam energy. Each shower position in the forward calorimeter at least 3 cm from the detector edge. No particle in the non-triggering charged particle detector just downstream of the target. Successful reconstruction of a track which extrapolates to within 1 cm of the target. A probability or confidence level greater than 1.5% for a kinematic fit to the reaction hypothesis, using the energy predicted by the tagging system.

At this stage we had 1908 Compton candidates and 578 quasi-elastic ω candidates. For the Comptons, we used the data with the charged particle bank to correct for background processes (predominantly $\gamma \to e^+e^-$ followed by ep elastic scattering) which satisfied all our criteria except for the forward showering particle being charged. We needed to extract our ω signal from a significant background which was smooth in histograms of the multi-shower invariant mass. So for each bin in the kinematic variables, we estimated the quasi-elastic ω contribution as the number of events in a mass window containing the peak (640-920 MeV/c^2) minus the number in two adjacent windows of half that width. Dividing our observed cross sections by 8.8% (the $\omega \to \pi^0\gamma$ branching ratio) gave us our final quasi-elastic ω photoproduction cross sections, which we quote <u>without</u> correction for interference with ρ decay. We determined our efficiencies using Monte Carlo simulation techniques.

Our results are the differential cross sections for these two reactions. The following functional forms give equally good fits to our data:

I. Ae^{bt}

II. $Ae^{8.9t + 2.2t^2}$, which corresponds to the sum of the π^+p and π^-p elastic cross sections as expected from applying the additive quark model to the Vector Meson Dominance model. Our differential cross section values for three beam energy intervals are shown in Figure 1a for the Compton and in Figure 1b for the quasi-elastic omega photoproduction. Superimposed are curves given by fits of I to our data summed over all beam energies. Fit I gives A = 541 ± 53 nb/GeV^2 and b = 6.9 ± 0.3 GeV^{-2} for the Compton and A = 6.9 ± 1.3 μb/GeV^2 and b = 6.9 ± 0.6 GeV^{-2} for the quasi-elastic ω photoproduction. Figure 1c and Figure 1d show the beam-averaged differential cross sections for Compton scattering and quasi-elastic ω photoproduction with curves from II normalized to our data. We find that function II gives a good fit to both the Compton (A = 726 ± 38 nb/GeV^2) and the quasi-elastic ω photoproduction (A = 9.6 ± 0.7 μb/GeV^2). Thus, both processes agree with Vector Meson Dominance model predictions for the shape of the differential cross section.

The lower bound on the differential cross section from the

Optical Theorem (using the γp total cross section) is 681 nb/GeV^2 ± 12 nb/GeV^2, shown at t = 0 GeV^2, where I and II give 541 nb/GeV^2 ± 53 nb/GeV^2 and 726 nb/GeV^2 ± 38 nb/GeV^2, respectively. If either function I or II describe the differential cross section at t = 0 GeV^2, the corresponding amplitude is mainly imaginary.

Summing our beam averaged data (for $|t| \geq .07$ GeV^2) with our fit to II (for $|t| < .07$ GeV^2) gives 88 ± 4 nb for the elastic γp cross section, roughly .75% of the total photon cross section.

Similarly for the quasi-elastic ω, summing our data from all beam energies (for $|t| \geq .1$ GeV^2) with our fit to II (for $|t| < .07$ GeV^2) gives 1.16 ± .08 μb for the quasi-elastic ω photoproduction cross section. The Vector Meson Dominance model relates this cross section to the photon-vector meson coupling constant, $\gamma_\omega{}^2/4\pi$, for which we find 5.12 ± 0.35 (possibly up to 5.6 with maximal ρ-ω interference), which compares well with the value from $e^+e^- \rightarrow \omega$, 4.6 ± 0.5. We find that the shape and magnitude of our differential cross sections agree with another independent measurement (R.M. Egloff et al., PRL 43, 1545 (1979), PRL 44, 690 (1980)) made at our beam energies for $|t| < 0.5$ GeV^2.

Thus, for the energy range 50-130 GeV, we find Compton scattering and quasi-elastic ω photoproduction to be consistent with each other in the framework of the Vector Meson Dominance model.

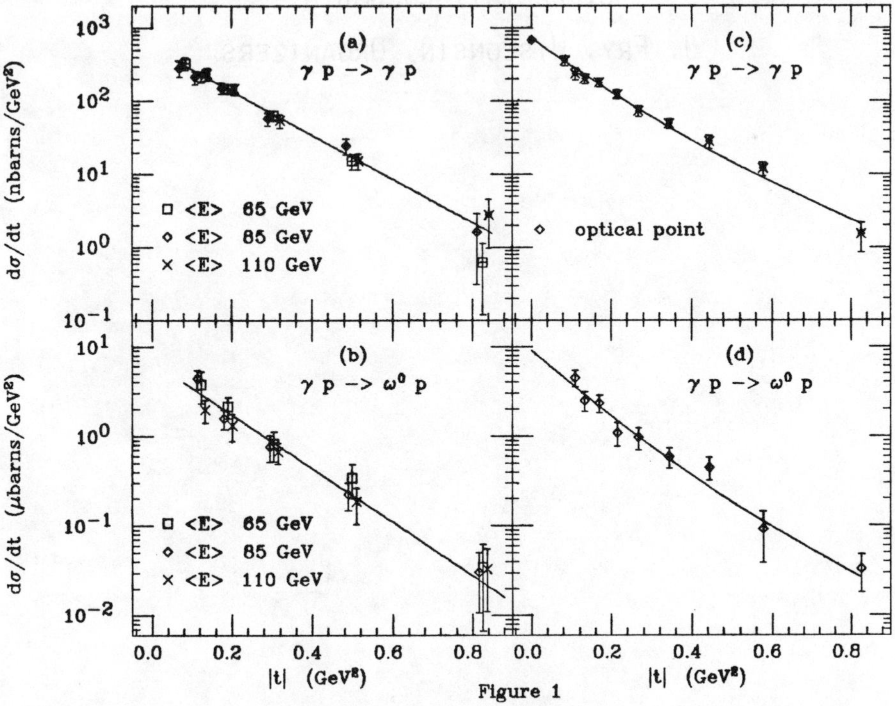

Figure 1

NEW PARTICLE PRODUCTION II, EXPERIMENT

C. Baltay, Columbia

J. Fry, Wisconsin, Organizers

MEASUREMENT OF PROMPT NEUTRINO FLUXES IN A BEAM DUMP EXPERIMENT

CERN-Dortmund-Heidelberg-Saclay-Collaboration[1)]

presented by K. Kleinknecht
Institut für Physik, Universität Dortmund, Dortmund, W.-Germany

ABSTRACT

A prompt ν_μ flux from a 400 GeV proton beam interacting in a copper dump is observed. The flux of prompt $\bar{\nu}_\mu$ in the forward direction is smaller than the ν_μ flux. The rate of prompt ν_e production confirms results from the earlier CDHS beam dump experiment. The ratio of prompt ν_e and ν_μ fluxes is compatible with unity. Observed energy spectra of the prompt events are in qualitative agreement with charm production and decay.

INTRODUCTION

The production of prompt neutrinos in collisions of 400 GeV protons with nuclei was first found in the 1977 beam dump experiments at CERN by the BEBC[2)], GGM[3)] and CDHS[4)] collaborations. In these experiments a flux of prompt ν_e was established, but the question of a prompt ν_μ flux had not been settled.

EXPERIMENT

We report here on a second experiment of this type. A total of $6.61\cdot10^{17}$ protons were dumped into a copper dump of full density (length 200 cm) and $2.48\cdot10^{17}$ protons impinged on a copper dump of 1/3 density, 236 cm long. The fiducial mass of the CDHS neutrino detector[5)], situated at 890 m downstream of the dump in the exact forward direction, was 492 tons for the analysis of single muon (charged current, CC) neutrino events, and 465 tons for analyzing muonless events.

Single muon events

For events with one muon, the shower energy E_{sho} and the muon momentum were measured. For $E_{vis}=E_{sho}+E_\mu>20$ GeV, a total of 1270 μ^- events and 251 μ^+ events was found. The distribution in the Bjorken scaling variable $y=E_{sho}/E_{vis}$ is shown in fig.1 for the full density dump data. The distributions agree perfectly with the expected ones from muon neutrino interactions, viz. $\nu_\mu + Fe \rightarrow \mu^- + X$ and $\bar{\nu}_\mu + Fe \rightarrow \mu^+ + X$; we therefore identify the μ^- events with charged current ν_μ events ($CC\mu^-$) and the μ^+ events with charged current $\bar{\nu}_\mu$ interactions ($CC\mu^+$). The visible energy distributions of $CC\mu^-$ events for the full density ($\rho=1$) and $\rho = 1/3$ density dumps are given in figs. 2 and 3.

In order to extract the contribution in this event sample of neutrinos from very short-lived sources, where the decay path of the parent is short compared to the nuclear interaction length in copper (12 cm) we employ two different methods: i) extrapolation method: the $\rho=1$ and $\rho=1/3$ density data are used to extrapolate to infinite density to get the prompt signal, ii) subtraction method: the non-prompt

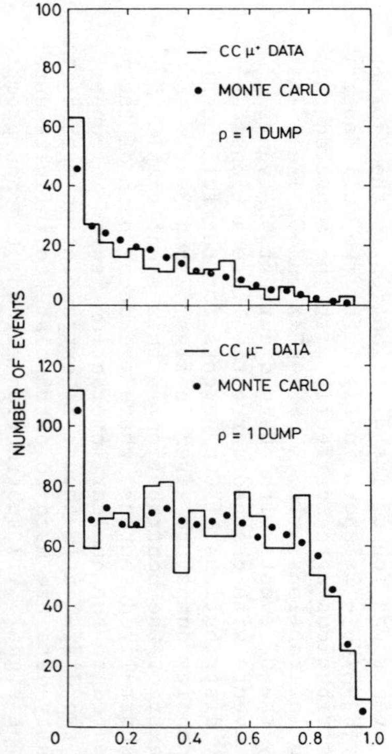

Fig. 1: Distribution of single μ events in the scaling variable y

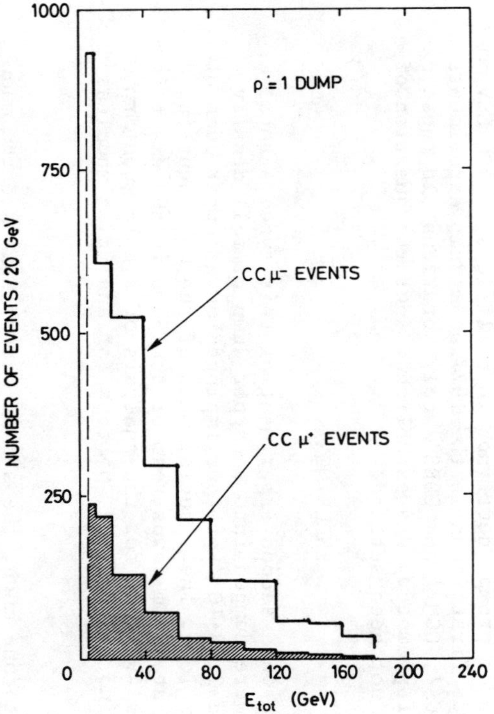

Fig. 2: Visible energy spectrum of charged current events; ρ = 1 dump

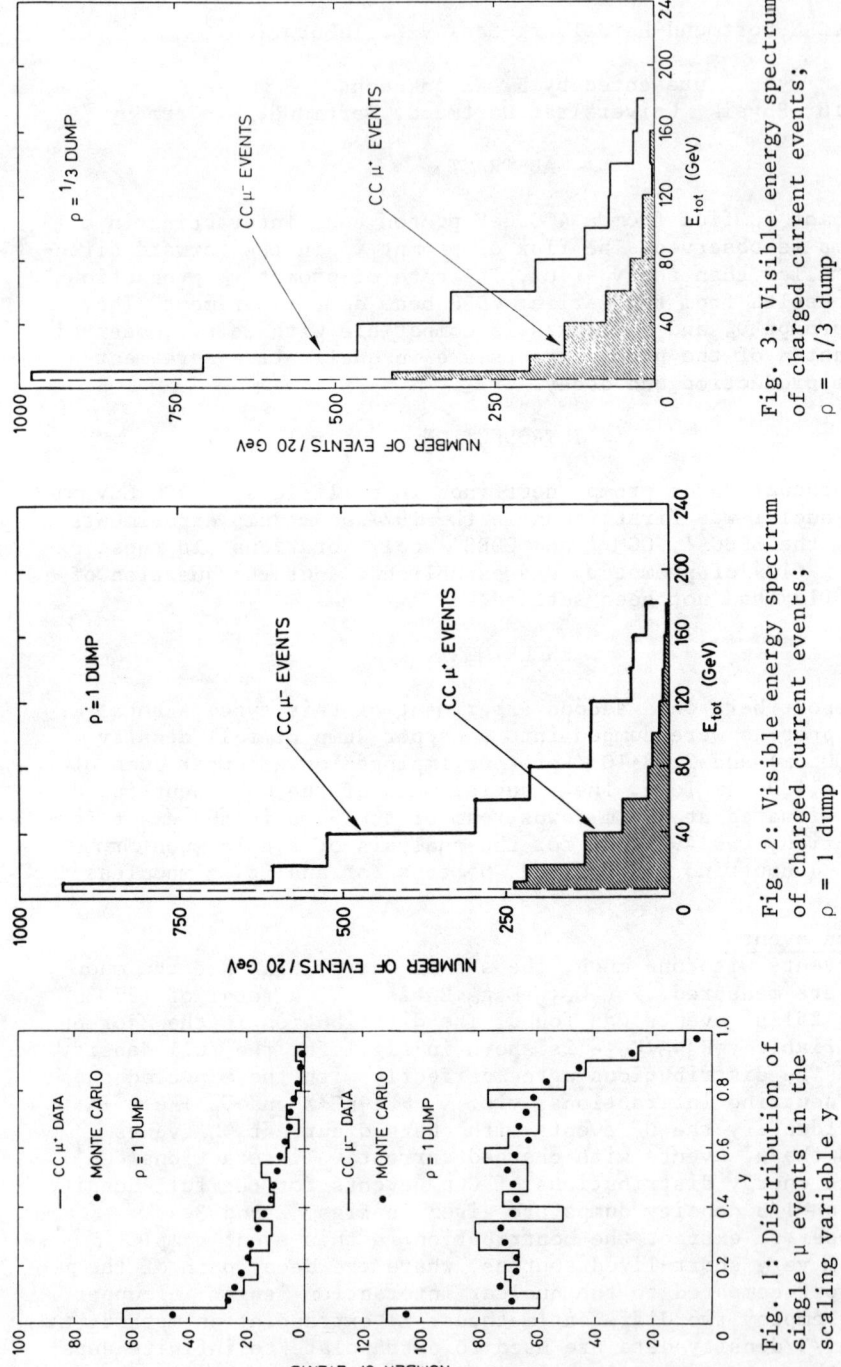

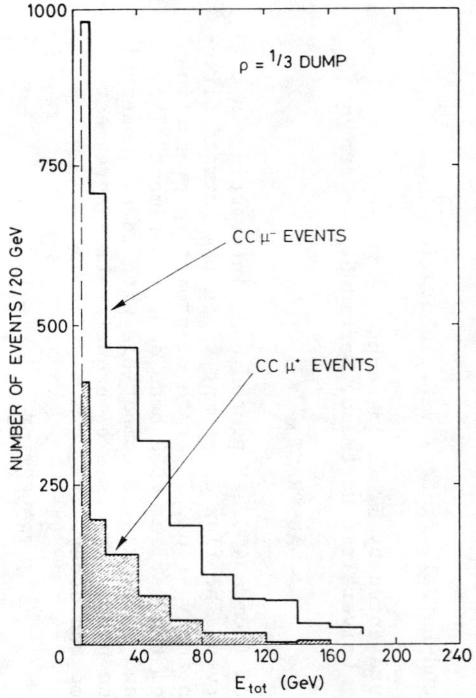

Fig. 3: Visible energy spectrum of charged current events; ρ = 1/3 dump

contributions of all conventional sources of neutrinos (π, K, hyperon decays) are computed using experimental production cross-sections[6]. The result of method i) can be seen in fig.4. With a linear density extrapolation, a definite prompt $CC\mu^-$ signal of 459±83 events is observed. The corresponding number for $CC\mu^+$ is 38±39 events, giving no evidence for a prompt $\bar{\nu}_\mu$ flux. The subtraction method, on the other hand, gives for the difference between observed events and the calculated non-prompt events 537±44 prompt $CC\mu^-$ events and 126±20 $CC\mu^+$ events.

While the results of the two methods agree for $CC\mu^-$ events, the calculation gives a smaller non-prompt $CC\mu^+$ rate than the one given by the density extrapolation. The rates of prompt events per ton of detector and 10^{18} protons are given in table 1.

Table I Prompt $CC\mu$ rates / $(t \cdot 10^{18} p)$		
	$CC\mu^-$	$CC\mu^+$
Extrapolation method	1.23 ± 0.25 ± 0.32	0.07 ± 0.12 ± 0.07
Subtraction method	1.35 ± 0.13 ± 0.48 (stat) (syst)	0.38 ± 0.06 ± 0.08 (stat) (syst)

For both methods, the prompt $CC\mu^+$ rate is smaller than the $CC\mu^-$ rate.

Muonless events

Events of this type are treated analogously to our neutral-current analysis[7]. Using the length of the event in the iron detector, the number of genuine 0μ events is obtained by subtracting cosmic ray background and CC events with muon hidden in the hadron shower. The remaining 0μ event sample consists of events induced by the new source of ν_e and $\bar{\nu}_e$, events from conventional ν_e and $\bar{\nu}_e$ sources (hyperon β decays, K_{e3} decays: "standard $\overset{(-)}{\nu}_e$ background") and the background from neutral current events induced by $\overset{(-)}{\nu}_\mu$. The "standard ν_e background" can be calculated, and the NC background is obtained from the measured $CC\mu$ rates and the measured NC/CC ratio. Subtracting the background, an excess of muonless events is found, whose distribution in the shower energy E_{sho} is shown in fig.5.

The rate of these excess events above E_{sho}=20 GeV is (1.21±0.11(stat)±0.11(syst))/(ton·10^{18} protons) for the full density dump. In order to obtain the rate of prompt CC events induced by ν_e or $\bar{\nu}_e$, this number has to be multiplied by 0.83±0.02, such that the prompt CCe^-+CCe^+ rate is (1.00±0.09±0.09)/ton·10^{18} protons).

ν_e/ν_μ ratio

Using the numbers given above, we obtain for the ratio r=(prompt CCe^-+CCe^+)/(prompt $CC\mu^-+CC\mu^+$) the numbers quoted in table 2.

Table II ν_e/ν_μ Flux ratio	
	Ratio $r = \phi(\nu_e)/\phi(\nu_\mu)$
Extrapolation ($CC\mu$); E_{sho}>20 GeV	0.77±0.18±0.24
Subtraction ($CC\mu$); E_{sho}>2o GeV	0.58±0.07±0.19
Extrapolation ($CC\mu$); E_{sho}>8o GeV	0.80±0.20

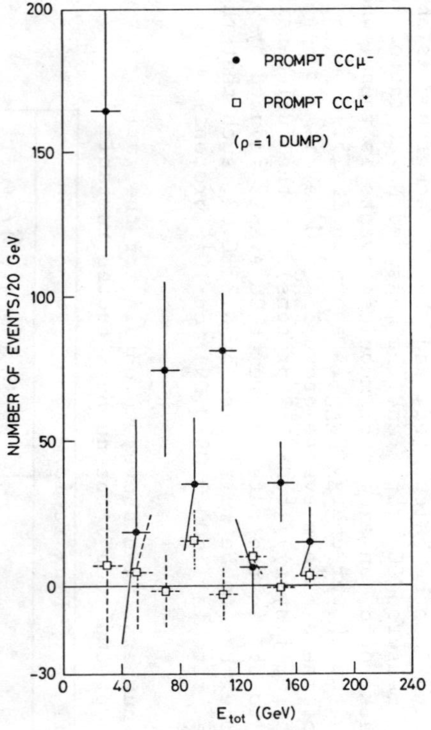

Fig. 4: Prompt $CC\mu^-$ and $CC\mu^+$ events vs. visible energy

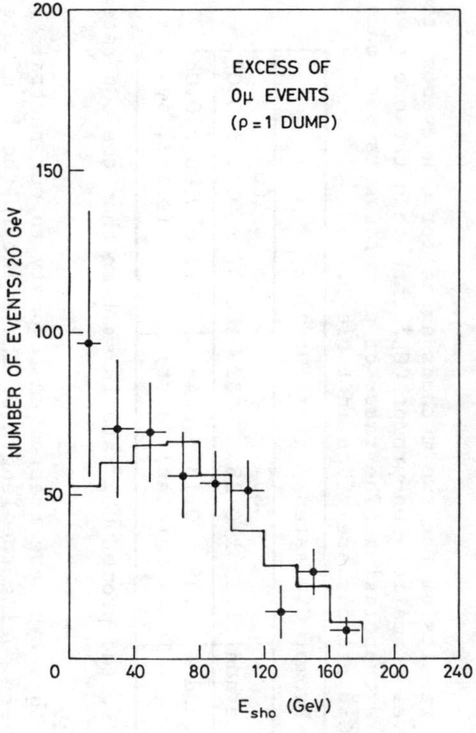

Fig. 5: Excess of muonless events vs. shower energy E_{sho}

If the neutrinos come from the weak decay of a heavy particle mass with $m \gg m_\mu$, μ-e universality would predict r=1. The data are compatible with this expectation, and there is no significant energy dependence of r.

CONCLUSIONS

i) The asymmetry between prompt $\bar{\nu}_\mu$ and ν_μ fluxes could indicate a contribution of charmed hadron production in the forward direction near X_F=1.
ii) The energy spectra of the prompt events are in qualitative agreement with charm production and decay.
iii) The ratio of fluxes $\phi(\nu_e)/\phi(\nu_\mu)$ is compatible with μ-e universality.

REFERENCES

1) The members of this collaboration are: H.Abramovicz, J.G.H.de Groot, J.T.He, J.Knobloch, J.May, P.Palazzi, A.Para, F.Ranjard, T.Z.Ruan, A.Savoy-Navarro, D.Schlatter, J.Steinberger, H.Taureg, W.von Rüden, H.Wahl, J.Wotschak, W.M.Wu (CERN); F.Eisele, H.Klasen, K.Kleinknecht, H.Lierl, B.Pszola, B.Renk, H.J.Willutzki (Universität Dortmund); I.Becker. G.Conforto, F.Dydak, T.Flottmann, C.Geweniger, V.Hepp, R.Herden, J.Królikowski, K.Tittel (Universität Heidelberg); P.Bloch, C.Guyot, S.Loucatos, J.P.Merlo, B.Peyaud, J.Rander, J.Rothberg, J.P.Schuller, R.Turlay (CEN Saclay)
2) P.C.Bosetti et al., Phys.Lett. 74 B (1978) 143.
3) P.Alibran et al., Phys.Lett. 74 B (1978) 134.
4) T.Hansl et al., Phys.Lett. 74 B (1978) 139.
5) M.Holder et al., Nucl.Instr.Meth. 148 (1978) 235.
6) H.Wachsmuth, Neutrino and Muon fluxes in the CERN 400 geV proton beam dump experiment, report CERN-EP/79-125.
7) M.Holder et al., Phys.Lett. 71 B (1977) 222.

EXPERIMENTAL STUDY OF PROMPT NEUTRINO PRODUCTION IN 400 GeV PROTON-NUCLEUS COLLISIONS

(The CHARM Collaboration)

M. Jonker, J. Panman and F. Udo
NIKHEF, Amsterdam, The Netherlands

J.V. Allaby, U. Amaldi, G. Barbiellini[1], A. Baroncelli, V. Blobel[2], W. Flegel, W. Kozanecki[3], K.H. Mess, M. Metcalf, J. Meyer, R.S. Orr, F. Schneider, V. Valente[1], A.M. Wetherell and K. Winter
CERN, Geneva, Switzerland

F.W. Büsser, P.D. Gall, H. Grote, B. Kröger, E. Metz, F. Niebergall, K.H. Ranitzsch and P. Stähelin
II. Institut für Experimentalphysik, Universität Hamburg[4], Hamburg, Germany

P. Gorbunov, E. Grigoriev, V. Kaftanov, V. Khovansky and A. Rosanov
Institute for Theoretical and Experimental Physics, Moscow, USSR

R. Biancastelli[5], B. Borgia[6], C. Bosio[5], A. Capone[6], F. Ferroni[6], E. Longo[6], P. Monacelli[6], F. de Notaristefani[6], P. Pistilli[6] and C. Santoni[5]
Istituto Nazionale di Fisica Nucleare, Rome, Italy

ABSTRACT

Results are reported from a proton beam-dump experiment performed at the 400 GeV CERN SPS using the CHARM neutrino detector. Prompt electron-neutrinos and prompt muon-neutrinos, produced by decays of short-lived parents, have been observed. The ratio of the fluxes of $(\bar{\nu}_e + \nu_e)$ and of $(\bar{\nu}_\mu + \nu_\mu)$, measured by the rates of charged-current interactions with $E_{vis} > 20$ GeV, is 0.48 ± 0.12 (statistical) $\pm$ 0.10 (systematic). The ratio of $\bar{\nu}_\mu$ and ν_μ fluxes is $1.3^{+0.6}_{-0.5}$. At low shower energies, $2 < E_{vis} < 20$ GeV, we observe 54 ± 19 (statistical) $\pm$ 9 (systematic) prompt muonless events in excess of electron- and muon-neutrino interactions expected from standard $D\bar{D}$ production and decay.

* * *

The 400 GeV proton beam of the CERN Super Proton Synchrotron (SPS) has been directed onto a 2 m long copper target which was segmented to obtain different mean densities by suitably spacing the disks. Two densities, 3 g/cm^3 and 9 g/cm^3, called 1/3 and 1 respectively, were used during data taking, allowing the prompt neutrino flux to be determined directly by extrapolation of the event rates to infinite density.

---

1 On leave of absence from the Laboratori Nazionali dell'INFN, Frascati, Italy.
2 On leave of absence from II. Institut für Experimentalphysik, Universität Hamburg, Fed. Rep. Germany.
3 Now at University of California, Riverside, California, USA.
4 Supported by the Bundesministerium für Forschung und Technologie, Bonn, Fed. Rep. Germany.
5 Istituto Superiore di Sanità, Roma and INFN Sez. Sanità, Rome, Italy.
6 Istituto di Fisica, Università di Roma and INFN Sez. di Roma, Italy.

The muon flux was measured[1] in the first four gaps of the iron shield, using targets of three mean densities (3, 4.5, and 9 g/cm^3). Extrapolation to infinite density allows the flux of prompt and non-prompt decay muons to be determined. The muon fluxes, monitored throughout the experiment, were used to determine the flux of non-prompt neutrinos from the decay of pions and kaons in the dump target, which were then subtracted from the observed neutrino event flux to obtain the contributions of prompt neutrinos. Both methods -- extrapolation and subtraction -- are subject to different uncertainties, and a comparison of their results is therefore an important check of consistency.

The fine-grain target calorimeter of the CHARM Collaboration[2] was located 910 m from the target. The apparatus consists of a segmented calorimeter surrounded by a magnetized iron frame, and of a muon spectrometer. It has high efficiency for detecting showers with energy $E_{sh} > 1.5$ GeV, good energy resolution, and high power of pattern recognition, allowing, for example, the detection of muons with momenta as low as 1 GeV/c. With a fiducial target mass of 100 tons, it combines many of the advantages of a visual detector and a high-mass calorimeter.

Data were taken with the full density beam-dump for 6.96×10^{17} protons on target (abbreviated as POT) and with the 1/3 density beam-dump for 2.60×10^{17} POT.

Neutrino interaction candidates were selected in a fiducial volume of 100 tons if the energy deposited was at least 2 GeV.

Events were classified as "one-muon" events if they had at least one primary track of at least 1 GeV/c momentum which shows no interaction and extrapolates to the vertex. All other events are called "muonless".

Figure 1 shows the visible energy spectrum of prompt muonless events corrected for NC interactions of muon-neutrinos, and for neutral-current (NC) and charged-current (CC) interactions of non-prompt electron-neutrinos.

In Table 1 we list, for the range in visible energy from 2 GeV to 20 GeV, the raw event numbers and corrections applied to the data, including the systematic errors. The contribution of NC interactions of muon-neutrinos was calculated using the corrected number of single-muon events and the average ratio of NC to CC cross-sections for neutrinos and antineutrinos, $\langle R \rangle = 0.324$.

The full points in Fig. 1 represent an estimate of the spectrum of prompt NC and CC electron-neutrino interactions, calculated under the assumption that they are due to pair production and subsequent semileptonic decay of charmed D mesons[3]. Normalizing to the number of events observed with $E_{vis} > 20$ GeV, we obtain for the product of total cross-section and branching ratio, assuming that the cross-section is proportional to the number of nucleons,

$$\sigma_{tot}(pp \to D\bar{D} + X) \cdot BR(D \to \nu_e eX) = 1.5 \pm 0.5 \ \mu b \ . \qquad (1)$$

The error includes uncertainties of the production model.

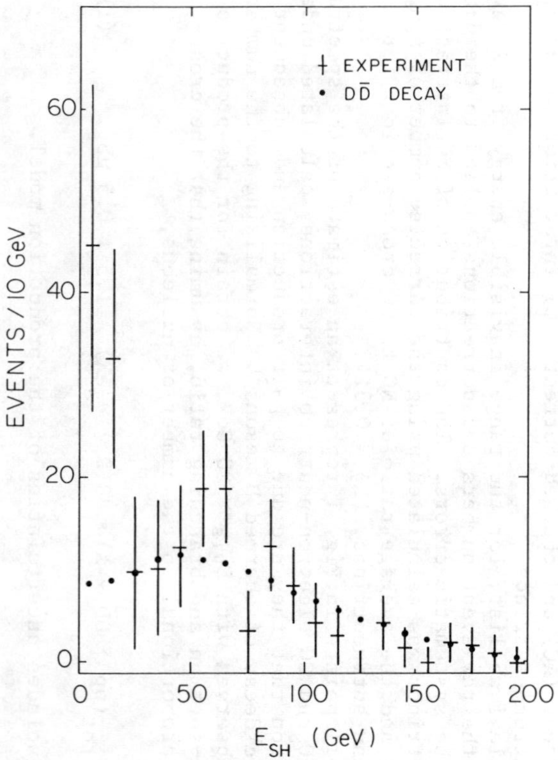

Fig. 1 Shower energy distribution of prompt muonless events in excess of muon-neutrino NC for ρ = 1 and 1/3 interactions. The points show the distribution of electron-neutrino interactions expected from standard $D\bar{D}$ production and decay.

<u>Table 1</u>

Event summary ($2 < E_{sh} < 20$ GeV).
Combined data from density-1 and density-1/3 runs.

	Muonless events	One-muon events	Systematic error
Raw event numbers	291	469	
Corrections from:			
πK decay muons	+ 3.7 ± 0.2	- 3.7 ± 0.2	±1.6
soft-muon loss	-23.3 ± 1.7	+23.3 ± 1.7	±4.0
side-muon loss	-11.5 ± 0.5	+11.5 ± 0.5	±2.0
cosmic background	- 9.5 ± 1.5	0	0
Corrected event numbers	250.4 ± 17.2	500.1 ± 21.7	±4.7
Expected contributions:			
Muon-neutrino NC	162 ± 7		±3.0
Electron-neutrino (non-prompt)	19.7 ± 0		±4.0
Prompt events:			
by subtraction	69 ± 19		±8
by extrapolation	43 ± 38		
Prompt excess: over $D\bar{D}$	54 ± 19		±9

Below E_{vis} = 20 GeV we observe 54 ± 19 (statistical) ± 9 (systematic) prompt muonless events in excess of ν_e and ν_μ interactions expected from standard $D\bar{D}$ production and decay.

To search for prompt muon-neutrinos, we have used the sample of one-muon events obtained with the full density target. For events with p_μ > 4 GeV/c, the charge of the muons can be determined by the muon spectrometer. Table 2 gives the results of the analysis after corrections for the momentum cut. We find evidence for prompt ν_μ and $\bar{\nu}_\mu$ production.

Table 2

Prompt one-muon event rates (E_ν > 20 GeV)

Reaction	Rate/(ton × 10^{17} POT)	
	Subtraction of conventional background (from density-1 runs)	Extrapolation to infinite density
$\nu_\mu N \to \mu^- X$	0.107 ± 0.027 (stat.) ± 0.033 (syst.)	0.106 ± 0.054
$\bar{\nu}_\mu N \to \mu^+ X$	0.069 ± 0.015 (stat.) ± 0.006 (syst.)	0.091 ± 0.029

The flux ratio of prompt electron- and muon-neutrinos can be evaluated by the number of their CC interactions. We restrict this analysis to data obtained with full dump density in order to minimize the systematic error due to the subtraction of non-prompt background. We find 138 ± 20 (statistical) ± 29 (systematic) prompt events of the reaction $\nu_\mu,\bar{\nu}_\mu N \to \mu^-,\mu^+ X$ with E_{vis} > 20 GeV. The systematic error is due to an estimated 11.5% uncertainty[1] on the rate of non-prompt muon-neutrino events. The number of NC and CC electron-neutrino interactions has been determined by two different methods. In one method we used the spectrum of prompt neutrinos from $D\bar{D}$ production and semileptonic decay to describe the observed spectrum of visible energy of prompt muonless events for E_{vis} > 20 GeV. The ratio of fluxes above E_{vis} = 20 GeV is then

$$R_\phi = \frac{\phi(\bar{\nu}_e+\nu_e)}{\phi(\bar{\nu}_\mu+\nu_\mu)} = \frac{N(\nu_e,\bar{\nu}_e N \to e^-,e^+X)}{N(\nu_\mu,\bar{\nu}_\mu N \to \mu^-,\mu^+X)} = 0.48 \pm 0.12\ (\text{stat.}) \pm 0.10\ (\text{syst.})\ . \tag{2}$$

By extrapolation to infinite density we find

$$R_\phi = 0.49 \pm 0.21\ . \tag{3}$$

The systematic error in Eq. (2) includes the uncertainties in the corrections applied to the raw data, the uncertainties in the fluxes[1] of non-prompt muon-neutrinos (±11.5%) and electron-neutrinos (±20.8%), and the uncertainty in the production mechanism of D mesons[3] by varying the exponent describing the dependence on $x = p_L/p_L^{max}$ from n = 3 to 5.

Evaluating R_ϕ as a function of the visible energy above 20 GeV we find that R_ϕ is energy-independent within the errors quoted in Eq. (2).

In the other method, we identify electron-neutrino CC events directly. This method is based on three characteristic features of electromagnetic showers: their small width, their regular longitudinal profile, and the strong correlation between the total shower energy and the energy detected at the shower maximum. The energy detected in a road of scintillators, one counter (15 cm) wide, starting at the vertex and crossing the shower maximum, has been calibrated using isolated electron showers obtained in a test beam. Neutral-current events with π^0's of high energy may simulate ν_e CC events. We determine this contribution using muonless events obtained in the wide-band beam, and subtract it from the beam-dump data.

For $E_{vis} > 20$ GeV and $\rho = 1$ we find, after subtraction of non-prompt contributions[1], 60.7 ± 13 (statistical) ± 5.3 (systematic) CC events induced by electron-neutrinos. The flux ratio,

$$R_\phi = 0.44 \pm 0.11 \text{ (statistical)} \pm 0.10 \text{ (systematic)} , \qquad (4)$$

thus directly determined, agrees with the model-dependent result in Eq. (2).

The flux ratio is significantly different from 1, in disagreement with expectations from the $D\bar{D}$ production model, whereas the shape and magnitude of the measured spectrum of CC ($\nu_e + \bar{\nu}_e$) events agree reasonably well with the predictions from this model.

REFERENCES

1. H. Wachsmuth, Neutrino and muon fluxes in CERN 400 GeV proton beam-dump experiment, preprint CERN-EP/79-125 (1979).
2. A.N. Diddens et al., CHARM Collaboration, A detector for neutral-current interactions of high-energy neutrinos, preprint CERN-EP/80-63 (1980), to be published in Nucl. Instrum. Methods.
3. We have used a parametrization for the pair production of $D\bar{D}$

$$\frac{d^2\sigma}{dx dp_T} = a(1 - |x|)^4 \frac{p_T}{E_D^*} \exp(-2p_T) , \quad \text{with } x = p_L/p_L^{max} ,$$

which gives reasonable agreement with the observed visible energy spectrum above 20 GeV.

STUDY OF PROMPT NEUTRINO PRODUCTION IN PROTON-Cu INTERACTIONS USING BEBC IN CONNECTION WITH NARROWBAND ν_e RESULTS

Aachen-Bonn-CERN-IC London-Oxford-Saclay Collaboration
Presented by: P.O. Hulth, CERN, Geneva, Switzerland

Abstract

Evidence for prompt ν production in proton-Cu interactions at 400 GeV/c is presented using BEBC as detector. An upper limit on the oscillation probability $P(\nu_e \rightarrow \nu_\tau)$ of 0.35 at 90% CL is given. The hypothesis that the τ lepton has the electron lepton number is ruled out to 90% CL. The interpretation of ν oscillations from the prompt $e^\pm/\mu^\pm$ ratio seems to be ruled out from the determination of $P(\nu_e \rightarrow \nu_e) = 1.04 \pm 0.15$ obtained from a narrowband beam experiment performed by the same collaboration.

INTRODUCTION

This report is a summary of the talk given at the XXth International Conference on High-energy Physics, University of Wisconsin - Madison 1980 describing the BEBC beam dump experiment and a measurement of the probability $P(\nu_e \rightarrow \nu_e)$ obtained from a narrowband beam experiment made by the same collaboration at BEBC.

A bubble chamber filled with a heavy Ne/H_2 mixture is excellent to detect electrons which are identified via bremsstrahlung. This feature is important when studying rare processes like ν_e interactions. The aim of a beam dump experiment is to study ν (or any new weakly interacting neutral particle) produced by short-lived particles (here called prompt) in interactions between protons and the material in the dump. The conventional ν_e and ν_μ from π and K decays are suppressed because these long-lived particles will mainly interact in the dump before they decay. During 1977 and 1979, two sets of beam dump experiments were performed at CERN [1-4]. In this report combined results of the two BEBC experiments are given. Ref. [5] gives a description of the experimental set-up.

In BEBC the observed events are classified into five classes. Events with a μ^- or μ^+ > 5 GeV/c identified in the External Muon Identifier (EMI) are classified as charged current muon ν events, $CC\mu^-$ or $CC\mu^+$. Events with an electron or positron identified via bremsstrahlung are called CCe^- or CCe^+ events. If there is no identified lepton in the event, it is called a neutral current candidate (NC).

In table 1 the observed number of events with E_{vis} > 10 GeV from the 1977-1979 full density dump and the 1979 1/3 density dump is presented.

TABLE 1

Observed number of events with E_{vis} > 10 GeV

	Full density (1977 + 1979)		1/3 density (1979)	
Type	Total/t and 10^{18} protons		Total/t and 10^{18} protons	
NC	47	3.2 ± 0.5	38	8.9 ± 1.4
$CC\mu^-$	85	5.7 ± 0.6	53	12.4 ± 1.7
$CC\mu^+$	24	1.6 ± 0.3	13	3.0 ± 0.8
CCe^-	17	1.1 ± 0.3	10	2.3 ± 0.7
CCe^+	6	0.4 ± 0.2	3	0.7 ± 0.4

These numbers of events have to be corrected for: (i) muon events, the EMI inefficiency (2%) and the 5 GeV/c momentum cut; (ii) electron events for the detection efficiency (90 ± 5%) and the probability to observe an electron from the interaction vertex (96 ± 2%) and (iii) NC events for the contamination of unobserved $CC\mu^{\pm}$ and $CCe^{\pm}$ events, where the lepton is not recognized.

The corrected number of events is shown in table 2 together with the number of prompt events/t of detector and 10^{18} protons at 820 m distance from the dump. The expected number of events have been calculated from ref. [5] estimating the ν flux from the conventional decays of π, K, etc. in the dump. The method of extrapolating to infinite density to get the prompt signal gives the same result but with larger errors.

TABLE 2

Summary of corrected number of events with $E_{vis} > 10$ GeV

Type	Total	Expected	Prompt signal	prompt signal ev/t . 10^{18} p
$NC^{(a)}$	30.2 ± 7.3	17.0	13.2 ± 7.5	0.89 ± 0.51
$CC\mu^{-(a)}$	99.4 ± 10.7	69.6	29.8 ± 13.2	2.01 ± 0.89
$CC\mu^{+(a)}$	25.3 ± 5.2	14.0	11.3 ± 5.4	0.76 ± 0.36
CCe^-	31.4 ± 5.9	7.6	23.8 ± 6.0	1.25 ± 0.32
CCe^+	10.0 ± 3.5	2.4	7.6 ± 3.5	0.40 ± 0.18

(a) Only full density.

SOME FEATURES OF THE DATA

(a) The ratio of the prompt CCe^+ to CCe^- is 0.32 ± 0.17, 0.36 ± 0.22 for the $CC\mu^+/CC\mu^-$ and 0.35 ± 0.15 for the combined $CC(e^+ + \mu^+)/CC(e^- + \mu^-)$. The expected ratio is 0.48 if the prompt flux is equal in ν and $\bar{\nu}$. We find the data compatible with equal fluxes.

(b) The ratio of prompt $CCe^\pm$ to prompt $CC\mu^\pm$ events is $0.59^{+0.35}_{-0.21}$ (68% CL) $^{+0.85}_{-0.29}$ (90% CL). For any known short-lived source (charm) we expect unity for this ratio. This experiment alone cannot claim any asymmetry (see below).

(c) Since we observe CCe^+ and CCe^- events, it is possible to determine the ratio of NC/CC for the electron and muon ν together (prompt + conventional) and we get NC/CC = 0.28 ± 0.07 ($E_{had} > 10$ GeV) when 0.32 is expected (assuming μ-e universality). Subtracting the expected number of NC events from the $\overset{(-)}{\nu}_\mu$ interactions we obtain $NC(\nu_e + \bar{\nu}_e + \text{anything})/CC(\nu_e + \bar{\nu}_e) = 0.1 \pm 0.5$.

TAU-NEUTRINOS

A beam dump experiment is well suited to study ν_τ interactions because of the good signal to background ratio. The ν_τ are assumed to be produced: (i) as prompt ν via F meson decay in the dump or (ii) from a possible $\nu_e \rightarrow \nu_\tau$ oscillation ($\nu_\mu \rightarrow \nu_\tau$ mixing is known to be very small). The τ lepton produced in BEBC will normally decay too fast to be detected. We have found that the best way of detecting the presence of ν_τ is from the NC/CC ratio. Since the charged current τ interaction will show up only as "CC($e^\pm$, $\mu^\pm$)" in 35% of the true CCτ there will be an apparent NC/CC ratio of $\simeq$ 2.5 compared with the normal 0.32. Taking into account that the charged current τ cross section is suppressed by a factor of 2 for our energy spectra compared with $\nu_\mu \rightarrow \mu$ because of threshold effects, the estimated number of events which could be due to ν_τ interactions in the sample are -6.8 ± 11.3 events. This corresponds to < 10% (90% CL) of the prompt events. Assuming that ν_e oscillates into ν_τ, an upper limit of $P(\nu_e \rightarrow \nu_\tau) < 0.35$ (90% CL), for our distance and energy spectra, is obtained.

The τ lepton has been assumed to have its own lepton number. However, the hypothesis that it has the same lepton number as the electron has not been ruled out [6]. Using the NC/CC ratio we can rule out this hypothesis at 90% CL.

THE INTERPRETATION OF NEUTRINO OSCILLATIONS AND THE NEW RESULT FROM THE BEBC NARROWBAND EXPERIMENT

The three CERN beam dump experiments [7] give, using the subtraction method, a prompt $e^\pm/\mu^\pm$ ratio which could be less than unity. For any known prompt source this ratio should be unity. It is important to emphasize that all three experiments deal with some systematic errors in common and that the interpretation of this results as evidence for an asymmetry has to be taken with caution. The interpretation that the ν_e oscillates to another ν (not ν_μ) has been proposed. It was reported in ref. [7] that no significant energy dependence was observed for the ratio. This implies that in the oscillation hypothesis $\delta m^2 > 100\ eV^2$ (when assuming $\sin^2 2\theta = 1.0$).

New data from a different ν experiment made by the same BEBC collaboration [8] are relevant to the oscillation hypothesis. If the "asymmetry" in the beam dump exists and is due to $\nu_e \rightarrow \nu_x$ oscillations, this effect should also be visible in other ν beams at high energies, which always contain some fraction of ν_e mainly coming from $K^{\pm}_{e3}$ and K^{0}_{e3} decays. A problem with most ν beams is that the flux of ν_e is badly known (± 10-30% only). However, in a narrowband beam, where the π and K are momentum selected before they enter the decay tunnel, the ν_e flux is known within a few percent. Knowing the number of $K_{\mu 2}$ decay we can directly get the number of K_{e3} from the branching ratios. We have studied the amount of e^- events in our narrowband experiment (200 GeV/c transport momentum). In a sample of 3200 $CC\mu^-$ we found 63 e^- events with $E_{vis} > 20$ GeV. After corrections for detection efficiency they become 76 ± 9. Subtracting 3 ± 3 events from ν_e produced outside the decay tunnel, we get 73 ± 10 events when 70 are expected. This gives $P(\nu_e \rightarrow \nu_e) = 1.04 \pm 0.15$ corresponding to a $\delta m^2 < 50\ eV^2$ at 90% CL ($\sin^2 2\theta = 1.0$). It seems that this result rules out the hypothesis that ν oscillations are responsible for a prompt $e^{\pm}/\mu^{\pm}$ "asymmetry" in the beam dump experiments.

REFERENCES

[1] P.C. Bosetti et al., Phys. Lett. 74B (1978) 143.
[2] P. Alibran et al., Phys. Lett. 74B (1978) 134.
[3] T. Hansl et al., Phys. Lett. 74B (1978) 139.
[4] H. Wachsmuth, Proc. 1979 Int. Symp. on Lepton and Photon Interactions at High Energies, Fermilab, 541.
[5] H. Wachsmuth, CERN/EP 79-125 (1979).
[6] J. Kirkby, Proc. 1979 Int. Symp. on Lepton and Photon Interactions at High Energies, Fermilab, 107.
[7] F. Dydak, Neutrino Conference 1980, Erice.
[8] H. Deden et al., subm. to Phys. Lett. B, (CERN/EP 80-164).

28 GeV/c BEAM DUMP EXPERIMENTS

John M. LoSecco
University of Michigan, Ann Arbor, Michigan 48109

ABSTRACT

Three experiments that were run concurrently at the Brookhaven AGS are reviewed and compared. The experiments searched for prompt neutrinos and penetrating neutral particles from a beam dump exposed to 4.2×10^{18} 28 GeV/c protons. Some indications of unusual production mechanisms have been reported.

INTRODUCTION

Beam dump experiments are a good way to search for evidence for new particle states. The initial proton beam is incident on a dense large block of material. The beam produces large numbers of conventional hadrons, such as π's and K's. The high density of the dump rapidly absorbs these before they can decay and produce neutrinos. The prompt signal can come from the decay of short lived states, such as charm, or from the production of penetrating neutral states such as the hypothetical axion. The advantage of beam dump studies at 28 GeV/c is that charm production is known to be highly suppressed, by 10^3 or more, over its production rate at 400 GeV/c. So this interesting, but known, background is removed. Another advantage is the high sensitivity that can be obtained since the accelerator can produce more protons and the experiments can be located much closer to the dump.

The three experiments to be considered are: the Rutgers, Stevens, Columbia group[1] that made use of the 7' bubble chamber situated 43 meters from the dump (Figure 1), the Columbia, Illinois,

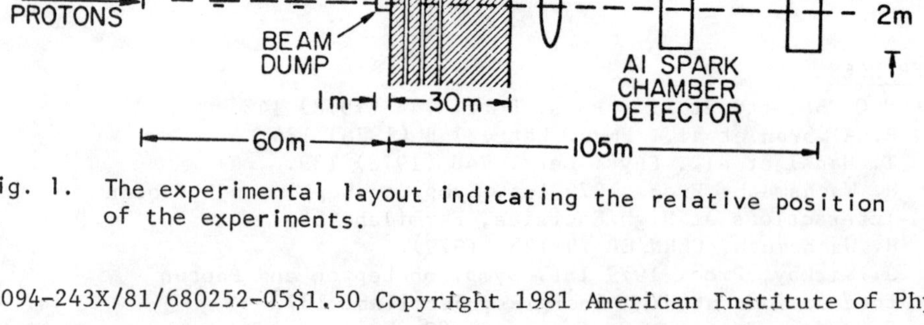

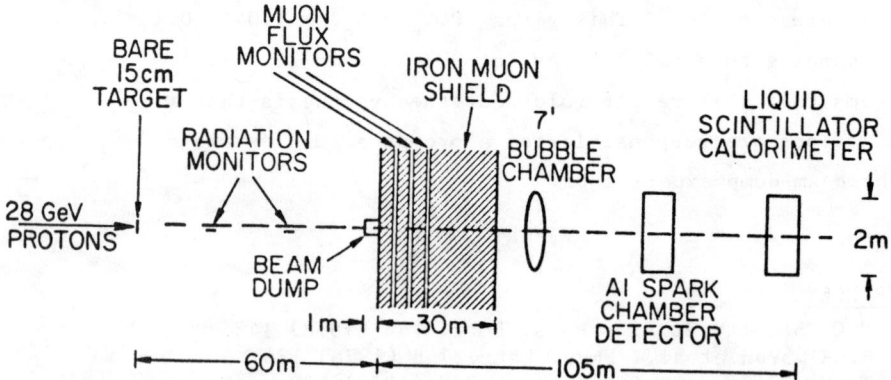

Fig. 1. The experimental layout indicating the relative position of the experiments.

0094-243X/81/680252-05$1.50 Copyright 1981 American Institute of Physics

Brookhaven group[2] that made use of an aluminum spark chamber detector situated 76 meters from the dump and the Harvard, Pennsylvania, Brookhaven, Oak Ridge group[3] that made use of a liquid scintillator detector situated 105 meters from the dump.

In addition to the dump two other targets were studied. A sample of conventional neutrino interactions were generated from a 15 cm thick bare target followed by a 60 meter decay space. Losses were studied with a series of transmission targets positioned at 5 points along the beam transport system.

THE EXPERIMENTS

The bubble chamber has reported on a sample of 147,000 pictures randomly selected from 700,000 taken during the dump. This represents an exposure of 1.3×10^{18} protons. The chamber was filled with 62% neon hydrogen mixture and had an effective fiducial mass of 2.8 tons. The scanning efficiency was 75 ± 10%.

The pictures were scanned for μ^-, μ^+, $e^\pm$, neutral current and e^+e^- candidates. To remove cosmic ray background the events were required to have $E_{visible} > .5$ GeV and a total momentum within 30° of the beam axis. Their experience with wide band neutrino beam exposures suggests that these cuts retain 98% of the beam associated signal. Their results after background subtraction and correction for scanning efficiency and misidentification are reported in Table I. The expected rate is calculated from wide band running correcting for the decay space and the effects of focusing. The event candidates have the expected properties of muon neutrino induced events.

Table I

7 Foot Bubble Chamber Event Summary

	Observed	Corrected	Expected
μ^-	10	12	11
μ^+	6	4	2
N.C.	6	6	4
$e^\pm$	0	0	0.1
Total	22	22±6	17±6

The Columbia, Illinois, Brookhaven group used their 4.5 ton detector to study neutrals from the dump and compared them with events observed from the bare target. Timing was used to eliminate background and indicated that cosmic ray induced or beam associated background was very small.

Their results are reported in Table II. A 15 to 20% correction for detection efficiency of low multiplicity events is needed, but has not been applied.

Table II

Columbia, Illinois, Brookhaven Event Summary

	Bare Target	Dump
Protons	1.9×10^{17}	4.7×10^{18}
C.C.	53	49
N.C.	9	14
C.C. > 3 Prong	24	29
N.C. > 2 Prong	9	14
$e^{\pm}$	1	1
μ^{+}	4	6
μ^{-}	19	19
$N.C./_{C.C.}$ > 2,3 Prong	.38 ± .15	.48 ± .16
Total	62 ± 8	63 ± 8

The observed suppression factor for the dump is Target Rate/ Dump Rate = 24±4. The group concludes that there are no obvious differences between dump and bare target events.

The Harvard, Pennsylvania, Brookhaven, Oak Ridge group used an 11 ton liquid scintillator detector to study dump and bare target events. The timing technique previously mentioned was also used. Background was very small. Their results are reported in Table III.

Table III

Harvard, Penn, BNL, ORNL Event Summary

	Bare Target	Dump	
Protons	1.97×10^{17}	4.87×10^{18}	
C.C.	115	90	
N.C.	24	14	
$N.C./_{C.C.}$	0.21±0.05	0.16±0.04	
Total	139	104	
Event Rate	70.6×10^{-17}	2.14×10^{-17}	events/$_{proton}$

The Rate Target/Rate Dump = 33±4. No major differences between dump and bare target events were reported. This group also reported an analysis of the transport loss study. Less than 21 of their 104 beam dump events are attributable to beam losses at 90% confidence level.

COMPARISON

In Table IV we compare the 3 experiments. We have standardized by scaling all experiments to an exposure of 10^{19} protons and a 10 ton detector at 100 meters.

Table IV

Comparison of Different Experiments

Group	7'	CIB	HPBO
Mass (tons)	2.8	4.55	11
ℓDump (m)	43	76	105
Protons	1.3×10^{18}	4.7×10^{18}	4.87×10^{18}
Events	22±6	63±8	104±10
Standard Events	112±21	204±26	214±21

The CIB group standard value has been corrected (+20%) for detection efficiency. The difference between the 7' group and the HPBO group is statistically significant.

$$\text{HPBO} - 7' = 102\pm30 \tag{1}$$

Since the HPBO detector can only see events out to 10 mrad and the bubble chamber extends out beyond 20 mrad a flux fall off with angle could explain the difference since the standard value was obtained by scaling by ℓ^2. Such a rapid fall off is not typical of hadronic showers at this energy.

CONCLUSIONS

All groups agree that there is no striking and unique signal coming from the dump. An analysis of the conventional π decay signal from the dump can be done and an excess searched for. This has been done by each group.

As seen from Table I the 7' bubble chamber group claims agreement between their predicted and observed event rate. The CIB group has done two calculations. One predicts half of the observed rate and the other predicts the rate observed. They do not trust the calculations to within a factor of 2 and do not draw any conclusions.

The HPBO group has done two separate calculations with different systematic errors. The first calculation scales the bare target event rate so it is insensitive to questions of absolute flux normalization or detection efficiency. The most critical parameter in the extrapolation is the effective pion absorption length in the dump. An absorption length of 29cm is indicated by experiments and includes the effect of hadron showers and secondary interactions.

The second calculation is a detailed hadron cascade Monte Carlo that follows the protons and secondaries in the dump. As a check on the Monte Carlo it was used to calculate the bare target rate and gave the observed value.

The detailed Monte Carlo calculation and the scaling calculation are in agreement. They predict 56 events for the HPBO group. The group has concluded that they have an excess of 48±10 events with a systematic error of ±12 events. Their studies indicate that it is unlikely that transport losses could account for the excess.

The calculations can be applied to the other two experiments. They indicate that the CIB group has a 45% excess. No excess is indicated for the 7' bubble chamber since it has a lower observed event rate.

I would like to thank M. Kalelkar and R. Fine for discussions and clarification of their experiments.

REFERENCES

1. P. F. Jacques, M. Kalelkar, P. A. Miller, R. J. Plano, P. Stamer, E. B. Brucker, E. L. Koller, S. Taylor, C. Baltay, H. French, M. Hibbs, R. Hylton, K. Shastri, and A. Vogel, Phys. Rev. D21, 1206 (1980).

2. P. Coteus, M. Diesburg, R. Fine, W. Lee, P. Sokolsky, R. Brown, S. Fuess, P. Nienaber, T. O'Halloran, and Y. Y. Lee, Phys. Rev. Lett. 42, 1438 (1979).

3. A. Soukas, P. Wanderer, W. T. Weng, M. Bregman, M. Claudson, J. LoSecco, L. Rivkin, J. Roeder, S. Russek, L. Sulak, P. Timbie, M. Yudis, T. A. Gabriel, R. S. Galik, J. Horstkotte, J. Knauer, M. Levine, and H. H. Williams, Phys. Rev. Lett. 44, 564 (1980).

HADRONIC PRODUCTION OF PROMPT SINGLE MUONS

B.C. Barish, A. Bodek,[1] K.W.B. Merritt,[2] M. Shaevitz, E. Siskind
California Institute of Technology, Pasadena, Ca. 91125

A. Diamant-Berger,[3] J.P. Dishaw,[4] M. Faessler,[5] J. Liu,[6]
F. Merritt,[2] S. Wojcicki
Stanford University, Stanford, Ca. 94305

ABSTRACT

Results from a prompt single muon search at Fermilab are reported. Prompt $1\mu^+$ and $1\mu^-$ production from 400 GeV protons on iron was observed at approximately equal levels. Charm models of $D\overline{D}$ production predict $1\mu^-/1\mu^+ \simeq 0.9$, from differences in trigger efficiency. The observed ratio was 1.3±0.3. Total charm cross sections of 20-90 μb/nucleon were calculated, assuming linear A dependence, for correlated and uncorrelated $D\overline{D}$ production. Upper limits on diffractive production of Λ_c-D pairs were set.

INTRODUCTION

This experiment triggered on a muon of momentum greater than 20 GeV emerging from a 400 GeV proton interaction in an iron target-calorimeter. One-muon and two-muon events were separated by counting tracks in spark chambers after 3m of iron (see Fig. 1). The triggering muon penetrated an iron toroidal spectrometer which measured its momentum and sign. The relative rate of prompt and non-prompt 1μ events was measured by varying the effective density of the upstream 1.2m of iron in the target. Data were collected at three effective densities.

In the analysis, the data at each density were separated into 1μ and 2μ categories and plotted versus inverse density. The 2μ rate should not vary with density; the 1μ rate should vary linearly with an intercept at zero inverse density that measures the prompt single muon rate. Figure 2 contains the extrapolation plots for triggering μ^+ and μ^- events. The 2μ rates are indeed flat; and both the $1\mu^+$ and $1\mu^-$ samples show a significant prompt signal, of approximately equal magnitude. The sizes of the single prompt muon signals after background subtraction are also shown on the plots.

---

1. Present address: University of Rochester, Rochester,N.Y. 14627.
2. Present address: University of Chicago, E.F.I., Chicago, IL 60637.
3. Permanent address: Departement de Physique des Particules Elèmentaires, Saclay, France.
4. Present address: Intel Corp., Santa Clara, CA 95051.
5. Present address: CERN, Geneva, Switzerland.
6. Present address: Global Union Bank, Wall Street Plaza, New York, NY 10005.

0094-243X/81/680257-05$1.50

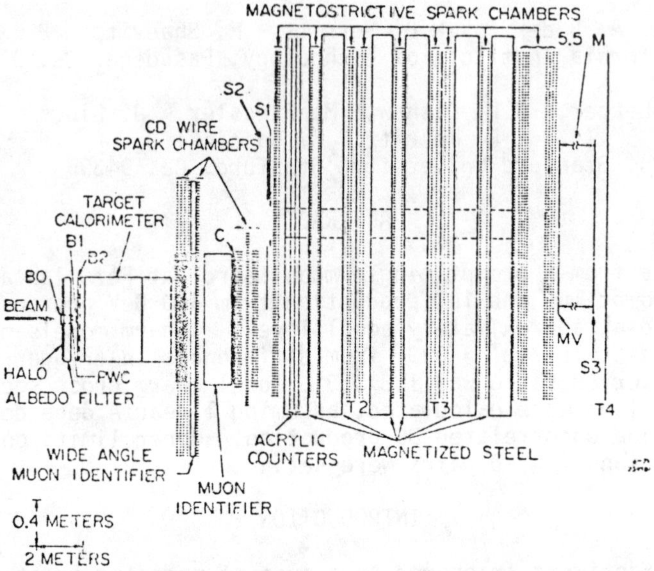

Fig. 1. The apparatus in the configuration used in the high p data taking.

Distributions in momentum p and transverse momentum p_t of the muon were obtained by a density extrapolation in each bin of p or p_t.

Two backgrounds had to be subtracted from the p and p_t distributions. The first background came from π and K decays occurring downstream of the expanded region of the calorimeter, where no density change was made. This source accounted for 14% of the $1\mu^+$ signal and 5% of the $1\mu^-$ signal, with a 30% systematic error. The background was calculated with a detailed Monte Carlo of the hadron cascade. The second background was 2μ events which were counted in the 1μ sample because the second muon either ranged out in the 3m of steel ($E_\mu < 4$ GeV) or had a very large angle ($\theta_{lab} > 150$ mrad) so that its trajectory missed the spark chambers downstream of the muon identifier. Again, a Monte Carlo of the hadronic cascade was used to calculate the background, including 2μ production from resonance decay, the Drell-Yan mechanism, and Bethe-Heitler production. The subtraction was 25% of the $1\mu^+$ signal and 21% of the $1\mu^-$ signal. A 10% systematic error was assigned to this calculation. Background from halo muons was completely eliminated with hardware and software requirements on incident charged particles. Hyperon decays would be a potential background, because their low values of $\gamma \cdot c\tau$, the decay length, result in nonlinear density dependence. However, their low cross section and branching ratio to muons and their low efficiency for producing muons of high momentum meant that the trigger rate for muons from hyperon decay was negligible.

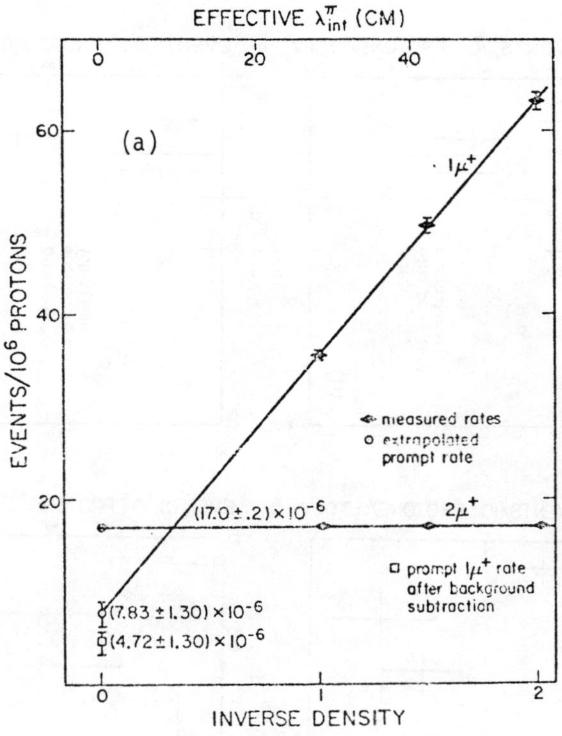

Fig. 2(a). The measured $1\mu^+$ and $2\mu^+$ rates from the high p trigger versus effective inverse density.

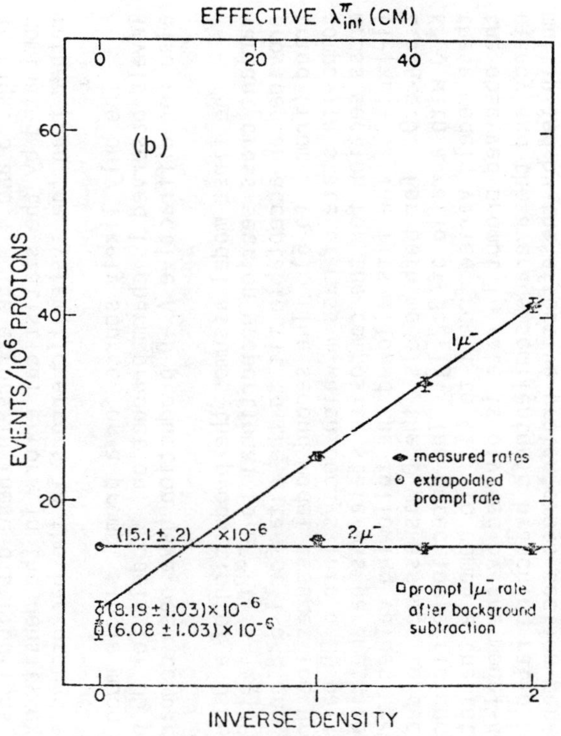

Fig. 2(b). The measured $1\mu^-$ and $2\mu^-$ rates from the high p trigger versus effective inverse density.

Fig. 3

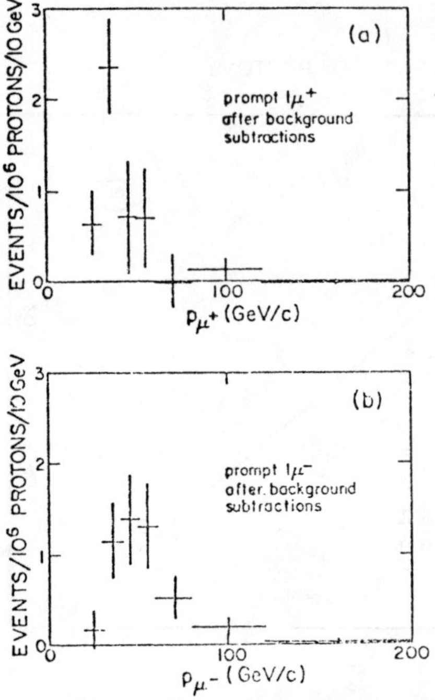

Fig. 4

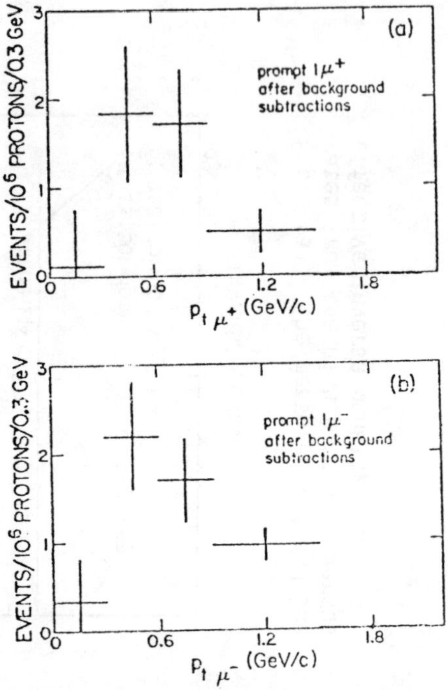

The distributions in p and p_t of the prompt $1\mu^+$ and $1\mu^-$ signal, after subtraction of the two significant backgrounds, are presented in Figs. 3 and 4. The errors in these distributions are completely dominated by the statistical errors in the density extrapolations rather than the systematic errors in the background subtractions.

The only likely source for a prompt single muon signal at the levels observed is charm production. Models for $D\bar{D}$ production and also for diffractive Λ_c-$\bar{D}$ production have been compared to the data.

The first model assumes the production of a D meson with an invariant cross section proportional to $e^{-\alpha p_t}(1-|x_F|)^{\beta}$. This model provided an acceptable fit to the data for $1<\alpha<3$ and all values of β tried (from 2 to 6). The second model assumes the production of a composite state of mass m which decays into a $D\bar{D}$ pair. The invariant cross section for the composite state was parametrized as $e^{-\gamma m}e^{-\alpha p_t}(1-|x_F|)^{\beta}$. The fits allowed the following values: $\gamma>0.5$, $\alpha>0.4$, $2.5<\beta<5.0$. For both models, the D was assumed to decay into $K\mu\nu$ and $K^*\mu\nu$ with a ratio of 3:2[1]. The detection efficiency obtained for these models varied from 1 to 4%. To compute the total cross section, the observed prompt 1μ rate is divided by the model-dependent efficiency and the average semileptonic branching ratio of 0.08[1], and multiplied by the effective proton cross section per nucleon (13 mb if linear A dependence is assumed for charm production). The cross sections obtained range from 20 to 90 μb/nucleon, in agreement with our previous results on single prompt muon production at high p_t [2], and with preliminary results from a new data run [3].

Our results may be used to predict the flux of prompt neutrinos observed in the CERN beam dump experiments, assuming a model. Figure 5 shows a comparison of the p distribution from the BEBC group [4] with a range of models normalized to our data. There is good agreement with the shape of the BEBC distribution for all the models. Our observed ratio of $1\mu^-/1\mu^+$ of 1.3±0.3 implies a ratio for prompt $\bar{\nu}_\mu/\nu_\mu$ of the same value over the same kinematic region. Neither the $D\bar{D}$ nor the Λ_c-D models produce a sign asymmetry at low p_t which would not be observed over the p_t range of this experiment.

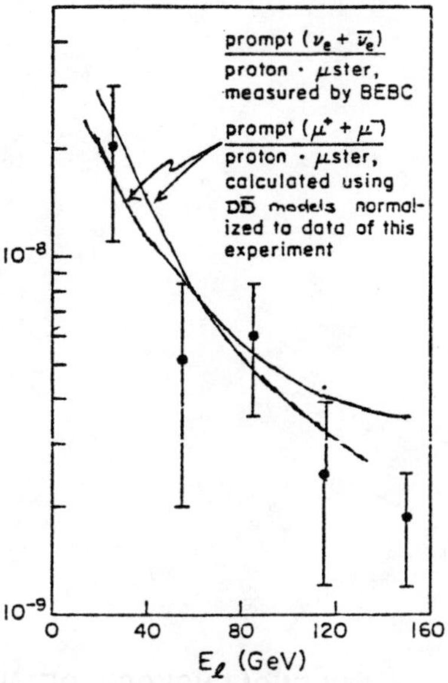

Fig. 5. Comparison with BEBC prompt νe rates.

A very simple diffractive model assuming production of Λ_c and $\bar{D}$, each with an e^{-6m_t} distribution and a flat x_F distribution out to the kinematic limit, was compared to the data. [5] The predicted p distributions are not a good fit to the data, but a strict upper limit on the diffractive cross section at $\sqrt{s}$ = 27 GeV is obtained by assuming that all of our signal is due to diffractive production. This 95% CL limit is 66 μb/nucleon, assuming $A^{2/3}$ dependence for the diffractive cross section. Extrapolating this limit to $\sqrt{s}$ = 60 GeV [6] gives σ_{diff}<165 μb/nucleon at 95% CL.

The first conclusion of this experiment is that there exists a significant prompt single muon signal for both μ^+ and μ^- of approximately equal magnitude, for p > 20 GeV. Models for $D\bar{D}$ production can be fit to the data and indicate cross sections in the range 20 - 90 μb/nucleon. The data are used to set upper limits on the diffractive cross section.

1. J. Kirkby, Proc. Lepton-Photon Symposium, Fermilab, eds. T.B.W. Kirk and H.D.I. Abarbanel, p. 107 (1979).
2. K. W. Brown et al., Phys. Rev. Letters 43, 410 (1979).
3. J. L. Ritchie et al., Phys. Rev. Letters 44, 230 (1980).
4. H. Wachsmuth, Proc. Lepton-Photon Symposium, Fermilab, eds. T.B.W. Kirk and H.D.I. Abarbanel p. 541 (1979).
5. $m_t = \sqrt{p_t^2 + m^2}$ for the Λ_c or for the $\bar{D}$.
6. An energy dependence of $e^{-10.8m/\sqrt{s}}$ was used.

PHENOMENOLOGY OF NEW PARTICLE PRODUCTION

E. REYA, DORTMUND, ORGANIZER

LEPTOPRODUCTION OF HEAVY QUARKS

J. P. Leveille
Physics Department, University of Wisconsin, Madison, WI 53706

T. Weiler
Physics Department, Northeastern University, Boston, MA 02115

ABSTRACT

The photon-gluon fusion model of heavy quark production is reviewed. New, untested predictions are presented, and comments are made concerning the next order calculation.

This talk presents recent theoretical progress on the dynamics of heavy quark production in photon (real or virtual)-nucleon scattering. Experimental production agrees remarkably well with the QCD motivated model to be presented here. For detailed comparison between model and data, the reader is referred to experimental talks presented at this conference,[1] and references therein.

Consider the dissociation of a gluon into a heavy quark-antiquark pair inside a nucleon. The tenets of asymptotic freedom justify the single gluon approximation when the relevant mass scale of the process, here $2m_Q$, is large relative to the hadronic scale, Λ or M_N. Thus for $2m_Q \gtrsim M_N$ the $Q\overline{Q}$ arise as fluctations in the one-gluon propagator and according to the uncertainty principle are correlated in space with length scale $(2m_Q)^{-1}$. The picture of $Q\overline{Q}$ pairs statically situated in the nucleon's sea is not valid for heavy quarks.

A complementary argument views the $Q\overline{Q}$ production as arising from the dissociation of the photon. Parton calculations are performed in the photon-incident parton cms (energy $\sqrt{\hat{s}}$). In this frame, the photon of invariant mass squared $-Q^2$ has three momentum $(\hat{s}+Q^2)/2\sqrt{\hat{s}}$. From the uncertainty principle, the inverse of this is the distance scale for photon dissociation. An implication is that real photoproduction of charm ($\hat{s} \simeq m_\psi^2$) has the same short distance properties as inclusive deep inelastic scattering of a $Q^2 = 2.4$ GeV^2 virtual photon. The factorization theorem for QCD ensures that every short distance process has a Feynman diagram expansion in terms of quark and gluon fields, with vertex strengths given by the renormalization group improved "running" coupling $\sqrt{4\pi\alpha_s}$.

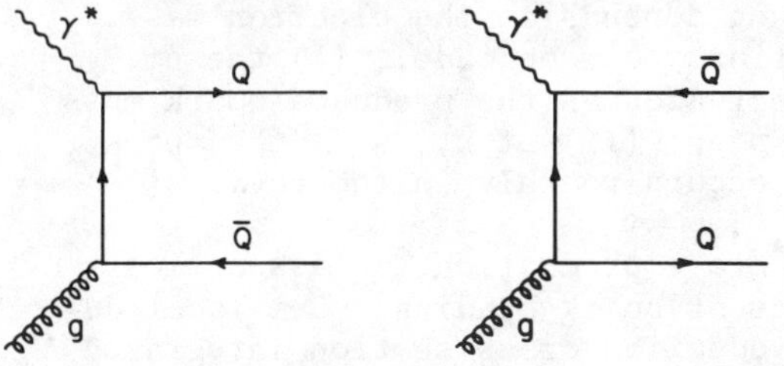

Fig. 1. Photon-gluon fusion diagrams.

The $O(\alpha_s)$ diagrams for heavy quark leptoproduction are shown in Fig. 1. A gluon and photon fuse[2,3] to create the heavy $Q\overline{Q}$ pair in a color octet. It is a model assumption that final state interactions, e.g. color rearrangement leading to color singlet heavy hadrons, do not affect production rates. To predict characteristics of associated production, the stronger assumption, that hadronization does not appreciably alter the Q or $\overline{Q}$ four momentum, is needed.

Several noteworthy predictions can be inferred from the $O(\alpha_s)$ diagrams. First of all, the dynamics implies correlations among observables. The momentum of Q transverse to the photon direction is balanced by that of the $\overline{Q}$, i.e. $p_\perp(Q) = -p_\perp(\overline{Q})$. The lab energy of the photon is shared between the Q and $\overline{Q}$, i.e. $Z(Q) + Z(\overline{Q}) = 1$, where $Z(Q) \equiv E_{lab}(Q)/\nu$. Quantitative evaluation of the correlations, expressed as distributions in $p_\perp$, Z, and rapidity, or as $\langle p_T \rangle$ or $\langle Z \rangle$, are given in Ref. 2. We reproduce in Fig. 2 the Z-distribution of charm production. Notice that as $m_Q^2/(Q^2 \text{ or } 2M\nu) \to 0$, the Z-distribution moves toward δ-fcns at Z = 0 and 1, i.e., current and hole fragmentation regions develop. Thus, as the resolving power of the photon is increased beyond the $Q\overline{Q}$ correlation length, the model approaches the uncorrelated static sea of the old parton model.

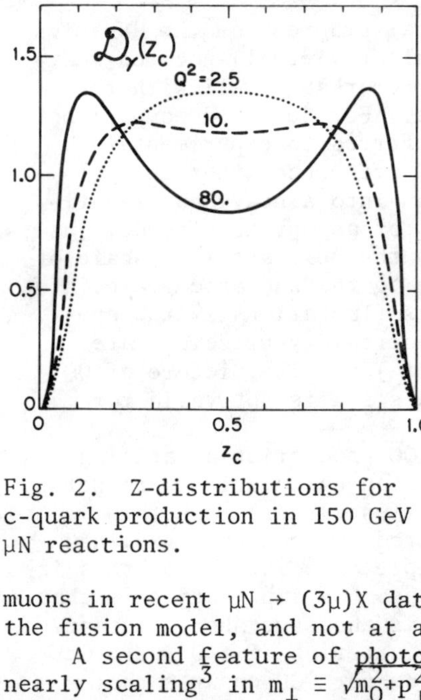

Fig. 2. Z-distributions for c-quark production in 150 GeV μN reactions.

However, at present laboratory energies, photon-gluon fusion predicts strong correlations between c and $\bar{c}$ quarks ($m_c = 1.5$ GeV). This contrasts markedly with the static sea model, where Q and $\overline{Q}$ are completely uncorrelated so that $p_\perp(Q) \simeq p_\perp(\overline{Q}) \simeq 0$ and one of $Z(Q)$, $Z(\overline{Q})$ is unity, the other being zero. Correlations among muons in recent $\mu N \to (3\mu)X$ data[1] are described remarkably well by the fusion model, and not at all by the static sea model.

A second feature of photon-gluon fusion is a cross section nearly scaling[3] in $m_\perp \equiv \sqrt{m_Q^2 + p_\perp^2}$. Of course, production of strange, charmed and beautiful hadrons will not show clean $m_\perp$ scaling since the fragmentation functions $\mathcal{D}_{H/Q}(E_H/E_Q)$ will depend on the quark masses, just as QED brehmsstrahlung depends on the electron mass. However, some qualitative predictions[2] can be made: (i) the $p_\perp^2$ distribution flattens and $\langle p_\perp^2 \rangle$ increases as the produced quark mass increases, thus $\langle p_\perp^2 \rangle_K < \langle p_\perp^2 \rangle_D < \langle p_\perp^2 \rangle_B$; (ii) at $x_{Bj} \lesssim .1$ the $\langle p_\perp^2 \rangle_D$ resulting from γ, W and Z fusion occurs roughly in the ratio 1.6 : 1.0 : 1.6.

To predict heavy-onia (ψ,Υ, etc.) production, an assumption must be made to model the dynamics of quark binding. Semilocal duality holds that the continuum production cross section integrated through the resonance region ($4m_Q^2 \le \hat{s} \le 4m_H^2$, where m_H is the mass of the lightest hadron containing Q) equals the cross section for resonance production. $\hat{s} \geq 4m_H^2$ is the condition for associated production. This division corresponds to a charm quark cms escape velocity of $\beta_{cr} = .6$; below this value the $c\bar{c}$ are bound together by the strong color force. The range $\beta < .6$ is consistent with charmonia potential models, where $\langle\beta\rangle \simeq .5$. If a single resonance production cross section is sought, a weight f, defined formally as the ratio of the single resonance production to all resonance production, is introduced. A simple statistical guess is that f^{-1} equals the number of resonances (8 for charm and 18 for beauty according to potential models). Notice the freedom to make a different choice for the interval of bound production with a compensating change in the parameter f.

An important feature of photon-gluon fusion is the proportionality of the cross section to the gluon component of the nucleon wavefunction. Thus heavy quark production data can be inverted[4] to yield the probability distribution $G(\eta)$ for finding a gluon in the nucleon with momentum fraction η. Fitting $\mu N \to \mu\psi N$ data[1] to a functional form $\eta G(\eta) = (1-\eta)^{\alpha}$ yields[4] $\alpha = 5.6^{+0.8}_{-1.2}$. (Reference 5 omits some dubious threshold data and finds an exponent of 4.4). This power agrees well with counting rule predictions of 4 or 5, especially when allowance is made for steepening due to $m_\psi^2 \simeq 10$ GeV^2 scale breaking.

In addition to the shape of the ψ excitation curve, valuable information is contained in the normalization. Relating the model to the experiment one has[4] $\alpha_s f\tilde{G}(2) = .038 \pm .011$, where G(2) is the fraction of the nucleon's momentum carried by gluons, known from deep inelastic scattering to be about 50%. Allowing for some heavier resonances cascading to ψ, a reasonable range for f^{-1} is 4 to 8. Then there results $\alpha_s(m_\psi^2)/\pi = .15 \pm .04$. The agreement between α_s/π just deduced and the four active flavor QCD value $\alpha_s(m_\psi^2)/\pi = 6/25\ell n(m_\psi/.5) = .13$ must be regarded as more than fortuitous! Consistency of the perturbative model is now ensured since the expansion parameter α_s/π is seen to be small compared to unity. The color degree of freedom makes an essential contribution to the agreement: although the non-Abelian aspects of color are not tested, color averaging of the diagrams of Fig. 1 yields an important factor of ½. One can also compare this determination of $\alpha_s(m_\psi^2)/\pi$ with the recent determination[6] of $\alpha_s(Q^2)/\pi$ from $e^+e^- \to$ 3-jet data. There it was found that $\alpha_s(900\ \text{GeV}^2)/\pi = .07 \pm .01$, which scaled to the m_ψ^2 mass becomes $.16 \pm .02$, in excellent agreement with the value deduced here (we have assumed a common value of $\Lambda = .5$ GeV for both processes). Since the normalization of ψ production agrees so well with the model prediction, we may predict that the $\mu N \to \mu\Upsilon N$ cross section is within an order of magnitude of the present experimental upper bound.[1]

The shape and normalization of associated charm production data also test photon-gluon fusion. Agreement is generally excellent.[1] Only in the double differential $d\sigma/\gamma^* N \to D\bar{D}X)/dQ^2 d\nu$ is there the slightest discrepancy. We now show that in fact this "discrepancy" is well within the tolerance of a leading ℓn calculation. The Feynman rules for short distance QCD determine the dependence of the cross section on Q^2 and $\hat{s}$ everywhere except in $\alpha_s(\mu^2)$. In lowest order, $\alpha_s(\mu^2)$ corresponds to a leading ℓn summation with an inherent ambiguity of argument which is resolved only by a next order (next-to-leading ℓn) calculation. We present two plausible arguments for a proper scale μ^2. Recall that the mass scale associated, via the uncertainty principle, with virtual photon dissociation into $Q\bar{Q}$ is $(\hat{s}+Q^2)/2\sqrt{\hat{s}}$. This characteristic mass scale, then, is a reasonable choice for the argument of α_s. A second argument comes from ladder summation techniques[7] which generate $\alpha_s(\mu^2)$. The scale μ is just the inverse distance of the off-shell leg of the strong interaction vertex from its mass shell. In photon-gluon fusion the dominant distance is $|\hat{t}_{min}-m_Q^2| = (\hat{s}+Q^2)(1-\beta)/2$, $\beta = \sqrt{1-4m_Q^2/\hat{s}}$ being the heavy quark cms velocity. For charmonia production, $\langle\beta\rangle \simeq .5$ and therefore $\mu^2 \simeq (m_\psi^2+Q^2)/4$; for open charm, $\langle\beta\rangle \simeq .7$ and $\mu^2 \simeq (\hat{s}+Q^2)/7$. Figure 3 presents model predictions normalized to double differential data.

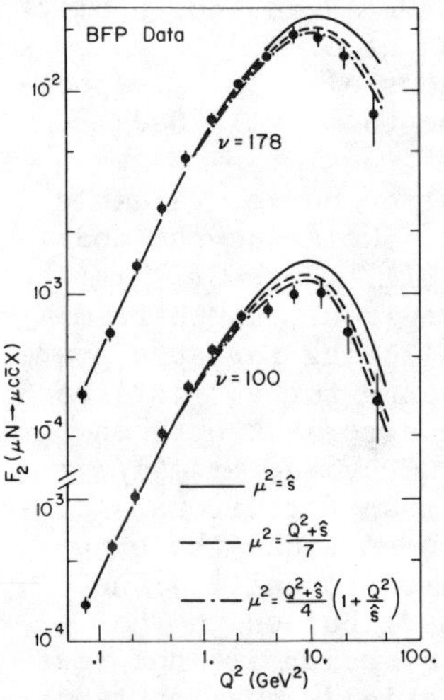

Fig. 3. Model predictions of F_2 for charm production with various scales for the strong coupling constant. In each case Λ = 500 MeV.

Next we consider the dependence of $O(\alpha_s)$ heavy quark leptoproduction on the gluon spin and parity. As pointed out in Ref. 8, the ratio of the longitudinal to transverse cross sections for the process $\gamma^* N \to (\psi$ or $D\bar{D})$ + X is very sensitive to gluon J^P. Experimental determination of this ratio is difficult. Perhaps a more accessible observable sensitive to J^P is the azimuthal asymmetry in $\gamma^* N \to D\bar{D}X$. Defining ϕ as the angle between the lepton scattering plane and the $c\bar{c}$ production plane, one has $(2\pi/\sigma)d\sigma(\gamma^* N \to c\bar{c}X)/d\phi = 1+\underline{a}\cos(2\phi)$. Calculation[9] reveals that sign $(\underline{a})$ is positive (negative) {zero} for gluon $J^P = 1^-(0^+,1^+$ with $Q^2 \lesssim \hat{s})\{0^-\}$. Thus the <u>sign of the asymmetry alone</u> differentiates among gluon spin-parities. Furthermore, the asymmetry parameter $\underline{a}$ is moderately large, $\simeq$ 15% for the QCD case.[2]

Not yet discussed is the fact that photon-gluon fusion is a diffractive model in the sense that

i) $d\sigma/dt \sim \delta(t)$ since $t \equiv (q_\gamma - P_{c\bar{c}})^2 = k^2(\text{gluon}) = 0$ (non-perturbative effects create an e^{bt} smear);

ii) $c\bar{c}$ rapidity in the hadronic rest frame is positive definite ($\frac{1}{2}\ell n\, W^2/\hat{s}$);

iii) $\sigma \xrightarrow{s\to\infty}$ constant for $\hat{s} + Q^2$ bounded;

iv) $[E(c\bar{c})/\nu]_{Lab} \sim 1$ and

v) no flavor quantum numbers are exchanged.

Nondiffractive $Q\bar{Q}$ production occurs first at $O(\alpha_s^2)$. Of the twelve diagrams at this order, three are completely finite, four have divergences which are to be absorbed into heavy quark fragmentation functions, four have divergences to be absorbed into the gluon distribution function, and one remains with a divergence related to the hadronic content of a real photon. In Ref. 10 this latter divergence was investigated and regulated via a vector meson dominance/Drell-Yan hybrid model. The result was an $O(\alpha_s^2)$ cross section for real photons comparable in magnitude to the $O(\alpha_s)$ contribution. The complete set of $O(\alpha_s^2)$ diagrams is currently under investigation.[11]

T. W. expresses appreciation to the Aspen Center for Physics, where the written version of this talk was drafted.

This work was supported in part by the National Science Foundation, the University of Wisconsin Research Committee with funds granted by the Wisconsin Alumni Research Foundation, and in part by the Department of Energy under contract DE-AC02 76ER00881-170.

REFERENCES

1. A. R. Clark, Berkeley-Fermilab-Princeton collaboration; E. Mount, EMC collaboration, XX International Conference on High Energy Physics, Madison, Wisconsin (1980).
2. J. P. Leveille and T. Weiler, Nucl. Phys. B147, 147 (1979).
3. L. M. Jones and H. W. Wyld, Phys. Rev. D17, 759, 2332 (1978); J. Babcock et al., Phys. Rev. D18, 162 (1978); M. Gluck and E. Reya, Phys. Lett. 79B, 453 (1978); 83B, 98 (1979); V. Barger et al., Phys. Rev. D20, 630 (1979), and references therein.
4. T. Weiler, Phys. Rev. Lett. 44, 304 (1980).
5. V. Barger et al., Phys. Lett. 91B, 253 (1980).
6. A. Ali et al., Phys. Lett. 93B, 155 (1980).
7. C. H. Llewellyn-Smith, private communication; Acta Physica Austriaca, Suppl. XIX, 331 (1978).
8. G. R. Farrar, Nucl. Phys. B77, 429 (1974); J. F. Gunion, Phys. Rev. D10, 24 (1974).
9. J. P. Leveille and T. Weiler, in preparation.
10. J. P. Leveille and T. Weiler, Phys. Lett. 86B, 377 (1979).
11. E. Braaten et al., in progress; J. Owens et al., in progress.

QCD FUSION MODELS FOR HEAVY QUARK PRODUCTION

M. Glück
Institut für Physik , Universität Dortmund
4600 Dortmund 50 , West - Germany

ABSTRACT

The applications of the locally dual QCD fusion model for heavy quark production as calculated with a recently obtained broad gluon distribution are compared with data on hidden and open heavy quark production as well as with other theoretical models.

The basic idea of the perturbative QCD fusion model for heavy quark production is summarized by the following expression

$$\sigma(AB \to Q\bar{Q} + X) = \int_a^b d\hat{s} \int dx_1\, dx_2\, \delta(x_1 x_2 S - \hat{s})$$

$$\sum_{i,j} q_A^i(x_1,\hat{s})\, q_B^j(x_2,\hat{s})\, \sigma^{QCD}_{q_i q_j \to Q\bar{Q}}(\hat{s}) \quad (1)$$

with q_A^i - the density of partons of type i in A and $\sigma^{QCD}_{q_i q_j \to Q\bar{Q}}(\hat{s})$ the perturbatively calculated cross-section for the subprocess $q_i q_j \to Q\bar{Q}$. For hidden charm (charmonium) production

$$a = 4\, m_c^2 \le \hat{s} \le b = 4\, m_D^2 \quad , \quad (2)$$

while for open charm $(D,D^*,...)$ production

$$a = 4\, m_D^2 \le \hat{s} \le b = \infty \quad . \quad (3)$$

Originally[1] only quark-antiquark fusion was considered but eventually it was realized[2] that data on beam ratios and x_F distributions imply a non-negligible gluon fusion subprocess $gg \to Q\bar{Q}$ as well. Straightforward generalization to heavy quark photoproduction with the basic subprocess $\gamma g \to Q\bar{Q}$ was soon suggested[3].

Present calculations have following uncertainties:

1. The normalization presciption[4] for calculating specific quarkonium states may be wrong by factors of two to three.
2. The precise value of m_Q to be used in the calculation.
3. The lower limit in eq. (3) as well as the limits in eq. (2), though plausible, are not compelling.

4. Non leading corrections to σ^{QCD} are not available at present. Hence the arbitrary choice of the mass scale $\hat{s}$ in the parton distributions and the running coupling constant is liable to induce large non-leading corrections.

For all these reasons it seems unsafe to attempt a determination of the gluon distribution from data on J/ψ production combined with eq.(1) to lowest order. Even the shape of the gluon distribution obtained this way is unreliable. Rather a careful analysis of scaling violations in deep inelastic scattering should be preferred for this purpose.

Recently a broad gluon

$$G(x,Q^2 = 4\ \text{GeV}^2) = 0.93(1 + 8.56x + 53.57x^2)(1 - x)^6 \qquad (4)$$

was extracted[5] by this method and compared with data on hidden [Fig. 1] and open [Fig.2] charm production. Note the discrepancy with the ISR data[6] at $\sqrt{s}$ = 53 GeV.

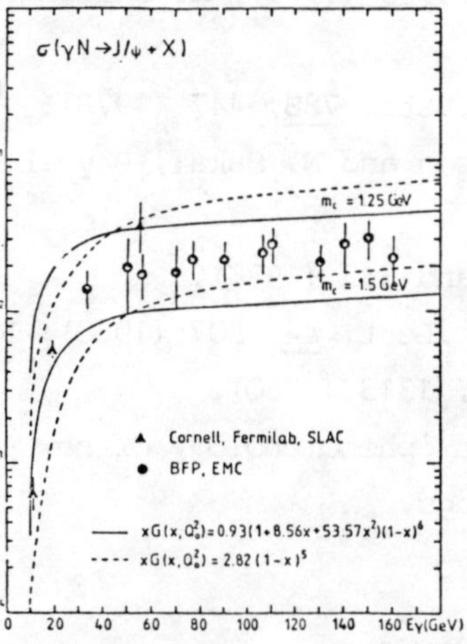

Fig. 1 Predictions[5] of eq.(1) for J/ψ photoproduction for broad (full-line) and conventional (dashed-line) gluon inputs (Q_0^2 = 4 GeV^2) continued to the appropriate $\hat{s}$.

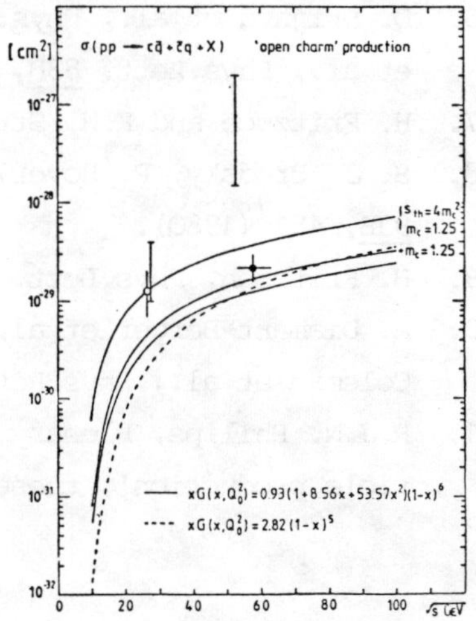

Fig. 2 Same as in Fig.1 for open charm production for different masses and $\hat{s}$ integration thresholds. The lowest full and dashed lines correspond to m_c = 1.5 GeV.

Some non-perturbative[7,8] models were suggested to account for this, seemingly diffractive, production. However ref. 7 also predicts[9] a rather high rate for B meson production in conflict with recently published upper bounds[10] which are in better agreement with the expectations of the perturbative fusion model. Recent data on the charm structure functions seem to be in conflict[11] with the charm constituent model of ref. 8. A further critical evaluation of the problematical ISR data at $\sqrt{s}$ = 53 GeV can be found in ref. 11.

REFERENCES

1. H. Fritzsch, Phys.Lett. 67B, 217 (1977).
2. M. Glück, J.F. Owens and E. Reya, Phys.Rev. D17, 2324 (1978).
3. L.M. Jones and H.W. Wyld, Phys.Rev. D17, 2332 (1978).
4. M. Glück and E. Reya, Phys.Lett. 79B, 453 (1978).
5. M. Glück, E. Hoffmann and E. Reya, Univ. of Dortmund report DO-TH 80/13.
6. D. Drijard et al., Phys.Lett. 81B, 250 (1979); A. Chilingarov et al., Phys.Lett. 83B, 136 (1979).
7. H. Fritzsch and K.H. Streng, Phys.Lett. 78B, 447 (1978).
8. S. J. Brodsky, P. Hoyer, C. Peterson and N. Sakai, Phys.Lett. 93B, 451 (1980).
9. H. Fritzsch, Phys.Lett. 86B, 164 and 343 (1978).
10. A. Diamant-Berger et al., Phys.Rev.Lett. 44, 507 (1980); R.N. Coleman et al., Phys.Rev.Lett. 44, 1313 (1980).
11. R.J.N. Philips, Plenary Session on 'phenomenology of new particle production', these proceedings.

GLUON FUSION

V. Barger and W. Y. Keung
Physics Department, University of Wisconsin, Madison, WI 53706

R. J. N. Phillips
Rutherford Laboratory, Chilton, Didcot, Oxon, England

ABSTRACT

The production of bound heavy flavors is successfully described by the mechanisms of (i) photon-gluon fusion in photoproduction and muoproduction, (ii) gluon-gluon fusion and quark-antiquark fusion in hadroproduction. ψ and Υ cross section data specify x-dependences of the gluon distributions in hadrons in agreement with power counting. ψ-production by neutrino beams is predicted from $Z°$-gluon fusion.

Heavy quark $Q\bar{Q}$ pair production rates are calculated from convolutions of the QCD subprocess cross sections $\sigma(\gamma g \to Q\bar{Q})$, $\sigma(gg \to Q\bar{Q})$, $\sigma(q\bar{q} \to Q\bar{Q})$ with the distributions of the gluons g, quarks q, or antiquarks $\bar{q}$ involved. Using semilocal duality, the cross section for production of $c\bar{c}$ ($b\bar{b}$) bound states is given by the $c\bar{c}$ ($b\bar{b}$) cross section with the pair mass m integrated from $2m_c$ to $2m_D$ ($2m_b$ to $2m_B$). A fixed fraction F of quark cross section is attributed to $\psi(\Upsilon)$ production.

The window of integration is sufficiently narrow that the gluon distribution G(x) can be evaluated at the integration midpoint and factored out of the convolution, to an excellent approximation. This yields a point-to-point correspondence between the measurable cross section and the gluon distribution.

Figure 1a shows the gluon distribution extracted from ψ-photoproduction and muoproduction data with $m_c = 1.5$ GeV and $\alpha_s = 0.375$. A best fit to the data in Fig. 1a assuming the standard form $G_i(x) = \frac{1}{2}(n_i+1)(1-x)^{n_i}/x$ (for i = N,π,K) yields $n_N \simeq 5$ and $F \simeq 1/6$. A similar analysis of ψ-hadroproduction data with the $q\bar{q}$ fusion background removed yields $n_N \simeq 5$, $n_\pi \simeq n_K \simeq 3$ and $F \simeq 1/12$. The fact that the F-values are different for photon and hadron reactions may be associated with the different color rearrangements in the final states. Data on Υ-production by nucleons also yields $n_N \simeq 5$. The power-law dependence of the gluon distributions are in close agreement with naive counting rule results, as illustrated by the solid curves in Fig. 1a-e.

Predictions for neutrino production of ψ via neutral currents based on an analogous $Z°$-gluon fusion approach are given in Fig. 1f. [Supported in part by the Department of Energy under contract DE-AC02 76ER00881-168.]

REFERENCES

See V. Barger, W. Y. Keung and R. J. N. Phillips, Phys. Lett. 91B, 253; 92B, 179 (1980); Z. Phys. (in press) and references therein.

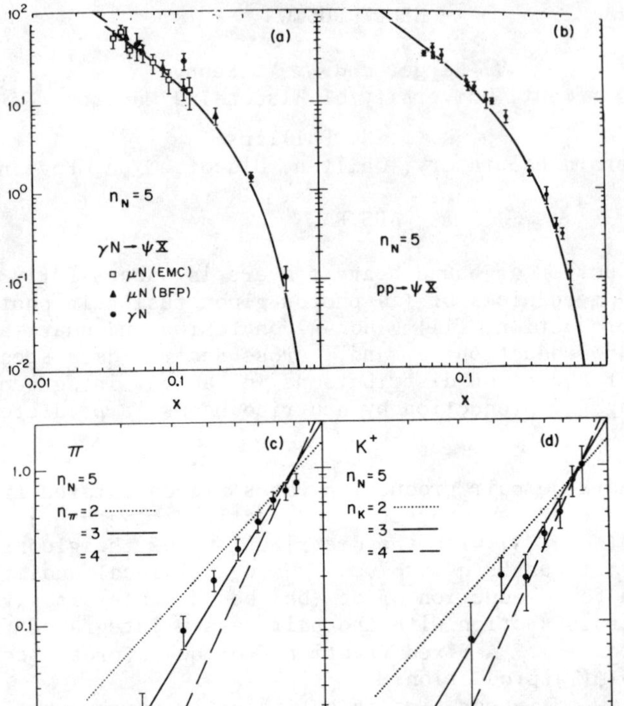

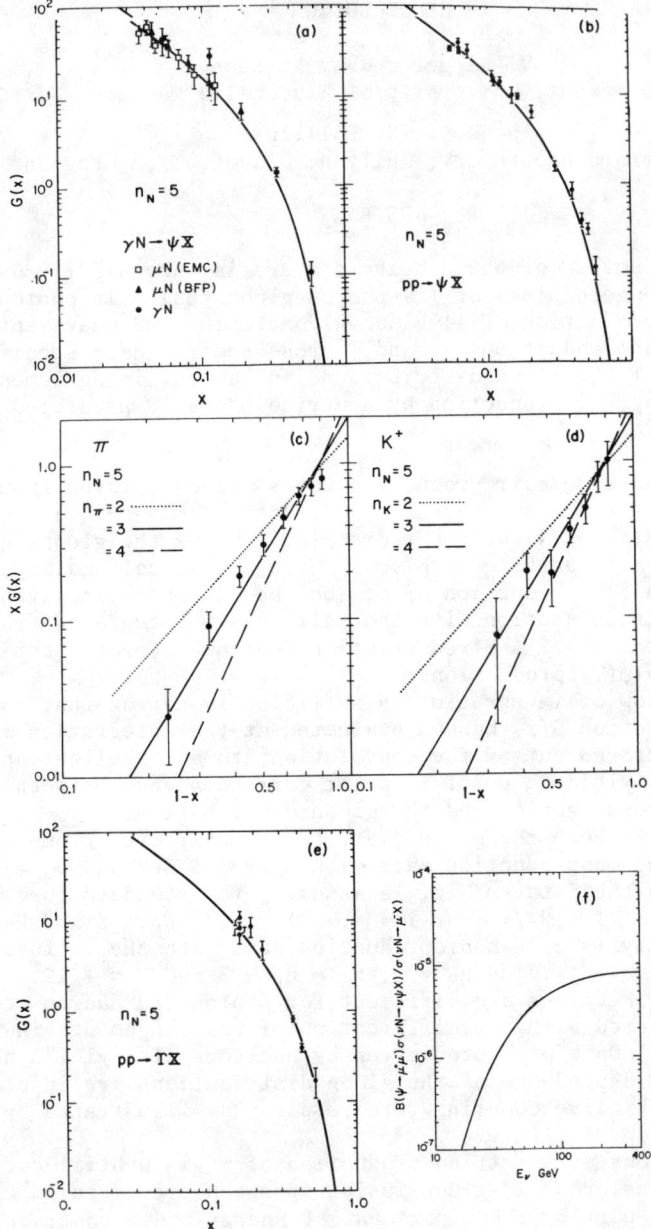

Fig. 1. Results of gluon fusion analyses.

NEW QUARK AND WEAK BOSON SIGNATURES AT $\bar{p}p$ COLLIDERS

F. Halzen and D. M. Scott
Physics Department, University of Wisconsin, Madison, WI 53706

ABSTRACT

We discuss leptonic signatures of charm and bottom, the discovery of the top, and the shape of the lepton spectrum from W decay in $\bar{p}p$ interactions at $\sqrt{s} \simeq 540$ GeV.[1]

LEPTONS FROM HEAVY QUARKS, Q = c,b,(t?)

The large semileptonic branching ratios of heavy quarks make them important sources of leptons in hadronic interactions. We calculate c and b cross sections in $O(\alpha_s)$ QCD.[2] This may be an underestimate, as experimentally there seems to be a large diffractive component of charm production.[3] Calculating the decay by a two-step collinear fragmentation first into a meson, followed by the meson's semileptonic decay (calculated at the quark level), we find[4] the single and double lepton cross sections shown in Figs. 1 and 2. Note that the contribution from b's is bigger than from c's. This is because b's are heavier than c's, and have weak isospin-1/2 as opposed to + 1/2. Both these properties make leptons from b's harder than those from c's. The magnitude depends crucially on the gluon density G(x). Two choices of parton densities lead to the bands in Fig. 2. We have used $B(Q \to \ell + x) = 10\%$, though early data from CESR suggest the bottom B may be bigger.[5]

In Fig. 1 we have also plotted an estimate of the cross section from the semileptonic decay of top for quark mass $m_t = 20$ GeV: it will be hard to find tops this way. In Fig. 2 we compare the dilepton cross section from $Q\bar{Q}$ with that from Drell-Yan: the $Q\bar{Q}$ source dominates to high masses. It has, however, two distinct characteristics: $\mu^+\mu^- = e^+\mu^-$, and each lepton is contained in a hadron jet. As it seems unlikely that top will be found via the customary $V(t\bar{t}) \to \ell^+\ell^-$,[4] correlation measurements will be required. Three possibilities are multileptons,[6] same sign dileptons, and same side high p_T $e^{\pm}\mu^{\mp}$.[4]

THE JACOBIAN PEAK OF THE W

Whereas the Z^0 should be discovered by its $\ell^+\ell^-$ decay, the W presents more problems. Much study[8] has centered on the Jacobian peak in the charged lepton spectrum in its decay to $\ell\nu$ (it may be, though, that the missing neutrino momentum can be constrained by measurement of hadron momentum). Using the Drell-Yan model for W production, the peak has two important properties: it is above background (see Fig. 3) and as noted by Paige,[9] it is forward, because $d\sigma/dp_T d\Omega = (p_T/\sin^2\theta) E d\sigma/d^3p$ in the collider frame.

The Drell-Yan model gives $<p_T>_W \simeq 0$. $0(\alpha_s)$ QCD calculations[10] predict $<p_T^2>_W = 100\text{-}200\ GeV^2$ (see Fig. 4), where p_T is generated by $q\bar{q} \to Wg$, $gq \to Wq$ ($q \equiv$ quark, $g \equiv$ gluon). The leading logarithm form factor[11] $S = \exp[-(2\alpha_s/3\pi)\ell n^2(s/p_T^2)]$ may be important. The dotted line in Fig. 4 is $<p_T^2>_W$ calculated with the $0(\alpha_s)$ p_T distribution multiplied by S.

The $0(\alpha_s)$ contribution gives a high p_T tail to the single lepton cross section, and surely affects the Jacobian peak. Different phenomenological schemes to regularize the $p_T \simeq 0$ divergence of the W cross section lead to different conclusions, but Aurenche and Lindfors[12] have given a solution to $0(\alpha_s)$ in QCD. They use the phase space integral of the ν to integrate over the W's p_T distribution. This enables them to cancel divergences and reallocate mass singularities as scaling violations in the factorization theorem way. The result to $0(\alpha_s)$ is the dotted line in Fig. 3. It is possible, though, that $S(s/p_T^2)$ may affect this. The Jacobian peak has been softened and moved to a slightly lower p_T, but notice that the structure is still significantly above backgrounds, which are indicated by the hatched area. Note that the lepton cross section has been calculated[12] without requiring the intermediate step of finding a regularized cross section for the W itself. It may be possible to check the experimental relevance of the calculation by studying the p_T distribution of single leptons originating from lepton pairs of given mass in existing data.[13]

ACKNOWLEDGEMENTS

We thank M. Dechantsreiter and S. Pakvasa for collaboration on many aspects of this paper. Work supported in part by the University of Wisconsin Research Committee with funds granted by the Wisconsin Alumni Research Foundation, and in part by the Department of Energy under contract DE-AC02 76ER00881.

REFERENCES

1. An extended version of this paper is available as UW-Madison report DOE-ER/00881-161.
2. See, e.g., B. Combridge, Nucl. Phys. B151, 429 (1979).
3. See talks by R.J.N. Phillips, S.Wojicki, this conference.
4. S. Pakvasa, M. Dechantsreiter, F. Halzen and D.M. Scott, Phys. Rev. D20, 2862 (1979); D. M. Scott, Workshop on the Production of New Particles in Super High Energy Collisions, Madison, 1979.
5. See talk by R. Berkelman, this conference.
6. N. Cabibbo and L. Maiani, report TH.2726-CERN
7. M. Abud, R. Gatto and C. A. Savoy, Phys. Lett. 79B, 435 (1978).
8. See A. Astbury et al., CERN report SPSC/78-06.
9. F. Paige, Workshop on the Production of New Particles in Super High Energy Collision, Madison, 1979.

10. F. Halzen and D.M. Scott, Phys. Lett. 78B, 318 (1978).
11. G. Parisi and R. Petronzio, Nucl. Phys. B154, 427 (1979).
12. P. Aurenche and J. Lindfors, report TH.2877-CERN
13. F. Halzen, D.M. Scott and P.M. Stevenson, investigation in progress.

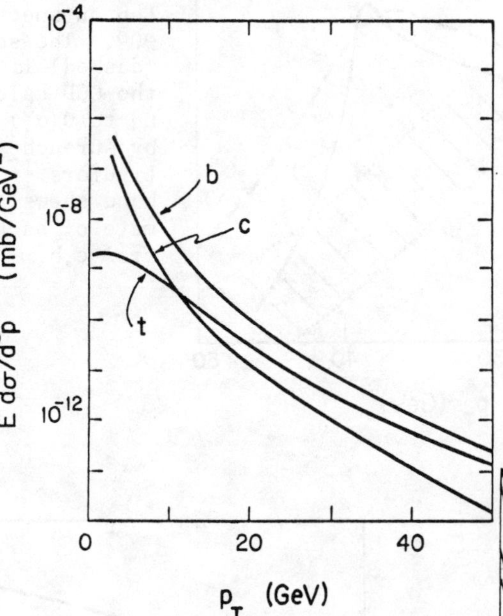

Fig. 1

Lepton spectrum at 90° from b,c,t production and semileptonic decay. $m_t = 20$ GeV.

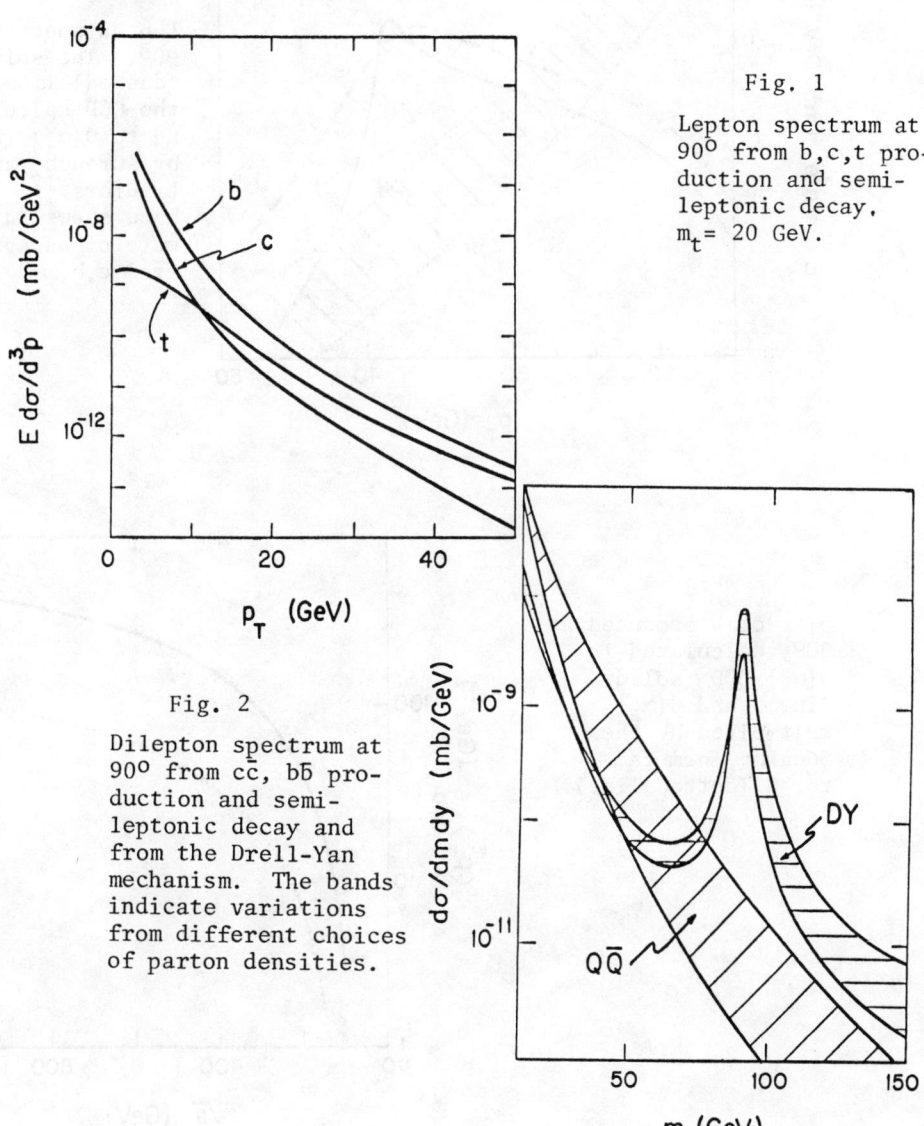

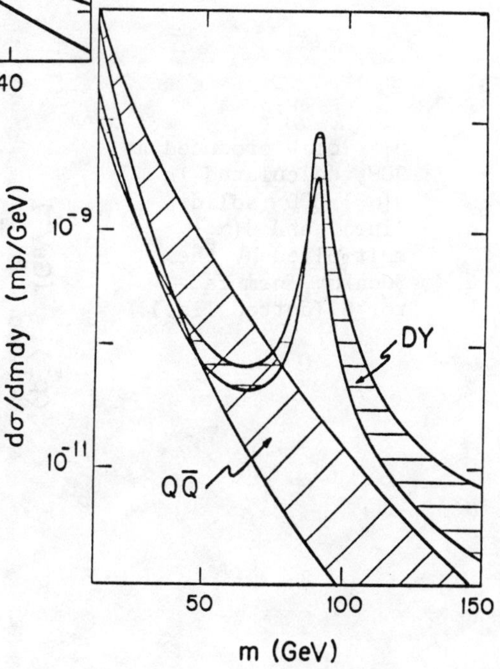

Fig. 2

Dilepton spectrum at 90° from $c\bar{c}$, $b\bar{b}$ production and semileptonic decay and from the Drell-Yan mechanism. The bands indicate variations from different choices of parton densities.

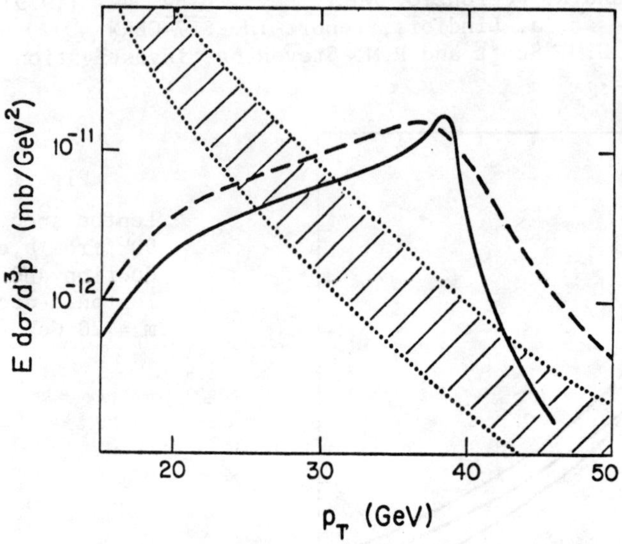

Fig. 3

Lepton spectrum at 90°. The solid (dashed) line is the QCD calculation up to $0(\alpha_s^0)$ $(0(\alpha_s))$ by Aurenche and Lindfors.[12] The band shows an estimate of backgrounds from c,b.

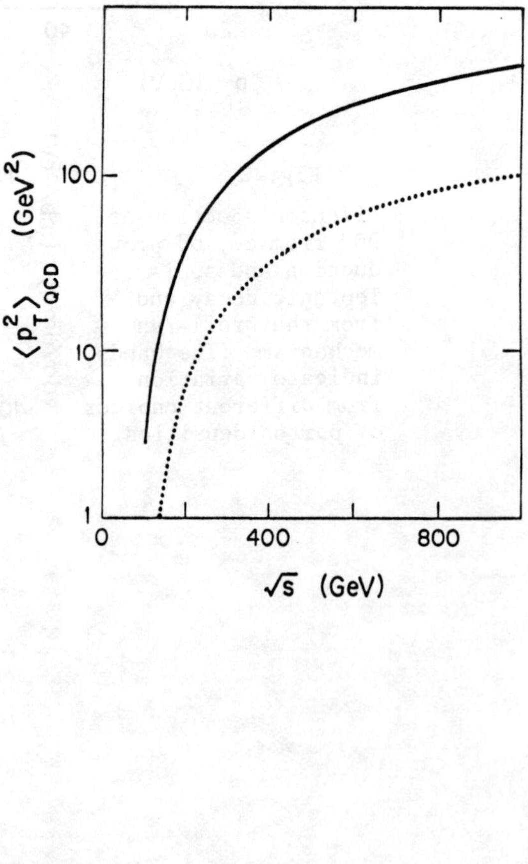

Fig. 4

$\langle p_T^2\rangle$ of W produced at 90°, calculated in $0(\alpha_s)$ QCD (solid line), and $0(\alpha_s)$ multiplied by the Sudakov form factor S (dotted line).

THE INTRINSIC CHARM OF THE PROTON*

P. Hoyer
Nordita, Copenhagen

ABSTRACT

Recent data give unexpectedly large cross-sections for charmed particle production at high x_F in hadron collisions. This may imply that the proton has a non-negligible uud$c\bar{c}$ Fock component.

The data on hadronic production of charmed particles (Λ_c,D) at high (ISR) energies[2,3] has several surprising features: (i) The cross-sections are large (several 100 μb). (ii) The produced particles are fast (the x_F-distributions may even be flat). This also applies to the $D^+ = (c\bar{d})$, which shares no valence quark with the proton. (iii) Part of the charm is produced diffractively.[3]

These features cannot be described by perturbative mechanisms ($gg \to c\bar{c}$, etc.), which give an order of magnitude smaller cross-sections, and predict all charm to have small x_F. In particular, the existence of diffractive (coherent) production suggests that the charm was not created in the collision (this would be a short-distance process), but was present in the incoming state. We are thus led to consider the existence of a uud$c\bar{c}$ component in the proton wave function. As we shall next discuss, such a (non-perturbative, long time-scale) component will in fact lead to the characteristic charm signals seen in the experiments.

In a frame where the proton has a large momentum, the u, u, d, c and $\bar{c}$ constituents all move with similar velocities. By virtue of their mass, this implies that the c,$\bar{c}$ quarks carry most of the momentum. One can estimate the momentum distribution of the quarks from the behavior of energy denominators in the infinite-momentum frame - we find[1] $< x_c > = 2/7$, $< x_u > = < x_d > = 1/7$. Hence the charmed particles - including the D^+ - will be fast.

If the proton has a uud$c\bar{c}$ component, then this can be excited in a diffractive collision. Thus diffractive charm production is an almost inevitable consequence of our picture. We may use it to estimate the magnitude of the charm component. If σ_{Diff}(charm) = 100...300 μb then, given that σ_{Diff}(tot) ≈ 8 mb, we get a 1...3% admixture of uud$c\bar{c}$ states in the proton.

The existence of multi-quark Fock components of the type described above leads to many interesting and observable consequences. In particular, (i) The charm quarks should manifest themselves through charm production at moderate x_{Bj} ($x_{Bj} \approx 0.3$) in deep inelastic $\mu p \to \mu X$ and $\nu p \to \nu X$ reactions. However, threshold effects require large W, and hence also large Q^2. (ii) If the hadronic production of charm is dominantly diffractive, the cross-

* This work was done[1] in collaboration with S. Brodsky, C. Peterson and N. Sakai.

section would increase dramatically as the energy becomes large enough to allow high-mass diffraction. This may partly explain the apparent rise in $\sigma(c\bar{c})$ from Fermilab-SPS to ISR energies. The effective threshold would be higher for nuclear targets, because of the stricter coherence requirements.

The existence of multi-quark components in hadrons is of basic importance for our understanding of the spectrum and dynamics. The study of heavy quark production in hadron collisions may provide new insights into this fascinating question.

REFERENCES

1. S. Brodsky, P. Hoyer, C. Peterson and N. Sakai, Phys. Letters 93B, 451 (1980).
2. D. Drijard et al., Phys. Lett. 81B, 250 (1979) and 85B, 452 (1979). W. Lockman et al., Phys. Lett. 85B, 443 (1979).
3. K. L. Giboni et al., Phys. Lett. 85B, 437 (1979).
   J. Eickmeyer et al., paper #512 submitted to the XX International Conference of High Energy Physics, Madison (1980).
   L. J. Koester et al., paper #618, ibid.

Heavy Flavor Production With Final State Interaction

B. Margolis
McGill University, Montreal, Quebec, Canada H3A 2T8

Introduction

I will discuss work done with Y. Afek and C. Leroy[1]. We calculate $D\bar{D}$ ($B\bar{B}$) production as the rate for $c\bar{c}$ ($b\bar{b}$) production by gluon-gluon (in p-p interaction) or photon-gluon fusion (in γ-p interaction) followed by the capture of light sea quarks. We calculate quarkonium production assuming that a produced $Q\bar{Q}$ pair (Q is a heavy quark) becomes a bound state after single hard gluon emission. There is expected to be a high probability for soft gluon emission and absorption by the heavy quarks before light quark capture or hard gluon emission in the case of D(B) and $\psi,\psi'...(\Upsilon,\Upsilon'...)$ production respectively. The soft gluon emission is assumed to create a random phase situation so that relative rates are determined by intensity ratios.

Calculations and Results

The rate for quarkonium production, R_1, is calculated using a dipole matrix element and non-relativistic quarkonium wave functions.[1]

$$R_1 = |\langle f|V'|i\rangle|^2 \; 2\pi\delta(E_i E_f - \omega) \; d^3k/2\omega \; (2\pi)^3 \qquad (1)$$

where $V' = g\Lambda_a \vec{\nabla}\cdot\vec{A}_a$. Here $\vec{A}_a$ is the color gauge field and Λ_a the color matrices for Q and $\bar{Q}$. $|i\rangle$ is the initial plane wave $Q\bar{Q}$ wave function and $|f\rangle$ the final wave function taken to be S.H.O. in form. To get the rate for light quark capture, R_2, to form D,B,.. we assume the existence of a thermodynamic sea of light quarks created in the interaction in the rest frame of Q and similarly of $\bar{Q}$. We have then the rate for formation of D etc.

$$R_2 = \frac{\int \frac{N}{V} e^{-E_q/T} \sigma_{q\bar{Q}}(\bar{s}) \sqrt{(p_q \cdot P_Q)^2 - m_q^2 m_Q^2} \,/E_q E_Q \, d^3 p_q}{\int e^{-E_q/T} d^3 p_q} \qquad (2)$$

where N is a normalization factor for the thermodynamically distributed light quarks ($T \sim m_\pi$), $\sigma_{q\bar{Q}}$ is the light on heavy quark total cross section which we get from the usual quark model considerations and photoproduction data. In general $\sigma_{q\bar{Q}} \sim 1/M_Q^2$ above threshold for open Q production. The factor N is found to be $\sim .2$ and is kept fixed for all processes.

The produced $Q\bar{Q}$ pairs capture each a light quark for the most part, to produce $D\bar{D}$ or $B\bar{B}$ (and $K\bar{K}$ if we extrapolate down to strange quarks). We identify such production then with heavy $Q\bar{Q}$ production in lowest order QCD. However we take the threshold for production as twice the mass of the produced quark and not twice the mass of the produced mesons. The remaining energy comes from the captured light quarks. The capture rate of light quarks is not important since $D\bar{D}$, $B\bar{B}$ (and $K\bar{K}$) production is dominant. The production of $Q\bar{Q}$ quarkonium states in competition with the dominant processes of course depends on the capture rate R_2.

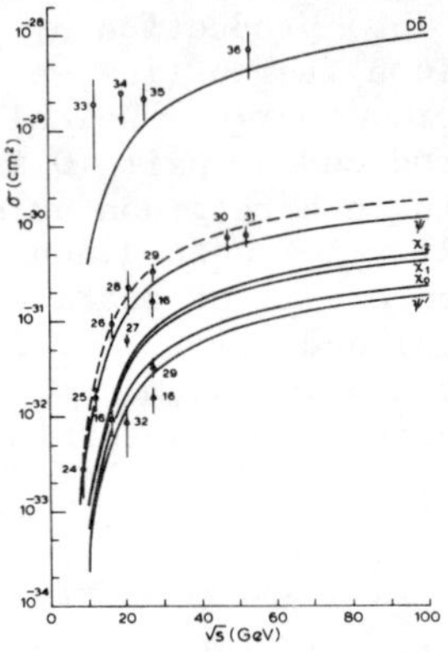

Fig. 1. p-p charm production

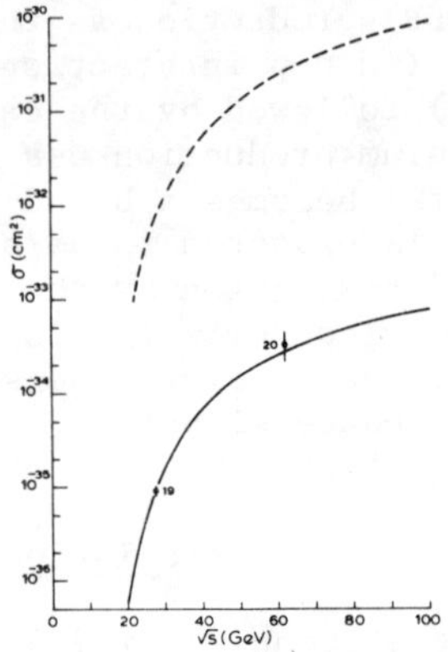

Fig. 2. p-p bottom production

Figures 1 and 2 show results for our calculation of $D\bar{D}$ and $B\bar{B}$ production in p-p interaction with $\alpha_s = 0.3,\ 0.23$ respectively. $m_c = 1.2$ GeV and $m_b = 4$ GeV. We also have calculated transverse momentum distributions for D and B mesons(1).

Following the above we calculate bound state production as the QCD cross section for gluon-gluon (p-p) or photon-gluon (γ-p) fusion times the appropriate branching ratio.

$$\sigma_{BS}(\hat{s}) \cong \sigma_{QCD}(\hat{s}) \frac{R_1}{R_2} \tag{3}$$

We convolute (3) with the appropriate gluon structure function. We use a Q^2 independent structure function with n=5. ψ production it is found gets an additional 20% contribution from χ states and a further 30% from the Carlson-Suaya mechanism. The results for ψ, ψ' and χ_0, χ_1, χ_2 production (p-p) are shown in figure 1. The dashed curve adds the Carlson-Suaya contribution to ψ production. Figure 2 shows the cross section for Υ production in p-p interaction. The η_c, η_b production should be dominated by the Carlson-Suaya mechanism.

We go from p-p to γ-p interaction replacing one gluon by a photon. Figures 3 and 4 show photoproduction cross sections for $D\bar{D}$, ψ, ψ' and $B\bar{B}$, Υ respectively. Here we need a more detailed Q^2 dependent structure function(1). Otherwise the parameters are as in p-p interaction.

<u>Vector Dominance and Total Cross Sections in γ-p Interaction.</u>

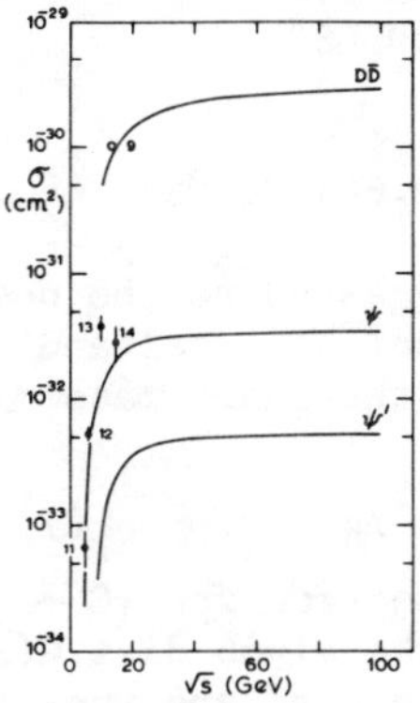

Fig. 3. γ-p charm production

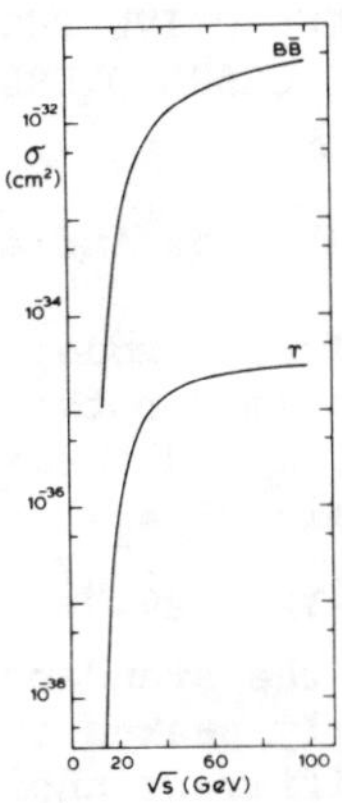

Fig. 4. γ-p bottom production

Figures 3 and 4 show that $\sigma(\Upsilon)/\sigma(B) \ll \sigma(\psi)/\sigma(D)$ which in fact is smaller than $\sigma(\phi)/\sigma(K)$ calculated in the same way. The latter ratio is slightly less than unity. Form factors inhibit heavy quarkonium production. If we could calculate well for light quarks we might expect $\sigma_{\gamma\rho}(\rho)$ to be very large i.e. there is vector dominance for light quarks. Lowest order QCD diagrams in fact fail for light quark production[1] as one might expect. However $\sigma^{tot}_{\gamma p} - \sigma^{QCD}_{\gamma p}(D\bar{D}) - \sigma^{QCD}_{\gamma p}(K\bar{K})$ leads to the full curve of figure 5 which we compare with $1/216\ \sigma^{tot}_{\rho p}$ (dot-dash curve). Vector dominance is good for light quarks. The rising σ^{tot} is due in part to rising $\sigma^{tot}_{\rho p}$ and in part to threshold effects in D and K production.

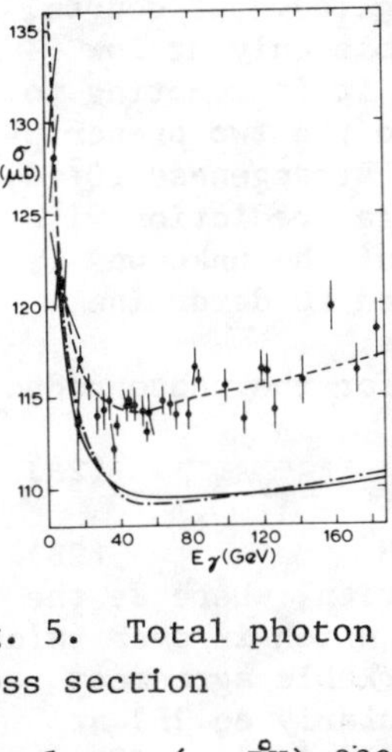

Fig. 5. Total photon cross section

The Production of D*(2010) in p-p and γ-p Interaction

We calculate D* production as a three step process. (1) $c\bar{c}$ production using lowest order QCD. (2) c captures a light quark q creating a charmed fireball. (3) This charmed fireball follows a cascade decay in general following the statistical bootstrap model and at some point a D or D* materializes. At $P_{lab} \sim 100$ GeV we calculate $\sigma(\gamma p \to D^0 X) = 230$nb and $\sigma(\gamma p \to D^{*+} X) = 125$nb. to be compared with FNAL measurements $\sigma(\gamma p \to D^0 X) = 295 \pm 130)$nb and $\sigma(\gamma p \to D^{*+} X) = (118 \pm 49)$nb.

References

1. Y. Afek, C. Leroy and B. Margolis, Phys. Rev. 22B, 86 (1980), ibid 22B, 93 (1980).

DIFFRACTIVE PHOTOPRODUCTION OF STRANGENESS AND CHARM-OVERLAP OF QCD AND REGGE MODELS*

D.P. Roy
TIFR, Bombay-5, India and ANL, Argonne, Ill.60439, USA

The γG fusion model of Fig.1a has been suggested as the dominant QCD mechanism for the diffractive photoproduction (real and virtual) of charm and possibly strangeness, with the quark mass m_q setting the scale of α_s.

$$d\sigma/dM^2 \; (\gamma p \to q\bar{q}(M^2)X) = \bar{\sigma}(M^2)G(x)/s; \; x = M^2/s \qquad (1)$$

where $\bar{\sigma}(M^2)$ is the standard Bethe-Heitler cross-section for $\gamma G \to q\bar{q}$ which is strongly peaked at low $M^2(\approx M^2_{th})$. For the gluon dist.G(x) three widely different choices, which have been in use, are the so called flat, power counting and steep solutions

(I) $(1-x)^5(1.03+13.5x)/x$ (II) $3(1-x)^5/x$ (III) $5.5(1-x)^{10}/x$.

The model further assumes colour compensation via soft gluon exchanges (collectively denoted by the dashed line) involving negligible energy transfer back from the excited $q\bar{q}$ to the recoil system X. This means that the hadronic system $H_{q\bar{q}}$ resulting from $q\bar{q}$ carries its energy ($\approx E\gamma$) and invariant mass ($\approx M^2_{th}$), where as the recoil system is dominated by the lowest energy colour singlet state (the proton).

All the three features above are characteristics of the Regge model for hadron diffraction (Pomeron exch.), which is linked to photoproduction of specific flavours via VMD (Fig.1b). Of course, the VMD and Regge model are quantitatively reliable only at low Q^2 and m_q and the QCD model at large Q^2 and/or m_q. It is tempting to speculate, nonetheless, an overlap region between the two prescriptions, hopefully covering the photoproduction of strangeness ($Q^2=0$, $m_q=m_s$). This can be tested by comparing the Regge prediction with the QCD (eq.1) at fixed x, which is independent of the unknown quantity G(x). If they agree then one can proceed to determine G(x), by comparing the two predictions at fixed s.

Fig.2 shows the VMD and Regge prediction for the photoproduction of strangeness, i.e.

$$\frac{d\sigma}{dM^2}\,(\gamma(\phi)p \to H(M^2)p) = \begin{cases} .2\mu b/M^2 & (2A) \\ .2\mu b/M^2 + 0.5\ \mu b\ GeV^2/M^4 & (2B) \end{cases}$$

The normalisation follows from Pomeron factorisation, where as the shapes correspond to the extrema $0 < \sigma^{PPR} < \sigma^{PPP}$ which is over twice the typical latitude of σ^{PPR}. Fig.2a shows remarkable agreement between the QCD and the Regge prediction (particularly eq.2B) at fixed x. Then, a comparison between the two predictions at fixed s (Fig.2b) clearly favours the flat gluon distribution (soln. I). The analogous comparison between the two models for charm production are shown in Fig.3. Although the VMD and Regge predictions are less reliable for charm, it is reassuring that one gets identical results.

Finally it is evident that the s-independence of the Regge prediction (eq.2) corresponds to a flat xG(x) (eq.1) over its range of validity ($x \lesssim .1$), as in solution I. Since the Regge model is quantitatively reliable for strangeness production, the appropriate mass scale for the predicted flat gluon distribution should be $Q^2 \sim m_\phi^2$.

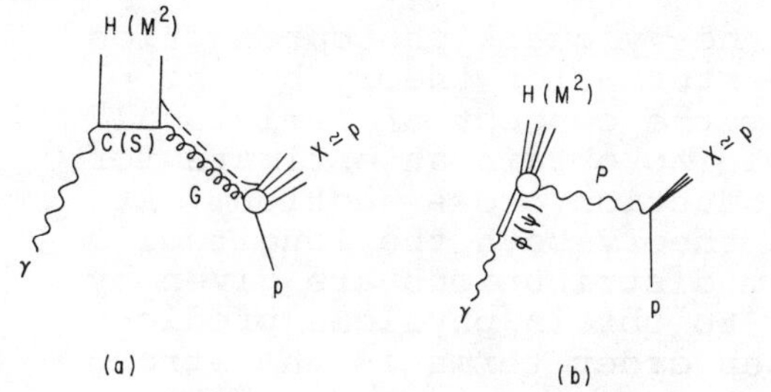

Fig.1. (a) γG fusion model and (b) VMD and Regge model for diffractive photoproduction of strangeness and charm.

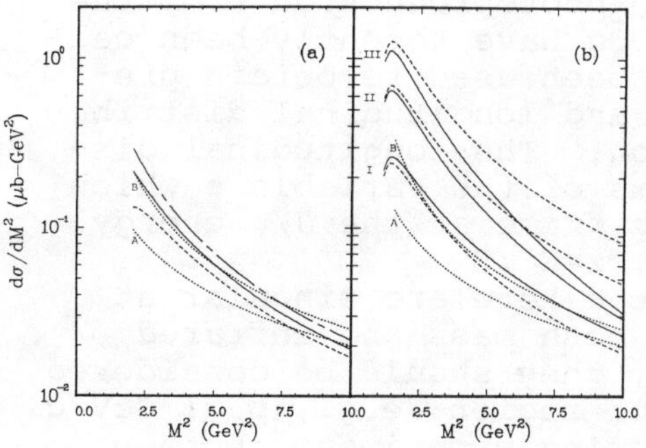

Fig.2. Regge prediction (dotted) for $\gamma p \rightarrow H_{s\bar{s}}(M^2)p$ compared with the γG fusion model ($\alpha_s \approx .3$). (a) fixed x; $m_s \approx .5$ (solid), .6 (dashed), .4 (chain dashed) (b) fixed s = 100 (solid), 1000 (dashed).

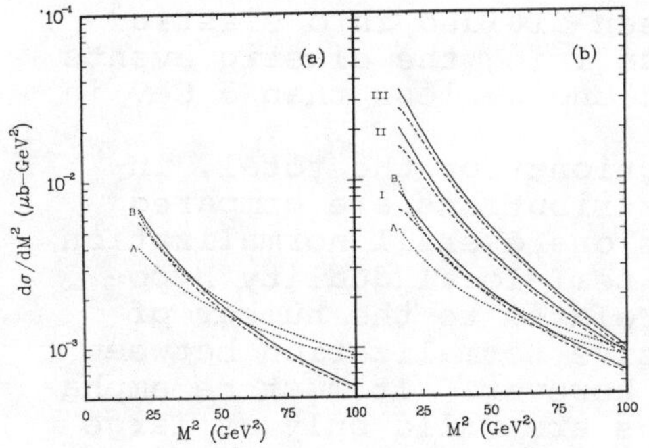

Fig.3. Regge prediction for $\gamma p \rightarrow H_{c\bar{c}}(M^2)p$ compared with γG fusion (a) fixed x (b) fixed s. $m_c \approx 1.5$ (solid), 1.65 (dashed).

*For details see ANL-HEP-PR-80-49, to be published in Phys. Letters B.

HIGHER ORDER CORRECTIONS FOR J/ψ PHOTOPRODUCTION*

J.F. Owens

Physics Department, Florida State University
Tallahassee, FL 32306

The cross section for heavy quark photoproduction is given in lowest order perturbation theory by the subprocess $\gamma g \rightarrow Q\bar{Q}$. By using the concept of semi-local duality[1] it is also possible to obtain an estimate for the total J/ψ or ϒ photoproduction cross section. At this order in perturbation theory both the longitudinal and transverse momentum distributions are given by delta functions. In order to obtain physical predictions, one must go to higher order terms in the strong coupling constant.

The contributions to $Q\bar{Q}$ photoproduction from the subprocesses $\gamma g \rightarrow Q\bar{Q}g$ and $\gamma q \rightarrow Q\bar{Q}q$ have recently been calculated[2]. The results have been used to obtain predictions for the transverse and longitudinal distributions for J/ψ photoproduction. The longitudinal distributions are given in terms of the variable z which is defined in the laboratory frame as the J/ψ energy scaled by the photon energy, $z=E_{\psi}/E_{\gamma}$.

The predictions presented here are singular at $p_T=0$ due to the presence of both mass and infrared singularities. Accordingly, they should be considered reliable only in the large p_T region, e.g., $p_T \gtrsim 1$ GeV/c.

In a recent muon scattering experiment, both p_T and z distributions for J/ψ photoproduction have been measured. The data have been divided into elastic[3] and inelastic[4] samples by defining the elastic events to be those with a hadronic energy less than 5 GeV in the laboratory.

In figure 1 the predictions for the total, inelastic, and elastic p_T distributions are compared with the data[3,4]. There is one overall normalization parameter which, using the semi-local duality hypothesis[1], is approximately related to the number of bound $c\bar{c}$ states. The relative normalization between the three curves is fixed, however. It must be emphasized that these predictions are valid only at large p_T. Nevertheless, the agreement, both in normalization and shape, between the predictions and the data is quite striking.

*Supported in part by the U.S. Department of Energy.

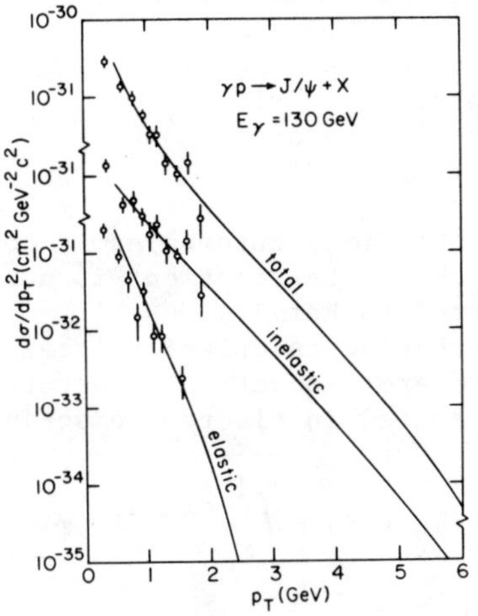

Fig. 1 Comparison between the data[3,4] and the theoretical predictions.

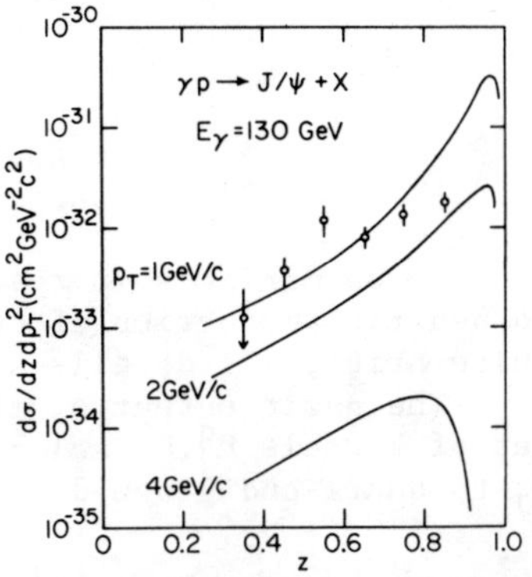

Fig. 2 Comparison between the shape of the experimental z distribution[4] and the theoretical predictions.

The predictions for the z distributions at several values of p_T are shown in figure 2. As yet the experimental data are not available in bins of both z and p_T. However, in the middle portion of the z region the slopes of the z distributions are approximately independent of p_T. In this same region the data[4], integrated over p_T, show a similar z slope.

There is encouraging agreement between the available data and the predictions presented here. The addition of new data at higher values of p_T should soon allow more detailed tests of these predictions.

REFERENCES

1. H. Fritzsch, Phys. Lett. 67B, 217 (1977).
2. D.W. Duke and J.F. Owens, FSU-HEP-800709 (1980), to be published in Phys. Lett.
3. J.J. Aubert et al., Phys. Lett. 89B, 267 (1980).
4. J.J. Aubert et al., CERN-EP/80-84 (1980).

QCD CORRECTIONS TO NEUTRINO CHARM PRODUCTION

Thomas Gottschalk
High Energy Physics Division
Argonne National Laboratory
Argonne, Illinois 60439

We summarize some results of the $O(\alpha_s)$ chromodynamic corrections to neutrinoproduction of heavy quarks. The treatment is necessarily quite brief; all details are deferred to Ref.1.

The basic output of the perturbative calculation itself is the set of kernels K_j^q, K_j^G relating the charm production structure functions F_j to quark and gluon densities measured in electroproduction, e.g.:

$$F_2(x) = 2x\left\{\hat{q}(x) + \int_x^1 \frac{dz}{z} K_2^q(\lambda,z)\hat{q}(x/z) + \int_x^1 \frac{dz}{z} K_2^G(\lambda,z)G(x/z)\right\} \quad (1)$$

where $\lambda \equiv Q^2/(Q^2+M^2)$. The kernels are given by differences in parton-level terms

$$K_j^q(\lambda,z) \equiv H_j(W^*q \to cX) - H_2(\gamma^*q \to qX) \quad (2)$$

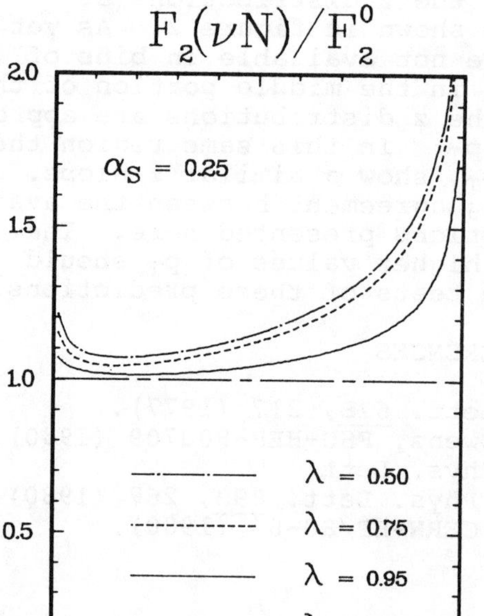

Fig.(1) Fractional corrections to the charm production structure function F_2.

The cancellation of perturbative singularities in Eq.(2)--mandated by factorization--<u>requires</u> a slow rescaling parton model. (This means, for example, that presently observed dimuon rates do <u>not</u> conclusively demonstrate suppression of the strange sea.) Uncertainties in the kernels arise from (i) choice of subtraction scale and (ii) treatment of the charmed sea.

Fig.(1) gives the ratio of corrected to lowest order results for $F_2(\nu N \to \mu cX)$. The region $x' \lesssim 0.3$ is somewhat sensitive to subtraction scheme ambiguities. The divergences for $x' \to 1$ can be traced to $\ell n(P/E)$ phase space factors and thus should be softened by summing higher orders. For $\lambda \to 1$, the corrections vanish, as required by the subtraction scheme in Eq.(2).

In Fig.(2) we show fractional corrections to the total charm production rates $R_c \equiv \sigma_{charm}/\sigma_{Total}$. The different curves are for (I) full perturbative subtraction and

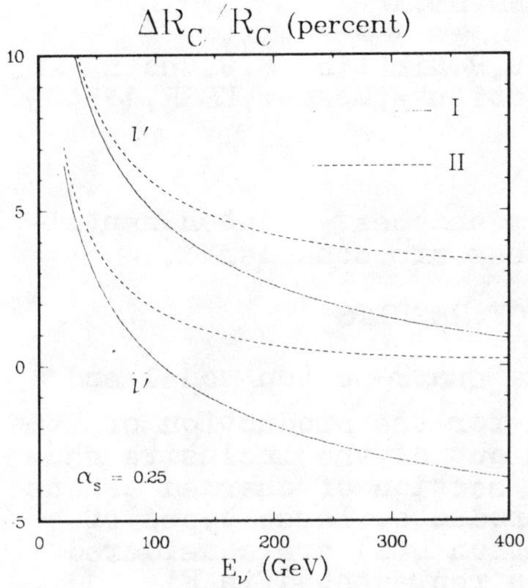

Fig.(2) Fractional corrections to the total charm production rates R_c.

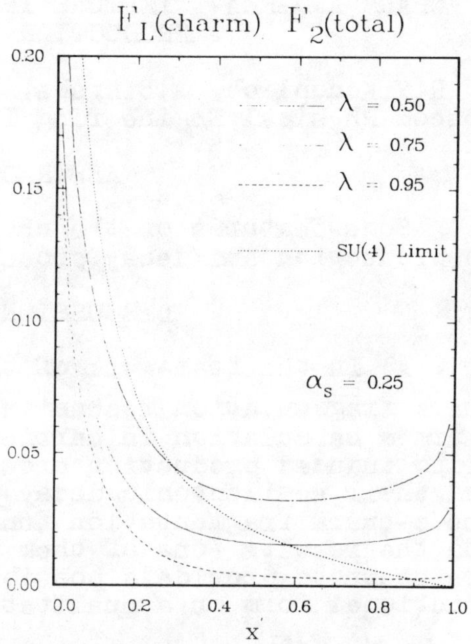

Fig.(3) Charm contributions to the longitudinal structure function F_L.

(II) no $\gamma^*G \to c\bar{c}$ subtraction (i.e., assuming no intrinsic charmed sea). In both cases, the corrections to R_c are modest. The small differences between I,II are significant in another regard: the smallness of $\gamma^*G \to c\bar{c}$ suggests that fusion model calculations based on $W^*G \to \bar{c}s$ are dominated by $\ell n(Q^2/m_s^2)$ mass singularities, and thus probably are not perturbatively reliable.

Finally, Fig.(3) shows charm contributions to the longitudinal structure function F_L for inclusive νN scattering. The dotted curve in Fig.(3) is the total F_L/F_2 signal expected for four massless quarks. Except for very large Q^2, charm provides a substantial fraction of the total F_L signal. Such effects must be taken into consideration in present experimental attempts to measure F_L/F_2.

REFERENCE

T. Gottschalk ANL-HEP-PR-80-35 (1980) and references therein.

CHARM AND HEAVY LEPTONS IN NEUTRINO INTERACTIONS: PRODUCTION AND DECAY

R.V.Konoplich, A.B.Krebs, Yu.P.Nikitin, S.G.Rubin
Moscow Physical Engineering Institute, Moscow, USSR, 115409

ABSTRACT

Some features of the charm and heavy lepton neutrino production and decay processes are considered.

DISCUSSION OF RESULTS

A) In the framework of the quark-parton model and quark fragmentation mechanism[1] for the production of charm a calculation is carried out of the inclusive neutrino induced production cross section of charmed mesons and their semileptonic decay products. Three types of the c-quark fragmentation function D(z) are considered and the results (one of them is represented in Fig. 1) are shown to provide a possibility to establish its functional form on a qualitative level. A condition

$$(qp_c) \leq 0 \qquad (1)$$

("fragmentation condition") with q and p_c – the momenta of W-boson and charm respectively is used to exclude the target fragmentation region. This condition is argued[2] to be more correct than the commonly used condition

$$E_c/\nu \geq 0.1 \div 0.2 \qquad (2)$$

and is shown to produce a considerable effect on the resulting charm and its decay products spectra. A crucial point is that it excludes some additional area in the (x,y) plane, interpreted in terms of the other dinamical mechanisms (quasielastic, diffractive) for charm production. The new definition of the fragmentation variable

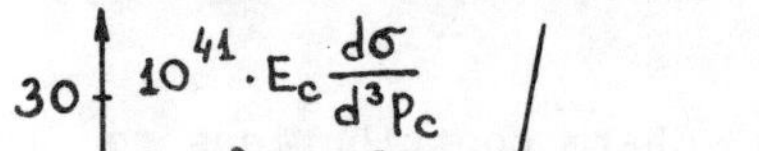

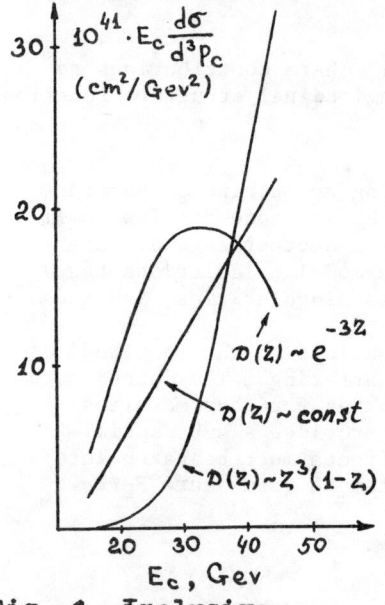

Fig. 1. Inclusive cross section for D-meson production (E_ν = 50 Gev, θ_c = = 0.1°).

$$z = (p^B_{cL} - p^B_{c\,min})/(p^B_{c\,max} - p^B_{c\,min}) \qquad (3)$$

is introduced which is more adequate to the accepted fragmentation condition and provides the exact normalization of the D(z) function (superscript "B" denotes the Breit reference frame and subscript "L" - the longitudinal impulse component).

B) We also consider heavy lepton decays through the neutral currents. The unknown constants for the $Z \leftrightarrow h^0$ transitions λ^0_h are obtained (Z - neutral intermediate boson, $h^0 - \pi^0$, K^0, etc.). With λ^0_h fixed all the main partial widths may be determined. For example,[3]

$$\lambda^0_\pi = 0.9\, m_\pi \frac{f^u_A - f^d_A}{\sqrt{2}\, f^{ud}_A}\,; \quad \lambda^0_\rho = \frac{m^2_\rho}{g_\rho}(f^u_V - f^d_V)\,;$$

$$\lambda^0_\eta = 0.9\, m_\pi \frac{f^u_A + f^d_A - 2f^s_A}{\sqrt{6}\, f^{ud}_A}\,; \quad \lambda^0_\omega = \frac{m^2_\rho}{g_\rho}(f^u_V + f^d_V). \qquad (4)$$

In above quark neutral currents are assumed to have the form $\bar{u}_1 \gamma_\mu (f_V + f_A \gamma_5) u_2$. Constants f_V , f_A are fixed by the choice of the particular model of weak and electromagnetic interaction. In this case the lifetime of the definite heavy lepton may be calculated.

C) Both charm and heavy lepton production have some kinematic features providing the possibility of their effective detection and evaluation of their parametres [4,5]. The simple rules are: 1) the most probable transverse momentum of a secondary particle from the two-particle charm or heavy lepton decay in neutrino induced reaction is $p_\perp = m_p/2$; 2) the same for a secondary particle from three-particle decay is $p_\perp = m_p/4$, where m_p is a parent particle mass and the products are assumed to be much lighter than the parents.

REFERENCES

1. M.Gronau, F.Ravndal, V.Zarmi, Nucl.Phys. B51, 611 (1973).
2. R.V.Konoplich, A.B.Krebs, Yu.P.Nikitin, Yad.Fiz. 31 , 713 , (1980).
3. V.I.Konjushko et al , Preprint IHEP 79-174 (1979).
4. Yu.P.Nikitin, S.G.Rubin, Yad.Fiz. 23, 1020 (1976).
5. V.A.Golubkov et al , Pisma JETP 17, 159 (1973).

SUPPRESSED PRODUCTION OF F, F* AND Σ_c

P. Mukhopadhyay

Department of Physics, Jadavpur University,
Calcutta 700032, India

This paper offers a model-independent explanation for the suppressed production of F, F* and Σ_c. It also predicts that the B and B* mesons will be suppressed if they are interpreted as $(c\bar{b})$ systems. This paper also demands that the production of the so far unobserved spin-half charmed baryons S and A (in the notations of Gaillard et al.) will be suppressed.

We assume that the hadron symmetry is broken in such a way that the SU(2) i.e. the isospin invariance still remains intact. If we denote isospin by I and if K is a c-number in isospace, then, the quantity $(-1)^{KI}$ is rotationally invariant in isospace. It is well known that the isospace and the actual spin space are completely disjoint and as such the actual spin J is a c-number in isospace. This implies that $(-1)^{LJ}$ will behave as a constant in isospace, L being a c-number in isospace. For reasons discussed above, the following quantity is invariant under rotations in isospace.

$$(-1)^{KI} + (-1)^{LJ} \equiv \text{invariant} \qquad (1)$$

If we set L = 2, then, $(-1)^{LJ}$ remains real for the bosons and fermions as well. If, however, we set K = 2 also, then, the invariant defined by relation (1) becomes trivial for the non-ordinary hadrons as it does not yield any vital information for them. As B (baryon number), S (strangeness), C (charm), B' (bottom/beauty) and T (top/truth) do not respond to rotations in isospace, therefore, we can set

$$K = B + S + C + B' + T + X \qquad (2)$$

where X is a c-number in isospace and it is necessary for a non-zero value of K. The values of the scalar quantum numbers of the non-ordinary hadrons suggest that the reality for the quantity $(-1)^{KI}$ is ensured for

$$X = 2,4,6\ldots\ldots\ldots\ldots \text{for iso-bosons} \qquad (3a)$$

and, $$X = 1,3,5\ldots\ldots\ldots\ldots \text{for iso-fermions} \qquad (3b)$$

0094-243X/81/680290-02$1.50

We rewrite relation (1) in its explicite form given below.

$$(-1)^{(B + S + C + B' + T + X)(I)} + (-1)^{2J} \equiv \text{invariant} \quad (4)$$

Obviously, the numerical value of this invariant can be either 0 or 2. Without any loss of generality we may choose the value 0 for the invariant concerned. Therefore, we can write

$$(-1)^{(B + S + C + B' + T + X)(I)} + (-1)^{2J} = 0 \quad (5)$$

For the moment we confine our attention to the non-ordinary hadrons the productions of which are not suppressed. For these hadrons the numerical value of the invariant is 0 when

$$X = \pm 2, \pm 6, \pm 10 \ldots\ldots \text{for iso-bosons} \quad (6a)$$

and $$X = \pm 1, \pm 5, \pm 9 \ldots\ldots \text{for iso-fermions} \quad (6b)$$

where + sign holds for the particles and - sign for the antiparticles. We can recast eqn. (5) in the following form

$$(-1)^{(B + S + C + B' + T + X)(I)} = (-1)^{(2J \pm 1)} \quad (7)$$

where for the sake of generality we have used the relation $(-1) = (-1)^{\pm 1}$. The value of X occurring in eqn. (7) may be uniquely fixed by demanding that

$$|(B + S + C + B' + T + X)|(I) = |(2J \pm 1)| \quad (8)$$

where X must also satisfy the relation $(B + S + C + B' + T + X) \neq 0$ since $K \neq 0$. The quantity X appearing in eqn. (8) can take one of the values given by relations (6a) and (6b). At this point we demand that a non-ordinary hadron can be produced unsuppressed if for it eqn. (8) is found to hold true. It is easy to check that eqn. (8) holds true for the charmed particles D, D* and Λ_c but not for the F, F* and Σ_c (which, therefore, must be suppressed). Eqn. (8) is also not satisfied for the spin-half charmed baryons S(C = + 1, S = - 1, I = ½) and A(C = + 1, S = - 1, I = ½) apart from Σ_c. It is interesting to note that B and B* mesons must also be suppressed, according to eqn. (8), if they are assinged the values B' = + 1 and C = + 1 i.e. if they are treated as $(c\bar{b})$ systems instead of $(u\bar{b})$ systems (for which B' = + 1, C = 0).

QUARK SEARCHES, EXPERIMENT AND THEORY

A. Zichichi, CERN

G. Morpurgo, INFN-Genoa, Organizers

QUARK SEARCH IN HIGH ENERGY EXPERIMENTS

G. Valenti
Istituto Nazionale di Fisica Nucleare, Bologna, Italy.

ABSTRACT

Results on free quark searches, in cosmic rays ($|Q| < 1$), proton-nucleon interactions ($|Q|$ 1/3, 4/3) and neutrino (antineutrino)-nucleon interactions ($|Q|$ 1/3) presented to this conference are reviewed.

INTRODUCTION

Since many years indirect experimental evidence supports the idea that nucleons are made of super-elementary point-like spin 1/2 fractionally charged objects: "quarks". They have never been observed as such. Experimental searches reported to this conference confirm and extend production upper limits.

COSMIC RAYS

A cosmic ray telescope[1] was operated at sea level for a few thousand hours. The apparatus comprises a vertical telescope with six scintillator planes, two wide gap spark chambers and steel absorbers. The charge is determined, with a resolution of 0.05 Q_e, by measuring the dE/dx in the scintillation counters, the same counters also provide a measurement of $\beta = v/c$ with a time-of-flight path of 2 mt. and a resolution of 0.4 nsec.. The mass is finally determined with a range-dE/dx method to about half proton mass (mp). Data were collected with a delayed trigger selecting $\beta < 1/2$.

Three candidates are reported with charges: 0.75 ± 0.05, 0.7 ± 0.05 and 0.89 ± 0.06, and masses > 5.3 mp, > 2.8 mp and (9.3 ± 3.0) mp. The first two candidates appear to be complicated multiparticle events, while the third is not incompatible with $|Q| = 1$. At this level these candidates correspond to the rejection limit of the experiment.

PROTON-NUCLEON COLLISIONS

A search for long lived particles[2] with charge $|Q| = \pm 2/3, \pm 1, \pm 4/3, \pm 2$ produced at zero degrees in proton-beryllium and proton alluminum at beam energies of 200, 210 and 240 GeV has been conducted at the CERN SPS. The apparatus consisted of 8 scintillation counters for dE/dx and time-of-flight measurements, threshold and differential Cherenkov counters. The charge and mass of momentum selected particles was determined by measuring dE/dx in the 8 scintillation counters, the threshold was set at $|Q| > 2/3$, and by determining $\beta = v/c$ with time-of-flight on a base of 200 mt. and with differential Cherenkov counters.

No candidate s were found for $|Q| = 2/3$ and 4/3, the 95% confidence level flux limits agree with limits obtained in similar searches.

NEUTRINO AND ANTINEUTRINO NUCLEON INTERACTIONS

Asearch for free fractionally charged quarks ($|Q| = 1/3$) has been performed in the wide band neutrino and antineutrino beam of the CERN SPS[4] , preliminary analysis shows no candidate. The experiment was prompted by the idea that a single high momentum transfer is needed to free the proton constituents, and the neutrino is the natural point-like particle capable of this. The apparatus, shown in fig. 1, consists of:

i) $200\cdot60\cdot170$ cm^3 lead target.
ii) Scintillation counters V, P_1, P_2, P_3, P_4 and S_1 to tag the $\nu(\bar{\nu})$-interaction inside the target.
iii) A 0.4 tesla magnet, B, to sweep away low momentum particles.
iv) A $235\cdot60\cdot125$ cm^3 avalanche chamber, AC-235, to measure the specific ionization of all particles independently of multipli city; events are recorded by three cameras equipped with light amplifiers.
v) Two double layer hodoscopes, ST1, ST2, to select particles according to their dE/dx.
vi) Scintillation counters, C, V_2, to select μ of specific momentum for calibration purposes.
vii) Four proportional wire chambers, W_1, W_2, with multihits capa bility for geometric reconstruction and track timing.

The reported results are relative to "isolated" (i.e. quarks accompaigned by low multiplicity) quarks, $|Q| = 1/3$. Events have been selected according to their dE/dx in the scintillation hodoscopes ST1, ST2, all candidates have then been scanned visually.

The efficiency of such analysis was determined via Monte Carlo assuming for the produced quark an angular distribution similar to the μ ("leptonic") or to the hadrons ("hadronic") produced in neutrino interactions. The acceptance so determined is 44% for the "leptonic" case, 28% for the "hadronic".

The total analysis efficiency was 0.285. The results, summari zed in table I, are expressed in terms of 90% confidence level flux upper limit where the flux is defined by (quark/ $\nu(\bar{\nu})$-interactions).

Narrow band (high ν energy) and high multiplicity analysis are in progress and should push these limits, if no candidate is found, down more than one order of magnitude.

Table I Summary of results.

beam type	$\nu(\bar{\nu})$-interactions	90% confidence level upper flux limit "leptonic"	"hadronic"
ν wide band	$2.4\cdot10^5$	$< 7.64\cdot10^{-5}$	$< 1.2\cdot10^{-4}$
$\bar{\nu}$ wide band	$1.04\cdot10^5$	$< 1.77\cdot10^{-4}$	$< 2.78\cdot10^{-4}$

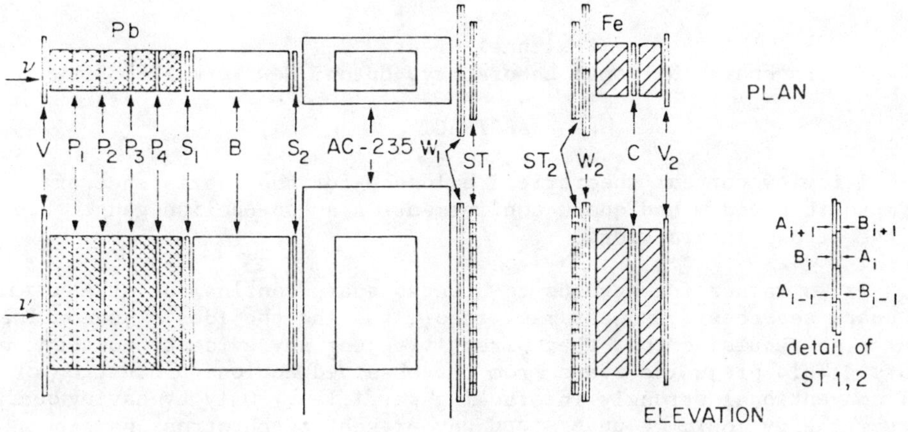

Fig. 1

REFERENCE

1) P. C. M. Yock, "Heavy particles in the cosmic radiation?" Phys. Rew. D (in press), Phys. Rew. D18, 641 (1978).
2) A. Bussiere et al. Nuclear Physics B (in press), A. Bozzoli et al. Nuclear Physics B159 (1979) 363, B144 (1978) 271, B140 (1978) 271.
3) L. W. Jones, Rew. Mod. Phys. 49 (1977) 713.
4) M. Basile et al. Nuovo Cimento 45A, 281 (1978).
5) M. Basile et al. "Primary ionization measurements in a large avalanche chamber for free quarks detection". Presented at the International Conference on experimentation at LEP, Uppsala, Sweden 1980.

QUARK CONFINEMENT

Michael Creutz
Brookhaven National Laboratory, Upton, New York 11973

ABSTRACT

I review current theoretical evidence for the coexistence of asymptotic freedom and quark confinement in a non-Abelian gauge theory of the strong interaction.

It is rather incongruous to discuss quark confinement at a session on quark searches. Indeed, many theorists find the idea of unconfined quarks so repulsive that they directly reject any evidence to the contrary. This prejudice stems from the observed copious production of all conventional strongly interacting particles. Only by having quarks be very heavy could we understand why present synchrotrons have not provided us with separated quark beams. However, a large quark mass would represent an extravagent new parameter in hadronic physics and would make it difficult to understand the remarkable successes of the naive non-relativistic quark model. It has become easier to imagine exact quark imprisonment rather than an approximate confinement that breaks down at some unknown energy.

An underlying gauge field forms the basis of most theoretical models of confinement. Quarks are coupled to a vector "gluo-electric" field through a generalization of Gauss's law $\vec{\nabla} \cdot \vec{G} = \rho$ where ρ is the quark density. In theories with a non-Abelian symmetry this equation is embelished with internal symmetry indices and extra source terms from the charged gluon field. Confinement in an automatic consequence of Gauss's law if the theory does not have massless gauge bosons in its spectrum. Without a massless field to support a Coulombic spreading the conserved flux of $\vec{G}$ must form itself into "flux tubes" which can only end on the other sources.[1] The only finte energy states are neutral clusters of quarks joined by these tubes with finite energy per unit length. At large separation this yields the conventional linearly rising interquark potential.

The behavior of hypothetical magnetic monopoles in a superconducting medium represents a simple example of this phenomenon. Here $\vec{G}$ represents the magnetic field and can pass through the medium only in the quantized form of "vortices" or "tubes" of magnetic flux which only end on sources carrying monopole charge. Modules of a strong interaction dynamics based on this idea have been proposed, although in complexity they appear somewhat contrived.[2]

The desire for an economical theory has led to an essentially universal enthusiasm for a Yang Mills theory of non-Abelian gauge mesons interacting with the quarks. This elegant generalization of electrodynamics endows the gauge bosons with an internal symmetry, and these "gluons" are charged with respect to each other. The confinement conjecture presumes an inherent instability of a theory of massless gluons. Consequently, a spreading gluonic field will automatically draw itself into the flux tubes necessary for confinement. A semiclassical analysis based on classical solutions to Euclidian

0094-243X/81/680296-06$1.50

Yang-Mills equations has suggested a possible mechanism for this flux collapse.[3]

If it occurs in four-dimensional space-time, confinement must be a non-perturbative phenomenon. Indeed, simple renormalization group arugments show that K, the energy per unit length of the flux tube, must exhibit an essential singularity in the bare coupling constant

$$K \sim \frac{1}{a^2} (g_0^2)^{(-\beta_1/\beta_0^2)} \exp\left(-\frac{1}{\beta_0 g_0^2}\right) \qquad (1)$$

Here g_0 is the base coupling constant, defined with a cutoff of length a imposed to remove ultraviolet divergences. The numerical constants β_0 and β_1 are the first terms in a perturbative expansion of the Gell-Mann Low function

$$a \frac{\partial}{\partial a} g_0 = \beta_0 g_0^3 + \beta_1 g_0^5 + O(g_0^7) \qquad (2)$$

Equation (1) should be valid in the limit of cutoff removal by taking a to zero. The important consequence of Eq. (1) is the impossibility of any perturbation calculation of K.

A non-perturbation treatment requires an ultraviolet cutoff that is not based on the Feynman expansion. The most extensively studied such regulator is the lattice proposed by Wilson.[4] Here the cutoff parameter a is the lattice spacing, which is to be taken eventually to zero. Before this is done, the path integral defining the quantum theory is formally equivalent to a partition function for a system of variables on a four dimensional crystal. In this analogy the bare coupling constant squared corresponds to temperature. Applying high temperature series techniques to this system, Wilson showed that in the strong coupling limit the theory describes quarks connected by strings with a finite energy per unit length. In other words, confinement is automatic in the lattice theory for strong enough coupling. However, a low temperature expansion at weak coupling reproduces conventional Feynman perturbation theory. This series is at best asymptotic, but its existance suggests a possible low temperature phase of free quarks and massless gluons. As this is a behavior qualitatively distinct from confinement, one expects at least one phase transition separating the high and low temperature domains, if the free quark phase exists. Balian, Drouffe, and Itzzkson have argued that such a confinement-nonconfinement phase transition will occur in large space-time dimensionality.[5]

Ultimately we are interested in the continuum limit of the theory. In the language of solid state physics, this requires taking the bare coupling constant to a critical value so that correlation scales, i.e., physical Compton wavelengths, become large relative to the cutoff represented by the lattice spacing. The perturbative renormalization group indicates one such critical point at vanishing bare coupling. A continuum limit at this point yields the phenomenon of asymptotic freedom; the effective renormalized coupling will go to zero when defined on decreasing length scales. This phenomenon allows perturbative predictions of scaling phenomena in high momentum transfer processes.

To have asymptotic freedom in the same phase that Wilson's expansion demonstrates confinement, four dimensional space time must be inadequate to exhibit the deconfining phase transition mentioned above. Based on an approximate analysis of Migdal and Kadanoff, current lore is that four dimensions are critical for gauge theories.[6] In more than four dimensions all gauge groups should exhibit a spin-wave phase transition whereas for less than four dimensions any continuous gauge group will always confine. In exactly four dimensions only Abelian groups should show a non-trivial phase structure; indeed, this is necessary if Wilson's formalism is to describe quantum electrodynamics, the prototype of all gauge theories.

Recently from two rather different techniques strong evidence has appeared supporting the "standard" picture of the phase structure of lattice gauge theory. The first technique is to extrapolate the strong coupling series into a region where weak coupling predictions should apply.[7] In being able to smoothly join the weak and strong coupling behavior, one obtains evidence for the lack of a phase transition separating these regimes. These methods have been quite successful in identifying the parameters characterizing this matching; in particular they agree to the Λ parameter discussed below.

The other technique supporting the standard picture is Monte Carlo simulation. Considering the path integral as a partition function for a statistical system at a given temperature, a Monte Carlo procedure generates a sequence of configurations which are typical of an ensemble in thermal equilibrium. This is done by making random changes in the gauge fields in such a way that the probability of obtaining any configuration C is proportional to the Boltzmann factor

$$P(c) \sim e^{-\beta S(c)} \tag{3}$$

where S(c) is the action associated with the gauge field configuration and β is the inverse temperature or inverse coupling squared.

As the entire lattice is stored in the computer memory, one can measure any desired correlation function. One is effectively doing experiments on a four dimensional crystal. For SU(2) gauge theory my crystals have been up to 10^4 sites in size while for SU(3) I have so far been limited to 6^4.

In Figure 1 I show the results of thermal cycles on the internal energy of several of the models.[8] The internal energy P is the expectation value of the action density and is normalized so that at infinite temperature it has value 1 and at zero temperature it vanishes. By slowly increasing the temperature from cold to hot and then reducing it, regions of slow convergence will appear as hysteresis effects. These are hints of phase transitions, where the convergence time should diverge on an infinite lattice. The figure shows SU(2) gauge theory in both four and five dimensions as well as the SO(2) theory of electrodynamics in four dimensions. I include SU(2) in five dimensions to illustrate the criticality of four dimensions. The signals of phase transitions in the four dimensional SO(2) and the five dimensional SU(2) models are clear whereas the four dimensional SU(2) model appears much smoother.

To provide more support for the lack of a transition in the four dimensional non-Abelian case, I have studied the interaction between external sources with quark quantum numbers. This is done by measuring

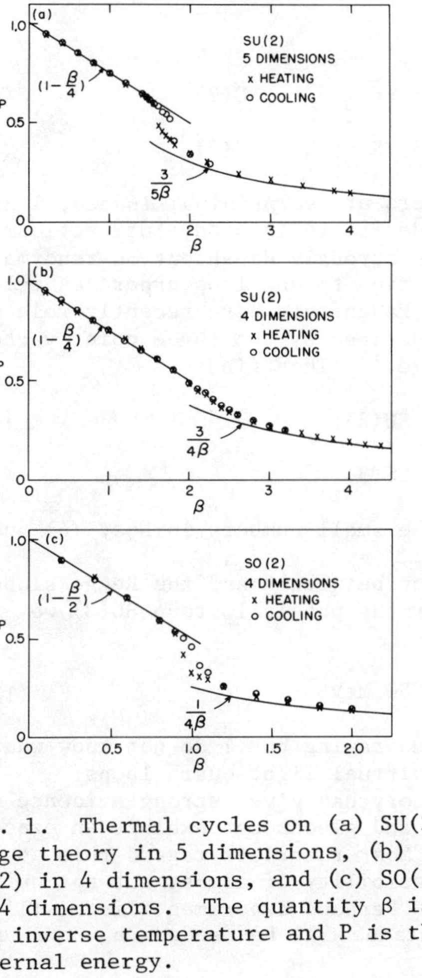

Fig. 1. Thermal cycles on (a) SU(2) gauge theory in 5 dimensions, (b) SU(2) in 4 dimensions, and (c) SO(2) in 4 dimensions. The quantity β is the inverse temperature and P is the internal energy.

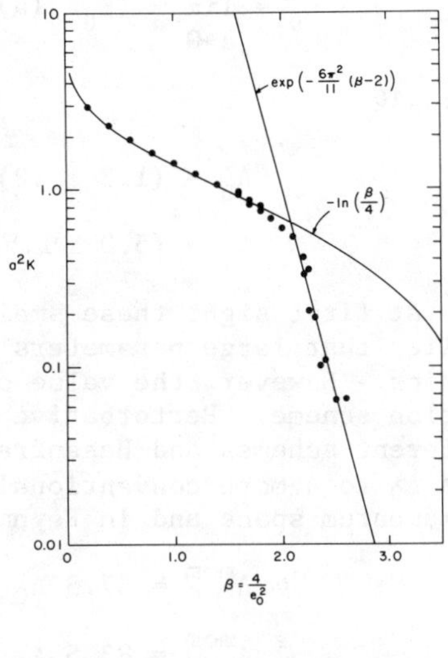

Fig. 2. The cutoff squared times the string tension as a function of β. The solid lines are the strong and weak coupling limits.

expectation values of Wilson loops, exponentials of the gauge field integrated about closed curves in the lattice. From this analysis I have extracted the coefficient K of the linear potential between widely separated quarks.[9] In Figure 2 I plot the measurements of a^2K versus the inverse temperature $\beta = \frac{4}{g_0^2}$. Also plotted are the first term in the strong coupling expansion

$$a^2K = -\log\frac{\beta}{4} + O(\beta^2) \tag{4}$$

and the asymptotic freedom prediction of Eq. (1) with the prefactor neglected and an arbitrarily chosen normalization. If a^2K for large β does indeed follow the asymptotic freedom prediction, then the linear potential will survive the continuum limit and confinement is the consequence.

One previously unknown number follows from this analysis. This is the overall normalization in Eq. (2) and is a parameter relating the short distance asymptotic freedom behavior to the long distance confining potential. Defining the parameter Λ_0 by

$$\Lambda_0 = \lim_{a\to 0} \frac{1}{a} \left(\beta_0 g_0^2(a)\right)^{(-\beta_1/2\beta_0^2)} \exp\left(\frac{-1}{2\beta_0 g_0^2(a)}\right) \tag{5}$$

I find[10]

$$\Lambda_0 = (1.3 \pm .2) \times 10^{-2} \sqrt{K} \qquad SU(2) \tag{6}$$

$$\Lambda_0 = (5.0 \pm 1.5) \times 10^{-3} \sqrt{K} \qquad SU(3) \tag{7}$$

At first sight these small numbers are surprising; indeed, I argued earlier that large parameters are undesirable in stong interaction physics. However, the value of Λ_0 is strongly dependent on renormalization scheme. Perturbative calculation to one loop order can relate different schemes and Hasenfratz and Hasenfratz have recently related this Λ_0 to a more conventional Λ^{mom} defined by the three point vertex in momentum space and in Feynman gauge.[11] They find

$$\Lambda^{mom} = 57.5\ \Lambda_0 \qquad SU(2) \tag{8}$$

$$\Lambda^{mom} = 83.5\ \Lambda_0 \qquad SU(3) \tag{9}$$

These large factors largely cancel the small numbers in Eqs. (6) and (7).

Using the string model connection between K and the Regge slope gives $\sqrt{K} \sim 400$ MeV; consequently, for the physical group SU(3) we obtain

$$\Lambda^{mom} = 170 \pm 50 \text{ MeV} \tag{10}$$

This number is phenomenologically encouraging but I do not know what corrections arise from inclusion of virtual light quark loops.

In conclusion, lattice gauge theory has given strong evidence that a non-Abelian gauge theory of quarks and gluons can exhibit an exact confinement of quarks into hadrons. Most theorists regard this as a more aesthetic situation than the possibility of almost but not quite confined constituents of the proton. Perhaps experimentalists will prove this position wrong, but persistance will be needed to persuade us.

REFERENCES

1. J. Kogut and L. Susskind, Phys. Rev. D9, 3501 (1974).
2. Y. Nambu, in Proceedings of Johns Hopkins Workshop on Current Problems in High Energy Particle Theory (Baltimore, 1974); M. Creutz, Phys. Rev. D10, 2696 (1974); G. Parisi, Phys. Rev. D11, 970 (1975).
3. C.G. Callan, R.F. Dashen and D.J. Gross, Phys. Rev. D19, 1826 (1979).
4. K. Wilson, Phys. Rev. D10, 2445 (1975).
5. R. Balian, J.M. Drouffe, and C. Itzykson, Phys. Rev. D10, 3376 (1974); D11, 2098 (1975); D11, 2104 (1975).
6. A.A. Migdal, Zh. Eksp. Teor. Fiz. 69, 810 (1975); 69 1457 (1975) [Sov. Phys. - JETP 42, 413 (1975); 42,743 (1975)]; L.P. Kadanoff, Rev. Mod. Phys. 49, 267 (1977).

7. J. Kogut, R.B. Pearson and J. Shigemitsu, Phys. Rev. Lett. 43, 484 (1979); J.B. Kogut and J. Shigemitsu, Preprint (1980); G. Münster, DESY Preprint 80/44 (1980); G. Münster and P. Weisz, Preprint (1980).
8. M. Creutz, Phys. Rev. Lett. 43, 553 (1979).
9. M. Creutz, Phys. Rev. D21, 2308 (1980).
10. M. Creutz, Preprint BNL-27752 (1980).
11. A. Hasenfratz and P. Hasenfratz, Preprint TH. 2727-CERN (1980).

ADDITIONAL EVIDENCE FOR FRACTIONAL CHARGE OF 1/3 e ON MATTER

G. S. LaRue, J. D. Phillips and W. M. Fairbank
Stanford University, Stanford, Ca. 94305

ABSTRACT

We report more data consistent with the existence of fractional charge of $\pm$ 1/3 e on matter. We have measured fractional charges on several new niobium balls and on balls previously measured. We continue to see fractional charge changes of $\pm$ 1/3 e, indicating that the fractionally charged atoms are either being added to or removed from the surface of the balls. We have made 21 more measurements on five different balls and have seen four fractional charges consistent with + 1/3 e, four with - 1/3e, and thirteen with zero. The residual charge values consistent with $\pm$ 1/3 e and their statistical errors in units of electric charge are -0.376 $\pm$ 0.045, 0.277 $\pm$ 0.026, -0.371 $\pm$ 0.025, 0.364 $\pm$ 0.023, 0.357 $\pm$ 0.026, -0.324 $\pm$ .014, -0.385 $\pm$ 0.032, 0.370 $\pm$ 0.056. We have measured one ball 14 times during a 3½ year period (every measurement consistent with $\pm$ 1/3 e or 0 e) and have seen nine charge changes of 1/3 e. Extensive measurements demonstrate that the fractional charges cannot be explained by background forces.

INTRODUCTION

In previous publications,[1-4] we presented results of a superconducting magnetic levitation experiment giving evidence for the existence of fractional charge of ~ 1/3 e on matter. Since then we have continued to modify the experiment and have more than doubled the experimental data. We report here 21 new independent measurements on 5 niobium balls which yield 4 values consistent with -1/3 e, 4 values consistent with +1/3 e, and 13 values consistent with 0 e. This brings the total number of measurements to 39. All known dipole effects have been considered and could not have accounted for the fractional charges observed.

APPARATUS

The apparatus is kept at 4° K and consists of a $\leq 9 \times 10^{-5}$ gm. niobium ball suspended magnetically between two horizontal capacitor plates. The plates are made by sputtering titanium or copper onto 15 cm. diameter 2.5 cm. thick optically flat quartz, and are separated uniformly by 1 cm. quartz spacers. The ball's position is sensed by a SQUID magnetometer. Its charge can be changed at will with movable β+ and β- emitters. The plates can be accurately moved with respect to the ball while a fixed separation is maintained between them.

By measuring the response of the ball to an alternating electric field $\vec{E}_A$ applied to the capacitor plates, the force applied to the ball can be measured. The alternating force on the ball in the vertical z direction is

$$F_A = (q_r + ne)E_A - \vec{P}_A \cdot \vec{\nabla}E_F - \vec{P}_F \cdot \vec{\nabla}E_A + F_M + F_Q \quad (1)$$

0094-243X/81/680302-06$1.50

where q_r is the residual charge, n is the integer charge number, R is the radius of the ball, E_A is the z component of $\vec{E}_A$, $\vec{P}_F$ is the permanent electric dipole of the ball, $\vec{P}_A$ is the induced electric dipole of the ball, F_M is the magnetic force, F_Q is the quadrupole force, and E_F is the z component of the fixed electric field arising from contact potential variations on the plates near the ball.

$\vec{E}_A$ is established by applying 3000V across the plates in the form of a square wave in phase with the velocity of the ball. Every 50 cycles the phase of the square wave with respect to the ball is reversed. The difference between the time rate of change of the envelope of the ball's oscillation before and after reversal is independent of the damping and proportional to F_A.

The measured residual force F_A^r is defined as the value of F_A for n = 0 in (1). It can be measured to within 0.01 eE_A with this technique in several hours. To determine the fractional charge, we need to accurately determine all of the background forces.

BACKGROUND FORCES

We have shown [1-4] by measuring $\vec{P}_F$ and $\partial E_A/\partial z$ that all known electromagnetic multipole forces are negligible at z_o (where $\partial E_A/\partial z = 0$) except for the force due to d.c. electric field gradients from the plates.

Marinelli and Morpurgo[5,6] have recently reported a charge mimicking effect which they conclude must arise from an electrically induced magnetic moment on the surface of their iron balls.

This magnetic electric force cannot account for our observed fractional charge of $\pm 1/3$ e for the following three reasons: 1) A reversal of the magnetic field would change the sign of the effect. When we reversed the magnetic field when measuring ball 6,[2] the residual charge remained consistent with +1/3 e and did not change to -1/3 e. 2) Marinelli and Morpurgo have reported that the magnitude of this effect drifts with time. We have never seen such drifts with time. 3) If such an effect were present in a magnitude sufficient to mimic a force 1/3 eE_A, then random values of the residual charge would be expected. Every measurement we have obtained for which $\partial E_F/\partial z$ was constant has been consistent with 0 e and $\pm$ 1/3 e.

Eliminating the negligible forces discussed above, (1) becomes

$$F_A^r / E_A = q_r - R^3 \partial E_F/\partial z(z) - P_z \partial E_A/\partial z(z) \qquad (2)$$

We measure F_A^r as a function of each ball's position in order to measure $\partial E_F/\partial z(z)$. For balls of identical radius, differences in F_A^r/E_A at z_o are differences in q_r. This only determines the residual charges up to an additive constant. By measuring balls of several radii it is possible to determine the constant as well. If all the balls have $q_r = 0$ then $F_A^r/E_A = -R^3 \partial E_F/\partial z(z_o)$ so that a plot of $F_A^r(z_o)/E_A$ vs R^3 would yield a straight line through zero with a slope of $-\partial E_F/\partial z(z_o)$. If one of the balls does not have $q_r = 0$, then its residual charge is the intercept of a line drawn through its point on this plot with slope $-\partial E_F/\partial z(z_o)$. The error in determining R^3 $\partial E_F/\partial z(z_o)$ must be added in quadrature to the error in F^r_A to get the errors in q_r.

Comparing the measured residual forces is valid only if $\partial E_F/\partial z$ remains constant throughout the measurements. This constancy is determined by measuring F^r_A as a function of position for every ball and by repeating these measurements on the same ball.

DATA ANALYSIS

If $\partial E_F/\partial z$ remains constant for a set of measurements, the data can be fit to the form in (2) where $\partial E_A/\partial z$ is determined experimentally. For each levitation of each ball (i.e., for each measurement) in the set, a vertical dipole term corresponding to P_z and a constant term are fit to the data. A function is chosen which best fits $\partial E_F/\partial z(z)$ with as few parameters as possible. It is required to be the same for all measurements within a set. This function is used to interpolate between the points to get $F^r_A(z_o)$.

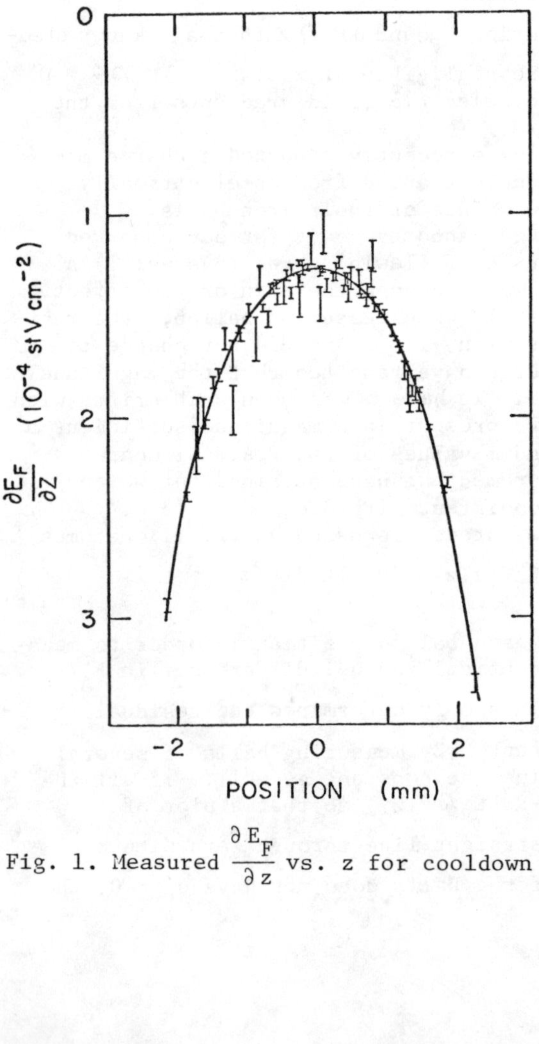

Fig. 1. Measured $\frac{\partial E_F}{\partial z}$ vs. z for cooldown 1.

In the first cooldown 11 measurements were alternately made on balls 6 and 10. R^3 of ball 10 is .352 times that of ball 6. $\partial E_F/\partial z$ changed after the first two measurements. The change in $\partial E_F/\partial z$ was evident from a change in F^r_A/E_A measurements which could not be accounted for by a constant q_r term or a vertical dipole term in (2). The constancy of $\partial E_F/\partial z$ for set 2 is shown in Fig. 1. Fig. 2 illustrates $F^r_A(z_o)/E_A$ for each ball plotted against R^3.

It is essential to note that even without considering the measurements on ball 10, those on ball 6 are not all consistent with the same residual charge, and differ in q_r by ~1/3 e.

For m measurements $\{F^r_{A_i}\}$, $i = 1, \ldots, m$, there are $m + 1$ unknowns $\{\partial E_F/\partial z(z_o), q_{r_1}, \ldots, q_{r_m}\}$. To find the residual charges, we must make an assumption. We assume that one of the three values obtained for ball 6 is 0 e. This gives three possibilities for $\partial E_F/\partial z(z_o)$, depending on which measurement of ball 6 has $q_r \approx 0$ e. One and only one of these is consistent with $q_r = 0$ e or $\pm$ 1/3 e for ball 10. We choose this slope $\partial E_F/\partial z(z_o)$, which is consistent with $q_r = 0$ e for ball 10. The slope is determined by minimizing the sum of the squares of the distances of each point from the line nearest it.

In the second cooldown, 10 measurements were made on 5 balls. $\partial E_F/\partial z$ remained constant for the first 5 measurements, changed to a new function for the next two measurements, and then changed to a still different function for the last three measurements. For each set of measurements the data were fit to (2). $\partial E_F/\partial z$ was different for each set but within each set $\partial E_F/\partial z$ was required to be the same for each ball. Because of the shape of the graph of $\partial E_F/\partial z$ vs. z, errors in the position of one ball relative to another for this cooldown could lead to residual charge errors $\lesssim .02$ e E_A.

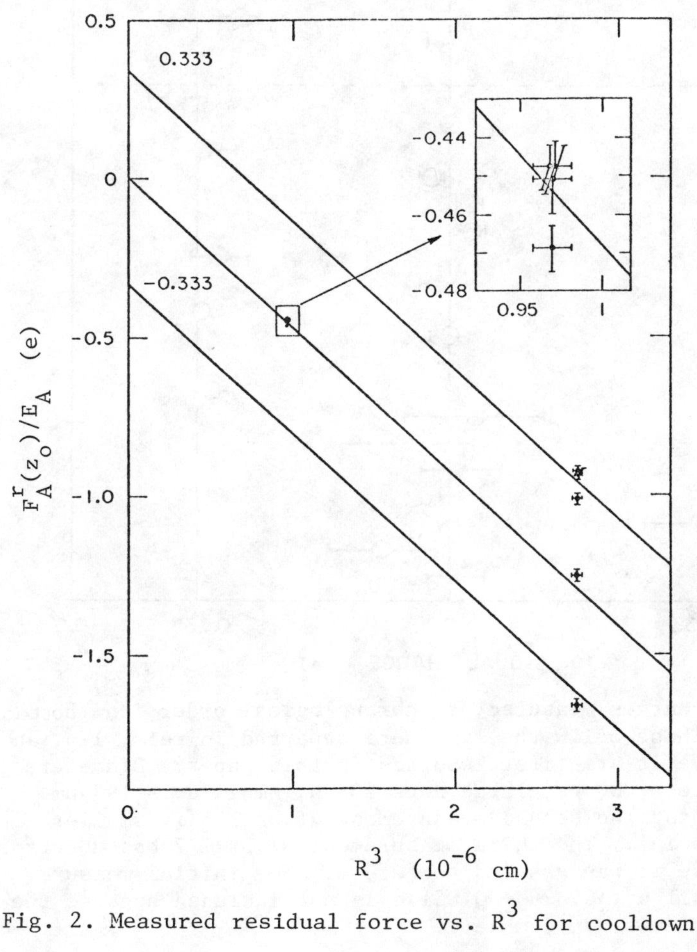

Fig. 2. Measured residual force vs. R^3 for cooldown 1.

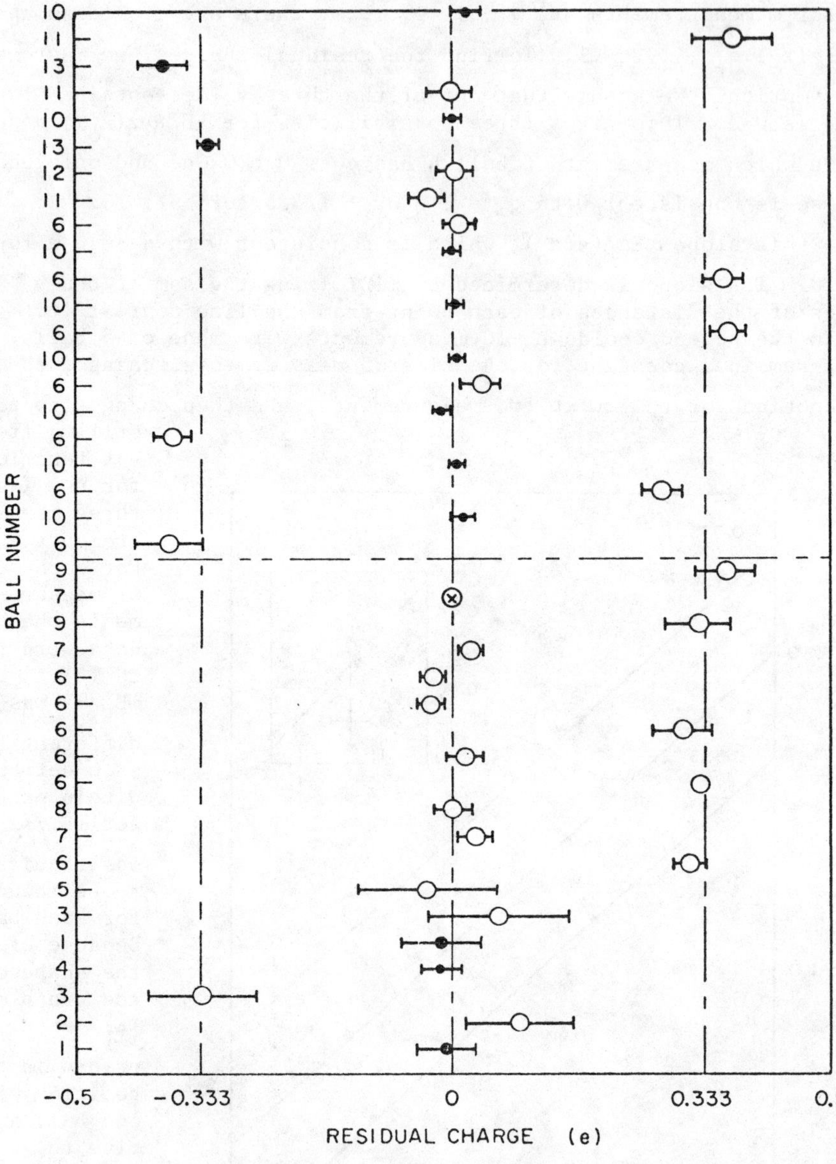

Fig. 3. Residual charges measured, in chronological order from bottom to top. Those below the line were reported in refs. 1-4, and those above it are first reported in this paper. Diameters: open circles 280 μm, large dots 233 μ, small dots 197 μm. The error for the second measurement of ball 6 is smaller than the point. The third measurement of ball 7 has no error bar because it was assumed to be 0 e. The initial measurement of ball 6 ($+0.326 \pm 0.051$)e is not included because the measurement was not complete.

RESULTS

We have made 39 independent measurements on 13 balls. The results are shown in Fig. 3. The last 21 measurements have been made since our last publication.[3,4] The statistical errors represent 1 standard deviation. The weighted averages are (-0.348 ± 0.009)e [5], (+0.001 ± 0.003)e [26] and (+0.330 ± 0.006)e [8]where the numbers in brackets are the number of measurements. For the 21 new measurements the ball stayed at low temperatures during any one cooldown. All of the balls were heat treated on a tungsten plate at 1800° C for 17 hours in a vacuum of 10^{-9} torr.

Out of 26 repeat measurements, we have observed 11 residual charge changes, in each case of ± 1/3 e. These changes have occurred only when a ball was brought into contact with other surfaces. We have made 14 independent measurements on ball 6 over a 3½ year period, every measurement consistent with 0 e or ± 1/3 e, and have observed 9 changes of 1/3 e.

Our apparatus measures the residual force on the ball to ~ 0.01 eE_A. We have shown by measurement of the relevent parameters that all electromagnetic multipole forces are negligible except for that force due to patch effect fields from the plates. We have measured the patch effect fields and have shown that we can take them into account and so obtain the true residual charge.

In order for a measurement to be valid, the patch effect field must not change with time. In all cases where this was true, the value of the residual charge was 0 e or ± 1/3 e.

We have shown that the new electromagnetic force reported by Marinelli and Morpurgo[5,6] cannot account for the fractional charges we have observed.

REFERENCES

1. G. S. LaRue, Ph.D. Thesis, Stanford University 1978(unpublished).
2. G. S. LaRue, W. M. Fairbank, and A. F. Hebard, Phys. Rev. Lett. 38, 1011 (1977).
3. G. S. LaRue, W. M. Fairbank, and J. D. Phillips, Phys. Rev. Lett. 42, 142 (1979).
4. Erratum to Ref. 3, Phys. Rev. Lett. 42, 1019 (1979).
5. M. Marinelli and G. Morpurgo, Phys. Letters B (in course of publication).
6. M. Marinelli and G. Morpurgo, Phys. Letters B (in course of publication).

SEARCH FOR QUARKS IN MATTER

M. Marinelli and G. Morpurgo(*)
Istituto di Fisica dell'Università and Sezione I.N.F.N.
Genova (Italy)

ABSTRACT

A magneto electric force acting on the levitating spheres in the experiment of search for quarks by the magnetic levitation electrometer is identified. All the 47 steel spheres (23 with diameter 0.2 mm. and 24 with diameter 0.3 mm. totalling a mass of 3.4 mg.) examined after the elimination of the effects of such force produce the result: $N_{quarks}/N_{nucleons} \leqslant 5.10^{-22}$.

Due to reasons of time this report will be very concise. It will deal with our work in Genoa on which we presented two papers at this conference[1,2] and two preprints[3,4] are circulating.

All this series of experiments with the magnetic levitation electrometer started in Genoa in 1965 after a theoretical paper of mine(*) containing the basis and several results of what is now called the "naive" or "non relativistic" quark model[5]. I was so impressed by the successes of that simple model that I concluded with the following statement: "Finally, if the present ideas are valid the quarks should exist; they should not be only mathematical entities... So one should finally discover the quarks. Which are the most appropriate conditions for this should be investigated. I hope to come back to this point in the future".

I certainly have kept this promise, although so far we have not discovered the free quark. Indeed I can anticipate our results: in 3.4 milligrams of steel there is no quark . To show you how we did reach this conclusion will be the main aim of this talk.

Before entering the subject I add a remark: it is important to realize that looking for quarks using accelerators or searching them in matter are complementary

(*) presented by G. Morpurgo.

approaches. The accelerator way is limited by the mass of the free quark, the search in matter by its abundance (the mass here is irrelevant).

I now proceed to describe the essentials of our experiments; due to time I have to be very schematic, omitting all the details of the instrument. The ball levitates in an evacuated chamber ($\sim 7\ 10^{-6}$ torr) at the center of two circular plates (compare the fig. 1 of ref. 1) and the idea of the method consists in measuring the oscillation amplitude A of the shadow of the ball on a differential photodiode under the action of an oscillating square wave electric field at the resonance frequency of the magnetic valley. The amplitude A, measured by a lock-in, (phase locked to the electric field) is proportional to the force F_x acting on the ball. How should we write this force?

Ideally one expects: $F_x = q\ E_x$ where q is the true charge of the ball and E_x the x component of the electric field; but, in practice, the situation is not so simple. Therefore, if we wish to measure the charge of the ball (or better the minimum charge q that the ball has when an appropriate number of electrons is added or subtracted from it; this q should be zero if there are no quarks inside the ball) we must write, first of all, a correct expression of the force F_x. Up to some time ago[6] this was:

$$F_x = q\ E_x + \alpha\ E_x \frac{\partial E_x^V}{\partial x} + \underset{\sim}{d} \cdot \frac{\partial \underset{\sim}{E}}{\partial x} \tag{1}$$

where the symbols will be explained in a moment. Our finding of this year has been that another term must be added to eq. (1); we shall show that after this new term was discovered all the difficulties in the interpretation of our results disappeared. But, before doing this, let me first clarify the various terms in (1). The second term is the Volta or "patch" effect term due to the fact that the potential on the surfaces of the plates is not uniform; it may differ from region to region typically by 20 to 50 millivolts, thus producing a patch or Volta electric field $\underset{\sim}{E}^V$. We thus have an apparent charge $\alpha\, \partial E_x^V/\partial x$ where $\alpha = 4\pi\mathcal{E}_o r^3$ is the polarizability of the sphere (r is the radius of the sphere).

The third term in (1) is due to the non uniformity

of the applied field on the permanent electric dipole moment $\underset{\sim}{d}$ of the sphere. This term is totally negligible in our experimental conditions because of the parallelism of the plates. We can therefore rewrite (1) as: $F_x = Q\ E_x$ where the apparent charge Q is given by:

$$Q = q + \alpha \frac{\partial E_x^V}{\partial x} \qquad (2)$$

We call Q_R (R means residual) the minimum value of Q for each object ($-0.5\,e \leqslant Q_R \leqslant +0.5\,e$). Q_R is the quantity that we measure on each ball. Below I will report, in the fig. 1, the results of the measurements of Q_R on a large number of different spheres.

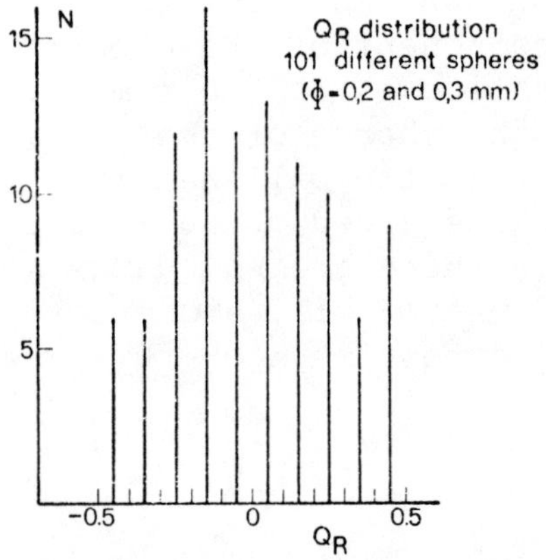

Fig. 1 - The distribution in Q_R for 101 different spheres before the comparison procedure (comparing the Q_R's of different spheres with the same radius without opening the chamber) was started.

I recall that whereas the first five measured objects[7] passed through $Q_R = 0$ (to $\approx 1/10\,e$) the subsequent ones failed to behave so "simply". They gave rise to the broad distribution in Q_R of the fig. 1. Should we attribute this distribution to the patch effects? Or do we have true continuous fractionary charges? Or something else? To solve this we decided to work, possibly, at "constant patches". In other words whereas the distribution of fig. 1 was obtained opening the levitation chamber to levitate each sphere (thereby changing presumably the patches by the influx of air) we modified the instrument to be able to levitate, deposit and rilevitate up to four different spheres in succession <u>without opening the chamber</u>.

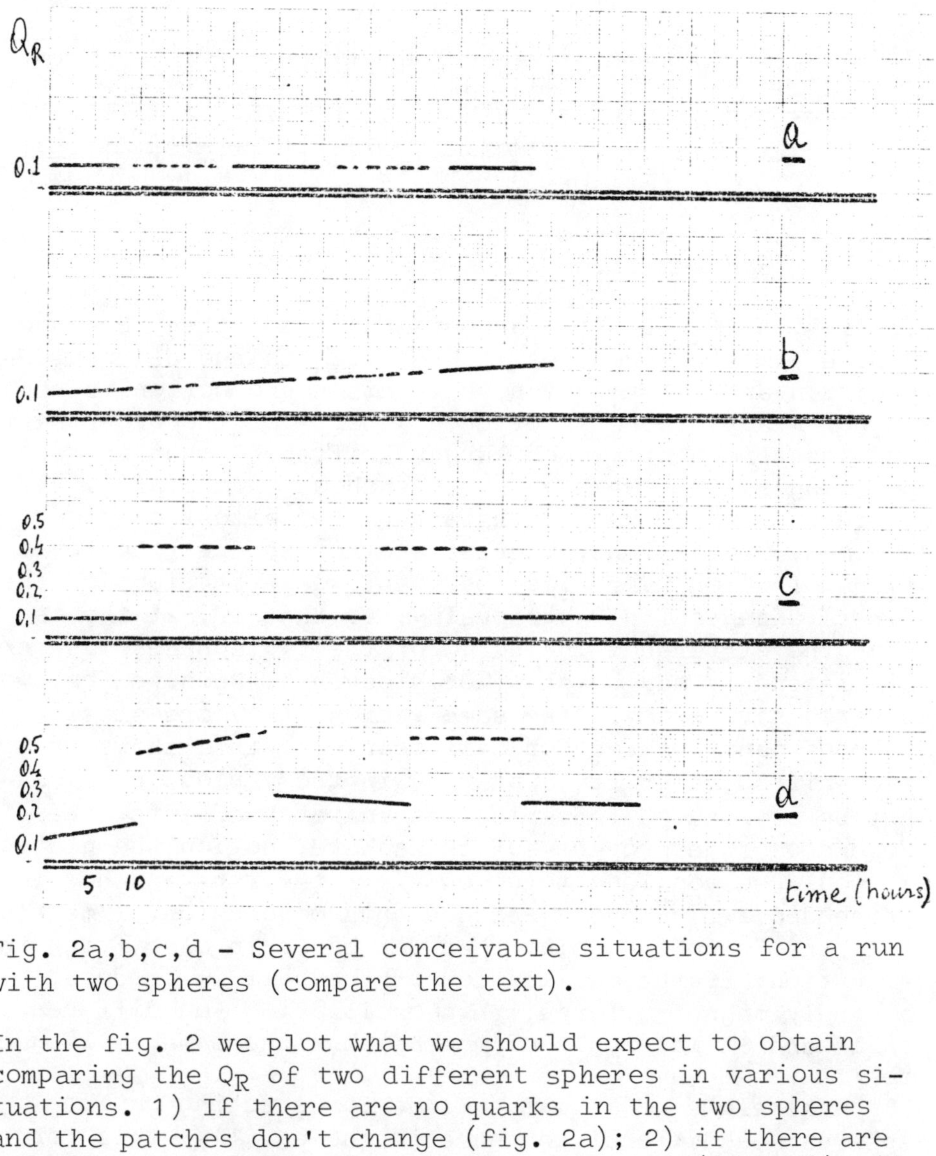

Fig. 2a,b,c,d - Several conceivable situations for a run with two spheres (compare the text).

In the fig. 2 we plot what we should expect to obtain comparing the Q_R of two different spheres in various situations. 1) If there are no quarks in the two spheres and the patches don't change (fig. 2a); 2) if there are no quarks and the patches change with time (fig. 2b); if one sphere contains a quark and the other does not (figs. 2c and 2d). Finally I plot a typical real "run" (fig. 3); and I mention that all the runs performed are shown schematically in Fig. 3 of ref. 1. (It should be pointed that a "run", consisting usually of many levitations and re-depositions, takes from 4 to 10 days or so of uninterrupted operation). From the fig. 3 of this paper and from

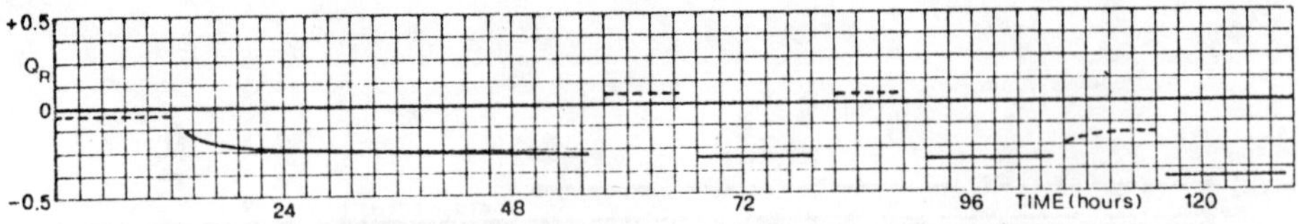

Fig. 3 - A typical run with two spheres showing drifts and shifts.

the fig. 3 of ref. 1 it appears that: a) often the same sphere almost reproduces its residual charge after a deposition and relevitation, but this reproduction is usually not exact: there is often a "shift" of the order e/10 and less frequently much larger shifts; b) what is more puzzling is that we have runs (such as that of fig. 3) where one sphere keeps constant Q_R (or almost constant Q_R) for many levitations; whereas another sphere (the levitations of which take place between two levitations of the first) changes its Q_R gradually; so that if, at the beginning the difference in the Q_R of the two spheres was e/3, at the end it may be zero; or viceversa. Because the two spheres have exactly the same radius (they are steel spheres manufactured for ball bearings) this shows that the shifts and drifts appearing in the various runs cannot be attributed to variations in patch effects. They depend on some property of the sphere, not on the plates. It would be too long to discuss all the runs and how we proceeded after having reached this conclusion. I mention only the following point that was basic in clarifying the situation: if the oscillation of the ball is produced by a purely magnetic force, this oscillation (on different balls and at intervals of months) has the same amplitude to 1 %.

Therefore if it exists some force acting on the spheres - additional to those in the eq.(1)- this force is either purely electric (and in perfect phase with $\underset{\sim}{E}$, as the lock-in shows) or due to the simultaneous presence of the electric and magnetic fields. In view of the fact that the so called magneto-electric effect was known in some substances (compare ref. 2), we decided to explore the latter assumption. A magneto-electric force amounts to add to the energy a (non time-reversal invariant) term of the form: $\mathcal{E} = \beta \, \underset{\sim}{E} \cdot \underset{\sim}{H}$ where β is a coefficient depending on the sphere under study. In a uniform electric field this term produces a force:

$$F_x^{\text{magneto-electric}} = \beta E_x \frac{\partial H_x}{\partial x} \equiv \beta E_x H_{xx} \qquad (3)$$

so that, on inserting in (1) and in (2), we get for Q_R:

$$Q_R = q + \alpha \frac{\partial E_x^V}{\partial x} + \beta \frac{\partial H_x}{\partial x} \qquad (4)$$

How can we measure the last term in (4)?

The best way would be to reverse $\underset{\sim}{H}$ thus changing sign to the last term in (4) while keeping the others unchanged. This is not feasible without depositing the sphere. What we could do and we did with the sphere constantly levitating, was to change H_{xx} from its normal value H°_{xx} to the values $H^{\circ}_{xx} + \Delta H_{xx}$ and $H^{\circ}_{xx} - \Delta H_{xx}$. By measuring then Q_R we get the two equations:

$$Q_R^{\pm} = q' + \beta \left(H^{\circ}_{xx} \pm \Delta H_{xx}\right) \qquad (5)$$

where we abbreviated with q' the quantity:

$$q' = q + \alpha \frac{\partial E_x^V}{\partial x}$$

Eliminating β from the two equations (5) and introducing the quantities:

$$Y = (Q_R^{+} + Q_R^{-})/2 \qquad\qquad X = Q_R^{-} - Q_R^{+} \qquad (6)$$

and calling $k = - H^{\circ}_{xx}/2\,\Delta H_{xx}$ (= 1.67 in our experimental conditions) we get:

$$q' = Y - kX \qquad (7)$$

Therefore if the true residual charge as well as the patch effects are zero, all the data points should be along the line:

$$Y = k\,X \qquad (8)$$

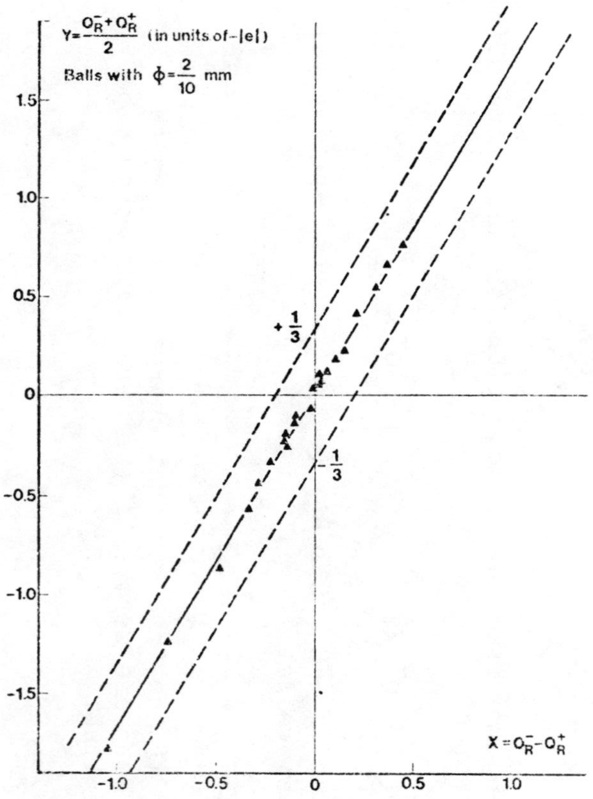

Fig.4 – All the 23 spheres with diameter 0.2 mm. measured after the identification of the magneto electric force.

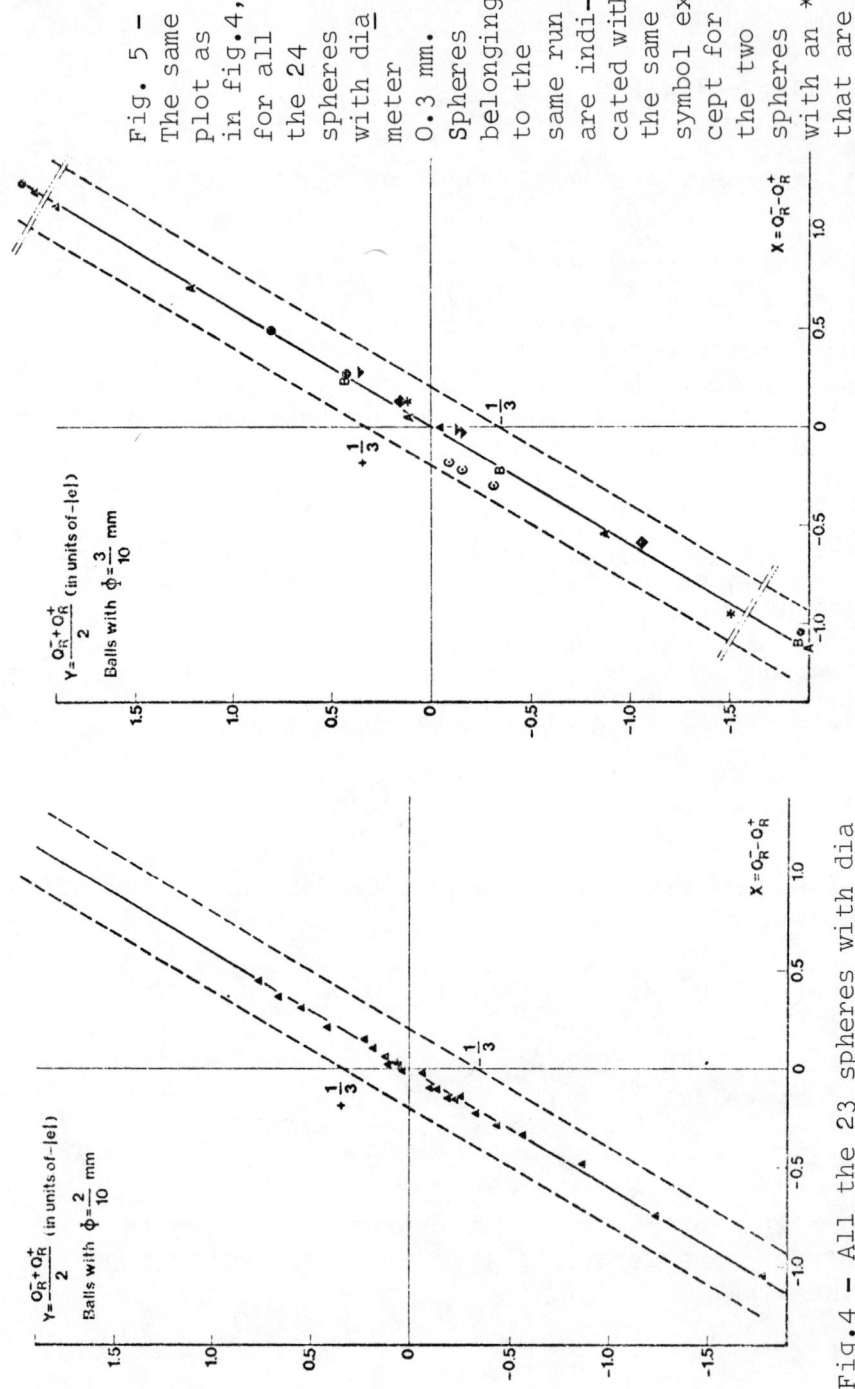

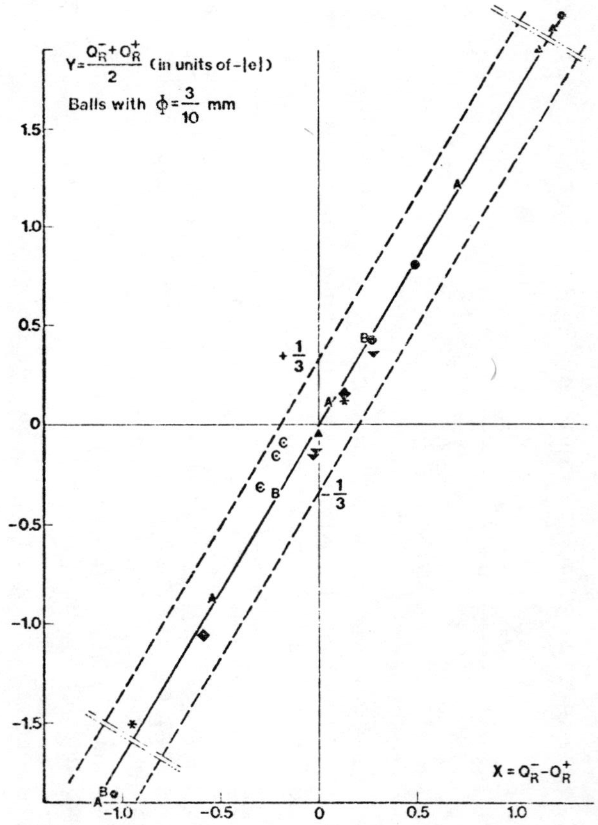

Fig. 5 – The same plot as in fig.4, for all the 24 spheres with diameter 0.3 mm. Spheres belonging to the same run are indicated with the same symbol except for the two spheres with an * that are individual measurements (not belonging to a run).

The two plots (figs. 4 and 5) show the situation for 23 spheres with diameter 0.2 mm. (fig. 4) and for 24 spheres with diameter 0.3 mm. (fig. 5). Recall that the patch effect is proportional to the volume of the sphere; this explains the larger deviations from the "standard line" for the 0.3 mm. spheres with respect to the 0.2 spheres. We stress however that for the 0.3 spheres belonging to the same run the deviations are equal (within the errors) for all the spheres.

In conclusion no quark was found in the 3.4 mg. of steel explored; the errors are mostly 0.05 e (or better), in some noisy cases they are 0.10 e.

Before concluding this report on our results (I have to skip entirely, for reasons of time the description of a method[3] by which we obtained a totally independent confirmation of the magneto electric force) I will make some comments on the magneto electric force. First, just to give a simple example, I will show how a magneto electric force might arise (but does not arise in our case). Clearly a magneto electric force arises if the oscillating $\underset{\sim}{E}$ produces an oscillating $\underset{\sim}{\mu}$ of the sphere; indeed this oscillating $\underset{\sim}{\mu}$ determines in a non uniform magnetic field an oscillation of the center of mass of the sphere. How can an oscillating $\underset{\sim}{E}$ produce an oscillating $\underset{\sim}{\mu}$? For a non spinning object the answer is trivial. The oscillating $\underset{\sim}{E}$ produces a periodic tilting of the permanent electric dipole moment $\underset{\sim}{d}$ of the sphere. If the sphere has also a permanent magnetic dipole moment $\underset{\sim}{\mu}$ any tilting of $\underset{\sim}{d}$ entails a tilting of this magnetic moment. This is why we always operated with spinning spheres; the gyroscopic stability prevents the tilting of d_z and, therefore, also of the permanent magnetic moment. That the gyroscopic stability does indeed its job can be clearly seen stopping the sphere. But as we stress, with the sphere spinning, (at 400 Hz.) this is not the mechanism determining the magneto electric force. Some ideas on the origin of this force (to be associated to a spontaneous breaking of time reversal invariance related a non vanishing correlation between the directions of the internal microscopic electric and magnetic fields near the surface of our sphere) have been presented in a short note[4].

REFERENCES

1. M. Marinelli and G. Morpurgo: New results in the search of quarks in matter by the magnetic levitation electrometer (in course of publication in Phys.Lett.B).
2. M. Marinelli and G. Morpurgo: Confirmation of the electric neutrality of matter at the milligram level (in course of publication in Phys. Lett. B).
3. M. Marinelli and G. Morpurgo: A new independent determination of the magneto electric force in the experiment of search for quarks by the magnetic levitation electrometer (Genoa preprint, June 28, 1980 - to be published).
4. M. Marinelli and G. Morpurgo: An example of spontaneous breaking of time reversal invariance (Genoa preprint, July 15, 1980 - to be published).
5. G. Morpurgo, Physics (N.Y.) 2, 95 (1965).
6. G. Gallinaro, M. Marinelli and G. Morpurgo - Proc. of the 4th EPS Gen. Conf. (York, Sept. 1978), Chapter 8, p. 562 (Adam-Hilger (London) 1979).
7. G. Gallinaro, M. Marinelli and G. Morpurgo, Phys. Rev. Lett. 38, 1255 (1977).

Chapter 3

Low Energy Weak Interactions and Weak Decays

LOW ENERGY WEAK INTERACTIONS, EXPERIMENT

F. REINES AND

J. SCHULTZ, UC-IRVINE, ORGANIZERS

PARITY VIOLATION IN PROTON-NUCLEUS SCATTERING AT 6 GeV/c

N. Lockyer and T. A. Romanowski, Department of Physics, Ohio State University, Columbus, Ohio 43210

J. D. Bowman, C. M. Hoffman, R. E. Mischke, D. E. Nagle, J. M. Potter, and R. L. Talaga, University of California, Los Alamos Scientific Laboratory, Los Alamos, New Mexico 87545

E. C. Swallow, Department of Physics, Elmhurst College, Elmhurst, Illinois 60126

D. M. Alde, Department of Physics, University of Illinois, Urbana, Illinois 60801

D. R. Moffett, Argonne National Laboratory, Argonne, Illinois 60439

J. Zyskind, Department of Physics and Astronomy, California Institute of Technology, Pasadena, California 91109

ABSTRACT

We have measured a parity-violating asymmetry in the total cross section for 6 GeV/c polarized protons on a water target. The asymmetry $A_L = (\sigma_+ - \sigma_-)/(\sigma_+ + \sigma_-)$, defined as the fractional difference of of the total cross sections for positive and negative helicity protons on an unpolarized target, was found to have the value $A_L = (2.61 \pm 0.55) \times 10^{-6}$. The quoted error includes both statistical and systematic contributions.

We have performed a search for parity violation in proton-nucleus scattering at 6 GeV/c by studying the dependence of the total cross section on the helicity of the incident polarized proton beam. This dependence is expressed by the parity-violating asymmetry $A_L = (\sigma_+ - \sigma_-)/(\sigma_+ + \sigma_-)$, where $\sigma_+(\sigma_-)$ is the total cross section for positive (negative) helicity protons on a water target. At present, investigation of parity-violating phenomena is the only way to gain experimental knowledge of the strangeness conserving hadronic weak interaction.

A parity-violating asymmetry A_L can arise from the interference of the strong and weak interactions between nucleons. The magnitude of A_L from a dimensional estimate is $A_L \simeq 10^{-6}$.

The vertically polarized proton beam from the Argonne National Laboratory Zero Gradient Synchrotron (ZGS) used in this experiment had an average polarization $|\overline{P}| = 0.71 \pm 0.03$. The polarization direction was reversed at the source between ZGS pulses. An upward bend of the beam rotated the polarization to a longitudinal direction. The horizontal position of the beam centroid at the target was stabilized with the aid of a feedback loop controlling the current in

0094-243X/81/680318-04$1.50

an upstream bending magnet.

A schematic representation of the experiment is shown in Fig. 1. The transmission Z through an 81-cm-long water target was measured by two independent detector systems which have been described previously.[1] One consisted of a pair of identical plastic scintillation counters, each viewed by four phototubes. The second detection system was a set of three identical ion chambers. The output currents from ion chambers upstream and downstream of the target were subtracted before amplification. The difference signal was normalized to the beam intensity with a third ion chamber. Integral counting techniques allowed high beam intensities so that the desired level of sensitivity could be achieved in a reasonable time. Auxiliary scintillation detectors shown in Fig. 1 monitored various beam properties. The data include 184 data runs and 54 control runs each consisting of $\sim$1600 pulses for a total of 10^{14} incident protons. The average fractional change in transmission Z with helicity reversal $\Delta Z/2Z = (Z_+ - Z_-)/(Z_+ + Z_-)$ was computed for each run. Errors were calculated from pulse to pulse fluctuations in the change of transmission and were typically 3 times as large as fluctuations in the finite number of incident protons. The measured $\langle \Delta Z/2Z \rangle = A_L |\bar{P}| \ln Z$ + BACKGROUND TERMS. A small reduction of the statistical fluctuations in $\langle \Delta Z/2Z \rangle$ was obtained by a linear regression analysis of $\langle \Delta Z/2Z \rangle$ for each run against the variables for horizontal and vertical beam position, beam position squared, and the beam intensity. The regression coefficients were determined by forming polarization independent linear combinations of data from sets of four consecutive pulses. The noise reduction process did not appreciably change the value of $\langle \Delta Z/2Z \rangle$ indicating that the regression variables are independent of polarization.

To take into account systematic contributions, the observed asymmetry for each data run was treated as the sum of a parity violating signal (proportional to A_L) and terms from known background processes: residual transverse polarization and beam matter interactions. Each background term is the product of the polarization correlated change in one of these variables and the sensitivity of the $\langle \Delta Z/2Z \rangle$ to that variable. The variables were monitored each beam pulse and the sensitivities were measured in control runs in which the effect of the relevant variable was enhanced. In the control runs measuring the effects of transverse polarization, the horizontal and vertical components were increased at the target with solenoids in the beam line. In the runs measuring the interaction of the vertically polarized beam with matter in the beam channel (beam-matter interaction), the beam halo was increased by adding material upstream of magnet B1. During both types of control runs the beam position and angle were varied. Beam-matter interactions contributed the largest term to the transmission asymmetry $\langle \Delta Z/2Z \rangle = \gamma \Delta H d$. In this expression ΔH is the average change of the transverse asymmetry with polarization reversal normalized to the incident beam intensity and measured by the X detector, $\gamma(\mathrm{cm}^{-1})$ is the measured sensitivity of $\langle \Delta Z/2Z \rangle$ to beam position (angle) in the presence of a finite ΔH and d(cm) is the distance of the beam position (angle) from the measured symmetry axis. The value of γ and d was determined from a least square fit of the measured values of $<\Delta Z/2Z>$ in the data and control runs to the sum of a beam-matter interaction, transverse polarization and a parity violating term. The data is consistant with this model yielding a $\chi^2/df \sim 1.17$ for both detector systems. In the nominal data taking runs the proton beam was positioned to minimize d and solenoids were adjusted to null the transverse polarization.

A correlation between transverse polarization and position in the beam can lead to a transmission asymmetry even when the average transverse component is zero. We measured the average transverse polarization for the upper, lower, left, and right portions of the beam. Using the measured sensitivity of $\Delta Z/2Z$ to transverse polarization and beam displacement, we determined the correction from this effect to be $-0.40 \pm 0.35 \times 10^{-6}$.

Decays of polarized hyperons produced in the target could be a possible source of a parity signal. The magnetic spectrometer and collimator, after the target, suppressed any hyperon background by at least a factor of 2000, and the corresponding correction to $<\Delta Z/2Z>$ is negligible.

Consecutive data runs were started on alternate polarization states to avoid effects from slowly varying drifts. These drifts were small and essentially random over the entire experiment. The background contributions in the determination of A_L are summarized in Table I.

TABLE I

Summary of Backgrounds in Determination of A_L Results in Units of 10^{-6}

Stage of Analysis	Scintillation Counters	Ionization Chambers	Combined
Raw Asymmetry	4.26±0.62	5.33±0.70	4.73±0.46
After Regression	4.04±0.55	4.98±0.67	4.42±0.42
Corrected for:			
Transverse Polarization	3.80±0.54	4.91±0.66	4.26±0.42
Beam-Matter Scattering	2.34±0.64	3.97±0.79	2.98±0.51
Polarization Correlation with Phase Space	1.94±0.71	3.57±0.85	2.61±0.55
Final Results	1.94±0.71	3.57±0.85	2.61±0.55

Our final result for A_L is $(1.94\pm0.71) \times 10^{-6}$ for the scintillation counters and $(3.57\pm0.85) \times 10^{-6}$ for the ion chambers including all corrections. The correlation coefficient between the values of A_L from the two detector systems is 0.2. The weighted average gives $A_L = (2.61\pm0.55) \times 10^{-6}$. The quoted error is the root-mean-square-value. From experiments at low energies and first calculation for higher energies,[2] we expected $A_L \sim 10^{-7}$.

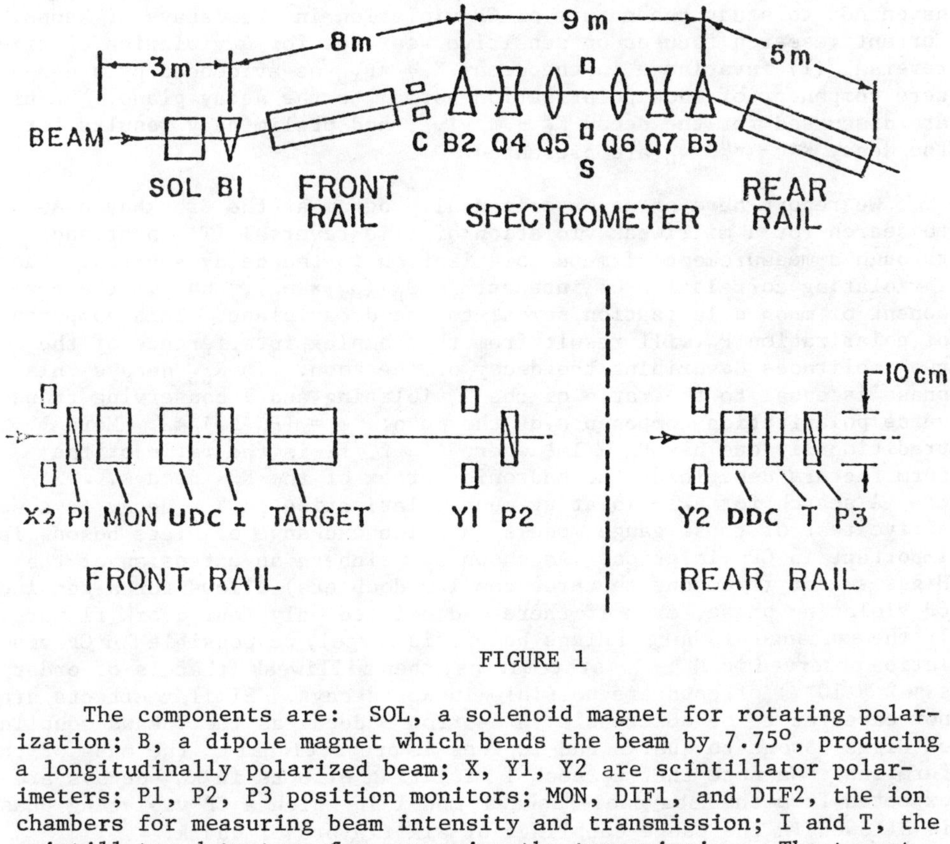

FIGURE 1

The components are: SOL, a solenoid magnet for rotating polarization; B, a dipole magnet which bends the beam by 7.75°, producing a longitudinally polarized beam; X, Y1, Y2 are scintillator polarimeters; P1, P2, P3 position monitors; MON, DIF1, and DIF2, the ion chambers for measuring beam intensity and transmission; I and T, the scintillator detectors for measuring the transmission. The target was 81 cm of water. C is a brass collimator with 5.5-cm-diameter aperture. S is a beam halo counter telescope.

REFERENCES

1. J. D. Bowman, et al., Phys. Rev. Lett. 34, 1184 (1975).
2. E. M. Henley and F. R. Krejs, Phys. Rev. D11, 605 (1975).

EXPERIMENTAL CONSTRAINTS ON MODELS OF CP VIOLATION

M.P. Schmidt, R.K. Adair, J.K. Black,
S.R. Blatt, M.K. Campbell and H. Kasha
Yale University, New Haven, CT 06520

R.C. Larsen, L.B. Leipuner and W.M. Morse
Brookhaven National Laboratory, Upton, NY 11973

ABSTRACT

A review is presented of the experimental efforts at the Brookhaven AGS to study the nature of CP violation in the decays of kaons. Current research focuses on sensitive searches for a violation of time-reversal (T) invariance in the decay $K \rightarrow \pi\mu\nu_\mu$ as evidenced by a non-zero component of muon polarization normal to the decay plane. Results are discussed for the decay $K_L^0 \rightarrow \pi^-\mu^+\nu_\mu$, and preliminary results for the decay $K^+ \rightarrow \pi^0\mu^+\nu_\mu$ are presented.

We report here on an experimental program at the Brookhaven AGS to search for a milliweak violation of time reversal (T) invariance through a measurement of muon polarization in the decay $K \rightarrow \pi\mu\nu$. The T violating correlation of interest is $\vec{S}_\mu \cdot (\vec{p}_\pi \times \vec{p}_\mu)$, that is the component of muon polarization normal to the decay plane. This component of polarization P_n will result from the complex interference of the two amplitudes describing the decay of the kaon. In $K_{\mu 3}$ decays this phase is equal to the ratio of the T violating and T conserving transverse polarization components of the muon: $\phi = (P_n/P_t)_{cms}$. More traditionally one has $P_n \propto \mathrm{Im}\xi$ where $\xi \equiv f_-/f_+$ is the ratio of the form factors describing the hadronic vertex of the $K_{\mu 3}$ decays.

A search for a T violating muon polarization in $K_{\mu 3}$ decay is a sensitive test of those gauge models in which exchange of Higgs bosons is important in CP violation. As shown by Weinberg[1] an extension of the Higgs sector (from one to three complex doublets) allows for a genuine CP violating phase, even if there had existed only four quark flavors. If the exchange of charged Higgs bosons is largely responsible for CP violation observed in $K^0 - \overline{K}^0$ systematics, then milliweak (that is of order $\varepsilon \sim 2 \times 10^{-3}$) effects are possible in $K_{\mu 3}$ decays. Similar effects are not expected to be observable in neutron β decay as the Yukawa coupling of Higgs bosons to quarks and leptons is proportional to the mass of the fermions. We note that no such T violating effects in $K_{\mu 3}$ decays are expected from the Kobayashi-Maskawa[2] model in which a CP violating phase is allowed in the gauge couplings of six flavors of quark.

The $K_{\mu 3}^0$ experiment was conducted in a 6° neutral beam which traveled through the center of a cylindrically symmetric detector (Fig. 1). Positive muons from decays ($K_L^0 \rightarrow \pi^-\mu^+\nu_\mu$) occuring in the 5 meter drift space upstream of the detector were focused by the steel toroidal magnet and brought to rest in the aluminum polarimeter. The path of the muons was determined from the coincidence of scintillation counter pulses from the hodoscopes labeled A-B, M, F and G. Field Programmable Logic Arrays (FPLA's) were used to determine the azimuthal position of stopping muons in the segmented polarimeter and to abort events for which one unambiguous muon stop was not found.

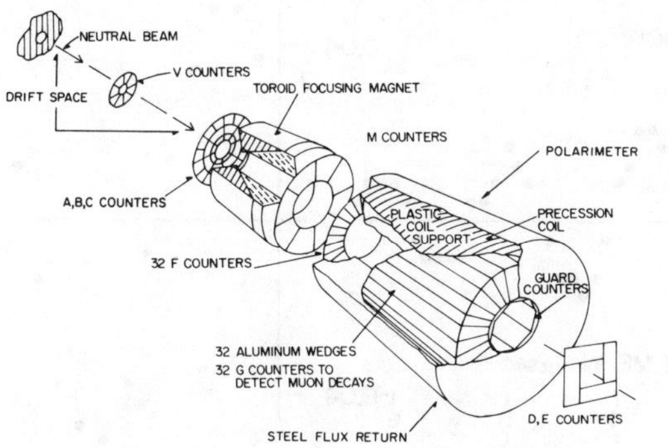

Figure 1.

In order to maximize the sensitivity of the detector to the T violating component of polarization, events were selected such that the K_L^0 laboratory momentum was very nearly in the $K_{\mu 3}$ decay plane. For such events the T violating component of polarization in the laboratory is normal to the plane defined by the laboratory momenta of the kaon and the muon, that is $P_n \alpha \vec{S}_\mu \cdot (\vec{P}_K \times \vec{P}_\mu)$. The event selection was accomplished by fast trigger logic which required, in coincidence with the muon, a pion hit in the D-E hodoscope or in an A-B counter adjacent to that hit by the muon. Monte Carlo calculations shows that about 6% of all $K^0_{\mu 3}$ decays with muons stopping in the polarimeter would fulfill the event selection criteria yielding $\langle P_n \rangle^{Lab} = 1.8 \times 10^{-3}$ for Imξ = 0.01 and Re$\xi \sim 0$.

After coming to rest in the aluminum polarimeter, the muon spin precessed with a period of 1.2 μsec in a 60 G axial magnetic field. The positron from the muon decay was detected by one of the two G counters flanking the muon stop position. Each G counter was associated with a "clock" which was gated on for 6.4 μsec by the fast trigger. The detection of the positron thus recorded the time and direction of the muon decay.

The geometry of the polarimeter and the applied magnetic field allowed a measurement of two components of the muon polarization in the laboratory: P_t, the T conserving transverse polarization, and P_n, the T violating component. For the ensemble of muons the polarization was determined from the amplitude of the asymmetry, $A = (R - L)/(R + L)$, in the number of positrons detected to the right (R) or the left (L) of the muon stop position. Reversing the direction of the precession field before each beam pulse allowed an independent determination of the P_t and P_n components. This is understood by noting that the contribution to the asymmetry from the P_t component reverses under a reversal of the field direction, whereas the contribution from the P_n component is invariant.

The curves of Fig. 2 shows the measured asymmetry, A, plotted as a function of time for the 12 million events collected. The upper curves show the expected sinusoidal dependence of A_t, the asymmetry due to the T conserving polarization P_t. The lower curves (note the change in the vertical scale) show the time dependence of the T violating asymmetry A_n with representative errors. The curves on the left display the data from the on-line analysis. The "damping" of the sinusoid in the A_t plot is due to random "clock" stops caused by neutron fluxes in the detector cave. The curves on the right show a simple Fourier analysis fit to the same data with the period fixed and the background subtracted. The least squares fit to the amplitude $A_t(0)$ yields a T conserving polarization of $P_t = 0.42 \pm 0.06$, which is consistent with Monte Carlo expectations and serves to calibrate the de-

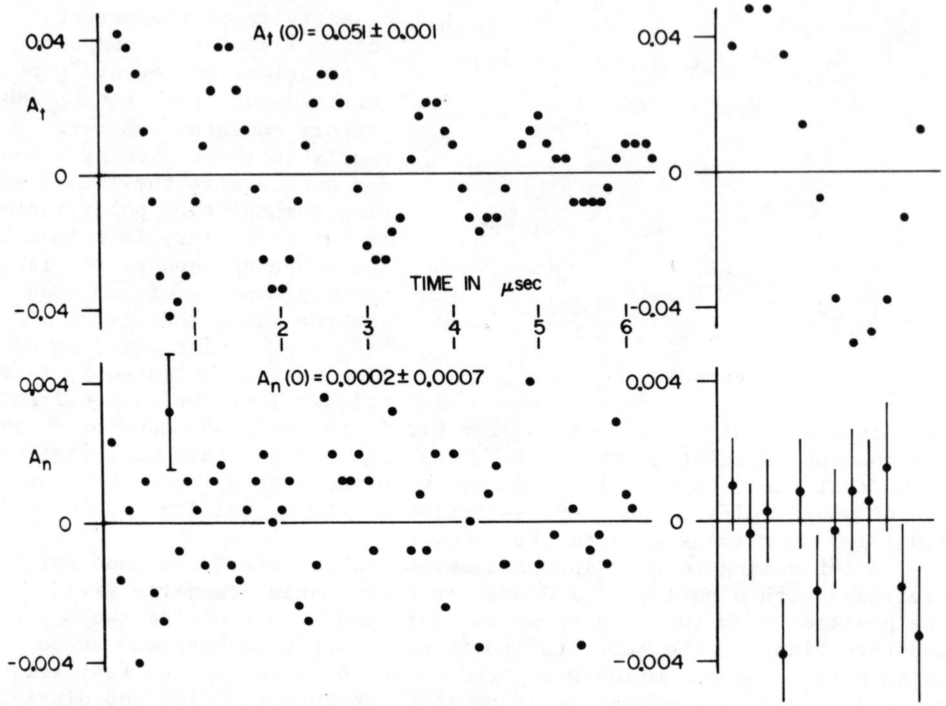

Figure 2.

tector. For the T violating data the fit implies a value for the polarization $P_n = 0.0021 \pm 0.0048$ which is consistent with zero.[3]

Using the calculated sensitivity of the detector this value of P_n yields a value for the T violating phase $\phi = (P_n/P_t)_{cms} = 0.004 \pm 0.009$ which is consistent with time reversal invariance, and should be compared with the value 0.002 that we suggest as a central value to be expected from milliweak theories. In terms of the traditional kaon form factor ratio ξ our result corresponds to a value $\text{Im}\xi = 0.012 \pm 0.026$, which is not significantly different than the value 0.008 expected from the electromagnetic final state interactions.

The completed $K^0_{\mu 3}$ experiment is considered the first of a program to detect milliweak CP violation in $K_{\mu 3}$ decays, and Brookhaven is well suited for such precision measurements. This summer we have begun collecting data on muon polarization in the decay $K^+ \to \pi^0 \mu^+ \nu_\mu$. The experiment is performed in a monochromatic (4 GeV/c) beam and we have replaced the D-E hodoscope (Figure 1) with a lead glass array ($\sim$ 8 radiation lengths) in order to detect γ-rays from π^0 decays. The requirement of a high energy γ-ray ($E_\gamma > 1.2$ GeV) enhances the selection of events where the π^0 is emitted forward in the kaon rest frame, thus increasing the sensitivity by about 25% compared to the $K^0_{\mu 3}$ experiment. We note that final state effects are negligible in $K^+_{\mu 3}$ decays. In a preliminary run ($\sim$ 100 hrs) we have found with $\sim$ 1 million $K^+_{\mu 3}$ events $\text{Im}\xi = 0.052 \pm 0.066$.

It is worth noting that we are employing a prototype of the Brookhaven Fast Bus[4] for data acquisition which will facilitate our expectation of collecting events at a rate almost an order of magnitude greater than that of the $K^0_{\mu 3}$ experiment. The collection of 150 million events would allow a measure of the CP violating phase ϕ to $\pm$ 0.003, well within the range expected from milliweak models of CP violation.

Further improvements of the limits on CP violating effects outside the $K^0 - \overline{K}^0$ system will be fundamental in reaching an understanding of the nature of CP violation. Competitive with and complimentary to the $K_{\mu 3}$ measurements are the continued searches for an electric dipole moment of the neutron. Present limits ($D^e_n < 3 \times 10^{-24}$ ecm)[5] already approach the sensitivity required to distinguish between the milliweak predictions ($D^e_n \sim 10^{-24} - 10^{-25}$ ecm)[1] of the extended Higgs models and the superweak predictions ($D^e_n \lesssim 10^{-32\pm 2}$ ecm)[6] of the Kobayashi-Maskawa 6 quark model of CP violation. Also of great interest are the renewed efforts to detect milliweak effects of direct CP violation in the $K^0 - \overline{K}^0$ system. We note here the proposed experiments[7] for a precise (1%) measurement of $|\eta_{00}|/|\eta_{+-}|$, the ratio of the CP violating amplitudes for the decays of neutral kaons into neutral and charged two pion final states. These measurements should be sensitive to the deviations (on the order of a few percent) of this ratio from unity expected from both the extended Higgs and the 6 quark models of CP violation.[8] In the not too distant future we may expect to see a resolution of the question as to the milliweak vs. superweak nature of CP violation.

REFERENCES

1. S. Weinberg, Phys. Rev. Lett. 37, 657 (1976). See also, A.A. Anslem, and D.I. D'Yakonov, Nucl. Phys. B145, 271 (1978).
2. M. Kobayashi and T. Maskawa, Prog. of Theor. Phys. 49, 652 (1973).
3. Results have been published: M.P. Schmidt, et al., Phys. Rev. Lett. 43, 556 (1979), and W.M. Morse, et al., Phys. Rev. D21, 1750 (1980).
4. See the paper presented by L. Leipuner, et al., in these proceedings.
5. W.B. Dress, et al., Phys. Rev. D15, 9 (1977).
6. E.P. Shabalin, ITEP Preprint 131 (1979) and references therein.
7. B. Winstein, et al., Fermilab Proposal, R.K. Adair, et al., Brookhaven Proposal; see also G. Thomson, et al., Fermilab Proposal for a 25% measurement of η_{+-0}.
8. L. Wolfenstein, AIP Conference Proceedings, Particles and Fields, 1979, Montreal, pg. 365, and references therein.

NEUTRINO INSTABILITY

Henry W. Sobel, Frederick Reines, Elaine Pasierb

[Presented by Henry W. Sobel]

Department of Physics, University of California, Irvine, CA 92717

INTRODUCTION

We have obtained indications of neutrino instability from our data[1] on the processes:

$$\bar{\nu}_e + d \rightarrow n + n + e^+ \quad \text{(charged current)}$$
$$\phantom{\bar{\nu}_e + d} \rightarrow n + p + \bar{\nu}_e \quad \text{(neutral current)}$$

Although the deuteron reaction is not the ideal way to search for neutrino oscillations, it has a number of attractive features which allow us to side step many of the problems currently inherent in reactor neutrino studies. The reactor is a good source since it provides a very intense ($2 \times 10^{13}\ \bar{\nu}_e \text{cm}^{-2}\text{sec}^{-1}$) pure source of $\bar{\nu}_e$'s at low energies (<10 MeV). The difficulty has been the uncertainty in the $\bar{\nu}_e$ spectrum.

The antineutrino spectrum itself is calculated by adding up the beta decay spectra of all the fission products. Unfortunately, approximately 30% of the fission products involve unknown decay schemes, and as a consequence the neutrinos from these decays require extrapolation and modeling. The two newest theoretical predictions[2,3] for the antineutrino spectrum are shown in Fig. 1. They disagree to a level of about 30%, varying with $\bar{\nu}_e$ energy.

$\bar{\nu}_e + p \rightarrow n + e^+$ EXPERIMENTS

The inverse beta decay reaction has been studied in several different experiments and new experiments are being built. Data from the experiments can be used in several ways.

1. Since the e^+ takes essentially all of the $\bar{\nu}_e$ energy in the reaction, a measurement of the e^+ spectrum is a measurement of the $\bar{\nu}_e$ spectrum. This neutrino spectrum can be used as input in other experiments.

2. The process itself can be used to study neutrino oscillations.

   a. Experimental results can be compared with theoretical prediction: This technique suffers from the uncertainty in the predicted neutrino spectrum in that any observed discrepancies can be attributed to that source.

   b. We can compare the results of different detectors at various source to detector distances: These data are just beginning to be available with the required precision, and analysis is incomplete.[4]

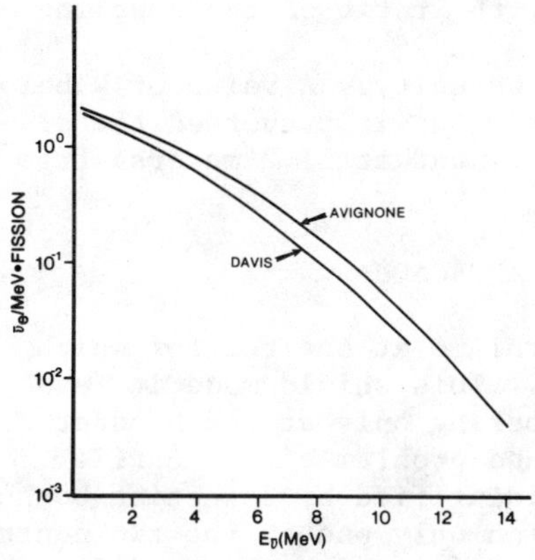

Fig. 1. Predicted $\bar{\nu}_e$ spectrum.

c. We can compare the results of the same detector taken at different distances: This clearly is the experiment of choice and two groups are actively pursuing this technique.[5,6]

There are three data sets available at this time. They are the 6.5 meter data of Nezrick and Reines,[7] the 11.2 meter data of Reines, Gurr, and Sobel,[8] and the 8.7 meter data of the C.I.T.-Grenoble-Munich group.[9]

THE DEUTERON EXPERIMENT

We have recently realized that the deuteron experiment could be used as a neutrino oscillation test. The neutral current branch (ncd) is independent of ν type, while the charged current branch (ccd) will only occur for incident $\bar{\nu}_e$'s. In addition, while the predicted rates of the individual branches are sensitive to the predicted neutrino spectrum the ratio of the predicted rates is not (Table I).

Table I. Predicted cross sections for charged current and neutral current reactions, and the ratio of these predictions.

Cross section (cm^2/fission)	Davis Spectrum	Avignone Spectrum
ncd	2.87×10^{-44}	3.73×10^{-44}
ccd	1.21×10^{-44}	1.64×10^{-44}
Ratio $\frac{ccd}{ncd}$	0.42	0.44

We choose to define the quantity:

$$R = \frac{\left(\frac{ccd}{ncd}\right)_{experiment}}{\left(\frac{ccd}{ncd}\right)_{predicted}}$$

The denominator of this quantity has the following features:

1. It is independent of the reactor neutrino absolute normalization.
2. It is insensitive to the precise shape of the reactor neutrino spectrum.
3. The ncd process is independent of neutrino type.
4. The ccd process only occurs with $\bar{\nu}_e$'s.

5. Assuming the standard model, the ratio of the coupling constants is known to $\sim 5\%$.[10]

The quantity, R, is expected to be unity. A value of R below unity would signal the instability of $\bar{\nu}_e$ as it traversed the distance (centered in this deuteron experiment at 11.2 meters) from its origin to the detector.

EXPERIMENTAL APPROACH

We have constructed a shielded volume at the reactor which has a greatly reduced neutron background. This shield made it feasible to search for the n.c. reaction by looking only at the product neutron, so avoiding the proton background problem of the earlier approaches.[11] The c.c. reaction is identified by detecting both product neutrons. Those cases in which only one of the two neutrons from the c.c. reaction was detected represented background for the n.c. reaction.

DETECTOR DESCRIPTION

The target consists of 268 kg. of D_2O. Immersed in the D_2O are 10 cylindrical, helium-3 filled, neutron proportional counters. The target is enclosed in 10.2 cm of lead, 0.1 cm of cadmium and immersed in a 2200 liter anticoincidence detector.

NEUTRON DETECTION EFFICIENCY

The neutron detection efficiency was determined using Monte Carlo techniques and a ^{252}Cf source.

The efficiency for single neutrons produced uniformly throughout the D_2O by the neutral current and charged current processes is:

$$\bar{\eta}_{n.c.} = \bar{\eta}_{c.c.} = .32 \pm .02$$

The efficiency for detecting two neutrons in the c.c. reaction is $\overline{\eta^2} = 0.112 \pm 0.009$ where the two neutron efficiency is averaged over the D_2O volume.

During single neutron analysis, we use $\overline{\eta'} = .89\ \bar{\eta}$, the efficiency loss due to a background reduction, cut.

DATA

We report the results of two data sets. Each data set consists of a number of reactor on and reactor off sequences alternating in time. Several time groupings were made from these sequences, each consisting of reactor on and off data. Reactor associated (reactor on minus reactor off) single and double neutron rates were obtained for each group. Some of the groups in data set 1 are shown in Table II.

From this, we see that we have a reactor associated signal from both single and double neutrons.

Table II. Sample neutron rates in data set 1.

	Group 1	Group 2	Group 3 . . .		Group 7	Weighted Mean
		Single Neutron Data				
Reactor on (day^{-1})	386.90 ±11.83	387.27 ±11.90	406.31 ± 6.72	. . .	439.95 ± 6.98	
Reactor off (day^{-1})	323.73 ± 7.32	320.01 ± 5.09	333.92 ±12.92	. . .	385.61 ± 5.68	
On-Off (day^{-1})	63.17 ±13.49	67.26 ±12.94	72.39 ±14.56		54.34 ± 8.90	68.26 ±4.11
		Double Neutron Data				
Reactor on (day^{-1})	53.51 ±4.40	54.12 ±4.45	58.54 ±2.55	. . .	51.85 ±2.40	
Reactor off (day^{-1})	47.67 ±2.81	51.55 ±2.04	55.87 ±4.96	. . .	46.81 ±1.98	
On-Off (day^{-1})	5.84 ±5.22	2.57 ±4.90	2.67 ±5.58	. . .	5.04 ±3.11	3.66 ±1.52

BACKGROUND TESTS

1. Neutrons

Our neutron background was established by completely surrounding our detector with an additional neutron shield. The observed change in our signal implied a neutron background of 0.7 ± 0.14 day^{-1}.

2. Gammas

The reactor associated gamma ray spectrum was measured with a 300 kg NaI detector. The background due to the (γ,n) reaction on the deuteron was calculated to be 0.05 day^{-1}.

3. $\bar{\nu}_e$ Background

a. Our D_2O is not pure. Since the inverse beta process has a relatively large cross section, this small contaminant gives a neutron background of 1.6 ± 0.1 day^{-1}.

b. The liquid scintillator anticoincidence detector consists of $CH_{1.8}$. Most of the neutrons which are produced by inverse beta decay are thermalized in the scintillator, and captured on hydrogen or our cadmium shield. The residual background is 7.9 ± 0.7 day^{-1}.

The total background to the single neutron signal from these sources is therefore 10.2 ± 0.7 day^{-1}.

DEUTERON RATES

We can now calculate the c.c.d. rate (R^{ccd}) and the n.c.d. rate (R^{ncd}) from the observed one neutron and two neutron signals (S_{1N}, S_{2N}).

$$R^{ccd} = \frac{S_{2N}}{\eta^2} \text{ and } R^{ncd} = \frac{S_{1N}^{ncd}}{\eta_{1N}}$$

where $S^{ncd}_{1N} = S_{1N} - S_{1N}(\text{background}) - S^{ccd}_{1N}$

and $S^{ccd}_{1N} = 2(\eta_{1N})(1-\eta_{1N})R^{ccd}$

The experimental ratio of ccd to ncd is therefore:

$$r_{exp} = \frac{R^{ccd}}{R^{ncd}} = \frac{\overline{\eta'} S_{2N}}{\overline{\eta^2}\,(S_{1N}-S^{BKGND}_{1N})-2(.89)(\bar{\eta}-\overline{\eta^2})S_{2N}}$$

and the error in r_{exp} is calculated from

$$\sigma^2_{r_{exp}} = \left(\frac{\partial r}{\partial S_{1N}}\right)^2 \sigma^2_{S_{1N}} + \left(\frac{\partial r}{\partial S_{2N}}\right)^2 \sigma^2_{S_{2n}} + \ldots$$

r_{exp} from data sets 1 and 2, combine to be $r_{exp} = 0.167 \pm 0.093$.

We had that $R \equiv \dfrac{r_{exp}}{r_{theory}}$

so, $R_{\substack{\text{Avignone}\\ \text{spectrum}}} = \dfrac{0.167 \pm .093}{0.44} = 0.38 \pm 0.21$

and, $R_{\substack{\text{Davis}\\ \text{spectrum}}} = \dfrac{0.167 \pm .093}{0.42} = 0.40 \pm 0.22$

These represent a 3.0 to 2.7 standard deviation departure from unity, assuming that the σ_r calculated above is representative of a normal distribution.

OSCILLATION PARAMETERS

We assume, for simplicity, a neutrino with two base states. The allowed values of $\Delta(|m_1^2-m_2^2|c^4)$ and $\sin^2 2\theta$ (θ is the mixing angle) for R = .38 ± .21 are plotted in Fig. 2.

CONSISTENCY CHECKS

1. We have mentioned another experiment at the 11.2 meter position which measured the ccp process.[8] This positron spectrum has been used to obtain a $\bar{\nu}_e$ spectrum for $E_{\bar{\nu}_e} \geq 4$ MeV (Fig. 3). The value for R deduced using this spectrum, extrapolated below 4 MeV, is 0.47 ± .24, a 2.2 standard deviation effect. If neutrino oscillations occur, with the parameters implied by the deuteron experiment, then the extrapolation of the neutrino spectrum to lower energies would be in error and the value of R reduced.

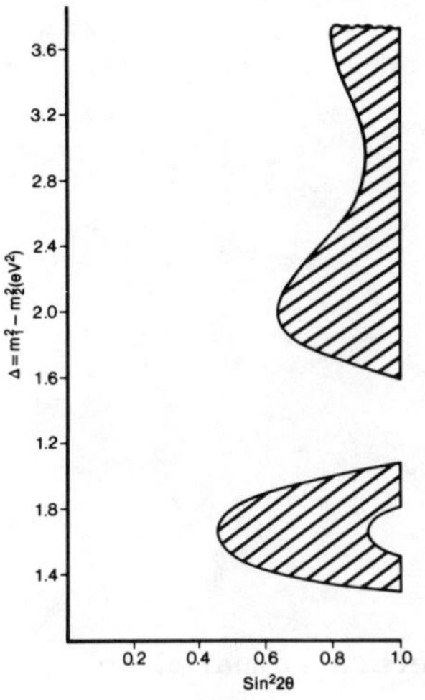

Fig. 2. $\Delta(eV^2)$ vs. $\sin^2 2\Theta$ for $R = 0.38 \pm 0.21$.

2. In Table III we list the individual ccd and ncd rates, and other reactor results compared to the predicted rates using the Avignone spectrum, the Davis spectrum and the measured $\bar{\nu}_e$ spectrum at 11.2 meters. Unlike the insensitivity of R to the reactor neutrino spectrum, all other ratios of experimentally determined rates to predicted rates are markedly dependent on the spectrum and normalizations.

a. We note that since our measurement of the neutrino spectrum is only sensitive to $\bar{\nu}_e$ it should enable us to correctly predict the ratio for the charged current branch. Table III indicates that the prediction for this ratio using the measured spectrum is 1.3 standard deviations from the expected value of unity. If the difference can be attributed to a normalization error between the two experiments it would have no effect on the ratio R. If, however, the difference is due to a statistical fluctuation and we therefore choose for the charged current the most likely value consistent with the two experiments, then the ratio R would become 0.62 ± 0.16. We note in this case that whereas R has increased, its error has diminished reflecting the greater precision of the prediction based on the measured $\bar{\nu}_e$ spectrum.

b. Allowed regions of Δ vs. $\sin^2 2\Theta$ can be drawn for each of the ratios listed in Table III. For the Avignone spectrum there is an overlapping region consistent with all the experiments. This yields, at one standard deviation,

$$0.5 \leq \sin^2 2\Theta \leq 0.8 \, (32^\circ > \Theta > 22^\circ)$$

and

$$0.7 \leq \Delta(eV^2) \leq 1.0.$$

We find that the Davis spectrum yields no overlapping region at the level of one standard deviation.

c. If oscillations occur with these approximate parameters, then the observed spectrum at 11.2 meters should show evidence of spectral changes. In Fig. 4 we plot the ratio of the observed 11.2 meter data to the Avignone prediction as a function of neutrino energy. If oscillations do not exist and further if the Avignone spectrum is the correct one, then the ratio should be 1.0 independent of energy. For comparison, the predicted ratio for $\Delta = 1\ eV^2$ and $\sin^2 2\Theta = 1$ is also plotted.

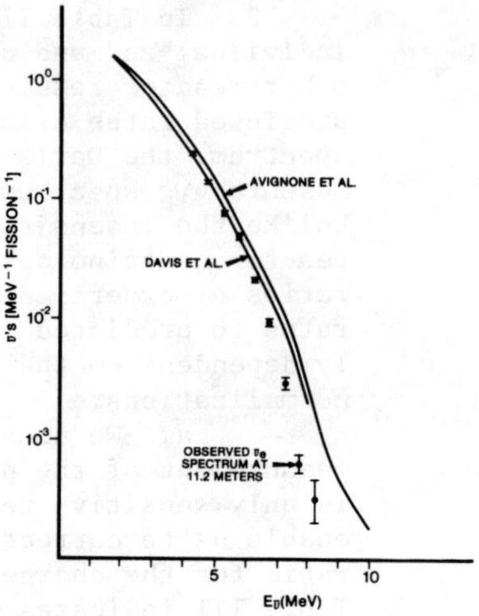

Fig. 3. The observed $\bar{\nu}_e$ spectrum at 11.2 meters compared to Avignone and Davis predictions.

Table III. Summary of Results for the Ratio $\frac{\bar{\sigma}_{expt.}}{\bar{\sigma}_{th.}}$

Distance from Core Center (Meters)	Reaction	Neutrino Detection Threshold (MeV)	Ratio Avignone Spectrum	Davis Spectrum	Measured $\bar{\nu}_e$ Spectrum (preliminary)
11.2	ncd	2.2	.83 ± .13	1.10 ± .16	1.3 ± .22
11.2	ccd	4.0	.32 ± .14	.44 ± .19	.61 ± .29
11.2	ccp	4.0	.68 ± .12	.88 ± .15	≡ 1.0
11.2	ccp	6.0	.42 ± .09	.58 ± .12	≡ 1.0
6	ccp	1.8	.65 ± .09	.84 ± .12	-
6	ccp	4.0	.81 ± .11	1.02 ± .15	1.19 ± .27
8.7	ccp	3.0	.68 ± .15	.87 ± .14	-

CONCLUSIONS

The results of a reactor experiment comparing the observed rates of the charged current and neutral current interactions of reactor neutrinos with deuterons gives an indication of neutrino instability at the 2 to 3 standard deviation level.

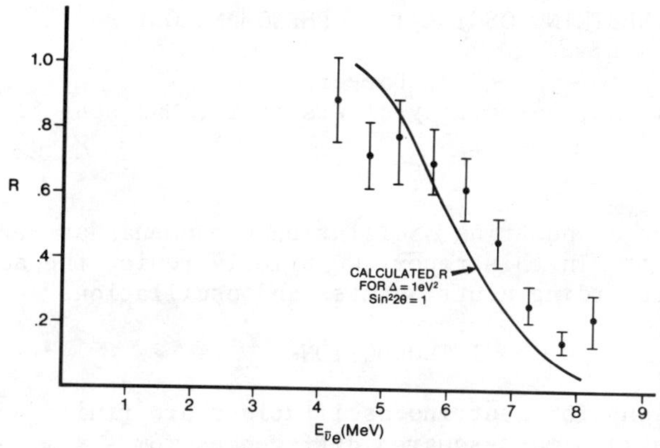

Fig. 4. The ratio of 11.2 meter ccp data to Avignone predicted spectrum.

REFERENCES

1. E. Pasierb, H.S. Gurr, J. Lathrop, F. Reines and H.W. Sobel, Phys. Rev. Lett. 43, 96 (1979).
2. F.T. Avignone III and Z.D. Greenwood (University of South Carolina preprint Feb. 1980).
3. B.R. Davis, P. Vogel, F.M. Mann, and R.E. Schenter, Phys. Rev. C19, 2259 (1979).
4. A. Soni and D. Silverman, private communication.
5. UCI internal reports - UCI-10P19-141 (1979), UCI-10P19-132 (1978), UCI-10P19-126 (1977).
6. The California Institute of Technology-Grenoble-Munich group plan measurements with the same detector at larger distances from a more powerful reactor, F. Boehm, private communication.
7. F. Nezrick and F. Reines, Phys. Rev. 142, 852 (1966).
8. F. Reines, H.S. Gurr, and H.W. Sobel, Phys. Rev. Lett. 37, 315 (1976).
9. F. Boehm, private communication.
10. R.M. Ahrens and L. Gallaher, Phys. Rev. D20, 2714 (1979) and private communication (1980). S.L. Glashow, J. Iliopoulos and L. Maiani, Phys. Rev. D2, 1285 (1970).
11. J.H. Munsee and F. Reines, Phys. Rev. 177, 2002 (1969).

NEUTRINO OSCILLATION PHENOMENOLOGY

V. Barger
Physics Department, University of Wisconsin, Madison, WI 53706

ABSTRACT

New evidence of neutrino oscillation phenomena has recently been presented.[1,2] In this report we briefly review the accumulated evidence regarding neutrino mass and oscillations.

INTRODUCTION

The conditions for neutrino oscillations are finite neutrino masses (m_i), non-zero mass-squared differences ($\delta m_{ij}^2 \equiv m_i^2 - m_j^2 \neq 0$ for $i \neq j$), and mixing (charge current eigenstates ν_α are superpositions of mass eigenstates ν_i). For a ν_α state of momentum p at time t = 0 given by $|\nu_\alpha\rangle = \sum_i U_{\alpha i}|\nu_i\rangle$, the time evolution is $|\nu_\alpha\rangle = \sum_i e^{-iE_it}U_{\alpha i}|\nu_i\rangle = \sum_{i,\beta} e^{-iE_it}U_{\alpha i}U_{\beta i}^*|\nu_\beta\rangle$ where $E_i \simeq p + m_i^2/(2p)$. Thus the $\nu_\alpha \to \nu_\beta$ transition probability at a distance $L \simeq t$ from a ν_α source is

$$P(\alpha\to\beta) = |\sum_i e^{-iE_iL} U_{\alpha i}U_{\beta i}^*|^2 . \tag{1}$$

The oscillations arise from the $\cos\Delta_{ij}$ dependence of interference terms in $P(\alpha\to\beta)$, with arguments $\Delta_{ij} = \frac{1}{2}\delta m_{ij}^2(L/E)$. The oscillations are periodic in L/E, with frequencies proportional to (mass)^2 differences and amplitudes determined by the mixing matrix U.

The leading oscillation - the first to occur as L/E increases from zero - has a simple form if one δm^2 dominates (i.e., $|\delta m_{in}^2| >> |\delta m_{ij}^2|$ for $i,j \neq n$). The transition probabilities for the leading oscillation are given by[3]

$$\begin{aligned} P(\alpha\to\alpha) &= 1-4(|U_{\alpha n}|^2 - |U_{\alpha n}|^4)\sin^2(\frac{\delta m^2}{4}\frac{L}{E}) \\ P(\alpha\to\beta) &= 4|U_{\alpha n}|^2|U_{\beta n}|^2 \sin^2(\frac{\delta m^2}{4}\frac{L}{E}) . \end{aligned} \tag{2}$$

For a single channel (e.g. $e \to e$ or $e \to \mu$), this is equivalent to the two-neutrino form often used in oscillation analyses.

If L/E is small compared to $1/\delta m^2$, oscillations have little effect. The amplitude and phase of the oscillations can be most readily mapped out for L/E of the order of $1/\delta m^2$. At L/E values much greater than $1/\delta m^2$ only average probabilities can be measured: see Fig. 1. If the mixing is small, oscillations are hard to detect anywhere. The ranges of δm^2 to which various experiments are sensitive are given in Fig. 1.

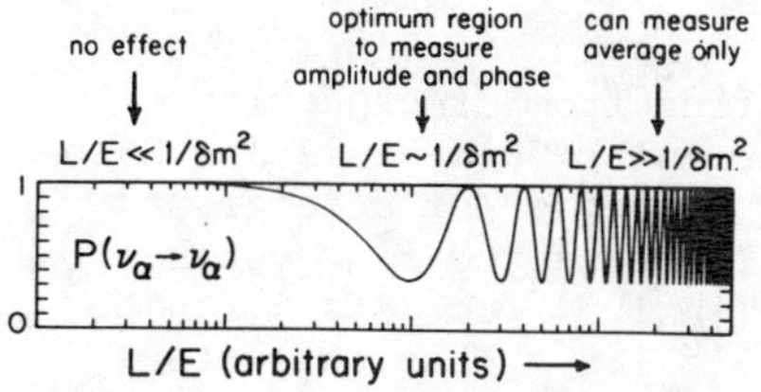

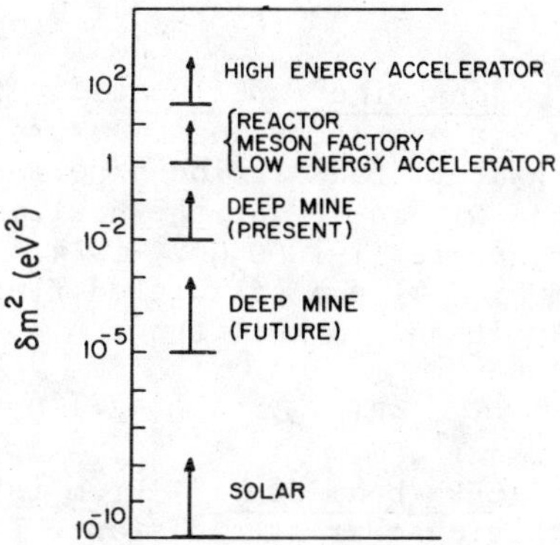

Fig. 1. Sensitivity of experiments to δm^2.

CLASSES OF OSCILLATIONS

With the usual SU(2)×U(1) assignments, (ν_{eL}, e_L^-) form a doublet and η_{eL}, e_L^+ are singlets. First class oscillations mix doublet flavors $(\nu_{eL} \leftrightarrow \nu_{\mu L} \leftrightarrow \nu_{\tau L})$ and second class oscillations mix doublet and singlet neutrinos[4] (e.g., $\nu_{eL} \leftrightarrow \eta_{eL}$). We contrast the predictions below for oscillations of two neutrinos:

1st class ($\nu_e \to \nu_\tau$)		2nd class ($\nu_e \to \eta_e$)	
$P(e\to e)$	$= 1-\sin^2 2\alpha \sin^2 \frac{\Delta}{2}$	$P(e\to e)$	$= 1-\sin^2 2\alpha' \sin^2 \frac{\Delta'}{2}$
$P(e\to\tau)$	$= \sin^2 2\alpha \sin^2 \frac{\Delta}{2}$	$P(e\to\eta)$	$= \sin^2 2\alpha' \sin^2 \frac{\Delta'}{2}$
σ_{CC}	oscillates	σ_{CC}	oscillates
σ_{NC}	immune	σ_{NC}	oscillates
σ_{CC}/σ_{NC}	oscillates	σ_{CC}/σ_{NC}	immune

Here CC denotes the charged current and NC the neutral current. Since η_L has only Higgs couplings, it is effectively non-interacting and the transition probability $P(e\to\eta)$ is lost. In general, oscillations of both classes could be simultaneously present.

POSSIBLE EVIDENCE OF NEUTRINO MASS

<u>β-spectrum of $^3H^0 \to {}^3H_e^+ \bar{\nu}_e e^-$ near endpoint</u>: Preliminary results of ITEP measurements[5] of the high energy part of the tritium β-spectrum give a neutrino mass in the interval $14 \le m_\nu \le 46$ eV at the 99% confidence level; assuming the atomic structure of the H_e^+ excited states, the most probable mass is $m_\nu = 34$ eV.

<u>solar neutrinos</u>:[6] The $\nu_e + {}^{37}Cl \to e^- + {}^{37}Ar$ experiment of Davis et al. measures the ν_e flux from the interior of the sun over the energy range 0.8-14. MeV. The discrepancy between the observed neutrino capture rate of 2.2 ± 0.4 SNU (captures/sec/10^{36} nuclei) and the standard solar model prediction of 7.5 ± 1.5 SNU suggests a neutrino oscillation effect with an average probability

$$\langle P(e \to e) \rangle = 0.3 \pm 0.1 \quad \text{at} \quad L/E \sim 10^{10} \text{ m/MeV} .$$

deep mine experiments:[7] The ν_μ-flux from atmospheric $\pi, K \to \mu \to e$ decays was measured in the Kolar Gold Field and Case-Withwaterstrand-Irvine experiments by detection of underground muons within 40° of the horizontal. The measured flux, with ν_μ-energies 1-1000 GeV, is 4.4 ± 0.4 (in $10^{-13}/cm^2/s/sr$), to be compared with a calculated flux of 7.5 ± 1.8. This suggests a transition probability

$$\langle P(\mu \to \mu) \rangle = 0.58 \pm 0.15 \quad \text{at} \quad L/E \sim 50 \text{ m/MeV} .$$

CERN beam dump:[8] From interactions of prompt neutrinos in a bubble chamber at a distance $L \simeq 800$ m from the source, an electron to muon ratio $R(e/\mu) = N(e^\pm)/N(\mu^\pm) = 0.59^{+0.35}_{-0.21}$ is observed. These prompt neutrinos presumably originate from charm particle decays which give equal numbers of ν_e and ν_μ. The deviation of $R(e/\mu)$ from unity can be interpreted[9] as ν_e oscillations with mean value

$$\overline{P}(e \to e) \sim 0.56^{+0.38}_{-0.24} \quad \text{at} \quad L/E \sim 0.01 \text{ m/MeV} .$$

Note that limits from other experiments preclude ν_μ-oscillations[3] at this L/E: see Fig. 2.

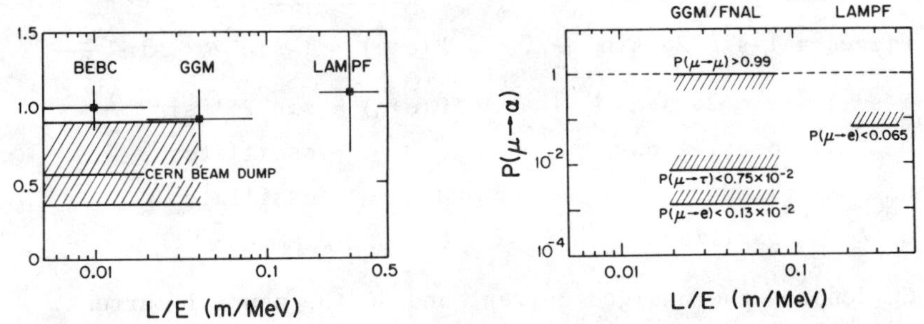

Fig. 2. Accelerator limits.

A corresponding depletion of the known ν_e flux in a narrow-band beam should be observed; however, a BEBC experiment finds

$$\langle P(e \to e) \rangle = 1.00 \pm 0.14 \quad \text{at} \quad L/E \simeq 0.01 \text{ m/MeV} .$$

It is still possible that an oscillation effect exists here, with an $\langle P(e \to e) \rangle$ intermediate between the two different measurements.

reactor data (proton target): The ν_e flux from nuclear reactor fissions was measured by Reines et al.[10] at L = 6 m and 11.2 m through the reaction $\overline{\nu}_e p \to e^+ n$. Taking the ratio of measured to calculated flux,[11] we deduced[1] that $P(\overline{e} \to \overline{e})$ shows an oscillation down to $P \sim \frac{1}{2}$, for $L/E \sim$ 1-3m/MeV. This conclusion is affirmed by new data from the L = 11.2 m experiment.[2] Preliminary Grenoble data by Boehm et al. at L = 8.7 m indicate a smaller oscillation effect than the Reines et al. data (see comparisions in Sobel report[2]).

reactor data (deuteron target): The ratio of CC to NC cross sections for deuteron break-up by reactor antineutrinos, $\sigma(\bar{\nu}_e d \to e^+ nn)/\sigma(\bar{\nu} d \to \bar{\nu} pn)$, was recently measured by Reines et al.[2] The NC channel is blind to first class oscillations and thus monitors the initial $\bar{\nu}_e$ flux. The average first-class oscillation probability determined from these data is[2]

$$\bar{P}(\bar{e}\to\bar{e}) = \frac{[\sigma(CC)/\sigma(NC)]\text{experiment}}{[\sigma(CC)/\sigma(NC)]\text{theory, no osc}} = 0.39 \pm 0.22 .$$

astrophysics: If the dark matter surrounding galaxies is a cloud of gravitationally bound massive neutrinos, a lower bound $m_\nu \geq 20$ eV can be placed on the neutrino mass.[12]

OSCILLATION INTERPRETATIONS OF DATA

solar: The solar neutrino experiment measures the average $\langle P(e\to e)\rangle$. With three neutrinos (ν_e, ν_μ, ν_τ), the minimum value is $\langle P\rangle = 1/3$ for first class oscillations with all $\delta m^2 \gg 10^{-10}$ eV^2. With both first and second class oscillations, the corresponding minimum is $\langle P\rangle = 1/6$. To achieve a near minimal $\langle P\rangle$, the mixing must be large, and hence sizeable oscillation effects are to be expected elsewhere.

deep mine: Here we need all $\delta m^2 > 10^{-2}$ eV^2 and large mixing to achieve significant flux reduction.

accelerator limits: The experimental limits shown in Fig. 2 require ν_μ to be essentially decoupled from oscillations at $L/E \sim 0.01$. This can be realized with either (i) all $\delta m^2 \ll 50$ eV^2 or (ii) a very small $U_{\mu n}$ matrix element, where n denotes the neutrino of large mass: see Eq. (2). In either case ν_μ oscillations will occur at larger L/E where non-leading δm^2 enter.

CERN beam dump: An e/μ ratio significantly less than unity requires $\delta m^2 > 50$ eV^2 and large mixing. Oscillation predictions[3] versus the average probability $\bar{P}(e\to e)$ are shown in Fig. 3 for the

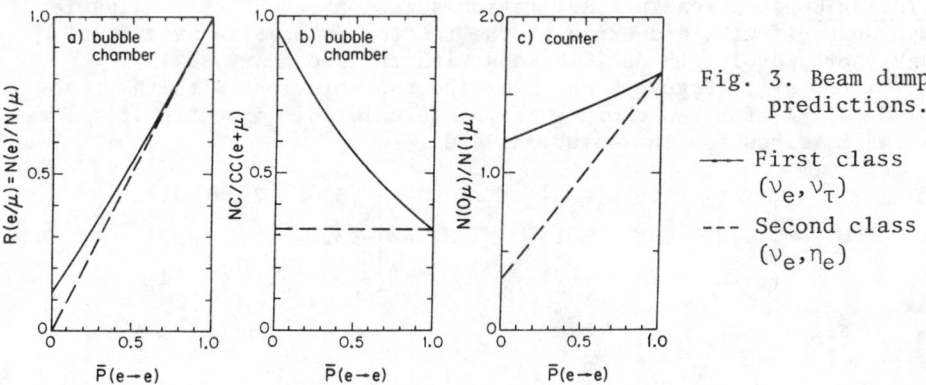

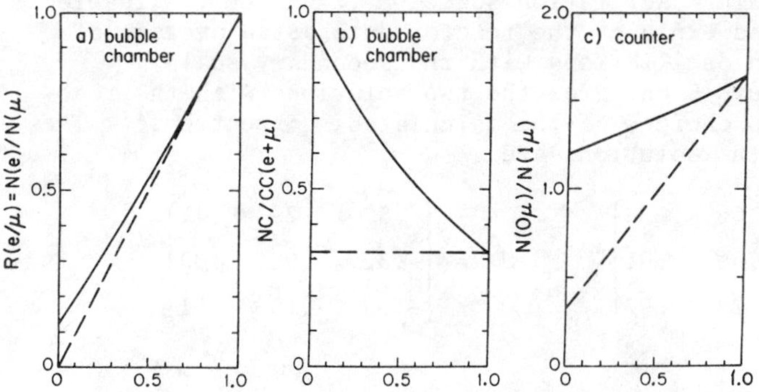

Fig. 3. Beam dump predictions.

—— First class (ν_e, ν_τ)

--- Second class (ν_e, η_e)

quantities measured in the bubble chamber and counter experiments.

reactor data: The observed flux depletions can be explained with first class oscillation parameters $\delta m^2 \sim 0.7\text{-}0.9\ eV^2$ and $\sin^2 2\alpha \sim 0.4\text{-}0.8$.

OSCILLATION SOLUTIONS

We now briefly discuss two solutions which accommodate the solar, deep mine, and reactor observations. Solution A has a leading oscillation $\delta m^2_{31} = 0.9\ eV^2$, matches the reactor data, but has no beam dump effects; the secondary oscillation scale is taken to be $\delta m^2_{21} = 0.05\ eV^2$. Figure 4 shows the propagation probabilities versus L/E.

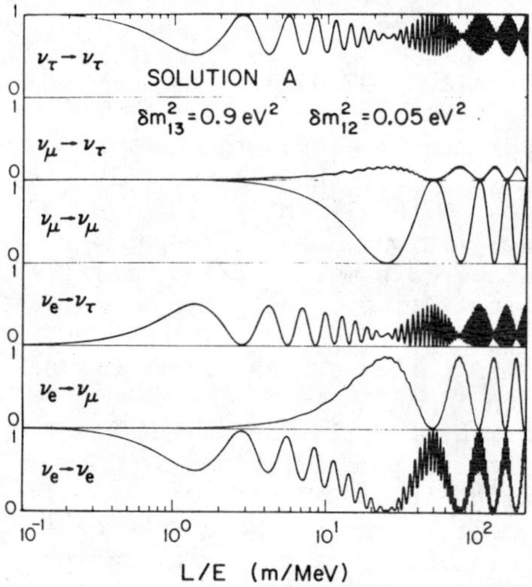

Fig. 4. Vacuum oscillations of solution A.

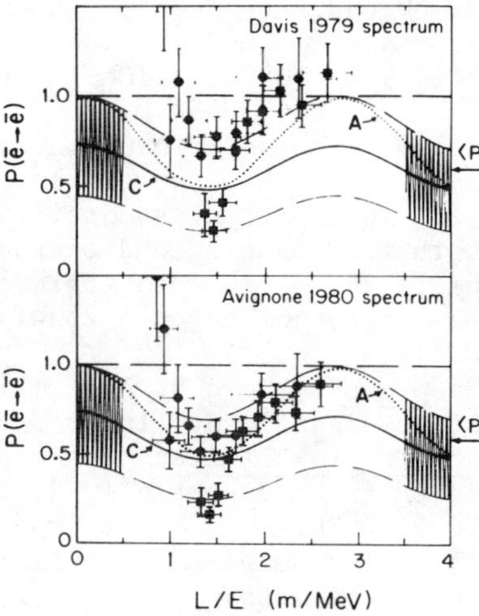

Fig. 5. Oscillation descriptions of reactor data.

Solution C has a leading oscillation scale $\delta m^2_{31} = 50\ eV^2$, sizeable beam dump effects, and explains the reactor data as an average of many short wavelength oscillations with the secondary scale $\delta m^2_{21} = 0.9\ eV^2$. Figure 5 compares the two solutions with the reactor data,[2,10] for two choices of the calculated ν_e spectrum.[11] The mixing matrices for the solutions are

$$U^A = \begin{pmatrix} .64 & .66 & .38 \\ -.72 & .69 & .01 \\ -.26 & -.28 & .92 \end{pmatrix} \qquad U^C = \begin{pmatrix} .87 & .29 & .41 \\ -.32 & .95 & .02 \\ -.38 & -.15 & .91 \end{pmatrix} .$$

The values for the element $U_{\mu 3}$ are well within the experimental limits, $|U_{\mu 3}| < 0.95$ for A and $|U_{\mu 3}| < 0.07$ for C. The solutions yield the following average probabilites:

solar	$\langle P(e \to e) \rangle$	0.4(A)	0.6(C)
deep mine	$\langle P(\mu \to \mu) \rangle$	0.5(A)	0.8(C)
beam dump	$\overline{P}(e \to e)$	1.0(A)	0.8(C)

SOME FUTURE TESTS

$\nu e \to \nu e$ and $\overline{\nu} e \to \overline{\nu} e$ cross sections: For an initial ν_μ source, oscillations amplify[13] $\sigma(\nu e)$; for an initial $\overline{\nu}_e$ source, oscillations depress[1,13,14] $\sigma(\overline{\nu} e)$, except in a limited kinematic region. Figure 6 illustrates effects on $\sigma(\overline{\nu} e)$ averaged over reactor spectra; the results are insensitive to the choice of calculated spectra.

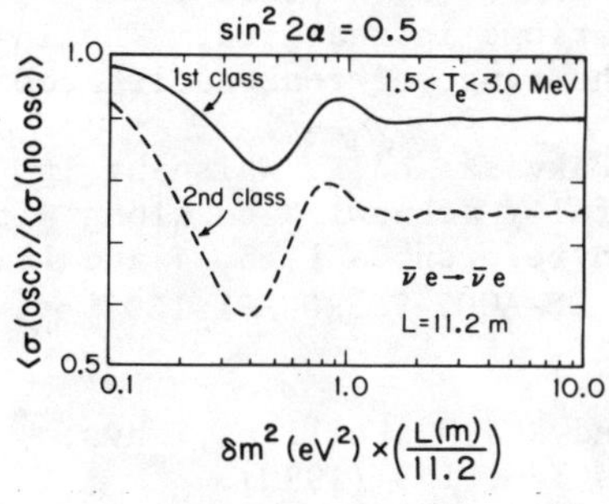

Fig. 6. The $\overline{\nu}$ - e cross section.

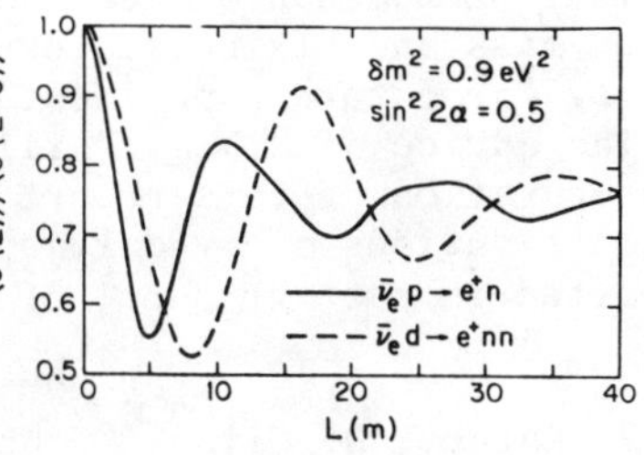

Fig. 7. Reactor cross sections versus distance.

reactor cross sections versus distance: Oscillations can be traced out versus L, as illustrated in Fig. 7. The oscillation patterns are different for proton and deuteron targets, due to the difference in threshold energies.

meson factory and accelerator experiments at favorable L/E: To confirm oscillations of mass scale $\delta m^2 \sim 1\ eV^2$, future experiments should be designed to make measurements at $L/E \sim 1.5 m/MeV$. In accelerator experiments, searches for τ-leptons produced via $\nu_e \to \nu_\tau$ oscillations are important to test flavor-changing transitions.

matter oscillations:[15] Vacuum oscillations are modified in matter by coherent ν_e-e charged current scattering. In deep mine experiments these modifications are significant if (i) the neutrinos traverse more than 0.2 of the earth's radius and (ii) the neutrinos have energy $E_\nu(MeV) \gtrsim 5 \times 10^5\ \delta m^2 (eV^2)$, with δm^2 the minimum mass-squared difference. Figure 8 illustrates matter oscillations for a two-neutrino case; the matter corrections depend sensitively on the sign of δm^2.

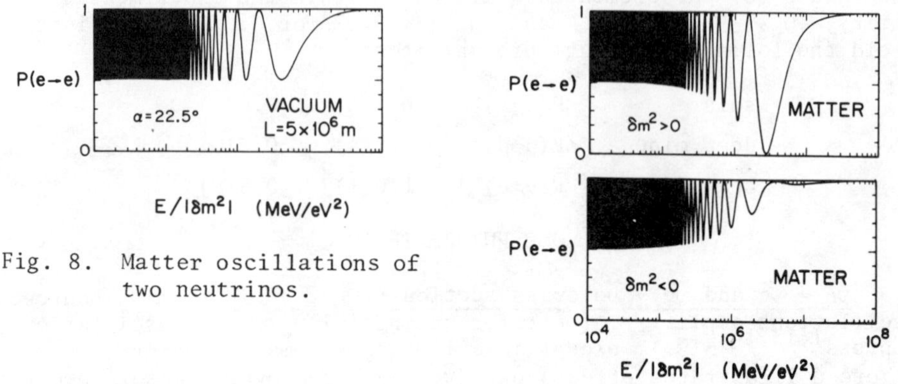

Fig. 8. Matter oscillations of two neutrinos.

A half-dozen experiments have yielded tantalizing evidence for finite neutrino masses; thus we have reason to be hopeful that the beautiful phenomenon of neutrino oscillations is realized. The indicated mass and mixing scales are in the range of reasonable accessibility for future experiments.

The advice of R.J.N. Phillips, S. Pakvasa and K. Whisnant in the preparation of this report is gratefully acknowledged, along with the contributions of my collaborators in references 1, 3, 4 and 15. [Supported in part by Dept. of Energy: DE-AC02 76ER00881-169.]

REFERENCES

1. V. Barger, D. Cline, K. Whisnant and R.J.N. Phillips, Phys. Lett. 93B, 194 (1980); UW preprints 146, 148 (1980).
2. H. W. Sobel, report in these proceedings; F. Reines, H. W. Sobel, and E. Pasierb, Irvine preprint (1980).
3. V. Barger, K. Whisnant, R.J.N. Phillips, Phys. Rev. (in press).
4. See, e.g., report in these proceedings by V. Barger, P. Langacker, J. Leveille, S. Pakvasa, and references therein.
5. V. A. Lyubimov et al., Phys. Lett. 94B, 266 (1980).
6. J. Bahcall et al., IAS preprint (1980); R. Davis, Jr., Proc. of Brookhaven Solar Neutrino Conference BNL 50879, 1, 1 (1978).
7. M. R. Krishaswamy et al., Proc. Roy. Soc. London A323, 489 (1971); M. F. Crouch et al., Phys. Rev. D18, 2239 (1978). See analysis of combined event sample by C. Rubbia, CERN-EP/80-108.
8. P. O. Hulth, K. Kleinknecht and F. Niebergall, reports in these proceedings.
9. A. de Rújula et al., Nucl. Phys. B168, 54 (1980).
10. F. Nezrick and F. Reines, Phys. Rev. 142, 852 (1966); F. Reines et al., in Proc. of Fermilab Ben Lee Memorial Conf. (1978).
11. F. T. Avignone III and Z. D. Greenwood, Phys. Rev. C (in press); B. R. Davis et al., Phys. Rev. C19, 2259 (1979).
12. E. Witten, HUTP-80/A031; S. Tremaine and J. E. Gunn, Phys. Rev. Lett. 42, 407 (1979).
13. B. Kayser and P. Rosen, Purdue-NSF preprint (1980).
14. B. Halls and B.H.J. McKellar, Melbourne preprint (1980).
15. L. Wolfenstein, Phys. Rev. D17, 2369 (1978); V. Barger, K. Whisnant, S. Pakvasa, R.J.N. Phillips, UW preprint 152 (1980).

EMULSION EXPERIMENTS ON SHORT-LIVED PARTICLES, EXPERIMENT

J. PRENTICE, TORONTO, ORGANIZER

OBSERVATION OF CHARMED PARTICLES PRODUCED BY HIGH-ENERGY PHOTONS IN NUCLEAR EMULSIONS COUPLED WITH A MAGNETIC SPECTROMETER

(Photon-Emulsion Collaboration)

(Bologna[1*]-CERN[2]-Florence[3*]-Genova[4*]-Madrid (JEN)[5]-Moscow (LPI and GCPP)[6]-Paris VI[7]-Santander[8]-Valencia[9])

M.I. Adamovich[6], Y.A. Alexandrov[6], J.M. Bolta[9], L. Bravo[8], A.M. Cartacci[3], B. Conforto[3], A. Conti[3], M.M. Chernyavski[6], M.G. Dagliana[3], M. Dameri[4], G. Diambrini-Palazzi[4], G. Di Caporiacco[3], A. Forino[1], R. Gessaroli[1], E. Higon[9], S.P. Kharlamov[6], V.G. Larionova[6], R. Llosa[5], J. Lory[7], A. Mattei[3], C. Meton[7], G.I. Orlova[6], B. Osculati[4], G. Parrini[3], N.G. Peresadko[6], A. Quareni-Vignudelli[1], A. Ruiz[8], K.M. Romanovskaya[6], M. Sannino[4], D. Schune[7], G. Tomasini[4], M.I. Tretyakova[6], Tsai Chu[7], G. Vanderhaeghe[2], E. Villar[8], B. Willot[7],

and

(Omega-Photon Collaboration)

(Bonn[10]-CERN[2]-Glasgow[11]-Lancaster[12]-Manchester[13]-Rutherford[14]-Sheffield[15])

M. Atkinson[14], T. Brodbeck[12], G.R. Brookes[15], P.J. Bussey[11], M. Davenport[14], J.-P. Dufey[2], R.J. Ellison[13], W. Galbraith[15], K. Heinloth[10], J.S. Hutton[14], R.E. Hughes-Jones[13], M. Ibbotson[13], B.R. Kumar[14], G.D. Lafferty[12], J.B. Lane[13], J.-C. Lassalle[2], R. McClatchey[15], D. Mercer[13], J.V. Morris[14], D. Newton[12], G.N. Patrick[11], C. Raine[15], A. Schlösser[10], K.M. Storr[12], A.P. Waite[13]

(Presented by G. Diambrini-Palazzi)

The charmed events we describe in the present paper were observed in nuclear emulsion pellicles exposed at CERN to the tagged photon beam from the SPS in the Omega prime spectrometer (experiment WA58). Six thousand BR2 emulsion pellicles of 20 cm × 5 cm × 600 μm (36 litres) were used**). The grain density of minimum ionization tracks in the various batches of emulsion ranges from 32 to 40 per 100 μm.

A mechanical device brought single pellicles one at a time to the target position. They were put at an angle of 5° to the beam axis, so that their effective thickness was about 6 mm. Each pellicle was exposed to a dose of 10^6 tagged photons with energies ranging from 20 to 70 GeV. Secondary particles from photon interactions in the emulsion were detected and analysed by the spectrometer. This special method of exposure was already used in a previous experiment (WA45) in which a $\bar{D}^0$ meson was observed[1].

During the exposure the trigger logic efficiently excluded electromagnetic background. Only events with a multiplicity greater than or equal to three were recorded. These events were reconstructed using the CERN geometry programme TRIDENT[2]. About 75% of them were

---

*)Istituto di Fisica dell'Università and Sezione INFN.

**)The pellicles were produced by the State Research Institute for Photo-chemical Projects (GCPP) of Moscow.

found by scanning the emulsion over areas of up to 75 mm^2 (15 mm parallel to the beam direction and 5 mm perpendicular to it) around the positions predicted by TRIDENT. Events are classified as "matching" if there is a good correspondence between the angles measured in the emulsion and the predicted ones (the mean absolute difference between measured and predicted angles is $\sim 0.5°$ in azimuth and $\sim 1.5°$ in dip).

Up to now, about 1400 events out of 160,000 have been found and measured. Among the matching events, three have been interpreted as double production and decay of charmed particles. Two more possible charmed pairs have been observed among the non-matching events.

It should be stressed that the presence of two decay-like vertices at short distances from a primary interaction is a strong indication of double production of charmed particles. The probability of finding a background event (secondary interaction or strange particle decay) with such a topology can be estimated at about 3×10^{-3} in the whole sample of 1400 events. Hereafter, we describe briefly the five events found and we give the results of their analysis; Table I summarizes their main features.

Event No. 1

A photon of 25 GeV makes a primary interaction of the type 7b + 0g + 5s *) (Fig. 1). At a distance of 50 μm from the primary vertex one of the shower tracks (No. 4) undergoes a deflection of 9° and, at a distance of 124 μm, there is a vertex with four relativistic tracks, known as a W^0 vertex (No. 6.1-4). The four tracks from the W^0 vertex match four of the tracks predicted by TRIDENT. The resultant transverse momentum of the four particles with respect to the direction of flight of the decaying particle is $p_T = (0.04 \pm$ $\pm\ 0.06)$ GeV/c, consistent with zero. If we assign the K mass to one of the positive particles and the π mass to the other three particles, the resulting effective mass is 1.847 GeV/c^2, compatible with the D mass**). The W^0 can therefore be interpreted as the decay of a $\bar{D}^0 \to K^+\pi^+\pi^-\pi^-$.

Track 4.1 (after the large deflection of track 4) does not match any prediction of TRIDENT; it was presumably lost in the plexiglas support of the emulsion. On the other hand, a Λ^0 has been recorded in the spectrometer but its decay products were not seen in the emulsion. However, its reconstructed line of flight is coplanar with tracks 4 and 4.1 within the errors (mixed scalar product = = 0.007 ± 0.100). One can therefore assume a two-body decay of particle 4 into $\Lambda^0\pi^+$. The computed effective mass of this particle is (2.33 ± 0.05) GeV/c, which is compatible with the Λ_c mass. The resulting π^+ momentum is (3.96 ± 0.24) GeV/c, in agreement with the

---

*) b stands for "black" tracks, g for "grey" tracks, and s for "shower" or relativistic tracks.

**) The error on the mass, as computed from TRIDENT, is 0.003 GeV/c^2, but there are good reasons to believe that the values obtained would be affected by systematic errors, because at the present time the program coefficients are not fully optimized.

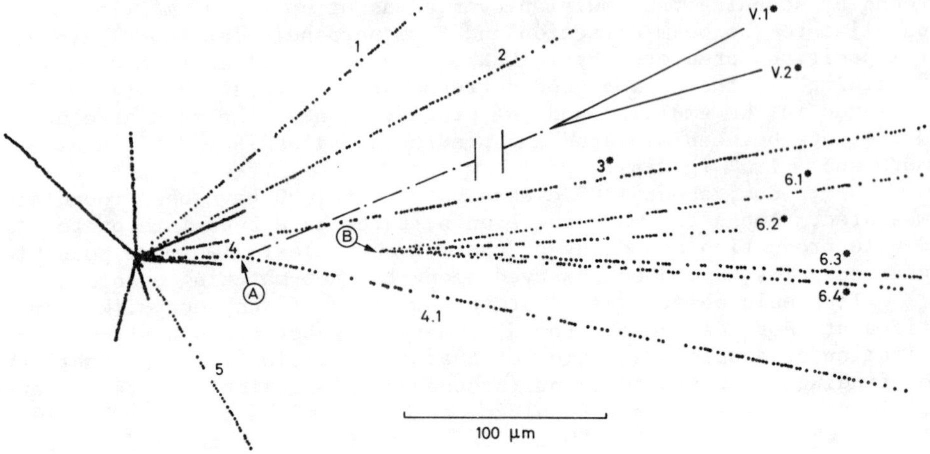

Fig. 1

Track numbers with an asterisk are tracks recorded in the spectrometer

value derived from multiple scattering measurements (see Table I). The decay can therefore be interpreted as $\Lambda_c^+ \to \Lambda^0\pi^+$.

This event is the first fully reconstructed example of associated photoproduction of a charmed baryon and a charmed meson. The computed decay times for the two charmed particles are

$$\tau(\bar{D}^0) = (0.86 \pm 0.01) \times 10^{-13}\ \text{s}; \quad \tau(\Lambda_c^+) = (0.57 \pm 0.02) \times 10^{-13}\ \text{s}\ .$$

Event No. 2

A photon of 60 GeV makes a primary interaction of the type 4b + 0g + 4s. At a distance of 94 μm from the primary vertex one of the shower tracks gives rise to a wide angle trident (three-prong event), and at a distance of 267 μm there is a V^0 vertex (Fig. 2). All eight outgoing low-ionization tracks (three from the primary interaction, three from the trident and two from the V^0) match tracks

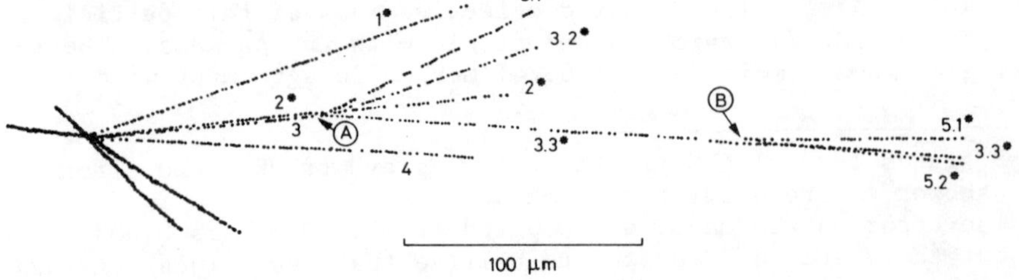

Fig. 2

Table 1

Event No.	Emulsion						Spectrometer			Nature of particles and decay mode	Decay time (10^{-13} s)
	Topology (N_{ch})	Track No.	Length (μm)	Azimuth (degrees)	Dip (degrees)	pβ (GeV/c)	Azimuth (degrees)	Dip (degrees)	p (GeV/c)		
1	1	4	50 ± 1	-1.8	4.7					$\Lambda_c^+ \to \Lambda^0\pi^+$	
		41		9.1	0.8	3.1 ± 0.8				π^+	0.57 ± 0.02
	4	(6)	124 ± 1	-2.7	0.1					$\bar{D}^0 \to K^+\pi^+\pi^-\pi^-$	
		61		-7.7	-2.5		-8.7	-3.9	-2.35	π^-	
		62		-4.0	-10.3		-3.9	-10.7	1.02	K^+	0.86 ± 0.01
		63		-0.2	3.2		-0.7	2.3	-4.47	π^-	
		64		0.5	9.4		0.2	9.5	1.21	π^+	
2	3	3	94 ± 1	-4.5	4.1					$D^- \to \pi^+\pi^-\pi^-(K^0)$	
		31		-24.1	0.7		-23.4	0.2	0.88	π^+	0.57 or 0.88
		32		-14.2	4.6		-13.8	5.2	-2.38	π^-	
		33		4.0	4.1		3.9	4.0	-1.65	π^-	
	2	(5)	267 ± 2	0.9	3.1					$D^0 \to > 2$ bodies	
		51		-0.2	3.1		-0.2	3.6	14.32		0.45-0.85
		52		5.1	2.9		5.1	3.1	-5.67		
3	3	1	685 ± 5	-1.3	20.8					? → > 3 bodies	
		11		1.0	11.3		0.9	11.1	0.97		
		12		6.4	8.5						7.1-14.2
		13		32.0	25.6		32.2	24.4	-0.98		
	3	2	980 ± 5	0.2	4.1					? → ≥ 3 bodies	
		21		-9.2	4.6		-9.1	4.1	3.69		
		22		1.4	2.6						3.8-7.6
		23		9.8	6.8						
4	2	(7)	44 ± 1	1.0	2.5					$D^0 \to \geq 2$ bodies	
		71		1.0	0.0	≥ 1.0					0.04-0.13
		72		5.0	3.0	≥ 1.0					
	1	4	260 ± 2	-2.0	0.0					? → ≥ 2 bodies	
		41		13.0	-39.0						1.0-5.4
5	3	6	32 ± 1	12.5	-2.4					? → ≥ 3 bodies	
		61		0.2	-2.9	1.8 ± 0.2					
		62		32.0	35.8						0.23-0.66
		63		35.0	0.0	1.2 ± 0.3					
	1	3	1900 ± 6	-1.0	7.7					? → ≥ 2 bodies	
		31		1.5	3.7						1.7-40

recorded by the spectrometer. The charge of the trident is negative and its transverse momentum is unbalanced: $p_T = (0.442 \pm 0.028)$ GeV/c. Ionization measurements on the positive track rule out the proton assignment. Taking it as a positive pion, and assuming the presence of an extra K^0 particle in order to balance the transverse momentum, the hypothesis of a $D^- \to \pi^+\pi^-\pi^-K^0$ gives two solutions for the D momentum, i.e. 6.63 and 10.33 GeV/c. No other hadronic decay mode is consistent with the data.

The hypothesis of a two-body decay of a D^0 at the V^0 vertex is not compatible with the measured parameters. However, a decay of a D^0 into three or more particles can describe the data and gives D^0 momenta between 20 and 37.6 GeV/c.

In conclusion, the trident can be interpreted as a $D^- \to \pi^+\pi^-\pi^-K^0$ with a decay time $\tau(D^-) = 0.57 \times 10^{-13}$ s or 0.88×10^{-13} s, and the V^0 as a $D^0 \to 2$ charged particles + neutral(s) with the following decay time limits: $(0.45 \leq \tau(D^-) \leq 0.85) \times 10^{-13}$ s.

Event No. 3

A photon of 65 GeV makes a primary interaction of the type 1b + 0g + 3s. Two of the shower tracks give rise to tridents, one after 685 μm and the other at 980 μm (Fig. 3). Two tracks of the first trident and only one track of the second match with tracks detected by the spectrometer. The unmatched tracks are too short for evaluating their momentum from multiple scattering measurements in the emulsion. The transverse momentum of the first trident is clearly unbalanced. Several Cabibbo-allowed decay modes in more than three bodies are possible, giving a maximum value of the decaying particle momentum of 6.0 GeV/c. A minimum value of about 3 GeV/c can be inferred from the fact that the decaying particle track is at minimum ionization. Limits for the momentum of the second trident can be inferred from the maximum opening angle of the charged tracks. The distribution of this angle as a function of the incident momentum, given by a Monte Carlo calculation, indicates that the most probable limits for the momentum are 8 and 16 GeV/c.

The decay time limits are therefore

$$(7.1 \leq \tau_1 \leq 14.2) \times 10^{-13} \text{ s} \quad \text{and} \quad (3.8 \leq \tau_2 \leq 7.6) \times 10^{-13} \text{ s} .$$

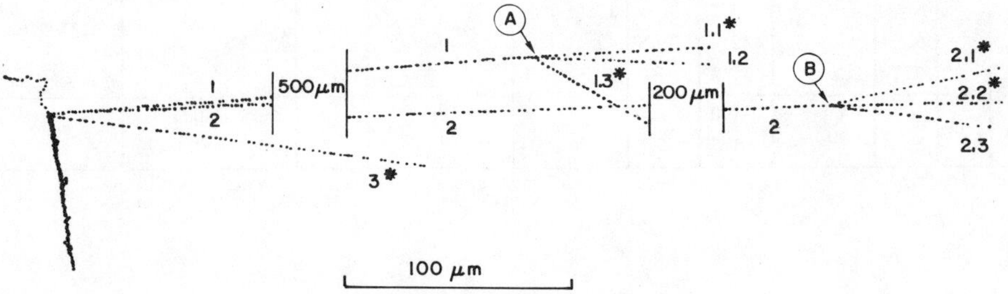

Fig. 3

Events Nos. 4 and 5

The topology of these non-matching events is shown in Figs. 4 and 5; their main features are given in Table I. The limits on their decay times are estimated on the basis of multiple scattering measurements in emulsion (whenever possible), event topology, and upper limit of the primary photon energy.

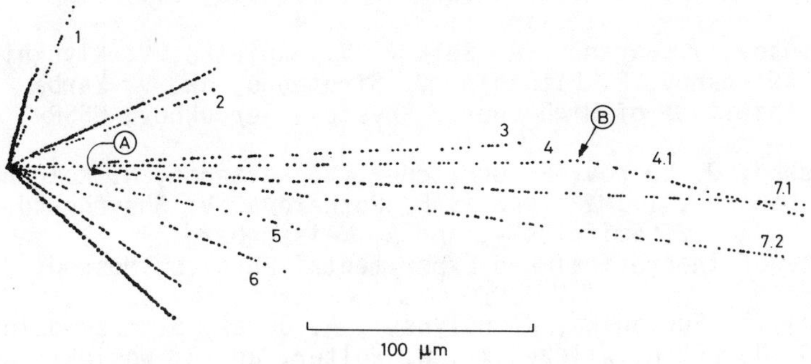

Fig. 4

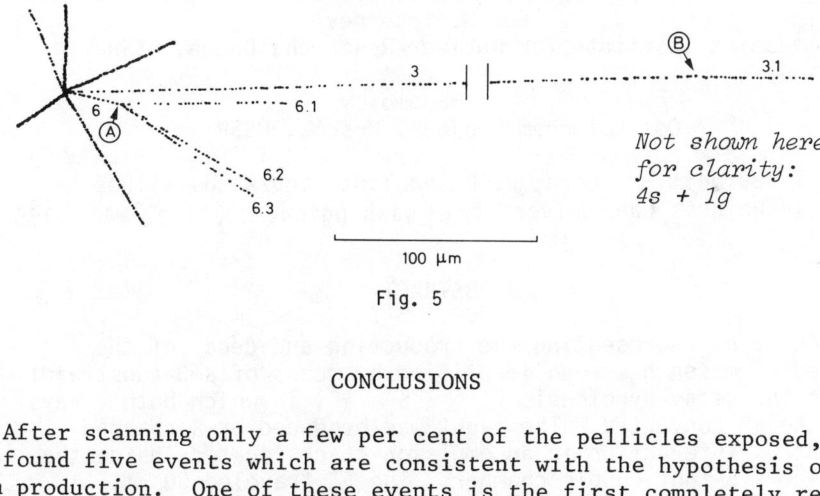

Not shown here for clarity: 4s + 1g

Fig. 5

CONCLUSIONS

After scanning only a few per cent of the pellicles exposed, we have found five events which are consistent with the hypothesis of charm production. One of these events is the first completely reconstructed example of associated photoproduction of a charmed baryon and a charmed meson.

The average charged multiplicity for charged and neutral decays is found to be 2.1 and 2.7, respectively.

REFERENCES

1. Proc. 19th Int. Conf. on High-Energy Physics, Tokyo, 1978, Physical Society of Japan, Tokyo (1979), p. 297, and M.I. Adamovich et al., Phys. Lett. 89B, 427 (1980).
2. J.C. Lassalle et al., Internal report CERN-DD/EE/79-2 (1979).

OBSERVATION OF CHARMED F^+ MESON PRODUCED BY NEUTRINO INTERACTION IN EMULSION

R. Ammar, D. Coppage, R. Davis, N. Kwak, R. Riemer, and R. Stump
University of Kansas, Lawrence, Kansas 66045

H. Kautzky, W. Smart, S. Velen, and L. Voyvodic
Fermi National Accelerator Laboratory, Batavia, Illinois 60510

V. Ammosov, V. Baranov, A. Belkov, V. Gapienko, V. Klyukhin,
V. Koreshev, P. Pitukhin, V. Sirotenko, and V. Yarba
Institute of High Energy Physics, Serpukhov, USSR

V. Efremenko, O. Egorov, P. Goritchev, V. Kaftanov, N. Kolganova,
M. Kubantsev, I. Makhlyueva, E. Pozharova, V. Shevchenko,
V. Smirnitsky, and A. Weissenberg
Institute of Theoretical and Experimental Physics, Moscow, USSR

J. Babecki, B. Furmanska, R. Holynski, A. Jurak, S. Krzywdzinski,
G. Nowak, H. Wilczynski, W. Wolter, and B. Wosiek
Institute of Nuclear Physics, Krakow, Poland

S. Bunyatov, M. Ivanova, O. Kuznetsov, V. Lyukov, V. Sidorov,
and H. Tchernev
Joint Institute for Nuclear Research, Dubna, USSR

K. Bogomolov
Gos Fotochem Project, Moscow, USSR

T. Burnett, J. Lord, R. Rosenbladt, and R. J. Wilkes
Visual Techniques Lab, University of Washington, Seattle, WA 98195

ABSTRACT

An event representing the production and decay of the charmed F^+ meson has been identified by means of a 3-constraint fit to the decay hypothesis $F^+ \to \pi^+\pi^+\pi^-\pi^\circ$, in which both γ-rays from the π° converted. The event was produced by a charged current ν_μ interaction in an emulsion stack located inside the Fermilab 15-foot bubble chamber. The F^+ traveled 50 μm, corresponding to a proper time of 1.4×10^{-13} seconds, before decaying in flight. Its mass was determined to be 2017 $\pm$ 25 MeV.

INTRODUCTION

Evidence for the charmed meson F(2030) was first presented by the DASP collaboration[1] which detected the decay $F \to \pi\eta$ following its production in e^+e^- annihilations. To date, however, there has been no convincing confirmation of this result from similar experiments at SLAC.[2] Another matter of considerable

0094-243X/81/680348-04$1.50

current interest is the determination of this, and other, charmed particle lifetimes which theoretical estimates place in the range 10^{-12} to 10^{-14} seconds, and which require high spatial resolution to detect the track of the decaying particle.[3] In this report we present a brief discussion of an event which represents F^{+} production and decay in an emulsion stack located inside the Fermilab 15-foot bubble chamber exposed to a neutrino beam; a more detailed discussion may be found elsewhere.[4]

EXPERIMENTAL DETAILS

About 320,000 pictures were taken in the 15-foot chamber, equipped with two-plane external muon identifer (EMI), exposed to neutrinos from the wideband ν_{μ} beam with single horn. The machine energy was 350 GeV with an average of 1.6 x 10^{13} protons per pulse on target. Twenty-two 1-liter stacks of Cryogenic Sensitized BR2 emulsions were contained in 2 stainless steel boxes mounted on the nose-cone flange just above and below the median plane of the chamber.

The bubble chamber pictures were scanned for tracks which leave the emulsion boxes and enter the chamber. These tracks were then measured and projected back to predict their common origin in the emulsion; events were subsequently found by scanning in the vicinity of the predicted vertex. The emulsion extends $\lesssim$ 5 cm along the beam direction and only 0.6 cm of steel, comprising the front face of the emulsion box, lies between the emulsion and the chamber liquid.

DISCUSSION OF THE EVENT

Fig. 1(a) shows the event as it appears in the emulsion. The particle of interest is produced at point A and travels 50 μm to point B. The vertex B is clean, with no sign of neclear excitation or recoil, as expected for a decay.

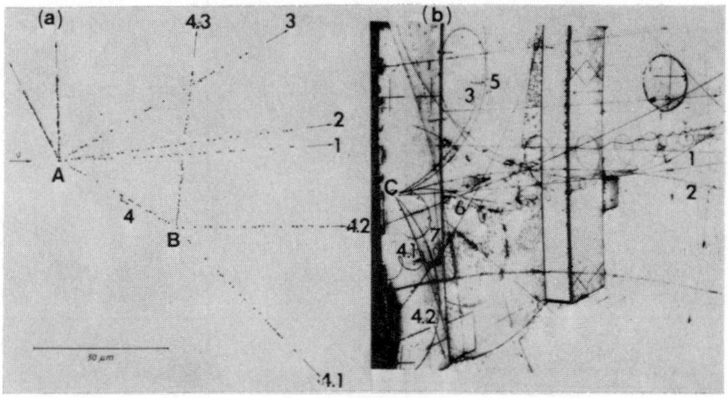

Fig. 1. Event as seen (a) in emulsion and (b) in bubble chamber

The two unnumbered tracks at the primary vertex A are nuclear fragments each of which travels less than 500 μm before stopping in the emulsion. All other tracks from the primary vertex are consistent with minimum or plateau ionization and enter the bubble chamber; they are shown, appropriately labeled, in Fig. 1(b). Of these, track 1 is due to a 12 GeV/c negative particle identified as a μ^- on the basis of the EMI data.

Of the tracks at the secondary vertex B, numbers 4.1 and 4.2 (both consistent with minimum/plateau ionization) are also seen in the bubble chamber but track 4.3 is not. Ionization measurements on track 4.3 in the emulsion yielded $\beta = 0.65 \pm 0.06$. In addition, if the vertex B represents a decay then the charge of track 4.3 must be negative.

There are 3 additional tracks which emerge from the box at point C but which are not seen in the emulsion. Of these, tracks 5 and 6 are consistent with an e^-e^+ pair[5] from a γ conversion (γ_1) while track 7 may be interpreted as an e^+ from another γ conversion (γ_2) in which the corresponding e^- does not enter the bubble chamber.

Two important features of the event should be noted: (a) the transverse momentum of track 4.1 (relative to the direction of track 4) is ∿600 MeV/c which is much higher than corresponding momenta encountered in the decays of non-charmed "stable" particles, and (b) it is not possible to balance transverse momentum using only the charged particles observed at vertex B. These expectations are confirmed by kinematic fitting; the 3-constraint fits obtained are shown in Table I.

Table I. 3-C Fits Obtained

Hypothesis			χ^2	Probability
$F^+(2030)$	→	$\pi^+\pi^+\pi^-\pi^\circ$, $\pi^\circ \to \gamma_1\gamma_2$	3.77	29%
		$K^+\pi^+\pi^-\pi^\circ$, $\pi^\circ \to \gamma_1\gamma_2$	12.7	0.5%
		$\pi^+\pi^+\pi^-\gamma_1$	16.7	0.08%
$\Lambda_c^+(2257)$	→	$p\pi^+\pi^-\pi^\circ$, $\pi^\circ \to \gamma_1\gamma_2$	11.7	0.9%

The only one of these fits with an acceptable χ^2 is $F^+ \rightarrow \pi^+\pi^+\pi^-\pi^\circ$; the results of this fit are presented in Table II.

Table II. Fit Results for $F^+(2030) \rightarrow \pi^+\pi^+\pi^-\pi^\circ$

TRACK	IDENTITY	AZIMUTH (deg.)	DIP (deg.)	MOMENTUM (GeV/c)
4	F^+	-28.7±0.2	5.1±0.4	2.37±0.03
4.1	π^+	-45.2±0.2	11.5±0.4	2.01±0.01
4.2	π^+	-2.3±0.3	41.3±0.3	0.299±0.003
4.3	π^-	86.4±0.4	-62.4±0.4	0.133±0.018
	π°	25.0±2.2	-28.2±1.6	0.568±0.028

We conclude that the event represents the production of F^+ by ν_μ in a charged current reaction, with the subsequent decay $F^+ \rightarrow \pi^+\pi^+\pi^-\pi^\circ$ occuring after 1.4×10^{-13} sec. Treating the F^+ mass as unknown a 2-C fit found its value to be 2017 ± 25 MeV. The invariant mass of the $\pi^+\pi^-\pi^\circ$ system, using the slower of the two π^+, is 808 ± 20 MeV, not inconsistent with the ω mass.

REFERENCES

1. R. Brandelik et al., Phys. Lett. 80B, 412 (1979).
2. V. Lüth, Proceedings of the Ninth International Symposium on Lepton and Photon Interactions at High Energies, 1979, Ed. H. Abarbanel, T. Kirk; J. Kirby, Ibid.
3. There are a number of experiments utilizing emulsions to study such decays. See, for example: review by L. Voyvodic, Proceedings of the Ninth International Symposium on Lepton and Photon Interactions at High Energies, 1979, Ed. H. Abarbanel, T. Kirk; J. D. Prentice, Ibid; N. Ushida et al., submitted to Phys. Rev. Letters.
4. R. Ammar et al., "Production and Decay of $F^+(2030)$ Observed in ν_μ Interactions in Emulsion", Phys. Lett., to be published.
5. Note that track 6 has a very large δ-ray.

CHARMED PARTICLE PRODUCTION AND DECAY LIFETIMES AND A NEUTRINO OSCILLATION TEST

K. Niu
Nagoya University, Nagoya 464, Japan

ABSTRACT

This paper reports Fermilab Experiment 531, "A study of Weak Decay Lifetimes of Neutrino Produced Particles in a Tagged Emulsion Spectrometer". Lifetimes deduced from 81% of the materials scanned so far are presented for D^0, $D^{\pm}$, and $F^{\pm}$ charmed mesons, and Λ_c^{+} charmed baryons. Some of physical speculations are made basing on those observed lifetimes. Upperlimit of ν_μ-ν_τ oscillation rate is also estimated depending to the present statistics.

INTRODUCTION

A large collaboration of United States, Canadian, Korean and Japanese scientists in the list[1] attached to the end of this report, constructed and operated Fermilab Experiment 531, in which a hybrid emulsion spectrometer shown in Fig. 1. was exposed to a wide-band neutrino beam to study lifetimes of charmed particles.

While many of the first generation experiments with a hybrid emulsion counter apparatus[2] have suffered grievously lacking the efficient cooperation of emulsion with counter, we have obtained a large sample of charm decays applying a new emulsion technology developed for the purpose in addition to a complete counter effort.

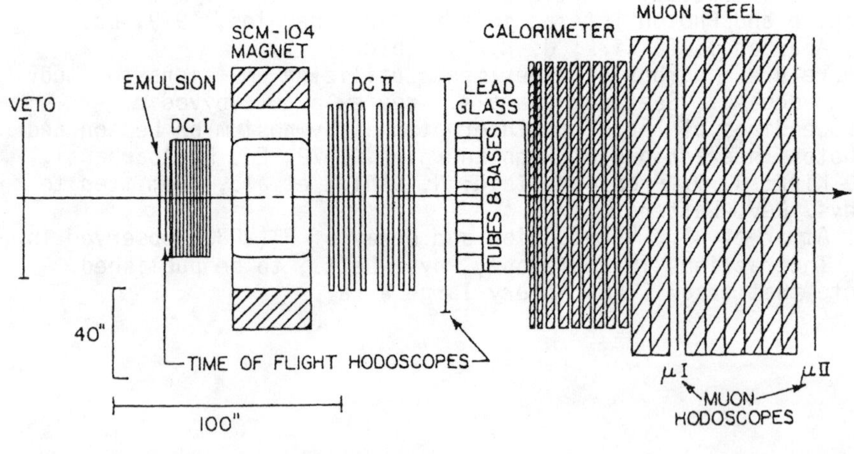

Fig. 1. Plan view of the experiment

EXPERIMENTAL PROCEDURE

As we wished to measure lifetimes of charmed particles, we used neutrinos for producing charm, emulsions to see the topology of the resulting decay with short flight length, and an electronic detector to specify event location and identify decay products. A magnet surrounded by drift chambers provides momentum analysis and aids in finding events within the emulsion. A time-of-flight system separates pions and kaons to 2.5 GeV and protons to 5 GeV. A wall of lead glass detectors is used to identify electromagnetic showers, and muons are identified by passage through 1.6 and/or 3 GeV equivalent of steel. Non-electromagnetic hadronic energy was recorded in a conventional iron calorimeter. Information on resolution of spectrometer elements is shown in Table I.

Table I Resolution of spectrometer elements

Quantity measured	Error (σ)
Drift chamber position resolution	120 microns
Charged particle momenta for Tracks passing through magnet	$\frac{\delta P}{P} = 0.13 + .005P$
Tracks analyzed only by fringe field	$\frac{\delta P}{P} \simeq 0.3P$
Gamma ray energy	$\frac{\delta E}{E} = \frac{0.3}{\sqrt{E(GeV)}}$
Gamma ray position	5 cm
Time-of-flight	120 picoseconds
Calorimeter	$\frac{\delta E}{E} = \frac{1.0}{\sqrt{E(GeV)}}$

This apparatus was exposed to the single horn focussed neutrino beam at Fermilab. A total of 7×10^{18} protons of 350 GeV/c momentum were incident on the production target. A simple neutrino trigger required only no charged particle be incident and that two or more charged particles exit the emulsion and pass through the magnet. Charged and neutral current events were accepted equally.

As for the performance of the counter system of our apparatus, it has been reported repeatedly in each symposium and conference[3]. Using counter data only, E-531 exhibits features typical of most bulk neutrino experiments in terms of the variables such as E_{vis}, x, and y. However, description of these distributions is skipped because of space-limitation.

NEW EMULSION TECHNIQUES

Before designing emulsion target which will be combined with counter spectrometer, our Nagoya Group has analysed the cause of low locating rate of tagged events in emulsion in the preceeding experiments with hybrid system[2]. The conclusion was that the deficit was inevitably connected with the use of volume scanning method to search events in a conventional pellicle stack. To overcome the deficit, we have developed new method as an extension of the emulsion chamber technique which has been successfully applied to the study of charmed particles in cosmic ray and accelerator energies since 1971[4]. There are 4 points in this new method.

The first is, adoption of thick emulsion films which are fablicated by coating both sides of a 70 micron polystyrene sheet with 330 micron of emulsion. Modules of 68 layers of such films stacked on four lucite posts are exposed perpendicularly to the beam as is shown in Fig. 2. Unlikely to the case of pure pellicle stack, doubly coated emulsion films are used as a position detector of tracks traversing the film base as well as analyser of 3 dimensional topology of the events in thick emulsion layers. By analogical expression, this is compared with a supermulti-layer counter assembly with spatial resolving power of submicron. This type of detector could have a better matching with counter array than a pellicle stack does.

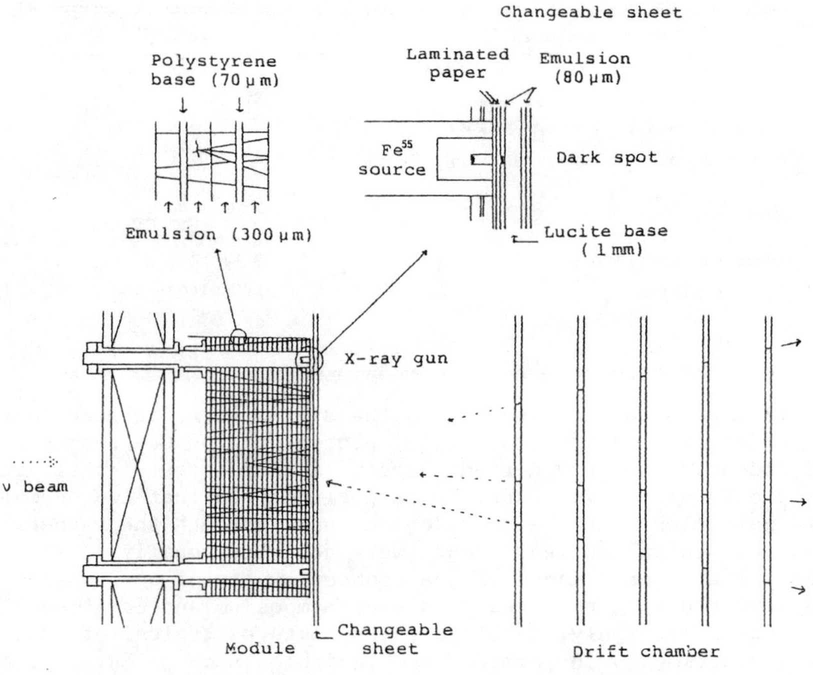

Fig. 2. Emulsion film chamber with changeable sheet

The second one is the tracing back method of secondary tracks predicted by the counters. The idea of the method is shown in Fig.2. Once an individual track from an event is found at the down stream surface of a module, it could be extrapolated to the interaction vertex employing doubly coated emulsion films. Such a technique is highly efficient because it doesn't rely on the existence of a large number of black tracks from nuclear breakup as does the volume scanning method.

This trace back method is also applicable for search of neutral decay away from original interaction vertex. Picking up at the downstream face of a module such a track that not seen at interaction vertex but tracked in a drift chamber spectrometer, one can easily trace back it to a neutral decay vertex no matter how far is it from the interaction point. This method is, however, could not be successfull without suppressing background density in the emulsion.

To realize the tracing back method under actual condition with high muon background, so called changeable sheet is introduced as the third new technique. That is an 800 micron sheet of lucite coated on both sides with a thin layer of emulsion placed immediately downstream of the emulsion target. An important function of this fiducial sheet is to serve as a low background interface to couple the drift chamber tracks to those seen in the emulsion module. As is shown in Fig. 2, the marks left on this sheet by the Fe^{55} sources imbedded in each mounting post record the relative positions of all modules. By changing this sheet many times during the experiment, muon and other backgrounds are kept sufficiently low to pick up a single minimum track predicted by drift chambers on this sheet.

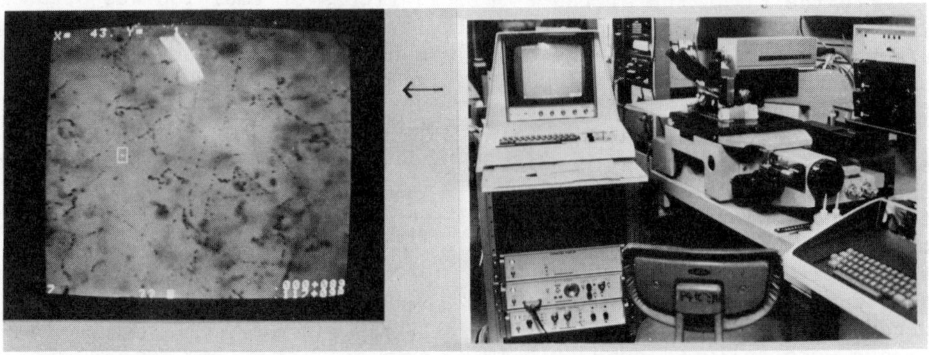

Fig. 3. Measuring microscope

The fourth one is using an epoch-making computer-controlled semiautomatic scanning technique to speed up location time of an event. It consists of CPU and several peripheral units. Among them there is a measuring microscope. This unit consists of a microscope, moving stage driven by a signal from CPU, ITV camera and console with CRT. This CRT plays a role of a video monitor of emulsion image as well as a character display as shown in Fig. 3. Measurement of a track co-ordinate is carried out on the CRT screen adjusting a measuring window by a joy stick. Relative co-ordinate of the most dark point inside the window to one of the corner of the TV image is measured electrically and the absolute co-ordinate is calculated by the CPU. Accuracy of this measurement is 0.1 μm or 1 μm over the working range of 10 cm × 10 cm.

X and Y coordinates of a grain spot is measured in a few seconds. Measuring coordinates of a track at 4 different depth, one can get x, y, θ and ϕ of a track in about 10 seconds, and those values are displayed on the TV screen to be compared with the predicted ones. Using this system, scanning of a track on a fiducial sheet is carried out in an area of 2.4 mm × 2.4 mm around the point predicted by the drift chambers. It takes less than 10 minutes per one predicted track. Efficiency of this scanning method is very high and about 96%. The found track is then followed into the most down-stream emulsion plane in the module using same measuring stage. It also takes 10 minutes. Then the track is extrapolated back to the interaction vertex employing highly precise emulsion techniques on ordinary microscope. It takes only each one minute to trace a track plate by plate. Mean location time of one neutrino event is about one hour. This machine has changed completely the image of emulsion work. Style of new emulsion work is more or less likened to the bubble chamber film analysis.

SCANNING SUMMARY

In the first run of the E-531, 23 liters of emulsion was organized into 27 modules of perpendicular type and 12 modules of pellicle stacks mounted on posts edge on to the beam. Up dated event finding summary is shown in Table II, disposing the data according to the finding method.

853 located events have been searched for charm decay candidates. Neutral decays were looked for by following back tracks into the emulsion as well as by volume scan in a cylinder, 1 mm long with a radius of 300 μm. For charged decays, all minimum-ionizing tracks were followed down to 6 mm.

In Table II, number of found multi-prong decays are also disposed according to the method of searching. Location efficiency, ratio of charms over found events and also charms over searched events by track follow back method are much higher than that by volume scanning method. To see the situation clearer, N_h distribution of the found events by the track follow method is shown in Fig. 4. Content of white stars is rather high about 27%, and in this part, 50% of charm decays were found associated. The white star is very hard to find by the volume scanning method. Now the cause

Table II Event finding summary

Group		I	II	III
Method		Track follow	Mix	Volume scan
Predicted		952	199	1046
After cut		750	161	832
Searched	A	548	151	718
Found	B	449	102	302
Multi-prong charm	C	22	1	7
B/A		0.82	0.68	0.42
C/B		0.049	0.010	0.023
C/A		0.040	0.007	0.010

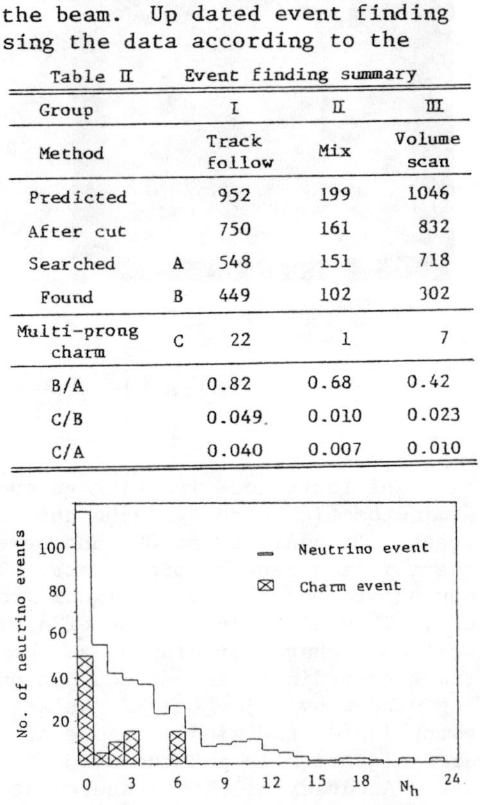

Fig. 4. N_h distribution

of very low locating efficiency of the neutrino events and low charm content in conventional pellicle stacks is made clearer.

MASS FITTING OF CHARMED PARTICLES

We have found so far 29 multiprong charm decay and about 30 kinks. These candidates have been fit to various decay hypothesis. Uniquely fitted results are listed in the following Table III. When a daughter has been identified by one of the particle identification systems with a confidence level of 95%, it appears in the Table underlined. Unseen particles added to balance momentum transverse to the direction of the parent particle are indicated in brackets.

Table III-a D^0 Fits

Event	Decay length (μm)	P_μ (GeV/c)	Hypothesis	P (GeV/c)	Mass (MeV)	Decay time ($\times 10^{-13}$ sec)
493- 177	326	-18.7	$D^0 \to \pi^+ \underline{\pi^-} (k^0)$	19.27		1.05
478-2638	126	- 4.8	$D^0 \to \underline{\pi^-} \pi^- \underline{\pi^+} \underline{\pi^+} \pi^- \pi^+ (\pi^0)$	9.1		0.86
486*-6857	256	not seen	$D^0 \to k^- \underline{\pi^+} \pi^- \pi^+ (\pi^0)$	12.8		1.24
513-8010	27.2	+11	$D^0 \to k^+ \pi^+ \underline{\pi^-} \underline{\pi^-} \underline{\pi^0}$	9.2	1766± 48	0.18
518-4935	116	- 4	$D^0 \to \pi^+ k^- \underline{\pi^0} \underline{\pi^0}$	30.1	1935±132	0.24
556*-152	41	-10	$D^0 \to \pi^- k^- \underline{\pi^+} \underline{\pi^+} \underline{\pi^0}$	15.4	1855± 43	0.17
577*-5409	67	-30	$D^0 \to \pi^+ \underline{\pi^-} (k^0)$	11.3		0.37
654-3711	6.5	- 4	$D^0 \to \pi^+ \underline{\pi^+} k^- \underline{\pi^-} \underline{\pi^-} \underline{\pi^+}$	19.2	1923± 46	0.021
661-6517	2647	-26	$D^0 \to k^- \underline{\mu^+} (\nu)$	22.8 32.7		7.20 4.24
670*-7870	187	+34	$D^0 \to k^+ \underline{\pi^-} (\pi^0)$	6.8 9.5		1.71 1.22

* These events have a D* with mass ∿ 2008 ± 3 MeV.

For the neutral decays, 10 events among 12 are consistent with D^0 decays. One of the remaining 2 decays took place when the magnet was off and could not be fitted now. The last one has a proton as a daughter of a neutral parent which decayed at 4.4 mm down, and a neutral charmed baryon hypothesis has been examining.

Results of fitting for 19 charged multiprong decays and 2 of kinks are as follows.

Two events are consistent only with F decays. Three semileptonic decays are consistent only with the $D^\pm$ hypothesis. Two events are somewhat ambignous in interpretation. One of these two fits both D and F with similar likelihood, but the event is used in the fit for the $D^\pm$ lifetime on the basis that F decay is Cabibbo Unfavored. For another event a Λ_c^+ hypothesis is possible but the confidence level is only 3%, and this decay is also used in the $D^\pm$ lifetime fit.

Five events have an identified baryon, either by time of flight or by reconstructions of a Λ^0, and fit a Λ_c^+ decay hypothesis. The remaining 9 events are not listed yet. Among them 5 are ambiguous between $F^+/D^+/\Lambda^+$ and are not used in the lifetime fit. Three events are not yet fully analysed. The last one event with a possible charge 2 baryon decay is also recorded during magnet was off and is still on the way of detailed analysis.

Table III-b F± Fits

Event	Decay length (μm)	P_μ (GeV/c)	Hypothesis	P (GeV/c)	Mass (MeV)	Decay time ($\times 10^{-13}$ sec)
527-3682	670	+30	$F^- \to \pi^- \pi^- \pi^+ \pi^0$	12.25	2026± 56	3.70
597-1851	130	not seen	$F^+ \to \underline{k^+} \underline{\pi^-} \underline{\pi^+} k^0$	9.70	2089±121	0.91

Table III-c D± Fits

Event	Decay length (μm)	P_μ (GeV/c)	Hypothesis	P (GeV/c)	Mass (MeV)	Decay time ($\times 10^{-13}$ sec)
512-5761	457	>150	$D^+ \to k^- \pi^+ \underline{\pi^+} \underline{\pi^0}$	10.1	1829± 35	2.82
546-1339	2145	- 7	$D^+ \to k^- \underline{\pi^+} \underline{\mu^+} (\nu)$	16.1		8.33
580-4508	2307	+ 7	$D^- \to \pi^- \underline{k^+} \underline{e^-} (\nu)$	9.4		15.3
598-1759	1802	-11	$D^+ \to k^- k^+ \pi^+ \pi^0$	17.0	1860± 25	6.60
663-7758	13000	>150	$D^+ \to k^- \pi^+ \underline{e^+} (\nu)$	118		6.86

Table III-d Λ_c^+ Fits

Event	Decay length (μm)	P_μ (GeV/c)	Hypothesis	P (GeV/c)	Mass (MeV)	Decay time ($\times 10^{-13}$ sec)
476-4449	27.7	-59	$\Lambda_c^+ \to \underline{P} \underline{\pi^+} \pi^- (k^0)$	2.9 5.0		0.73 0.42
549-4068	20.6	-11	$\Lambda_c^+ \to k^- \underline{P} \underline{\pi^+} (\pi^0)$	2.2		0.70
567-2596	175	- 5.4	$\Lambda_c^+ \to \underline{P} (k^0)$	6.5		2.10
610-4088	221	- 8	$\Lambda_c^+ \to \underline{\pi^+} \pi^- \pi^+ \underline{\Lambda^0}$	4.7	2382 ± 90	3.58
650-6003	40.6	-15	$\Lambda_c^+ \to \underline{\pi^+} \pi^- \pi^+ \underline{\Lambda^0}$	5.7	2192 ± 92	0.54

LIFE TIME

The lifetime for each charmed particle is obtained by maximizing a likelyhood function which is weighted by the finding efficiency dependent to the decay length. Fig. 5 shows finding efficiencies for neutral and charged particles. For decays which occur at distances less than 16 micron, tracks from the primary vertex obscure the decay point,

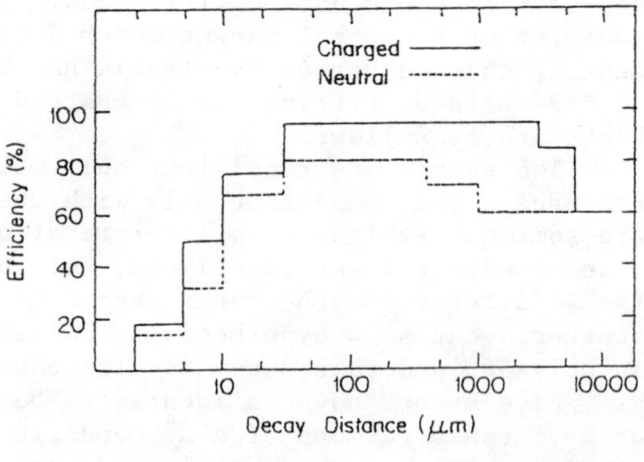

Fig. 5. Charm finding efficiency

thus reducing the efficiency. For neutral decays the finding efficiency inside the scan volume is approximately 80%. This is based on comparing found γ-conversions inside the scan volume to a Monte-Carlo prediction, and by a rescan of 420 events which had a track in the drift chambers not seen at the primary vertex. The latter method yielded 3 neutral events. Since all decay candidate could have been found by this technique, the finding efficiency outside the scan volume is 60%.

Table IV Lifetime summary

Particle	Lifetime ($\times 10^{-13}$ sec)	Events
D^0	$1.01^{+0.43}_{-0.27}$	10
$D^\pm$	$10.3^{+10.5}_{-4.1}$	5
$F^\pm$	$2.2^{+2.8}_{-1.0}$	2
Λ_c^+	$1.36^{+0.84}_{-0.46}$	5

The results of the lifetime fit are presented in Table IV, with $\pm 1\ \sigma$ limit. These values are still preliminary but are not expected to change beyond the $1\ \sigma$ limit. For D^0 statistic has increased and one lifetime has become assured. From the table we note that D^0, $F^\pm$ and Λ_c^+ are significantly shorter-lived than $D^\pm$. This suggests that process other than the simple radiative decay of the free charmed quark may be important in the calculation of decay rates for these particles.

SPECULATIONS AND A NEUTRINO OSCILLATION TEST

Basing on those lifetimes, we can make some of interesting estimations. For example, using lifetime of charged D meson obtained by our experiment and branching ratio of $D^+ \to X^0 e^+ \nu_e$ obtained at SLAC[5], one can estimate partial decay width of $D^+ \to X^0 e^+ \nu_e$ as $1.9^{+1.4}_{-1.0} \times 10^{11}$ /sec. Assuming that the D^+ and D^0 have the same semileptonic partial widths, and using the D^0 lifetime, we obtain the branching ratio of $D^0 \to X^- e^+ \nu_e$ as about 2%. If we relate pure loptonic width of $F \to \tau\nu$ to $\pi \to \mu\nu$, we can estimate branching ratio of $F \to \tau\nu$ as 1.6%.

Finally, depending of our statistics, we can estimate upperlimit of $\nu_\mu - \nu_\tau$ oscillation rate. So far we have observed 600 charged current events but no tau decays. As a signature of τ decays, we take events with no muon from vertex, and with charged trident decay or kink with P_T higher than 50 MeV/c and not with identified daughter as a proton. We observed no such signal. Correcting scanning efficiency (0.55) and considering cross-section ratio of ν_τ and ν_μ (0.6 for our beam energy), we obtain rate of $(\nu_\mu \to \nu_\tau)$ as less than 1.1% with 90% C.L. Taking average pass length (790m) and average neutrino energy (24GeV), we obtain the upper limit of $\delta m^2_{\mu\tau} = (m^2_{\nu_\tau} - m^2_{\nu_\mu})$ as a function of mixing angle $\theta_{\mu\tau}$ as shown in the Fig. 5. Any topology of τ decay could be observed in our experiment. So, this is the first result based on direct observation, and this indicates that the mixing of ν_μ and ν_τ may be very small.

In concluding, we will finish the analysis of the material from the first run in a few months, and are expecting still more 10 - 20 decays at that time. A second run will start in Dec. of this year and expect to find 70~100 more decays with improved particle identification. Therefore, new data expected before the next conference should further elucidate these questions and have promise of producing many interesting surprise.

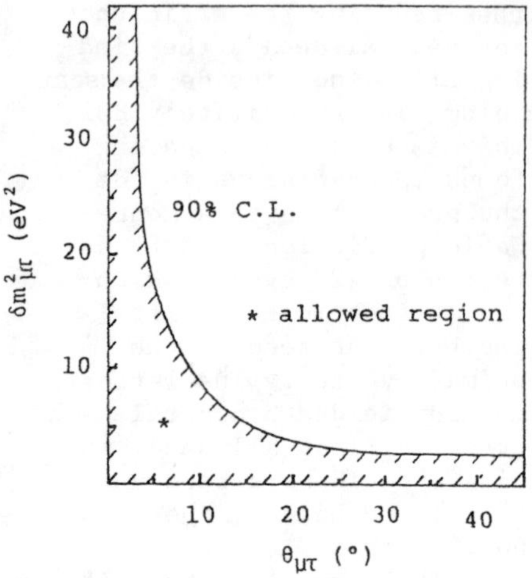

Fig. 6. Upper limit of $\delta m^2_{\mu\tau}$

ACKNOWLEDGMENT

This work has been supported in part by the National Science Foundation and Department of Energy of the United States, the Natural Sciences and Engineering Research Council of Canada, the Department of Education of the Province of Quebec, Canada, the Ashigara Research Laboratories of the Fuji Photo Film Co.,Ltd., the Mitsubishi Monsant Chemical Co.,Ltd., the Mitsubishi Foundation, the Nishina Memorial Foundation, the Nissan Scince Foundation, the Ministry of Education of Japan, the Japan Society for Promotion of Science, and finally, by the Japan-United States Cooperative Research Program.

REFERENCES

1. The list of E531 collaborators is as follows:
N.Ushida, T.Kondo, G.Fujioka, H.Fukushima, Y.Takahashi, S.Tatsumi, C.Yokoyama, Y.Homma, Y.Tsuzuki, S.Bahk, C.Kim, J.Park, J.Song, D.Bailey, S.Conetti, J.Fischer, J.Trischuk, H.Fuchi, K.Hoshino, M.Miyanishi, K.Niu, K.Niwa, H.Shibuya, Y.Yanagisawa, S.Errede, M.Gutzwiller, S.Kuramata, N.Reay, K.Reibel, T.Romanowski, R.Sidwell, N.Stanton, K.Moriyama, H.Shibata, T.Hara, O.Kusumoto, Y.Noguchi, M.Teranaka, H.Okabe, J.Yokota, J.Harnois, C.Hebert, J.Herbert, S.Lokanathan, B.McLeod, S.Tasaka, P.Davis, J.Martin, D.Pitman, J.Prentice, P.Sinervo, T.Yoon, H.Kimura, Y.Maeda.
2. L. Hand et al., Nucl. Inst. and Method 167, 261 (1979).
A. L. Read et al., Phys. Rev. 19, 1287 (1979).
D. Allasia et al., to be published in Nucl. Phys. (1980).
3. N. W. Reay, XXth Int. Cosmic Ray Conf, (Kyoto, 1979).
J. Prentice, Lepton Photon Symp, (Fermilab, 1979), 563.
N. W. Reay, Montreal Meeting of the APS (Div. of Particle and Field, 1979).
J. Trischuk, XVth Recontre de Moriond, Les Arcs, France (1980) and else where.
4. H. Fuchi et al., AIP conf. Proc. No49, Particles and Fields No.16 Cosmic Rays and Particle Physics-1978 (Bartol Conf.) 49 (1979).
K. Niu, ibid, 181.
5. $BR(D^+ \to X^0 e^+ \nu_e) = 22^{+4.4}_{-2.2}\%$ (DELCO), $16.8 \pm 6.4\%$ (MARK II).

PRODUCTION OF CHARM PARTICLES IN PROTON-EMULSION INTERACTIONS AT 400 GeV/c

Bombay-Chandigarh-Delhi-Jammu Collaboration

Presented by
P.K. Malhotra
Tata Institute of Fundamental Research, Homi Bhabha Rd., Bombay-400005

ABSTRACT

A total of 12 three-prong charm candidates have been recorded by following secondaries to one mm length in a total of 2500 proton-emulsion interactions at 400 GeV/c. The estimated background is 4. The signal implies an observed cross section of 43 ± 20 μb/nucleon. Attributing this to Λ_c^+, we obtain 120 ± 55 μb/nucleon for production cross section of Λ_c^+. Five associated events were also recorded.

STACK DETAILS

The Bombay-Chandigarh-Delhi-Jammu Collaboration has exposed three emulsion stacks to hadron beams at Fermilab and CERN to study production and decay of short-lived particles in hadron interactions. The details of the stacks are (a) 400 GeV/c protons, 600 μm Ilford K5 emulsion, 70 pellicles each of 15 cm x 10 cm dimensions (Fermilab), (b) 340 GeV/c π^-, 600 μm Fuji emulsion, 70 pellicles, 8 cm x 8 cm (CERN) and (c) 340 GeV/c π^-, 600 μm Ilford G5 emulsion, 30 pellicles, 19 cm x 5 cm (CERN).

We report here preliminary results from a total of 2500 proton-emulsion interactions at 400 GeV/c which have been scrutinised so far.

NEUTRAL CANDIDATES

The search for neutral charm candidates was carefully carried out partly under 600 X and partly under 1500 X magnification. The search extended to a distance of 1 mm downstream, 75 microns upstream, and a width of 150 microns throughout the depth of the emulsion pellicle containing the primary interaction. The candidates were checked for association with the primary interaction and whether a low energy electron was associated with the vertex of the candidate. For two-prong events, a cut of a minimum of 10 mrad opening angle was imposed to minimise contamination due to gamma rays. Furthermore, the prongs were followed to see if they could be due to e^+ or e^-. In this way a total of 38 candidates were recorded. The topological break-up is as follows:

No. of prongs	2	4	6	8
No. of candidates	35	1(+e)	1(+e?)	1

These events are being subjected to further scrutinisation and measurements.

CHARGED CANDIDATES

The search for charged charm candidates was carried out carefully following the shower (minimum ionising tracks) and grey tracks under 1500 X magnification. In this way a total of 44 charm candidates were recorded with the toplogical break-up being as follows:

No. of prongs	1	3	5
No. of candidates	30	12	2

The one-prong (kink) candidates require further work. The two five-prong candidates are consistent with the estimated background.

Regarding the 12 three-prong candidates, we estimate that they include a background of 4 due to three-prong secondary interactions and gammas overlapping a charged track. We are therefore left with a signal of 8 charm events associated with 2500 primary interactions. Assuming that the charm production in hadron-nucleus collisions is proportional to A, we estimate that this signal corresponds to an observed cross section, for charged charm particles decaying to three-prong topology, of

$$\sigma_{obs}(\text{charm-three-prongs}) = 43 \pm 20\ \mu b/\text{nucleon} \qquad (1)$$

Using a branching ratio of 0.57 for decay into three-prong toplogy, which is strictly valid only for $D^{\pm}$, we obtain

$$\sigma_{obs}(\text{charged charm particles}) = 74 \pm 34\ \mu b \qquad (2)$$

This has to be corrected for decay probability. The results on identified charm particles reported at this Conference imply mean life-times of $\tau(D^{\pm}) = 9 \times 10^{-13}$ s (6 events) and $\tau(\Lambda_c^+) = 1.4 \times 10^{-13}$ s (6 events). Since in our experiment the search is restricted to only 1 mm, it appears that our signal is mainly due to Λ_c^+(and possibly Σ_c^+). If we assume this, then we obtain the following value for the Λ_c^+ production cross section in 400 GeV/c proton-nucleus interactions

$$\sigma(\Lambda_c^+) = 120 \pm 55\ \mu b/\text{nucleon}. \qquad (3)$$

We have recorded a total of 5 associated events so far. These are (a) one three-prong associated with a one-prong (kink), (b) one three-prong associated with a V^0, (c) one kink associated with another kink and (d) two events in which two V^0's are associated.

The experiment is still in progress. Work is also continuing on the π^--emulsion stacks.

WEAK DECAYS, EXPERIMENT

G. Goldhaber and

G. Abrams, UC-Berkeley, Organizers

MEASUREMENT OF THE VECTOR AND AXIAL VECTOR COUPLING CONSTANTS IN $\Lambda^{0} \rightarrow p + e^{-} + \bar{\nu}$

D.A. Jensen, M.N. Kreisler, F. Lomanno,[a] R. Poster,[a]
M.S.Z. Rabin, K. Raychaudhuri, M. Way,[b] and J. Wise[c]
University of Massachusetts, Amherst, MA 01003

J. Humphrey
Brookhaven National Laboratory, Upton, NY 11973

ABSTRACT

Preliminary results based on 10^4 Λ_β events (the full sample contains 10^5 events!) are presented. The BR$(\Lambda \rightarrow pe\bar{\nu})$/BR$(\Lambda \rightarrow p\pi)$ is determined as $(1.318 \pm 0.024) \times 10^{-3}$ and $|g_1|/|f_1|$ is determined from a projection of the Dalitz Plot onto the $\cos\theta_{e\nu}$ axis to be $|g_1|/|f_1| = 0.734 \pm 0.031$. These values together with known factors yield $|f_1| = 1.229 \pm 0.034$ and $|g_1| = 0.903 \pm 0.046$.

A precise study of the process $\Lambda^{0} \rightarrow p + e^{-} + \bar{\nu}$, Λ_β decay, results in a detailed probe of the structure of the strangeness changing weak hadronic current thus complimenting the precise studies of neutron beta decay. This provides strong tests of the validity of the Cabibbo model[1] and allows a search for the existence of second class currents. Four form factors, defined below, determine the dynamics of Λ_β decay. The measurement of two of those form factors is the subject of this presentation.

The experiment was performed at Brookhaven National Laboratory with an apparatus that has been described elsewhere[2] and will therefore not be discussed here.

The event trigger was Λ_β + 2T/N1 + $K\pi\pi$/N2 where the 2T trigger was the fast electronics trigger requiring two tracks in the spectrometer, Λ_β was a trigger formed by requiring two threshold electron Čerenkov counters in coincidence and two pion threshold Čerenkov counters in the proton arm of the spectrometer in veto, and the $K\pi\pi$ trigger was a 2T trigger in coincidence with one of the pion threshold Čerenkov counters. N1 and N2 are trigger prescale factors. The $K\pi\pi$ triggers, along with other special runs, were used to study the efficiencies of all of the Čerenkov counters.

The analysis, after track reconstruction, proceeded as follows. First, the sample of $\Lambda p\pi$ events was studied in great detail and compared with a Monte Carlo simulation of the experiment. Because the $\Lambda p\pi$ events are overconstrained, they are particularly useful as a probe of one's understanding of the apparatus. Second, a sample of Λ_β events was studied. These too were compared with the Monte Carlo simulation of the apparatus. From that comparison, it

a) Brandeis University, Waltham, MA.
b) 321 Conestoga Road, Wayne, PA.
c) Computerized Biomechanical Analysis, Inc., Amherst, MA.
Presented by D.A. Jensen

was determined that there was a residual contamination of $\Lambda p\pi$ events of ~(2.5 ± 0.5)% and Ke3 of ~(0.7 ± 0.3)% in the Λ_β sample. These $\Lambda p\pi$ and Λ_β data were then compared to obtain the relative branching ratios for the two modes. That ratio is:

$$R = \frac{\Gamma(\Lambda^0 \to pe\overline{\nu})}{\Gamma(\Lambda \to p\pi^-)} = \frac{\Lambda_\beta(\text{Observed})}{\Lambda_{p\pi}(\text{Observed})} \times \frac{A_{p\pi}}{A_\beta} \times \frac{1}{\text{Prescale}}; \quad (1)$$

where $A_{p\pi}/A_\beta$ is the ratio of the spectrometer acceptances for the two decays as determined from the Monte Carlo. "Prescale" is the average prescale factor for the 2T trigger and equals 735.8. The result is

$$R = \frac{10{,}039}{25.450} * 2.45 * \frac{1}{735.8} = (1.318 \pm 0.024) \times 10^{-3}. \quad (2)$$

The 1.8% error in the result is determined from the statistics of the data (1.2%) and the Monte Carlo (0.6%), the uncertainty in the Čerenkov counter efficiencies (1.1%), and the uncertainties in the residual background subtractions (0.5 and 0.3%).

Using the current world average[3] for $\Gamma(\Lambda^0 \to p\pi)/\Gamma(\Lambda^0 \to \text{all}) = 0.642 \pm 0.005$, we present this measurement as a branching ratio:

$$BR = \Gamma(\Lambda^0 \to pe\overline{\nu})/\Gamma(\Lambda^0 \to \text{all}) = (0.846 \pm 0.017) \times 10^{-3}. \quad (3)$$

Finally, using the world average[3] for the Λ^0 lifetime of $(2.632 \pm 0.02) \times 10^{-10}$ sec, the absolute rate for Λ_β decay is calculated to be $(3.215 \pm 0.068) \times 10^6 \text{ sec}^{-1}$. This result represents a significant improvement over previous determinations[4,5,6,7] which were primarily limited due to statistics.

As a third and final phase of analysis (at this time), the ratio $|g_1|/|f_1|$ was determined. This was carried out as follows: the weak hadronic current for the Λ_β decay process may be written as:

$$J_\mu^{\ h} = \overline{\psi}_p \left\{ f_1(q^2)\gamma_\mu + \frac{i}{M} f_2(q^2)\sigma_{\mu\nu}q_\nu + g_1(q^2)\gamma_\mu\gamma_5 + \frac{i}{M} g_2(q^2)\sigma_{\mu\nu}q_\nu\gamma^5 \right\} \psi_\Lambda$$

where terms of order m_e/M in the amplitude have been dropped. In this expression the vector coupling constant is $f_1(q^2)$, the axial vector coupling is $g_1(q^2)$, the weak magnetism term is $f_2(q^2)$, and the "second class" current is $g_2(q^2)$. In the simplest form of the Cabibbo model,[1] $f_1(0) = \sqrt{3/2}$, $f_2(0)/f_1(0) = 0.97$, and $g_2(0) = 0$. Using standard techniques, one can derive[8,9] an expression for the differential rate as a function of the electron energy, Y, and the angle, $\theta_{e\nu}$, between the electron and the neutrino in the Λ^0 rest frame, $d^2W/(dY\, d(\cos\theta_{e\nu}))$.

The form factors used in the expression $d^2W/dY\, d(\cos\theta_{e\nu})$ were modified by radiative corrections following the prescription of Yokoo et al.[10] and the expected dipole q^2 dependence.[11] (A future paper will describe these calculations in more detail.) This differential decay rate was then used to generate Λ_β Monte Carlo events. The χ^2 comparison of Monte Carlo and data as a function of $|g_1|/|f_1|$ then yields a value for that ratio. This comparison of data vs. Monte Carlo in the center of mass was made after summing the Dalitz plot over Y to obtain the projection onto the cos $\theta_{e\nu}$ axis.

To extract values for the form factors, the events must be transformed to the Λ center of mass. This transformation is ambiguous since a kinematic analysis yields two solutions for the magnitude of the (unknown) Λ momentum. There are several possibilities for dealing with this problem but if Monte Carlo events and data are treated in the same way, the comparisons are valid and the extraction of the form factors will be correct. However, some techniques will yield more statistical power than others. The techniques chosen, based on Monte Carlo studies where the thrown Λ momentum is available, were to: a) choose the solution closest to the peak of the momentum spectrum, and for comparison b) take the average. In choice a) the correct solution is obtained ~70% of the time.

The value obtained for $|g_1|/|f_1|$ is 0.734 ± 0.020 (statistical error only) using method a) for obtaining p_Λ and 0.746 ± 0.026 using technique b). (Note that the statistical power of technique a) is ~1.7 times that of technique b).) The χ^2 = 17.2/19 DOF. The quoted result is the value obtained by method a).

Other sources of error (which will be discussed in greater detail in a later publication) include 1) effects due to $\Lambda_{p\pi}$ and Ke3 contamination (0.007), 2) sensitivity to the peak of the momentum spectrum used in deciding the best solution (0.001), 3) sensitivity to the momentum spectrum (0.004) and 4) Čerenkov efficiency uncertainty (0.002). Adding these statistical errors in quadrature yields a statistical uncertainty of 0.022.

There is one more possible source of systematic error. The effects of Λ polarization[12] have not yet been fully addressed. A simple test - that of dividing the data into a low momentum region where the Λ polarization is negligible, and a high momentum region where there is ~20% polarization - should be sensitive to polarization. These two samples yield $|g_1|/|f_1|$ values that differ by 2.3 standard deviations - a marginally significant effect. To be safe, we multiply the quoted error by $\sqrt{2}$ to yield a total error of 0.031. This systematic error will clearly be resolved by further analysis.

This result represents a significant improvement over previous excellent work which has been, in general, limited by statistics. For comparison, we mention that these results include values of $|g_1|/|f_1|$ of 0.73 +.1/-.08 (645 events)[13], 0.63 ± 0.06 (817 events,[14] and 0.79 +.15/−.13 (409 events).[15] A future paper will discuss overall comparisons in detail.

Using this result and our previous measurement of the absolute decay rate for Λ_β decay, we can extract the values of $|g_1|$ and $|f_1|$ themselves. To lowest order in $\Delta(\equiv (M_\Lambda - M_p)/M_\Lambda)$, the absolute decay rate, $R_{\Lambda\beta}$, can be expressed as:

$$R_{\Lambda\beta} = 1.45 \times 10^7 \sin^2\theta_c \quad (|f_1|^2 + 3\,|g_1|^2),$$

where θ_c is the Cabibbo angle ($\theta_c = 0.239 \pm 0.005$).[3] We then calculate the values of the form factors and find:

$$|f_1| = 1.229 \pm 0.035$$

$$|g_1| = 0.903 \pm 0.046$$

It is particularly interesting to compare the value of $|f_1|$ with the prediction of the simplest Cabibbo model in which $|f_1| = \sqrt{3/2} = 1.22$. The agreement is surprisingly good lending strong support to the supposition that the Λ_β decay process is well described by the current model of weak interactions.

We would like to thank F. Shoemaker of Princeton University for the loan of a large amount of equipment, the Director and staff of the A.G.S. for their assistance and continual encouragement, and B. Holstein for many discussions. Special thanks are due to M. Pieraccini and P. Smart for their technical assistance and to the University of Massachusetts Computer Center for their invaluable help. This work was supported in part by the National Science Foundation under grant NSF PHY78 22997.

REFERENCES

1. N. Cabibbo, Phys. Rev. Letters 10, 531 (1963).
2. J. Wise et al., Phys. Letters 91B, 165 (1980).
3. Particle Data Group, Reviews of Modern Physics, 75B, 79 (1978).
4. J. Lindquist et al., Phys. Rev. D16, 2104 (1977).
5. K.H. Altoff et al., Phys. Letters, 37B, 531 (1971).
6. J. Canter et al., Phys. Rev. Letters, 26, 868 (1971).
7. C. Katz, Ph.D. Thesis, University of Maryland, (1973) Unpublished.
8. B. Holstein and S. Treiman, Phys. Rev. C, 3 , 1921 (1971).
9. R. Abrams et al., ZGS Workshops, 369 (1971) (ANL/HEP-7208).
10. Y. Yokoo et al., Prog of Theoretical Phys., 50, 1894 (1973).
11. W.A. Mann et al., Phys. Rev. Letters 13, 844 (1973).
12. F. Lomanno et al., Phys. Rev. Letters 43, 1905 (1980).
13. C. Katz, Ph.D. Thesis, University of Maryland (1973) Unpublished.
14. K. Altoff et al., Phys. Letters 37B, 531 (1971).
15. J. Lindquist et al., Phys. Rev. D16, 2104 (1977).

MEASUREMENT OF THE DECAYS $\tau^- \to \rho^- \nu_\tau$ and $\tau^- \to K^{*-}(892)\nu_\tau$ USING THE MARK II DETECTOR AT SPEAR*

Jonathan Dorfan
Stanford Linear Accelerator Center
Stanford University, Stanford, California 94305

ABSTRACT

Measurement of the branching fractions for the Cabibbo favored decay $\tau^- \to \rho^- \nu_\tau$ and the Cabibbo suppressed decay $\tau^- \to K^{*-}(892)\nu_\tau$ are presented. The energy dependence of the $\tau^+\tau^-$ production cross section is measured using the decay $\tau^- \to \rho^- \nu_\tau$ which yields a measurement $M_\tau = (1790 \pm 40)$ MeV. A 2σ upper limit for the forbidden decay $\tau^- \to K^{*-}(1430)\nu_\tau$ is also presented.

INTRODUCTION

The heavy lepton is now a well established particle. Leptonic and hadronic branching fractions of the τ have been measured and agreement between the experimental determinations and the theoretical calculations is good.[1] This agreement supports the notion that decays of the τ are mediated by the W boson. We present herein measurements of the branching fractions for $\tau^- \to \rho^- \nu_\tau$ and $\tau^- \to K^{*-}(892)\nu_\tau$[2] which probe respectively the Cabibbo favored and Cabibbo suppressed weak hadronic coupling in τ decays. The data were obtained by the SLAC-LBL group using the Mark II detector at SPEAR.[3] We reported recently[4] a measurement of the branching fraction for the decay $\tau^- \to \rho^- \nu_\tau$ and the results presented herein represent our final measurement of $B(\tau^- \to \rho^- \nu_\tau)$ based on approximately four times as much data. Using eleven events (of which two are background) containing $K^{*\pm}(892)$ -- $\ell^\pm$ coincidences (the symbol ℓ stands for either an electron or a muon) we obtain the first measurement of a τ Cabibbo suppressed branching fraction.

MEASUREMENT OF THE BRANCHING FRACTION FOR $\tau^- \to \rho^- \nu_\tau$

The Mark II solenoidal detector is fully described in the articles listed under Ref. 5. The data used for this analysis come from all Mark II running at SPEAR with the exception of the ψ and ψ'.[6] These data span the center-of-mass energy range $3.52 \leq E_{c.m.} \leq 6.7$ GeV, and correspond to an integrated luminosity of 21,000 nb^{-1} or, using the theoretical $\tau^+\tau^-$ production cross section, 58,000 produced $\tau^+\tau^-$ pairs. In searching for the decay $\tau^- \to \rho^- \nu_\tau$, we attempt to isolate events arising from the decay sequence:

0094-243X/81/680368-05$1.50

---

* Work supported primarily by the Department of Energy under contracts DE-AC03-76SF00515 and W-7405-ENG-48.

$$e^-e^+ \to \tau^-\tau^+$$
$$\tau^+ \to \ell^+\nu_\ell\bar{\nu}_\tau$$
$$\tau^- \to \rho^-\nu_\tau$$
$$\rho^- \to \pi^-\pi^o$$
$$\pi^o \to \gamma\gamma$$

which results in two charged tracks and two photons in the detector. The lepton tag (ℓ^+) helps provide a clean signature for events containing τ's. The selected events have two oppositely charged tracks, one a pion and the other a lepton, and two photons with $E_\gamma \geq 100$ MeV. For the purposes of this analysis a pion is any charged particle which is not positively identified as a kaon, proton, electron or muon. The invariant mass of the two photons ($M_{\gamma\gamma}$) is formed and the photon energies of the π^0 candidates (those for which $80 \leq M_{\gamma\gamma} \leq 200$ MeV) are adjusted using the π^0 mass as a single constraint. These π^0's are then combined with the charged pion to form the $\pi^\pm\pi^0$ invariant mass ($M(\pi^\pm\pi^0)$) as shown in Fig. 1. An impressive $\rho^\pm$ signal is seen with very little background. Defining the signal region as $M(\pi^\pm\pi^0) \leq 1.25$ GeV/c^2 there are 215 ρe events and 137 $\rho\mu$ events. There are fewer $\rho\mu$ events because of the 700 MeV/c threshold of the muon tagging system. Background contributions of 12% due to hadronic events and 14% due to the decays $\tau \to A_1\nu_\tau$ and $\tau \to (4\pi)\nu_\tau$ are removed. In order to obtain $B(\tau^- \to \rho^-\nu_\tau)$ we need to know $B(\tau^- \to \ell^-\nu_\ell\bar{\nu}_\tau)$. This has been measured using 294 events containing an electron, a muon and no photons ($e\mu$ events). A small (12%) correction is made to these events for backgrounds arising from QED and hadronic processes. A Monte Carlo simulation program has been used to generate the detection efficiencies $\varepsilon_{\rho e} = 3.4\%$, $\varepsilon_{\rho\mu} = 2.1\%$ and $\varepsilon_{e\mu} = 7.1\%$. Combining these efficiencies with the number of corrected events and the number of produced $\tau^+\tau^-$ pairs yields:

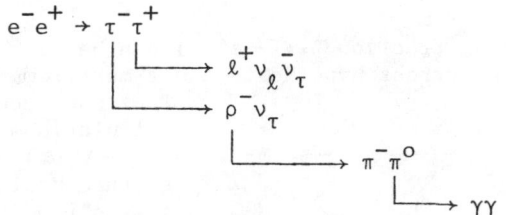

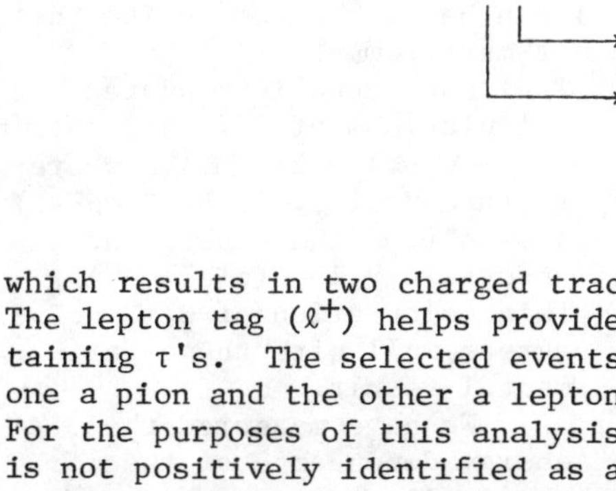

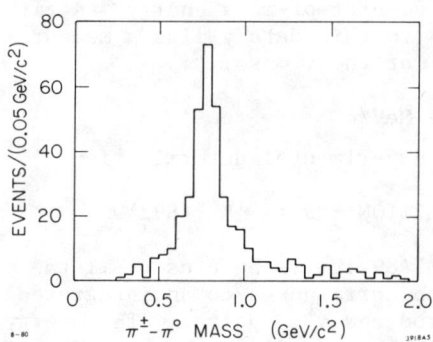

Fig. 1. The $\pi^+\pi^o$ invariant mass spectrum for $\ell^\pm\pi^\mp\pi^o$ events.

$$B(\tau^- \to \rho^-\nu_\tau)\, B(\tau^+ \to e^+\bar{\nu}_\tau\nu_e) = 0.034 \pm 0.004 \pm 0.007 \quad (1)$$

$$B(\tau^- \to \rho^-\nu_\tau)\, B(\tau^+ \to \mu^+\bar{\nu}_\tau\nu_\mu) = 0.041 \pm 0.005 \pm 0.007 \quad (2)$$

$$B(\tau^- \to e^-\nu_\tau\bar{\nu}_e)\, B(\tau^+ \to \mu^+\bar{\nu}_\tau\nu_\mu) = 0.030 \pm 0.002 \pm 0.007 \quad (3)$$

where the first error is statistical and the second is systematic.

Assuming μ-e universality in τ decays (3) can be combined with (1) and (2) to yield

$$B(\tau^- \to \rho^-\nu_\tau) = (21.6 \pm 1.8 \pm 3.2)\%$$

The branching fraction $B(\tau^- \to \rho^- \nu_\tau)$ can be calculated[7] using the conserved vector current hypothesis and a measurement of $e^+e^- \to \rho^0$. Tsai has recently[8] updated his calculations of 1971 and obtains $B(\tau^- \to \rho^- \nu_\tau) = 21.5 \pm 1.5$ where the uncertainly in $B(\tau^- \to \rho^- \nu_\tau)$ arises from the experimental uncertainly in $\Gamma(e^+e^- \to \rho^0)$. This calculation of $B(\tau^- \to \rho^- \nu_\tau)$ agrees well with the experimental result.

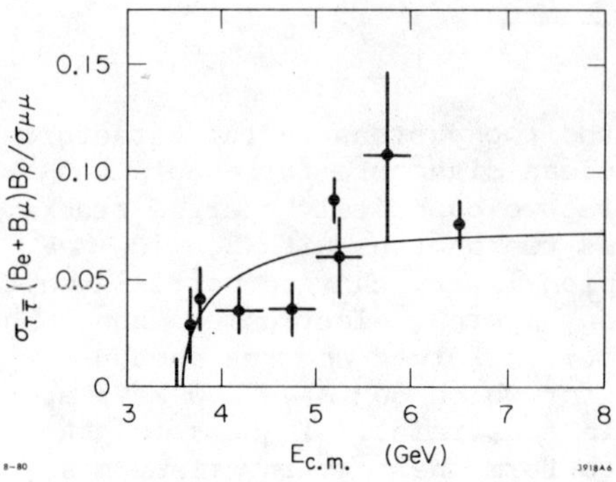

Fig. 2. The $\tau\bar{\tau}$ production cross section, normalized to the mu-pair cross section, as a function of $E_{c.m.}$.

We have measured the energy dependence of the $\tau^+\tau^-$ production cross section ($\sigma_{\tau\bar{\tau}}$) by studying the quantity $\sigma_{\tau\bar{\tau}} B\rho (B_e + B_\mu)/\sigma_{\mu\mu}$, where $\sigma_{\mu\mu}$ is the point cross section. Figure 2 shows a plot of this quantity as a function of center-of-mass energy. A fit to this data yields a measure of the τ mass:

$$M_\tau = (1790 \pm 40)\ \text{MeV}/c^2$$

which is in good agreement with other experimental determinations.[1]

MEASUREMENT OF THE BRANCHING FRACTION FOR $\tau^- \to K^{*-}(892)\nu_\tau$

For the study of the decay $\tau^- \to K^{*-}(892)\nu_\tau$ we have used all the Mark II data with $E_{c.m.} > 4.2$ GeV, which corresponds to an integrated luminosity of 14,600 nb^{-1} or 40,200 produced $\tau^+\tau^-$ pairs.[9] The energy threshold of 4.2 GeV is chosen to avoid those regions where production of D mesons could constitute a significant background. The decay $\tau^- \to K^{*-}\nu_\tau$ is identified via the topology:

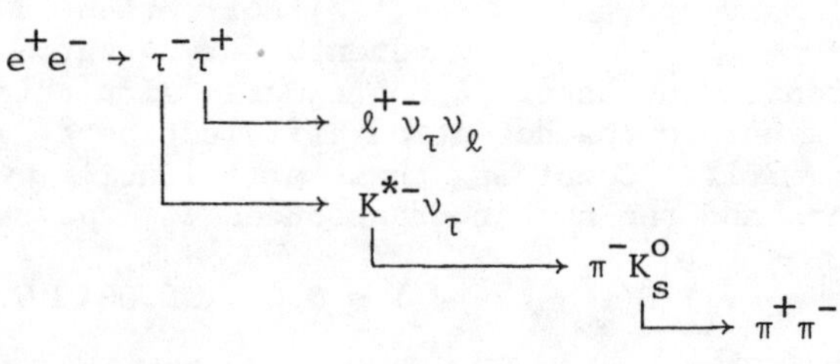

which results in four charged particles and no photons in the detector. Events are required to have four charged particles, two positive and two negative, and no photons with $E_\gamma \geq 100$ MeV. Two of the charged particles are required to form a secondary vertex, distant from the primary vertex by at least 1 cm, and have a mass consistent with the K^0_S. We further require that one of the particles at the primary vertex be a lepton; the other is hitherto referred to as a $\pi^\pm$.

Figure 3 shows the $K_S^0\pi^\pm$ invariant mass spectrum for the events selected as described above. A clear excess of events is present at the $K^*(892)$ mass and we attribute these $K^*(892)$-lepton coincidences to the decay topology outlined at the beginning of this section.

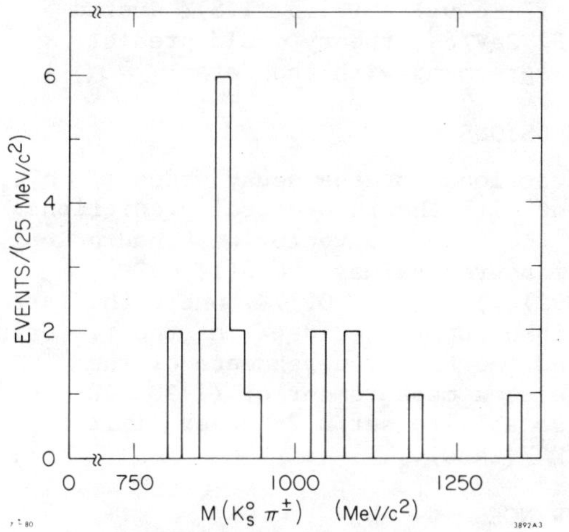

Fig. 3. $K_S^0\pi^\pm$ invariant mass spectrum for $K_S^0\pi^\pm\ell^\mp$ events.

From the eleven events in the five bins from 925-950 MeV/c^2 we have subtracted a background of two events obtained from the ten surrounding bins. The resulting signal is 9.0 ± 3.6 events. Of the eleven events, seven have an electron tag and four a muon tag which is consistent with the relative electron/muon tagging efficiency of 1.6. The momentum spectrum of the leptons is "hard" -- nine out of eleven leptons have momenta above 700 MeV/c. This momentum spectrum and that of the K^*'s is well reproduced by the Monte Carlo simulation program assuming a $\tau^+\tau^-$ production process.

Since D mesons are both a source of leptons and strange particles it is important to eliminate them as a source of the K^*-ℓ-no photon events. Charm events have a relatively high charged and neutral multiplicity and this, combined with the good solid angle coverage of the Mark II detector, make it very difficult for charmed events to populate the low multiplicity $K_S^0\pi^\pm\ell^\mp$ topology. A Monte Carlo program has been used to estimate the D background. Charmed particle production contributes at most 0.1 events to the $K^{*\pm}\ell^\mp$ sample. The leptons which come from semi-leptonic D decays, as simulated by the Monte Carlo program, have a momentum spectrum which is much "softer" than that observed in the $K^*\ell$ events. Almost all these leptons would have momenta below 700 MeV/c, in strong contrast to the data.

The detection efficiencies for the K^*-electron and K^*-muon events are 2.1% and 1.3% respectively. These efficiencies are combined with the number of produced $\tau^+\tau^-$ pairs and the number of signal events to yield

$$B(\tau^- \to K^{*-}(892)\nu_\tau)\ B(\tau^+ \to \ell^+\nu_\ell\bar{\nu}_\tau) = 0.0031 \pm 0.0013 \quad .$$

Using the Mark II measurement[4] of

$$B(\tau^- \to \ell^-\bar{\nu}_\ell\nu_\tau) = (18.5 \pm 1.5)\%$$

and assuming electron-muon universality in τ decays, we obtain

$$B(\tau^- \to K^{*-}(892)\nu_\tau) = (1.7 \pm 0.7)\% \quad .$$

We observe no $K^{*\pm}\ell^\mp$ events in the region of the $K^*(1430)$ which allows us to set a 2σ upper limit $B(\tau^- \to K^{*-}(1430)\nu_\tau) < 0.9\%$. This decay is, of course, forbidden in any V/A theory.

A detailed calculation by Tsai[7] yields

$$B(\tau \to K^{*} \nu_\tau) = \tan^2\theta_c \, B(\tau \to \rho \nu_\tau) \, F(M_\tau, M_{K^*}, M_\rho) \quad .$$

The function F accounts for phase space effects and is numerically equal to 0.93. Using the value $B(\tau^- \to \rho^- \nu_\tau) = (21.5 \pm 1.8)\%$ quoted earlier, $\tan^2\theta_c = 0.05$ and $M_\tau = 1.782$ GeV/c^2, theory would predict $B(\tau^- \to K^{*-} \nu_\tau) = (1.0 \pm 0.1)\%$ in good agreement with the data.

CONCLUSIONS

We find that the branching fractions for the decay modes $\tau^- \to \rho^- \nu_\tau$ and $\tau^- \to K^{*-} \nu_\tau$ are in good agreement with the theoretical predictions and hence support the notion that the standard vector weak hadronic coupling mediates τ decays. The measured values are $B(\tau^- \to \rho^- \nu_\tau) = (21.6 \pm 1.8 \pm 3.2)\%$ and $B(\tau^- \to K^{*-}(892) \nu_\tau) = (1.7 \pm 0.7)\%$, where the latter is the first measurement of a Cabibbo suppressed decay of the τ. Using the decay $\tau^- \to \rho^- \nu_\tau$ we have explored the energy dependence of the $\tau^+\tau^-$ production cross section which yields a measurement of (1790 ± 40) MeV/c^2 for the τ mass. We are also able to set a 2σ upper limit of 0.9% for the forbidden decay $\tau^- \to K^{*-}(1430) \nu_\tau$.

REFERENCES

1. See for instance G. Feldman, Proc. of the 19th International Conference on High Energy Physics, Tokyo (1978), p. 777.
2. The notation $\tau^- \to \rho^- \nu_\tau$, $\tau^- \to \ell^- \nu_\tau \bar{\nu}_\ell$ and $\tau^- \to K^{*-} \nu_\tau$ imply also the charge conjugate reactions.
3. Members of the SLAC-LBL collaboration: G. S. Abrams, M. S. Alam, C. A. Blocker, A. M. Boyarski, M. Breidenbach, D. L. Burke, W. C. Carithers, W. Chinowsky, M. W. Coles, S. Cooper, W. E. Dieterle, J. B. Dillon, J. Dorenbosch, M. W. Eaton, G. J. Feldman, M. E. B. Franklin, G. Gidal, G. Goldhaber, G. Hanson, K. G. Hayes, T. Himel, D. G. Hitlin, R. J. Hollebeek, W. R. Innes, J. A. Jaros, P. Jenni, A. D. Johnson, J. A. Kadyk, A. J. Lankford, R. R. Larsen, V. Lüth, R. E. Millikan, M. E. Nelson, C. Y. Pang, J. E. Patrick, M. L. Perl, B. Richter, A. Rousarrie, D. L. Scharre, R. H. Schindler, R. F. Schwitters, J. L. Siegrist, J. Strait, H. Taureg, M. Tonutti, G. H. Trilling, E. N. Vella, R. A. Vidal, I. Videau, J. M. Weiss and H. Zaccone.
4. G. S. Abrams <u>et al</u>., Phys. Rev. Lett. <u>43</u>, 1555 (1979).
5. The most thorough review of the detector appears in Vera Lüth's talk in the Proc. of the 1979 International Symposium on Lepton and Photon Interactions at High Energy, Batavia (1979), p. 78. Also G. S. Abrams <u>et al</u>., Phys. Rev. Lett. <u>43</u>, 477 (1979).
6. The results presented in this section are more fully discussed in Craig Blocker's Ph.D. Thesis, LBL-10801, April 1980.
7. M. B. Thacker and J. J. Sakurai, Phys. Lett. <u>36B</u>, 103 (1971); Y.-S. Tsai, Phys. Rev. <u>D4</u>, 2821 (1971).
8. Y.-S. Tsai, SLAC-PUB-2450 (to be published in the Proc. of the Guangzhou Conference on Theoretical Particle Physics, January 5-14, 1980).
9. This analysis is more fully described in SLAC-PUB-2566 (submitted to Phys. Rev. Lett.).

RECENT RESULTS ON DECAYS OF D MESONS FROM MARK II

SLAC-LBL Mark II Collaboration*

Presented by A. J. Lankford
Lawrence Berkeley Laboratory
University of California
Berkeley, CA 94720

ABSTRACT

Recent results on the decay of charmed D mesons obtained with E_{cm} = 3.771 GeV at SPEAR using the Mark II detector are presented. Results on semileptonic branching fractions and decays into two-body final states are emphasized and suggest enhancement in nonleptonic D^0 decays relative to nonleptonic D^+ decays.

INTRODUCTION

In order to study decays of charmed D mesons, an integrated luminosity of 2850 nb^{-1} was accumulated on the $\psi''(3770)$ resonance[1] at $E_{cm} = 3.771 \pm 0.001$ GeV with the Mark II detector at SPEAR. The ψ'' resonance provides enhanced charm production above the thresholds for $D^0\bar{D}^0$ and D^+D^- production but below the thresholds for production of excited D mesons, providing quasi-two-body production of $D\bar{D}$ with well-defined momenta. The momenta involved allow excellent particle identification by time-of-flight techniques, and kinematic constraints further reduce backgrounds. The Mark II detector and particle identification techniques are described in Ref. 2.

Results of studies of exclusive final states of Cabibbo-favored decays of both D^0 and D^+ are presented in Table I. Exclusive channels with the greatest statistics ($D^0 \to K^-\pi^+$ or $K^-\pi^+\pi^-\pi^+$, and $D^+ \to K^-\pi^+\pi^+$) along with constraint to the beam energy provide improved mass values for D^0 (1863.8 ± 0.5 MeV) and D^+ (1868.4 ± 0.5 MeV), where both masses have an additional uncertainty of ± 2.45 MeV due to uncertainty in the beam energy. The mass difference ($D^+ - D^0$) is (4.7 ± 0.3) MeV. Observed D^0 and D^+ decays in modes with low background ($D^0 \to K^-\pi^+$ or

---

*The analysis presented in this talk is predominantly the work of R. H. Schindler. The members of the SLAC-LBL Collaboration are: G. S. Abrams, M. S. Alam, C. A. Blocker, A. P. Blondel, A. M. Boyarski, M. Breidenbach, D. L. Burke, W. C. Carithers, W. Chinowsky, M. W. Coles, S. Cooper, W. E. Dieterle, J. B. Dillon, J. Dorenbosch, J. M. Dorfan, M. W. Eaton, G. J. Feldman, M. E. B. Franklin, G. Gidal, G. Goldhaber, G. Hanson, K. G. Hayes, T. M. Himel, D. G. Hitlin, R. J. Hollebeek, W. R. Innes, J. A. Jaros, P. Jenni, A. D. Johnson, J. A. Kadyk, A. J. Lankford, R. R. Larsen, M. E. Levi, V. Lüth, R. E. Millikan, M. E. Nelson, C. Y. Pang, J. F. Patrick, M. L. Perl, B. Richter, A. Roussarie, D. L. Scharre, R. H. Schindler, R. F. Schwitters, J. L. Siegrist, J. Strait, H. Taureg, M. Tonutti, G. H. Trilling, E. N. Vella, R. A. Vidal, I. Videau, J. M. Weiss, and H. Zaccone.

Table I. Branching fractions for Cabibbo-favored D^0 and D^+ decays.

Decay mode	Signal	$\sigma \cdot B$ (nb)	B (%)
$K^-\pi^+$	263.0 ± 17.0	0.24 ± 0.02	3.0 ± 0.6
$\bar{K}^0\pi^0$	8.5 ± 3.7	0.18 ± 0.08	2.2 ± 1.1
$\bar{K}^0\pi^+\pi^-$	32.0 ± 7.7	0.30 ± 0.08	3.8 ± 1.2
$K^-\pi^+\pi^0$	37.2 ± 10.0	0.68 ± 0.23	8.5 ± 3.2
$K^-\pi^+\pi^+\pi^-$	185.0 ± 18.0	0.68 ± 0.11	8.5 ± 2.1
$K^{*-}\pi^+$		0.26 ± 0.08	3.3 ± 1.2
$\bar{K}^{*0}\pi^0$		$0.11^{+0.16}_{-0.11}$	$1.4^{+2.1}_{-1.4}$
$K^-\rho^+$		$0.58^{+0.21}_{-0.22}$	$7.2^{+2.9}_{-3.0}$
$\bar{K}^0\rho^0$		$0.01^{+0.03}_{-0.01}$	$0.1^{+0.3}_{-0.1}$
$\bar{K}^0\pi^+$	35.7 ± 6.7	0.14 ± 0.03	2.3 ± 0.7
$K^-\pi^+\pi^+$	239.0 ± 17.0	0.38 ± 0.05	6.3 ± 1.5
$\bar{K}^0\pi^+\pi^0$	9.5 ± 5.5	0.78 ± 0.48	12.9 ± 8.4
$\bar{K}^0\pi^+\pi^+\pi^-$	21.0 ± 7.0	0.51 ± 0.18	8.4 ± 3.5
$K^-\pi^+\pi^+\pi^0$	6.4 ± 3.9	1.2 ± 0.9	20 ± 15
$K^-\pi^+\pi^+\pi^-\pi^+$	<11.5	<0.23	<4.1 at 90% C.L.
$\bar{K}^{*0}\pi^+$		<0.26	<4.6 at 90% C.L.

$K^-\pi^+\pi^-\pi^+$ and $D^+ \rightarrow K^-\pi^+\pi^+$ or $\bar{K}^0\pi^+$) provide 536 D^0 tags and 330 D^+ tags with 11% background for inclusive studies of D decays. Studies of charged multiplicity in D^0 and D^+ decay have been reported previously.[3]

SEMILEPTONIC D^+ AND D^0 DECAYS

D^+ and D^0 semileptonic branching fractions were directly measured by studying inclusive electron production in the tagged D^+ and D^0 event sample. Electrons were identified using time-of-flight and liquid argon shower counter information. Residual hadronic background (about half the raw electron sample) was accurately estimated from the observed spectrum of candidate tracks and the hadron misidentification rates measured using pions from K^0_S and $\psi(3095)$ decays. Residual electromagnetic background was determined by observed electrons with charge opposite that expected from the tag. After background subtraction, significantly more electrons are observed in tagged D^+ events (23.3±6.7

e^+ in 295 ± 19 D^+ tags) than in the D^0 sample (12.3 ± 7.6 e^+ in 477 ± 24 D^0 tags). The measured D^+ and D^0 semileptonic branching fractions are $B^+(D^+ \rightarrow Xe^+\nu) = (16.8 \pm 6.4)\%$ and $B^0(D^0 \rightarrow Xe^+\nu) = (5.5 \pm 3.7)\%$. The associated multiplicity distribution in the D^+ sample determines that semileptonic D^+ decays are composed of $(68 \pm 28)\%$ $D^+ \rightarrow \bar{K}^0 e^+\nu$. Since the $\Delta I = 0$ nature of semileptonic D decays dictates equal semileptonic decay rates for D^+ and D^0 decays, the ratio of semileptonic branching fractions equals the ratio of D^+ and D^0 lifetimes. The result from a maximum likelihood fit:

$$\frac{\tau^+}{\tau^0} = \frac{B^+}{B^0} = 3.1^{+4.6}_{-1.4}$$

is two sigma different from equal D^+ and D^0 lifetimes and is consistent with the D^+ lifetime being significantly longer than the D^0 lifetime.[4]

The D^+ semileptonic decay rate $(1.7^{+1.0}_{-1.6}) \times 10^{11}$ sec, deduced from the semileptonic branching fraction reported here and the D^+ lifetime measured in emulsion,[5] is consistent with naive and QCD corrected spectator quark ideas. The much smaller D^0 semileptonic branching fraction and shorter D^0 lifetime suggest enhancement of nonleptonic D^0 decays. Decays to two-body and Cabibbo-forbidden final states probe the dynamics of this enhancement. (For theoretical discussion, see N. Cabibbo, "Decay of Charm," Session B-8 at this Conference.)

DALITZ PLOT ANALYSIS OF THREE-BODY D^0 AND D^+ DECAYS

Dalitz plots for the three-body decays $D^0 \rightarrow \bar{K}^0\pi^+\pi^-$, $D^0 \rightarrow K^-\pi^+\pi^0$, and $D^+ \rightarrow K^-\pi^+\pi^+$ are shown in Fig. 1a-c without efficiency corrections. The three-body decay $D^+ \rightarrow \bar{K}^0\pi^+\pi^0$ has poor statistics; consequently, that Dalitz plot has not been studied. The data on each Dalitz plot was fit using maximum-likelihood techniques to an efficiency-corrected density function representing allowed final states involving K^* and ρ Breit-Wigner amplitudes, nonresonant three-body decays, and background whose distribution and magnitude were estimated from a low-mass sideband. Figures 1d-f show projections of the Dalitz plots on dominant axes. The solid curves are the smoothed solutions to the fits.

The $D^0 \rightarrow \bar{K}^0\pi^+\pi^-$ decay mode (52 events with $33 \pm 9\%$ background) is dominated by the $K^{*-}\pi^+$ contribution ($70^{+14}_{-16}\%$). The $\bar{K}^0\rho^0$ contribution ($2^{+9}_{-2}\%$) is very small, and the nonresonant contribution is $30^{+17}_{-14}\%$. The $D^0 \rightarrow K^-\pi^+\pi^0$ mode (56 events with $43 \pm 9\%$ background) is dominated by the $K^-\rho^+$ contribution ($85^{+10}_{-15}\%$). Small contributions arise also from $\bar{K}^{*0}\pi^0$ ($11^{+16}_{-11}\%$), from $K^{*-}\pi^+$ ($7^{+8}_{-4}\%$), and from nonresonant ($6^{+8}_{-6}\%$) final states. Branching fractions for these two-body D^0 modes are included in Table I.

The Dalitz plot of the $D^+ \rightarrow K^-\pi^+\pi^+$ decay mode (292 events with $12 \pm 2\%$ background) is neither uniformly populated[6] nor dominated by a $\bar{K}^{*0}\pi^+$ contribution. Figure 1f shows the superposition of the two $M^2_{K^-\pi^+}$ projections along with results of unsuccessful fits to pure three-body phase space (solid curve) and to phase space plus 15% $\bar{K}^{*0}\pi^+$ (dashed curve). The assumption that all mass combinations in the $\bar{K}^{*0}$ mass band arise from $\bar{K}^{*0}$ final states leads to the very conservative upper limit of 0.39 at 90% C.L. (with ± 0.06 additional systematic error)

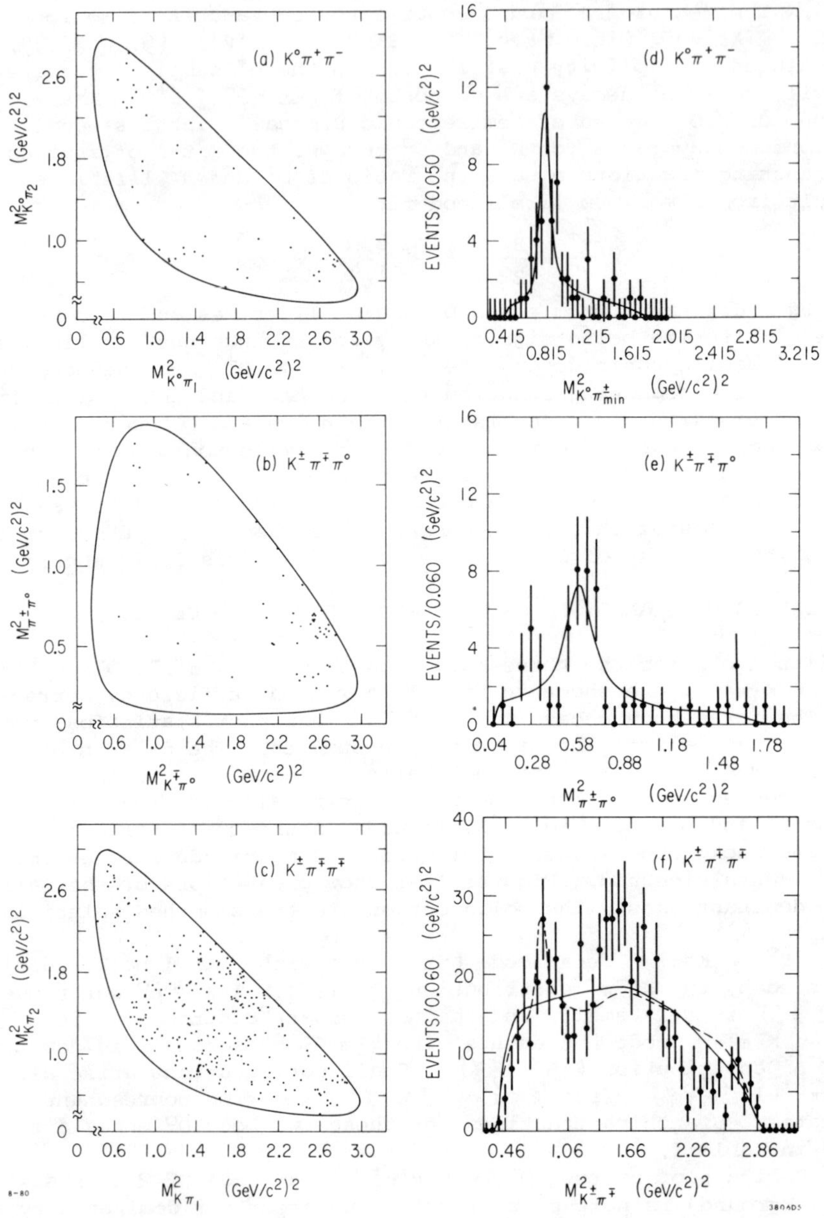

Fig. 1. Dalitz plots (without efficiency corrections) for final states (a) $K^0\pi^+\pi^-$, (b) $K^-\pi^+\pi^-$, and (c) $K^-\pi^+\pi^+$, and (d-f) projections onto the axes of dominant channels. See text for explanation of curves.

for the fractional contribution of the decay $D^+ \to \bar{K}^{*0}\pi^+$ to the observed three-body decay $D^+ \to K^-\pi^+\pi^+$.

The decay rates for D^0 and D^+ decays into two-body final states, deduced from studies of Dalitz plots and of branching fraction measurements for three-body decays, are expressed as ratios in Table II.

Table II. Two-body D^0 and D^+ decay rates.

Relative D^0 rates	Relative D^+ and D^0 rates
$\frac{\Gamma(D^0 \to \bar{K}^0\pi^0)}{\Gamma(D^0 \to K^-\pi^+)} = 0.75 \pm 0.34$	$\frac{\Gamma(D^+ \to \bar{K}^0\pi^+)}{\Gamma(D^0 \to K^-\pi^+)} = (0.77 \pm 0.28)\frac{\tau^0}{\tau^+}$
$\frac{\Gamma(D^0 \to \bar{K}^{*0}\pi^0)}{\Gamma(D^0 \to K^{*-}\pi^+)} = 0.44^{+0.65}_{-0.44}$	$\frac{\Gamma(D^+ \to \bar{K}^{*0}\pi^+)}{\Gamma(D^0 \to K^{*-}\pi^+)} < 1.7\frac{\tau^0}{\tau^+}$ at 90% C.L.
$\frac{\Gamma(D^0 \to \bar{K}^0\rho^0)}{\Gamma(D^0 \to K^-\rho^+)} = 0.01^{+0.04}_{-0.01}$	$\frac{\Gamma(D^+ \to \bar{K}^0\rho^+)}{\Gamma(D^0 \to K^-\rho^+)}$ --- see text

The $K\pi$ and $K^*\pi$ systems show (1) enhancement of D^0 rates relative D^+ comparable to the ratio of total widths and (2) ratios of $\bar{K}^0\pi^0$ ($\bar{K}^{*0}\pi^0$) to $K^-\pi^+$ ($K^{*-}\pi^+$) final states consistent with 1/2 and with no color suppression. These observations are in general agreement with enhancement of $I = 1/2$ final states relative to $I = 3/2$. On the other hand, the $K\rho$ system appears to be distinctly different from the $K\pi$ and $K^*\pi$ systems. The ratio of $\bar{K}^0\rho^0$ to $K^-\rho^+$ final states is much smaller than 1/2. The $\Delta I = 1$ nature of the weak-hadronic current along with the measured branching fraction $(12.9 \pm 8.4)\%$ for $D^+ \to \bar{K}^0\pi^+\pi^0$ implies that (1) this decay must be dominated by $D^+ \to \bar{K}^0\rho^+$, and (2) the decay rate for $D^+ \to \bar{K}^{*0}\pi^+$ must be much less than the rate for $D^+ \to \bar{K}^0\rho^+$.

CABIBBO-SUPPRESSED D^0 AND D^+ DECAYS

Studies of inclusive kaon rates in D^0 and D^+ decays using tagged events are summarized in Table III. In exclusive channels, Cabibbo-

Table III. D^0 and D^+ branching fractions to Cabibbo-favored ($D \to K^-$) and Cabibbo-suppressed ($D \to K^+$) charged kaons and to neutral kaons (indistinguishable strangeness).

	D^0		D^+	
Decay mode	#(events-bkgd)	BR (%)	#(events-bkgd)	BR (%)
$D \to K^-$	113 - 4	56 ± 11	26 - 3	19 ± 5
$D \to K^+$	25 - 9	8 ± 3	12 - 5	6 ± 4
$D \to \bar{K}^0$ or K^0	17 - 4	29 ± 11	18 - 3	52 ± 18

suppressed decays of $D^o \rightarrow K^-K^+$ and $\pi^-\pi^+$ have been previously reported.[7] Table IV summarizes these results and similar studies of D^o and D^+ decays of other Cabibbo-suppressed final states. Cabibbo-suppressed decays of both D^o and D^+ into "wrong sign" kaons are consistent with expectations of $\sim 5\%$ ($\tan \theta_c = 0.05$), as are limits on relative rates for three-body Cabibbo-suppressed D^+ decays to $K^-K^+\pi^+$ and $\pi^-\pi^+\pi^+$ final states. A fraction larger than $\sim 5\%$ is allowed for Cabibbo-suppressed decays containing no kaon, particularly for D^+ decays, and for two-body final states in D^+ decays; however, large statistical errors prevent firm conclusions.

Table IV. Relative branching fractions for Cabibbo-suppressed D^o and D^+ decays. (All limits are at 90% C.L.)

Decay mode	Events	Relative efficiency	Relative branching fraction
$D^o \rightarrow K^-\pi^+$	234.5 ± 15.8	1.00	1.00
$\rightarrow K^-K^+$	22.1 ± 5.2	0.84	0.113 ± 0.030
$\rightarrow \pi^-\pi^+$	9.3 ± 3.9	1.19	0.033 ± 0.015
$D^+ \rightarrow \bar{K}^o\pi^+$	31.5 ± 5.8	1.00	1.00
$\rightarrow K^oK^+$	5.6 ± 3.0	0.71	0.25 ± 0.15
$\rightarrow \pi^o\pi^+$	< 7.5	1.03	<0.30
$D^+ \rightarrow K^-\pi^+\pi^+$	239.0 ± 17.0	1.00	1.00
$\rightarrow K^-K^+\pi^+$	8.7 ± 4.0	0.56	<0.140
$\rightarrow \pi^-\pi^+\pi^+$	8.3 ± 9.2	1.12	<0.084
$D^o \rightarrow K^-\pi^+\pi^+\pi^-$	185.0 ± 18.0	1.00	1.00
$\rightarrow \pi^-\pi^+\pi^+\pi^-$	14.1 ± 8.3	1.28	<0.210

SUMMARY

Table I presents branching fractions for Cabibbo-favored D^o and D^+ decays. Including semileptonic decays, a considerable fraction of all D decays, particularly D^+ decays, have now been identified. Measurement of the individual semileptonic branching fractions of D^o and D^+ demonstrates a difference in D^o and D^+ lifetimes, which suggests enhancement of nonleptonic D^o decays. Measurements of Cabibbo-suppressed decays allow a similar enhancement of nonleptonic Cabibbo-suppressed D^+ decays but do not demonstrate such an enhancement. Studies of two-body $K\pi$ and $K^*\pi$ systems of final states suggest enhancement of $I = 1/2$ final states in D decay relative to $I = 3/2$; however, the $K\rho$ system appears substantially different. We look forward to an understanding of the dynamics of charmed quark decays and to future studies of D decay with substantially greater statistics.

This work was supported primarily by the U. S. Department of Energy under Contract numbers W-7405-ENG-48 and DE-AC03-76SF00515.

REFERENCES

1. R. H. Schindler et al., Phys. Rev. D21, 2716 (1980).
2. G. S. Abrams et al., Phys. Rev. Lett. 43, 477 (1979).
3. V. Lüth, Proceedings of the IX International Symposium on Lepton and Photon Interactions at High Energies, Batavia, IL, 1979.
4. W. Bacino et al., SLAC-PUB-2500 (1980); submitted to Phys. Rev. Lett.
5. N. Ushida et al., DOE/ER/01545-278 (1980).
6. Acceptance over this Dalitz plot (Fig. 1c) is uniform, except at the corners of the plot where momenta are less than 100 MeV/c.
7. G. S. Abrams et al., Phys. Rev. Lett. 43, 481 (1979).

MEASUREMENT OF INCLUSIVE η PRODUCTION IN e^+e^- INTERACTIONS NEAR CHARM THRESHOLD*

F. C. Porter
(Representing the Crystal Ball Collaboration)[1]
Stanford Linear Accelerator Center
Stanford University, Stanford, California 94305
and
California Institute of Technology
Pasadena, California 91125

ABSTRACT

We have measured the inclusive cross section for η production in e^+e^- interactions near charm threshold using the Crystal Ball detector at SPEAR. By comparing the inclusive η production above and below charm threshold we obtain the limits: $R(e^+e^- \rightarrow F\bar{F}X)\,BR(F \rightarrow \eta x) < 0.3$ (90% C.L., $E_{c.m.} < 4.5$ GeV); $BR(D \rightarrow \eta x) < 0.13$ (90% C.L., averaged over charged and neutral D components of the ψ'').

INTRODUCTION

It is expected that the charmed-strange F-meson has a significant branching fraction to η's.[2] As early as 1977, the DASP collaboration found a strong threshold in inclusive η production in e^+e^- interactions at $E_{c.m.} \sim 4.4$ GeV, which they interpreted as evidence for production of the F meson.[3,4] In this paper, we present a higher statistics measurement of inclusive η production as a function of $E_{c.m.}$ at similar energies using the Crystal Ball detector at SPEAR. No evidence for strong thresholds in η production is found.

METHOD

The Crystal Ball detector has been described elsewhere,[5] so only a summary of the relevant parameters is presented here. This detector consists primarily of a segmented array of NaI(Tℓ) crystals for high-resolution ($\sigma_E/E \sim 2.6\%/[E(\text{GeV})]^{1/4}$) measurement of electromagnetic showers. The solid angle coverage is 94% of 4π steradians with the main array, which is extended to 98% of 4π with crystals in the endcap regions. The angular resolution for photons is $1-2^{\circ}$, depending on energy. In addition, there are spark and proportional chambers for charged particle detection and tracking (no magnetic field).

Table I summarizes the data used in the present inclusive measurement. The data for $E_{c.m.} = 3.878$ to 4.500 GeV are divided into bins in $E_{c.m.}$ as motivated by the structure in R (Fig. 1). The hadron selection efficiency is estimated at $(93 \pm 5)\%$.

The number of η's at each energy was extracted by performing fits to the $m_{\gamma\gamma}$ mass plots for the $\eta \rightarrow \gamma\gamma$ decay. To be included in the mass plots, photons had to satisfy the following conditions:

---

* Supported in part by the Department of Energy, contracts DE-AC03-76SF00515 and DE-AC03-79ER0068, and by a Chaim Weizmann Fellowship.

Table I

Data Sample for Inclusive η Measurement

$E_{c.m.}$ GeV	$\int$Ldt nb^{-1}
Fixed Points	
J/ψ (3.095) [a]	110
3.670	500
ψ' (3.684) [a]	290
ψ" (3.77)	1700
4.028	840
5.2	6700
Scan Data	
3.878-4.004	370
4.005-4.082	850
4.083-4.142	1700
4.143-4.225	2100
4.226-4.300	1100
4.301-4.364	820
4.365-4.500	1500

[a]The full data samples at J/ψ, ψ' are considerably larger.

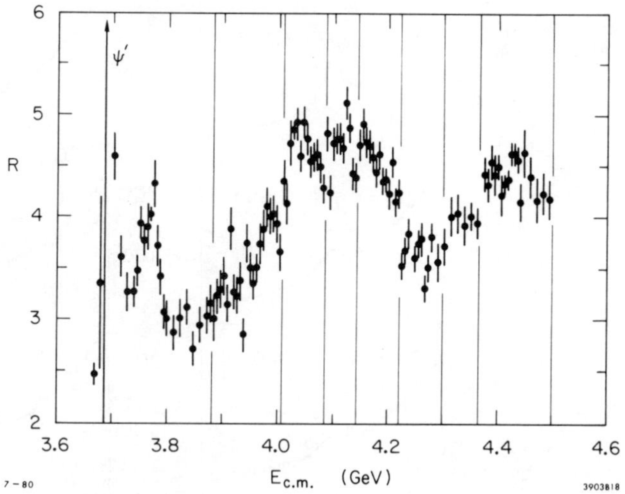

Fig. 1. Crystal Ball measurement of R (1979 data), showing the center-of-mass binning used for the inclusive η analysis (light vertical lines). [Contributions from the τ are excluded, but no radiative corrections have been made. The errors are statistical only.][6]

i) $|\cos\theta_{\gamma\cdot beam}| < 0.85$; ii) $\cos\theta_{\gamma t} < 0.9$, where $\theta_{\gamma t}$ is the opening angle between the γ and any other track in the event; iii)$E_\gamma > 30$ MeV (irrevelant for $m_{\gamma\gamma}$ in the η region, but can affect π^0 subtraction). Optionally, the elimination of γ's which form a π^0 with another γ ("π^0 subtraction"), and/or the elimination of γ's which do not exhibit a reasonable lateral shower distribution ("pattern cut") has been used. For example, Fig. 2(a-m) shows the $m_{\gamma\gamma}$ mass distributions after application of the above cuts, including π^0 subtraction but not the pattern cut.

A fit is performed to each $m_{\gamma\gamma}$ distribution assuming a Gaussian shape for the η peak, the width ($\sigma \sim 4\%$) corresponding to our resolution. Various checks have been made to estimate the sensitivity of the number of η's found to the width and mean of the peak and the shape of the background assumed in the fits. The background form used in the fits shown in Fig. 2(a-m) is a quadratic times $m_{\gamma\gamma}^{-2.7}$.

RESULTS

Figure 3 shows f_η, the number of η's per hadronic event, as a function of the center-of-mass energy. The error bars do not include the estimated uncertainty in absolute normalization (~25%). It should be noted that an η signal is observed everywhere, including at the off-resonance point below charm threshold (3.67 GeV).

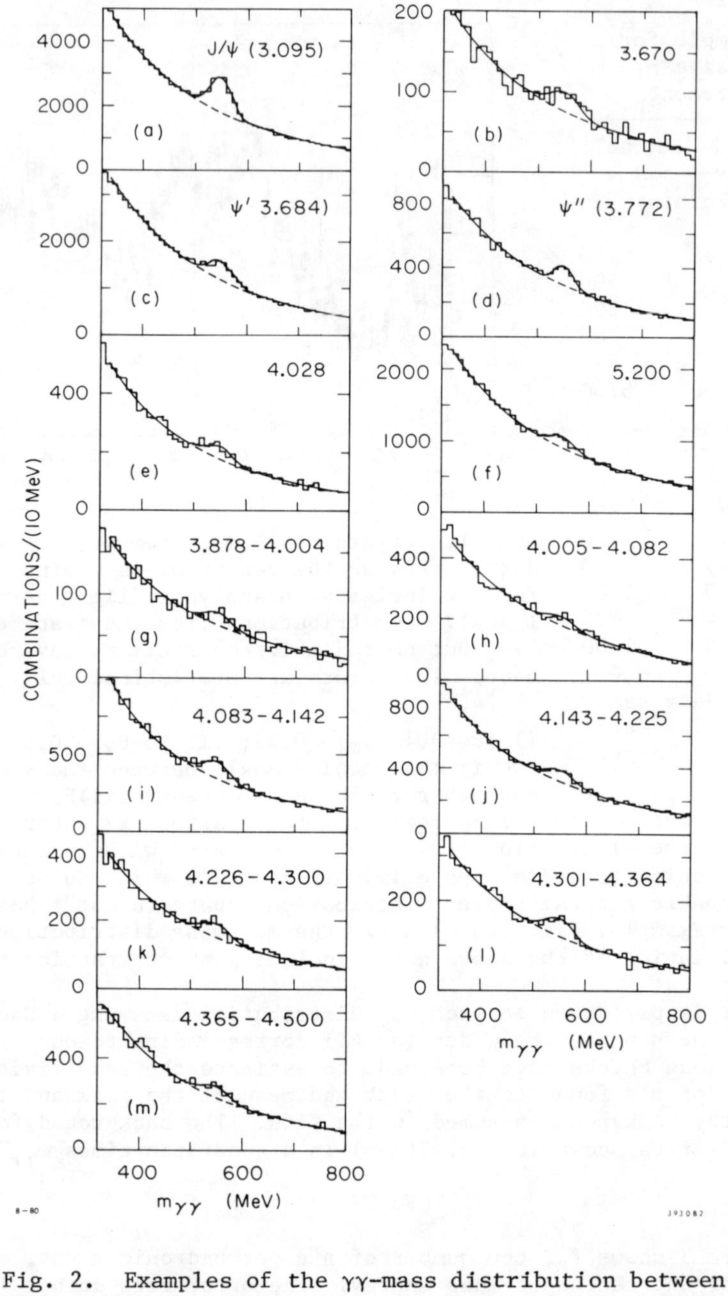

Fig. 2. Examples of the γγ-mass distribution between 320 and 800 MeV for the center-of-mass energies used. The curves are fits to the data (dashed is background, solid is background plus η) as described in the text.

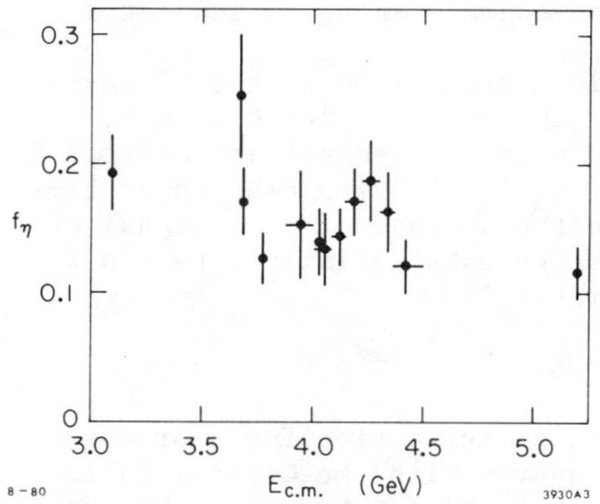

Fig. 3. f_η, the number of η's per hadronic event, versus $E_{c.m.}$ The first four points are (in order) J/ψ, 3.67 GeV, ψ' and ψ".

In Figure 4(a) we plot the inclusive cross section for η production in the form of $R_\eta = \sigma(e^+e^- \to \eta X)/\sigma(e^+e^- \to \mu^+\mu^-)$. Because R_η at the J/ψ and ψ' resonance is large (off-scale), the contributions from the radiative tails of these resonances to the other points has been subtracted (maximum correction = 0.08, at the ψ").

Figure 4(a) indicates that the cross section for η production does not vary much in the region above charm threshold from the 3.67 GeV value below threshold. This may be contrasted with Fig. 4(b), in which we have divided the published DASP cross sections[4] by $\sigma(e^+e^- \to \mu^+\mu^-)$

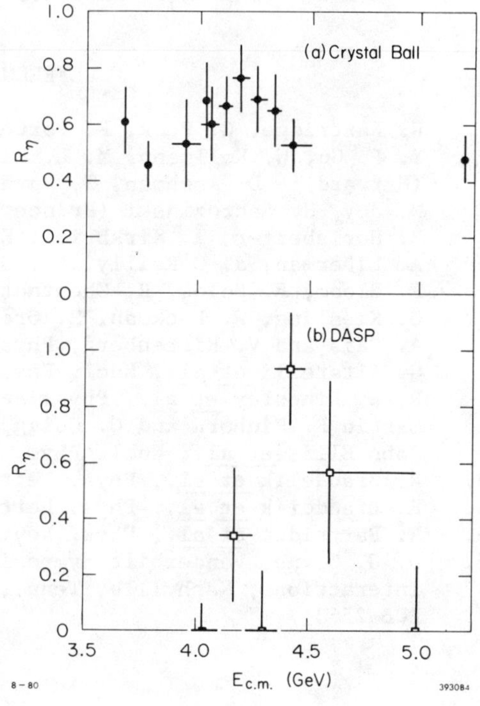

Fig. 4. R_η, the cross section for η production divided by the QED point cross section, versus $E_{c.m.}$: a) Crystal Ball experiment (first two points are 3.67 GeV and ψ"), b) DASP experiment.[4]

to obtain R_η. The DASP experiment reported η signals only at 4.17, 4.42, and possibly at 4.60 GeV, the 4.42 GeV point implying a strong jump in η production. A number of multiplicity and energy cuts were made in the DASP experiment, but it appears difficult to explain the discrepancy in terms of these cuts, as we have been unable to generate substantial energy-dependence in the 3.9-4.5 GeV region with similar cuts. It should be cautioned, however, that the two detectors are very different, and hence it is impossible to mimic the DASP experiment exactly.

Using the highest point, at 4.18 GeV (Fig. 4(a)), we may set an upper limit on the product of the cross section for $F\bar{F}$ production (including excited F's) times the branching ratio for F decay to η: $R(e^+e^- \to F\bar{F}X)\ BR(F \to \eta x) < 0.31$ (90% C.L.).

In this limit, we have corrected for the smaller η-detection efficiency in $F\bar{F}$ events, compared with non-charm events, as indicated by Monte Carlo simulations.

Because we observe no increase in η production at the ψ'' resonance, we may also set an upper limit on the inclusive branching ratio for D decay into η's (averaged over the neutral and charged D components of ψ''): $BR(D \to \eta x) < 0.13$ (90% C.L.). The DASP upper limit is 0.02 (unstated C.L.),[4] based on the absence of an η signal at 4.03 GeV. Note, however, that we observe substantially more η's in the 4.03 GeV region than the DASP limit.

CONCLUSIONS

In a measurement of the inclusive cross section for η production as a function of center-of-mass energy with the Crystal Ball, no substantial increase above charm threshold is observed. The implications for the charmed D and F mesons may be summarized in the limits:

$BR(D \to \eta x) < 0.13$ (90% C.L., averaged over neutral and charged D components of ψ'')

$R(e^+e^- \to F\bar{F}X)\ BR(F \to \eta x) < 0.3$ (90% C.L., $E_{c.m.} < 4.5$ GeV) .

REFERENCES

1. R. Partridge, C. Peck, F. Porter (Caltech); D. Antreasyan, Y. F. Gu, W. Kollmann, M. L. Richardson, K. Strauch, K. Wacker (Harvard); D. Aschman, T. Burnett, M. Cavalli-Sforza, D. Coyne, M. Joy, H. Sadrozinski (Princeton); R. Hofstadter, R. Horisberger, I. Kirkbride, K. Königsmann, H. Kolanoski, A. Liberman, J. O'Reilly, A. Osterheld, J. Tompkins (Stanford); E. Bloom, F. Bulos, R. Chestnut, J. Gaiser, G. Godfrey, C. Kiesling, W. Lockman, M. Oreglia (SLAC).
2. A. Pais and V. Rittenberg, Phys. Rev. Lett. 34, 707 (1975); G. Altarelli et al., Nucl. Phys. B88, 285 (1975); R. L. Kingsley et al., Phys. Rev. D11, 1919 (1975); Martin B. Einhorn and C. Quigg, Phys. Rev. D12, 2015 (1975); John Ellis et al., Nucl. Phys. B100, 313 (1975).
3. R. Brandelik et al., Phys. Lett. 70B, 132 (1977).
4. R. Brandelik et al., Phys. Lett. 80B, 412 (1979).
5. R. Partridge et al., Phys. Rev. Lett. 44, 712 (1980).
6. D. G. Coyne, Vanderbilt Symposium on High Energy e^+e^- Interactions, Nashville, Tenn., May 1-3, 1980; (also SLAC-PUB-2563).

WEAK DECAYS, THEORY

M. Suzuki, UC-Berkeley, Organizer

PENGUINS AND THE $\Delta I=1/2$ RULE

Christopher T. Hill
Fermi National Accelerator Laboratory, Batavia, Illinois 60510

ABSTRACT

We discuss the role of "penguin" operators in the weak decays of light particles, including an extra operator, O_7, entering in two loops. The evidence for large penguin contributions is reviewed and found to be uncompelling.

INTRODUCTION

The connection between QCD and the $\Delta S=1$ nonleptonic weak decays was first explored in 1974,[1] and is conventionally discussed within the context of the Wilson expansion and the renormalization group. A "standard" operator basis may be summarized as follows:

$$(I=\tfrac{1}{2})\quad O_1^q = \bar{s}\gamma_\mu q_L \bar{q}\gamma^\mu d_L - \bar{s}\gamma_\mu d_L \bar{q}\gamma^\mu q_L \; (q=u \text{ or } c) \qquad d_1^q=+4$$

$$(I=\tfrac{1}{2})\quad O_2^q = \bar{s}\gamma_\mu q_L \bar{q}\gamma^\mu d_L + \bar{s}\gamma_\mu d_L \bar{q}\gamma^\mu q_L + 2\delta_{qu}(\bar{s}\gamma_\mu d_L \bar{d}\gamma^\mu d_L + \bar{s}\gamma_\mu d_L \bar{s}\gamma^\mu s_L) \qquad d_2^q=-2$$

$$(I=\tfrac{1}{2})\quad O_3 = \bar{s}\gamma_\mu d_L \bar{u}\gamma^\mu u_L + \bar{s}\gamma_\mu u_L \bar{u}\gamma^\mu d_L + 2\bar{s}\gamma_\mu d_L \bar{d}\gamma^\mu d_L - 3\bar{s}\gamma_\mu d_L \bar{s}\gamma^\mu s_L \qquad d_3=-2$$

$$(I=\tfrac{3}{2})\quad O_4 = \bar{s}\gamma_\mu d_L \bar{u}\gamma^\mu u_L + \bar{s}\gamma_\mu u_L \bar{u}\gamma^\mu d_L - \bar{s}\gamma_\mu d_L \bar{d}\gamma^\mu d_L \qquad d_4=-2 \tag{1}$$

with bare coefficients $c_1^u=-c_1^c=-1$, $c_2=1/5$, $c_3=2/15$, $c_4=2/3$, and the weak $\Delta S=1$ Hamiltonian is given by:

$$H_{wk} = \sqrt{2}\, G_F \sin\theta_c \cos\theta_c \sum_i \tilde{c}_i \theta_i . \tag{2}$$

Our problem is the exact calculation of the renormalized $\tilde{c}_i$ in QCD and, particularly the estimation of the operator matrix elements. In the absence of penguins we have $\tilde{c}_i=c_i(\bar{g}^2(m_W^2)/\bar{g}^2(\mu^2))^{-d_i/b_0}$ where d_i is the anomalous dimension ($\times\ 16\pi^2/g^2$) of Eq. (1) and $b_0=11 - 2/3\ n_f$.

PENGUINS

In 1977 Shifman, Vainshtein and Zakharov[2] (SVZ) pointed out that the diagrams of Fig. 1 are potentially important as the GIM cancellation is only logarithmic. Direct computation yields the operator:

$$\frac{g}{16\pi^2}\int_0^1 dx\; 4(x-x^2)\log\left(\frac{q^2(x-x^2)-m_u^2}{q^2(x-x^2)-m_c^2}\right)\bar{s}\gamma_\mu \frac{\lambda^A}{2} d\left(D_\nu^{AB} G^{\mu\nu B}\right) , \quad (3)$$

By use of the operator equation of motion[3]

$$\left(D_\mu G^{\mu\nu}\right)^A = g\sum_q \bar{q}\gamma^\nu \frac{\lambda^A}{2} q \quad (4)$$

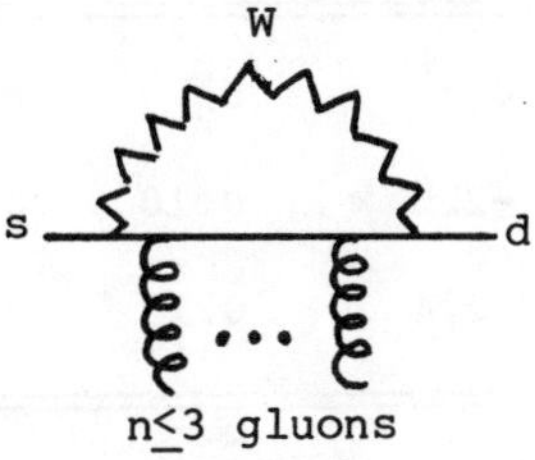

Fig. 1. Penguin Diagram.

we have:

$$= \frac{g^2}{16\pi^2}\,\frac{2}{3}\left(\log\frac{m_c^2}{\mu^2}\right)\bar{s}\gamma_\mu \frac{\lambda^A}{2}\; d\Big(\sum_q \bar{q}\gamma\, \frac{g^A}{2}\, q\Big) \;. \quad (5)$$

By color + Dirac Fierz rearrangement Eq. (5) can be rewritten in terms the preceding O_1^7, O_2^9 and the new ops:

$$O_5 = (\bar{s}\gamma_\mu \frac{\lambda^A}{2} d_L)(\bar{u}\gamma^\mu \frac{\lambda^A}{2} u_R + \ldots + \bar{c}\gamma^\mu \frac{\lambda^A}{2} c_R)$$

$$O_6 = \bar{s}\gamma_\mu d_L(\bar{u}\gamma^\mu u_R + \ldots + \bar{c}\gamma^\mu c_R) . \quad (6)$$

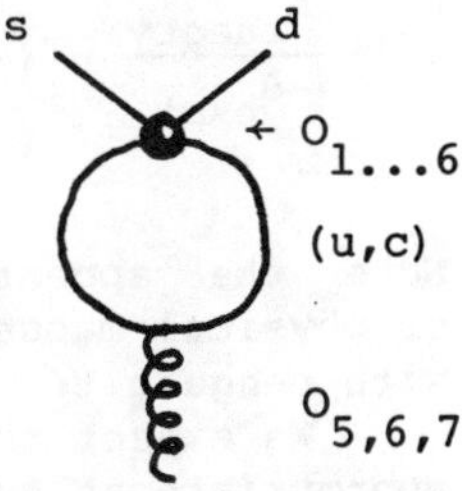

Fig. 2. Operator Mixing Diagram.

In two loops we further encounter[4]

$$O_7^{(-)} = m_s \bar{s}_R \sigma_{\mu\nu} \frac{\lambda^A}{2} d_L G^{\mu\nu A} - m_d \bar{s}\sigma_{\mu\nu} \frac{\lambda^A}{2} d_R G^{\mu\nu A} , \quad (7)$$

with coefficient: $3/8\; g^3/(16\pi^2)^2\; G_F \cos\theta_c \sin\theta_c\; (\log\; m_c^2/\mu^2)$. No further d=6 operators can occur (5).

The operators $O_1^u \ldots O_6$ mix into O_5, O_8 and O_7 through the diagram of Fig. (2) (O_7 requires two loops). An anzätz for treating this expanded operator mixing problem, with inherent subtleties due to the GIM cancellation, is discussed in Ref. (2). The mixing generated by

Fig. (2) leads to mixed leading logs like $(\log M_w^2)^p(\log m_c^2)^q$. But is this the exact log structure of the actual Feynman diagrams? In Ref.(6) this is answered in the negative for two reasons: (a) Large $(\log m_c^2)^{p+q}$ are present which are leading and not properly summed by this ansatz (b) the $(\log M_w^2)^p(\log m_c^2)^q$ were found to cancel in some diagrams in which they should be present. To get an idea as to the potential size of these effects we compare in Table I the results for $\tilde{c}_1,\ldots\tilde{c}_7$ computed by the anzatz of Ref. (2) and by an anzatz in Ref. (6) in which only $(\log m_c^2)^{p+q}$ terms are summed for $\tilde{c}_5$, $\tilde{c}_6$, $\tilde{c}_7$.

Table I. Comparison of operator coefficients
(Λ=0.5 GeV, μ=1, , m_c=2)

$\tilde{c}_1^u$	$\tilde{c}_2^u$	$\tilde{c}_4$	$\tilde{c}_5$	$\tilde{c}_6$	$\bar{g}(\mu^2)\tilde{c}_7$	
-2.5	0.10	0.41	-0.05	-0.01	---	SVZ
-2.4	0.13	0.41	-0.032	-0.006	-0.003	Ref. (4,6)

AMPLITUDES

We give a parameterization[2] for the $\Delta I=1/2$, $K_s\to\pi^+\pi^-$ decay in the valence quark approximation[4] (VQA), ignoring effects of $O_7^{(-)}$ which we expect to be small (the MIT bag model (7) is expected to give roughly similar results):

$$\frac{A_{theory}}{A_{expt}} = (-0.087)\tilde{c}_1^u + (\tilde{c}_5+\frac{3}{16}\tilde{c}_6)\frac{(-0.292)m_\pi^2}{(m_u+m_d)(m_s-m_d)} \quad . \tag{8}$$

Note the appearance of the renormalization group noninvariant ratio of physical masses to quark masses $(m_\pi^2/(m_u+m_d)(m_s-m_d))$ associated with penguin terms. What quark masses are to be used here?

We expect that the usual PCAC masses $m_u:m_d:m_s$=5:10:150 MeV are appropriate at energy scales ∽ few GeV where PCAC sum rules saturate. For our problem we must evolve (hopefully) down to mass scales of order m_K,m_π where the quark masses increase. In the MIT bag, for example, we have[7] m_u=10, m_d=20, m_s=300 (MeV). With these masses and c_1^u=-2.5 and c_5 + 3/16 c_6 = -0.06 we find A_{thy}/A_{expt} = 0.26; the O_1^u contribution is 87% here where the penguin is only 13%. In Ref. (2) the (extremely optimistic) values m_u=m_d=5 MeV, m_s=150 are used with c_5 + 3/16 c_6 = -0.25 (five times greater than theory predicts) to obtain A_{thy}/A_{expt} = 1.13. In this estimate the "reliable" short distance calculation of c_5 and c_6 has been abandoned and the penguin contribution is greatly enhanced by the "unreliable" matrix element estimate. Since the matrix element of O_1^u is expected to be accurate only to within a factor of 2∽3, we believe that the former estimate more accurately reflects the real situation than the latter. Here penguins account for ∽ 1/5 of the total amplitude.

If we simply fit hyperon s-waves by pure penguins in the VQA we deduce an effective c_5' and c_6' to compare with theory. With the bag model quark masses one obtains $c'_{5,6}/\tilde{c}_{5,6} \sim 30$. With a pure O_1^u fit we obtain $c_1'/c_1 \sim 3$. Hence, one must conclude that the penguin coefficients are far too small to allow a large contribution to the observed amplitudes with reasonable PCAC quark masses.

The hyperon processes have been reexamined by Finjord & Gaillard[8] and do not agree with theory in any approximation. Also, the $\Delta I=3/2$ decays involve only O_4 and are overestimated by ~ 2 in theory.

Clearly, external input is needed to determine whether penguins are really important or not. If they have a substantial contribution here then they may be isolated in CP-violation measurements of $|\varepsilon'/\varepsilon|$[9]. Also, such measurements as $\Omega^- \to \Xi^- \pi / \Omega^- \to \Xi^* \pi$[8] are sensitive to penguins. We expect penguin contributions to be $\sim 1/5$ to $1/10$ the contribution of other operators.[6]

REFERENCES

1. M.K. Gaillard, B.W. Lee, Phys. Rev. Letters <u>33</u> (1974) 108. G. Altarelli, L. Maiani, Phys. Lett. <u>52B</u> (1974) 351.
2. M.A. Shifman, A.I. Vainshtein, V.I. Zakharov, Sov. Phys. JETP <u>45</u> (1977) 670.
3. For discussion of off-shell structure see W. Deans, J. Dixon, Phys. Rev. <u>D18</u> (1978) 1113 and classic references therein. By explicit calculation the null operator $(\bar{s}\gamma_\mu \lambda^A/2\, d)((D_\nu G^{\mu\nu})^A - \bar{q}\gamma^\mu \lambda^A/2\, q)$ has an α (gauge) -dependent anomalous dimension (C. Hill, G. Ross unpublished), hence is unphysical.
4. C.T. Hill, G.G. Ross, Fermilab-Pub-79/82-THY, Dec. 1979 (to be published in Nucl. Phys. B).
5. See R.K. Ellis, Nucl. Phys. <u>B106</u> (1976) 239; except possible Higgs induced. For full treatment of op mixing with Higgs see C.T. Hill, Caltech PhD. Thesis, (unpublished).
6. C.T. Hill, G.G. Ross, Univ. of Oxford Preprint, Ref. 75/79 (to be published in Physics Letters).
7. J. Donoghue, E. Golowich, B. Holstein, Phys. Rev. <u>D21</u> (1980) 186.
8. J. Finjord, M.K. Gaillard, Bern Preprint, March 1980.
9. F. Gilman, M. Wise, Physics Letters, <u>83B</u> 83, (1979).

THE ROLE OF NON-SPECTATOR INTERACTIONS IN CHARM AND BOTTOM DECAYS

V. Barger, J. P. Leveille and P. M. Stevenson
Physics Department, University of Wisconsin, Madison, WI 53706

ABSTRACT

The role of annihilation and exchange diagrams in charmed meson, Λ_c^+ and B-meson decays is analyzed.

INTRODUCTION

Recent measurements[1] of charmed particle lifetimes and semileptonic branching fractions have made it necessary to revise our theoretical description of heavy particle decays. The conventional quark model for these decays assumes that the heavy quark decays freely with light quark constituents acting merely as spectators: see Fig. 1. This spectator model predicts equal lifetimes and semileptonic branching ratios for Λ_c^+ and the charmed mesons D^0, D^+, F^+, whereas recent data indicate that $\tau(D^+)/\tau(D^0) \sim \tau(D^+)/\tau(\Lambda_c^+) \sim 10$, $\tau(D^+)/\tau(F^+) \sim 5$, and $B(D^+ \to eX)/B(D^0 \to eX)$ of order 3 or larger. Although preliminary, the data strongly suggest that D^0, F^+ and Λ_c^+ are shorter-lived than the D^+.

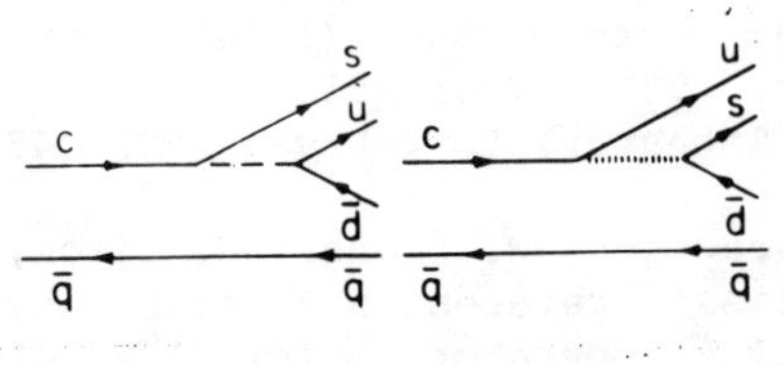

Fig. 1. Spectator diagrams ($D^0, D^+, F^+, \Lambda_c^+$ decays)

Additional contributions to the decay rates have been known for some time. These are of the non-spectator type, e.g. $c\bar{u} \to s\bar{d}$ in case of the D^0. However, these contributions vanish in Born approximation with vanishing light quark masses, e.g. Γ(non-spect.)/Γ(spectator) $\sim m_s^2 |(\psi(0)|^2/m_D^5$ where $|\psi(0)|^2$ is the probability of finding the two constituent quarks at the origin. This vanishing of lowest order annihilation and exchange contributions with light quark masses is due to helicity suppression. It is analogous to the suppression of $\pi^- \to e^-\bar{\nu}_e$. In the rest frame of the decaying 0^- meson the outgoing quark and antiquark are collinear, so their spins must be antiparallel. Hence they must be either both right-handed or both left-handed. Since V-A theory wants a left-handed quark and a right-handed antiquark at the production vertex, chirality must be flipped at the cost of a power of a light quark mass in the amplitude.

It was recently realized[2] that helicity suppression could be avoided in nonleptonic decays. A hard gluon can be radiated perturbatively from the light quark prior to the interaction allowing the quark-antiquark pair to interact in a spin one state: see Fig. 2. Alternatively, a gluon may be present in the D^0 wavefunction. In

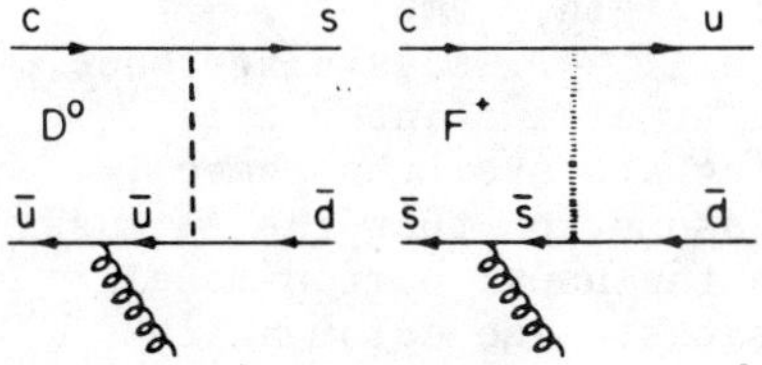

Fig. 2. Single gluon emission diagrams.

the following we shall only consider the perturbative radiation of a gluon and refer to it as the <u>one-gluon model</u>. As we will show in the sequel, this model is in fair agreement with lifetimes and decay patterns observed experimentally. Most of our predictions, however, transcend the one-gluon model and remain valid provided that some mechanism enhances non-spectator contributions.

For convenience we summarize our results here:

(i) We find a substantial enhancement for the Λ_c^+ decay rate from the interaction between the c and d quarks. The lifetime relation $\tau(\Lambda_c^+)/\tau(D^+) \simeq [1 + \tau(D^+)/3 \times 10^{-13}\ s]^{-1}$ is derived, based on the assumption that the spectator model is valid for the D^+.

(ii) For charmed mesons, given $\tau(D^+)/\tau(D^0) = 5 - 10$, we predict: (a) $\tau(F^+)/\tau(D^+) = .7 - .4$; (b) an enhanced $\Delta S = 0$ rate for D^+, (c) a substantial multipion branching fraction for the F^+, (d) an $F^+ \to \tau^+ \nu$ branching fraction of order 10-20%, and (e) restrictive bounds on the $D^0 \to \bar{K}^0 \pi^0$ mode.

(iii) We give formulae for the B mesons lifetimes in the one-gluon model, which suggest that non-spectator effects are not as significant for B mesons.

Our analysis is based on the effective weak interaction Lagrangian $L_{eff} = (G_F/\sqrt{2})[f_1(\bar{u}d')(\bar{s}'c) + f_2(\bar{s}'d')(\bar{u}c)]$, where $(\bar{q}Q)$ denotes a color singlet V-A current. The short distance enhancement factors f_1 and f_2 due to gluon exchanges are given numerically by $f_+ \equiv f_1 + f_2 = .69$, $f_- = f_1 - f_2 = 2.09$. In the absence of strong interactions $f_+ = f_- = 1$. Here $d' = d \cos\theta + s \sin\theta$ and $s' = s \cos\theta - d \sin\theta$, where θ is the Cabibbo angle, $\sin^2\theta = 0.05$.

Λ_c^+ LIFETIME[3]

As a first example of non-spectator contributions, we consider the effect of cd interactions for the Λ_c^+ lifetime: see Fig. 3.

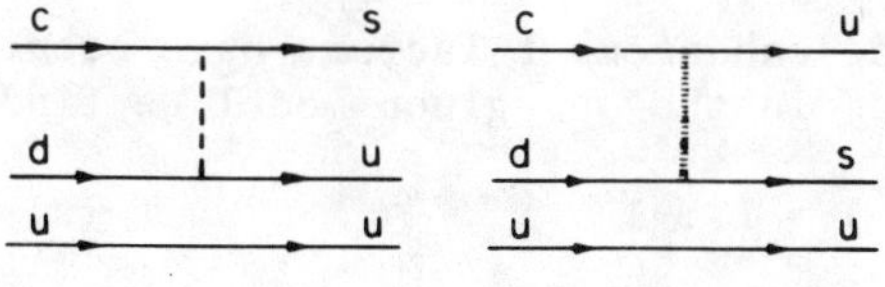

Fig. 3. Non-spectator diagrams of Λ_c^+ decay. Note: --- and ┅ correspond to the coefficients of f_1 and f_2 respectively in the effective Lagrangian.

Helicity suppression does not apply here since the initial state contains two quarks. Our approach is to directly calculate the non-spectator contribution to the Λ_c^+ rate in the free quark model, including short-distance enhancement. We use a non-relativistic approximation for the quarks in the Λ_c^+. The non-spectator contribution is proportional to the square modulus of the wavefunction for two quarks at the origin, $|\psi(0)|^2 \equiv \langle\psi|\delta^3(\vec{r}_1-\vec{r}_2)|\psi\rangle$ which we estimate from the $\Sigma_c^+ - \Lambda_c^+$ mass difference, using the QCD analogue of the Fermi-Breit hyperfine interaction. To obtain the inclusive decay rate we integrate over the phase space of the final state quarks, on the assumption that the Λ_c^+ mass is large enough for this to represent (in the usual parton-model fashion) the sum of all hadronic final states. The color antisymmetry of the wavefunction leads to a large enhancement factor $f_-^2 \approx 4$. We find a non-spectator rate $\Gamma(cd \to su) \simeq .32 \times 10^{13}\ s^{-1}$ to be compared with the estimated spectator rate $\Gamma_{spect.} \simeq .2\times10^{13}\ s^{-1}$. To avoid the uncertainties in the latter estimate, we use $\Gamma_{spect.} \simeq 1/\tau(D^+)$, i.e., we assume that the conventional dogma (charmed hadron decay = charmed quark decay) is correct for the D^+, where all non-spectator interactions are Cabibbo-suppressed. We thus obtain the lifetime relation $\tau(\Lambda_c^+) = \tau(D^+)[1 + \tau(D^+)/(3 \times 10^{-13}\ s)]^{-1}$. With input $\tau(D^+) \simeq 10 \times 10^{-13}$ s, we obtain $\tau(\Lambda_c^+) \sim 2.3 \times 10^{-13}$ s in fair agreement with recent data[1] $\tau(\Lambda_c^+) = 1.14^{+.90}_{-.44} \times 10^{-13}$ s. The poor statistics of the present data preclude serious quantitative comparisons at this point, but the trend of the data is nicely reproduced.

CHARMED MESON DECAYS AND LIFETIMES[4]

(a) Lifetimes: The common spectator contributions to the decays of the D^+, D^0 and F^+ are of two types. The nonleptonic contributions $c \to s u\bar{d}, \ldots$ including all radiative corrections will be denoted by N. The semileptonic decays to electrons plus muons, with gluon corrections, will be denoted by L. We shall parametrize these spectator contributions rather than calculate them directly, thereby avoiding the various ambiguities inherent in the quark masses and parton model assumptions.

The contributions of non-spectator diagrams are parametrized as

$$\begin{aligned}
\Gamma_g(D^0(c\bar{u}) \to s\bar{d}; d\bar{d}; s\bar{s}; d\bar{s}) &= (C^4; C^2S^2; C^2S^2; S^4)\ f_1^2 G(D^0) \\
\Gamma_g(D^+(c\bar{d}) \to u\bar{d}; u\bar{s}) &= (C^2S^2, S^4)\ f_2^2 G(D^+) \\
\Gamma_g(F^+(c\bar{s}) \to u\bar{d}, u\bar{s}) &= (C^4, C^2S^2)\ f_2^2 G(F^+)\ ,
\end{aligned} \tag{1}$$

where $C = \cos\theta$, $S = \sin\theta$. The strong enhancement factors have been isolated in Eq. (1) for convenience. In the one-gluon model we find

$$G(D^0) : G(D^+) : G(F^+) = 1 : 1 : r \tag{2}$$

where $r \simeq .55$ is an SU(3) breaking factor. An important contribution to the F^+ rate is the $\tau\nu$ decay which is $\Gamma_{F\to\tau\nu} = C^2G_F^2 f_F^2\, m_F m_\tau^2 (1-m_\tau^2/m_F^2)^2/8\pi$, where $f_F^2 = 12|\psi(0)|_F^2/m_F$ in the non-relativistic approximation. The f_F, f_D values are uncertain; for numerical estimates we use $f_F = f_D \simeq 430$ MeV as suggested by experimental $D^{*+,0}$ and $D^{+,0}$ mass splittings. Neglecting other purely leptonic decay modes, we obtain the total decay rates $\Gamma(D^0) = L + N + f_1^2 G(D^0)$,

$\Gamma(D^+) = L + N + f_2^2S^2G(D^+)$ and $\Gamma(F^+) = L + N + f_2^2C^2G(F^+) + \Gamma_{F\to\tau\nu}$. In the one-gluon model, Eq. (2) leads to the lifetime results of Fig. 4. For $\tau(D^+)/\tau(D^0) \approx 10$ we find $\tau(F^+)/\tau(D^+) \simeq .4$-$.5$, in good agreement with present data.

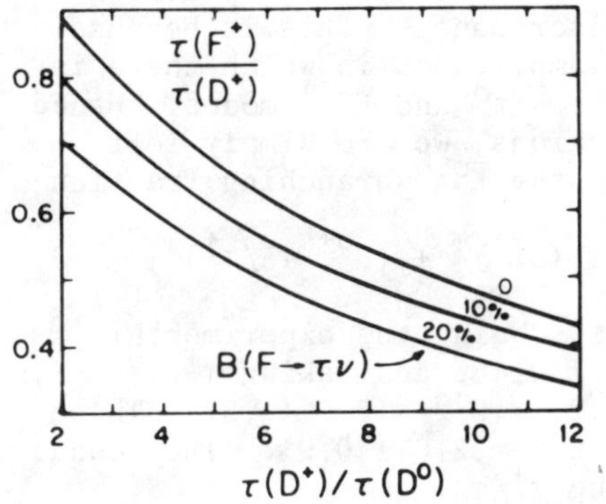

Fig. 4. F^+ lifetime predictions.

We note that since one-gluon diagrams do not contribute to decays to leptons, the semileptonic branching fractions into electrons are in the same ratios as the lifetimes in the one-gluon model. However, if multiple gluon emission is important, then the semileptonic branching fraction of the F^+ could be enhanced by the annihilation interaction $c\bar{s} \to e^+\nu_e$ + gluons.

(b) Inclusive ΔS=0 decays: We now turn to inclusive Cabibbo suppressed $\Delta S = 0$ decays; i.e., transitions in which the final state has the same strangeness as the initial state. In the spectator model the branching fractions for these decays are the same for the D^0, D^+ and F^+, with $B(\Delta S = 0) \approx 8\%$. Including exchange interactions, we find that $\Delta S = 0$ branching ratios are slightly increased to 8-10% for the D^0, whereas for the F^+, $B(\Delta S{=}0)$ is constrained to lie in the range 4-8%.

For the D^+, non-spectator interactions contribute only to Cabibbo suppressed decays, and can lead to a significant enhancement of the $\Delta S = 0$ branching fraction. In the one-gluon model with $B_e(D^+) = 0.22$ and $\tau(D^+)/\tau(D^0) = 5$ and 10, we find $B(D^+,\Delta S = 0) =$ 12% and 17% respectively. There is some indication from the measured $B(D^+ \to \bar{K}X)$ that the $\Delta S = 0$ branching fraction for the D^+ is enhanced.

(c) F^+ multipion decay modes: In F^+ decays the spectator transition $c\bar{s} \to s\bar{s}u\bar{d}$ will give final states with either overt or hidden strangeness (e.g., $F^+ \to K^-K^+X^+$ or $F^+ \to \eta X^+$). Non-resonant multipion final states are not expected at a significant level. The situation is markedly different with the non-spectator transition $c\bar{s} \to u\bar{d}$ + gluon(s), where states of three or more pions are anticipated. The branching fraction for this transition with no s or $\bar{s}$ quarks originating from the weak interaction can be estimated in the one-gluon model. We obtain $B(F^+ \to \text{no } s \text{ or } \bar{s}) \simeq (25$-$30)\%$. Since $\bar{u}u$ or $\bar{d}d$ pair creation is favored over $\bar{s}s$, the $u\bar{d}g$ final state will evolve mainly into multipion or η plus pions modes. Although we are unable to separate the relative proportion, phase space considerations suggest that the multipion states will predominate. In this connection it is interesting to note that several $F^- \to \pi^-\pi^+\pi^-\pi^0$ events have now been observed.

(d) The "color-suppression" puzzle: The spectator model predicts $B(D^0 \to \overline{K}^0\pi^0)/B(D^0 \to K^-\pi^+) = 0.05$, whereas observed rates for these modes are $B^{00} = 2.0 \pm .9\%$, $B^{-+} = 2.8 \pm .6\%$. Non-spectator interactions offer a way out of the discrepancy. This is because gluon-enhanced exchange diagrams give amplitudes in which there is no color suppression factor between the $\overline{K}^0\pi^0$ and $K^-\pi^+$ modes. Hence adding spectator and exchange contributions, we are simply left with an isospin triangle inequality on the $\overline{K}^0\pi^0$ branching fraction

$$\tfrac{1}{2}[(B^{-+})^{\frac{1}{2}} - (B^{0+}\ \tau^0/\tau^+)^{\frac{1}{2}}]^2 \le B^{00} \le \tfrac{1}{2}[(B^{-+})^{\frac{1}{2}} + (B^{0+}\ \tau^0/\tau^+)^{\frac{1}{2}}]^2 \qquad (3)$$

where τ^0,τ^+ denote the D^0,D^+ lifetimes. Using the experimental values $B^{-+} = 2.8 \pm 0.6\%$ and $B^{0+} = 2.1 \pm 0.4\%$, and taking $\tau^+/\tau^0 = 5$, we find the numerical bounds $0.5 \pm 0.2\% \le B^{00} \le 2.7 \pm .5\%$. These bracket nicely the experimental value $B^{00} = 2.0 \pm 0.9\%$. The bounds become more restrictive with increasing τ^+/τ^0.

(e) Distinctive decay modes: Special decay modes in which the so-called spectator quark is absent in the final state are of special interest since they can occur only through the non-spectator interaction. An example is the transition $c\bar{u} \to s\bar{d}g$ with an $s\bar{s}$ pair created by the gluon. This would give rise to $D^0 \to \overline{K}^oK^o\overline{K}^o$ or $D^0 \to \overline{K}^0\phi$ decays. A more exotic example is $F^+ \to p\bar{n}$; although certain to be rare, such a decay has a spectacular signature!

B MESON LIFETIMES AND DECAYS

We give a brief summary of the situation for bottom mesons $B_u(b\bar{u})$, $B_d(b\bar{d})$, $B_s(b\bar{s})$ and $B_c(b\bar{c})$. The net spectator contribution with strong enhancement factors included (note for B mesons $f_+ = .80$, $f_- = 1.56$) is:

$$\Gamma_{spect.}(B) = \gamma_b[7.69|U_{bu}|^2 + 3.07|U_{bc}|^2] \qquad (4)$$

where $\gamma_b^{-1} \equiv (G_F^2\ m_b^5 b/192\pi^3)^{-1} \simeq .16 \times 10^{-13}$ s and $U_{bu}(U_{bc})$ are the $b \to u$ $(b \to c)$ transition elements. Adding all possible contributions (spectator and non-spectator), we find the decay rates:

$$\begin{aligned} \Gamma(B_d) &= \gamma_b[10.41|U_{bu}|^2 + 4.95|U_{bc}|^2] \\ \Gamma(B_s) &= \gamma_b[\ 8.67|U_{bu}|^2 + 3.75|U_{bc}|^2] \\ \Gamma(B_u) &= \gamma_b[\ 9.67|U_{bu}|^2 + 3.07|U_{bc}|^2] \\ \Gamma(B_c) &= \gamma_b[\ 7.69|U_{bu}|^2 + 5.17|U_{bc}|^2]\ . \end{aligned} \qquad (5)$$

These rates are not dramatically different from the spectator rate. The role of non-spectator diagrams in these rates are as follows:

(a) Charged mesons: Annihilation diagrams into final states containing a charmed quark or a τ lepton are neither helicity- nor color-suppressed. However, gluon radiation diagrams are color-suppressed by a factor $(f_+ - f_-)^2/4 \simeq .14$.

(b) Neutral mesons: Exchange diagrams to final states with heavy quarks or leptons are not helicity-suppressed but are strongly color-suppressed by a factor $(2f_+ - f_-)^2 = 0.002$. Gluon radiation diagrams are enhanced by $(f_+ + f_-)^2/4 \simeq 1.4$.

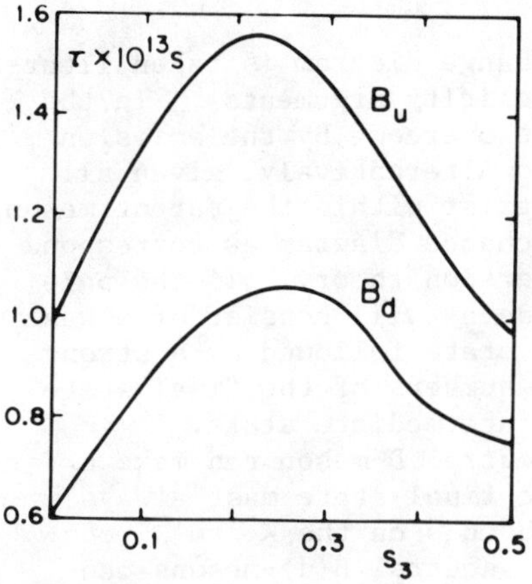

Fig. 5. B-lifetime predictions vs. the KM parameter s_3, based on Eq. (6) and the analysis of Ref. 5 with B = 0.4 and $\delta < \pi/2$.

For the expected ranges[5] of U_{bc} and U_{bu} we find 6×10^{-14} s $< \tau(B) < 2\times10^{-13}$ s: see Fig. 5.

The semi-leptonic branching fractions for B_d and B_u mesons are $B_{e+\mu}(B_d) = \gamma_b[2.11|U_{bu}|^2 + 0.92|U_{bc}|^2]/\Gamma(B_d)$ and $B_{e+\mu}(B_u) = \gamma_b[2.25|U_{bu}|^2 + 0.92|U_{bc}|^2]/\Gamma(B_u)$. For (i) $U_{bu} = 0$, (ii) $|U_{bu}| = |U_{bc}|$, the predicted semi-leptonic rates are (i) 19% (ii) 20% for B_d and (i) 30%, (ii) 25% for B_u.

The expected mean multiplicities of kaons per B-meson arising from strange and charm quarks created in the weak interaction are $\langle n_K(B_d)\rangle = \gamma_b[1.9|U_{bu}|^2 + 5.7|U_{bc}|^2]/\Gamma(B_d)$ and $\langle n_K(B_u)\rangle = [4.3|U_{bu}|^2 + 3.8|U_{bc}|^2]/\Gamma(B_u)$ which yield (i) 1.1, (ii) 0.5 for B_d and (i) 1.2, (ii) 0.6 for B_u. We expect these multiplicities to be augmented by 15-20% from $s\bar{s}$ pair creation from the vacuum.

We are grateful to S. Pakvasa and M. Suzuki for discussions. This research was supported in part by the Department of Energy under contract DE-AC02 76ER00881-158.

REFERENCES

1. See, e.g., reports by R. Ammar, G. Diambrini-Palazzi and K. Niu in these proceedings.
2. M. Bander, D. Silverman and A. Soni, Phys. Rev. Lett. 44, 4 and 962 (E) (1980); V. Barger, R. N. Cahn, Y. Kang, J. P. Leveille, S. Pakvasa and M. Suzuki (unpublished); H. Fritzsch and P. Minkowski, Phys. Lett. 90B, 455 (1980).
3. V. Barger, J. P. Leveille and P. M. Stevenson, Phys. Rev. Lett. 44, 226 (1980); S. P. Rosen, ibid. 44, 4 (1980).
4. V. Barger, J. P. Leveille and P. M. Stevenson, Phys. Rev. D22, 693 (1980).
5. V. Barger, W. F. Long and S. Pakvasa, Phys. Rev. Lett. 42, 1585 (1979).

COMMENT ON W-EXCHANGE IN B-DECAY

S. P. Rosen
Purdue University, West Lafayette, Indiana 47907

ABSTRACT

Tests of the W-exchange mechanism in the nonleptonic decays of neutral B-mesons are briefly discussed.

It is well-known that if the W-exchange diagram is taken literally, then it is highly suppressed by helicity arguments[1]. In the literal picture, this suppression can be overcome by the emission of a gluon from one of the quark legs.[2] Alternatively, given all the long-range complications that must exist within the parent meson state-vector, one can think of the W-exchange diagram as corresponding to a pole term in the sense of dispersion theory. If the pole term dominates the amplitude, then the decay will consist of a weak transition to some intermediate excited state followed by a strong decay to the final state[3]. The quantum numbers of the final state are therefore the same as those of the intermediate state.

In charm decay, for example, the neutral D-meson can make a weak transition to an excited $\bar{K}^o$ and the final state must always have $T = 1/2$. This can be used to set bounds on the K^- to $\bar{K}^o$ content of the decay modes[3]. In a similar, neutral B(d)-mesons can make weak transitions to excited charm states, and the final state becomes the $T = -T_3 = 1/2$ member of a triplet. The ratio of D^+ to D^o final states is then bounded by[4]

$$0 \leqslant \Gamma(D^+ n\pi)/\Gamma(D^o n\pi) \leqslant 2$$

$$0 \leqslant \Gamma(D^+X)/\Gamma(D^oX) \leqslant 1.5$$

Similar results hold for $B^o(s)$.

Other tests for the dominance of W-exchange are: (i) $\tau(B^o) << \tau(B^-)$; (ii) $BR(B^o \to \psi K\pi) << BR(B^- \to \psi K\pi)$; (iii) $BR(B^o \to \ell\nu X) << BR(B^- \to \ell\nu X)$. Data presented to this conference by two groups at CESR hint that the semi-lepton branching ratio of B-mesons may be significantly smaller than theoretical expectations[5]. Should this trend continue, it will present a problem for the standard model[1] and give support to W-exchange dominance.[6]

1. See talk by N. Cabibbo in this session for complete references.
2. M. Bander, D. Silverman, and A. Soni, Phys. Rev. Lett. 44, 7 (1980). M. Suzuki (unpublished).
3. S. P. Rosen, Phys. Rev. Lett. 44, 4 and Phys. Lett. 89B, 246 (1980)
4. S. P. Rosen, Phys. Lett., 93B, 492 (1980)
5. See talks in sessions C6 and W3.
6. The dynamics of W exchange are discussed by J.P. Leveille preceeding talk.

Chapter 4

Quantum Flavor Dynamics and Unified Theories

QFD I

R. PECCEI, MAX-PLANCK INSTITUTE, ORGANIZER

CP VIOLATION IN THE SIX-QUARK MODEL*

Mark B. Wise†
Stanford Linear Accelerator Center
Stanford University, Stanford, California 94305

ABSTRACT

Some of the recent work on CP violation in the six-quark model is reviewed.

INTRODUCTION

CP violation has been observed in the $K^0 - \bar{K}^0$ system. Nonzero values for the quantities η_{+-} and η_{00}, defined by

$$\eta_{+-} \equiv \frac{\langle\pi^+\pi^-|H_{eff}^{|\Delta S|=1}|K_L\rangle}{\langle\pi^+\pi^-|H_{eff}^{|\Delta S|=1}|K_S\rangle} \quad \text{and} \quad \eta_{00} \equiv \frac{\langle\pi^0\pi^0|H_{eff}^{|\Delta S|=1}|K_L\rangle}{\langle\pi^0\pi^0|H_{eff}^{|\Delta S|=1}|K_S\rangle} , \qquad (1)$$

are an indication of CP violation. Experimentally[1] $|\eta_{+-}| = (2.274\pm.022)\times10^{-3}$ and $|\eta_{00}| = (2.32\pm.09)\times10^{-3}$. In the standard model of strong, weak and electromagnetic interactions based on the gauge group $SU(3) \otimes SU(2) \otimes U(1)$ the quarks get their masses through Yukawa couplings to the Higgs fields. The resulting mass matrices for the quarks can be made diagonal, with real positive elements, by performing unitary transformations on the left-handed and right-handed quark fields. In the minimal model, which contains only one Higgs doublet, CP violation can then appear in the Lagrangian in only two ways.

In the strong interaction part of the Lagrangian density there is a term $\theta\varepsilon^{\mu\nu\lambda\sigma}G^a_{\mu\nu}G^a_{\lambda\sigma}$, where $G^a_{\mu\nu}$ is the gluon field strength tensor, $a\varepsilon\{1, \ldots 8\}$, and $\mu,\nu\varepsilon\{0,1,2,3\}$. Such a term violates both P and CP invariance. An electric dipole moment for the neutron also violates both P and CP. The stringent experimental upper limit, $|D_n| \lesssim 10^{-24}$ cm, on the electric dipole moment of the neutron[2] gives rise to an upper bound on θ which shows that strong interaction violation of CP is much too small to explain the CP violation observed in the kaon system.[3,4]

In the six-quark model CP violation can also occur in the weak interaction portion of the Lagrangian through the coupling of the quarks to the W-bosons.[5] This part of the Lagrangian density has the form

$$\mathscr{L}_I = \frac{g}{2\sqrt{2}} \; J_\mu^{(+)} \; W^{(-)\mu} + \text{h.c.} \quad , \qquad (2)$$

where $W_\mu^{(-)}$ is the charged W-boson field, g is the gauge coupling of the weak SU(2) group and $J_\mu^{(+)}$ is the charged weak current. In the

---

* Supported by the Department of Energy, contract DE-AC03-76SF00515.
† Present address: Department of Physics, Harvard University, Cambridge MA 02138.

six-quark model

$$J_\mu^{(+)} = (\bar{u}\,\bar{c}\,\bar{t})\,\gamma_\mu(1-\gamma_5)\,\mathcal{U}\begin{pmatrix} d \\ s \\ b \end{pmatrix} \quad . \tag{3}$$

$\mathcal{U}$ is a 3×3 unitary matrix which arises from the diagonalization of the quark mass matrices. u, c and t denote the quark fields with charge $+2/3$ and d, s and b the quark fields with charge $-1/3$. In general a 3×3 unitary matrix is specified by nine independent real parameters. However, five of the parameters used to specify $\mathcal{U}$ can be absorbed into the phases of the quark fields. Consequently $\mathcal{U}$ can be written in terms of only four real quantities. Three of these are Euler-type angles denoted by θ_1, θ_2 and θ_3 and the fourth is a phase denoted by δ. With the standard choice of quark field phases[5]

$$\mathcal{U} = \begin{pmatrix} c_1 & -s_1c_3 & -s_1s_3 \\ s_1c_2 & c_1c_2c_3 - s_2s_3e^{i\delta} & c_1c_2s_3 + s_2c_3e^{i\delta} \\ s_1s_2 & c_1s_2c_3 + c_2s_3e^{i\delta} & c_1s_2s_3 - c_2c_3e^{i\delta} \end{pmatrix} \quad , \tag{4}$$

where $s_i \equiv \sin\theta_i$ and $c_i \equiv \cos\theta_i$ for $i\varepsilon\{1,2,3\}$. The signs of the quark fields are chosen so that θ_1, θ_2 and θ_3 all lie in the first quadrant. Then the quadrant of the phase δ has physical significance and cannot be specified by convention. Experimental information from β-decay give $s_1^2 \approx 0.05$. Combining this with experimental information on semileptonic hyperon decays provides the limit $s_3 \lesssim 0.5$ on violations of universality.[6]

By readjusting the phases of the quark fields the phase δ can be moved from one part of the matrix $\mathcal{U}$ to another. However, it is impossible to render the matrix $\mathcal{U}$ real by readjusting the phases of the quark fields. Consequently there is CP violation when δ differs from zero. In the next two sections the phenomenology of weak interaction CP violation in the $K^0 - \bar{K}^0$ and $B_d^0 - \bar{B}_d^0$ systems is discussed.

THE $K^0 - \bar{K}^0$ SYSTEM

Within the phase convention where the $K \to \pi\pi$ $(I=0)$ amplitude, A_0, is chosen real, the quantities η_{+-} and η_{00} are approximately given by $\eta_{+-} = \varepsilon + \varepsilon'$ and $\eta_{00} = \varepsilon - 2\varepsilon'$. The quantity ε determines the eigenstates K_S and K_L in terms of the K^0 and $\bar{K}^0$ states. Explicitly

$$K_S = \frac{1}{\sqrt{2(1+|\varepsilon|2)}}\left[(1+\varepsilon)K^0 + (1-\varepsilon)\bar{K}^0\right] \tag{5a}$$

$$K_L = \frac{1}{\sqrt{2(1+|\varepsilon|2)}}\left[(1+\varepsilon)K^0 - (1-\varepsilon)\bar{K}^0\right] \quad . \tag{5b}$$

To first order in CP violating quantities

$$\varepsilon = \frac{i\left(\mathrm{Im}\,\Gamma_{12}/2 + i\mathrm{Im}\,M_{12}\right)}{\tfrac{1}{2}\left(\Gamma_S-\Gamma_L\right) + i\left(m_S-m_L\right)} \quad , \tag{6}$$

where M_{12} and Γ_{12} are the $K^0 - \bar{K}^0$ mass and width transition matrix elements. In the phase convention A_0 real

$$\varepsilon \approx \frac{e^{i\pi/4}}{2\sqrt{2}}\left(\frac{\mathrm{Im}\,M_{12}}{\mathrm{Re}\,M_{12}}\right) \quad , \tag{7}$$

where the experimental values of the K_S and K_L masses and widths have been used. In addition $m_S - m_L = 2\,\mathrm{Re}\,M_{12}$ was used. The quantity ε' measures the deviation of η_{+-}/η_{oo} from unity and is defined by

$$\varepsilon' = \frac{i}{\sqrt{2}}\, e^{i(\delta_2-\delta_0)}\, \frac{\mathrm{Im}\,A_2}{A_0} \quad , \tag{8}$$

where A_2 is the $K \to \pi\pi$ $(I = 2)$ amplitude and δ_2 and δ_0 are the $\pi\pi$ $(I = 2)$ and $\pi\pi$ $(I = 0)$ phase shifts.

The CP violation parameters ε and ε' can be computed from the matrix elements of the effective Hamiltonians for $\Delta S = 2$ $K^0 - \bar{K}^0$ mixing and $\Delta S = 1$ weak nonlepontic decays. These effective Hamiltonians are derived by a four step process in which the W-boson t-quark, b-quark and c-quark are treated as heavy and their fields removed from explicitly appearing in the theory. The choice of quark field phases made in Eq. (4) puts the CP violating phase only in the couplings of the heavy quarks to the W-bosons. Therefore CP violation in the effective Hamiltonian for $\Delta S = 1$ weak nonleptonic decays only occurs from so called penguin-type diagrams which contain a heavy quark loop. These diagrams are purely $I = \tfrac{1}{2}$ and may be responsible for the $\Delta I = \tfrac{1}{2}$ rule in nonleptonic kaon and hyperon decays.[7] The penguin-type diagrams give an imaginary CP violating part to the $K \to \pi\pi$ $(I = 0)$ amplitude A_0. Therefore, the choice of quark fields made in Eq. (4) does not correspond to the phase convention where A_0 is real. Transforming the strange quark field, $s \to e^{i\xi}s$, to make A_0 real causes the amplitude A_2 to pick up on imaginary part. Consequently

$$\varepsilon' \approx \frac{1}{20\sqrt{2}}\; e^{i\pi/4}(-\xi) \quad , \tag{9}$$

where the experimental relation $\mathrm{Re}\,A_2/A_0 \approx 1/20$ has been used. The phase which has been approximated by $\pi/4$ follows from the $\pi\pi$ phase shifts and has the experimental value $(37\pm6)^o$.

The other CP violation parameter ε arises from the matrix elements of the effective Hamiltonian for $K^0 - \bar{K}^0$ mixing.[8,9] The six-quark model parameters θ_2, θ_3 and δ may be fit to the experimental values of the $K^0 - \bar{K}^0$ mass difference and the CP violation parameter ε;

however, no prediction can be made for these quantities. The measured phase of ε forces δ to lie in the upper half plane for small s_3.[10] For s_3 near the universality bound of .5 there is also a small region of allowed angles for which δ lies in the lower half plane.[11,12] The region with δ in the lower half plane exists only for $\cos\delta<0$ while the region with δ in the upper half plane exists for both $\cos\delta<0$ and $\cos\delta>0$.[11,12,13,14]

Both ε and ε' are proportional to the combination of angles $c_2s_2s_3\sin\delta$ and their ratio ε'/ε is fairly insensitive to the values of the six-quark model parameters. There have been two approaches to estimating ε'/ε. In order to understand these computations it is necessary to know a few facts about the effective Hamiltonian for $\Delta S = 1$ weak nonleptonic decays. In the leading logarithmic approximation the effective Hamiltonian is a sum of Wilson coefficients multiplied by local four-quark operators:[7,15,16] $\mathcal{H}_{eff}^{|\Delta S|=1} = \sum_{i=1}^{6} C_i Q_i +$ h.c.. The operator Q_6 has a chiral structure $(V-A)\otimes(V+A)$ and is induced by penguin-type diagrams with a heavy quark loop.[17] The Wilson coefficient C_6 is small in magnitude compared with those of the $(V-A)\otimes(V-A)$ operators. However, the $(V-A)\otimes(V+A)$ chiral structure of Q_6 leads to enhanced matrix elements and this operator may make important contributions to nonleptonic decay amplitudes. If this is the case then an understanding of the $\Delta I = \frac{1}{2}$ rule is possible since Q_6 is purely $I = \frac{1}{2}$. The ratio $\mathrm{Im}\, C_6/\mathrm{Re}\, C_6$ is of order $c_2s_2s_3\sin\delta$ while the ratio of imaginary to real parts of the Wilson coefficients of the familiar $(V-A)\otimes(V-A)$ operators are of order $10^{-2}\ c_2s_2s_3\sin\delta$. Thus it is the matrix elements of Q_6 which contribute the largest CP violating imaginary part to the amplitude A_o. Let f be the fraction of the $K\to\pi\pi$ $(I=0)$ amplitude that arises from the matrix elements of Q_6. Then the isospin zero amplitude has a phase ξ where

$$\xi = \mathrm{Im}\, C_6 \, \frac{\langle \pi\pi(I=0)|Q_6|K^o\rangle}{A_o \quad e^{i\delta_o}} \tag{10a}$$

$$= f\, \frac{\mathrm{Im}\, C_6}{\mathrm{Re}\, C_6} \quad . \tag{10b}$$

Recall that ε' is proportional to ξ (cf., Eq. (9)). One approach to estimating ε' assumes that penguin-type diagrams are responsible for the $\Delta I = \frac{1}{2}$ rule and therefore chooses a large value for f. ε' is then calculated from Eqs. (9) and (10b) using a leading logarithmic calculation of C_6. Of course, the value of f is renormalization point dependent. It is not known what value of the renormalization point mass, μ, corresponds to the value of f used. Therefore several different values of μ are used in the computation of C_6 to get an idea of the uncertainties involved. This approach typically finds ε'/ε to be of order a fraction of a percent although the uncertainties are large.[15] The second approach to estimating ε' recognizes that a leading logarithmic calculation of $\mathrm{Re}\, C_6$ is very uncertain since it depends on integrations over virtual momenta primarily in the range $\mu^2 \lesssim p^2 \lesssim m_c^2$. In this approach Eq. (10a) is used to estimate ξ. The amplitude A_o is taken from experiment and a quark-model-type or vacuum insertion estimate of the matrix element $\langle 2\pi(I=0)|Q_6|K^o\rangle$ is used. This approach also involves an implicit choice of μ, namely that for which the matrix element estimate is correct. Predictions for ε' are, however, now not as sensitive to the value of the renormalization mass, μ, used to compute C_6 since $\mathrm{Im}\,C_6$ is less sensitive to variations in μ than $\mathrm{Re}\,C_6$. In general the values of ε'/ε obtained in this way are somewhat smaller than those obtained by the first approach.[16,18,19]

The present experimental limit is $|\varepsilon'/\varepsilon| \lesssim 1/50$, however, upcoming experiments should be able to determine ε'/ε to the fraction of a percent level.[20] The measurement of a nonzero value for ε'/ε in these experiments would be qualitative evidence that penguin-type diagrams are responsible for the $\Delta I = \frac{1}{2}$ rule. In addition information on the six-quark model parameters would be obtained from a measurement of ε'/ε. The theoretical estimates of ε'/ε predict that it is almost real and has the same sign as $\sin\delta$. If ε'/ε is measured to be negative then we would have very tight constraints on the angles θ_2, θ_3 and δ because the allowed region for these angles is very small when δ lies in the lower half plane.[11,12]

CP violation in low energy systems is characterized by the combination of angles $s_2 s_3 \sin\delta$. The experimental value of ε implies that $s_2 s_3 \sin\delta$ is of order 10^{-3}. Estimates of D_n, the electric dipole moment of the neutron,[21] indicate that $|D_n| \approx 10^{-30}$ cm when $s_2 s_3 \sin\delta \approx 10^{-3}$. The electric dipole moment is very small because first order weak diagrams do not contribute to it. Therefore, unlike strong interaction CP violation, CP violation in the weak couplings of the quarks to the W-bosons can be responsible for the observed values of η_{+-} and η_{oo} without giving too large an electric dipole moment to the neutron.

THE $B_d^0 - \bar{B}_d^0$ SYSTEM

At the present time observation of CP violation has been confined to the neutral kaon system. It may be possible to also observe CP violation in B meson decays. The analysis of the $B_d^0 - \bar{B}_d^0$ system is similar to that of the neutral kaon system. The eigenstates are

$$B_1 = \frac{1}{\sqrt{2(1+|\varepsilon|^2)}} \left[(1+\varepsilon)B_d^0 + (1-\varepsilon)\bar{B}_d^0\right] \tag{11a}$$

$$B_2 = \frac{1}{\sqrt{2(1+|\varepsilon|^2)}} \left[(1+\varepsilon)B_d^0 - (1-\varepsilon)\bar{B}_d^0\right] \quad . \tag{11b}$$

Since CP $B_d^0 = \bar{B}_d^0$ and CP $\bar{B}_d^0 = B_d^0$ the eigenstates B_1 and B_2 would also be CP eigenstates if $\varepsilon = 0$. To lowest nontrivial order in CP violating quantities

$$\varepsilon = \frac{i\left(\mathrm{Im}\,\Gamma_{12}/2 + i\,\mathrm{Im}\,M_{12}\right)}{\frac{1}{2}\left(\Gamma_{B_1} - \Gamma_{B_2}\right) + i\left(m_{B_1} - m_{B_2}\right)} \quad , \tag{12}$$

where Γ_{12} and M_{12} are the $B_d^0 - \bar{B}_d^0$ width and mass transition matrix elements. Recall that in low energy systems CP violating quantitites are characterized by the combination of angles $s_2 s_3 \sin\delta$. For heavier systems like the B mesons this is no longer true. To leading order in the large W-boson and t-quark masses the box diagram for $B_d^0 - \bar{B}_d^0$ mixing gives rise to a mass transition matrix element which is proportional to the combination of angles $s_1^2 s_2^2 (c_1 s_2 s_3 - c_2 c_3 e^{i\delta})^2$.

Then, for small s_3, $\mathrm{Im}\,M_{12}/\mathrm{Re}\,M_{12} \approx \tan 2\delta$, which can be large if δ is not small.[22] Unfortunately a calculation of the absorbtive part of the box diagram reveals that the leading contribution to the width transition matrix element is proportional to the same combination of angles so that ε is almost pure imaginary in the region where $B_d^0 - \bar{B}_d^0$ mixing is large.[23] A purely imaginary ε can be transformed away by readjusting the quark field phases so that CP violating physical quantities, like the charge asymmetry in the number of same sign dilepton events from semileptonic $B_d^0 - \bar{B}_d^0$ decays

$$\frac{\ell^{++} - \ell^{--}}{\ell^{++} + \ell^{--}} = \frac{-4\mathrm{Re}\,\varepsilon\left(1+|\varepsilon|^2\right)}{\left(1+|\varepsilon|^2\right)^2 + 4\left(\mathrm{Re}\,\varepsilon\right)^2}, \quad (13)$$

vanish when ε has no real part. It may be better to look for CP violation in processes where CP violation coming from the decay amplitudes also plays a role.[24]

REFERENCES

1. Particle data group, Phys. Lett. 75B, 1 (1978).
2. W. B. Dress et al., Phys. Rev. D15, 9 (1977).
3. V. Baluni, Phys. Rev. D19, 2227 (1979); R. J.Crewther et al., Phys. Lett. 88B, 123 (1979).
4. See also M. A. Shifman et al., Nucl. Phys. B166, 494 (1980).
5. M. Kobayashi and T. Maskawa, Prog. Theor. Phys. 49, 652 (1973).
6. R. E. Shrock and L. L. Wang, Phys. Rev. Lett. 41, 1692 (1978).
7. A. I. Vainshtein et al., Zh. Eksp. Teor. Fiz. Pisma Red 22, 123 (1975) [JETP Lett. 22, 65 (1975)]; M. A. Shifman et al., Nucl. Phys. B120, 316 (1977) and ITEP-63, ITEP-64 (1976) unpublished.
8. J. Ellis et al., Nucl. Phys. B109, 213 (1976).
9. For QCD corrections to the effective Hamiltonian in the six-quark model see: F. J. Gilman and M. B. Wise, Phys. Lett. 93B, 129 (1980).
10. F. J. Gilman and M. B. Wise, Phys. Lett. 83B, 83 (1979).
11. J. S. Hagelin, Harvard University Preprint HUTP-80/A018 (1980) unpublished.
12. B. D. Gaiser et al., SLAC-PUB-2523 (1980) unpublished.
13. V. Barger et al., Phys. Rev. Lett. 42, 1585 (1979).
14. R. E. Shrock et al., Phys. Rev. Lett. 42, 1589 (1979).
15. F. J. Gilman and M. B. Wise, Phys. Rev. D20, 2392 (1979).
16. B. Guberina and R. D. Peccei, Nucl. Phys. B163, 289 (1980).
17. Here the notation of Ref. 15 is used.
18. V. V. Prokhorov, Yad. Fiz 30, 111 (1979).
19. J. S. Hagelin, Harvard University Preprint HUTP-79/A081 (1979) unpublished.
20. R. Bernstein et al., Fermilab experiment E-617 and R. K. Adair et al., Brookhaven experiment.
21. D. V. Nanopoulos et al., Phys. Lett. 87B, 53 (1979); B. F. Morel, Nucl. Phys. B157, 23 (1979).
22. J. Ellis et al., Nucl. Phys. B131, 285 (1977).
23. J. Hagelin, Phys. Rev. D26, 2893 (1979); E. Ma et al., Phys. Rev. D26, 2888 (1979).
24. For some recent work on this see: M. Bander et al., Phys. Rev. Lett. 43, 242 (1979); A. B. Carter and A. I. Sanda, Rockefeller University Preprints DOE/EY/2232B-205 and DOE/EY/2232B-203 (1980).

PARITY VIOLATION IN NUCLEI

Dubravko Tadić
Zavod za teorijsku fiziku, Prirodoslovno-matematički fakultet,
University of Zagreb, Croatia, Yugoslavia

INTRODUCTION

Any weak interaction Hamiltonian can be divided into leptonic, semi-leptonic, strangeness-violating non-leptonic and strangeness conserving non-leptonic sectors

$$H_W = H_L + H_{SL} + H_{NL}(\Delta S=1) + H_{NL}(\Delta S=0) . \tag{1}$$

In the case of the Weinberg-Salam (WS-GIM) model the last term in (1) can give us useful information about neutral hadronic currents. Its existence which is responsible for parity-violating (PV) nuclear processes seems to be well established experimentally, as described in numerous extensive reviews of the subject.[1,2,3,4] We shall particularly stress theoretical difficulties encountered when dealing with the $n + p \to d + \gamma$ and $n + Sn^{117} \to n + Sn^{117}$ experiments. The general scheme of our paper is shown in Fig. 1.

WEAK PV POTENTIAL

The PV nucleon-nucleon scattering amplitude is usually approximated by OBE contribution as illustrated in Fig. 2. The PV amplitude $n \to p + \pi^-$ ($A(n_-^o)$) can be calculated in the same way as the hyperfine non-leptonic decay amplitudes. Recent investigations[5,6,7] based on the WS-GIM model with the inclusion of QCD corrections concluded that penguin four-quark operators, in which left-handed and right-handed currents are mixed, play the dominant role. They appear already in the "bare" WS Hamiltonian where neutral currents contain both left-handed and right-handed pieces. QCD renormalization enhances the contribution of the penguin operators. The final result (in $\hbar = c = 1$ units)

$$A(n_-^o) = (4\text{-}12)\cdot 10^{-8} \tag{2}$$

shows an enhancement between 5-15 times in comparison with the old value which was found by using charged currents only (i.e., Cabibbo-model) and by connecting the $A(n_-^o)$ amplitude to the hyperon non-leptonic decay amplitudes through sum rule which is a natural generalization of the Lee-Sugawara sum rule.

The vector-meson exchange contribution (we denote those generically by the ρ-meson) can be approximated by the vacuum-saturated diagrams (separable contributions), i.e.,

$$\langle N_1' N_2' | H_W | N_1 N_2 \rangle \sim \text{const.}(\langle N_1' | A_\lambda | N_1 \rangle \langle N_2' | V_\lambda \; N_2 \rangle + \langle N_2' | \tilde{A}_\lambda | N_1 \rangle \langle N_1' | \tilde{V}_\lambda | N_2 \rangle) . \tag{3}$$

The second term is obtained by Fierz-transforming products of currents. As current form factors are dominated by vector mesons one can easily interpret the axial vector current matrix element as a source of the ρ-meson.

The PV NNπ and NNρ amplitudes can be connected through a sum rule based on $SU(6)_W$ symmetry.[8] It has been found that the theoretical calculations (2) and (3) do not satisfy that sum rule and that additional terms might be needed in which a $1/2^-$ baryon resonance is exchanged.[9] Direct calculations based on quark models can fix absolute signs of all amplitudes.[10,11] Obtained results appear to be in good agreement with the $SU(6)_W$ sum rule. PV V_W has a complex structure:

$$V_W = V_1(\Delta I=0) + V_2(\Delta I=0)\vec{\tau}_1\cdot\vec{\tau}_2 + V_3(\Delta I=2)(3\tau_{1Z}\tau_{2Z}-\vec{\tau}_1\cdot\vec{\tau}_2)$$
$$+ V_4(\Delta I=1)(\tau_1+\tau_2)_Z + V_5(\Delta I=1)(\tau_1-\tau_2)_Z + iV_\pi(\vec{\tau}_1\times\vec{\tau}_2)_Z \ . \quad (4)$$

We have displayed here only isospin (τ) dependence of V_W. As an illustration we show explicitly only the pion exchange term

$$V_\pi = C_\pi(\vec{\sigma}_1+\vec{\sigma}_2)\ [\vec{p}_{12},\exp(-m_\pi r)/r]_- \ . \quad (5)$$

The form of V_W is quite general, what changes is its strength, i.e., C_π in (5) which is proportional to the A(n$^{\underline{o}}$) amplitude. There were also attempts to change potential strengths in (4) in such a way as to fit all, then existing, experimental data.[12,13]

The importance of the $1/2^-$ state contributions is illustrated in the following Table I.

Table I. Potential strengths[3] in 10^{-8} fm^3

	V_1	V_2	V_3	V_4	V_5
Approx. by formula 3	0.48	-1.69	0.28	0.50	0.53
$1/2^-$ intermediates or $SU(6)_W$	3.13	3.99	0.28	0.81	0.73

HEAVY NUCLEI AND p-p EXPERIMENTS

The theoretical potential, corresponding to the last row of Table I, is in reasonable agreement with the heavy nuclei PV experiments (O^{16},F^{18},F^{19},Ne^{21},K^{41},Lu^{175},Ta^{181}) and with PV p-p scattering. This can be illustrated for Ta^{181} where experiment found significant circular polarization of the emitted γ's: $P_\gamma = -(5.2\pm0.5)\cdot10^{-6}$. Theoretical predictions for various V_W are summarized in Table II. In the scattering of the longitudinally polarized protons there are two existing measurements of the cross-section asymmetry Az; Az(15 MeV) = $(-1.7\pm0.8)\cdot10^{-7}$ and Az(45 MeV)[14] = $(-3.2\pm1.1)\cdot10^{-7}$. The relative magnitudes are consistent with theoretical expectation that Az(15 MeV) $\cong$ 1/2 Az(50 MeV). As PV π-exchange cannot contribute to the p-p scattering the theoretical determination of Az tests ρ-exchange PV potential. One finds[3] for the 15 MeV protons: Az(formula 3) = $+0.36\cdot10^{-7}$ while Az($1/2^-$ or $SU(6)_W$) = $-3.3\cdot10^{-7}$.

Table II. Ta^{181}

V_W	$P_\gamma(TH)\cdot 10^6$
ρ formula 3 π Cabibbo	+1.71
ρ $SU(6)_W$ π Cabibbo	-6.72
ρ formula 3 π penguin	-0.42
ρ $SU(6)_W$ π penguin	-8.85

The reproduction of the correct sign is a decisive test for the theory. The methods used in calculating V_W and its effects could not be expected to predict magnitudes very accurately.

$$n + p \rightarrow d + \gamma$$

Polarized neutrons lead to the asymmetry A_γ in the photon emission with respect to the neutron spin orientation. The published result $A_\gamma = (6\pm21)\cdot10^{-8}$ is too inaccurate for any firm conclusion. A_γ is completely determined by the π-exchange. If this is dominated by the penguin terms, one expects $A_\gamma(TH) \sim (2\text{-}6)\cdot10^{-8}$ (otherwise it would be 5 - 15 times smaller). The theoretical uncertainty is due to the treatment of the penguin terms, as the nuclear physics calculations can be performed quite accurately.

The other reaction, with randomly oriented neutrons, leads to the relatively large circular polarization of the emitted photons $P_\gamma = -(130\pm45)\cdot10^{-8}$. This effect is completely determined by the ρ-exchange contribution. On the basis of Ref. 15 we can estimate: P_γ(formula 3, no neutral currents) = $2.3\cdot10^{-8}$; P_γ(formula 3, WS model) = $5.9\cdot10^{-8}$ and $P_\gamma(SU(6)_W$, WS model) = $4.8\cdot10^{-8}$. The last result corresponds to the combination which worked well for complex nuclei and for p-p scattering. In the more empirical approach, potential strengths are used to fit all experiments mentioned so far.[12,13] Alternatively one might speculate about the relativistic effects.[16,17,18] The admixture of the small components in the nuclear wavefunctions destroys the isospin selection rules and P_γ can receive contributions from the penguin enhanced π-exchange also.[18]

$$n + Sn \rightarrow n + Sn$$

Experiment[19] has found much greater n-spin rotation in Sn than envisaged by naive theoretical estimates. The numbers in Table III show that effect obviously depends on nuclear structure. Neutron

Table III. n-spin rotation

	J^P	%	$\phi/10^{-6}$ rad cm^{-1}
Sn natural		100	$-(4.95 \pm 0.93)$
${}_{50}Sn^{124}_{74}$	0^+	5.9	$-(0.63 \pm 0.95)$
${}_{50}Sn^{117}_{67}$	$1/2^+$	7.6	$-(38 \pm 5)$

scattering on the atomic electrons must be weak.[20,21] In the first theoretical description we may thus safely limit ourselves to the study of the PV non-leptonic interactions.

Spin of thermal neutrons traveling through a medium will rotate about the direction of propagation by angle ϕ.[20,22] In the first order of weak interactions effect gets contributions from two diagrams shown in Fig. 3. Direct scattering (DS) (Fig. 3a) was investigated first.[20,22] It is essentially determined by Born term[20]

$$f_W^{DS} = -(M/2\pi) \langle \phi_f \Omega_o | V_W | \Omega_o \phi_i \rangle \tag{6}$$

Here ϕ and Ω_o are neutron and nucleus wavefunctions, respectively. A p-wave resonance in the n-nucleus channel can greatly enhance the value of the matrix element (6), even when scattering happens at energies far from the resonance. This effect has been theoretically confirmed through the study of solvable models.[24,25] A related strong nuclear structure enhancement appears also in the reaction $n + Sn^{117} \to S^{118} + \gamma$. In the indirect scattering (IS, Fig. 3b) PV effects appear because non-leptonic weak interactions mix nuclear states of the opposite parity.[26] Nuclear state is no longer Ω_o but a mixture

$$|\Omega\rangle = |\Omega_o\rangle + \sum_n |\Omega_n\rangle\langle\Omega_n|V_W|\Omega_o\rangle (E_o-E_n)^{-1} = |\Omega_o\rangle + \mathcal{F}|\tilde{\Omega}\rangle . \tag{7}$$

Scattering is strong, but between nuclear states of the opposite parity

$$((-)2\pi/M) f_W^{IS} = \langle \phi_f \Omega | T_S | \Omega \phi_i \rangle = \mathcal{F}(\langle \phi_f \tilde{\Omega} | T_S | \Omega_o \phi_i \rangle + \langle \phi_f \Omega_o | T_S | \tilde{\Omega} \phi_i \rangle) \tag{8}$$

First calculation[21] of such a contribution in Bi^{209}, which did not include any resonance effects, concluded that $f_W^{IS} < f_W^{DS}$. However a resonance can lead to a considerable enhancement of f_W^{IS}.[27] The whole scattering amplitude can be expressed using the two-channel reaction matrix formalism

$$f_W \sim (1-iK_o)^{-1}_{s,p} K_W (1-ik_o)_{p,s} \qquad K = K_o(\text{strong}) + K_W . \tag{9}$$

A general expression of the form (9) contains in itself actually both DS(6) and IS(8) contributions. This is easily seen by noting that the outgoing scattering state $\psi^{(+)}(\phi_1\Omega)$ is connected to the principal value state ψ_p, which in turn determines Born approximation for the K-matrix. The formulae

$$\psi_\ell^{(+)} = \exp(i\delta_\ell)\cos\delta_\ell \psi_{p\ell} ; (1-iK_o)^{-1}_\ell = \exp(i\delta_\ell)\cos\delta_\ell ; K_W = \langle \psi_p | V_W | \psi_p \rangle \tag{10}$$

speedily establish connection between (6) and (9). If there is a p-wave resonance in n-Sn^{117} scattering at $E = 1.3$ eV the measured neutron width Γ_p for Sn^{117} and the largest Γ_s found in that energy range appear in the approximation of the formula (9):

$$f_W^{IS} \sim \text{Re}[p^{-1}(p/p_o)^2 \exp(i\delta_s + i\delta_p)\cos\delta_p \cos\delta_s \cdot \mathcal{F} \cdot (E-E_o)^{-1}\sqrt{\Gamma_p\Gamma_s}] \tag{11}$$

The obtained enhancement[27] might be sufficient to explain the experimental result.

In all detailed theoretical calculations one should use the most sophisticated V_W available. But neither such potential nor the semiempirical potentials[12,13] can explain the experimental result without the enhancement produced by nuclear resonance. In order to illustrate that point we have roughly estimated ϕ using (6) and reducing V_W to an effective single particle potential.[22] While the simple estimate of Ref. 20 gave $\phi = 2.43$ (in 10^{-8} rad cm^{-1}) a semiempirical potential with large isotensor component[12] gives $\phi = -55.6$ in the same units. The sophisticated theoretical potential predicts that $\phi = -8.9$. The result changes sign if V_π is multiplied by 1/4.

Full explanation of the Sn^{117} effect will probably combine both the nuclear resonance and the correct V_W. Obviously the prediction of the correct sign will depend on the latter.

CONCLUSION

Theoretically derived V_W based on WS model and QCD is in rough qualitative agreement with the majority of the nuclear PV experiments. It is an open question whether this agreement, rough as it is, is really statistically significant.[28] Two experiments, $n + p \rightarrow d + \gamma$ and $n + Sn^{117}$, still require a theoretical explanation. In the first case theoretical confusion might be cleared up if the reverse reaction is performed. Spin precession experiments in other nuclei than Sn will be also very helpful.

The author wishes to express appreciation for the hospitality of the Department of Physics, Purdue University and to thank Prof. E. Fischbach for kindly reading the manuscript.

REFERENCES

1. E. Fischbach and D. Tadić, Phys. Rep. 6C, 123 (1973).
2. M. Gari, Phys. Rep. 6C, 317 (1973).
3. B. Desplanques, invited talk at the 8th International Conference of High Energy Physics and Nuclear Structure, Vancouver (1979), Orsay preprint.
4. D. Tadić, Rep. Progr. Phys. 43, 67 (1980) and reviews listed therein.
5. J. G. Körner, G. Kramer and J. Willrodt, Phys. Lett. 81B, 365 (1979).
6. B. Guberina, D. Tadić and J. Trampetić, Nucl. Phys. B152, 429 (1979).
7. F. Bucella, M. Lusignoli, L. Maiani, and A. Pugliese, Nucl. Phys. B152, 461 (1979).
8. B. H. J. McKellar and P. Pick, Phys. Rev. D6, 2184 (1972); D7, 260 (1973); D8, 265 (1973).
9. B. Desplanques, J. F. Donoghue and B. R. Holstein, Ann. of Phys. 124, 449 (1980).
10. B. Desplanques and J. Micheli, Phys. Lett. 68B, 339 (1977).

11. D. Palle, I. Picek, D. Tadić, and J. Trampetić, Nucl. Phys. B166, 149 (1980). I. Picek, D. Tadić and J. Trampetić, IRB-TP-6-1980 preprint.
12. M. A. Box, A. J. Gabric, K. R. Lassey, and B. H. J. McKellar, J. Phys. G.:Nucl. Phys. 2, L107 (1976).
13. B. Desplanques and J. Missimer, Nucl. Phys. A300, 286 (1978); Phys. Lett. 84B, 363 (1979).
14. R. Balzer, et al., Phys. Rev. Lett. 44, 699 (1980).
15. B. A. Craver, E. Fischbach, Y. E. Kim, and A. Tubis, Phys. Rev. D13, 1376 (1976).
16. V. M. Dubovik and A. P. Kobushkin, preprint JINR, Dubna, USSR (1978).
17. S. Morioka and T. Ueda, Prog. Theor. Phys. 60, 299 (1978).
18. V. B. Kopeliovich, Phys. Lett. 78B, 529 (1978).
19. W. Forte, et al., Abstract APS Meeting, Washington, April 1980; Washington meeting talk by B. Heckel.
20. L. Stodolsky, Phys. Lett. 50B, 352 (1974).
21. D. Tadić and A. Borroso, Nucl. Phys. A294, 376 (1978).
22. F. C. Michel, Phys. Rev. 133, B329 (1964).
23. M. Forte, ILL Research Proposal 03-05-002 (1976); Inst. Phys. Conf. Ser. 42, 86 (1978).
24. G. Karl and D. Tadić, Phys. Rev. C16, 1726 (1977); C20, 1959 (1979).
25. A. Barroso and F. M. Margaça, J. Phys. G.:Nucl. Phys. 6, 657 (1980).
26. R. J. Blin-Stoyle, private communication, 1977.
27. L. Stodolsky, SLAC-PUB-May 1980 preprint.
28. J. D. Bowman, B. F. Gibson, P. Herczeg, and E. M. Henley, LA-UR-79-1423 preprint, 1979.

PV THEORY AND EXPERIMENT

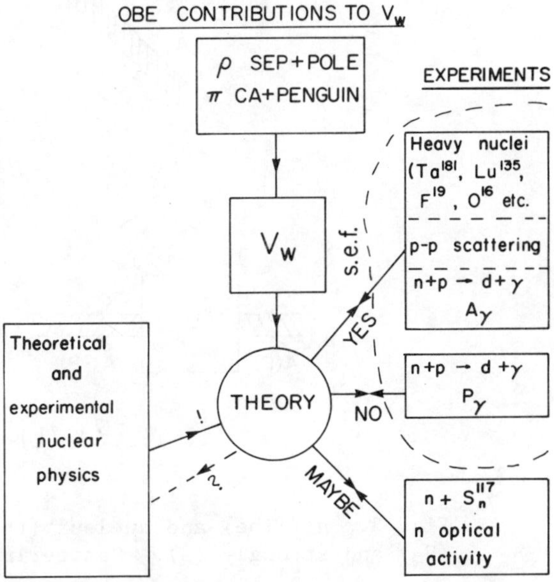

Fig. 1. Here s.e.f. indicates experiments used in the semi-empirical fit for V_W.

PV N-N SCATTERING

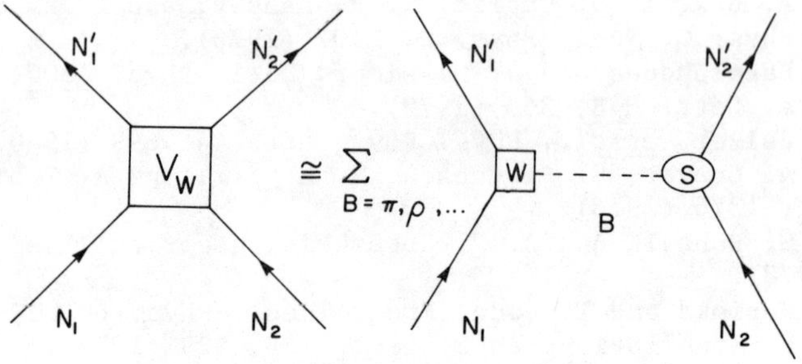

Fig. Weak W and strong S vertices are indicated.

n-NUCLEUS SCATTERING

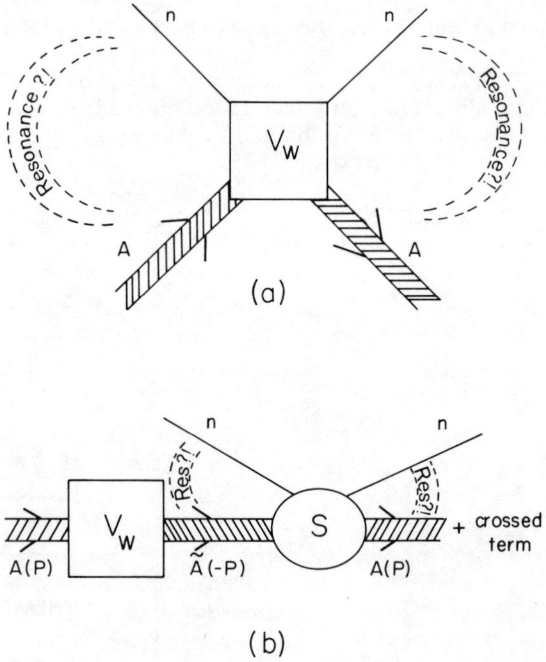

Fig. 3. n (line) and nucleus (thick line) can interact weakly (V_W) and strongly (S). Scattering states might resonate.

HORIZONTAL INTERACTIONS

R. D. Peccei
Max-Planck-Institut für Physik und Astrophysik, Munich (Fed.Rep.Germany)

ABSTRACT

The use of Horizontal symmetries to constrain the fermion mass matrix is discussed, with particular attention payed to a gauged $SU(2)xU(1)xG_HxU(1)_{B-L}$ group, with G_H = SU(2) or SU(3). The feebleness of generation changing transitions is related to the near masslessness of the neutrinos.

INTRODUCTION

The W-S-GIM SU(2)xU(1) model[1] is by now well established[2]. Nevertheless there are features which are not well understood, like the pattern of fermion masses and Cabibbo angles or the phenomena of family replication. In the standard model fermions are organized in left-handed doublets and right-handed singlets and the fermion masses are arbitrary since they are related to arbitrary Higgs-fermion Yukawa couplings. To determine the fermion mass matrices, and hence the Cabibbo angles, one must find a symmetry reason to restrict the Yukawa couplings. This suggests imposing an intrafamily-Horizontal-interaction, which could be either discrete or continuous. Discrete Horizontal symmetries have the advantage of not giving rise to Goldstone bosons under spontaneous breakdown and have been the subject of numerous papers recently[3]. The "typical" result of these papers are formulas for the Cabibbo angles in terms of quark masses. Discrete Horizontal symmetries have, however, some disadvantages:

(1) To obtain reasonable results one often needs to use a larger group than SU(2)xU(1). Indeed, under certain conditions, one can prove[4] that no useful formulas can arise from SU(2)xU(1) for the Cabibbo angles.

(2) No good criterion exists for choosing a particular discrete symmetry.

(3) If the Higgs fields were to be replaced by a dynamical spontaneous breakdown mechanism, it is difficult to see how discrete symmetries could ensue.

Here I will focus only on continuous Horizontal symmetries, which must be gauged to avoid unwanted Goldstone bosons. Clearly these symmetries are more dynamical, but one has to be sure to respect the known experimental bounds for intrafamily transitions. One can consider very large groups which have mixed Horizontal and Vertical interactions[5]. I will consider here, however, only the case in which the Horizontal group G_H is separate from the vertical, SU(2)xU(1), group. In this case the Horizontal gauge bosons are neutral. Recent papers on Horizontal interactions have used G_H=U(1)[6] or G_H=SU(2)[7]. I want to discuss particularly the phenomenology of a model based

0094-243X/81/680411-04$1.50

on the group $SU(2)xU(1) \times G_H xU(1)_{B-L}$, with $G_H=SU(2)$, or $SU(3)$, which I studied in collaboration with Y. Chikashige, G. Gelmini and M. Roncadelli[8]. Very similar ideas are also due to Yanagida[9]

$SU(2) \times U(1) \times G_H \times U(1)_{B-L}$ MODEL

The model is based on two primary assumptions:

(1) There are 3 generations of fermions and as far as G_H goes the fermions of a given charge and helicity are treated symmetrically. Thus $G_H=SU(2)$ or $SU(3)$.Clearly, if there are more generations this assumption can be trivially modified.

(2) Only Higgs mesons with quantum numbers of bound states of the fermion-(anti)fermions in the theory will be allowed. This last assumption is motivated by the hope of one day replacing the Higgs bosons by fermion condensates.

If $G_H=SU(2)$ there are no problems with anomalies[10]. For $G_H=SU(3)$, if one wants to cancel anomalies, one must introduce a triplet of right handed neutrinos. Then anomalies can be cancelled by either making the Horizontal interaction totally vector-like or by the unconventional assignments

$$3:\quad (u,c,t)_L;\quad (\nu_e,\nu_\mu,\nu_\tau)_L;\quad (d,s,b)_R;\quad (e,\mu,\tau)_R \tag{1a}$$

$$\bar{3}:\quad (u,c,t)_R;\quad (\nu_e,\nu_\mu,\nu_\tau)_R;\quad (d,s,b)_L;\quad (e,\mu,\tau)_L \tag{1b}$$

The allowed Higgs fields that contribute to the fermion mass matrix are all doublets under $SU(2)xU(1)$ and transform under G_H as follows

$$G_H = SU(2) \qquad \Phi \sim 3 \times 3 = 5 + 3 + 1 \tag{2a}$$

$$G_H = SU(3)_{conv} \qquad \Phi \sim \bar{3} \times 3 = 8+1 \tag{2b}$$

$$G_H = SU(3)_{unconv} \qquad \Phi \sim 3 \times 3 = 6 + \bar{3} \tag{2c}$$

For a given fermionic charge the mass matrix takes the form

$$M_{ij}(Q) = \sum_a c_a(Q) < \Phi_{ij}(a)> \tag{3}$$

Here a runs over the various Higgs representations. Clearly (3) is too general to be of use. To get some results one must assume that some representation(s) dominate and that the resultant breaking is along some preferred direction. In particular, the large disparity between the masses of leptons and quarks suggests that as a lowest approximation one should have

$$M_{diag}(Q) \cong m(Q) \begin{pmatrix} 0 & & \\ & 0 & \\ & & 1 \end{pmatrix} \tag{4}$$

Such a pattern is only possible for $G_H=SU(2)$ or $SU(3)_{conv}$ if a delicate cancellation between two Higgs representations is arranged. However, as first emphasized by Gell Mann[11], if one has a sextuplet of Higgs, as in our $SU(3)_{unconv}$ case, then (4) emerges directly.

Two questions emerge immediately: How does one go from this zero-order pattern to the real life pattern and, if neutrinos have a mass, why is their mass so much less than that of the charged fermions. The answer to the first question is not known, although it is presumed that perhaps the remaining small masses can arise out of radiative corrections.[12] The second question has an easy answer if one assumes that the right-handed neutrinos have a large Majorana mass, M. Then the physical neutrino masses, neglecting intra generation mixing, come from diagonalizing the 2x2 matrix

$$M_{\nu} = \begin{pmatrix} 0 & m \\ m & M \end{pmatrix} \tag{5}$$

where m is the neutrino ordinary Dirac mass. For $m \ll M$ one of the eigenvalues is M while the other is $m^2/M \ll m$. We see thus that the physical neutrinos, even if m is comparable to the charged fermion masses, could naturally have small masses.

The Majorana mass term for the neutrinos arises from a Higgs field $(\Phi_{Maj})_{ij}$ coupled to $\nu_{Ri}\nu_{Rj}$. Such a field is necessarily an SU(2)xU(1) singlet but carries Horizontal quantum numbers. The Majorana mass is proportional to $\langle\Phi_{Maj}\rangle$ while the Dirac mass is proportional to $\langle\Phi\rangle$. To obtain light neutrinos we require $\langle\Phi_{Maj}\rangle \gg \langle\Phi\rangle$. This requirement, is precisely what one needs <u>also</u> to guarantee that the Horizontal interactions do not violate any known experimental bounds on generation changing processes, like $\mu \to e\gamma$. If $\langle\Phi_{Maj}\rangle \gg \langle\Phi\rangle$ the effective Fermi constant for Horizontal interactions is much less than G_F

$$G_H \sim \frac{1}{\langle\Phi_{Maj}\rangle^2} \ll \frac{1}{\langle\Phi\rangle^2} \sim G_F \tag{6}$$

Recent analysis of generation changing processes[8,13] give the bound $G_H/G_F \leq 10^{-4} - 10^{-6}$. If one uses the plausible formula

$$m_{\nu_e} = m\left(\frac{m}{M}\right) \cong m\left(\frac{\langle\Phi\rangle}{\langle\Phi_{Maj}\rangle}\right) \tag{7}$$

taking $m \cong 1$ Mev and using $m_{\nu_e} \cong 30$ ev, as indicated by the ITEP experiment[14], one deduces $G_H \cong 10^{-9}\ G_F$ which is well within the bounds.

Because Φ_{Maj} carries lepton number, the assumption that $\langle\Phi_{Maj}\rangle \neq 0$ would imply an unwanted Goldstone boson unless lepton number is gauged. In fact, what must be gauged is B-L so as to avoid anomalies. Thus the final local group for the model is SU(2)xU(1) x G_HxU(1)$_{B-L}$. We note that the effective Fermi coupling for the B-L interactions is also small, being of the same order as G_H. Because of the lepton number violating Majorana mass in the neutrino sector, neutrinoless double β-decay is in principle allowed. However, in the present model one is well below the bounds found in the analysis of Halprin et al.[15]. These bounds are only threatened if there is a large $\nu_e - \nu_\tau$ mixing and m_{ν_τ}/m_{ν_e} scales as m_τ/m_e.

REFERENCES

1. S. Weinberg, Phys.Rev.Lett. 19, 1264 (1967); A. Salam, Proceedings of the 8th Nobel Symposium, ed. by N. Svartholm (Almquist and Wiksells, Stockholm 1968), p. 367; S.L. Glashow, J. Iliopoulos and L. Maiani, Phys.Rev. D2, 1285 (1970).
2. J.E. Kim et al., Contributed paper (# 478) to this Conference.
3. Some representative references include, S. Weinberg, Trans N.Y. Acad. of Sciences II, 38 (1977); H. Fritzsch, Phys.Lett. 70B, 43 (1977) and 73B, 602 (1978); F. Wilczek and A. Zee, Phys. Lett. 70B, 418 (1977); G. Ecker et al., Contributed paper (# 604) to this Conference.
4. R. Barbieri, R. Gatto and F. Strocchi, Phys.Lett. 74B, 344 (1978).
5. See for example the talk of I. Bars at this Conference.
6. A. Davidson, M. Koca and K.C. Wali, Phys.Rev. D20, 1195 (1979); A.K. Ray and M.M. Kundu, Contributed paper (# 744) to this Conference.
7. F. Wilczek and A. Zee, Phys.Rev.Lett 42, 421 (1979); C.L. Ong, Phys.Rev. D19, 2739 (1979); J. Chakrabarti, Phys.Rev. D20, 2411 (1979).
8. Y. Chikashige, G. Gelmini, R.D. Peccei and M. Roncadelli, MPI preprint MPI-PAE/PTh 17/80, to be published in Phys.Lett. B.
9. T. Yanagida, Tohoku University preprint TU/80/208.
10. H. Georgi and S.L. Glashow, Phys.Rev. D6, 429 (1972).
11. M. Gell Mann, private communication.
12. M. Gell Mann, P. Ramond and R. Slansky in "Supergravity", ed. by P. Van Nieuwenhuizen and D. Freedman (North Holland, Amsterdam 1979) p. 315.
13. G. Kane and R. Thun, Univ. of Michigan preprint UM HE 80-8; R.N. Cahn and H. Harari, Berkeley preprint, LBL-10823.
14. V.A. Lyubimov et al., ITEP preprint ITEP-62.
15. A. Halprin et al., Phys.Rev. D13, 2567 (1976).

OPERATOR ANALYSIS OF NEW PHYSICS

H. Arthur Weldon
University of Pennsylvania, Philadelphia, Pa. 19104

ABSTRACT

A clear signal for the existence of new physics beyond M_W would be the non-conservation of either the separate lepton numbers L_e, L_μ, L_τ, of total lepton number L, or of baryon number B. At low energy these violations are described by an effective theory containing only the known quarks and leptons with vertices of dimension five or higher that are SU(3)xSU(2)xU(1) invariant. Here we discuss some of the relations imposed by this invariance on the numerous processes which might violate L_e, L_μ, L_τ, L, or B.

INTRODUCTION

Experiments at presently available energies all confirm the standard SU(3)xSU(2)xU(1) theory of strong, weak, and electromagnetic interactions.[1] Many theorists expect that this is not the ultimate theory but that at higher energy scales there are new types of interactions, e.g. $SU(2)_L \times SU(2)_R \times U(1)$, horizontal gauge interactions between generations, super-strong hypercolor interactions that dynamically break SU(2) x U(1), SU(5) or SO(10) gauge interactions that unify the standard interactions, and many others.

The obvious phenomenological question that we wish to discuss here is how can we learn about physics beyond the standard SU(3) x SU(2) x U(1) theory? The answer is provided by the technique of effective Lagrangians[2]: Given the full Lagrangian containing all the new interactions we must integrate out the superheavy fields to obtain an effective field theory of ordinary quarks and leptons. If the new interactions occur at a mass scale M*, the result of this computation will be

$$\mathcal{L}_{(\text{eff})} = \mathcal{L}^R_{(\text{standard})} + \sum_{n=5}^{\infty} \left(\frac{1}{M^*}\right)^{n-4} \mathcal{O}^{(n)}$$

where the running coupling constants of $\mathcal{L}^R_{(\text{standard})}$ depend logarithmically on M^* and where $\mathcal{O}^{(n)}$ are dimension n products of ordinary quark and lepton fields that are invariant under SU(3) x SU(2) x U(1) and Lorentz transformations. These new operators are non-renormalizable (i.e. $n \geq 5$) but constrained by the low energy invariance group. This is precisely analogous to Fermi's analysis in 1934 of beta decay: $U(1)_{EM}$ and Lorentz invariance there reduced the possible dimension six operators to S, V, T, A, P.

To learn about new physics we must construct the operators $\mathcal{O}^{(n)}$. Unfortunately, many of the $\mathcal{O}^{(n)}$ are not really indicative of new physics because they will also arise as radiative corrections to $\mathcal{L}_{(\text{standard})}$. For example, the dimension five operator $(\phi^+\tau_n\phi)W_{n,\mu\nu} \times B^{\mu\nu}$, which modifies $M_W = M_Z \cos\theta$, will be generated by $\mathcal{L}_{(\text{standard})}$ even without any new interactions.

The resolution of this dilemma lies in a remarkable property of $\mathcal{L}$(standard): Even though the standard SU(3) x SU(2) x U(1) theory contains all possible operators of dimension four or less it automatically conserves L_e, L_μ, L_τ, and B. Hence any operator $\mathcal{O}^{(n)}$ that violates one of these quantum numbers cannot come from a radiative correction to $\mathcal{L}$(standard). It therefore signals the existence of new physics beyond M_W. To learn something about this new physics we must separate the constraints of SU(3) x SU(2) x U(1) x Lorentz invariance from the dynamics of the new interactions. A systematic catalogue of all operators of low dimension (five, six, seven,...) that violate L_e, L_μ, L_τ, L, or B has been constructed with A. Zee.[3] Here we will discuss only a portion of that catalogue. The most important fact that should impress the reader is that for each type of process the list of allowed operators is fairly short. This should make a phenomenological determination of the operator coefficients possible. As an added bonus, we occasionally find relations between processes that are independent of the coefficients and rely only on the SU(3) x SU(2) x U(1) framework.

The notation we employ is $\ell_{eL} = (\nu_e, e_L)$, $\ell_{\mu L} = (\nu_\mu, \mu_L)$, $\ell_{\tau L} = (\nu_\tau, \tau_L)$, $q_L = (u_L, d_L\cos\theta + s_L\sin\theta)$, $\phi = (\phi^+, \phi^o)$ for SU(2) doublets (with Y/2 = -1/2, -1/2, 1/6, 1/2, respectively), and $(\tilde{\ell}_L)_i = \varepsilon_{ij}(\ell_L)_j$, $(\tilde{q}_L)_i = \varepsilon_{ij}(q_L)_j$, $(\tilde{\phi})_i = \varepsilon_{ij}(\phi)_j$ for their SU(2) duals.

$\Delta L = 0$ OPERATORS WITH $\Delta L_e = -\Delta L_\mu \neq 0$

Two Fermions: Of the operators that violate L_e and L_μ separately but preserve $L_e + L_\mu$, the simplest contain only one e and one $\bar{\mu}$:

$$A_1 = (\bar{\ell}_{\mu L}\, \phi\, \sigma_{\alpha\beta}\, e_R)\, B^{\alpha\beta}$$

$$A_2 = (\bar{\ell}_{\mu L}\, \tau_n\, \phi\, \sigma_{\alpha\beta}\, e_R) W_n^{\alpha\beta} \quad ,$$

These are the dimension six possibilities. Their coefficients would necessarily contain a factor $(M^*)^{-2}$. With ϕ replaced by its vacuum expectation value it is easy to see that each operator induces both $Z^o \to \mu^- e^+$ and $\mu^- \to e^- \gamma$. Thus a measurement of $BR(Z^o \to \mu^- e^+)/BR(\mu^- \to e^- \gamma)$ would determine the relative strengths of A_1 and A_2.

Four Fermions, semi-leptonic: Within this class the most interesting are those containing two neutrinos:

$$B_1 = (\bar{\ell}_{\mu L}\, \tau_n\, \gamma_\alpha\, \ell_{eL})\, (\bar{q}_L\, \tau_n\, \gamma^\alpha\, q_L)$$

$$B_2 = (\bar{\ell}_{\mu L}\, \gamma_\alpha\, \ell_{eL})\, (\bar{q}_L\, \gamma^\alpha\, q_L)$$

$$B_3 = (\bar{\ell}_{\mu L}\,\gamma_\alpha\,\ell_{eL})\,(\bar{u}_R\,\gamma^\alpha\,u_R)$$

$$B_4 = (\bar{\ell}_{\mu L}\,\gamma_\alpha\,\ell_{eL})\,(\bar{d}_R\,\gamma^\alpha\,d_R)$$

These operators are analogous to those used in the model independent analysis of neutrino neutral current interactions. Because they change lepton flavor they all induce $\nu_\mu q \to \nu_e q$ oscillations in matter. Wolfenstein[4] showed that any such coupling with strength $G_F \sin^2\alpha/\sqrt{2}$ will have an oscillation length $\lambda \approx 10^4\text{km}/\sin^2\alpha$. Here $\sin^2\alpha \approx (M_w/M^*)^2$. Unfortunately, B_1 and B_2 also produce $K_L^0 \to \mu^- e^+$ so that we expect $M_w/M^* \leq 10^{-2}$ and consequently $\lambda \geq 10^8\text{km}$.

Four Fermions, purely leptonic: In this class we can relate such processes as $\nu_\mu e^- \to \nu_e e^-$ to $\mu^- \to e^- e^+ e^-$.

$\Delta L = 2$ OPERATORS

Two Fermions: The lowest possible dimension here is five. There are only two operators:

$$D_{ab}^{(1)} = (\tilde{\ell}_{aL}\,C\ell_{bL})(\tilde{\phi}\phi') \qquad (a \neq b)$$

$$D_{ab}^{(2)} = (\tilde{\ell}_{aL}\,C\,\tau_n\,\ell_{bL})(\tilde{\phi}\,\tau_n\,\phi)\ ,$$

where $a,b = e,\mu$ independently. $D^{(1)}$ requires the presence of two Higgs doublets. When the Higgs are replaced by their vacuum expectation value only $D^{(2)}$ survives. It produces neutrino oscillations in vacuum (i.e. Majorana masses) $\nu_e \to \bar{\nu}_e$ or $\bar{\nu}_\mu$ and $\nu_\mu \to \bar{\nu}_e$ or $\bar{\nu}_\mu$. The magnitude of the mass is M_w^2/M^*, so that either a small coefficient or a large $M^* \approx 10^{12}$ GeV is required. Precisely the latter occurs in a specific model of Witten[5].

Four Fermions, semileptonic: The simplest operators within this class are of dimension seven:

$$E_{ab}^{(1)} = (\tilde{\ell}_{aL}\,C\,\ell_{bL})(\bar{q}_L\,\phi\,u_R)$$

$$E_{ab}^{(2)} = (\tilde{\ell}_{aL}\,C\,\tau_n\,\ell_{bL})(\bar{q}_L\,\tau_n\,\phi\,u_R)$$

$$E_{ab}^{(3)} = (\tilde{\ell}_{aL}\,\phi\,\gamma_\alpha\,\ell_{bR})(\bar{d}_R\,\gamma^\alpha\,u_R)$$

$$E_{ab}^{(4)} = (\tilde{\ell}_{aL}\,C\,D_\alpha\,\ell_{bL})(\bar{d}_R\,\gamma^\alpha\,u_R)\ ,$$

where D_α is the gauge-covariant derivative. Note that all operators contain a u_R. Hence we have the rather surprising result that, regardless of the coefficients, $\sigma(\nu_a d \to \bar{\nu}_b d) = 0$ for $a, b = e$ or μ.

ΔB = 1 OPERATORS

$\Delta B = \Delta L$ (e.g. $p \to e^+\pi^0$): These processes were historically the first to which the technique of operator analysis based on SU(3) x SU(2) x U(1) was applied.[6,7] For completeness, we display the allowed operators:

$$G_1 = (\tilde{\ell}_{eL} C q_L)(u_R C\ d_R)$$

$$G_2 = (e_R\ C\ u_R)(\tilde{q}_L\ C\ q_L)$$

$$G_3 = (\tilde{\ell}_{eL}\ C\ q_L)(\tilde{q}_L\ C\ q_L)$$

$$G_4 = (\tilde{\ell}_{eL}\ \tau_n\ C\ q_L)(\tilde{q}_L\ \tau_n\ C\ q_L')$$

$$G_5 = (e_R\ C\ u_R)(u_R\ C\ d_R)$$

$$G_6 = (e_R\ C\ d_R)(u_R\ C\ u_R')$$

where primes denote second generation quarks, viz. $q_L' = (c_L, s_L \cos\theta - d_L \sin\theta)$ and $u_R' = c_R$. Various selection rules, rate relations, and polarization tests may be deduced from this list.[6,7] Note that all these operators are of dimension six.

$\Delta B = -\Delta L$ (e.g. $p \to e^-\pi^+\pi^+$): The analysis of this and the following $\Delta B \neq 0$ processes was done simultaneously by ourselves[3] and by S. Weinberg.[8] The simplest $\Delta B = -\Delta L$ operators are of dimension seven. (For example, the dimension six combinations $\bar{\ell}_L q_L d_R d_R$ and $\bar{e}_R d_R d_R d_R$ are allowed by SU(3) x SU(2) x U(1) but cannot be made Lorentz invariant.) The allowed operators are of two types: those containing a covariant derivative, e.g. $(\bar{e}_R \gamma_\alpha d_R)(d_R\ C\ D^\alpha\ d_R)$, and those containing a Higgs field. Since the first type is suppressed by a factor M_N/M_W relative to the second, we shall only discuss the latter. With only one generation of quarks there are only two allowed operators:

$$H_1 = (\bar{\ell}_{eL}\ \tilde{\phi}^+\ d_R)(d_R\ C\ u_R)$$

$$H_2 = (\bar{\ell}_{eL}\ \tilde{\phi}^+\ d_R)(\tilde{q}_L\ C\ q_L).$$

When ϕ is replaced by its vacuum expectation value a curious thing happens. Both operators collapse to $(\bar{\nu}_e d_R)(d_R\ C\ u_R)$ and contain no e^-. To produce an e^- we must examine the operators allowed when more quark generations are present:

$$H_3 = (\bar{e}_R\ \phi^+\ q_L)(d_R\ C\ s_R)$$

$$H_4 = (\bar{\ell}_{eL}\ \tau_n\ \tilde{\phi}^+\ d_R)(\tilde{q}_L\ \tau_n\ C\ q_L')$$

$$H_5 = (\bar{\ell}_{eL}\ \tilde{\phi}^+\ d_R)(\tilde{q}_L\ C\ q_L')$$

$$H_6 = (\bar{\ell}_{eL} \tilde{\phi}^+ s_R)(\tilde{q}_L \, C \, q_L)$$

$$H_7 = (\bar{\ell}_{eL} \, \phi \, d_R)(d_R \, C \, s_R)$$

$$H_8 = (\bar{\ell}_{eL} \tilde{\phi}^+ d_R)(s_R \, C \, u_R)$$

$$H_9 = (\bar{\ell}_{eL} \tilde{\phi}^+ s_R)(d_R \, C \, u_R)$$

$$H_{10} = (\bar{\ell}_{eL} \tilde{\phi}^+ u_R)(d_R \, C \, s_R).$$

When ϕ is replaced by its vacuum expectation value only H_3, H_4, and H_7 can produce an e^- (or a μ^-). Of these, H_3 and H_7 always contain an s quark but H_4 is a Cabibbo mixture of $\Delta S = 1$ and $\Delta S = 0$ pieces. We thus obtain the rather surprising rule that for nucleon decays into negatively charged leptons the $\Delta S = 1$ processes (e.g. $p \to e^- K^+ \pi^+$ and $n \to e^- K^+$) dominate by a Cabibbo factor or more over the $\Delta S = 0$ processes (e.g. $p \to e^- \pi^+ \pi^+$ or $n \to e^- \pi^+$).

<u>$\Delta B = -\Delta L/3$ (e.g. $p \to \nu\nu\nu\pi^+$)</u>: This decay was first discussed in the Pati-Salam model[9], which is not SU(3) x SU(2) x U(1) at low energy, but it may be analysed more generally. The simplest operators in this class have dimension ten (three quarks, three leptons, and one Higgs field) but all produce charged leptons (e.g., $p \to e^- \nu\nu\pi^+\pi^+$, $n \to e^- \nu\nu\pi^+$). To obtain the original Pati-Salam decay into three neutrinos one must actually go to dimension twelve.

<u>$\Delta B = \Delta L/3$ (e.g. $p \to e^+ \nu\nu$)</u>: The simplest operators of this class that induce nucleon decay have dimension eleven. Some of these produce quite spectacular decays such as $p \to e^+ e^+ \bar{\nu} \pi^-$ or $p \to e^+ \mu^+ \bar{\nu} \pi^-$.

CONCLUSIONS

We have here highlighted some of the results of Ref. 3, which is, in itself, only a first step toward a model independent analysis of a new frontier. Much more can be done. One could relax the SU(3) x SU(2) x U(1) framework slightly by allowing right handed neutrinos or perhaps disallowing elementary Higgs fields. Ultimately, one would like a catalogue of cross sections parametrized by the operator coefficients in order to fit data. Of course, a complete determination of the coefficients is not a theory. It is essential to have specific models. They suggest which operators to expect and provide an interpretation of the measurements. In this sense the two approaches, operator analysis and model building, are entirely complementary.

REFERENCES

1. S. L. Glashow, Nucl. Phys. 22, 579 (1961); S. Weinberg, Phys. Rev. Lett. 19, 1264 (1967); A. Salam, in Elementary Particle Physics, ed. N. Svartholm (Almquist and Wicksells, Stockholm, 1968) p. 367.
2. B. Ovrut and H. J. Schnitzer, Phys. Rev. D21, 3369 (1980) and Brandeis preprint (February, 1980); S. Weinberg, Phys. Lett. 91B, 51 (1980); Y. Kazama and Y. P. Yao, Phys. Rev. D22, 514 (1980).
3. H. A. Weldon and A. Zee, Univ. of Pennsylvania preprint UPR-0153T (to be published in Nucl. Phys.).
4. L. Wolfenstein, Phys. Rev. D17, 2369 (1978).
5. E. Witten, Phys. Lett. 91B, 81 (1980).
6. S. Weinberg, Phys. Rev. Lett. 43, 1566 (1979).
7. F. Wilczek and A. Zee, Phys. Rev. Lett. 43, 1571 (1979).
8. S. Weinberg, Harvard preprint HUTP-80/A023.
9. J. Pati and A. Salam, Phys. Rev. Lett. 31, 661 (1973).

QFD II

E. MA, HAWAII, ORGANIZER

RADIATIVE CORRECTIONS IN THE STANDARD MODEL

Ernest Ma
Department of Physics and Astronomy
University of Hawaii at Manoa, Honolulu, Hawaii 96822

ABSTRACT

The effects of radiative corrections on the masses of the standard W and Z bosons as well as their decay widths and induced neutral-current couplings are briefly reviewed. In contrast, the model of Berezinsky and Smirnov, where all weak gauge boson masses can be substantially heavier than 100 GeV, is also described.

Several papers[1-7] were submitted to this conference, dealing with the effects of radiative corrections to processes involving the W and Z bosons of the standard electroweak gauge model.[8] In this talk, for lack of time, I shall only be able to mention some of the results. At the end, I shall also describe briefly a model due to Berezinsky and Smirnov,[9] which allows all weak gauge bosons to be substantially heavier than those of the standard model.

The mass of the W boson is related to the Fermi weak coupling constant by the well-known formula

$$\frac{G_F}{\sqrt{2}} = \frac{e^2}{8M_W^2 \sin^2\theta_W} . \tag{1}$$

However, the experimental value of $\alpha = e^2/4\pi$ is obtained at zero momentum transfer, whereas G_F and $\sin^2\theta_W$ are determined more or less at the mass of the W boson. To correct for this effect, Marciano[10] used the renormalization behavior of the fine-structure "constant" as a function of q^2, i.e.

$$\alpha^{-1}(M_W) \simeq \alpha^{-1}(0) - \frac{2}{3\pi} \sum_f Q_f^2 \ln \frac{M_W}{m_f} , \tag{2}$$

where the sum is over all fermions of charge Q_f, to estimate the numerical shift in M_W. Since $\alpha^{-1}(M_W)$ must be less than $\alpha^{-1}(0)$, the corrected value of M_W must be greater than the uncorrected value. Numerically,

$$\begin{aligned} M_W &= \frac{37.32 \text{ GeV}}{\sin\theta_W} \quad \text{[uncorrected]} \\ &\to \frac{38.5 \text{ GeV}}{\sin\theta_W} \quad \text{[corrected]} . \end{aligned} \tag{3}$$

0094-243X/81/680422-05$1.50

For specific values of $\sin^2\theta_W$, those of M_W and M_Z are given below.

$\sin^2\theta_W$	M_W (GeV)	M_Z (GeV)
0.21	84.0	94.5
0.22	82.1	93.0
0.23	80.3	91.6
0.24	78.6	90.2
0.25	77.0	88.0

(4)

Similar results were obtained by Sakakibara,[6] Antonelli, Consoli, and Corbo,[11] and Veltman.[12]

It has also been pointed out by Albert, Marciano, Wyler, and Parsa[1] that since the decay widths of W and Z are proportional to $G_F M^3_{W,Z}$, the 3% (or so) shift in mass amounts to a 10% (or so) shift in decay rate. For $\sin^2\theta_W = 0.23$, the leptonic decay rates of W and Z are then given by

$$\begin{aligned} \Gamma(W^- \to \ell^- + \bar{\nu}_\ell) &= 0.226 \text{ GeV}, \\ \Gamma(Z^o \to \ell^- + \ell^+) &= 0.084 \text{ GeV}, \\ \Gamma(Z^o \to \nu_\ell + \bar{\nu}_\ell) &= 0.168 \text{ GeV}. \end{aligned} \tag{5}$$

For the hadronic decay rates, there is an additional correction due to QCD, amounting to an overall factor of $1 + \alpha_s(M_Z)/\pi$. If $\alpha_s(M_Z)$ is estimated to be about 0.15, then this is a 5% effect. Numerically,

$$\begin{aligned} \Gamma(W^- \to \text{hadrons}) &= 2.13 \text{ GeV}, \\ \Gamma(Z^o \to \text{hadrons}) &= 2.09 \text{ GeV}, \end{aligned} \tag{6}$$

which would have been 2.03 and 1.99 GeV respectively without the QCD correction. [In the above estimates, all fermion masses are set equal to zero. If they are not, then there may be important QCD corrections for the axial-vector couplings.[13]]

The number of neutrino species N_ν can be determined from Z^o decay by

$$\begin{aligned} N_\nu\, \Gamma(Z^o \to \nu + \bar{\nu}) = \Gamma(Z^o \to \text{all}) &- \Gamma(Z^o \to \ell^- + \ell^+) \\ &- \Gamma(Z^o \to \text{hadrons}). \end{aligned} \tag{7}$$

In principle, all quantities on the right-hand side are experimentally measurable, so that only $\Gamma(Z^0 \to \nu + \bar{\nu})$ would have to be calculated precisely in order to determine N_ν.[14] However, since neutral hadrons are often not detected, $\Gamma(Z^0 \to \text{hadrons})$ may still have to be estimated.

In the leptonic decay of the W boson, the energy of the charged lepton can only be determined within a certain energy resolution ΔE, hence there may be important radiative corrections due to the emission of photons with energy less than ΔE. Specifically,

$$\frac{\Gamma(W^- \to \ell_1 + \bar{\nu}_1 + (\gamma),\ \Delta E)}{\Gamma(W^- \to \ell_2 + \bar{\nu}_2 + (\gamma),\ \Delta E)} \simeq 1 - \frac{\alpha}{2\pi}\,\ell n\,\frac{m_{\ell_2}}{m_{\ell_1}}\left[4\,\ell n\,\frac{M_W}{2\Delta_E} - 3\right], \tag{8}$$

where it has been assumed that ΔE is the same for both processes. This result was also obtained by Inoue, Kakuto, Komatsu, and Takeshita.[4] For $\ell_1 = e$, $\ell_2 = \mu$, $\Delta E = 100$ MeV, and $M_W = 80$ GeV, the above ratio is 0.87, resulting in an apparent violation of e-μ universality.

In the work of Sakakibara,[6] the weak mixing angle θ_W is defined by the mass eigenstates of the renormalized vector gauge bosons A_3^R and B^R, i.e.

$$Z = A_3^R\cos\theta_W + B^R\sin\theta_W, \qquad A = -A_3^R\sin\theta_W + B^R\cos\theta_W. \tag{9}$$

The semiweak coupling g is then defined by $g = e/\sin\theta_W$. In this scheme, the effective neutral-current interaction for neutrinos and quarks is given by

$$\mathcal{H}_{eff} = \frac{G_F}{\sqrt{2}}\,\rho\bar{\nu}\gamma^\alpha(1-\gamma_5)\nu\bar{q}\,\gamma_\alpha[T_3(1-\gamma_5) - \frac{c}{a}(2Q)\sin^2\theta_W]q, \tag{10}$$

where radiative corrections result in

$$\rho = (1-q^2/M_Z^2)^{-1} + 0.001, \qquad 1 - c/a = 0.02. \tag{11}$$

The mass of W is now determined by

$$\frac{G_F}{\sqrt{2}} = \frac{g^2}{8M_W^2}\,(1 + 0.94), \tag{12}$$

and for specific values of $\sin^2\theta_W$, those of M_W and M_Z are given below.

$\sin^2\theta_W$	M_W (GeV)	M_Z (GeV)
0.225	82.2	93.4
0.230	81.3	92.7
0.235	80.4	92.0

(13)

It should be kept in mind that because of the different ways used in defining θ_W (which is only unique in the tree approximation), the $\sin^2\theta_W$ in (13) should be corrected by (11) before comparing with (4). Similar work on radiative corrections to effective current-current interactions in the standard model has also been done by Sirlin,[15] and by Marciano and Sirlin.[16]

Let me now describe briefly a model[9] which makes it possible for all weak gauge bosons to be substantially heavier than those of the standard model, and could therefore escape experimental detection even at the next generation of machines. Consider the gauge group $SU_2\times U_1\times U_1'$ with couplings g, g_B, and g_B'. Let the electric charge be given by $Q = T_3 + \frac{1}{2}Y + \frac{1}{2}Y'$, then

$$e^{-2} = g^{-2} + g_B^{-2} + g_B'^{-2}. \tag{14}$$

Let $Y = Y'$ for fermions, and assume $g_B' \ll g_B$, g. Then from Eq. (14), it is clear that $g_B' \underset{\sim}{\sim} e$, and that the photon is contained mostly in U_1'. The effective neutral-current interaction at low energies is now approximately given by

$$\mathcal{H}_{eff} \simeq \frac{4G_F}{\sqrt{2}} \{(J_3 - \sin^2\theta_W J_{em})^2 + \frac{1}{4} x \cos^4\theta_W Y^2\} , \tag{15}$$

where $\tan^2\theta_W = g_B m_{B3}^2/g m_B^2$, and $x = (g_B/g)^2(m_B^2 m_3^2 - m_{B3}^4)/m_B^4$, which must of course be small in order to fit the data. The mass of the W boson in this model is related to the other parameters by

$$\frac{m_3^2}{m_W^2} = \frac{1}{\cos^4\theta_W} (1 + \frac{1}{x} \tan^4\theta_W). \tag{16}$$

Since m_3 remains arbitrary even if x and θ_W are fixed, m_W can be set greater than, say, 200 GeV. For some typical values, let $x = 0.1$, $\sin^2\theta_W = 0.2$, and $m_B = m_W$, then $g_B = 0.064\ g$, $m_{Z_1} = 0.72\ m_W$

and $m_{Z_2} = 1.74\ m_W$. Therefore, it is still conceivable, within the context of a gauge model, to have all weak gauge bosons at much greater masses than are expected.

I acknowledge the support of the U.S. Department of Energy under Contract DE-AC03-76ER00511.

REFERENCES

1. D. Albert, W.J. Marciano, D. Wyler, and Z. Parsa, Nucl. Phys. B166, 460 (1980).
2. B. Humpert and W.L. van Neerven, Phys. Lett. 93B, 456 (1980).
3. T. Inami and C.S. Lim, Univ. of Tokyo-Komaba report.
4. K. Inoue, A. Kakuto, H. Komatsu, and S. Takeshita, Kyushu Univ. report no. KYUSHU-80-HE-4 (1980).
5. K. Aoki, Z. Hioki, R. Kawabe, M. Konuma, and T. Muta, Kyoto Univ. report no. RIFP-395 (1980).
6. S. Sakakibara, Univ. of Dortmund report no. DO-TH80/01 (1980).
7. R. Rodenberg, Phys. Inst. Aachen report.
8. S. Weinberg, Phys. Rev. Lett. 19, 1264 (1967); Phys. Rev. D5, 1412 (1972); A. Salam, in Elementary Particle Theory: Relativistic Groups and Analyticity, Nobel Symposium No. 8, edited by N. Svartholm (Almqvist & Wiksell, Stockholm, 1968), p. 367; S.L. Glashow, Nucl. Phys. 22, 579 (1961); S.L. Glashow, J. Iliopoulos, and L. Maiani, Phys. Rev. D2, 1285 (1970).
9. V.S. Berezinsky and A. Yu. Smirnov, Inst. for Nucl. Res., Moscow report.
10. W.J. Marciano, Phys. Rev. D20, 274 (1979).
11. F. Antonelli, M. Consoli, and G. Corbo, Phys. Lett. 91B, 90 (1980).
12. M. Veltman, Phys. Lett. 91B, 95 (1980).
13. K.J.F. Gaemers, remark at this conference.
14. N. Cabibbo, remark at this conference.
15. A. Sirlin, Phys. Rev. D, to be published.
16. W.J. Marciano and A. Sirlin, Rockefeller Univ. report no. COO-2232B-206 (1980).

MULTI-W AND Z BOSONS

V. Barger and W. Y. Keung
Physics Department, University of Wisconsin, Madison, WI 53706

E. Ma
Physics Department, University of Hawaii, Honolulu, HI 96822

ABSTRACT

The standard SU(2)×U(1) model is extended to include multi-W and Z bosons without changing the structure of its low energy phenomenology. The lightest W and Z in these models must be lighter than the W and Z of the standard model.

The standard SU(2)×U(1) model explains all low energy data, but it is not completely tested until the gauge bosons W and Z and the Higgs bosons are discovered. We have examined[1] a possible extension of the standard model which has multi-W and Z bosons.[2] The simplest extension is based on the electroweak gauge group SU(2)×U(1)×SU(2)'. The charge operator is $Q = T_3 + Y/2 + T_3'$ where T, Y and T' are the generators of the subgroups SU(2), U(1), SU(2)', respectively. All quarks and leptons are assumed to transform with respect to the SU(2)×U(1) subgroup only, as in the standard model. In addition to the usual Higgs scalar doublet $\phi = (\phi^+, \phi^0)$ in the representation (½,1,0), we need a quartet $\eta = (\eta^0, \eta^-, \chi^+, \chi^0)$ in the representation (½,0,½) linking SU(2) and SU(2)' to generate masses. Spontaneous symmetry breakdown occurs with non-vanishing vacuum expectation values $\langle\phi^0\rangle$ and $\langle\eta^0\rangle = \langle\chi^0\rangle$. The latter equality is guaranteed by discrete symmetries $\eta \to i\eta$ and $\eta \to \tau_2\eta^*\tau_2$ in the Higgs potential. This pattern of symmetry breaking reproduces the standard model results for the ratio of neutral current to charged current coupling strengths at low energy.

The salient features of this extension are as follows:

(i) There exists a pair of left-handed W bosons as well as a pair of Z bosons. The interaction Hamiltonian in the fermion sector is

$$H = \sum_{i=1}^{2} n_i (j_\nu^{(3)} - M_{Z_i}^{-2} M_Z^2 \sin^2\theta_W\, j_\nu^{em}) Z_{i\nu} + \sum_{i=1}^{2} N_i (j_\nu^{(+)} W_{i\nu}^{+} + \text{h.c.}) \quad * (1)$$

where the currents are defined as

$$j_\nu^{em} = \overline{\psi}\gamma_\nu Q\psi$$
$$\vec{j}_\nu = \overline{\psi}\gamma_\nu[(1-\gamma_5)/2]\vec{T}\psi \quad (2)$$
$$j_\nu^{(\pm)} = j_\nu^{(1)} \pm i\, j_\nu^{(2)} .$$

0094-243X/81/680427-04$1.50

The mixing coefficients n_i and N_i satisfy the sum rules:

$$\begin{aligned}\sum_{i=1}^{2} n_i^2 M_{Z_i}^{-2} &= 8G_F/\sqrt{2}\\ \sum_{i=1}^{2} n_i^2 M_{Z_i}^{-4} &= 8G_F\, M_Z^{-2}/\sqrt{2}\\ \sum_{i=1}^{2} N_i^2 M_{W_i}^{-2} &= 4G_F/\sqrt{2}\\ \sum_{i=1}^{2} N_i^2 M_{W_i}^{-4} &= 4G_F\, M_W^{-2}/\sqrt{2}\,.\end{aligned} \quad * \ (3)$$

Here M_W and M_Z are the gauge boson masses of the standard model

$$\begin{aligned}M_W &= e(G_F^{\frac{1}{2}}\, 2^{5/4}\, \sin\theta_W)^{-1}\\ M_Z &= M_W/\cos\theta_W\,.\end{aligned} \quad (4)$$

(ii) One W must be lighter than M_W of the standard model, $M_{W_1} < M_W$. This is a result of sum rules in Eq. (3). Similarly, one Z must be lighter than M_Z of the standard model, $M_{Z_1} < M_Z$. If these low mass gauge bosons exist, they could be observed at the higher energies that will soon be reached in e^+e^- and $\bar{p}p$ colliders.

(iii) Only 2 free parameters, e.g. M_{Z_1} and M_{Z_2}, are present in the model. Other masses are given by the following relations from the traces and determinants of the mass matrices of the gauge fields

$$\begin{aligned}\sum_{i=1}^{2} M_{W_i}^2 &= \sum_{i=1}^{2} M_{Z_i}^2 - M_Z^2 \sin^2\theta_W\\ \prod_{i=1}^{2} M_{W_i} &= \prod_{i=1}^{2} M_{Z_i} \cos\theta_W\,.\end{aligned} \quad * \ (5)$$

(iv) The effective current-current interaction has a new piece $C(j_\mu^{em})^2$ in the neutral current sector:

$$H_{eff} = (4G_F'/\sqrt{2})[(j_\nu^3 - \sin^2\theta_W\, j_\nu^{em})^2 + C(j_\nu^{em})^2]\,. \quad (6)$$

Such an expression is generally true in most models as long as the fermions only transform under the subgroup SU(2)×U(1) as in the standard model. However, G_F' may be different from G_F defined in the charged current sector

$$H_{eff} = (4G_F/\sqrt{2})(j_\nu^+ j_\nu^- + \text{h.c.})\,. \quad (7)$$

The symmetry breaking pattern $\langle\eta^0\rangle = \langle\chi^0\rangle$ mentioned above guarantees

$$G'_F = G_F \ . \tag{8}$$

The C coefficient in Eq. (6) is non-negative:

$$C = \sin^4\theta_W \left(\frac{8G_F}{\sqrt{2}}\right)^{-1} \sum_{i=1}^{2} \frac{n_i^2}{M_{Z_i}^2} \left(\frac{M_Z^2}{M_{Z_i}^2} - 1\right)^2 \geq 0 \ . \qquad * \ (9)$$

In terms of M_{Z_1} and M_{Z_2}

$$C = \sin^4\theta_W (M_{Z_2}^2 - M_Z^2)(M_Z^2 - M_{Z_1}^2) M_{Z_1}^{-2} M_{Z_2}^{-2} \ . \tag{10}$$

The additional term $C(j_\mu^{em})^2$ is not stringently tested because it has to compete with the much stronger electromagnetic interaction at these energies.

In this model the muon anomalous magnetic moment $a_\mu = (g_\mu - 2)/2$ is given by

$$a_\mu = a_\mu^{S.M.} + \sqrt{2} G_F m_\mu^2 C/(3\pi^2) \tag{11}$$

where $a_\mu^{S.M.}$ is the standard model value, $a_\mu^{S.M.} \simeq 2\times10^{-9}$. The discrepancy between the latest experimental value[3] for a_μ and the theoretical electromagnetic contribution through eighth order is $(4\pm22) \times 10^{-9}$. Allowing for one standard deviation from the mean experimental value, the resulting bound $C < 3.87$ is not very restrictive.

Limits on the deviation from QED of $e^+e^- \to \mu^+\mu^-$ and $e^+e^- \to e^+e^-$ cross sections provide[1,4] a much more stringent bound on C. The most recent data[5] yield the bound $C \leq 0.017$. Figure 1 shows the corresponding restriction on the masses M_{Z_1}, M_{Z_2}. For $M_{Z_2} \gg M_Z$, the mass of the light Z boson is restricted to the range $76 \leq M_{Z_1} \leq 89$ GeV.

Generalization[1] of the model to N W and N Z bosons based on the gauge group $SU(2)\times U(1)\times\Pi_{j=1}^{N-1}[SU_j(2)]$ is straightforward. A sequence of Higgs quartets η_i linking two SU(2) subgroups are introduced for symmetry breaking. Discrete symmetries are imposed such that $\langle\eta_i^0\rangle = \langle\chi_i^0\rangle$ as before. The equations above with asterisks still hold true provided that the summations $\Sigma_{i=1}^{2}$ and products $\Pi_{i=1}^{2}$ are replaced by $\Sigma_{i=1}^{N}$ and $\Pi_{i=1}^{N}$, respectively. For the case of unequal number of W's and Z's, we need a gauge group $\Pi_{j=1}^{M-N}[U_j(1)'] \times SU(2)\times U(1)\times\Pi_{j=1}^{N-1}[SU_j(2)']$. This model[1] provides N W's and M Z's. Equations (6), (7) and (8) for the low energy phenomenology remain the same. Among these sequential W and Z bosons, one W and one Z must be lighter than those in the standard model.

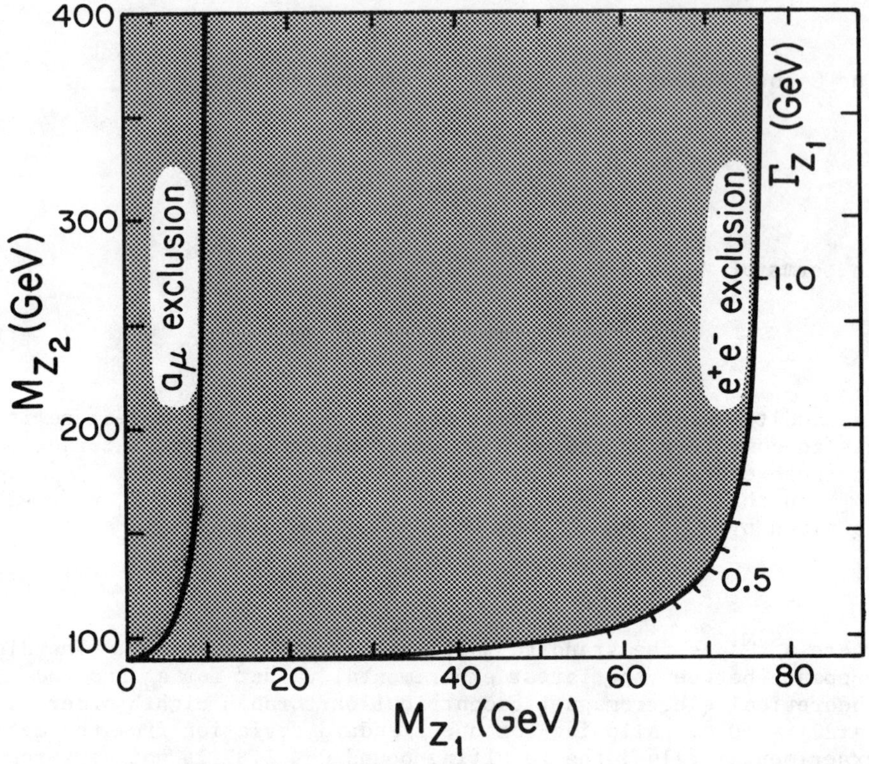

Fig. 1. Excluded regions (to left of curves) for Z_1 and Z_2 masses from a_μ and e^+e^- measurements.

This work was supported in part by the Department of Energy under contracts DE-AC02 76ER00881-166 and DE-AC03 76ER00511.

REFERENCES

1. V. Barger, W. Y. Keung and E. Ma, Phys. Rev. Lett. 44, 1169 (1980); Phys. Rev. D22, 727 (1980), Phys. Lett. 94B, 377 (1980).
2. Extensions to multi-Z bosons only have been considered by E. H. DeGroot, G. J. Gounaris and D. Schildknecht in Phys. Lett. 85B, 399 (1979); Zeit. Phys. C5, 127 (1980).
3. J. Bailey et al., Phys. Lett. 67B, 225 (1977); J. Calmet et al., Rev. Mod. Phys. 49, 21 (1976).
4. E. H. DeGroot, G. J. Gounaris and D. Schildnecht, Phys. Lett. 90B, 427 (1980).
5. See report in these proceedings by B. Wiik; also, M. Chen (private communication).

DEPARTURE FROM WEINBERG-SALAM MODEL AND GRANDUNIFICATION

N.G. Deshpande
University of Oregon, Eugene, OR. 97403

ABSTRACT

The spontaneous breaking of grandunified groups like SO(10) or SU(8) x SU(8) can lead to an extra U(1) group beyond Weinberg-Salam (W-S) SU(2) x U(1). Neutral current data is now shown to depend on two more parameters. We examine the data and put limits on the mass of the extra Z boson. Need for more experiments on parity violation effects in atoms is stressed.

INTRODUCTION

Is proton decay the only direct experimental test for grandunification? The answer in the context of the SU(5) model is unfortunately yes. This is not necessarily true in other models of grandunification. Several models permit low energy group $SU(3)_L \times SU(2)_L \times U(1) \times U(1)$ with two Z-bosons. There should be a characteristic deviation from W-S model in the low energy data which can be saught for. Here we shall examine a class of models that have $SU(4) \times SU(2)_L \times SU(2)_R$ as a subgroup. Some models that have this subgroup are SO(10),[1] $SU(8)_L \times SU(8)_R$[2] and SU(4)[4] of Pati and Salam.[3] By suitable choice of Higgs bosons the low energy group could be (a) $SU(3)_C \times SU(2)_L \times SU(2)_R \times U(1)$, (b) $SU(3)_C \times SU(2)_L \times (T_3)_R \times U(1)$ or (c) $SU(3)_C \times SU(2)_L \times U(1)$ of W-S. The possibility (a) has been extensively studied in the literature,[4] but recent determination[5] of $x \equiv \sin^2 \theta_W$ at a value of $x = .23 \pm .02$ seems to rule this out. The value of x in possibility (a) is restricted to $1/4 < x < 3/8$ and should be close to .28. The possibility (b) has the same constraints as (c) i.e., $1/6 < x < 3/8$, and is consistent with data. The right-handed charged bosons, $W_R^{\pm}$ can be shown to be at least as heavy as 10^9 GeV for this reason. An intermediate scale between the low-energy scale of 10^2 GeV and unification scale at 10^{15} GeV seems to help in raising x from 0.20 to 0.22 thus improving the agreement with experiment. The possibility that the low energy group (b) is realized will be pursued in this talk.

Another way of generating additional U(1) group has recently been pointed out by Barr and Zee.[6] A group SU(N) can break through Higgs in adjoint representation into several extra U(1) groups. The case we consider is a special case of their general case of $SU(N) \to SU(5) \times U(1)$, with known quarks and lepton having appropriate quantum numbers under U(1).

MODEL

The model we consider,[7] based on $SU(2)_L \times (T_3)_R \times U(1)$ can be considered in its own right. It is the only model other than W-S

0094-243X/81/680431-05$1.50

model which is free of anomalies and has natural flavor conservation.[8] Right-handed neutrinos play a central role in the model, so if the neutrino oscillations are ever verified, this model could be an interesting alternative. The subgroup $SU(2)_L$ is identical to W-S, $(T_3)_R$ corresponds to the third component of the $SU(2)_R$ and U(1) corresponds to (B-L)/2 where B is the baryon number, and L is the lepton number. The charge Q is given by $Q = T_{3L} + T_{3R} + (B-L)/2$. The neutral current sector is described by the Lagrangian (suppressing the Lorentz index)

$$L_{int} = g_L L\cdot J_L + g_R R\cdot J_R + g_B B\cdot J_B \tag{1}$$

From normalization of the currents, at grandunification scale we find $g_L = g_R$ and $g_B = (3/2)^{\frac{1}{2}} g_R$. The first relation changes at lower scale because the couplings evolve differently, while the second relationship between two U(1) groups is <u>true at all scales</u>. By a change of basis we can rewrite Eq. (1) in the form

$$L_{int} = eA\cdot Q + g_L(\sec\theta_W)Z\cdot J_Z + g_R C\cdot(2J_R - 3J_B)(10)^{\frac{1}{2}} \tag{2}$$

where A is the photon, $\tan\theta_W \equiv (3/5)^{\frac{1}{2}}(g_R/g_L)$, $e = g_L \sin\theta_W$, J_Z is the neutral current W-S model. Z and C are linear combinations of the original gauge fields L, R and B. The first two terms belong to SU(5) while the third is the extra U(1). We now consider the most general mass-mixing of Z-C system

$$M^2 = \begin{matrix} M_Z^2 & M_{Z-C}^2 \\ M_{Z-C}^2 & M_C^2 \end{matrix} \tag{3}$$

It is convenient to define three equivalent parameters $\rho = M_W^2 + /M_Z^2 \cos^2\theta_W$, $\alpha^2 = M_{Z-C}^4/\det M^2$ and $\beta = -(5\sin\theta_W/\sqrt{6})(M_Z^2/M_{Z-C}^2)$. We shall see that ρ and β take simple values when the Higgs structure is simple. The low energy Lagrangian now takes the form[7]

$$L_{eff} = -(8G_F/\sqrt{2})\ \{J_Z^2 + \alpha\ [J_Z + (\beta/5)(2J_R - 3J_B)]^2\} \tag{4}$$

Low energy neutral current phenomena can be described by ten parameters. We adopt the standard definition where $u_{L,R}$ and $d_{L,R}$ stand for left and right handed couplings of u and d quarks to the left-handed neutrinos. The neutrino-electron vector and axial couplings are described by g_V and g_A respectively. For parity violation electron-quark coupling we follow Bjorken convention[9] which differs in sign from the convention used in Ref. 5.

Neutrino-Scattering	Electron=Scattering
$u_L = (1/2x/3)\rho_1 - z$	$\varepsilon_{VA}(e,) = (1/2-2x)\rho_2 - z'$
$u_R = (-2x/3)\rho_1 + z$	$\varepsilon_{VA}(e,d) = -\varepsilon_{VA}(e,u)$
$d_L = (-1/2+x/3)\rho_1 - z$	$\varepsilon_{VA}(e,u) = (1/2-4x/3)\rho_2$
$d_R = (x/3)\rho - 3z$	$\varepsilon_{AV}(e,d) = (-1/2+2x/3)\rho_2 - z'$
$g_V = (-1/2+2x)\rho_1 + 4z$	where $x = \sin^2\theta_W$
$g_A = (-1/2)\rho_1 + 2z$	$\rho_2 = \rho[1+\alpha^2(1-2\beta/5]$
where $x = \sin^2\theta_W$	$z' = 2\rho\alpha^2\beta(1-2\beta/5)/5$
$\rho_1 = \rho[1+\alpha^2(1+3\beta/5)]$	
$z = \rho\alpha^2\beta(1+3\beta/5)/10$	

We note that if symmetry is broken only by Higgs doublets and singlets, $\rho=1$. Further if $\beta=5/2$ there is deviation only in neutrino-scattering. If $\beta=-5/3$, there is deviation only in electron scattering. We consider these two cases separately.

SPECIAL CASES

Case I. $\rho=1$ and $\beta=5/2$

This can be accomplished by choosing a doublet and a singlet of Higgs, with the doublet having (B-L)=0. The electron-quark sector is the same as W-S. In the neutrino sector we have two parameters, α^2 and x. From the analysis of Ref. 5, we have $u_L = .351 \pm .034$, $u_R = -.180 \pm .028$, $d_L = -.415 \pm .028$, $d_R = -.011 \pm .046$, $g_V = .043 \pm .066$ and $g_A = -.545 \pm .045$. We then find

$$x = .24 \pm .02$$

$$\alpha^2 = .016 \pm .012$$

The central value of α^2 corresponds to $M_{Z_1} = .99$. $M_W\sec\theta_W$ and $M_{Z_2} = 2.6\ M_W\sec\theta_W$. From $\alpha^2 < .028$ we have $M_{Z_2} > 1.6\ M_W\sec\theta_W$.

Case II. $\rho=1$, $\beta=-5/3$

The Higgs here must satisfy the Georgi-Weinberg condition, namely the doublet with large expectation value must be neutral under $(T_3)_R$ as in ν_L. We find for the electron-deuteron asymmetry

$$A^{ed}/Q^2 = -3G_F/(10\sqrt{2}\ \pi\alpha)\{(3/2-10x/3)+\delta^2(1/2-2x)$$
$$+\ 3f(y)[(1/2-2x) + \delta^2(7-12x)/10]$$

where $\delta^2 = \alpha^2\beta^2$. We note that for $x \approx .25$, y independent part is the same as W-S theory. Further y dependent part is small if $\delta^2 < .4$. For $\delta^2 = .4$ we have compared the theory with experiment, and find the fit as good as the S-W model (see Fig. 1). The expression for the parity violation in atoms is rather sensitive to δ^2. Thus for Bismuth $Q_W = -126+192\delta^2$. If $\delta^2 = .4$, we have for $Q_W = -49$ which is a value much smaller than given by W-S model. Similarly parity violation in Hydrogen will also be different from the standard model. Another test is the elastic scattering of electrons off deuterium. For details of these processes see Ref. 7 This value of δ^2 corresponds to $M_{Z_1} = .9\ M_Z$ and $M_{Z_2} = 1.78\ M_Z$, where $M_Z = M_W \sec\theta_W$.

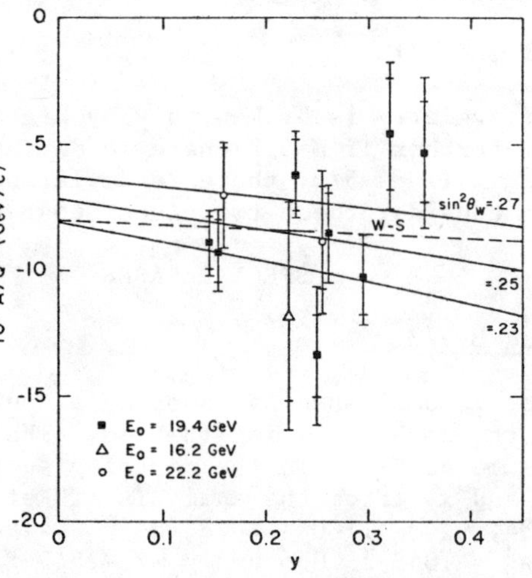

Fig. 1. The asymmetries in the SLAC e–d deep-inelastic scattering as a function of *y*. The predictions of WS model (dashed lines) as well as the predictions of our model for $\delta^2 = 0.4$ and various values of $\sin^2\theta_W$ are shown.

GENERAL CASE

Barr and Zee[6] have observed that if the combination $R=2U_L-U_R+2d_L$ and $S=U_L+2U_R+d_L+5d_R$ are plotted on y and x axis respectively, W-S model is at the origin. All SU(5)xU(1) models lie on straight lines through the origin. The model we have considered lies on the line $R = S/3$. A general SU(N) $\rightarrow$ SU(5)xU(1) breaking will lie on $R = 2S$. The data favors $R = S/3$ to W-S model, though the errors have to be considerably reduced for definitive conclusion (see Fig. 2).

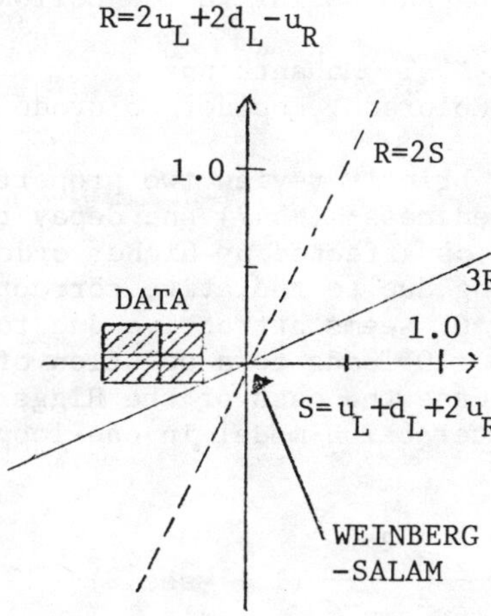

Figure 2

REFERENCES

1. H. Fritzsch and P. Minkowski. Ann. of Physics. 93. 193 (1974) and Nuclear Phys. B103, 61 (1976).
2. N.G. Deshpande and P. Mannheim. Phys. Lett. (1980) and University of Oregon preprint OITS 143.
3. J. Pati and A. Salam. Phys. Rev. D10, 275 (1974).
4. R.N. Mohapatra and J.C. Pati. Phys. Rev. D11, 566, 2558 (1975) G. Senjanovic and R.N. Mohapatra. Phys. Rev. D12, 1502 (1975).
5. P. Langacker et al. Univ. of Pennsylvania preprint COO-3071-243, to be published in Neutrino-79.
6. S.M. Barr and A. Zee. Phys. Lett. 92B 297 (1980).
7. N.G. Deshpande and D. Iskandar. Nuclear Physics B167 223 (1980) and Phys. Lett 87B, 383 (1979).
8. M.S. Chanowitz, J. Ellis and M.K. Gaillard, Nuclear Physics B128, 506 (1977).
9. J.D. Bjorken. Phys. Rev. D18 3239 (1978).

HIGGS MESON AND RADIATIVE CORRECTIONS

K. T. Mahanthappa
University of Colorado, Boulder, Colorado 80309

In this talk I will briefly review two properties of Higgs meson, radiatively induced mass[1] (#547) and decay rate into hadrons[2] (#117 and #247) as affected by higher order corrections.

The symmetry breaking due to radiative corrections[3] (with bare Higgs mass parameter $\mu = 0$) seems attractive due to the work of Weinberg[4] showing that $\mu = 0$ leads to a solution of the gauge hierarchy problem. In this case the mass of the Higgs meson in the standard electro-weak interaction model in one-loop approximation is given by

$$\frac{m_s^2}{m_W^2} = \frac{3e^2}{16\pi^2 \sin^2\theta} (1 + \tfrac{1}{2}\sec^4\theta)$$

which gives $m_s = 10.4$ GeV for $\sin^2\theta = 0.2$. Because of the overlap with upsilon states, the phenomenology becomes complicated and the detection techniques differ according to the location of the Higgs meson relative to upsilon states[5]. Some clue as to its relative position could be obtained by using the precisely determined value of $\sin^2\theta$ in SU(5) in the higher order mass formula for m_s.

With this in view, we calculate the $O(e^4)$ terms of the scalar-vector mass ratio. For this purpose we need the two-loop effective potential, the one-loop vector propagator, the one-loop scalar propagator and renormalization constants associated with coupling constants g and g' of SU(2) and U(1) respectively.

The most general form of the effective potential through two loops is

$$\begin{aligned} V(\phi) = {} & \tfrac{1}{4}\lambda\phi^4 + \frac{1}{16\pi^2}\tfrac{1}{4}\Big[a_1\lambda^2 + a_2\lambda g^2 + a_3\lambda g'^2 + a_4 g^4 + a_5 g^2 g'^2 \\ & + a_6 g'^4\Big]\phi^4 \ln(g^2\phi^2/4M^2) + \frac{1}{(16\pi^2)^2}\tfrac{1}{4}\Big[a_7\lambda^2 + a_8\lambda g^2 + a_9\lambda g'^2 \\ & + a_{10}g^4 + a_{11}g^2g'^2 + a_{12}g'^4\Big]\phi^4 + \frac{1}{16\pi^2)^2}\tfrac{1}{4}\Big\{\Big[b_1\lambda^3 + b_{21}\lambda^2 g^2 \\ & + b_{22}\lambda^2 g'^2 + b_{31}\lambda g^4 + b_{32}\lambda g^2 g'^2 + b_{33}\lambda g'^4 + b_4 g^6 + b_5 g^4 g'^2 \\ & + b_6 g^2 g'^4 + b_7 g'^6\Big]\phi^4 + \Big[c_1\lambda^3 + c_{21}\lambda^2 g^2 + c_{22}\lambda^2 g'^2 + c_{31}\lambda g^4 \\ & + c_{32}\lambda g^2 g'^2 + c_{33}\lambda g'^4 + c_4 g^6 + c_5 g^4 g'^2 + c_6 g^2 g'^4 + c_7 g'^6\Big] \end{aligned}$$

$$\times\ \phi^4 \ln (g^2\phi^2/4M^2) + \Big[d_1\lambda^3 + d_{21}\lambda^2 g^2 + d_{22}\lambda^2 g'^2 + d_{31}\lambda g^4$$

$$+ d_{32}\lambda g^2 g'^2 + d_{33}\lambda g'^4 + d_4 g^6 + d_5 g^4 g'^2 + d_6 g^2 g'^4 + d_7 g'^6\Big]$$

$$\times\ \phi^4 \ln^2 (g^2\phi^2/4M^2)\Big\}$$

where a_i, b_i, c_i and d_i are coefficients to be determined. Most of the two-loop integrals must be done numerically. The vector-meson and scalar-meson propagators can be written as

$$G^{-1}_{\mu\nu}(p^2) = -ig_{\mu\nu}\left\{p^2 - \tfrac{1}{4}g^2\langle\phi\rangle^2 - \frac{1}{16\pi^2} g^2(yp^2 - z\langle\phi\rangle^2) + O(\lambda p^2, g^2p^4)\right\}$$

and

$$G^{-1} = i\left\{p^2 - \frac{1}{16\pi^2} g^2p^2x - \frac{d^2V}{d\phi^2}\bigg|_{\phi=\langle\phi\rangle} + O(\lambda p^2, g^2p^4)\right\}$$

where x, y and z are coefficients to be determined. The particle masses are given by the zeroes of the inverse propagators. We express the results in terms of the renormalized coupling constants. We define

$$g^2 = g_R^2\left(1 - \frac{1}{16\pi^2} g_R^2 s\right); \quad g'^2 = g_R'^2\left(1 - \frac{1}{16\pi^2} g_R'^2 r\right)$$

where s and r are to be determined. Minimizing the potential and finding the poles of the propagators yields, dropping the subscript R,

$$\frac{m_s^2}{m_W^2} = \frac{1}{2\pi^2}\left\{a_4g^2 + a_5g'^2 + a_6g'^4/g^2\right\} + \frac{1}{32\pi^4}$$

$$\times\Big\{(a_4x - a_4s + c_4 + d_4 - a_4y - 4a_4z - a_2A_4)g^4$$

$$+ (a_5x - a_5y - 4a_5z + c_5 + d_5 - a_3A_4 - a_2A_5)g^2g'^2$$

$$+ (a_6x - a_5r - a_6s - a_6y - 4a_6z + c_6 + d_6 - a_3A_5 - a_2A_6)g'^4$$

$$+ (-2a_6r + c_7 + d_7 - a_3A_6)g'^6/g^2\Big\}$$

The relevant quantities have been calculated. The two-loop potential was calculated in the Feynman gauge using the method of Ref. 6. We use the renormalization procedure in the Feynman gauge of Ref. 7.

Taking into account the uncertainties in the calculation of $\sin^2\theta$ in SU(5) due to uncertainties in the QCD parameter Λ_{QCD} and threshold effects, we find $m_s = 10.8 \pm .3$ GeV.

Before I leave this subject, I would like to make a brief remark concerning the cosmological implication of $\mu = 0$. It has been shown very recently[8] that for $\mu = 0$, the phase transition $SU(2) \times U(1) \rightarrow U(1)_{em}$ takes place at extremely low temperatures (~300 eV). It has been pointed out by Guth and Weinberg that this may lead to n_B/n_r smaller than $10^{-8\pm1}$, and it is prevented by having a lower bound on m_s. Witten has pointed out that the chiral symmetry breaking may rescue the situation and the phase transition can take place at temperatures of ~300 MeV.

Braaten and Leveille[2] have calculated the lowest order QCD corrections involving gluons to the decay rate of a Higgs boson into $q\bar{q}$. After renormalization, in the limit $m/M \rightarrow 0$ (m = quark mass; M = Higgs mass), they find the correction to be $-(8/7)$ of the zeroth order rate. Thus they are led to sum the leading logarithms to all order in perturbation theory. This is done using Kinoshita's theorem, the renormalization properties of the Higgs vertex and the renormalization group equation for α_s. They find that, in the leading log approximation, the decay rate can be written as (with $m/M \rightarrow 0$)

$$\Gamma = (3/8\pi)(G_F \sqrt{2}\ \tilde{m}(M)\ )\ M$$

where $\tilde{m}(M)$ is the mass parameter which occurs in mass independent renormalization group procedure of t Hooft and Weinberg. Note that the only difference between the above expression and the zeroth order expression for decay rate is the physical mass m of the quark is replaced by running mass $\tilde{m}(M)$.

Non-leading terms could be important. These have been calculated by Inami and Kubota[2] using the operator product expansion of the product of currents that occur in the polarization of Higgs meson, the imaginary part of which gives decay rate. They construct and solve the RGE's for the coefficients in the expansion. For one flavor, their result is

$$\Gamma = (3/8\pi)(G_F\sqrt{2}\ \tilde{m}\ (M)\ M \left\{1 + \frac{b_1 c_m^{(0)}}{b_0^3} \frac{\ln\ln(M^2/\Lambda^2_{QCD}}{\ln\ (M^2/\Lambda^2_{QCD})} \right.$$

$$\left. + [10/3 - c_m^{(0)}(2-\gamma_E+\ln 4\pi) + \frac{1}{b_0}\ (\frac{b_1}{b_0}\ c_m^{(0)} - c_m^{(1)}\ )]\ \frac{1}{b_0 \ln(M^2/\Lambda^2_{QCD})} \right\}$$

where b_0 and b_i are coefficients in the expansion of the β-function and $c_m^{(0)}$ and $c_m^{(1)}$ are coefficients in the expansion of the anomalous mass dimension. Note that the first log-term is of $O(\alpha_s(M) \ln \alpha_s(M))$ and the second one of $O(\alpha_s(M))$. Very recently this result has been verified by Sakai[8] who calculates the non-leading contribution to be 35% of the leading one for Higgs mass of 50 GeV in a model with six flavors.

REFERENCES

1. K. T. Mahanthappa and M. A. Sher, Colorado preprint COLO-HEP-18 (to be published in Phys. Rev. D).
2. E. Braaten and J. P. Leveille, Wisconsin preprint COO-881-127; T. Inami and T. Kubota, Tokyo preprint.
3. S. Coleman and E. Weinberg, Phys. Rev. D7, 1888 (1973).
4. S. Weinberg, Phys. Lett. 82B, 387 (1979).
5. J. Ellis et al., Phys. Lett. 83B, 339 (1979).
6. S. Y. Lee and A. M. Sciaccaluga, Nucl. Phys. B96, 435 (1975).
7. D. A. Ross and J. C. Taylor, Nucl. Phys. B51, 125 (1973); S. Sakakibara, Aachen preprint 79/17 (1979).
8. A. D. Linde, Phys. Lett. 92B, 119 (1980); A. H. Guth and E. Weinberg, SLAC-PUB-2525; E. Witten HUTP-80/A040; Gale Cook and K. T. Mahanthappa (unpublished).
9. N. Sakai, Fermilab - PUB-80/51-THY.

POSSIBLE SPIN-ONE RESONANCES IN THE STRONGLY COUPLED HIGGS SECTOR

Makoto Kobayashi and Takayuki Matsuki
National Laboratory for High Energy Physics, Oho-machi
Tsukuba-gun, Ibaraki-ken, 305 Japan

The Weinberg-Salam model[1] has successfully explained the bulk of weak-interaction phenomena. Gauge couplings in this theory are controlled by a clear principle, i.e., the gauge invariance. In contrast, much is yet to be known about the Higgs sector. Especially important and interesting to be studied is the case that coupling constants in the Higgs sector become very strong when a mass of the Higgs boson becomes very large (∿1TeV) or when there are much heavy quarks and/or leptons. In that case, a perturbation calculation is not applicable and there will happen the strong coupling phenomena such as resonances. Though these problems have already been discussed in Refs.2 and 3, we shall investigate here especially the spin-one resonance states when a mass of the Higgs boson is very large.

It is well-known[3,4] that in the high energy limit scattering amplitudes of longitudinal components of weak bosons coincide (up to order α) with those of the Nambu-Goldstone bosons in the Higgs sector without gauge bosons. The following Lagarangian in which gauge coupling is turned off will be enough to see the consequences derived from a strongly coupled Higgs sector.

$$L = \partial^{\mu}\phi^{\dagger}\partial_{\mu}\phi + \mu^{2}\phi^{\dagger}\phi - \lambda(\phi^{\dagger}\phi)^{2} \quad . \qquad (1)$$

One should add to (1) the heavy fermion terms if they exist. Writing the field ϕ as

$$\phi = \frac{1}{\sqrt{2}}\begin{vmatrix} i\Pi_1 + \Pi_2 \\ \sigma_0 - i\Pi_3 \end{vmatrix} \quad , \qquad (2)$$

one can easily notice that the Lagrangian (1) is completely equivalent to the chiral σ-model and is $SU(2)\otimes SU(2)$ invariant. Correspondence between the values of parameters of two systems is given as follows.

W-S model	chiral σ-model
$v = \langle\sigma_0\rangle = \dfrac{1}{2^{1/4}G_F^{1/2}} \sim 248$ GeV	$v = f_\pi \sim 100$ MeV
$\lambda = m_H^2/2v^2$	$\lambda = m_\sigma^2/2f_\pi^2$

In the case that the value of λ is equal to each other, the difference between the two systems is only in the mass scale.

In the quark model, hadronic resonances are usually considered to be bound atates of quarks and/or anti-quarks. It is, however, known that resonance states may be constructed from self-interactions among π and σ fields. In fact, by utilizing the first diagonal Padé approximant for the perturbation series, the s- and p-wave phase shifts are computed and have a good agreement with the experimental data.[5] The ρ-meson pole appears in the p-wave as a resonance state of the pions. The above parallelism between the two systems naturally enforces it on us that there should be also, in the strongly coupled Higgs sector, an iso-triplet resonance state with spin one like the ρ-meson in the chiral σ-model, together with the neutral Higgs boson. Since v is fixed by the Fermi coupling constant, the mass of such a resonance depends on essentially one parameter m_H. Using the [1,1] Padé approximant, we have $m_V \sim 2.5$ TeV for $m_H \sim 1.5$ TeV, as a typical result of the numerical estimation.

To see the behavior of the resonance state, we explicitly introduce an effective field V_μ^i (i=1,2,3) corresponding to the spin-one resonance and derive an effective Lagrangian for V_μ^i and the Higgs sector. It is convenient to introduce the following 2x2 matrix instead of the field ϕ in order to see manifestly the transformation property under $SU(2)_L \otimes SU(2)_R$.

$$M = |\tilde{\phi}\ \phi| = \frac{1}{\sqrt{2}}\begin{vmatrix} \sigma_0+i\Pi_3 & i\Pi_1+\Pi_2 \\ i\Pi_1-\Pi_2 & \sigma_0-i\Pi_3 \end{vmatrix} = \frac{1}{\sqrt{2}}\left(\underline{1}\sigma_0 + i\sum_i \tau_i \Pi_i \right) \quad . \tag{3}$$

The transformation property of the field M under $SU(2)_L \otimes SU(2)_R$ is as follows.

$$\begin{aligned} U_L \in SU(2)_L = SU(2)_{WS} &: \quad M \to U_L M \\ U_R \in SU(2)_R \qquad\qquad &: \quad M \to M U_R^\dagger \quad . \end{aligned} \tag{4}$$

Matrix notations are also introduced for the $SU(2)_{WS} \otimes U(1)_{WS}$ gauge fields:

$$A_\mu = \frac{1}{2}\sum_i \tau_i A_\mu^i \quad , \quad B_\mu = \frac{1}{2}\tau_3 B_\mu^0 \quad ; \tag{5}$$

Then, gauge coupling being turned on, the Lagrangian for the Higgs sector is given by

$$\mathcal{L}_M = \frac{1}{2}\,\mathrm{Tr}[(D^\mu M)^\dagger (D_\mu M)] + \frac{1}{2}\mu^2 \mathrm{Tr}(M^\dagger M) - \frac{1}{4}\lambda[\mathrm{Tr}(M^\dagger M)]^2 \quad , \tag{6}$$

where

$$D_\mu M = \partial_\mu M - igA_\mu M + ig' M B_\mu \quad . \tag{7}$$

Next we must know the transformation properties of V_μ^i under the $SU(2)_{WS} \otimes U(1)_{WS}$ gauge transformation in order to determine the interactions among V_μ^i, A_μ^i, and B_μ^0. Following the conventional method,[5] which appears in treating the chiral Lagrangian, we introduce the following 2x2 unitary matrix.

$$\mathcal{M} = \left(\frac{2}{\sigma_0^2+\Pi^2}\right)^{1/2} \mathrm{M} \quad , \quad \mathcal{M}^\dagger \mathcal{M} = 1 \quad . \tag{8}$$

Here we introduce the following auxiliary fields.

$$V_{L\mu} = \sqrt{\mathcal{M}}\, V_\mu \sqrt{\mathcal{M}^\dagger} \quad , \quad V_{R\mu} = \sqrt{\mathcal{M}^\dagger}\, V_\mu \sqrt{\mathcal{M}} = \mathcal{M}^\dagger V_{L\mu} \mathcal{M} \quad . \tag{9}$$

The transformation properties of V_L and V_R under $SU(2)_L \otimes SU(2)_R$ are $(\underline{3},\underline{1})$ and $(\underline{1},\underline{3})$, respectively. How to embed $SU(2)_{WS} \otimes U(1)_{WS}$ in $SU(2)_L \otimes SU(2)_R$ can be seen from Eqs.(4) and (7). Therefore, the covariant derivatives of these fields are defined as

$$\begin{aligned} D_\mu V_{L\nu} &= \partial_\mu V_{L\nu} - ig\,[A_\mu, V_{L\nu}] \quad , \\ D_\mu V_{R\nu} &= \partial_\mu V_{R\nu} - ig'[B_\mu, V_{R\nu}] \quad . \end{aligned} \tag{10}$$

Then, the Lagrangian for V_μ is given by

$$\mathcal{L}_V = -\frac{1}{4}\,[\, \mathrm{Tr}(F^{L\mu\nu}F^L_{\mu\nu}) + \mathrm{Tr}(F^{R\mu\nu}F^R_{\mu\nu}) \,] - m_V^2\, \mathrm{Tr}(V^\mu V_\mu) \quad , \tag{11}$$

where

$$F^L_{\mu\nu} = D_\mu V_{L\nu} - D_\nu V_{L\mu} \quad , \quad F^R_{\mu\nu} = D_\mu V_{R\nu} - D_\nu V_{R\mu} \quad . \tag{12}$$

We may allow the additional self-interaction terms of V_μ in Eq.(11) which are invariant under $SU(2)_L \otimes SU(2)_R$ in the case of $g=g'=0$.

The matrix $\mathcal{M}$ can be expanded around a vacuum expectation value $v = \langle\sigma_0\rangle$ as

$$\mathcal{M} = 1 + \frac{i}{v} \sum_i \tau_i \Pi_i + \cdots \quad . \tag{13}$$

The Lagrangian $\mathcal{L}_V$ is not renormalizable because $\mathcal{L}_V$ includes interactions of infinitively higher orders. We do not, however, worry about non-renormalizability of the Lagrangian $\mathcal{L}_V$ since our aim is to find an effecitve Lagrangian at low energies ($\lesssim$ 1 TeV). Now, setting $\mathcal{M}$=1 in Eq.(11), we find

$$L_V \sim -\frac{1}{4}\{\mathrm{Tr}[\,(\partial_\mu V_\nu-\partial_\nu V_\mu-ig[A_\mu,V_\nu]-ig[A_\nu,V_\mu])^2\,]$$
$$+\,\mathrm{Tr}[\,(\partial_\mu V_\nu-\partial_\nu V_\mu-ig'[B_\mu,V_\nu]-ig'[B_\nu,V_\mu])^2\,]\}$$
$$-m_V^2\,\mathrm{Tr}(V^\mu V_\mu)\quad, \tag{14}$$

which determines the gauge coupling of V_μ. Here one should notice that V_μ is not a vector representation of $SU(2)_{WS}$.

The interaction of VMM type can be described by the following interaction Lagrangian:

$$L_{VMM} = i\,\frac{f}{m_V^2}\,\mathrm{Tr}[\,D^\mu M(D^\nu M)^\dagger(D_\mu V_{L\nu} - D_\nu V_{L\mu})$$
$$+\,(D^\mu M)^\dagger D^\nu M(D_\mu V_{R\nu} - D_\nu V_{R\mu})\,]\quad. \tag{15}$$

In conclusion, it should be noted that we might have resonance phenomena in TeV region, even if the Higgs boson is an elementary object, in contrast to the techniquark picture. The phenomenological Lagrangian approach will be useful in both cases.

REFERENCES

1. S. Weinberg, Phys. Rev. Lett. 19, 1264 (1967); A. Salam, Proc. 8th Nobel Symp., ed., N. Svartholm (Almqvist and Wiksell, Stockholm, 1968), p.367.
2. M. Veltman, Acta Phys. Polon. B8, 475 (1977).
3. B. W. Lee, C. Quigg, and H. B. Thacker, Phys. Rev. Lett. 38,883 (1977); Phys. Rev. D16, 1519 (1977). See also, D. A. Dicus and V. S. Mathur, Phys. Rev. D7, 3111 (1973).
4. J. M. Cornwall, D. N. Levin, and G. Tiktopoulos, Phys. Rev. D10, 1145 (1974).
5. J. L. Basdevant and B. W. Lee, Phys. Rev. D2, 1680 (1970); K. S. Jhung and R. S. Willey, Phys. Rev. D9, 3132 (1974); and references cited therein.
6. See for example, K. Hiida, Y. Ohnuki, and Y. Yamaguchi, Prog. Theor. Phys. Suppl. Ext. p.337 (1968).

QFD III

R. Slansky, Los Alomos, Organizer

DECOUPLING THEOREMS AND EFFECTIVE FIELD THEORIES

Burt A. Ovrut†
and
Howard J. Schnitzer†
Department of Physics
Brandeis University, Waltham, MA 02254

ABSTRACT

The status of decoupling theorems and effective field theories is reviewed. It is emphasized that decoupling is valid for minimal subtraction if an effective field theory with appropriate effective couplings is constructed.

The basic decoupling theorem is that of Appelquist and Carazzone,[1] which states that in renormalizable field theories, without broken symmetry, the effects of heavy particles on low-energy Green's functions for purely light-particle processes are either suppressed by inverse powers of M (where M is a characteristic heavy mass) or are absorbed into the renormalization of effective light-particle couplings. Recent work has shown that the restriction to theories without broken symmetry can be removed.[2] The often forgotten alternative that allows heavy particle effects to be absorbed into the redefinition of effective couplings is basic to the extension of decoupling to theories with broken symmetry, and to effective field theories renormalized by minimal subtraction (MS).

An important application of decoupling[3] is the evolution of the gauge couplings of the $SU(2)xU(1)xSU(3)_c$ subgroups of a grand unified theory from the unification mass M to much lower mass-scales m. Such relations are schematically of the form

$$\frac{1}{g_i(m)^2} - \frac{1}{g_i(M)^2} = b_i \ln\left(\frac{M^2}{m^2}\right) + O(1) \tag{1}$$

Decoupling implies that the constants b_i are obtained from β-functions which refer to light particles only. A systematic discussion of the corrections to Eq. (1) is now possible due to the improved understanding of the role that MS plays in the decoupling of effective field theories.

It should be emphasized that although the β-function, $\beta_{TOT}(g(\mu))$, for the renormalization group (RG) equations of the complete theory is mass-independent, decoupling still takes place. That is, although

$$\mu \frac{dg}{d\mu} = \beta_{TOT}(g) \tag{2}$$

---

†Research supported in part by the DOE under contract EY-76-S-02-3230-A002.

nevertheless

$$\mu \frac{d g_{eff}}{d\mu} = \beta_{light}(g_{eff}) \tag{3}$$

a result which has been verified explicitly to 2-loops for both QED and QCD, and to 1-loop in scalar theories with broken symmetry.[4] One can derive g_{eff} from g by "integrating out" the heavy fields of the theory, with the result schematically written as[2,4,5]

$$g_{eff}(\mu) = g(\mu) + C g^3 \ln \frac{M}{\mu} + D g^3(\mu) + \ldots \tag{4}$$

The $\ln(M/\mu)$ terms are an essential part of the definition of g_{eff}, and guarantee the correct scaling and decoupling. The boundary condition for (3), $g_{eff}(\mu=M)$, is obtained from (4). It contains information dependent on the heavy sector, and is sensitive to subtle non-local effects[2,4] involving overlapping heavy and light exchanges. It should be emphasized that as a result there is <u>no decoupling for the boundary condition</u>, as is already anticipated in the original statement of the Appelquist-Carazzone theorem.

The effective field theory is easily defined by the methods of functional integration.[2] Consider the renormalizable Lagrangian

$$\mathcal{L} = \mathcal{L}(\phi) + \mathcal{L}(\chi) + \mathcal{L}(\phi;\chi) \tag{5}$$

where ϕ represents the set of fields (of any spin) with characteristic mass-scale m, and χ those fields with mass-scale M, where $m \ll M$. The generating functional for ϕ processes is

$$\begin{aligned} Z[j] &= \int [d\phi]\, e^{i\int \mathcal{L}(\phi)+j\phi} \int [d\chi]\, e^{i\int \mathcal{L}(\chi)+\mathcal{L}(\phi;\chi)} \\ &= \int [d\phi]\, e^{i\int \mathcal{L}(\phi)+j\phi}\, e^{iW(\phi)} \\ &= \int [d\phi]\, e^{i\int \mathcal{L}_{eff}(\phi)+j\phi} \end{aligned} \tag{6}$$

where the effective Lagrangian $\mathcal{L}_{eff}(\phi)$ is defined by

$$\int \mathcal{L}_{eff}(\phi) = \int \mathcal{L}(\phi) + W(\phi) \tag{7}$$

At this stage Eqs. (6) and (7) are exact. Note that $W(\phi)$ is the connected vacuum functional for the χ-field theory interacting with a classical ϕ background field. In order to maintain the gauge invariance of $\mathcal{L}_{eff}(\phi)$, one should use a variant of the background gauge in integrating out the heavy fields,[5] which then permits a numerical evaluation of the O(1) corrections to Eq. (1).[6] In general $\mathcal{L}_{eff}(\phi)$ is non-local and an infinite polynomial in ϕ. However if $m \ll M$ and $|k_i| \ll M$ for all k_i, where the k_i are the external momenta of the ϕ lines, then $\mathcal{L}_{eff}$ can be expressed as an infinite series of local terms, where

$$\mathcal{L}_{eff}(\phi)_{local} = \underbrace{[\text{renormalizable}]}_{\dim \le 4} + \underbrace{[(\tfrac{1}{M})^n \text{suppressed}]}_{\dim \ge 5} \tag{8}$$

The program for computing low-energy light-particle Green's functions is easily formulated in the context of the above discussion.

(I) Evaluate $W(\phi)$, defined by (6) and (7), correct to N heavy-loops (including certain mixed graphs) in perturbation theory.

(II) Define $g_{eff}(\mu)$, as in Eq. (4). Here μ is $O(M)$ for the heavy loop-expansion to be valid. The presence of $\log(M/\mu)$ in (4) is not a violation of decoupling.

(III) $$g_{eff}(M) = g(M) + D\, g^3(M) + \ldots \tag{9}$$
is the boundary condition at the heavy mass-scale.

(IV) Use the R.G. to scale $g_{eff}(\mu \simeq M) \xrightarrow{R.G.} g_{eff}(\mu \simeq m)$, where

$$\mu \frac{dg_{eff}}{d\mu} = \beta_{light}(g_{eff}) \qquad \text{[decoupling]}$$

and

β_{light} is the (N+1) loop β-function of $\mathcal{L}_{eff}$

(V) Go to the local limit, using the above g_{eff}, and evaluate $\int[d\phi]\exp i\int\mathcal{L}_{eff}(\phi)_{local}$ to N-loops in ϕ-perturbation, using a MS subtraction scheme. This gives the low-energy ϕ-Green's functions. At no stage of this calculation are there large logarithms, and the ϕ-Green's functions exhibit decoupling and correct threshold behavior.

One can illustrate how decoupling takes place with minimal subtraction by considering the quantum electrodynamics of n_H muons of mass M and n_L electrons of mass m, with $n_T = n_L + n_H$ the total number of fermions. One inte grates out the muon fields Ψ, obtaining

$$\int[d\Psi]\exp i\int\bar{\Psi}[\gamma^\mu(\partial^\mu - g a^\mu) + M]\Psi$$
$$\longrightarrow \exp i\int g^2\Big[C_1 ff + \frac{C_2}{M^2} f\Box f + \ldots\Big] + \frac{g^4}{M^2}\Big[C_3 (ff)^2 + \frac{C_4}{M^2}(ff)(f\Box f) + \ldots\Big] + \ldots \tag{10}$$

with the C_i pure numbers, and $f_{\mu\nu} = \partial_\mu a_\nu - \partial_\nu a_\mu$. To first order in the heavy loop-expansion we require the graph illustrated in Fig. 1.

$$= g^2 C_1 f_{\mu\nu} f^{\mu\nu}$$

Fig. 1 One loop vacuum polarization

From this contribution to the vacuum polarization one obtains[4] to this order

$$g_{eff}(\mu) = g(\mu) + \frac{g^3}{24\pi^2}\, n_H \ln\frac{M^2}{\mu^2} \tag{11}$$

Then

$$
\begin{aligned}
\mu \frac{dg_{eff}(\mu)}{d\mu} &= \mu \frac{dg(\mu)}{d\mu} - \frac{g^3 n_H}{12\pi^2} \\
&= \frac{g^3}{12\pi^2}(n_T - n_H) = \frac{g^3}{12\pi^2} n_L \\
&= \frac{g_{eff}^3}{12\pi^2} n_L = \beta_{light}(g_{eff})
\end{aligned} \tag{12}
$$

In (12) we have used the fact that $g(\mu)$ satisfies (2), and have replaced g by g_{eff} on the right-side of (12) correct to the order we are considering. The final result is the decoupling described in (1)-(3).

If one goes to the local limit in (9) before considering certain mixed or overlap graphs contributing to the vacuum polarization, then one will <u>incorrectly</u> obtain Eq. (11) as an exact result. Rather, at the 2-loop level one must consider the contribution of Fig. 2 to g_{eff}, and similar effects in higher orders.

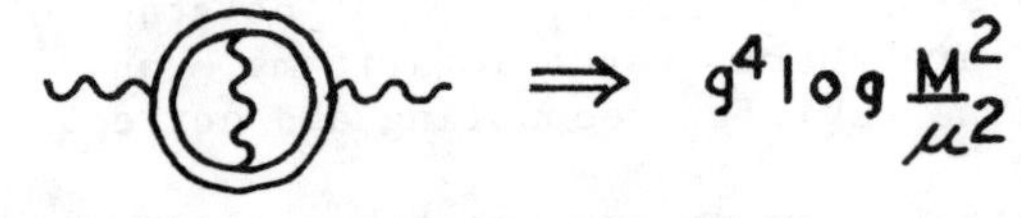

Fig. 2 Two-loop vacuum polarization

The correct relation is given by Eq. (4), with the boundary condition (9). The terms $D\,g^5 \ln (M^2/\mu^2)$ and $E\,g^5$, and similar ones in higher orders, would be omitted by a naive application of (10). These additional terms are essential for decoupling and the correct boundary conditions. The decoupling has been verified <u>explicitly</u> to 2-loop order in QED and QCD for MS, using the ideas sketched above.

The problem of the contribution of mixed graphs to the definition of g_{eff} occurs in scalar theories with broken symmetry even at the <u>one-loop</u> level.[2] In this sense decoupling is more subtle for scalar theories than gauge theories. A typical example is shown in Fig. 3.

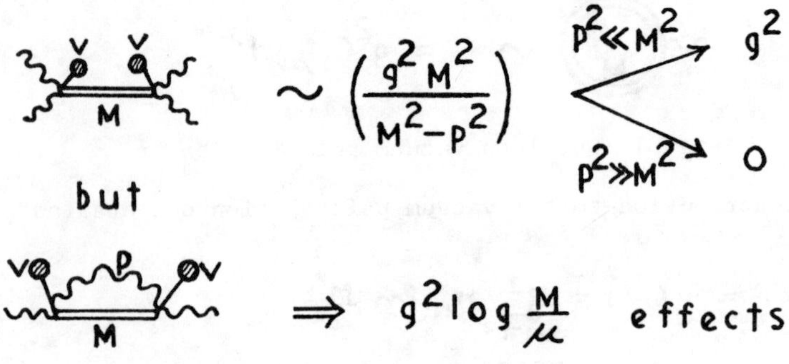

Fig. 3 Mixed graph in scalar theories

Both $p^2 \ll M^2$ and $p^2 \gg M^2$ contribute to the one-loop mixed graph. All such one-loop (and higher order) mixed graphs must be computed if the correct g_{eff} is to be obtained. Nevertheless, decoupling for effective scalar field theories, renormalized by MS schemes has also been demonstrated for these cases even when there is a broken symmetry linking the heavy to the light sector.[2] The final result is again of the form of Eqs. (1)-(3) for all effective parameters of the effective field theory, with all non-local or overlapping effects absorbed into the definition of g_{eff}, Eq. (4).

In summary, we believe the results sketched here for effective field theories defined by MS schemes are likely to be quite general, and valid to all orders. The reader should consult Table I for a comparison of earlier work on decoupling based on momentum space subtraction, with decoupling using MS subtraction.

ACKNOWLEDGEMENTS

We are indebted to Professor L.F. Abbott for his contributions to our understanding of this problem. Many of the issues reported here were anticipated by Witten[9] in his work on weak interactions.

REFERENCES

1) T. Appelquist and J. Carazzone, Phys. Rev. D11, 2856 (1975).
2) B.A. Ovrut and H.J. Schnitzer, Phys. Rev. D21, 3369 (1980) and "Decoupling Theorems for Effective Field Theories," Brandeis preprint (1980).
3) H. Georgi, H. Quinn, and S. Weinberg, Phys. Rev. Lett. 33, 451 (1974); A. Buras, et al, Nucl. Phys. B135, 66 (1978).
4) B.A. Ovrut and H.J. Schnitzer, Brandeis reports in preparation.
5) S. Weinberg, Phys. Lett. 91B, 51 (1980). See also P. Binétruy and T. Schücker, CERN preprints TH-2802 and TH-2857; and N.P. Chang, A. Das, and J. Mercader, City College preprints (1979) and (1980).
6) L. Hall, Harvard preprint HUTP-80/A024. See also G. Cook, K.T. Mahanthappa, and M. Sher, Colorado preprint.
7) For a discussion of such effects with BPHZ subtraction see C.K. Lee, Nuc. Phys. B161, 171 (1979); Y. Kazama and Y-P Yao, Phys. Rev. Lett. 43, 1562 (1979); Phys. Rev. D21, 1116 (1980); ibid. D21, 1136 (1980). These elegant papers derive Callan-Symanzik like equations for the coefficient functions in Γ_{eff}, the 1PI functions of the light fields.
8) H. Georgi and H. D. Politzer, Phys. Rev. D14, 1829 (1976); D. A. Ross, Nucl. Phys. B140, 1 (1978); T. J. Goldman and D. A. Ross, Phys. Lett. 84B, 208 (1979); W. Marciano, Phys. Rev. D20, 274 (1979).
9) E. Witten, Nucl. Phys. B122, 109 (1977).

TABLE I

Table I: Comparison of decoupling for momentum space and MS subtraction.

Momentum Space Subtraction	MS Subtraction
Refs. 7, 8.	Refs. 2, 4, 5.
Consider full theory	Construct effective field theory for light particle processes
Neglect heavy particles in low-energy Green's functions of light particles due to Q^2/M^2 suppression	Integrate out heavy particles in functional method
Work with couplings of full theory	Define effective couplings $g_{eff}(\mu)$ for the effective field theory
logM/m subtracted away	$g_{eff}(\mu) = g(\mu) + Cg^3 \ln M/\mu + \ldots$, $\ln M/\mu$ essential part of definition of g_{eff}. Guarantees correct scaling and decoupling
Decoupling intuitive, but R.G. complicated since β -function mass-dependent	$\beta_{TOT}(g)$ = mass-independent, where $\mu \frac{dg}{d\mu} = \beta_{TOT}(g)$ but $\mu \frac{dg_{eff}}{d\mu} = \beta_{light}(g_{eff})$ [Decoupling]

NEUTRAL LEPTON MASS MATRIX

J. A. Harvey, P. Ramond*, and D. B. Reiss
California Institute of Technology, Pasadena, California 91125

Presented By
P. Ramond

For historical reasons, the neutral lepton masses and mixing angles have not been analyzed in the literature in the same depth as those of charged particles. As long as the only neutral leptons were massless neutrinos there was little need for an extensive analysis. However, with the advent of grand unified theories it was realized[1] that neutrinos were likely to acquire masses and that new extra neutral leptons were a real possibility. Although none of the above have been experimentally confirmed, there are strong theoretical prejudices for their manifestation. Thus we feel that it is worthwhile to present a kinematical discussion of the neutral lepton mass sector.

Spinor particles can be characterized by their transformation properties under the Lorentz group. The Lorentz group is algebraically equivalent to $SU_2 \times SU_2$ which leads to the classification of its representations by two numbers (s_1, s_2) where s_1 and s_2 denote the spin of each SU_2. However, since the Lorentz group is non-compact, these two SU_2's are not unitarily realized with the result that they can be related to one another by a conjugation operation, as well as by a parity flip. Indeed if $\vec{J}$ and $\vec{K}$ are the rotation and boost generators respectively, the two SU_2's are generated by $\vec{J} + i\vec{K}$ and $\vec{J} - i\vec{K}$; in this form it is evident that they are related by conjugation as well as by parity.

The simplest non-trivial representations are realized by complex two-component spinors (Weyl spinors). They are

$$\psi_L \sim (\tfrac{1}{2},0); \; \psi_R \sim (0,\tfrac{1}{2}).$$

Physically $\psi_{L(R)}$ represents a left(right)-handed massless particle with L(R) denoting its helicity. Parity flips ψ_L with ψ_R and vice-versa

$$P: \; \psi_{L(R)} \to \psi_{R(L)}. \tag{1}$$

Thus a parity conserving theory must involve both ψ_L and ψ_R.

Under the Lorentz group

$$\psi_{L(R)} \to e^{i \frac{\vec{\sigma}}{2} \cdot (\vec{\omega} \pm i\vec{\nu})} \psi_{L(R)} \tag{2}$$

---

* Address after Sept. 1, 1980: Physics Dept. University of Florida, Gainesville, Florida 32611.

where $\vec{\omega}(\vec{\nu})$ are the rotation (boost) parameters. It follows from (2) that the new field

$$\sigma_2 \psi_L^* \text{ transforms as } (0,\tfrac{1}{2})$$

and

$$\sigma_2 \psi_R^* \text{ transforms as } (\tfrac{1}{2},0).$$

We therefore identify the transformations

$$C: \begin{cases} \psi_L \to \sigma_2 \psi_R^* \\ \psi_R \to \sigma_2 \psi_L^* \end{cases} \tag{3}$$

with the operation of charge conjugation C, which changes a left(right)-handed particle into a left(right)-handed antiparticle. Note that this operation requires both ψ_L and ψ_R. However, the operation

$$CP: \begin{cases} \psi_L \to \sigma_2 \psi_L^* \\ \psi_R \to \sigma_2 \psi_R^* \end{cases} \tag{4}$$

change a left-handed particle into a right-handed antiparticle, and is to be identified with CP. As an example, ψ_L can represent a left-handed neutrino and $\sigma_2 \psi_L^*$ a right-handed antineutrino. Contrast this situation with that of an electron which is respresented by both e_L and e_R; e_L is the weak interaction eigenstate, but since the electron has mass it no longer corresponds to a definite helicity state but rather to a handedness.

Kinetic terms for ψ_L and ψ_R can be constructed and are of the form

$$\psi_L^+ \sigma_\mu \partial^\mu \psi_L , \quad \psi_R^+ \bar{\sigma}_\mu \partial^\mu \psi_R , \tag{5}$$

where $\sigma_\mu = (1, \vec{\sigma})$, $\bar{\sigma}_\mu = (1, -\vec{\sigma})$. Thus in this notation a Dirac spinor is described by both ψ_L and ψ_R assembled in a four component field, (here shown in the Weyl representation).

$$\psi_D = \begin{pmatrix} \psi_L \\ \psi_R \end{pmatrix} . \tag{6}$$

A Majorana spinor in turn is built out of ψ_L only arranged in a four component form

$$\Psi_M = \begin{pmatrix} \psi_L \\ -\sigma_2 \psi_L^* \end{pmatrix} . \tag{7}$$

It is then clear that any theory involving spin ½ particle can be interpreted as a theory involving many left-handed spinors ψ_{La} a = 1, ..N. Their physical interpretation will depend on the peculiarities of their mass matrix.

With spinors one can construct two types of masses. Starting with ψ_L alone one can build the Majorana mass corresponding to $[(\frac{1}{2},0) \otimes (\frac{1}{2},0)]_A = (0,0)$. It is algebraically written in the form

$$i\, \psi_L^T\, \sigma_2 \psi_L \,.$$

When more than one ψ_L is present one can form the Dirac mass which is nothing but an off-diagonal Majorana mass. It is traditionally written as $i\, \psi_L^{+} \psi_R$, but it is always possible to interpret ψ_R as another ψ_L by means of the C operation (3). For instance in a theory with the left-handed spinors ψ_{L1} and ψ_{L2}, the most general mass matrix is

$$i\, \psi_{La}^T\, M_{ab}\, \psi_{Lb} + c.c. \,, \tag{8}$$

where M is a symmetric 2 x 2 matrix. By identifying

$$\psi_{L2} = \sigma_2 \psi_{R1}^{*} \,, \tag{9}$$

it is clear that the M_{12} component will give rise to the Dirac mass while M_{11} and M_{22} give Majorana masses. Thus we lose no generality in discussing only the Majorana mass term (9). We also note that a mass term be it Majorana or Dirac, will flip helicities. Hence, ψ_L for a massive Weyl spinor (Majorana) is a linear superposition of the spin states of a self-conjugate fermion. When both ψ_L and ψ_R are linked by a Dirac mass, ψ_L is a linear superposition of distinct particle and antiparticle states.

The most general non-interacting theory with N Weyl spinors ψ_{La}, a = 1, ..N is described by the Lagrangian

$$\mathcal{L} = \psi_{La}^{+} \sigma\cdot\partial\, \psi_{La} + \frac{1}{2}\, \psi_{La}^{T} \sigma_2 M_{ab}\, \psi_{Lb} + \frac{1}{2}\, \psi_{La}^{+} \sigma_2 M_{ab}^{*}\, \psi_{Lb}^{*} \,, \tag{10}$$

where M_{ab} is a symmetric N x N matrix. One can define a discrete transformation under which

$$\psi_{La} \rightarrow \sigma_2 \psi_{La}^{*} \,, \tag{11}$$

which we identify with CP. It leaves the kinetic term invariant, but not the mass term unless M has real matrix element. We now proceed to diagonalize the mass matrix.

It is not generally known[2], but nevertheless true, that a symmetric complex matrix can be cast in diagonal form by means of a unitary congruence (Schur's[3] theorem), i.e. we can always set

$$M = U^T D U, \tag{12}$$

where T stands for transposition and U is a unitary matrix

$$U^+U = UU^+ = 1, \tag{13}$$

and D is a diagonal matrix with real positive entries. Then we define the mass eigenstates

$$\chi_L \equiv U\psi_L, \tag{14}$$

in terms of which we rewrite the Lagrangian (10) in diagonal form

$$\mathcal{L} = \chi^+_{La}\sigma\cdot\partial\, \chi_{La} + (i\, \chi^T_{La}\sigma_2 M_a \chi_{La} + \text{c.c.}), \tag{15}$$

where M_a are the entries of D. If by chance some degeneracies occur in D, say $M_1 = M_2$, then we can always set $\chi_{L2} = \sigma_2\chi^*_{R1}$ and interpret the resulting Lagrangian for χ_{L1} and χ_{L2} as describing a neutral Dirac spinor with distinct particle and antiparticle. The M_a entries can be obtained as the square root of the eigenvalues of the hermitian matrix MM^*, since

$$MM^* = U^+D^2U, \tag{16}$$

which has manifestly positive eigenvalues. A similar treatment can be carried out for the charged and colored sectors of the theory. However, since QED and QCD are both vector-like theories the mass matrix in these sectors will preserve C^4.

REFERENCES

1. M. Gell-Mann, P. Ramond, R. Slansky, unpublished,
   P. Ramond, Sanibel Symposia Talk, Feb. 1979, CALT 68-709, (unpublished),
   H. Georgi, D. V. Nanopoulos, Nucl. Phys. B159 (1979) 16,
   E. Witten, Harvard Preprint (1980),
   J. A. Harvey, P. Ramond, D. B. Reiss, Phy. Lett. B92, 309 (1980).
2. We are indebted to our mathematical colleagues Profs. Taussky Todd and A. Schepp for guidance.
3. A. I. Shur, Amer. Journal Math. V.67, 472 (1945).
4. M. Gell-Mann, R. Slansky, unpublished.

CALCULABILITY OF THE N-P MASS DIFFERENCE IN GAUGE THEORIES

Joe Kiskis
Theoretical Division, Los Alamos Scientific Laboratory
University of California, Los Alamos, New Mexico 87545

ABSTRACT

The requirement of a calculable N-P mass difference leads to a consideration of unified gauge theories. Future developments in grand unified models may provide a realistic framework for the calculation of the N-P mass difference. The possibility that the relatively soft ultraviolet behavior of QCD softens the divergence in the lowest order electromagnetic mass shift is considered in detail. It is shown that if the bare mass and QCD coupling are constrained to be independent of the electromagnetic coupling, as is natural, then the lowest order electromagnetic shifts of the renormalized mass and QCD coupling are infinite.

I. UNIFIED GAUGE THEORIES

The universe would be a very different place if the neutron were lighter and thus very much more stable than the proton. Let us take the position that a splitting of this importance does not get its observed value by accident, unrelated to other properties of the "complete theory". Thus, we begin with the assumption that the N-P mass difference is calculable.

The further assumption that the mass difference is mainly electromagnetic (EM) leads, with rough calculation, to the conclusion that the neutron is lighter. However, it is now commonly held that the nucleons are made of u and d quarks. So the u-d mass difference is a required imput ot any N-P mass difference calculation. Unfortunately, indications are that the d quark is heavier; the problem has simply been passed down a level. From now on, we will discuss the problem at this level: the u-d mass difference (Δud).

Since the charges of the u and d quarks are different, there is certainly an EM contribution to Δud. In lowest order, this contribution diverges, and we are faced with an ultraviolet (UV) problem. From now on, the discussion will concentrate upon the calculability of the UV contribution to Δud. Can we get a finite result, or equivalently, a result that is not a free parameter after renormalization?

We have seen that the EM contribution is not calculable. A related observation is that EM explicitly breaks flavor symmetry and not in a soft way. Something to soften or cancel the EM UV contribution is needed.

One might hope that the relatively tame UV behavior of QCD could soften the UV divergence. We will analyze this point in some detail in Section II and show that it does not work out. Composite quarks might have softer EM properties in the UV. However, resorting to this argument simply passes the problem down yet another level (a ploy that did not work last time). We will insist upon facing the problem at the quark level.

Consider divergent contributions to Δud that could cancel the EM one. QCD, being flavor symmetric, cannot cancel the nonsymmetric EM contribution. However, the weak interactions are not symmetric and could work. But to do so they would have to have the same strength as the EM contributions in the UV. Thus, we are led to consider electroweak unification!

The standard $SU(2)_L \times U(1)$ electroweak theory, while unified enough to be renormalizable, is not unified enough to make Δud calculable. There are gauge and Yukawa couplings that explicitly break flavor symmetry. Δud is a free parameter.

Electroweak unification schemes with tighter structure could fix the free parameters in the standard model and give a calculable Δud. Unfortunately models of this type seem to have phenomenological difficulties.[1]

We are left with the possibility of including $SU(3)_{color}$ and unifying further: grand unified theories (GUT). The general structure that a GUT must have to give a calculable Δud can be deduced from Weinberg's work.[2] In short: the symmetry and representation content of the fundamental fields must conspire with the requirement of renormalizability in such a way as to rule out a counterterm for Δud. Δud is zero at the tree level for any[2] choice of the Lagrangian parameters. But the residual unbroken symmetry does not prevent the appearance of (necessarily finite) contributions in higher order.

The SU(5) model in its usual form gives a zeroth order relation $m_e = m_d$ and a calculable Δed. Δud remains a free parameter. SO(10) allows one to put all the first generation fermions in a single irreducible representation. This is more interesting. A number of schemes[3] for Higgs scalars, symmetry breaking, and fermion masses have been discussed. With few enough scalars, there are many zeroth order mass relations. However, my impression is that schemes that are restrictive enough to leave Δud calculable are not realistic.

Nevertheless, this work is moving in a promising direction, and future developments may yield a framework for a Δud calculation.

II. $SU(3)_{color} \times U(1)_{EM}$

We return now to the idea of softening rather than canceling the EM UV divergence. Recently Brodsky, Schmidt and de Téramond[4] (BST) suggested that the relatively soft UV behavior of QCD may give lowest order EM self-energy integrals that converge on the scale of an eleventh quark flavor mass (if such exists) rather than diverge or converge on some grand unification scale as we have been discussing. This is based on the observation that the integral over the running mass

$$\bar{m}(q^2) = \bar{m}(q_0^2)\left(\frac{\ln q_0^2}{\ln q^2}\right)^{\gamma} , \qquad (1)$$

$$\int d^4q\, q^{-4}\, \bar{m}(q^2) \qquad (2)$$

will converge if $\gamma > 1$. This condition implies in turn that the number of flavors n_f exceed ten. Dine[5] has analyzed their argument carefully. Craigie, Narison, and Riazuddin[6] have also discussed the subject.

On the other hand, Collins[7] and West[8] have argued generally, using the Cottingham approach, the operator product expansion (OPE), and the conservation of the stress energy tensor that the photon loop integral

must diverge. The two arguments are rather different, but both claim to include QCD to all orders. We have looked at this more carefully.[9]

Consider the $SU(3)_{color} \times U(1)_{EM}$ theory dimensionally regulated ($d=4-\varepsilon$) with minimal subtraction. Let g_B be the dimensionless bare coupling. So $\mu^{\varepsilon/2}\, g_B$ appears in the Lagrangian. When the EM coupling e is zero the connection between the bare (g_B, m_B) and renormalized (g_S, m_S) parameters is

$$g_B = Z_g^S(g_S,\varepsilon)g_S \qquad m_B = Z_m^S(g_S,\varepsilon)m_S \quad . \qquad (3)$$

If the β function and mass anomalous dimensions are

$$\beta_S(g_S) = -bg_S^3 + \ldots \qquad \gamma_m^S(g_S) = -ag_S^2 + \ldots \quad , \qquad (4)$$

then a renormalization group analysis[9] shows that at fixed g_S and $\varepsilon \to 0$

$$\lim_{\varepsilon \to 0} g_B = \left(\frac{\varepsilon}{2bg_S^2}\right)^{1/2} g_S \qquad \lim_{\varepsilon \to 0} m_B = \left(\frac{\varepsilon}{2bg_S^2}\right)^{\frac{a}{2b}} m_S \quad .$$

Now turn on EM. It is natural to ask if the renormalized g and m that parameterize the theory remain finite when EM is turned on without changing the bare parameters g_B and m_B. The Cottingham-OPE approach effectively does this and answers "no".[9] We will now give another analysis which does not give the photon loop such a special role. We proceed <u>as if</u> the intent is to develope the theory to all orders in e. The new relationship between the bare and renormalized parameters is

$$g_B = Z_g(g,e,\varepsilon)g \qquad m_B = Z_m(g,e,\varepsilon)m \quad . \tag{5}$$

Also

$$\beta(g,e) = \beta_S(g) + e^2(cg^3 + \ldots) + \ldots \tag{6}$$

$$\gamma_m(g,e) = \gamma_m^S(g) + e^2(D + dg^2 + \ldots) + \ldots \tag{7}$$

Another renormalization group analysis shows that if we write

$$Z_g = Z_g^S\left[1 + e^2 z_g\right] \quad \text{and} \quad Z_m = Z_m^S\left[1 + e^2 z_m\right] \quad ,$$

then

$$\lim_{\varepsilon\to 0} z_g(g,\varepsilon) = \frac{1}{4}\frac{c}{b} \quad \text{and} \quad \lim_{\varepsilon\to 0} z_m(g,\varepsilon) = \frac{D}{\varepsilon} - \frac{1}{2b}\left(d - a\frac{c}{b}\right)\ell n\varepsilon .$$

As discussed, we now require that g_B and m_B be independent of e and chosen so that g and m are finite and equal to g_S and m_S when $e=0$. In equations:

$$Z_g^S(g_S,\varepsilon)g_S = Z_g(g,e,\varepsilon)g$$

$$Z_m^S(g_S,\varepsilon)m_S = Z_m(g,e,\varepsilon)m \quad .$$

So g and m are fixed in terms of g_S, m_S, e, and ε. Further analysis gives

$$\lim_{\varepsilon\to 0} g = g_S\left[1 + e^2\frac{1}{\varepsilon}\left\{\frac{1}{2}\frac{c}{b}\,\beta_S(g_S)\,\frac{1}{g_S}\right\}\right]$$

$$\lim_{\varepsilon\to 0} m = m_S\left[1 + e^2\frac{1}{\varepsilon}\left\{\frac{1}{2}\frac{c}{b}\,\gamma_m^S(g_S) - D\right\}\right] \quad .$$

Thus the lowest order EM shifts of g and m diverge. We feel that this calculation most accurately expresses the intuitive concept of "electromagnetic mass shift".

However, there is a different approach that is equivalent to that of BST. Rather than ask for the EM shifts in the renormalized parameters with the bare parameters fixed, one calculates the EM shifts in the bare parameters when the renormalized parameters are held fixed. Then g_B and m_B certainly have an e dependence, and we find

$$\lim_{\varepsilon \to 0} g_B = \left(\frac{\varepsilon}{2bg^2}\right)^{1/2} g \left[1 + e^2 \frac{1}{4}\frac{c}{b}\right] \longrightarrow 0$$

$$\lim_{\varepsilon \to 0} m_B = \left(\frac{\varepsilon}{2bg^2}\right)^{\frac{a}{2b}} m \left[1 + e^2 \frac{D}{\varepsilon}\right] .$$

Thus the shift in the bare mass will be finite only if $\frac{a}{2b} \geqslant 1$, and as BST have observed, this requires $n_f \geqslant 11$.

REFERENCES

1. H. Harari, Phys. Rep. 42, 235 (1978).
2. S. Weinberg, Phys. Rev. Lett. 29, 388 (1972); Phys. Rev. D7, 2887 (1973).
3. For example:
   H. Fritzsch and P. Minkowski, Ann. Phys. 93, 193 (1975);
   H. Georgi and D. Nanopoulos, Nuc. Phys. B155, 52 (1979);
   M. Chanowitz, J. Ellis, and M. Gaillard, Nuc. Phys. B129, 506 (1977);
   M. Gell-Mann, P. Ramond, and R. Slansky, unpublished;
   E. Witten, Phys. Lett. 91B, 81 (1980).
4. S. Brodsky, G. de Téramond, and I. Schmidt, Phys. Rev. Lett. 44, 557 (1980).
5. M. Dine, SLAC-PUB-2502.
6. N. Craigie, S. Narison, and Riazuddin, Trieste Preprints.
7. J. Collins, Nucl. Phys. B149, 90 (1979).
8. G. West, Los Alamos Preprint LA-UR-79-1690 (1979).
9. J. Kiskis and G. West, Los Alamos Preprint LA-UR-80-1125.

SEQUENTIAL INTERNAL SUPERSYMMETRY

Y. Ne'eman[†*+"] and S. Sternberg[**]
Tel Aviv University, Tel Aviv, Israel

ABSTRACT

The supergroups SU(2/1), SU(5/1) and SU(5=k/1) provide fitting classifications, unifying weak-electromagnetic, color, and sequential flavor respectively.

SU(2/1): ELECTROWEAK DYNAMICS

Experience with old-fashioned SU(3) has shown that a good classification group should be studied, even if neither its dynamics nor the quantum-statistics of its representations are understood at the time. It has recently been suggested[1,2] that the simple supergroup $SU(2/1) \supset SU(2) \times U(1)$, applied as an <u>internal</u> supergauge, provide a highly-restricted Salam-Weinberg model.

We have proved elsewhere[3] that the standard-model matrices, when applied to leptons and quarks sequentially, obey a condition of supertracelessness when left and right chiral fields are assigned to different gradings. This is because SU(2) is traceless, and acts only within the left-chiral sector; and because electric charges Q have to match in both sectors. The U(1) is then a supertraceless matrix U, since it is a linear combination, $Q = I_3 + \frac{1}{2}U$.

The defining representation is $(\nu^0_L, e^-_L / [e^-_R]_L)$ with the quantum numbers $(-\frac{1}{2}, \frac{1}{2})$ in the SNR notation[4] $(\frac{1}{2}u_{mid}, i_{3max})$ characterizing the eigenvalues of the highest I_3 state. To commute with the Lorentz group, this is a left-chiral spin-half multiplet. The boson $[e^-_R]_L$ has the internal quantum numbers of e^-_R. Several suggestions have been made with respect to the "wrong" statistics and chirality

a) adding dimensions to space-time[2], preferably chiral-fermionic[5]

b) treating the entire system as a classification with no statistics implications. For example, a study of anomaly restrictions relating constituents and their composites[6] has shown that the SU(n/m) groups fulfill these conditions

c) we have shown[7] that the renormalization-ghosts include an effective field, with "wrong" statistics, for every matter field in the theory. Our [ ] would be the parity-transforms of these ghosts. One of us (Y.N.) may have apreference for this interpretation, but the results of this article do not depend on that choice, except where stated.

Note that one might be tempted to forego the complications of a supergroup, and replace it by SU(3) with $(\nu^0_L, e^-_L, (\overline{e^-_R})^+_L)$ and $|\Delta\ell|=2$ generators and currents, and the same $\Theta_W = 30^o$. However, SU(2/1) scores highly in providing us with the <u>4-dimensional fundamental representation</u>[4,8] $(u^{2/3}_L, d^{-1/3}_L / [d^{-1/3}_R]_L, [u^{2/3}_R]_L)$, (1/6,1/2) in SNR notation. Indeed, the <u>4</u> representation is characterized[9] by

† also Center for Particle Theory, U. of Texas, Austin
* Wolfson Chair Extraordinary of Th. Physics
+ supported in part by U.S.-Israel Binational Science Foundation
" " " " " " D.O.E.contract EY-76-S-05-3992
**also Dept. of Math., Harvard University

$b = \frac{1}{2} u_{mid} + \frac{1}{2}$; b= 2/3,1/3 respectively for quarks and antiquarks, and b= 0,1 respectively for leptons and antileptons. In the system of representations of SU(n/m) that we use[9], integer values for b imply reducibility. SU(2/1) indeed predicts the decoupling of ν_R^0 due to the integer charges of the leptons!

SU(2/1) also predicts the $I = \frac{1}{2}$, $U = \mp 1$ assignments of the Goldstone-Higgs multiplet[1,2]. In the ghost interpretation these scalar fields make a perfect 8, together with the Faddeev-Popov ghosts of SU(2) x U(1). Indeed, this restriction could be taken as a definition of the theory. In that picture, the Φ^4 coupling is determined and the mass of the Higgs field ∿ 250 GeV. Treating the full SU(2/1) as a local gauge makes the basic set of leptons or quarks have masses of the order of the W mass. This looked surprising at the time but it is precisely the picture of "technicolor" or "hypercolor" etc.[10].

Local SU(2/1) has one main difficulty, relating to the Killing metric $g_{AB} = \mathrm{str}\,(M_A M_B)$ which is non-positive-definite for the 8th direction $g_{UU} = \mathrm{str}\,(M_U)^2 < 0$. This precludes the straightforward construction of a second-quantized theory. Either we break the symmetry (by using the metric of SU(3), with a certain justification in the ghost picture[8]) or we treat SU(2/1)/SU(2) x U(1) globally only (this also cancels the heavy fermion masses).

SU(5/1): COLOR ADDED

In SU(5/1), taking for $I_3 = \mathrm{diag}\,((\frac{1}{2}, -\frac{1}{2}, 0, 0, 0/0))$ there are two choices of U. Previous authors[11,12] adopted $U \subset SU(5) \subset SU(5/1)$ i.e. the Georgi-Glashow choice

$$\tfrac{1}{2}N = \mathrm{diag}\ \tfrac{1}{2}\,((1,\ 1,\ -2/3,\ -2/3,\ -2/3\ /\ 0)) \qquad (1)$$

but this clashes with $SU(2/1) \subset SU(5/1)$. We have selected[9]

$$\tfrac{1}{2}U = \mathrm{diag}\ \tfrac{1}{2}\,((0,\ 0,\ 1/3,\ 1/3,\ 1/3\ /\ 1)) \qquad (2)$$

which have $SU(5/1) \supset SU(3)_{color} \times SU(2/1)$. The b=2 representation is 32 dimensional and contains precisely two generations (or a generation and its ghosts). It is reducible, and one may select [11,12] 16 or 31 (dropping the ν_R).

SU(5+k/1): SERIALITY ADDED

The representations of SU(n/m) have recently been described[9,11,13]. They follow the same Clifford-algebra pattern

$$|\,"0"\,\rangle \oplus \square\,|\,"0"\,\rangle \oplus \begin{smallmatrix}\square\\ \square\end{smallmatrix}\,|\,"0"\,\rangle \oplus \begin{smallmatrix}\square\\ \square\\ \square\end{smallmatrix}\,|\,"0"\,\rangle \oplus \ldots$$

which we used in exploring extended supergravity and discovering the N=8 limit. Here too, we have a restriction on the range of U going from +2 to -2, with ±4/3, ±1, ±2/3, ±1/3 as the only allowed intermediate values. We have proved[14] that these values restrict the color group uniquely to SU(3), or no color at all as an alternative.

Our representations $b = \frac{4+k}{2}$, with $u_{max} = 2$, for

$$\tfrac{1}{2}U = \mathrm{diag}\ \tfrac{1}{2}\,\left(\left(\tfrac{-k}{4+k},\ \tfrac{-k}{4+k},\ \left(\tfrac{4}{4+k}\right)_{k\ \text{times}},\ \left(\tfrac{4-2k}{3(4+k)}\right)_{3\ \text{times}}\ /\ \tfrac{4}{4+k}\right)\right) \qquad (3)$$

$$I_3 = \mathrm{diag}\,((\tfrac{1}{2},\ -\tfrac{1}{2},\ 0_{(k+3)\ \text{times}}\ /0)) \qquad (4)$$

are 2^{5+k} dimensional. They contain 2^{k+1} generations, or 2^k generations plus their ghosts, depending upon the interpretation (see discussion of ε in ref[8]). Thus k=2 (or k=3) predicts 8 generations. We have provided the exact construction elsewhere [9,3]. Half of the generations have inverted physical chiralities.

Note that for $k \geq 1$, we now have $g_{UU} > 0$, so that a quantized local supergauge theory becomes possible.

The defining representation describes the primitive constitutents of the theory. For k=2 they have charges,

$$Q = \text{diag}\ ((1/3,-2/3,1/3,1/3,0,0,0/\ 1/3)) \qquad (5)$$

REFERENCES

1. Y. Ne'eman, Phys. Lett. 81B, 190 (1979).
2. D. B. Fairlie, Phys. Lett. 82B, 97 (1979).
3. Y. Ne'eman, Proc. of Europhysics Conference on Grand Unification Theories and Supergravity (Erice 1980), to be pub.
4. M. Scheunert, W. Nahm and V. Rittenberg, J. Math. Phys. 18, 155 (1977).
5. E. J. Squires, Phys. Lett. 82B, 395 (1979).
   J. G. Taylor, Phys. Lett. 83B, 331 (1979).
   P. H. Dondi and P. D. Jarvis, Phys. Lett. 84B, 75 (1979).
6. T. Banks, S. Yankielowicz and A. Schwimmer, report WIS-80/19/5-Ph (to be pub.).
7. Y. Ne'eman and J. Thierry-Mieg, PNAS (USA) 77, 720 (1980).
8. Y. Ne'eman and J. Thierry-Mieg, Proc. (Salamanca, 1979) Int.Conf. on Diff. Geom. Methods in Phys., to be pub. by Springer Verlag (Math. Series) (also rep. TAUP 727-79).
9. Y. Ne'eman and S. Sternberg, PNAS (USA) 77, 3127 (1980).
10. See papers by M.A.Beg, H. R. Pagels etc. in the Dynamical Symmetry Breaking session of these Proceedings.
11. P. H. Dondi and P. D. Jarvis, Z. Physik C, Particles and Fields 4, 201 (1980).
12. J. G. Taylor, Phys. Rev. Lett. 43, 826 (1979).
13. I. Bars and Balantekin, Proc. 9th Int. Conf. on Applications of Group Theory to Physics (Cocoyoc, Mexico 1980), to be pub.
14. Y. Ne'eman and S. Sternberg, ibid. (also report TAUP 133-80).

QFD IV

J. Pati, Maryland, Organizer

PROBING THE HIERARCHY OF GRAND UNIFICATION THROUGH CONSERVATION LAWS

Jogesh C. Pati*
International Centre for Theoretical Physics, Trieste, Italy,
and
Department of Physics, University of Maryland, College Park,
Maryland 20742, USA.

1. The hypothesis of grand unification [1-3] stands at present primarily on its aesthetic merits. It gives the flavour of synthesis in that it provides a rationale for the existence of quarks and leptons by assigning the two sets of particles to one multiplet of a gauge symmetry G. It derives their forces - weak, electromagnetic as well as strong - through one principle - gauge unification. It provides a reason for the quantization of electric charges.

With quarks and leptons in one multiplet F of a local spontaneously broken gauge symmetry G, baryon and lepton number conservations cannot be absolute. To see this simply, take for illustration a multiplet F of three quarks of red, yellow and blue colours plus one lepton (i.e. $F = \{q_r, q_y, q_b, \ell\}$) and gauge the maximal symmetry $SU(4)_{colour}$ in this space[2]. The gauge particle V_{15} associated with the 15th generator of this SU(4) symmetry couples to the current $\left[\sum_{r,y,b} \bar{q}_i\gamma_\mu q_i - 3\bar{\ell}\gamma_\mu \ell\right](2\sqrt{6})^{-1}$.

The corresponding charge is proportional to $(B_q-3L) = 3(B-L)$; here B_q denotes quark number, which is three times the baryon number B. This example serves to demonstrate the general result that once quarks and leptons are put in one multiplet and their maximal symmetry is locally gauged, some linear combination of baryon and lepton numbers (which in this example is B-L) must be among the generators of the local symmetry; it is thus conserved in the basic Lagrangian. Given that no massless vector particle can remain coupled to such a generator due to limits from the Eötvös-type experiments, however, the associated gauge particle (in this example V_{15}) must acquire a mass through spontaneous breakdown of the local symmetry. Thereby the associated charge (in this example B-L) must be violated spontaneously.

Instead of gauging the maximal symmetry SU(4), one might have chosen to gauge a subgroup of the maximal symmetry SU(4). In this illustrative example the subgroup might have been SU(2), which treats (q_r, q_y) and (q_b, ℓ) as doublets. In this case, the gauge interactions of the basic Lagrangian would break B and L explicitly even prior to spontaneous symmetry breaking because one and the same gauge particle would couple for example to $\bar{q}_r\gamma_\mu q_y$ as well as to $\bar{q}_b\gamma_\mu \ell$ currents. An analogous situation is in fact what happens in some realistic models of grand unification (see elaborations later). The point of the remark made above, however, is that even if baryon and lepton numbers are conserved in the basic Lagrangian (and this holds automatically if we

*Supported in part by the National Science Foundation.
On leave of absence from University of Maryland till 1 January 1981.

gauge the maximal symmetry of the quark-lepton-multiplet), either B or L or both must be violated spontaneously as the gauge particles of the quark-lepton symmetry acquire masses; this is in order that the theory may be compatible with the empirical fact based on Eötvös-type experiments. This line of reasoning had led Salam and myself to suggest in 1973 that the lightest baryon - the proton - must ultimately decay into leptons[2]. Theoretical considerations in the context of a number of models suggest a lifetime for the proton in the range of 10^{28}-10^{33} years[2-6].

Experiments are now underway to test proton stability to an accuracy one thousand times higher than before. In view of this, I shall concentrate primarily on the question of expected proton decay modes within the general hypothesis of quark-lepton unification and on the question of intermediate mass scales filling the grand plateau between 10^2 and 10^{15} GeV, which influence proton decay. The main point of my talk would be to stress that "maximal" symmetries such as SU(16) or its family extensions permit as a rule the existence of several intermediate mass scales lying within the grand plateau; consequently they allow four major decay modes* for proton some of which can even coexist[7]. These are:**

(i) $p \to 3$ leptons + (mesons) ($\Delta F = 0$; $\Delta(B-L) = -4$)
(ii) $p \to$ lepton + (mesons) ($\Delta F = -2$; $\Delta(B-L) = -2$)
(iii) $p \to$ antilepton + (mesons) ($\Delta F = -4$; $\Delta(B-L) = 0$)
(iv) $p \to 3$ antileptons + (mesons) ($\Delta F = -6$; $\Delta(B-L) = +2$) . (1)

It is furthermore noted[8] that within such symmetries not only proton decay but also $\Delta B = 2$ $n \leftrightarrow \bar{n}$ oscillation[9] and $\Delta L = 2$ neutrinoless β decay can in general coexist with measurable strengths.

2. Much of what I have to say arises in the context of those unifying symmetries for which violations of baryon and lepton numbers and in general also of fermion number $F \equiv B_q + L = 3B + L$ arises only spontaneously rather than explicitly in the manner discussed in Sec.I. Examples of this kind of symmetry are (i) the left-right symmetric subunification symmetry[2] $SU(2)_L \times SU(2)_R \times SU(4)_{L+R}$ and its unifying extensions, (ii) $[SU(4)]^4$, which operates on four flavours and four colours and (iii) $[SU(6)]^4$, which operates on six flavours and six colours,etc. For these symmetries (B-L) is locally gauged,but fermion number F is still at best a global symmetry. Further extensions which put n left-handed fermions F_L and their n antiparticles F_L^c in one multiplet and gauge the "maximal" symmetry*** SU(2n), gauge not only

*That the proton may in general decay via all these four modes satisfying $\Delta F = 0, -2, -4$ and -6 was first noted in Ref.7.
**Here F stands for fermion number, which is +1 for a quark or a lepton; leptons comprise ($e^-, \nu_e, \mu^-, \nu_\mu$) and mesons stand for π's and K's, etc.
***Such symmetries generate triangle anomalies, which are avoided by postulating that there exist a conjugate mirror set of fermions $F^m_{L,R}$ supplementing the basic fermions $F_{L,R}$ with the helicity flip coupling represented by the discrete symmetry $F_{L,R} \longleftrightarrow F^m_{R,L}$. Thus by "maximal" symmetry we mean a symmetry which is maximal upto discrete symmetries such as the mirror symmetry.

(B-L) but also fermion number F as local symmetries. As an example, for a single family of two flavours and four colours including leptonic colour, n = 8, and thus the "maximal" symmetry G is SU(16), which treats the 16 left-handed fields $\{u_{r,y,b}, \nu_e; d_{r,y,b}, e^- | d^c_{r,y,b}, e^+; u^c_{r,y,b}, \nu^c_e\}_L$ as members of a single 16-plet.

To include the e, μ and τ families, while attributing the distinction, which each family deserves in the sense of maximal gauging, one would need to gauge $[SU(16)]^3$ or the still extended symmetry* SU(48). Spontaneous symmetry breaking can permit the descent of these gigantic symmetries SU(48) or $[SU(16)]^3$ to the familiar low energy symmetry $SU(2)_L \times U(1) \times SU(3)^c$ via for example the diagonal symmetry** $SU(16)_{e+\mu+\tau}$ so that interfamily universality $e \leftrightarrow \mu \leftrightarrow \tau$ appears only below some energy scale M_f; M_f need not be any higher than about 10^5 GeV. It turns out that these extended maximal symmetries such as $[SU(16)]^3$ or SU(48) permit[10] signals for grand unification at low and intermediate mass scales ($\sim 10^4$-10^5 GeV and 10^8-10^{10} GeV) and thereby richer experimental possibilities[8] than for example SU(5) or SO(10). In the discussions to follow, SU(16) will be used for simplicity as a language for maximal symmetry, though it may be viewed ultimately as part of the extended maximal symmetries such as $[SU(16)]^3$ or SU(48).

As noted before B-L as well as F are locally gauged within SU(16) and are therefore conserved in the basic Lagrangian. They are violated in two distinct ways: (a) through spontaneously induced mixings of gauge particles carrying different B, L and F and (b) through spontaneously induced Yukawa transitions of the type $q \to \ell + \phi$, $q \to \bar{\ell} + \phi'$, or even $q \to \bar{q} + \phi''$, where ϕ's denote appropriate Higgs fields. In either case the violation is only spontaneous. Two illustrative examples for case (a) and two for case (b) are shown in Figs.1 and 2 respectively.

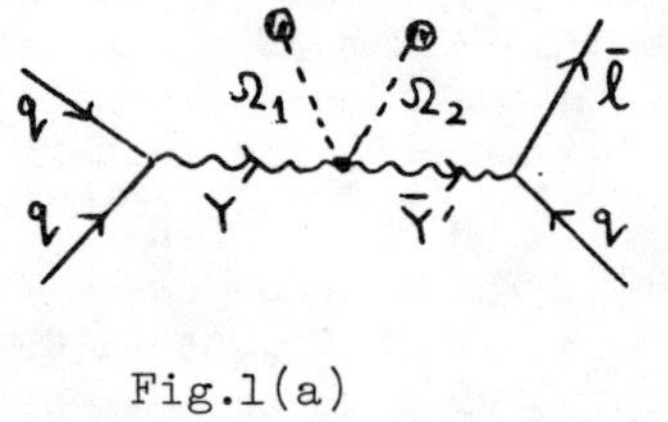

Fig.1(a)

($\Delta F = -4$) $p \to \bar{\ell}$ + mesons

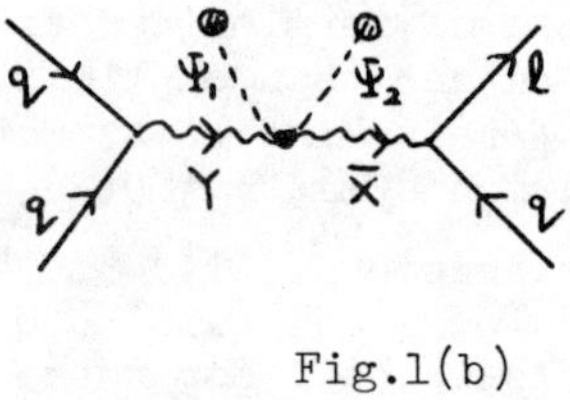

Fig.1(b)

($\Delta F = -2$) $p \to \ell$ + mesons

*These symmetries are no doubt gigantic, but if quarks and leptons are proliferated why not the associated gauge particles? As expressed elsewhere, the answer to proliferation must come by viewing quarks, leptons and the associated gauge particles as composites of more elementary objects - preons. From this point of view $[SU(16)]^3$ and [SU(48)] are only effective gauge symmetries generated from a more economical basis of preons.

**Note that for this diagonal symmetry $B = B_e + B_\mu + B_\tau$ and likewise for L and F.

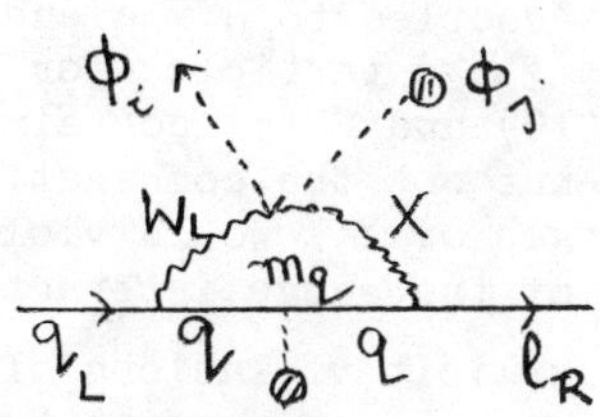

Fig.2(a)

$q \to \ell + \phi$

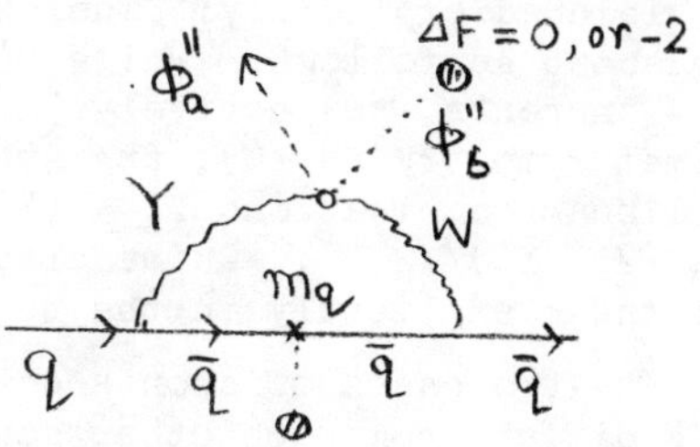

Fig.2(b)

$q \to \bar{q} + \phi''$ with $\Delta F = 0$ or -2

Fig.1(a) shows the mixing of the diquark gauge particle Y coupled to the $\bar{q}^c_L \gamma^\mu q_L$ current with the $\bar{Y}'$ gauge particle coupled to the $\bar{\ell}^c_L \gamma_\mu q_L$ current; such a mixing induces the $\Delta F = -4$ (i.e. $\Delta(B-L) = 0$) transitions:

$$\text{proton} \to (e^+ \text{ or } \bar{\nu}_e) + \text{mesons etc.} \qquad (2)$$

The mixing of Y with the leptoquark gauge particle $\bar{X}$ coupled to the $\bar{\ell}_L \gamma_\mu q_L$ current shown in Fig.1(b) induces the $\Delta F = -2$ (i.e. $\Delta(B-L) = -2$) transitions:

$$\text{proton} \to (e^- \text{ or } \nu_e) + \text{mesons etc.} \qquad (3)$$

The effective Yukawa transition $q \to \ell + \phi$ shown in Fig.2(a) taken in third order[11] induces the $\Delta F = 0$ (i.e. $\Delta(B-L) = -4$) transitions

$$\text{proton} \to 3 \text{ leptons} + \text{mesons} \quad . \qquad (4)$$

A similar mechanism involving $q \to \bar{\ell} + \phi'$ taken in third order would induce $\Delta F = -6$ decays (i.e. proton $\to 3\bar{\ell}$ + mesons), while $q \to \bar{q} + \phi''$ (shown in Fig.2(b)) taken in third later (see remarks later) induces $\Delta B = 2$ $n \longleftrightarrow \bar{n}$ transitions. It may be noted incidentally, that none of these mechanisms inducing the various proton decay modes and the $n \longleftrightarrow \bar{n}$ transition depend in any way upon the nature of the quark charges.

Instead of gauging the maximal symmetry SU(16), for which the violations of B,L and F are only spontaneous, one might have gauged one of its two special subgroups SO(10)[12] or SU(5),which preserve the spirit of grand unification. These two subgroups violate B, L and F explicitly just as the illustrative example - the SU(2) subgroup - does in Sec.1 and for the same reason. For SU(5) neither B-L nor F is locally gauged and B,L as well as F are violated explicitly*. For SO(10), B-L is locally gauged, but not F and here too B,L and F

*For the <u>minimal</u> SU(5) model, B-L turns out to be a global symmetry even though B,L and F are violated explicitly.

are violated explicitly. The explicit violations for SU(5) or SO(10) come about as follows. While Y's and Y''s coupled to $\bar{q}^c \gamma_\mu q$ and $\bar{\ell}^c \gamma_\mu \ell$ currents, respectively, are distinct gauge particles for the maximal symmetry SU(16), the subgroups SO(10) and SU(5) contain only the linear combinations $Y_s \equiv (Y + \bar{Y}')/\sqrt{2}$ but not the combinations $Y_a \equiv (Y - \bar{Y}')/\sqrt{2}$. Quite clearly exchanges of Y's would violate B, L and F explicitly in the second order of the gauge interactions.

The two cases of spontaneous versus explicit violations of B, L and F differ from each other conceptually as well as physically. Spontaneous violation would vanish at high temperatures exceeding the masses of the relevant gauge particles, where both Y and Y' or equivalently Y_s and Y_a would exist as degenerate massless particles. Explicit violation, on the other hand, would acquire its maximum strength at such high temperatures, since in this case only the Y_s gauge particles would exist as massless particles with the Y_a's being still absent. This distinction would play its most obvious role in the early stage of the Universe.

It is of interest to see the alternative routes for the spontaneous descent of SU(16) down to the low energy symmetry $SU(2)_L \times U(1) \times SU(3)_C$. The Higgs structure permitting the descents via such alternative routes is given in detail in a recent paper by Salam, Strathdee and myself.[8] The three most obvious descents go via 1) SO(10) with respect to which the 16-plet remains irreducible, 2) the maximal chiral symmetry $SU(8)_F \times SU(8)_{\bar{F}} \times U(1)_F$ where the two SU(8)'s operate in the spaces of eight fermions (F_L) and the eight antifermions (F_L^c) respectively, and $U(1)_F$ represents fermion number and 3) $SU(12)_q \times SU(4)_\ell \times U(1)_{|B|-|L|}$, where $SU(12)_q$ operates on (6 quarks + 6 antiquarks), $SU(4)_\ell$ on (2 leptons + 2 antileptons) and U(1) denotes $|B_q| - 3|L| = 3(|B| - |L|)$.

Note that in contrast to the first two routes, the third route separates quarks from leptons at the very first stage of symmetry breaking. Complexions for proton decay as well as the strength of $n \leftrightarrow \bar{n}$ oscillation depend crucially upon the route for spontaneous descent, which in turn depends upon the pattern of VEV of the Higgs multiplets. For instance, an adjoint $\underline{255}$ of Higgs by itself can take SU(16) to $SU(8)_F \times SU(8)_{\bar{F}} \times U(1)_F$, while a four index multiplet $\Phi^{\{AB\}}_{\{CD\}} \sim \{16 \times 16\}_{symm} \times \{16^* \times 16^*\}_{symm}$ by itself can break SU(16) in general to $SU(2)_L \times SU(2)_R \times SU(4)^c_{L+R}$ or for a special case[8] into SO(10). If $\langle \underline{255} \rangle$ far exceeds $\langle \Phi^{\{AB\}}_{\{CD\}} \rangle$, the descent via $SU(8) \times SU(8) \times U(1)_F$ would be prominent; in this case proton can decay via a variety of channels satisfying $\Delta F = -4$, $\Delta F = -2$ as well as $\Delta F = 0$; here $\Delta F = -2$ (i.e. $p \to e^- +$ mesons) and/or $\Delta F = 0$ (i.e. $p \to 3\ell +$ mesons) can even far supercede $\Delta F = -4$ mode (i.e. $p \to e^+ +$ mesons). If on the other hand $\langle \Phi^{\{AB\}}_{\{CD\}} \rangle$ far exceeds $\langle \underline{255} \rangle$ and in particular the special SO(10) chain materializes, then only the $\Delta F = -4$ mode ($p \to e^+ +$ mesons etc.) would be dominant. Thus predictions of SO(10) (or SU(5)) can emerge as special cases within those of the maximal symmetry SU(16). Finally, the third route with the intermediate $SU(12)_q \times SU(4)_\ell \times U(1)$ symmetry can result from an alternative pattern of VEV of the adjoint 255 plus $\Phi^{\{AB\}}_{\{CD\}}$. The third route has the potentiality of yielding $\Delta F = -4$ proton decay ($p \to e^+ +$ mesons) coexisting[8] with $\Delta B = 2$ $n \leftrightarrow \bar{n}$ oscillation with observable strength. This is because, for this route, the leptoquark X and the lepto-antiquark Y'gauge particles coupled respectively to the currents $\bar{q}\gamma_\mu \ell$ and $\bar{q}^c\gamma_\mu \ell$ acquire masses at the first stage of spontaneous symmetry breaking, while the diquark Y gauge particles coupled to the currents $\bar{q}^c\gamma_\mu q$ acquire their masses only at a secondary stage of SSB. This generates the possibility (for $m_Y \sim 10^4$-10^5 GeV, $m_{Y'} \sim 10^4$ GeV and Y-$\bar{Y}'$ mixing masses $\sim m_Y \sim 10^4$-10^5 GeV) that both $\Delta F = -4$ proton decay (i.e. $p \to e^+\pi^0$ etc.) induced via Fig.1(a) and $\Delta B = 2$ $n \leftrightarrow \bar{n}$ oscillation (induced via effective third order iteration of Fig.2(b) followed by a quartic $\lambda\phi^4$ interaction) can occur with observable strength.

It would be only premature to speculate on which of these routes if any is preferred by Nature. One must wait for the forthcoming experiments searching for proton decay as well as $n \leftrightarrow \bar{n}$ oscillation to provide guidance in this matter. Further elaborations on these questions may be found in Refs.8 and 13.

REFERENCES

1. J.C. Pati and Abdus Salam, "Lepton hadron unification" (unpublished) reported by J.D. Bjorken in the Proceedings of the 15th High Energy Physics Conference held at Batavia, Vol.2, p.304, September (1972); J.C. Pati and Abdus Salam, Phys. Rev. D8, 1240 (1973).
2. J.C. Pati and Abdus Salam, Phys. Rev. Letters 31, 661 (1973); Phys. Rev. D10, 275 (1974); Phys. Letters 58B, 333 (1975).
3. H. Georgi and S.L. Glashow, Phys. Rev. Letters 32, 438 (1974).
4. J.C. Pati, Proc. Seoul Symposium (1978).
5. H. Georgi, H. Quinn and S. Weinberg, Phys. Rev. Letters 33, 451 (1974).
6. T. Goldman and D. Ross, Cal. Tech. preprint (1979).
7. J.C. Pati, Abdus Salam and J. Strathdee, Il Nuovo Cimento 26A, 77 (1975); J.C. Pati, Proceedings of the Second Orbis Scientiae, Coral Gables, Florida, p.253, January (1975); J.C. Pati, S. Sakakibara and Abdus Salam, ICTP, Trieste, preprint IC/75/93 (unpublished).
8. J.C. Pati, Abdus Salam and J. Strathdee, forthcoming preprint, ICTP, Trieste.
9. V.A. Kuzmin, Pisma Zh. E Ksp. Teor. Fiz. 13, 335 (1970); S.L. Glashow, Cargese Lectures (1979); R.N. Mohapatra and R.E. Marshak, VPI-HEP-80/1 and 80/2; L.N. Chang and N.P. Chang, CCNY-HEP-80/5 (1980). J.C. Pati, Abdus Salam and J. Strathdee, work reported at 20th Int. Conf.on H.E. Physics, Madison, Wisconsin (July, 1980) and preprint to appear.
10. B. Deo, J.C. Pati, S. Rajpoot and Abdus Salam, preprint to appear.
11. B. Deo, J.C. Pati, S. Rajpoot and Abdus Salam, preprint to appear.
12. H. Fritzsch and P. Minkowski, Ann. Phys. (N.Y.) 93, 193 (1975); H. Georgi, Proceedings of the Williamsburg Conference (1974).
13. J.C. Pati, ICTP, Trieste, Preprint IC/80/178, Proc. of 1980 ν-Conference held at Erice (to appear).

ASPECTS OF UNIFIED GAUGE THEORIES

Q. Shafi
Theoretical Physics Division, CERN
1211 Geneva 23, Switzerland

ABSTRACT

The talk is divided into two parts. The first part outlines the phenomenology of an SO(10) model which breaks down to $SU(3)_c \times U(1)_{em}$ via $SU(4)_c \times SU(2)_L \times SU(2)_R$. The second part describes how a strong first order phase transition in unified gauge theories may suppress the superheavy magnetic monopole density at the time of nucleosynthesis.

1. - MOTIVATION FOR GOING BEYOND MINIMAL SU(5)

Let us list some reasons which motivate us to go beyond SU(5) [1] in search of a unified gauge model of strong, weak and electromagnetic interactions.

a) Fermions of each generation belong to a reducible representation, i.e., to $\bar{5}_L + 10_L$ of SU(5).

b) Since the maximal electroweak subgroup of SU(5) is $SU(2)_L \times U(1)$, the observed parity violation in weak interactions is put in by hand.

c) Mass relations exist in SU(5) that involve charge $-\frac{1}{3}$ quarks and charged leptons. However, nothing can be said about charge $\frac{2}{3}$ quarks.

d) SU(5) predicts $(m_d/m_s) = (m_e/m_\mu)$ in the asymptotic limit [2]. This is known to be badly violated.

e) Although B+L is violated, B−L remains as an accidental global symmetry of SU(5). The apparent masslessness of the neutrino is accidental in SU(5) which is not very satisfactory.

f) The baryon asymmetry in the Universe as predicted by minimal SU(5) is several orders of magnitude too small.

g) Last, but by no means least, the experimental value of $\sin^2\theta_W$ may be somewhat larger than the SU(5) prediction [3] of 0.2.

A MINIMAL SO(10) MODEL

The fermions of each generation belong to the $\underset{\sim}{16}$ dimensional irreducible spinor representation of SO(10) [4]. The maximal electroweak subgroup is $SU(2)_L \times SU(2)_R \times U(1)$ [5] which is left-right symmetric and is superstrongly broken down to $SU(2)_L \times U(1)$. Points (a) and (b) listed above are therefore easily taken care of. The resolution of the remaining points requires a more detailed analysis [6] which we briefly describe. One is led to what one may call a minimal SO(10) model.

SYMMETRY BREAKING PATTERNS IN SO(10)

By employing a minimal Higgs system ($\underset{\sim}{45}$, $\underset{\sim}{126}$ and $\underset{\sim}{10}$ or $\underset{\sim}{54}$, $\underset{\sim}{126}$ and $\underset{\sim}{10}$) the following symmetry breaking patterns are allowed :

i) $SO(10) \to SU(3)_c \times SU(2)_L \times U(1) \to SU(3)_c \times U(1)_{em}$

ii) $SO(10) \to SU(5) \to SU(3)_c \times SU(2)_L \times U(1) \to SU(3)_c \times U(1)_{em}$

iii) $SO(10) \to SU(4)_c \times SU(2)_L \times SU(2)_R \to SU(3)_c \times SU(2)_L \times U(1)$
$\qquad\qquad\qquad\qquad\qquad\qquad\qquad\qquad\qquad\downarrow$
$\qquad\qquad\qquad\qquad\qquad\qquad\qquad\qquad SU(3)_c \times U(1)_{em}$

Scheme iii) :

Predicts $\sin^2\theta_W > 0.2$ (see (g)),

Keeps the "good" SU(5) predictions,

Overcomes (c), (d), (e) and (f),

Predicts the top quark mass plus the masses of the particles of a possible fourth generation,

Is likely to avoid the cosmological monopole problem since it necessarily involves a strong first order phase transition [7].

PROTON LIFETIME

Consider the breaking

$$SO(10) \underset{M_x}{\overset{54}{\to}} SU(4)_c \times SU(2)_L \times SU(2)_R \underset{M_R}{\overset{126}{\to}} SU(3)_c \times SU(2)_L \times U(1)$$
$$10 \downarrow M_L$$
$$SU(3)_c \times U(1)_{em}$$

Using standard renormalization group arguments [8]

$M_x \sim 10^{15}$ GeV

$M_R \sim (1\text{-}2)\times 10^{13}$ GeV

with $\sin^2\theta_W \simeq 0.23$ and $\alpha_s(M_L) \simeq 0.11$

The proton lifetime τ_p is estimated to be [9]

$\tau_p \simeq (1\text{-}20)\times 10^{31}$ yr.

Note that B-L violating processes in SO(10) are strongly suppressed. It would be difficult to distinguish between SU(5) and this SO(10) model on the basis of proton decay[10].

CHARGED FERMION MASSES

The main breaking of $SU(2)_L \times U(1)$ down to $U(1)_{em}$ is achieved by the $SU(2)_L$ doublets contained in $\underset{\sim}{10}$. However, due to mixing, the $SU(2)_L$ doublets in $\underset{\sim}{126}$ acquire an induced vacuum expectation value v where[6]

$$v \simeq (M_R/M_x)^2 <\underset{\sim}{10}> \quad << \quad <\underset{\sim}{10}>$$

This has the following effect on the fermion mass relations obtained by considering only the $\underset{\sim}{10}$.

i) The bad relation $m_d = m_e$ involving the first generation is substantially altered. Indeed, in contrast to minimal SU(5), there is enough freedom in this minimal SO(10) model to arrange matters so that the first generation masses and the Cabibbo angle come out right.

ii) The mass relations involving the second generation are altered by about 10%. i) and ii), when taken together, take care of (d).

iii) Mass relations involving the third and possible fourth generation are essentially unchanged.

Examples

$$m_b(M_R) = m_\tau(M_R) \pm 1\%$$

$$\rightarrow m_b(10 \text{ GeV}) \simeq 4.6 - 5.5 \text{ GeV}$$

$$m_s(M_R) = m_\mu(M_R) \pm 10\%$$

$$\rightarrow m_s(1 \text{ GeV}) \simeq (370\text{-}500) \text{ MeV} \pm 10\%$$

Top quark mass

$$m_t(M_R) = m_c(M_R)\ (m_\tau(M_R)/m_\mu(M_R)) \pm 10\%$$

$$\rightarrow m_t(40 \text{ GeV}) \simeq 20 \pm 2 \text{ GeV}$$

Fourth generation

It is possible to have a fourth generation $\nu_{\tau'}\tau' t' b'$ in the SO(10) scheme discussed here. Their masses can be predicted as follows.

One expects that

$$17 \text{ GeV} \lesssim m_{\tau'} \lesssim 20 \text{ GeV}$$

where the lower bound is experimental and the upper bound comes from the requirement [11] that the Yukawa couplings should not become strong anywhere in the range between M_L and M_x. We then predict

$$m_{t'}(2m_{t'}) \simeq 170 - 200 \text{ GeV}$$

$$m_{b'}(2m_{b'}) \simeq 38 - 45 \text{ GeV}$$

NEUTRINO MASSES [12]

The decomposition of 16 under $SU(4)_c \times SU(2)_L \times SU(2)_R$ is as follows :

$$\underset{\sim\sim}{16} \to (4,2,1) + (\bar{4},1,2)$$

One has both left-handed (ν_L) and right-handed (ν_R) neutrino fields. Let us consider one generation. Then

ν_R acquires a Majorana mass term of order M_R (through $\underset{\sim\sim\sim}{126}$)
ν_L acquires an induced Majorana mass term of order M_L^2/M_R (through $\underset{\sim\sim\sim}{126}$)
One also has $m_\nu^{\text{Dirac}} = m_{\frac{2}{3}}$ (from $\underset{\sim\sim}{10}$; see point c) above).

The neutrino mass matrix is as follows :

$$\frac{1}{2}(\bar{\nu}_L^c \; \bar{\nu}_R) \begin{pmatrix} c & b \\ b & a \end{pmatrix} \begin{pmatrix} \nu_L \\ \nu_R^c \end{pmatrix} + \text{h.c.}$$

where

$$a \sim M_R >> b \sim m_{\frac{2}{3}} >> c \sim (M_L^2/M_R)$$

Diagonalization leads to two Majorana mass eigenstates with eigenvalues M_R and $c - b^2/a$. The "heaviest" light neutrino mass can lie in the 1-10 eV range (point e)).

BARYON ASYMMETRY

An interesting source of generating baryon asymmetry in the Universe within the SO(10) model may be the decay of superheavy right-handed neutrinos at temperatures below M_R [13]. With suitable Yukawa couplings and the above value for the Majorana mass, it seems possible to account for a baryon asymmetry of order $10^{-8\pm1}$ (point f)).

2. - FIRST ORDER PHASE TRANSITION AND MAGNETIC MONOPOLE SUPPRESSION

Our reasons for supposing that a strong first order [14] phase transition can suppress the 't Hooft-Polyakov [15] density differ somewhat from the arguments given in Ref. 16 [17]. Let us briefly summarize the scenario for the SU(5) case [7] with a Higgs potential involving one 24-plet.

At very high temperatures ($T>T_{c1}' \simeq 10^{16}$ GeV) the SU(5) symmetry is fully operative and no monopoles are present. At a temperature $T < T_{c1}'$, the effective Higgs potential develops a local minimum corresponding to the $SU(4)\times U'(1)$ phase which, to start with, is thermodynamically unfavoured compared to the SU(5) phase. However, the former phase quickly overtakes the latter in being the lower minimum and, after further cooling, the SU(5) phase is not even a local minimum of the potential. The system then undergoes a rapid transition (at temperature T_{c1}) to the $SU(4)\times U'(1)$ phase. Superheavy magnetic monopoles carrying SU(4) and U'(1) magnetic charges are produced.

As the Universe cools further in the $SU(4)\times U'(1)$ phase, the Higgs potential develops a local minimum corresponding to the $SU(3)_c\times SU(2)_L\times U(1)$ phase which becomes degenerate with the former phase at some critical temperature T_c. After more cooling, the $SU(3)_c\times SU(2)_L\times U(1)$ phase has lower free energy than the $SU(4)\times U'(1)$ phase. For reasonable choice of Higgs parameters a large barrier separates the two minima. The transition from the false to the true vacuum proceeds through barrier penetration (bubble formation) which, at least to start with, is a rather slow process. Therefore, the Universe will supercool in the false ($SU(4)\times U'(1)$) phase until a temperature T_{c2} is reached at which the rate of barrier penetration becomes large. Indeed, at temperature T_{c2}, the false phase in our case becomes a local maximum in its SU(4) directions. As soon as this happens, the false vacuum decays almost instantaneously into the true vacuum. Some comments are in order :

1) If the Universe stays in the false ($SU(4)\times U'(1)$) phase for a sufficiently long time, it undergoes an exponential expansion until the temperature T_{c2} is reached. The formation of bubbles of the true ($SU(3)_c\times SU(2)_L\times U(1)$) vacuum in the early stages of the phase transition can be neglected since they can, at most, only expand with the speed of light.

2) The probability of converting the false ($SU(4)\times U'(1)$) monopoles to $SU(3)_c\times SU(2)_L\times U(1)$ monopoles is expected to be very low. They will mostly be annihilated.

3) The release of latent heat at T_{c2} warms the Universe back to $T_r < T_c$. Typically, $T_r \simeq 0.45\ T_c$. The monopole density is expected [15] to be proportional to $\exp-(m(T_r)/T_r)$, with $(m(T_r)/T_r) \simeq 70$-80. Thus, as desired, monopole production is strongly suppressed by a first order transition.

ACKNOWLEDGEMENT

George Lazarides collaborated on all the aspects of the work described here.

REFERENCES

1. H. Georgi and S.L. Glashow - Phys.Rev.Letters 32 (1974) 438.

2. A. Buras, J. Ellis, M.K. Gaillard and D. Nanopoulos - Nuclear Phys. 135B (1978) 66.

3. H. Georgi, H. Quinn and S. Weinberg - Phys.Rev.Letters 33 (1974) 451.

4. H. Fritzsch and P. Minkowski - Ann.Phys. 93 (1975) 193 ;
H. Georgi - Particles and Fields, Ed. C.E. Carlson (AJP) (1975) ;
M.S. Chanowitz, J. Ellis and M.K. Gaillard - Nuclear Phys. B128 (1977) 506.

5. J. Pati and A. Salam - Phys.Rev. D10 (1974) 275.

6. G. Lazarides, Q. Shafi and C. Wetterich - Univ. of Freiburg Preprint 80/2 (1980).

7. For a discussion on monopoles in SO(10), see : G. Lazarides, M. Magg and Q. Shafi - CERN Preprint TH. 2856 (1980).

8. H. Georgi and D. Nanopoulos - Harvard Preprint HUTP 79/A039 (1979) ;
Q. Shafi, M. Sondermann and C. Wetterich - Phys.Letters 92B (1980) 304 ;
S. Rajpoot - Imperial College Preprint (December 1980).

9. J. Ellis, M.K. Gaillard and D. Nanopoulos - CERN Preprint TH. 2749 (1979) ;
C. Jarlskog and F.J. Yndurain - Nuclear Phys. B149 (1979) 29 ;
A. Din, G. Girardi and P. Sorba - LAPP Preprint TH. 08 (1979) ;
B. Machacek - Nuclear Phys. B169 (1979) 37 ;
M.B. Gavela, A. Le Yaouanc, L. Oliver, O. Peire and J.C. Raynal - Orsay Preprint LPTHE 80/6 (1980).

10. See, however : A. De Rujula, H. Georgi and S.L. Glashow - Harvard Preprint

11. N. Cabibbo, L. Maiani, G. Parisi and R. Petronzio - Nuclear Phys. B158 (1979) 295.

12. M. Gell-Mann, P. Ramond and R. Slansky - unpublished.
Also, see : R. Barbieri, D.V. Nanopoulos, G. Morchio and F. Strocchi - Phys.Letters 90B (1980) 91 ;
M. Magg and C. Wetterich - CERN Preprint TH 2829 (1980).

13. T. Yamagida and M. Yoshimura - KEK Preprint TH. 8 (1980) ;
J. Harvey, P. Ramond and D. Reiss - Caltech Preprint CALT 68-758 (1980).

14. For estimates based on second order transitions, see :
J.P. Preskill - Phys.Rev.Letters 43 (1979) 1365 ;
M.B. Einhorn, D.L. Stein and D. Toussaint - Phys.Rev. D12 (1980) 3295 ;
T.W.B. Kibble - J.Phys. A9 (1976) 1387.

15. G. 't Hooft - Nuclear Phys. B79 (1974) 276 ;
A.M. Polyakov - JETP Letters 20 (1974) 194.

16. A.H. Guth and S.H. Tye - Phys.Rev.Letters 44 (1980) 631.

17. For details, see : A. Kennedy, G. Lazarides and Q. Shafi - to be published.

FIGURE CAPTION

Plot of SU(5) phase diagram with one 24 plet ϕ of Higgs fields. The Higgs potential is

$$V = \frac{a}{4}(\mathrm{Tr}\phi^2)^2 + \frac{b}{2}\mathrm{Tr}\phi^4 + \frac{c}{3}\mathrm{Tr}\phi^3 - \frac{\mu^2}{2}\mathrm{Tr}\phi^2 .$$

Define $\eta = a/b > -\frac{7}{15}$ and $\xi = -\mu^2 b/c^2$ $(b > 0)$. To the right of A only the SU(5) phase exists. Between A and A' a local minimum corresponding to SU(4)×U'(1) appears. To the left of A' the SU(4)×U'(1) phase is the absolute minimum and becomes the sole minimum between $\xi = 0$ (where SU(5) disappears) and B. To the left of B there appears a local minimum corresponding to $SU(3)_c \times SU(2)_L \times U(1)$. Between C and D the SU(4)×U'(1) phase is the false minimum. To the left of D the SU(4) directions develop instabilities. The false vacuum then decays instantaneously to the true $SU(3)_c \times SU(2)_L \times U(1)$ vacuum.

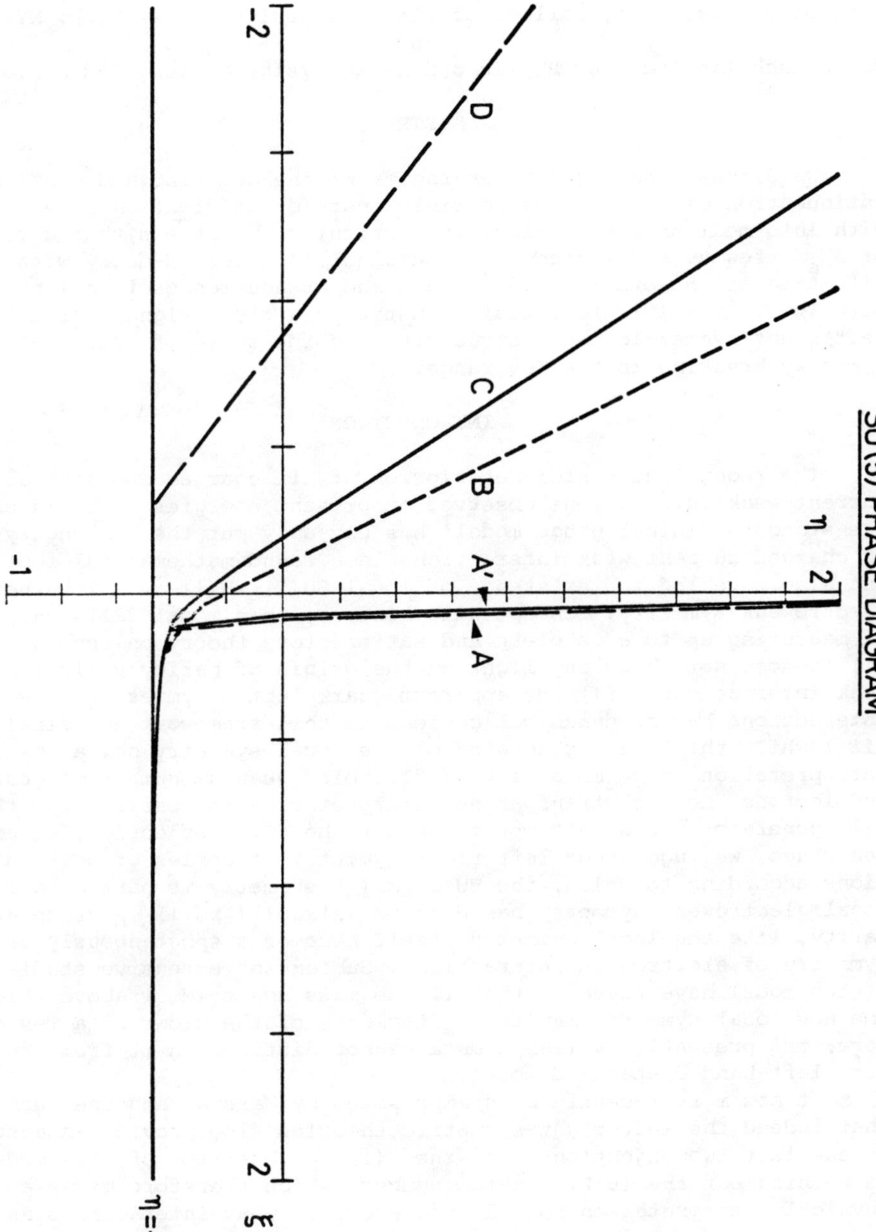
SU(5) PHASE DIAGRAM
η
2
-1
-2
2
ξ
A
A'
B
C
D
η=-7/15

MAJORANA MASSES FOR NEUTRINOS AND NEUTRON OSCILLATIONS $(N \leftrightarrow \bar{N})$ AS TESTS OF UNIFICATION MODELS WITH INTERMEDIATE MASS SCALES

Rabindra N. Mohapatra
Dept.of Physics, City College of City University of New York, NY-10031
and
Max-Planck-Institut für Physik und Astrophysik, Munich, Fed.Rep.Germany

ABSTRACT

We discuss the Majorana character of the neutrino and neutron-antinueutron oscillation as possible tests of unification models with intermediate mass scales. In particular, we stress that a value of $m_{\nu e}$ few ev's, existence of neutrinoless double β-decay with life-time of the order of 10^{24} years and/or neutron oscillation with τ_{N-N} 10^{6}-10^{10} sec. would constitute a clear signal for a left-right symmetric gauge structure with the scale of local B-L symmetry breaking in the Tev range.

INTRODUCTION

The recent successful description of all charged and neutral current weak interactions observed at present energies in terms of the standard minimal gauge model[1] has not only put the V-A theory[2] of charged current weak interactions in a sound mathematical footing but also revealed the existence of local $SU(2)_L xU(1)$ as a low energy electroweak symmetry. Conceptually however, this model falls short of measuring up to a complete and satisfactory theory on several counts: (i) It does not throw any light on the origin of parity violation in weak interactions; (ii) the apparent quark-lepton symmetry[3] of weak interactions has no dynamical content in this framework and finally, (iii) while the $SU(2)_L$ generator of the local symmetry has a physical interpretation in terms of a weak "Isospin" quantum number of quarks and leptons, no such transparent interpretation is possible for the U(1)-generator. In an attempt to answer the first objection, several years ago, we suggested[4] left-right symmetric theories of weak interactions according to which, the $SU(2)_L xU(1)$ symmetry is part of a larger local electroweak symmetry based on $SU_L(2)xSU(2)_R xU(1)_{L+R}$ group and parity, like the local symmetry itself becomes a spontaneously broken symmetry of electroweak interactions. Subsequent extensive studies[4] of the model have revealed that if the mass scale (M_{W_R}) above which the new local symmetry manifests itself is of the order of a Tev or more, the presently available data cannot distinguish it from the pure left-handed standard model.

It has more recently been emphasized by Marshak and the author[6] that indead the left-right symmetric theories also provide an answer to the last two objections i.e. the $U(1)_{L+R}$ generator of this model is nothing but the (B-L)-quantum number, which therefore gives a physical interpretation to all generators of weak interactions and in a way,gives a dynamical meaning to the quark-lepton symmetry. It has been further noted[7] that the smallness of the neutrino mass in these models is linked to the dominant V-A nature of low energy charged current weak interactions.

In view of the immensely satisfactory framework that the left-right symmetric theories provide for the description of electroweak interactions, it is important to point out the various tests of this model and furthermore, it is also important to compare and contrast its predictions with those of other unification models that coincide the $SU(2)_L xU(1)$ model in the low energy regime. In this talk, we discuss the neutrino mass, lifetimes for ν-less doulde β-transitions, rare decays such as $\mu \to e\gamma$ and the phenomenon of neutron oscillation as tests between left-right symmetric and other unification models with intermediate mass scales.

MASSIVE NEUTRINOS

Even though the V-A theory was motivated by the considerations based on a massless neutrino, the present terrestrial experiments give only the following bounds on the various neutrino masses: 16 ev $< m_{\nu_e} <$ 46 ev (shape of H^3 β-spectrum[11]), $m_{\nu_\mu} <$ 0.57 Mev. (Muon Range in $\pi^+ \to \mu^+ + \nu_\mu$) and $m_{\nu_\tau} <$ 250 Mev (τ-decay). There are also indications of neutrino oscillations in recent reactor experiment of Reines et al.[12] which would be consistent with δm_{12} (i.e. mass difference between any two types of neutrinos) few ev. There also exist certain persuasive astrophysical arguments in favor of a massive neutrino[13]. Should these preliminary indications of non-vanishing neutrino mass be confirmed in the future, electroweak models will have to be constructed to accommodate this. Below we present a brief overview of the various options for gauge models with massive neutrinos.

There are two distinct possibilities: (i) Dirac Neutrinos, (ii) Majorana Neutrinos. A characteristic experimental signature of case (ii) is the existence of neutrinoless double β-decay, which we discuss shortly.

(a) <u>$SU(2)_L xU(1)$ model:</u> First, we note that in the standard $SU(2)_L xU(1)$ model absence of ν_R and the existence of conserved global leptonic quantum number lead to a massless neutrino. Though not a very natural possibility, a massive neutrino can be accommodated in $SU(2)_L xU(1)$ model in two ways. Adding a ν_R allows the following invariant Yukawa coupling $h_\nu(\bar{\nu}_L \nu_R \Phi_0^* - \bar{e}_L \nu_R \Phi^-)$ which on substituting $\langle\Phi_o\rangle = \kappa$ gives a Dirac mass to the neutrino. The other possibility is not to add ν_R but instead a triplet Higgs meson (1, 2) which then allows for the following gauge-invariant Yukawa coupling: $h_\Delta(\nu_L{}^T c^{-1} \nu_L \Delta^o + \nu_L{}^T c^{-1} e_L^- \Delta^+ + e_L^{-T} c_L e^{-T} \Delta^{++})$. On substituting for $\langle\Delta^o\rangle = v$ leads to a Majorana mass for the neutrino. In both these cases, however, the smallness of neutrino mass remains a mystery.

(b) $SU(2)_L xSU(2)_R xU(1)_{B-L}$ model: In contrast with the $SU(2)_L xU(1)$ model, in the left-right symmetric model, the neutrino mass arises naturally. This is because, left-right symmetry demands that, there must be a right-handed doublet $\Psi_R \equiv (\nu_R, e^-_R)$ accompanying the standard left-handed doublet $\Psi_L \equiv (\nu_L, e^-_L)$. Thus, the very Higgs multiplet $\Phi(1/2,1/2,0)$, which gives mass to the electron via the Yukawa coupling $\bar{\Psi}_L \Phi \Psi_R$, also gives a Dirac mass to the neutrino. This point was emphasized repeatedly earlier[4,5]. In a recent paper, Senjanoviĉ and this author[7] have argued that, a more physically appealing way to have a small m_{ν_e} exists in this model. We note in this model, due to the relation $Q = I_{3L}+I_{3R}+\frac{B-L}{2}$ we find that, $\Delta I_{3R} = -(1/2)\Delta(B-L)$ or the scale of parity violation (i.e. m_{W_R}) and the scale of (B-L) breaking are identical. To realize the above idea, we choose the Higgs multiplets $\Delta_L(1,0,2)$ and $\Delta_R(0,1,2)$ which carry non-vanishing B-L quantum number and implement parity breakdown by the choice of vev $<\Delta_R> \neq 0$ and $<\Delta_L> = 0$. One then obtains a mass matrix for neutrinos of type[14]

$$\begin{array}{c c} & \begin{array}{cc} \nu_L & \nu_R \end{array} \\ \begin{array}{c} \nu_L \\ \nu_R \end{array} & \begin{pmatrix} 0 & h<\Phi> \\ h<\Phi> & h'<\Delta_R> \end{pmatrix} \end{array} \tag{1}$$

As a result, ν_L and N_R become two two-component Majorana leptons with $m_{\nu_L} \approx \frac{m^2}{m_{N_R}} \approx \frac{m^2}{h'<\Delta_R>}$ (and $m_{N_R} \gg m_L$). It is clear from this that for $<\Delta_R> \to \infty$, $m_{\nu_L} \to 0$ thus relating the smallness of m_ν to the suppression of V+A charged currents. A choice of the right-handed gauge boson mass, m_{W_R} in the Tev range leads to neutrino masses well within the present terrestrial upperbounds.[7] In particular, for the electron neutrino, this model can accommodate a value of several electron volts. The main point however is that, due to the existence of a heavier mass scale above m_{W_L} (i.e. m_{W_R}) the smallness of the neutrino mass is no more a mystery. In fact, similar results also follow in various grandunified models which generally contain heavier mass-scales.

(c) Grandunified models: While the minimal SU(5) model of Georgi and Glashow does not allow for a neutrino mass, the inclusion of a {15}-dim Higgs multiplet in this model leads to a Majorana neutrino[10] with $m_{\nu_e} \approx 10^{-3}$ ev. Similar small values of m_ν also arise in the grandunified SO(10) model[14] if a Higgs multiplet belonging to the {126}-dim representation is included in the theory. These models, however, lead to $m_{\nu_e} \approx 10^{-3}$ ev and therefore can be distinguished from a low mass W_R theory of type described in sec. (2b)[15].

Aside from measuring the m_{ν_e}, m_{ν_μ} etc. from direct terrestrial experiments, another way to discriminate between the above options (a), (b) and (c) is to look for neutrinoless double β-transitions and $\mu \to e\gamma$ decay. Due to the existence of a low mass Majorana lepton N (with $m_N \approx 100$ Gev) in the left-right symmetric model, it predicts $(\beta\beta)_0$ transition lifetimes for ^{82}Se of order 10^{24} years[10] and also branching ratios for the $\mu \to e\gamma$ decay of order 10^{-9} to 10^{-11}. These processes are extremely suppressed in the models of type (a) or (c).

NEUTRON OSCILLATIONS ($N \leftrightarrow \bar{N}$): PHENOMENOLOGY AND THEORY

It was observed by Marshak and this author[6] that the identification of the U(1) electroweak generator with (B-L) quantum number allows for the existence of $\Delta B = 2$ transitions (without $\Delta B = 1$ transitions) of typ $N \leftrightarrow \bar{N}$ (Neutron Oscillations) and $N_1 + N_2 \to \pi$'s with characteristic N-$\bar{N}$ transition times $\tau_{N-\bar{N}} \gtrsim 10^6$ sec. (corresponding to $\tau_{N_1+N_2 \to \pi's} \gtrsim 10^{30}$sec) provided $m_{W_R} \gtrsim 10^3$ to 10^4 Gev. Such transitions were also discussed from purely phenomenological standpoint independently in ref. 9. Detailed phenomenology and theoretical predictions concerning N-$\bar{N}$ transitions have been discussed earlier[16]. Below we summarize the main points.

If we define an effective N-$\bar{N}$ transition operator $_{\Delta B=2} \cong \delta m\, N^T C^{-1} N$ + h.c., we find $\delta m \approx 10^{-21}$ ev from the relation $(\delta m)^2 \approx (\Gamma_{N+N \to \pi's} \cdot M)$ and the present upperlimits on nuclear stability of 10^{30} years. Defining $\tau_{N-\bar{N}} \simeq (\hbar/\delta m)$, we find the corresponding $\tau_{N-\bar{N}} \gtrsim 10^6$ sec. To show why such a small $\Delta B=2$ transition time does not lead to instantaneous annihilation of all baryonic matter in the universe, we have noted earlier[8] that coherent evolution of $\bar{N}$ from an ensemble of neutrons is very sensitive to the existence of external fields. For example, if there exists an external field that discriminates between a neutron and antineutron i.e. $(V_N \neq V_{\bar{N}})$, then, transition probability to $\bar{N}$, $P_{\bar{N}}(t)$ at time t starting with pure neutrons at time t=0, is given by

$$P_{\bar{N}}(t) = \left(\frac{\delta m}{V_N - V_{\bar{N}}}\right)^2 \sin^2((V_N - V_{\bar{N}})t) \tag{2}$$

$V_N - V_{\bar{N}} \approx 2\mu B_{Earth} \approx 10^{-11}$ ev for a beam of neutrons in the laboratory. In nuclear matter, $V_N - V_{\bar{N}} \approx 100$ Mev. So, we see that in each case, for reasonable flight times, t, $P_{\bar{N}}(t)$ is extremely tiny. This is therefore the reason why catastrophy is avoided even with $\tau_{N-\bar{N}}$ of about a month. The same reason could however make observation of N-$\bar{N}$ transition difficult. However, in a number of proposed experiments,[17,18,19,20] it is planned to de Gauss the experimental tube to an extent that, one has $V_N - V_{\bar{N}} \approx 10^{-14}$ ev and then, for reasonable transit times $\frac{(V_N - V_{\bar{N}})}{\hbar} \ll 1$ so that, one obtains $P_{\bar{N}}(t) \approx (\delta m \cdot t)^2$. For typical reactor fluxes of 10^{13} N/sec, this could lead to $\approx 10^4$ counts a year if $\delta m \approx 10^{-21}$ ev and would constitute a clear signal above the background. Turning the question around, no event in such an experiment running once a year would reduce the upperband on δm by two orders of magnitudes (corresponding $\tau_{\Delta B=2}$ Nuc.Transition $\gtrsim 10^{34}$ years).

On the theoretical side, the only[21] model which leads to a clear cut prediction with such a small $\tau_{N-\bar{N}}$ is the partial unification model of Marshak and the author based on the group $SU(2)_L \times SU(2)_R \times SU(4')$ gauge group, provides one accepts an $m_{W_R} \gtrsim 10^3$ to 10^4 Gev, (the same values which also lead to $m_{\nu_e} \lesssim$ few ev.). Embedding this model in SO(10) grandunified group tends to suppress the prediction for δm unless small values for Higgs scalar self couplings are chosen. Another possibility is to embed the partial unification group in grandunified groups of type $[SU(2N)]^4$, $(N > 2)$[22] in which case, due to much lower grandunification mass scales, a large value of δm[23] can be obtained[16]. The observation of neutron oscillation with $\tau_{N-\bar{N}}$ 10^6-10^8 sec. will, therefore, be crucial to the future direction of grand unification.

I would like to acknowledge useful discussions with R.E. Marshak, J.C. Pati, Riazuddin, G. Senjanivoč, R. Wilson and P.K. Kabir.

REFERENCES

1. S. Weinberg, Phys,Rev.Lett. 19, 1264 (1967); A. Salam, "Elementary Particle Theory", ed. by N. Svartholm (1968); S.L. Glashow, Nucl.Phys. 22, 579 (1961).
2. R.E. Marshak and E.C.G. Sudarshan, Proc.of the Padua-Venice Conf. on "Mesons and Newly discovered Particles" (1957); R.P. Feynman and M. Gell Mann, Phys.Rev. 109, 193 (1958).
3. A. Gamba, R.E. Marshak and S. Okubo, Proc.Nat.Acad. of Sci. 45, 881 (1959); B.J. Bjorken and S.L. Glashow, Phys.Lett. 11, 255 (1964).
4. J.C. Pati and A. Salam, Phys.Rev. D10, 275 (1974); R.N. Mohapatra and J.C. Pati, Phys.Rev. D11, 566,2558 (1975); G. Senjanovič and R.N. Mohapatra, Phys.Rev. D12, 1502 (1975).
5. G. Costa, M. d'Anna and P. Marcolungo, Nuovo Cimento 50A, 177 (1979); G. Senjanovič, Nucl.Phys. B153, 334 (1979). For a review, see R.N. Mohapatra, "New Frontiers in High Energy Physics", edited by A. Perlmutter and L. Scott, Plenum 1968 , p. 33.
6. R.E. Marshak and R.N. Mohapatra, Phys.Lett. 91B, 222 (1980); R.N. Mohapatra and R.E. Marshak, Phys.Rev.Lett. 44, 1316 (1980).
7. R.N. Mohapatra and G. Senjanovič, Phys.Rev.Lett 44, 912 (1980) and FermiLab Preprint (1980).
8. R.N. Mohapatra and R.E. Marshak, ref. 6 and Phys.Lett. B (to appear).
9. V.A. Kuzmin, K.G. Chetrkin, M.A. Kazarnovsky and M.E. Shaposhmikov, P-0161, Moscow (1980); See also, S.L. Glashow, HUTP-79/A029(1979).
10. For a more detailed exposition of the results reported in this talk, see ref. 6,7,8 and R.E. Marshak, R.N. Mohapatra and Riazuddin, VPI-HEP-80.
11. M. Tretyakov et al., ITEP preprint, Moscow (1980).
12. F. Reines, H.W. Sobel and E. Pasierb, Univ. of Irvine Preprint (1980).
13. See E. Witten, HUTP-80/A031 for a review.
14. M. Gell Mann, P. Ramond and R. Slansky, (unpublished); E. Witten, Phys.Lett. 91B, 81 (1980).
15. For other models that lead to small m_ν , see for example, T. Yanagida, KEK lectures (1979); Y. Chikashige, R. Peccei, G. Gelmini and M. Roncadelli, Phys.Lett. B (to appear).
16. R.N. Mohapatra, Invited talk at the "First Workshop on Grand Unification", held in Durham, Newhampshire (to be published in the proceedings).
17. R. Wilson, Harvard Preprint (1980).
18. M. Baldoceolin et al., Padova Preprint (1980).
19. T. Pinelli, P. Trower and P. Ratti, Private communication.
20. W.C. Sauder, Private communication.
21. L.N. Chang and N.P. Chang, Phys.Lett. 93B, (1980) argue that a complicated SU(5) model with a {45} -dim Higgs could also lead to similar values for $\tau_{N-\bar{N}}$.
22. V. Elias, J.C. Pati and A. Salam, Phys.Rev.Lett. 40, 920 (1978).
23. There of course exist interesting models where δm is unobservably small: see T.K. Kuo and S. Love, Phys.Rev.Lett. 45, 93 (1980).

NEUTRINO OSCILLATIONS OF THE SECOND CLASS

V. Barger, J. P. Leveille
Physics Department, University of Wisconsin, Madison, WI 53706
P. Langacker*
Stanford Linear Accelerator, Stanford, CA 94305
S. Pakvasa
Physics Department, University of Hawaii, Honolulu, HI 96822

ABSTRACT

We consider a second class of neutrino oscillations which can arise when both Majorana and Dirac neutrino mass terms are present in the Lagrangian. These oscillations mix neutrino members of weak current doublets with singlets of the same chirality. A depletion of a neutrino beam would result, with apparent non-conservation of probability. Possible relevance to current oscillation experiments is discussed.

The possibility that neutrinos have mass has received experimental and theoretical support in recent months. The experimental evidence[1,2] is not yet conclusive, but is tentalizing nevertheless. On the theoretical side many grand unified theories3 require non-vanishing neutrino masses. In the following we shall assume that neutrinos are massive. For convenience, we shall work in an SU(2)×U(1) framework with the usual lepton assignment:

$$\begin{pmatrix}\nu_e\\ e^-\end{pmatrix}_L \quad \begin{matrix}\eta_{eL}\\ e^+_{\ L}\end{matrix} \quad ; \quad \begin{pmatrix}\nu_\mu\\ \mu^-\end{pmatrix}_L \quad \begin{matrix}\eta_{\mu L}\\ \mu^+_{\ L}\end{matrix} \quad ; \quad \begin{pmatrix}\nu_\tau\\ \tau^-\end{pmatrix}_L \quad \begin{matrix}\eta_{\tau L}\\ \tau^+_{\ L}\end{matrix} \quad \ldots \tag{1}$$

The number of families is arbitrary. We assume that the mass matrix is already diagonal in the sector with electric charge $Q = \pm 1$. For convenience we have chosen to use left-handed singlet fields which are conjugate fields of the more familiar right-handed singlets, e.g., $e^-_R = C\gamma^0(e^+_L)^\dagger$ and $\eta^c_{eR} = C\gamma^0(\eta_{eL})^\dagger$ with $C = i\gamma^2\gamma^0$. In Eq. (1) we have also assumed only one neutral singlet per family, although more could be added. As defined above, ν_{eL} is the field associated with the electron neutrino produced in the inverse β-decay reaction $e^-p \rightarrow \nu_{eL}n$, whereas $\nu^c_{eR} \equiv C\gamma^0(\nu_{eL})^\dagger$ is the field of the electron anti-neutrino created in β-decay $n \rightarrow pe^-\nu^c_{eR}$. Note that the singlet fields η_{iL}, $i = e,\mu,\tau$ ... are not coupled to the $W^\pm$ and Z bosons since they are electrically neutral isospin singlets. Because the singlets interact with other fermions via Higgs bosons only, they are effectively decoupled from light fermions.

Two choices are possible for the mass terms in the Lagrangian.[3,4] With the usual Higgs doublet only, a Dirac-type mass term

*On leave from the Univ. of Pennsylvania, Philadelphia, PA 19174.

0094-243X/81/680483-05$1.50

can be constructed

$$L_D = -\sum_{i,j=e,\mu,\tau} (d_{ij} \overline{\nu_{iL}} \eta^c_{jR} + h.\ c.)\ . \qquad (2)$$

In Eq. (2), we have omitted by fiat possible bare mass terms of the form $\overline{\eta_{iL}} \eta^c_{jR}$. Such terms can be dismissed by imposing a discrete symmetry on the Lagrangian:

$$\left[\begin{pmatrix}\nu_e\\ e^-\end{pmatrix}_L, \begin{pmatrix}\nu_\mu\\ \mu^-\end{pmatrix}_L, \begin{pmatrix}\nu_\tau\\ \tau^-\end{pmatrix}_L\right] \to e^{+i\alpha}\left[\begin{pmatrix}\nu_e\\ e^-\end{pmatrix}_L, \begin{pmatrix}\nu_\mu\\ \mu^-\end{pmatrix}_L, \begin{pmatrix}\nu_\tau\\ \tau^-\end{pmatrix}_L\right].$$
$$[e^+{}_L, \eta_{eL}, \mu^+{}_L, \eta_{\mu L}, \tau^+{}_L, \eta_{\tau L}] \to e^{-i\alpha}[e^+{}_L, \eta_{eL}, \mu^+{}_L, \eta_{\mu L}, \tau^+{}_L, \eta_{\tau L}] \qquad (3)$$

Equation (2) is invariant under the transformation Eq. (3), while Majorana mass terms are not: $\overline{\eta_{iL}} \eta^c_{jR} \to e^{2i\alpha} \overline{\eta_{iL}} \eta^c_{jR}$. The invariance of the Lagrangian Eq. (2) under Eq. (3) leads to lepton number conservation, with all doublet fields chosen to carry lepton number $\ell = +1$, all left-handed singlets $\ell = -1$. Majorana mass terms change lepton number by two units. With the discrete symmetry of Eq. (3), electron, muon and τ numbers are not separately conserved.

In the case where lepton number is conserved, Eq. (2) is the most general mass Lagrangian, and the mass eigenstates are three (Dirac) four-component neutrinos, in the case of three families. The weak interaction eigenstates ν_{eL}, $\nu_{\mu L}$, $\nu_{\tau L}$ are linear superpositions of these Dirac mass eigenstates. Through the usual formalism, one is led to oscillation among the three flavors, $\nu_{eL} \leftrightarrow \nu_{\mu L} \leftrightarrow \nu_{\tau L}$. Similarly, one has oscillations $\eta_{eL} \leftrightarrow \eta_{\mu L} \leftrightarrow \eta_{\tau L}$ which are undetectable, however, since the singlets are not coupled to the gauge bosons. We call these oscillations among flavors (without chirality change) first class oscillations.

If we do not impose lepton number conservation, then Eq. (2) is not the most general Lagrangian mass term. Including all allowed Majorana couplings, we obtain

$$L = -\tfrac{1}{2} \sum_{i,j=e,\mu,\tau} a_{ij} \overline{\nu_{iL}} \nu^c_{jR} + d_{ij} (\overline{\nu_{iL}} \eta^c_{jR} + \overline{\eta_{jL}} \nu^c_{iL}) + s_{ij} \overline{\eta_{iL}} \eta^c_{jR} + h.c. \qquad (4)$$

Here we have used the identity $\overline{\nu_L} \eta^c{}_R = \overline{\eta_L} \nu^c{}_R$ to reduce the numbers of independent constants. Within an SU(2)×U(1) context additional Higgs fields would be needed to generate the extra terms: $a_{ij} \neq 0$ requires a Higgs triplet whereas a singlet Higgs or simply bare mass terms will allow non-vanishing s_{ij}'s.

Diagonalization[3,4] of Eq. (3) reveals that the mass eigenstates are 6 Majorana (i.e., self-conjugate) neutrinos in the case of 3 families. The weak eigenstates ν_{iL} and η_{iL}, $i = e,\mu,\tau$, are linear superpositions of these six states. Besides the first class oscillations $\nu_{iL} \leftrightarrow \nu_{jL}$, $\eta_{iL} \leftrightarrow \eta_{jL}$, one can now have lepton-number changing oscillations involving singlet-to-doublet transitions, $\nu_{iL} \leftrightarrow \eta_{jL}$. We call these oscillations of the second class. We note that these oscillations do not flip chirality. Chirality flip oscillations are suppressed by powers of m_ν/E_ν and are negligible.

We consider in detail the consequences of having both Majorana and Dirac neutrino mass terms in a single family. Defining the doublets $\omega^\alpha{}_L \equiv (\nu_L, \eta_L)$, $\omega^{\alpha c}{}_R \equiv (\nu^c{}_R, \eta^c{}_R)$, the Lagrangian mass term from Eq. (3) can be cast in the form

$$L_{mass} = -\tfrac{1}{2}\overline{\omega^\alpha{}_L}\, M^{\alpha\beta}\, \omega^{\beta c}{}_R + \text{h.c.} \tag{5}$$

with mass matrix

$$M = \begin{pmatrix} a & d \\ d & s \end{pmatrix} . \tag{6}$$

The diagonalized mass matrix is $M_D = U_L^\dagger M U_R$ where U_L and U_R are unitary transformations of the ω_L and $\omega^c{}_R$ fields. Since M is symmetric, $U_R = U_L^* K^\dagger$ with K a unitary matrix. The relation of mass eigenstates ν_{iL} to $\omega^\alpha{}_L$ is $\omega^\alpha{}_L = U_L^{\alpha i}\, \nu_{iL}$ (i = 1,2). The corresponding right-handed transformation is $\omega^{\alpha c}{}_R = C\overline{(\omega^\alpha{}_L)}^T = U_R^{\alpha i}\, K_{ij}\, \nu^c_{jR} \equiv U_R^{\alpha i}\, \nu^{\tilde c}_{iR}$ where $\nu^{\tilde c}_{iR} \equiv K_{ij}\nu^c_{jR} = K_{ij}\, C\overline{(\nu_{jL})}^T$. The free Lagrangian for the neutral leptons is diagonal in the basis $\nu_i = \nu_{iL} + \nu^{\tilde c}_{iR}$. We find $\nu^{\tilde c}{}_i = \nu_i$, where $\nu^c{}_i \equiv K_{ij}\, C(\overline{\nu}_j)^T$. Hence, the ν_i are Majorana neutrino fields since they are self-conjugate. The combined Dirac and Majorana mass terms in the Lagrangian produce two Majorana eigenstates which in general have different masses m_1 and m_2. When $m_1 \neq m_2$, there is no conserved lepton number. The weak eigenstates ν_L and η_L are linear superpositions of the two Majorana mass eigenstates

$$\nu_L = \cos\alpha\ \nu_{1L} + \sin\alpha\ \nu_{2L}\ ,\quad \eta_L = -\sin\alpha\ \nu_{1L} + \cos\alpha\ \nu_{2L} \tag{7}$$

where $\cos\alpha = (U_L)^{11}$, $\sin\alpha = (U_L)^{12}$. The doublet member ν_L has the usual charged and neutral current couplings to gauge bosons. In the mass eigenstate basis the neutral current is non-diagonal, leading to the possibility of decays of the mass eigenstates by neutral currents, but with lifetimes which are much longer than the age of the universe.

Our primary considerations are for the logical possibility in which both m_1 and m_2 are small compared to the electron mass. This possibility has interesting implications for neutrino oscillations. Since the mass eigenstates propagate differently in time, second class oscillations $\nu_{eL} \leftrightarrow \eta_{eL}$ which conserve chirality can occur. At a distance L from a source of ν_{eL}, the probability (for energy $E \gg m_1, m_2$) of finding ν_{eL} is

$$P(\nu_{eL} \to \nu_{eL}) = 1 - \sin^2 2\alpha\ \sin^2(\tfrac{1}{2}\Delta) \tag{8}$$

where the oscillation argument is $\tfrac{1}{2}\Delta = 1.27\ \delta m^2\ L/E$, with $\delta m^2 = m_1^2 - m_2^2$ in eV^2 units and L/E in m/MeV units. The oscillations result in a depletion of an electron neutrino beam, or equivalently a deviation from a $1/r^2$ law for a point ν_{eL} source. Moreover, since η_{eL} is effectively non-interacting, probability conservation would appear to be violated by an amount $P(\nu_{eL} \to \eta_{eL}) = 1 - P(\nu_{eL} \to \nu_{eL})$, in contrast to first class oscillations where a depletion in $\nu_{eL} \to \nu_{eL}$ coincides with $\nu_{eL} \to \nu_{\mu L}, \nu_{\tau L}, \ldots$ transitions which are in principle observable.

In second class oscillations, both the charged current (CC) $\nu_{eL}p \to e^- X$ and neutral current (NC) $\nu_{eL}p \to \nu_{eL}X$ cross sections oscillate, $\sigma(L)/\sigma(L=0) = P(\nu_{eL} \to \nu_{eL};\ L/E)$, and the ratio σ_{NC}/σ_{CC} is unaffected in the one-family case. This should be contrasted with first class oscillations where σ_{CC} and σ_{NC}/σ_{CC} oscillate, but σ_{NC} does not. Corresponding statements apply to ν^c_{eR} cross sections.

We now turn to possible phenomenological implications of second class oscillations for current experiments.

Solar: Lepton number violating oscillations have the capability of explaining the deficiency in the ratio of observed to expected solar neutrinos.[5] With first and second class oscillations among three families, the minimum probability for $\nu_e \to \nu_e$ transitions is 1/6.

Reactor: The cross sections for an initial ν^c_{eR} beam scattering on proton and deuteron targets indicate depletions[1] in $\sigma_{CC}(p)$, $\sigma_{CC}(d)$ and $\sigma_{CC}(d)/\sigma_{NC}(d)$ but not (at the $\simeq$ 20% uncertainty level) in $\sigma_{NC}(d)$. To explain both the σ_{CC} and σ_{CC}/σ_{NC} results, first class oscillations are required with $\delta m^2 \simeq 1\ eV^2$.

Beam dump: Charged and neutral current events are produced by prompt neutrinos created in the dump. Since the prompt neutrinos originate from decays of charmed particles, identical ν_e and ν_μ spectra and numbers are generated. The charged and neutral current interactions of the prompt neutrinos are measured in bubble chamber and counter experiments at CERN at a distance L $\simeq$ 800-900 m downstream.

In the bubble chamber experiment, the measured e/μ ratio[6] is $R(e/\mu) = 0.59^{+0.35}_{-0.21}$. Such deviations of the e/μ ratio from unity may indicate a $P(\nu_e \to \nu_e)$ depletion arising from oscillations.[2,7] For the CERN beam dump $L/E \simeq 0.01$ m/MeV, so the mass scale of the oscillations would be $\delta m^2 \simeq 100\ eV^2$. To discuss such oscillations we assume a prompt neutrino beam with equal parts of ν_{eL} and $\nu_{\mu L}$, neglecting any ν^c_{eR} and $\nu^c_{\mu R}$ contributions for simplicity.

For second class oscillations of the ν_e family alone, the e/μ ratio is given by

$$R(e/\mu) = [\langle P(\nu_e \to \nu_e)\sigma_{CC}\rangle]/\langle\sigma_{CC}\rangle \qquad (9)$$

where σ_{CC} is the inclusive production cross section for e or μ and < > denotes a spectrum average. For first class oscillations $\nu_e \to \nu_e$, $\nu_e \to \nu_\tau$ (stringent experimental limits exist on $\nu_\mu \to \nu_e$ and $\nu_\mu \to \nu_\tau$ oscillations in this L/E range), the corresponding prediction is

$$R(e/\mu) = \frac{\langle P(\nu_e \to \nu_e)\sigma_{CC}\rangle + 0.17\langle P(\nu_e \to \nu_\tau)\sigma^\tau_{CC}\rangle}{\langle\sigma_{CC}\rangle \qquad + 0.17\langle P(\nu_e \to \nu_\tau)\sigma^\tau_{CC}\rangle} \qquad (10)$$

where $_{CC}$ is the inclusive cross section. For comparable mixing in the two classes, the predictions in Eqs. (9) and (10) are similar. One can discriminate experimentally between the classes of oscillations by ascertaining whether is produced and whether $_{NC}/_{CC}$ changes.

The beam dump counter experiments measure the ratio N(0)/N(1) of muonless to single muon events. With second class oscillations of the $_e$ family the prediction is

$$N(0\mu)/N(1\mu) = [\langle(1+P(\nu_e\to\nu_e))\sigma_{NC}\rangle + \langle P(\nu_e\to\nu_e)\sigma_{CC}\rangle]/\langle\sigma_{CC}\rangle \qquad (11)$$

in the limit of perfect acceptance. The corresponding prediction for first class oscillations is

$$\frac{N(0\mu)}{N(1\mu)} = \frac{2\langle\sigma_{NC}\rangle + \langle P(\nu_e \to \nu_e)\sigma_{CC}\rangle + 0.83\ \langle P(\nu_e \to \nu_\tau)\sigma^\tau_{CC}\rangle}{\langle\sigma_{CC}\rangle + 0.17\ \langle P(\nu_e \to \nu_\tau)\sigma^\tau_{CC}\rangle}\ . \qquad (12)$$

Taking comparable mixing in the two classes (and hence similar R(e/) predictions), the value of N(0)/N(1) is significantly lower for second class oscillations. A detailed analysis with experimental cuts could thereby differentiate between first and second class oscillations in this L/E range on the basis of measured R(e/) and N(0)/N(1) values. Still other alternatives are simultaneous first and second class oscillations or first class oscillations involving additional families.

This work was supported in part by the Department of Energy under contract DE-AC02 76ER00881-167.

REFERENCES

1. See, e.g., reports by H. W. Sobel and V. Barger in these proceedings.
2. V. Barger, K. Whisnant and R.J.N. Phillips, UW-Madison report, DOE-ER/00881-151, Phys. Rev. (in press) and references therein.
3. V. Barger, P. Langacker, J. P. Leveille and S. Pakvasa, UW-Madison report COO-881-149, May 1980, Phys. Rev. Lett. (in press).
4. J. Schechter and J.W.F. Valle, Syracuse University preprint SU-4217-167, June 1980; T. P. Cheng and L. F. Li, Carnegie-Mellon preprint COO-3066-152, June 1980; Dan-di Wu, Harvard preprint HUTP-80/A032, June 1980; S. M. Bilenky, J. Hosek and S. T. Petcov, Dubna preprint (1980); T. Yanagida and M. Yoshimura, KEK preprint TH-14 (1980); M. Gell-Mann, R. Slansky and G. Stephenson, unpublished.
5. J. Bahcall et al., IAS preprint (1980).
6. P. O. Hulth, report in these proceedings.
7. A. de Rújula et al., Nucl. Phys. B168, 54 (1980).

DYNAMICAL SYMMETRY BREAKING, THEORY

M.A.B. BÉG, ROCKEFELLER, ORGANIZER

DYNAMICAL SYMMETRY BREAKING AND HYPERCOLOR*

M.A.B. Bég
The Rockefeller University, New York, N.Y. 10021

The use of hypercolor, for implementing the Higgs mechanism and establishing gauge hierarchies in a dynamical way, is described and discussed. Phenomenological implications of the scheme are examined and its experimental signatures are contrasted with those of the canonical theory with elementary spin-0 fields.

1. To establish the motivation for hypercolor, it is necessary to start with a declaration of an article of faith. We hold it to be self-evident that all physics stems from a hierarchy of nested gauge groups:

$$\begin{aligned} U(1)_Q \otimes SU(3)_C &\subset U(1) \otimes SU(2)_L \otimes SU(3)_C \\ &\subset \cdots\cdots\cdots\cdots \\ &\subset \cdots\cdots\cdots\cdots \\ &\subset \text{Grand Unification Group} \end{aligned} \qquad (1)$$

Here $U(1)_Q$ and $SU(3)_C$ are the unbroken gauge groups of QED and QCD respectively, $U(1) \otimes SU(2)_L$ is the gauge group of QFD at least in the low energy ($\lesssim 10^2$GeV) sector. The Grand Unification Group presumably describes physics at energies $\gtrsim 10^{14}$GeV; the groups denoted by dots bridge the gap between 10^2GeV and 10^{14}GeV and thereby make Glashow's desert bloom.

To establish the gauge hierarchy of Eq.(1) within the framework of the canonical methodology, one is obliged to introduce vast multitudes of Higgs fields with judiciously chosen couplings and carefully rig the Lagrangian so it would yield the right number of would-be Goldstone modes. There is a surfeit of parameters and the elegance of the gauge theoretic approach is irretrievably lost. The alternative path, which we discuss, is to opt for dynamical symmetry breaking and generate the Higgs fields as bound states of fermion -- anti-fermion pairs, $f_1\bar{f}_2$; this we attempt to implement by introducing particles with extra-strong interactions, interactions generated by gauging a new degree of freedom hereinafter called hypercolor.[1,2]

2. The strategy of hypercolor is based on the assumption that chiral symmetry in QCD-like theories with zero current quark-mass is realized in the Nambu-Goldstone way. A feature of this realization is the spontaneous generation of a mass scale which manifests itself via dynamical masses of quarks (~ 300 MeV for QCD) and non-vanishing values of the Goldberger-Treiman constants of Goldstone bosons such as the pion (f_π ~ 100 MeV). Quite obviously the mass-scale of QCD is not adequate for endowing the weak bosons, $W^\pm$ and Z, with masses ~ 100 GeV. One therefore postulates hyperfermions with the attribute of hypercolor; gauging of this hypercolor degree of freedom

leads to QC´D, a QCD-like theory with a natural mass scale such that for hyperpions $f_{\pi'} \lesssim 250$ GeV. Such hyperpions, if they are true Goldstone bosons, can indeed play the role of unphysical Higgs fields, and thereby get the physical Higgs mechanism off the ground.

3. In the first step of the hypercolor scenario one starts with a gauge group $G = U(1) \otimes SU(2)_L \otimes SU(3)_C \otimes G_{C'}$, $G_{C'}$ being the unbroken hypercolor group. The hyperquarks, presumed to come in $N_{C'}$ colors and an <u>even</u> number of flavors, $N_{F'}$, are assigned weak iso-spins so that there are $N_{F'}/2$ left-handed doublets and $N_{F'}$ right-handed singlets. For at least one linear combination of doublets it is assumed that there is no mechanism for current mass, so we start out with at least a massless π'-like triplet and an η'-like singlet. The η' acquires mass by virtue of the 't Hooft mechanism, whereas the π' is absorbed in the Higgs mechanism. [We use the generic symbol π' to indicate any particle, massless or otherwise, transforming as a hyperpion under the <u>electroweak</u> group.] Consideration of π' poles in the vacuum polarization functions of $W^{\pm}$ and Z fields yields the relationships

$$m_z \cos\xi = m_w = (e/2\sin\xi) f_{\pi'} (N_{F'}/2)^{1/2} \tag{2}$$

where e is the positron charge and ξ is the Glashow-Weinberg-Salam angle ($\sin^2\xi \cong 0.22$). Thus, from the known value of the Fermi constant in β-decay,

$$f_{\pi'} \cong 250 \, (N_{F'}/2)^{-1/2} \text{ GeV} \tag{3}$$

The mass scale of QC´D is rendered manifest by Eq.(3). Note the tacit assumption

$$f_{\pi'^+} = \sqrt{2}\, f_{\pi'^o} \tag{4}$$

The final step is to introduce effective local Higgs fields as "phenomenological props"; thus Eq.(2) would follow if we introduce the Higgs doublet

$$\Phi = \begin{pmatrix} i\,\pi'^+ \\ \dfrac{\sigma' - i\,\pi'^o}{\sqrt{2}} \end{pmatrix} \tag{5}$$

with the identification

$$f_{\pi'} = \langle \sigma' \rangle (N_{F'}/2)^{-1/2} \tag{6}$$

in the effective Lagrangian

$$L = (D_\mu \Phi)^+ (D^\mu \Phi) + \ldots . \tag{7}$$

where D_μ is a gauge-covariant derivative

It seems therefore that we have recovered the Weinberg-Salam model in the low-energy limit in a dynamical way, but read on.

4. We proceed to review the balance sheet on hypercolor.

On the credit side we have a "natural" explanation for the so-called $\Delta I_{weak} = 1/2$ rule as well as the aesthetic appeal of a notion which promises a dynamical resolution of the gauge-hierarchy problem.

At this time, however, the debit side is much longer; we are faced with problems that may well prove insurmountable. To review but a few: (a) The repetition of fermion representations implies unwanted global symmetries and Goldstone bosons. One may get rid of these bosons, by gauging these horizontal symmetries, and absorbing them in the Higgs mechanism. The relevant gauge fields must however become so massive that these horizontal interactions do not disturb the phenomenology of weak interactions -- reproduced so well by the Weinberg-Salam model. The problem of unwanted Goldstone bosons is therefore resolved by bracketing it under the heading of another unsolved problem, the gauge hierarchy problem. Another approach is to envisage only one electroweak uhr doublet with other doublets appearing as composite fermions; this approach, however, appears to create more problems than it solves. (b) The second serious problem facing the scheme stems from the failure of the Φ's to couple directly to fermions without the attribute of hypercolor. There is thus no mechanism for generation of lepton mass or of current quark mass. The notion of "extended hypercolor", proposed to resolve the mass problem[1], is plagued by its own problems; if one tries to ensure that $m_c \neq m_s$ (say), one can in fact jeopardize Eq.(4) and the only significant triumph of the scheme, to wit the $\Delta I_{weak} = 1/2$ rule. This point will be discussed in more detail by Dr. Sikivie. Finally, (c) there is the problem of calculational intractability, common to all theories in which symmetry-breaking is dynamical; the use of phenomenological Higgs fields Φ is a start towards a resolution of this problem. To summarize: the subject of hypercolor is afflicted with many problems which make it an arena of opportunity, with room for many gladiators!

5. The phenomenological implications[1,2,3] and the experimental signatures of hypercolor-based schemes are very different from those of schemes which utilize the canonical methodology. To bring out the distinction we try to be as model-independent as possible, using only the following input in the hyperfermion sector: (a) The possibility of Grand Unification which implies $11N_{C'} - 2N_{F'} > 21$. (b) N_C-scaling, which quantifies the notion that QC$'$D is a scaled-up version of QCD and implies $(f_{\pi'} / \Lambda_{C'}\sqrt{N_{C'}}) = (f_\pi / \Lambda_C \sqrt{N_C})$. (c) Current algebra and PCAC. With $\Lambda_C \approx 300$ MeV, it is evident from (b) that $\Lambda_{C'} < 750$ GeV. In the following, we take $\Lambda_{C'} \sim 0.5$ TeV.

In the trans-TeV region, we expect hyperbaryons (defined as objects transforming according to the completely anti-symmetric representation of the hypercolor group) at an average mass[1] $m_B \approx 1.8 (N_{C'}/N_{F'})^{1/2}$TeV; some of these may be stable and lead to new islands of stability in the TeV region (Puzzle for cosmologists?). There are, in the meson sector, the ρ', ω' and the σ' at masses ~1TeV; the last is interesting in that it is the precise analogue of the left-over Higgs of the Weinberg-Salam model. Unlike the Higgs, the σ' would decay rapidly into $2\pi'$ and would be too broad to be recognizable as a particle.

More interesting, at this time, is hyperhadron spectroscopy in the 10-100 GeV region, the pseudo-Goldstone sector. These pseudo-Goldstones acquire mass from sub-electroweak (horizontal?) interactions. For comparable current masses for quarks and hyperquarks we expect[2] $m_{\pi'} \lesssim$ 7GeV, to be compared with $m_\phi >$ 7GeV for the Weinberg-Salam Higgs. The mass ranges of π' and ϕ could thus overlap and π'^o could mock the ϕ. An experimental handle is afforded, however, by parity[2]: whereas ϕ couples to $\bar{f}f$, π'^o couples to $\bar{f}\, i\gamma_5 f$. Thus $\phi \to D\bar{D}$ and $\pi'^o \to \pi\, D\bar{D}$ are allowed, whereas $\phi \to \pi\, D\bar{D}$ and $\pi'^o \to D\bar{D}$ are forbidden. In the canonical methodology, there will always be one or more ϕ-like scalar states, in the hypercolor scenario all low-lying states are pseudoscalar.

The first hyperpions accessible to observation, however, would very likely be charged; the distinction between these and canonical Higgses is regrettably somewhat blurred. It would not be uninteresting however to investigate the reaction[1] $e^+e^- \to \pi'^+ \pi'^- \to \tau^+ + \nu_\tau$ + Hadron Jet $\to \mu^+$ + Hadron Jet + ν's at energies at which the π''s are point-like ($\sqrt{s} << 1$ TeV) and contribute an amount $(1-4m^2_{\pi'}/s)^{3/2}/4$ to R; the hypercolor scheme dces make very specific predictions for this reaction which can emerge from the canonical methodology only by adding extra unmotivated constraints.

The principal moral to be drawn from the above discussion is that experiment could pinpoint the nature of the Higgs mechanism before theorists have resolved the underlying problems -- a conclusion with which Dr. Kane, the next speaker, happens to be in agreement.

* Work supported in part by the U.S. Department of Energy under Contract Grant Number DE-AC02-76ER02232B.

1. M.A.B. Bég, Proceedings of Orbis Scientiae, 1980, Coral Gables (Plenum, New York, 1980). This paper contains an extensive list of references to the literature up to January 1980; for a listing of more recent papers, see ref. 3 below.

2. M.A.B. Bég, D. Politzer and P. Ramond, Phys. Rev. Lett. 43, 1701 (1979).

3. S. Dimopoulos, S. Raby and G. L. Kane, University of Michigan preprint (1980).

TESTING TECHNICOLOR THEORIES

G. L. Kane
Randall Laboratory of Physics
University of Michigan, Ann Arbor, MI 48109

ABSTRACT

We provide improved estimates of the masses, decay modes and widths, and production cross sections of the physical particles expected in theories with dynamical symmetry breaking. The most important results are charged pseudo-Nambu-Goldstone-bosons $a_T^{\pm}$ with $m_{\pm} \approx 8$ GeV (and thus detectable at PETRA/PEP), two neutral pseudoscalars with $m_0 \lesssim 2.5$ GeV, and the colored technieta (m=240 GeV) with observable production cross sections at the Tevatron Collider and at Isabelle. The calculations were done with S. Dimopoulos and S. Raby.

INTRODUCTION

There is not yet a dynamical symmetry breaking model which could be fully realistic. As a result, any calculations must be done in models which might not be generally applicable. Nevertheless, many features are expected to hold in any reasonable Technicolor theory,[1,2] such as the techniquarks being colored, and the existence of interactions which couple quarks to leptons. In our calculations we use an SU(N) model which has such general features and we avoid particular assumptions which might have less generality. We give numerical results for N=4. Our results are given in detail in ref. 3.

There are many Goldstone bosons which arise in such a theory from breaking the original chiral symmetry. We will mention 12 of them here, the technieta color octet η_T^a, and the color-singlet light pseudoscalars ("pseudos") $a_T^{\pm}$, a_T^{0}, $\tilde{a}_T^{0}$. There are no light scalars in a Technicolor theory, an important prediction. The pseudo-Nambu-Goldstone bosons get mass from color, electroweak, and extended technicolor interactions[4,5].

Particles come in 3 mass scales. Resonances such as the techirho, ρ_T, will occur on the mass scale of the theory, about 1 TeV. Pseudos that get mass from color, etc., will have $m^2 \sim \alpha_C m_{TC}^2 \sim (300\ \text{GeV})^2$. Those which still get no mass from color interactions will have $m^2 \lesssim \alpha m_Z^2 \sim (\text{few GeV})^2$.

ρ_T We find the technirho will have mass of about 900 GeV. It will be produced in $\bar{p}p$ collisions, as shown,

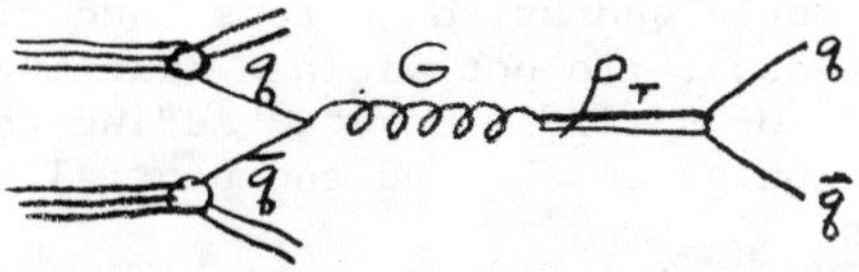

with $(d\sigma/dy)_{y=0} \simeq 6\text{x}10^{-36}$ cm^2 at $\sqrt{s}$ = 2000 GeV. This rate scales as $u(m_{\rho T}/\sqrt{s})/m^4_{\rho T}$ so it would increase considerably if $m_{\rho T}$ went down. The total width of ρ_T is $\Gamma \simeq 4$ GeV, with dominant decay modes being $\rho_T \rightarrow GG, q\bar{q}, W^+_L W^-_L$.

η^a_T A clear test of technicolor theories will come from production of η^a_T since a lower limit in its production cross section can be computed from the triangle contribution. Since chiral symmetry should be as good an approximation here as for a pion, this should be reliable. We find $M(\eta^a_T) \simeq 240$ GeV. It is produced via

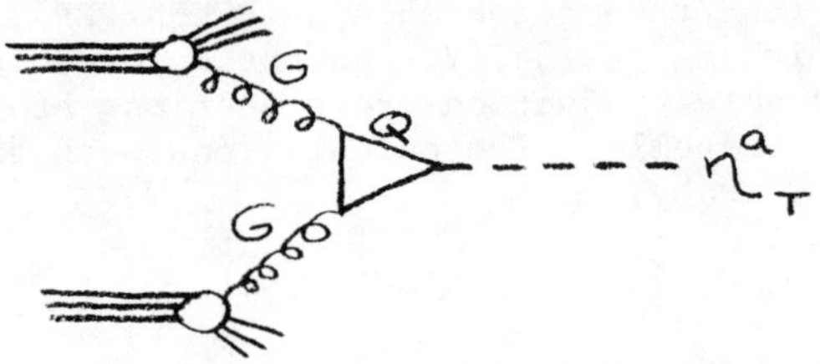

and gives a large cross section because the technifermions are assumed to carry color and couple to gluons. We find

$\sqrt{s}$(GeV)	$(d\sigma/dy)_{y=0}$(cm^2)
500	$2\text{x}10^{-37}$
800	$4.5\text{x}10^{-36}$
2000	$44\text{x}10^{-36}$

This cross section has been computed in ref. 6 also and they agree with us. Note that this result is about $8\text{x}N^2 = 128$ times larger than the cross section for producing a fundamental Higgs of similar mass, since η^a_T is a color octet and there are N=4 technifermions in the loop to sum over.

The important η^a_T decays are $\eta^a_T \rightarrow GG$ (opposite to the production) with $\Gamma(GG) \simeq 60$ MeV, and $\eta^a_T \rightarrow f\bar{f}'$ with f,f' heavy fermions. The effective coupling to the fermions is $(m_f+m_{f'})/F_T$, from models or from a Goldberger-Treiman type argument. For m_f=25 GeV, $\Gamma(f\bar{f}') \simeq 1$ GeV and will dominate. Various modes are $\eta^a_T \rightarrow t\bar{t}$, $b\bar{b}$, GG, Gγ, GZ°, GGG,---. Other pseudos will have modes such as GW^+, $t\bar{b}$.

a_T The charged light pseudos get about 7.7 GeV of mass from electroweak interactions, while the neutrals remain massless. All get some mass from the extended technicolor interaction[4,5]. While this contribution cannot be calculated reliably, we can put a limit on it. We make the important assumption that the same leptoquark bosons that couple technifermions to technileptons (and give mass contributions) also couple quarks to leptons (and give flavor charging neutral currents). To not violate existing limits on $K_L \rightarrow \mu e$, this implies $g^2/M^2 < (1/310 \text{ TeV})^2$. Adding this to the electroweak contribution gives charged and neutral Higgs-like particles with masses

$$m(a_T^{\pm}) \simeq 8 \text{ GeV}$$

$$m(a_T^{\circ}) \lesssim 2.1 \text{ GeV}$$

$$m(\tilde{a}_T^{\circ}) \lesssim 2.5 \text{ GeV}$$

The $a_T^{\pm}$ are charged pseudoscalars and can be found in $e^+e^- \rightarrow a_T^+a_T^-$ at PETRA or PEP. They are produced with a β^3 threshold behavior, 1/4 unit of R, $\sin^2\theta$ production distribution. They decay dominantly into $\tau\nu_\tau$ (about 40%), $c\bar{s}$ (like a heavy $F^{\pm}$; about 40%), and $c\bar{b}$ (about 20%), and $\mu\nu_\mu$ (about 0.1%). About 60% of the events have 4 strange quarks.

Interestingly, if $a_T^{\pm}$ exists, the mode $t \rightarrow ba_T^{+}$ dominates t decay since it is semiweak, and t decays are not as in the standard model.

The neutrals $a_T^{\circ}, \tilde{a}_T^{\circ}$ can be produced in Drell-Yan reactions (K beams are best for good signal/noise), or in decay of heavier states such as ψ, Υ. Their main modes are a_T°, $\tilde{a}_T^{\circ} \rightarrow \mu^+\mu^-$ (about 1/3), $K^*\bar{K}$, $\phi\phi$, $\Lambda\bar{\Lambda}$, $K\bar{K}\pi$, and perhaps the parity violating mode $K\bar{K}$. They are pseudoscalars that may have parity violating couplings to fermions but not to $\gamma\gamma$, GG.

The above predictions can be tested, and guarantee that soon (finally) there will be experimental input into understanding the origin of spontaneous symmetry breaking.

REFERENCES

1. See M. A. B. Bég, Proceedings of Orbis Scientiae, 1980, Coral Gables, (see ref. 2, for a recent review of the theory).

2. L. Susskind, these proceedings.

3. S. Dimopoulos, S. Raby, and G. L. Kane, Michigan preprint UM HE 80-22.

4. S. Dimopoulos and L. Susskind, Nuc. Phys. B1555 237 (1979).

5. E. Eichlen and K. Lane, Phys. Lett. 90B 125 (1980).

6. F. Hayot and O. Napoly, Saclay preprint.

FERMION MASSES AND WEAK ISOSPIN IN TECHNICOLOUR MODELS

P. Sikivie
CERN, CH-1211 Geneva 23, Switzerland

ABSTRACT

Isospin breaking is needed in models of dynamical symmetry breaking to generate up-down splittings in the quark and lepton mass matrices. It is shown how this isospin breaking can be introduced without upsetting the relation $M_W = M_Z \cos\theta_W$.

In their seminal papers [1] on dynamical symmetry breaking, Susskind and Weinberg pointed out that the $M_W = M_Z \cos\theta_W$ relation (also called the "weak $\Delta I = \frac{1}{2}$ rule") would follow from isospin conservation by the "technicolour" (or "hypercolour") forces that break the electroweak gauge group. On the other hand, it is clear that isospin breaking must be present in the mechanism [2] through which quark and lepton masses are generated. The question thus arose whether these two requirements can be made compatible. The answer is simply yes. The general conditions under which one can have both $M_W = M_Z \cos\theta_W$ and $m_u \neq m_d$ will be found below. I follow here the discussion given in Ref. 3 although the question has also been discussed by several other authors [4].

The general requirement for an interaction to respect the weak $\Delta I = \frac{1}{2}$ rule is the following:

1. that it conserves electric charge
2. that it conserves a "custodial" SU(2), which is defined as a symmetry of both the Lagrangian and the vacuum, under which the generators (or gauge fields) of $SU_L(2)$ transform as a triplet.

It is easy to show [5] that the contributions to the $SU_L(2) \times U_Y(1)$ vector boson $(\text{mass})^2$-matrix from any interaction that obeys these two requirements are such that: $\delta(M_Z)^2 = \delta(M_W)^2 \cos^2\theta_W$. The custodial symmetry must be SU(2) because SU(2) is the only group with a real representation of dimension three. The custodial SU(2) can but need not be isospin.

In the standard $SU_L(2) \times U_Y(1) \times SU^c(3)$ model with one elementary Higgs doublet, the strong colour interactions and the Higgs self-interactions do conserve "custodial" SU(2)'s, whereas the electroweak gauge interactions and the Yukawa interactions do not. Thus:

$$M_W = M_Z \cos\theta_W \left(1 + O(\alpha) + O\,\alpha\left(\frac{m_u^2 - m_d^2}{M_W^2}\right)\right) \qquad (1)$$

These corrections to the weak $\Delta I = \frac{1}{2}$ rule have been calculated and should soon be tested experimentally. The custodial SU(2) for the strong interactions is of course isospin. The custodial SU(2) for the Higgs self-interactions is defined in Ref. 3. Note that if there are several Higgs doublets, the conservation of a custodial SU(2) is not automatic, and unless some special care is taken to impose one, Eq. (1) will be violated in higher orders of the Higgs self-interactions.

To assure the $\Delta I = \frac{1}{2}$ rule in technicolour models, we must require that the technicolour (TC) interactions conserve a custodial SU(2). We assume that the quarks and leptons acquire their masses from the TC condensates through broken "Extended Technicolour" (ETC) gauge interactions as described in Ref. 2. Clearly, if the custodial SU(2) is TC isospin, then ETC must violate isospin in order that we may have $m_u \neq m_d$, $m_c \neq m_s$, ... This strategy can in general be implemented since the ETC flavour symmetry is always a subgroup of (and therefore in general smaller than) the TC flavour symmetry (see Fig. 1a). As an example, consider the gauge group $SU_L(2) \times U_Y(1) \times SU^c(3) \times SU^{ETC}(3) \times Sp^{T'C}(6)$ and the (left-handed) fermion representation content:

$$\begin{aligned} \begin{pmatrix} U \\ D \end{pmatrix} &= (2,\ \tfrac{1}{6},\ 3,\ 3,\ 1) \\ \bar{U} &= (1,\ +\tfrac{2}{3},\ \bar{3},\ \bar{3},\ 1) \\ \bar{D} &= (1,\ +\tfrac{1}{3}.\ \bar{3},\ 3,\ 1) \\ \mathcal{F} &= (1,\ 0,\ 1,\ \bar{3},\ 6) \end{aligned} \tag{2}$$

T'C and the fermions $\mathcal{F}$ have been introduced to break $SU^{ETC}(3) \to SU^{TC}(2)$ at a scale $\Lambda^{T'C} \simeq 30$ TeV through the condensate:

$$<\mathcal{F}^a_i(-i\sigma_2)\ \mathcal{F}^b_j\ \eta^{ij}\ \varepsilon_{abc}>_0\ \sim \delta_{c3} \tag{3}$$

which is a Lorentz scalar, T'C singlet and ETC triplet. The quark multiplets break up into techniquarks (TC doublets) and ordinary quarks (TC singlets):

$$\begin{aligned} \begin{pmatrix} U \\ D \end{pmatrix} &= \left(\begin{array}{cc|c} U_1 & U_2 & u \\ D_2 & D_2 & d \end{array}\right) = (2,\ \tfrac{1}{6},\ 3,\ 2{+}1,\ 1) \\ \bar{U} &= (\bar{U}_1\ \ \bar{U}_2\ |\ \bar{u}) = (1,\ -\tfrac{2}{3},\ \bar{3},\ 2{+}1,\ 1) \\ \bar{D} &= (\bar{D}_1\ \ \bar{D}_2\ |\ \bar{d}) = (1,\ +\tfrac{1}{3},\ \bar{3},\ 2{+}1,\ 1) \end{aligned} \tag{4}$$

of $SU_L(2) \times U_Y(1) \times SU^c(3) \times SU^{TC}(2) \times SU^{T'C}(6)$. The TC flavour symmetry includes the chiral isospin group $SU_L(2) \times SU_R(2)$ which gets broken spontaneously down to TC isospin $SU_{L+R}(2)$ when the techniquarks condense:

$$<\bar{U}U + \bar{D}D>_0\ \simeq (\Lambda^{TC})^3 \simeq (\tfrac{1}{2}\ \text{TeV})^3 \tag{5}$$

TC isospin is a custodial symmetry and therefore the weak $\Delta I = \frac{1}{2}$ rule will be assured to all orders of the strong TC interactions. On the other hand, we will have $m_u \neq m_d$ since ETC violates isospin ($U_R \sim 3^{ETC}$ whereas $D_R \sim \bar{3}^{ETC}$). Indeed we find in lowest order of ETC/TC vector boson exchange:

$$m_u \simeq \frac{g^2_{ETC}}{m^2_{ETC}} \langle \bar{U}U \rangle_0 \ , \quad m_d = 0 \tag{6}$$

The corrections to the weak $\Delta I = \frac{1}{2}$ rule due to the isospin violation in the ETC interactions are of the same order of magnitude as those due to the isospin violation in the Yukawa couplings of an elementary Higgs doublet. They are acceptably small provided $m_u^2 - m_d^2 \lesssim M_W^2$ (see Eq. (1)).

In general a gauge interaction is isospin symmetric if its gauge group has real fermion representation content [3]. It is then interesting to note that if ETC "tumbles" [7] down to TC, ETC will automatically violate isospin whereas TC will conserve isospin. See Ref. 8 for an attempt to combine the ideas of "custodial SU(2)" and "tumbling" into a realistic model.

If ETC conserves isospin, TC will conserve isospin as well and the only way in which one can have $m_u \neq m_d$ is by having the TC condensates break isospin spontaneously. However, this will upset $M_W = M_Z \cos\theta_W$ unless we can impose a custodial SU(2) other than isospin. This can be done provided the flavour symmetry of the TC interactions is sufficiently large (see Fig. 1b). An example of such a model was constructed in Ref. 3.

Finally we note that $\nu_\ell - \ell$ splittings can be obtained in the lepton mass matrix by giving large Majorana masses to the right-handed neutrinos [9].

In summary, Fig. 1 represents the two different ways in which $M_W = M_Z \cos\theta_W$ can be made compatible with $m_u \neq m_d$. In the first mechanism (Fig. 1a), isospin is conserved by the TC interactions and the TC condensates but is not a symmetry of ETC. In the second mechanism, isospin is a symmetry of both the ETC and TC interactions but is violated spontaneously by the TC condensates. The custodial symmetry is a SU(2) other than isospin.

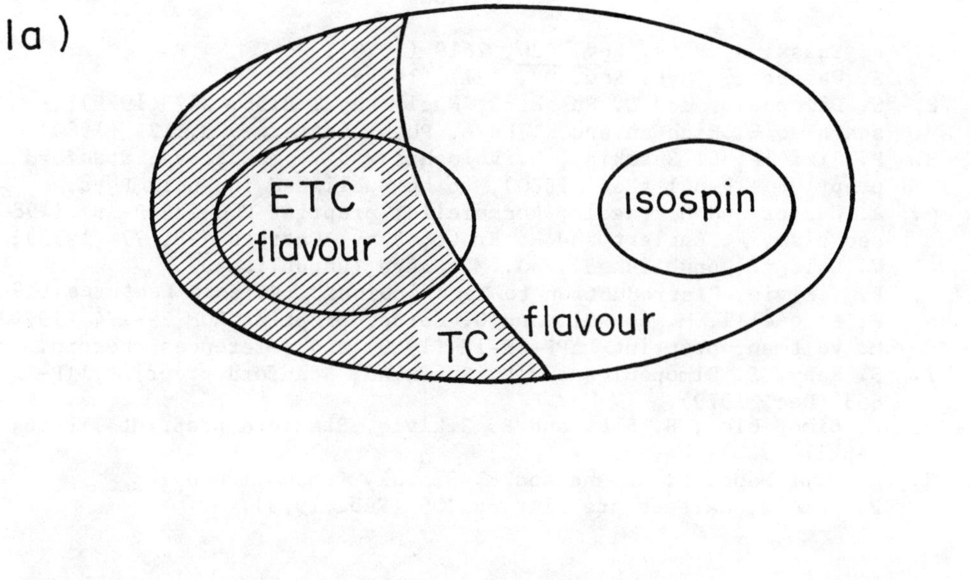

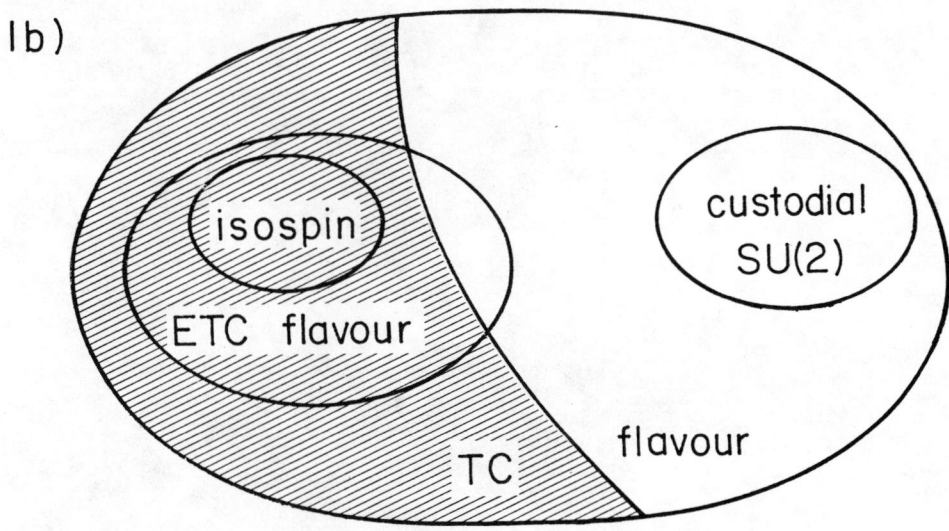

Fig. 1 : The two flavour symmetry geographics that assure simultaneously $M_W = M_Z \cos \theta_W$ and $m_u \neq m_d$. The shaded areas represent the flavour symmetries which are spontaneously broken by the TC condensates.

REFERENCES

1. L. Susskind, Phys. Rev. D20, 2619 (1979);
   S. Weinberg, Phys. Rev. D9, 1277 (1979).
2. S. Dimopoulos and L. Susskind, Nucl. Phys. B155, 237 (1979); see also E. Eichten and K. Lane, Phys. Letters 90B, 125 (1980).
3. P. Sikivie, L. Susskind, M. Voloshin and V. Zakharov , Stanford preprint ITP-661 (Feb. 1980), to be published in Nucl. Phys.
4. A. Carter and H. Pagels, Rockefeller preprint COO-2232B-187 (1980); see also, F. Englert and R. Brout, Phys. Letters 49B, 77 (1973); V. Baluni (unpublished); J.-M. Frère (unpublished).
5. P. Sikivie, "Introduction to Technicolour", Varenna Lectures (1980).
6. F. Antonelli, M. Consoli and G. Corbo, preprint ROM 79-174 (1979); M. Veltman, preprint LAPP-TH-12 (1980) and references therein.
7. S. Raby, S. Dimopoulos and L. Susskind, Stanford preprint ITP-653 (Dec. 1979).
8. S. Dimopoulos, S. Raby and P. Sikivie, Stanford preprint ITP-664 (April 1980).
9. M. Gell-Mann, P. Ramond and R. Slansky (unpublished);
   P. Ramond, Caltech preprint 680709 (Feb. 1979).

DYNAMICALLY BROKEN GAUGE THEORIES

Heinz R. Pagels
The Rockefeller University, New York, NY 10021

ABSTRACT

We describe several of the attractive features of the dynamically broken gauge theories (DBGT) and calculate the pion decay constant, pion electromagnetic form factor and the light current quark masses in QCD. A dynamically broken $SU_L(2) \times U(1)$ model is examined and the mass ratios of the W and Z gauge bosons, leptons and bound state Higgs meson are calculated in terms of the weak angle θ.

FEATURES OF DYNAMICALLY BROKEN GAUGE THEORIES[1]

In the standard $SU_C(3) \times SU_L(2) \times U(1)$ model with three generations of quarks and leptons there are 16 free parameters. DBGT offer the hope that the number of free parameters can be reduced to zero. If a single simple group with just gauge fields and fermions describes nature then there is only one coupling constant and this can be eliminated, by dimensional transmutation, to set the mass scale. All other constants are determined — in particular the fine structure constant $\alpha = 1/137$ is in principle fixed.

If a DBGT has chiral invariance then there are no bare fermion mass terms. Since mass divergences are proportional to such terms there are no mass divergences — all masses are finite, calculable numbers.

This means that the bare current quark masses in QCD vanish — the current quark masses are weakly momentum dependent for momenta on the scale of hadron energies but vanish at high momenta. This observation implies the strong P and T violating term θF^*F can be removed by a chiral rotation[2,3]. However in a DBGT all P and T violation can be removed to every finite order in perturbation theory. This undesired feature may not actually be realized because non-perturbative effects can generate P and T violations[2].

Conceivably even the gauge fields and fermions like quarks and leptons are boundstates of yet more fundamental fields. This could solve the notorious gauge hierachy problem providing that the observed masses of fermions and gauge fields are small perturbations of some very large mass scale.

CHIRAL BREAKING IN QCD

The renormalized quark self energy $\Sigma(p)$ for a single flavor has a dynamically generated and explicit component $\Sigma(p) = \Sigma_D(p) + \Sigma_E(p)$ with the asymptotic behavior[4]

$$\Sigma_D(p) \to \frac{4M_D^3}{p^2} \ell n^{d-1}(p^2) \qquad \Sigma_E(p) \to m \ell n^{-d}(p^2) \tag{1}$$

0094-243X/81/680501-04$1.50

where $d = 12/(33-2n_f)$ and m_D = dynamical quark mass ≃ 270 MeV, m = current quark mass. In QCD an exact integral representation for the pion decay constant f_π can be obtained and which in an approximation becomes[5]

$$f_\pi^2 = \frac{3}{(2\pi)^2} \int_0^\infty \frac{dp^2 \; p^2[\Sigma_D^2(-p^2) - \frac{1}{2} p^2 \Sigma_D'(-p^2)\Sigma_D(-p^2)]}{[p^2 + \Sigma_D(-p^2)]^2} \tag{2}$$

Using this Pagels and Stokar[5] estimate the integral and find

$$f_\pi^2 = \frac{\sqrt[3]{2}}{2\pi\sqrt{3}} m_D^2 \qquad\qquad f_\pi \simeq 83 \text{ MeV} \tag{3}$$

in good agreement with the experimental value $f_\pi \simeq 93$ MeV.

The pion electromagnetic form factor may be estimated for large momentum transfers and it is given by an expansion $q^2 \to -\infty$.

$$-q^2 F_\pi(q^2) = c_0 + c_1\alpha_s(q^2) + c_2\alpha_s^2(q^2) + \ldots \tag{4}$$

where $\alpha_s(q^2) \to (\ln q^2)^{-1}$ is the strong coupling and the c_i are constants. The non perturbative term c_0 has been estimated[5] to be $c_0 \simeq (4 \ln 2/\sqrt{3}\sqrt[3]{2}\pi) m_D^4/f_\pi^2$ and c_1 has been computed exactly[6]. Together these leading terms give a good fit to the observed form factor; the dominant term is c_0.

MAGNITUDE OF THE LIGHT CURRENT QUARK MASSES

The light current quark mass ratios are renormalization group invariant and these have been estimated from current algebra phenomenology (including lending order symmetry breaking effects) to be[7]

$$\frac{m_u}{m_d} = 0.38 \pm 0.13 \qquad \frac{m_d}{m_s} = 0.045 \pm 0.011 \tag{5}$$

The absolute magnitude of a quark mass is not a renormalization group invariant and depends on its definition which can be made precise in terms of QCD Green's functions. If we use conventional definitions of Green's functions normalized by lending order perturbation theory at the renormalization point $\mu \simeq 1$ GeV where $\alpha_s \simeq 0.2$. Then current algebra implies

$$f_\pi^2 m_\pi^2 = \langle 0|m_u\bar{u}u + m_d\bar{d}d|0\rangle \simeq (m_u + m_d)\langle 0|\bar{u}u|0\rangle \tag{6}$$

while the operator product expansion implies for small α_s, $\langle\bar{u}u\rangle_0 = 12\, m_D^3(4\pi\alpha_s)^{-d}(2/b)^{d-1}$ where $b = (33-2n_f)/24\pi^2$. With $n_f = 3$, we obtain[8], using the above values of the quark mass ratios,

$$m_u = (1.8 \pm 0.7)\text{MeV} \quad m_d = (4.3 \pm 1.7)\text{MeV} \quad m_s = (112 \pm 66)\text{MeV} \tag{7}$$

which compare reasonably with other estimates[9].

DYNAMICALLY BROKEN $SU_L(2) \times U(1)$

The weak $\Delta I = 1/2$ rule, $m_W^2/m_Z^2 = \cos^2\theta + O(g^2)$, follows in the conventional Weinberg-Salam model because the elementary Higg's fields are in the (1/2,1) representation. If the Higg's fields are dynamically generated boundstates in the extended technicolor model with isospin breaking one loses the $\Delta I = 1/2$ rule because large isospin breaking in the fermion sector implies the same in the gauge boson sector. However using the irregular solution for the boundstate Higgs[10] (or a solution that behaves like the irregular solution for a large subasymptotic energy regime) one retains the $\Delta I = 1/2$ rule[11,12]. Assuming the Higg's field is a boundstate of mass m_H of a fermion with large mass m_E and a massless neutrino, the $SU_L(2) \times U(1)$ model has only two parameters g and θ. All mass ratios of the standard model can be calculated in the weak coupling limit with the result obtained by A. Carter and H. Pagels[12],

$$m_W^2:m_Z^2:m_E^2:m_H^2 = 1:\cos^{-2}\theta:6\tan^2\theta:12\tan^2\theta \tag{8}$$

suggestive of supersymmetry. This model illustrates an important feature of DBGT — the parameters of the theory are reduced in number. This model implies the existence of heavy quarks or leptons with mass $\sim$ 100 GeV.

REFERENCES

1. See M. A. B. Beg, Proceedings of this conference; S. Weinberg, Phys. Rev. D13, 974 (1976).
2. H. Pagels, Phys. Lett. 87B, 222 (1979).
3. S. Susskind and S. Dimopolis, Stanford University preprint (1979).
4. K. Lane, Phys. Rev. D10, 2605 (1974); H. D. Politzer, Nuc. Phys. B117, 397 (1976); H. Pagels, Phys. Rev. D19, 3080 (1979).
5. H. Pagels and S. Stokar, Phys. Rev. D20, 2947 (1979).
6. A. V. Efremov and A. V. Radyuskin, Dubna preprint E2-11983 (1978); D. R. Jackson, Thesis CalTech, Pasadena (1977); S. J. Brodsky and G. P. Lepage, SLAC-PUB 2294 (1979); A. Duncan and A. Muehler, Columbia University preprint (1979).
7. These estimates are those of P. Langacker and H. Pagels, Phys. Rev. D19, 2072 (1979). Other estimates are S. Weinberg, Festschrifft for I. I. Rabi, edited by L. Motz, Academy of Sciences, NY (1977); J. Gasser and H. Leutwyler, Nucl. Phys. B94, 269 (1975); J. Gunion, Phys. Rev. D8, 517 (1973); C. A. Dominguez and A. Zepeda, Phys. Rev. D18, 884 (1978).
8. H. Pagels and S. Stokar, Phys. Rev. D (to be published 1980).
9. H. Leutwyler, Nucl. Phys. B76, 413 (1974); Phys. Lett. 48B, 431 (1973); M. Testa, Phys. Lett. 56B, 53 (1975); E. Gava, F. Legovini, N. Paver, Nuovo Cim. Lett. 14, 41 (1975); G. Furlan, N. Paver, C. Verzegnassi, Nuovo Cim. 32A, 75 (1976); S. Weinberg, ref. 7.
10. A. A. Migdal and A. M. Polyakov, Sov. Phys. JETP 24, 91 (1967); R. Jackiw and K. Johnson, Phys. Rev. D8, 2386 (1973); J. M.

10. (cont.) Cornwall and R. E. Norton, Phys. Rev. D8, 3338 (1973); H. Pagels, Phys. Rev. D21, 2336 (1980).
11. S. Englert and R. Brout, Phys. Lett. 49B, 77 (1973).
12. A. Carter and H. Pagels, Phys. Rev. Lett. 43, 1845 (1979).

SUPPRESSION OF SUPERHEAVY MAGNETIC MONOPOLES IN GRAND UNIFIED THEORIES*

So-Young Pi
Stanford Linear Accelerator Center
Stanford University, Stanford, California 94305

ABSTRACT

The superheavy magnetic monopoles predicted by grand unified theories would not be produced in significant numbers if electromagnetic gauge invariance is spontaneously broken when the temperature T is greater than $T_c \gtrsim 1$ TeV.

Grand unified theories predict the existence of superheavy magnetic monopoles.[1-7] These monopoles are of the type discovered by 't Hooft and Polyakov.[6] They exist if a semi-simple group is broken down to a subgroup which contains U_1 factor. The monopole mass M_m is of order M_X/α, where $\alpha = g^2/4\pi$, g is a gauge coupling and M_X is a typical mass of a gauge boson associated with a broken generator. For example, in Georgi-Glashow model $M_X \simeq 10^{14}$ GeV and $M_m \simeq 10^{16}$ GeV.

The problem of monopole production and their subsequent annihilation, in the context of a second order or weakly first order phase transition, was analyzed by Zeldovich and Khlopov[1] and by Preskill.[2] In preskill's analysis, it was found that relic monopoles would exceed present bounds by roughly 14 orders of magnitude. Since it seems difficult to modify the estimated annihilation rate, one must find a way which suppresses the production of these monopoles.

One interesting solution to this problem, suggested by Preskill,[2] Einhorn et al.,[3] and Guth and Tye[4] is that the phase transition at which the U_1 factor occurs is strongly first order. The problem of monopole production in a strongly first order phase transition was treated in detail by Guth and Tye.

In this talk I will describe an alternative scenario for the suppression of monopoles, developed in collaboration with P. Langacker,[7] in which the universe undergoes two or more phase transitions (which can be second order)

$$G \xrightarrow[T_1]{} H_1 \xrightarrow[T_2]{} H_2 \;\cdots\; \xrightarrow[T_n]{} H_n \xrightarrow[T_c]{} SU_3^c \times U_1^{EM} \quad , \tag{1}$$

where U_1^{EM} is not a subgroup of H_n. The critical temperature at which U_1^{EM} appears is $T_c \gtrsim 1$ TeV. For example, in SU_5 model,

$$SU_5 \xrightarrow[T_1 \lesssim M_X]{} SU_3^c \xrightarrow[T_c \gtrsim 1 \text{ TeV}]{} SU_3^c \times U_1^{EM} \quad . \tag{2}$$

Since $T_c \lll M_m \simeq 10^{16}$ GeV no monopoles will be produced.

We consider a model which at T = 0 is the standard SU_5 model with symmetry breaking

$$SU_5 \rightarrow SU_3^c \times SU_2 \times U_1 \rightarrow SU_3^c \times U_1^{EM} \tag{3}$$

---

* Work supported by the Department of Energy, contract DE-AC03-76SF00515.

by an adjoint Higgs representation and three five Higgs representations. (It turns out that this is the minimum number of five Higgs representations required for our purpose.) The Higgs potential at $T=0$ is

$$V = V_\Phi + V_{\Phi\phi} + V_\phi \tag{4a}$$

$$V_\Phi = -\frac{1}{2}m^2\,\mathrm{Tr}\Phi^2 + \frac{1}{4}a(\mathrm{Tr}\,\Phi^2)^2 + \frac{1}{2}b\,\mathrm{Tr}\,\Phi^4 \tag{4b}$$

$$V_{\Phi\phi} = \sum_{i=1}^{3}\left[\alpha_i\,\phi_i^\dagger\phi_i\,\mathrm{Tr}\Phi^2 + \beta_i\,\phi_{ia}^\dagger\,\Phi_{ab}^2\,\phi_{ib}\right] \tag{4c}$$

$$V_\phi = \sum_{i=1}^{3}\left[-\mu_i^2\,\phi_i^\dagger\,\phi_i + \lambda_i(\phi_i^\dagger\phi_i)^2\right] + \sum_{i<j}\left[\sigma_{ij}(\phi_i^\dagger\phi_i)(\phi_j^\dagger\phi_j)\right.$$
$$\left. + \rho_{ij}(\phi_i^\dagger\phi_j)(\phi_j^\dagger\phi_i) + \eta_{ij}(\phi_i^\dagger\phi_j)^2 + \eta_{ij}^*(\phi_j^\dagger\phi_i)^2\right] \tag{4d}$$

where Φ is an adjoint Higgs representation and ϕ_i are five Higgs representations. We have imposed discrete symmetries $\Phi\to-\Phi$ and $\phi_i\to-\phi_i$ for simplicity.

For $0\le T \ll M_X$ we need only to consider the $SU_2\times U_1$ part of the model. (We assume that SU_3^c is never broken.) Therefore let us first consider $SU_2\times U_1$ part of the model with Higgs potential V_ϕ, Eq. (4d), in which ϕ_i are SU_2 doublets.

At $T=0$ we choose the parameters in the potential such that the vacuum expectation values (VEV) of the Higgs fields are $\langle\phi_1(0)\rangle = (0\ v_1)^T/\sqrt{2}$ and $\langle\phi_2(0)\rangle = \langle\phi_3(0)\rangle = 0$. $SU_2\times U_1$ symmetry is broken down to U_1^{EM}. We also require the parameters satisfy the sufficient conditions for V_ϕ to be bounded below.[7] We also take $\rho_{ij} > 2|\eta_{ij}|$ so that when two fields ϕ_i and ϕ_j develop VEV they want to be orthogonal, i.e., $\phi_i = (0\ v_i)^T/\sqrt{2}$ and $\phi_j = (v_j\ 0)^T/\sqrt{2}$. We want U_1^{EM} to be unbroken at $T=0$ but broken for $T>T_c$.

At high temperatures, we have to calculate the finite temperature effective potential to study the symmetry behavior of the system.[8-10] For sufficiently high T, the ensemble averages $\langle\phi_i(T)\rangle$ can be obtained by minimizing the effective potential[9,10]

$$V_\phi(T) = V_\phi(0) + \sum_{i=1}^{3}\frac{1}{2}T^2 F_i\,\phi_i^\dagger\,\phi_i \quad . \tag{5}$$

The functions F_i are given by[7]

$$F_i = (3g^2+g'^2)/8 + \lambda_i + \sum_{j\neq i}\left[\frac{\sigma_{ij}}{3} + \frac{\rho_{ij}}{6}\right] + \text{Yukawa terms} \ . \tag{6}$$

For small fermion masses Yukawa terms are negligible. The effective mass terms at high temperature will be $M_i^2(T) = \mu_i^2 - \frac{1}{2}F_iT^2$. Therefore, if $F_i>0$ then for $T^2 \gtrsim 2\mu_1^2/F_1$ $SU_2\times U_1$ will be restored. However, if $F_i<0$ the symmetry will stay broken[11,12] or may be further broken down to a lower symmetry[9] at high temperature. We choose parameters so that $F_{1,2}<0$. This turns out to require $F_3>0$ so that for sufficiently high T, we may have a phase transition to a phase where $SU_2\times U_1$ is completely broken.

We have found a range of parameters such that at high temperature $\langle\phi_1(T)\rangle = (0\ v_1(T))^T/\sqrt{2}$, $\langle\phi_2(T)\rangle = (v_2(T)\ 0)^T/\sqrt{2}$, $\langle\phi_3(T)\rangle = 0$ is (at least) a local minimum of $V_\phi(T)$.[7] These parameters satisfy

$$\begin{aligned} &\lambda_1 \simeq \lambda_2 \simeq \lambda >> g^4\ ,\ |\rho_{ij}| \\ &-\sigma_{13} \simeq -\sigma_{23} \simeq \sigma > 3\lambda + \sigma_{12} + 3X \\ &|\mu_2^2| > \mu_1^2\ ,\ 2 > \sigma_{12}/\lambda > -1\ ,\ \lambda_3 > \sigma^2/\lambda \quad . \end{aligned} \tag{7}$$

where $X = (3g^2+g'^2)/8 \simeq 0.16$. The condition $\lambda << g^4$ allows us to neglect radiative corrections to V_ϕ. For a typical set of numbers, choose $\lambda \simeq -\sigma_{12} \simeq g^2 \simeq 0.4$, $\sigma \gtrsim 1.3$, $\lambda_3 \gtrsim 4.1$. We see that there is a range of parameters which satisfy the above conditions, but a rather large value for λ_3 is required. The second order phase transition occurs at T_c such that $v_2(T_c) = 0$. T_c is given by

$$T_c = A\mu_1/\sqrt{\lambda_1} = (246\ \text{GeV})A \tag{8}$$

where A is a function of the parameters in the potential and is typically of order unity, but can be made much larger or smaller by adjusting parameters. We will assume $T_c \gtrsim 1$ TeV. We have therefore demonstrated the existence of a phase transition in which $SU_2 \times U_1$ is broken to U_1^{EM} at $T = 0$ and $SU_2 \times U_1$ is completely broken for $T > T_c$.

Now I would like to describe how to embed our scheme to SU_5. We study the complete SU_5 potential, Eqs. (4a)-(4d). The conditions that V is bounded below and have symmetry breaking Eq. (3) at $T = 0$ are

$$\begin{aligned} &b > 0,\quad 15a + 7b > 0,\quad \beta_i < 0,\quad 5\alpha_i + 4\beta_i > 0 \\ &\lambda_i > 0,\quad \sqrt{\lambda_i\lambda_j} + \sigma_{ij} > 0\ . \end{aligned} \tag{9}$$

For $T > M_X$ we have to consider the heavy particle contributions to V(T). The effective potential at high temperature, $T > M_X$, are given by

$$V(T) = V_\Phi(0) + V_\phi(0) + V_{\Phi\phi}(0) + \frac{1}{2}GT^2\mathrm{Tr}\Phi^2 + \sum_{i=1}^{3}\frac{1}{2}F_i'T^2\phi_i^\dagger\phi_i \tag{10}$$

where

$$G = \frac{1}{60}\left[13a + 94b + 75g^2 + \sum_{i=1}^{3}(50\alpha_i + 10\beta_i)\right]$$

$$F_i' = 2\lambda_i + \sum_{j\neq i}\left[\frac{5}{6}\sigma_{ij} + \frac{1}{6}\rho_{ij}\right] + \frac{6}{5}g^2 + \text{Yukawa terms}$$

G is always positive for the parameters satisfying Eq. (9). Therefore, at sufficiently high T, the effective mass of Φ, $-\frac{1}{2}m^2 + \frac{1}{2}GT^2 > 0$ so that VEV $\langle\Phi(T)\rangle$ will vanish. The parameters λ_i, σ_{ij} and ρ_{ij} have been already chosen as Eq. (7) for the phase transition at $T_c \gtrsim 1$ TeV. For those λ_i, σ_{ij} and ρ_{ij} and any α_i and β_i in Eq. (9) F_3 will be always positive. F_1 and F_2 may be positive or negative depending upon the values of α_i and β_i. It is very likely that SU_5 symmetry have been restored in the very early universe. In this case α_i and β_i should be chosen such that F_1 and F_2 are also positive.

It is a difficult problem to study the effective potential near $T \lesssim M_X$. There may be intermediate phases between SU_5 and SU_3^c phases (for example, $SU_3^c \times SU_2 \times U_1$):

$$SU_5 \xrightarrow[T_1]{} \text{intermediate phases} \xrightarrow[T_n]{} SU_3^c \xrightarrow[T_c]{} SU_3^c \times U_1^{EM} \quad . \qquad (11)$$

There should be essentially no magnetic monopoles in our model. Any monopoles produced during intermediate phases at $T \lesssim M_X$ will become unstable once the SU_3^c phase is entered. They would presumably either decay or be confined in pairs which could subsequently annihilate. Stable monopoles of mass $M_m \approx 10^{16}$ GeV could, in principle, exist for $T < T_c$, but the number $r \simeq \exp(-M_m/T_c)$ expected from thermal fluctuations when $T \simeq T_c$ is extremely small.

For $T_n > T > T_c$ the U_1^{EM} is spontaneously broken. During this period the photon has a mass and electric charge is violated. Charge violating reactions are in equilibrium for $T \gtrsim T_c$.[7] If there is a net charge density in the present universe left over from fluctuations from equilibrium as $T > T_c$ it is far smaller[7] than the observational limit[13,14] from galaxies and cosmology.

REFERENCES

1. Ya. B. Zel'dovich and M. Y. Khlopov, Phys. Lett. 79B, 239 (1979).
2. J. P. Preskill, Phys. Rev. Lett. 43, 1365 (1979).
3. M. B. Einhorn, D. L. Stein and D. Toussaint, Phys. Rev. D21, 3295 (1980).
4. A. H. Guth and S.-H. H. Tye, Phys. Rev. Lett. 44, 631, 963 (1980).
5. H. Georgi and S. L. Glashow, Phys. Rev. Lett. 32, 438 (1974).
6. G. 't Hooft, Nucl. Phys. B79, 276 (1974); A. M. Polyakov, Pis'ma Eksp. Teor. Fiz. 20, 430 (1974) [JETP Lett. 20, 194 (1974)]. For an introduction, see S. Coleman in New Phenomena in Subnuclear Physics, Part A, ed. A. Zichichi (Plenum, N.Y., 1977), p. 297.
7. P. Langacker and S.-Y. Pi, Phys. Rev. Lett. 45, 1 (1980).
8. D. A. Kirzhnits and A. D. Linde, Phys. Lett. 42B, 471 (1972) and Ann. Phys. 101, 195 (1976).
9. S. Weinberg, Phys. Rev. D9, 3357 (1974).
10. L. Dolan and R. Jackiw, Phys. Rev. D9, 3320 (1974).
11. R. N. Mohapatra and G. Senjanovic, Phys. Rev. Lett. 42, 1651 (1979), Phys. Rev. D20, 3390 (1979), and CCNY preprint HEP-7916 (1979).
12. A. Zee, Phys. Rev. Lett. 44, 703 (1980).
13. R. A. Lyttleton and H. Bondi, Proc. R. Soc. Lond. A252, 313 (1959).
14. A. Barnes, Astron. J. 227, 1 (1979).

WEAK DECAYS, THEORY

D. NANOPOULOS, CERN, ORGANIZER

QUARK FLAVOR MIXING AND ITS PHYSICAL IMPLICATIONS

Ling-Lie Chau Wang
Brookhaven National Laboratory, Upton, New York 11973

This is a very fast developing field and the amount of literature on the subject is vast. It is impossible not to miss quoting some papers. I intend to inform you of the scope of the developments on the subject of flavor mixing, especially the Kobayashi-Maskawa model, since the Tokyo conference in 1978.

The history of the development of our understanding the weak interaction quark states is certainly an amazing one. Cabibbo in '63 observed that a single parameter θ_c can account for all hyperon semileptonic decays. That led to the weak-interaction quark structure $(u,d')_L$, s_L, where $d' = \cos\theta_c d + \sin\theta_c s$. Due to the presence of the strangeness-changing current of $\sin\theta_c \cos\theta_c (d\bar{s})$ from $(d'\bar{d}')$, the small branching ration of $B(K_L \to \mu\bar{\mu}) = (9.1 \pm 1.8) \times 10^{-9}$ could not be explained in this scheme, especially since the renormalizability[1] of the weak interaction theories was at hand. That led to the conjecture of the existence of a new quark in '70 by Glashow, Iliopoulos and Maiani, the charm c, and the two doublet quark structure $(u,d')_L$, $(c,s')_L$, where $d' = \cos\theta_c d + \sin\theta_c s$, and $s' = -\sin\theta_c d + \cos\theta_c s$. Now there is no strangeness changing neutral current $(d\bar{s})$ due to the cancellation in $(d'\bar{d}') + (s'\bar{s}')$. It is now history that charm particles have been found, though after a long wait since the discovery of its implicit state J. In '73 in a then quite unnoticed paper by Kobayashi and Maskawa,[2] a third pair of quarks was introduced in order to incoroporate CP violation[3] effects from the complexity of the coupling of the $W^{\pm}$'s to the quarks.[4] Now the doublets are $(u,d')_L$, $(c,s')_L$ and $(t,b')_L$, where

$$(d',s',b') = (d,s,b) \begin{pmatrix} V_{ud} & V_{cd} & V_{td} \\ V_{us} & V_{cs} & V_{ts} \\ V_{ub} & V_{cb} & V_{tb} \end{pmatrix}, \tag{1}$$

here V is unitary $V^{+}V = 1$. In general for n doublets, the number of physically significant parameters in V is equal to the number of parameters for a nxn unitary matrix minus the relative phases of the doublets, i.e. $n^2 - (2n-1)$. An orthogonal matrix can be characterized by $\frac{1}{2}n(n-1)$ angles, thus the rest of the parameters $[n^2-(2n-1)] - \frac{1}{2}n(n-1) = \frac{1}{2}(n-1)(n-2)$ has to be charaterized by phases. For n=2, V can be characterized by an angle θ_c and no phase. For n=3, V is characterized by three angles and one phase. It is this complexity in V that provides the CP violation. Thus the salient feature of the K-M model is that the CP violation effect is tied with the nonvanishing of some of the matrix elements in the third row or third column, which means that the b and the t flavored particles must have pure hadronic decays. Models with CP violation coming from the Higgs couplings, by having more Higgs doublets than the standard $SU(2)_L \times U(1)$ model, have no such correlation.[5] Actually in many of these models, the b-flavored particles have only semileptonic decays though this is not imposed on by any first principles.

<u>The determination of the mixing matrix V</u>: The four parameters of the V matrix have been so far determined from four sets of experimental informations. The $0^+ \to 0^+$ nuclear β decay rates comparing to that of μ decay (assuming no effects from the mixing of the leptons) determines $|V_{ud}|$, and the hyperon semileptonic decays determines $|V_{us}|$. The result of shrock and Wang's analysis[6] in '78, $|V_{ud}| = .9737 \pm .0025$ $|V_{us}| = .219 \pm .003$, giving $|V_{ud}|^2 + |V_{us}|^2 = .996 \pm .004$. The important point of the result is that the central value of $|V_{ud}|^2 + |V_{us}|^2$ is less than one, indicating that the old Cabibbo theory was not exactly true and there is "leakage" from the first two doublets.[6,7] It allows the third doublet to decay, i.e. the b can decay unto u. The K_L, K_S mass difference $M_L - M_S$ and the CP violation parameter ε determine the rest of the mixing matrix. As an example we give the central values of the V matrix determined in Ref. (8)

$$V = \begin{array}{c} \\ \\ \\ \end{array} \begin{matrix} u & c & t \\ \end{matrix}$$

$$V = \begin{pmatrix} .97 & -.20 & -.11 \\ .22 & .95-.75\times10^{-3}i & .20+1.3\times10^{-3}i \\ .068 & -.22+2.4\times10^{-3}i & .97-4.1\times10^{-3}i \end{pmatrix} \begin{matrix} d \\ s \\ b\,. \end{matrix} \qquad (2)$$

(columns labelled u, c, t)

It is interesting to note that the matrix elements in the V of Eq. (2) get smaller as they move away from the diagonal, i.e. though there are flavor mixing, the flavors like to keep their original identity. Also it implies that b and t prefer to decay in a cascade fashion, $t \to b \to c \to s$. Here it is important to keep in mind that the theoretical estimates of $M_S - M_L$ and ε depend upon an estimate of the matrix element $\langle \bar{K}^0 | \bar{s}\Gamma_\mu d \bar{s}\Gamma_\mu d | K^0 \rangle$, which depends upon various dynamical schemes. Though the central value of the V matrix, Eq. (2), has not been challenged by various considerations, it is of importance to have independent determinations of V_{cs}, V_{cd} in a more model-independent[9] way similar to the determination of V_{ud}, V_{us}. Here I list a few of such possibilities:

(1) Obtain V_{cs} from $D \to \ell\bar{\nu}_\ell X$ (with K), and V_{cd} from $D \to \ell\bar{\nu}_\ell X$ (without K). It is desirable to study such decay rates in $e^+e^- \to \psi(3770) \to D\bar{D}$ with one D or $\bar{D}$ explicitly selected from its exclusive decays.

(2) From the results of Ref.(10) $\Gamma(D^+ \to \pi^+\pi^0)/\Gamma(D^+ \to \bar{K}^0\pi^+) = \frac{1}{2}|V_{cd}/V_{cs}|^2$, which, in addition, has the nice feature that both final states $\pi^+\pi^0$, $\bar{K}^0\pi^+$ are exotic, thus free from possible complications of final state interactions.

(3) Comparing the decays $b \to cW^+ \hookrightarrow c\bar{s}$ and $b \to cW^+ \hookrightarrow \mu\bar{\nu}_\mu$, ought to give information[11] about V_{cs}.

It is interesting to note that if $V_{ud}V_{us} \neq -V_{cs}V_{cd}$, i.e. if the strangeness neutral current is not cancelled in the first two doublet then the t quark that so far eludes observation is needed.[12] If $|V_{cs}|^2 + |V_{cd}|^2 < 1$, the b flavored particle must decay into charm.

Nonleptonic Decays: Since the establishment of asymptotic freedom[13] of the non-Abelian gauge theory (QCD), ambitous attempts have been made, using perturbative QCD, to explain various pure hadronic phenomena including the nonleptonic decays[14] of the Kaons and the charm particles.

a) The $\Delta I = \frac{1}{2}$ rule of K decays: In '77 Shifman et al[15] noted that the enhancement of $\Gamma(K^0 \to \pi^+\pi^-)/\Gamma(K^+ \to \pi^+\pi^0) \simeq 670$, the $\Delta I = \frac{1}{2}$ rule, could be explained by the so-called "Penguin" diagram. It was subsequently observed by Gilman and Wise[16] that if the Penguin diagram is dominating in the K-M model, the parameter ε' in K decays can be non-negligible due to the complexity introduced in the K-2π amplitudes from the V_{ij}'s. The expression for ε' is

$$\varepsilon' \equiv \frac{1}{3}(\eta_{++} - \eta_{00}) = \sqrt{2}e^{i(\delta_2 - \delta_0 + \pi/2)} \mathrm{Im}\frac{A_2}{A_0} , \qquad (3)$$

where $\eta_{++} = A(K_L \to \pi^+\pi^-)/A(K_S \to \pi^+\pi^-) = \varepsilon + \varepsilon'$

$\eta_{00} = A(K_S \to \pi^0\pi^0)/A(K_S \to \pi^0\pi^0) = \varepsilon - 2\varepsilon'$.

and δ_2, δ_0 are respectively the I = 2, I = 0 phase shifts of scattering amplitudes. In the superweak model $\mathrm{Im}(A_2/A_0) = 0$. In the K-M model the "Penguin" diagram gives a phase to A_0, $\mathrm{Im}(A_2/A_0) = \mathrm{Im}(V_{ts}V_{td}^*)N \cdot \mathrm{Re}A_2/|A_0|$, where N is a number from the gluon exchange effect. The estimate of ε' varies, depending upon the QCD calculations,[17] ranging from a few times 10^{-2} to 10^{-3}. There are indications from experiments that the central value of $|\varepsilon'/\varepsilon|$ is not zero, however the errors are still rather large that with two standard deviations it is still consistent with zero.[18] It was commented by Li and Tuan[19] that the phase $\pi/2 + \delta_2 - \delta_0$ measured in the latest experiment is consistent with that obtained from other sources only if two standard deviation of error is taken. All these certainly give new imputs to measure ε' more accurately.

b) Charm Nonleptonic Decays: As we mentioned before we can also learn the mixing matrix in the nonleptonic decays of charm. Actually there is already one interesting result that the experimental observation $\Gamma(D^0\ K^+K^-)/\Gamma(D^0 \to \pi^+\pi^-) \gtrsim 1$ can be easily accommodated in the K-M model with $V_{ud}/V_{us} \neq V_{cs}/V_{cd}$ even in the case of SU(3) symmetry.[10,20] Further the twenty plus decays of D^+, D^0, F^+ into two-meson final states can be written in terms of five independent amplitudes[10,21] and the mixing matrices. They will provide valuable information in the amplitude and mixing matrix analysis. There are already indications[22] from the data, $\tau(D^+)/\tau(D^0) = 3.1^{+4.1}_{-1.3}$ and $\Gamma(D^0 \to K^-\pi^+)/\ (D^0 \to \bar{K}^0\pi^0) = 1.6 \pm .9$, that the W exchange diagrams dominate. This actually challenges the simple perturbative QCD calculations.

The Neutron Electro-Dipole Moment: There are three form factors for the neutron, $<n|J_\mu^{e.m.}(0)|n> \sim \bar{u}(p')[F_1(q^2)\gamma_\mu - F_2(q^2)\sigma_{\mu\nu}\gamma_\nu + F_3(q^2)i\gamma_5\sigma_{\mu\nu}q^2]u(p)$, where $F_1(0) = 0$ the charge form factor,

$F_2(0) = \mu_n$ the magnetic moment and $F_3(0) = d_n$ the electron dipole moment. Again the complexity in V_{ij} can give d_n of the neutron via the diagrams of Fig.(1a) with a photon attached in all possible ways. It was first

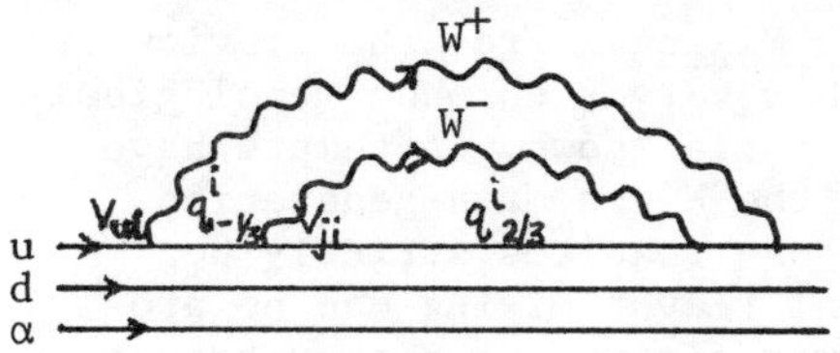

Fig. (1a)

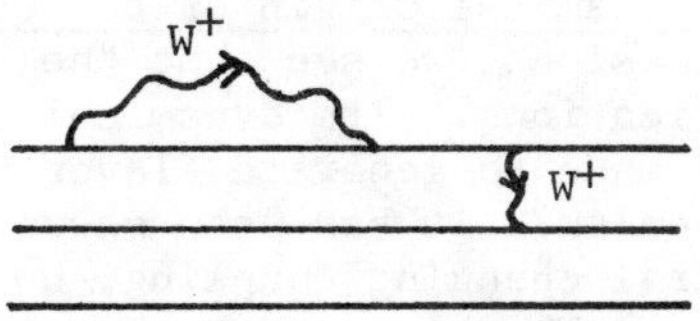

Fig. (1b)

Diagrams considered for the neutron electro-dipole moment, where $q^i_{-1/3}$, $q^j_{2/3}$ are the quarks of charge of -1/3 and 2/3 respectively.

estimated by Ellis, Gaillard, Nanopoulos[7] in '76, $d_n \sim 10^{-30}$ cm. Then Shabalin[23] showed that actually the sum of graphs in Fig. (1a) gives $d_n = 0$. Calculations have also been done including strong interactions[24] and interquark exchange forces[25] Fig. (1b). The results are quite model dependent but they all gave very small d_n in contrast to the close-to-the-experimental limit $d_n \leq 1.6 \times 10^{-24}$ cm of Ramsey from the Higgs CP violation.[5]

Decay Interference from CP Violation: As in the K^0 system, due to the $K^0 \rightleftarrows \bar{K}^0$ transition and CP violation, where K_S, K_L are composed of unequal amounts of K^0 and $\bar{K}^0$, the $D^0{}_S$, $D^0{}_L$, and $B^0{}_S$, $B^0{}_L$ will be composed of unequal amounts of D^0, $\bar{D}^0$, and B^0, $\bar{B}^0$ respectively. That leads to the following asymmetries.[26]

(1) The asymmetry δ of same sign double-lepton final state in $e^+e^- \to D^0\bar{D}^0X^0$(or $-B^0\bar{B}^0X^0$) $\to \ell^+\ell^+X^{--}, \ell^-\ell^-X^{++}$ is $\delta \equiv (N_{++}-N_{--})/(N_{++}+N_{--}) = 4\mathrm{Re}\varepsilon$, where ε is the CP violation parameter for D^0, or B^0 system. It was estimated[27] to be small, ($\delta \sim 10^{-3}$) for the K-M model, but bigger ($\delta \sim 10^{-2}$) for the Higgs CP violation.[5] Thus a large double charge asymmetry in e^+e^- experiment can rule out the K-M model. However such double lepton charge asymmetry has severe contamination from the chain semileptonic decays of quarks.

(2) Recently Carter and Sanda[28] have observed an interesting interference effect in the weak interaction tree diagrams. That gives an asymmetry effect a from the B^0, B^0 decay in e^+e^- interaction, $e^+e^- \to B^0\bar{B}^0X$, where $B^0\bar{B}^0$ must be in s-wave,

$$a = \frac{\sigma(K^+K_SX^-) - \sigma(K^-K_SY^+)}{\sigma(K^+K_SX^-) + \sigma(K^-K_SY^+)}$$

The estimate of a depends sensitively on the values of the mixing matrix. For the values of V_{ij} determined from Ref. (8) $a \sim 1\%$, which is very difficult to observe. The restriction that $B^0\bar{B}^0$ must be in the s-wave state may put very serious limitation on the feasibility of the observation of such an effect.

(3) CP violation can also lead to unequal decay rates in decays

related by charge conjugation. Bander, Silverman and Soni[29] estimated that such asymmetry may reach observable values in some of the decay channels of $B^{\pm}$.

Dynamical Origin of the Quark Flavor Mixing: From the previous discussions, we see that the K-M model has very rich phenomenological implications. Its dynamical origin is little known. Attempts have been made to generate flavor mixing via the Higgs mass-generation mechanism. It has been shown[30] that if one requires strictly no neutral changing coupling, no non-trivial flavor mixing can be produced. Thus to produce K-M type of flavor mixing one either has to introduce large Higgs masses or more complicated Higgs structures in order to suppress the neutral flavor changing coupling. Another interesting fact is that for the three-doublet model, there are four parameters in V but only three independent quark masses. So if we want to express them in terms of physical masses only, there is always a relation among them.

Alternatives: So far the most serious question to the three doublet model is that the t quark is still missing (experiments at PETRA limits $M_t > 18$ GeV). Various alternatives[31] have been offered. However, most of these models have been ruled out by the recent results from CESR,[32] analyzing the decay products in the $\Upsilon(4s)$ region, that the B meson does have appreciable pure hadronic decays and also favors the chain decay $b \to c \to s$. The Georgi-Glashow[31] and the Georgi-Machacek[31] models are ruled out because their lonely b can only decay semileptonically. The Georgi-Pais[31] model is ruled out because their b quark prefers to decay into u quark rather then the c quark.[4]

Here I want to emphasize an interesting phenomenological model that accommodates all existing data. It is essentially the Cabibbo world enlarged by one doublet: the quark states are $(u,d')_L$, $(c,s')_L$, b_L, where $d' = V_{ud}d + V_{us}s + V_{ub}b$, $s' = V_{cd}d + V_{cs}s + V_{cb}b$, and, in addition to the orthogonality and unitarity, the constraint of no strangeness changing neutral current is imposed, $V_{ud}V_{us} + V_{cd}V_{cs} = 0$. However the b-flavor changing neutral currents $d\bar{b}$, $s\bar{b}$ are allowed. Such a model was discussed by Barger and Pakvasa.[33] Taking the known value of $V_{ud} = .97$, $V_{us} = .23$, one can actually solve the mixing elements. One of the results is $V_{ub} = .1$, $V_{cd} = .89$, $V_{cs} = -.25$, $V_{cb} = .39$. They calculated the $s\bar{b}$ flavored neutral current decay of the b and gave a branching ratio $B(b \to se^+e^-) \approx 2\%$. Also $B^0_{(bs)} \to \tau^+\tau^-$ should exist. This shows the importance of studying the existence of $(b\bar{s})$ flavor changing neutral current.

To end the lecture, I would put these challenges to the experimentalists:

(1) "Direct" measurements of V_{cs}, V_{cd}: Inclusive and semileptonic decays of charm and B decays, $\Gamma(D^+ \to \pi^+\pi^0)/\Gamma(D^+ \bar{K}^0\pi^+)$.

(2) B decay properties: Does B decay only semileptonically? Which decay of B is favored $b \to c \to s$ or $b \to u$? For these CESR will soon have a definite answer. Is there b-changing neutral current, $b \to q\, \ell\bar{\ell}$, $B \to \ell\bar{\ell}$?

(3) CP properties of the charm and the B system: ε', $\frac{N^{++}-N^{--}}{N^{++}+N^{--}}$, ...

(4) Better neutron electrodipole moment measurements.

The real challenge that confronts us is the "family" problem. How many generations of quarks are there? How does the mixing come about? What is the origin of CP violation? It is likely that the current distinction between the K-M origin and complex-Higgs origin may turn out to be a superfluous one.

Acknowledgments: I would like to thank Thomas Rizzo for many helpful discussions in preparing this talk.

REFERENCES

1. G. 't Hooft and M. Veltman, Nucl. Phys. B50, 318 (1972); B. W. Lee and J. Zinn-Justin, Phys. Rev. D7, 1049 (1973); for a review, see E. Abers and B. W. Lee, Phys. Reports, 9C, 1 (1973).
2. M. Kobayashi and T. Maskawa, Prog. Theor. Phys. 49, 652 (1973).
3. J. H. Christenson, J. W. Cronin, V. L. Fitch and R. Turlay, Phys. Rev. Lett. 13, 138 (1964).
4. For a recent review on the subject and references, see L.-L. Chau Wang, "Flavor Mixing and Quark Decay", talk presented at the VI Int. Conf. on Meson Spectroscopy, BNL, April 24-25, 1980.
5. For a review of CP violation from other mechanisms, see G. Senjanović's talk in this parallel session of the conference.
6. R. Shrock and L.-L. Wang, Phys. Rev. Lett. 41, 1692 (1978).
7. The previous result was $|V_{ud}|^2 + |V_{us}|^2 = 1.004 \pm .004$, see M. Roos, Nucl. Phys. B77, 420 (1974). The earliest attempt to evaluate this leakage was given by J. Ellis, M. K. Gaillard, and D. V. Nanopoulos, Nucl. Phys. B109, 213 (1976).
8. R. E. Shrock, S. B. Treiman and L.-L. Chau Wang, Phys. Rev. Lett. 42, 1589 (1979). See also V. Barger, W. F. Long, and S. Pakvasa, ibid, 42, 1585 (1979).
9. See such a constraint from $K_L \to \mu\bar{\mu}$, R. E. Shrock, M. B. Voloshin, Phys. Lett. 87B, 375 (1979).
10. L.-L. Chau Wang and F. Wilczek, Phys. Rev. Lett. 43, 816 (1979).
11. F. Wilczek, private communication.
12. See also contributed paper G. L. Kane, University of Michigan preprint, UMHE 80-18 (1980).
13. D. J. Gross and F. A. Wilczek, Phys. Rev. Lett. 30, 1343 (1973), H. D. Politzer, ibid 30, 1346 (1973).
14. M. K. Gaillard, B. W. Lee, Phys. Rev. Letters 33, 108 (1974); G. Altarelli and L. Maiani, Phys. Letters, 32B, 351 (1974). For QCD calculation of charm decays, see N. Cabibbo and L. Maiani, Phys. Lett. 73B, 418 (1973), B. Stech, Nucl. Phys. B133 315 (1978).
15. M. A. Shifman, et al., JETP Lett. 22, 55 (1975); Nucl. Phys. B120, 316 (1977); Sov. Phys. JETP 45, 670 (1977); J. F. Donoghue and B. R. Holstein, "Dynamical Effects in Two Body Decay", MIT and NSF Preprint (1979).

16. F. J. Gilman and M. B. Wise, Phys. Rev. Lett. 83B, 83 (1979).
17. L. Wolfenstein's talk in AIP Conference Proceedings No. 59 Particle and Fields Subseries No. 19, (Montreal 1979), edited by M. Margolis and D. Stairs; B. Guberina, R. D. Peccei, Nucl. Phys. B163, 289 (1980).
18. See the latest experimental results of J. H. Christenson et al., Phys. Rev. Lett. 43, 1209 (1979); and review of previous results, K. Kleinknecht, Proceedings of the 17th Int. Conf. on H.E. Physics, London, England '75.
19. Li, Xiaoyuan and S. F. Tuan, "Is the CP Puzzle Resolved?" Institute of Theoretical Physics, Beijing, preprint ASITP-79-007, '79.
20. M. Suzuki, Phys. Rev. Lett. 43, 818 (1980); C. Quigg, Zeit Phys. C4, 55 (1980).
21. T. Rizzo and L.-L. Chau Wang, "The Quark-Diagram Classification of Charm Decays", Brookhaven preprint BNL 27950 '80. Ref. (3) of this paper and references therein. See contributed paper by S. P. Rosen, W-Exchange Dominance in Neutral B-Decay, Purdue University preprint '80.
22. See A. Lankford's talk in this conference.
23. E. P. Shabalin, Sov. J. Nucl. Phys. 28, 75 (1978).
24. J. Ellis, M. K. Gaillard, Nucl. Phys. B150, 141 (1979), E. P. Shabalin, preprint ITEP-131-1979
25. B. F. Morel, Harvard preprint HUTP-79/A009 (1979), D. V. Nanopoulos, A. Yildiz, and P. H. Cox, Harvard and University of New Hampshire preprint, HUTP-79/A024, E. P. Shabalin, Inst. of Theo. and Exp. Phys. preprint, ITEP-131 (1979).
26. A. Pais and S. B. Treiman, Phys. Rev. 12, 2744 (1975); L. B. Okun V. I. Zakharov, and B. M. Pontecorvo, Nuovo Cimento Lett. 13 218 (1975)
27. A. Ali, Z. Z. Aydin, Nucl. Phys. B148, 1651 (1979); J. S. Hagelin, Phys. Rev. D20, 2893 (1979).
28. A. B. Carter and A. I. Sanda, Rockefeller University preprint, DOE/EY/2232B-203 (1980), to be published in the Phys. Rev. Lett.
29. M. Bander, D. Silverman and A. Soni, Phys. Rev. Lett. 43, 242 (1979).
30. R. Gatto, G. Morchio, F. Strocchi, Phys. Lett. 83B 348 (1979). The literature on the subject is vast, see the recent paper on the subject G. Segre, H. A. Weldon, and J. Weyers, Phys. Lett. 83B, 351 (1979).
31. See the discussions in the parallel sessions (1-4 of this conference).
32. See E. Thorndike's talk at this conference.
33. H. Georgi and S. Glashow, Nucl. Phys. B167, 173 (1980). H. Georgi and M. Machacek, Phys. Rev. Lett. 43, 1639 (1979). H. Georgi and A. Pais, Phys. Rev. D19, 2746 (1979). V. Barger and S. Pakvasa, Phys. Lett. 81B, 195 (1979).
34. See also contributed paper, G. L. Kane, R. Thun, University of Michigan preprint, UM HE 80-8, 1980.

CP VIOLATION AT HIGH ENERGIES

Mary K. Gaillard
Fermi National Accelerator Laboratory, Batavia, Illinois 60510
and
LAPP, Annecy-le-Vieux, France

ABSTRACT

The high energy sector of gauge theories with hard CP violation is discussed with emphasis on "strong" CP violation and baryon number generation.

INTRODUCTION

Existing data are compatible with the "standard" GWS[1]-GIM[2]-KM[3] model of weak interactions: an $SU(2)_L \otimes U(1)$ electroweak gauge theory spontaneously broken via the introduction of (minimally) one Higgs doublet whose Yukawa couplings to fermions are responsible for quark masses and their generalized, complex Cabibbo angles which are in this model the only source of CP violation. The low energy phenomenology of this model has been reviewed by Lin-Li Wang; I shall instead discuss ways of probing the high energy sector using the standard model and its minimal extension to a unified theory of strong and electroweak interactions as a reference point.

One low energy probe of the high energy sector is provided by the experimental limit on the "strong" CP violation parameter θ which appears[4] via non-perturbative topological effects in the effective QCD Lagrangian:

$$\mathscr{L}_{QCD} \ni \theta \frac{\alpha_s}{\pi} \tilde{F}_{\mu\nu}{}^{i} F_{i}{}^{\mu\nu} \tag{1}$$

the most stringent bound on θ is provided by the low experimental limit[5] on the neutron dipole moment which leads to the estimate[6]

$$\theta < \text{a few} \times 10^{-9} \ . \tag{2}$$

Since in the standard model defined above CP violation is "hard," θ is infinitely renormalized and the cut-off Λ which must be introduced to render it finite might indicate an energy level at which new physics should intervene. However, the GWS-GIM-KM model requires only

$$\Lambda < \exp(10^{25}) \text{ GeV} \quad , \tag{3}$$

while we expect new physics at considerably more modest energies: at least at 10^{19} GeV where gravitational effects become important, and more probably at 10^{14} GeV as suggested by grand unified theories (GUTs).

The "minimal" GG[7]-BEGN[8] model for grand unification is a straightforward extension of the "standard" electroweak model, namely SU(5) with the Higgs sector restricted to an adjoint to provide the

0094-243X/81/680517-07$1.50

initial breaking to $SU(3)_c \otimes SU(2)_L \otimes U(1)$ and (minimally) a single 5-plet containing the electroweak doublet whose complex Yukawa couplings again provide quarks masses and mixing angles and the only source of CP violation. One might hope that further unification might ease the "strong" CP problem, but it turns out that this minimal version slightly aggravates it. More generally, the concept of grand unification leads naturally to further probes of the high energy sector: the low energy probe provided by baryon decay as well as a high energy probe provided by cosmological CP violation which we hope will account for the observed baryon-to-photon density ratio:

$$n_B/n_\gamma \simeq 10^{-9\pm 1} \quad . \tag{4}$$

Efforts to explain the number (4) suggest that the minimal GUT is in fact insufficient; this result feeds back to the renormalization of θ in an essentially model-independent way and renders the constraint (2) more interesting.

θ-RENORMALIZATION

We define the strong interaction Lagrangian by taking the quark mass matrix to be real and diagonal, and—on the grounds that it is experimentally tiny—the θ parameter to vanish in the absence of weak radiative corrections. The CP violating property of the latter will induce a non-vanishing θ; in particular the renormalied mass matrix $M_{ren.}$ must be rediagonalized at each order in electroweak perturbation theory. Since CP violation introduces complex matrix elements which are logarithmically divergent, a redefinition of the quark basis so as to give a real, diagonal mass matrix involves a chiral transformation which, because of the anomolous divergence of the flavor singlet axial current, induces[9] a correspondingly divergent renormalization of θ:

$$\delta\theta = \mathrm{Arg}\,\det M_{ren.} \equiv a \ln \Lambda/m_N \tag{5}$$

The bound (2) then implies a limit on the cut-off parameter:

$$\Lambda < m \exp(10^{-9}/a) \quad , \tag{6}$$

which might be interpreted as the energy where new physics must come into play. This is analogous to the cut-off of a few GeV required to make, say, the quadratically divergent amplitude for $K_L \to \mu\mu$ compatible with experimental data in pre-GIM days, assuming vanishing strangeness changing neutral currents in lowest order weak interactions. In that case new physics—namely charm—did indeed appear to provide the needed cut-off. A more modern formulation would be to express a "running" $\theta(m) = a \ln\mu/m$ in terms of the scale[10] μ where an as yet unknown symmetry principle requires it to vanish.

In the "standard" model, and in the quark basis where electroweak gauge couplings are flavor diagonal, the Yukawa couplings are specified by two arbitrary complex coupling matrices:

$$\mathscr{L}_Y = (\bar{\psi}_L \cdot \phi) G_k \psi_R^k + (\bar{\psi}_L \cdot \tilde{\phi}) G_a \psi_R^a + \text{h.c.} \tag{7}$$

where $\psi_L^T = (\psi_L^a, \psi_L^k)$ is an electroweak doublet, the indices a and k denote "anoquark": $\psi_a^T = (u,c,t,...)$ and "cathoquark": $\psi_k^T = (d,s,b,...)$ vectors in flavor space, and $\tilde{\phi} = -i\,\tau_2 \phi^*$. The mass matrices are generated by the usual shift $\phi = H + \langle\phi\rangle$ of the Higgs doublet with $\langle\phi^T\rangle = (0, v/\sqrt{2})$. A general complex square matrix can be written in terms of a diagonal matrix multiplied on the left and on the right by independent unitary matrices:

$$G_k = \frac{\sqrt{2}}{v} U_1^\dagger U_c M_k U_2 \quad , \quad G_a = \frac{\sqrt{2}}{v} U_1^\dagger M_a U_3 \quad . \tag{8}$$

Since U_i, i=1,2,3 can be absorbed into the definitions of the quark fields while leaving the gauge couplings invariant, the only observables in (8) are the mass matrices $M_{k,a}$ and the Cabibbo matrix U_c which appears in the charged current matrix as well as the (unphysical) charged Higgs coupling after redefinition of ψ_L^k so as to give a diagonal zeroth order mass matrix M_0. The renormalized mass matrix[11] is given by the sum of radiative corrections to the quark propagator with one mass insertion and is in general complex:

$$\mathscr{L}_{mass} = \bar{\psi}_R M_{ren.} \psi_L + \text{h.c.} \quad ; \quad M_{ren.} \equiv M_0(1+C) \quad , \tag{9}$$

and the change in θ induced[10] by the chiral transformation which makes M_{ren} real and diagonal is

$$\delta\theta = \text{Arg Det } M_{ren.} = \text{Arg Det}(1+C) = \text{ImTr}\ln(1+C) = \text{ImTr}C + \ldots \tag{10}$$

The multiple GIM-type cancellations associated with CP violation in the KM model require quark mass factors to appear in ImC. Since C is at most logarithmially divergent any additional mass insertions render it finite so that the divergent part can arise only from multiple (physical or unphysical in a renormalizable gauge) Higgs exchange. Inspection shows[10] that the first non-vanishing divergent contribution appears for six Higgs loops (plus one U(1) loop without which the ano- and catho-quark contributions would cancel identically), e.g.:

$$\delta\theta_{inf} \propto \text{ImTr}(U^\dagger M_a^4 U M_k^4 U^\dagger M_a^2 U M_k^2) \quad . \tag{11}$$

Finite contributions occur in a lower order giving[10]

$$\delta\theta = \delta\theta_{inf} + \delta\theta_{fin} \quad , \quad \delta\theta_{fin} \sim \left(\frac{a}{\pi}\right)^2 \left(\frac{m_q}{m_W}\right)^4 \sim 10^{-16}$$

$$\delta\theta_{inf} \sim \left(\frac{\alpha}{\pi}\right)^7 \left(\frac{m_q}{m_W}\right)^{12} \ln(\Lambda/m) \sim 10^{-34} \ln(\Lambda/m) \quad , \tag{12}$$

so that $\delta\theta < 10^{-9}$ as long as $\ln(\Lambda/m) \lesssim 10^{25}$. Other sources of CP-violation such as renormalization[12] of the di-gluon operator in

Eq. (1) involve similar traces of quark loops with Higgs exchange and give similar results.[13]

THE MINIMAL GUT

In the GG-BEGN model there are again two independent Yukawa coupling matrices:

$$\mathscr{L}_Y = 10_L^T\, C\, G_a\, 10_L\, H + \bar{5}_R\, G_k\, 10_L\, \bar{H} + \text{h.c.} \qquad (13)$$

where 10 and 5 denote the conventional quark multiplets in SU(5). Again expressing the coupling matrices in terms of real, diagonal mass matrices

$$G_k = W\, M_k\, U_c^\dagger V \quad ; \quad G_a \equiv G_a^T = V^T S\, M_a V \quad , \qquad (14)$$

we see that the unitary matrices W and V may be absorbed in the definition of the 5-plet and 10-plet, respectively, while the symmetry of G_a

$$G_a G_a^\dagger = G_a^T G_a^* = (G_a^\dagger G_a)^* \qquad (15)$$

requires that S commute with M_a, so it is diagonal in the same basis:

$$S M_a^2 S^\dagger = M_a^2 \quad ; \quad S_{ab} = e^{i\phi_a}\delta_{ab} \quad . \qquad (16)$$

Since an overall common phase may again be reabsorbed, we are left with[8,14,15] ($N_{generation}$ - 1) observable phases in addition to the Cabibbo parameters which include one observable phase for three generations.

Immediate consequences of the above analysis are that the Cabibbo angles for baryon decay are completely determined[8,14,16] in the minimal model, and that no CP violation can occur in lowest order since the effective Lagrangian is simply multiplied by an overall phase:[14] $e^{i\phi_1}$. Thus detailed studies of nucleon decay can in principle provide a test of this model, although deviations from its predictions are unfortunately expected to be small in many more general models.

A further consequence is that infinite θ renormalization will occur in a lower order[10] because a) the six-quark model contains two new phases which are unconstrained by low energy data, b) the presence of only the four heaviest quarks need be felt to generate CP violating effects and c) the presence of a new vertex, namely $q+H\rightarrow\bar{q}$ permits more complicated structures for the relevant trace; one finds,[15] e.g.:

$$\delta\theta_{inf} \propto \mathrm{ImTr}(M_a^3 S U M_k^2 U^\dagger M_a S^\dagger U^* M_c^2 U^T) \quad . \qquad (17)$$

Inserting the appropriate masses and Cabibbo parameters, one sees that the θ renormalization is enhanced relative to the previously evaluated K-M case by a factor

$$\left(\frac{\delta\theta_{KM}}{\delta\theta_{GUT}}\right)_{inf} \backsim \theta_c^2\left(\frac{M_t M_c M_s^2}{M_W^4}\right)\left(\frac{\alpha}{\pi}\right)^3 (3)^8 \backsim 10^{-12} \quad . \qquad (18)$$

where the factor $(3)^8$ takes into account quark mass renormalization relative to their low energy values which were used in the estimate (12). A plausible guess is that Λ lies in the expected range of new physics $\Lambda \backsim (10^{15}-10^{19})$GeV, so that $\delta\theta_{GUT}^{inf.} \backsim 10^{-20}$ which is still safe.

BARYON NUMBER GENERATION

As discussed by a number of authors[17] GUTs contain the three ingredients necessary for generating a calculable net baryon number as first enumerated by Andrei Sakharov:[18] CP violating interactions, baryon number violating interactions, and a non-equilibrium epoch for the latter. Within the context of present theories, the dominant mechanism is believed to be the decay of superheavy Higgs bosons ($M_H \backsim 10^{0\pm2}$ M_X where $M_X \backsim 10^{15}$ GeV are superheavy GUTs gauge vectors) at a temperature $T \lesssim 10^{14}$ GeV where baryon number violating forces drop from thermal equilibrium. As a first approximation, the baryon-to-photon density ratio (4) is simply given by

$$(n_B/n_\gamma)_{H\text{-decay}} \backsim \left(\frac{g_H}{g_{tot}}\right)(\Delta B)_H \quad , \qquad (19)$$

where g_H and g_{tot} are (essentially) the numbers of helicity states of scalars and of all particles in the theory, and $(\Delta B)_H$ is the baryon asymmetry intrinsic to Higgs decay. Since in the quark basis where gauge couplings are diagonal CP violation is confined to the Yukawa couplings, the leading order contribution to $(\Delta B)_H$ can be obtained by considering only Higgs-exchange radiative corrections to the $H \to f_1 f_2$ vertex. Summation over fermion final states reduces the calculation to the trace of a fermion loop with all possible scalar vertices inserted, just as in the calculation of $\delta\theta_{inf}$, and in the minimal GG-BEGN GUTs model the leading contribution is just proportional to the expression (17). Then one finds[15]

$$\Delta B = \frac{\Gamma(H \to B)-\Gamma(H \to -B)}{\Gamma(H)} \backsim 10^{-1}\alpha^4\left(\frac{M_b^4 M_t^3 M_c}{M_W^8}\right)\Big/\alpha(M_t/M_W)^2 \backsim 10^{-14}, \qquad (20)$$

where the factor 10^{-1} arises from Cabibbo angles. Since one expects $g_H/g_{tot} \backsim 10^{-2}$, the experimental number (4) requires a much larger value: $\Delta B \backsim 10^{-7\pm1}$.

In fact, there are various effects which tend to decrease n_B/n_γ relative to the simple estimate (19). Baryon number violating fermion-fermion scattering can wash out[19] the ΔB generated by Higgs decay in the non-equilibrium epoch; if the mass of the GWS Higgs doublet is sufficiently close to the Coleman-Weinberg value[20] of about 10 GeV, the breakdown of $SU(2)_L \otimes U(1)$ to $U(1)_{em}$ will be associated with a reheating which will further dilute[21] n_B/n_γ. So we probably should require $(\Delta B)_H \gtrsim 10^{-5\pm1}$, which is nine orders of magnitude larger than the value (20) obtained in the minimal GG-BEGN model.

There are various possibilities for increasing the minimal model prediction (20). One would be simply to add arbitrarily heavy fermion generations. However, this is disfavored by a number of arguments. The successful calculation[8,22] of (M_b/M_c) and astrophysical arguments[23] on the number of neutrinos favor three or at most four generations of not-too-heavy ($M \lesssim M_W$) fermions, while very heavy fermions ($M \gtrsim M_W$) with the usual $SU(3)_c \otimes SU(2)_L \otimes U(1)$ quantum numbers would induce unacceptable radiative corrections to the relative strengths of the neutral- and charged-current fermi coupling constants[24] and to the Higgs potential,[25] rendering the observed vacuum unstable. A more acceptable modification is simply to add more super-heavy scalars. These need not be "Higgs" scalars in that they could have vanishing vacuum expectation values, but would have arbitrary complex Yukawa couplings, unconstrained by low energy phenomenology, i.e, unrelated to quark masses and mixing angles. On general grounds, a non-vanishing ΔB can be induced[26] only at 4th order in the Yukawa couplings. Systematic analysis[26] of possible decay channels shows that with the range of masses and couplings expected in a general GUT framework one can obtain $(\Delta B)_H = 10^{-10} - 10^{-6}$, by adding more scalars to the minimal SU(5) GUT, and $(\Delta B) = 10^{-6} - 10^{-2}$, if a more complex structure for the gauge group is introduced.

The conclusion is that GUTs do indeed provide the possibility of a quantitative understanding of the density ratio (4). Recall, however, that the calculation of $\delta\theta_{inf}$ is essentially the same as that of $(\Delta B)_H$. The first is the imaginary part of scalar-exchange radiative corrections to a mass insertion on a fermion loop divided by the uncorrected mass insertion; the second is the imaginary part of the same corrections to a Yukawa coupling divided by the uncorrected Yukawa coupling. If we complicate the scalar sector of the theory so as to jack up ΔB by nine orders of magnitude relative to (20), we expect that in the same theory we should find a corresponding increase in (18), giving $\delta\theta_{inf} \sim 10^{-11\pm?}$, for a cut-off ($\Lambda \sim 10^{15} - 10^{19}$)GeV. Since there are undoubtedly errors of an order of magnitude or so in this estimate and in the "experimental" limit (2), the latter constraint starts to become interesting.

The analysis of both phenomena could be complicated by the presence of super-heavy fermions carrying exotic $SU(3)_c \otimes SU(2)_L \otimes U(1)$ quantum numbers which are to be expected in the framework of a truly unified, minimal parameter theory. Hopefully these would provide the desired "initial condition" $\theta(\mu \sim 10^{15} - 10^{19}$ GeV$) \equiv 0$, while dropping from equilibrium sufficiently soon to have no effect on (or perhaps give an additional contribution to) the value of ΔB generated by scalar decays. An alternative picture is one in which CP conservation is broken only "softly", as discussed in the following talk.

REFERENCES

1. S.L. Glashow, Nucl. Phys. 22, 579 (1961); S. Weinberg, Phys. Rev. Lett. 19, 1264 (1967); A. Salam, Proc. 8th Nobel Sympos. ed. N.Svartholm (Almqvist and Wiksells, Stockholm, 1968) p. 367.
2. S.L. Glashow, et al., Phys. Rev. D2, 1258 (1970).
3. M. Kobayashi and K. Maskawa, Progr. Theor. Phys. 652 (1969).
4. A.A. Belavin, et al., Phys. Lett. 59B, 85 (1975); R.D. Peccei and H.R. Quinn, Phys. Rev. Lett. 38, 40 (1977).
5. W.B. Dress, et al., Phys. Rev. D15, 9 (1977).
6. V. Baluni, Phys.Rev. D19, 2227 (1979); R. Crewther, et al., Phys. Lett. 88B, 123 (1979) and 91B, 487 (1980).
7. H. Georgi and S.L. Glashow, Phys. Rev. Lett. 32, 438 (1975).
8. A.J. Buras, et al., Nucl. Phys. B135, 66 (1977).
9. S. Weinberg, Phys.Rev. Letters 40, 22 (1978).
10. J. Ellis and M.K. Gaillard, Nucl. Phys. B150, 141 (1979).
11. Since the Feynmann integrals relevant to CP violation are governed by large cut-offs and/or large masses we neglect further non-perturbative effects.
12. F. Wilczek, Phys.Rev. Letters 40, 279 (1978)
13. C.G. Han, Phys. Rev. D20, 996 (1979).
14. J. Ellis, et al., Phys. Lett. 88B, 320 (1979).
15. J. Ellis et al., Phys. Lett. 80B, 360 and 82B, 464 (1979).
16. R.N. Mohapatra, Phys. Rev. Lett. 43, 893 (1979).
17. For a recent review and references see J. Ellis, et al., preprint CERN-TH-2858/LAPP-TH-19 in "Unification of the Fundamental Interactions," Proc. Europhysics Study Conference, Erice (Plenum Press, 1980).
18. A.D. Sakharov, Pis'ma Zh. Theor. Fiz. 5, 32 (1967).
19. E.W. Kolb and S. Wolfram, Phys. Lett. 91B, 217 (1980); J.N. Fry, et al., Enrico Fermi Inst. preprint 80-07(1980); S.B. Treiman and F. Wilczek, Princeton preprint (1980). (See also Ref. 17).
20. S. Coleman and E. Weinberg, Phys. Rev. D7, 1888 (1973); J. Ellis, et al., Phys. Lett. 83B, (1979).
21. A. Guth and E. Weinberg, SLAC preprint SLAC-PUB-2525 (1980); E. Witten, Harvard preprint HUTP-80/A040 (1980); M. Sher, U.C. Santa Cruz preprint UCSC 80/132 (1980).
22. D.V. Nanopoulos and D.A. Ross, Nucl. Phys. B157, 273 (1979).
23. J. Yang, et al., Ap. J. 227, 697 (1979).
24. M. Veltman, Nucl. Phys B123, 89 (1977); M.S. Chanowitz, et al., Nucl. Phys. B153, 402 (1979).
25. N.V. Kraznikov, Sov. J. Nucl. Phys. 28, 549 (1978); P.Q. Hung, Phys. Rev. Lett. 42, 873 (1979); H.D. Politzer and S. Wolfram, Phys. Lett. 82B,, 242 and 83B, 421 (1979).
26. D.V. Nanopoulos and S. Weinberg, Phys. Rev. D20, 2484 (1979); S. Barr, et al., Phys. Rev. D20, 2494 (1979).

SOFT CP VIOLATION: PRESENT STATUS*

Goran Senjanović**
Department of Physics, University of Maryland
College Park, Md. 20742

ABSTRACT

The theories in which CP nonconservation in weak interactions originates due to spontaneous symmetry breaking, are reviewed. Special classes of such models are shown to lead to natural suppression of strong interactions induced CP phase, i.e. resolution of the so-called axion problem. The high temperature behavior of such theories is investigated. It is argued that soft CP violation does remain at $T \sim 10^{15}$ GeV, as to lead to the production of matter-antimatter asymmetry in the early universe. The characteristic properties of such models, such as $|\eta_{+-}/\eta_{oo}|$ predictions and the value of the electric dipole moment of the neutron being typically $10^{-27} - 10^{-24}$ ecm are stressed.

INTRODUCTION

Almost two decades after the discovery of CP violation in neutral K meson system,[1] we still do not have an accepted theory of CP nonconservation in weak interactions.[2] The current belief (prejudice) is that gauge models of weak and electromagnetic interactions will provide such a theory. In this talk, I will describe a particular class of gauge models in which CP violation originates due to the spontaneous symmetry breaking of an originally CP conserving theory.[3]

There are basically two different ways of introducing CP nonconservation. Let us, for the purpose of discussion, write the weak Lagrangian as $L_w = L_w^{(+)} + L_w^{(-)}$, where $L_w^{(+)}$ is CP even and $L_w^{(-)}$ is CP odd. Then, by the way they treat $L_w^{(-)}$, we have the following two types of the theories of CP violation:

(i) HARD CP VIOLATION. In this case the canonical dimension of $L_w^{(-)}$ is: $d(L_w^{(-)}) \geq 4$ (in gauge theories d=4, since renormalizability forbids higher dimensions).

Examples: Yukawa interactions $h\bar{\psi}\psi\phi$ or scalar self-couplings $\lambda\phi^4$, with h, λ complex numbers. Kobayashi-Maskawa[4] extension of the standard model[5] (KM) is based on complex h.

(ii) SOFT CP VIOLATION. In these models $d(L_w^{(-)}) \leq 3$. An important feature of such theories is that if an operator O satisfies $d(O) > 3$, then CP odd piece of O will be a finite an calculable quantity in perturbation theory. This will turn out to be crucial in solving

*Invited talk given at parallel session B9 of the XXth International Conference on High Energy Physics, held at Madison, Wisconsin in July, 1980 (to appear in the Proceedings).

**Address as of September, 1980: Dept. of Physics, Brookhaven National Laboratory, Upton, L.I., N.Y. 11973.

the axion problem (section III).

Example: CP invariance broken spontaneously. Hereafter, I will concentrate on exactly such a case and by soft CP violation, I will mean models of CP violation being due to the spontaneous symmetry breaking.

II. SOFT CP VIOLATION: A MINIMAL SCHEME

The model is based on $SU(2)_L \times U(1)$ gauge group with the standard fermionic assignment and the Higgs fields being in doublet representation (it preserves all the successful features of the standard theory). Unlike in the KM case, all the couplings of the symmetric theory are assumed real, so that CP is conserved prior to the symmetry breaking.

Now, if there is a single Higgs doublet ϕ, it is well known that $<\phi>$ can be made real by terms of $SU(2)_L \times U(1)$ rotation. In this case: no CP violation (recall that the KM scheme needs complex Yukawa couplings).

Therefore, a minimum of two Higgs doublets is needed:[6] ϕ_1, ϕ_2. One can show, consistent with the minimization of the potential, that $<\phi_1> = \begin{pmatrix} 0 \\ v_1 \end{pmatrix}$, $<\phi_2> = \begin{pmatrix} 0 \\ e^{i\alpha}v_2 \end{pmatrix}$, with v_1, v_2 and α real numbers. It turns out that quark mass matrices are most general complex matrices. It follows, then, that the Cabibbo rotation is complex and can be put in the standard KM form. Therefore, as far as the gauge meson sector is concerned, this model is equivalent to KM extension of the standard theory.

Higgs sector. Enriched Higgs sector leads to a plethora of new interactions. There are 3 neutral Higgs particles $H_i^0 (i=1,2,3)$ and a charged Higgs meson $H^{\pm}$. Their interactions with fermions violate CP.

(a) Neutral Higgs. Their exchange, at the tree level, leads to the K^0-$\bar{K}^0$ mixing. One gets for the CP even and odd amplitudes, respectively

$$S_H^{(+)} \simeq G_F \frac{m_d m_s}{m_{H^0}^2}, \quad S_H^{(-)} \simeq S_H^{(+)} \sin\delta \tag{2.1}$$

where δ is the CP phase in Cabibbo rotation. Using $\sin\delta = 1/10$, requirement $S_H^{(-)} \lesssim 10^{-9} G_F$ leads to the constraing $m_{H0} \gtrsim (100\text{-}300)$GeV, in which case $S_H^{(+)} \simeq 10^{-8}\ G_F$. Therefore, we expect $S_H^{(+)} << S_W^{(+)}$ and $S_H^{(-)} \simeq S_W^{(-)}$, where S_W is the usual GIM induced amplitude through the gauge currents.

Due to the large mass of H_i^0, direct $\Delta S=1$ $K\to 2\pi$ decays will obviously be strongly suppressed ($\sim 10^{-8} - 10^{-9}\ G_F$).

(b) Charged Higgs. From the bound on its mass: $m_{H^+} \gtrsim 2$ GeV (from τ decay), we can conclude that for the one-loop induced $\Delta S=2$ amplitude which involves charges Higgs: $S^{(+)} \lesssim 10^{-9}\ G_F$ and $S^{(-)} \lesssim 10^{-10}\ G_F$, which is negligible.

In direct $K\to 2\pi$ decays the situation is somewhat different. It is easy to estimate $S_H^{(-)}(\Delta S=1) = G_F \frac{m_d m_s}{m_{H^+}^2} \sin\delta \lesssim 10^{-5}\ G_F$. Therefore, if H^+ is light ($\sim$ few GeV) $\Delta S=1$ amplitudes could in principle compete with the KM contribution, and somewhat change the usual estimates for $\eta_{ij} \equiv \frac{S(K_L \to \pi_i \pi_j)}{S(K_S \to \pi_i \pi_j)}$ (experimentally[7] $|\eta_{+-}/\eta_{oo}| \simeq 1.05 \pm 0.46$). One should add, that since m_{H^+} and m_{H^0} originate from basically the same couplings in the Lagrangian, it is more likely that H^+ is heavy ($m_{H^+} \gtrsim 10$-100 GeV), in which case KM contribution dominates.

The electric dipole moment of the neutron (d_n^e). The value of d_n^e is, as we shall see, the most characteristic feature of this model. The main contribution comes from the charged Higgs exchange (Fig. 1). One gets

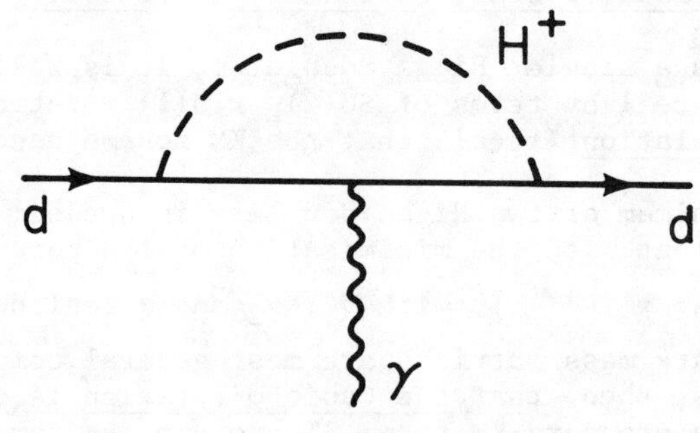

Fig. 1. Charged Higgs contribution to d_n^e

$$d_n^e = e\, G_F\, m_d \sin\delta\, \xi^2 \begin{Bmatrix} 1 & : & m_t > m_{H^+} \\ \left(\frac{m_t}{m_{H^+}}\right)^2 & : & m_{H^+} > m_t \end{Bmatrix} \quad (2.2)$$

where ξ is the mixing between d and t quarks. Using $\sin\delta = 1/10$, $m_d = 10^{-2}$ GeV and taking $\xi = 1/10$, results in

$$d_n^e = 10^{-24} \begin{Bmatrix} 1: m_t > m_{H^+} \\ (m_t/m_{H^+})^2: m_{H^+} > m_t \end{Bmatrix} \text{ecm} \quad (2.3)$$

Now, unitarity constraints (i.e. validity of perturbation theory) lead to upper bounds on Higgs bosons mass: $m_{H^+} \lesssim 1000$ GeV. That in turn restricts d_n^e to be in the range

$$10^{-27}\ \text{ecm} \lesssim d_n^e \lesssim 10^{-24}\ \text{ecm}: \quad (2.4)$$

This is an important feature of the model and serves as the distinction from KM and superweak theories[8] (experimental[9] upper bound on d_n^e is $(.4 \pm 1.5) \times 10^{-24}$ ecm).

In summary, main features of the model are:

- o $\eta_{ij} \simeq (\eta_{ij})_{KM}$
- o $d_n^e \simeq (10^{-27} - 10^{-24})$ ecm
- o existence of $\Delta S=2$ transitions at the tree level, which lead to

the limit: $m_{H^0} \gtrsim 300$ GeV.

It should be added that $\Delta S=2$ tree level transitions can be forbidden[10] by coupling ϕ_1 only to up quarks and ϕ_2 only to down quarks. It turns out that CP nonconservation then requires the existence of at least three Higgs doublets.[11] We only list the predictions of such models

(i) Cabibbo rotation is real, independently of the number of flavors, so that CP violation is possible only through the Higgs interactions.[11]

(ii) $|\eta_{+-}/\eta_{oo}| \simeq 1.06$[12]

(iii) $d_n^e \simeq 10^{-25}$ ecm[13]

III. NATURAL SUPPRESSION OF STRONG P ANT T NONINVARIANCE

It has been shown,[14] that nonperturbative effects in QCD induce effective, P and T violating interaction of the form

$$L_{eff} \propto e^{i\theta} \det |\bar{Q}_L Q_R| + h.c. \tag{3.1}$$

where Q denotes up and down quark flavors. When weak and electromagnetic interactions are switched on, they induce further contribution to θ, through the complex rotations needed to diagonalize quark mass matrices. One obtains the effective $\bar{\Theta}$ (M is the quark mass matrix)

$$\bar{\Theta} = \Theta + i\, \ell n \frac{\det M}{\det M^+} \tag{3.2}$$

The analysis[15] of the $\bar{\Theta}$ contribution to the electric dipole moment of the neutron leads to the constraint

$$\bar{\Theta} \lesssim 10^{-8} \tag{3.3}$$

It is a challenge, then, to understand why $\bar{\Theta}$ has to be so small (setting $\bar{\Theta}$ tree = 0 does not work in general, since higher orders in perturbation theory induce infinite Θ[16]). That is the essence of so-called strong CP problem.

There were several resolutions suggested:

(i) Existence of extra axial U(1) symmetry[17] sets $\bar{\Theta}=0$. However, $U_A(1)$ gets broken, which leads to the existence of light pseudoscalar meson ($m_a \simeq 10$-100 keV), the axion.[18] That was ruled out experimentally, at least in the context of the standard theory.

(ii) The idea that $m_u=0$ is in serious disagreement with current algebra and is therefore, a rather unappealing possibility.

(iii) Use soft CP violation to have $\bar{\Theta}$ finite and small in perturbation theory.[19] I will describe this approach in the context of left-right symmetric gauge theories,[20] which achieve it naturally The strategy is very simple:

(1) Left-right symmetry dictates: $\Theta^{tree} = 0$

(2) By use of symmetries, construct manifest left-right symmetric[21] model, characterized by $J_{\mu L} = J_{\mu R}$ (charged weak

currents). That is achieved by having hermitean quark mass matrices: $M = M^{+}$. Consequence: $\bar{\Theta}^{tree} = 0$.

(3) Higher orders, due to the lack of counterterms of the symmetric theory, will be finite. The analysis shows: $\bar{\Theta}^{1\text{-loop}}=0$; $\bar{\Theta}^{2\text{-loop}} \lesssim 10^{-10}$, clearly in agreement with experiment.

The predictions of the model, as far as the weak CP violation is concerned, are the same as in the minimal scheme.

IV. SOFT CP VIOLATION AT HIGH TEMPERATURE AND BARYON PRODUCTION IN THE EARLY UNIVERSE

The small observed excess of matter over antimatter in the universe has been an outstanding puzzle for generations of physicists. Recently, progress[22] has been made by attributing the baryon number of the observable universe $n_B/n_\gamma = 10^{-10}$–10^{-8} to a product of baryon and CP violating decays of superheavy bosons of grand unified theories during the early stages of universe.

Quantitatively, the baryon number can be derived to be[20]

$$\frac{n_B}{n_\gamma} = \frac{N_X}{N}(r-\bar{r}) \qquad (4.1)$$

where N_X is the number of superheavy particle species, N is the number of light particle species and $r(\bar{r})$ are the branching ratios in the decays of superheavy bosons to produce the excess of baryons (antibaryons). Now, CP invariance implies $r=\bar{r}$, so in order to have a nonvanishing baryon number we clearly need CP to be broken at $T \sim m_X \simeq 10^{14}$ GeV, when n_B was created.

If CP noninvariance is of the hard origin, i.e. through the complex couplings of the basic Lagrangian, then, of course, it will remain at all temperatures. However, in the case of soft CP violation the situation is more complicated and it depends on the high temperature behaviour of spontaneously broken gauge theories. By the analogy with the ferromagnetic systems, one would expect symmetry restoration at high T. This conjecture has been shown[23] to be true in the simplest Higgs models and it has been argued that the symmetry restoration occurs at $T \sim m_W/g \sim 300$ GeV. Now, if this was true in general, it would imply that soft CP violation is to be ruled out or we would have to give up our understanding of matter-antimatter asymmetry of the universe.

It was shown, however, that the symmetry does not have to be restored at high temperature.[24] A simple example is provided by an SU(2) x U(1) gauge theory with two Higgs doublets. One can show, that there is solution which ensures that for $T > T_c \simeq 300$ GeV $\langle\phi_1\rangle = 0$, but $\langle\phi_2\rangle \neq 0$. That means that SU(2) x U(1) symmetry is broken down to $U_{em}(1)$ at $T \gg 300$ GeV.

A realistic grand unified theory in which CP is broken at $T \sim m_x$ is then easy to construct.[24] It is based on SU(5) gauge group and requires the existence of three $\underline{5}$ dimensional Higgs multiplets ϕ_1, ϕ_2 and ϕ_3. At high T, the pattern of symmetry breaking is the following one: $\langle\phi_3\rangle = 0$, $\langle\phi_{1,2}\rangle \propto T$ with $\langle\phi_2\rangle$ being a complex number. The analysis[25] of the induced baryon number gives

$$\frac{n_B}{n_\gamma} \simeq 10^{-2}\ h^2(T_X) \tag{4.2}$$

where $h(T_X)$ is the value of Yukawa couplings at $T \sim m_X$. Agreement with the observation is achieved for $h(T_X) \simeq 10^{-4} - 10^{-3}$.

I should mention that in the minimal SU(5) scheme, $n_B/n_\gamma \lesssim 10^{-18}$ which tends to suggest[26] that we require more than one Higgs 5. That provides a rationale for the ideas presented in this talk.

Finally, it is easy to show that the low temperature predictions of the above model are the same as in the minimal scheme.

I would like to end with an amusing comment. The suggestive eventual restoration of symmetry at $T > m_X$, predicts the existence of the domains with positive and negative CP phase, i.e. the domains with matter and antimatter.[27] Therefore, either universe is globally baryon symmetric with random domains of matter and antimatter or soft CP models are not a viable alternative in understanding the origin of CP nonconservation in nature.

I wish to thank Rabi Mohapatra for discussions.

REFERENCES

1. J. H. Christensen, J. Cronin, V. L. Fitch and R. Turlay, Phys. Rev. Letters 13, 138 (1964).
2. For a recent review and references to various models of CP violation, see R. N. Mohapatra, Proceedings of the XIX International Conference on High Energy Physics, Tokyo (1978), ed. S. Homma et al., Physical Society of Japan, 1979.
3. T. D. Lee, Phys. Rev. D8, 1226 (1973); Phys. Reports C9, 148 (1974); R.N. Mohapatra & J.C. Pati, Phys. Rev. D11, 566 (1975).
4. M. Kobayashi and T. Maskawa, Progr. Theor. Physics 49, 642 (1973).
5. S. L. Glashow, Nuc. Phys. 22, 579 (1961); S. Weinberg, Phys. Rev. Lett. 19, 1264 (1967); A. Salam, in Elementary Particle Theory, ed. N. Svartholm (Wiley, N.Y. (1969)).
6. T. D. Lee, ref. 2; P. Sikivie, Phys. Lett. 65B, 141 (1976).
7. For a review and references on the experiments on CP violation, see E. Paul: Springer Tracts in Modern Physics 79, 53 (1976). See also, K. Kleinknect: Ann. Rev. Nucl. Sci. 26, 1 (1976).
8. L. Wolfenstein, Phys. Rev. Lett. 13, 562 (1964).
9. N. F. Ramsey, W. Dress, P. Miller, P. Perrin and J. Pendelbury, Phys. Rev. D15, 9 (1977); V. Lobashev et al., Leningrad Institute of Nuclear Physics preprint (1978).
10. S. L. Glashow and S. Weinberg, Phys. Rev. D15, 1958 (1977).
11. S. Weinberg, Phys. Rev. Lett. 37, 657 (1976).
12. A. A. Ansel'm and N. G. Uraltsev, Sov. J. Nucl. Phys. 30, 240 (1979); G. C. Branco, Phys. Rev. Lett. 44, 504 (1980) and Carnegie-Mellon preprint COO-3066-150 (1980); K. Shizuya and S.-H. H. Tye, Cornell preprint CLNS80/458 (1980).
13. S. Weinberg, ref. 11; N. G. Deshpande and E. Ma, Phys. Rev.

13. D16, 1583 (1977); A. A. Ansel'm and D. I. D' Yakonov, Nucl. Phys. B145, 271 (1978); A. B. Lahanas and N. J. Papadopoulos, Lett. Nuovo Cimento 18, 123 (1977); A. B. Lahanas and C. E. Vayonakis, Phys. Rev. D19, 2158 (1979).
14. G. 't Hooft, Phys. Rev. D14, 3432 (1976); C. Callan, R. Dashen and D. Gross, Phys. Lett. 63B, 334 (1976); R. Jackiw and C. Rebbi, Phys. Rev. Lett. 37, 172 (1976).
15. V. Baluni, Phys. Rev. D19, 2227 (1979); R. Crewther, P. Di Vecchia, G. Veneziano and E. Witten, Phys. Lett. 88B, 123 (1979).
16. For an example of infinities in higher orders in perturbation theory, see J. Ellis and M. K. Gaillard, Fermilab preprint (1978).
17. R. Peccei and H. Quinn, Phys. Rev. D16, 1791 (1977).
18. S. Weinberg, Phys. Rev. Lett. 40, 223 (1978); F. Wilczek, ibid. 40, 279 (1978).
19. M. A. B. Beg and H. S. Tsao, Phys. Rev. Lett. 41, 278 (1978); R. N. Mohapatra and G. Senjanović, Phys. Lett. 79B, 283 (1978); H. Georgi, Hadr. J. 1, 155 (1978). See also, G. Segrè and H. A. Weldon, Phys. Rev. Lett. 42, 1191 (1979); S. M. Barr and P. Langacker, Phys. Rev. Lett. 42, 1654 (1979). For alternative attempts in the context of technicolor gauge theories, see E. Eichten, K. Lane and J. Preskill, Phys. Rev. Lett. 45, 225 (1980) and references therein.
20. J. C. Pati and A. Salam, Phys. Rev. D19, 275 (1979); R. N. Mohapatra and J. C. Pati, Phys. Rev. D11, 566; 2588 (1975); G. Senjanović and R. N. Mohapatra, Phys. Rev. D12, 1502 (1975).
21. M. A. B. Beg, R. V. Budny, R. N. Mohapatra and A. Sirlin, Phys. Ref. Lett. 38 1252 (1977).
22. M. Yoshimura, Phys. Rev. Lett. 41, 381 (1978); A. Yu. Ignatiev, N. Y. Krasnikov, V. A. Kuzmin and A. N. Tavkhelidze, Phys. Lett. B76, 436 (1978); S. Dimopoulos and L. Susskind, Phys. Rev. D18, 4300 (1978); D. Touissant, S. Treiman, F. Wilczek and A. Zee, Phys. Rev. D19, 1036 (1979); S. Weinberg, Phys. Rev. Lett. 42, 850 (1979); J. Ellis, M. Gaillard and D. V. Nanopoulos, Phys. Lett. 80B, 360 (1978), E. W. Kalb and S. Wolfram, Phys. Lett. 91B, 217(1980) and Caltech preprint (1979).
23. D. A. Kirzhnitz and A. D. Linde, Phys. Lett. 42B, 471 (1972); S. Weinberg, Phys. Rev. D9, 3537 (1974), L. Dolan and R. Jackiw, Phys. Rev. D9 2904 (1974); C. Bernard, Phys. Rev. D9, 3312(1979).
24. R. N. Mohapatra and G. Senjanović, Phys. Rev. Lett. 42, 1651 (1979); Phys. Rev. D20, 3390 (1979) and also Phys. Lett 89B, 57 (1979).
25. R. N. Mohapatra and G. Senjanović, Phys. Rev. D21, 3470 (1980).
26. S. M. Barr, G. Segrè and H. A. Weldon, Phys. Rev. D20, 2494 (1979); D. V. Nanopoulos and S. Weinberg, Phys. Rev. D20, 2484 (1979); A. Yildiz and P. Cox, Phys. Rev. D21, 906 (1980).
27. R. W. Brown and F. Stecker, Phys. Rev. Lett. 43, 315 (1979); G. Senjanović and F. Stecker, Univ. of Maryland preprint (1980). This is true even if superheavy Higgs sector is responsible for CP breaking at $T \simeq M_x$, as in J. A. Harvey, P. Ramond and D. B. Reiss, Caltech preprint (1980).

A QUANTUM STRUCTUREDYNAMIC MODEL OF QUARKS, LEPTONS, WEAK VECTOR BOSONS, AND HIGGS MESONS*

O. W. Greenberg and Joseph Sucher
Center for Theoretical Physics
Department of Physics and Astronomy
University of Maryland
College Park, Maryland 20742

ABSTRACT

We propose a model in which quarks, leptons, and weak vector bosons are two-body composites and Higgs mesons are two- or many-body composites confined by an SU(N) local gauge "quantum structure-dynamics" interaction, quark-lepton parallelism holds, and the strong and weak interactions are on a parallel footing -- both are residual effects of a flavor-independent local gauge interaction.

Composite structure of quarks, leptons, and other observed or hypothesized particles may be necessary to understand quark-lepton parallelism and repetition of generations. Our model[1] differs from most composite models by having a confining $SU(N)_S$ "quantum structuredynamics" (QSD) local gauge interaction which binds the constituents. If the QCD and QSD couplings coincide at a very small distance, we expect that for some N>3 QSD will form composites of size $r_0 <<$ typical hadronic sizes. The fundamental fields are a spin- 1/2 flavor doublet $F = (F_u, F_d)$, a spin-0 color antitriplet C, and a spin-0 "leptoscalar" S, with electric charges (2/3,-1/3), 0, and 2/3, respectively. (A more general possible charge pattern is (q+1,q), q+ 1/3, and q+1, respectively.) We assume F, C, and S transform as the fundamental representation N of $SU(N)_S$. The Lagrangian for QSD is the sum of the $SU(N)_S$ and $SU(3)_c$ gauge Lagrangians, the Lagrangian for the massless spinors (F_u, F_d), and the Lagrangians for the massive scalars C and S, with gauge couplings appropriate to their $SU(N)_S$ and $SU(3)_c$ representations as given above. We assign the first generation of quarks u,d, and leptons ν_e,e to the lowest J= 1/2 states of $(F_u\bar{C})$, $(F_d\bar{C})$ and $(F_u\bar{S})$, $(F_d\bar{S})$, respectively, and higher generations to J= 1/2 excited states of these systems. Confinement implies infintely many generations. In addition to the local $SU(3)_c \times SU(N)_S$ symmetry, L has global $G = SU(2)_L \times SU(2)_R \times U(1)_L \times U(1)_R$ flavor symmetry which gives a chiral classification of the $\bar{F}F$ bound states. We identify the scalar composites $\psi = \bar{F}_L F_R$, and vector composites $W^i_{L\mu} = \bar{F}_L \gamma_\mu \tau^i F_L$ and $W^i_{R\mu} = \bar{F}_R \gamma_\mu \tau^i F_R$, i = 0,2,3,4, as Higgs bosons and massless gauge bosons (with $V \mp A$ couplings) of the effective low-energy theory, respectively. There will also be N-body scalar composites which we identify as additional Higgs bosons. According to the conjecture[2] that when the size r_0 of the composites is much

*Talk delivered by O. W. Greenberg

less than their Compton wavelength, the effective low-energy Lagrangian for the composite fields L_{eff} is renormalizable and local, L_{eff} is a spontaneously broken, G gauge-invariant Lagrangian with left-right symmetry, whose reduction to the standard $SU(2)_L \times U(1)$ theory has been discussed earlier. All the two-body $SU(N)_S$ singlets in our model, except $\bar{C}C$, $\bar{S}S$, $\bar{S}C$, and $\bar{C}S$, which we assume lie high in mass, correspond to quarks, leptons, weak vector bosons, or Higgs mesons. The photon and/or the $SU(3)_c$ gluons can be taken either to be elementary or composite. If the photon is elementary, electromagnetism can break global flavor symmetry in the usual way.

A major feature of our type of model is that weak interactions are residual effects of flavor-independent QSD, just as strong interactions are residual effects of QCD, with analogous constituent line diagrams in each case. This parallelism supplements the quark-lepton parallel counting in each generation.

A major problem in our model is the mass spectrum, which includes the following issues: why are quarks, leptons, weak vector bosons, etc., very light compared to r_0^{-1} or to grand unified mass scales, why are mass spacings between quarks and leptons much smaller than r_0^{-1}; why are quarks and leptons much less massive than weak vector bosons; why do quark and lepton spacings increase rapidly with generation; why do quarks and leptons have spin 1/2; why do neutrinos have small masses; why are electromagnetic mass splittings much less than e^2/r_0?

The constraint of g-2 requires a size $\lesssim (5\ \text{TeV})^{-1}$; the constraint of $\mu \not\to e\gamma$ requires a size $\lesssim (10^8\ \text{TeV})^{-1}$ or $(10^2\ \text{TeV})^{-1}$, if we estimate the matrix element as $em_\mu r_0$ or $e(m_\mu r_0)^2$, respectively.

With some additional assumptions, the Cabibbo structure and GIM mechanism follow from the global $SU(2)_L$ algebra associated with the phenomenological flavor currents constructed from the F's.

Naive potential models with $V \propto r^\alpha$ can have many radial excitations below the first orbital excitation for $\alpha \to -2$. If the QSD confining gauge flux forms a rigid string or "rod," then longitudinal excitations lie below orbital excitations if $\pi MLv \ll 25\hbar$, where M, L, and v are the mass, length, and velocity of longitudinal sound waves associated with the rod, respectively.

We are greatly indebted to Jogesh Pati for detailed discussions, as well as to Haim Harari, Tony Kennedy, and Shmuel Nussinov. This work was supported in part by the National Science Foundation.

REFERENCES

1. O. W. Greenberg and J. Sucher, University of Maryland Physics Publication No. 81-026, 1980. This article contains more relevant references than space allows to give here.
2. J. C. Pati, Bhubaneswar and Trieste preprint, 1980, and private communication; M. Veltman, quoted in J. Ellis, et al., LAPP preprint LAPP-TH-15, 1980.

A SOLUTION TO THE PROBLEM OF THE FERMION MASSES *)

D.V. Nanopoulos
CERN, Geneva, Switzerland

A B S T R A C T

A new mechanism for giving masses to the first generation (u,d, e,ν_e) through radiative corrections at the two-loop level is presented.

In this nano-comment I would like to suggest a new mechanism [1] for generating masses for the first generation (u,d,e,ν_e) through radiative corrections at the two-loop level.

Let me briefly recall that in the framework of grand unified theories the following "mysteries" of the fermion mass spectrum have been explained [2] :

1) Quark-lepton mass differences are due [3] to the simple fact that quarks have strong interactions while the leptons have not.

2) The lightness of the neutrinos (or the neutrino-charged lepton mass differences) is due [4] to the fact that neutrinos, being chargeless may get Majorana masses. That is, m_{ν_R} may be superlarge ($\sim$ M, the superunification mass scale), while $m_{\nu_L} \sim m^2/M$ (m the "low energy" scale).

3) We assume that the third generation (t,b,τ,ν_τ) may get a "direct" mass at the tree-level, while the first and second generations remain massless at that level because of some "forbidden" symmetry. The underlying philosophy is explained in full detail in Ref. 2. Then the second generation (c,s,μ,ν_μ) gets "naturally" masses of order $\sim \alpha M_W$ through the one-loop diagrams [5]

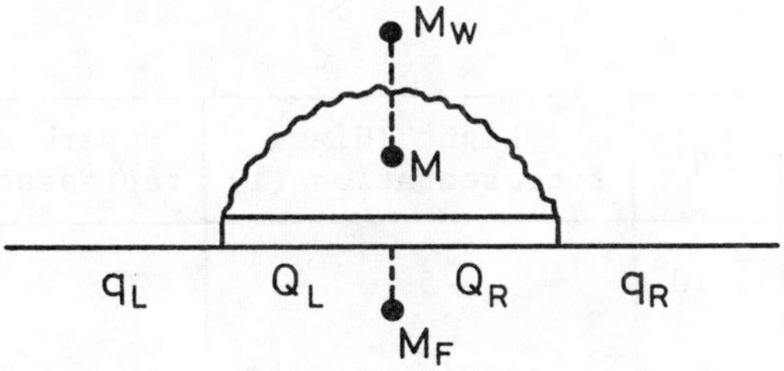

(every body in the loop is superheavy) which give

$$\delta m \propto \left(\frac{\alpha}{\pi}\right) \cdot \frac{M_W}{M} \cdot M_F \simeq \left(\frac{\alpha}{\pi}\right) M_W \qquad (1)$$

More detailed calculations [5, 6] show that there is a tendency that $m_c > m_s \simeq m_\mu$ and that the first generation remains massless even at the one-loop level. But then who gives masses to the first generation ?

*) Talk presented at the XXth International Conference on High Energy Physics, Madison, 17-23 July 1980.

Here comes our new mechanism [1] to fill the last gap by giving masses to the first generation.

4) The first generation (u,d,e,ν_e) "naturally" gets masses of order $\sim \alpha^2 M_W$ through the two-loop diagrams

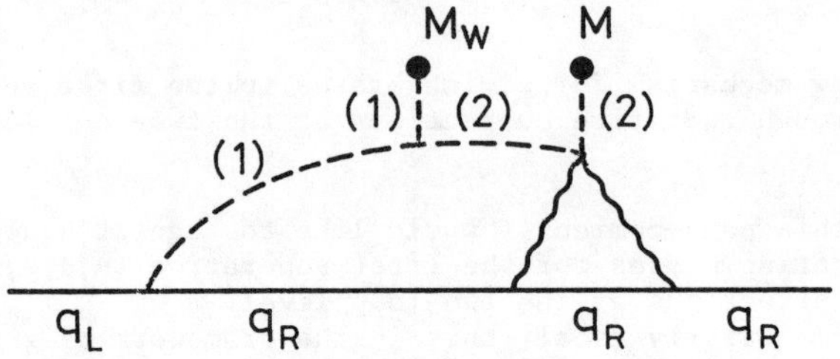

(the wavy lines represent as usual gauge boson)

which give [1]

$$\delta m \propto \left(\frac{\alpha}{\pi}\right)^2 M_W , \qquad (2)$$

our new and highly desirable result.

The two-loop diagram [7] has the following attractive features:

i) It is of the correct order to magnitude ($\sim$ few MeV) to be interpreted as a "mass" term for the first generation. The other two generations, being much heavier, are insensitive to such "miniature" mass terms.

ii) It is present in all leading grand unified models with the following identification:

	q	"Light" Higgs representation (1)	"Superheavy" Higgs representation (2)
SU(5)	$\bar{5}$,10	5	24
O(10)	16	10	45
E_6	27	27	351

iii) The "effective" Higgs fields which couple to the first generation, i.e., the ones contained in the product (1) $\otimes$ (2), will in general give different contributions to u,d,e and so provide the "highly" desirable d-e and u-d mass splitting. The fact that within our framework m_d/m_e and m_u/m_d are "calculable", is of great importance. Hopefully they will turn out to have the correct magnitude [1].

We are definitely not far from a satisfactory understanding of the fermion mass spectrum. The strategy is clear. All that remains is to fill in the details.

REFERENCES

1. R. Barbieri and D.V. Nanopoulos, CERN preprint, in preparation.
2. For a recent review see:
   D.V. Nanopoulos, CERN preprint TH.2866 (1980).
3. A.J. Buras, J. Ellis, M.K. Gaillard and D.V. Nanopoulos, Nucl. Phys. B135, 66 (1978);
   D.V. Nanopoulos and D.A. Ross, Nucl. Phys. B157, 273 (1979).
4. H. Georgi and D.V. Nanopoulos, Nucl. Phys. B155, 52 (1979) and unpublished;
   M. Gell-Mann, P. Ramond and R. Slansky, in "Supergravity", Proc. of the Supergravity Workshop at Stony Brook, Ed. by P. Van Nieuwenhuizen and D.Z. Freedman (North Holland, Amsterdam, 1979), p. 315;
   R. Barbieri, D.V. Nanopoulos, G. Morchio and F. Strocchi, Phys. Letters 90B, 91 (1980).
5. R. Barbieri and D.V. Nanopoulos, Phys. Letters 91B, 369 (1980) and CERN preprint TH.2870 (1980), Phys. Letters B in press.
6. R. Barbieri, A. Masiero and D.V. Nanopoulos, CERN preprint, in preparation.
7. Similar two-loop diagrams have been used by:
   E. Witten, Phys. Letters B91, 81 (1980), to give superheavy masses to ν_R.

Chapter 5

e^+e^- Physics and Electromagnetic Properties, Experiment and Theory

QED AND ELECTROMAGNETIC PROPERTIES OF PARTICLES

T. DEVLIN, RUTGERS, ORGANIZER

RADIATIVE WIDTHS OF K* AND ρ MESONS

D.Berg, C.Chandlee, S.Cihangir, T.Ferbel, J.Huston, T.Jensen, F.Lobkowicz, C.A.Nelson jr.,[a] T.Ohshima,[b] P.Slattery, M.McLaughlin, P.A.Thompson[c]
University of Rochester, Rochester, N.Y. 14627

J.Biel, T.Droege, A.Jonckheere, P.F.Koehler
Fermi National Accelerator Laboratory, Batavia, IL. 60510

B.Collick, S.Hepplemann, T.Joyce, Y.Makdisi,[c] M.Marshak, E.Peterson, K.Ruddick, M.Shupe, T.Walsh
University of Minnesota, Minneapolis, MN. 55455

ABSTRACT

The dissociation of K and π mesons in the nuclear Coulomb field bas been studied. Results are: $\Gamma(\rho \to \pi\gamma)=67\pm7$ KeV., $\Gamma(K^*(890) \to K\pi)=60\pm15$ KeV; and the observation of Coulomb excitation of the following states: $K^*(1420,1700)$, A_2, A_1, B. The measured widths are in rough agreement with a simple quark model.

Accurate measurements of the radiative decay widths of mesons provide a particularly simple test of quark models. In a simple model, the transition rate is proportional to magnetic moments of the quarks. Because of competition from other decay modes, the inverse process[1] ($M\gamma \to M^*$) is used in these experiments.

Data have been taken at FNAL with mixed K and π beams of 156, 200, and 250 GeV/c. For studying the excited kaon states, 2 m of beryllium were placed in the beam to filter out protons. With this arrangement, the kaon fraction of the beam was 15%. The experimental apparatus consisted of a target (typically Pb), a charged particle spectrometer (drift chambers and a 37 Kg-m magnet), and a gamma ray calorimeter (1.4 x 0.7 m^2 liquid argon detector). An important source of background is the strong interaction contribution. At the beam momenta used this becomes significant at $t \sim 10^{-3}$ GeV^2, whereas the coulomb part is at $t < 10^{-4}$,and the most recent data have a resolution of 1.5 x 10^{-4}. Thus the strong interference amounts to a few percent of the coulomb cross section and the uncertainties are small compared to the statistical ones.

Figure 1 shows a preliminary mass plot for the excited states of the K^+. This plot represents ∿15% of the data. Equivalent data were also taken for other decay modes.(particularly $\pi^+\pi^-\pi^+$)

Table 1 summarizes the results of this and other experiments together with calculated values from a quark model[2]. We expect to have useable statistics for several high mass states, whose dynamics are significantly different and thus provide additional information. The third column in the table lists the results of a non-relatavistic,SU3 model. In the next column the quark magnetic moments are changed to those suggested by Teese et al.[3] Considering the simplicity of the model, the agreement is good except for the K^{*0}. It is difficult to see how this discrepancy can be removed by straightforward changes in the model.(eg. relatavistic phase space,non-ideal mixing angles).

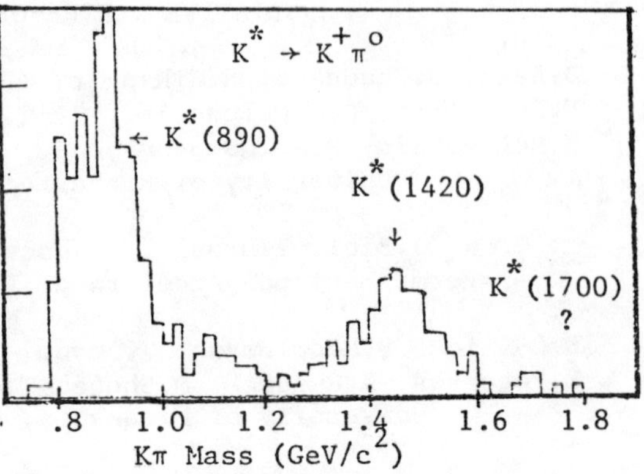

Fig. 1 Kπ mass spectrum.

TABLE I: Summary of Radiative Widths

Meson	M^2	Γ(SU3)	Γ(u,d,s)†	Γ(exp)
ρ-	$(\mu_d+\mu_u)^2$	84	76	67± 7 KeV[4]
ω^0	$(\mu_u-\mu_d)^2$	783	798	789±92 KeV[5]
$K^{*\pm}(890)$	$(\mu_s+\mu_u)^2$	47	83	60±15 Kev[6]
$K^{*0}(890)$	$(\mu_s+\mu_d)^2$	188	137	75±35 KeV[7]

†SU3 symmetry except following values used for magnetic moments:[3] μ_u=1.852, μ_d=-0.971, μ_s=-0.614.

Preliminary data: $K^{*+}(1420)$ ∿ 1000 events, Γ ∿ 600 KeV, $K^{*+}(1700)$ ∿ 100 events, A_2,B ∿500 events, A_1 ∿ 10^6 events.

REFERENCES

a)Presently at Fermi National Accelerator Laboratory,Batavia, IL. 60510
b)Presently at the Institute for Nuclear Studies, University of Tokyo, Tanashi-City, Tokyo 188, Japan
c)Presently at Brookhaven National Laboratory, Upton, NY. 11973
1. H.Primakoff, Phys. Rev. 81, 899 (1951).
2. C.Becchi and G. Morpurgo, Phys. Rev. 140B, 687 (1965).
3. R.B.Teese and R.Settles, Phys. Lett. 87B, 111 (1979).
L.Schachinger et al., Phys. Rev. Lett. 41, 1348 (1980).
4. D.Berg et al., Phys. Rev. Lett. 44, 706 (1980).
5. T.Ohshima, University of Rochester Report UR-733 (1980), a revision fo the particle group value.
6. D.Berg et al., University of Rochester Report UR-751 (1980).
7. W.C.Carithers et al., Phys. Rev. Lett. 35,349 (1975).
Research supported by the U.S. Department of Energy, the National Science Foundation and funds from the Universities of Minnesota and Rochester.

MEASUREMENT OF THE Ξ° AND Ξ^- MAGNETIC MOMENTS

R. Handler, R. Grobel, R. March, L. Pondrom,
M. Sheaff, C. Wilkinson, B. Lundberg
University of Wisconsin, Madison, Wisconsin 53706

P.T. Cox, J. Dworkin, O.E. Overseth, C. Dukes
University of Michigan, Ann Arbor, Michigan 48109

L. Deck, T. Devlin, B. Luk, G. Ramieka, P. Skubic
Rutgers University, Piscataway, New Jersey 08902

G. Bunce
Brookhaven National Laboratory, Upton, L.I. New York 11973

K. Heller
University of Minnesota, Minneapolis, Minnesota 55455

Experiments were performed in the M2 line of FNAL to determine if Ξ° and Ξ^- particles inclusively produced by 400 GeV protons off a Be target were polarized in the parity allowed direction for production angles greater than four millirads. If so a precise measurement of their magnetic dipole moments was possible.

The decay chains studied were $\Xi^\circ \rightarrow \pi^\circ + \Lambda$, $\Lambda \rightarrow p + \pi^-$, $\pi^\circ \rightarrow \gamma + \gamma$ and $\Xi^- \rightarrow \pi^- + \Lambda$, $\Lambda \rightarrow p + \pi^-$. Multi wire proportional chambers in conjunction with a spectrometer magnet were used to measure final state charged particle momenta. A lead glass array gave γ ray energy and position coordinates. The polarization of the parent particles was found by measuring the difference in the decay asymmetry of the daughter Λ for cascades produced at opposite production angles. Equipment biases cancelled out of this difference and the accuracy of the measurement was limited only by statistics.

The observed magnitude of the polarization of the cascades was similar to that previously measured for lambdas.[1,2] The values found for the magnetic moments, expressed in proton nuclear magnetons were

$$\mu_{\Xi^\circ} = -1.253 \pm .014$$

$$\mu_{\Xi^-} = -0.75 \pm .06$$

---

[1] K. Heller, et al., Phys. Rev. Lett. 41, 607 (1978).
[2] L. Schachinger, et al., Phys. Rev. Lett. 41, 1348 (1978).

MAGNETIC MOMENTS OF QUARKS IN BARYONS AND MESONS

Jonathan L. Rosner
School of Physics and Astronomy, University of Minnesota
Minneapolis, Minnesota 55455

ABSTRACT

Recent measurements of hyperon magnetic moments and of magnetic dipole transitions in meson radiative decays are discussed in the context of the nonrelativistic quark model. It is shown that this model works surprisingly well. Various "fine-tuning effects" are discussed, including configuration mixing in the baryons, quark anomalous moments, and SU(3) breaking.

INTRODUCTION

We shall compare some new measurements of hyperon magnetic moments[1-4] and meson radiative decays[5,6] with quark model predictions. We find that:

1. A naive quark model[7] works surprisingly well[8] if SU(3) breaking is introduced through an enhanced strange quark mass (and a correspondingly reduced magnetic moment).[1,8-10]
2. The data are consistent with small isoscalar anomalous magnetic moments of quarks,[11,12] corresponding to a few percent reduction in μ_u and increases in $|\mu_d|$, $|\mu_s|$ from their naive values.
3. Small "configuration mixing" effects[12-15] may alter the symmetry of the ground state baryon wave functions.
4. The present value of the Σ^+ magnetic moment,[4] $\mu_{\Sigma^+}=2.30\pm0.14$ n.m.,[4] is too low for most models, even with the fine-tuning mentioned above.
5. The measured rate $\Gamma(K^{*0}\to K^0\gamma) = 75 \pm 35$ keV[16] is too low to be accommodated by most models.

The magnetic moment of a quark of flavor i may be expressed in terms of its mass m_i and charge Q_i by $\mu_i=|e|Q_i/2m_i$. This moment enters into both the magnetic moment of a baryon and the rate for a magnetic dipole transition in a meson.

We can construct a crude estimate of quark masses by taking into account hyperfine splittings in hadrons[10] and writing the hadron mass M as

$$M = \sum m_i + \left\{ \begin{matrix} a \\ 2b \end{matrix} \right\} \sum_{i<j} \frac{\vec{\sigma}_i \cdot \vec{\sigma}_j}{m_i\, m_j} \quad ; \qquad (1)$$

where a,b are constants appropriate for $\frac{1}{2}^+$, $\frac{3}{2}^+$ baryons or 0^-, 1^- mesons, respectively.

BARYONS

We fit four $1/2^+$ baryons (N, Λ, Σ, Ξ) and four $3/2^+$ baryons (Δ, Σ^*, Ξ^*, Ω) with three parameters in Eq. (1):

Presented at parallel session C-8, 20th Int. Conf. on High Energy Physics, Madison, Wisc., July 17-23, 1980.

$m_u = m_d = 363$ MeV; $m_s = 538$ MeV; $a/m_u^2 = 50$ MeV. (2)

If all the quarks in a baryon are in a relative S-wave, one can show that

$$p = \frac{4}{3} u - \frac{1}{3} d \text{ (similar expressions for n, } \Sigma^{\pm}, \Xi) \quad (3)$$

$$\Lambda = s \quad (4)$$

$$\Sigma \to \Lambda = (d-u)/\sqrt{3} \quad . \quad (5)$$

where we shall use the symbol A for μ_A. The masses (2), when combined with quark charges and formulae (3) - (5), then lead to the "naive model" predictions of Table 1. The agreement is not bad for a zero-parameter fit. Cohen and Lipkin have given some arguments why this should be so.[8]

Several attempts to "fine-tune" the model are also shown in Table 1. No simultaneous agreement is possible for all baryons. Those models which accomodate the Σ^+ moment[4] have trouble with something else (e.g., Ref. 18 with Ξ^- or Ref. 19 with Λ and Ξ^0.) We are partial to the model in column (b), as it incorporates effects that ought to be present at some level. These are: 1) configuration mixing, and 2) small anomalous moments of quarks.

A simple model of configuration mixing envisions the two like quarks (e.g. uu in a proton) as being in a spin-1 state $\cos^2\theta$ of the time and a S=0, $\ell_1=1$ state $\sin^2\theta$ of the time. When $\ell_1=1$, the odd quark is also taken to have $\ell_2=1$, and we take $\vec{\ell}_1 + \vec{\ell}_2=0$. Then (3) is replaced by

$$p = \left(\frac{4}{3} u - \frac{1}{3} d\right) \cos^2\theta + d \sin^2\theta \quad , \quad (6)$$

(4) becomes

$$\Lambda = s \cos^2\theta + \left[\frac{2}{3} (u + d) - \frac{1}{3} s \right]\sin^2\theta \quad , \quad (7)$$

and (5) remains unchanged. If we use p, n, Λ, and Ξ^0 moments as input, we get the predictions of Column (b) of Table 1 when u = 2.14 n.m., d = -1.26 n.m., s = -0.693 n.m., and $\cos^2\theta = 0.894$. Thus only a 10% mixing is implied. The quarks have anomalous moments since $u \neq -2d$. However these are small effects. If the anomalous moment is isoscalar, as suggested by a quark loop model,[12]

$$u = u^0 - \delta, \qquad d = d^0 - \delta \quad , \quad (8)$$

then the fitted values of u, d, and the ratio $u^0 = -2d^0$ imply $\delta/u^0 \simeq 6\%$.

Other mixing models[14,15] do not relax the condition u = -2d, but treat the possible configurations with which mixing can occur in more detail.

In some SU(3) breaking treatments, the effective mass of a quark in a hadron may depend on the mass of the hadron. (See Refs. 9, 18-20 for details.)

If configuration mixing is universal for octet baryons composed of quarks AAB, their magnetic moments can be expressed as $\alpha A + \beta B$ (α, β arbitrary), and one finds[8,9] for any u,d,s:

Table 1. Magnetic moments of baryons, nuclear magnetons, in various models.

Baryon	Expt.[(a)]	Naive Model	Mixing[12-15] (b)	(c)	(d)	SU(3) Breaking[18-20] (e)	(f)
p	2.793	2.79	2.79*	2.85	2.79*	2.79*	2.79*
n	-1.913	-1.86	-1.91*	-1.85	-1.91*	-1.86	-1.91*
Λ	-0.6138 ±0.0047[(g)]	-0.58	-0.61*	-0.61	-0.61*	-0.61*	-0.80
$\Sigma^0 \to \Lambda$	$-[1.82^{+.25}_{-.18}]$[(h)]	-1.61	-1.96	-1.51	-1.68	-1.46	-1.35
Σ^+	2.30 ±0.14[(i)]	2.68	2.68	2.54	2.74	2.46	2.20
Σ^-	-1.48 ±0.37[(j)]	-1.05	-1.37	-1.00	-1.21	-0.92	-0.69
Ξ^0	-1.237 ±0.016[(k)]	-1.40	-1.24*	-1.20	-1.46	-1.20*	-1.37
Ξ^-	-0.75 ±0.06[(l)]	-0.47	-0.58	-0.43	-0.52	-0.45	-0.63

*Input
a) Ref. 17 unless indicated
b) Model based on Ref. 12; $u \neq -2d$
c) Ref. 14; $u = -2d$
d) Ref. 15; $u = -2d$
e) Ref. 18
f) Ref. 19
g) Ref. 1
h) Ref. 21
i) Ref. 4
j) Ref. 22
k) Ref. 2
l) Ref. 3

$$(-4.21\pm 0.07 \text{ n.m.}) = n-p+\Xi^- - \Xi^0 = \Sigma^- - \Sigma^+ = (-3.78 \pm 0.40 \text{ n.m.}). \quad (9)$$

If Σ^-[23] has a normal moment (≈ -0.8 n.m.), as suggested by analogy with Ξ^-, Eq. (9) is violated. (In the absence of SU(3) breaking, all negative hyperons Y have an approximately normal moment, since all their quarks have the same charge -1/3 and mass $\approx m_Y/3$.)

Further relations of this model include

$$\frac{\Sigma^+ - \Sigma^- + \Xi^0 - \Xi^-}{p+n} = \frac{u-d}{u+d} = \begin{array}{ll} 3 \text{ if } u = -2d^{24} & (10a) \\ 3.77 \pm 0.45 \text{ (expt.)}, & (10b) \end{array}$$

$$(-1.03 \text{ n.m.}) = 2n+p = \Xi^0 - \Xi^- = (-0.49 \pm 0.07 \text{ n.m.}) \quad (11)$$

(implied by (9) and u = -2d in (10); see also Ref. 9), and

$$\Sigma^+ = [\, \Xi^0(2n+p) + 2(p^2 - n^2)]/(2p+n) \quad (12)$$
$$= 2.60 \text{ n.m.(pred.)}; = 2.30 \pm 0.14 \text{ (expt.)}$$

(again only for u = -2d). Unless configuration mixing varies throughout the baryon octet,[14] the failure of (12) and especially of (11) suggests $u \neq -2d$.

MESONS

The masses of 0^- and 1^- mexons are fit by Eq. (1) and $m_u = m_d$ = 310 MeV, m_s = 483 MeV, b/m_u^2 = 80 MeV. The predicted radiative decay rates for $1^- \to 0^-\gamma$ and $0^- \to 1^-\gamma$ are then compared with experiment in Table 2. Here I = $\langle f|i \rangle$ describes the overlap of spatial wave functions, which may not be complete.[25]

The widths are nearly all consistent with $|I|^2 = 1/2$, favoring an SU(3) motivated fit.[28] However, those experimental widths proportional to $(u \pm d)^2$ all have $|I|^2 \lesssim 1/2$. A small isoscalar anomalous moment in the direction suggested earlier for baryons $\delta/u^0 \approx$ (few %) can remove this small discrepancy.[12] However, this would raise the predicted $K^{*0} \to K^0\gamma$ rate since it would contribute negatively to d and presumably also to s, both of which already are negative. We await with interest the remeasurement of the $K^{*0} \to K^0\gamma$ rate at Fermilab.[30]

Heavy quarks may have larger anomalous moments due to light quark loops,[12] just as the muon anomalous moment is particularly sensitive to the electron loop. If $c = c^0$ - (few %)u $\simeq 0.7\, c^0$, predicted rates for such M1 processes as $J/\psi \to \gamma\eta_c$ could be cut in half. Limits on M2 effects in $\psi' \to \gamma\chi \to \gamma\gamma\psi$ indicate that $|c| \lesssim |c^0| \equiv |e Q_c/2m_c|$,[31] and presumably these can be improved. The rates for $\psi' \to \gamma\chi_{0,1,2}$ also are able to display effects of quark anomalous moments.[32] For b quarks, if $b = b^0$ - (few %)u $\simeq$ 1.5b, M1 rates could be doubled with respect to naive expectations [such as $B(\Upsilon \to \gamma\eta_b) \lesssim 10^{-3}$.[33]]

CONCLUSIONS

The quark model provides a good zeroth-order description of magnetic transitions in mesons and baryons. Fine tuning takes lots of work and the payoff is marginal. However, anomalous quark magnetic moments may explain small discrepancies if a) the rest of the light-quark pattern emerges as predicted [see Column (b) of Table 1 for the Σ^- moment], b) the effects turn out to be calculable, e.g., in QCD, and c) the c and b quark magnetic moments are really found to be diminished and augmented in magnitude,[12] respectively. Finally, the ground state baryons probably have some admixture of more complex states -after all, so does the deuteron!

Table 2. Magnetic dipole (M1) transitions for mesons composed of light quarks.

Process	Rate; units of $\frac{\omega^3}{3\pi}$ $\|I\|^2$	Prediction, keV	Experiment, keV	$\|I\|^2$
$\omega \to \pi^0\gamma$	$(u-d)^2$	1390 $\|I\|^2$	889 ± 50[a)]	0.64 ± 0.04
$\rho \to \pi\gamma$	$(u+d)^2$	148 $\|I\|^2$	67 ± 7[b)]	0.45 ± 0.05
$\omega \to \eta\gamma$[c)]	$(u+d)^2/2$	11 $\|I\|^2$	$3 {}^{+2.5}_{-1.8}$ [a)]	$0.27 {}^{+0.23}_{-0.16}$
$\rho \to \eta\gamma$[c)]	$(u-d)^2/2$	92 $\|I\|^2$	50 ± 13[a)]	0.54 ± 0.14
$\eta' \to \omega\gamma$[c)]	$3(u+d)^2/2$	17 $\|I\|^2$	7.6 ± 3[a,d)]	0.45 ± 0.18
$\eta' \to \rho\gamma$[c)]	$3(u-d)^2/2$	171 $\|I\|^2$	83 ± 30[a,d)]	0.49 ± 0.18
$\phi \to \eta\gamma$[c)]	$2s^2$	110 $\|I\|^2$	62 ± 9[a)]	0.56 ± 0.08
$\phi \to \pi^0\gamma$[e)]	0	0	5.7 ± 2[a)]	-
$K^{*+} \to K^+\gamma$	$(u+s)^2$	153 $\|I\|^2$	60 ± 15[f)]	0.39 ± 0.10
$K^{*0} \to K^0\gamma$	$(d+s)^2$	224 $\|I\|^2$	75 ± 35[a)]	0.34 ± 0.16

a) See Refs. 12, 17 for further refs. b) Ref. 5
c) Here $\eta = (uu+dd)/2 - ss/\sqrt{2}$, $\eta' = (uu+dd)/2 + ss/\sqrt{2}$ (an octet-singlet mixing angle of 10°) as suggested in Ref. 26.
d) Total η' width of Ref. 27 used: $\Gamma(\eta' \to all) = 0.28 \pm 0.1$ MeV.
e) $\phi \equiv s\bar{s}$.
f) Ref. 6.

ACKNOWLEDGMENTS

I thank Jay Dworkin, Gina Ramieka, and my colleagues at Minnesota for many fruitful discussions. This work was supported in part by the U. S. Department of Energy under Contract No. EY-76-C-02-1764.

REFERENCES

1. L. Schachinger, et al., Phys. Rev. Lett. 41, 1348 (1978).
2. G. Bunce, et al., Phys. Lett. 86B, 386 (1979); R. Handler, talk at parallel session C8, this conference; P.T. Cox, thesis, Univ. of Michigan, 1980 (unpublished).
3. R. Handler, ref. 2; R. Ramieka, et al. (to be published).
4. R. Settles, et al., Phys. Rev. D20, 2154 (1979).
5. D. Berg, et al., Phys. Rev. Lett. 44, 706 (1980).
6. B. Collick, et al., "New Results on Radiative Meson Decays," presented by K. Ruddick at Experimental Meson Spectroscopy Conf., Brookhaven Nat. Lab., 1980 (to be published); P. Thompson, talk at parallel Session C8, this conference.

7. M.A.B. Bég, B.W. Lee, and A. Pais, Phys. Rev. Lett. 13, 514 (1964); G. Morpurgo, Physics (N.Y.) 2, 95 (1965); W. Thirring, Acta Physica Austriaca Supp. II, 205 (1965); H. Rubinstein, F. Scheck, and R. H. Socolow, Phys. Rev. 154, 1608 (1967).
8. I. Cohen and H. J. Lipkin, Phys. Lett. 93B, 56 (1980); H. J. Lipkin, Phys. Rev. Lett. 41, 1629 (1978); Phys. Lett. 74B, 399 (1978); Phys. Lett. 89B, 358 (1980); in Common Problems in Low- and Medium-Energy Nuclear Physics, edited by B. Castel, B. Goulard, and F. C. Khanna (Plenum Publishing Co., New York, 1979); and in Proc. of the IV Int. Conf. on Baryon Resonances, Toronto, July 14-16, 1980 (to be published).
9. J. Franklin, paper no. 531, this conference; Phys. Rev. 172, 1807 (1968), 182, 1607 (1969), D20, 1742 (1979).
10. A. De Rújula, H. Georgi, and S. L. Glashow, Phys. Rev. D12, 147 (1975).
11. A. Grunberg and F. M. Renard, Nuov. Cim. 33A, 617 (1976); A. Bohm and R. B. Teese, Phys. Rev. D18, 330 (1978) and references therein; A. N. Kamal, Phys. Rev. D18, 3512 (1978).
12. D. A. Geffen and Warren Wilson, Phys. Rev. Lett. 44, 370 (1980).
13. H. J. Lipkin, Phys. Lett. 35B, 534 (1971).
14. N. Isgur and G. Karl, Phys. Rev. D21, 3175 (1980).
15. M. Bohm, R. Huerta, and A. Zepeda, paper no. 819, this conference.
16. W. C. Carithers, et al., Phys. Rev. Lett. 35, 349 (1975).
17. Particle Data Group, Rev. Mod. Phys. 52, S1 (1980).
18. Ramesh C. Verma, "Magnetic Moments of Baryons in Broken Spin-Unitary-Spin Symmetry," Univ. of Alberta report Thy-5-80, unpublished.
19. S. Oneda, et al., paper no. 772, this conference.
20. See also Y. Tomozawa, Phys. Rev. D19, 1626 (1979); R. B. Teese and R. Settles, Phys. Lett. 87B, 111 (1979).
21. F. Dydak, et al., Nucl. Phys. B118, 1 (1977).
22. B. L. Roberts, et al., Phys. Rev. D12, 1232 (1975).
23. R. Ramieka, private communication and Ref. 3.
24. R. G. Sachs, private communication.
25. J. S. Kang and J. Sucher, Phys. Rev. D18, 2698 (1978).
26. N. Isgur, Phys. Rev. D13, 122, 129 (1976); D. Ebert and M. Volkov, paper no. 76, this conference; T. N. Pham, paper no. 731, this conference.
27. D. M. Binnie, et al., Phys. Lett. 83B, 141 (1979); G. S. Abrams, et al., Phys. Rev. Lett. 43, 477 (1979); B. Wiik, this conference.
28. T. Ohshima, Univ. of Rochester report, 1980 (unpublished), and Ref. 34.
29. T. N. Pham, paper no. 731, this conference.
30. B. Winstein, private communication.
31. T. Burnett, talk at parallel session C5, this conference.
32. G. Karl, S. Meshkov, and J. Rosner, Phys. Rev. Letters 45, 215 (1980).
33. See, e.g., J. Rosner in Particles and Fields - 1979, ed. B. Margolis and D. Stairs, American Institute of Physics, New York, 1980, p. 325.
34. N. Isgur, Phys. Rev. Lett. 36, 1262 (1976).

OBSERVATION OF A $\phi'(1.65)$ VECTOR MESON IN E+E- ANNIHILATION AT DCI

J-C Bizot, D. Bisello, J. Buon, A. Cordier, B. Delcourt, I. Derado, L. Fayard, F. Mane, J-L Bertrand, P. Eschstruth, J. Jeanjean, M. Jeanjean, J-C Parvan, M. Ribes, F. Fumpf.

Laboratoire de l'Accélérateur Linéaire, ORSAY 91405

and J. Layssac

Laboratoire de Physique Mathématique, USTL, 34060 Montpellier, France

The e^+e^- annihilation into hadrons has been studied with the magnetic detector DM1 in the 1.4-2.2 GeV range. Accurate magnetic analysis of charged particle momenta allowed constrained kinematical reconstruction of events with at most one neutral missing.

The $2\pi^+2\pi^-$ production is dominated by $\rho^\circ\pi^+\pi^-$. A fit of both Novosibirsk[1] and our results gives for the ρ' the parameters shown below

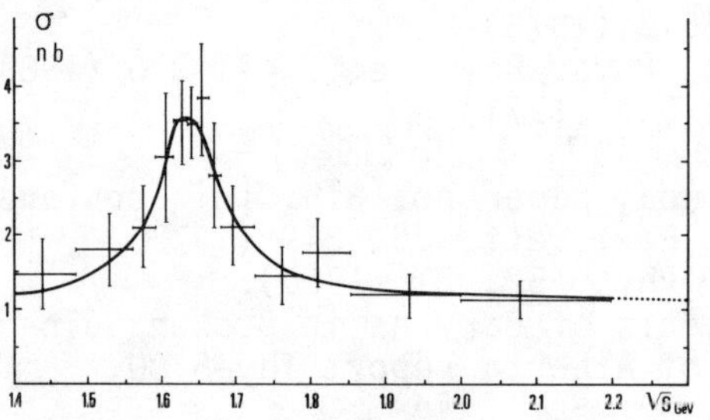

The $2\pi^+2\pi^-\pi^\circ$ production is dominated by $\omega\pi^+\pi^-$ below 1.9GeV. The cross section for events compatible with $\omega\pi^+\pi^-$ (Fig) shows a bump at 1.65 GeV with 6.7 sd confidence level

The K^+K^- and $K^\circ_S K^\circ_L$ cross sections require an isoscalar amplitude interfering with the $\rho\omega\phi$ tail and the ρ'.

The $K^\circ_S K^\pm \pi^\mp$ production is dominated by KK^* and shows also a bump just above the KK^* threshold. The cross section is rather large : 4.5 nb which would give ≃ 13 nb for the whole KK^* production. From the Dalitz plot we can extract the isoscalar and the isovectorial amplitudes and their relative phase. The isoscalar cross section shows a bump around 1.6 GeV. Its absolute phase is obtained by adding to the measured relative phase the isovector one computed from the ρ'. It increases fastly and goes through 90° near 1.67 GeV.

We conclude that we have observed a new $\phi'(s\bar{s})$: the large KK^* production, the small $\omega\pi\pi$ cross section and the narrowness of the bump support this conclusion. One model to interpret our results is a new 1^{--} nonet with an ω' degenerate with the ρ' and then difficult to observe. The $\omega\pi\pi$ production would then be due to an interference between this ω' and the ϕ'. Within this model, the parameters of the ρ' and ϕ' are :

	M(MeV)	Γ(MeV)	$\Gamma_{ee} B_{K\bar{K}}$	$\Gamma_{ee} B_{KK^*}$	$\Gamma_{ee} B_{\rho\pi\pi}$	$\Gamma_{ee} B_{\omega\pi\pi}$
ρ'	1588± 4	444±18	35±29ev	305±71ev	3510±90ev	0
ϕ'	1647±10	152±19	53±35ev	413±33ev	0	≃ 17ev

1. L-M. Kurdadze et al. Preprint 79-69. Institute of Nuclear Physics Novosirbirsk.

ELECTROMAGNETIC FORM FACTORS OF HADRONS

B. T. Chertok*
American University, Washington, DC 20016

INTRODUCTION

There has been significant progress in experiment and theory of the elastic electromagnetic form factors since the 18th International Conference in Tbilisi in 1976. Elastic form factors have been measured and extended to larger q^2 for fixed target scattering $-q^2 \equiv Q^2 = 4\ E_i E_f \sin^2\theta/2$ and for the annihilation channel $q^2 = s > 0$. Meson form factor determinations have been reported for π^-, K^-, $\eta \rightarrow \mu^+\mu^-\gamma$, and $\eta' \rightarrow \mu^+\mu^-\gamma$. New measurements of the baryon form factors are the neutron and proton [$p\bar{p} \rightarrow e^+e^-$(1)]. Larger Q^2 data for the deuteron and ^{3}He and ^{4}He (2) have been reported.**

These experiments on mesons, baryons and light nuclei are challenging for a number of essential reasons. Principally, [1] the cross sections are small because of the e-m coupling constant $\alpha^2 \sim 5 \times 10^{-5}$ and the form factors $F_H^2(q^2)$ which fall rapidly, especially for $Q^2 > M_T^2$ (3); [2] some of the most interesting elementary particles are only found in sufficient density bound inside matter, i.e., u and d quarks and the neutron, or they decay too rapidly to be probed at large Q^2, as with the π^0.

In spite of these and other problems, the deep structure determined from elastic form factors is certainly one of the key ingredients in understanding elementary particles. Impressive support comes almost annually now from e^+e^- storage rings that the electron-photon vertex is still point-like (see paper in this section by A. Böhm) and therefore an ideal tool for strong interaction research.

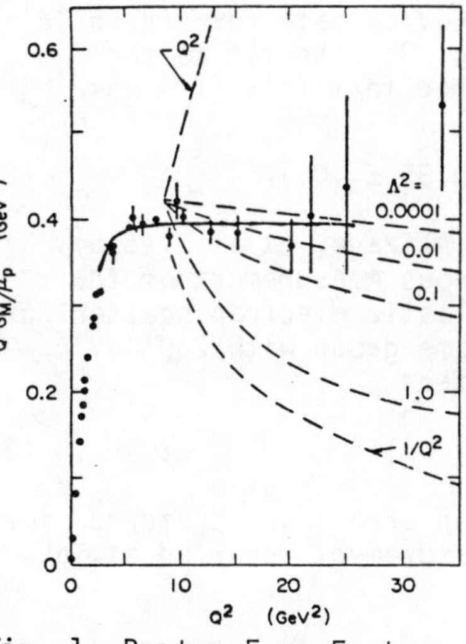

Fig. 1--Proton Form Factor with QCD Predictions

THEORY

Theoretical development of the exclusive scaling laws for the general reaction a + b → c + d have led to many successful predictions. The principal modification from simple QCD arguments is the appearance of the quark-gluon running coupling constant, $\alpha_s = \frac{4\pi}{\beta} / \ln\left(\frac{Q^2}{\Lambda^2}\right)$ leading to the asymptotic prediction for electromagnetic form factors of hadrons

$$F_H \rightarrow [\alpha_s(Q^2)/Q^2]^{n_H - 1} \quad (1)$$

where n_H is the number of valence quarks.

A calculation by Lepage and Brodsky (4) of the proton's magnetic form factor in perturbative QCD is compared with experiment in Fig. 1. The data are consistent with $n_H = 3$ and $\Lambda < .4$ GeV. A more exact determination of Λ is possible if a precision experiment were performed for $Q^2 \gtrsim 15$ $(GeV/c)^2$.

A non-perturbative calculation by A. Kaidalov (#818) of the meson and proton form factor uses a color tube model for confinement, connecting these form factors with the intercepts of Regge Trajectories. Results are included as the solid line in Fig. 1.

The world's data on the proton's form factors (except for G_{mp}-- see Fig. 1) are displayed in Fig. 2 along with fits of Bardek and Zovko. They fit both space like and time like data using an Extended (8 resonance) Vector Meson Dominance model (Paper #270).

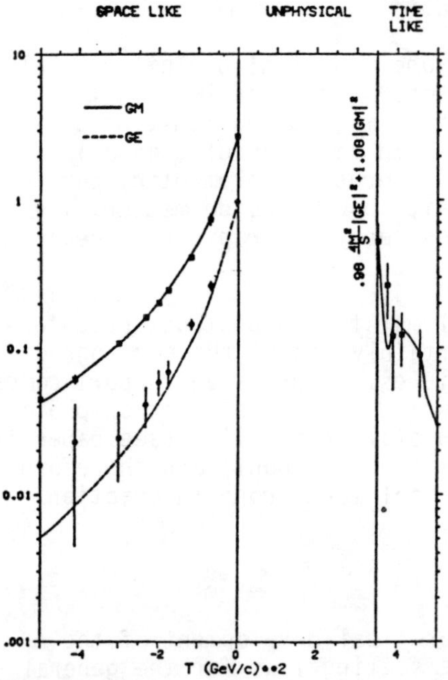

Fig. 2--Proton Form Factors

Another theoretical calculation, inspired by the development of intensely polarized electron sources, is the description of polarization transfer in $\vec{e}\vec{T}$ elastic scattering for T(argets) = n, p or d (5).

EXPERIMENT

New data reported to this conference were tabulated in the Introduction. The US-USSR Collaboration (Papers 964, 965, Dally, et al.) reported 250 GeV π^- scattering data with the combined fit to the 50,100 and 250 GeV/cπ data for $e(\pi, \pi')e'$ as in Fig. 3. The fit to the constrained form $F_\pi^2 = [1 - \frac{q^2 \langle r^2 \rangle}{6}]^{-2}$ gives

$$\langle r_\pi^2 \rangle = (0.39 \pm .04) F^2 \quad (2)$$

See also Alizade, et al. (#615). Simultaneous measurement of the π^- and K^- elastic electron scattering by the same group with 250 GeV/c mesons at the Fermilab gives the 3 - σ effect

$$\langle r_K^2 \rangle - \langle r_\pi^2 \rangle = (-.18 \pm .06) F^2 \quad (3)$$

This result should be free of normalization errors and preferred for the Kaon size rather than the absolute measurement reported at Tokyo (XIX ICHEP-1978).

The reports from Serpukhov of Dzhelyadin, et al. (Papers 600, 601, 602) present measurements for the rare e-m decays of the η and η' produced by $\pi^- p \to \eta(\eta')n$. The B.R. for $\eta \to \mu^+\mu^-$ is 6.5 ± 2.1(-6) and

$\eta' \to \mu^+\mu^-\gamma$ is 8.9 ± 2.4(-5). The measured form factors for $\eta(\eta') \to \mu^+\mu^-\gamma$ can be fitted with monopoles $F(m^2_{\mu\mu}; 0) = (1 - m^2_{\mu\mu}/\Lambda^2)^{-1}$ in rough agreement with VMD.

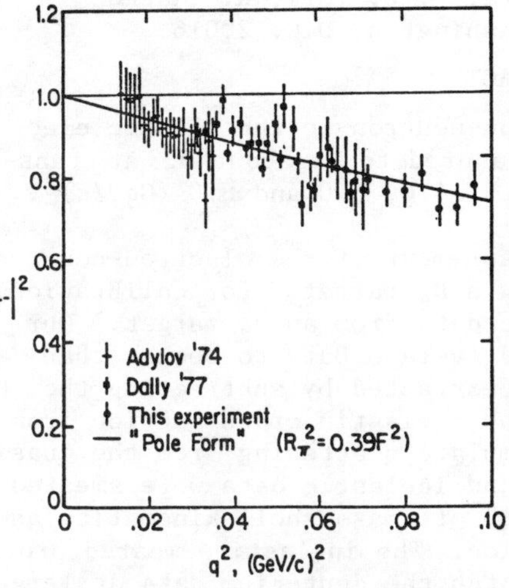

Fig. 3--Elastic π^--e^- Scattering

The principal new result on baryon form factors is the determination at SLAC of the electron-neutron elastic cross section for $2.5 \leq Q^2 \leq 10$ $(GeV/c)^2$ reported by Prof. S. Rock in the next paper. A measurement of D(e, e') X at $Q^2 = 8$ near $x_D = 1$ is reported in Paper #592 leading to an indirect determination of the elastic deuteron form factor. The results continue to favour the quark-like interpretation of the deuteron (3).

CONCLUSION

As reported by Prof. C. Llewellyn-Smith in Plenary Session T4, there is evidence accumulating that meson and baryon elastic form factors may be a highly important testing ground toward a viable theory of the Strong Interactions, e.g. Quantum Chromodynamics. We note that the range of interest appears to be

$$10 < Q^2 < 100 \ (GeV/c)^2 \qquad (4)$$

The electron accelerator required to perform fixed target experiments in this Q^2 range is not under active discussion (however, see Prof. A. Skrinsky's presentation in M1). The energy requirement would be $\underset{\sim}{\sim}$100 GeV at currents of $\sim 10^{14}$ e^-/sec and duty factor $\sim$3 X that of the SLAC accelerator of .03%. In this time of great enthusiasm to reach for very large values of C.M. energy by Storage Rings and Colliders, it is important not to lose sight of significant research to be carried out in the fixed target-Super Luminosity mode.

REFERENCES

*Research supported principally by the National Science Foundation, PHY78-09378.
**See Conference Abstracts 270, 579, 592, 600, 601, 602, 615, 964, 965.

1. G. Bassompiere, et al., Phys. Lett., 68B, 477 (1977), and B. Delcourt, et al., ibid, 86B, 395 (1979).
2. R. Arnold, et al., Phys. Rev. Lett., 40, 1429 (1978).
3. S. Brodsky and B. Chertok, Phys. Rev., D14, 3003 (1976).
4. G.P. Lepage and S. Brodsky, SLAC-PUB-2478, Phys. Rev. D (1980).
5. R. Arnold, C. Carlson and F. Gross, SLAC-PUB-2551, Phys. Rev. C (1980).

MEASUREMENT OF THE ELASTIC ELECTRON-NEUTRON CROSS SECTION AT HIGH Q^2*

S. Rock, R. G. Arnold, B. T. Chertok, B. A. Mecking,†
I. A. Schmidt, Z. M. Szalata, R. C. York, R. Zdarko
American University, Washington, D.C. 20016

ABSTRACT

The ratio of the elastic electron-neutron to the elastic electron-proton cross section was measured to be about 0.3 at transverse momenta squared (Q^2) of 2.5, 4.0, 6.0 and 8.0 $(GeV/c)^2$.

We have made a single-arm measurement of the electron-neutron elastic cross section at SLAC using a D_2 target. For calibration, we used the electron-proton elastic data from an H_2 target. Our cross section agrees with the World Average Data to better than 2%.

The neutron cross section was extracted by subtracting the elastic proton cross section and the inelastic cross section from the deuteron cross section. To simulate scattering from the quasi-free nucleons, the proton elastic and inelastic data were smeared, using the impulse approximation with off mass shell kinematics and a model of the deuterium fermi motion. The inelastic smeared proton cross sections were scaled up to match the deuterium data at large missing mass squared (W^2) ($\sim 2.0\{GeV/c\}^2$) where the elastic contribution is small. The fermi smeared ratio σ_n/σ_p is independent of W^2 in the region near the quasi-elastic peak, indicating that the models of inelastic and elastic cross sections are basically correct.

Our preliminary results on the ratio of elastic cross sections σ_n/σ_p is shown in Fig.1. The final analysis should have substantially reduced errors. This ratio agrees with previous data at low Q^2 [1] and may point toward $F_{1n}^2/F_{1p}^2 = \frac{1}{4}$ at high Q^2. This rules out a completely symmetric wave function for the 3 quarks in the neutron ($F_{1n}=0$). The reduced form factor $Q^4\sqrt{\sigma_n}$ is almost constant as a function of Q^2 as predicted by Dimensional Scaling.

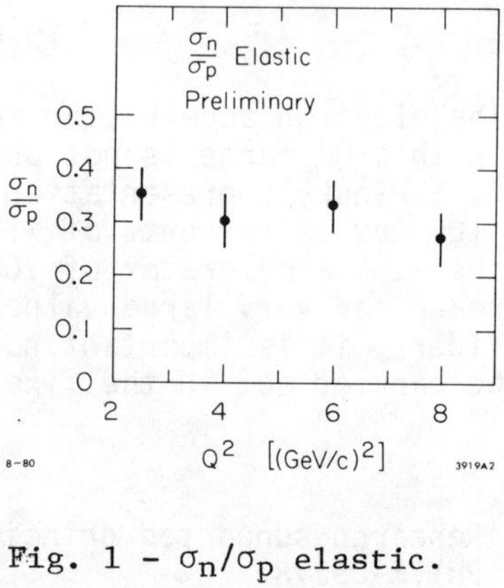

Fig. 1 - σ_n/σ_p elastic.

We acknowledge the staff at SLAC for their support of our investigations and especially Dr. David Sherden and SFG for help with the new on-line computer.

*Work supported in part by the DOE (DE-AC03-765F00515) and in part by the NSF (Grant PHY78-09378).

†On leave from Universität Bonn, Bonn, Germany.

REFERENCES

1. W. Albrect et al, Phys. Lett. 26B, 642 (1968).

TEST OF ELECTRO-WEAK THEORIES AT PETRA

A. Böhm

III. Phys. Institut, RWTH Aachen, D51 Aachen, W.-Germany

ABSTRACT

Measurements on the reactions $e^+e^- \rightarrow e^+e^-$, $\gamma\,\gamma$, $\mu^+\mu^-$ and $\tau^+\tau^-$ at PETRA are compared with theoretical predictions from QED and electroweak theories. Limits are derived for the weak coupling constants in theories of neutral current interactions including models of multi gauge bosons.

INTRODUCTION

Tests of electro-weak theories in e^+e^- reactions can be divided into two classes: tests[1] of QED and the search for weak interaction effects. A test of QED is usually done neglecting weak effects, since they are small in the standard model with $\sin^2\Theta_W = 0.23$ and within measurement errors. The complementary approach is to assume the validity of QED and test[2] the theory of neutral current interactions. The sensitivity of such tests strongly increases with energy. To be precise, weak effects increase faster than the c.m. energy squared, s, due to the Z^0 propagator. The high energies, which are now available at PETRA up to 36 GeV, enable a search for weak interaction effects in e^+e^- reactions to be made for the first time. Despite the fact that we do not yet see weak effects we can put strong limits on the coupling parameters of neutral currents.

In this talk the reactions $e^+e^- \rightarrow e^+e^-$, $\gamma\,\gamma$, $\mu^+\mu^-$ and $\tau^+\tau^-$ will be discussed. They are measured by four experiments, JADE, MARK J, PLUTO and TASSO. Limited space does not allow to discuss the detection of these reactions, but in most cases the methods are already published[3-6].

TEST OF QED

The theoretical cross-sections of QED reactions have a simple and compact form only in lowest order of α. Higher orders lead to radiative corrections[7] which usually depend on the detector and the selection of events. We therefore adopt the following procedure for testing QED: a measured cross-section, say $d\sigma/d\Omega$, is corrected for radiative effects δ_r and hadronic vacuum polarisation δ_h and then compared to the lowest order QED cross-section $d\sigma QED/d\Omega$

$$\frac{d\sigma}{d\Omega}\,(1 - \delta_r - \delta_h) = \frac{d\sigma_{QED}}{d\Omega}\,(1 + \delta) \qquad (1)$$

A deviation δ can have two sources: a violation of QED or weak interaction effects.

A violation of QED can have the following origins:

1. A deviation from the pointlike nature of leptons.
2. The exchange of heavy photonlike object.
3. The existence of a heavy lepton e^* with quantum numbers like the electron.

These sources will modify vertices and/or propagators. Deviations from QED are generally parametrized[8] in terms of a form factor $F(q^2)$

$$F(q^2) = 1 \mp \frac{q^2}{q^2-\Lambda_\pm^2} \approx 1 \pm \frac{q^2}{\Lambda_\pm^2}$$

The first expression is used by JADE, MARK J and TASSO, while PLUTO uses the second expression. The opposite sign convention of JADE will be taken into account by exchanging the limits on Λ_+ and Λ_-. To shorten the list of results we will not apply different cut-off parameters for the form factors in the spacelike and timelike regions.

Before we look at the data, we should get an idea of the deviations resulting from a cut-off parameter with a typical value of 100 to 150 GeV. Figure 1 demonstrates the effect for Bhabha scattering at the c.m. energy of 35 GeV in an experiment which does not measure the charge of the scattered particles and therefore measures scattering angles only between 0° and 90°.

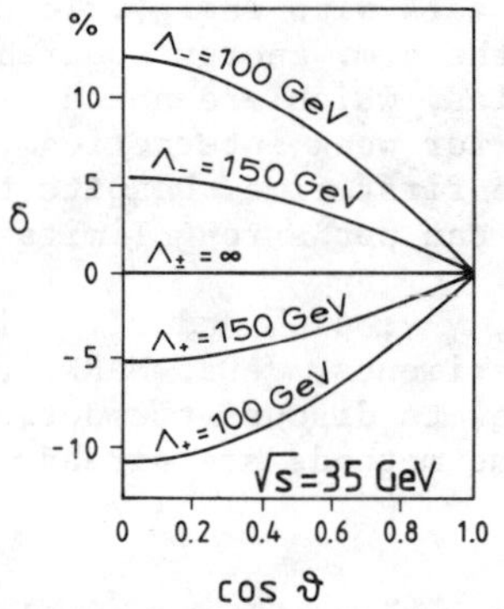

Fig. 1

Relative deviation δ of the angular distribution of Bhabha scattering at 35 GeV from QED for different values of the cut-off parameters Λ_+ and Λ_-.

The deviation from the QED cross-section increases with the scattering angle and amounts to about ± 5% for $\Lambda_\pm$ = 150 GeV at large angles.

RESULTS ON QED TESTS

At PETRA a large number of results exists on the reactions $ee \to ee$, $\gamma\gamma$, $\mu\mu$ and $\tau\tau$. In this talk it is only possible to show a small selection of data obtained by the experiments JADE, MARK J, PLUTO and TASSO. All four groups measure Bhabha scattering and the annihilation into two photons $ee \to \gamma\gamma$. One important reason for studying Bhabha scattering is to measure the luminosity. For a test of QED the luminosity has to be measured at small scattering angles, where cut-off parameters do not modify the cross-section. Otherwise one can only test the form of the angular distribution.

As examples for the high quality of the data a selection of measurements will be presented here. Figure 2 shows the angular distribution of Bhabha scattering and annihilation into two photons measured by JADE. As the cross-sections are multiplied by the c.m. energy squared s, the measurements at different energies can be compared to the same QED prediction. The experiments JADE, PLUTO

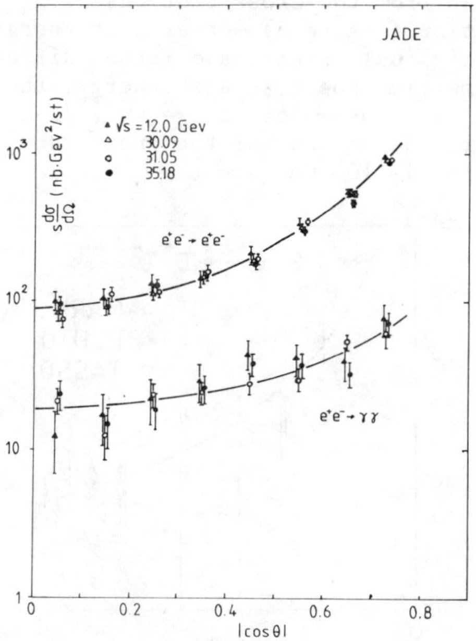

Fig. 2

Angular distribution of Bhabha scattering and ee → γγ measured by JADE. The solid lines display the QED cross-sections.

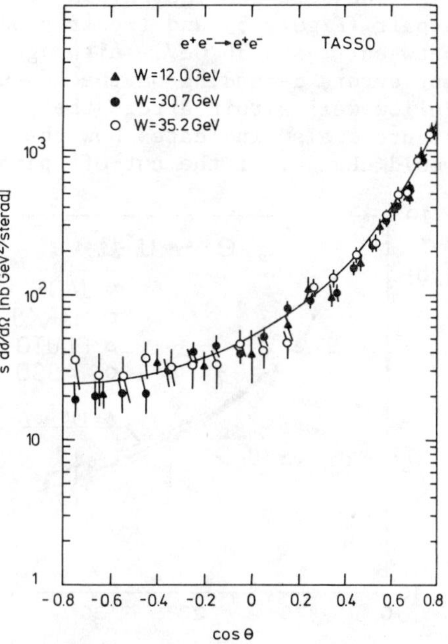

Fig. 3

Angular distribution of Bhabha scattering obtained by TASSO. The solid line is the QED prediction.

and TASSO are able to measure the charge of the scattered electrons. Then the angular distribution of Bhabha scattering extends to scattering angles larger than 90°, as shown in Figure 3 in a measurement by TASSO.

The PLUTO group gives in Figure 4 the cross-section for ee → γγ integrated over $|\cos\Theta| < 0.75$ which falls like 1/s as expected from QED. One point in Figure 4 is measured directly on the Υ resonance and, as expected, it does not increase. In summary all the measurements shown compare well with the QED prediction.

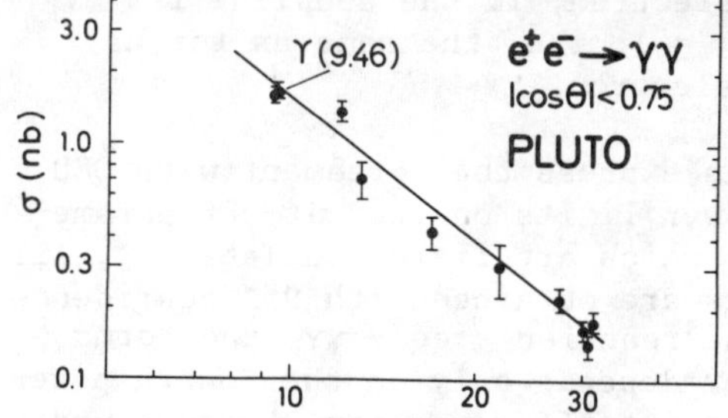

Fig. 4

The cross-section ee → γγ integrated over $|\cos| < 0.75$. The solid line indicates the 1/s behaviour of the QED cross-section.

The same conclusion can be drawn from the cross-sections of μ-pair (Figure 5) and τ-pair production (Figure 6) measured at energies between 12 and 36 GeV. Although individual points have rather different errors depending on the measuring time spent at each energy, they follow well within errors the pointlike cross-section $\sigma = 4\pi\,\alpha^2/3s$. Figure 5 also indicates how the energy dependence of the cross-section would change, if the cut-off parameter is 100 GeV.

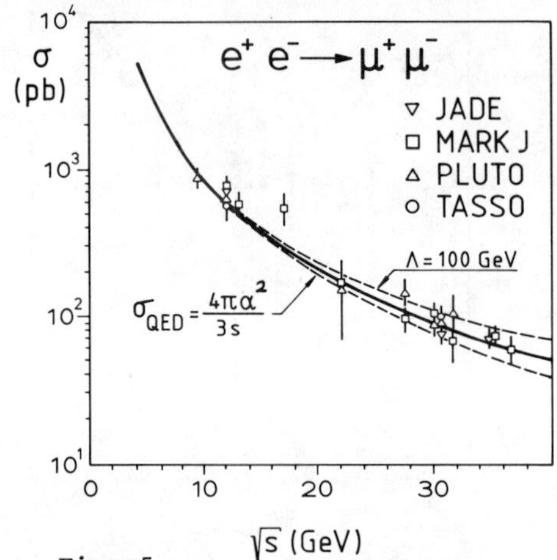

Fig. 5

μ-pair cross-section as a function of energy. Also indicated is the QED cross-section (solid line) and the limits (dashed line) corresponding to a cut-off parameter of 100 GeV.

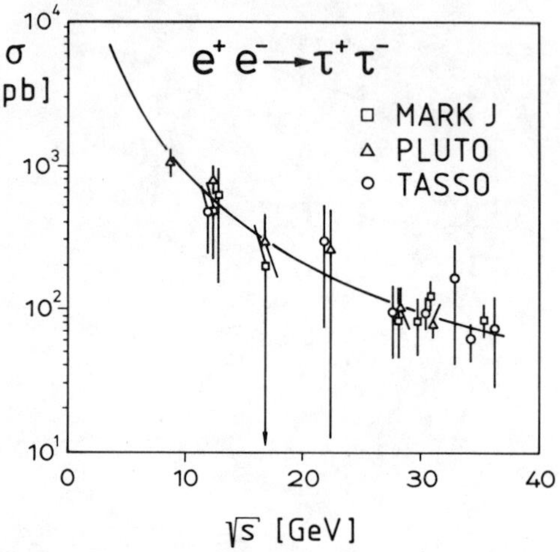

Fig. 6

Energy dependence of the τ-pair cross-section compared with the QED prediction.

	JADE	MARK J	PLUTO	TASSO
ee → ee				
Λ_+	112	91	80	150
Λ_-	106	142	234	136
ee → γγ : $F(q^2) = 1 \pm q^4/\Lambda_\pm^4$				
Λ_+	—	44	46	—
Λ_-	—	34	36	—
heavy electron : $\Lambda_+^2(e^*) = \frac{M_{e^*}^2}{\lambda}$				
$\Lambda_+(e^*)$	47	55	46	34
$\Lambda_-(e^*)$	44	38	—	42
ee → μμ				
Λ_+	137	123	116	80
Λ_-	96	142	101	118
ee → ττ				
Λ_+	—	76	74	115
Λ_-	—	154	65	76

Table I

Limits on the cut-off parameters in GeV with 95% confidence. In the reaction ee → γγ the cut-off parameter $\Lambda_+(e^*)$ can be interpreted[10] as a limit on the mass M_{e^*} of the exchanged heavy electron, if the coupling is $\lambda = 1$, i.e. the same as for an electron.

We can express the agreement with QED by lower limits on the cut-off parameters, which are listed in Table I. All limits are obtained with 95% confidence. In the reaction ee → γγ the form-factor depends only on the fourth power of the cut-off parameter, because modifications on the electron propagator cancel[9] due to current conservation. Limits are also given on the mass of an exchanged heavy lepton with the quantum numbers of the electron[10].

Summarizing the test of QED we can say that QED is valid up to $s = 1225\ GeV^2$ and $q^2 \approx 1000\ GeV^2$. The leptons are pointlike particles with a limit on their charge radius of $R < 2 \cdot 10^{-16}$ cm. This result is especially remarkable for the τ-lepton having a mass about twice as heavy as the proton.

WEAK INTERACTION EFFECTS

A realistic deviation from QED is expected from the weak neutral current interaction. The main effect comes from an interference between photon and Z^0 exchange. Within the standard SU(2) x U(1) theory the change of a cross-section, e.g. for μ-pair production, is small, as it depends mainly on the vector coupling constant which is practically zero for $\sin^2\Theta_w = 0.23$. Therefore, one should test the shape of the angular distribution.

As an example of the sensitivity of the test Figure 7 shows the deviation from QED of the angular distribution for Bhabha scattering measured by MARK J. It is compared with predictions of the standard weak interaction theory for $\sin^2\Theta_w = 0.25, 0.01, 0.55$. Clearly the curve $\sin^2\Theta_w = 0.55$ can be excluded by this measurement. The largest relative effect of weak interaction is expected as a forward-backward asymmetry in the angular distribution of the reaction $ee \to \mu\mu$. Figure 8 shows the individual measurements of JADE, MARK J, PLUTO and TASSO and Figure 9 the combined angular distribution for energies between 27 and 35 GeV. Unfortunately, the angular distribution is not measured at values near $|\cos\Theta| \approx 1$, where it is most sensitive to the shape $1 + \cos^2\Theta$ predicted by QED.

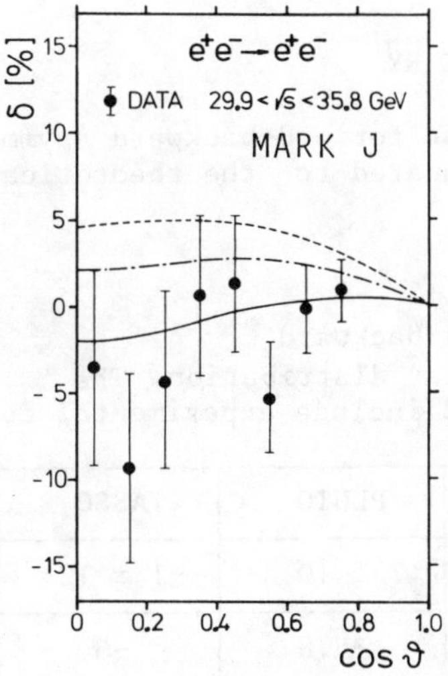

Fig. 7

Deviation of Bhabha scattering from lowest order QED prediction as measured by MARK J. The solid, dash-dotted and dashed curves are the predictions of the standard electroweak theory with $\sin^2\Theta_w$ = 0.25, 0.01 and 0.55 respectively.

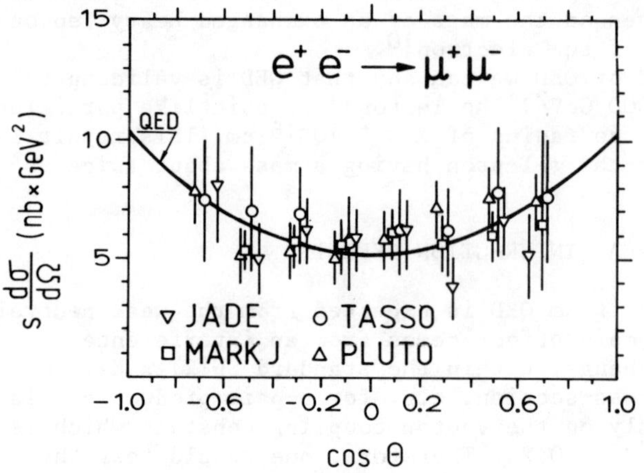

Fig. 8

Angular distribution of μ-pair production measured by PETRA experiments at energies between 27 and 35 GeV.

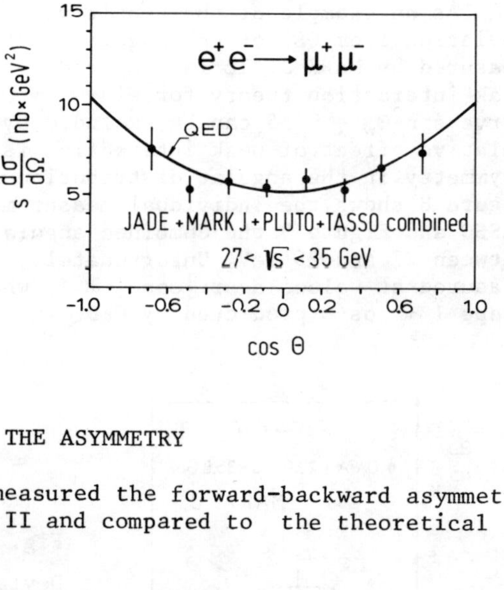

Fig. 9

Combined angular distribution of μ-pair production from PETRA experiments.

RESULTS ON THE ASYMMETRY

All four experiments have measured the forward-backward asymmetry. The results are listed in Table II and compared to the theoretical expectations.

Table II

PETRA results on the forward-backward asymmetry $A_{\mu\mu}$ of the μ-pair angular distribution. The expected values are from SU(2) x U(1) and include experimental cuts.

	JADE	MARK J	PLUTO	TASSO
$A_{\mu\mu}$ in %	-8 ± 9	0 ± 9	7 ± 10	-1 ± 12
EXPECTED	-6	-6	-5.8	-6

Unfortunately, the data have large errors due to a lack of statistics. We therefore combine all PETRA measurements and obtain an asymmetry $A_{\mu\mu} = (-0.9\pm4.9)\%$, which should be -6% in the standard SU(2) x U(1) theory. The asymmetry depends essentially on the axial vector coupling and to second order on the mass m_z of the Z^0 boson[11].

$$A_{\mu\mu} \approx \frac{3G_F}{4\sqrt{2}\pi\alpha} g_A^2 \cdot \frac{s\cdot m_z^2}{s-m_z^2} = \frac{2.7\cdot 10^{-4}}{GeV^2} g_A^2 \frac{s\cdot m_z^2}{s-m_z^2}$$

With 95% confidence the combined asymmetry is larger than -9.0%. From this value the following limits can be determined again with 95% confidence:

1) If we set $g_A^2 = \frac{1}{4}$, as predicted by the standard SU(2) x U(1) theory, we find $m_z > 49$ GeV.
2) Reversing the argument, we find $|g_A|<0.65$, where $g_A = -0.5$ is expected in the standard SU(2) x U(1) theory.
3) Instead of g_A^2 we really measure the product of axial coupling of the electron and the muon, $g_A^e \cdot g_A^\mu$. If we again set $g_A^e = -0.5$ we can give a limit of the axial coupling of the neutral current interaction for the muon without the use of μ-e universality

$$|g_A^\mu| < 0.85 .$$

The limits for 2) and 3) are obtained with the assumption of $m_z = \infty$, which gives the weakest constraints of the coupling constants.

LIMITS ON THE WEAK NEUTRAL COUPLING

The previous results suggest that one should use all the reaction ee → ee, ee → μμ, and ee → ττ to determine the leptonic coupling of the weak neutral current interaction. The cross-sections for Bhabha scattering, μ-pair and τ-pair production[12] depend in general on four parameters[13] m_z, h_{VV}, h_{AA} and h_{VA}. In models with a single Z^0 the parameter h_{VA} factorizes $h_{VA}^2 = h_{VV}\cdot h_{AA}$.

We start our discussion with the simplest model, the standard SU(2) x U(1) electroweak theory, where $\sin^2\Theta_w$ is the only parameter. It gives the relations

$$h_{VV} = g_V^2 = \frac{1}{4}(1-4\sin^2\Theta_w)^2 \qquad h_{AA} = g_A^2 = \frac{1}{4}$$

$$m_z = \frac{37.4 \text{ GeV}}{\sin\Theta_w \cos\Theta_w}$$

Comparing the theoretical cross-section with the data, the following limits for the weak mixing angle have been determined at PETRA.

Table 3: Results on $\sin^2\Theta_W$ for $27 < \sqrt{s} < 31$ GeV. MARK J also includes 35 GeV Data.

group	limit on $\sin^2\Theta_W$ with 95% confidence: lower	upper	final states used	$\sin^2\Theta_W$ with one standard deviation error
JADE[14]	-	0.55	ee,μμ	0.25 ± 0.18
MARK J[15]	0.07	0.42	ee,μμ,ττ	0.24 ± 0.11
PLUTO	-	0.57	ee,μμ	0.23 ± 0.17
TASSO	-	0.52	ee	-

Next we discuss a more general case still with a single Z^o, but do not restrict the vector and axial coupling constants. We would have 3 parameter m_z, g_V and g_A, but setting m_z to infinity leads to the most pessimistic limits. In fig. 10 the 95% confidence contours of JADE[14], MARK-J and PLUTO are plotted for positive values of $g_V{}^2$ and $g_A{}^2$.

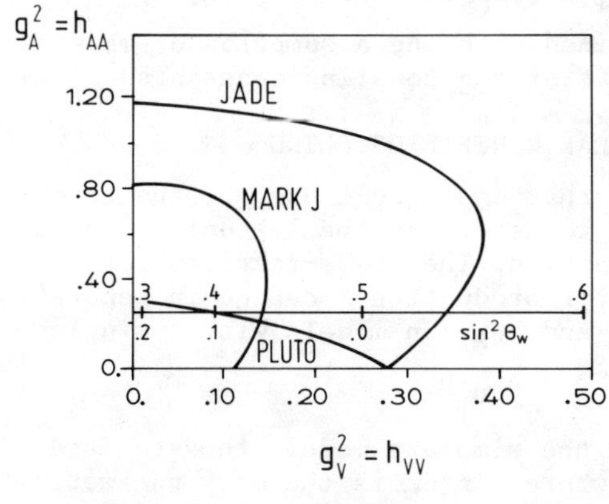

Fig. 10:
95% confidence contours on $g_V{}^2$ and $g_A{}^2$ obtained from the reactions ee → ee and ee → μμ.

The MARK J group[15] has compared their limits on g_V and g_A with the measurements of the neutrino-electron scattering[16], which is also a purely leptonic interaction (fig. 11). Neutrino electron scattering limits the values of g_V and g_A to two regions in the $g_V - g_A$ plane. To solve this ambiguity one previously had to consider lepton-hadron scattering with the inherent complication of hadronic targets. Now the MARK J data determine[15] a unique solution for the weak neutral coupling constants from purely leptonic interactions. While the MARK J result is obtained by combining Bhabha scattering, μ-pair and τ-pair production, the group can show that Bhabha scattering alone is already sufficient to resolve the ambiguity, as it gives the main constraint on the vector coupling.

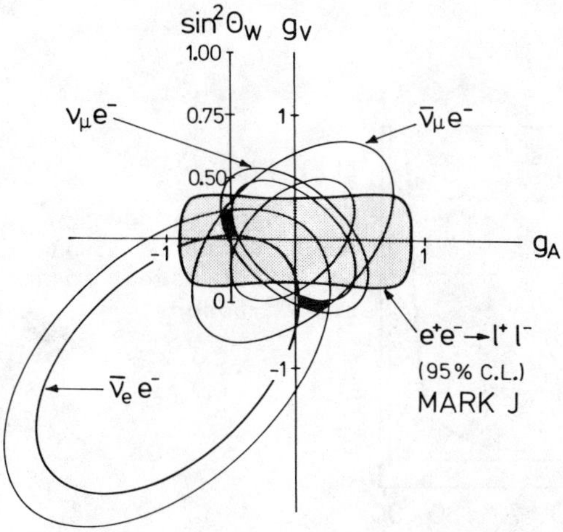

Fig. 11:

Results obtained from neutrino[16] experiments and the MARK-J experiment expressed in terms of limits on g_V and g_A. The regions in between the concentric ellipses correspond to 1σ limits from the neutrino-electron scattering experiments. The two black areas indicate the two allowed regions for g_V and g_A from the combined neutrino data. The shaded area represents the 95% confidence limit contour from the MARK-J experiment.

MULTI GAUGE BOSON THEORIES

Now we abandon the restriction to one neutral gauge boson Z^0 and allow[17] for a larger group SU(2) x U(1) x G, but keep predictions for neutrino and polarized electron-deuteron scattering at low momentum transfer identical to those of SU(2) x U(1). In this case[2] the parameter h_{VV}, h_{VA} and h_{AA} turn out to be

$$h_{VV} = \frac{1}{4}\,(1 - 4\sin^2\Theta_w)^2 + 4\,C$$

$$h_{VA} = \frac{1}{4}\,(1 - 4\sin^2\Theta_w) \qquad h_{AA} = \frac{1}{4}$$

The parameter C contains the couplings of the gauge bosons and their masses.

To reproduce the low energy neutrino data we use $\sin^2\Theta_W = 0.23$ and determine a limit on C. From measurements of Bhabha scattering and μ-pair production at 30 and 35 GeV JADE and MARK J obtain with 95% confidence the following limits:

JADE	$-\,0.059 < C < +\,0.033$
MARK J	$-\,0.097 < C < +\,0.027$

In the case[18] of G = U'(1) with one $W^{\pm}$, but two neutral bosons the parameter C can be expressed by the masses m_1 and m_2 of the Z^0-bosons:

$$C = \cos^4\Theta_w \left(\frac{m_z^2}{m_1^2} - 1\right)\left(1 - \frac{m_z^2}{m_2^2}\right)$$

For G = SU'(2) with two charged and neutral bosons[19] the factor $\cos^4\Theta_W$ is replaced by $\sin^4\Theta_W$. With these expressions we can convert the limits on C into limits on the two neutral boson masses (fig. 12).

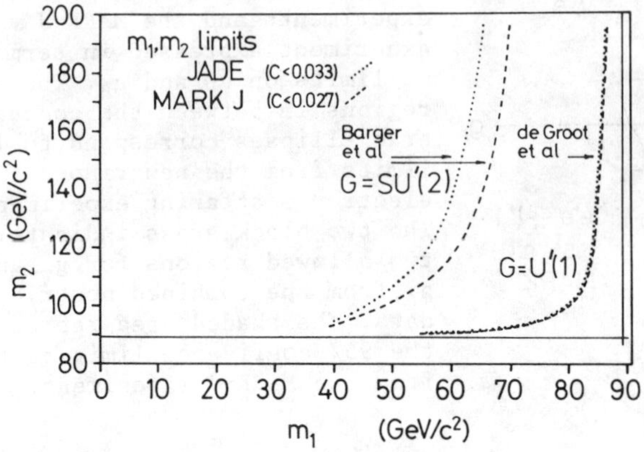

Fig. 12:
Limits on m_1 and m_2, deduced from a limit on C, for models with two Z^0-bosons.

Due to the factor $\sin^4\Theta_W$ the model of Barger et al. is generally less restricted, but for both models a large range of masses m_1 is excluded if $m_2 \gtrsim 100$ GeV.

CONCLUSION

The tests of electro-weak theories performed at PETRA lead to the following conclusions:

1. QED is valid up to $s = 1225$ GeV^2 and $q^2 \approx 1000$ GeV^2.
2. Leptons, including the τ-lepton, are point-like particles with a charge radius less than $2 \cdot 10^{-16}$ cm.
3. The search for weak interaction effects not only leads to strong limits on the parameter of neutral currents, but also tests the theories at very large s and q^2.
4. The vector and axial-vector couplings of the weak neutral currents can be uniquely determined by purely leptonic interactions.
5. For multi gauge boson models tight restrictions are put on coupling constants and neutral boson masses.

REFERENCES

1. S.J. Brodsky and S.D. Drell, Annu. Rev. Nucl. Sci. 20 (1970) 147; R. Hofstadter, Proc. 1975 Int. Symp. on Lepton and Photon Interactions at High Energies, Stanford 1975, p. 869.
2. N. Wright and J.J. Sakurai, Phys. Rev. D22 (1980) 220; E.H. de Groot and D. Schildknecht, Bielefeld preprint, BI-TP 80/08, May 1980.
3. JADE: W. Bartel et al., Phys. Lett. 92B (1980) 206 and private communication R. Marshall, B. Naroska and Y. Totsuka.

4. MARK J: D.P. Barber et al., Phys. Rev. Lett. 42 (1979) 1110; Phys. Rev. Lett. 43 (1979) 1915 and The MARK J Collaboration, Physics Reports 63 (1980) 337.

5. PLUTO: Ch. Berger et al., Z. Physik C4 (1980) 269; Phys. Lett. 94B (1980) 87 and private communication V. Hepp, W. Lührsen and O. Meyer.

6. TASSO: R. Brandelik et al., Phys. Lett. 94B (1980) 259; Phys. Lett. 92B (1980) 199 and private communication U. Martyn.

7. F.A. Berends, K.J.F. Gaemers and R. Gastmans, Nucl. Phys. B57 (1973) 381; Nucl. Phys. B63 (1973) 381; Nucl. Phys. B68 (1974)541; F.A. Berends and G.J. Komen, Phys. Lett. 63B (1976) 432 and private communication F.A. Berends and R. Kleiss.

8. H. Salecker, Zeits. Naturforsch. 8a (1953) 16 and 10a (1955) 349; S.D. Drell, Ann. Phys. (N.Y) 4 (1958) 75; T.D. Lee and G.C. Wick, Phys. Rev. D2 (1970) 1033.

9. J.A. McClure and S.D. Drell, Nuovo Cim. 37 (1965) 1638; N.M. Kroll, Nuovo Cim. 45A (1966) 65.

10. A. Litke, Harvard University thesis 1970 (unpublished).

11. J. Ellis and M.K. Gaillard, CERN yellow report CERN 76-18, p. 21.

12. R. Budny, Phys. Lett. 45B (1973) 340; Phys. Lett. 55B (1975) 227.

13. P.Q. Hung and J.J. Sakurai, Phys. Lett. 69B (1977) 323; Phys. Lett. 88B (1979) 91.

14. R. Marshall, Rutherford Lab. report RL-80-29, May 1980.

15. D.P. Barber et al., Aachen report PITHA 80/8, July 1980.

16. H. Faissner, New Phenomena in Lepton-Hadron Physics (1979), ed. D.E. Fries and J. Wess (Plenum Publishing Corp., New York),p.371; H. Reithler, Phys. Blätter 35, 630 (1979); F.W. Bullock, Proc. of Neutrino 79, Bergen 1979, p. 398; R.H. Heisterberg et al., Phys. Rev. Lett. 44 (1980) 635; L.W. Mo, Contribution to Neutrino 80, Erice (1980), and private communication H. Faissner and H. Reithler.

17. H. Georgi and S. Weinberg, Phys. Rev. D17 (1978) 275; J.D. Bjorken, Phys. Rev. D19 (1979) 335.

18. E.H. de Groot, G.J. Gounaris and D. Schildknecht, Phys. Lett. 85B (1979) 399; Phys. Lett. 90B (1980) 427 and Z. Physik C5 (1980)127.

19. V. Barger, W.Y. Keung and E. Ma, Wisconsin-Hawai reports UW-COO-881-126 (1980); UW-COO-881-133 (1980) and UW-COO-881-138 (1980).

E^+E^- PHYSICS I

W. FRAZER, UC-SAN DIEGO, ORGANIZER

TWO-PHOTON REACTIONS IN $e^+ e^-$ COLLIDING BEAMS

William R. Frazer
University of California, San Diego, La Jolla, CA. 92093

Since this is the first in this series of conferences to have an entire session devoted to two-photon physics, I think it is appropriate to begin with a brief theoretical review.[1] The process we are discussing is shown in Fig. 1. The areas of greatest current interest are:

1. Tests of QED
2. Resonance production
3. Tests of QCD
   a. Jet production
   b. Photon structure
   c. Exclusive final states

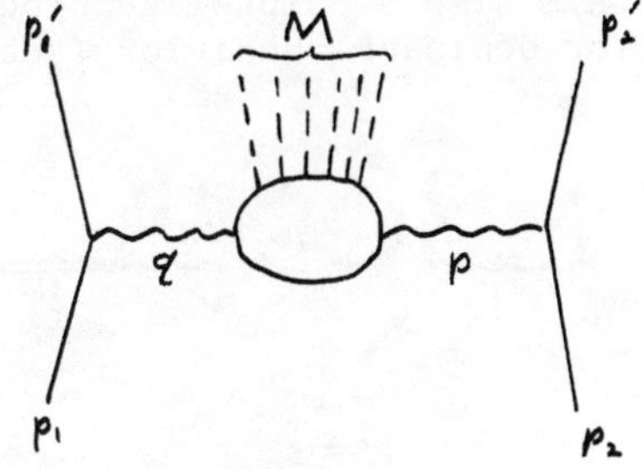

Fig. 1. Two-photon reaction

Since there is very little new theory about tests of QED I leave this topic to the experimental talks. Tests of QCD in exclusive final states will be the subject of the following talk, by Brodsky. After a few brief remarks about resonance production I shall concentrate on tests of QCD in jet production and by measurement of the photon structure function.

RESONANCE PRODUCTION

Resonance production, wherein the entire mass M produced (see Fig. 1) consists of a single resonance R (neutral, C=+), provides the opportunity for measurement of two-photon decay widths $\Gamma(R \to \gamma\gamma)$. If neither of the scattered leptons is detected ("tagged" is the jargon in this field) the two photons can to good approximation be taken to be real, with the resulting cross section[2]

$$\sigma(e^+e^- \to e^+e^- R) = \left(4\alpha \log \frac{E}{M_e}\right)^2 f\left(\frac{M_R}{2E}\right) \frac{(2J_R+1)\Gamma(R \to \gamma\gamma)}{M_R^3} \tag{1a}$$

where

$$f(x) = -(2 + x^2)^3 \log x - (1 - x^2)(3 + x^2) \tag{1b}$$

Gilman[3] has given a complete tabulation of cross sections expected for all the known low-lying C = + states. At beam energies of 15 GeV the cross sections are a respectable few nanobarns up through the f region, but fall to an estimated .05 nb for the η_c and drop to a discouraging 4×10^{-5} by the η_b. The factor M_R^{-3} in (1) will make it difficult to wander very far up the mass scale.

TESTS OF QCD

Two-photon reactions offer an excellent testing ground for QCD, because QCD makes especially simple and definitive predictions with photon targets as compared to hadron targets. Whereas the fragmentation of a hadron into quarks and gluons is a bound state problem outside the domain of perturbative QCD, the photon can directly produce a quark-antiquark pair. In order to take advantage of this simplicity, however, one must look for regions of the variables in which this simple point-like production dominates over the hadron-like vector dominance behavior (see Fig. 2).

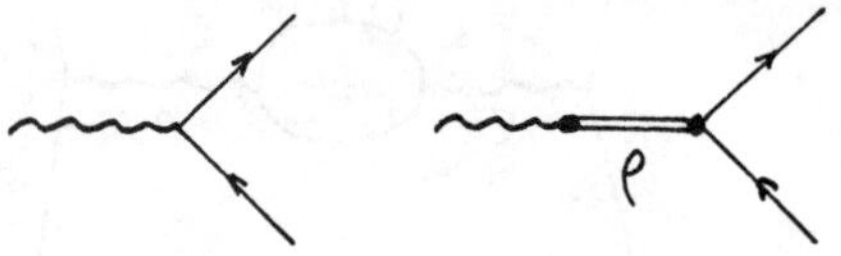

Fig. 2. Point-like vs hadron-like quark production by a photon.

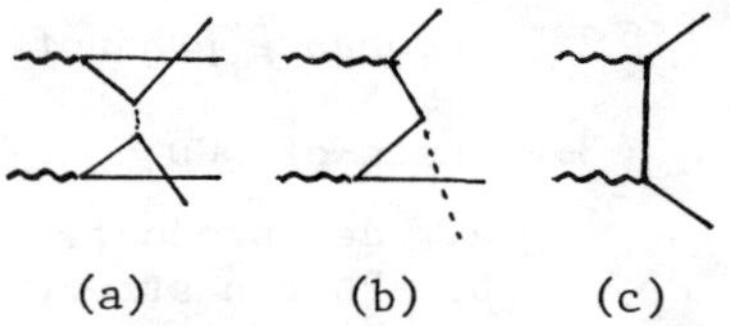

Fig. 3. Some diagrams leading to high-$p_\perp$ jets.

a. Jets at high transverse momentum: Jets at high $p_\perp$ are the simplest of QCD predictions.[4,5] Simplest of all is the two-jet process in Fig. 3c. The predicted rate is most succinctly given in terms of a ratio $R_{\gamma\gamma}$, the ratio of two-jet production to two-muon production:

$$R_{\gamma\gamma} \equiv \frac{d\sigma(e^+e^- \to e^+e^- \, q\bar{q} \to e^+e^- + 2 \text{ jets})}{d\sigma(e^+e^- \to e^+e^- \, \mu^+\mu^-)} \tag{2a}$$

$$= \sum_{\substack{\text{color,}\\ \text{flavor}}} Q_i^4 \left(1 + O[\alpha_s(4p_\perp^2)]\right) \tag{2b}$$

This process is a source of most remarkable events, in which all the hadrons are emitted at large $p_\perp$. Another remarkable prediction, which holds also for the three- and four-jet processes shown in Fig. 3, is that all these processes obey exact Bjorken scaling, free from even logarithmic violations.

b. Photon structure: Looking back at Fig. 1, events in which q^2 is large but $p^2 \approx 0$ can be thought of as deep inelastic scattering off a nearly-real photon. The cross section can be expressed in terms of the structure photons of the photon. For unpolarized leptons one can measure three structure functions,[6]

$$\frac{d\sigma}{dE_1{}' dE_2{}' d\cos\theta_1 d\phi} = \frac{4\alpha^2 EE_1{}'}{\nu^2 Q^2} N_2(y_2{}^2 - 2y_2 + 2)$$

$$\times (W_1 + \varepsilon_2 W_L + \varepsilon_1\varepsilon_2 W_3 \cos 2\phi) \qquad (3)$$

where $\nu = p\cdot q$, $y_1 = \nu/p_2\cdot q$, $y_2 = \nu/p_1\cdot q$,

$$\varepsilon_i = \frac{1 - y_i}{1-y_i + \frac{1}{2}\, y_i{}^2} .$$

The angle ϕ is the angle between the planes of scattering of the leptons, and N_2 is the equivalent-photon density. The structure functions $W_i(x,Q^2)$ depend on Q^2 and $x \equiv Q^2/2p\cdot q$.

In the parton model there is a point-like contribution to the photon structure from the box diagram, in which the target photon produces a quark-antiquark pair, one of which is then struck by the other photon.[7] The QCD results are summarized in the following table.

	Parton model (Box diagram)	QCD (leading order)	
W_1	$= f_1^{BOX}(x) \log\left(\frac{Q^2}{m_q^2}\,\frac{1-x}{x}\right)$	$f_1(x) \log \frac{Q^2}{\Lambda^2}$	(4a)
W_L	$= f_L^{BOX}(x)$	$f_L(x)$	(4b)
W_3	$= f_3^{BOX}(x)$	$f_3^{BOX}(x)$	(4c)

Note that the QCD result, which (in physical gauges) comes from ladder graphs, preserves the Q^2 dependence of the box diagram. The x-dependence of the box diagram result for W_1 and W_L is modified in a calculable way, as was first shown by Witten.[8] The functions $f_i(x)$ can be found in Ref. 8,9,10. On the other hand, the third structure function W_3 is given, to leading order, by the unmodified box diagram result.[11]

The QCD result described above is only the point-like piece; there is also a hadron-like piece which one can estimate by a vector-meson-dominance model.[5,6] The result in Fig. 4 shows that even at $Q^2 = 2$ GeV^2 the vector-meson background can be made negligible by avoiding the region $x \lesssim 0.4$. It should be emphasized, however, that the VMD model is just a model; at small x where its contribution cannot be neglected there is no unambiguous QCD prediction of the structure functions.

A second caveat concerns the region $x \approx 1$. Bardeen and Buras[12]

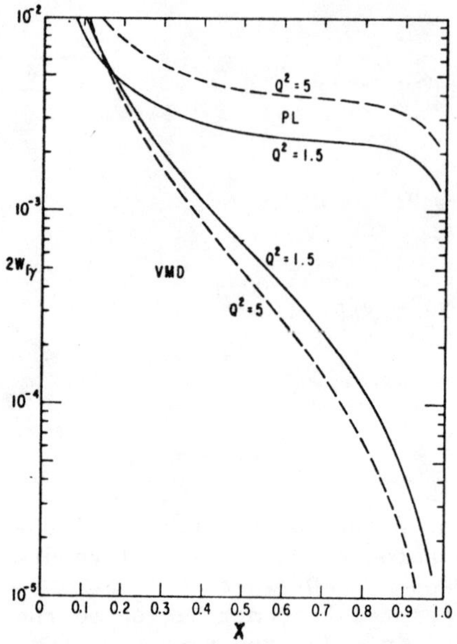

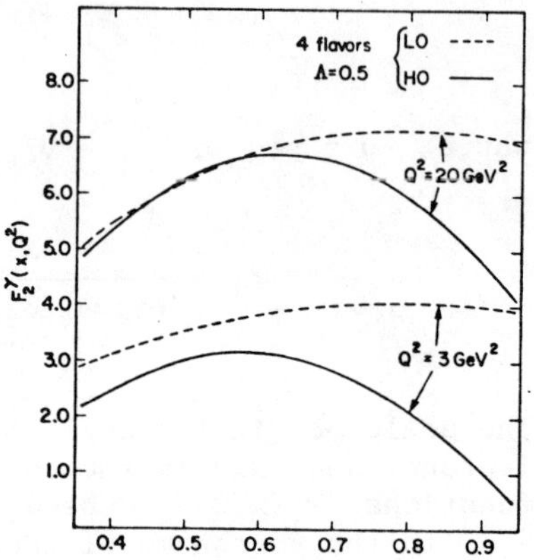

Fig. 4. Point-like leading order QCD result compared to vector dominance contribution (Ref. 9)

Fig. 5. Leading order QCD result compared to result of first two orders (labeled HO). From Bardeen and Buras, Ref. 12

have found that the higher-order QCD contributions become large in this region, as can be seen in Fig. 5. It is widely believed that this is a kinematic effect of the phase-space boundary, shown in Eq. (4a) by the factor 1 - x in the argument of the logarithm. Such problems have received a gread deal of attention recently, and plausible prescriptions exist for handling the $x \approx 1$ region, but we are still lacking a definitive demonstration by means of a higher-order calculation in the two-photon process.[13]

For the time being, the photon structure functions definitively test QCD only for mid-range values of x; for example, at $Q^2 = 3$ one should trust only $0.3 \lesssim x \lesssim 0.7$. At higher values of Q^2 the region expands. It is not that we have any evidence of a failure of convergence of QCD perturbation, it is just that the convergence is not uniform in x.

In summary, the two-photon process provides, because of the simple point-like fragmentation of the photon into quarks, definitive tests of QCD in high-$p_\perp$ jet production and in measurement of the photon structure.

REFERENCES

1. In these brief remarks I cannot give a complete historical review. See H. Terazawa, Rev. Mod. Phys. 45, 615 (1973); V. M. Budner et al., Phys. Rep. 15C, 181 (1974).
2. F. E. Low, Phys. Rev. 120, 582 (1960).
3. F. J. Gilman, Proceedings of the 1979 International Conference on Two-Photon Interactions (J. F. Gunion, ed.).
4. C. H. Llewellyn Smith, Phys. Lett. 79B, 83 (1978).
5. S. Brodsky, T. De Grand, J. Gunion, and J. Weis, Phys. Rev. Lett. 41, 672 (1978); Phys. Rev. D19, 1418 (1979).
   K. Kajantie, Phys. Scripta 29, 230 (1979);
   K. Kajantie and R. Raitio, Nucl. Phys. B159, 528 (1979).
6. R. P. Worden, Phys. Lett. 51B, 57 (1974).
7. T. Walsh and P. Zerwas, Nucl. Phys. B41, 551 (1972);
   R. L. Kingsley, Nucl. Phys. B60, 45 (1973).
8. E. Witten, Nucl. Phys. B120, 189 (1977).
9. W. R. Frazer and J. F. Gunion, Phys. Rev. D20, 147 (1979).
10. R. J. DeWitt, L. M. Jones, J. D. Sullivan, D. E. Willen, and H. W. Wyld, Phys. Rev. D19, 2046 (1979).
11. W. Frazer and G. Rossi, Phys. Rev. D21, 2710 (1980);
    K. Sasaki, contribution 333 to this Conference;
    C. Peterson, T. F. Walsh, and P. M. Zerwas, preprint NORDITA-80/13.
12. W. A. Bardeen and A. J. Buras, Phys. Rev. D20, 166 (1979).
13. G. Parisi, Phys. Lett. 90B, 295 (1980);
    D. Amati, A. Bassetto, M. Ciafaloni, and G. Marchesini, CERN preprint Th. 2831 (1980);
    M. Ciafaloni, contribution to this Conference;
    S. Brodsky, contribution to this Conference.

STRUCTURE FUNCTIONS AND HIGH TWIST CONTRIBUTIONS IN PERTURBATIVE QUANTUM CHROMODYNAMICS*

Stanley J. Brodsky
Stanford Linear Accelerator Center
Stanford University, Stanford, California 94305
and
G. Peter Lepage†
Laboratory of Nuclear Studies
Cornell University, Ithaca, New York 14853

ABSTRACT

Perturbative QCD predictions are presented for the x near 1 behavior of hadronic structure functions. The available energy xW^2 is shown to control structure function evolution. In the case of meson structure functions, the $x \sim 1$ behavior is dominated by a high twist contribution to the longitudinal structure function $F_L \sim Cx^2/Q^2$ which can be rigorously computed and normalized.

One of the most important areas of study of perturbative quantum chromodynamics is the behavior of the hadronic wavefunctions at short distances or at far off-shell kinematics. This behavior can be tested not only in exclusive reactions such as form factors at large momentum transfer but also in deep inelastic scattering reactions at the edge of phase space. In this talk we will review the QCD predictions for the behavior of the hadronic structure functions $F_i(x,Q)$ in the end-point $x_{Bj} \sim 1$ region.[1] The endpoint region is particularly interesting because one must understand in detail (a) the contributions of exclusive channels, (b) the effect of high twist terms (power-law scale-breaking contributions) which can become dominant at large x, and (c) the essential role of the available energy W in controlling the logarithmic evolution of the structure functions. Note that as $x \sim 1$, essentially all of the hadron's momentum must be carried by one quark (or gluon), and thus each propagator which transfers this momentum becomes far-off shell: $k^2 \sim -(\vec{k}_\perp^2 + \mathcal{M}^2)/(1-x) \to -\infty$ [see Fig. 1]. Accordingly, if the spectator mass $\mathcal{M}$ is finite the leading power-law behavior in (1-x) is determined by the minimum number of gluon exchanges required to stop the hadronic spectators, and only the valence Fock states, $|qqq\rangle$ for baryons, and $|q\bar{q}\rangle$ for mesons, contribute to the leading power behavior. If one simply computes the connected tree graphs, as in Fig. 2, one finds that the results depend in an essential way on the quark's helicity:

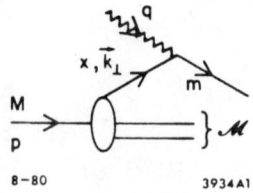

Fig. 1. Kinematics for inelastic structure functions.

$$G_{q/p}(x) \sim \begin{cases} (1-x)^3 & \text{parallel quark/nucleon helicity} \\ (1-x)^5 & \text{anti-parallel quark/nucleon helicity} \end{cases}$$

* Work supported by Department of Energy, contract DE-AC03-76SF00515.
† Work supported by the National Science Foundation.

and $G_{q/\pi} \sim (1-x)^2$. If the nucleon wavefunctions satisfy the standard SU(6) spin-flavor symmetry, then the above results imply[2]

$$\frac{G_{u/p}}{G_{d/p}} \underset{x\to 1}{=} 5 \quad .$$

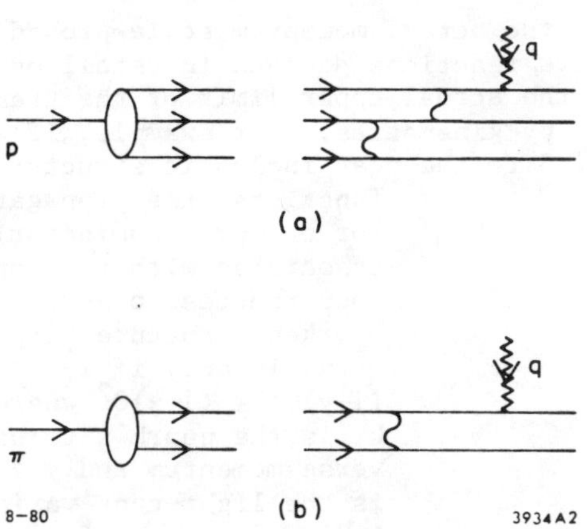

Fig. 2. Perturbative QCD tree diagrams for computing the $x \sim 1$ power behavior of baryon and meson structure functions.

Let us now consider how these results for the power-law behavior emerge within the complete perturbative structure of QCD. Including corrections from gluon radiation, vertex and self-energy corrections, and continued iteration of the gluon-exchange kernel, one finds for the nucleon's quark distribution[1]

$$G_{q/p}(x,Q) \underset{x\to 1}{=} (1-x)^3 \alpha_s^4(k_x^2) \left| \sum_{j=0}^{\infty} b_j \left(\log \frac{k_x^2}{\Lambda^2} \right)^{-\gamma_j^N} \right|^2 P_q(x,Q)$$

$$\times \left[1 + \mathcal{O}\left(\alpha_s(k_x^2), 1/Q \right) \right] \quad . \qquad (1)$$

The powers of α_s and $(1-x)$ reflect the behavior of the hard scattering amplitude at the off-shell value $k_x^2 = (\langle k_\perp^2 \rangle + \mathcal{M}^2)/(1-x)$ where $\langle k_\perp^2 \rangle$ is set by the spectator transverse momentum integrations. The anomalous dimensions γ_j^N are the anomalous dimensions of the nucleon's valence Fock state wavefunction at short distances. Their contribution to $G_{q/p}(x,Q)$ are due to the evolution of the wavefunction integrated up to the transverse momentum scale $\vec{\ell}_\perp^2 < k_x^2$ as in the corresponding exclusive channel analyses.[1] The last factor $P_q(x,Q)$ represents the target-independent evolution of the structure function due to gluon emission from the struck quark: $(C_F = 4/3)$

$$P_q(x,Q) \sim (1-x)^{4C_F \xi(Q)} \qquad (2)$$

where

$$\xi(Q) = \int_{Q_0^2}^{Q^2} \frac{dj_\perp^2}{j_\perp^2} \alpha_s(j_\perp^2) \sim \log\left(\frac{\log Q^2/\Lambda^2}{\log Q_0^2/\Lambda^2} \right) \quad . \qquad (3)$$

The lower limit Q_0^2 of the gluon's transverse momentum integration is set by the mean value of the spectator quark's transverse momenta and masses. This hadronic scale sets the starting point for structure functions evolution.[3] Equation (1) then gives the light-cone momentum distribution for parallel-helicity quarks with x near 1 at the transverse momentum scale Q.

It should be emphasized that the actual momentum scale probed by various deep inelastic inclusive reactions depends in detail on the process under consideration; the actual upper limit of the transverse momentum integration is set by kinematics. For example, if we consider the contribution of Fig. 3 to the deep inelastic structure functions, the propagator (or energy denominator) associated with the top loop reduces to the usual Bjorken structure $2q\cdot p - Q^2/x + i\varepsilon$ only if $k_\perp^2 << (1-y)Q^2 \le (1-x)Q^2$ where $k_\perp$ is the quark's transverse momentum and $y \ge x$ is the light-cone variable indicated in the figure. The remaining structure factorizes into a form which defines $G_{q/p}(x/y, k_\perp)$. Thus the actual relation between the structure function and the momentum distribution for $x \sim 1$ is[1,4]

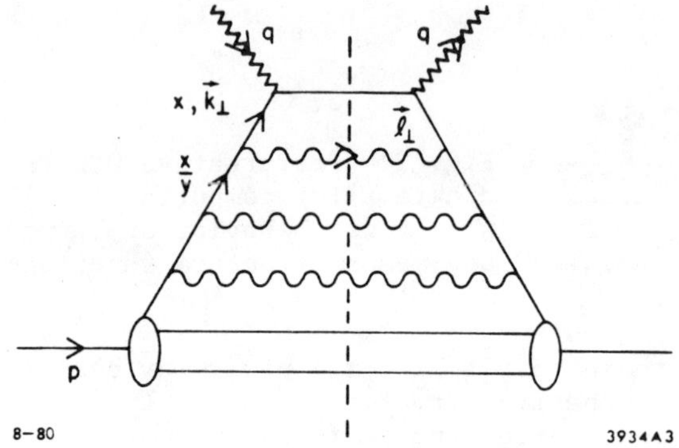

Fig. 3. Perturbative QCD diagrams for structure function evolution.

$$F_2(x,Q) = \sum_i e_i^2 \, x_{Bj}\left[G_{q_i/p}(x_{Bj},Q) + \delta G_{q_i/p}(x_{Bj},Q)\right] \qquad (4)$$

where

$$\delta G_{q/p}(x,Q) = -2C_F \int_0^1 dy \, \frac{1+y^2}{1-y} \int_{(1-y)Q^2}^{Q^2} \frac{dk_\perp^2}{k_\perp^2} \frac{\alpha_s(k_\perp^2/y)}{4\pi}$$

$$\times \left\{ G_{q/p}(x/y,k_\perp) \, \frac{\theta(y>x)}{y} - G_{q/p}(x,k_\perp) \right\} \qquad (5)$$

corrects for the fact that the top loop is integrated to $k_\perp^2 < (1-y)Q^2$ ($\le x_{Bj}W^2$) not Q^2. [The argument of α_s is also crucial here.] Other inclusive reactions have to be individually examined: in the case of the Drell-Yan process $q_a\bar{q}_b \to \mu^+\mu^-$, the structure functions evolve to $(1-y_a)Q^2$ and $(1-y_b)Q^2$ not $Q^2 = (p_+ + p_-)^2$.

The actual evolution of structure functions in deep inelastic lepton scattering is thus controlled by the available energy $x_{Bj}W^2$, and is more moderate at $x_{Bj} \sim 1$ than would be expected from lowest order expectations. Analytic forms for the (1-x) behavior are readily computed.[1] The most important features are the following: (1) The $\delta G/G$ correction to leading order in α_s reproduces the critical $2C_F(\alpha_s(Q^2)/4\pi)\log^2 n$ terms in the structure function moments as calculated using the operator product expansion and renormalization group. In our analysis a series of terms of all orders in $(\alpha_s \log^2(1/1-x))^p$ or $(\alpha_s \log^2 n)^p$ arises simply from the fact that the natural evolution parameter for the structure functions $F_i(x,Q)$ and moments $\mathcal{M}_n(x,Q)$ is controlled by $(1-y)Q^2 < (1-x)Q^2$ and not Q^2; the basic momentum distributions $G(x,Q)$ do not contain the anomalous double-log terms and have a straightforward perturbative evolution. (2) The extended evolution equations based on Eqs. (4) and (5) have a number of phenomenological advantages. After taking into account the appropriate evolution limits, each deep inelastic process can be related to the basic distributions $G_q(x,Q)$, avoiding large kinematic corrections. The scale parameter Λ_n which has been introduced to eliminate the strong n-dependence of the higher order corrections to the moments is unnecessary. The fact that xW^2 controls the evolution suggest its use in structure function parameterizations and studies of moment factorization in fragmentation processes. A study of the application of this method to photon structure functions is in progress. (3) The exclusive-inclusive connection fails in QCD.[1,5] At fixed but large W^2, $F_{2N}(x,Q)$ falls as $(1-x)^{3+\delta}$ where $\delta > 0$, whereas, modulo logarithmic factors, exclusive channels in QCD give contributions $\sim(1-x)^3$ from the Q^{-4} scaling of the leading nucleon form factors. Thus exclusive channels will eventually dominate the leading twist contributions to inclusive cross sections at fixed W^2, $Q^2 \to \infty$.

A complete treatment of the hadron structure functions must take into account higher twist contributions. Although such contributions are suppressed by powers of $1/Q^2$, they can have fewer powers of (1-x) and, accordingly, may be phenomenologically important in the large x domain.[6] In the case of nucleons, the $\ell + qq \to \ell' + qq$ subprocess (in which the lepton recoils against two quarks) leads to a structure function contribution $\sim(1-x)/Q^4$ since only one quark spectator is required. A large longitudinal structure function is also expected.[6,7,8] Although complete calculations of such terms have not been done, the presence of such terms can reduce the amount of logarithm scale-violation required from the leading twist contributions in phenomenological fits.[6,9]

The analysis of meson structure functions at $x \sim 1$ is similar to that of the baryon, with two striking differences: (1) The controlling power behavior of the leading twist contribution is $(1-x)^2$ from perturbative QCD.[2,10] The extra factor of (1-x) -- compared to what would have been expected from spectator counting -- can be attributed to the mismatch between the quark spin and that of the meson. (2) The longitudinal meson structure function has an anomalous nonscaling component[7,11] which is finite at $x \to 1$: $F_L(x,Q) \sim Cx^2/Q^2$. This high twist term, which comes from the lepton scattering off an instantaneous fermion-line in light-cone perturbation theory, can be rigorously computed and normalized in perturbative QCD. The crucial fact is that the wavefunction evolution and spectator transverse momentum integrations in Fig. 4 can be written directly in terms of a corresponding calculation of the meson form factor. The result for the pion structure function to leading order in $\alpha_s(k_x^2)$ and $\alpha_s(Q^2)$ is[11,12]

$$F_L^\pi(x,Q) = \frac{2x^2}{Q^2} C_F \int_{m^2/(1-x)}^{Q^2} dk^2\ \alpha_s(k^2)\, F_\pi(k^2) \qquad (6)$$

which numerically is $F_L \sim x^2/Q^2$ (GeV^2 units).

The dominance of the longitudinal structure functions in the fixed W limit for mesons is an essential prediction of perturbative

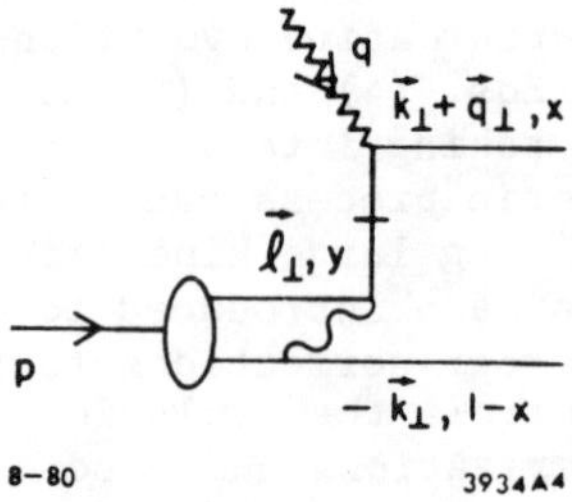

Fig. 4. Perturbative contribution to the meson longitudinal structure function $F_L \sim C/Q^2$.

QCD. Perhaps the most dramatic consequence is in the Drell-Yan process $\pi p \to \ell^+\ell^- X$; one predicts[7] that for fixed pair mass Q, the angular distribution of the ℓ^+ (in the pair rest frame) will change from the conventional $(1+\cos^2\theta_+)$ distribution to $\sin^2(\theta_+)$ for pairs produced at large x_L. A recent analysis of the Chicago-Illinois-Princeton experiment[13] at FNAL appears to confirm the QCD high twist prediction with about the expected normalization. Striking evidence for the effect has also been seen in a Gargamelle analysis[14] of the quark framentation functions in $\nu p \to \pi^+\mu^- X$. The results yield a quark fragmentation distribution into positive charged hadrons which is consistent with the predicted form: $dN^+/dzdy \sim B(1-z)^2 + (C/Q^2)(1-y)$ where the $(1-y)$ behavior corresponds to a longitudinal structure function. It is also crucial to check that the $e^+e^- \to MX$ cross section becomes purely longitudinal $(\sin^2\theta)$ at large z at moderate Q^2. The implications of this high twist contribution for meson production at large p_T will be discussed elsewhere.[12]

REFERENCES

1. S. J. Brodsky and G. P. Lepage, SLAC-PUB-2447, Proc. of SLAC 1979 Summer Inst. on Particle Physics, and references therein.
2. G. R. Farrar and D. R. Jackson, Phys. Lett. 35, 1416 (1975). Recent BEBC data appear to be consistent with this result. See F. Sciulli, this meeting.
3. S. J. Brodsky, G. P. Lepage and T. Huang, SLAC-PUB-2540 (1980).
4. See also G. Curci and M. Greco, Phys. Lett. 92B, 175 (1980); D. Amati *et al.*, CERN-TH-2831 (1980); and M. Ciafaloni, this meeting.
5. See also G. Parisi, Phys. Lett. 84B, 225 (1979).
6. R. Blankenbecler and I. Schmidt, Phys. Rev. D16, 1318 (1977); R. Blankenbecler *et al.*, Phys. Rev. D12, 3469 (1975).
7. E. L. Berger and S. J. Brodsky, Phys. Rev. Lett. 42, 940 (1979); E. L. Berger, Phys. Lett. 89B, 241 (1980).
8. L. Abbott *et al.*, Phys. Lett. 88B, 157 (1979).
9. L. Abbott *et al.*, Phys. Rev. D22, 582 (1980).
10. Z. F. Ezawa, Nuovo Cimento 23A, 271 (1974). This result assumes the mass of the spectator is finite. If $\mathcal{M} \to 0$, perturbative and even-non-perturbative calculations allow $F_{2\pi}(x) \sim (1-x)$. See A. DeRujula and F. Martin, MIT-CTP-851 (1980); F. Martin, SLAC-PUB-2581, this conference; and S. J. Brodsky *et al.*, Ref. 3.
11. Tests of this result will be discussed in E. L. Berger, S. J. Brodsky and G. P. Lepage (in preparation).
12. See also A. Duncan and A. Mueller, Phys. Lett. 90B, 159 (1980).
13. K. J. Anderson *et al.*, Phys. Rev. Lett. 43, 1219 (1979).
14. CERN-MILAN-ORSAY Collaboration, CERN-EP/80-124 (1980); C. Matteuzzi *et al.*, contribution to this meeting.

TWO-PHOTON RESULTS FROM SPEAR*

A. Roussarie
Stanford Linear Accelerator Center
Stanford University, Stanford, California 94305

ABSTRACT

I report results obtained by two experiments at SPEAR on the two-photon production of lepton pairs and resonances. Both experiments find agreement of lepton production with QED expectations, and observe an enhancement of the $\pi\pi$ mass spectrum in the 1250 MeV/c^2 region. The Mark II finds this enhancement not consistent with the decay of the f(1270 MeV/c^2) alone. The $\gamma\gamma$ partial width of the η' has also been measured by the Mark II.

INTRODUCTION

Two experiments, performed at SPEAR, have measured dilepton and resonance production via the two photon mechanism: $e^+e^- \rightarrow e^+e^-X$. In addition to the $\gamma\gamma \rightarrow X$ cross section, the total cross section for these processes contains the probability for the incident electrons to emit the bremsstrahlung photons. This gives the kinematic peculiarity of these events:

(1) The X system has a low mass and is confined at a very low transverse momentum with respect to the beams.

(2) The cross sections rise logarithmically with the beam energy.

SP 14: U.C. SAN DIEGO[1]

A luminosity of 10 pb^{-1} was accumulated for a center-of-mass energy between 6 and 7.2 GeV/c^2. The apparatus consists of 3 parts:

(1) A small angle detector (polar angle θ from 55 to 180 mrad) tags the outgoing e^+e^-. Doubly tagged events are selected for which one can compute the squared mass of the $\gamma\gamma$ system (M_X^2).

(2) An inner detector ($15^o \leq \theta \leq 30^o$) is able to separate e, μ and hadrons. 28 events with a μ and 30 events with a π were recorded, showing that in this solid angle, π and μ rate are comparable. Figures 1a and 1b display the M_X^2 spectrum for these events. On Fig. 1a the line gives the normalized QED expected spectrum which agrees very well with the data.

(3) A crude central detector provides, in a wide solid angle ($22^o \leq \theta \leq 158^o$) information on the azimuth of charged particles. It is used to search for two body decay of resonances. Figures 1c to 1f give the M_X^2 distributions for two prong events. Figure 1c shows all the 189 events while Fig. 1e is restricted to tagged events which have both electrons on the same side of an horizontal plane. This configuration cannot be faked by the Bhabha annihilation events that contaminate Fig. 1c in the zero mass region. The ee and $\mu\mu$

---

* Work supported by Department of Energy, contract DE-AC03-76SF00515.

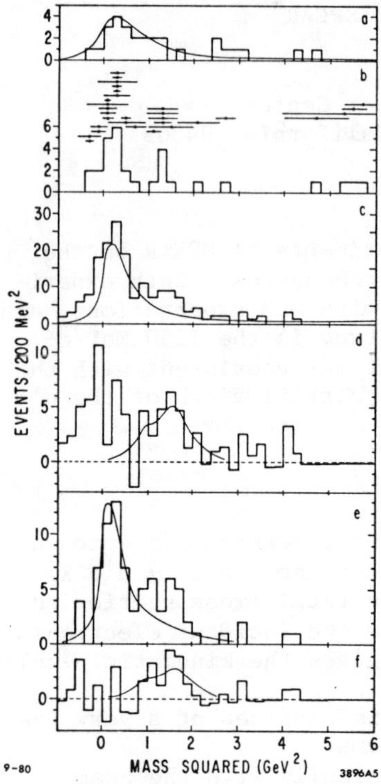

Fig. 1. SP 14 — histograms of the $\gamma\gamma$ mass squared (see text).

event rate can be computed from QED (solid lines) and subtracted from Figs 1c and 1e to get Figs. 1d and 1f respectively. Both distributions display, above the QED continuum, an enhancement around 1.5 $(GeV/c^2)^2$ of 26 events. The process $\gamma\gamma \to \eta' \to \rho\gamma$ is expected to contribute 5 events to this region (computed from the known $\gamma\gamma$ width of the η').[2] This leaves 21 events that can be f(1270), ε(1100 to 1300) or A_2(1310) decay. The result is that if the resonance responsible for these events has a spin 2, its total $\gamma\gamma$ decay width is $9.5 \pm 3.9 \pm 2.4$ keV.

SP 29 (MARK II): SLAC-LBL

A luminosity of 14 pb^{-1} was obtained between 4.4 and 7.2 GeV/c^2. The MARK II detector has no tagging system. Two-photon events are selected by requiring the observation of exclusive channels of low invariant mass (≤ 2 GeV/c^2) produced at low transverse momentum (≤ 250 MeV/c).

(1) η' production: I will just recall briefly this published result.[2] From 61 $\eta' \to \gamma\gamma$ events observed, the $\gamma\gamma$ partial width of the η' was determined to be

$$\Gamma_{\eta \to \gamma\gamma} = (5.8 \pm 1.1 \pm 1.2) \text{ keV}$$

which together with the known branching ratio gives a total width for the η' in very good agreement with a recent missing mass measurement.[3]

(2) 2 prong production: In addition to the M_X and p_T cut, the 2 tracks are required to be well inside the liquid argon detector solid angle ($|\cos\theta| \leq 0.6$).

— 2 prong ee: For the events in which the two particles are identified by the liquid argon as electrons, the mass plot of Fig. 2 is obtained. The data (dots) are in very good agreement with a complete QED computation

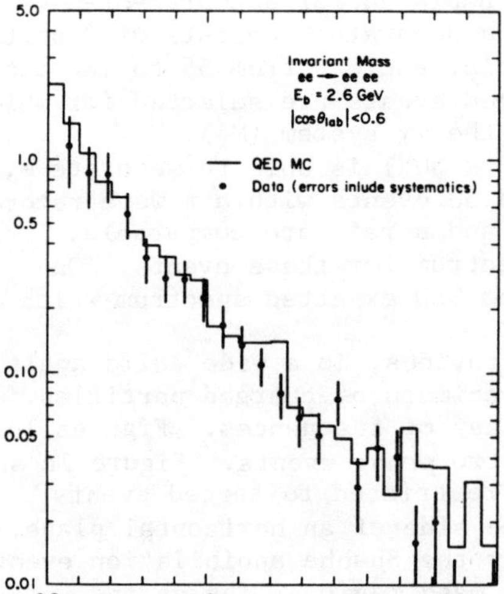

Fig. 2. MARK II — M_{ee} (GeV).

using the Vermaseren program.[4] Not shown here is the very good agreement with the two photon process of the total p_T and center-of-mass energy of these events.

— 2 prong all: Using for each particle the π mass, the mass distribution of all the 2 prong events was obtained (not shown here). Below 900 and above 1400 MeV/c^2 its shape is perfectly similar to the computed QED distribution (ee and $\mu\mu$ pairs) but 8% higher. The QED continuum is normalized to the data in the 700-900 MeV/c^2 region. The subtracted distribution is shown in Fig. 3. Possible sources for the excess of 540 events between 950 and 1500 MeV/c^2 are the decay into $\pi\pi$ of the C even resonances: S*(980), f(1270) and ε(between 1100 and 1300). If all these events are f (mass = 1270 MeV/c^2, width = 180) the mass distribution is expected to be the line drawn on Fig. 3. If this assumption were true, the $\gamma\gamma$ partial width of the f(spin and helicity 2) would be 3.5 ± 0.6 keV. But the fit is bad; the mass spectrum is not consistent with the decay of the f alone. An upper limit of the f to $\gamma\gamma$ partial width is

$$\Gamma_{f \to \gamma\gamma} \leq 4.7 \text{ keV (95\% confidence level)} \quad .$$

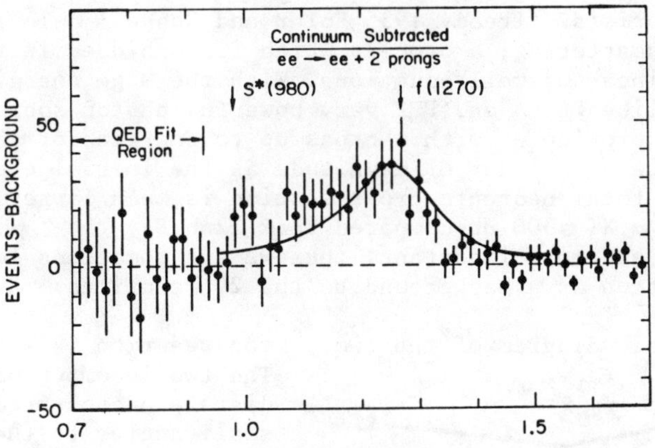

Fig. 3. MARK II — $M_{\pi\pi}$ (GeV).

REFERENCES

1. C. J. Biddick et al., Paper #583, submitted to this conference.
2. G. S. Abrams et al., Phys. Rev. Lett. 43, 477 (1979).
3. D. M. Binnie et al., Phys. Lett. 83B, 141 (1979).
4. J. Smith, J.A.M. Vermaseren and G. Grammer, Phys. Rev. D15, 3220 (1977).

Two Photon Processes at PETRA

W. Wagner
Deutsches Elektronen-Synchrotron DESY, Notkestr. 85, Hamburg

ABSTRACT

The analysis of two photon physics at the e^+e^- storage ring PETRA is reviewed. Higher order QED processes $e^+e^- \rightarrow e^+e^-e^+e^-$ $(e^+e^-\mu^+\mu^-)$ have been measured in good agreement with the QED calculations. The production of the f^0 resonance has been investigated and the radiative width has been determined. The total cross section $\sigma(Q^2,W)$ of the inelastic electron photon scattering has been studied in some detail. Evidence for the production of large P_T jets in the reaction $\gamma\gamma \rightarrow q\bar{q}$ is reported and a first measurement of the photon structure function in deep inelastic $e\gamma$ scattering is discussed.

INTRODUCTION

Long before the two photon reactions were investigated experimentally the scattering of light by light attracted the attention of many theorists. Already 1935 Euler and Kockel[1] calculated the elastic $\gamma\gamma$ scattering, a process which is forbidden in the classical notion of linear maxwell equations. With the high energy e^+e^- storage rings like PETRA and PEP very powerful photon sources are available which provide us with photons up to energies of 15 GeV and fluxes of the same order of magnitude as the initial e^+e^- beams. Because the total hadronic cross section is much larger for $\gamma\gamma \rightarrow X$ than for $ee \rightarrow X$ ($\approx$300 nb compared to $R\cdot 22nb/E^2$) the 2 photon processes become more and more important, thus we are sometimes looking at the 1γ annihilation as a background to the 2γ reactions.

The basic diagram of the two photon reaction is shown in fig. 1. The two incoming particles radiate a photon predominantly at small angles to the beam and with small energies. These two photons react and produce a final state. Typically the 2γ events are measured in the following way: the final state is detected in a central detector, and the scattered electrons are detected in a separate device, a forward spectrometer, covering the range of $1^o - 10^o$ and $170^o - 179^o$ respectively.

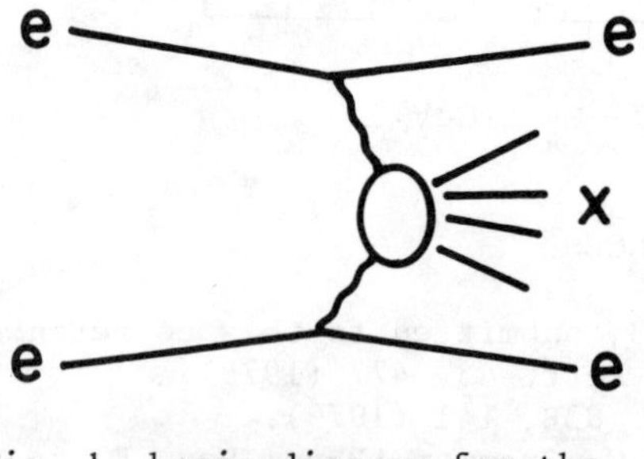

Fig. 1 basic diagram for the reaction $ee \rightarrow eex$

None of the PETRA detectors has a 0° tagging system. We distinguish between three experimental conditions:

1. 'double tag': both of the scattered electrons are detected, both photons are virtual: $Q^2 > 0.1$ GeV^2 for a typical PETRA detector ($-Q^2$ is the invariant mass squared of the photon). Though this condition was considered to be the only clean way of measuring 2γ phy-

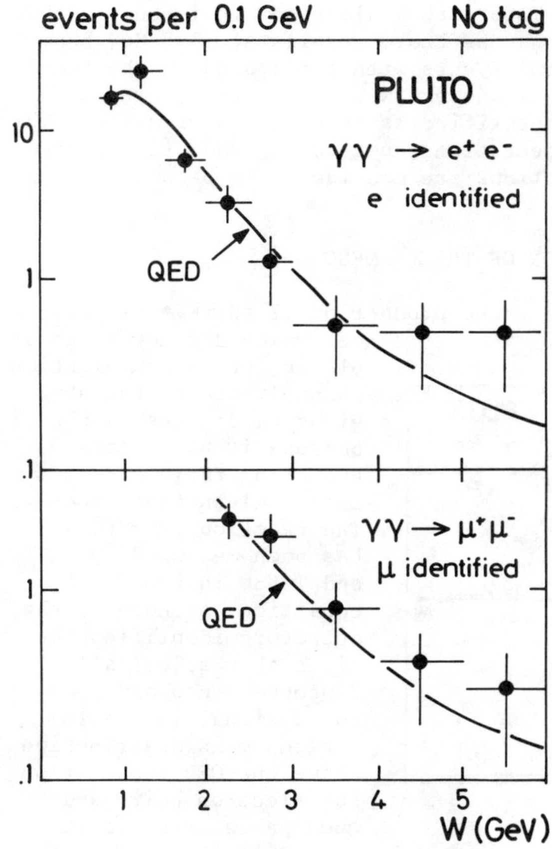

Fig. 2 Invariant mass distribution of the electron pairs and the μ pairs from the reaction $\gamma\gamma \to ee(\mu\mu)$

sics a couple of years ago, there are no experimental results up to now. The double tag event rate is very small, and the interpretation is complicated, because in general 8 structure functions can contribute to the inclusive cross section.

2. 'single tag': only one of the electrons is measured and the other electron stays in the beam pipe, radiating an almost real photon ($< Q^2 > < 0.01$ GeV^2). Besides the fact that the event rate is higher than in the double tag condition, the interpretation is much clearer. We can regard this process as electroproduction off an almost real photon target. It should be noted that for this interpretation of the single tag events it is essential that one has a complete coverage of electron detection in θ between 1° and 179° to make sure that the second electron is scattered at small angles.

3. 'no tag': both electrons stay in the beam pipe, both photons are almost real. This is the (theoretically) most easy case, only one structure function contributes, and the measured cross section is only a function of W, the invariant mass of the final state X.

The common signature of all 2γ reactions is the small fraction of the total e^+e^- energy carried by the photons. This leads to only little energy in the central detector, which allows a clean separation from 1γ annihilation events.

TEST OF THE QED IN THE REACTIONS $\gamma\gamma \to ee, \mu\mu$

The reaction ee → ee + lepton pairs can be completely calculated in QED. The measurement of a higher order QED process with an amplitude $\sim e^4$ as an isolated reaction (and not as usual as a radiative correction) is important by itself, although the Q^2 values involved are quite small. The QED reactions have been measured by four PETRA groups MARK J, JADE, PLUTO, TASSO, they all find consistency with QED calculations[2]. Fig. 2 shows the PLUTO results at 15.5 GeV beam energy in the no tag condition. Events with two prongs in the central detector and no additional showers were selected, where at least one of the tracks was identified as an electron or a muon. The curve drawn shows the QED expectations, neglecting radiative corrections. These radiative corrections are considered to be small (1-2%)[3].

PRODUCTION OF THE f^o MESON

Two photon reactions allow the production of the C = +1 resonances which are not accessable in the 1γ annihilation channel. One of the obvious candidates is the f^o because it has a large branching ratio of 55% into a clean final state. The reaction $\gamma\gamma \to f^o \to \pi^+\pi^-$ has been measured by PLUTO and TASSO in the no tag condition. As none of the detectors identifies the pions they select all 2 prong events and look for a signal in the invariant mass distribution above the QED expectation for electron pairs and muon pairs. Fig. 3a shows this distribution for the PLUTO detector with the QED subtracted data as an insert and Fig. 3b shows the TASSO data after QED subtraction. Both experiments show a clean peak around 1250 MeV. In order to determine the radiative width $\Gamma^{f^o}_{\gamma\gamma}$ from the number of events in the peak one has to make an assumption

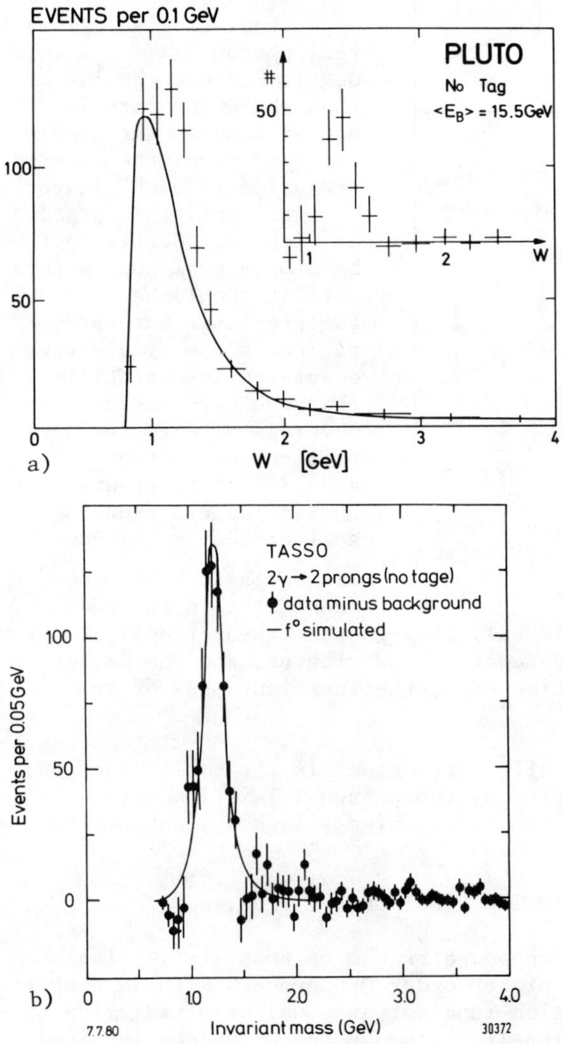

Fig. 3 Invariant mass distribution of the two prong events. The curve in 3a is an absolute QED prediction. The insert and the diagram from TASSO show the difference between data and QED background.

about the helicity amplitude for the f^o production[4] . Two amplitudes are possible, $\lambda = 0$ and $\lambda = 2$, which lead to about 50% different trigger efficiencies. Assuming $\lambda = 2$ we get:

$$\begin{aligned} \text{PLUTO} \quad \Gamma^{f^o}_{\gamma\gamma} &= (2.3 \pm .5 \text{ (stat.)} \pm .35 \text{ (syst.)) keV} \\ \text{TASSO} \quad \Gamma^{f^o}_{\gamma\gamma} &= (4.1 \pm .4 \text{ (stat.)} \pm .6 \text{ (syst.)) keV} \end{aligned} \tag{1}$$

Stretching the systematic errors the data agree, but one should note that the TASSO results are still preliminary.[11]

TOTAL CROSS SECTION FOR $\gamma\gamma$-HADRONS

The total cross section for multihadron production via two photons has been measured by PLUTO and TASSO with quite considerable statistics (ca. 1000 events). The data have been taken in the single tag mode allowing a study of the dependence of σ on W and Q^2, where Q^2 is determined from the scattered electron and W from the final state hadrons observed in the central detector. In the Vector Dominance Model (VDM) the Q^2 dependence of the cross section for transverse polarized photons is mainly given by a ρ pole form factor

$$\sigma_t(W,Q^2) = \sigma_{\gamma\gamma}(W) \left(\frac{m_\rho^2}{m_\rho^2 + Q^2}\right)^2 \tag{2}$$

A second term $\varepsilon\sigma_l$ for the longitudinal polarized photons is considered to be small[5]. $\sigma_{\gamma\gamma}(W)$, the cross section for two real photons can be estimated using Pomeron factorization and Resonance Regge duality[6]

$$\sigma_{\gamma\gamma}(W) = \sigma^{VDM}(W) = 240\text{nb} + \frac{270\text{nb.GeV}}{W} \tag{3}$$

Fig. 4 shows the PLUTO results for $\sigma_{\gamma\gamma}$ as a function of the visible invariant mass, W_{vis}, at 15.5 GeV beam energy. Above 3 GeV the shape agrees with the VDM predictions, where at small W the data are substantially higher. This could indicate the presence of non Regge terms[6] like the quark box diagram (Fig. 5) which would lead to the typical $1/W^2$ behaviour of point-like processes, although this assumption is debated[7]. Fig.6 shows the Q^2 behaviour for two

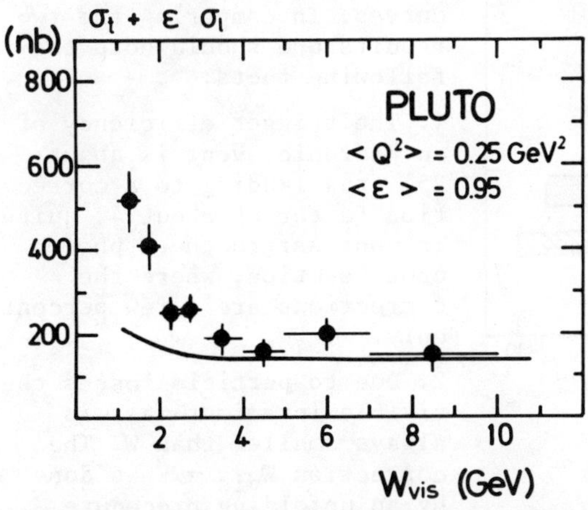

Fig.4 Total cross section vers. W_{vis}

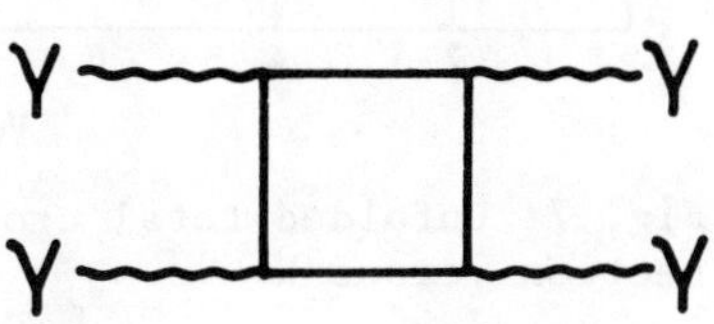

Fig.5 Quark box diagram

different invariant mass bins, both being consistent with the ρ pole form factor. For $W_{vis} > 3.5$ GeV one can assume that the total cross section is dominated by the VDM contribution. A separate fit of this

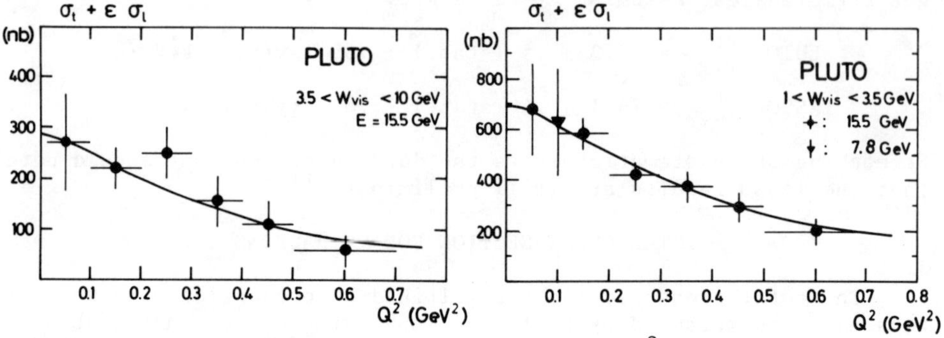

Fig. 6 Total cross section as a function of Q^2 for two different W_{vis} bins

W region results in $\sigma(W) = (1.21 \pm .13) \cdot \sigma^{VDM}(W)$. Finally the whole data sample is fit by the ansatz:[5]

$$\text{PLUTO:} \qquad \sigma(W) = A \cdot \sigma^{VDM}(W) + B/W^2 \qquad (4)$$

in order to account for the enhancement at low W_{vis}. The data from TASSO are fit by a slightly different ansatz:[11]

$$\text{TASSO:} \qquad \sigma(W) = A + B/W \qquad (5)$$

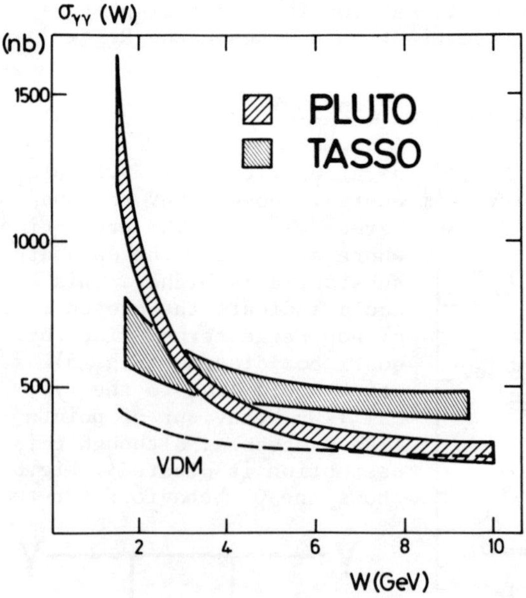

Fig. 7 Unfolded total cross section versus W

whereas both groups assume the same Q^2 behaviour (Eq.2). The fit results are represented by the hatched bands in Fig. 7 with an additional systematic error of 25%. The two experiments agree on average but differ very clearly in the shape of the curves. In comparing the two results one should note the following facts:

1. The trigger efficiency of an hadronic event is about 25% thus leading to a correction factor of about 4, quite in contrast to the 1 photon cross section, where the corrections are a few percent only.

2. Due to particle losses the visible invariant mass is always smaller than W. The correction $W_{vis} \rightarrow W$ is done by an unfolding procedure

which needs a specific ansatz. (The PLUTO data can also be fit by an ansatz A + B/W, resulting in a somewhat flatter cross section.)

3. There are two major differences in the analysis of the two experiments: a) PLUTO uses single tags only where TASSO includes the double tags, b) TASSO uses charged particles only where PLUTO includes the neutrals, leading to a smaller correction $W_{vis} \to W$.

HARD SCATTERING PROCESSES

Though the gross features of the total cross section are well described by the VDM picture there is certainly room for additional processes; especially the PLUTO data seem to suggest a contribution from the pointlike photon quark coupling (Fig. 5). A possibly better way to look for this pointlike coupling is the study of hard scattering processes (Fig. 8).

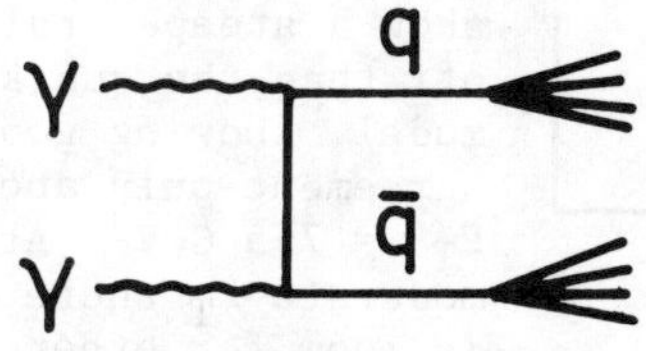

Fig. 8 pointlike photon quark coupling

There are two cases of interest:

1. the exchanged quark is highly virtual. This leads to the production of two coplanar but non collinear jets with large transverse momenta, a process which cannot occur in the VDM picture, but which can easily be related to the μ pair production:

$$R_{\gamma\gamma} = \frac{\sigma\ \gamma\gamma \to q\bar{q}}{\sigma\ \gamma\gamma \to \mu\mu} = 3\ \Sigma\ e_q^4\ (1 + O(\alpha_s)) \approx \frac{34}{27} \text{ for } q = udsc \qquad (6)$$

2. one of the photons is highly virtual. The process can be understood as deep inelastic electron photon scattering, thus probing the hadronic structure of the photon.

The part of the photon structure function, resulting from the pointlike γq coupling can be calculated from first principles in QCD[8,9].

PRODUCTION OF LARGE P_T JETS

A good candidate for a two photon jet event, seen in the PLUTO detector, is shown in Fig. 9. Similar events are also seen by CELLO and TASSO. For a quantitative analysis PLUTO selected all events (single tag + no tag) with 3 GeV < W_{vis} < 9 GeV. The P_T^2 distribution of the charged particles of these events shows a very pronounced tail (Fig. 10), which can be partly explained by the process $\gamma\gamma \to q\bar{q}$ including charm quarks (full curve). In a further step the particles are ordered into two non collinear jets by maximizing the twoplicity T_2 (=Thrust along 2 different axis). Fig. 11 shows that the tail in the P_T distribution is entirely due to the jetlike events (T_2 > .75), whereas the transverse momenta for the more isotropic events (T_2 < .75) are limited.

A crucial test of the production mechanism is the P_T behaviour of the partons in the underlying hard scattering process which can be studied by the P_T distribution of the jets. For the process $\gamma\gamma \to q\bar{q}$ we expect P_T^{-4} whereas higher twist terms[9] like $\gamma\gamma \to q\bar{q}$ meson, where a meson is involved in the basic hard process should be $\sim P_T^{-6}$. Fig. 11 shows the P_T distribution of the jets, compared with the Monte Carlo expectation for $\gamma\gamma \to q\bar{q}$ including u d s c quarks. Obviously the data show a steeper fall off than the quark model, showing good agreement only above $P_T^2 = 7.5\ GeV^2$. At moderate P_T there is room for other contributions (like $q\bar{q}$ meson) but up to now there is no positive evidence for these processes because for invariant masses below 9 GeV it seems to be very

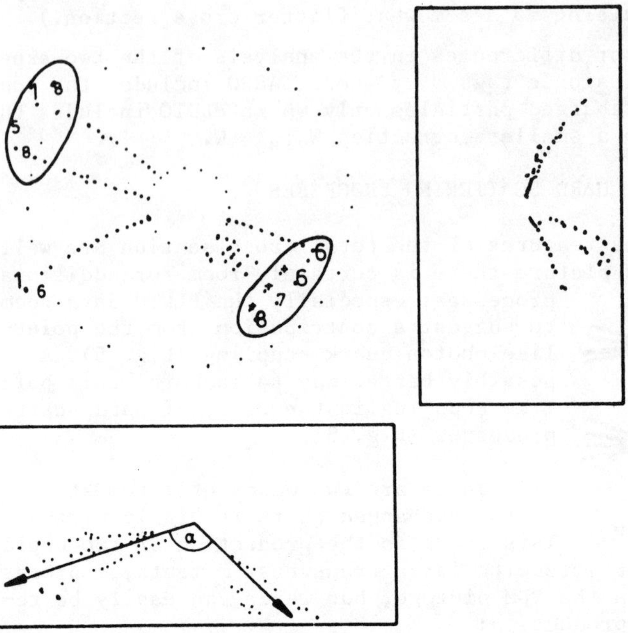

Fig. 9 example of a two photon initiated two jet event

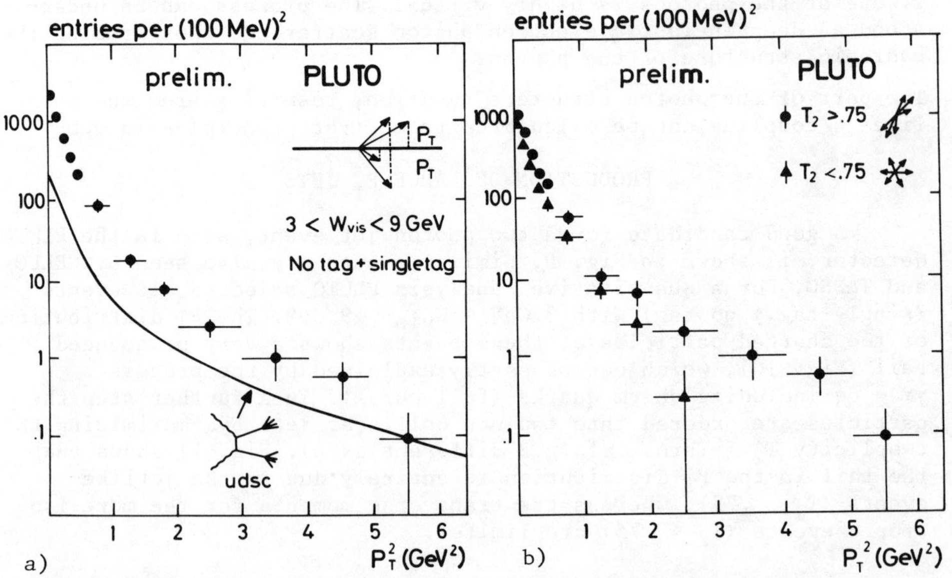

Fig. 10 transverse momenta of all charged particles with respect to the beam (a), and for different twoplicity values (b)

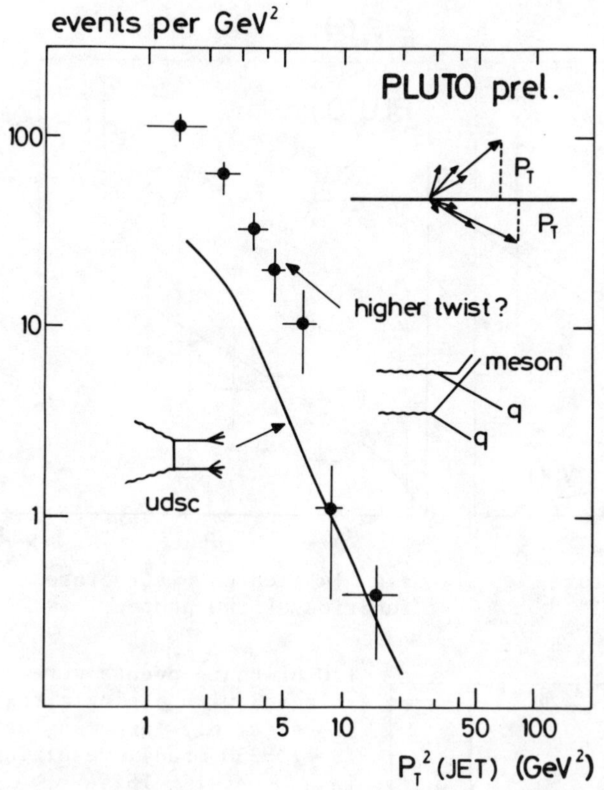

hard to distinguish between a $q\bar{q}$ and a qq meson final state. A Monte Carlo analysis, including higher twist terms and diffractive hadron production is under work.

Fig. 11 transverse momenta of the jets

DEEP INELASTIC eγ SCATTERING

The precise measurement of the photon structure function might turn out to be one of the best testing grounds of QCD. The situation is quite different from nucleon scattering where the structure function cannot be calculated but only the (small!) QCD corrections can. In the case of the photon the structure function is completely dominated by the perturbative part. The Born term as well as the higher order corrections can be calculated in QCD. In the leading log approximation we get

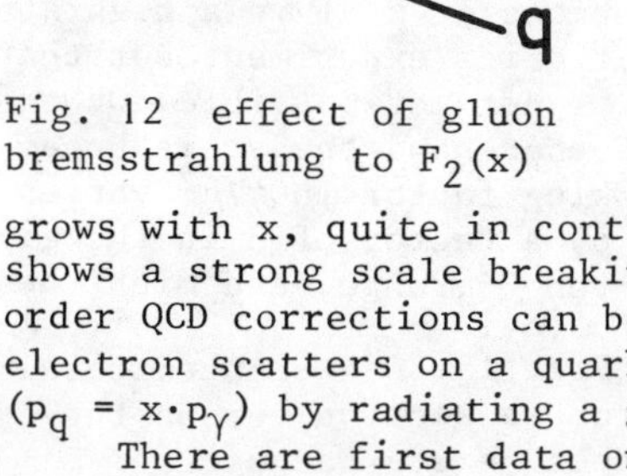

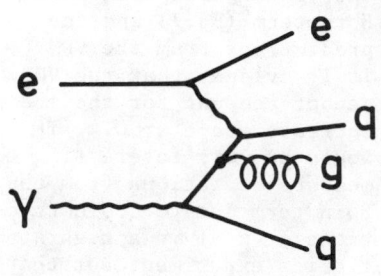

Fig. 12 effect of gluon bremsstrahlung to $F_2(x)$

$$F_2^{Born} = 2xF_T = \frac{\alpha}{\pi}\,\Sigma e_q^4\, x\,(x^2+(1-x)^2)\,\ln\frac{Q^2}{\Lambda^2} \quad (7)$$

F_2^{Born} shows two interesting features: it grows with x, quite in contrast to the VDM expectation ($F_2^{VDM} \sim 1-x$) and shows a strong scale breaking effect $\sim \ln Q^2/\Lambda^2$. The effect of higher order QCD corrections can be easily understood in a simple picture: The electron scatters on a quark which has lost part of it's momentum ($p_q = x\cdot p_\gamma$) by radiating a gluon. This cancels part of the increase of F_2.

There are first data on the photon structure function from PLUTO.

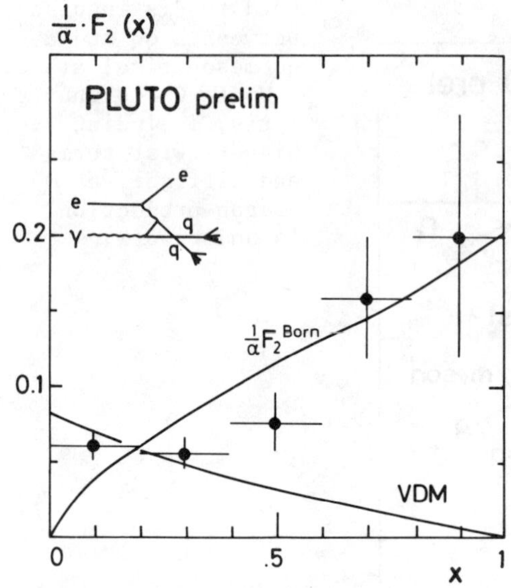

Fig. 13 hadronic structure function of the photon

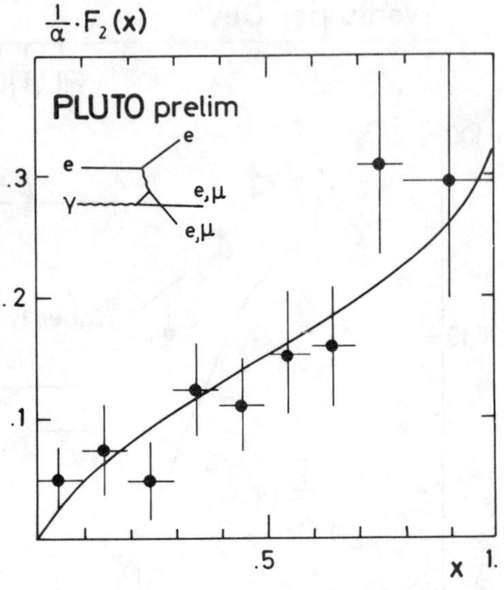

Fig. 14 leptonic structure function of the photon

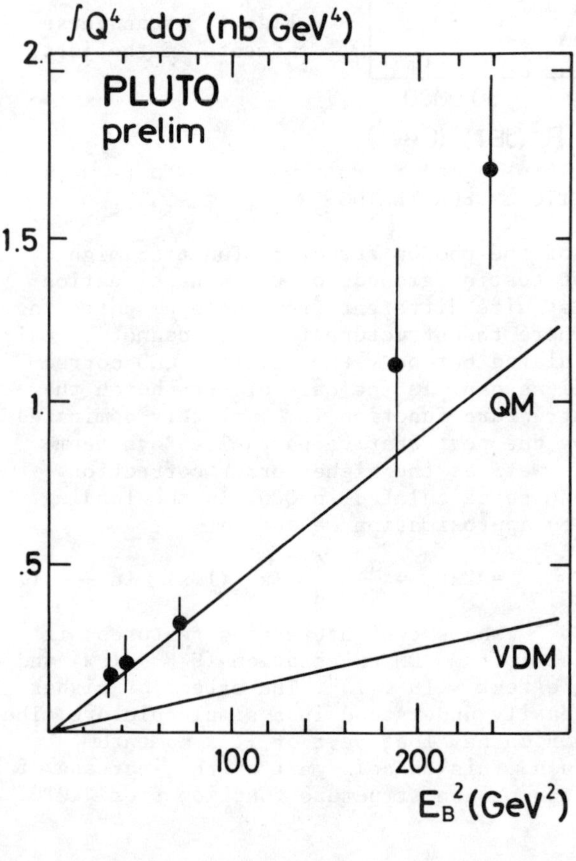

120 hadronic events were selected with a single tag at relatively large angles ($\theta = 70$-250 mrad), resulting in a Q^2 of 1 - 15 GeV^2 ($\langle Q^2 \rangle = 5$ GeV^2). This allows to measure the structure function up to x = 0.95. The result is shown in Fig.13, and is compared with the Born term (Eq.7) and the predictions from the VDM. It is evident that the VDM cannot account for the steep increase above x = 0.4. It would be very interesting to measure deviations from the Born term due to QCD corrections at $x \to 1$ in a high statistics experiment but that is not yet possible. On the other hand the scale breaking factor $\ln Q^2/\Lambda^2$ varies by a factor of 3 in the given Q^2 range, and might be

Fig. 15 Q^4 weighted total cross section versus the beam energy acquired visible even with low statistics. As a test of the method, the 'leptonic structure function' was determined with the same procedure but using only two prong final states (ee,$\mu\mu$). The result in Fig.14 shows good agreement with the QED expectation. It is quite exciting to see that hadronic final states with low mass and low multiplicity seem to be produced by the same mechanism as lepton pairs.

One can finally do a global scaling test by integrating over the structure function. If F_2 is a function of x only one can easily show:

$$\int Q^4 \frac{d\sigma}{dxdy} \, dxdy \sim E^2 \qquad (8)$$

Scaling is predicted in VDM and (approximately) in the quark model, but with different slopes. Fig.15 shows the PLUTO data compared to the two models (m_g = 300 MeV was used in the QM). At large beam energies, corresponding to large Q^2, the data seem to exceed the QM predictions. This could already indicate the presence of scale breaking effects, but with the present statistics this is not conclusive.

CONCLUSIONS

An enormous progress has been made in two photon physics in the last two years. The photon appears, if we don't look too deep inside, as a vector meson. But the VDM obviously doesn't tell us the whole story: if we investigate the photon at very short distances it exhibits a 'hard component'. The analysis of the hard scattering processes with high statistics will be a good testing ground of QCD.

REFERENCES

1. H. Euler and B. Kockel, Naturwiss. 23, 246 (1935).
2. D.P. Barber et al., Phys. Rev. Lett. 43, 1915 (1979).
   PLUTO Collaboration, Ch. Berger et al., Phys. Lett. 94B, 254 (1980).
3. G. Cochard, Invited Talk given at the International Workshop on $\gamma\gamma$ Collisions, Amiens (1980).
4. J. Babcock, J.L. Rosner, Phys. Rev. D14, 1286 (1976).
5. Ch. Berger, Invited Talk given at the International Workshop on $\gamma\gamma$ Collisions, Amiens (1980), PITHA 80/07 (1980).
   PLUTO Collaboration, Ch. Berger et al., DESY 80/94 (1980), to be published in Phys. Lett..
6. M. Greco, Y. Srivastava, Nuovo Cim. 43A, 88 (1978).
7. J. Gunion, Invited Talk given at the International Workshop on $\gamma\gamma$ Collisions, Amiens (1980), SLAC-PUB-2503 (1980).
8. E. Witten, Nucl. Phys. B120, 189 (1977).
   C.H. Llewellyn Smith, Phys. Lett. 79B, 83 (1978).
9. S.J. Brodsky et al., Phys. Rev. D19, 1418 (1979).
10. P. Zerwas, Phys. Rev. D10, 1485 (1974).
11. E. Hilger, Invited Talk given at the International Workshop on $\gamma\gamma$ Collisions, Amiens (1980), DESY 80/75 (1980).

Rho Rho Production by Two Photon Scattering

TASSO Collaboration
Aachen, Bonn, DESY, Hamburg, Imperial College, Oxford, Rutherford, Weizmann, Wisconsin*

(presented by E. Hilger)

We present here the first observation of the reaction

$$\gamma\gamma \rightarrow \rho^{\circ}\rho^{\circ} \ . \tag{1}$$

The experiment[1] was carried out with the TASSO detector[2] at the storage ring PETRA using the information on charged particles from the central detector. The data were taken at beam energies between 15.0 and 18.3 GeV with a total integrated luminosity of 3710 nb^{-1}.

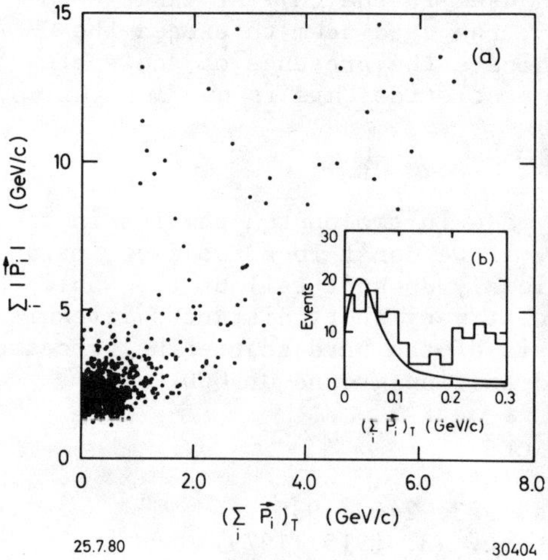

Fig. 1

Candidates for reaction (1) were required to have two positive and two negative tracks, originating from the interaction point, with transverse momenta greater than 0.2 GeV/c with respect to the e^+e^- beam direction. Fig. 1a gives the distribution of the sum of the particle momenta, $\Sigma|\vec{p}_i|$, versus the net transverse momentum of the four particle system, $p_T = (\Sigma\vec{p}_i)_T$. There is an isolated cluster of about 800 events at low $\Sigma|\vec{p}_i|$ (1 - 3 GeV/c). This region corresponds to the kinematical configuration expected for events of the type $e^+e^- \rightarrow e^+e^- + 4$ charged particles + undetected particles, where the incident e^+ and e^- are scattered at such small angles that they are not detected. We estimated the background from radiative 1γ annihilation to be about 10 events. The contamination from beam-gas scattering was negligible.

Events of the type $\gamma\gamma \rightarrow \rho^{\circ}\rho^{\circ}$ are expected to cluster at small p_T. Fig. 1b shows the distribution of p_T for events with $1.5 < W_{\gamma\gamma} < 2.3$ GeV where $W_{\gamma\gamma}$ is the cm-energy of the 4 particles assumed to be pions. The curve in Fig. 1b shows the expected p_T distribution for $\gamma\gamma \rightarrow 4\pi$ events where the effects of resolution and $\gamma\gamma$ kinematics have been taken into account. A cut at $p_T = 0.15$ GeV/c selects nearly all the $\gamma\gamma \rightarrow 4\pi$ events. These requirements are satisfied by 89 events. We estimated the background from states with additional unobserved particles to be about 15 events, and from radiative 1γ annihilation to be less than 1 event. The 89 events were therefore assumed to be

$$\gamma\gamma \rightarrow \pi^+\pi^+\pi^-\pi^- \tag{2}$$

in the rest of the analysis.

*for a list of the current members of the TASSO Collaboration see the paper by Sau Lan Wu in these proceedings.

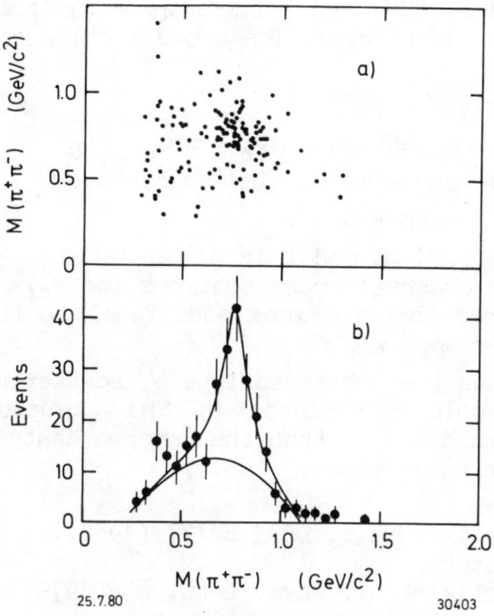

Fig. 2

Fig. 2a shows the 2-dimensional mass distribution of one $\pi^+\pi^-$ combination versus the other one for $1.5 < W_{\gamma\gamma} < 2.0$ GeV (2 entries /event). The pronounced enhancement when both mass values are near the ρ mass is evidence for the final state $\rho^0\rho^0$. Fig. 2b gives the projection onto the $\pi^+\pi^-$ mass axis (4 entries/event). The number of $\rho^0\rho^0$ events was determined with a maximum likelihood fit. A sum of noninterfering contributions from $\rho^0\rho^0$ production ($N_{\rho\rho}$), ρ^0 + nonresonant $\pi^+\pi^-$ production ($N_{\rho\pi\pi}$), and phase space (N_{PS}) was fitted simultaneously to the two density distributions $M_{\pi_1^+\pi_2^-}$ vs. $M_{\pi_3^+\pi_4^-}$ (Fig. 2a) and $M_{\pi^+\pi^+}$ vs. $M_{\pi^-\pi^-}$. The fit yielded the number of events shown in Table 1. The results indicate that the four charged pion state results predominantly from the production of $\rho^0\rho^0$ at low $W_{\gamma\gamma}$. The curves in Fig. 2b represent the sum of all fitted contributions and the background of all fitted nonresonant $\pi^+\pi^-$ combinations.

$W_{\gamma\gamma}$(GeV)	$N_{\rho\rho}$	$N_{\rho\pi\pi}$	N_{PS}	$\sigma(\gamma\gamma \to \rho^0\rho^0)$ (nb)
1.50 - 1.75	40 ± 6	-6 ± 11	8 ± 8	77 ± 12
1.75 - 2.0	14 ± 6	-4 ± 12	18 ± 9	35 ± 14
2.0 - 2.3	4 ± 3		15 ± 3	10 ± 7

The reaction $e^+e^- \to e^+e^- + \rho^0\rho^0$ was simulated in our detector using a Monte Carlo method to calculate the cross section for $\gamma\gamma \to \rho^0\rho^0$. The virtual photon fluxes were computed according to Ref. 3. For the production angular distribution $f(\cos\Theta^*)$ we tried reasonable assumptions such as a constant and $1 + \cos^2\Theta^*$; for the distribution of the polar helicity decay angle, $W(\cos\Theta^H)$, we tried a constant and $\sin^2\Theta^H$. The results were insensitive to these assumptions.

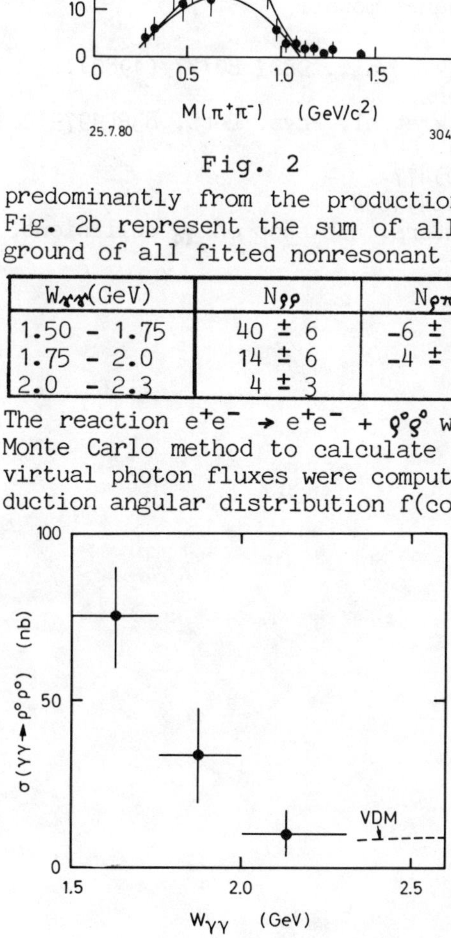

Fig. 3

Averaged over $1.5 < W_{\gamma\gamma} < 2.3$ GeV the acceptance was found to lie between 5 and 7%. The cross sections are given in Table 1. The errors are statistical only. A systematic error of ±25%, mainly due to the uncertainty in the acceptance, has to be added. The cross section values are shown in Fig. 3. Radiative corrections have not been included; they may be very small[5].

In the vector meson dominance model (VDM) the process $\gamma\gamma \to \rho^0\rho^0$ proceeds predominantly via elastic $\rho^0\rho^0$ scattering. The cross section can then be given by

$$\sigma(\gamma\gamma \to \rho^0\rho^0) = g^4_{\rho\gamma} \cdot \sigma(\rho^0\rho^0 \to \rho^0\rho^0) , \qquad (3)$$

where $g_{\rho\gamma}$ is the photon-rho coupling constant, $g^2_{\rho\gamma} = \alpha\pi/\gamma_\rho^2$ ($\gamma_\rho^2/4\pi$ = 0.5). At high energies the additive quark model (QM) suggests

$$\sigma(\rho^0\rho^0 \to \rho^0\rho^0) = (4/9)^2 \cdot \sigma(pp \to pp) . \qquad (4)$$

This prediction of the asymptotic VDM-QM model is shown in Fig. 3. The substantial enhancement of the observed cross sections for $W_{\gamma\gamma} < 2$ GeV indicates the presence of other contributions additional to the Pomeron in the threshold region for $\gamma\gamma \to \rho^0\rho^0$.

In summary the reaction $\gamma\gamma \to \rho^0\rho^0$ has been observed in e^+e^- scattering and its cross section near threshold determined. In this region $\sigma(\gamma\gamma \to \rho^0\rho^0)$ has been found to be much larger than the Pomeron contribution predicted by the vector dominance model.

References

1. TASSO Collaboration, R. Brandelik et al., DESY 80/77 (1980), to be published in Physics Letters
2. TASSO Collaboration, R. Brandelik et al, Phys. Lett. 83B(1979)261 and Z.Physik C, 4(1980)87
3. J.H.Field, Nucl.Phys., B168(1980)477
4. J.D.Jackson, Nuovo Cimento 34(1964)1644
5. G.Cochard, Proceedings of the Intern. Workshop on $\gamma\gamma$ collisions, Amiens, April 1980

HIGH ENERGY e^+e^- REACTIONS
(R, INCLUSIVE DISTRIBUTIONS, JETS)
EXPERIMENT

H. SPITZER, DESY, ORGANIZER

MEASUREMENTS OF R AND SEARCH FOR NEW THRESHOLDS AT PETRA

Dieter Cords
Deutsches Elektronen-Synchrotron DESY, Hamburg, Germany

ABSTRACT

Data are presented from five experiments at PETRA and combined where possible. The measured R values show a constant behaviour for the centre of mass energies between 17 and 36.5 GeV, and are compatible with expectations from the quark parton model including five quarks u,d,s,c,b and QCD corrections. The results of two energy scans are reported. The data exclude any $t\bar{t}$ threshold below 36.5 GeV of 2/3 charged t quarks decaying into many hadrons. The search for superheavy sequential leptons and scalar electrons yields negative evidence and lower mass limits are given.

INTRODUCTION

The reported results were made possible by the efficient operation of the PETRA storage ring, which during its first year of operation supplied a luminosity of about 5500 nb^{-1} for each of the four interaction regions. Each of the experiments JADE, PLUTO, MARK J, and TASSO collected about 2500 multihadronic events; CELLO replaced PLUTO at the beginning of this year.

MEASUREMENTS OF R

The ratio $R = \sigma(e^+e^- \rightarrow \text{hadrons})/\sigma(e^+e^- \rightarrow \mu^+\mu^-)$ tests the pointlike nature of quark pair production and counts the number of quarks involved. In order to discriminate hadronic annihilation events from two-photon and beam-gas background, the most important cut - employed by all PETRA experiments [1] - requires that at least 50% of the total centre of mass energy be visible in the detector (25% for the case of only charged particle detection). In addition each event is required to have at least four charged tracks and $\tau^+\tau^-$ topologies are explicitely removed from the event sample with four tracks. These cuts lead to a clean sample of hadronic annihilation events losing only a few good events. The missing fraction mainly depends on how well the detectors cover the full solid angle. All quoted cross sections are corrected for acceptance and radiative effects which in all cases include hadronic vacuum polarization of about 5%. In Fig. 1a the measured R values are given for five PETRA experiments in the energy range from 12 to 36.5 GeV and a few measured points at lower energies [2] for comparison. The most obvious feature is that from 17 GeV onwards the R values seem to be constant and compatible with what one expects for five quarks plus QCD corrections.

According to the quark-parton model the expected R value is obtained by adding the squares of the quark charges Q_i and multiplying this sum by QCD corrections up to second order. In the version of Dine and Sapirstein[3] R reads:

$$R = 3\sum_{i=1}^{N_f} Q_i^2 \left\{ 1 + \frac{\alpha_s(s)}{\pi} + (1.98 - 0.115\, N_f)\left(\frac{\alpha_s(s)}{\pi}\right)^2 \right\} \qquad (1)$$

0094-243X/81/680590-06$1.50 Copyright 1981 American Institute of Physics

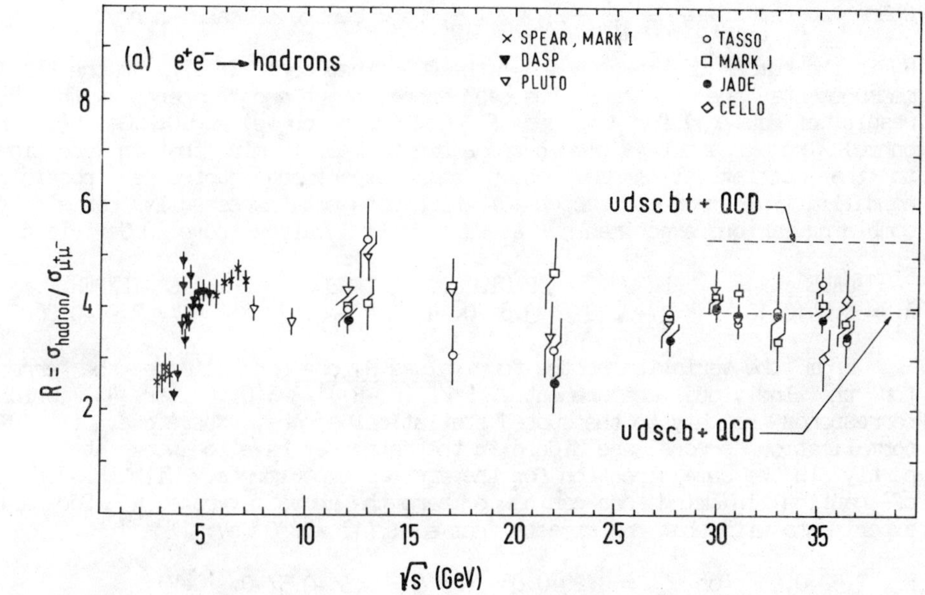

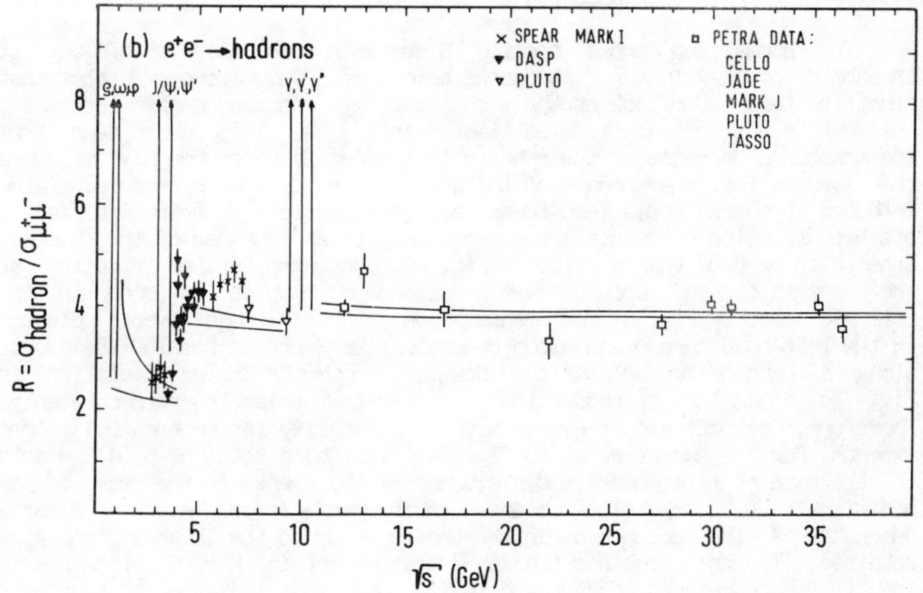

Fig. 1: Ratio of hadronic to $\mu^+\mu^-$ cross section as function of centre of mass energy $\sqrt{s}$. Measurements at PETRA are shown (a) separate and (b) combined. The curves are calculated from equ. (1) with Λ = 1.0/0.2 GeV for upper/lower line.

where $\alpha_s(s) = \frac{12\pi}{(33-2N_f)\ln(s/\Lambda^2)}$, $\sqrt{s}$ = centre of mass energy.

N_f is the number of flavours and Λ the QCD parameter. In Fig. 1b the PETRA measurements are averaged for each energy point and compared with the result of equ. (1) for Λ values of 1 GeV (upper curve) and 0.2 GeV (lower curve). The agreement is good but one has to keep in mind that in addition to the statistical errors shown each experiment quotes a possible normalization error of about 10% which does not necessarily cancel by combining various experiments. Averaging all R values above 20 GeV yields:

	JADE	MARK J	PLUTO	TASSO	ALL 4 EXPERIM.
R_i =	3.84+0.10	4.17+0.10	3.82+0.14	4.0+0.13	R_o = 3.97+0.06

From the variation of the four values R_i one can estimate the error for the individual measurement $\Delta = \sqrt{\Sigma(R_o-R_i)^2/3} = 0.16$ (or 4%) which corresponds roughly to the quoted statistical errors. Therefore, if 10% normalization errors are hidden in the data they have to work at least partly in the same direction for the various experiments. With this note of caution in mind one can now compare the above average R_o for all experiments with what one expects from equ. (1) at 30 GeV:

R = 3.92+0.06 for α_s = 0.20+0.04 (or Λ = 0.5 +0.5/−0.3 GeV)

SEARCH FOR $t\bar{t}$ THRESHOLDS

The normalized cross section R suggests the absence of any $t\bar{t}$ threshold of 2/3 charged t quarks below 36.5 GeV. However, the most sensitive test is to look at the event topology. As the energy increases, the events are more and more collimated into jets. This can be seen from the monotonic decrease of the sphericity [4] (Fig. 2) which roughly measures the average of the square of the jet cone half opening angle. Pair-produced t quarks would lead to an isotropic particle distribution for a sizable fraction of events and consequently to an increase of the average sphericity by 0.08 (dashed line in Fig. 2). One step further is to look at the sphericity and aplanarity [4] simultaneously which is done in the triangle plot of Fig. 3a for events above 35 GeV. Two-jet events cluster in the left-hand corner and gluon bremsstrahlung spreads some of the events along a band of low aplanarity. However, the Monte-Carlo simulation in Fig. 3c shows that $t\bar{t}$ production populates the triangle rather evenly. Therefore, selecting only events with an aplanarity larger than 0.15, one expects for the energy range in Fig. 3a (Fig. 3b) a total of 5 (1) events in the case of five quarks u,d,s,c,b and 57 (9) events in the case of an additional t quark. The observed number of 2 (0) events clearly demonstrates the absence of any $t\bar{t}$ threshold up to the highest energies obtained. If one combines the TASSO measurement in Fig. 3 with nearly identical measurements of JADE and MARK J, one finds that the $t\bar{t}$ threshold is excluded with 12 s.d. at 35.3 GeV and with 5 s.d. at 36.5 GeV. This statement holds if the t quark decays via the sequence $t \to b \to c \to s$ into many hadrons, as suggested by the Kobayashi-Maskawa generalized Cabibbo matrix [5]. Another consequence of this scheme is that the t decay is a rich source of

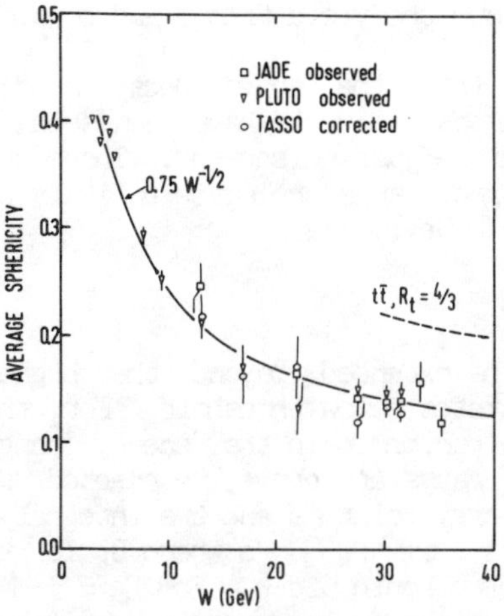

Fig. 2: Average shpericity versus centre of mass energy W.

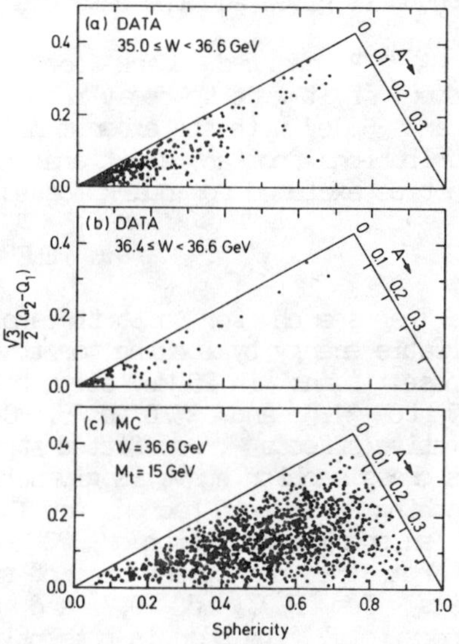

Fig. 3: Distribution of events as a function of sphericity and aplanarity A. (a) and (b) show TASSO data for two energy regions. (c) is a Monte-Carlo simulation for $e^+e^- \longrightarrow t\bar{t} \longrightarrow$ hadrons.

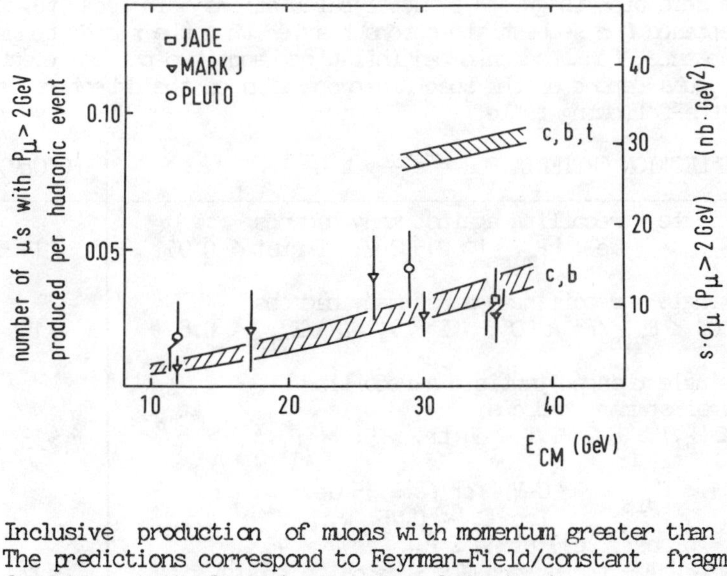

Fig .4: Inclusive production of muons with momentum greater than 2 GeV. The predictions correspond to Feynman-Field/constant fragmentation functions at the lower/upper edge of the bands.

leptons. In Fig. 4 the fraction of events with a muon of at least 2 GeV momentum is given and the data clearly favour the yield from c and b quarks only.

A 1/3 charged quark does not fit into the above decay scheme. However, if it were to decay into many hadrons, it would have been seen at 30 and 35 GeV with the experimental technique just discussed. Therefore, contributions from any additional 1/3 charged quarks are quite unlikely but cannot be excluded for quark masses below 10 GeV.

ENERGY SCAN

The search for a $t\bar{t}$ threshold can be extended beyond the highest available energy by looking for $t\bar{t}$ bound states below threshold. For this purpose, scans in 20 MeV steps have been performed in the energy ranges 29.9 to 31.5 and 35.0 to 35.6 GeV. A gaussian curve, corrected for radiative effects[6], was fitted at each energy point $\sqrt{s}$ and the integral of this cross section curve is given by $\int \sigma \, d\sqrt{s} = 6\pi^2 B_h \Gamma_{ee} / s$ where B_h is the branching ratio into hadrons and Γ_{ee} the electronic width. For $\Gamma_{ee} = 5$ KeV and an energy resolution of 20 MeV one expects an increase in R of about 7. From the non-observation of any significant peak one obtains upper limits for $B_h \Gamma_{ee}$ of 0.7 and 0.4 KeV in the above energy ranges respectively. The scan is presently continued from 35 GeV downwards.

SEARCH FOR SUPERHEAVY SEQUENTIAL LEPTONS

When searching for final states including neutrinos, which escape detection, one has to employ a cut on the visible energy which is quite different from the case of selecting purely hadronic final states. In fact, the selection is guided by model calculations[7] in order to maximize the acceptance for sequential leptons heavier than the τ and to minimize the acceptance for hadronic annihilation and two-photon events. The detailed cuts depend on the specific properties of the detectors and are given in the following table.

EXPER.	SELECTION CRITERIA FOR $e^+e^- \to L^+L^-$	M_L (GeV)
PLUTO	single recoiling against many hadrons $E_{vis} > 3$ GeV; $\lvert\vec{P}_{miss}\rvert > 2.5$ GeV; Thrust < 0.95	> 14.5
MARK J	single recoiling against many hadrons $10\% < E_{vis}/\sqrt{s} < 50\%$; $(\Sigma \vec{P})(-\vec{\mu})/E_{vis} < 0.8$	> 15
TASSO	single charged particle A recoiling against many hadrons $\Sigma \lvert\vec{P}_i\rvert \geq 8$ GeV, i = ch.tr. ≥ 5; $\alpha(\vec{P}_i, \vec{A}) \geq 90^\circ$	> 13
JADE	$11 < E_{vis} < 32$ GeV for $\sqrt{s} = 35$ GeV $\lvert\cos\Theta_T\rvert < 0.75$ w.r.t. e^+ axis (plane($\vec{T}$, $\vec{e^+}$), plane($\Sigma \vec{P}_{OTS}$, $\vec{e^+}$)) $> 45^\circ$ where T = thrust and OTS = opposite thrust side	> 17

No events have been found by the four experiments and lower mass limits for superheavy sequential leptons are given in the right-hand column of the table.

SEARCH FOR SCALAR ELECTRONS

Supersymmetric theories require all particles to have partners with spin different by half a unit[8]. Therefore, the partner of the electron would be a scalar electron which can be pair-produced in e^+e^- annihilation and which decays into an electron and a neutrino-like particle. The signature of such a process would be two electrons, which do not form a plane containing the beam axis, and no additional signal in the detector. PLUTO and JADE searched for these final states by demanding an acoplanarity angle of more than $15^\circ/10^\circ$ and electron energies of at least 20%/30% of the beam energy. Because no candidates were observed at centre of mass energies of 31 GeV for PLUTO and 35 GeV for JADE, the lower mass limits for scalar electrons are 13 and 16 GeV respectively.

CONCLUSIONS

Measurements of the normalized cross section R show a constant behaviour over the energy range from 17 to 36.5 GeV. They agree well with the quark-parton model with five quarks plus QCD corrections but are fairly insensitive to variations of the strong coupling constant α_s. Additional processes contributing to the multihadronic final state are not completely excluded due to possible normalization errors of 10% but they are certainly limited to less than half a unit of R.

Lower mass limits for top quarks, superheavy sequential leptons, and scalar electrons are 18, 17, and 16 GeV respectively.

REFERENCES

1. JADE Collab., W.Bartel et al., Phys.Lett. 88B,171 (1979)
MARK J Collab., D.Barber et al., Phys.Lett. 43,901 (1979)
PLUTO Collab., C.Berger et al., Phys.Lett. 81B,410 (1979)
TASSO Collab., R.Brandelik et al., Phys.Lett. 83B,261 (1979)
2. MARK I Collab., J.E.Augustin et al., Phys.Rev.Lett. 34,76 (1975)
DASP Collab., R.Brandelik et al., Nucl.Phys. B148,189 (1979)
PLUTO Collab., C.Berger et al., Phys.Lett. 82B,449 (1979)
3. M.Dine and J.Sapirstein, Phys.Rev.Lett. 43,668 (1979)
4. Sphericity, thrust and aplanarity are explained by S.L.Wu, "Results from TASSO on jets", these proceedings.
5. M.Kobayashi and T.Maskawa, Prog.Theor.Phys. 49,652 (1973)
6. J.D.Jackson and D.L.Scharre, Nucl.Instr.Meth. 128,13 (1975)
7. Y.S.Tsai, Phys.Rev. 4D,2821 (1971)
8. G.R.Farrar and P.Fayet, Phys.Lett. 89B,191 (1980)

FEATURES OF INCLUSIVE HADRON PRODUCTION IN e^+e^- ANNIHILATION AT PETRA

D.Pandoulas

Blackett Laboratory, Imperial College, London SW 7 2AZ, UK

ABSTRACT

The gross features of multihadron final states produced by e^+e^- annihilation at high energy are discussed and their dependence on the center of mass energy is examined. Topics include the fraction of c.m. energy carried by neutrals, charged particle multiplicity, inclusive hadron spectra with and without particle identification.

INTRODUCTION

Hadron production in e^+e^- annihilation at high energies proceeds primarily through the formation of two back to back jets. At sufficiently high energies, a process leading to the production of three-jet events, presumably through hard gluon bremsstrahlung has also been observed[1,2]. We report here on the gross features of hadronic final states produced by e^+e^- annihilation and discuss their dependence on the center of mass energy W. The data presented were obtained with the JADE, PLUTO, and TASSO detectors at the DESY storage ring PETRA at energies W between 12 and 36.6 GeV. Another review is given in Ref. 3.

ENERGY CARRIED BY NEUTRALS

The fraction of the energy W carried by photons (either from π^0, η, ... decay or from direct production) and by neutrals has been measured by the JADE collaboration[4]:

$$f_\gamma = \sum_i E_{\gamma_i} / W \quad \text{and} \quad f_{neutral} = 1 - \sum_i E_{CH_i} / W,$$

where E_{CH_i} is the energy of charged particle i. The contribution from K^o_s and $\Lambda(\bar{\Lambda})$ particles is included in $f_{neutral}$ but not in f_γ. The results are given in Table I for energies W between 12 and 35 GeV.

TABLE I - Fraction of energy carried by photons, neutrals and neutrinos

W (GeV)	f_γ	$f_{neutral}$	f_ν
12.00	0.21 ± 0.07	0.31 ± 0.04	-0.02 ± 0.08
30.35	0.26 ± 0.06	0.38 ± 0.04	-0.01 ± 0.07
34.93	0.31 ± 0.06	0.442± 0.04	-0.02 ± 0.08

Furthermore, as also seen in Table I, a two-standard deviation upper limit of about 15% is found for the fraction of energy f_ν carried by neutrinos. In a model where free quarks with unit charge are produced before they fragment into hadrons[5] f_ν is expected to lie between 20% and 30%, a prediction disfavoured by the JADE results.

CHARGED PARTICLE MULTIPLICITY

Fig. 1 shows the average charged particle multiplicity* $\langle n_{CH} \rangle$ as a function of the c.m. energy[6-12]. Above W $\gtrsim$ 7 GeV, $\langle n_{CH} \rangle$ is seen to rise (logarithmically) faster than at lower energies. The broken curves represent the data for pp collisions[13] and $\bar{p}p$ annihilation[14]. They bracket the e^+e^- data at high energies**.

The prediction $\langle n_{CH} \rangle = a + b \ln s$, resulting from the simplest form of scaling of charged particle production, is clearly at variance with the data when the full energy range is considered. The onset of $b\bar{b}$ production is expected to contribute only ∿0.2 units to $\langle n_{CH} \rangle$ and thus cannot explain the observed rise of the multiplicity. QCD leads to an increase of $\langle n_{CH} \rangle$ above the scaling curve through the additional contribution from gluon fragmentation, but the exact form of the energy dependence is not yet clear. In the limit of infinitely heavy quarks, one expects[16]

$$\langle n_{CH} \rangle = n_0 + a \exp(b \sqrt{\ln(s/\Lambda^2)}) \quad (1) .$$

Fits of this form (solid curve in Fig. 1) reproduce the trend of the data[11,12], but this should not be interpreted as evidence for hard gluon effects. Indeed the $q\bar{q}$ model with Field-Feynman fragmentation functions but without hard gluon contributions, $e^+e^- \to q\bar{q} \to$ hadrons, accounts well for the rise in $\langle n_{CH} \rangle$, while the inclusion of hard gluons has a negligible effect for W $\lesssim$ 10 GeV and adds only 0.8 units at W = 35 GeV. We conclude therefore that the rapid rise of $\langle n_{CH} \rangle$ with W is mostly due to the growing phase space.

The PLUTO group[12] has measured the ratio $\langle n_{CH} \rangle / D_{CH}$, where D_{CH} is the dispersion of the multiplicity distribution, as a function of W and they have compared it to data for pp collisions[13], $\bar{p}p$ annihilations[14], and νp interactions[17]. The e^+e^- data are in agreement with the QCD prediction of 3.2[18] and also agree within errors with the $\bar{p}p$ and νp results. The pp data show a larger dispersion.

INCLUSIVE SPECTRA WITHOUT PARTICLE IDENTIFICATION

We discuss next the energy dependence of the scaling cross section $s\, d\sigma/dx$ in terms of the scaled particle momentum $x = p/p_{beam}$. At particle energies E large enough that particle masses (m) can be neglected, the scaling cross section is related to the structure functions $\bar{W}_1$ and $\bar{W}_2$,

---

* $\langle n_{CH} \rangle$ includes the contribution from $K^0_S \to \pi^+\pi^-$ decay. This amounts to 0.4, 0.6, and 1 unit at W = 7.4, 12, and 30 GeV respectively. For the pp and $\bar{p}p$ data shown by the curves the K^0_S contribution is excluded.

** In Ref. 15 the c.m. energy available for hadronization in pp collisions is redefined by subtracting the energy carried by the leading protons. Good agreement between pp and e^+e^- data is then obtained.

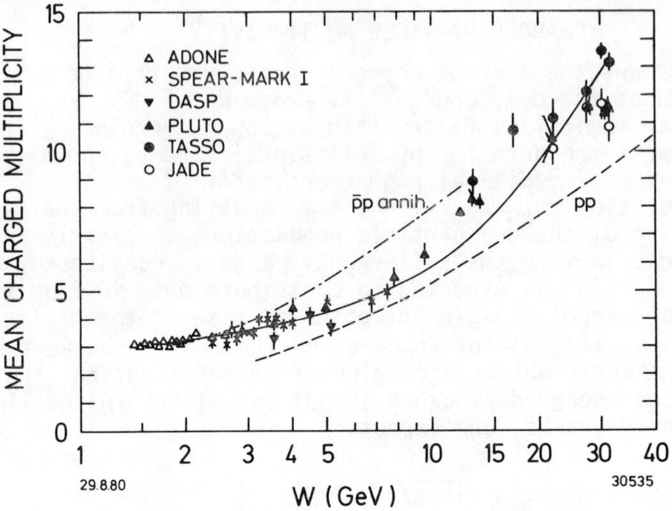

Fig. 1 - Average charged particle multiplicity as a function of c.m. energy.

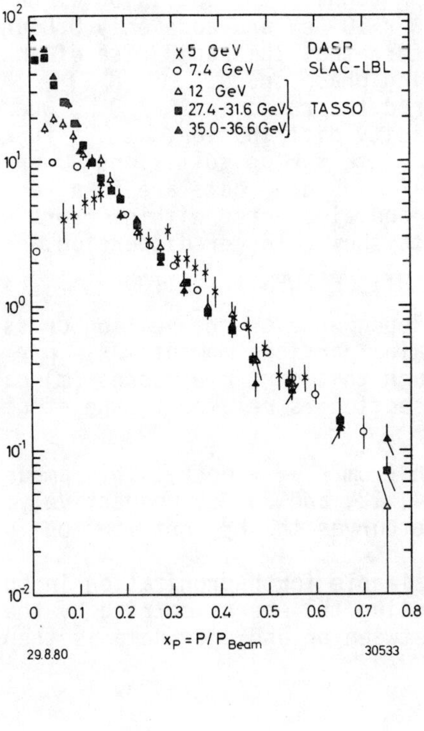

Fig. 2

The scaling cross section $s\ d\sigma/dx$ $(x = p/p_{beam})$ for inclusive charged particle production.

$$s \frac{d\sigma}{dx} \approx 4\pi \alpha^2 x \{ m \bar{W}_1 + \frac{1}{6} x \nu \bar{W}_2 \} \qquad (2)$$

where $\nu = \sqrt{s}\,E/m$ is the virtual photon energy in the particle rest system. The structure functions in general depend on two variables, e.g. x and s. If scale invariance holds $\bar{W}_1$ and $\nu\bar{W}_2$ are functions of x alone and s dσ/dx is energy independent.

Fig. 2 shows the data on s dσ/dx measured by TASSO[19] for W between 12 and 36 GeV. For x > 0.2 they are the same within errors and agree with the low energy data from DASP[20] (5 GeV) and SLAC-LBL[21] (7.4 GeV) to within 30%. At low x values the particle yield rises dramatically as the c.m. energy increases from 5 to 36 GeV. Thus the rise of the multiplicity with W discussed above is associated with the increase in yield of low momentum particles.

Scale breaking effects should result from the emission of hard gluons as the primary momentum is shared by quark and gluon leading to a depletion of particles at high x and to an excess at low x. The effect becomes more pronounced as the energy rises, e.g. the 30 GeV data are predicted to be higher by ~10% at x = 0.2 and lower by ~20% at x = 0.7 than the 5 GeV data[29]. The measurements are not accurate enough to test this prediction.

π, K AND p($\bar{p}$) CROSS SECTIONS

Inclusive $\pi^\pm$, $K^\pm$, $K^0(\bar{K}^0)$, and p($\bar{p}$) cross sections were measured at PETRA by the experiments listed in Table II.

TABLE II - Experiments measuring particle separated cross sections

type of particle	experiment	technique	momentum range(GeV/c)	remark
$\pi^\pm$	JADE	dE/dx	< 0.7, 2-7	preliminary
	TASSO	TOF[23]	< 1.1	
		Cerenkov	< 5	preliminary
$K^\pm$	JADE	dE/dx	< 0.7	preliminary
	TASSO	TOF[23]	< 1.1	
		Cerenkov	< 5	preliminary
$K^0,\bar{K}^0$	PLUTO	$K^0_S \to \pi^+\pi^-$	all p	preliminary
	TASSO	$K^0_S \to \pi^+\pi^-$[24]	all p	
$\bar{p}$	JADE	dE/dx	< 0.9	preliminary
p,$\bar{p}$	TASSO	TOF[23]	< 2.2	
		Cerenkov	< 4	preliminary

Fig. 3a shows the scaling cross section s/β dσ/dx as a function of x = 2E/W for the sum of $\pi^+ + \pi^-$ production for W = 5.2[20], 12, and 30 GeV. The 12 and 30 GeV points agree with each other, but seem to

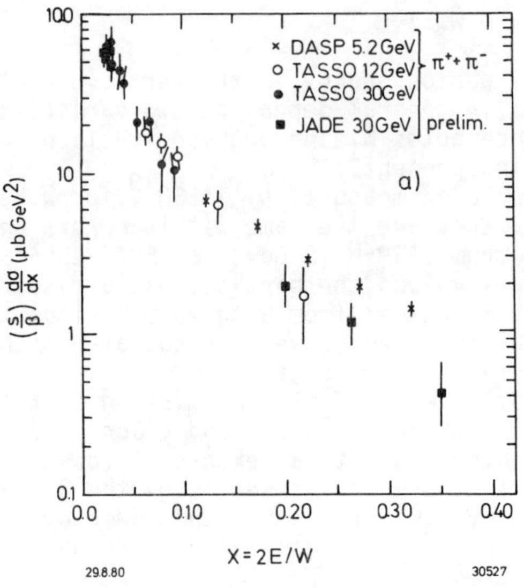

Fig. 3 - The cross section $s/\beta \frac{d\sigma}{dx}$ ($x = E/E_{beam}$) for
a) the sum of $\pi^+ + \pi^-$ production,
b) $K^+ + K^-$ and $K^0 + \bar{K}^0$ production, and
c) $p + \bar{p}$ production

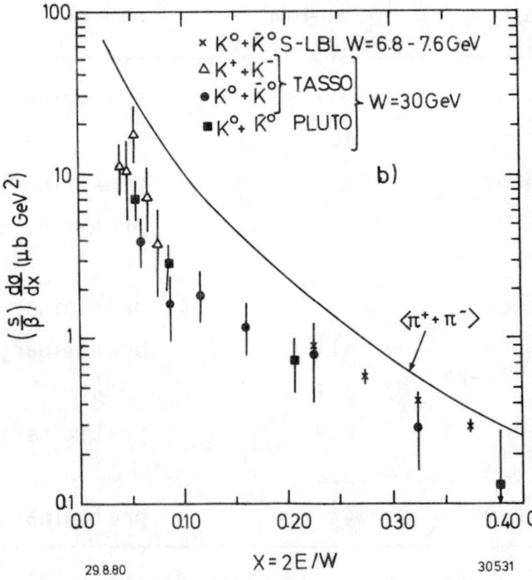

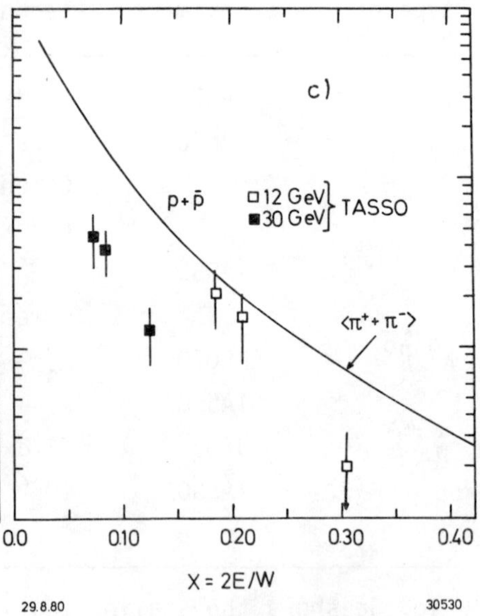

be lower than the 5.2 GeV data by $\approx$ 30% for $x \gtrsim 0.2$. There is a break in slope near $x = 0.1$, the data at lower x values having a steeper slope.

Fig. 3b displays the same quantity for $K^+ + K^-$ and $K^0 + \bar{K}^0$ production, while in Fig. 3c the corresponding measurements for $p + \bar{p}$ production are plotted. In Fig. 3b results for $K^0 + \bar{K}^0$ at W around 7.4 GeV[25] are shown, in addition to the PETRA data at W = 30 GeV. For $x < 0.1$ and W = 30 GeV, where $K^\pm$ data are available, the $K^\pm$ and $K^0, \bar{K}^0$ yields agree within errors. The $K^0, \bar{K}^0$ points indicate a break in slope similar to that seen in the $\pi^\pm$ data. The curves in Figs. 3b and 3c represent a hand drawn average through the high energy $\pi^\pm$ points (Fig. 3a). The K yield is a factor of 2 to 4 lower than the π yield, but the difference becomes smaller at high x values. Within the large errors the p, $\bar{p}$ and K^o, $\bar{K}^0$ yields appear to be equal.

Fig. 4 shows the relative fraction of $\pi^\pm$, $K^\pm$, and p, $\bar{p}$ at W = 30 GeV as a function of particle momentum p. The low momentum particles are essentially all pions. As the momentum increases, the fractions of $K^\pm$ and p, $\bar{p}$ rise and for $p \approx 4$ GeV/c the $K^\pm$ and $\pi^\pm$ yields appear roughly equal. This trend is also seen in a $q\bar{q}g$ model for hadron production, $e^+e^- \to q\bar{q}g \to$ hadrons (curve in Fig. 4).

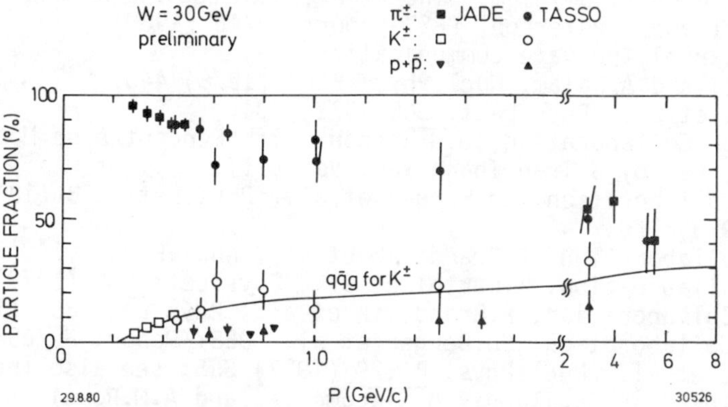

Fig. 4 - Charged particle fractions as a function of momentum

An average event at W = 30 GeV has approximately 11 $\pi^\pm$, 1.4 $K^0 + \bar{K}^0$, 1.4 $K^\pm$, and 0.4 $p + \bar{p}$ in the final state. Assuming that $n(\bar{n})$ production equals $p(\bar{p})$ production and that no more than one baryon pair per event is produced, we conclude that about 40% of the events contain a baryon-antibaryon pair in the final state. The yield of $\sim$ 1.4 $K^0, \bar{K}^0$ per event is 2-3 times larger than that observed in pp final states, where at W = 24 GeV there are on the average 0.5 K^0, $\bar{K}^0$ per event[26]. The excess of kaons in e^+e^- annihilation is likely to be due to c and b quark contributions.

CONCLUSIONS

Multihadron final states from e^+e^- annihilation show a rapid (logarithmic) rise of the charged particle multiplicity for c.m. energies above $\sim$ 7 GeV. This rise is mainly due to a strong increase of the yield of low momentum particles. The cross section s $d\sigma/dx$ scales to within 30% for $x = p/p_{beam} \gtrsim 0.2$. The scaling cross sections for pions and kaons show similar behaviour, including a break in the exponential slope near x = 0.1. The kaon and p,$\bar{p}$ yields are similar and lower by a factor of 2-4 than the $\pi^{\pm}$ yield depending on x. The relative large number of baryons produced in high energy e^+e^- annihilations is surprising. At high particle momenta the pion and kaon yields appear to be equal. The large fraction of kaons compared to pp collisions may be an indication of the different primary quark flavours.

REFERENCES

1. TASSO Collaboration, R.Brandelik et al., Phys.Lett.86B(1979) 243.
2. MARK J Collaboration, D.P.Barber et al., Phys.Rev.Lett. 43 (1979) 830; PLUTO Collaboration, Ch.Berger et al., Phys.Lett. 86B (1979) 418; JADE Collaboration, W.Bartel et al., Phys.Lett. 91B (1980) 142.
3. G.Wolf, Talk at the XI International Symposium on Multiparticle Dynamics, Bruges, Belgium, June 22-27,1980, DESY Report 80/85 (1980)
4. W.Bartel and A.Peterson, DESY-Report 80/46 (1980) and W.Bartel, private communication.
5. J.C.Pati and A.Salam, Nucl.Phys. B144 (1978) 445.
6. C.Bacci et al., Phys.Lett. 86B(1979) 234.
7. SLAC-LBL Collaboration, G.G.Hanson, 13th Rencontre de Moriond (1978), ed. by J.Tran Thanh Van, Vol. III.
8. PLUTO Collaboration, Ch.Berger et al., Phys.Lett.81B (1979) 410, 78B (1978) 176.
9. DASP Collaboration, R.Brandelik et al., Nucl.Phys. B148(1979)189.
10. JADE Collaboration, W.Bartel et al., Phys.Lett.88B (1979) 171.
11. TASSO Collaboration, R.Brandelik et al., Phys.Lett.89B (1980) 418.
12. PLUTO Collaboration, Ch.Berger et al., DESY-Report 80/69 (1980)
13. W.Thomé et al., Nucl.Phys. B 129 (1977) 365; see also the review by E.Albini, P.Capiluppi, G.Giacomelli, and A.M.Rossi, Nuov.Cim. 32A (1976) 101.
14. R.Stenbacka et al., Nuov.Cim. 51A (1979) 63; S.Barshay, A.Fridman, and P.Juillot, Phys.Rev.D15 (1977) 2702.
15. M.Basile et al., CERN preprint (1980)
16. J.Ellis (private communication), motivated by W.Furmanski, R.Petronzio, and S.Pokorski, Nucl.Phys. B155 (1979) 253.
17. N.Schmitz, Review Talk at the Intern.Symp. on Lepton and Photon Interactions at High Energies, FNAL (1979) and MPI preprint MPI-PAE/Exp. 80 (1979)
18. S.Wolfram, Caltec Preprint CALT-68-778 (1980)
19. TASSO Collaboration, R.Brandelik et al., Phys.Lett.89B (1980) 418.
20. DASP Collaboration, R.Brandelik et al., Nucl.Phys.B148(1979) 189.

21. G.J.Feldman and M.L.Perl, Physics Reports 33 (1977) 285
22. R.Baier, J.Engels, and B.Peterson, Univ. of Bielefeld Report BI-TP 79/10 (1979);
W.R.Frazer and J.F.Gunion, Phys.Rev. D20 (1979) 147.
23. TASSO Collaboration, R.Brandelik et al., Phys.Lett. 94B (1980) 444.
24. TASSO Collaboration, R.Brandelik et al., Phys.Lett. 94B (1980) 91.
25. V.Lüth et al., Phys.Lett. 70B (1977) 120.
26. Dao et al., Phys.Rev.Lett. 30 (1973) 1151;
Sheng et al., Phys.Rev. D11 (1975) 1733.

TASSO RESULTS ON JETS AND QCD

TASSO-Collaboration[1]

Aachen[§], Bonn[§], DESY, Hamburg[§], Imperial College[§§], Rutherford[§§], Oxford[§§], Weizmann[§§§], Wisconsin[§§§§]

Presented by Sau Lan Wu,University of Wisconsin,Madison,WI 53706,USA

ABSTRACT

Recent results obtained with the TASSO detector are presented, including in particular the values of jet fragmentation parameters found from the two-jet events, accurate determination of the quark-gluon strong coupling constant α_s to be $0.17 \pm 0.02 \pm 0.03$, comparison with QCD, evidence that the gluon spin is 1, and the first observation of a long-range charge correlation in opposite jets from e^+e^-.

This is a report on the recent experimental results obtained by the TASSO Collaboration at PETRA. Since there is a large amount of data, I have chosen to concentrate on those that are obtained recently, including some that are preliminary. Some of the TASSO publications are listed in references 2-10.

The TASSO detector employs a solenoid with cylindrical proportional and drift chambers plus two hadron spectrometer arms with three types of Cerenkov counters each (one with aerogel of 12 m^2 coverage and two with gases) to identify charged particles up to the highest momenta. These counters are now operative and preliminary results have presented to this conference by D.Pandoulas (parallel session C2). Photons are detected in scintillation and liquid argon shower counters. The latter provides good energy and precise spatial resolution. An intriguing although less typical event observed at c.m. energy of 30 GeV is displayed in Fig. 1. It shows an isolated photon of 9.3 GeV which is separated from any charged particles by more than 90^0. Such an

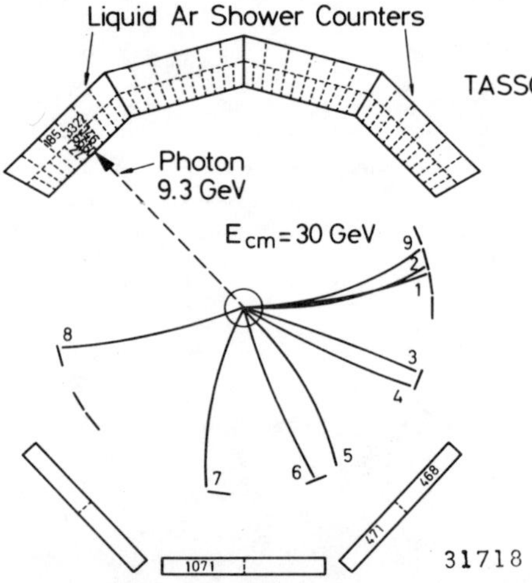

Fig. 1 - An event with isolated hard photon.

---

§ supported by the Bundesministerium for Forschung u.Technologie
§§ supported by the UK Science Research Council
§§§ supported by the Minerva Gesellschaft f.Forschung mbH., München
§§§§ supported in part by the US Department of Energy, contract WY-76-C-02-0881

event is likely to come from e^+e^- initial state bremsstrahlung.

1. J E T S

We shall be concerned mostly with the high-energy data where the center-of-mass energy W is between 27 and 37 GeV. At these energies, most of the hadron events consist of two back to back jets, and are interpreted as the production of a quark-antiquark pair: $e^+e^- \rightarrow q\bar{q}$. Last summer, shortly after PETRA reached 27 GeV, the TASSO collaboration, using a method of generalized sphericity[11], found the first few events with a distinctly different shape. These events, with three jets instead of two, were reported[2,3] just over a year ago. Very soon after, the number of observed three-jet events had increased rapidly[4,12-14]. By now, there are several hundred such events from TASSO alone; the precise number is not well defined and depends on the somewhat artificial definition, because a three-jet event gradually changes over to a two-jet event if the angle between two of the jets is reduced. In spite of the ambiguity, the number of the observed three-jet events is roughly five times that expected from statistical fluctuations in the fragmentation process of the above mentioned process $e^+e^- \rightarrow q\bar{q}$. Instead, these three-jet events are most naturally explained by hard non-collinear gluon bremsstrahlung[15] $e^+e^- \rightarrow q\bar{q}g$, where the quark, the antiquark and the gluon each materializes as a jet of hadrons with limited transverse momentum.

A recent example of a three-jet event is shown in Fig. 2. This example is especially interesting because, as seen on the right-hand side of Fig. 2, there is a 1.67 GeV/c positive track and a 1.32 GeV/c negative track, both entering one of the hadron arms of the TASSO detector and hence identified as K^+ and K^-, respectively. Since the invariant mass is 1.01 GeV, what is seen here is likely to be $\phi \rightarrow K^+K^-$.

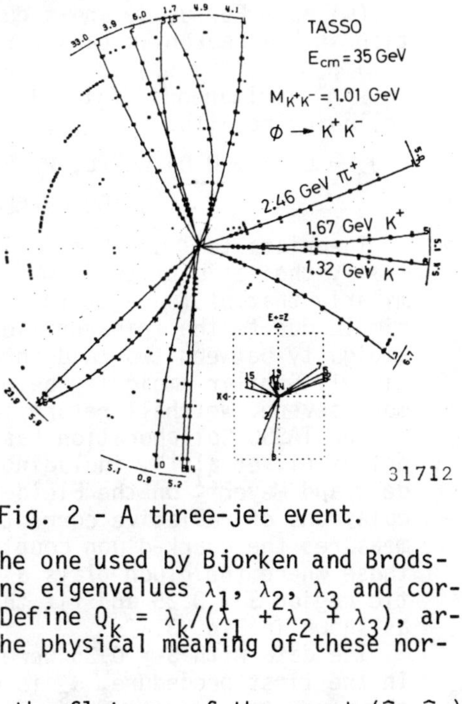

Fig. 2 - A three-jet event.

For later purposes it is useful to review the quantities used in the method of generalized sphericity. Let $\vec{p}_j$, j=1..N be the hadronic momenta and $p_{j\alpha}$, α = 1,2,3, their rectangular components. From the momentum tensor $\Sigma_j p_{j\alpha} p_{j\beta}$, which is similar to the one used by Bjorken and Brodsky[16], and Hanson et al.[17], one obtains eigenvalues λ_1, λ_2, λ_3 and corresponding eigenvectors $\hat{n}_1$, $\hat{n}_2$, $\hat{n}_3$. Define $Q_k = \lambda_k/(\lambda_1 + \lambda_2 + \lambda_3)$, arranged such that $0 \leq Q_1 \leq Q_2 \leq Q_3$. The physical meaning of these normalized eigenvalues is such that

$Q_1 = \min_{\hat{n}} \Sigma_j (\vec{p}_j \cdot \hat{n})^2/\Sigma_j \vec{p}_j^{\,2}$ gives the flatness of the event ($\hat{n} = \hat{n}_1$),

$Q_2 = \min_{\hat{n}\perp\hat{n}_1} \Sigma_j(\vec{p}_j \cdot \hat{n})^2/\Sigma_j \vec{p}_j^{\,2}$ gives the width of the event ($\hat{n}=\hat{n}_2$), and

$Q_3 = \max_{\hat{n}} \Sigma_j(\vec{p}_j \cdot \hat{n})^2/\Sigma_j \vec{p}_j^{\,2}$ gives the length of the event ($\hat{n}=\hat{n}_3$).

Sphericity S and aplanarity A are respectively $3(Q_1 + Q_2)/2$ and $3Q_1/2$; the triangular plot is a two-dimensional plot with axes S and $Y = \sqrt{3}\,(Q_2-Q_1)/2$. TASSO event distribution at 30 GeV is shown in Fig.3.

2. PARAMETERS OF JET FRAGMENTATION

At present, there is at a fundamental level no theory that starts from first principles and gives quantitatively the fragmentation of a quark into a jet of hadrons. Nevertheless, there are good models for this process, for example the model of Field and Feynman[18]. The specific models that we use combine the ideas of Field and Feynman with the quark-gluon coupling of QCD, and are due to Hoyer et al.[19] and Ali et al.[20]. Similar results are obtained from both models. The results we quote are from the second model which include second-order terms in α_s. The parameters used by Field and Feynman are

(A) σ_q = The distribution of the transverse momentum k_T of the quarks in the jet cascade is assumed to be proportional to $\exp(-k_T^2/2\sigma_q^2)$.

(B) P/(P+V) - Here P/V is the ratio of primordial pseudoscalar mesons to vector mesons produced in the fragmentation process.

(C) a_F - For u, d, and s quarks, the primordial quark fragmentation into a hadron is parametrized as $1-a_F+3a_F(1-z)^2$, where $z = (E + p_\parallel)_h \,/\, (E + p_\parallel)_q$.

A simultaneous fit[10] to all data with sphericity S<0.25 yields the following results:

$$\sigma_q = 0.32 \pm 0.04 \text{ GeV/c},\quad P/(P+V) = 0.56 \pm 0.15,\quad a_F = 0.57 \pm 0.20.$$

3. QUARK-GLUON STRONG COUPLING CONSTANT

Theorists tell us that it is very easy to determine α_s: it is essentially the ratio of the numbers of three-jet to two-jet events, or similarly that of four-jet to three-jet events. Actually it is not so simple due to the fragmentation of quarks and gluons and the resulting ambiguity between two- and three-jet events already mentioned in section 1. Similar ambiguity between three- and four-jet events is even more severe. We shall return to this subject later.

The TASSO Collaboration has determined α_s using again the QCD model of Ali et al.[20], including in particular the improvements by Sander and Meyer[21] on the Field-Feynman fragmentation and the recent calculations of radiative corrections by Berends and Kleiss[22]. Since α_s measures the quark-gluon coupling, the experimental data used must be those where the gluon plays a significant role. We choose the kinematic region S > 0.25 and fit the two dimensional S vs A distribution as shown in Fig. 3.

The data with S > 0.25 were analyzed by two different procedures. In the first procedure, α_s is determined without any assumptions on the parameters of jet fragmentation. As shown in Table 1[10], we found that the value of α_s is totally insensitive to the values of the fragmentation parameters, and the result[10] is α_s=0.16±0.04 (statistical).

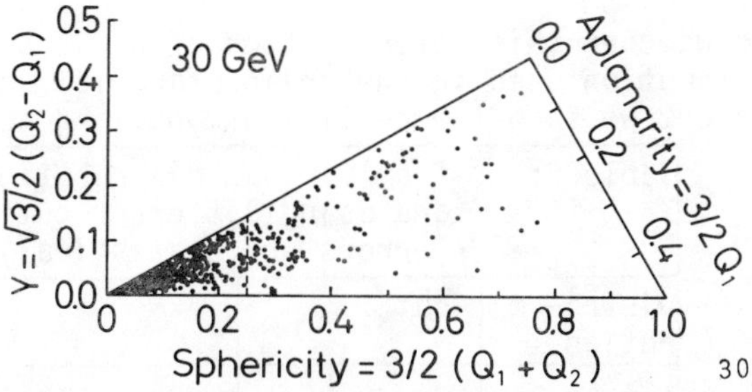

30112

Fig. 3 - Distribution of the events as a function of sphericity and aplanarity.

Table 1: Fitted values of α_s and σ_q for different input values of a_F and P/(P+V)

a_F \ P/(P+V)	0.1	0.3	0.5	0.7	0.9
0.1	$\alpha_s=0.17\pm0.03$ $\sigma_q=0.44\pm0.11$	0.17 ± 0.03 0.46 ± 0.10	0.17 ± 0.03 0.47 ± 0.10	0.16 ± 0.03 0.48 ± 0.09	0.16 ± 0.03 0.48 ± 0.08
0.3	$\alpha_s=0.17\pm0.04$ $\sigma_q=0.42\pm0.12$	0.16 ± 0.04 0.44 ± 0.11	0.16 ± 0.04 0.46 ± 0.10	0.15 ± 0.04 0.47 ± 0.09	0.15 ± 0.04 0.48 ± 0.08
0.5	$\alpha_s=0.17\pm0.04$ $\sigma_q=0.35\pm0.12$	0.16 ± 0.04 0.38 ± 0.12	0.16 ± 0.04 0.41 ± 0.12	0.15 ± 0.04 0.43 ± 0.10	0.14 ± 0.04 0.44 ± 0.09
0.7	$\alpha_s=0.17\pm0.03$ $\sigma_q=0.28\pm0.09$	0.17 ± 0.04 0.30 ± 0.10	0.16 ± 0.04 0.33 ± 0.10	0.15 ± 0.05 0.36 ± 0.10	0.14 ± 0.05 0.39 ± 0.09
0.9	$\alpha_s=0.17\pm0.03$ $\sigma_q=0.21\pm0.08$	0.17 ± 0.04 0.23 ± 0.08	0.16 ± 0.04 0.26 ± 0.08	0.15 ± 0.04 0.30 ± 0.09	0.14 ± 0.05 0.33 ± 0.08

To further reduce this error, in the second procedure, the result in section 2 on the parameters of jet fragmentation is used. The result is consistent with the previous one, but of improved statistical accuracy:

$$\alpha_s = 0.17 \pm 0.02 \text{ (statistical)} \pm 0.03 \text{ (systematical)}.$$

It should be pointed out that in the QCD model of Ali et al., up to order of α_s^2, all the bremsstrahlung diagrams are included but not the virtual correction diagrams. Hence we should emphasize that this value of α_s should be taken within the context of the QCD model used. With this α_s and the values of the fragmentation parameters determined in section 2, various comparisons of the experimental distributions with model calculations are shown in Fig. 2 of ref.10, both at E_{cm} = 12 GeV (using α_s = 0.21 for the running coupling constant) and at 30 GeV. Since the 12GeV data were not used in determining any of the parameters, the good agreement shows that we have indeed succeeded in separating the perturbative effects which are strong at 30 GeV and hardly observable at 12GeV, from the nonperturbative effects which are generally assumed to be independent of energy.

The question may be raised whether this determination of α_s is sensitive to the particular distribution used in the triangular plot and the cut in sphericity. In Table 2 we show the results of applying the second procedure for various distributions and cuts. It is seen that the values of α_s obtained in this way are always between 0.16 and 0.19. This investigation, as well as the comparison with the model of Hoyer et al.[19] (α_s =0.19±0.02), led to the estimate of the systematic error given above.

Table 2: α_s from various distributions and using different cuts (errors are statistical)

Distribution	cut	α_s
triangular plot (S vs.A)	S > 0.25 All S	0.17 ± 0.02 0.18 ± 0.01
S	All S S > 0.2 S > 0.25	0.18 ± 0.01 0.17 ± 0.01 0.18 ± 0.01
A	All A A > 0.050 A > 0.060	0.16 ± 0.01 0.16 ± 0.01 0.17 ± 0.02
Q_2	All Q_2 $Q_2 > 0.12$ $Q_2 > 0.14$ $Q_2 > 0.16$	0.19 ± 0.01 0.18 ± 0.01 0.17 ± 0.02 0.17 ± 0.02
Y	Y > 0.10 Y > 0.15	0.18 ± 0.02 0.19 ± 0.02
Thrust T	All T T < 0.95 T < 0.90	0.19 ± 0.01 0.19 ± 0.01 0.18 ± 0.01

In QCD, α_s is expressed as

$$\alpha_s = \frac{12\pi}{(33-2N_f)\ln(Q^2/\Lambda^2)}$$

where N_f is the number of flavors, five for the energy range under consideration. It is less clear what Q^2 should be. With our value of S>0.25, and with Q identified with the center-of-mass energy (E_{cm} = 30 GeV), then Λ = 0.24 GeV. However, one could define alternatively by identifying Q^2 with the average mass squared of the virtual quark or antiquark. In this case $\langle Q^2 \rangle$ is found to be 140 GeV^2 and accordingly Λ = 0.095 GeV.

With α_s determined, we return briefly to four-jet events. A few clear four-jet events have been seen, with one example shown in Fig.4. However, the ambiguity between three- and four-jet events is more severe: for example Fig. 5 shows a plot of the 191 events with S > 0.25 together with the Monte Carlo result, normalized to the total number of events with no S or A cut, as a function of the aplanarity A. In this and in a number of similar plots, the number of three-jet events is everywhere larger than that of four-jet events. Nevertheless subtracting from the experimental data the $q\bar{q}$ and the $q\bar{q}g$ contributions according to the QCD model of Ali et al.[20] there remains a total of 36 four-jet events, in good agreement with the expectation for four-jet events from the QCD model calculation.

4. THREE-JET DISTRIBUTION AND THE GLUON SPIN

As already mentioned in section 1, three-jet events were first seen at PETRA over a year ago. We discuss a few features of these three-jet events. Note that the method of generalized sphericity gives all three jet axes together with the partitions of the hadrons into the three jets.

(A) In Fig. 6 the transverse momentum distribution of the hadrons with

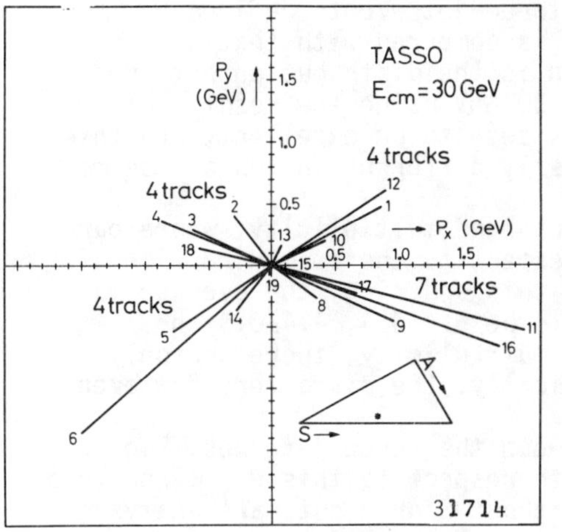

Fig. 4 - A four-jet event.

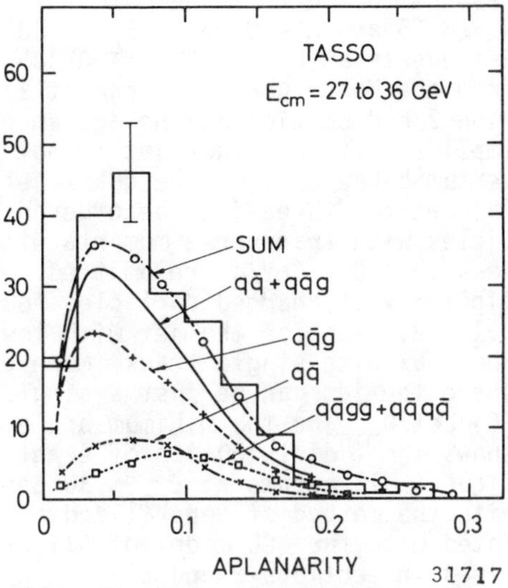

Fig. 5 - Distribution of events as a function of aplanarity for S > 0.25.

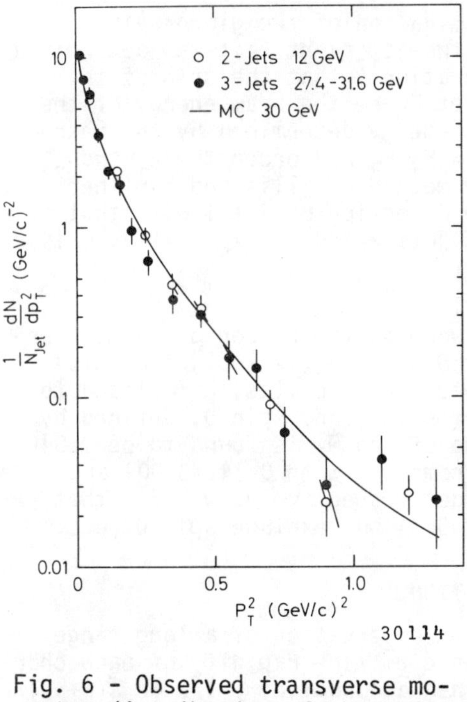

Fig. 6 - Observed transverse momentum distribution of the hadrons from the planar region (S > 0.25, A < 0.08) with respect to the three axes at E_{cm} = 30 GeV (●). It is compared with the transverse momentum distribution relative to the sphericity axis for all events (no S or A cut) at E_{cm}=12 GeV, analyzed as two-jet events (○). It is also compared with the result from the QCD model at 30 GeV (curve).

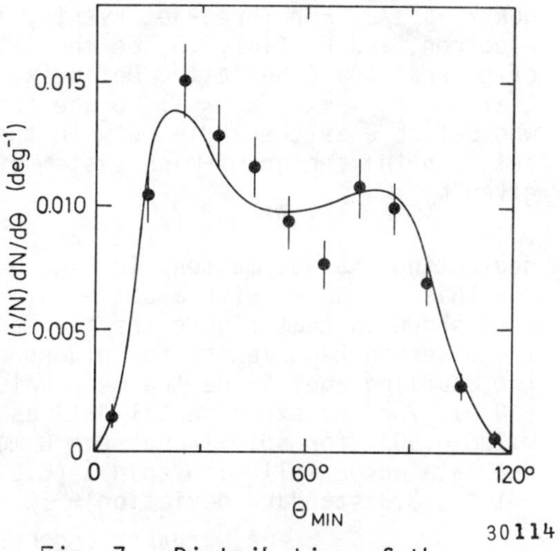

Fig. 7 - Distribution of the smallest angle between any of the three jets, when all events (no S or A cuts) at E_{cm} = 30 GeV are analyzed with the generalized sphericity method. The curve show the result from the QCD model.

respect to the three jet axes of the three-jet events defined by $S > 0.25$ and $A < 0.08$ at $E_{cm} \sim 30$ GeV is compared with that of two-jet events at $E_{cm} = 12$ GeV. Also shown is the distribution obtained from the Monte Carlo program at $E_{cm} = 30$ GeV using the results of section 2 and section 3. The agreement is seen to be excellent, and this implies that the gluon jet is not grossly different in transverse momentum behavior from the quark jet.
(B) Let the three-jets be ordered by charged multiplicity, where particles with transverse momenta with respect to the beam direction of less than 0.1 GeV/c are omitted. With this ordering, the average multiplicity of charged particles found to be 6.4±0.4, 4.4±0.3 and 3.1±0.2. Even for the jet with lowest multiplicity, there are only about 5% with single track. More explicitly, there are very few events where the jet can be just a single π.
(C) Let θ_{min} be the minimum angle between the three jet axes. Fig. 7 shows the distribution of events with respect to this θ_{min}. For this figure, all events at $E_{cm} \sim 30$ GeV, without S or A cut, are analyzed with the method of generalized sphericity. The solid curve is calculated with the QCD model of Ali et al., using the parameters determined in sections 2 and 3.
(D) Finally, we report a recent determination of the gluon spin. From the angular distribution of the two-jet events, it was determined some time ago by the SLAC-LBL Collaboration[17] that the spin of the quark is 1/2. For three-jet events, let E_B be the beam energy of the electron, and E_i $(i=1,2,3)$ be the jet energy determined by the method of generalized sphericity. Define $x_i = E_i/E_B$ and order the x_i such that $0 \leq x_3 \leq x_2 \leq x_1 \leq 1$. We use the method of Ellis and Karliner[23] who define $\tilde{\theta}$ as the angle between the direction of jet 1 with that of jet 2 in the center-of-mass system of jets 2 and 3. Explicitly, $\tilde{\theta}$ is given by

$$|\cos\tilde{\theta}| = (x_2 - x_3) / x_1$$

neglecting the jet masses. In Fig. 8 we plot the number of events versus this variable, with a cut of $x_1 < 0.9$ to remove two jet events. Also shown in that figure are the Monte Carlo results, normalized to the observed 248 events, for gluons of spin 1 and spin 0, defined by the coupling $\phi q\bar{q}$. Since the mean value of $|\cos\tilde{\theta}|$ is found to be 0.349 ± 0.013 for the experimental data as compared with 0.341±0.004 and 0.298±0.003, for spin 1 and spin 0 model respectively, we find that the data agree well with spin 1 (C.L.=55%) but exclude spin 0 (C.L. = 10^{-4}, 3.8 standard deviation)[24].

5. CHARGE CORRELATION

In this section, we report the first observation of a long range charge correlation in opposite jets in e^+e^-. The rapidity of each charged hadron in a two-jet event is defined as usual by $y=1/2 \ln [(E+p_\parallel)/(E-p_\parallel)]$, where $p_\parallel$ is the component of the momentum parallel to the jet axis. If n is the charged multiplicity of the event, then we define a charge correlation function, the compensating charge flow $\bar{\phi}$ by

$$\bar{\phi}(y,y') = -\frac{1}{\Delta y \Delta y'} \left\langle \frac{1}{n} \sum_{k=1}^{n} \sum_{i \neq k} e_i(y)\, e_k(y') \right\rangle \qquad (1)$$

where, for example, for the ith particle with rapidity y_i, $e_i(y) = +1$ or -1 according to the charge of this particle if y_i is inside an interval Δy around y and $e_i(y) = 0$ otherwise. In (1) $< >$ means averaging over all events.

This charge correlation function is to be compared with the corresponding particle density function defined by

$$\bar{\rho}(y,y') = \frac{1}{\Delta y \Delta y'} < \frac{1}{n(n-1)} \sum_{k=1}^{n} \sum_{i \neq k} |e_i(y)|\ |e_k(y')| > \qquad (2)$$

where the sum is over all charged particles. The denominators n and n(n-1) in (1) and (2) are chosen such that

$$\int dy \int dy'\ \bar{\phi}(y, y') = \int dy \int dy'\ \bar{\rho}\ (y,y') = 1$$

because $\sum_{i \neq k} e_i e_k = -n$ and $\sum_{i \neq k} |e_i||e_k| = n^2 - n$

Let $P(y_+, y'_-)$, for example, be the probability of having a positively charged particle with rapidity y and a negatively charged particle with rapidity y'. In terms of this probability, $\bar{\rho}$ and $\bar{\phi}$ are

$$\bar{\rho}(y,y') \propto P(y_+, y'_-) + P(y_-, y'_+) + P(y_+, y'_+) + P(y_-,y'_-)$$

and $$\bar{\phi}(y,y') \propto P(y_+, y'_-) + P(y_-, y'_+) - P(y_+, y'_+) - P(y_-, y'_-)$$

For practical purposes, it is more convenient to plot instead

$$\rho(y, y') = \bar{\rho}(y, y') \Big/ \int dy\ \bar{\rho}(y, y')$$

and $$\phi(y, y') = \bar{\phi}(y, y') \Big/ \int dy\ \bar{\phi}(y, y')$$

such that $$\int dy\ \rho(y, y') = \int dy\ \phi(y, y') = 1.$$

Thus $\rho(y, y')$ is the probability that a charged particle with rapidity y' finds another charged particle with rapidity y, while $\phi(y, y')$ is the probability that the charge of a particle with rapidity y' is compensated by another particle of opposite charge with rapidity y.

Using the TASSO data with the additional cut that the total observed charge is 0 or ± 1, $\rho(y,y')$ is plotted in Fig. 9 for (a) $-0.75 \leq y' < 0$ and (b) $-5 \leq y' \leq -2.5$. Also $\phi(y,y')$ is plotted in Fig.9 for (c) $-0.75 \leq y' < 0$ and (d) $-5 \leq y' \leq -2.5$. A comparison of these figures shows the following features.

(A) For small value of y', Fig. 9a shows a broad distribution while Fig. 9c shows a narrower distribution which peaks at $y \sim y'$. This is the evidence for a short range charge correlation. Moreover as y' becomes larger, there exhibits a peak adjacent to y' as shown in Fig.9d but not so in Fig.9b. This further supports the evidence for short range charge correlation.

(B) As shown in Fig. 9(d), there is a noticable rise in the ϕ distribution near large positive y when y' is large and negative. The area beyond $y \geq 2.5$ is 0.101 ± 0.033 from Fig. 9(d) compared with the corresponding value of 0.011 ± 0.014 from Fig. 9(c) for small y'. This is the evidence of a long-range charge correlation in opposite jets from

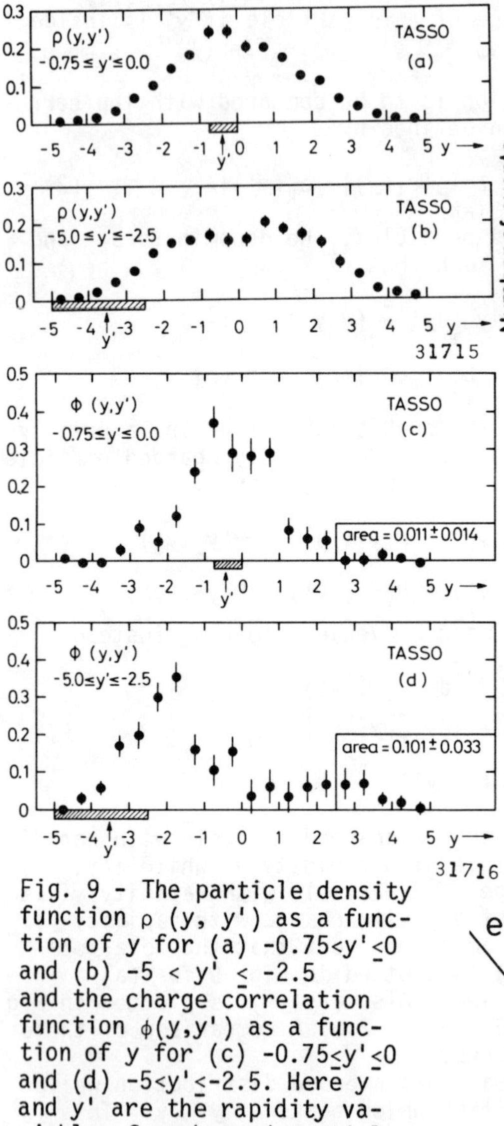

Fig. 9 - The particle density function ρ (y, y') as a function of y for (a) $-0.75<y'\leq 0$ and (b) $-5 < y' \leq -2.5$ and the charge correlation function $\phi(y,y')$ as a function of y for (c) $-0.75\leq y'\leq 0$ and (d) $-5<y'\leq -2.5$. Here y and y' are the rapidity variables for charged particles. The data are from center of mass energies of 27 to 36 GeV.

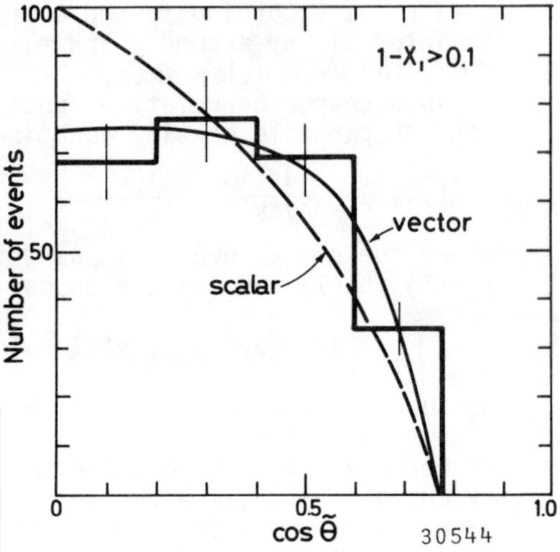

Fig. 8 - Observed distribution of the data in the region $1-x_1>0.10$, as a function of the cosine of the Ellis-Karliner angle $\tilde{\theta}$. The solid line shows the QCD prediction, the dotted line the prediction for scalar gluons, both normalized to the number of observed events. The data are from center of mass energies of 27 to 36 GeV.

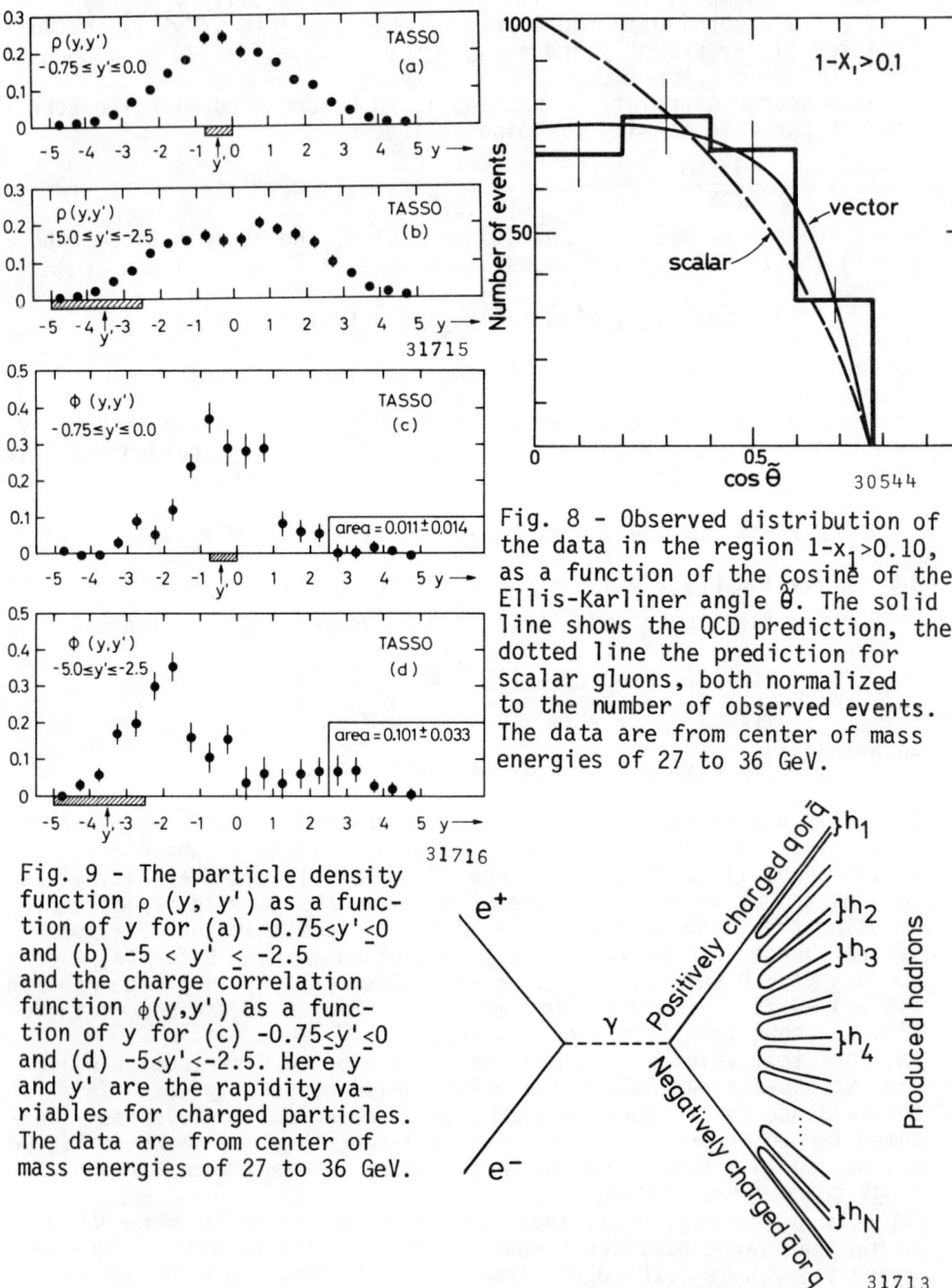

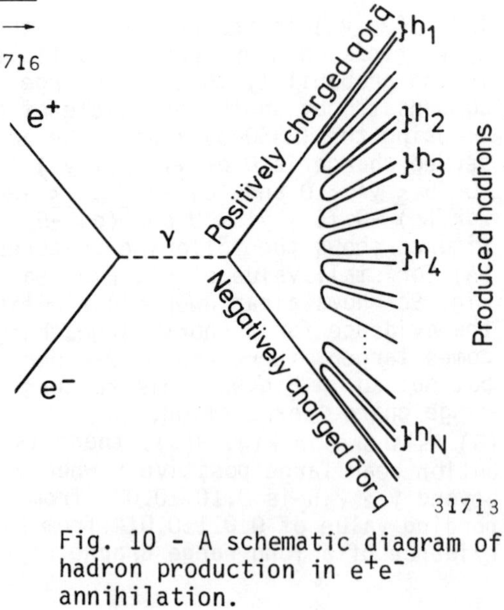

Fig. 10 - A schematic diagram of hadron production in e^+e^- annihilation.

e^+e^- annihilation.

The result of long range correlation may be qualitatively understood as follows. With the schematic diagram of hadron production of Fig.10, let the produced charged hadrons be ordered as shown there. Since the quark charges are ±1/3 and ±2/3, we have with this ordering $e_1 = 1$, $e_2 = -1$, $e_3 = 1$, $e_4 = -1$ etc. Although this ordering cannot be determined experimentally, it is strongly correlated with the rapidity ordering. Accordingly, the presence of a negative charge for large negative y increases the probability of the presence of a positive charge for large positive y', and vice versa. Thus our observation of a long range charge correlation in opposite jets gives independent corroboration of the picture of jet formation via a $q\bar{q}$ pair as shown in Fig. 10.

REFERENCES

1. TASSO-Collaboration:
R.Brandelik, W.Braunschweig, K.Gather, V.Kadansky, F.J.Kirschfink, K.Lübelsmeyer, H.-U.Martyn, G.Peise, J.Rimkus, H.G.Sander, D.Schmitz, A.Schultz von Dratzig, D.Trines, W.Wallraff, I.Physikalisches Institut der RWTH Aachen, Germany; H.Boerner, H.M.Fischer, H.Hartmann, E.Hilger, W.Hillen, G.Knop, L.Koepke, H.Kolanoski, P.Leu, B.Löhr, R.Wedemeyer, N.Wermes, M.Wollstadt, Physikalisches Institut der Universität Bonn, Germany; H.Burkhardt, D.G.Cassel, D.Heyland, H.Hultschig, P.Joos, W.Koch, P.Koehler, U.Kötz, H.Kowalski, A.Ladage, D. Lüke, H.L.Lynch, P.Mättig, G.Mikenberg, D.Notz, J.Pyrlik, R.Riethmüller, M.Schliwa, P.Söding, B.H.Wiik, G.Wolf, DESY, Hamburg, Germany; R.Fohrmann, M.Holder, G.Poelz, O.Römer, R.Rüsch, P.Schmüser, II.Institut für Experimentalphysik der Universität Hamburg, Germany; I.Al-Agil, D.M.Binnie, P.J.Dornan, N.A.Downie, D.A.Garbutt, W.G. Jones, S.L.Lloyd, D.Pandoulas, J.Sedgbeer, R.A.Stern, S.Yarker, C. Youngman, Depart. of Physics, Imperial College, London, England; R.J.Barlow, I.C.Brock, R.J.Cashmore, R.Devenish, P.Grossmann, J. Illingworth, M.Ogg, B.Roe, G.L.Salmon, T.R.Wyatt, Depart. of Nuclear Physics, Oxford University, England; K.W.Bell, B.Foster, J.C. Hart, J.Proudfoot, D.R.Quarrie, D.H.Saxon, P.L.Woodworth, Rutherford Laboratory, Chilton, England; E.Duchovni, Y.Eisenberg, U.Karshon, D.Revel, E.Ronat, A.Shapira, Weizmann Institute, Rehovot, Israel; T.Barklow, J.Freeman, P.Lecomte, T.Meyer, G.Rudolph, E.Wicklund, Sau Lan Wu, G.Zobernig, Department of Physics, University of Wisconsin, Madison, Wisconsin, USA.
2. B.H.Wiik Proceedings of the Intern.Neutrino Conference,Bergen, Norway, 18-22 June 1979, p. 113
3. P.Söding, Proceedings of the European Phys.Society Intern.Conf. on High Energy Phys.Geneva, Switzerland, 27 June - 4 July,1979,p.271
4. TASSO-Collaboration,R.Brandelik et al.,Phys.Lett.86B(1979)243
5. TASSO-Collaboration,R.Brandelik et al.,Phys.Lett.88B(1979)199
6. TASSO-Collaboration,G.Wolf,Proceedings of the 1979 Intern.Symp. on Lepton and Photon Interactions at High Energies, FNAL,Batavia, Illinois 23-29 August 1979
7. TASSO-Collaboration,R.Brandelik et al.,Phys.Lett.89B(1980)418

8. TASSO-Collaboration, R.Brandelik et al., Phys.Lett. 92B(1980)199, Phys.Lett. 94B(1980)91, Phys.Lett.94B(1980)259 and Phys.Lett.94B(1980)444
9. H.Boerner et al., DESY 80/27 (1980)
10. TASSO-Collaboration, R.Brandelik et al., Phys.Lett.94B(1980)437
11. S.L.Wu and G.Zobernig, Particle and Fields(Z.Phys.C), 2(1979)107
12. MARK-J.Collaboration, D.P.Barber et al., Phys.Rev.Lett.43(1979)830
13. PLUTO-Collaboration, Ch.Berger et al., Phys.Lett.86B(1979)418
14. JADE-Collaboration, W.Bartel et al., Phys.Lett.91B(1980)142
15. J.Ellis, M.K.Gaillard and G.G.Ross, Nucl.Phys.B111(1976)253; T.A.DeGrand, Y.J.Ng, and S.-H.Tye, Phys.Rev.D16(1977)3251; A.DeRujula, J.Ellis, E.G.Floratos and M.K.Gaillard, Nucl.Phys.B138(1978)387
16. J.D.Bjorken and S.J.Brodsky, Phys.Rev.D1(1970)1416
17. G.Hanson et al., Phys.Rev.Lett. 35(1975)1609
18. R.D.Field and R.P.Feynman, Nucl.Phys. B136(1978) 1
19. P.Hoyer, P.Osland, H.E.Sander, T.F.Walsh, and P.M.Zerwas, Nucl.Phys. B161(1979)349
20. A.Ali, E.Pietarinen, G.Kramer, and J.Willrodt, Phys.Lett. 93B(1980) 155
21. Private communications, H.E.Sander and Tom Meyer of TASSO-Col.
22. F.A.Berends and R.Kleiss, DESY-Report 80/73 (1980)
23. J.Ellis and I.Karliner, Nucl.Phys. B148(1979)141
24. TASSO-Collaboration, R.Brandelik et al., DESY Report 80/80 (1980)

HIGH ENERGY e^+e^- INTERACTIONS

S. Orito, DESY, Organizer

RECENT RESULTS FROM THE JADE COLLABORATION*

presented by S.Yamada
LICEPP, University of Tokyo, Hongo, Bunkyo-ku Tokyo

ABSTRACT

An upper limit on the b-meson lifetime is obtained to be 3 x 10^{-11}sec. Hadron events are analysed in the QCD scheme and the strong coupling constant is determined together with the fragmentation parameters. The best fit of α_s is 0.18±0.03± 0.03. A possible difference of gluon and quark fragmentation is seen in three-jet events.

INTRODUCTION

Recent results from the JADE collaboration on the b-meson life time and the QCD analysis are reported. For further results from JADE refer to the talks by the subject speakers. The detector is described in detail[1,2] elsewhere. It consists of a cylindrical drift chamber with 48 sampling layers enabling the ionization loss measurement, a solenoid of 4.5kgauss, TOF/trigger counter, lead-glass shower counter arrays, five layers of muon drift chambers embedded between the hadron absorbers, and forward tagging counters.

UPPER LIMIT ON B-MESON LIFETIME

There are good reasons to assume that 1/11 of the observed multihadron events are initiated by $e^+e^- \to b\bar{b}$ and contain b-mesons.[3] If the b-meson liftime (τ_b) is longer than a few nsec, their tracks can be identified in the drift chamber by means of the momentum(p) and the energy loss (dE/dx) measurement.[2] We searched for massive particles in the 1668 hadron events between 27 and 35GeV[4]. Fig.1 shows dE/dx vs p of the long tracks. Pions, Kaons and protons are seen. No heavily ionizing particle is observed for a mass bigger than 4GeV/c^2 and an upper limit on their lifetime is calculated. In order to estimate the number of produced particles in the sensitive momentum range the following spectra are used.

*Members
DESY: W.Bartel, T.Canzler, D.Cords,P.Dittmann, R.Eichler, R.Felst, D.Haidt, S.Kawabata, H.Krehbiel, B.Naroska, L.H.O'Neill, J.Olsson, P.Steffen, W.L.Yen. Hamburg: E.Elsen, M.Helm,A.Petersen,P.Warming, G.Weber. Heidelberg: H.Drumm, J.Heintze, G.Heinzelmann, R.D.Heuer, J.von Krogh, P.Lennert, H.Matsumura, T.Nozaki, H.Rieseberg, A.Wagner. Lancaster: D.C.Darvill, F.Foster, G.Hughes, H.Wriedt. Manchester: J.Allison, J.Armitage, A.H.Ball, I.Duerdoth, J.Hassard, F.Loebinger, H.McCann, B.King, A.Macbeth, H.Mills, P.G.Murphy, H.Prosper, K.Stephens. Rutherford: D.Clarke, M.C.Goddard, R.Hedgecock, R.Marshall, G.F.Pearce. Tokyo: M.Imori, T.Kobayashi, S.Komamiya, M.Koshiba, M.Minowa, S.Orito, A.Sato, T.Suda, H.Takeda, Y.Totsuka, Y.Watanabe, S.Yamada, C.Yanagisawa.

0094-243X/81/680616-06$1.50

$$d\sigma/dp \propto (4\pi p^2/E) \exp(-3.5E) \quad (1)$$

$$d\sigma/dp \propto 1 \quad (2)$$

For b-mesons of mass 5GeV the 90% C.L. limit for the case (1) is 1.7 $\times 10^{-9}$sec and for the case (2) it is 2.2×10^{-9}sec.

Once we know the above limit, the decay vertex distribution can be used to obtain a better limit on τ_b. Secondary vertices with ≥4 prongs at a distance of ≥5cm from the beam are looked for. where the detection efficiency is verified to be better than 70%. Background from nuclear interactions is removed by rejecting those vertices which contain protons among their products or which are in dense material such as the beam pipe. There is no decay vertex of ≥4 prongs. Hence, we get a 90% C.L. limit of 10^{-10}sec for (1) and 3x 10^{-11}sec for (2).

The secondary vertex can be more efficiently found in the vicinity of the primary vertex if one knows which tracks are produced by the decay. For this purpose muons are used. A total of 29 muon candidates of above 1.4GeV/c are found among 265 multihadron events taken with $0.92\times4\pi$ str muon acceptance, of which (17±6) are estimated to be genuine signals; the number is in good agreement with the model preditions including c and b quarks.[5] Using 10% for the $b\to\mu x$ branching ratio[6], we expect 4.6 muons from b-mesons. All the muons are extrapolated to the initial vertex. They originate within 3mm from the beam and no secondary vertex is seen. The 90% C.L. upper limit from this analysis is 3×10^{-11}sec for (1) and 10^{-11}sec for (2).

In the standard six quark model the recently predicted τ_b is

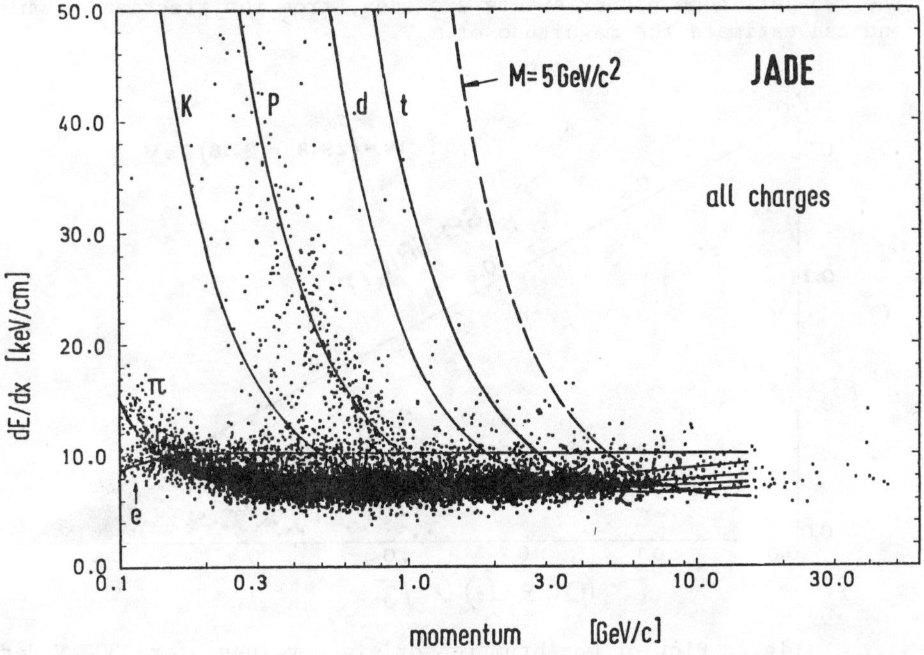

Fig.1 dE/dx vs momentum of long tracks of multitrack events.

shorter than the limit[7], while it puts some constraint on the b-lifetime expected by the five quark model[8].

DETERMINATION OF α_s

Three-jet events which cannot be explained by statistical fluctuations of two jets have been observed at PETRA energies.[9,10] Since the two jets are well described by the quark parton model, it is natural to interpret the phenomenon as a higher order QCD effect[11], $e^+e^- \to q\bar{q}g$, where quarks and gluon fragment into three jets. It was shown that by taking the strong coupling constant α_s to be 0.17 at 30GeV a good agreement was obtained between the QCD prediction and observation.[10] A systematic study has been done using 1002 hadron events collected around 30GeV. The event selection criteria are published elsewhere.[1]

The jet analysis is done by means of the normalized momentum tenser defined by

$$T_{\alpha\beta}=\sum_{i=1}^{n} P_\alpha P_\beta \Big/ \sum_{i=1}^{n}|P_i|^2, \tag{3}$$

where α,β=x,y and z, and i goes over all charged and neutral particles. The eigen values of the tensor, $Q_1 \leq Q_2 \leq Q_3$, characterize the event shape. The corresponding eigen vectors $\vec{n}_1$, $\vec{n}_2$ and $\vec{n}_3$ give the orientation of the momentum ellipsoid. Typically two-jet events have $Q_1;Q_2 \sim 0$ and $Q_3 \sim 1$, and $\vec{n}_3$ is parallel to the jet axis. For three-jet events planarity, Q_2-Q_1, is large and ($\vec{n}_2$, $\vec{n}_3$) defines the event plane. The Q plot of the data is shown in Fig.2. Besides the two-jets some planar events are seen, from the fraction of which one can estimate the magnitude of α_s.

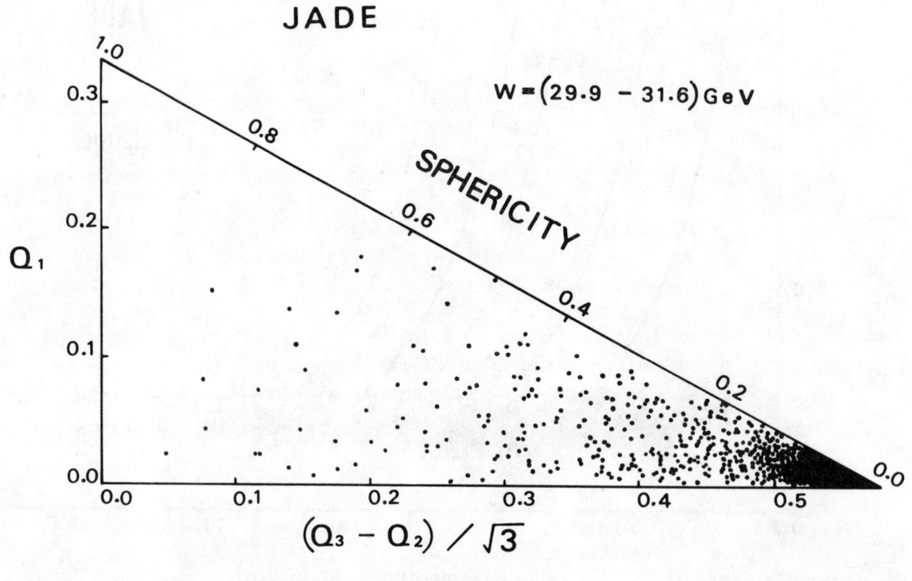

Fig.2 Plot of momentum tensor eigen values for 30GeV data.

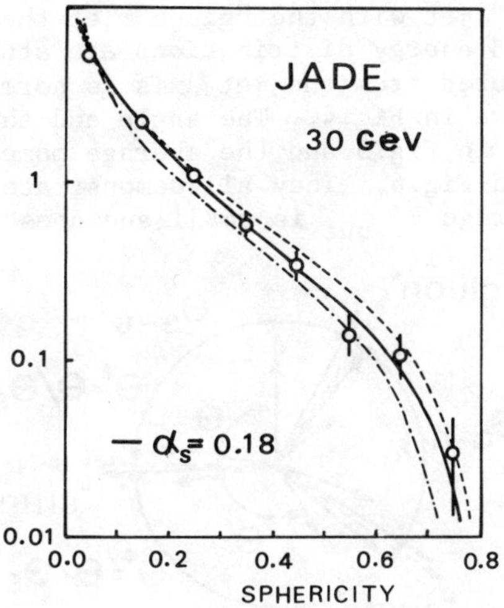

Fig.3 Sphericity distribution. The curves are expectations by the QCD model of Hoyer et al; ——— α_s=0.18, ----α_s=0.26 and -·-·- α_s=0.12.

To compare the QCD prediction with the data we employed the model by Hoyer et al.,[12] which takes into account $O(\alpha_s)$ effect and the fragmentation of quarks and gluons according to Field and Feynman.[13] The fragmentation parameters are also determined. The primordial fragmentation function is expressed as

$$f_q(z)=1-a+3a(1-z)^2, \quad (4)$$

$$z=\frac{(E+P_{11})\text{hadron}}{(E+P_{11})\text{quark}}, \quad (5)$$

where a is an adjustable parameter for u,d,s quarks, while for heavy quarks and gluons it is fixed to 0 and 1, respectively. The relative production ratio of secondary u,d, and s quarks is assumed to be 2:2:1. Both the parameter a and the fraction of pseudoscalar mesons among the produced mesons are determined from the inclusive hadron spectrum to be 0.5±0.1 and 0.5±0.1. The transverse momentum of hadrons w.r.t. the jet axis is distributed as

$$\frac{1}{\sigma}\frac{d\sigma}{dp_t^2} = \frac{1}{2\sigma_q^2}\exp\left(-\frac{P_t^2}{2\sigma_q^2}\right). \quad (6)$$

The effect of heavy quark decay is taken into account as well as hard photon radiation in the initial state.

α_s and σ_q are obtained by simultaneously fitting the sphericity and Q_1 distributions. The best values are α_s=0.18±0.03 and σ_q= 0.34±0.03. The errors are statistical. The result is the same as our earlier number within the error.[10] A systematic error is estimated to be 0.03 by comparing different method of fitting.

We repeated the same analysis by using other QCD models. One is by Ali et al.[14] which includes $O(\alpha_s^2)$ effects and the other is the Lund model[15] which uses the string model for fragmentation along the two colour lines. They give consistent values of α_s. Similar numbers are obtained by other PETRA experiments.[16]

STUDY OF THE THREE-JET EVENTS[17]

Planar events mostly consist of a slim and a broad jet. In its rest system the broad jet can be decomposed into two jets[10], of which the lower energy one in the lab. system is denoted as the "gluon" jet and the other is called the "quark" jet. According to the QCD prediction the gluon tends to have the lowest energy among $q\bar{q}g$ and the method identified the original gluon jet for about 50% of the case when applied to simulated events.

In order to compare the "quark" jet with the "gluon" jet the angle dependence of multiplicity and energy distributions are studied in the event plane. The angle measured from the jet axis is normalized by the jet opening angle as shown in Fig.4. The angle and the energy flow distributions are shown in Fig.5 and the average momentum in and out of the plane are shown in Fig.6. They all demonstrate clear three jet structure. The average $P_{t,out}$ is small and does not indicate angle dependence. A difference can be seen in the relative depths of the multiplicity and energy flow distributions between the slim-quark jet interval and the slim-gluon jet interval. It is not explained by the reflection of the angle normalization or the "gluon" selection criteria. In the frame of the above described QCD models the difference is reproducible only when the gluon fragments softer than quarks. An example

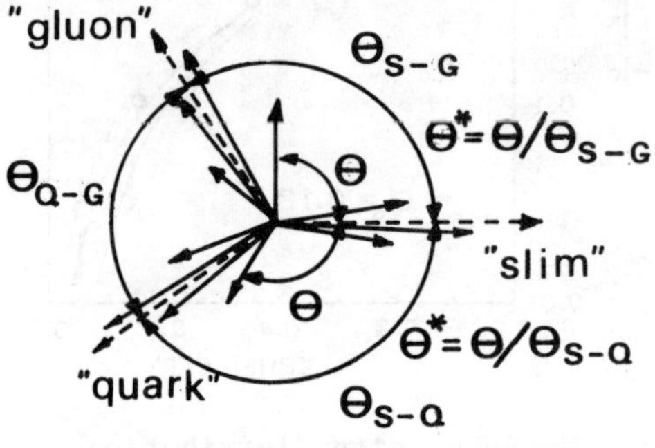

Fig.4. Definition of Normalized angles in the event plane.

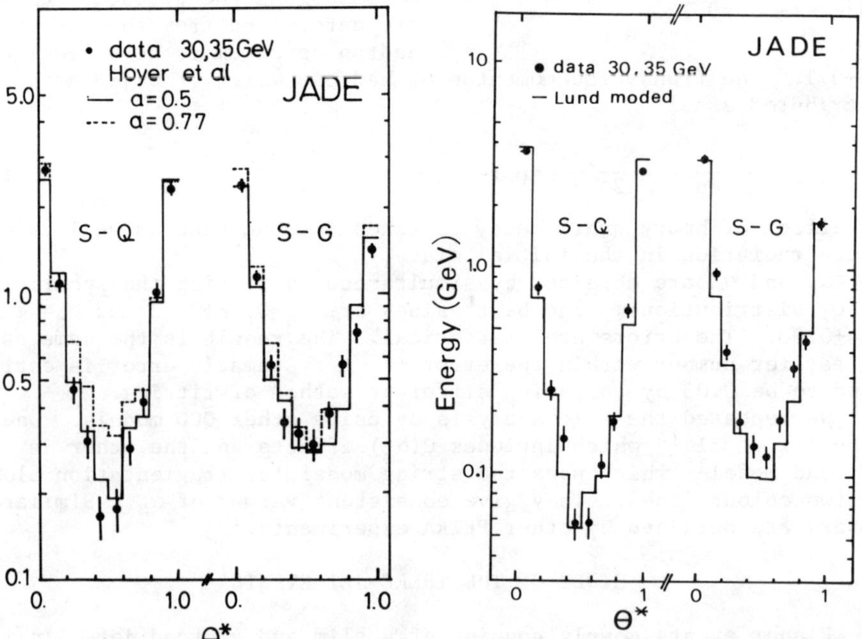

Fig.5a,b Multiplicity and energy distribution in the event plane for the three-jet events, respectively.

is shown is Fig.5-a using Hoyer et al.'s model. The energy flow distribution compares with the models in a similar way. In Fig.5-b a prediction of Lund model is shown. The observation may indicate the softness of gluon fragmentation in contrast to quark fragmentation.

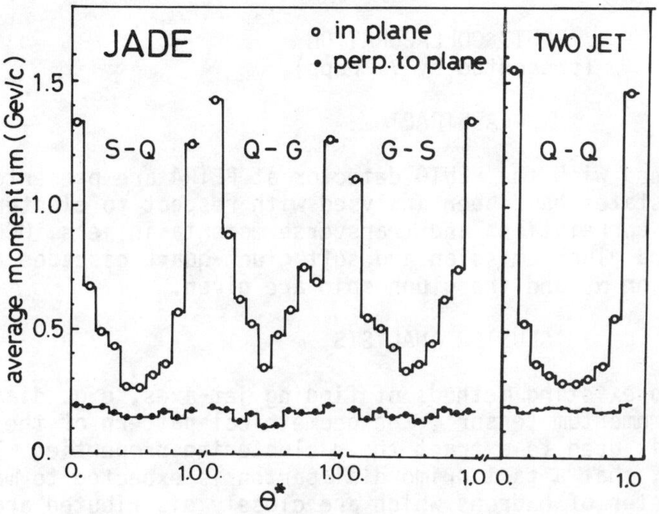

Fig.6 Average momentum in and out of the event plane for the three-jet events. The same is shown for the two-jet events w.r.t. a plane containing the jet axis for comparison.

REFERENCES

1 Jade collaboration. Phys.Lett.88B 171(1979)
2 H.Drumm et al.DESY 80/38
3 talk by D.Cord, this conference
4 JADE collaboration, DESY 80/71; talk by J.von Krogh, this conference
5 A.Ali, Zeitschrigt f.Physik C1 25 (1979)
6 talks by J.Yoh and E.Thornadike, this conference. CLEO Collaboration, CLNS 80/464
7 R.E.Shrock, S.B.Treiman and L.L.Wang, Phys.Rev.Lett. 42 158(1979)
8 H.Georgi and S.L.Glashow,HUTP 79/A073
9 TASSO Collaboration, Phys.Lett.86B 243 (1979)
MARK j Collaboration, Phys.Rev.Lett. 43 830 (1979)
Pluto Collaboration, Phys.Lett. 86B 418 (1979)
10 JADE Collaboration, Phys.Lett. 91B 142 (1980)
11 J.Ellis,M.K.Gaillard and G.G.Ross, Nucl.Phys. B111 253 (1976)
T.A.Degrand,Y.J.Hg and S.H.H.Tye, Phys.Rev.D16 3251 (1977)
12 P.Hoyer et al., Nucl.Phys.B161,349 (1979)
13 R.D.Field and R.P.Feynman, Nucl.Phys. B136 1 (1978)
14 T.Sjostrand and B.soderberg, LU TP 78-18 (1978)
T.Sjostrand, LU TP 79-8 (1979)
15 A.Ali, E.Pietarinen, G.Kramer and J.Willrodt, DESY 79/86
16 TASSO Collaboration, DESY 80/40
talks by S.L.Wu, V.Hepp, H.Newman, this conference
17 Earlier results are presented at the 15th Rencontre de Moriond 1980, DESY 80/46

PLUTO RESULTS ON JETS AND QCD

PLUTO COLLABORATION
(presented by V. Hepp)

ABSTRACT

Results obtained with the PLUTO detector at PETRA are presented. Multihadron final states have been analysed with respect to clustering, energy-energy correlations and transverse momenta in jets. QCD predictions for hard gluon emission and soft gluon-quark cascades are discussed. Results on α_s and the gluon spin are given.

CLUSTER ANALYSIS

In contrast to existing methods of finding jet-axes, e.g. diagonalization of the momentum tensor[1], the geometrical pattern of the final state hadrons is used to extract their clustering properties. The idea is, of course, that a fast primordial parton is expected to materialize into a cluster of hadrons which are closely distributed around the original parton axis at high energies. In a first step, we use only the directions of charged and neutral particles to define preclusters by requiring any two of them to be within a 'collecting angle' α. Then preclusters are merged into clusters, if the momentum vectors of any two of them subtend an angle smaller than β. A cluster is called a jet, if the energy contained in it exceeds typically 2 GeV, as motivated by low energy results. For α and β values around 30^0 were used, corresponding to ~ 7 % of 4 π. With this method the experimental number of jets in every event was determined. This number is not necessarily equal to the number of fast primordial partons. The efficiency of the collecting algorithm was determined with $q\bar{q}$, $q\bar{q}g$ and ggg Monte Carlo studies including complete simulation of the detector and radiation. The unbiased reconstruction of the parton quantities based on zero mass kinematics was verified.

The PLUTO data at cms energies between 27 and 32 GeV were passed through the cluster program. For each event the number of jets/event (n_j) was determined. In the sample of 860 selected hadronic events ~ 250 were found with n_j = 3. Fig. 1 shows the observed n_j distribution (full circles) compared with the expected one for the $q\bar{q}$ and $q\bar{q}g$ Monte Carlo (triangles and rectangles, resp.). We find good agreement with the $q\bar{q}g$ prediction.

The n_j distribution for isotropic phase space has a maximum at n_j = 5 (not shown). $t\bar{t}$ production with Q = 1/3 would predict an excess of entries for n_j > 3 and is excluded by more than 7 s.d.. Since all events in the n_j = 4 class can be explained by the 1st order QCD predictions, we do not see any sizeable production of 4 energetic jets in our data.

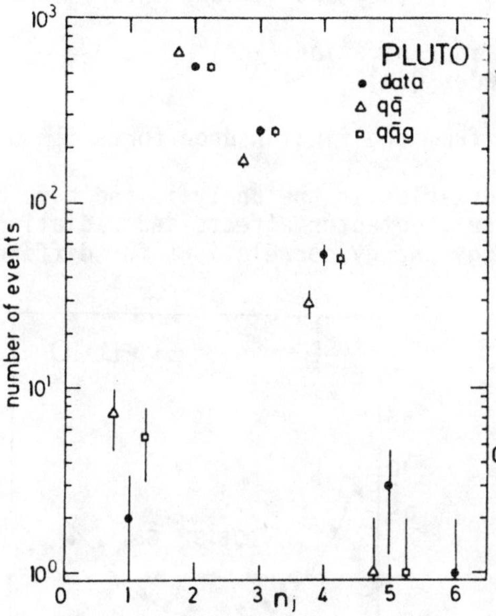

Fig. 1 n_j distributions (see text)

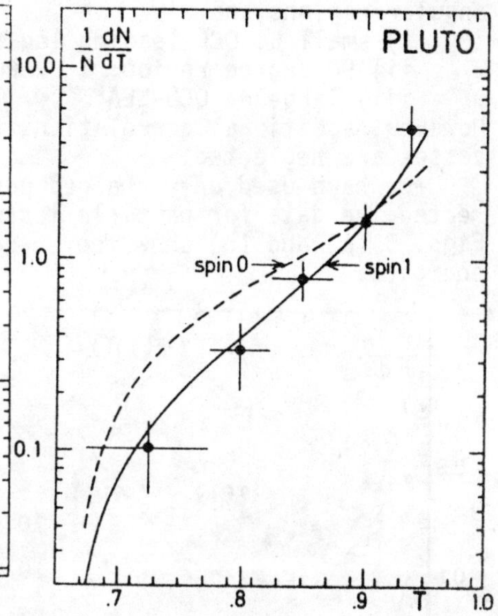

Fig. 2 Parton thrust distribution solid (dashed) curves: spin = 1 (0) predictions

From the n_j = 2 class we determined the transverse spread σ_q in the quark fragmentation to be σ_q = (290 ± 30) MeV. From the number of events with n_j = 3 we obtained (after applying more severe cuts, e.g. thrust < 0.925) for the strong coupling constant: $\alpha_s = 0.16 \pm 0.03$ (stat. error) ± 0.03 (syst. error). Background due to hard photons or 2γ events was found to be small. The systematic error in α_s reflects the uncertainties introduced by the cuts and the correlations with σ_q.

The parton thrust distribution after subtraction of the $q\bar{q}$ background is displayed in Fig. 2. The solid curve in the figure is the absolute prediction[2] from 1st order QCD for gluon spin = 1. The dashed curve is the spin = 0 prediction[3] normalized to the number of events which is clearly disfavored (χ^2 is 28.7 for 4 degrees of freedom).

ENERGY-ENERGY CORRELATIONS

As a measure of parton cascading we studied energy-energy correlations of the form

$$\frac{1}{\sigma}\frac{d\sigma}{d\theta} = \sum_{a,\,b} \int \frac{1}{\sigma}\frac{d^3\sigma}{dz_a dz_b d\theta}\, z_a z_b dz_a dz_b$$

with $z_{a,\,b} = E_{a,\,b} / E_{cm}$, θ = angle between hadron a and hadron b.

The data are compared with theoretical QCD predictions which depend only on the scale parameter Λ. They are divided into 3 separate angular regions, namely:

i) small θ: QCD leading log approximation (LLA)[4]

ii) 90 degree region: 1st order QCD[5]

iii) large θ: QCD-LLA[6]

However, additional correlations from the final hadron formation processes are neglected.

We have used only charged particles in the analysis and have corrected the data for particle losses, detector effects and radiation. Figs. 3 (a) and (b) show the energy-energy correlations for different energies.

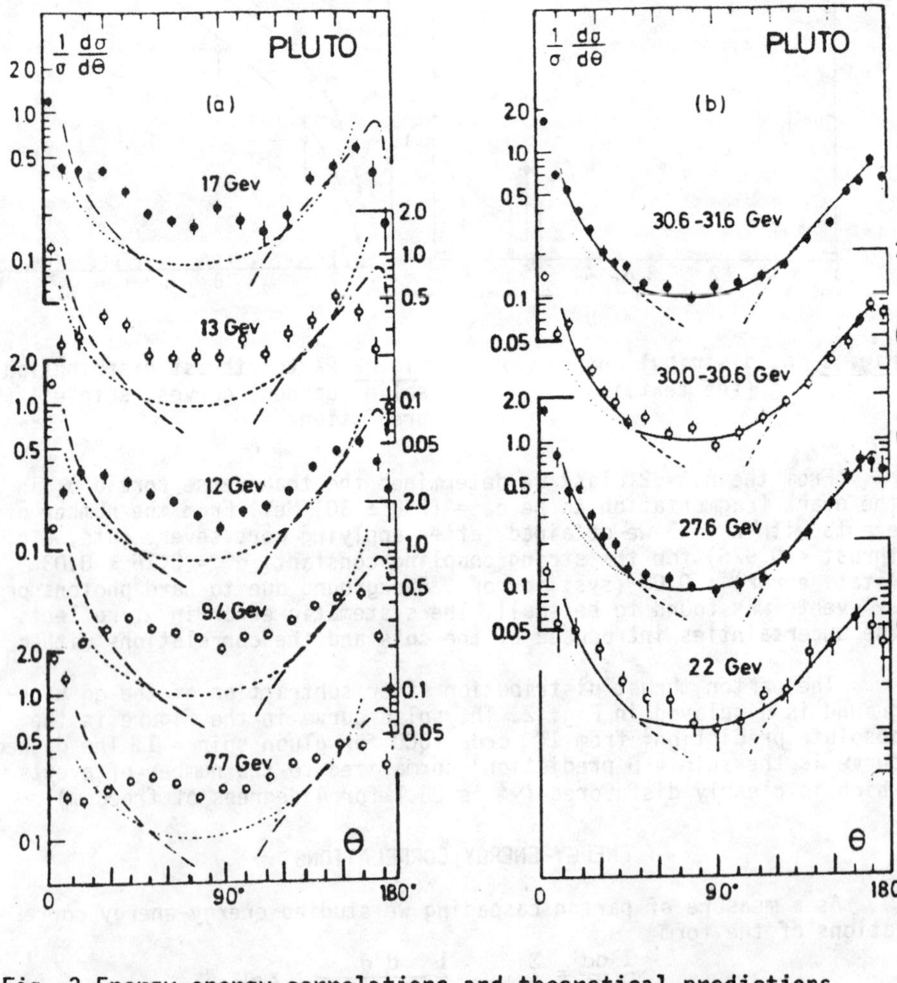

Fig. 3 Energy-energy correlations and theoretical predictions (see text). Dashed curves, ref. 4; dotted curves, ref. 5; dashed-dotted curves, ref. 6; full curves, fit to Λ.

In Fig. 3 (a) the theoretical expectations are given using Λ = 500 MeV. No attempt was made to fit the data, since large non-perturbative contributions are expected. The data above 20 GeV (Fig. 3 (b)) were fitted and an acceptable χ^2 was obtained for α_s = 0.26 ± 0.01 (stat. error) ± 0.06 (syst. error). This value is larger than α_s obtained from e.g. 3 jet studies indicating a non-negligible contribution due to fragmentation even at our highest energy points.

TRANSVERSE MOMENTUM IN JETS

The summed transverse momentum $K_\perp$ in jets is expected to be less sensitive to hadronization effects than single particle transverse momenta $p_\perp$.

We define: $K_\perp = |\Sigma_i \vec{p}_{\perp i}|$ where $\vec{p}_{\perp i}$ is transverse to the trust axis T and summation includes only particles on one side of an arbitrary plane containing T. The trust axis T is defined in the usual way with neutral and charged particles. We use, however, for $K_\perp$ only charged particles and average $K_\perp$ over several random planes to reduce fluctations. The data were fully corrected for detector- and radiation effects.

At low Q we find very good agreement with theoretical predictions[7] based on QCD-LLA, namely

i) the probability distribution $dP/dK_\perp$

ii) $x_\perp = K_\perp / < K_\perp >$ scales ("scaling in the mean")

iii) $< K_\perp^2 >$ is proportional to Q^2

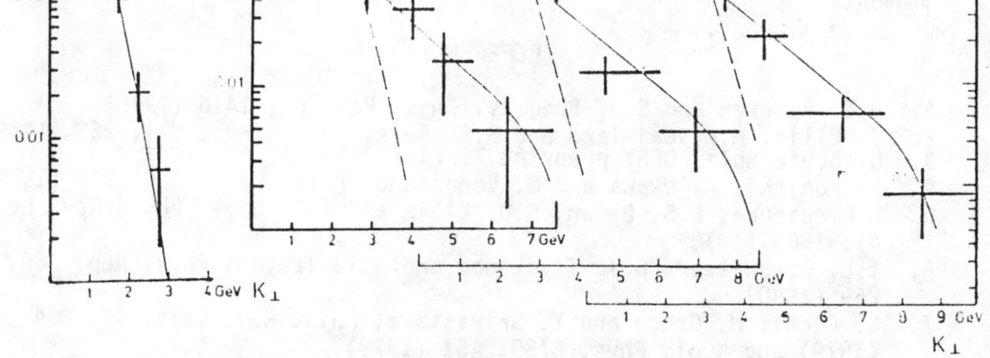

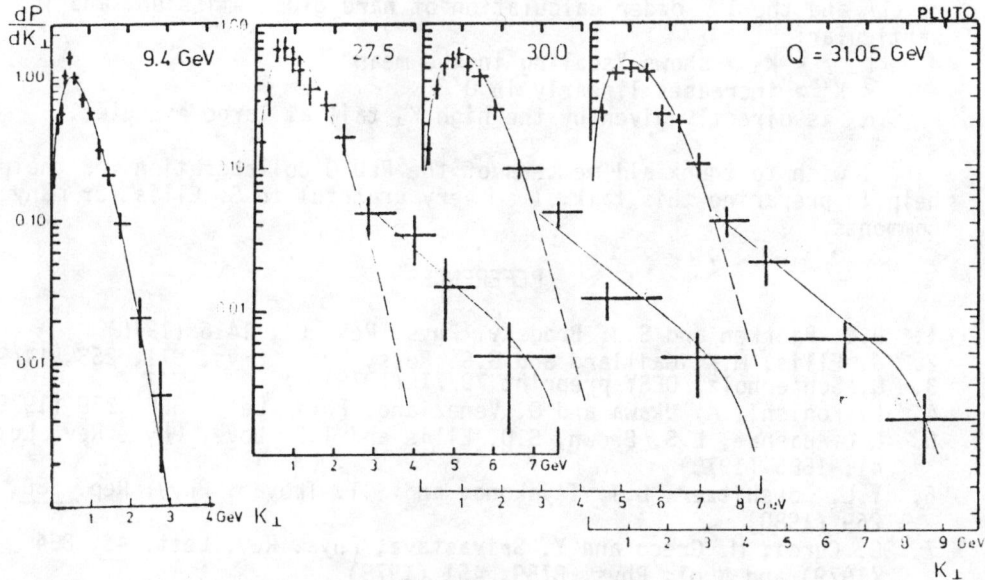

Fig. 4 Selected $K_\perp$ distributions (see text)

As an example we show in Fig. 4 some experimental $K_\perp$ distributions as function of Q. At low energies the QCD-LLA prediction i) describes the data very well. At high Q, however, we see clear deviations in the large $K_\perp$ tail of the distributions. We interpret the excess of data in the tails as due to the onset of hard gluon bremsstrahlung and fit that part of the distributions separately to 1st order QCD predictions. By comparing the two contributions we can determine α_s without recourse to Monte Carlo calculations involving fragmentation parameters or by explicitly counting jet numbers. Preliminarily we get: $\alpha_s = 0.19 \pm 0.02$ (stat. error) $\pm$ 0.04 (syst. error) at Q = 30 GeV. The systematic error is mainly due to the uncertainty in defining the high $K_\perp$ regime where $K_\perp$ is essentially given by the transverse momentum of the hard gluon.

SUMMARY

We have measured parton distributions in multi-jet events with a new cluster algorithm. We find excellent agreement with 1st order QCD predictions in the class of identified 3 jet events and determine: $\alpha_s = 0.16 \pm 0.03$ (stat. error) $\pm$ 0.03 (syst. error). From the 2 jet class we find: $\sigma_q = (290 \pm 30)$ MeV. The parton thrust distribution favors clearly gluon spin = 1.

We have measured the energy-energy correlations of any two particles in hadronic events. At high energies the data approach the predictions of perturbative QCD calculations.

We have measured the transverse momentum $K_\perp$ in jets as a function of energy. We find good agreement with various QCD predictions, based on LLA and the 1st order calculation of hard gluon emission, and in particular:

$K_\perp / < K_\perp >$ shows "scaling in the mean"

$< K_\perp^2 >$ increases linearly in Q^2

α_s is directly given by the high $K_\perp$ tail at large energies.

I wish to thank all members of the PLUTO collaboration for their help in preparing this talk. I am very grateful to S. Ellis for many comments.

REFERENCES

1. J.D. Bjorken and S.J. Brodsky, Phys. Rev. D1, 1416 (1970)
2. J. Ellis, M.K. Gaillard and G.G. Ross, Nucl. Phys. B11, 253,(1976)
3. G. Schierholz, DESY-preprint 79/71 (1979)
4. K. Konishi, A. Ukawa and G. Veneziano, Phys. Lett. 80B, 259 (1979)
5. C.L. Basham, L.S. Brown, S.D. Ellis and T.S. Love, Phys. Rev. Lett. 41, 1585 (1978)
6. Y.L. Dokshitzer, D.T. T'yakonov and S.T. Troyan, Phys. Rep. 58C, 269 (1980)
7. G. Curci, M. Greco and Y. Srivastava, Phys. Rev. Lett. 43, 834 (1979) and Nucl. Phys. B159, 451 (1979)

RESULTS ON JETS, QCD AND LEPTON PRODUCTION FROM THE MARK J

by Harvey B. Newman
Deutsches Elektronen-Synchrotron, Hamburg, West Germany

ABSTRACT

MARK J results on the properties of three jet events, on lepton pair production, and on a search for new heavy charged leptons in e^+e^- annihilation are presented. Analysis of the energy flow pattern of hadronic events shows that the production rate and the shape of flat three jet events are in excellent agreement with the predictions of QCD, and that the planar events in $e^+e^- \to$ hadrons have three well separated collimated jets. By use of the triplicity method we show that the energy of individual jets and the angles between the jets agree with the predictions of QCD. MARK J data on $e^+e^- \to e^+e^-$, $\mu^+\mu^-$ and $\tau^+\tau^-$ are also presented. A search for the production of a new heavy lepton and for the spin zero supersymmetric partners of the muon rules out the existence of a new heavy lepton below 16 GeV or a supersymmetric scalar particle below 15 GeV.

ENERGY FLOW ANALYSIS

Details of the apparatus and event selection procedures have been presented elsewhere[1,2,3]. The main criteria used to select hadronic events for jet analysis are $E_{vis} > 0.7\sqrt{s}$, $E_{//} < 0.5\sqrt{s}$ and $E_{\perp} < 0.5\sqrt{s}$, where E_{vis} is the total visible energy in the detector and $E_{//}$ and $E_{\perp}$ are the components of energy flow imbalance parallel and perpendicular to the beam line. The energy flow analysis uses the variables oblateness O, thrust T, F_{major} and F_{minor} and broad jet oblateness O_B to characterize the event shape (see Ref. 3). The comparison of the three jet structure observed in the data to the predictions of QCD makes use of the Monte Carlo model of Ali et al.[4] which has been extended to include an accurate representation of initial state radiative corrections[5].

0094-243X/81/680627-11$1.50

Figure 1a compares the broad jet oblateness distribution $\frac{1}{N}\frac{dN}{dO_B}$ seen in the data to the QCD prediction, and to the prediction of a $q\bar{q}$ model without gluon emission. Only the QCD model can explain the rate of production of planar events, which have large oblateness.

In Figure 1b we compare the energy flow pattern of planar events with $O_B > 0.3$ in the high energy ($\sqrt{s} \geq 27.4$ GeV) and the low energy ($\sqrt{s} \leq 22$ GeV) regions to the high energy predictions of QCD, and to the predictions of hadron production according to phase space. The pattern shown by the high energy data agrees with QCD but is quite distinct from the pattern given by phase space. The low energy data gives a pattern much closer to phase space as the jets at lower energy are less collimated.

To study the properties of the individual jets, we also divided the event plane into three subregions bounded by the angles corresponding to the minima in the energy flow diagram. The thrust distributions of the individual jets shown in Figure 1b are in excellent agreement with the QCD predictions, corresponding to the appearance of well collimated jets. The phase space model predicts distributions which are much broader in all three regions.

In Figure 1c we unfold the energy flow diagram of Figure 1a to see more clearly the comparison between the high energy planar events and the predictions of the QCD, $q\bar{q}$, and phase space models. All models in the figure are normalized to have equal area. Figure 1b demonstrates that at high energy, only QCD can explain the <u>shape</u> of the energy flow pattern of planar events.

Figure 1d shows the two dimensional distribution $d^2\sigma/d\theta_{12}d\theta_{23}$ for the planar events at high energy where θ_{12} is the angle between the first and second jet and θ_{23} is the angle between the second and the third jets. The three jets in this analysis are ordered according to the momentum projected along the major axis. The data are compared to the expectations of the QCD model, absolutely normalized. This figure demonstrates that the three jet kinematics seen in the data (which would be completely determined by θ_{12} and θ_{23} if the jets originated from massless constituents) agrees well with QCD predictions over the full range of θ_{12} and θ_{23}.

The structure of three jet events viewed on edge to the event plane is shown in Figures 2a-2c. The energy flow patterns of events with $O_B > 0.3$ viewed in the thrust-minor plane is given in Figure 2a, compared to the QCD and $q\bar{q}$ predictions. The individual jets, two of which overlap in this view, are shown separately. As seen in the figure, all three jets have approximately the same thickness, indicating that the P_t distributions relative to the axis of each jet are similar. This is seen more clearly in Figure 2b, where the energy flow diagram in Figure 2a is unfolded, and where the data are compared to the three afore-mentioned models. Figure 2b demonstrates that the individual jets are thinner than the expectations of the phase space model, and thicker than the $q\bar{q}$ model would predict. The data agrees well in all cases with the QCD predictions.

In Figure 2c we show the distribution in the fraction of the visible energy projected along the minor axis for the entire event (defined as F_{minor} in Ref. 3), compared to the model predictions. Figure 2c shows that if the source of low thrust oblate events were simply phase space production of hadrons, then the events would be thicker. The figure also shows that if the underlying process were $q\bar{q}$ production, then the events would be even flatter than what is actually observed. In this case the main contribution to the planar events would be from hard non-collinear photon emission.

TRIPLICITY ANALYSIS

Detailed properties of three jet events have also been analyzed using the triplicity[6] method to split the energy flow into three non-overlapping subregions. In each subregion a unit vector $\vec{n}_i$ giving the jet direction is constructed as the direction of the energy flow vector sum, where the division into subregions is chosen to maximize the triplicity. The total energies and directions of the three jets E_i and $\vec{n}_i$ $(i = 1,2,3)$ are then subjected to a kinematic fit satisfying energy - momentum conservation under the assumption that all jets are massless. The jets are ordered according to energy, with jet 1 the most and jet 3 the least energetic.

For events in which the jets are well separated from each other, the energy flow inside jets can be studied. This condition is fulfilled for jet 1 in all events at energies above 12 GeV, while jet 2 and jet 3 can only be reliably separated in clear 3-jet events ($O_B > 0.3$) at high energies ($\sqrt{s} \geq 27$ GeV).

Figure 3 shows the longitudinal energy flow $\frac{1}{E_j} dE_j/d(\cos\varepsilon)$ for the most energetic and least energetic jet at $\sqrt{s} \simeq 30$ and 35 GeV, where ε denotes the angle with respect to the corresponding jet axis $\vec{n}_j$. At high energies even the least energetic jets ($\langle E_3 \rangle$ = 6.4, 7.1 GeV respectively) show an exponentially falling distribution indicating their high degree of collimation around the jet axis. The longitudinal energy flow distribution of jet 3 in flat events at 30 GeV is nearly the same as that of jet 1 in all events at 12 GeV $\leq \sqrt{s} \leq$ 17 GeV, which has the same average energy, indicating that the fragmentation of gluons, forming $\sim$ 40% of the least energetic jets at high energies, is not substantially broader than that of quarks.

Figure 3 shows that about 90% of the jet energy is contained inside a cone of halfangle $\sim 20^o$ for a high energy jet and $\sim 40^o$ for a low energy jet. For a reliable separation of jet 2 and 3 in the high energy data it is necessary that they be more than about 60° apart. This requirement is naturally met by selecting events with $O_B > 0.3$, which have been shown earlier[1,2,3] to be of 3-jet nature, coming from the radiation of hard, non-collinear gluons. Figure 4a shows the double differential distribution $\sigma^{-1}d^2\sigma/dx_3d\theta_{23}$ for the whole hadron sample at high energies $27 \leq \sqrt{s} \leq 37$ GeV, which is dominated by 2-jet events up to an energy fraction $x_3 = E_3/\sqrt{s} \lesssim 25\%$ contained in the third "jet". Consequently, since the split up of one jet into "jet 2" and "jet 3" is artificial, the average opening angle between these "jets", $\langle\theta_{23}\rangle$ <u>decreases</u> with increasing energy fraction x_3, from about 65° at $x_3 \simeq 0$ to 55° at $x_3 = 10\%$. In contrast to this, genuine 3-jet events ($O_B > 0.3$, Figure 3b) show an opposite behaviour. When events with low x_3 or θ_{23} are cut out ($O_B \simeq 2x_3\sin\theta_{13}$), the average opening angle $\langle\theta_{23}\rangle$ <u>increases</u> with increasing x_3, being $\sim 70^o$ at $x_3 = 10\%$ and $\sim 90^o$ at $x_3 = 30\%$. The 3-jet events with $O_B > 0.3$

thus represent a sample where the two less energetic jets can be reliably separated and their properties can be studied.

Also shown in Figure 4 as solid curves are the double differential distributions obtained from the QCD model of Ali et al[4], which are in remarkable agreement with the data especially for the 3-jet sample in Figure 3b.

PRODUCTION OF CHARGED LEPTON PAIRS

Recent MARK J data on $e^+e^- \to e^+e^-$, $\mu^+\mu^-$ and $\tau^+\tau^-$ are shown in Figures 5 and 6, where the data are compared to the QED predictions after applying radiative corrections. Lower limits on the cut off parameters Λ_+ and Λ_- are given in A. Böhm's talk at this conference.

SEARCH FOR NEW SEQUENTIAL HEAVY LEPTONS

The result of a search for a new sequential heavy lepton[8] L produced in $e^+e^- \to L^+L^-$ is shown in Figure 7. Heavy lepton production is most easily recognized in the MARK J detector for events in which one lepton decays into a muon and neutrinos and the other lepton decays into hadrons and neutrinos[9]. Hadrons are detected by their energy deposit, E_{vis}, in the calorimeter. Heavy lepton candidates with masses greater than 6 GeV are selected by applying criteria described in Ref. 8, which distinguish $e^+e^- \to L^+L^-$ from $e^+e^- \to$ hadrons, $\mu^+\mu^-\gamma$, $\tau^+\tau^-$, and $e^+e^-\mu^+\mu^-$.

No event was observed in the data with $33 \leq \sqrt{s} \leq 36.7$ GeV, corresponding to a time integrated luminosity of 6.9 pb^{-1}. The number of events predicted by a Monte Carlo calculation as a function of the heavy lepton mass M_{HL} is shown in Figure 7 along with the 95% confidence level upper limit from the data. Figure 7 demonstrates that the existence of a sequential heavy lepton with a mass between 6 and 16 GeV is excluded.

Heavy leptons with $M_{HL} < 6$ GeV would decay into final states similar in appearance to those from τ decay, and would tend to be included in our sample of $e^+e^- \to \tau^+\tau^-$ events. The inset in Figure 7 shows the Monte Carlo prediction for the total number of events in the τ sample from $e^+e^- \rightarrow \tau^+\tau^-$ (52) and from heavy lepton production, as a function of M_{HL}. The inset demonstrates that we exclude the existence of a new heavy lepton with mass $M_{HL} < 6$ GeV with more than 95% confidence. We are thus able to rule out the existence of a new heavy lepton for $M_{HL} < 16$ GeV.

SEARCH FOR SCALAR SUPERSYMMETRIC PARTNERS OF THE MUON

In the framework of supersymmetric theories[10], spin zero partners of the muon are expected to decay only according to the reactions

$$s_\mu^- \rightarrow \mu^- + \text{photino (goldstino)} \qquad \text{(a)}$$

$$t_\mu^- \rightarrow \mu^- + \text{antiphotino (antigoldstino)} \qquad \text{(b)}$$

where s_μ and t_μ are the spin zero partners of the muon associated with the left and right handed parts of the muon field respectively, the photino is the spin 1/2 partner of the photon and the goldstino is the goldstone spinor associated with the breaking of supersymmetry.

Because of the uniqueness of the decay reactions (a) and (b), the extremely short lifetimes[10] of s_μ and t_μ and the prediction that the interaction cross sections of photino and goldstino are expected to be very small[11], only muon pairs are observed in the final state. Near threshold production of s_μ and t_μ the two residual muons would be produced isotropically in space. Data from SPEAR place a lower limit of 3.5 GeV[12] on the mass of s_μ and t_μ. Thus, over the PETRA energy range of 12 to 36.7 GeV, an increase in the production of acoplanar muon pairs should be observed if a new threshold is passed.

Further details on event selection and background suppression are given in Ref. 8.

Figure 8 shows the expected number of events predicted by a Monte Carlo calculation for the production of s_μ or t_μ pairs. We also indicate the number of events corresponding to the 95% confidence level.

Thus, the data show that no s_μ or t_μ type particle is produced with a mass between 3 GeV and 15 GeV. If s_μ and t_μ particles have the same mass and are both produced, the number of expected events is twice as large. In this case they can both be excluded with more than 99% confidence.

ACKNOWLEDGEMENT

I wish to thank all the members of the MARK J collaboration and the M.I.T. support staff for help in preparing this talk. The continuing support of the DESY directorate, and the hospitality of the organizers of this conference are gratefully acknowledged.

REFERENCES

1 H. Newman, Proc. 1979 Intern. Symp. on Lepton and Photon Interactions at High Energy Physics (Fermilab, Batavia, Illinois, 1979);
D.P. Barber et al., Phys. Rev. Lett. 43, 830 (1979).

2 D.P. Barber et al., Phys. Lett. 89B, 139 (1979).

3 D.P. Barber et al., Physics Reports 63, 337 (1980) and M.I.T./LNS Report No. 107 (1980).
D.P. Barber et al., Phys. Rev. Lett. 44, 1722 (1980).

4 A. Ali, E. Pietarinen, G. Kramer and J. Willrodt, DESY Report 79/86 (1979), and Phys. Lett. 93B, 155 (1980).

5 F.A. Berends and R. Kleiss, DESY Reports 80/66 and 80/73 (1980).

6 S. Brandt and H.D. Dahmen, Zeitschr. f. Physik C1, 61 (1979).

7 A. Böhm, talk presented at this conference, and Aachen Report PITHA 80/9 (1980).

8 D.P. Barber et al., M.I.T./LNS Report No. 113 (1980).

9 Heavy lepton decay branching ratios are taken from Y.S. Tsai, SLAC Preprint, SLAC-PUB-2450, December 1979, and the τ branching ratios are taken from the compilations of world data. The $\tau \to \mu\nu\nu$ branching ratio is set at 17%.

10 Yu. A. Gol'fand and E.P. Likhtman, JETP Letters 13, 323 (1971).
J. Wess and B. Zumino, Nucl. Phys. B70, 39 (1974).
For a review article, see:
P. Fayet and S. Ferrara, Phys. Report 32C, 249 (1977);
P. Fayet, Phys. Lett. 69B, 489 (1977);
G.R. Farrar and P. Fayet, Phys. Lett. 76B, 575 (1978); 79B, 442 (1978);
P. Fayet, Phys. Lett. 84B, 421 (1979);
G.R. Farrar and P. Fayet, Phys. Lett. 89B, 191 (1980).

References (cont'd)

11 P. Fayet, Phys. Lett. 86B, 272 (1979).

12 F.B. Heiel et al., Nucl. Phys. B138, 189 (1978).

FIGURE CAPTIONS

Figure 1 a) The broad jet oblateness distribution $\frac{1}{N}\frac{dN}{dO_B}$ compared with the QCD and $q\bar{q}$ models.

b) Energy flow diagram in the thrust-major plane for events with $O_B > 0.3$, for the high energy data ($\sqrt{s}$ = 27.4 to 36.7 GeV) and the low energy data ($\sqrt{s}$ = 12 to 22 GeV), compared to the predictions of QCD at high energy and of the phase space model. The thrust distribution $\frac{1}{N}\frac{dN}{dT}$ for each individual jet is shown beside the corresponding jet in the energy flow diagram.

c) The unfolded energy flow diagram of Figure 1a compared with the QCD , and phase space models.

d) The two-dimensional distribution in the opening angles between the jets θ_{12} and θ_{23}, compared with QCD predictions.

Figure 2 a) The energy flow diagram in the thrust-minor plane for the high energy events (27.4 to 36.6 GeV) with $O_B > 0.3$, compared with the predictions of the QCD, $q\bar{q}$ and phase space models.

b) The unfolded energy flow diagram of Figure 2a. Here the third jet is shown above the second jet at $\sim 180^o$.

c) The distribution $\frac{1}{N}\frac{dN}{dF_{minor}}$ in the fraction of the visible energy of the entire event projected along the minor axis (perpendicular to the event plane).

Figure 3 a) The energy flow density within a jet $\frac{1}{E}\frac{dE}{d(\cos\varepsilon)}$ vs. $\cos\varepsilon$, where ε is the angle from the individual jet axis. The lowest energy jet in the high energy data ($\sqrt{s} \simeq 30$ GeV) tends to be slightly broader than the highest energy jet from the low energy ($\sqrt{s} \simeq 13$ GeV) data. The leading jet at $\sqrt{s} \simeq 30$ GeV is well described by the QCD model.

Figure 3 b) The highest and lowest energy jets in events with $32 \leq \sqrt{s} \leq 37$ GeV are compared to the QCD model as in Fig.3a.

Figure 4 $\frac{1}{\sigma} \frac{d^2\sigma}{dx_3 d\theta_{23}}$ as observed after a kinematic fit for the three jet energies and directions, where x_3 is $E_3/\sqrt{s}$ and θ_{23} is the angle between the second and third jets, for:

a) all events satisfying the cuts described in the text,

b) planar events with $O_B > 0.3$.

The data points agree well with the predictions of QCD (solid line).

Figure 5 Measured cross sections for $e^+e^- \to \mu^+\mu^-$ (left scale) and $e^+e^- \to \tau^+\tau^-$ (right scale) corrected for radiative effects and acceptance, compared to the point-like QED cross section.

Figure 6 The corrected observed distribution $s \frac{d\sigma}{d(\cos\theta)}$ for $e^+e^- \to e^+e^-$ compared to the point-like QED distribution.

Figure 7 Number of events expected for the production of a new (sequential) heavy lepton as a function of mass. The inset shows the number of events expected in the τ sample from tau and heavy lepton production. We observe a total of 54 events. The dashed line corresponds to the 95% confidence level upper limits for τ events.

Figure 8 Number of events expected for the production of a spin zero partner s_μ or t_μ of the muon as a function of mass. The upper limit of events (95% confidence) and the mass range excluded are also indicated.

(a)

(b)

(c)

(d)

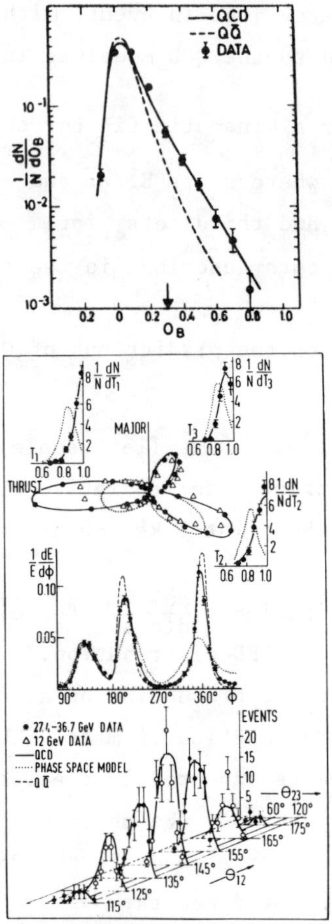

Fig. 1

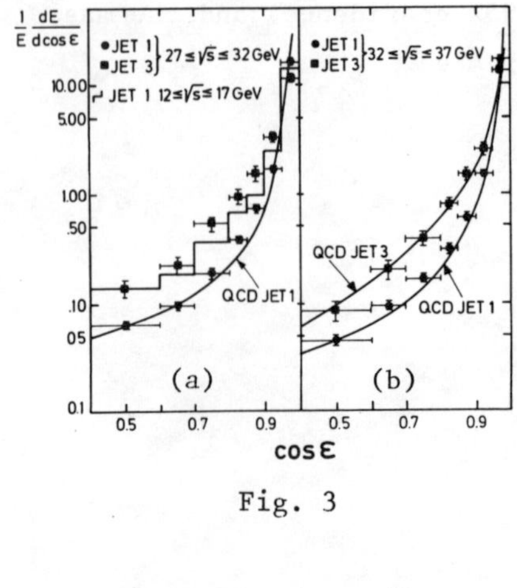

Fig. 3

PRELIMINARY

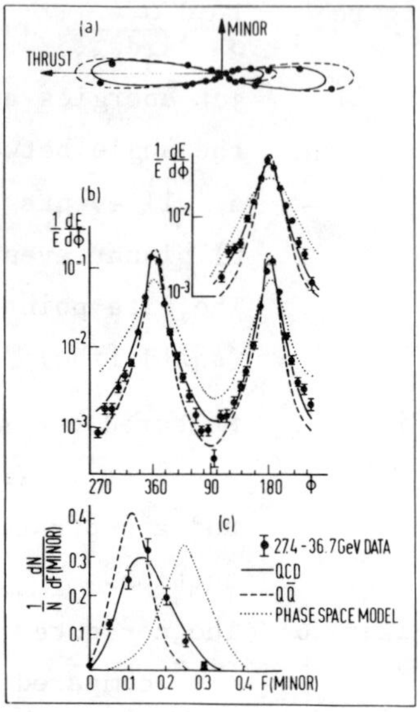

Fig. 2

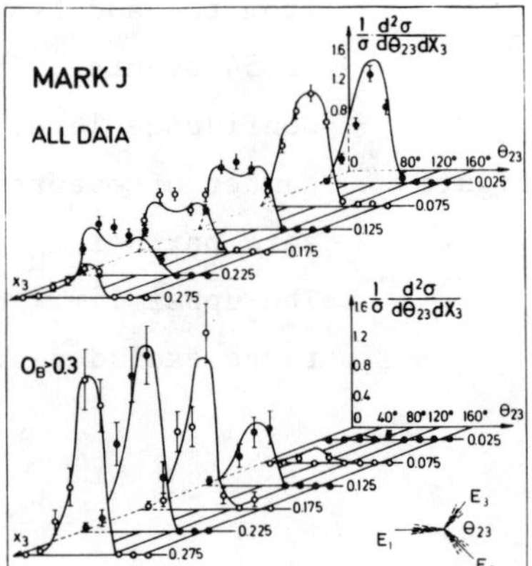

Fig. 4

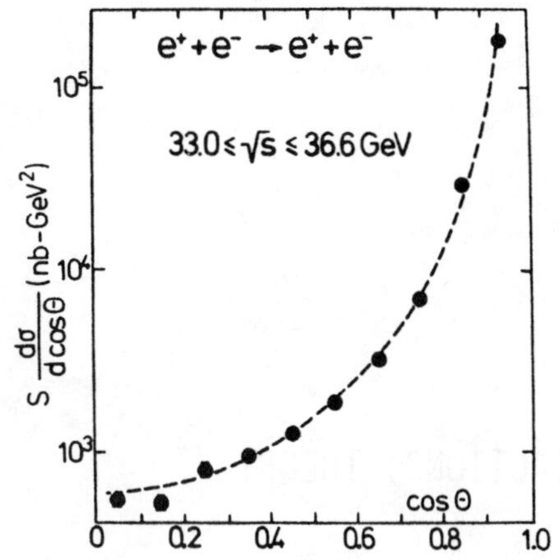

Fig. 5

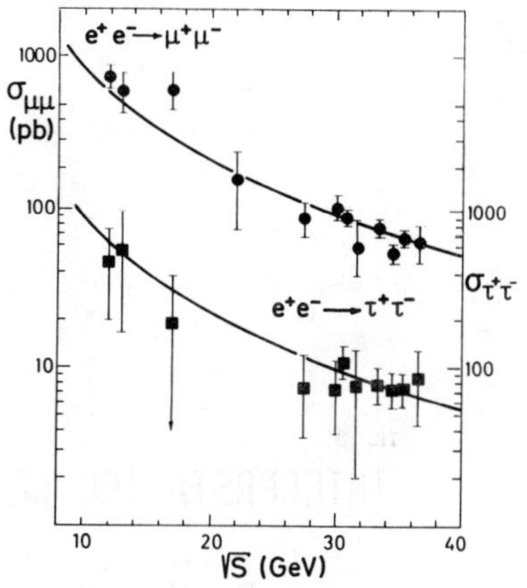

Fig. 6

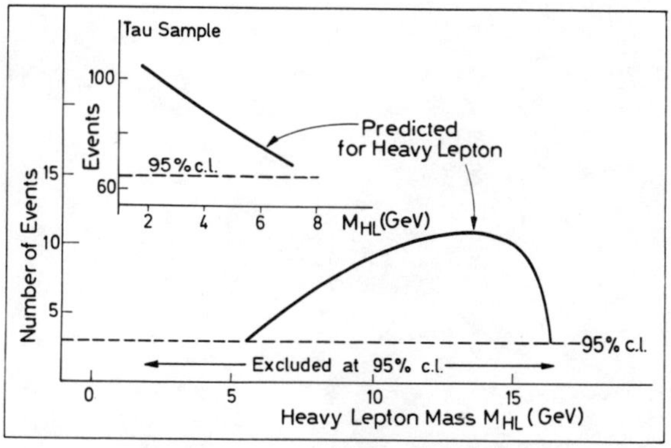

Fig. 7

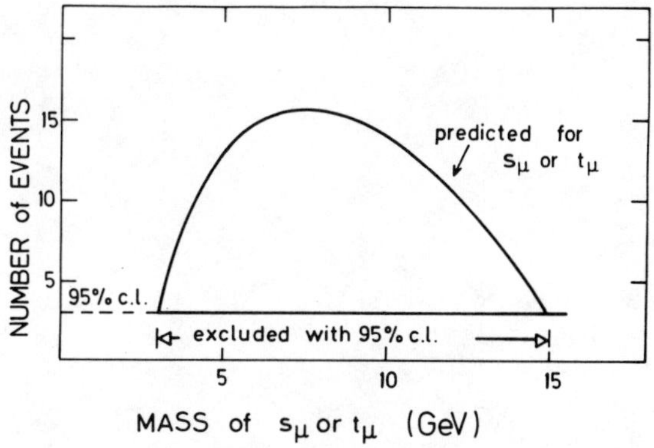

Fig. 8

INTERPRETATION OF e^+e^- REACTIONS, THEORY

P. Hoyer, Nordita, Organizer

USING e^+e^- CROSS-SECTIONS TO TEST QCD AND TO SEARCH FOR NEW PARTICLES*

R. Michael Barnett
Stanford Linear Accelerator Center
Stanford University, Stanford, California 94305

ABSTRACT

A careful analysis is presented of the most recent data for $R(e^+e^- \to \text{hadrons})$ using improved theoretical techniques. Recent calculations of higher-order corrections are discussed. It shown why R is potentially one of the best tests of QCD. For $\sqrt{s}$ near 7 GeV, the data lie about 16% above the theory; the experimental uncertainty is ±10% (dominated by systematics). While this discrepancy may well be due to experimental problems, we also consider the possibility that there is a threshold for new particles (at $\sqrt{s} \approx 6$ GeV) such as new quarks, Higgs bosons, heavy leptons, quixes and massive gluons.

Many process have been investigated as tests of QCD. For e^+e^- physics there has been considerable discussion concerning the use of jet phenomena as such a tool. At the same time it should be remembered that the total cross-section for e^+e^- annihilation to hadrons is also an excellent means of testing QCD. This cross-section is usually normalized to the muonic cross-section:

$$R \equiv \frac{\sigma(e^+e^- \to \text{hadrons})}{\sigma(e^+e^- \to \mu^+\mu^-)} \quad . \tag{1}$$

The magnitude of R is one of the best tests of QCD, because: (a) it is conceptually simple, (b) the <u>magnitude</u> of R at any <u>single</u> value of Q^2 is predicted by QCD (unlike deep-inelastic scattering), (c) Q^2 can (as a result of (b)) be chosen large in order to minimize non-perturbative effects such as higher-twist contributions, and (d) the α_s^2 term in R has been calculated[1] and is small ($\lesssim 1\%$ of the total R for $\sqrt{s} > 4$ GeV).

Neglecting masses, the perturbation expansion for R is

$$R = \sum_i 3Q_i^2 \left[1 + \sum_{n=1}^{\infty} C_n \left(\frac{\alpha_s}{\pi} \right)^n \right] \quad . \tag{2}$$

The calculation of the second-order term is very important since it provides some indication of how rapidly the perturbation series converges and since Λ is not well-defined without going to second-order. The calculation is most easily performed[2] by calculating the divergent part of the photon's vacuum polarization tensor. This is related to R through standard renormalization group and unitarity arguments. The calculation of C_2 has now been done by three groups using different methods, but with identical results. Dine and

* Work supported by the Department of Energy, contract DE-AC03-76SF00515.

Sapirstein[1] performed much of the calculation numerically with self-energy insertion diagram, treated analytically. Chetyrkin, Kataev and Tkachev[1] performed the calculation analytically in coordinate space while Celmaster and Gonsalves[1] used momentum space. The results depend, of course, on the renormalization procedure ($N_f \equiv$ number of flavors):

$$C_2 = 7.36 - 0.44\, N_f \quad \text{in MS} \equiv \text{minimal subtraction scheme}$$
$$C_2 = 1.99 - 0.12\, N_f \quad \text{in } \overline{\text{MS}} \text{ scheme (Bardeen } \underline{\text{et al.}}^{3})$$
$$C_2 = -2.19 + 0.16\, N_f \quad \text{momentum space scheme.}^{4}$$

The MS scheme appears to be an inappropriate scheme to use. In calculations of other processes, the MS scheme also gives larger corrections. Here the $\overline{\text{MS}}$ and momentum space schemes are smaller because

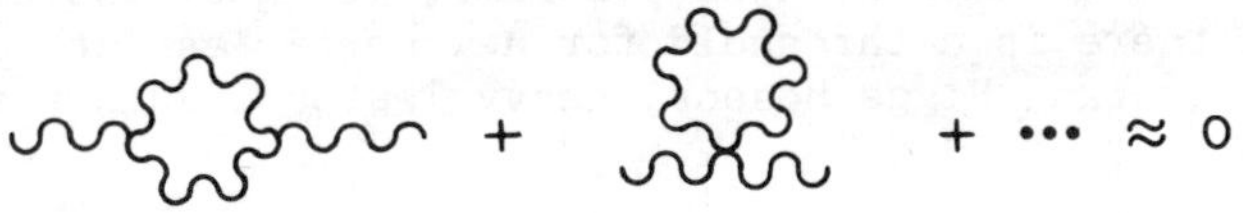

and

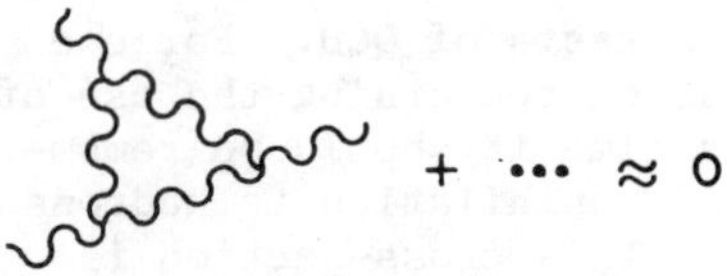

at a symmetric point with $q^2 = q_0^2 \equiv$ typical momentum. Therefore

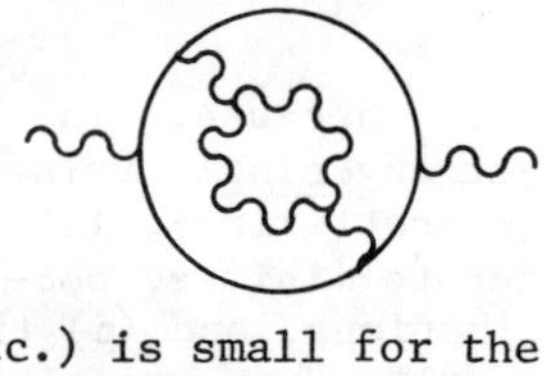

(etc.) is small for the $\overline{\text{MS}}$ and momentum space schemes. We may then conclude from the magnitude of C_2 in these two schemes that perturbution theory for R is reliable.

In work[5] I have done with Michael Dine and Larry McLerran we concentrated on the data above the charm resonance region with $5.5 \le \sqrt{s} \le 7.5$ GeV (where only the Mark I experiment has published significant data[6]), see Fig. 1.

To test QCD, we should smooth[7] the data and theory using an appropriate procedure. The smoothing assump-

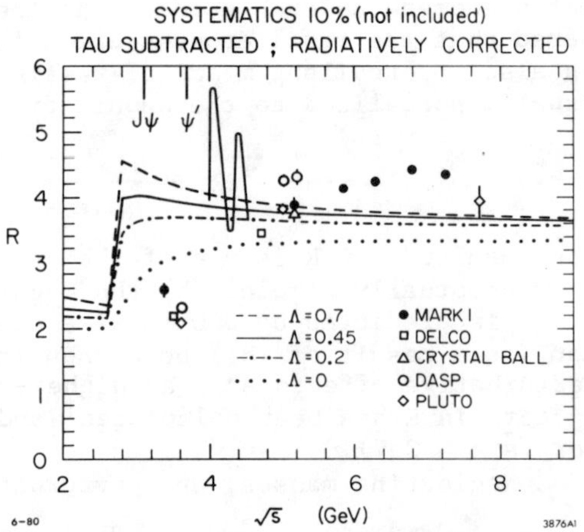

Fig. 1. Data for R from the SLAC-LBL COLLABORATION (Ref. 6) and from other collaborations (references in Ref. 5). The resonance region is shown schematically. The contributions of the τ have been subtracted, and radiative corrections have been applied. Only statistical errors are shown. The locations of J/ψ and ψ' have been indicated, since they are included in smoothing. The curves are the QCD predictions for R ($\Lambda = 0$ is the parton model).

tion is almost equivalent to the assumption of local duality. Dine, McLerran and I developed a general procedure[5] using

$$\bar{R}(s) = \int_{4m_\pi^2}^{\infty} ds' \, W(s,s',\Delta)\, R(s') \tag{3}$$

where W is a weight function such as

$$W \propto \exp\left[-\tfrac{1}{2}(s-s')^2/\Delta^2\right] \tag{4}$$

Figure 2 shows smoothed theory and data.

There is a discrepancy between theory and experiment of about 16%; systematic errors are reported to be 10%. From this discrepancy we can draw one of three conclusions:

(1) The experiment is inaccurate; the 10% systematic error is actually 16%. This is the most probable conclusion.
(2) QCD is wrong.
(3) There is a threshold for new particles.

The discussion[5] of what types of particles may have missed up to now is relevant not only at SPEAR but also at DORIS, CESR, PETRA and PEP.

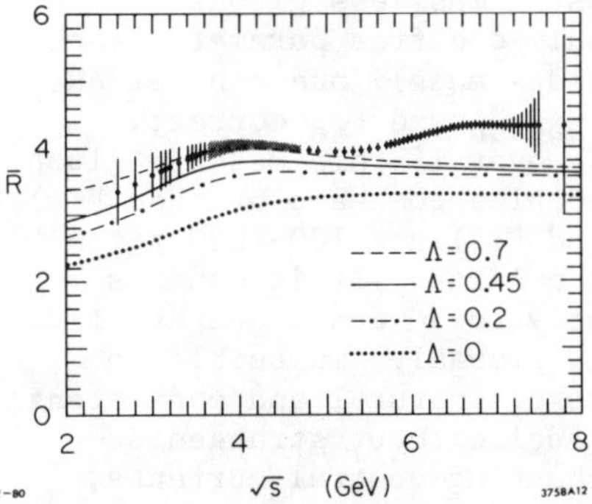

Fig. 2. The results of smoothing the theoretical and experimental values of R with $\Delta = 5$ GeV^2. All data were from Ref. 6. The error bars are statistical only. The curves are QCD. $\Lambda = 0$ indicates the parton model.

Is it possible that a quark of charge = −1/3 and mass ≈ 3 GeV has been missed? It would give an excellent fit to Mark I data[6] for R and cannot be ruled out by PETRA data. But the expected $Q\bar{Q}$ resonances have not been observed. Mark I data[8] give $\Gamma_{ee}(Q\bar{Q}) \lesssim 0.15$ keV (90% confidence) for $4.5 < \sqrt{s} < 7.5$ GeV, but we expect $\Gamma_{ee}(Q\bar{Q}) \approx$ 1 keV. Could the $Q\bar{Q}$ resonances be hidden by making them wide? They would have to be 100 MeV wide. Particles such as quix resonances are probably only about 3 MeV wide. Unless a mechanism to make $Q\bar{Q}$ resonances wide can be found, new quarks are ruled out.

The production of charged Higgs bosons cannot explain the data because their threshold has $\Delta R \propto (\text{velocity})^3$ unlike fermions which have $\Delta R \propto$ velocity. As a result R rises very slowly (asymptotically $\Delta R = 0.25$).

Ordinary charged heavy leptons, though consistent with Mark I data for R, are ruled out by examination of eμ, $eX^{\pm}$,... events at SPEAR and PETRA. However, consider

$$L^+ \to N_L^0 + (e\nu),\ (\mu\nu) \text{ or } (ud)$$

where mass $(N_L^0) \approx 2$ GeV and N_L^0 is relatively stable, or

$$L^+ \to \tau^+ \nu\nu$$

where this is the dominant decay. Then experimental cuts made at PETRA to eliminate backgrounds would also eliminate these events. At SPEAR these events would be counted, usually as 2-prong events (slow electrons and muons cannot be distinguished from hadrons).[9] It is possible (but not at all certain) the apparent rise in R is due mostly to a rise in the 2-prong cross-section, see Fig. 3.

Finally is it possible that the "rise" in R is not a threshold but is a 2 GeV wide resonance? This might correspond to an extra U(1) gluon separate from the usual massless gluons. With only one free parameter (besides mass), one can get ΔR, Γ_{hadron} and Γ_{ee} correct. However if this massive gluon couples to c and b, then $\Gamma(\psi)$ and $\Gamma(\Upsilon)$ are 100-1000 times too large. If it couples only to u and d quarks, it is probably impossible to make a natural and consistent model without strangeness-changing-neutral-currents, etc.

In conclusion, R is potentially one of the best tests of QCD. (There are also sum rule tests[10] of R not discussed here.) But currently R is 16% higher than theory for $\sqrt{s} = 6 - 7.5$ GeV although this is likely to be due to systematic error. The apparent rise may come from 2-prong events. Clearly more accurate (3-5% accuracy) experiments are needed, and already there are new data under analysis. It should be noted that some types of new particles may have eluded detection at present storage rings.

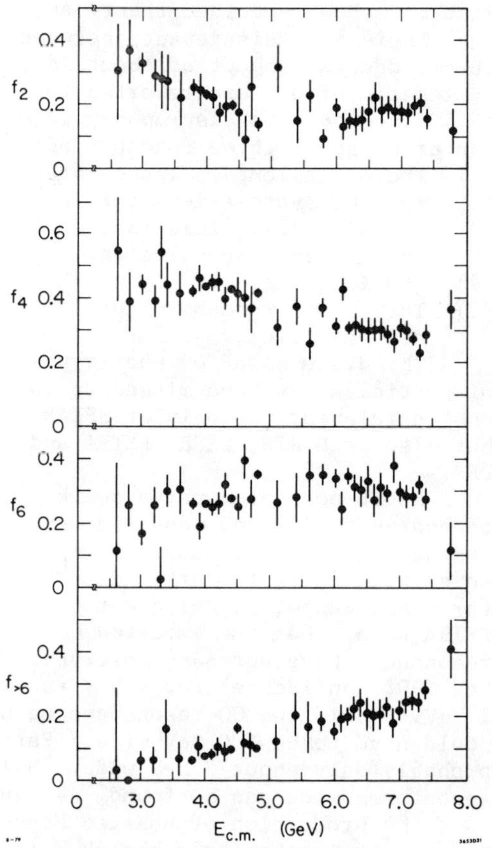

Fig. 3. The ratio of R for events with a given number of charged prongs to the total R. Data are from Ref. 6.

I would like to thank M. Dine and L. McLerran with whom this work was done for their help in preparing this talk. I also enjoyed the hospitality of the Aspen Center for Physics where I prepared the talk. This paper was supported by the Department of Energy under contract number DE-AC03-76SF00515.

REFERENCES

1. M. Dine and J. Sapirstein, Phys. Rev. Lett. 43, 668 (1979); K. G. Chetyrkin, A. L. Kataev and F. V. Tkachev, Phys. Lett. 85B, 277 (1979); W. Celmaster and R. J. Gonsalves, Phys. Rev. Lett. 44, 560 (1980).
2. T. Appelquist and H. Georgi, Phys. Rev. D8, 4000 (1973); A. Zee, Phys. Rev. D8, 4038 (1973).
3. W. A. Bardeen et al., Phys. Rev. D18, 3998 (1978).
4. W. Celmaster and R. J. Gonsalves, Phys. Rev. Lett. 44, 560 (1980).
5. R. M. Barnett, M. Dine and L. McLerran, Phys. Rev. D22, 594 (1980) (an analysis of the new heavy lepton hypothesis using earlier data appears in M. Biyajima and O. Terazawa, Prog. Theor. Phys. 60, 1240 (1978)).
6. J. Siegrist, Report No. SLAC-225 (1979) (Ph.D. Thesis).
7. E. C. Poggio, H. R. Quinn and S. Weinberg, Phys. Rev. D13, 1958 (1976); R. Shankar, Phys. Rev. D15, 755 (1978); R. G. Moorhouse, M. R. Pennington and G. G. Ross, Nucl. Phys. B124, 285 (1977); and Ref. 5.
8. Private communication with G. Feldman.
9. Similar arguments would also apply to fractionally-charged heavy leptons ("freptons"). In that case we could imagine a 2/3-charged lepton which decays quickly into a 2 GeV, -1/3-charged lepton. See Ref. 5.
10. T. Hagiwara and A. I. Sanda, Rockefeller Report No. COO-2232B-165 (1978); K. G. Chetyrkin, N. V. Krasnikov and A. N. Tavkhelidze, Phys. Lett. 76B, 83 (1978).

The Pertubative Calculation of Event Shapes in e^+e^- Annihilation

R. K. Ellis*, D. A. Ross**, and A. E. Terrano
California Institute of Technology, Pasadena, California 91125

Presented by A. E. Terrano

In this talk I describe the results of the calculation of event shapes in e^+e^- annihilation to order α_s^2(details and complete references have been presented elsewhere[1]). Our results are presented in terms of the tensor

$$\theta^{ij} = \sum_a \frac{P_a^i P_a^j}{|P_a|} \Big/ \sum_a |P_a| \quad ,$$

where P_a^i are the components of the center of mass three-momentum of hadron a and the sum runs over all hadrons. Since this tensor combines parallel momenta linearly, it is infra-red finite and calculable in perturbation theory. The eigenvalues of θ are determined by the characteristic equation

$$\lambda^3 - \lambda^2 + \frac{C}{3}\lambda - \frac{D}{27} = 0; \qquad 0 \le C, D \le 1 \quad .$$

The quantities C and D are symmetric functions of the eigenvalues

$$C = 3(\lambda_1\lambda_2 + \lambda_2\lambda_3 + \lambda_3\lambda_1), \; D = 27(\lambda_1\lambda_2\lambda_3) \quad ,$$

and provide a convenient measure of the shape of the event. Since the tensor has unit trace, the distributionin in the eigenvalues is fully determined by the distribution in C and D.

In QCD perturbation theory we can define an analogous tensor $\bar{\theta}$ by replacing the observed hadron momenta P by the parton momenta p. These two tensors are equal up to hadronization corrections which vanish like inverse powers of the total center of mass energy Q, but which at present energies play a significant role in the determination of the shape of the hadronic events. We have calculated the properties of θ for massless quarks and gluons in QCD and have taken no account of the effects of hadronization. Nevertheless, our calculation is a prerequisite for a more complete analysis.

For a two jet final state, only one of the eigenvalues of $\bar{\theta}$ will be nonzero, and both C and D will vanish; for a three jet final state, two of the eigenvalues will be nonzero with the result that C lies between 0 and .75 while D vanishes. Thus C measures the acollinearity of an event and D measures its acoplanarity.

---

*Address after Sept. 1980, Theory Division, CERN, Geneva, Switzerland
**Address after Oct. 1980, Physics Department, University of Southampton,SO9 5NH, United Kingdom

In Fig. 1 we display $\frac{C}{\sigma_0}\frac{d\sigma}{dC}$ in units of $\frac{\alpha_s}{2\pi}$ for the process $e^+e^- \to q\bar{q}G$ in the Born approximaiton. The distribution is completely specified by adding a term propotional to $\delta(C)$ such that the total cross-section is equal to $\sigma = \sigma_0(1 + \alpha_s/\pi)$. The determination of the $O(\alpha_s^2)$ correction to the distribution in Fig. 1 requires the calculation (for $C \neq 0$) of the cross-sections for $e^+e^- \to q\bar{q}GG$ and $q\bar{q}q\bar{q}$ in the Born approximation and $e^+e^- \to q\bar{q}G$ to one loop. The ultraviolet divergences are controlled using dimensional regularization and renormalization is performed according to the $\overline{MS}$ prescription. After renormalization the diagrams still contain mass singularities and infrared divergences. These are also controlled by dimensional regularization. Processes containing four partons in the final state masquerade as three jet events in the region in which one of the partons is soft and/or collinear with respect to another. The singularities present in this region are controlled by generalizing four particle phase space to n dimensions.

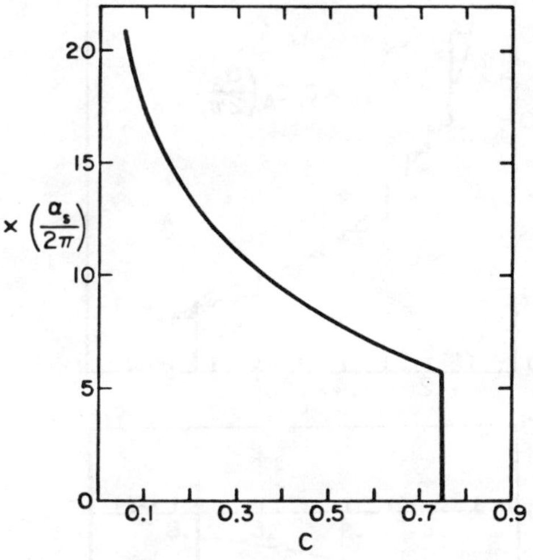

Fig. 1. $\frac{C}{\sigma_0}\frac{d\sigma}{dC}$ to $O(\alpha_s)$

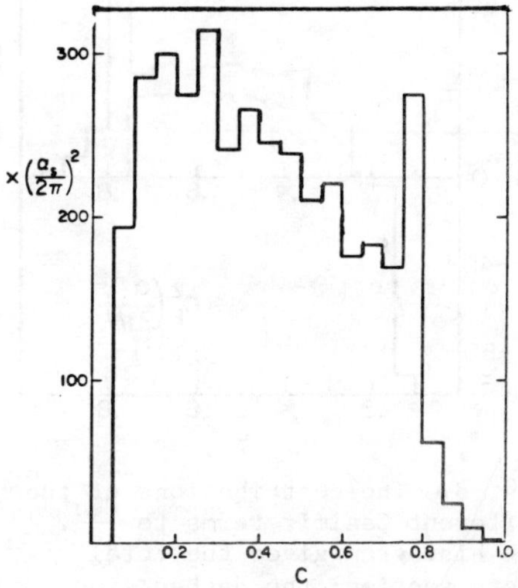

Fig. 2. $\frac{C}{\sigma_0}\frac{d\sigma}{dC}$ to $O(\alpha_s^2)$

In Fig. 2 we present our results for the $O(\alpha_s^2)$ contributions to $\frac{C}{\sigma_0}\frac{d\sigma}{dC}$ in units of $(\frac{\alpha_s}{2\pi})^2$ (the number of flavors is taken to be 5). An idea of the order of magnitude of the corrections with respect to the lower order term, Fig. 1, can be obtained by noting that for Q = 30 Gev, $\frac{\alpha_s(Q^2)}{2\pi} \sim \frac{1}{30}$ (Λ = 500 Mev); it is evident that they are comparable in size. Specifically, we may refer to all values of C greater than 0.5 as multijet events. Then, as an estimate of the size of the multijet fraction, we find

$$\int_{1/2}^{1} \frac{1}{\sigma}\frac{d\sigma}{dC}\, dC = 2.8\, \frac{\alpha_s(Q^2)}{2\pi}\left(1 + 37\, \frac{\alpha_s(Q^2)}{2\pi}\right)$$

At $Q^2 \simeq$ 30 Gev, the correction is 100%.

Some contributions to the cross section may be summed to all orders. The small C behavior has been shown to be

$$\frac{d\sigma}{dC} \sim \frac{\partial}{\partial C} \exp\left(-C_F \ln^2(\frac{C}{2})\, \frac{\alpha_s}{\pi}\right)$$

Furthermore, the continuation of Q^2 to positive values in the $\ln^2(-Q^2)$

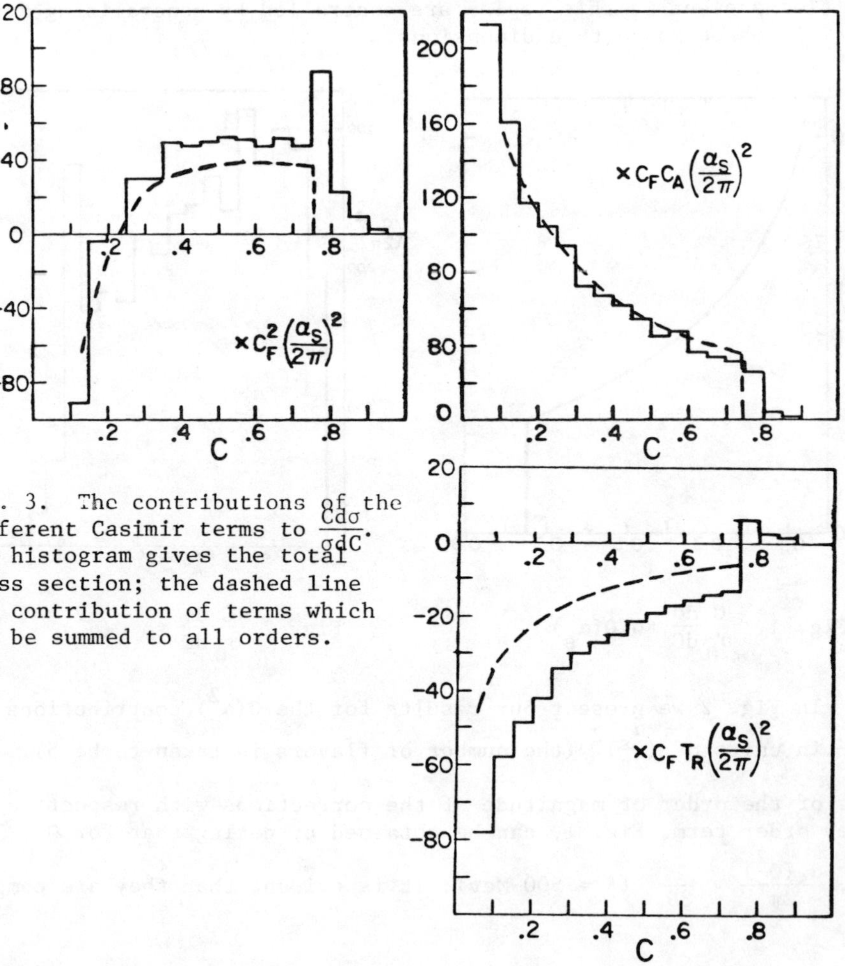

Fig. 3. The contributions of the different Casimir terms to $\frac{Cd\sigma}{\sigma dC}$. The histogram gives the total cross section; the dashed line the contribution of terms which may be summed to all orders.

terms from the loop integrations in the three parton final state diagrams gives rise to terms proportional to π^2 which, being associated with the leading poles, also may be exponentiated. Finally, the characteristic momentum of the interactions will be somewhat smaller than Q^2. The presence of the term

$$\beta_0 \ln\left(\frac{s_{13}s_{23}}{Q^2}\right) \sim \beta_0 \ln\left(\frac{C}{6}\right)$$

where s_{ij} is the invariant mass of partons i and j and β_0 is the coefficient of g^3 in the expansion of the beta function, suggests that the running coupling constant should be evaluated at $Q^2C/6$ rather than at Q^2. In Fig. 3 we have compared the sum of these terms with the $O(\alpha_s^2)$ contribution to the cross section. The terms proportional to C_F^2, C_FC_A, and C_FT_R are plotted seperately. The dashed lines give

$$\text{L.O.} \left(-2C_F \ln^2\left(\frac{C}{2}\right) + \left(C_F + \frac{C_A}{2}\right)\pi^2 - \left(\frac{11C_A}{6} - \frac{2}{3} T_R\right) \ln\left(\frac{C}{6}\right) \right) \frac{\alpha_s}{2\pi}$$

where L.O. is the lowest order cross section. The histogram gives the total cross section. As can be seen, the remaining $O(\alpha_s^2)$ corrections are small.

1) R. K. Ellis, D. A. Ross, and A. E. Terrano, Caltech preprint CALT 68-785 (to be published in Nucl. Phys. B)

A QCD ANALYSIS OF JETS IN E^+E^- ANNIHILATION

A. Ali
Deutsches Elektronen-Synchrotron DESY, Hamburg,
Federal Republic of Germany

Jets in e^+e^- annihilation are discussed in the context of perturbative Quantum Chromodynamics. Topics discussed include higher twist contribution, effects of quark masses and fragmentation on the 3 and 4 jet rates and some distributions bearing on the experimental verification of 4 jet events at the PETRA/PEP energies.

It is now almost five years that the idea of observing gluon jets in e^+e^- annihilation was first discussed [1]. Since then a lot of water has gone down the river Elbe. Several independent groups at DESY have presented independent and convincing evidence [2] bearing on the existence of gluon bremsstrahlung as a physical phenomenon. There is even a prima facie case [3] for the spin-1 nature of gluon [4] and indeed there is ample evidence [5] that the first estimates [6] of the effective quark gluon coupling, $\alpha_s(Q^2)$, are in conformity with the expectations of asymptotic freedom.

The triumph of the Born estimates, impressive as it is, however, by itself does not constitute a satisfactory quantitative description of the experimental data in terms of the underlying theory, QCD. The hurdles that prevent such a direct comparison are:

i) Non-perturbative effects, both related to confinement and to the breakdown of perturbation theory in certain kinematic domains.

ii) Influence of higher order effects on the normalization and shape of the lowest order distributions.

Clearly (i) is the more formidable of the two, since calculation of higher order effects in a theory does not involve any matter of principle once you admit the Born(term) philosophy. I would like to report some work related to points (i) and (ii) above from a practitioner's point of view. Most of it, if not all, is related to the determination of $\alpha_s(Q^2)$ from e^+e^- jet data.

First, let me discuss the effects of the higher twist terms in e^+e^- annihilation. In deep inelastic reactions, the term higher twist is almost synonymous with mass effects. In e^+e^- annihilation, there is no target mass and the final quark mass effects can perhaps be calculated in the context of perturbative QCD. What is generally understood by higher twist effects in e^+e^- annihilation are effects related to non-perturbative phenomenon, for example the process

$$e^+e^- \to q + \bar{q} + \text{large } p_T \text{ hadron(s)} \qquad (1)$$

which competes with the QCD process

$$e^+e^- \to q + \bar{q} + G \qquad (2)$$

The higher twist process (1) can be calculated in model field theories. A σ-model type estimate gives [7]:

$$\frac{1}{\sigma_{pt.}} \frac{d\sigma(1)}{dx_1 dx_2} \simeq \frac{\alpha_{HT}}{16\pi^2} \frac{1}{Q^2} \left(\frac{x_2^2}{(1-x_2)^2} + \frac{x_1^2}{(1-x_1)^2} \right)$$

$$\sigma_{pt.} = \frac{4\pi\alpha}{3s} \Sigma Q_i^2 \ , \ x_i = E_i/E_{beam} \tag{3}$$

where α_{HT} is the effective quark-meson coupling constant. This has to be compared with the QCD estimate:

$$\frac{1}{\sigma_{pt.}} \frac{d\sigma(2)}{dx_1 dx_2} = \frac{2}{3} \frac{\alpha_s(Q^2)}{\pi} \frac{x_1^2 + x_2^2}{(1-x_1)(1-x_2)} \tag{4}$$

(3) gives rise to a p_T-behaviour $\frac{d\sigma}{dp_T^2} \sim \frac{1}{p_T^4}$ as compared to the $O(\alpha_s)$ QCD prediction $\frac{d\sigma}{dp_T^2} \sim \frac{1}{p_T^2}$. Of course, intrinsic to both (1) and (2) is a non-perturbative p_T dependence, for which there is convincing experimental evidence that it is sharply peaked. Both the low-p_T hadron data[8] as well as the low energy e^+e^- data[9] suggest, though do not prove, that a Gaussian distribution

$$\frac{d\sigma}{dp_T^2} \sim e^{-bp_T^2} \tag{5}$$

with $b \cong 6\text{-}8\ \text{GeV}^{-2}$ describe the p_T distributions and $\langle p_T \rangle$ adequately. The question is how important could be the contribution from (3) or in other words how big is α_{HT}? It has been suggested[7] that the Spear p_T-distributions in the energy region $3.0\ \text{GeV}<Q<7.4\ \text{GeV}$ are well accounted for if one takes $b \cong 6\ \text{GeV}^{-2}$ in (5) and $\alpha_{HT} \cong 220\ \text{GeV}^2$. However, this analysis <u>ignores</u> the charm quark production and weak decay effects.

I have redetermined the strength of the effective coupling α_{HT} by taking into account charm production and decay at DORIS and SPEAR.[10] Using still $b = 6\ \text{GeV}^{-2}$, I get

$$\alpha_{HT} \leqq 10\ \text{GeV}^2 \tag{6}$$

Note that neither $\langle p_T \rangle$ nor the p_T-distributions can be explained in terms of the higher twist process (1) and the non-perturbative p_T distribution (5) in the entire DORIS/SPEAR, PETRA range. The multiplicity of the third jet also mutilates against a $q + \bar{q}$ + "meson" interpretation of these events. My contention here is that the bound (6) together with the $\frac{1}{Q^2}$ behaviour of (3) leads to the result that the contribution of the higher twist/CIM diagrams is $\leqq 1$ % of the* contribution due to the perturbative QCD process (2) at $Q \geqq 30$ GeV.

---

* This is correlated with the value of $\alpha_s(Q^2)$ and b. I have taken $\alpha_s(Q^2) = 0.15$ and $b = 6$ for this estimate.

Next, I would like to discuss the effects of quark masses on the 3 and 4 jet rates. Implicit in these estimates is the tacit assumption that the quark mass effects may be incorporated via perturbation theory. In $q\bar{q}$ production, these kinematic mass effects are not a faithful approximation since the resonances in the $q\bar{q}$ channel vitiate this simple quark model picture. One does not anticipate resonances in the $q\bar{q}G$ channel, more so when the three partons are so far apart in phase space. There is thus a hope that taking into account the quark mass effects in the 3 and 4 jet processes, using perturbation theory, may be a more meaningful exercise. I adopt this philosophy and present in tables I and II the effects of charm and bottom quark masses on the rates of 3 and 4 jet events. One could use the Sterman-Weinberg[11] variables to define finite 3 and 4 jet events, or equivalently the variable thrust, T[12] and acoplanarity, A,[13] to define the 3 and 4 jets respectively. The variables T and A are defined as:

$$T = \max(x_1, x_2, x_3)$$
$$A = 4 \min \frac{\Sigma p^i_{out}}{\Sigma |\vec{p}^{\,i}|} \tag{7}$$

The ratios $R_{3jet} \equiv \sigma(3jet, T<T_c)/\sigma_{total}$ and $R_{4jets} \equiv \sigma(4jet, A>A_c)/\sigma_{total}$ are presented in table I, where σ_{total} is the expression derived in ref. (14):

$$\sigma_{total} = \sigma_o\left(1 + \frac{\alpha_s(Q^2)}{\pi} + (1.98 - .116 n_f)\left(\frac{\alpha_s(Q^2)}{\pi}\right)^2\right) \tag{8}$$

and $\alpha_s(Q^2)$ is the quark gluon coupling constant in the so-called $\overline{MS}$ scheme. The entries in table II are obtained by imposing cuts on each invariant mass pair.

Table I. Fraction of 3 and 4 jet events with thrust and acoplanarity variables

Q (GeV)	T_c	R_{3jet} $n_f=3$	R_{3jet} $n_f=5$	A_c	R_{4jet} $n_f=3$	R_{4jet} $n_f=5$
20	.95	.30	.27	.05	.045	.035
	.90	.13	.12	.03	.085	.07
30	.95	.28	.26	.05	.04	.03
	.90	.12	.11	.03	.07	.06
40	.95	.26	.25	.05	.035	.03
	.90	.115	.11	.03	.065	.06

Table II. Fraction of 3 and 4 jet events with invariant mass cuts

Q (GeV)	M_{ij} (GeV)	R_{3jet} $n_f=3$	R_{3jet} $n_f=5$	R_{4jet} $n_f=3$	R_{4jet} $n_f=5$
20	5	.24	.28	4.5×10^{-3}	5×10^{-3}
	6	.155	.165	6.0×10^{-3}	7×10^{-3}
30	5	.445	.48	.038	.04
	6	.34	.36	.015	.018
40	5	.59	.62	.09	.10
	6	.48	.50	.05	.055

Note that the entries in tables I and II have been calculated for equal value of $\alpha_s(Q^2)$ for $n_f=3$ and $n_f=5$.* Also note that the quark mass effects are still $\sim$ 10–12 % in the range of PETRA jet experiments, and they introduce a substantial difference in the value of the QCD scale parameter, Λ.[15] I would also like to point out that only fixed angle type variables like the Sterman-Weinberg variables or T give cross sections proportional to $(\ln Q^2/\Lambda^2)^{-1}$ for the process $e^+e^- \rightarrow Q\bar{Q}G$, as opposed to the fixed p_T cuts, like the invariant masses, where the use of perturbation theory becomes suspect. This can be seen through the rapid growth of R_{3jets} and R_{4jets} with energy in table II. The entries for R_{4jets} are exact to $O(\alpha_s^2)$ but those for R_{3jets} are subject to $O(\alpha_s^2)$ virtual gluon corrections, whose exact contribution is controversial at the moment[16]. In any case, an $O(\alpha_s^2)$ calculation for 3jets including the quark masses has yet to be done.

The next point I would like to discuss concerns the effects of quark and gluon fragmentation. That there is a definite correlation between the fragmentation parameters and the value of $\alpha_s(Q^2)$ extracted from an experiment was first pointed out in ref. (6). Needless to say that the exact correlation is model dependent, and one should undertake a more detailed study of these matters than has been done so far. My purpose here is to show, by example of the model in ref. 6, the extent of one of such correlations namely between the intrinsic quark p_T parameter $\sigma_q \equiv \frac{1}{\sqrt{2b}}$ and the value of $\alpha_s(Q^2)$. The model described in ref. 6 is a modified Field and Feynman model[17] with a primordial p_T distribution as in (5), and a longitudinal fragmentation described in ref. (17) and specified through the parameter A_F. The fragmentation is then completely specified by stating the pseudoscalar, P, to vector meson, V, decay ratio from a $q\bar{q}$ pair. The residual $\sigma_q - \alpha_s(Q^2)$ correlation is shown in Fig. 1. The data comes from the TASSO collaboration[5,18]. I agree that the correlation is somewhat disconcerting in the absence of any other constraints. The fragmentation effects are important also for the so called "infra-red-safe fragmentation-insensitive variables" like thrust and acoplanarity. The distribution in acoplanarity is shown in Fig. 2. Note the effect of fragmentation on the 4jet distribution. Incidentally, Fig. 2 is a prediction of $O(\alpha_s^2)$ effects in e^+e^- annihila-

* The entries correspond to using $\Lambda_{\overline{MS}}$ = 350 MeV, $n_f=3$, $m_u=m_d=m_s=0$ m_c = 1.8 GeV and m_b = 4.5 GeV.

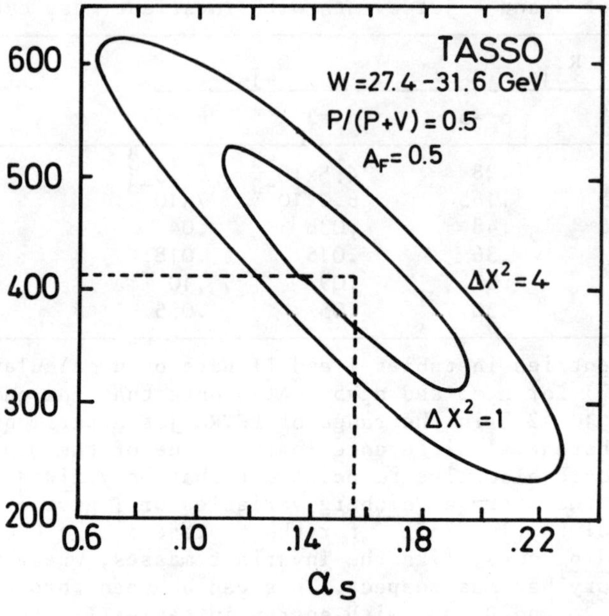

Fig. 1. The correlation between the intrinsic parameter σ_q and the quark gluon coupling constant $\alpha_s(Q^2)$ based on the model in ref. (6). The data includes events with sphericity $\geqq 0.25$. $\Delta\chi^2 = 1,4$ correspond to $\pm 1\sigma$ and $\pm 2\sigma$ respectively.

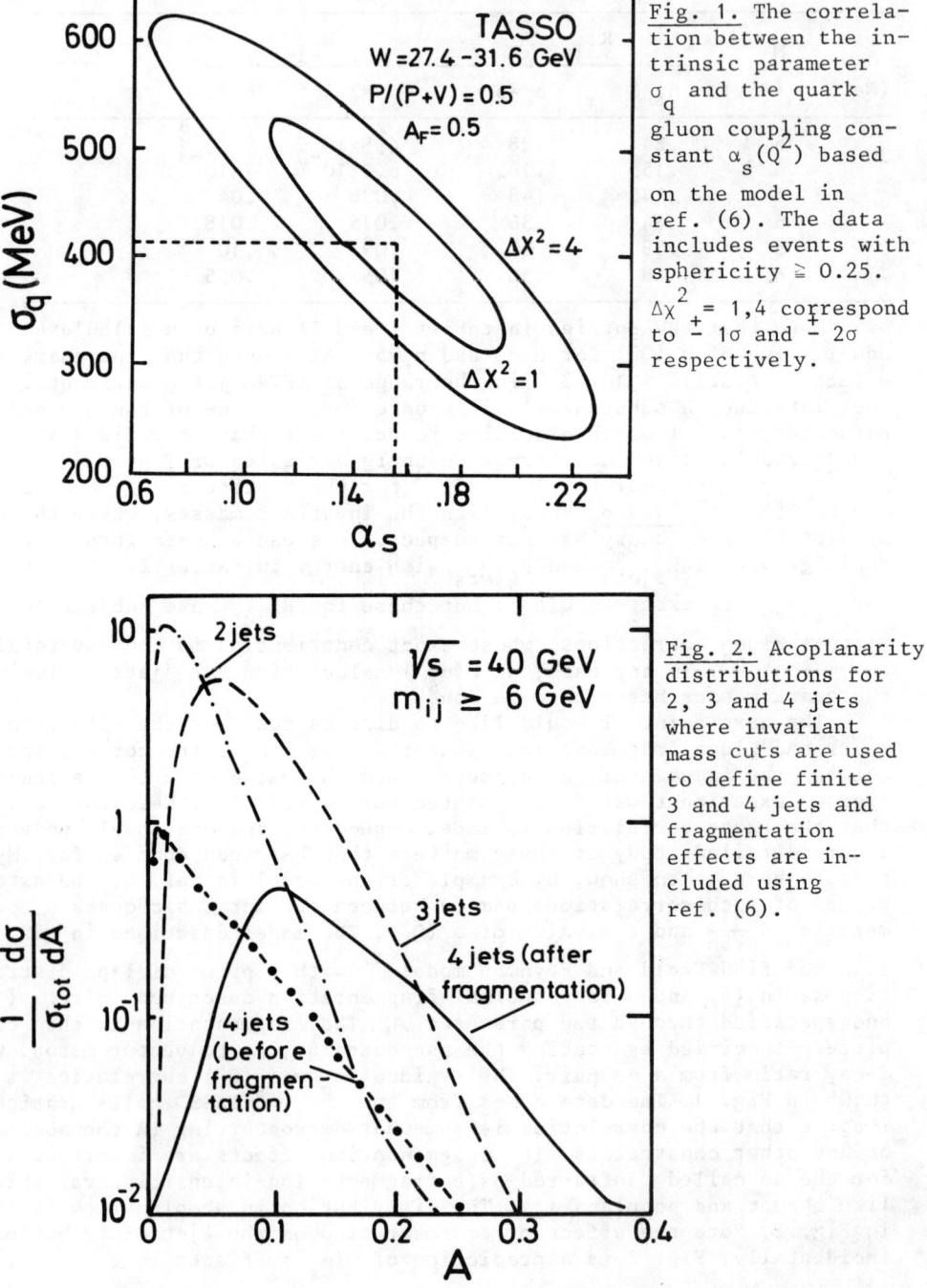

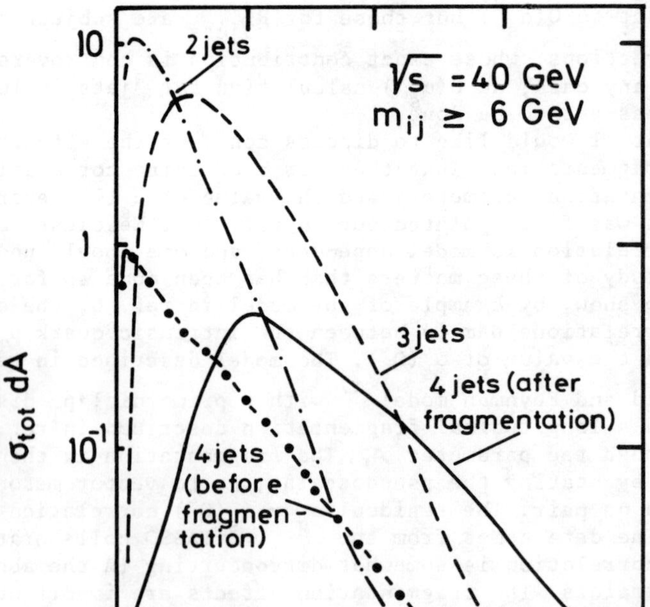

Fig. 2. Acoplanarity distributions for 2, 3 and 4 jets where invariant mass cuts are used to define finite 3 and 4 jets and fragmentation effects are included using ref. (6).

tion [19] -- perhaps the distributions $\frac{1}{\sigma}\frac{d\sigma}{dp^2_{out}}$ and $\frac{1}{\sigma}\frac{d\sigma}{dp^4_{out}}$ are more sensitive to the effects of 4jets. The rise of heavy quark multiplicity in e^+e^- annihilation [19] is another $O(\alpha_s^2)$ effect which might be observable at PETRA/PEP energies.

The extent of the non-perturbative effects at PETRA/PEP energies is a reminder that the determination of $\alpha_s(Q^2)$ from e^+e^- data is a somewhat model dependent enterprise.

I gratefully acknowledge useful and productive discussions with my colleagues at DESY. I would especially like to thank Gustav Kramer and Esko Pietarinen for sharing their insight with me.

REFERENCES

1. J. Ellis, M.K. Gaillard and G.G. Ross, Nucl. Phys. B111, 253 (1976).
2. For an exhaustive list of experimental references see the talk of B. Wiik in the Proceedings of this Conference.
3. V. Hepp, PLUTO Collaboration, DESY 80/84 (1980), TASSO Collaboration, R. Brandelik et al., DESY 80/80 (1980).
4. A. De Rujula, J. Ellis, E.G. Floratos and M.K. Gaillard, Nucl. Phys. B138, 387 (1978). J. Ellis and I. Karliner, Nucl. Phys. B148, 141 (1979). K. Koller and H. Krasemann, Phys. Lett. 88B, 119 (1979).
5. MARK J Collaboration, D.P. Barber et al., Phys. Lett. 89B, 139 (1979); TASSO Collaboration, R. Brandelik et al., Phys. Lett. 94B, 437 (1980).
6. A. Ali, E. Pietarinen, G. Kramer, J. Willrodt, Phys. Lett. 93B, 155 (1980) and A. Ali, E. Pietarinen and J. Willrodt, DESY Report T-80/01 (1980).
7. T.A. DeGrand, Y.J. Ng and S.-H.H. Tye, Phys. Rev. D16, 3251 (1977).
8. See, for example, L. Di Lella, CERN-EP/79-145 (1979).
9. See for example G.G. Hanson, SLAC-Reports SLAC-PUB-1814 (1976), SLAC-PUB-2118 (1978); PLUTO Collaboration, R. Brandelik et al., DESY 79/73 (1979).
10. A. Ali (unpublished calculation)
11. G. Sterman and S. Weinberg, Phys. Rev. Lett. 39, 1436 (1977).
12. E. Farhi, Phys. Rev. Lett. 39, 1587 (1977). See also S. Brandt et al., Phys. Lett. 12, 57 (1964).
13. A. De Rujula et al. in ref. (4).
14. M. Dine and J. Sapirstein, Phys. Rev. Lett. 43, 668 (1979); W. Celmaster and R.J. Gonsalves, Phys. Rev. Lett. 44, 560 (1980); K.G. Chetyrkin, A.L. Kataev and F.V. Tkachov, Phys. Lett. 85B, 277 (1979).
15. This for example is one of several differences between the model of ref. 6 and the one described in P. Hoyer, P. Osland, H.G. Sander, T.F. Walsh and P.M. Zerwas, Nucl. Phys. B161, 349 (1979).
16. R.K. Ellis, D.A. Ross and A.E. Terrano, Caltech Reports CALT-68-783 and CALT-68-785 (1980); K. Fabricius, I. Schmitt, G. Schierholz and G. Kramer, DESY Report 80/91(1980)
17. R.D. Field and R.P. Feynman, Nucl. Phys. B136, 1 (1978).
18. TASSO Collaboration and J. Freeman, Ph.D. thesis (under preparation), Univ. of Wisconsin. I am grateful to Jim Freeman for supplying fig. 1.
19. A. Ali, J.G. Körner, Z. Kunszt, J. Willrodt, G. Kramer, G. Schierholz and E. Pietarinen, Nucl. Phys. B167, 454 (1980).

JET ACOLLINEARITY AND QUARK FORM FACTORS

W. J. Stirling
University of Washington, Seattle, Wa. 98195

ABSTRACT

Perturbative Quantum Chromodynamic corrections involving the emission of gluons which are both soft and collinear are discussed for both hadronic production of lepton pairs and e^+e^- annihilation. The result is an exponential, double logarithmic quark "form factor". The effect of sub-leading corrections and the possible experimental observation of the form factor are discussed.

For many processes involving hadrons the parton model of non-interacting pointlike quarks is a remarkably good approximation. Furthermore, it can be shown that the general factorized picture of these interactions central to the parton model survives in the context of Quantum Chromodynamics (QCD)[1]. This result derives from the fact that the logarithmic singularities inherent in the theory can be factored in such a way as to associate them with the renormalized wave functions of quarks and gluons inside hadrons. The simplest perturbative results are, however, confined to processes characterized by a single large invariant Q^2 with all relevant "scaling" variables fixed. The perturbative expansion is a sum of terms of the form $\alpha_s^n \log^m Q^2$ with $m \leq n$, the logarithmic singularities arising from the emission of collinear gluons. The statement of factorization is that all such large logarithms can be reabsorbed into Q^2 dependent wave functions or running couplings $\alpha_s(Q^2)$.

It is of interest to extend the analysis into kinematic regions where there are two large invariants which nevertheless have a large ratio, thereby enlarging the region of phenomenological applicability of the theory. Specific examples of such a kinematic regime are the measurement in e^+e^- annihilation of energy-energy correlations with two calorimeters which are almost back to back, and the hadronic production of lepton pairs with $Q_T^2 \ll Q^2$. In such cases additional large logarithms occur e.g. $\log Q^2/Q_T^2$ and the leading corrections to the parton model take the form $\alpha_s^n \log^{2n} Q^2/Q_T^2$ arising from the emission of n soft <u>and</u> collinear gluons. The question of summing these leading corrections has been studied recently in various ways by several authors[2-5]. Presented here is a brief summary of the methods and results. In particular one finds an exponential quark form factor which leads to a vanishing rate in the limit $Q^2/Q_T^2 \to \infty$. The possibility of observing such form factors in experiments will be discussed.

Consider then the contribution to $pp \to \mu^+\mu^- X$ from the parton subprocess $q\bar{q} \to \gamma^* + ng$. It is convenient to use a non-covariant gauge so that the leading contribution comes from ladder diagrams

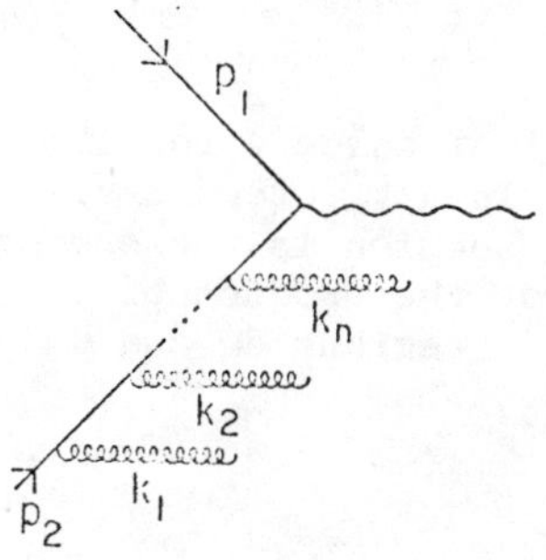

Fig. 1. A ladder diagram with n gluons.

An example is shown in Fig. 1. Define the kinematic variables $\hat{s}$, Q^2, p^2 to be the $q\bar{q}$ center of mass energy, the mass squared of the virtual photon and the (small) mass squared of the incoming quarks respectively, and parametrize the gluon momenta as $k_i = x_i p_2 + y_i p_1 + \vec{k}_{T_i}$ where $k_i^2 = 0$ implies $x_i y_i \hat{s} \approx k_{T_i}^2$. The leading contribution to $d\sigma/dQ^2$ comes from the region of integration where the transverse momenta are strong ordered $p^2 \ll k_{T_1}^2 \ll \ldots \ll k_{T_n}^2 \ll Q^2 \sim \hat{s}$. The nested k_{T_i} integrals each yield a large logarithm and the contribution is of the form $\alpha_s^n \log^n Q^2/p^2$ times a function of the scaling variable $Q^2/\hat{s}$ which comes from a convolution of the x_i integrals.

Now consider the cross section $d^2\sigma/dQ^2 dQ_T^2$ where $\vec{Q}_T$ is the transverse momentum of the lepton pair and $Q_T^2 \ll Q^2, \hat{s}$. The lower limit of each x_i integral is of order $k_{T_i}^2/\hat{s} \ll 1$ and the small x_i region yields an additional large logarithm. A typical factor is

$$\alpha_s \int_{p^2}^{Q_T^2} \frac{dk_T^2}{k_T^2} \int_{k_T^2/\hat{s}}^{1} \frac{dx}{x} = \frac{1}{2}\alpha_s (\ell n^2 \hat{s}/p^2 - \ell n^2 \hat{s}/Q_T^2) \quad . \tag{1}$$

Note that these additional logarithms are exactly cancelled in the integrated cross section $d\sigma/dQ^2$ by virtual gluons with contributions proportional to $\delta(Q_T^2)$. It is again possible to identify the region of integration which gives rise to these leading double logarithms[2]: the quantities $k_{T_i}^2/x_i$ are strong ordered from the outside to the inside of the ladder, as for $d\sigma/dQ^2$. The transverse momenta themselves are also ordered, but now <u>every</u> permutation gives a leading contribution:

$$k_{T_{i_1}}^2 \ll k_{T_{i_2}}^2 \ll \ldots \ll k_{T_{i_n}}^2 \sim Q_T^2 \quad . \tag{2}$$

In the double leading logarithm approximation (DLLA) these orderings simplify the momentum conservation δ-functions and decouple the integrand into nested soft gluon integrals (Eq. (1)). The gluons are thus effectively emitted independently and the cross section <u>exponentiates</u> i.e.

$$Q^2 \frac{d^2\sigma}{dQ^2 dQ_T^2} \underset{\text{DLLA}}{\simeq} \sigma_0 \delta(1-Q^2/\hat{s}) \frac{\partial}{\partial Q_T^2} F(Q_T^2/\hat{s}) \tag{3}$$

where

$$F(\eta) = \exp\left[\frac{-2\alpha_s}{3\pi}\,\ell n^2\eta\right], \quad \eta << 1 . \tag{4}$$

Note that this result disagrees with the original calculation of Ref. 5, due to an incorrect treatment of the gluon k_T orderings.

In e^+e^- annihilation the analogous cross section is the energy-energy correlation function defined in terms of the product of the energies dE and dE' which flow into two angular regions dΩ and dΩ' normalized to the total e^+e^- energy $W(= \sqrt{s})$[6]. Thus

$$\frac{d^2\Sigma}{d\Omega d\Omega'} = \frac{\sum_{\text{events}} (dEdE')}{\mathcal{L}\, TW^2 d\Omega d\Omega'} . \tag{5}$$

A calculation, almost identical to that described above, gives (in DLLA)

$$\frac{d^2\Sigma}{d\Omega d\Omega'} \underset{\text{DLLA}}{\sim} \frac{1}{8\pi}\frac{\partial}{\partial\eta} F(\eta = \frac{1+\cos\chi}{2})\,\frac{d\sigma}{d\Omega} \tag{6}$$

where $\cos\chi = d\hat{\Omega}\cdot d\hat{\Omega}'$. Again the cross section is proportional to a doubly logarithmic exponential quark form factor.

The characteristic feature of the quark form factor $F(\eta)$ is the suppression at small η. Since in effect quantities which are asymptotically large (for $\eta<<1$) at each order have been summed to give a quantity which is asymptotically small, it is important to estimate the effect of subleading corrections on the DLLA result. Subleading logarithms are generated by gluons which are not soft or not collinear, and by coupling constant corrections. Explicit second order calculations show that corrections due to soft non-collinear gluons are down by two logarithms[2]. In fact in the next-to-leading double logarithm approximation (NDLLA) the form factor is given by

$$F(\eta) = \exp\left[\frac{-4b}{3\pi}\left\{\left(\frac{3}{2} - \frac{b}{\alpha_s(\hat{s})}\right)\ell n\left[\frac{\alpha_s(\hat{s})}{\alpha_s(\eta\hat{s})}\right] + \ell n\eta\right\}\right] \tag{7}$$

where $\alpha_s(Q^2) = b/\ell n\, Q^2/\Lambda^2$. This expression is presumed to correctly include all even "less leading" logarithms of η which arise from all numbers of non-soft but collinear gluons. It does not however include contributions from non-collinear gluons which come, in general, from both ladder and non-ladder diagrams.

For values of $\eta \lesssim \Lambda^2/\hat{s}$, perturbation theory in $\alpha_s(\eta\hat{s})$ is inappropriate and non-perturbative effects due to the transverse momentum of quarks within hadrons are also important. However the position of the peak in $\partial F/\partial\eta$ is readily calculable using Eq. (7) and is found to be at $\eta \sim (\Lambda^2/\hat{s})^{25/41}$ well outside the non-perturbative region i.e. the peak in the cross sections as given by Eqs. (3,6,7) presumably survives the inclusion of non-perturbative smearing. It is an important question, therefore, whether the inclusion of further perturbative corrections to Eq. (7) "fills in" the dip in the cross section between the non-perturbative region $\eta \sim \Lambda^2/\hat{s}$ and the peak $\eta \sim (\Lambda^2/\hat{s})^{25/41}$.

Another approach used to deal with this question is the "impact parameter" method in which the two dimensional Fourier transform in $\vec{Q}_T$ of the cross section is calculated[4]. This approach has the advantage of having transverse momentum conservation built in, since the appropriate δ-function has a simple representation in transform space:

$$\delta^{(2)}(\vec{Q}_T - \Sigma\vec{k}_{T_i}) = \frac{1}{(2\pi)^2}\int d^2b \exp\left[i\vec{b}\cdot(\vec{Q}_T - \Sigma\vec{k}_{T_i})\right] . \quad (8)$$

At very large values of b the transform probes only those gluons with small k_T. In the soft gluon approximation, the gluon integrals factorize and b space exponentiation obtains. The positivity of the transform implies a cross section which decreases monotonically with increasing Q_T in apparent contradiction with the presence of a peak in the momentum space result. However, the inverse transform is dominated by values of b which are not asymptotically large and it is in precisely this region that the factorization of the gluon integrals breaks down. Once again therefore the precise behavior of the cross section is sensitive to the subleading corrections. However, it is possible that impact parameter space may be a better theoretical laboratory for computing such corrections and recent work in this direction has been reported at this conference[7].

The energy-energy correlation function has recently been measured by the PLUTO collaboration at PETRA[8]. In Fig. 2 the data are compared with the cross section of Eq. (6) using both the DLLA and NDLLA forms for the form factor. Note, in particular, the suggestion of a damping in the smallest

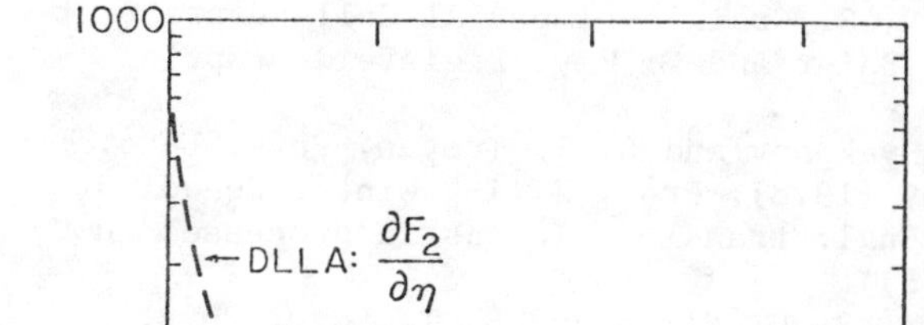

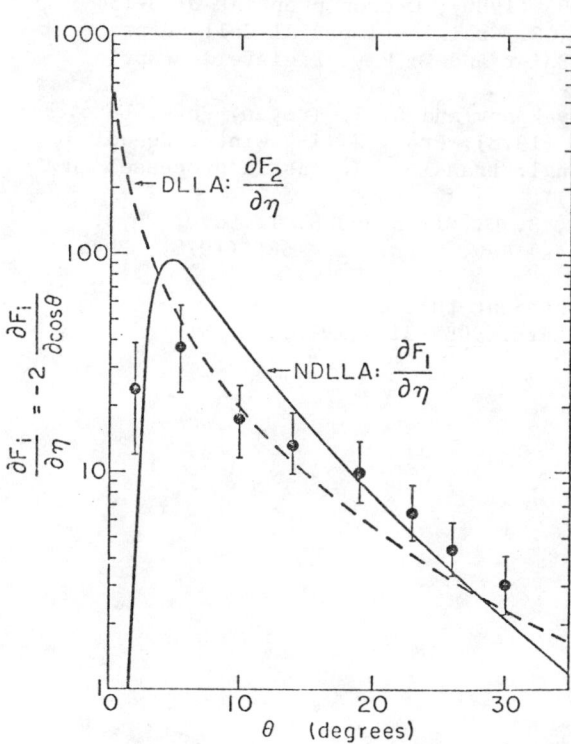

Fig. 2. Comparison of the energy weighted cross section in DLLA and NDLLA with data from Ref. 8.

angle bin and the general agreement in magnitude.

The form and observability of quark form factors is clearly an interesting question which deserves and presumably will receive further theoretical and experimental study.

REFERENCES

1. R. K. Ellis, H. Georgi, M. Machacek, H. D. Politzer and G. G. Ross, Nucl. Phys. B152, 285 (1979); S. B. Libby and G. Sterman, Phys. Rev. D19, 2468 (1979).
2. S. D. Ellis and W. J. Stirling, University of Washington preprint RLO-1388-821 (1980).
3. C. L. Basham, L. S. Brown, S. D. Ellis and S. T. Love, Phys. Lett. 85B, 297 (1979); G. C. Fox and S. Wolfram, Caltech preprint CALT 68-723 (1979); C. Y. Lo and J. D. Sullivan, Phys. Lett. 86B, 327 (1979); J. B. McKitterick, University of Illinois preprint ILL-(TH)-80-22 (1980); K. Kajantie and F. Pietarinen, DESY preprint 80/19 (1980); W. Marquardt and F. Steiner, DESY preprint 80/24 (1980).
4. G. Parisi and R. Petronzio, Nucl. Phys. B154, 427 (1979); D. F. Soper, Nucl. Phys. B163, 93 (1980); Oregon preprint OITS-134 (1980); H. F. Jones and N. S. Craigie, Imperial College preprint ICTP/79-80/21 (1980); R. Baier and K. Fey, Bielefeld preprint BI-TP 80/10 (1980).
5. Yu.L. Dokshitzer, D. I. Dyakonov and S. I. Troyan, Phys. Lett. 78B, 290 (1978); 79B, 269 (1978); Proc. XIIIth Winter School of LNPI (Leningrad, 1978) (Engl. transl.: Inelastic processes in QCD, SLAC-TRANS-183 (1978)).
6. C. L. Basham, L. S. Brown, S. D. Ellis and S. T. Love, Phys. Rev. D17, 2298 (1978); Phys. Rev. Lett. 41, 1585 (1978); Phys. Rev. D19, 2018 (1979).
7. J. C. Collins, talk presented at this conference.
8. Ch. Berger et al., Phys. Lett. 90B, 312 (1980).

MONTE CARLO JET GENERATION

R. Odorico
I.N.F.N. and Istituto di Fisica, Bologna

Abstract. Basic features of recent Monte Carlo calculations for jets in leading log QCD are reviewed.

A number of papers on the title subject have recently appeared[1,3-7]. The use of Monte Carlo (MC) techniques in this context is allowed by the probabilistic description underlying the leading logarithm approximation (LLA) for QCD. In the MC one has to follow the degradations in mass ($\sqrt{K^2}$) and longitudinal momentum or energy (described by a scaled variable x) along the LLA tree, down to a minimum off-shell mass Q_o below which the non-perturbative regime is supposed to start. The only dynamical input needed in the MC is the elementary LLA emission probability. E.g. for a quark leg:

$$dP_E = \frac{\alpha(K^2)}{2\pi}\frac{dK^2}{K^2}\int_{\varepsilon}^{1-\varepsilon} P_{q,qG}(z)\,dz \qquad (1)$$

Here $\varepsilon\equiv\varepsilon(K^2,K\eta)$ (where η is the gauge 4-vector) takes into account kinematic limits and absorbs with good approximation the gauge dependence of the splitting function (see DDT[8]). If desired, the form of dP_E can be changed, e.g. substituting $\alpha(K^2)$ with $\alpha(K_T^2)$ [9]. For a generic gauge, eq. 1 implies that the degradations in K^2 and x are strictly intertwined, which makes a calculationally viable MC cascade hard to find. But if one chooses η light-like and with space direction opposite to that of the jet axis, ε reduces to the kinematic limit $\varepsilon\simeq Q_o^2/K^2$, independent of x. In this case, given a parton leg of maximum allowed square mass (i.e. virtuality) K_J^2, from eq. 1 it is easy to calculate the probabilities that the parton leg in its evolution i) degrades down to Q_o without emission:

$$\Pi_{NE}(K_J^2,Q_o^2) = EXP(-\int_{4Q_o^2}^{K_J^2} dP_E) \qquad (2)$$

ii) undergoes the first branching in the neighbourhood of some K^2:

$$d\Pi_E(K_J^2,K^2) = \frac{d}{dK^2}(EXP(-\int_{K^2}^{K_J^2} dP_E))\,dK^2 \qquad (3)$$

Eq. 2 is easy to tabulate once for all, and with the same tabulation one can generate K^2 according to eq. 3. Under these conditions (i.e. ε independent of η) it is easy to write a MC cascade describing the LLA evolution of a parton leg: first decide whether the leg branches (according to eq. 2), if so generate the K^2 at which it branches (eq. 3), then generate z, finally assign virtualities to the two secondary legs (according to kinematic limits, see below) and iterate the cascade for each of them. If one takes Q_o very small and sets the virtualities of secondary legs equal to the K^2 of the parent (thus violating kinematic limits), this MC cascade provides an exact solution of the LLA evolution equations in their standard form[1].

Two crucial points to understand before applying the MC to event simulation (at the parton level) are i) the qualitative inconsistency of exact phase-space bounds with the LLA independent development of parton legs, ii) the fixing of the gauge in physical processes (e.g. e^+e^- annihilation), where more than one jet is present.

All phase-space bounds derive from the condition that in a branching ($K^2 \to K_1^2 + K_2^2$)

$$K_T^2 = z(1-z)(K^2 - K_1^2/z - K_2^2/(1-z)) \geqslant 0 \qquad (4)$$

where K_T is the relative transverse momentum between the secondary partons. This entails $K_1^2 \leqslant z(K^2 - K_2^2/(1-z))$, which is incompatible with the independent development of the two secondary legs. In analytic calculations[9,10], in order to cope with this problem the (less restrictive) approximate bound $K_1^2 \lesssim z\,K^2$ is used. In the MC one can enforce (or not, if one likes to) exact phase-space constraints, but in doing so it must be realized that there is no guidance from the LLA and that a specific recipe is as good as any other. (We find, however, that the procedure used in [6] unduly distorts the z distribution. Eq. 2, in fact, remains valid as far as K_1^2/K^2, $K_2^2/K^2 \ll 1$, which is reasonably well verified by the MC results. See fig. 1[1] and also cfr. [9-11]). Fortunately, results are largely recipe-independent. In the Table there are shown values of the (subleading) exponent γ appearing in $\langle n_{parton}\rangle = (\sqrt{\tau_2}/\sqrt{\tau_1})^{-\gamma}$ $\mathrm{EXP}(2\sqrt{C_A/\pi b}(\sqrt{\tau_2}-\sqrt{\tau_1}))$, $\tau = \log(Q^2/\Lambda^2)$, obtained from the growth of $\langle n_{parton}\rangle$ in various Q intervals as calculated from the MC with various recipes[1].

TABLE

$Q_1 - Q_2$ (GeV)	$z\,K^2$	Fastest first	Random	$\frac{1}{2}\,z\,K^2$
10 - 100	1.35	2.04	2.45	2.51
100 - 1000	1.27	2.07	2.71	2.58
10 - 1000	1.32	2.05	2.58	2.59

There is no big change when moving within exact phase-space recipes, whereas there is a sizable jump with respect to the approximation used in analytic calculations. The fact that γ is independent of the Q interval verifies the subleading nature of phase-space effects.

In e^+e^- annihilation, if one chooses η so as to have $\varepsilon(K^2,K\eta)$ independent of $K\eta$ for one "jet" (i.e. η light-like and running opposite to the jet direction), one has no emission at all for the other "jet", in the LLA. In [4], in order to ensure forward-backward symmetry, η is chosen with null space component in the c.m.s., so as to have equal radiation from both quarks. In doing that, however, the basic DDT result about the gauge dependence of splitting functions is ignored and it is replaced by model-dependent kinematic limits, meant to avoid the emission of backward radiation from the quarks. The MC cascade is correspondingly much more complicated than the one seen before. In [4,5] not enough details are given to understand its working and its degree of model dependence. The results quoted are at variance with those of [10,1]. Specifically, $<n_{parton}>$ is about twice that obtained analytically in [10], whereas it should actually be lower since, differently from [10], phase-space bounds are supposed to be exactly enforced (in [1] the result of[10] is correctly reproduced if the same phase-space approximations are used). In [6], although an application to e^+e^- annihilation is briefly considered, no details are given about the treatment of this point. If x is given a Lorentz invariant definition (i.e. fraction of K_0+K_L), even with light-like gauges the forward-backward symmetry of the results must follow as a consequence of the gauge invariance of the calculation[1]. In other words, in a light-like gauge a single MC cascade contains both jets in a symmetric way, the backward jet being generated by the small x quanta. Although this is far from being apparent in the calculational procedure, in [1] it is shown that it is nicely verified numerically (see e.g. the figure).

As to the hadronization of the quanta, necessary for phenomenological purposes, in [4-6] it is adopted a model which draws inspiration from the preconfinement idea of [12]. Eventually, however, the model has little to do with preconfinement, since gluons at the end of the perturbative cascade are forced to convert into $q\bar{q}$ pairs (whereas, with $Q_0 = 1$ GeV, colourless clusters consist on average of a $q\bar{q}$ pair and $\simeq 20$ gluons). In [3], instead, calculations are done keeping independent fragmentation of the quanta à la Feynman and Field, which as a matter of fact turns out to satisfy 4-momentum balance reasonably well, and incorporates (differently from the previous approach) standard QCD phenomenology at PETRA energies ($Q_0 \simeq$ 3 GeV must be assumed for this sake).

Among the new LLA results obtained by MC one should mention[1,3] i) the near equality of $<p_T^2>$ in gluon and quark jets, when no energy weight is applied, ii) the slower increase of $<p_T^2>$ with Q^2 ($\sim \alpha_s(Q^2)Q$ instead of $\sim \alpha_s(Q^2)Q^2$, as found in lowest perturbative order calculations), iii) a detailed calculation of the (sizable) delaying effects caused by phase-space constraints in reaching the asymptotic regime. But, of course, with the Monte Carlo one has full exclusive information about the final state, and therefore all desired LLA results about jets can be easily and exactly obtained.

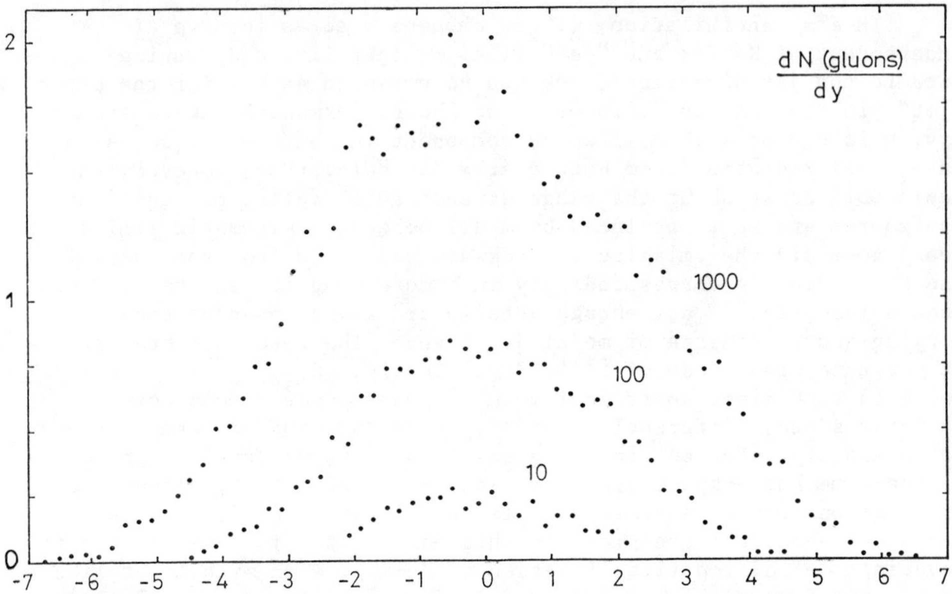

Fig. Rapidity distributions of gluons with respect to the sphericity axis for various e^+e^- c.m.s. energies (GeV). Q_o = 1 GeV.[1]

1. R. Odorico, paper No. 283 presented to the Conference, ECFA-LEP Note SSG/8/9 (1980), to be published in Nucl.Phys. B. Preliminary results of this work were reported at the Rome Meeting of the ECFA-LEP Working Group, September 1979, see ref. 2.
2. H. Bøggild, Status Report of SSG8, in ECFA-LEP Working Group 1979 Progress Report, ECFA/79/39, p. 97.
3. P. Mazzanti and R. Odorico, paper No. 590, Univ. of Bologna preprint IFUB 80/12 (1980), to be published in Phys. Lett. B; paper No. 806, Univ. of Bologna preprint IFUB 80/13 (1980).
4. G. C. Fox and S. Wolfram, Caltech preprint CALT-68-755 (1980).
5. S. Wolfram, Caltech preprint CALT-68-778(1980).
6. K. Kajantie and E. Pietarinen, DESY preprint 80/19 (1980).
7. C. H. Lai, J. L. Petersen and T. F. Walsh, paper No. 248, Niels Bohr Inst. preprint NBI-HE-80-8 (1980).
8. Yu. L. Dokshitzer, D. I. D'yakonov and S. I. Troyan, Proc. 13th Winter School of LNPI, Leningrad (1978).
9. D. Amati, A. Bassetto, M. Ciafaloni, G. Marchesini and G. Veneziano, CERN preprint TH-1831 (1980).
10. A. Bassetto, M. Ciafaloni and G. Marchesini, Nucl. Phys. B 163, 477 (1980).
11. K. Konishi, A. Ukawa and G. Veneziano, Nucl. Phys. B 157, 45 (1979).
12. D. Amati and G. Veneziano, Phys. Lett. 83B, 87 (1979).

E^+E^- PHYSICS IV

G. HANSON, SLAC, ORGANIZER

CRYSTAL BALL STUDIES OF THE REACTION $\psi' \to \gamma\gamma\psi$

T. H. Burnett
(Representing the Crystal Ball Collaboration[1])
Physics Department, University of Washington
Seattle, Washington 98195

INTRODUCTION

This talk is a summary of the analysis of 2048 events of the type

$$\psi' \to \gamma\gamma\psi \quad \psi \to \ell^+\ell^- \tag{1}$$

obtained from (776 ± 78)K ψ'decays detected by the Crystal Ball detector at SPEAR. A more complete description of the detector and details of the analysis can be found in reference 2.

The procedure is to first classify events as to whether the two photons originate from the cascade $\psi' \to \gamma\chi$ followed by $\chi \to \gamma\psi$, where χ represents an intermediate C-even charmonium state, or from the direct decay $\psi' \to \pi^0/\eta\psi$, the photons resulting from the π^0 or η decay. Events in the first category represent a search for χ states which cannot be formed directly by e^+e^- annihilation, allowing measurement of the product branching ratios, and determination of the spin of the χ state, statistics permitting. The spin analysis also allows study of the multipole structure of the decays, which we find to be predominantly dipole, and even a determination of the magnetic moment of the charmed quark. Determination of the branching ratio of $\psi' \to \pi^0\psi$ indicates isospin violation in the decay amplitudes.[3]

DETERMINATION OF BRANCHING RATIOS

The trigger is very efficient for hadron decays, >99%. but admits roughly equal numbers of Bhabhas which deposit twice the beam energy in the Ball, and beam-gas and cosmic rays, which contribute to a low energy tail and asymmetric distribution of the energy. Hadron events tend to deposit energy intermediate between the Bhabha and cosmic/beam-gas extremes, in a fairly symmetric manner. All triggers are classified according to the amount and symmetry of the energy deposited--this analysis results in the estimate of the number of psi decays in the sample, 776K ± 78K.

Further selection of the 2 gamma, 2 lepton topology is based on the response of the detector to the approximately 1.5 GeV $e^\pm$ or $\eta^\pm$ from the decay. An electron or positron deposits nearly all of its energy in 13 adjacent crystals. Muons, on the other hand, deposit 210±5 MeV in 1-4 crystals, corresponding to minimum ionization energy loss. Then a series of fiducial cuts are applied to require that the configuration be detectable with high efficiency and resolution.

Finally, a constrained fit to the reaction (1) hypothesis is performed, which is 5C for the electron final state, and 3C in the muon case. The requirement that the chi squared be reasonable, plus a final hand scan, results in the final sample of 2048 events. Figure 1 shows a Dalitz plot in the mass squared of the two photons vs. the mass squared of the more energetic photon and the psi. Bands corresponding to the known chi states at 3.51 and 3.55 GeV (which shall be referred to as χ_1 and χ_2, respectively, the subscripts corresponding to the expected, and with this analysis, confirmed, spins) are clearly visible, as are π^0 and η bands. Much less prominent, but still visible, is a band corresponding to the χ (3.41), or χ_0. The (product) branching ratios for these processes, as determined from fits to projections of fig. 1 are listed in Table I. With the exception of about 8 events from $\psi' \to \pi^0 \pi^0 \psi$, in which two photons either overlap or escape, the three chi cascades and π^0 and η decays account for all of the events. In particular, we set a limit of 0.04% (90% CL) for the product branching ratio of cascade via a state at 3.591 GeV, which is inconsistent with a previous measurement at DORIS of (0.18 0.06)%.[4]

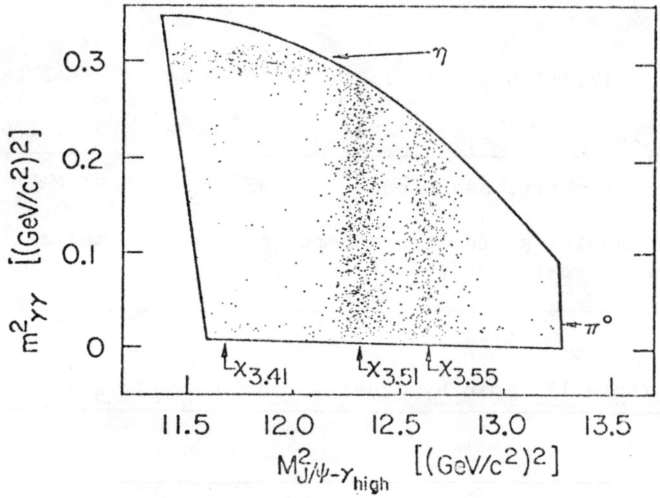

Figure 1

SPIN-MULTIPLE ANALYSIS OF THE CASCADES

Studying the angular correlations of the cascade decays allows determination of the chi spin, and relative strengths of the multipole amplitudes for the two transitions.[5] Cutting overlap regions in the Dalitz plot to avoid ambiguities, 442 χ_2 and 921 χ_1

events were isolated. A maximum-likelihood fit to all five angles describing the configuration of the two decays was performed, comparing the expected (using a Monte Carlo calculation) and actual contents of each of 486 bins in the five-dimensional space. Assuming, as predicted by theory, that the amplitudes are predominantly dipole, only diple and quadrupole terms were allowed, so there were just two parameters for a given spin, which took the values 0, 1 and 2. The fit confidence levels for the best fits are shown in Table II. The only acceptable fits correspond to the expected ordering of the triplet P-wave states.

Table I Branching Ratios determined from Dalitz plot

State	Mass[a]	Events	Eff.	(Product)B.R.(%)[b]
χ_2	3553.9±0.5±4	479	0.35	1.26±0.09
χ_1	3508.4±0.4±4	943	0.37	2.38± .12
χ_0	3413 (assumed)	17	0.28	0.59±0.015
η	547.3±1.0	386	0.44	2.18±0.14
π^0	135±3	23	0.26	0.09±0.02

a Determined from fit, assuming m_ψ = 3095, $m_{\psi'}$ = 3684 MeV

b Not including 16% systematic uncertainty from normalization. Assumes BR($\psi \to e^+e^-$) = BR($\psi \to \mu^+\mu^-$) = 6.9%

Table II Spin hypothesis confidence levels

	Spin		
State	0	1	2
χ_1	$<10^{-6}$	0.15	4×10^{-3}
χ_2	3×10^{-4}	10^{-2}	0.22

For the acceptable solutions, the quadrupole amplitudes, normalized so the sum of the squares of dipole and quadrupole is 1, are shown in Table III. All four are consistent with zero, justifying the decision to ignore higher multipoles. These quantities have been related to the charmed quark magnetic moment by Karl, Meshkov, and Rosner.[6] Assuming these relations, we have also performed a single parameter fit for both the χ_1 and χ_2 cascades obtaining values for the charmed quark magnetic moment shown.

Table III Fits to quadrupole coefficients and magnetic moment

State	Quadrupole coefficients $\psi' \to \gamma\chi_J$	$\chi_J \to \gamma\psi$	Magnetic Moment (in units of μ_{Dirac})
χ_1	$0.077^{+0.050}_{-0.045}$	$-0.002^{+0.008}_{-0.020}$	$0.93^{+0.72}_{-0.63}$
χ_2	$0.132^{+0.098}_{-0.075}$	$-0.333^{+0.116}_{-0.292}$	$5^{+2.5}_{-4.5}$

REFERENCES

1. R. Partridge, C. Peck, F. Porter (Caltech); D. Antreasyan, Y. F. Gu, W. Kollmann, M. Richardson, K. Strauch, K. Wacker (Harvard); D. Aschman, T. Burnett, M. Cavalli-Sforza, D. Coyne, H. Sadrozinski (Princeton); R. Hofstadter, R. Horisberger, I. Kirkbride, K. Konigsmann, H. Kolanoski, A. Liberman, J. O'Reilly, J. Tompkins (Stanford); E. Bloom, F. Bulos, R. Chestnut, J. Gaiser, G. Godfrey, C. Kiesling, W. Lockman, M. Oreglia (SLAC).
2. M. Oreglia, SLAC-PUB-2529, (presented at the XV Rencontre de Moriond, 1980); M. Oreglia et al., SLAC-PUB-2539 (submitted to Phys. Rev. Lett.)
3. N. Isgur, et al., Phys. Lett. 89B, 79(1979); P. Langacker, Phys. Let. 90B, 447 (1980).
4. W. Bartel et al., Phys. Lett. 79B, 492 (1978).
5. G. Karl, S. Meshkov and J. Rosner, Phys. Rev. D13, 1203 (1976).
6. G. Karl, S. Meshkov and J. Rosner, Phys. Rev. Lett. 45, 215 (1980).

RADIATIVE TRANSITIONS FROM THE $\psi(3095)$ AND $\psi'(3684)$ TO ORDINARY HADRONS*

Daniel L. Scharre
Stanford Linear Accelerator Center
Stanford University, Stanford, California 94305

ABSTRACT

Mark II results from SPEAR on radiative transitions from the $\psi(3095)$ and $\psi'(3684)$ to ordinary hadrons (i.e., hadrons which do not contain charmed quarks to first order) are reviewed.

I. INCLUSIVE ψ RADIATIVE TRANSITIONS

First order QCD calculations predict that a significant fraction of the hadronic decays of heavy quark-antiquark $^3S_1$ resonances (such as the ψ) result in the production of direct photons (i.e., photons not coming from secondary decays of π^o's and η's).[1] The direct hadronic decay of the ψ must proceed via an intermediate state consisting of at least three color-octet gluons. The dominant contribution to direct photon production arises by replacement of one of the outgoing gluon lines by a photon. It is expected that approximately 8% of all ψ decays should contain a direct photon. This direct photon production is expected to peak at $x = 1$, where x is the fraction of the beam energy taken by the photon, and hence should be experimentally observable.

Figure 1 shows the inclusive γ momentum distribution at the ψ (solid circles) compared with the γ distribution predicted from the measured π^o and η momentum distributions (open circles).[2] Whereas π^o and η decays can account for the measured γ spectrum at low x, there is clearly an excess in the spectrum for $x \gtrsim 0.5$. The direct photon contribution, which was obtained by subtracting the predicted distribution from the measured distribution, is shown in Fig. 2. Also shown in the figure is the theoretical distribution (convoluted with the Mark II photon energy resolution) from lowest order in QCD. The observed integrated rate of direct photon

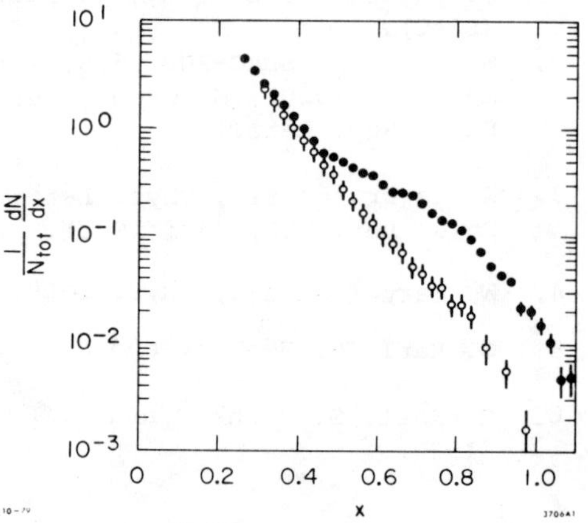

Fig. 1. Solid circles show the inclusive γ momentum distribution at the ψ, normalized to the total number of produced ψ events. Open circles show the expected distribution from π^o and η decays.

---

* Work supported by the Department of Energy, contract DE-AC03-76SF00515.

0094-243X/81/680668-07$1.50

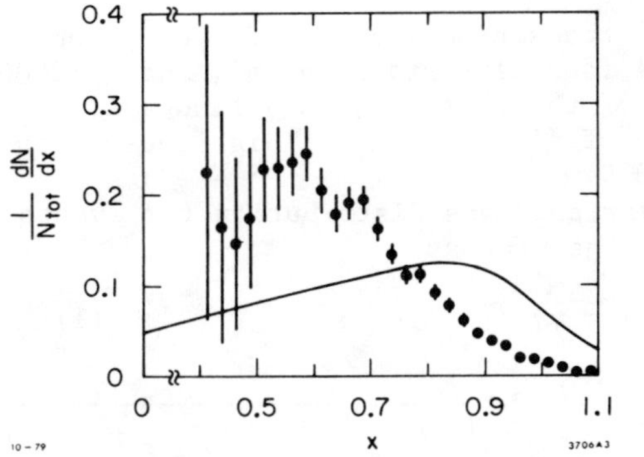

Fig. 2. Direct photon momentum distribution. Curve shows the leading-order QCD prediction.

production (4.1 ± 0.8% for x ≥ 0.6) agrees well with the theoretical estimate (5% for x ≥ 0.6), but the observed distribution is softer than expected. However, the leading-order calculation ignores the effects of the masses of the final-state hadrons and radiative and second-order QCD effects. These are expected to soften the momentum distribution.

We have compared some properties of the hadronic system recoiling against direct photons (presumably resulting from fragmentation of a 2-gluon system) and the hadronic system produced directly in e^+e^- annihilations (presumably resulting from fragmentation of a quark-antiquark system) at comparable invariant mass. Figure 3 shows the mean produced charged particle multiplicity as a function of the invariant mass of the hadronic system (M_X) from this experiment[2] and two low-energy e^+e^- annihilation experiments.[3] Also shown are mean multiplicities extrapolated from higher-energy e^+e^- annihilation data.[4] The data indicate that multiplicities from gluon and quark fragmentation are similar at these energies. A similar comparison of K^0 multiplicities in our data and e^+e^- annihilation data shows no evidence for an enhanced kaon yield from gluon fragmentation.[2]

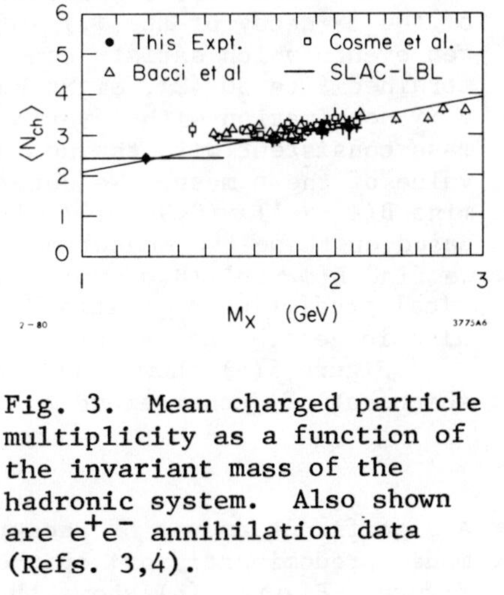

Fig. 3. Mean charged particle multiplicity as a function of the invariant mass of the hadronic system. Also shown are e^+e^- annihilation data (Refs. 3,4).

II. EXCLUSIVE ψ RADIATIVE TRANSITIONS

Transitions from the ψ to ordinary hadrons[5] are expected to proceed via emission of a photon from the initial-state charmed quark line as discussed in the previous section. The coupling to the ordinary quark-antiquark system is via a two-gluon system (in the case of the even C-parity states to be discussed here). Based on this model and the assumption of SU(3) invariance, a prediction for the relative widths of the γη and γη' transitions can be made. Unfortunately, this prediction does not agree with experimental measurements. However, if one allows for SU(3) symmetry breaking, these results are modified. Fritzsch and Jackson[6] have calculated the relative widths of the $\gamma\eta$ and $\gamma\eta'$ transitions by considering gluon-mediated mixing between the three isoscalar states η, η', and $\eta_c(2980)$. Based on the experimental masses of these states, they find a 1% admixture of η and a 2% admixture of η' in the η_c. This leads to the prediction $\Gamma(\psi\rightarrow\gamma\eta')/\Gamma(\psi\rightarrow\gamma\eta) \simeq 3.9$.

Figure 4 shows the $\pi^+\pi^-\gamma$ invariant mass distribution for events which satisfy constrained fits to the process

$$\psi \rightarrow \pi^+\pi^-\gamma\gamma \quad . \qquad (1)$$

We observe a peak at the η' mass and determine the branching fraction $B(\psi\rightarrow\gamma\eta') = (3.4 \pm 0.7) \times 10^{-3}$. In order to measure the η branching fraction, it was necessary to analyze the process

$$\psi' \rightarrow \pi^+\pi^-\psi \quad , \quad \psi \rightarrow 3\gamma \ . \qquad (2)$$

(The Mark II trigger requirement does not allow direct observation of the 3γ decay of the ψ.) Of the ten events which satisfy constrained fits to (2), eight have a $\gamma\gamma$ combination with invariant mass consistent with the nominal value of the η mass. We determine $B(\psi\rightarrow\gamma\eta') = (0.9 \pm 0.4) \times 10^{-3}$. Based on these two measurements, we find $B(\psi\rightarrow\gamma\eta')/B(\psi\rightarrow\gamma\eta) = 3.8 \pm 1.9$, which agrees with the theoretical prediction of Fritzsch and Jackson.[6] These measurements are also in general agreement with previous measurements.[5]

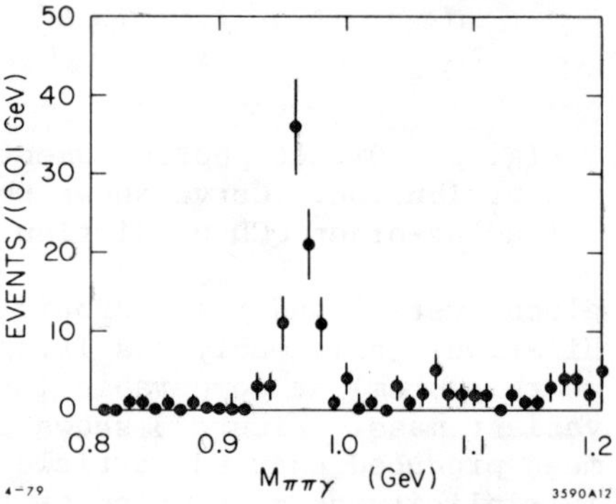

Fig. 4. $\pi^+\pi^-\gamma$ invariant mass distribution for events which satisfy (1).

Figure 5(a) shows the $\pi^+\pi^-$ invariant mass distribution for events which satisfy constrained fits to

$$\psi \rightarrow \pi^+\pi^-\gamma \quad . \qquad (3)$$

A significant amount of background is expected from the $\pi^+\pi^-\pi^0$ decay mode (predominantly $\rho\pi$) of the ψ, as shown by the histogram in the figure. Figure 5(b) shows the $\pi^+\pi^-$ mass distribution after subtraction of this background. The fitted parameters of the peak in this distribution (with mass $M = 1280 \pm 20$ MeV and width $\Gamma = 180 \pm 50$ MeV) are consistent with those of the f. We find $B(\psi\rightarrow\gamma f) = (1.3 \pm 0.3) \times 10^{-3}$. We have looked for evidence of the transition $\psi\rightarrow\gamma f'$ in the $K^+K^-\gamma$ final state, but observe no signal. We set a 90% confidence level (c.l.) upper limit of $B(\psi\rightarrow\gamma f') \times B(f'\rightarrow K\bar{K}) < 10^{-3}$. This limit is not inconsistent with the naive theoretical calculation $B(\psi\rightarrow\gamma f')/B(\psi\rightarrow\gamma f) = 0.5$ based on SU(3) invariance.

Figure 6(a) shows the $K_SK^\pm\pi^\mp$ invariant mass distribution for events which satisfy 5-constraint (5C) fits (a K_S mass constraint is imposed in addition to the normal energy-momentum constraints) to

$$\psi \rightarrow K_SK^\pm\pi^\mp\gamma \qquad (4)$$

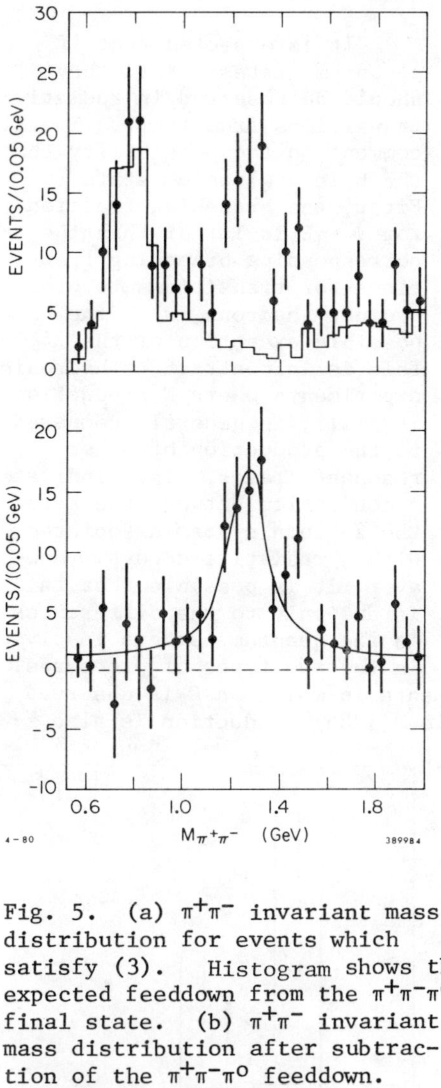

Fig. 5. (a) $\pi^+\pi^-$ invariant mass distribution for events which satisfy (3). Histogram shows the expected feeddown from the $\pi^+\pi^-\pi^0$ final state. (b) $\pi^+\pi^-$ invariant mass distribution after subtraction of the $\pi^+\pi^-\pi^0$ feeddown. Curve shows fit to data.

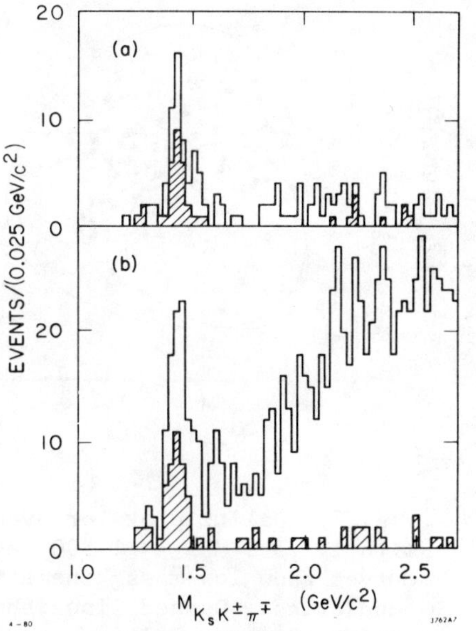

Fig. 6. $K_SK^\pm\pi^\mp$ invariant mass distributions for events which satisfy (a) 5C fits and (b) 2C fits to (4). Shaded regions have the additional requirement $M_{K\bar{K}} < 1.05$ GeV.

Figure 6(b) shows the same distribution for events which satisfy 2C fits (observation of the γ is not required) to (4). We observe a state near 1.4 GeV which we intepret as the E(1420).[7] From a fit to the distribution in Fig. 6(a), we determine $M = 1440^{+10}_{-15}$ MeV, $\Gamma = 50^{+30}_{-20}$ MeV, and $B(\psi \to \gamma E) \times B(E \to K\bar{K}\pi) = (3.6 \pm 1.4) \times 10^{-3}$. The Dalitz plot for the events shown in Fig. 6(b) with masses between 1.375 and 1.500 GeV (the signal region) is shown in Fig. 7. The data is inconsistent with a phase space distribution and we observe evidence for an excess of events in the region of the plot corresponding to low values of the $K_SK^\pm$ invariant mass. Figure 8 shows the $K_SK^\pm$ invariant mass projection for events in the signal region compared with the expected phase space distribution. There is an excess at low mass which we associate with the $\delta(980)$. As shown by the shaded region in Fig. 6 (which requires $M_{K\bar{K}} < 1.05$ GeV), this enhancement is associated with events in the signal region.

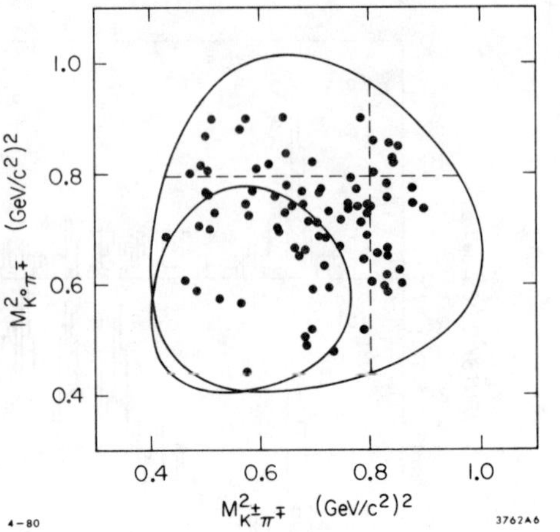

Fig. 7. Dalitz plot for events with $1.375 \le M_{K\bar{K}\pi} < 1.500$ GeV. Curves show low-mass kinematic boundaries. Dashed lines show nominal K^* mass values.

It is expected that if gluonium states exist, they should be observed in radiative transitions from the ψ. I will comment on the possibility that the E is a gluonium state. First, the branching fraction for $\psi \to \gamma E$ is larger than the corresponding branching fractions for transitions to other ordinary hadrons, with the possible exception of the η'. This is in contrast to hadronic experiments where E production is small, in general, compared to the production of other resonant states. This indicates a connection between the E and the 2-gluon system associated with ψ radiative decays. However, it is possible that this is due only to the differences in the quantum numbers involved in the two types of processes.

Second, in most hadronic experiments in which an E is observed in the $K\bar{K}\pi$ channel, roughly comparable D(1285) production is also observed. We see no evidence for the transition $\psi \to \gamma D$ with 90% c.l. upper limit

$$\frac{B(\psi \to \gamma D) \times B(D \to K\bar{K}\pi)}{B(\psi \to \gamma E) \times B(E \to K\bar{K}\pi)} < 0.2.$$

This might be interpreted as evidence for a large gluonium component in the E but not the D. However, if one hypothesizes that the E is predominantly an SU(3) singlet state and the D is predominantly an SU(3) octet state, the transition to the D could be greatly suppressed relative to the E.

Finally, Dionisi et al.[8] recently reported a determination of the spin-parity of the E from a partial wave analysis of the $K_S K^\pm \pi^\mp$ system produced in the reaction $\pi^- p \to K_S K^\pm \pi^\mp n$ at 3.95 GeV/c. The spin-parity $J^P = 1^+$ was

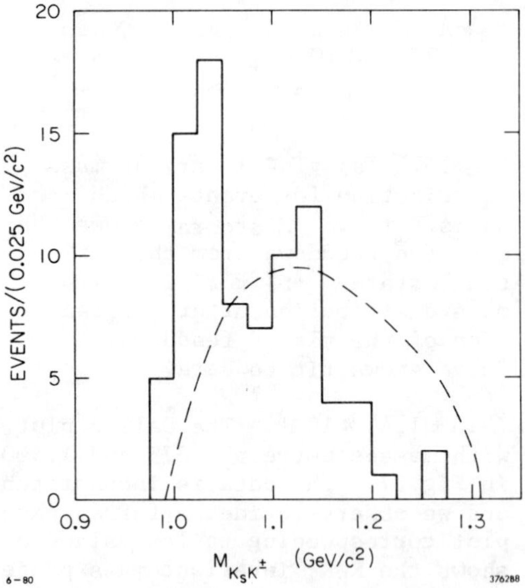

Fig. 8. $K_S K^\pm$ invariant mass distribution for events in the signal region. Curve shows expected phase space distribution.

determined from the $K^*\bar{K}$ (plus charge conjugate) decay mode of the E, which they find from their data to be the principal decay mode. This is inconsistent with Mark II results which find no evidence for a large $K^*\bar{K}$ decay mode, but rather require a significant fraction of the decay to go via $\delta\pi$. If the state observed by us and by Dionisi et al. are the same state, it is necessary to understand this inconsistency before the spin of the E can be considered as firmly established. (Unfortunately, our statistics are too limited to allow an independent determination of the spin.) If the E were finally established as an axial vector state, there would be no reason not to make the standard quark model interpretation and include it in the same nonet as the D(1285), A_1, and Q_A. If, on the other hand, the E were pseudoscalar, it would be difficult to interpret it within the standard quark model. The $J^P = 0^-$ nonet is complete, and one plausible interpretation of the E would be as a gluonium state.

III. EXCLUSIVE ψ' RADIATIVE TRANSITIONS

For a given exclusive decay of the ψ to some state X, the partial width is given by

$$\Gamma(\psi \to X) = |M_X|^2 |R(0)|^2 \quad ,$$

where M_X is the matrix element for the decay and R(0) is the wave function of the ψ at the origin. A similar expression can be written for the decay of the ψ' to some state X, where the wave function at the origin is expected to be different than for the ψ decay but the matrix element should be independent of the initial state. (All phase space factors are ignored in this discussion.) Thus, the ratio

$$R(X) = \frac{B(\psi' \to X)}{B(\psi \to X)} = \frac{B(\psi' \to e^+e^-)}{B(\psi \to e^+e^-)} \simeq 0.13$$

is expected to be independent of the state X. In addition to the e^+e^- final state (which was used in the determination of the value 0.13), the following measurements have been made:[9] $R(\pi^+\pi^-\pi^+\pi^-\pi^0) = 0.095 \pm 0.043$, $R(\pi^+\pi^-K^+K^-) = 0.19 \pm 0.08$, and $R(p\bar{p}) = 0.11 \pm 0.04$. These ratios are consistent with the expected value of 0.13.

Figure 9 shows the $\pi^+\pi^-\gamma$ invariant mass for events which satisfy constrained fits to the process

$$\psi' \to \pi^+\pi^-\gamma\gamma \quad . \qquad (5)$$

From the signal observed at the mass of the η', we determine $B(\psi' \to \gamma\eta') = (2.0 \pm 1.0) \times 10^{-4}$ and $R(\gamma\eta') = 0.06 \pm 0.03$. This value for R is somewhat smaller

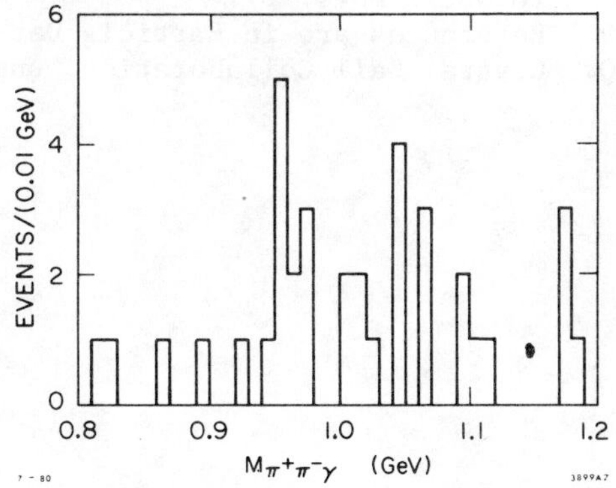

Fig. 9. $\pi^+\pi^-\gamma$ invariant mass distribution for events which satisfy (5).

than the expected value of 0.13.

No other radiative transitions from the ψ' have been observed, but 90% c.l. upper limits on two decay modes are significantly smaller than 0.13: $R(\eta\gamma) < 0.08$[10] and $R(\gamma E) < 0.03$[7]. However, since radiative transitions occur via photon emission from the charmed quark line, it is not correct to calculate partial widths in terms of the wave functions of the ψ or the ψ' at the origin. The calculation is more complicated because it involves intermediate $c\bar{c}$ states (resulting from the photon emission) which are different for each final state and also depend on whether the transition originated from the ψ or ψ'. To add confusion to the situation, we see no evidence for the decay $\psi' \to \rho^0\pi^0$ with 90% c.l. upper limit $R(\rho^0\pi^0) < 0.011$, which is an order-of-magnitude smaller than expected. Clearly, more ψ' data is required.

REFERENCES

1. S. J. Brodsky et al., Phys. Lett. 73B, 203 (1978) and references therein.
2. For details of the analysis, see G. S. Abrams et al., Phys. Rev. Lett. 44, 114 (1980) and D. L. Scharre et al., SLAC Report No. SLAC-PUB-2513, 1980 (submitted to Phys. Rev. D).
3. G. Cosme et al., Nucl. Phys. B152, 215 (1979); C. Bacci et al., Phys. Lett. 86B, 234 (1979).
4. J. L. Siegrist, SLAC Report No. SLAC-225, Ph.D. Thesis, Stanford University, 1979 (unpublished).
5. For a review of the subject of ψ radiative transitions, and references to previous experimental results, see D. L. Scharre, SLAC Report No. SLAC-PUB-2519, to be published in the Proceedings of the VI International Conference on Experimental Meson Spectroscopy, Brookhaven National Laboratory, Upton, Long Island, N.Y., April 25-26, 1980.
6. H. Fritzsch and J. D. Jackson, Phys. Lett. 66B, 365 (1977).
7. D. L. Scharre et al., SLAC Report No. SLAC-PUB-2514, 1980 (submitted to Phys. Rev. Lett.).
8. C. Dionisi et al., CERN Report No. CERN/EP 80-1, 1980 (submitted to Nucl. Phys. B).
9. References are in Particle Data Group, Phys. Lett. 75B, 1 (1978).
10. Crystal Ball Collaboration, unpublished results.

HADRONIC DECAYS OF THE η_c *

Presented by Kay Königsmann
(Representing the Crystal Ball Collaboration)[1]
Stanford Linear Accelerator Center
and
Physics Department and High Energy Physics Laboratory
Stanford University, Stanford, California 94305

ABSTRACT

Results on hadronic decays of the η_c candidate state are presented. A mass value of $M = (2978 \pm 9)$ MeV is obtained. The branching fraction for the decay into $\eta\pi^+\pi^-$ is presented and an upper limit for the decay into $\pi^0K^+K^-$ is given.

INTRODUCTION

Since the discovery of the charmonium states the question of the existence of the singlet partners of the triplet states J/ψ and ψ' have been of considerable interest. In particular its mass values, its widths, and branching ratios have been considered as crucial tests for potential and dispersion relation models.[2]

As the mass of the η_c is expected to lie between 20 MeV above and 100 MeV below the mass of the J/ψ, a detector with good photon resolution in that energy range is certainly of great help. The Crystal Ball[3] is a nonmagnetic detector consisting mainly of two elements: NaJ(Tℓ) for the detection of neutral particles and two spark chambers and one multiwire proportional chamber for tagging charged particles. Two hemispherical shells of NaJ subdivided into 672 modules cover about 94% of 4π. The three chambers surrounding the beam pipe cover 94%, 80%, and 71% of 4π respectively. In addition, two endcaps with two endcap spark chambers increase the total active volume of the Crystal Ball detector to 98% of 4π steradian.

The photon energy resolution is $\sigma_E/E = 2.6\%/\sqrt[4]{E}$ (this translates into a full width of 11 MeV for a photon of $E_\gamma = 110$ MeV). The direction of photons is calculated to $\sigma_\theta = (1 \text{ to } 2)^\circ$ depending on their energy. The resolution for charged particles is $\sigma_\theta = 1^\circ$.

One problem in selecting hadronic final states in the Crystal Ball arises when hadrons interact in the sodium iodine. Apart from the initial energy deposition at the place of the impact, additional energy traces show up. These would be recognized as being due to photons. To recognize these secondary energy traces special pattern recognition algorithms have been developed. When applied to the inclusive photon spectrum from the J/ψ (Fig. 1), we see that these algorithms are most effective at low energies; above 200 MeV hardly any fake photons are found.

---

* Supported in part by the Department of Energy, under contract DE-AC03-76SF00515 (SLAC), and by the National Science Foundation, under contract PHY78-00967 (HEPL).

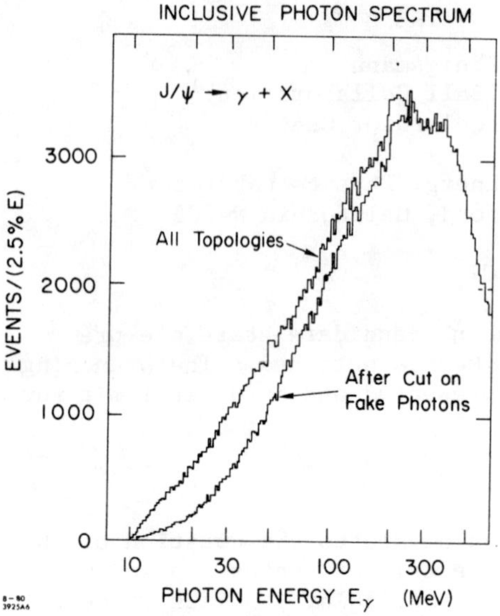

Fig. 1. Inclusive photon spectrum from J/ψ. Pattern recognition algorithms remove secondary energy depositions ('fake photons') from interacting charged particles.

RESULTS

Our experimental knowledge on the pseudoscalar state below the J/ψ is summarized in Fig. 2 and Fig. 3. An initial candidate for the η_c state, the X(2830) [4] had a mass lower than theoretically anticipated[2] (Fig. 2). This state was not confirmed in a more sensitive experiment by the Crystal Ball Collaboration.[3] In addition, this state was also not found in inclusive photon distributions[5] from the J/ψ; an upper limit was set below the expected branching fraction[2] (Fig. 3). The Crystal Ball Collaboration has observed[6] a state at a mass M = (2981 ± 15) MeV with

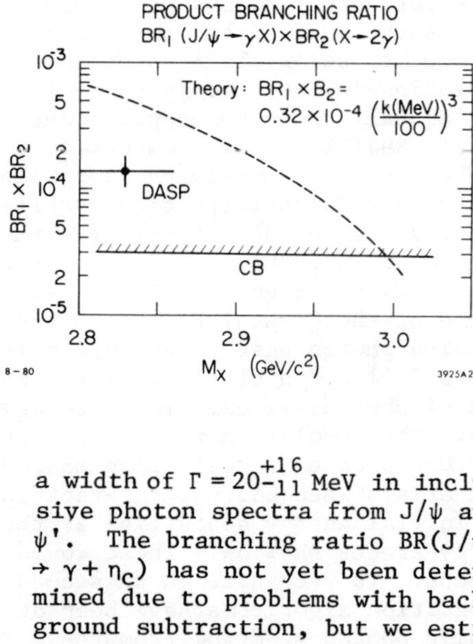

Fig. 2. Product branching ratio $J/\psi \to \gamma + X$, $X \to \gamma\gamma$. Measurements from DASP[4] and the Crystal Ball[3] are indicated. The theoretical prediction is taken from E. Eichten et al.[2]

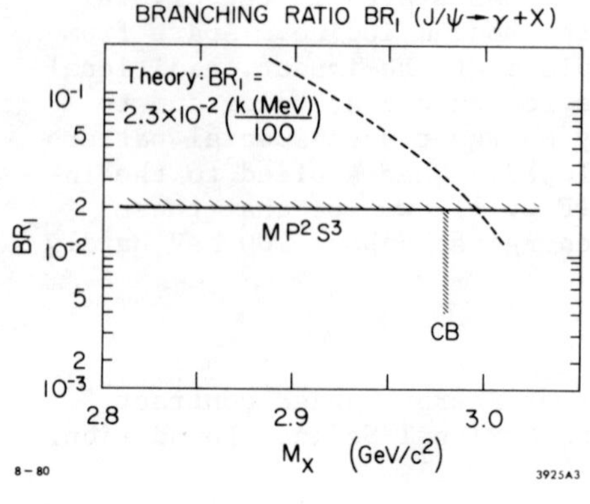

a width of $\Gamma = 20^{+16}_{-11}$ MeV in inclusive photon spectra from J/ψ and ψ'. The branching ratio BR($J/\psi \to \gamma + \eta_c$) has not yet been determined due to problems with background subtraction, but we estimate the ratio to be (0.4 to 4) * 10^{-2}.

Fig. 3. Inclusive branching ratio $J/\psi \to \gamma + X$. The data are from Maryland, Pavia, Princeton, San Diego, SLAC, Stanford[5] and from the Crystal Ball.[6]

We have also looked for hadronic decays of the η_c candidate state by performing 3-C kinematic fits to the final states:

$$J/\psi \to \gamma + \eta + \pi^+ + \pi^-, \quad \eta \to \gamma\gamma$$

$$J/\psi \to \gamma + \pi^0 + K^+ + K^-, \quad \pi^o \to \gamma\gamma \ .$$

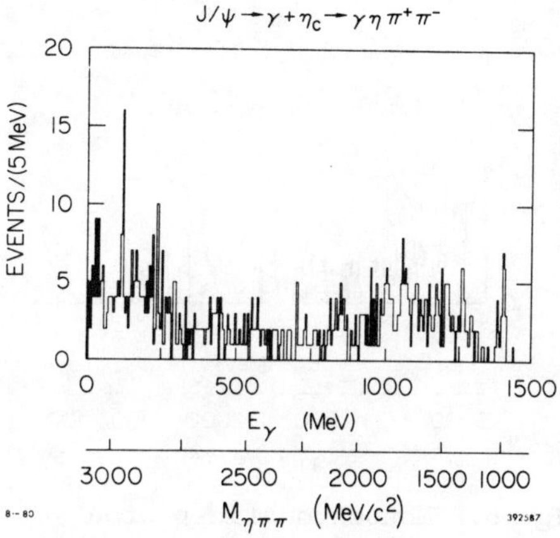

Fig. 4. Monochromatic photon energy of the decay $J/\psi \to \gamma + \eta + \pi^+ + \pi^-$.

The energy spectrum for the monochromatic photon is shown in Fig. 4 for events which pass the fit with a probability of χ^2 greater than 0.01. A clear signal is seen at a photon energy of 120 MeV. The region around the peak is enlarged in Fig. 5. A maximum likelihood fit gives an invariant $\eta\pi\pi$ mass of $M = (2974 \pm 2 \pm 9)$ MeV, where the first error is statistical and the second is an estimate of the systematical uncertainties. This mass value agrees well within errors with the value obtained from the inclusive spectra. Combining both mass values and the measurements of the widths[6] yields:

$$M = (2978 \pm 9) \text{ MeV}$$

$$\Gamma < 20 \text{ MeV (90\% C.L.)} \quad .$$

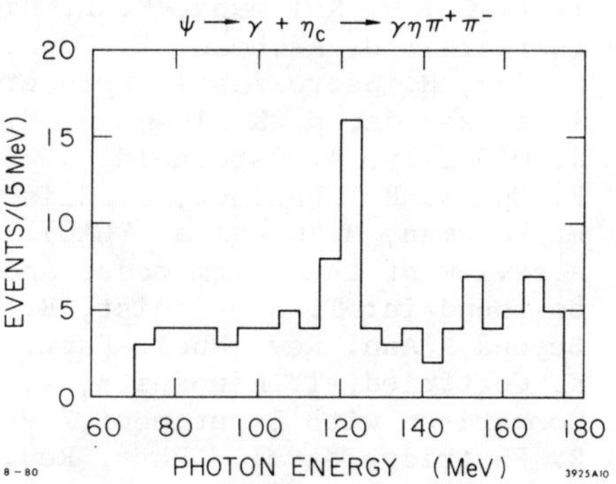

Fig. 5. Blowup of the region around $E_\gamma = 120$ MeV of Fig. 4.

A 16% detection efficiency was obtained with a Monte Carlo calculation where the η_c was assumed to have $J^P = 0^-$ and to decay with a phase space distribution. Our signal of 18 ± 6 events yields a product branching ratio:

$$BR(J/\psi \to \gamma + \eta_c) * BR(\eta_c \to \eta\pi^+\pi^-)$$

$$= (3.1 \pm 1.1 \pm 1.5) * 10^{-4} \quad ,$$

the errors being statistical and systematic respectively.

In the decay channel $\eta_c \to \pi^0 + K^+ + K^-$ we observe no signal (Fig. 6). The photon spectrum is flat in the energy range 70 MeV < 170 MeV. We deduce an upper limit on the product branching ratio of

$$BR(J/\psi \to \gamma + \eta_c) * BR(\eta_c \to \pi^0 K^+ K^-) < 1.5 * 10^{-4} \text{ (90\% C.L.)} \quad .$$

The Mark II collaboration at SPEAR has observed an enhancement[7] in $\psi' \to \pi^\pm + K^\mp K^o_S$ at an invariant mass of $M = (2980 + 8)$ MeV.

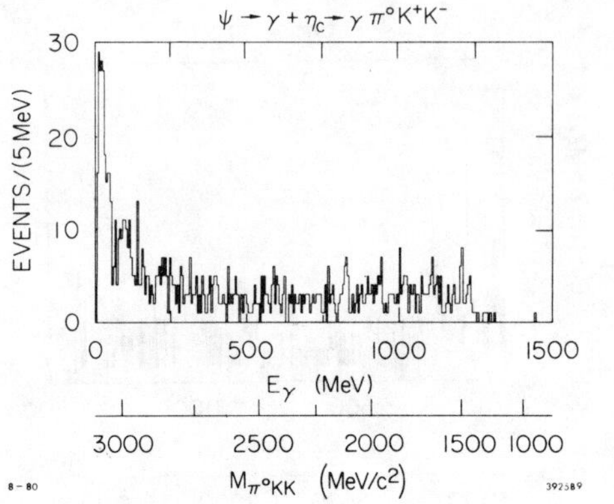

Fig. 6. Monochromatic photon energy of the decay $J/\psi \rightarrow \gamma + \pi^0 + K^+ + K^-$.

A comparison of these two results has to wait for an accurate determination of the branching ratio $J/\psi \rightarrow \gamma + \eta_c$

CONCLUSIONS

In summary, an η_c candidate state is observed at a mass of $M = (2978 \pm 9)$ MeV in both inclusive and exclusive radiative transitions from the J/ψ. The combined limit on the width is $\Gamma < 20$ MeV (90% C.L.) which is in line with theoretical expectations.[2] A final determination of this state as the pseudoscalar partner of the J/ψ has to wait for a determination of its spin-parity J^P.

REFERENCES

1. R. Partridge, C. Peck, F. Porter (Caltech); D. Antreasyan, Y. F. Gu, W. Kollmann, M. L. Richardson, K. Strauch, K. Wacker (Harvard); D. Aschman, T. Burnett, M. Cavalli-Sforza, D. Coyne, M. Joy, H. Sadrozinski (Princeton); R. Hofstadter, R. Horisberger, I. Kirkbride, K. Königsmann, H. Kolanoski, A. Liberman, J. O'Reilly, A. Osterheld, J. Tompkins (Stanford); E. Bloom, F. Bulos, R. Chestnut, J. Gaiser, G. Godfrey, C. Kiesling, W. Lockman, M. Oreglia (SLAC).
2. A review of the charm model and a comparison with experiment can be found in: T. Appelquist, R. M. Barnett, K. D. Lane, "Charm and Beyond," Ann. Rev. Nucl. Part. Sci. 28 (1978); and E. Eichten, K. Gottfried, T. Kinoshita, K. D. Lane and T. M. Yan, "Charmonium: Comparison with Experiment," Phys. Rev. D21, 203 (1980).
3. R. Partridge et al., Phys. Rev. Lett. 44, 712 (1980) and references therein.
4. W. Braunschweig et al., Phys. Lett. 67B, 243 (1977) and W. D. Apel et al., Phys. Lett. 72B, 500 (1978).
5. C. Biddick et al., Phys. Rev. Lett. 38, 1324 (1977).
6. R. Partridge et al., SLAC-PUB-2578, July 1980, submitted to Phys. Rev. Lett.
7. T. M. Himel et al., SLAC-PUB-2562, July 1980, submitted to Phys. Rev. Lett.

OBSERVATION OF $\eta_c(2980)$ IN $\psi'(3684)$ RADIATIVE DECAY

SLAC-LBL Mark II Collaboration*

In a study of $\psi'(3684)$ radiative decays with the Mark II detector at SPEAR, we observe the decay sequence $\psi' \rightarrow \gamma\eta_c(2980)$ with the $\eta_c(2980)$ decaying into several completely reconstructed hadronic modes. These results confirm the observations of the $\eta_c(2980)$ reported by the Crystal Ball Collaboration.[1]

The hadronic decay modes studied are listed in the first column of Table I, and are the simplest ones which can arise from the decay of a pseudoscalar particle with charged hadrons and K_S. The events were selected on the basis of appropriate charged particle topology and time-of-flight identification, small missing mass recoiling against charged particles and K_S, and the detection of a photon in the direction expected from measurements of the charged track momenta.

Table I. Branching ratios for the decay $\psi' \rightarrow \gamma\eta_c \rightarrow \gamma$ hadrons.

Hadronic mode	$B(\psi' \rightarrow \gamma\eta_c)B(\eta_c \rightarrow \text{hadrons})$
$K^{\pm}K_S\pi^{\mp}$	$(1.5^{+0.8}_{-0.6}) \times 10^{-4}$
$2\pi^+2\pi^-$	$(5.7^{+3.9}_{-2.4}) \times 10^{-5}$
$\pi^+\pi^-K^+K^-$	$(4.0^{+6.0}_{-2.5}) \times 10^{-5}$
$p\bar{p}$	$(8^{+8}_{-4}) \times 10^{-6}$
$\pi^+\pi^-p\bar{p}$	$< 5 \times 10^{-5}$ (90% C.L.)

Figure 1 shows the observed hadronic mass spectrum between 2.35 and 3.35 GeV/c^2 summed for all the final states studied. This spectrum incorporates a subtraction of the expected number of background events with the same hadronic final states accompanied by a π^0 rather than a photon. There is a clear peak in the bins between 2950 and 3000 MeV/c^2, and the probability that any of the 40 adjacent bin pairs provide a fluctuation as large or larger is 3×10^{-5}. Further analysis of the data shown in Fig. 1 yields the following parameters:

$$M(\eta_c) = 2980 \pm 8 \text{ MeV}/c^2 ,$$

$$\Gamma(\eta_c) < 40 \text{ MeV}/c^2 \text{ (90\% C.L.)} .$$

Estimates of the branching ratio products for η_c decay are given in Table I.

*The collaborators are listed in the talk by A. J. Lankford.

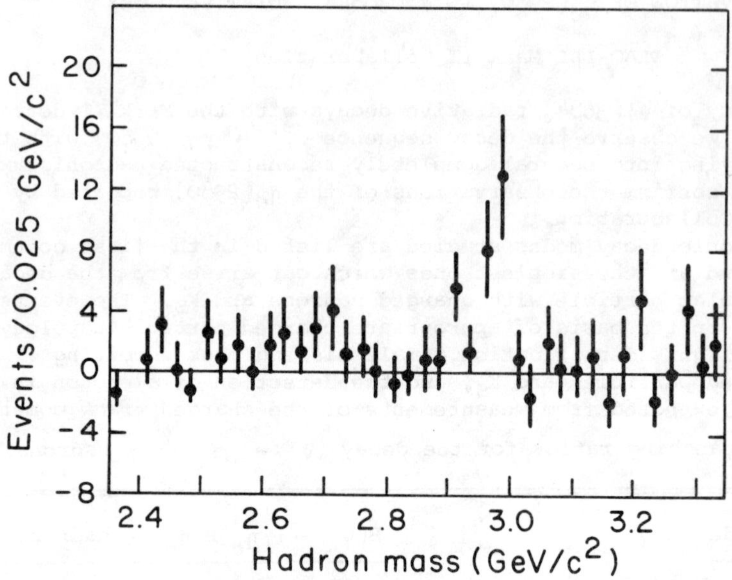

Fig. 1. Hadronic mass spectrum after π^0 background subtraction.

REFERENCE

1. R. Partridge et al., "Observation of an η_c Candidate State with Mass 2978 ± 9 MeV," SLAC-PUB-2578.

PRODUCTION AND DECAYS OF D* MESONS¶

Hartmut F. W. Sadrozinski†
Stanford Linear Accelerator Center
Stanford University, Stanford, California 94305
and
University of California, Santa Cruz, California 95064
(Representing the Crystal Ball Collaboration‡)

ABSTRACT

We report measurements of inclusive π^0 and γ production in e^+e^- annihilation at c.m. energy $E_{c.m.} = 4.028$ GeV with the Crystal Ball detector at SPEAR. The decays $D^* \to \pi^0 D$, $D^* \to \gamma D$ are observed and allow determination of the D^{*0}-D^0 mass difference, production ratio and γ/π^0 decay ratio's. In addition, the resonance parameters of the $\psi''(3770)$ resonance are given.

The total hadronic cross section in e^+e^- annihilation is dominated above open charm threshold by the $\psi''(3770)$ resonance, which decays into $D\bar{D}$ pairs, and resonances above 4 GeV of which the one centered at $E_{c.m.} = 4.028$ GeV decays into charmed meson pairs containing predominantly D^*. Comparison of inclusive π^0 and γ spectra taken at these two resonances yield information about the production of D^*'s and their decay into D's.

The apparatus and the selection procedure for hadronic events are described elsewhere.[1] Figure 1a shows the relative hadronic cross section R as function of c.m. energy: it is characterized by the radiative tails of the ψ and ψ' resonances above a non-charmed background and shows the ψ'' at $E_{c.m.}$ = 3770 MeV:

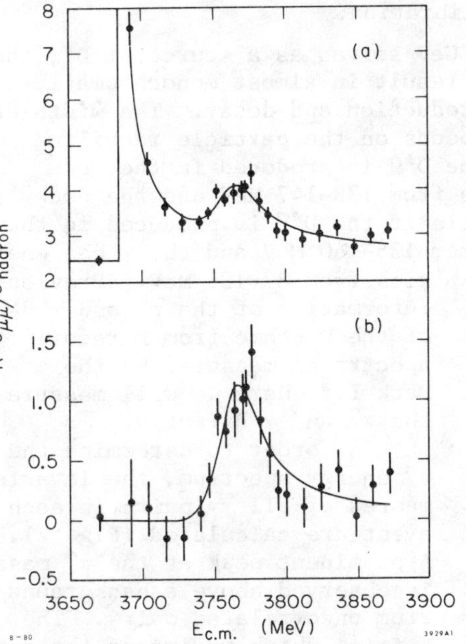

Fig. 1. (a) Relative hadronic cross section $R \equiv \sigma_{HAD}/\sigma_{\mu\mu}$ near the ψ''. The solid line is the fit to (1). (b) R of the ψ'' contribution.

¶ Work supported in part by the Department of Energy under contract DE-AC03-76SF00515 and by the NSF Grant PHY79-16461.

$$R = R_{\text{NON-CHARM}} + \psi_{\text{RAD TAIL}} + \psi'_{\text{RAD TAIL}} + \psi'' \qquad (1)$$

Assuming a non-relativistic Breit-Wigner resonance form of the ψ'' with an energy dependent width due to the closeness of the $D\bar{D}$ threshold,[2] the mass M, hadronic width Γ and leptonic width Γ_{ee} of the ψ'' can be determined. Figure 1b shows the ψ'' resonance after subtracting all other contributions. In Table I, the results of the fit are given and compared with previous measurements, with which they agree well.

TABLE I
ψ'' Resonance Parameters

Experiment	Mass* (MeV)	Γ_{ee} (eV)	Γ (MeV)
Crystal Ball	3768 ± 2	308 ± 56	36 ± 8
LGW[2]	3772 ± 3	345 ± 85	28 ± 5
Delco[3]	3770 ± 2	180 ± 60	24 ± 5
Mark II[4]	3764 ± 2	276 ± 50	24 ± 5

* The mass determination of all experiments have a common additional uncertainty of 4 MeV due to the absolute SPEAR energy calibration.

The resonance at $E_{c.m.} = 4.028$ GeV serves as a source of D^*, the decay of which into $D\pi^0$ or $D\gamma$ will result in almost monochromatic π^0 or γ's due to the low Q value of production and decay. The width of the π^0 or γ energy distribution depends on the particle recoiling against the D^*: for example, if the D^{*0} is produced in the mode $D^{*0}\bar{D}^{*0}$ the decay π^0's have energies from 138-147 MeV and the decay γ's have energies from 122-157 MeV, while if the D^{*0} is produced in the mode $D^{*0}\bar{D}^0$, the π^0 has energies from 135-160 MeV and the γ has energies from 87-185 MeV. Previous information of the π^0 and γ decay of the D^* came from D recoil spectra as measured by the Mark I.[5] Here we will measure the π^0 and γ directly.

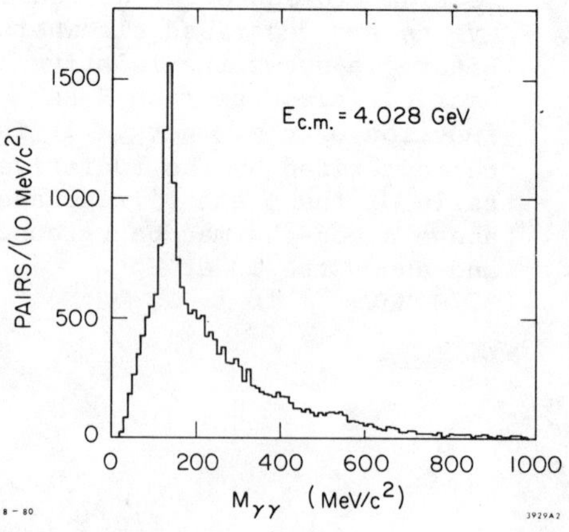

Fig. 2. Mass of all γ-γ pairs of 18k hadronic events at $E_{c.m.}$ = 4.028 GeV.

In order to determine the π^0 energy spectrum, the invariant masses of all $\gamma\gamma$ pairs in each event are calculated (Fig. 2). A prominent peak at the π^0 mass is observed above a background from uncorrelated pairs. The pairing which maximizes the number of π^0 is chosen and by constraining the masses to the π^0 mass, the energies of the photons $E_{\gamma 1}$, $E_{\gamma 2}$ are fitted and summed to form the π^0 energy $E_{\pi 0} = E_{\gamma 1} + E_{\gamma 2}$. Fig. 3a shows the low energy part

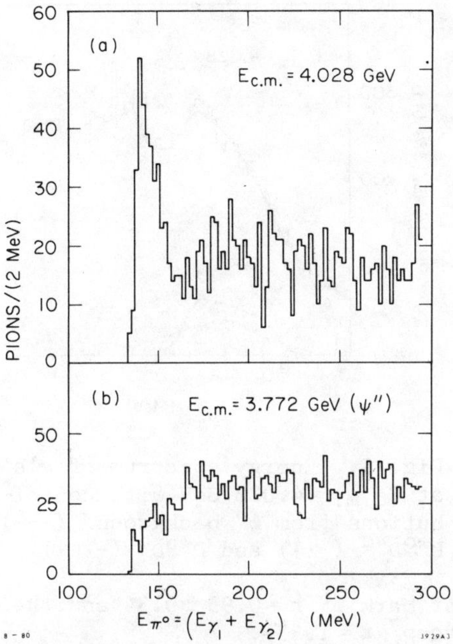

Fig. 3. (a) Energy spectrum of π^0's at $E_{c.m.} = 4.028$ GeV. (b) Energy spectrum of π^0's at $E_{c.m.} = 3.772$ GeV.

of the π^0 spectrum for 18k hadronic events at $E_{c.m.} = 4.028$ GeV and Fig. 3b the corresponding spectrum from 36k hadronic events at the ψ'', which is below the D^*D threshold. A distinct peak from the decay $D^* \to \pi^0 D$ is seen for $E_{c.m.} = 4.028$ GeV (Fig. 3a) above the background from D decays and non charmed background as determined at the ψ'' (Fig. 3b).

The π^0 spectrum is fitted with the smooth ψ'' background and the contribution from a Monte Carlo simulation of the following sources of π^0: (i) $D^{*0}\bar{D}^{*0} \to D^0\pi^0$; (ii) $D^{*0}\bar{D}^0 + \bar{D}^{*0}D^0 \to D^0\pi^0$; (iii) $D^{*+}D^{*-} \to \pi^0 D^+$; and (iv) $D^{*+}D^- + D^{*-}D^+ \to \pi^0 D^+$ as a function of the charmed meson masses $m(D^{*0})$, $m(D^0)$.[6] The fit gives the product of production cross section σ_j and decay branching ratios, BR, into π^0 for the different D^* modes (i)-(iv). The spectrum is sensitive to the $D^{*+}D^{*-}/D^{*0}\bar{D}^{*0}$ ratio, the D^{*0}-D^0 mass difference and the $D^*\bar{D}^*/D^*\bar{D}$ production ratio. The relative contribution of $D^{*0}\bar{D}^0$ and $D^{*+}D^-$ cannot be determined. A typical fit to the π^0 energy spectrum is shown in Fig. 4.

We determine the contribution from $D^{*+}D^{*-}$ production to be small and find

		Mark I[5]
$m(D^{*0}) - m(D^0)$	$= 142.2 \pm 0.5 \pm 1.5$	(142.7 ± 1.7)
$m(D^{*0})$	$= 2006 \pm 2 \pm 1.5$	(2006 ± 1.5)
$m(D^0)$	$= 1864 \pm 2 \pm 1.5$	(1863.3 ± 0.9)

where the first error is statistical and the second error an estimate of the systematical uncertainties. The previous measurements from Mark I[5] are in parenthesis. The production rate times branching ratios are:

$$\frac{\sigma(D^{*+}D^{*-}) \cdot BR(D^{*+} \to \pi^0 D^+)}{\sigma(D^{*0}\bar{D}^{*0}) \cdot BR(D^{*0} \to \pi^0 D^0)} < 0.11 \quad (90\%\ \text{C.L.})$$

$$\frac{\sigma(D^{*0}\bar{D}^0 + \bar{D}^{*0}D^0) BR(D^{*0} \to \pi^0 D^0) + (D^{*+}D^- + D^{*-}D^+) BR(D^{*+} \to \pi^0 D^+)}{\sigma(D^{*0}\bar{D}^{*0}) BR(D^{*0} \to \pi^0 D^0)} = 2.47 \pm 0.61$$

If we assume reasonable branching ratios[5] $BR(D^{*+} \to \pi^0 D^+) = 0.3$ and $BR(D^{*0} \to \pi^0 D^0) = 0.6$ and also $\sigma(D^{*0}\bar{D}^0) = \sigma(D^{*+}D^-)$ we get

$$r \equiv (\sigma(D^{*0}\bar{D}^0) + \sigma(\bar{D}^{*0}D^0))/\sigma(D^{*0}\bar{D}^{*0}) \approx 1.6$$

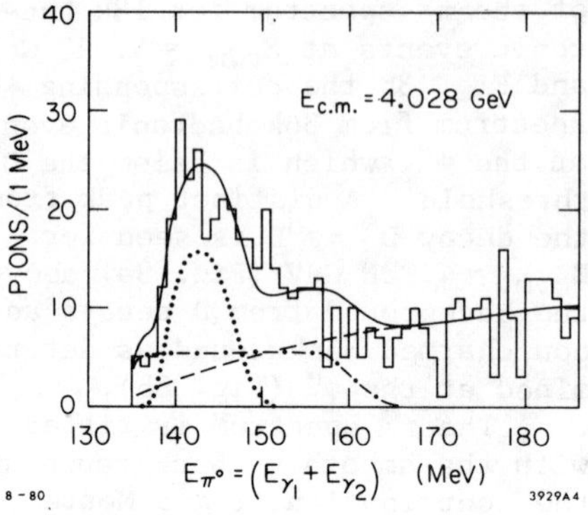

Fig. 4. Energy spectrum of π^0 at $E_{c.m.} = 4.028$ GeV with fitted contributions from ψ'' background (---), $D^{*0}\bar{D}^{*0}$ ($\cdots$) and $D^{*}D$ (-·-·). ($M(D^{*0}) = 2006$, $M(D^0) = 1863$.)

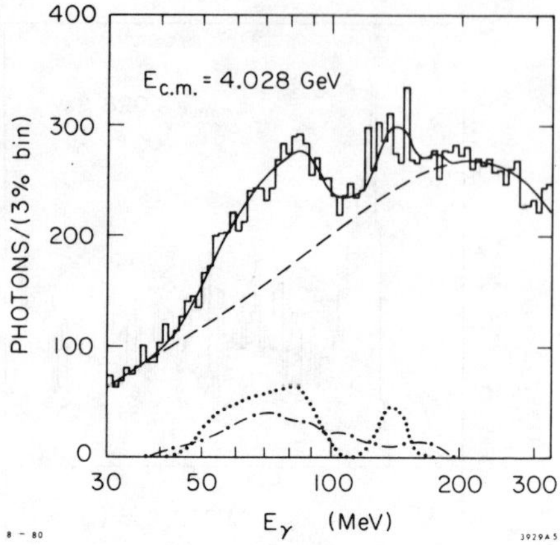

Fig. 5. Energy spectrum of γ's at $E_{c.m.} = 4.028$ GeV with contributions from ψ'' background (---), $D^{*0}\bar{D}^{*0}$ ($\cdots$) and $D^{*0}\bar{D}^{0}$ (-·-·).

(in agreement with the measurement of Mark I[5] $r = 0.95 \pm 0.34$ and theoretical prediction of the Cornell group[7] $r = 1.35$).

Additional information about D^* decays comes from the photon spectrum at $E_{c.m.} = 4.028$ GeV. Again the spectrum is fitted to a background from D decays and non-charmed background as determined at the ψ'' resonance and contributions from a Monte Carlo simulation of $D^*\bar{D}^*$ and $D^*\bar{D} + \bar{D}^*D$ production with subsequent decays into π^0 and γ's. Figure 5 gives the inclusive spectrum which shows (above the ψ'' background) contributions from $D^* \to D\pi^0 \to \gamma\gamma$ (centered at $E_\gamma = 70$ MeV) and $D^* \to D\gamma$ (centered at 140 MeV). The relative population of the two peaks allows to compute the branching ratio into γ's directly:

$$\frac{\sum \sigma_i(D^*)BR_i(D^* \to \gamma D)}{\sum \sigma_i(D^*)\left(BR_i(D^* \to \gamma D) + BR_i(D^* \to \pi^0 D)\right)} = 0.31 \pm 0.06$$

where the sum extends over production of $D^{*0}\bar{D}^{*0}$, $D^{*+}D^{*-}$, $D^{*0}\bar{D}^{0} + \bar{D}^{*0}D^{0}$, $D^{*+}D^{-} + D^{*-}D^{+}$, respectively. The error reflects the uncertainty in the background subtraction. In order to deduce the $BR(D^{*0} \to \gamma D^0)$, we make the following assumptions, as suggested by the fit to the π^0 spectrum, the Mark I measurements[5] and theoretical calculations[8]:

$$\sigma(D^{*+}D^{*-})BR(D^{*+} \to \pi^0 D^+) \ll \sigma(D^{*0}\bar{D}^{*0})BR(D^{*0} \to \pi^0 D^0)$$

$$\sigma(D^{*+}D^{*-})BR(D^{*+} \to \gamma D^+) \ll \sigma(D^{*0}\bar{D}^{*0})BR(D^{*0} \to \gamma D^0)$$

$$\sigma(D^{*+}D^{-})\,BR(D^{*+} \to \gamma D^+) \ll BR(D^{*0}\bar{D}^{0})BR(D^{*0} \to \gamma D^0)$$

$$\sigma(D^{*+}D^{-}) = \sigma(D^{*0}\bar{D}^{0})$$

$$BR(D^{*+} \to \pi^0 D^+) = 0.3$$

We get $BR(D^{*0} \to \gamma D^0) \approx 0.37$ in agreement with the Mark I measurement[5] $BR(D^{*0} \to \gamma D^0) = 0.45 \pm 0.15$.

ACKNOWLEDGEMENT

I would like to thank Donald Coyne, Boris Kayser, Robert Cahn, and Richard Partridge for discussions and help in the preparation of this talk. This work was partially supported by the Department of Energy under contract DE-AC03-76SF00515 and by the NSF Grant PHY79-16461.

REFERENCES

† On leave from Princeton University.
‡ Crystal Ball Collaboration: CALTECH: R. Partridge, C. Peck, F. Porter; HARVARD UNIVERSITY: D. Antreasyan, Y. F. Gu, W. Kollmann, M. Richardson, K. Strauch, K. Wacker; PRINCETON UNIVERSITY: D. Aschman, T. Burnett (visitor), M. Cavalli-Sforza, D. Coyne, M. Joy, H. Sadrozinski; SLAC: E. D. Bloom, F. Bulos, R. Chestnut, J. Gaiser, G. Godfrey, C. Kiesling, W. Lockman, M. Oreglia; STANFORD UNIVERSITY: R. Hofstadter, R. Horisberger, I. Kirkbride, H. Kolanoski, K. Königsmann, A. Liberman, J. O'Reilly, J. Tompkins.
1. J. C. Tompkins, "Recent Results from the Crystal Ball," SLAC Report 224, 578 (1980); E. D. Bloom, Proceedings of the 1979 International Symposium on Lepton and Photon Interactions at High Energies, August 23-29, 1979, Fermilab.
2. For a description of a similar analysis of the LGW data see P. Rapidis, SLAC Report 220, 53 (1979); P. Rapidis et al., Phys. Rev. Lett. 39, 5261 (1977).
3. W. Bacino et al., Phys. Rev. Lett. 40, 671 (1978).
4. G. Abrams et al., Phys. Rev. D21, 2716 (1980).
5. G. Goldhaber et al., Phys. Lett. 69B, 503 (1977); G. Feldman, SLAC-PUB-2068 (1977).
6. $m(D^{*+}) = 2008.6 \pm 1.0$ and $m(D^{+}) = 1868.3 \pm .9$ are taken from Ref. 5.
7. E. Eichten et al., Phys. Rev. D21, 203 (1980).
8. S. Ono, Phys. Rev. Lett. 37, 655 (1976).

E^+E^- PHYSICS V

B. GITTELMAN, CORNELL, ORGANIZER

PARALLEL SESSION ON UPSILON SPECTROSCOPY

B. Gittelman
Cornell Univeristy, Ithaca, N.Y. 14853

The parallel session on Υ spectroscopy consisted of 5 papers. The first three summarized the experimental results at the DORIS e^+e^- storage ring (DESY Laboratory) on the properties of Υ(9.4) and Υ'(10.0). The last two talks presented preliminary results of the groups at CESR (Wilson Laboratory).

C. Grupen, Experimental Evidence for Υ Decay into Three Gluons (PLUTO collaboration).
This was a detailed presentation comparing shape distributions (such as thrust, planarity, etc.) with Monte Carlo calculations for a 2 quark jet model, a phase space model, and a three gluon model. The data are best described by the latter. The angular distribution of the thrust axis direction relative to the electron beam favors a gluon spin equal 1.

W. Schmidt-Parzefall, Resonance Parameters of Υ and Υ' Measured by LENA, DASP2, and PLUTO plus Inclusive Spectra in the Υ Region Measured by DASP2
The speaker presented individual results and best values for the mass, electronic width, and total width of Υ(9.4) for the three DORIS experiments. The new data on $\Upsilon\to\mu^+\mu^-$ and its analysis exhibited substantial improvement over the situation of last year.

$$M(\Upsilon) = 9462\pm1\pm10 \text{ MeV}$$

$$\Gamma_{ee}(\Upsilon) = 1.29\pm0.09\pm0.13 \text{ keV}$$

$$\Gamma_{tot}(\Upsilon) = 43^{+20}_{-11} \text{ keV}$$

The mass and leptonic width of the Υ' were given

$$M(\Upsilon') - M(\Upsilon) = 553\pm1\pm10 \text{ MeV}$$

$$\Gamma_{ee}(\Upsilon')/\Gamma_{ee}(\Upsilon) = 0.45\pm0.05$$

Inclusive cross sections of $\pi^\pm$, $K^\pm$, and $\bar{P}$ were presented for Υ, Υ', and the nearby continuum. The experiment covers the momentum range about 200 MeV/c. The data show the fraction of baryons in the final state increases at the resonances.

F. Messing, New Results on Υ'(10.0) Hadronic Decay (LENA and DASP2 Collaborations.)
The sphericity distribution of Υ' decays were shown to be the same as those for Υ decays. From this information an upper limit of 21% for the branching fraction of the cascade decay, $\Upsilon'\to\gamma+(\text{Pstate})\to\gamma+(2\text{ gluons})$, was derived. The mean charged multiplicity of Υ' decays is found to be larger than that of Υ decays. By assuming this comes about because of the decay $\Upsilon'\to\pi\pi\Upsilon$ a value for the branching fraction of 27±9% was found for this channel.

In addition, the experimenters have 7 fitted events for $\Upsilon'\to\pi^+\pi^-\Upsilon\to\pi^+\pi^-$+leptons, thus giving a direct confirmation of the theoretical connection between Υ and Υ'.

J. Yoh, Recent Results from CUSB at CESR
A short description of the CUSB detector and their operating experience was presented. An analysis of ~400 nb^{-1} of data recorded on the $\Upsilon'(2S)$ has uncovered 3 events which fit the hypothesis $\Upsilon'\to\pi^+\pi^-\Upsilon\to\pi^+\pi^-e^+e^-$. They estimate the $\pi^+\pi^-\Upsilon$ branching fraction to be between 10 and 20%. The mass and width of the $\Upsilon'''(4S)$ were given

$$M(\Upsilon''') - M(\Upsilon) = 1114 \pm 2 \text{ MeV}$$

$$\Gamma_{ee}(\Upsilon''')/\Gamma_{ee}(\Upsilon) = 0.25 \pm 0.07$$

$$\Gamma(\Upsilon''') = 12.6 \pm 6.0 \text{ MeV}$$

The $\Upsilon'''(4S)$ was found to be a rich source of electrons in comparison with $\Upsilon'(2S)$or the continuum. Assuming $\Upsilon'''(4S)\to B\bar{B}$ the data is consistent with a branching fraction of 10 to 20% for $B\to De\nu$.

E. Thorndike, First Results on B Mesons from CLEO
Dr. Thorndike discussed what data is needed to establish that the $\Upsilon'''(4S)$ decays into mesons carrying a new quark flavor, stressing the importance of demonstrating that the mesons (hereafter called B) decay weakly. He then presented evidence for the semileptonic decays, $B\to e\nu X$ and $B\to\mu\nu X$. Preliminary values of the branching ratios were presented for the assumption that the b quark decays to the charm quark. This assumption was justified by data showing the Υ''' decays to final states containing 2.4±0.5±0.5 charged kaons. Although the above results are preliminary, they are consistent with general predictions of the K.M. model.

EXPERIMENTAL EVIDENCE FOR THE Υ-DECAY INTO 3 GLUONS
(Results From The PLUTO-Detector)

C.Grupen

Siegen University, 5900 Siegen 21, Federal Republic of Germany

ABSTRACT

A jet analysis of the Υ-direct decays is presented. The analysis involves both neutral and charged particles. The Υ-decay data are best described by a $\Upsilon \to ggg$ matrix element. A dominantly phase space like or 2 jet decay can be ruled out. A search for flatness of events does not show a significant difference for 'off'resonance or direct decay data. In this sense the Υ-direct decay events are not flat.

INTRODUCTION

The PLUTO collaboration* has measured the cross-section for $e^+e^- \to$ hadrons in the Υ-mass region[1]. Results on the Υ-decay have already been reported previously[2,3,4]. In this paper further experimental evidence for the Υ-decay into 3 gluons is presented. The additional evidence comes primarily from the careful and detailed inclusion of neutral particles into the analysis.

RESULTS

The data recorded on the Υ-resonance have to be corrected for the 'off'-resonance contribution and for vacuum polarisation. The 'off'-resonance contribution has been directly measured. The vacuum polarisation can be inferred from our measurement of the μ-pairs 'on' and 'off'-resonance[3,5]. The remaining signal contains ~1200 hadronic events.

If the Υ-decay proceeds dominantly via the 3 gluon decay one would naively expect the events to be planar. Indeed, Monte Carlo calculations have shown that there is a correlation between the 3 gluon input plane and the reconstructed plane. This correlation depends on the available center of mass energy. As expected it is pronounced at high energies (~30GeV), but it is also observed at the Υ-mass.

However, measures of flatness, such as $1 - Q_1/Q_2$, do not show within the errors a difference between phase space models, 2-jet models and the 3 gluon model. Here Q_1 and Q_2 are quantities derived from the sphericity tensor[3]. The same feature is also borne out by the 'on' and 'off'-resonance data. So, in this sense of flatness, the Υ-direct decay events are not flat.

To discriminate between various models one has to use more sophisticated topological quantities. We have decided to use thrust[6], triplicity[7], jet energies and the angles between jets[4].

We compare our direct decay data with 'off'-resonance data and to models, such as

* a complete list of the members of the PLUTO-collaboration is given in references 1,2,3,4

a) 3 gluon model (as predicted by quantum-chromodynamics)[8]
b) phase space model 1 (pseudoscalar mesons only, pion to kaon ratio 3:1)
c) phase space model 2 (pseudoscalar mesons and vector mesons in equal proportions)
d) 2-jet model (Field and Feynman)[9]

Firstly, we check the behaviour of the neutral particles. We reconstruct the thrust axis for charged and neutral particles separately. We find a strong correlation between these two reconstructed thrust axes, which is also given by a 3 gluon Monte Carlo calculation. A phase space model gives a weaker correlation. Secondly, we look at the thrust and triplicity distributions for neutrals only. These distributions are nicely reproduced by a corresponding 3 gluon Monte Carlo for neutrals only. The 2-jet model and phase space models do not describe the data well. Thirdly, we directly compare the thrust distributions for charged and neutral particles separately. We find agreement for the two distributions.

These consistency checks encourage us to look at the combined data using both charged and neutral particles. As an example the thrust distribution is shown in figure 1. The combined data clearly prefer the 3 gluon decay picture. Table 1 summarizes the results for average quantities used in this analysis. Here x_1, x_2, x_3 are the fractional energies of the 3-jets as determined by triplicity ($x_i = 2k_i/M_T$; k_i-jet energy) and Θ_i are the angles between the jet directions (Θ_1 is the angle between the jets with energies x_2 and x_3, etc.)

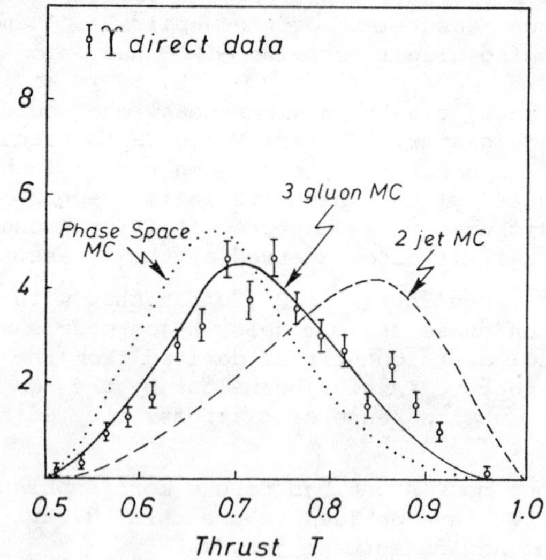

Fig.1 Thrust distribution of Υ-direct decay events in comparison to various models (The two phase space models give very similar results, so that only one common curve is shown)

The data are best described by the 3 gluon Monte Carlo. Phase space models and the two jet model are very well separated from the data. The 'off'-resonance data are described by the 2-jet picture.
We have also investigated whether a mixed model (50% phase space plus 50% two-jet) describes our data. We find that the 3 gluon decay picture of the Υ-decay is slightly favoured over such a mixed model.
We have also looked into the distribution of the thrust axis. The expected $1+0.39\cos^2\Theta$ distribution for vector gluons is consistent with the data. Scalar gluons do not describe the data.

Table 1: Average quantities for T-direct decay data, off resonance data and various theoretical models. (The errors for data are statistical. For Monte Carlo they contain the systematic uncertainties in the models)

	T-direct data	3 gluon Monte Carlo	phase space 1 Monte Carlo	phase space 2 Monte Carlo	2-jet Monte Carlo	'off'resonance data
<T>	0.732 ±0.004	0.72 ±0.01	0.69 ±0.01	0.69 ±0.01	0.80 ±0.01	0.808 +0.004
$<T_3>$	0.870 ±0.002	0.86 ±0.01	0.84 ±0.01	0.84 ±0.01	0.90 ±0.01	0.910 ±0.002
$<x_1>$	0.862 ±0.003	0.86 ±0.01	0.83 ±0.01	0.83 ±0.01	0.91 ±0.01	0.909 ±0.003
$<x_2>$	0.715 ±0.004	0.72 ±0.01	0.70 ±0.01	0.71 ±0.01	0.75 ±0.01	0.733 ±0.005
$<x_3>$	0.423 ±0.005	0.42 ±0.01	0.47 ±0.01	0.46 ±0.01	0.34 ±0.01	0.358 ±0.007
$<\Theta_1>$	82.7° ± 0.9°	84° ± 1°	91° ± 1°	90° ± 1°	70° ± 1°	68.8° ± 1.1°
$<\Theta_2>$	126.4° ± 0.6°	126° ± 1°	124° ± 1°	124° ± 1°	131° ± 1°	132.8° ± 0.8°
$<\Theta_3>$	151.0° ± 0.5°	151° ± 1°	146° ± 1°	146° ± 1°	159° ± 1°	158.5° ± 0.6°

CONCLUSIONS

The T-direct decay data are very well described by the T-decaying dominantly into 3 gluons. Dominantly phase space like or 2 jet models can be ruled out.

REFERENCES

1. PLUTO Collaboration, Ch.Berger et al. Phys.Lett.76B,243,(1978)
2. PLUTO Collaboration, Ch.Berger et al.Phys.Lett.78B,176,(1978)
3. PLUTO Collaboration, Ch.Berger et al.Phys.Lett.82B,449,(1978)
4. PLUTO Collaboration, Ch.Berger et al.(Talk by S.Brandt) Proceedings of the Int.Conf.on High Energy Physics, Geneva 1979, p.338 and DESY 79/43,(1979)
5. PLUTO Collaboration, Ch.Berger et al.Z.Physik C, Particles and Fields 1,343,(1979)
6. S.Brandt, Ch.Peyrou, R.Sosnowski, A.Wroblewski, Phys.Lett.12,57,(1964); E.Fahri, Phys.Rev.Lett.39,1587,(1977)
7. S.Brandt, H.D.Dahmen Z.Physik C, Particles and Fields 1,61,(1979)
8. K.Koller, T.F.Walsh, Phys.Lett.72B,227,(1977); 73B,504,(1978); Nucl.Phys.B140,449,(1978); K.Koller, H.Krasemann, T.F.Walsh, Z.Physik C, Particles and Fields 1,71(1979)
9. R.D.Field, R.P.Feynman, Nucl.Phys.B136,1,(1978)

RESONANCE PARAMETERS OF Υ AND Υ' AND INCLUSIVE SPECTRA MEASURED AT DORIS

W. Schmidt-Parzefall
Deutsches Elektronen-Synchrotron DESY, Hamburg

ABSTRACT

Recent results on measurements of the Υ and Υ' resonances by the DASP2, LENA and PLUTO collaborations obtained at the DORIS storage ring are reported. The combined result for the branching ratio for $\Upsilon \to \mu\mu$ and $\Upsilon \to ee$ is $B_{\mu\mu} = B_{ee} = (3.0 \pm 0.8)\%$. Thus the total width of the Υ state is determined. $\Gamma_{tot} = (43 {}^{+20}_{-11})$ keV. DASP2 studied inclusive particle production and observed an excess of antiprotons produced on the Υ.

The Υ and Υ' have first been seen in e^+e^- collisions at DORIS in summer 1978. Then DORIS had to serve PETRA as injector. After construction of the positron accumulation ring PIA, high energy physics could resume at DORIS. New data have been taken during a run from december 79 to easter 80. Fig. 1 shows the energy range covered by DORIS so far. Note the small statistical errors on the peaks of the resonances. The area under a resonance gives its electronic width. Γ_{ee} $(\Upsilon \to ee)$ is shown in Table I. The total width of the Υ can be determined indirectly using

$$\Gamma_{tot} = \frac{\Gamma_{ee}}{B_{ee}} = \frac{\Gamma_{ee}}{B_{\mu\mu}} \qquad (1)$$

The branching ratio $B_{\mu\mu}$ $(\Upsilon \to \mu\mu)$ has been measured by DASP2 [1] and LENA [2]. For rejection of a heavy background from cosmic muons, time of flight measuring equipment has been added to the detectors. Thus muon pairs could be safely identified over a large solid angle. The time of flight distributions are shown in Fig. 2.

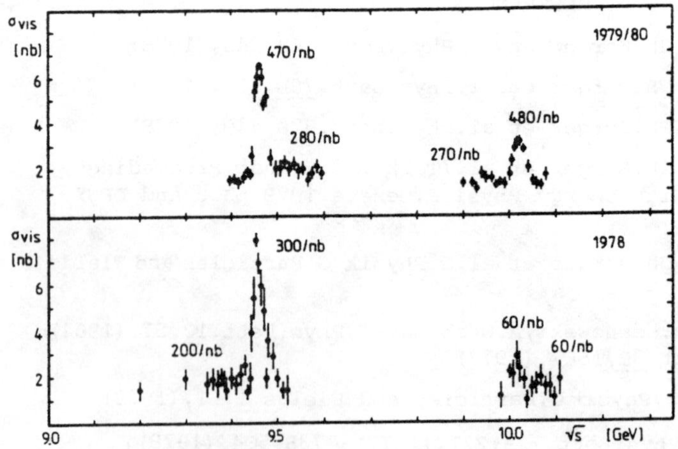

Fig. 1 DASP2 cross section and luminosity for all energies in the Υ and Υ' region scanned at DORIS.

TABLE I The electronic width of the Υ state

	PLUTO	LENA	DASP2	average
Γ_{ee} |keV|	1.33±0.14	1.23±0.10±0.14	1.35±0.11±0.22	1.29±0.09±0.13

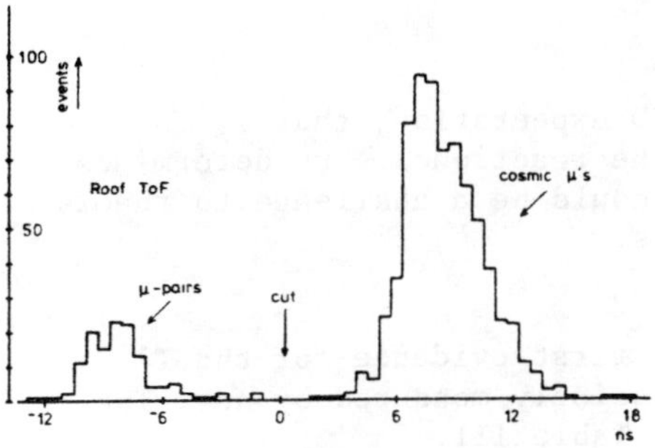

Fig. 2 LENA. Time of flight distributions for muon pairs and cosmic muons.

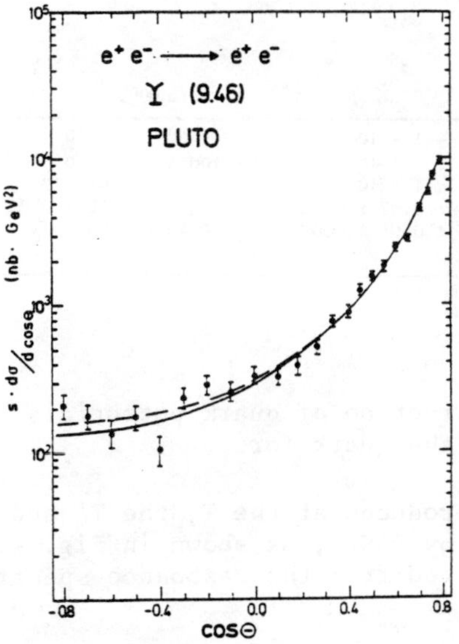

Fig. 3 Angular distribution of electron pairs. The solid line is the QED continuum, the dashed line is fitted to the data. The difference gives B_{ee}.

PLUTO [3] recently determined B_{ee} by studying the angular distribution of the reaction $\Upsilon \rightarrow ee$, as shown in Fig. 3. The results are shown in Table II. Assuming $B_{\mu\mu} = B_{ee}$, the average over all published results is $B_{\mu\mu} = (3.0 \pm 0.8)\%$. It gives a total width of the Υ state,

$$\Gamma^{\Upsilon}_{tot} = 43 \begin{matrix} +20 \\ -11 \end{matrix} \text{ keV}.$$

This is the first proof that the Υ resonance is a narrow state. According to QCD a Υ state decays into 3 gluons. The first order expression for the width Γ_{3g} $(\Upsilon \rightarrow 3g)$ is

$$\Gamma_{3g} = \frac{10(\pi^2-9)\alpha_s^3}{9\pi\,\alpha^2}\,\Gamma_{ee} \qquad (2)$$

where α is the electromagnetic and α_s the strong coupling constant. Denoting with R the ratio of the cross sections for hadron - over muon production, α_s is given by

$$\alpha_s = 0.0557\,\sqrt[3]{1/B_{\mu\mu}-(R+3)} \qquad (3)$$

which leads to

$\alpha_s^{\Upsilon} = 0.17 \pm 0.02$, to be compared with

$\alpha_s^{J/\psi} = 0.19 \pm 0.02$, obtained the same way.

TABLE II The branching ratio for Υ-decay into muon or electron pairs.

	DASP2	LENA	PLUTO	average
$B_{\mu\mu} B_{ee}$ |%|	$2.9 \pm 1.3 \pm 0.5$	$3.5 \pm 1.4 \pm 0.4$	5.1 ± 3.0	3.0 ± 0.8
1978		$1.0^{+3.4}_{-1.0}$	2.2 ± 2.0	

There is perfect agreement with the QCD expectation, that α_s falls with rising energies. Experimentally the reaction $\Upsilon \to \mu\mu$ determines α_s with a maximum of precision. This should be a challenge to theory to compute the higher order corrections.

Whereas in 1978 at DORIS just the first evidence for the Υ' resonance was obtained, it has been precisely measured by now. The resonance parameters are summarized in Table III.

TABLE III Υ' resonance parameters

	DASP2	LENA	Average	
M(Υ)	9463 ± 1 ± 10	9462 ± 1 ± 10	9462 ± 1 ± 10	MeV
M(Υ')	10017 ± 2 ± 10	10014 ± 1 ± 10	10015 ± 1 ± 10	MeV
M(Υ') - M(Υ)	554 ± 2 ± 10	553 ± 1 ± 10	553 ± 0.6 ± 10	MeV
$\Gamma_{ee}(\Upsilon')$	0.61 ± 0.12 ± 0.11	0.55 ± 0.07 ± 0.06	0.57 ± 0.06 ± 0.06	keV
$\Gamma_{ee}(\Upsilon)/\Gamma_{ee}(\Upsilon')$	0.45 ± 0.09 ± 0.05	0.45 ± 0.06 ± 0.02	0.45 ± 0.05	

This data is valuable for the construction of quark potentials and to test the flavour independence of the quark force.

Inclusive spectra of $\pi^{\pm}$, $K^{\pm}$ and $\bar{p}$ produced at the Υ, the Υ' and the nearby continuum have been measured by DASP2, as shown in Fig. 4. The continuum contribution has been removed from the resonance spectra. The pion spectra have been parametrized as

$$E \frac{d\sigma_\pi}{dp^3} \propto e^{-E/E_o}$$

and E_o was fitted to the data with a momentum cut-off p > 200 MeV/c.

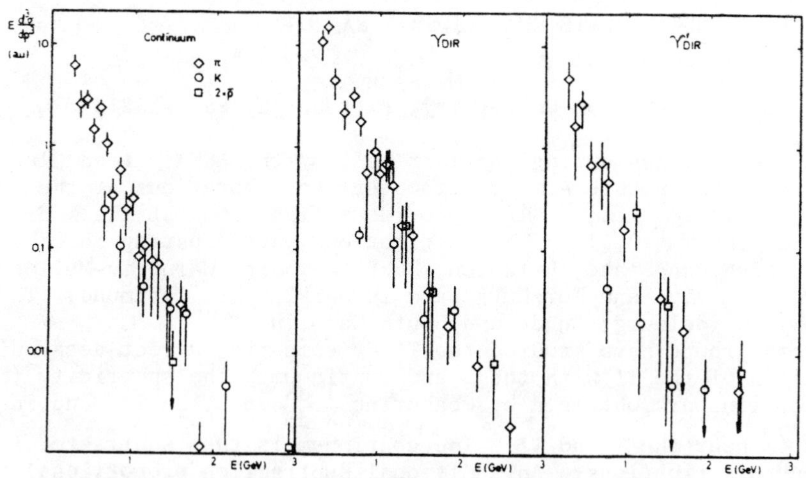

Fig. 4 DASP2. The invariant cross section plotted in arbitrary units versus the particle energy.

The results are E_0^c = (227 ± 16) MeV, E_o^{Υ} (230 ± 18) MeV, $E_o^{\Upsilon'}$ (199 ± 38) MeV. The difference between the Υ and Υ' spectra corresponds to a branching ratio around 30% for the reaction $\Upsilon' \rightarrow \pi\pi\Upsilon$. The absence of any noticeable difference between the pion spectra on the Υ and the nearby continuum remains still to be explained.

Table IV displays the measured particle ratios for momenta larger than 200 MeV/c.

TABLE IV Particle ratios

%	Continuum	Υ_{DIR}
$\pi^{\pm}$	87 ± 3	87 ± 3
$K^{\pm}$	10 ± 2	5 ± 2
$2\bar{p}$	3 ± 2	8 ± 2

The most striking result is the excess of antiprotons produced on the Υ-resonance. It suggests that some of the six quarks resulting from three gluons combine and facilitate the formation of baryons.

REFERENCES

1 H. Albrecht et al., Phys.Letters 93B (1980) 500
2 LENA Collaboration, B. Niczyporuk et al., DESY 80/53, June 1980
3 Ch. Berger et al., Phys.Letters 93B (1980) 497.

NEW RESULTS ON Υ'(10.01) HADRONIC DECAY

F. Messing
Carnegie-Mellon University, Pittsburgh, Pa. 15213

New data taken during the last year by the LENA[1] (Lead Glass NaI) and DASP2 (Double Arm Spectrometer) collaborations at the DESY e^+e^- storage ring, DORIS, provide a first look at the hadronic decays of the Υ'(10.01). The institutions participating in LENA are Cracow, Michigan State, Erlangen, DESY, Hamburg, Carnegie-Mellon, Saclay, Tel Aviv, and Würzburg, and in DASP2, DESY, Dortmund, ITEP (Moscow), Heidelberg, Lund, and South Carolina.

Both groups have studied the Υ' by comparing direct decay distributions of the Υ' with the Υ and continuum. The sphericity (S) distributions are obtained by measuring $\frac{1}{N}\frac{dN}{dS}$ on Υ, on Υ', and in the continuum near the Υ and Υ'. The continuum is then subtracted from the "on" distributions. An additional subtraction proportional to the continuum is made to account for the vacuum polarization (Fig.1) decays. This is done under the assumption that the virtual photon propagator is unaffected by the existence of the resonance. The amount of vacuum polarization is proportional to the mu-pair branching ratio of the resonance. For Υ, both groups have used their measured values: $B_{\mu\mu}(\Upsilon)=3.5\pm1.4\%$ (LENA),[2] and $B_{\mu\mu}(\Upsilon)=2.9\pm1.3\%$ (DASP2).[3] For the Υ' the LENA group has used a theoretical value of 2.2% and DASP2 has made no subtraction.

The standard interpretation of the direct decays is shown in Fig. 2 (minor decays, e.g. $P\to\gamma\Upsilon$, have been left off for clarity). The primary Υ decay is believed to be via three gluons. The Υ' is expected to decay via three primary modes. One mode is directly to three gluons. Another is the decay via pions or η to the Υ. The third is the radiative decay to a set of intermediate P states which would in turn decay primarily to two or three gluons depending on the total angular momentum. In Figs.3 & 4 the Υ' sphericity distribution is compared to the Υ. Within statistics the distributions agree. This supports the prediction that the Υ' decays into states consisting primarily of three gluons, and this is the first evidence that the Υ and Υ' are states with similar decay characteristics and are members of the same family of particles.

A quantitative measure of the similarity of the Υ and Υ' decays can be made within a simple model. The model is based on the observation that $[M(\Upsilon')-M(\Upsilon)]/M(\Upsilon')=.06$. This implies that the 3 gluon decays of Υ and Υ' should have similar characteristics. Furthermore, the pions from $\Upsilon'\to\Upsilon+X$ and the photons from $\Upsilon'\to P+\gamma$ will be of very low momentum and therefore unlikely to alter the sphericity distributions measureably. Monte Carlo studies support these assumptions. The only major Υ' decay expected to differ substantially from the 3 jets of the Υ is the decay $\Upsilon'\to\gamma+P(J\neq1)\to 2$ gluons. These would be expected to have a 2 jet sphericity distribution like the continuum ($e^+e^-\to 2$ quarks $\to 2$ jets) under the further assumption that quark and gluon jets are similar.

One can then make a one parameter fit of the Υ' sphericity distribution to a sum of the Υ (3 jet) and continuum (2 jet) sphericity distributions:

$$\frac{1}{N}\frac{dN}{dS}(\Upsilon') = (1-A)\underbrace{\frac{1}{N}\frac{dN}{dS}(\Upsilon)}_{3\text{ jet}} + A\underbrace{\frac{1}{N}\frac{dN}{dS}\text{ (continuum)}}_{2\text{ jet}}$$

The parameter A is the product of branching ratios of $\Upsilon' \to \gamma + P$ and $P \to 2$ gluons. The results are $A(\Upsilon' \to \gamma + P(J\neq 1) \to 2 \text{ gluons}) < 19\%$ for LENA. This is a 90% confidence level based on statistical errors only.

Further information of the Υ' hadronic decays can be obtained from the observed direct charged multiplicity distributions. The decays $\Upsilon' \to 3$ gluons and $\Upsilon' \to \gamma + P \to 3$ gluons should all have the same charged multiplicity distribution as the Υ due to the closeness of the masses. However, the decay $\Upsilon' \to \pi^+\pi^-(\pi^0)\Upsilon$ will have a multiplicity distribution shifted by 2 units. The branching ratio for the decay via $\pi^+\pi^-(\pi^0)$ can be measured by looking for this shift. A one parameter fit can be made of the Υ' multiplicity distribution to a sum of the Υ distribution and the Υ distribution shifted by 2 units:

$$\left.\frac{1}{N}\frac{dN}{dn_{ch}}\right|_{\Upsilon'} = (1-B)\left.\frac{1}{N}\frac{dN}{dn_{ch}}\right|_{\Upsilon} + B\left.\frac{1}{N}\frac{dN}{dn_{ch}}\right|_{\Upsilon\text{ shifted by two units}}$$

The use of the measured distributions, rather than Monte Carlo calculations, eliminates most of the systematic errors. The LENA result is $B(\Upsilon' \to \pi^+\pi^-(\pi^0)\Upsilon) = 27\pm 9\%$ (statistical error only). The contribution to this number induced by the larger mass of the Υ' can be estimated by applying the same fit to the continuum near the two resonances. The induced branching ratio obtained is $2\pm 4\%$, consistent with our assumption of a negligible contribution.

While the multiplicity distribution gives a good measure of the branching ratio to the Υ, it does not prove that such decays exist. To do this we have searched for $\Upsilon' \to \pi^+\pi^-\Upsilon$ followed by $\Upsilon \to e^+e^-$ or $\mu^+\mu^-$. We have accepted only those events in which the lepton is positively identified in our detector and the decay geometry is kinematically consistent with the hypothesis. LENA has a total of 7 candidates: 5 with e^+e^- and 2 with $\mu^+\mu^-$. The ratio of e^+e^- to $\mu^+\mu^-$ events comes from the different detector acceptances for muons and electrons and does not imply a violation of lepton universality. The background was estimated by searching for such events in the continuum and on the Υ resonance. We estimate the background at less than one event. The observation of the $\Upsilon' \to \pi^+\pi^-\Upsilon$ is the first evidence[4] directly linking the Υ' and Υ.

In summary, the Υ and Υ' sphericity distributions have been compared and found to support the belief that the Υ' decays primarily into 3 gluons. A model dependent limit on the decay of $\Upsilon' \to \gamma + P(J\neq 1)$ of 19% was found at the 90% confidence level. The measured Υ' charged multiplicity has been found to be higher than the Υ charged multiplicity. If we attribute this to the decays $\Upsilon' \to \pi^+\pi^-(\pi^0)\Upsilon$ we obtain a branching ratio of $B(\Upsilon'\to\pi^+\pi^-(\pi^0)\Upsilon)=27\pm9\%$. We have also found examples of this decay followed by $\Upsilon \to e^+e^-$ or $\mu^+\mu^-$ in which the leptons are positively identified by the detector. All of these results are consistent with standard quarkonium model predictions which interpret the Υ and Υ' as bound $b\bar{b}$ quark states.

REFERENCES

1. The detector was constructed by the DESY-Heidelberg collaboration and is described in W.Bartel et al., Phys. Lett. 77B, 331 (1978) and 66B, 483 (1976).
2. LENA collaboration, B.Niczyporuk et al., DESY report 80/53(1980).
3. DASP2 collaboration, C.W.Darden et al., DESY report 80/30(1980).
4. The CLEO and CUSB groups at CESR have both reported observing a few candidates for $\Upsilon' \to \pi^+\pi^-\Upsilon$. See talks presented by J. Yoh and E. Thorndike at this Conference.

FIGURE CAPTIONS

Fig. 1. Diagram for vacuum polarization decay of Υ or Υ'.

Fig. 2. Primary decay modes of the Υ'.

Fig. 3. Normalized sphericity distributions for Υ and Υ' measured by LENA. The dashed line is the measured continuum sphericity distribution. The average sphericities were $\langle S \rangle = .36\pm.012$ for Υ and $.367\pm.014$ for Υ'. 5247 events were collected on the Υ and 2882 events were collected on the Υ'.

Fig. 4. Normalized sphericity distributions for Υ and Υ' measured by DASP2. The solid line is a prediction from a 2 jet Monte Carlo. The average sphericities were $\langle S \rangle = .36\pm.02$ for Υ and $.32\pm.03$ for Υ'.

Fig. 5. Normalized observed charged multiplicity distributions for Υ and Υ' measured by LENA. The averages were $\langle n_{ch} \rangle = 6.25\pm.08$ for Υ and $6.45\pm.07$ for Υ'.

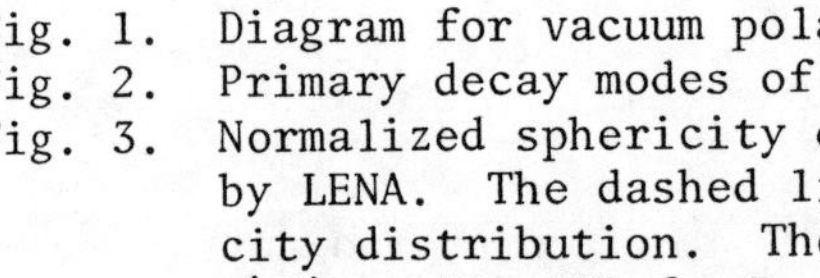

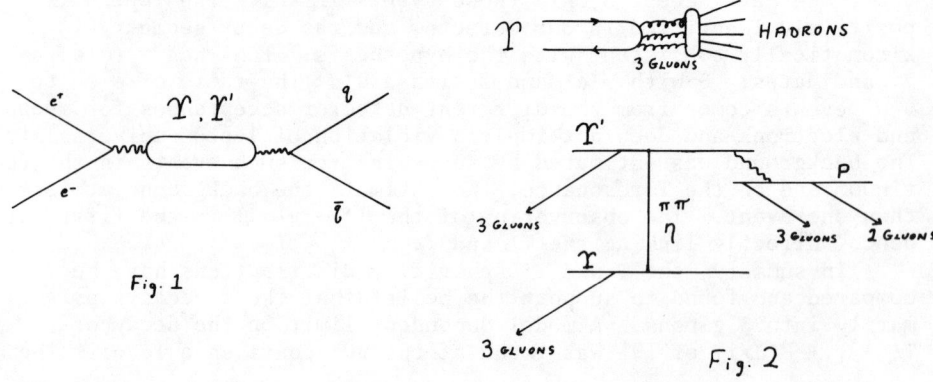

Fig. 1

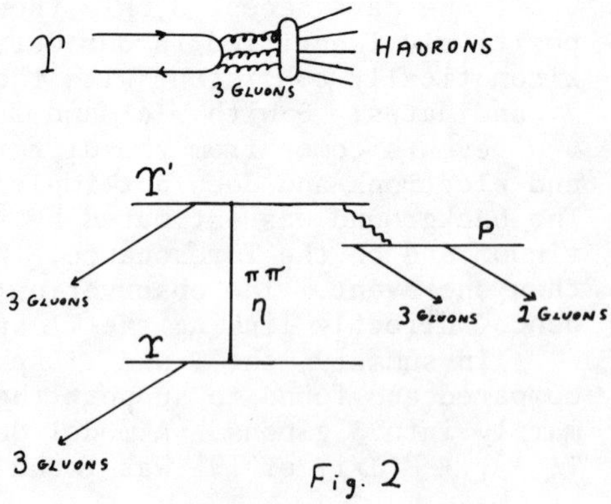

Fig. 2

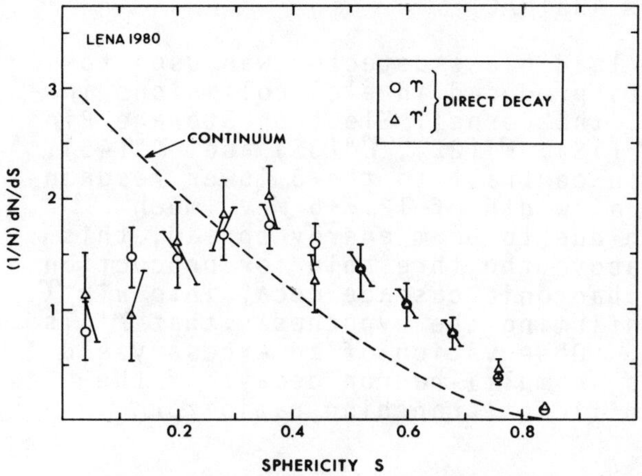
LENA 1980
Υ
Υ'
DIRECT DECAY
CONTINUUM
(1/N) dN/dS
SPHERICITY S
0.2
0.4
0.6
0.8
1
2
3

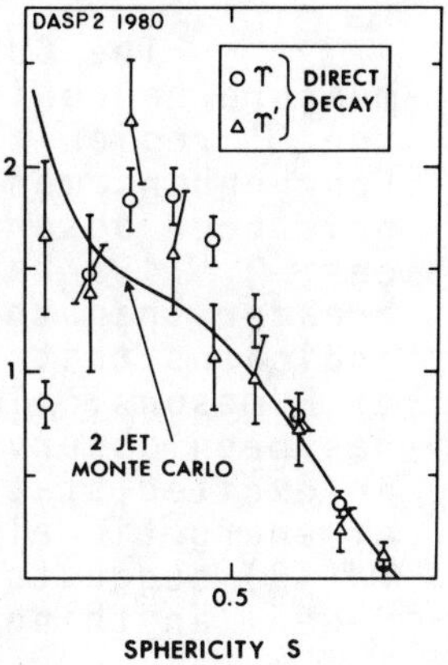
DASP 2 1980
Υ
Υ'
DIRECT DECAY
2 JET MONTE CARLO
(1/N) dN/dS
SPHERICITY S
0.5
1
2

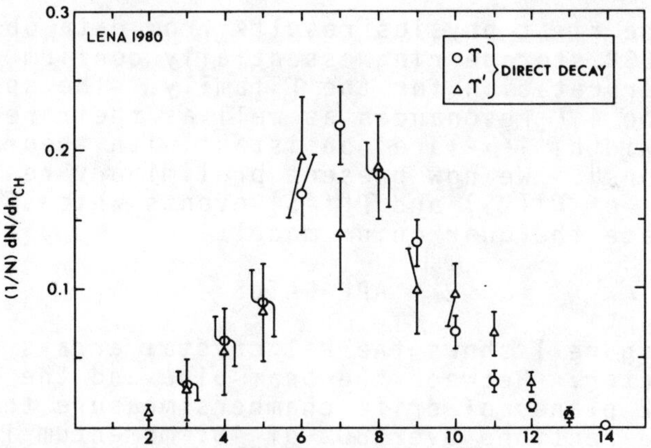
LENA 1980
Υ
Υ'
DIRECT DECAY
(1/N) dN/dn_CH
CHARGED MULTIPLICITY n_CH
0.1
0.2
0.3
2
4
6
8
10
12
14

RESULTS ON THE Υ RESONANCES FROM THE CUSB GROUP AT CESR

John K. Yoh
Columbia University, New York, NY 10027

ABSTRACT

The CUSB NaI/lead glass detector was used to measure hadronic events produced in e^+e^- collisions in the Υ region at CESR, the Cornell Electron Storage Ring. Four enhancements, Υ(1S), Υ'(2S), Υ''(3S) and Υ'''(4S), have been observed. In contrast to the 3 lower resonances, Υ'''(4S) has natural width of 12.6$\pm$6 Mev, much broader than the width due to beam energy spread; this indicates that it is above the threshold for production of B mesons. The Υ' hadronic cascade decay into $\pi^+\pi^-\Upsilon$ has been observed, confirming the hypothesis that Υ' is an excited state of Υ. Observation of an excess yield of energetic electrons in multi-hadron decays of the Υ'''(4S) suggests a significant branching ratio for $B \rightarrow e$ + anything.

INTRODUCTION

The first physics results from data obtained at the new CESR storage ring essentially confirm the quarkonium interpretation for the Υ family. The spacing between the 4 Υ resonances as well as their relative leptonic widths[1,2,3] are consistant with theoretical expectations[4]. We now present preliminary results on the decays of Υ'(2S) and Υ'''(4S) events which further substantiate the quarkonium model.

APPARATUS

Figure 1 shows the NaI crystal arrays of the CUSB detector. Between the beam pipe and the NaI blocks, 12 planes of drift chambers measure the charged particle directions over 80% of 4π; momentum is not determined since no magnetic field exists. The first 4 layers of NaI, each 1-1/8 r.l. thick, are interleaved with strip chambers (PWC with strip cathode readout); these chambers are used to record the position of particles which shower in the NaI. Leakages of electromagnetic showers beyond the 8.6 r.l. of NaI are measured by arrays of 256 lead glass blocks (7 r.l. thick).

DATA TAKING AND PERFORMANCE

The trigger requirement consists of at least .7 Gev energy deposit in the NaI crystals coincident with

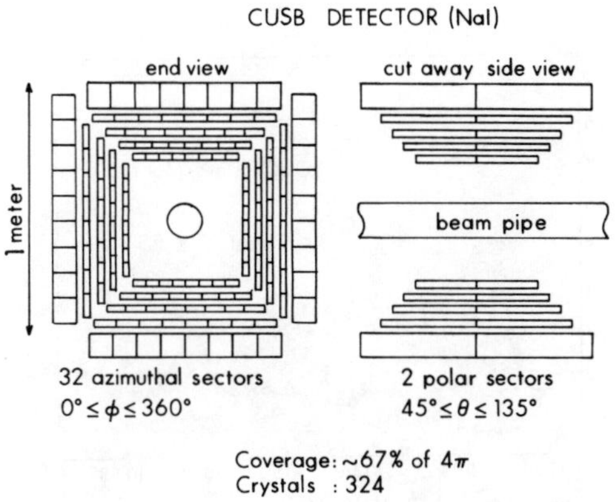

Figure 1 : The CUSB NaI crystal array

beam crossing. Roughly 1/4 of the triggers are beam-beam events--Bhabhas, multi-hadrons or 2-photon events. Multi-hadronic events are selected by various criteria; the most important criterium consists of the requirement of 3 or more particles penetrating the crystal array. Background from beam-gas , beam-wall or QED processes are at the few percent level. Details can be obtained from our publications[1,2].

In order to enhance the resonance signal relative to the continuum background, we have developed a thrust variable T. Data from the Υ'''(4S) are usually cut by requiring $T<.85$; this results in a 52% reduction in continuum events while only 26% of the resonance events are removed.

We have preliminary results of studies on the performance of our detector. The drift chamber track reconstruction efficiency is found to be about 95% for isolated tracks, such as from Bhabha events. Our preliminary calibration on the gain of the NaI and lead glass gives us an energy resolution of about 2.5%(rms) for electrons near 5 Gev.

THE UPSILON RESONANCES

Our results on the 4 Υ resonances have been reported[1,2]. Figure 2 shows the visible cross section of multi-hadron events for the 3S and 4S resonances. The resonance parameter for the Υ'''(4S) are :

$$M(\Upsilon''')-M(\Upsilon)=1.114 + .002 \text{ Gev}$$
$$\text{Natural width } (\Upsilon''')=12.6+6.0 \text{ Mev}$$
$$\Gamma_{ee}(\Upsilon''')/\Gamma_{ee}(\Upsilon) = .25 + .07$$

In addition to the data discussed above, we have recently taken data consisting of about 500 nb^{-1} at the

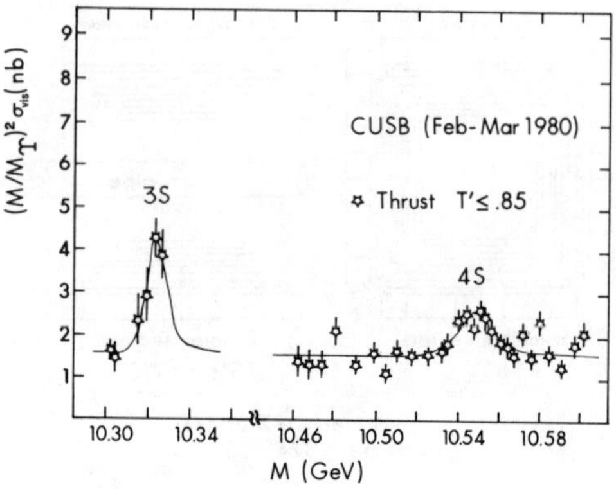

Figure 2 : Observed e^+e^- hadronic cross section

Υ', 1000 nb^{-1} at continuum near 10.4 Gev and 1300 nb^{-1} at or near the Υ'''(4S) peak. Preliminary analysis of the 4S data gives similar values to the above results. The natural width for the 4S from the improved statistics is now about 5 σ away from 1 Mev or below.

Υ' HADRONIC CASCADE INTO $e^+e^-\pi^+\pi^-$

About 2500 Υ' hadronic events have been taken. These events were scanned to look for the hadronic cascade $\Upsilon' \rightarrow \pi^+\pi^-\Upsilon \rightarrow \pi^+\pi^-e^+e^-$. Events were preselected requiring 2 energetic electrons and at least one more track; the two electrons are required to have observed mass exceeding 7 Gev and to be collinear to within 10^0 in azimuth.

3 events with 4 observed tracks (2 electrons and 2 other tracks) were found in our Υ' sample; these were fitted to $\Upsilon' \rightarrow \pi^+\pi^-e^+e^-$; the observed and fitted di-electron masses are shown in figure 3; the confidence levels for these fits are good. All the continuum and 4S data, with about 5 times the sensitivity, were also searched for these events; the only background candidate is also shown in figure 3; the fit is poor (C.L. of only .0002) and the fitted mass is inconsistant with the mass difference $\sqrt{S}-M_{ee}$ of $M(\Upsilon')-M(\Upsilon)$ = .56 Gev. Thus, there is no background to our 4 track cascade candidates.

Preliminary analysis on events with only 3 observed tracks (2 electrons + one other) gives 2 candidates from the Υ' data with fitted M_{ee} .71+.17 Gev and .47+.17 Gev below the mass of Υ'. Again, no background event was seen with the correct mass difference; one background event has $\sqrt{S}-M_{ee}$ of 1.02 +.25 Gev, indicating that the background level could be .1 to .2 event.

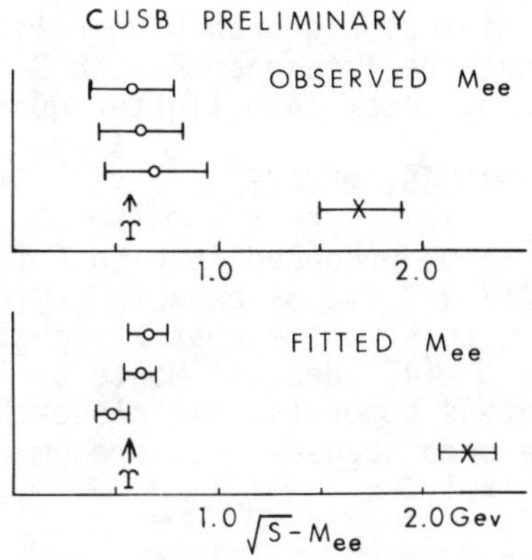

Figure 3 : a)observed and b)fitted M_{ee} for $ee\pi\pi$ events from Υ'(o) and continuum (X).

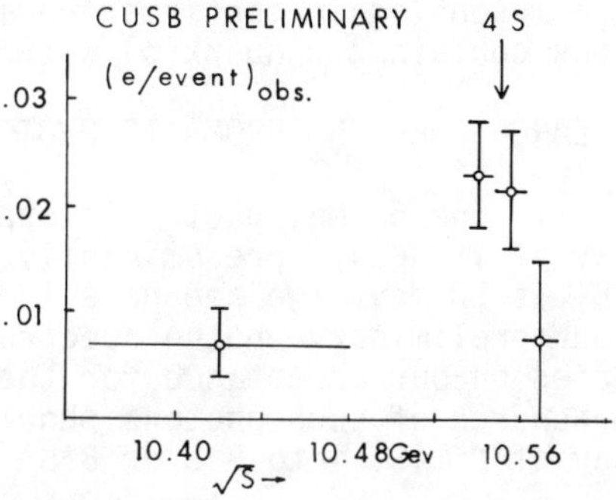

Figure 4 : Yield of energetic electrons per hadronic event with T<.85; see text for other cuts.

The efficiency for detecting 3 or 4 track events is approximately 40%; thus, the 5 events represent a branching ratio for $\Upsilon' \rightarrow \pi^+\pi^-\Upsilon$ of about 18%, assuming the recently reported value for BR($\Upsilon \rightarrow e^+e^-$) of 2.8%[5].

INCLUSIVE ENERGETIC ELECTRON YIELD IN B MESON DECAY

We have measured the yield of energetic (P>1Gev) electrons in multi-hadron events. The electron criteria have been developed using several hundred single-electron events from $e^+e^- \rightarrow e^+e^-e^+e^-$ in which only one of the 4 final state electrons hit our apparatus. Cuts are made on the agreement between track direction and shower centroid position and on the longitudional and transverse shower developments. We select only those events with high multiplicities (to reject electrons from taus) and T<.85 (to enhance the resonance/continuum ratio). We then normalize the electron candidates to either luminosity, number of hadronic events with T<.85, or the total number of tracks in these events.

Preliminary result on the electron yield at the 4S peak and at continuum is shown in figure 4. The excess at the 4S peak can not be accounted for by the 10% increase in charged tracks at the 4S peak; the significance of this excess is at the level of 3 standard deviations. Subtracting the continuum contribution and correcting for estimated efficiency gives a rough branching ratio of 10-20% for $B \rightarrow e$ + anything(assuming $4S \rightarrow B\bar{B}$). The energies of the electrons at the 4S average to 1.5 Gev and extend up to 2.3 Gev, consistant with the expected M.C. generated electrons from $4S \rightarrow B\bar{B}$ and $B \rightarrow De\nu$(although $B \rightarrow \pi e \nu$ can not be excluded).

The observation of a strong semi-leptonic branching ratio for B mesons again confirms the idea that the Υ resonances and B mesons contain new quark(s) without strong decay into lighter quarks.

SEARCH FOR MONOCHROMATIC PHOTONS IN Υ'''(4S) DECAYS

One 50 Mev photon is expected to be produced in each Υ'''(4S) decay if it decays predominantly into $B^*\overline{B}$ and if, as expected, $M(B^*)-M(B)$ is 50 Mev. We see no evidence for this monochromatic signal in our preliminary photon spectrum from Υ'''(4S) decays; Monte Carlo studies on our acceptance for these photons suggests that close to one hundred of such photons should have been seen. Thus, predominant decay of Υ'''(4S) into $B^*\overline{B}$ or $B^*\overline{B}^*$ is unlikely.

CONCLUSIONS

The observation of Υ' hadronic cascade decay and B meson semi-leptonic decay convincingly establishes the quarkonium interpretation for the Υ family and related particles. We are eagerly working on understanding other aspects of these particles. A wealth of other discoveries-- radiative transitions, new particles, other cascade decays, and even such esoteric possibilities such as CP violation ---may well be forthcoming.

Other members of the CUSB collaboration--Columbia, SUNY Stony Brook, Louisiana State Univ., and Max Planck Inst. (Munich)--contributed to all of the above results.

REFERENCES

1 T. Bohringer et. al., Phys. Rev. Lett. 44, 1111 (1980)
2 G. Finocchiaro et. al., Phys. Rev. Lett. 45,222 (1980)
3 D. Andrews et. al., Phys. rev. Lett. 44, 1108 (1980)
D. Andrews et. al., Phys. rev. Lett. 45, 219 (1980)
4 C. Quigg and J. Rosner, Phys. Rep. 56, No. 4 (1979)
E. Eichten et. al., Phys. Rev. D21, 203 (1980)
G. Bhanot and S. Rudaz, Phys. Lett. 78B, 119 (1978)
5 K. Berkelman, talk given at 20th HEP conf.,(Madison, Wisc),1980.

* This work is supported in part by National Science Foundation and by the Department of Energy.

FIRST RESULTS ON BARE b PHYSICS

E. H. Thorndike
University of Rochester, Rochester, NY 14627
(for the CLEO Collaboration)*

CLEO is the name given to the large magnetic detector in the main experimental area of the Cornell Electron Storage Ring (CESR). ("CLEO" doesn't stand for anything, but was picked as a name because we thought it went well with CESR).

CLEO was designed, built, and now is operated by a collaboration of six universities: Vanderbilt, Syracuse, Rutgers, Rochester, Harvard, and Cornell. It was first turned on in September, 1979, and was taking useful data by November. We still aren't exploiting the full capabilities of the instrument, but have nonetheless collected more interesting data in the past 7 months than I can cover in my allotted time. I will therefore concentrate on B meson physics, and omit our results on Υ and Υ' decays.

Consider first the total hadronic cross section as a function of center of mass energy. Motivated by the 1977 discovery of the Υ at Fermilab, physicists at DORIS, in 1978, found two narrow resonances, at 9.4 and 10.0 GeV. In our early running we confirmed the existence of these two resonances [hereafter called Υ(1S) and Υ(2S)]. At higher energy, we obtained our first new physics result, the Υ (3S) at 10.3 GeV. Like the two lower resonances, it is narrow, having an experimental width determined by the beam energy spread.

Moving up in energy, Fig. 1 shows the 3S and the 4S. The discovery of the 4S was our second new physics result. Note that it is noticeably wider than the 3S, i.e., it is measurably broad.

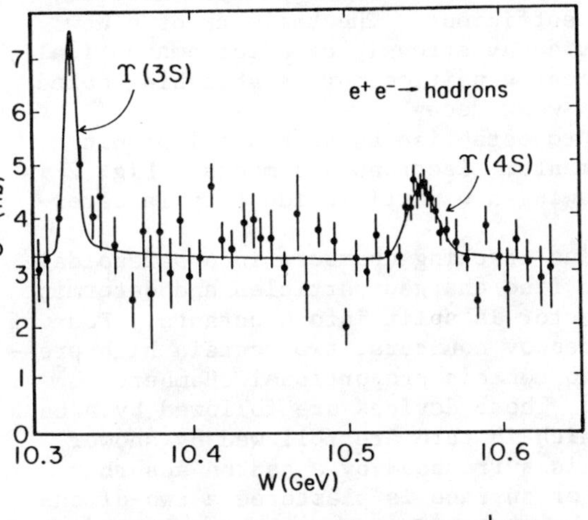

Fig. 1. Total cross section for $e^+e^- \to$ hadrons as a function of center of mass energy, showing the third and fourth Υ resonances.

Parameters of the four resonances are shown in Table I. I draw your attention to the bottom line of the table, the measured r.m.s. widths of the resonances. From the first 3 resonances, we confirm our understanding of the beam energy spread. In sharp contrast, the 4S is much broader than the beam energy spread.

From this result we can conclude that a

*Supported by the National Science Foundation and the Department of Energy.

Table I. Parameters for the First Four Υ Resonances

	(1S)	(2S)	(3S)	(4S)
Mass (MeV)	9433.6 ±0.2±28	9994.4 ±0.4±30	10323.1 ±0.4±30	10547.6 ±1.1±32
M-M(1S) (MeV)	0	560.8 ±0.4±3.0	889.5 ±0.5±4.0	1114.0 ±1.1±5.0
$\Gamma_{ee}\cdot(\Gamma_{had}/\Gamma_{TOT})$ $\equiv \Gamma$(KeV	0.925 ±.06±.14	0.468 ±.04±.07	0.288 ±.03±.05	0.221 ±.02±.03
Γ/Γ(1S)	1	0.51 ±.05±.05	0.31 ±.04±.03	0.24 ±.02±.03
Measured r.m.s. width (MeV)	3.14 ±0.20	3.53 ±0.22	3.76 ±0.24	9.00 ±0.82

threshold of some sort has been crossed between 10.3 GeV and 10.5 GeV. Conventional wisdom tells us that it is the threshold for producing naked beauty. If so, the mass of the lightest B meson has been determined to ±1%, i.e., it lies between 5.16 and 5.27 GeV.

Consider for a moment what need be done to establish that the threshold is indeed that for naked beauty. One's first reaction is to reconstruct masses and look for bumps in mass plots. However, this is neither necessary nor sufficient. The hallmark of a new naked flavor is that it cannot decay strongly or electromagnetically. Thus the necessary and sufficient condition for establishing naked beauty is the observation of a weak decay.

Since the easiest decays to establish as weak are leptonic decays, we looked for high momentum electrons and muons. Fig. 2 is a beam's eye view of CLEO, showing how particle identity is determined.

The central region contains tracking chambers in a solenoidal magnetic field, and is used to find charged particles and determine their momenta. The outer detector is split into 8 octants. Four of the 8 contain low pressure Cerenkov counters, two contain high pressure Cerenkov counters, and two contain proportional chambers measuring specific ionization. These devices are followed by a bank of time of flight counters, which in turn are followed by shower counters. The outer detector is surrounded by a hadron absorber (2-3 ft of iron), on whose outer surface is plastered a two-dimensional array of planar drift chambers.

Muons are identified by tracing an inner drift chamber track out through the iron, and then noting the distance from it to the nearest hit in the muon chambers. The distribution in the square of this distance is shown in Fig. 3. The conspicuous peak at small r^2 contains the muons, but also contains background from hadronic punch-thru and decay in flight. (A subtraction eliminates this background.)

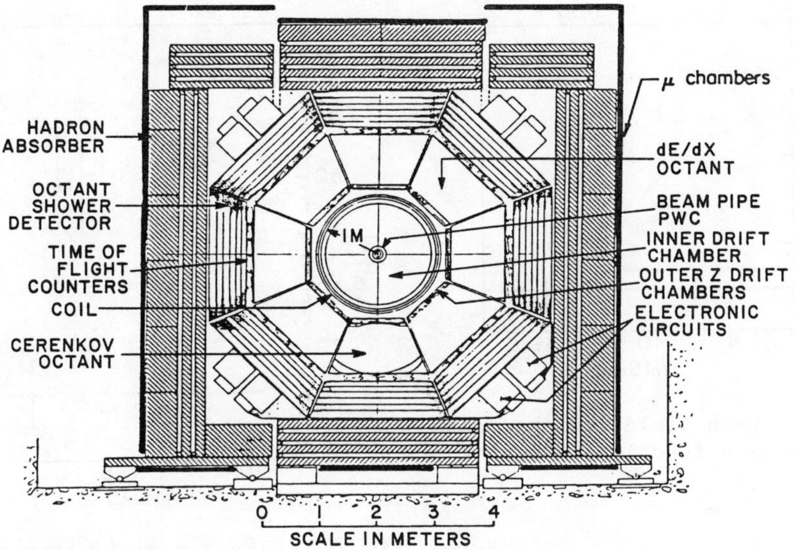

Fig. 2. Beam's eye view of the CLEO detector.

For electrons, we require a triple match of an inner drift chamber track with a momentum above 1 GeV, a Cerenkov counter hit, and a shower.

We searched our hadronic event sample for electrons and muons. To eliminate taus and 2γ events as background, we required that the event have a charged multiplicity of 5 or greater. The muon yield as a function of center of mass energy is shown in Fig. 4. There is a 3-standard deviation enhancement at the 4S. A similar plot for electrons is shown in Fig. 5. There is a 4-standard deviation enhancement at the 4S.

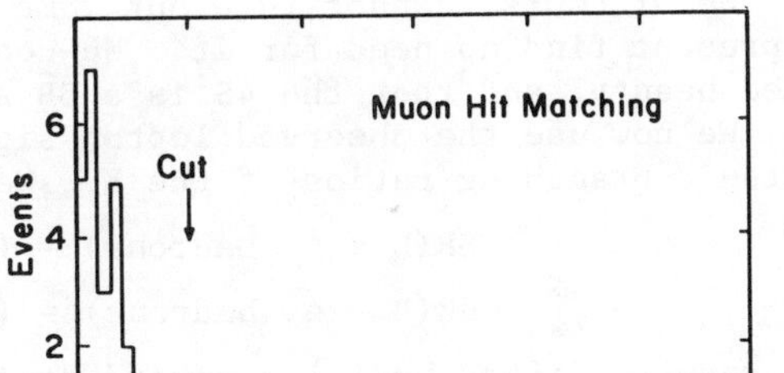

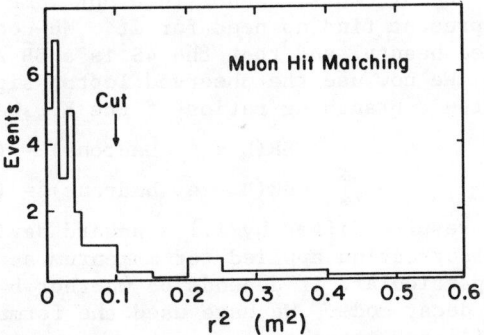

Fig. 3. Distribution in squared distance r^2 between tracks projected to the muon chambers and hits in the muon chambers.

Having established a yield of leptons from the 4S, we looked for an explanation other than a weak decay of B. Two possibilities were considered: strong decay of 4S to ψ + hadrons, followed by ψ→μμ; strong decay of 4S to D + hadrons, followed by D→μν hadrons. By pairing the detected lepton with all other charged tracks, and computing the pair invariant mass, we checked for and eliminated ψ's as the source of the leptons. Similarly, any plausible model of D

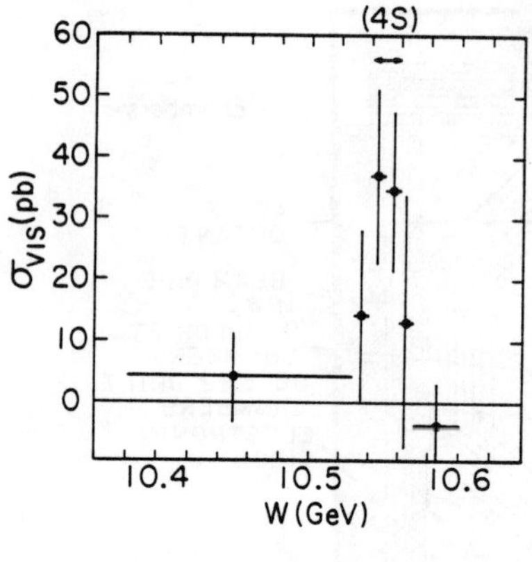

Fig. 4. Muon yield from hadronic events, as a function of center of mass energy.

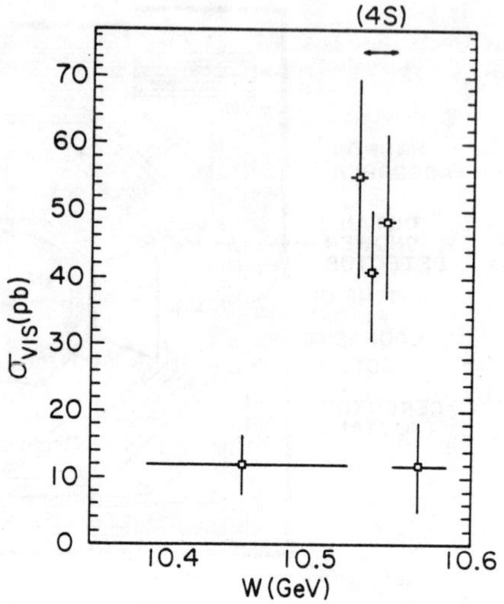

Fig. 5. Electron yield from hadronic events, as a function of center of mass energy.

production gives a rather soft lepton spectrum, which could account for at most 20% of the signal we see.

We of course cannot rule out some entirely new phenomenon, but at present find no need for it. We conclude that we are observing naked beauty, and that the 4S is a $B\bar{B}$ state.

We now use the observed lepton signal to determine the semileptonic branching ratios of the B. We find

$$BR(B \to \mu\nu \text{ hadrons}) = (7.5 \pm 3.1)\%$$

$$BR(B \to e\nu \text{ hadrons}) = (16 \pm 4 \pm 7)\%$$

The results differ by 1.1 standard deviations, and average to 12 ± 4%. The correction applied for momentum acceptance is model dependent. In particular, it depends on whether $b \to c\mu\nu$ or $b \to u\mu\nu$ is the dominant decay mode. We have used the former, on the basis of evidence I will now present.

Based on simple quark counting, one expects 1 s quark per $e^+e^- \to$ hadrons continuum event. If b decays to charm, one expects 3-1/3 s quarks per $B\bar{B}$ event, while if b decays to up, one expects 1-1/3 s quarks per $B\bar{B}$ events. A large increase in the number of strange particles on the 4S, as compared to the continuum, thus indicates a substantial decay of b to charm. So motivated, we looked for strange particles, in particular for charged kaons.

Recall what CLEO looks like (Fig. 2). We take those tracks found in the inner detector which project to a hit time of flight counter and a hit in the shower counter. (The shower counters detect minimum ionizing particles as well as showering ones.) We further require that the calculated π-K separation time for the track be greater than 1.0 ns (i.e., momentum less ∿.95 GeV). On the low momentum side, we quit at ∿0.7 GeV, because kaons begin ranging out in the coil. (Our TOF resolution is ∿0.4 ns.)

Results for tracks with a π/K separation time near 1.5 ns are shown in Fig. 6. Both $\pi^{\pm}$ and $K^{\pm}$ peaks are evident. (The $p/\bar{p}$ peak is smeared out by range losses.) The $K^{\pm}$ signal is extracted by fitting the distribution with two Gaussians. The same procedure is followed for other separation times, from 1.0 to 1.8 ns.

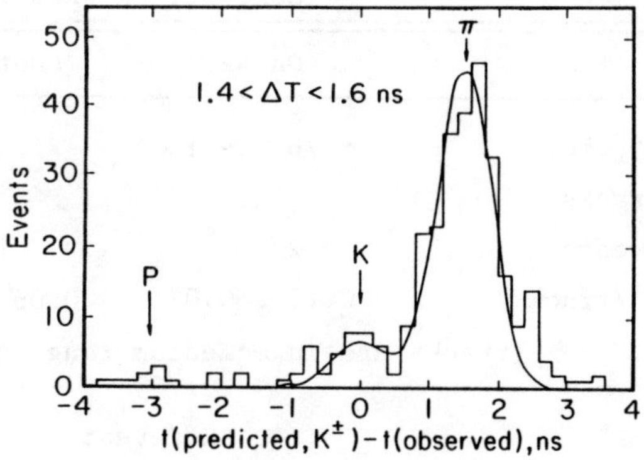

Fig. 6. Distribution of measured flight times (relative to the flight time calculated for the K hypothesis) for all tracks for which the calculated K/π flight time difference Δt lies in the range 1.4 < ΔT < 1.6 nsec.

Over a restricted solid angle, we can also identify π/K/p by specific ionization. In a sample of 113 tracks identifiable by both techniques, time of flight classified 100 tracks as $\pi^{\pm}$, 9 as $K^{\pm}$, and 4 as $p/\bar{p}$. Specific ionization differed only on two tracks, classifying one TOF pion as an electron, and one TOF pion as a kaon. The excellent agreement demonstrates the correctness of the TOF identification.

The K/π ratio as a function of center of mass energy is shown in Fig. 7. There is a 4 standard deviation enhancement at the 4S.

Table II gives our $K^{\pm}$ results in a numerical form. In the momentum range 0.7 to 0.9 GeV, the fraction of tracks that are K's is 7 times larger for $B\bar{B}$ events than for continuum events. Indeed, in this momentum range more than 1/3 of all tracks from B decays are K's.

To obtain the number of K's per <u>event</u>, one must apply two corrections. The first, geometrical acceptance and track handling efficiency, is model dependent and was found to be 25%. The second correction, for momentum acceptance, of course depends on the kaon momentum spectrum. From Monte Carlo calculations, and from comparisons with inclusive spectra, we obtain (16 ± 4)%. The corrected numbers of K's per event are given in Table II.

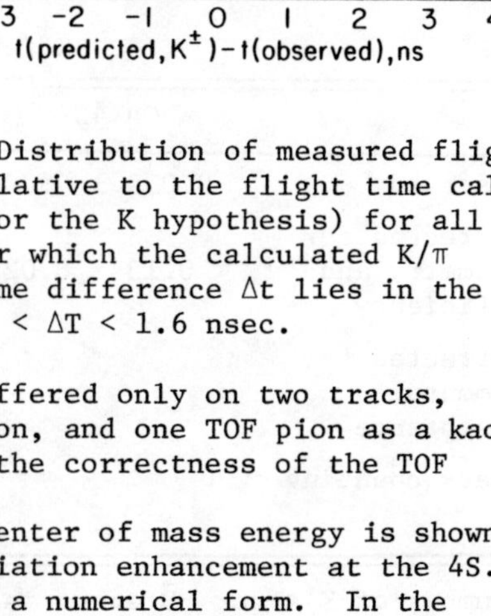

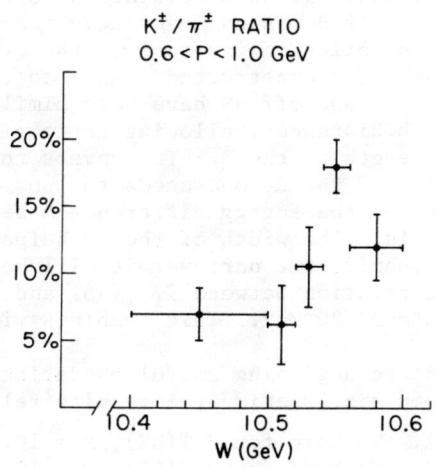

Fig. 7. $K^{\pm}/\pi^{\pm}$ ratio as a function of center of mass energy.

Table II. $K^{\pm}$ Results

	On 4S	Continuum	$B\bar{B}$
$K^{\pm}$ obs.	76 ± 9.1	21.4 ± 6.3	54.6 ± 11.1
tracks	501	369	132
Events	2276	1738	537
$K^{\pm}$/track	0.15 ± 0.02	0.06 ± 0.02	0.41 ± 008

(tracks are in momentum range of 0.7 - 0.9 GeV)

$K^{\pm}$/event

	On 4S	Continuum	$B\bar{B}$
observed	0.033 ± 0.004	0.012 ± 0.004	0.102 ± 0.021
corrected for geometry and efficiency	0.13 ± 0.02	0.05 ± 0.01	0.41 ± 0.08
corrected for momentum acceptance		0.31 ± 0.09 ±0.08	2.5 ± 0.5 ±0.6
Quark counting		.5	1.67 $b \to cW^-$
			0.67 $b \to uW^-$

The number of K's per $B\bar{B}$ event is large, as expected for a dominant decay of b to charm.

Finally, we have obtained an upper limit on the lifetime of the B meson. If B mesons moved away from the beam axis before decaying, we would notice an increase in the average distance of closest approach of reconstructed tracks to the beam line. However, the data on 4S and off 4S have very similar distributions in the closest approach distances, allowing us to place a limit of 2 mm on the mean decay length of the B. To convert this distance to a limit on the lifetime of the B, one needs to know the velocity of the B's, which depends on the energy difference Δ between the 4S mass and the $B\bar{B}$ threshold. The width of the 4S helps us here. The closer the 4S is to threshold, the narrower it will be. Using Eichten's prediction of the relation between $\Gamma_{TOT}(4S)$ and Δ, and our measured $\Gamma_{TOT}(4S)$ we conclude $\Delta \gtrsim 20$ MeV. This combines with $d \lesssim 2$ mm to yield $\tau_B < 1\times10^{-10}$ sec.

Since beginning useful operation last November, CLEO has obtained six experimental results relevant to bare-b physics:

1. We have found $\Upsilon(3S)$, M = 10.3 GeV, narrow.
2. We have found $\Upsilon(4S)$, M = 10.55 GeV, broad.
3. $(4S) \to \mu$ + hadrons.
4. $(4S) \to e$ + hadrons.

5. (4S) $\rightarrow K^{\pm}$ + hadrons, copiously.
6. (4S) decay path is less than 2 mm.

From these experimental results, we draw the following conclusions:

1. Naked beauty exists
2. $5.16 \gtrsim M_B < 5.27$ GeV.
3. BR ($B \rightarrow \mu\nu$ hadrons) + BR($B \rightarrow e\nu$ hadrons) = $23\frac{1}{2} \pm 7\frac{1}{2})$%.
4. $(b \rightarrow uW^-) << (b \rightarrow cW^-)$.
5. $\tau_B \rightarrow 1 \times 10^{-10}$ sec.

ONIUM THEORY AND SPECTROSCOPY, THEORY

C. QUIGG, FERMILAB, ORGANIZER

(QUARK)ONIUM THEORY AND SPECTROSCOPY

C. Quigg
Fermi National Accelerator Laboratory
Batavia, Illinois 60510

ABSTRACT

Introductory remarks to Parallel Session C7 at the XXth International Conference on High Energy Physics.

Now, nearly six years after the discovery of charmonium, the nonrelativistic theory of quarkonium spectroscopy has achieved a degree of maturity and permanence. For the most part, therefore, the contributions made to the subject in the past year represent extensions and refinements of calculations or analyses of new data, rather than novel theoretical initiatives.

The extensive results on the vector states in the Υ family obtained at CESR, together with new measurements of Υ and Υ' properties at DORIS, make it possible to further the program of unbiased (or "model-independent") determinations of the interquark interaction. Two facets of this line of investigation are reviewed in this session. What may be called the elementary quantum mechanics approach, in which the dependence of observables upon quark mass or principal quantum number is used to infer properties of the potential, is discussed in the contribution by André Martin. A complementary program in which bound-state properties are used to reconstruct the potential by means of inverse scattering techniques is the subject of Jonathan Rosner's report. Both methods provide support for the idea that the quark-antiquark interaction is flavor-independent. No direct evidence has been found for the one-gluon-exchange Coulomb potential at short distances; it appears that this awaits heavier quarkonia.

A key aspect of the nonrelativistic treatment is the need to make precise connections between properties of the Schrödinger wavefunction and observables. At issue are the importance of relativistic effects and of radiative (which is to say higher-order QCD) corrections to transition rates. Transcription of QED radiative correction formulae has led to the suspicion that such corrections may be large for quarkonium decays. Recent work, summarized here by W. Celmaster, suggests that at least for certain ratios of decay rates the corrections may not be uncontrollably large. The reliability of lowest-order expressions for absolute rates is still not established.

A worthy goal is the derivation of the interquark potential from a fundamental theory. Considerations of this sort lead to the expectations of Coulomb and linear potentials at very short and very long distances, and have yielded interesting suggestions for interpolations in the region of current experimental concern. QCD-inspired ruminations also point to the likelihood that non-quark degrees of freedom may manifest themselves as "extra" levels which cannot be reproduced by potential models. These may be regarded as string excitations or as $(Q\bar{Q}g)$ bound states. One view of this situation from the perspective of the MIT/Budapest bag model is presented by Julius Kuti.

Several topics on which there has been recent progress reported in contributions to this Conference could not be included in the presentations to this session. These include new work on hadronic transitions among quarkonium states, the influence of flavor threshold on upsilon line shapes, and theorems on bound-state properties. The Kepler problem of light quark-heavy quark bound states and extensions of sum rules for quarkonium states have also been omitted.

A final contribution to this session, by S. Meshkov, deals with quarkonium as an entrée to other new hadronic states. The specific issue addressed is whether the E(1440) meson observed in the decay $\psi \to \gamma$ + anything is to be interpreted as a glueball state in the process $\psi \to \gamma + (gg)$. More extended discussions of the characteristics of glueballs (gluonia?) have been presented by John Donoghue in parallel session A2 and by Carl Carlson in session D10.

A FIT OF HEAVY QUARKONIA

A. Martin
CERN, Geneva

The description of the J/ψ and T states as nonrelativistic $c\bar{c}$ and $b\bar{b}$ systems bound by a potential has had considerable success[1]. The nonrelativistic treatment can be a posteriori justified by evaluating the kinetic energy of the ground state from the formula[2]

$$\langle T\rangle_{1S} \simeq \frac{3}{4}\left(E_{1P}-E_{1S}\right)\left[1+\frac{7}{9}\left(\frac{E_{2S}+E_{1S}-2E_{1P}}{E_{2S}-E_{1S}}\right)^2\right] \qquad (1)$$

which gives for J/ψ, with the minimum C quark mass

$$\langle v^2/c^2\rangle \sim .25 \qquad (2)$$

Undoubtedly this value is not so small and there might be sizeable corrections if one inserts the potential in a Breit equation for instance[3]. However here we shall think of the potential as an effective potential to be put into the Schrödinger equation. We shall take this potential to be the <u>same</u> for J/ψ and T even though there might also be some corrections to a basically flavor independent quark-antiquark interaction.

Most of the time the potential chosen is inspired by theoretical prejudices according to which the short distance part is dominated by one gluon exchange and the long distance part is given by a linear confining potential. These prejudices may turn out to be correct, but, however, there is no reason to believe that the potential consists of the sum of these two contributions only or of a carefully chosen nice analytic interpolating formula. Nothing, in fact, is theoretically known about the intermediate range, which, as we shall see, is the really relevant fact in $c\bar{c}$ and $b\bar{b}$ spectroscopy. Under these conditions why not make the simplest possible Ansatz which is

$$V(r) = A + Br^{\alpha} \qquad (3)$$

and try to adjust A, B, α and the quark masses m_b and m_c. This has been done previously by Quigg and Rosner[4], and the various tests they have carried out point consistently towards values of α close to zero, i.e. toward a potential approximately logarithmic. The abundance of new data, from SLAC and CESR[5], has encouraged us to make a systematic fit with a potential of the form (3). For such a potential the energy levels are given by

$$M(n,\ell,m_q) = 2m_q + A + (B^2 m_q^{-\alpha})^{\frac{1}{\alpha+2}} C(\alpha,n,\ell) \qquad (4)$$

Fitting only the energy levels (but not the leptonic widths) of a single system will not allow to fix the quark masses since only the combinations $2m_q + A$ and $B^{2-\alpha} m_q$ appear. However a fit of both $c\bar{c}$ and $b\bar{b}$ will fix m_b and m_c.

The fit turns out to be very accurate so that hyperfine effects cannot be disregarded. These are removed by assuming a Fermi-type hyperfine splitting and identifying the state at 2.98 Gev discovered by Crystal Ball with the η_c. Using the WKB approximation, known to be good for potentials of type (3), we have for the $\ell = 0$ state:

$$\Delta_m(\text{triplet-singulet}) =$$

$$0.112\text{Gev} \left(\frac{m_q}{m_c}\right)^{-\frac{1+2}{2+\alpha}} \left(\frac{4n-1}{3}\right)^{2\,\frac{\alpha-1}{\alpha+2}} \qquad (5)$$

A preliminary estimate shows that α lies in the interval .1 to .2, allows to calculate (5) and to get masses for the center of gravity of the $\ell = 0$ states. This leads, using the 1S, 2S, 3S states to[6]

$$\alpha_{b\bar{b}} = 0.125(6) \quad \alpha_{c\bar{c}} = 0.104 \qquad (6')$$

However $\alpha = 0.104$ is acceptable for both systems (while $\alpha = 0$ is excluded). We then get a fit with the following parameters

$$\alpha = 0.104, \quad A = -7.15 \quad B = 6.58$$
$$m_c = 1.443 \quad m_b = 4.845 \qquad (7)$$

In Table 1 we give also the relative leptonic widths $\Gamma_{ee}(nS)/\Gamma_{ee}(1S)$ as calculated from the van Royen Weisskopf formula. We believe that the absolute leptonic widths are affected by too large corrections and are, in addition, not well known experimentally for the Υ system. We have adopted the energy scale of DESY.

Using the difference between the quark masses, rather insensitive to the choice of the potential[2], and the mass of the D mesons, flavor independence[7] leads to

$$M(b^-d)_{0^{-+}} = 5.285 \text{ Gev} \qquad (8)$$

compatible with the experimental facts presented at this conference:

$$\frac{10.58 - 0.05}{2} \leqslant M(b^-d)_{0^{-+}} \leqslant \frac{10.58}{2} \qquad (9)$$

The fit in Table 1 is good beyond all hopes. Notice the "prediction" of the Υ''' state and the mean P state energy of the $c\bar{c}$ system. The calculated relative leptonic widths are very reasonable

TABLE 1

	Energies		$\Gamma_{ee}(ns)/\Gamma_{ee}(1S)$		
	UPSILON				
	Experiment	Theory	Experiment		Theory
$^3 1S$	9.46	9.46*	1		1
$^3 2S$	10.02	10.02*	.44	.39	.43(R.45)
$^3 3S$	10.35	10.35	.35	.32	.28(R.32)
$^3 4S$	10.58	10.59 (R.10.60)	.27	.25	.20(R.26)
$^3 5S$		10.74			.16
$^3 1P$ Average		9.87			
$^3 2P$ Average		10.21			
	CHARMONIUM				
$^3 1S$	3.095	3.095*	1		1
$^3 2S$	3.684	3.684*	.45 ± .09		.35(R.45)
$^3 3S$	4.030	4.030*	.16 ± .02		.23(R.31)
$^3 4S$	{4.1, 4.4} ?	4.280			
$^3 1P$ Average	3.52	3.52			
$^3 1D$	3.77	"3.79" (No tensor force)			

* Input

TABLE 2

Hyperfine Splittings

$$M_{\Upsilon} - M_{\eta_b} = 0.056$$

$$M_{\Upsilon'} - M_{\eta_b'} = 0.027$$

$$M_{\Upsilon''} - M_{\eta''_b} = 0.018$$

$$M_{J\psi} - M_{\eta_c} = 0.112^*$$

$$M_{\psi'} - M_{\eta_c'} = 0.048$$

(10% to 20% deviation from the central experimental values). In parentheses we give the predictions of the Richardson potential[1] wherever they differ appreciably.

For completeness we give predicted hyperfine splittings in Table 2. Our conclusions are that

(1) Flavor independence is shown to hold almost too well.

(2) One cannot "prove" that the $b\bar{b}$ and $c\bar{c}$ potentials have a QCD shape from experimental data, and this presumably because the radii of $c\bar{c}$ and $b\bar{b}$, .3 and .2 Fermis respectively, are such that we explore neither the long range nor the short range region. Hopefully the $t\bar{t}$ will be discovered and with a predicted radius less than .1 Fermi (from $M_{t\bar{t}} > 36$ Gev and

$$\sqrt{\langle r^2 \rangle} < \frac{3}{m_t} (E_{1P}-E_{1S})^{(2)})$$

will allow an investigation of the one gluon exchange region. On the other hand, it will be very difficult to determine experimentally the nature of the confining part of the potential.

The hospitality of the Physics Department of the University of Washington, where this paper was written, is gratefully acknowledged.

REFERENCES

1. T. Applequist and H. D. Politzer, Phys. Rev. Lett. 34, 43 (1975).
   A. DeRujula and S. L. Glashow, Phys. Rev. Lett. 34, 46 (1975).
   T. Applequist and S. L. Glashow, Phys. Rev. Lett. 34, 46 (1975).
   E. Eichten et al, Phys. Rev. Lett. 34, 369 (1976).
   J. L. Richardson, Phys. Lett. 82B, 272 (1979).
   G. Bhanot and S. Rudaz, Phys. Lett. 78B, 119 (1978).
   H. Kraseman and S. Ono, Nucl. Phys. B154, 283 (1979).
   (The list is probably incomplete).
2. R. Bertlmann and A. Martin, Nucl. Phys. B168, 111 (1980).
3. H. J. Schnitzer, Phys. Rev. D18, 3482 (1978).
   D. Beavis, S. Y. Chu, B. R. Desai and P. Kaus, Phys. Rev. D20, 763 (1979).
4. C. Quigg and J. L. Rosner, Physics Reports 56, 167 (1979).
5. See K. Berkelmann, Rapporteur's talk, this conference.
6. A. Martin, CERN preprint TH 2843 (1980) to appear in Phys. Letters, and CERN preprint TH 2876.
7. A. Martin, Phys. Letters 88B, 133 (1979).

INVERSE SCATTERING AND THE Υ FAMILY

C. Quigg
Fermi National Accelerator Laboratory, Batavia, Illinois 60510

Jonathan L. Rosner[+]
School of Physics and Astronomy, Univ. of Minnesota, Minneapolis, MN 55455

ABSTRACT

A quarkonium potential is constructed with the help of masses and leptonic widths of the Υ(1S-4S) levels using the inverse scattering formalism. This potential agrees at all interquark separations beyond 0.06f with one constructed earlier from ψ and ψ', providing further evidence for flavor independence of the $Q\bar{Q}$ interaction. Comparison with other a priori potentials suggests that tests for a short-range Coulomb interaction (as predicted by QCD) will have to rely primarily on more precise values for $\Gamma(\Upsilon \to e^+e^-)$, on measurement of the 2S-2P spacing (predicted to be about 120 MeV for a short-range Coulomb-like interaction or in the inverse scattering formalism but about 150 MeV for an effective power-law potential), and on the discovery of heavier quarks.

INTRODUCTION

The nonrelativistic quark model, with Schrödinger dynamics, has some promise of describing the Υ family of resonances (reviewed in Ref. 1). A natural question is the form of the interquark interaction. The predictions of specific potentials with various degrees of theoretical motivation may be compared with spectroscopic data.[2-4] Scaling methods can show if a potential behaves like a power law $V(r) = A + Br^{\nu}$ in the region of interest.[5] The potential can be constructed directly from masses and leptonic widths with the help of the inverse scattering formalism.[6-8] A further step in this direction is described here.

We shall use the most recent parameters[1] of Υ(1S-4S) to construct a quarkonium potential which turns out remarkably similar to one constructed earlier from charmonium data.[6] Evidence for flavor independence of the quark-antiquark interaction is thereby extended beyond the range investigated earlier.[7] The predictions of this potential are compared with those of other a priori potentials.[2-4] It is found that most properties of the charmonium and Υ systems are sufficiently similar that tests for a QCD-motivated short-distance Coulomb singularity must be selected with some care. The masses of the lowest P-wave $b\bar{b}$ states, the leptonic width of the 1S Υ level, and properties of quarkonia heavier than the Υ family provide means for distinction among various potentials.

[+]Work supported in part by the U. S. Department of Energy under Contract No. EY-76-C-02-1764. Presented by J. Rosner.

METHODS AND RESULTS

A one-dimensional potential with a set of levels E_i may be constructed as follows. Choose an "ionization point" E_0 not far above the last level. If E_0 is chosen too high, a potential will be constructed with a large "energy gap"; it will consist of multiple buckets just as a solid (with energy gaps) also has multiple sites of attraction. Examples of this behavior have been demonstrated in some detail.[9] Let μ denote the reduced mass, and define

$$K_i = [2\mu(E_0-E_i)]^{1/2} \tag{1}$$

Then there is a unique symmetric, reflectionless potential $V(E_0, \{K_i\}, x)$, approaching E_0 at $x = \pm\infty$, with just the indicated bound states.

The potential is required to be reflectionless to avoid the need for phase-shift information in the inverse scattering formalism.[6,10] (We don't have quark-antiquark phase shifts.) Moreover the inverse problem for reflectionless potentials is an algebraic one, while for any other situation it involves the solution of an integral equation.

The potential is taken to be symmetric with an eye to the three-dimensional S-wave problem. For $\ell = 0$, the Schrödinger equation

$$[-\frac{\nabla^2}{2\mu} + V(r)]\ \Psi(\vec{r}) = E\ \Psi(\vec{r}) \tag{2}$$

becomes

$$[-\frac{1}{2\mu}\frac{d^2}{dr^2} + V(r)]\ [r\,\Psi(r)] = E[r\,\Psi(r)], \tag{3}$$

a one-dimensional Schrödinger equation for $r\,\Psi(r)$. Note that $r\Psi(r)$ vanishes at r=0. Hence it can be viewed as an odd-parity eigenfunction $\psi_{2n}(r) = r\,\Psi_n(r)$ (n=1,2,...) in a symmetric potential $V(-r) = V(r)$. Here n-1 labels the number of nodes between r=0 and $r=\infty$. The corresponding quarkonium masses may be identified with the odd-parity energy levels: $M(nS) = E_{2n}$ (n=1,2,...).

Let us solve the problem of N quarkonium levels. This leaves us with a collection of N unphysical levels $E_1, E_3, \ldots, E_{2N-1}$ and their corresponding even-parity wave functions $\psi_1, \psi_3, \ldots, \psi_{2N-1}$. In the absence of information about $E_1, E_3, \ldots$, we can specify instead the set of N values

$$|\psi'_{2n}(0)|^2 = |\Psi_n(0)|^2 = (16\,\alpha^2 e_Q^{\,2})^{-1}(M(nS))^2\,\Gamma(nS\to e^+e^-) \qquad (n=1,\ldots,N), \tag{4}$$

where the last equality is from Ref. 11. These are related to the $\{E_i\}$ by simple combinatorial identities.[8] For example, when N=1, we have

$$|\psi_2'(0)|^2 = \frac{K_2}{2}\ |K_1^2 - K_2^2|\ , \tag{5}$$

which may be used to solve for K_1.

The quarkonium information we use[1] is summarized in Table 1.

Table 1. Properties of $^3S_1$ quarkonium levels.

$c\bar{c}$ level	Γ_{ee}, keV	$b\bar{b}$ level	Γ_{ee}
		$\Upsilon_1(9.46)$	Γ_1 = 1.0 - 1.3 keV
$\psi(3.095)$	4.8 ± 0.6	$\Upsilon_2(10.02)$	(0.45 ± 0.07) Γ_1
		$\Upsilon_3(10.35)$	(0.32 ± 0.06) Γ_1
$\psi(3.684)$	2.1 ± 0.3	$\Upsilon_4(10.57)$	(0.25 ± 0.05) Γ_1

The ψ and ψ' information were used in Ref. 6 to construct a potential which gave the correct value of the $c\bar{c}$ spin-averaged P wave mass χ = 3.52 GeV, the correct Υ - Υ' spacing (0.56 GeV), and the correct Υ, Υ' leptonic widths. This was achieved with the choice E_0 = 3.8 GeV, m_c = 1.1 GeV. The potential is shown as the solid line in Fig. 1.

Previously[7] we used Υ, Υ' data to construct a potential which agreed with the solid curve in Fig. 1 out to the classical turning point of the Υ', but leveled off at 10.1 GeV for larger r. With four Υ levels it is now possible to extend the comparison to larger interquark separations. The results are the dashed and dotted curves in Fig. 1. The agreement with the charmonium potential is excellent. At very small distances ($r < 0.3$ GeV^{-1} = 0.06 fm) there is some sensitivity to the exact value of $\Gamma(\Upsilon \to e^+e^-)$. We expect that if the potential really has a short-distance Coulomb singularity its reconstruction using Υ levels will be deeper than that based on the ψ levels, since the heavier quarks in the Υ can probe shorter distances.

A check has been performed on the Υ potentials to see if they can provide the correct charmonium level spacings. The result of solving the Schrödinger equation is $\psi'-\psi$ = (576, 574) MeV, ψ' - χ = (154, 152) MeV for $\Gamma(\Upsilon \to e^+e^-)$ = (1,1.25)keV.

The Schrödinger equation also can be solved for other $b\bar{b}$ (P-wave and D-wave) levels. The results are shown as the solid lines in Fig. 2. Also shown are the predictions of a logarithmic potential[3] and of a QCD-inspired potential[4] whose form in momentum space is

$$V(\vec{q}^{\,2}) = -\frac{4}{3}\alpha_s(\vec{q}^{\,2})\frac{4\pi}{\vec{q}^{\,2}} \quad , \tag{7}$$

$$\alpha_s^{-1}(\vec{q}^{\,2}) = \frac{11-\frac{2}{3}n_f}{4\pi}\,\ln\left(1+\frac{\vec{q}^{\,2}}{\Lambda^2}\right) \quad , \tag{8}$$

where a fit to charmonium gave Λ = 398 MeV.

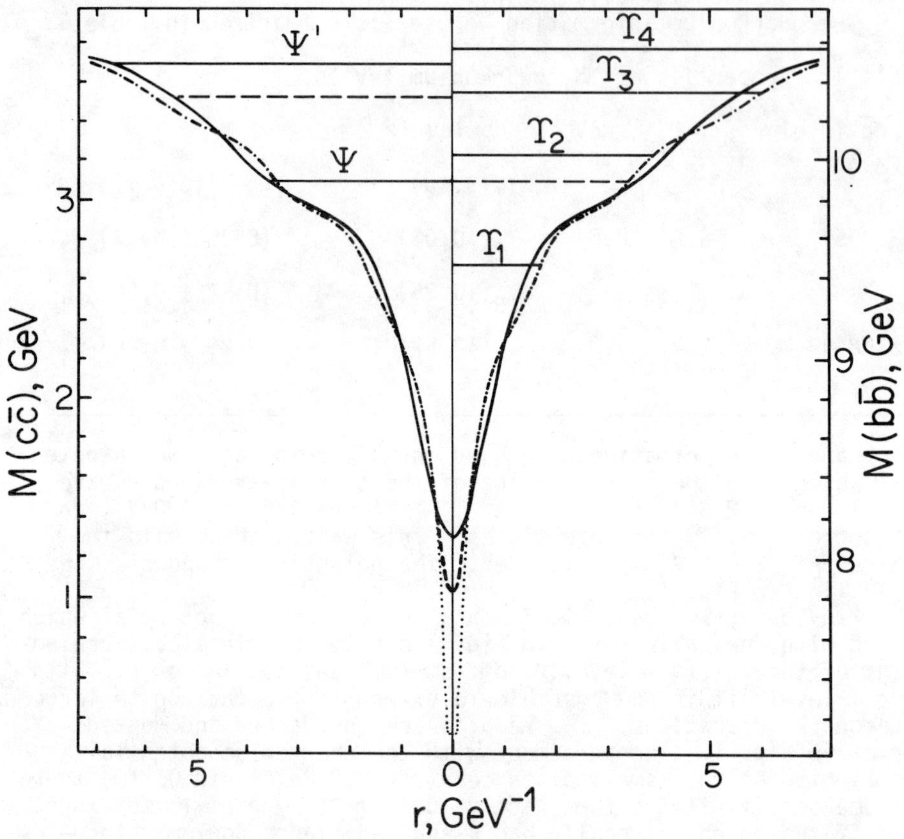

Fig. 1. Quarkonium potentials constructed by the inverse scattering method. Solid curve based on ψ, ψ' with E_0 = 3.8 GeV, m_c = 1.1 GeV. Dashed curve ($\Gamma(\Upsilon \to e^+e^-)$ = 1 keV) and dotted curve ($\Gamma(\Upsilon \to e^+e^-)$ = 1.25 keV) based on $\Upsilon_1, \ldots, \Upsilon_4$ with E_0 = 10.6 GeV, m_b = 4.5 GeV/c^2. Horizontal lines denote S wave levels (solid) and 2P levels (dashed) for charmonium (left) and upsilons (right).

The differences are not great. The Richardson potential is too "stiff" at large distances (as it is for charmonium) but this is not of serious concern to us. More important is the excess of predicted with respect to observed leptonic widths in the Richardson potential by factors of 1-1/2 to 2.[12] This would be evidence for important gluonic radiative corrections in the van Royen-Weisskopf formula (4). The expected form of these corrections if Coulomb binding dominates is

$$\Gamma_{ee} \sim |\Psi(0)|^2_{N.R.} \left[1 - \frac{16\alpha_s}{3\pi} \right], \qquad (9)$$

where N.R. denotes the non-relativistic limit, but opinions differ on

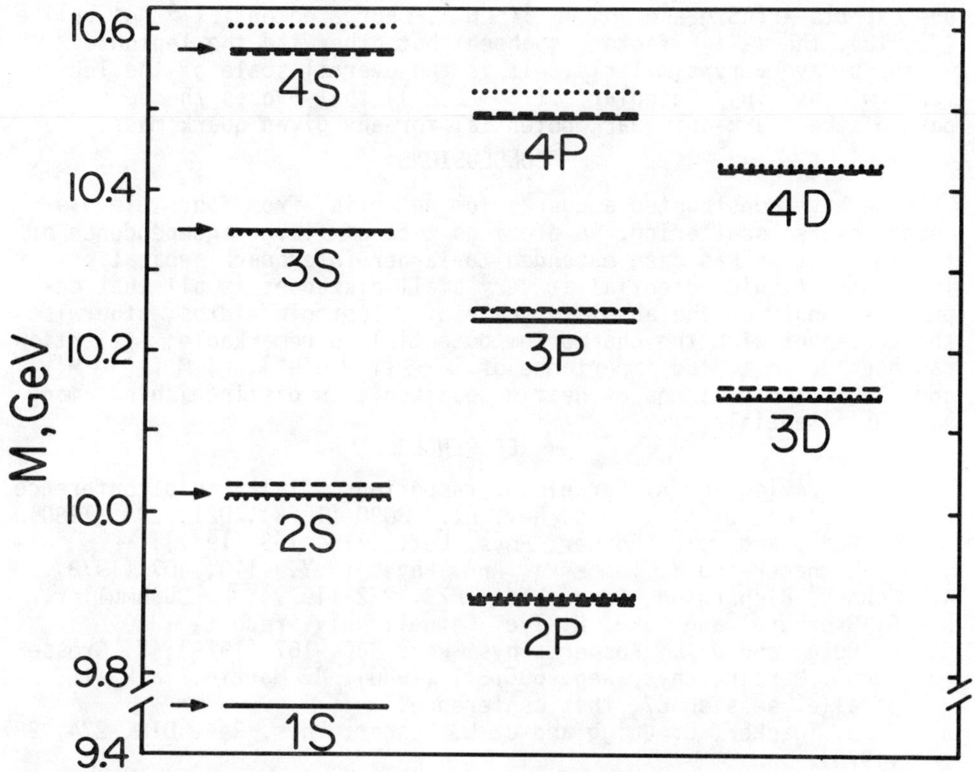

Fig. 2. Positions of S-, P-, and D-wave $b\bar{b}$ levels in several potentials. Solid lines: potentials constructed via inverse scattering (see Fig. 1). The average of predictions based on Γ_{ee} = 1.0, 1.25 keV is shown. Dashed lines: potential $V(r) = (0.715\ \text{GeV})\ \ell n(r/r_0)$. Dotted lines: Richardson potential (Eqs. (7), (8)). Arrows indicate positions of observed S wave Υ levels. The DESY mass scale (Υ = 9.46 GeV) is used.

---

the magnitude of this correction in the presence of a confining potential.[13]

The most important distinction among the predictions of Fig. 2 is the 2S-2P spacing:

$$2S - 2P = \begin{cases} 120\text{-}130\ \text{MeV (inverse)} \\ 150\ \text{MeV} \ (V \sim \ell n\ r) \\ 120\ \text{MeV (Eqs. (7), (8))} \end{cases} \qquad (10)$$

These differences may be crucial in observing monochromatic photons in $\Upsilon' \to \gamma\chi_b$. Even a smaller 2S-2P spacing is favored by a Coulomb + linear potential[2] in which the Coulomb strength does not decrease logarithmically at short distances.

We close by stressing the role of heavier quarks in deciding among potentials (e.g., $V \sim \ln r$ or Eqs. (7), (8).) For the former, one expects $\Gamma(nS \to e^+e^-) \sim m_Q^{-1/2}\, n^{-1}$. For a potential[14] not unlike (7), (8), the $m_Q^{-1/2}$ factor is absent but otherwise the leptonic widths behave very similarly. It is the overall scale of the leptonic widths (in particular $\Gamma(1S \to e^+e^-)$) that probes the deepest part of the quark-antiquark potential for any given quark mass.

CONCLUSIONS

We have constructed a quarkonium potential from four Υ levels using inverse scattering. A previous test of flavor independence of the interaction has been extended to larger interquark separations. The depth of this potential at very small distances is all that depends strongly on the absolute scale of Υ leptonic widths; otherwise the agreement with the charmonium potential is remarkable. Attention has been drawn to the importance of 1) $\Gamma(\Upsilon \to e^+e^-)$, 2) $M(\Upsilon') - M(\chi_b)$, and 3) leptonic widths of heavier quarkonia in distinguishing among various potentials.

REFERENCES

1. For a review see K. Berkelman, rapporteur's talk, this conference.
2. E. Eichten, et al., Phys. Rev. D17, 3090 (1978), D21, 203 (1980).
3. C. Quigg and J. L. Rosner, Phys. Lett. 71B, 153 (1977); M. Machacek and Y. Tomozawa, Ann. Phys. (N.Y.) 110, 407 (1978).
4. John L. Richardson, Phys. Lett. 82B, 272 (1979); W. Büchmuller, G. Grunberg, and S.-H. H. Tye, Cornell Univ. report, 1980.
5. C. Quigg and J. L. Rosner, Phys. Rep. 56C, 167 (1979); H. Grosse and A. Martin, Phys. Rep. 60C, 341 (1980); A. Martin, talk at parallel session C7, this conference.
6. H. B. Thacker, C. Quigg and J. L. Rosner, Phys. Rev. D18, 274, 287 (1978).
7. C. Quigg, H. B. Thacker, and J. Rosner, Phys. Rev. D21, 234 (1980).
8. J. F. Schonfeld, et al., Ann. Phys. (N.Y.) 127 (1980) (to be published).
9. W. Kwong, et al., Am. J. Phys., 1980 (to be published).
10. J. M. Gel'fand and B. M. Levitan, Am. Math. Soc. Trans. 1, 253 (1955); I. Kay and H. E. Moses, J. Appl. Phys. 27, 1503 (1956); A. C. Scott, F. Y. F. Chu, and D. W. McLaughlin, Proc. IEEE 61, 1443 (1973).
11. R. van Royen and V. Weisskopf, Nuovo Cimento 50, 617 (1967), 51, 583 (1967).
12. J. L. Richardson, private communication; Büchmuller, et al., Ref. 4.
13. R. Barbieri, et al., Phys. Lett. 57B, 455 (1975), Nucl. Phys. B105, 125 (1976); W. Celmaster, Phys. Rev. D19, 1517 (1979); E. Poggio and H. J. Schnitzer, Phys. Rev. D20, 1175 (1979); L. Bergström, H. Snellmann, and G. Tengstrand, Phys. Lett. 80B, 242 (1979); Ibid., 82B, 419 (1979); Royal Inst. of Technology (Stockholm) preprint TRITA-TFY-79-10, 1979 (unpublished).
14. M. Krammer, H. Krasemann, and S. Ono, DESY report 80/25, subm. to Z. Phys. C.

QCD CORRECTIONS TO QUARKONIA DECAYS

William Celmaster
Argonne National Laboratory, Argonne, IL 60439

ABSTRACT

A summary is given of recent work done on QCD corrections of hadronic decays of quarkonia. Particular attention is given to the decay $\eta_Q \to$ hadrons.

Some calculations[1,2] have been done of QCD corrections to quarkonia hadronic decays. Specifically, the branching ratios $P = \Gamma(Q \to e^+e^-)/\Gamma(Q \to \gamma\gamma)$ were computed beyond the leading order in α_s. These ratios may have the property of factorizing binding effects.[3] In particular, the wavefunction factors out in the ratio.

As of yet there is little data on these ratios, but the calculations are of great interest--not only because they are high-order QCD corrections--but because they serve as prototypes for the calculation of corrections to more easily measured processes such as $\Psi \to$ hadrons, etc.

I will mainly discuss the calculation of $P^{-+} = \Gamma(^1S_0 \to \text{hadrons})/\Gamma(^1S_0 \to \gamma\gamma)$. (An example of the $^1S_0$ is the η_c state.) The quoted result of this calculation was,[1] for 4 flavors

$$P^{-+} = C\alpha_s^2\left(1 + 22.14\,\frac{\alpha_s}{\pi}\right) \tag{1}$$

where

$$C = \frac{2}{9e_Q^4\alpha_{QED}^2} \tag{2}$$

If $\alpha_s \sim 0.2$ then the size of the correction is ~140%! This has caused a great deal of consternation since the calculation seems to suggest that a perturbation expansion of P will be unreliable.

What I wish to point out here is that the quoted result may be misleading in that the largeness of the coefficient is simply a consequence of the particular expansion parameter, α_s, used by Barbieri et al.[1] In fact, the definition of α_s depends on which method is used to renormalize the ultraviolet divergences encountered in the calculation of P^{-+}.[4-6] The α_s used by Barbieri et al. is known technically as $\alpha_{MS}(2m)$, where m is the quark mass and "MS" denotes "minimal subtraction"--an elegant mathematical trick for removing the divergences. A perhaps more physical technique[4,7] for eliminating divergences is "momentum space subtraction". This leads us to using the parameter $\alpha_{MOM}(m)$[8] which I call α_s' and is related to α_s by the equation

$$\alpha_s' = \alpha_s\left(1 + 10.18\,\frac{\alpha_s^2}{\pi}\right) \tag{3}$$

Substitution of Eq.(3) $\left(\alpha_s \sim \alpha_s' - 10.18\,\frac{\alpha_s^2}{\pi}\right)$ into Eq.(1) and Taylor expanding leads to

$$P^{-+} = C\alpha_s'^2\left(1 + 1.78\,\frac{\alpha_s'}{\pi} + \ldots\right) \tag{4}$$

Eq.(4) is equivalent, to this order of perturbation theory, to Eq.(1). The correction has been decreased at the expense of increasing the size of the coupling. A natural question is "Is the total value of P^{-+} the same in both Eqs.(1) and (4)?" I will show that it is not. Suppose $\alpha_s = 0.20$. (This is consistent with deep inelastic data.) Then from Eq.(3), $\alpha_s' = 0.33$. Fig.1 shows P^{-+}/C for Eq.(1) and for Eq.(4). The leading order results are very different in the two cases and are still far apart after including the first QCD correction but the α_s' method appears to be converging better. What is clear is that to this order of perturbation theory, the physical prediction is very different in the two schemes. Therefore, if we have reason to trust the α_s' perturbation expansion, then it is advisable to use Eq.(4) for comparison with experiment. In fact, α_s' may be trustworthy, for it leads to small corrections in other processes such as deep inelastic scattering.

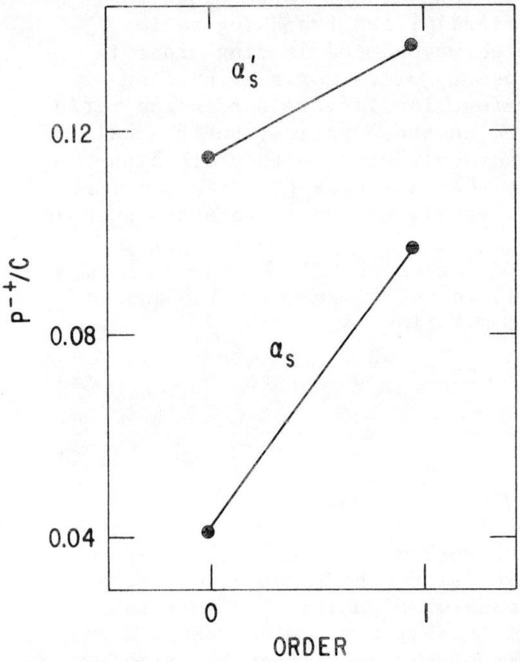

Fig.5

There is another argument due to Buras[6] which shows that the corrections to P^{-+} are probably small. He argues that in order to estimate the size of corrections to P^{-+} you can proceed as follows:

(1) Compute leading order P^{-+} and leading order deep inelastic scale violations (DIS).

(2) Use DIS data and the above leading order DIS result to obtain $\alpha_{(0)}$. Plug this into leading order P^{-+} to obtain $P^{-+}_{(0)}$.

(3) Repeat steps (1) and (2) but replacing "leading order" by "leading order + correction" and replacing $P^{-+}_{(0)}$ by $P^{-+}_{(1)}$.

(4) Finally, compare $P^{-+}_{(1)}$ and $P^{-+}_{(0)}$.

Buras[6] finds that the difference between $P^{-+}_{(0)}$ and $P^{-+}_{(1)}$ is not very large. Furthermore, this result is fairly insensitive to prescription. This analysis of Buras' agrees with our results expressed in terms of α_s'. That is to be expected simply because α_{MOM} leads to small corrections for both P^{-+} and for deep-inelastic scattering and therefore the correction must be small when comparing the two.

As a conclusion are the results of Barbieri et al.'s calculation of P-wave decays[2,9]

$$\frac{\Gamma(0^{++}\to \text{hadrons})}{\Gamma(0^{++}\to 2\gamma)} = C\alpha_s'^2\left(1+2.7\,\frac{\alpha_s'}{\pi}\right) \qquad (5)$$

and

$$\frac{\Gamma(2^{++}\to \text{hadrons})}{\Gamma(2^{++}\to 2\gamma)} = C\alpha_s'^2\left(1-3.8\,\frac{\alpha_s'}{\pi}\right) . \qquad (6)$$

They also predict

$$\frac{\Gamma(0^{++}\to \text{hadrons})}{\Gamma(2^{++}\to \text{hadrons})} = \frac{15}{4}\left(1+12\,\frac{\alpha_s}{\pi}\right) . \qquad (7)$$

This correction is, apparently, the only large correction of those computed by Barbieri et al. It also seems to be prescription independent, but one might question the wisdom of taking this ratio. In any case, if Eq.(7) is approximately valid, then it may well explain the experimental value of $\gtrsim 14$.[10]

REFERENCES

1. R. Barbieri, G. Curci, E. d'Emilio and E. Remiddi, Nucl. Phys. B154, 535 (1979).
2. R. Barbieri, M. Caffo, R. Gatto, and E. Remiddi, UGVA-DPT 1980/06-247.
3. A. Duncan and A. Mueller, CU-TP-175 (1980).
4. W. Celmaster and R. J. Gonsalves, Phys. Rev. Lett. 42, 1435 (1979) and Phys. Rev. D20, 1420 (1975).
5. P. M. Stevenson, Madison preprints DOE-ER/0881-153 (1980) and DOE-ER/0881-155 (1980).
6. A. Buras, Fermilab-Pub-80/43-THY (1980).
7. W. Celmaster and D. Sivers, ANL-HEP-PR-80-28 (1980).
8. W. Celmaster and D. Sivers, ANL-HEP-PR-80-29 (1980).
9. Actually it is not clear, from reading Ref.2, how many flavors of quarks are included in their calculations.
10. T. DeGrand, private communication and E. Bloom, in Proceedings of the 1979 International Symposium on Lepton and Photon Interactions at High Energy, T. Kirk and H. Abarbanel, editors.

QCD INSPIRED BAG MODEL OF QUARKONIUM

P. Hasenfratz, R.R. Horgan, J. Kuti[+] and J.M. Richard
CERN - Geneva

ABSTRACT

The QCD motivated bag model is applied to heavy quark - - antiquark systems. The effect of colored glue in the model is shown to explain the rapid cross-over of the static $Q\bar{Q}$ potential from the asymptotically free Coulomb region into the linear confinement regime. The spin-dependent force between static quarks is derived in Coulomb gauge from the exchange of a confined transverse gluon. The dimensional bag parameter $\Lambda_B = 235$ MeV and the quark-gluon coupling constant $\alpha = 0.38$ as defined at $r_{Q\bar{Q}} \simeq 0.2$ fermi are determined from a good fit of the $c\bar{c}$ and $b\bar{b}$ spectra. The fit is in serious disagreement with the widely accepted MIT parameters. As an important test of our model, we calculate the rich spectrum of $Q\bar{Q}$glue states. In Υ particle spectroscopy we predict a narrow $Q\bar{Q}$glue state with exotic quantum numbers $J^{PC} = 1^{-+}$ below the $B\bar{B}$ threshold. Its experimental confirmation would be the first direct evidence for colored glue in the hadron spectrum.

INTRODUCTION

This report is based on our two recent publications[1] where the bag picture[2] is assumed for the heavy $Q\bar{Q}$-system[3]. Accordingly, a bubble with well-defined surface is created around the heavy $Q\bar{Q}$-pair. Perturbation theory is valid inside with a chromoelectric permeability ϵ approximately one. Across the surface there is a rapid change of ϵ from one to zero. The 'ether' outside is a perfect diaelectric and paramagnetic medium against the gluon fields.

The energy required to create the bag phase is $\Lambda_B^4 V + \sigma A$ where V designates the volume of the bubble and A is the surface area. We ignore here the surface energy σA and only the volume energy is kept in the calculations. The volume energy density Λ_B is often denoted by $B^{1/4}$ in the bag literature.

In the adiabatic Born-Oppenheimer approximation[3] we proceed

[+]Speaker at the conference

in two steps. First, we solve the problem for fixed quark sources where the shape of the bag is determined by the minimum of the total energy with respect to the variation of the shape. In the second step the static energy of the $Q\bar{Q}$-system is used as the potential energy of the non-relativistic Schrödinger equation of the slowly-moving heavy quarks. With this method we generate the spectrum of $Q\bar{Q}$-states, the spin-dependent force between the $Q\bar{Q}$-pair, and a rich spectrum of $Q\bar{Q}$glue states where the transverse gluon adiabatically occupies the bag around the slowly moving quarks.

The surface of the bag is not a classical notion in our model, since the outside medium is pictured as a complicated quantum state. Whether the boundary of this medium exhibits some collective excitations remains an open theoretical problem. The latent dynamical degrees of freedom associated with the geometric bag variables are frozen out in our calculation.

$Q\bar{Q}$ SPECTROSCOPY

At short distances for $r < \Lambda_B^{-1}$ the chromoelectric Coulomb energy dominates and the bag is nearly spherical. For $r \sim \Lambda_B^{-1}$ we find a rapid crossover from approximately spherical shape into cylindrical shape. The radius of the chromoelectric vortex for large $Q\bar{Q}$-separation is given by

$$R_{vort} = \left(\frac{8\alpha}{3\pi}\right)^{1/4} \cdot \Lambda_B^{-1} \tag{1}$$

The energy per unit length (string tension) along the vortex tube is

$$\lambda = \left(\frac{32\pi}{3}\alpha\right)^{\frac{1}{2}} \cdot \Lambda_B^2 \tag{2}$$

The shape of the bag and the potential energy of the $Q\bar{Q}$-pair were determined accurately in a computer calculation. The Schrödinger equation was solved with our precise numerical potential which is Coulomb-like at short distances and linear at large distances. It is interesting to observe that the shape of the bag is almost spherical when the precocious confinement force sets in at a separation of $r_{Q\bar{Q}} \sim 0.6$ fermi.

In our numerical fit to charmonium and Υ-spectroscopy $\alpha = 0.4$ and $\Lambda_B = 235$ MeV were chosen. The calculated tension with these parameters is $\lambda = 0.2\ \mathrm{GeV}^2$ and the radius of the asymptotic flux tube is $R = 0.64$ fermi. In the Schrödinger equation $m_c = 1.35$ GeV and $m_b = 4.75$ GeV were taken for the masses of the c and b quarks, respectively. We find a good over-all fit to the heavy quarkonium spectra.

SPIN-DEPENDENT FORCE

We shall discuss now the exchange of a confined transverse gluon between the heavy $Q\bar{Q}$-pair. The level shift due to the exchange of the transverse gluon can be described in the second order of Schrödinger perturbation theory as

$$\Delta E_{spin} = \frac{g^2}{4M^2}\, \frac{1}{2}\lambda_1^a \sigma_1 \cdot G^{ab}_{mag}(\vec{r}_1, \vec{r}_2) \cdot \sigma_2 \frac{1}{2}\lambda_2^a \, , \tag{3}$$

where $G^{ab}_{mag} = \delta^{ab} \cdot G_m(\vec{r}_1, \vec{r}_2)$ is the static limit of the transverse chromomagnetic Green's function. In the spherical approximation, which is quite adequate for the ground state, $G_m(\vec{r}_1, \vec{r}_2)$ is calculable in closed form. The formula for the energy shift is given by

$$\Delta E_{spin}(r) = -\frac{2}{3}\vec{\mu}_1 \cdot \vec{\mu}_2\, \delta(\vec{r}) + \frac{1}{4\pi r^3}\left[\vec{\mu}_1 \vec{\mu}_2 - 3\,\frac{(\vec{\mu}_1 \cdot \vec{r})(\vec{\mu}_2 \cdot \vec{r})}{r^2}\right] +$$

$$+ \frac{R}{4\pi (R^2 + \frac{1}{4}r^2)^3}\left[\frac{1}{4}(\vec{\mu}_1 \cdot \vec{r})(\vec{\mu}_2 \cdot \vec{r}) - R^2 \vec{\mu}_1 \cdot \vec{\mu}_2\right] , \tag{4}$$

where $\vec{r} = \vec{r}_2 - \vec{r}_1$ is the quark-antiquark distance and $\vec{\mu} = \sqrt{\frac{4}{3}}\,(g/2M) \cdot \vec{\sigma}$ is the chromomagnetic moment of the quark. The second line of Eq.(4) displays the effect of confinement on the exchanged transverse gluon. For our charmonium ground state wave function the spin-dependent energy split

$$\Delta E_{spin} = 180 \text{ MeV}$$

is found. We expect a decrease of ΔE_{spin} due to asymptotic freedom effects at short distances.

$Q\bar{Q}$ GLUE STATES

In the Born-Oppenheimer approximation we predict a new set of unconventional $Q\bar{Q}$ glue states which bear some resemblance to diatomic molecules. The states are classified in Coulomb gauge where a transverse valence gluon is added to the heavy $Q\bar{Q}$-pair. The motion of the heavy $Q\bar{Q}$-pair is followed adiabatically by the constituent gluon which occupies a definite quantum state in the bag.

The lowest lying $Q\bar{Q}$glue states correspond to the TE gluon mode with eigenvalue $\lambda^{TE} = 2.74$. We find a $J^{PC} = 1^{-+}$ state which is exotic and has a mass of 10.5 GeV. The self mass of the gluon mode and spin-dependent level shifts in the $Q\bar{Q}$glue system are ignored in the calculation. That leads to an estimated $\pm$ 200 MeV ambiguity in our prediction for this $b\bar{b}$ glue state.

If our exotic and narrow $b\bar{b}$ glue state stays below $B\bar{B}$ threshold after the above mentioned corrections, its experimental confirmation would be the first direct evidence for gluon degrees of freedom in the hadron spectrum. It would also point, for the first time, to the predictive power of the QCD motivated bag model.

REFERENCES

1. P. Hasenfratz, R.R. Horgan, J. Kuti and J.M. Richard, Ref.TH. 2837-CERN and Ref.TH. 2838-CERN (1980) .
2. P. Hasenfratz and J. Kuti, Phys. Rep. 40C, 75 (1978) .
3. P. Hasenfratz, J. Kuti and A.S. Szalay, in Proc. 10th Rencontre de Moriond (1975) .

E(1440): GLUEBALL OR QUARKONIUM?

S. Meshkov
National Bureau of Standards, Washinton, DC 20234

ABSTRACT

The most likely assignment of the E(1440) is that of an ordinary $s\bar{s}$, L = 1, $J^{PC} = 1^{++}$ meson, rather than a flavor singlet meson or a $J^{PC} = 0^{-+}$ or 1^{++} glueball.

This talk is a report on work done with C. E. Carlson, J. J. Coyne, P. Fishbane and F. Gross. The E°(1440) has been with us for more than a decade and continues to be of experimental interest[1-3]; questions remain regarding its constitution and quantum numbers. There is even a suggestion[3] that the E is a glueball— a bound state of gluons with no valence quarks[4,5]. The more conventional description of the E is that it is a $J^{PC} = 1^{++}$ state belonging in the same nonet as the A_1 and the D°(1285). The usual quark content assignments are pure $s\bar{s}$ or pure flavor SU(3) singlet. In either case the D is an isospin zero state orthogonal to the E. We show that in the light of recent experimental data the interpretation of the E as a $J^{PC} = 1^{++}$ $s\bar{s}$ state is compelling, and that there are definite reasons for excluding both the SU(3) flavor singlet and glueball possibilities.

Consider the production cross sections of the D and the E. If their quark content is

$$E = s\bar{s} = \sqrt{\frac{1}{3}}\,\underset{\sim}{1} - \sqrt{\frac{2}{3}}\,\underset{\sim}{8} \quad \text{and} \quad D = \frac{1}{\sqrt{2}}(u\bar{u}+d\bar{d}) = \sqrt{\frac{2}{3}}\,\underset{\sim}{1} + \sqrt{\frac{1}{3}}\,\underset{\sim}{8} \qquad (1)$$

where $\underset{\sim}{1}$ and $\underset{\sim}{8}$ refer to flavor singlet and octet states, then qualitatively we expect the hadronic production cross section for the D to be larger than for the E. For most beams and all plausible targets, we make an E by fusing[6] two "ocean" quarks, whereas for the D at least one and (for the case of pions or nucleons) perhaps both of the quarks can be a valence quark. Valence quarks of a given flavor are more numerous than ocean quarks; ergo, it is easier to make Ds than Es. The production cross sections σ_E and σ_D calculated by this mechanism for the assignment of Eq. (1) are $\sigma_E = 2\,\mu b$, $\sigma_D = 15.5\,\mu b$. The same analysis repeated for the assignment

$$E = \underset{\sim}{1} = \sqrt{\frac{1}{3}}\,(u\bar{u}+d\bar{d}+s\bar{s}) \qquad D = \underset{\sim}{8} = \sqrt{\frac{1}{6}}\,(u\bar{u}+d\bar{d}-2s\bar{s})$$

yields $\sigma_E = 15\,\mu b$, $\sigma_D = 5\,\mu b$. Our result if the E were a glueball G depends on the quantum numbers.[4] If $J^{PC} = 1^{++}$, then the coupling of the glueball to massless gluons vanishes[4] and $\sigma_G = 0$; if $J^{PC} = 0^{-+}$, then using $\Gamma_G = .003$ GeV implies $\sigma_G = 7\,\mu b$.

Application of the fusion model down to m_E^2 may not be good to better than a factor of two, moreover the sea quark distributions are sufficiently uncertain so that in the $s\bar{s}$ assignment of the E, differing published distributions could lower σ_E to 1 μb or raise it to 7.5 μb.

Comparing the above predictions to the data must be done with care. Observations of the E, and often of the D, are made via $K\bar{K}\pi$

final states. It is necessary to understand what is implied by the experimental observation of a "δ"π decay mode, where "δ" $\to K\bar{K}$. The δ, mass 980 MeV (below $K\bar{K}$ threshold), has a width of 50 MeV and is an I=1, $J^{PC} = 0^{++}$ state composed of u and d quarks. We expect $\delta \to K\bar{K}$ to be quite rare.

For the E, we find that we can interpret the entire "δ"π signal as a kinematic reflection of $K^*\bar{K}$. In fact, if E decayed exclusively into $K^*\bar{K}$+c.c., the fraction of decays that would be called "δ" in ref. 1 would be 29%. This agrees with what is reported[1]. Using the data from ref.1 we find that $\sigma_E = 2.0 \pm 0.7$ μb and $\sigma_D = 5.4 \pm 3.0$ μb, consistent with the E being $s\bar{s}$ and in disagreement with the assignment E = $\underset{\sim}{1}$.

We turn to the decays of the J/ψ into γ+E and γ+D. From Eq.(1), $\frac{B(J/\psi \to \gamma D)}{B(J/\psi \to \gamma E)} = \frac{2}{1}$. On the other hand, if E is a flavor singlet and D is pure octet, then the preceding ratio would be zero. The E is clearly seen in the $K\bar{K}\pi$ channel, with the result $B(J/\psi \to \gamma E) \times B(E \to K\bar{K}\pi) = (3.6 \pm 1.4)\times 10^{-3}$ from Mark II data[3] and about 2×10^{-3} from Crystal Ball data[3,7]. For the D, an upper limit is quoted, $B(J/\psi \to \gamma D)\ B(D \to K\bar{K}\pi) \lesssim 0.7\times 10^{-3}$ from Mark II. With the $D \to K\bar{K}\pi$ branching ratio of 10%, there is no disagreement with either assignment for the E.

That the E might be a glueball is inconsistent with its width and observed branching fractions. The most likely choices for J^{PC} are 0^{-+} and 1^{++}. In Ref. 4 we show that for spin 0 and spin 1 glueballs, the expected width is rather narrow, of order 3 MeV. Also, a glueball would not be expected to decay so frequently into strange particles. For spin 0 glueballs, where helicity conservation arguments favor decays into heavy quarks[4,8], we expect the relative production of strange to non-strange quarks to be in the ratio $R = \frac{1}{2}\left[\frac{m_s}{m_u}\frac{\ell n(M_G/m_s)}{\ell n(M_G/m_u)}\right]^2$. The 1/2 takes into account that there are both u and d non-strange quarks. For a glueball mass 1.5 GeV, and with $\bar{m}_s = 0.5$ GeV, and $m_u \simeq m_d = 0.33$ GeV, we get R = 0.63. This is not compatible with the observed decays of the E.

REFERENCES

1. C. Bromberg et al., preprint CALT-68-747 (1980).
2. C. Dionisi et al., preprint CERN/EP 80-1 (1980).
3. D. L. Scharre et al., SLAC PUB-2514, 2538, 2519.
4. C. Carlson, et al., Glueballs and Oddballs: Their Experimental Signature, NBS preprint.
5. J. Bjorken, SLAC Summer Institute on Particle Physics, SLAC-PUB-2372; J. Coyne, P. Fishbane, and S. Meshkov, Phys. Lett.91B, 259 (1980). See references contained therein.
6. S. Ellis and M. Einhorn, Phys. Rev. D12, 2007 (1975); C. Carlson and R. Suaya, Phys. Rev. D14, 3115 (1976).
7. D. G. Aschman, to be published in the Proceedings of the 15th Rencontre de Moriond, Les Arcs, France, March 1980.
8. K. Ishikawa, Phys. Rev. D20, 2903 (1979); A. Soni, Nucl. Phys. B168, 147 (1980).

Chapter 6

Lepton-Nucleon Interactions

eN, μN, νN INTERACTIONS I

CROSS SECTIONS, SCALING, EXPERIMENT

A. Benvenuti, CERN, Organizer

NEW RESULTS ON INCLUSIVE νFe CHARGED CURRENT INTERACTIONS

CERN-Dortmund-Heidelberg-Saclay Collaboration *
Presented by
J.G.H. de GROOT
CERN, Geneva, Switzerland

INTRODUCTION

Based on 60.000(25.000) $\nu(\bar{\nu})$events obtained recently in the CERN 200 GeV Narrow Band Neutrino beam, we present new results on the nucleon structure functions $F_2(x,Q^2)$ and $xF_3(x,Q^2)$and on the Callan Gross relation. The observed scaling violations of F_2 and xF_3 are interpreted in terms of QCD. The strong coupling constant α_s is measured in agreement with the QCD predicted behaviour.

THE CALLAN GROSS RELATION

Previous analyses of deep inelastic processes have usually been performed assuming the validity of the Callan Gross relation [1] :

$$F_2(x) = 2x\ F_1(x)$$

Several effects [2] may give rise to a non-zero ratio of longitudinal to transverse structure function :

$$R = \frac{\sigma_L}{\sigma_t} = \frac{F_2\ (1+\frac{Q^2}{\nu^2}) - 2\ x\ F_1}{2\ x\ F_1}$$

The basic formula for the determination of R is :

$$\frac{d\sigma^{\nu}}{dy} + \frac{d\sigma^{\bar{\nu}}}{dy} \propto F_2\left[1 + (1-y)^2 - y^2R'\right] \text{ where } R' = \frac{F_2 - 2\ x\ F_1}{F_2} \qquad \text{I-1}$$

We fit expression I-1 to the data in ν-bins to avoid confusion with QCD-type scaling violations. The resulting values for R are shown in Fig. I-1 together with a first order QCD prediction [3] for Λ=.5GeV/c. No strong ν-dependence is observed. Averaged over ν we obtain :

$$R = .10 \pm .025(\text{stat}) \pm .07(\text{syst}).$$

Expression I-1 has also been fitted to the data in x-ν bins. over ν this gives the x-dependence of R as shown in Fig.I-2. Again the curve is a first order QCD prediction[3] for Λ=.5 GeV/c and a gluon distribution $\propto(1-x)^5$.

No strong x-dependence of R is observed. For this analysis, radiative corrections have been applied following a recent paper by A. De Rujula et al [4].

$F_2(x,Q^2)$, $xF_3(x,Q^2)$

The formulae used for the extraction of F_2 and xF_3 from the ν and $\bar{\nu}$ cross-section are :

$$q + \bar{q} = F_2 \propto \left[\frac{d^2\sigma^{\nu}}{dxdy} + \frac{d^2\sigma^{\bar{\nu}}}{dxdy}\right] / \left[1+(1-y)^2\right] \qquad \text{II-1}$$

$$q - \bar{q} = xF_3 \propto \left[\frac{d^2\sigma^{\nu}}{dxdy} - \frac{d^2\sigma^{\bar{\nu}}}{dxdy}\right] / \left[1-(1-y)^2\right] \qquad \text{II-2}$$

* H.Abramowicz, I.Becker, F.Dydak, F.Eisele, T.Flottmann, C.Geweniger, J.G.H.de Groot, C.Guyot, J.T.He, H.Klasen, K.Kleinknecht, J.Knobloch, J.Krolikowski, H.Lierl, J.May, J-P.Merlo, P.Palazzi, A.Para, B.Peyaud, B.Pszola, J.Rander, F.Ranjard, J.Rothberg, T.Z.Ruan, D.Schlatter, J.P.Schuller, J.Steinberger, K.Tittel, R.Turlay, W.von Rüden, H.Wahl, H.J.Willutzki, J.Wotschack and W.M.Wu.

Corrections are applied for radiative effects[4,5], and a nonzero value of R (R=.1). Fig. II-1 shows $xF_3(x,Q^2)$ as a function of Q^2 in x bins. Scaling violations are clearly present; xF_3 is rising with Q^2 for low values of x while xF_3 decreases with Q^2 for $x > .3$. In the QCD framework this feature may qualitatively be understood as due to the gluon radiation by quarks. Similar behaviour is observed for $F_2(x,Q^2)$ (Fig. II-2). The increase of F_2 with Q^2 at low x is more marked however. This is due to the combined effect of the sea and the onset of charm production, both of which should not be reflected in xF_3.

Both xF_3 and F_2 agree well with our earlier published measurements[6].

In first order QCD the Q^2 evolution of F_2 and xF_3 are given by the Altarelli-Parisi equations [7]:

$$Q^2 \frac{\partial x F_3}{\partial Q^2} = \frac{\alpha_s}{2\pi} \int_x^1 \frac{dy}{y}\, x F_3(y,Q^2)\, P_{qq}\left(\frac{x}{y}\right) \qquad \text{II-3}$$

$$Q^2 \frac{\partial F_2}{\partial Q^2} = \frac{\alpha_s}{2\pi} \int_x^1 \frac{dy}{y}\, F_2(y,Q^2)\, P_{qq}\left(\frac{x}{y}\right) + G(y,Q^2) P_{qg}\left(\frac{x}{y}\right) \qquad \text{II-4}$$

P_{qq} and P_{qg} are the splitting functions for gluon emission and quark pair production and α_s the strong coupling constant proportional to $[\ln Q^2/\Lambda^2]^{-1}$. In order to solve equations II-3 and II-4 numerically, we used a method developed by Abbot and Barnett[8], assuming a form $x^\alpha(1-x)^\beta(1+\gamma x)$ for F_3 and $(1-x)^\beta(1+\gamma x)$ for F_2 at a given $Q^2=Q_0^2$. A fit to the data then yields α,β,γ, and Λ. The results of the fit are shown in Figs. II-1 and II-2 in good agreement with the data. Our result for Λ determined from a fit to xF_3 for all x and F_2 for $x >.4$, $Q^2>5$ GeV^2/c^2 :

$\Lambda = 0.3\pm0.1(\text{stat}) \pm 0.1(\text{syst})$ GeV/c

THE STRONG COUPLING CONSTANT α_s

Measuring xF_3 and its Q^2 evolution for different values of Q^2 we can solve the Altarelli-Parisi equation for $x F_3$ (eq. II-3) to yield α_s. We used the form suggested by Abbot and Barnett to extrapolate to x = 1. The result for four values of Q^2 is shown in Fig.III-1 together with the expected QCD behaviour for Λ = .4 GeV/c.

REFERENCES

1. C.G. Callan and D.J. Gross, Phys.Rev.Lett. 22 (1969) 156.
2. R. Feynman, Photon-Hadron Interactions, (Benjamin New-York, 1972).
3. G. Altarelli and G. Martinelli, Phys.Lett. B76 (1978) 89.
4. A. de Rũjula, R. Petronzio and A. Savoy-Navarro, Nucl.Phys.B154(1979) 394.
5. J.Kiskis, Phys.Rev.D8 (1973), 2129
   R. Barlow and S. Wolfram, Oxford Report, OUNP 24/78 (1978).
6. J.G.H. de Groot et al.,Zeitschrift für Physik C1(1979),412.
   J.G.H. de Groot et al.,Phys.Lett. B82 (1979) 456.
7. G. Altarelli and G Parisi, Nucl.Phys. B126 (1977) 298.
8. L.F. Abbot and R.M. Barnett, Annuals of Physics 125 (1980) 276.

FIGURES

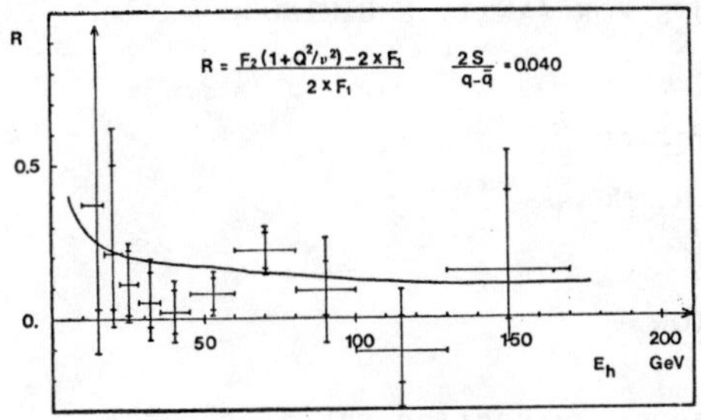

Fig.I-1 : ν-Dependence of R.

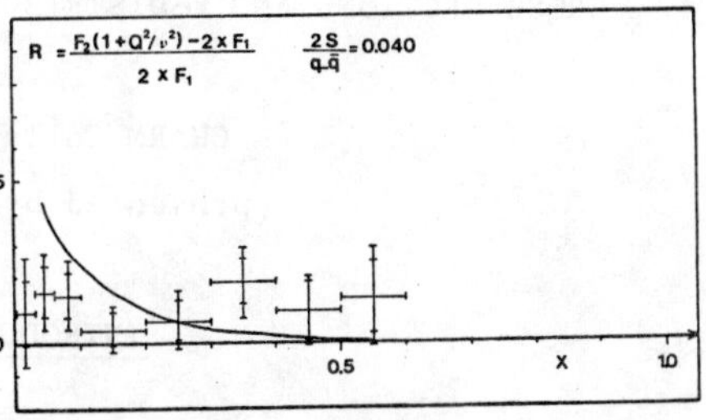

Fig.I-2 : x-Dependence of R.

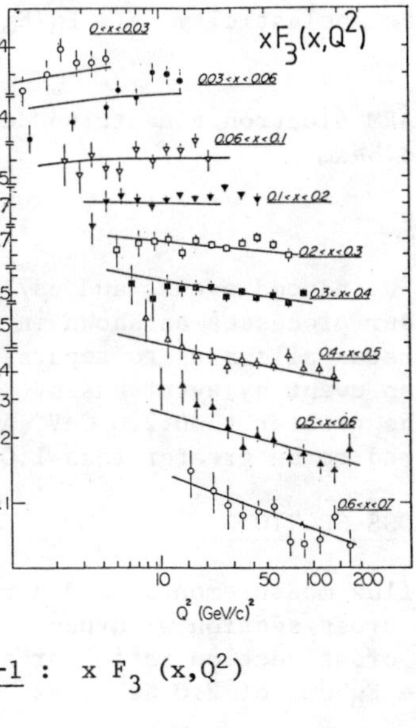

Fig.II-1 : $x F_3 (x,Q^2)$

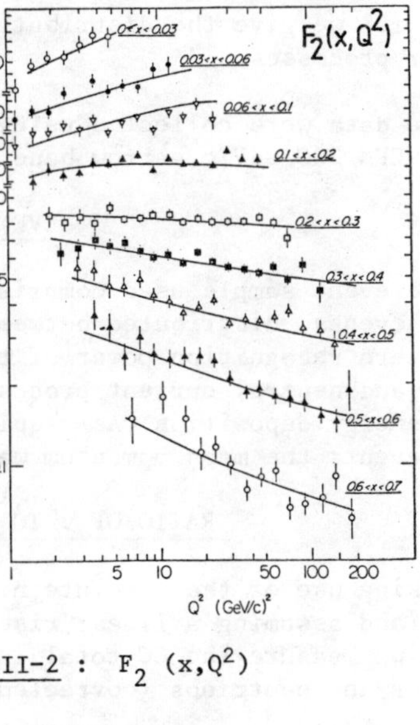

Fig.II-2 : $F_2 (x,Q^2)$

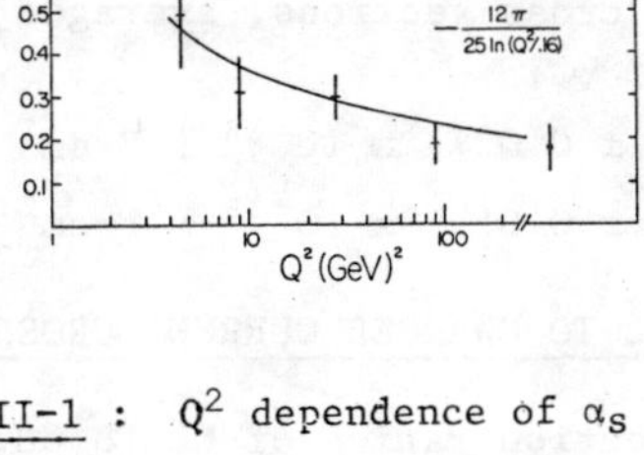

Fig.III-1 : Q^2 dependence of α_s

EXPERIMENTAL STUDY OF NEUTRAL AND CHARGED CURRENT CROSS SECTIONS AND Y-DISTRIBUTIONS FOR (ANTI) NEUTRINOS

CHARM Collaboration[1)]

presented by K.H. Mess

INTRODUCTION

We present here a measurement of the cross section ratio, r, of neutrino and antineutrino induced charged current events, and also the ratios, R and $\bar{R}$, of the neutral to charged current cross sections for $E_h > 2.0$ GeV. In addition we give the distribution in the inelasticity, $y(= E_h/E_\nu)$ for all four processes.

The data were collected using the CHARM electronic neutrino detector[1)] in the CERN 200 GeV/c narrow band neutrino beam.

EVENT SAMPLE

The event sample used comprises 8864 ν induced events and 3877 $\bar{\nu}$ induced events, distributed between the four processes as shown in Table I. The pattern recognition power of the apparatus allows us to separate the charged and neutral current processes on an event by event basis. The hadron energy deposition was required to be greater than 2.0 GeV, and, for CC events the muon momentum was required to be greater than 1.0 GeV/c.

RATIO OF ν TO ν CC CROSS SECTIONS

Making use of the absolute neutrino flux measurements as described in Ref. 2) and assuming a linear rise of the cross section with neutrino energy, we measure the CC total inelastic cross section ratio for antineutrinos and neutrinos (corrected for the E_h cut of 2.0 GeV), as

$$r = \frac{\sigma(\bar{\nu}N \to \mu^+X)}{\sigma(\nu N \to \mu^-X)} = 0.491 \pm 0.019$$

and the absolute inelastic cross sections, averaged over the neutrino energy range 20 to 200 GeV, as

$$\sigma(\nu N \to \mu^- X) = (0.594 \pm 0.027) \times 10^{-38} \times E_\nu \mathrm{cm}^2 \text{ (GeV nucleon)}^{-1}$$

$$\sigma(\bar{\nu}N \to \mu^+X) = (0.292 \pm 0.015) \times 10^{-38} \times E\text{-cm}^2 \text{ (GeV nucleon)}^{-1}$$

RATIO OF NEUTRAL TO CHARGED CURRENT CROSS SECTIONS

We obtain the cross section ratios of NC to CC, on an isoscalar target, for shower energies above 2 GeV.

$$R = \frac{\sigma(\nu N \to \nu X)}{\sigma(\nu N \to \mu^- X)} = 0.320 \pm 0.010$$

0094-243X/81/680738-03$1.50

$$\bar{R} = \frac{\sigma(\bar{\nu}N \rightarrow \bar{\nu}X)}{\sigma(\bar{\nu}N \rightarrow \nu^{+}X)} = 0.377 \pm 0.020$$

Using these ratios, together with the value of r for the same cut in E_h, in the Paschos-Wolfenstein formula[4] yields the following value of the electroweak mixing angle

$$\sin^2\theta = 0.230 \pm 0.023$$

FORM OF THE Y-DISTRIBUTIONS

Figures 1a) and 1b) show the acceptance corrected and resolution unfolded y-distributions. NC and CC events have been treated in an identical fashion. The muon information has been used only to classify each event as being an NC or a CC interaction. The CC distribution includes radiative corrections.

For both CC and NC the respective ν and $\bar{\nu}$ y distributions are consistent with being equal at y = 0. In the former case this indicates charge symmetry invariance between ν and $\bar{\nu}$; in the latter it indicates identical initial and final state neutrinos.

Parametrizing the ν and $\bar{\nu}$ distributions for both CC and NC as, in units of $\frac{G^2ME_\nu}{\pi}$

$$\frac{d\sigma^{\nu}}{dy} = ((1-\alpha) + \alpha(1-y)^2) \qquad \frac{d\sigma^{\bar{\nu}}}{dy} = (\alpha + (1-\alpha)(1-y)^2)$$

yields the following preliminary values:

$$\alpha_{CC} = 0.17 \pm 0.03$$

$$\alpha_{NC} = 0.23 \pm 0.04$$

The errors include systematic uncertainties added linearly to the statistical uncertainties.

In the limit of negligible strange sea α_{CC} is the total antiquark content of the nucleon. The present number agrees well with the other high precision determination[3], despite the completely different systematics in the two experiments.

The near equality of α_{CC} and α_{NC} indicates that the Lorentz structure of the neutral current is predominantly V-A.

REFERENCES

1) A.N. Diddens et al., A detector for neutral current interactions of high energy neutrinos. CERN-EP/80-63. To be published in Nucl. Inst. and Meth.

2) M. Jonker et al., Experimental study of neutral and charged current neutrino cross sections. Contribution to the International Conference on Neutrino Physics and Astrophysics, Erice, Italy. 1980.

3) J.G.H. de Groot et al., Z. f. Physik C1, 143-162 (1979).

4) E.A. Paschos and L. Wolfenstein, Phys. Rev. D7 (1973) 91.

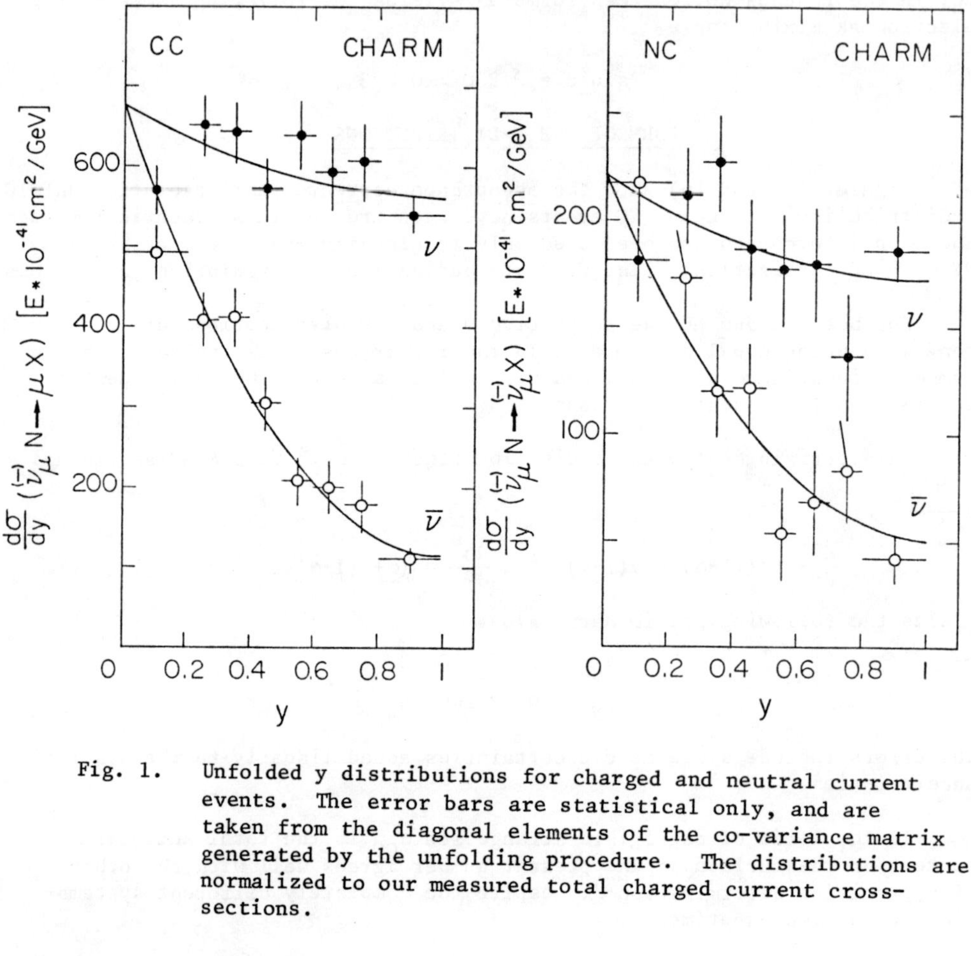

Fig. 1. Unfolded y distributions for charged and neutral current events. The error bars are statistical only, and are taken from the diagonal elements of the co-variance matrix generated by the unfolding procedure. The distributions are normalized to our measured total charged current cross-sections.

RECENT RESULTS FROM THE CFRR NEUTRINO EXPERIMENT AT FERMILAB

B. Barish, B. Blair, Y. Chu, B. Jin, D. MacFarlane, R. Messner, J. Lee, J. Ludwig, D. Novikoff, M. Purohit, F. Sciulli, and M. Shaevitz,
California Institute of Technology, Pasadena, California 91125

D. Edwards, E. Fisk, Y. Fukushima, D. Yovanovitch, Q. Kerns, T. Kondo, P. Rapidis, S. Segler, R. Stefanski, and D. Theriot,
Fermi National Accelerator Lab, Batavia, Illinois 60510

A. Bodek, R. Coleman, and W. Marsh
University of Rochester, Rochester, New York 14627

O. Fackler and K. Jenkins
Rockefeller University, New York, New York 10021

Talk Presented By
M. Shaevitz

ABSTRACT

Preliminary results from a new high-statistics study of neutrino and antineutrino interactions are presented. The full data sample includes 130,000 charged-current neutrino and 23,000 antineutrino events detected in the large acceptance Lab E neutrino target at Fermilab with the N-30 dichromatic beam. Results based on a preliminary sample of one-third of the data are given for normalized cross sections and neutral currents.

INTRODUCTION

Deep inelastic neutrino scattering has long been a source not only of weak interaction study, but also nucleon structure and new flavor production. Over the past five years, the experiments built to study this process have become much larger, yielding increased data samples for analysis. For these new data samples, systematic uncertainties dominate the statistical errors. In this paper, we report preliminary results from a new high statistics experiment to measure neutrino interactions with good understanding of systematic effects.

DETECTOR

The Lab E neutrino detector is a separated function device with a 690 ton target-calorimeter and 420 ton muon spectrometer. The target is composed of non-magnetized steel with transverse dimensions of 3m x 3m and is interspersed with scintillation counters every 10cm of steel and spark chambers every 20cm. The muon spectrometer is made of solid iron toroids 12 feet in diameter giving a total Pt kick of 2.4 GeV.

Absolute calibrations of the energy scale for the calorimeter and muon spectrometer have been performed using momentum tagged hadron and muon beams coming into Lab E.

NEUTRINO BEAM AND FLUX MONITORING

The experiment uses the new N-30 dichromatic train at Fermilab. In this setup, a 400 GeV primary proton beam impinges on a BeO target. Secondary particles are momentum and sign selected and then directed into a 340m evacuated decay pipe. The neutrino detector is located in Lab E, 910 meters downstream of the decay pipe. The flux and energy spectrum of neutrinos at the neutrino detector are calculated from measurements of the number of pions and kaons in the decay pipe along with their energies and spatial distribution.

A combination of detectors is used to measure the flux of secondaries. The total flux (independent of particle type) is measured by ion chambers placed at two locations along the decay pipe. These chambers are calibrated using foil activation techniques. This technique ultimately relies on the foil activation cross section. We have measured this cross section several times using either a beam current transformer, an RF cavity, or single particle counting as the primary normalization. All measurements agree to better than two percent.

Particle fractions are measured with a differential Cerenkov counter using a He radiator which integrates over the beam spill. The counter has been calibrated by a mono-energetic 200 GeV primary proton beam coming directly from the accelerator. This calibration provides a measurement of the gas index of refraction to better than .1% and also maps the counter response to a mono-energetic beam. The particle fractions are measured for each secondary momentum setting. (For example, the K/π ratio at +200 GeV and -200 GeV is $.145 \pm .006$ and $.0484 \pm .0025$ respectively, the π/P ratio at +200 GeV is $.223 \pm .013$.)

The flux monitoring system is designed to have many cross checks for the study of systematic errors. A segmented ion chamber behind the secondary dump measures the number and spatial distribution of decay muons and, therefore, the neutrino flux. At present this device has not been absolutely calibrated and can only be used to check the relative flux between different momenta and polarities. The pion fluxes obtained with this chamber agree with the above measurements using the Cerenkov counter and ion chamber to better than one percent for negative settings and five percent for positives.

For the preliminary results presented here, we estimate the systematic error in the neutrino flux to be below 10%. In the future, we should eventually reduce these errors to 2-3% using the additional information available from the calibrated muon chamber and the RF cavity placed in the decay pipe.

Knowledge of the energy spectrum of secondaries in the decay pipe is crucial for inferring the energy and flux of neutrinos at the detector. From measurements of particle yields for pions, kaons, and protons, energy spectra for each of the momentum settings are calculated. These spectra can be checked in two ways. First, for protons and kaons, the mean pressure observed in the Cerenkov counter is related to the mean momentum by

$$\text{Mean}\left(\frac{1}{(\text{Momentum})^2}\right)=\left(\frac{m^2_{K\text{ or }P}}{\langle\theta_c^2\rangle - 2K\langle\mathbb{P}\rangle}\right)^{-1}$$

where K and $\langle\theta_c^2\rangle$ are measured with the 200 GeV mono-energetic proton calibration and $\langle\mathbb{P}\rangle$ is the mean pressure. This procedure determines the energy to better than 1% and agrees with the value predicted from the particle yields at that level. An additional check using the measured energy of observed neutrino events also agrees to 1%.

TOTAL CROSS SECTION RESULTS

Neutrino and antineutrino data were recorded at five secondary energies (250,200,169,140,120 GeV) over a period of eight months ending January 1980. The total sample includes 130,000 charged current neutrino interactions and 23,000 antineutrinos. The cross section results shown here are preliminary and correspond to about one-third of the above sample. All events are reconstructed by computer with fiducial and reconstruction cuts applied to limit the sample to regions with good acceptance and small background. With these cuts, the fiducial volume of the target is reduced to 430 tons. Corrections are made for wide band background, cosmic ray contamination (<1%), geometrical efficiency, and the unsampled region at high x and high y (typically <3%). Cross sections for the different energy settings are then formed using the neutrino fluxes described previously.

For the final results shown in Figure 1, the different energy settings are combined; checks made in regions where different settings overlap show the cross sections to agree within statistical errors.

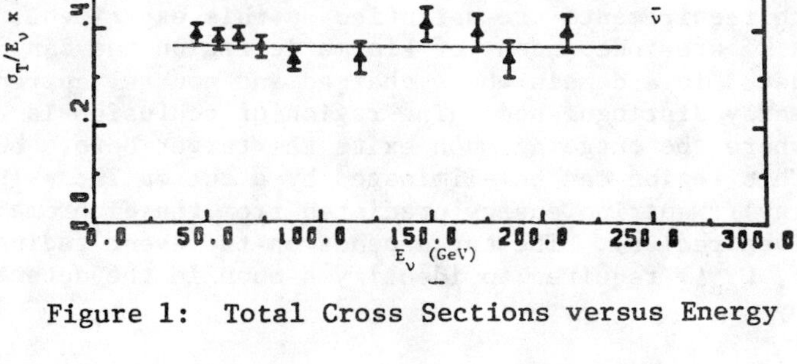

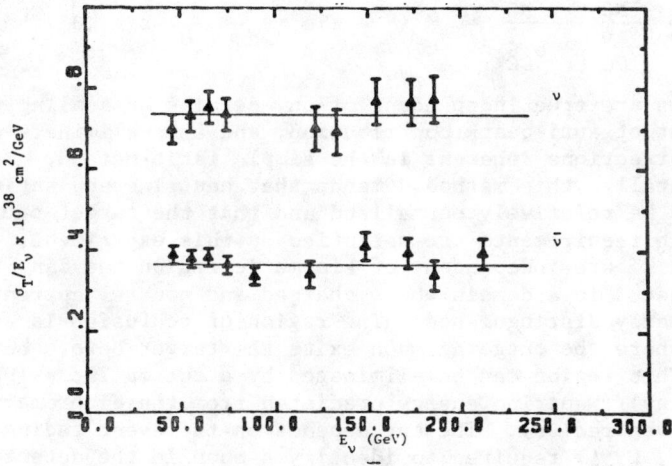

Figure 1: Total Cross Sections versus Energy

The average total cross sections for the range from 40-200 GeV are:

$$\sigma_T^{\nu}/E_{\nu} = .733 \pm .005 \pm .073 \times 10^{-38}\ \mathrm{cm}^2/\mathrm{GeV}$$

$$\sigma_T^{\bar{\nu}}/E_{\bar{\nu}} = .371 \pm .004 \pm .037 \times 10^{-38}\ \mathrm{cm}^2/\mathrm{GeV}$$

where the first error is statistical and the second systematic. The results agree with our previous measurement[1] for neutrinos of (.700 ± .015 ± .035) taken with this beam and detector but are 15-20% higher compared to the world average of .63 ± .02 for neutrinos and .30 ± .01 for antineutrinos.

NEUTRAL CURRENTS AND SIN $^2\theta_W$

The most precise measurements of sin $^2\theta_W$ are from high statistics studies of deep inelastic neutrino scattering. The errors in these measurements are dominated by systematic uncertainties. Experimentally, large backgrounds exist from wide band events, charged-current events identified as neutral currents, and electron-neutrino contamination in the beam. Theoretically, if one extracts sin $^2\theta_W$ from the ratio of neutral to charged currents, corrections must be applied for scaling violations, anti-quarks, neutron-proton excess, and the experimental cut on the minimum detectable hadronic energy. We have chosen to measure sin $^2\theta_W$ from the data by using the Paschos-Wolfenstein relations[2] with additional kinematic cuts in an effort to minimize some of the above uncertainties. In this method,

$$R- = \frac{\sigma_{NC}^{\nu} - \sigma_{NC}^{\bar{\nu}}}{\sigma_{CC}^{\nu} - \sigma_{CC}^{\bar{\nu}}} = \rho\,(\tfrac{1}{2} - \sin^2\theta_W)$$

and

$$R+ = \frac{\sigma_{NC}^{\nu} + \sigma_{NC}^{\bar{\nu}}}{\sigma_{CC}^{\nu} + \sigma_{CC}^{\bar{\nu}}} = \rho\,(\tfrac{1}{2} - \sin^2\theta_W + \frac{10}{9}\sin^4\theta_W)$$

These relations are true independent of any details of scaling violations or amount of anti-quark contribution, and thus eliminate some theoretical corrections inherent in the simple ratio method.

Experimentally, this method demands that neutrino and antineutrino cross sections be relatively normalized and that the target nucleus be isoscalar; both requirements are satisfied by this experiment. The ratios, R^+ and R^- are independent of kinematic region and can, therefore, be evaluated in a domain where charged and neutral current events are clearly distinguished. The region of confusion is at high x and y where the outgoing muon exits the target before being identified. This region can be eliminated by a cut on Ync = (Observed hadron energy)/(Neutrino energy predicted from the dichromatic beam at the given radius). The cut depends on the event radius, r, and the length, L_{cut}, required to identify a muon in the detector and is given by

$$Ync < Ycut = \frac{E_\nu \theta^2_{max}}{2M_p + E_\nu \theta^2_{max}} \quad \text{where } \theta_{max} = \tan^{-1} \frac{60''-r}{L_{cut}}$$

This procedure reduces the charged current contamination in the neutral current sample to 4% for neutrinos and 2% for antineutrinos.

The preliminary results given here are for the 200 GeV momentum setting (about 1/3 of the total data) for neutrinos and antineutrinos from pion decay with Lcut=210cm and Ehadron > 20 GeV. Small corrections (7.7% for neutrinos and 6.8% for antineutrinos) have been applied for the backgrounds mentioned previously giving the results:

$$R- = .258 \pm .035$$

$$R+ = .326 \pm .015$$

Additional corrections for strange/charm quarks, charm quark threshold effects, kaon neutrino contamination, and deviations of the target from pure isoscalarity amount to +3.4% for R- and +3.9% for R+. Converting these ratios to measurements of $\sin^2\theta_w$ gives the results:

Two Parameter Fit $\rho = 1.01 \pm .10$

$\sin^2\theta_w = .246 \pm .056$

One Parameter Fit $\sin^2\theta_w = .239 \pm .023$ with $\rho = 1.0$

(The errors shown are statistical only).

REFERENCES

1. J. Lee, Calif. Inst. of Tech., Thesis (unpublished).

2. E. A. Paschos and L. Wolfenstein, Phys. Rev. D7, 91, (1973). P. Q. Huns and J. J. Sukurai, Phys. Lett. 63B, 295, (1976).

NEUTRAL CURRENT INTERACTIONS IN ν N AND ν P COLLISIONS AND STUDY OF THE PRODUCED HADRONS

H. Yuta
IIT-Maryland-Stony Brook-Tohoku-Tufts Collaboration
(E545)

ABSTRACT

From an analysis of ν_μ interactions in deuterium we report the ratio of the neutral to charged current events for the neutron and proton targets and differences in P_t^2 distributions of the struck quark and of the recoiling diquark.

INTRODUCTION

We present recent results[1-4] from a study of ν_μ D interactions obtained from an exposure of the Fermilab 15' bubble chamber with the two plane EMI to a wide-band single-horn focused neutrino beam produced by 350 GeV/c protons. A total of 328,000 pictures were taken corresponding to 4.9×10^{18} protons on target. The data used for the present analysis consist of about 50% of the total exposure.

The film was scanned twice for neutral induced interactions with two or more charged particles with visible energy greater than 2 GeV. For events reconstructed with the geometry program TVGP, we require that the sum of the visible longitudinal momenta be greater than 5 GeV/c to remove neutral hadron induced background and that charge balance be satisfied. Multiplicity dependent corrections for scan-measuring losses were applied to the data to obtain an unbiased inclusive sample. The neutron and proton target events were first separated by the presence or absence of a spectator proton(including invisible) in the events. Then, a correction was applied to account for events converted from the neutron to proton target events due to rescattering in the deuteron[5].

NEUTRAL CURRENT

The neutral current events (NC) and the charged current events (CC) are separated by using a kinematic method[6] and the EMI information. Corrections for the EMI inefficiency and its limited acceptance are made for both NC and CC samples. Backgrounds from antineutrino CC and NC events are also corrected. The largest background in our neutrino NC sample comes from hadron induced events. To remove the hadron background, we use the similar procedure as used by Blietschau et al[7]; we first select from the above sample those events with the total visible hadron longitudinal momentum, $P_\ell^h > 5$ GeV/c. Then, we examine variation of the NC/CC ratio with the total visible hadron transverse momentum, P_t^h and take the ratio as one free from the hadron background if the ratio becomes independent of the cut value of P_t^h.

Using the procedure described above, we obtain the ratios of the NC and CC events with $P_t^h > 1.75$ GeV/c on the deutron target to be $R^d = 0.29 \pm 0.03$, which is to be compared with the world average for

Table I   Ratios of NC to CC events

	This experiment	Previous measurement
R^d	0.29 ± 0.03	0.301 ± 0.007 (Ref. 8)
$R^d(V^o)$	0.27 ± 0.06	———
R^p	0.50 ± 0.08	0.48 ± 0.17 (Ref. 9)
		0.51 ± 0.04 (Ref. 7)
R^n	0.21 ± 0.03	———

the ratio on isoscalar targets, 0.301 ± 0.007[8]. Similarly, we obtain the NC/CC ratios for events with V^o, $R^d(V^o)$, the proton target, R^p and the neutron target, R^n separately and give them in Table I. The ratio, R^n is the first measurement for such quantity. For comparison, we also list the previously measured ratios. The ratio of NC cross sections on neutron and proton targets is calculated to be

$$\sigma(\nu_\mu+n \to \nu_\mu+X)/\sigma(\nu_\mu+p \to \nu_\mu+X) = 1.02 \pm 0.19.$$

This value is consistent with the previous measurement of this ratio in the neutrino-neon experiment, 1.27 ± 0.36[10].

Using the ratios, R^p and R^n obtained from this analysis, we have determined the neutral current coupling constants, u_L^2 and d_L^2, where we use the expressions of Sehgal[11] to describe the NC cross sections for neutron and proton targets. The results are displayed in Fig. 1 with the prediction of the standard SU(2) x U(1) model[12]. The Weinberg angle is determined to be $\sin^2\theta_W = 0.17 \pm 0.07$ and 0.30 ± 0.09 from the proton and neutron target measurements, respectively, which are compatible with the SU(2) x U(1) prediction of $\sin^2\theta_W = 0.23$. The point of intersection of the two lines in Fig. 1 gives a measurement of the coupling constants:

$$u_L^2 = 0.20 \pm 0.08 \text{ and}$$
$$d_L^2 = 0.11 \pm 0.07.$$

These values are consistent with the previous measurement of $u_L^2 = 0.15 \pm 0.05$ and $d_L^2 = 0.17 \pm 0.07$ obtained by combining results of

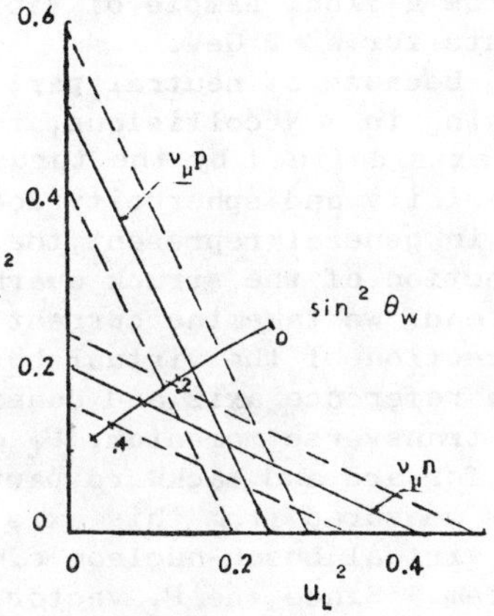

Fig. 1 Straight lines in the u_L^2-d_L^2 plane corresponding to R^p and R^n with the SU(2)xU(1) predictions. The dashed lines indicate the error sizes.

two different experiments; one using an isoscalar target[13] and the other using a hydrogen target[7].

TRANSVERSE MOMENTUM OF HADRONS

In this sections, we present results obtained from a study of jets produced in high energy ν-deuterium interactions. In particular, we analyze the data in terms of the transverse momentum, P_t, and the hadronic energy, W, in the rest frame of the virtual boson-nucleon system. The variable, W, has been suggested by Mazzanti et al[14] to be useful in avoiding complications due to finite experimental energies. We examine whether the current axis (direction of the virtual boson) is the appropriate axis for the jet analysis of ν-nucleon collisions and compare the size of the outgoing single quark jet with that of the recoiling diquark jet in the energy $2 < W < 18$ GeV and find differences in $\langle P_t^2 \rangle$ between these two jets. The differences in $\langle P_t^2 \rangle$ between these two jets have been suggested by Stevenson [15] and Ranft et al[16].

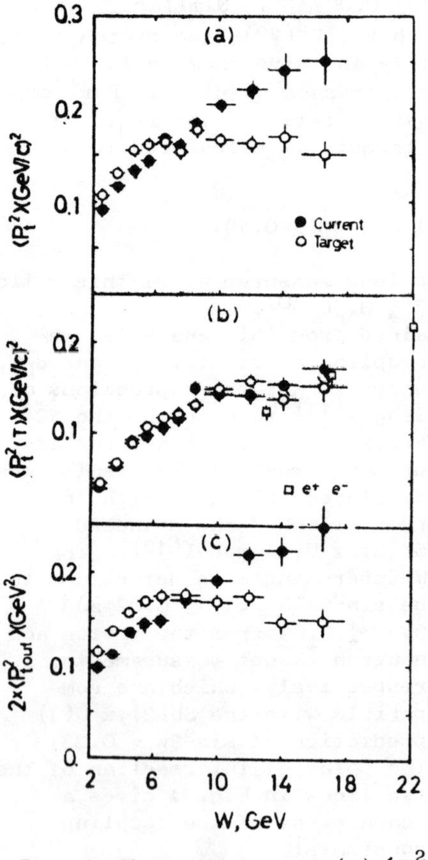

Fig. 2 W-dependence of (a) $\langle P_t^2 \rangle$, (b) $\langle P_t^2(T) \rangle$ and (c) $\langle P_{t\,out}^2 \rangle$. The dark and open circle points are for the current and target fragmentations.

The charged current events are selected by applying the kinematic method[17] with additional requirements of $E_\nu > 10$ GeV, Bjorken $y < 0.9$ and hadron multiplicity ≥ 4. This selection yields a final sample of 3162 events for $W > 2$ GeV.

Because of neutral particle missing in ν N collisions, the jet axis defined by the thrust, sphericity and spherocity does not in general represent the direction of the struck quark. Instead, we take the current axis (direction of the virtual boson) as a reference axis and consider the transverse momentum, P_t of the forward and backward particles measured from this axis in the virtual boson-nucleon c.m. system. Since the P_t vector sums for all particles should balance between the forward and backward hemispheres, any difference in P_t^2 between the two hemispheres reflects the difference in the jet properties between the single quark and the recoiling diquark system.

Fig. 2 (a) displays the $\langle P_t^2 \rangle$ as a function of W in the forward (current) and backward (target) hemispheres, defined by the plane perpendicular to the current axis. The data points are corrected for the prong-dependent scan-measuring efficiency. A Monte Carlo simulation has been used to correct for effects arising from our kinematic selection of the charged current events; the largest correction is for contamination of antineutrino charged current events which becomes important in the forward hemisphere for hadronic energies $W > 12$ GeV.

The most striking feature in Fig. 2 (a) is that both $\langle P_t^2 \rangle$ distributions have approximately the same magnitudes and rise together with W upto 8 GeV and then start deviating from each other for $W > 9$ GeV: the current $\langle P_t^2 \rangle$ keeps increasing, whereas the target $\langle P_t^2 \rangle$ levels off at about 0.15 GeV^2, indicating different jet properties of the single quark and the diquark system. The similar deviations in $\langle P_t^2 \rangle$ have been observed in the Ne-H_2 mixture experiment at CERN(18).

Fig. 2 (b) shows the W-dependences of $\langle P_t^2(T) \rangle$ for the forward and backward hemispheres, where $P_t(T)$ is the transverse momentum measured from the thrust axis, indicating no deviation as observed in Fig. 2 (a). This is due to the fact that the procedure used to determine the thrust axis tends to balance the transverse momenta between the forward and backward particles. Thus, analysis using the thrust or sphericity axis is not a sensitive way to detect small differences between the two hemispheres. The open square points in Fig. 2 (b) are the $\langle P_t^2(T) \rangle$ from the e^+e^- experiment(19) and plotted for comparison. Fairly good agreement is seen between the ν-D and the e^+e^- data for $W > 13$ GeV.

The inaccuracy of the energy calculation comes primarily from the missing neutral particles which causes deviations in the current axis. Consequently, this might give possible error in P_t^2. The transverse momentum, $P_{t\,out}$ measured perpendicular to the plane containing the incident ν and the outgoing μ^- is not influenced by the deviation of the current axis and is related to $\langle P_t^2 \rangle$ according to $\langle P_{t\,out}^2 \rangle = \langle P_t^2 \rangle / 2$ if the P_t^2 is correctly measured. Fig. 2 (c) shows the $\langle P_{t\,out}^2 \rangle$ vs. W, demonstrating that $\langle P_{t\,out}^2 \rangle$ agrees with the $\langle P_t^2 \rangle / 2$ within error for both hemispheres over the whole energy range $2 < W < 18$ GeV.

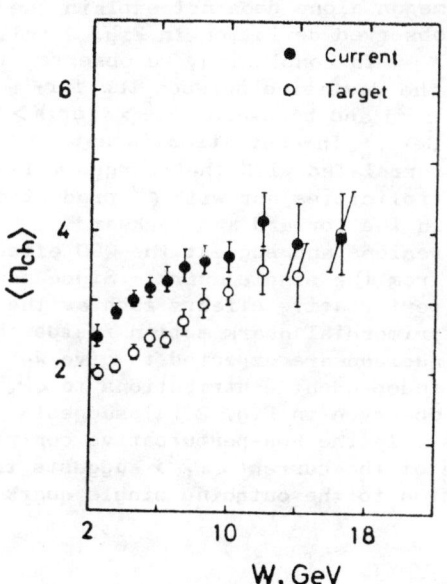

Fig. 3 W-dependence of the average charged hadron multiplicity.

Fig. 3 (a) shows the averaged charged multiplicities $\langle n_{ch} \rangle$, for the forward and backward hemispheres to see possible correlation of $\langle n_{ch} \rangle$ with the observed $\langle P_t^2 \rangle$ deviation. The $\langle n_{ch} \rangle$ for the target hemisphere in general gives smaller values than the current $\langle n_{ch} \rangle$. Both target and current $\langle n_{ch} \rangle$ rise together with W, but no significant $\langle n_{ch} \rangle$ behaviour which explaines the $\langle P_t^2 \rangle$ deviation is observed for $W > 9$ GeV.

We have also examined contributions of vector meson production to the forward and backward $\langle P_t^2 \rangle$. Since the ρ^o signal is the only one which is significant in the $\pi^+\pi^-$ mass distribution, we calculate the $\langle P_t^2 \rangle$ for π^+ and π^- tracks whose pair forms the $\pi^+\pi^-$ mass in the ρ^o mass range between 0.64 and 0.88 GeV. This is calculated for the forward and backward particles separately.

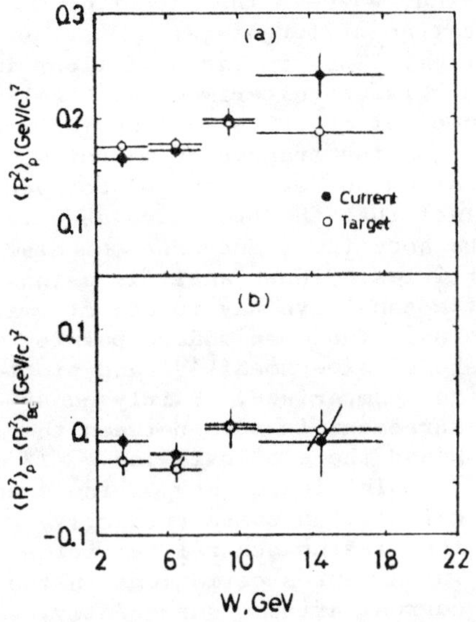

Fig. 4 W-dependence of (a) $\langle P_t^2 \rangle_\rho$ for the ρ^o mass region and (b) the differences of $\langle P_t^2 \rangle$ between the ρ^o and the control mass regions.

Fig. 4 (a) shows the plot of $\langle P_t^2 \rangle$ vs W indicating a similar deviation between the forward and backward $\langle P_t^2 \rangle$ observed in Fig. 2 (a). Fig. 4 (b) shows the differences between the $\langle P_t^2 \rangle_\rho$ in the ρ^o mass region and the $\langle P_t^2 \rangle_{BG}$ in the control mass region, where the control mass regions are taken to be from 0.40 to 0.60 GeV and from 0.92 to 1.16 GeV. The differences of $\langle P_t^2 \rangle_\rho - \langle P_t^2 \rangle_{BG}$ show no significant deviation between the forward and backward regions, indicating that the ρ^o meson alone does not explain the observed deviation in Fig. 2 (a).

In conclusion, we observe the deviation between the forward and backward $\langle P_t^2 \rangle$ for $W > 9$ GeV. The deviation is not correlated with the charged multiplicities nor with ρ^o production in the forward and backward regions and suggest the QCD effect from the single quark. Since non-perturbative effects such as the primordial quark motion inside the nucleon are expected to give W-independent contributions to $\langle P_t^2 \rangle$ the flattening of the target $\langle P_t^2 \rangle$ observed in Fig. 2 (a) suggests that the recoiling diquark receives only the non-perturbative contribution for $W > 9$ GeV, while the rise of the current $\langle P_t^2 \rangle$ suggests the presence of an additional contribution to the outgoing single quark due to the gluon emission.

REFERENCES

1. C. C. Chang et al., Paper No. 519 submitted to this conference.
2. T. Kitagaki et al., Paper No. 700 submitted ot this conference.
3. T. Kitagaki et al., Paper No. 738 submitted to this conference.
4. S. Sommars et al., Paper No. 539 submitted to this conference.
5. J. Hanlon et al., Neutrino '79 Conference, Bergen, June 1979.
6. G. Myatt, CERN/ECFA/72-4, Vol. II, p. 117.
7. J. Blietschau et al., Phys. Letters 88B, 381 (1977).
8. C. Baltay, "Proceedings of the 19th International Conference on High Energy Physics, Tokyo, Aug. 1978", Phys. Soc. of Japan (1979).
9. F. A. Harris et al., Phys. Rev. Letters 39, 437 (1977).
10. J. Marriner, Ph. D. Thesis, Univ. of California, Berkeley (1979).
11. L. M. Sehgal, Phys. Letters 71B, 99 (1979).
12. S. Weinberg, Phys. Rev. Letters 19, 1264 (1967); A. Salam, "Elementary Particle Physics", Almquist and Wiksells, Stockholm (1968).
13. H. Deden et al., Nucl. Phys. B149, 1 (1979).
14. P. Mazzanti et al., Phys. Letters 81B, 219 (1979).
15. P. M. Stevenson, Nucl. Phys. B150, 357; ibid, B156, 43 (1979).
16. J. Ranft and G. Ranft, Phys. Letters 82B, 129 (1979).
17. J. Bell et al., Phys. Rev. D19, 1 (1979).
18. H. Deden et al., CERN/EP-66, 13 May 1980.
19. G. Wolf, Proceeding of the 1979 International Symposium on Lepton and photon Interactions at High Energies, Aug. 23-29, FNAL.

EVIDENCE FOR GLUON RADIATION IN HIGH ENERGY NEUTRINO INTERACTIONS

Presented by V.J. Stenger for the Berkeley, Fermilab,
Hawaii, Seattle, Wisconsin Collaboration*

The hypothesis that gluon bremsstrahlung is a major mechanism in the production of high p_T hadrons in e^+e^- interactions seems now to be strongly confirmed by several experimental groups at PETRA/1/. While the high cm energy available for e^+e^- collisions makes this a good place to observe gluon effects, there are some advantages for studying these phenomena in lepton-hadron collisions. For example, the direction of the current vector $\vec{Q}$ is known whereas the "thrust" axis must be determined in e^+e^- experiments. This is particularly important since detailed QCD calculations predict a strong forward backward asymmetry in high p_T hadron production.

In this report evidence is presented for gluon radiation effects in ~4000 high energy charged current νN interactions in the Fermilab 15-foot bubble chamber operated in the Quad Triplet Beam. A portion of the sample was selected with a muon momentum transverse to the beam direction of greater than 4 GeV/c. The resulting sample has $\langle E_\nu \rangle = 112$ GeV, $\langle Q^2 \rangle = 26.9$ GeV^2/c^2 and $\langle W^2 \rangle = 67.3$ GeV^2. We present results on the p_T of charged hadrons, measured relative to the visible $\vec{Q}$ vector direction, and preliminary results on angular energy flow.

In Fig. 1 the p_T^2 distribution is shown for two regions of W^2. We see a high p_T^2 tail which increases with increasing W^2. The data are compared to a QCD calculation of Halzen and Scott/2/ in which no parameters were adjusted to fit our data, α_s being determined from PETRA data.

In Fig. 2 we present $\langle p_T^2 \rangle$ as a function of W^2 and Q^2 for forward and backward charged hadrons separately. We see that the high p_T tail is associated entirely with the forward hadrons, as predicted by QCD. The forward data are systematically high compared to a Monte Carlo based on Longitudinal Phase Space (the curves labelled LPS--an average of only slightly different forward and backward predictions).

If gluon bremsstrahlung is the process responsible for high p_T phenomena, events with high p_T should also tend to be planar. We define a quantity $\pi_N = A \sum_F (p_T - 0.320)/\sqrt{n_F}$ as a measure of excess forward high p_T in the event as a whole (constants chosen to give a mean ~0 and a width ~1 independent of multiplicity). To obtain a quantity which measures the planar nature of the event we perform an event-by-event rotation about $\vec{Q}$ to maximize the "planarity" $P = \Sigma(p_{T1}^2 - p_{T2}^2)/\Sigma(p_{T1}^2 + p_{T2}^2)$. The angular energy flow (AEF), defined as the fraction of energy radiated into angular regions in the plane which maximizes planarity, is shown in Fig. 3 for the region of high π_N and planarity ($\pi_N > 2.4$, $P > 0.6$). It is seen to be consistent with the configuration expected from a qqG process.

We have compared these data with a Monte Carlo based on the

Feynman-Field parameterization of quark fragmentation/3/ with no QCD effects/4/. We find that the Monte Carlo events in the high π_N, high P region have a forward dip in the AEF which is characteristic of this kinematic region. However, the Monte Carlo predicts many fewer events in that region, as seen in Table I. For a given π_N region, the Monte Carlo gives approximately equal numbers of events in the high P and low P regions, in sharp disagreement with the data. We are currently developing this Monte Carlo further to include gluon effects and preliminary results indicate that, indeed, they lead to an excess of events at high π_N and P.

In summary, hard gluon bremsstrahlung such as observed at PETRA, or something quite similar, is manifested by the following features of our data: i) the high p_T tail associated with forward hadrons, increasing with Q^2 and W^2, ii) the forward high p_T events being mostly highly planar and occurring at a rate much greater than predicted by Monte Carlos without gluon effects. In addition, the events with high forward transverse momenta and planarity exhibit the expected 3-jet structure. Taken as a whole, these results cannot be explained by any non-ad-hoc models involving simple assumptions such as scaling or LPS.

This work was supported, in part, by the U.S. Department of Energy and the National Science Foundation.

---

*H.C. Ballagh, H.H. Bingham, W.B. Fretter, T. Lawry, G.R. Lynch, J. Lys, J. Orthel, M.D. Sokoloff, M.L. Stevenson, G.P. Yost, University of California and Lawrence Berkeley Laboratory, Berkeley, CA 94720; D. Gee, F.R. Huson, E. Schmidt, W. Smart, E. Treadwell, J.Wolfson, FERMILAB, P.O. Box 500, Batavia, IL 60510; R.J. Cence, C.H. Fujiyoshi, F.A. Harris, M.D. Jones, A. Koide, S.I. Parker, M.W. Peters, V.Z. Peterson, V.J. Stenger, G.N. Taylor, Department of Physics, University of Hawaii at Manoa, Honolulu, HI 96822; T.H. Burnett, L. Fluri, H.J. Lubatti, K. Moriyasu, D. Rees, G.M. Swider, E. Wolin, Visual Techniques Lab, Department of Physics, University of Washington, Seattle, WA 98195; U. Camerini, W. Fry, R.J. Loveless, P. McCabe, M. Ngai, D.D. Reeder, Department of Physics, University of Wisconsin, Madison, WI 53706.

REFERENCES

1. R. Brandelik et al.,Phys. Lett. 83B,261(1979);D.P. Barber et al., Phys. Rev. Lett. 43,830(1979);Ch. Berger et al.,Phys. Lett.86B, 418(1979);N. Bartel et al.,Phys. Lett. 91B,142(1980).

2. F. Halzen and D.M. Scott,Univ. of Hawaii preprint UH-511-386-80 (1980),DOE-ER/0081-157(1980),and private communication.

3. R.D. Field and R.P. Feynman, Nucl. Phys. B136,1(1978).

4. C. Day, private communication.

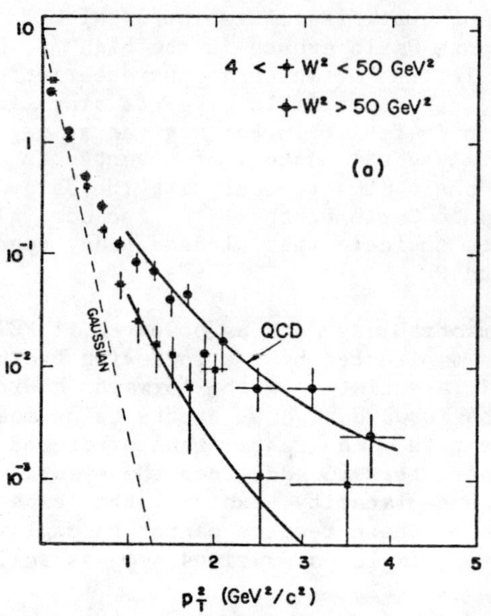

Fig. 1. Distribution in p_T^2 of charged hadrons for two ranges of W^2.

Table I. Number of events with P > 0.6 and high π_N, compared

	$\pi_N > 1$	$\pi_N > 2.5$
Data	84	20
Monte Carlo	25	2

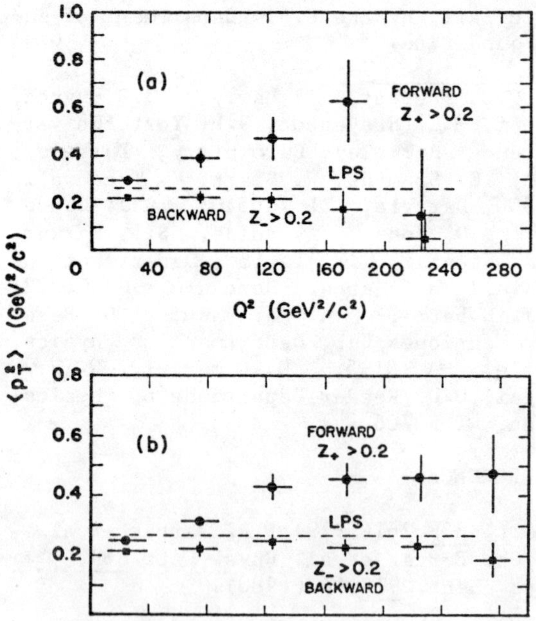

Fig. 2. $\langle p_T^2 \rangle$ vs. Q^2 and W^2 for forward and backward charged hadrons.

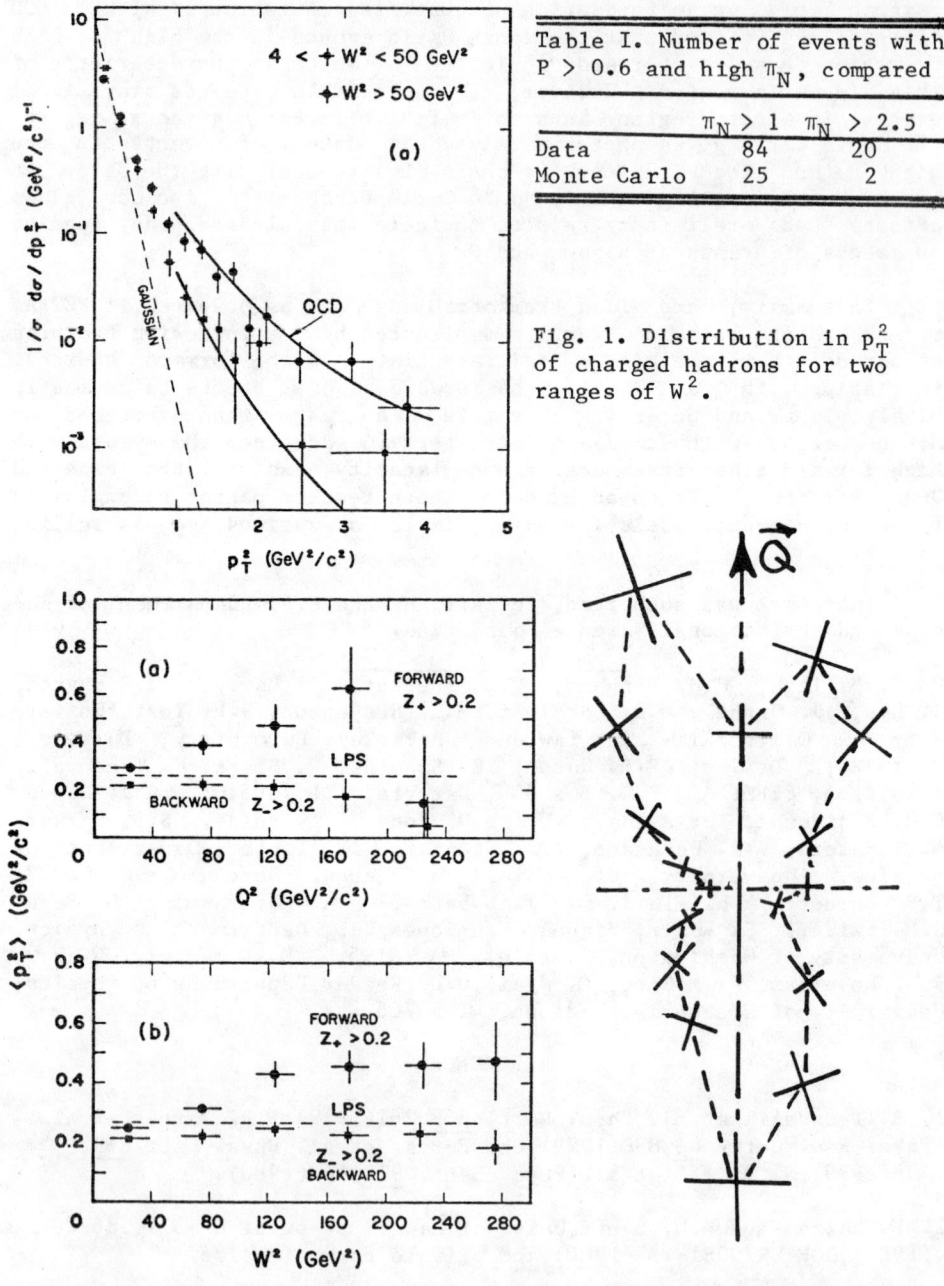

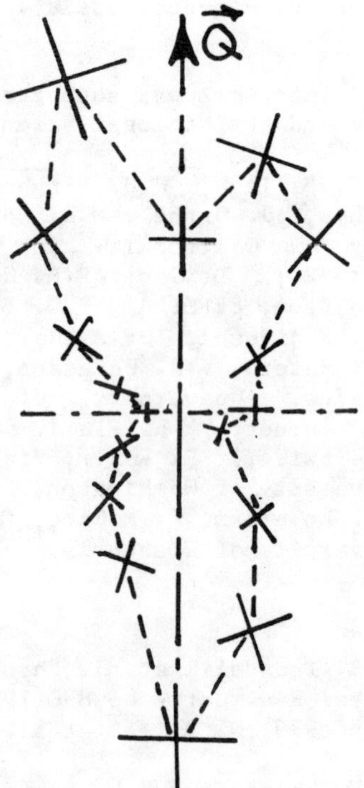

Fig. 3. Angular energy flow in the plane of maximum planarity for events with P > 0.6 and π_N > 2.5.

ENERGY DISTRIBUTION AND AVERAGE TRANSVERSE MOMENTUM OF PRODUCED HADRONS IN DEEP INELASTIC MUON SCATTERING

The European Muon Collaboration,
presented by F.W. Brasse

Energy distributions of charged and neutral hadrons in the forward direction and the average transverse momentum squared of charged hadrons have been determined in the interactions of 280 GeV muons with protons. The muons were incident on a 6m long hydrogen target, of which the downstream 2m only have been used for this study of hadron production. Charged hadrons as well as the scattered muons were measured in momentum and angle by a forward spectrometer described elsewhere[1]. The muon was identified after approximately 2.5m of iron which absorbed all the hadrons. Neutral hadrons were measured by a lead-scintillator calorimeter sitting in front of the absorber. The space resolution of this calorimeter however did not allow to identify two photons from π^0 decay. The momentum of each incident muon was measured with an accuracy of ±0.3%. The data presented here correspond to a total flux of $3 \cdot 10^{11}$ muons and are derived from ∿23,000 events, satisfying the conditions $20 < \nu < 260$ (240) GeV, $W^2 > 40$ GeV^2, $E_\mu' > 20$ GeV and $Q^2 > 1$ (5) GeV^2. The values in brackets correspond to the energy distributions only.

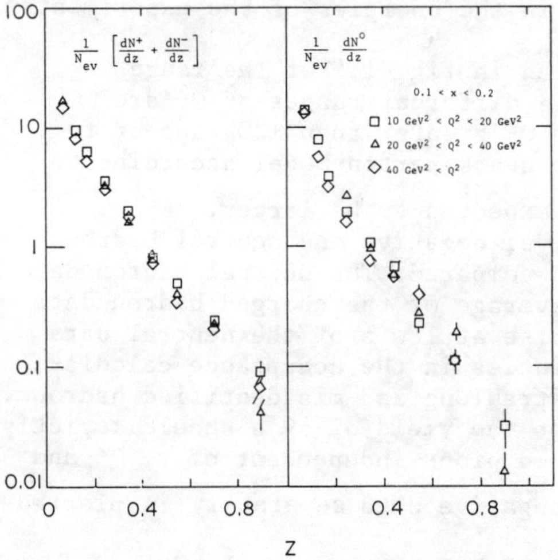

Fig. 1 – Q^2-dependence of energy distributions of charged and neutral hadrons.

RESULTS ON ENERGY DISTRIBUTIONS

The data are presented in terms of normalized $z = E_h/\nu$ distributions, $\frac{1}{N_{ev}}\frac{dN}{dz}$, where N_{ev} is the number of muon scatters in a range of Q^2 and x.

The z distributions for $Q^2 > 5$ GeV^2, integrated over x for charged particles have been compared with previous high energy muon scattering experiments[2,3]. Agreement between the experiments is about ±20%, which is our present systematic error. Since there is no equivalent neutral hadron data comparison was made with low energy π^0 electroproduction data[4,5].

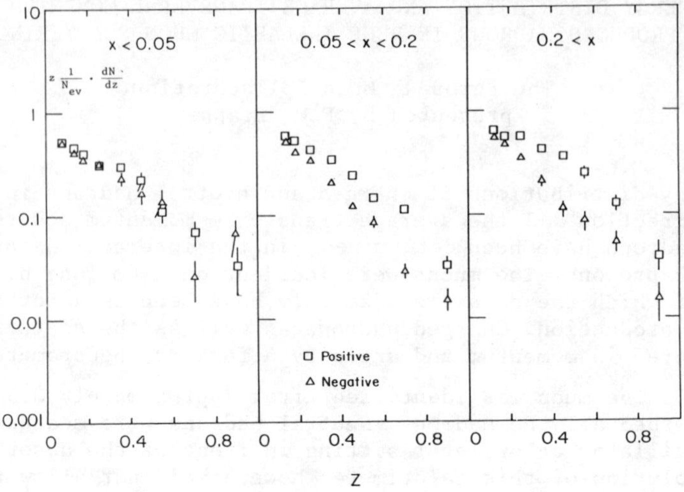

Fig. 2 - Comparison of energy distributions of hadrons with positive and negative charge.

In view of the extreme differences in the energies of the experiments the agreement is remarkable.

The Q^2 dependence is illustrated in fig. 1. For the range $0.1 < x < 0.2$ the data for three different ranges of Q^2 are plotted. Systematic variations with Q^2 greater than ±20% appear to be excluded. QCD corrections to the quark parton model according to recent numerical calculations[6] are expected to be larger.

The z distributions for positive, negative and neutral hadrons for $Q^2 > 5$ GeV^2 and all x have been compared. The neutral hadron data seem to represent rather well the average of the charged hadron data in the range $z > 0.3$. An apparent rise at low z of the neutral data may be related to residual uncertainties in the acceptance calculation and to backgrounds from Bremsstrahlung and misidentified hadrons. The quark parton model predicts that the yield of π^0's should strictly equal half the total yield of charged pions independent of x, Q^2 and z. $\frac{z}{N_{ev}} \cdot \frac{dN}{dz}$ for the positive and negative data separately is plotted for three different x ranges in fig. 2. For $x < 0.05$ the data do not depend on the charge whereas for the higher x regions $\frac{dN^+}{dz} > \frac{dN^-}{dz}$, the difference between the charges increasing with z. The quark model interpretation of this behaviour is that at low x the muon scatters from the charge symmetric quark-antiquark sea whereas at higher x scattering from thr valence u quarks dominates. This charge asymmetry then appears at high z, a region qualitatively free of target

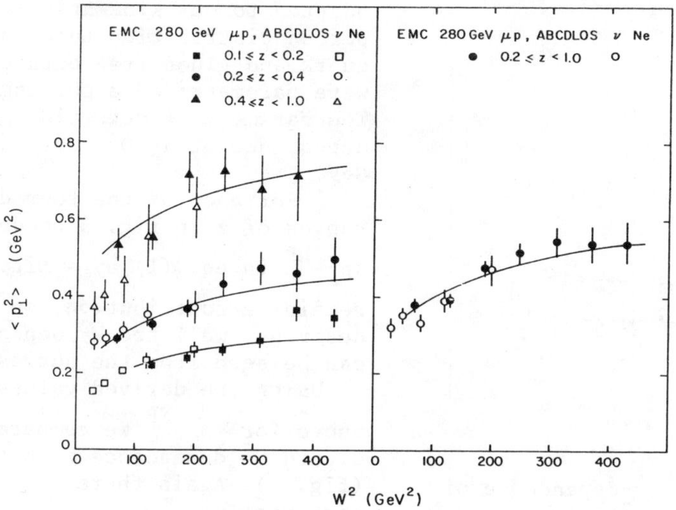

Fig. 3 - W^2-dependence of $\langle p_\perp^2 \rangle$ of charged hadrons. Curves correspond to QCD prediction.

fragments, if the faster hadrons contain the struck quark.

RESULTS ON THE AVERAGE TRANSVERSE MOMENTUM

Averaging over all events with $Q^2 > 5\ GeV^2$ we show in fig. 3 the W^2 dependence of $\langle p_\perp^2 \rangle$ for four different ranges of z. The mean accepted Q^2 of the individual points is in the region of 20 GeV^2 but drops to 11 GeV^2 at the highest W^2. The mean accepted z falls with increasing W^2 from 0.35 to 0.28 and 0.56 to 0.52 for $0.2 \leq z < 1$ and $0.4 \leq z < 1$ respectively. Also shown are the data from a neutrino experiment[7]. A clear rise of $\langle p_\perp^2 \rangle$ with W^2 is visible. The excellent agreement with the neutrino data at low W is remarkable as the relative contribution of the individual quark flavours and the spectrum of virtual bosons is much different.

Contrary to the simple quark parton model, QCD predicts a rise of $\langle p_\perp^2 \rangle$ with W^2 [8] as a consequence of gluon emission. We therefore make the simple ansatz

$$\langle p_\perp^2 \rangle = \langle p_\perp^2 \rangle^{NP} + \langle p_\perp^2 \rangle^{QCD} \quad (1)$$

assuming that in the kinematic range covered the non-perturbative part, $\langle p_\perp^2 \rangle^{NP}$, is a function of z only. The perturbative part $\langle p_\perp^2 \rangle^{QCD}$ is calculated to first order in $\alpha_s(Q^2)$, the effective strong coupling constant. We used for this calculation a program of Kroll[9], based on

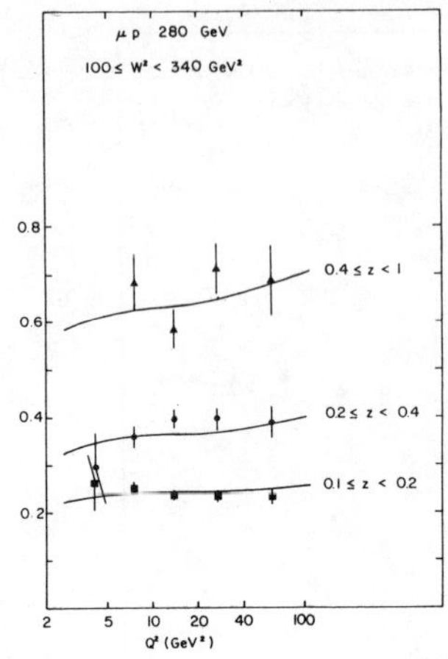

Fig. 4 – Q^2-dependence of $\langle p_\perp^2\rangle$ of charged hadrons. Curves correspond to QCD prediction.

Altarelli and Martinelli and others[8], adapted to our kinematic range. The parton distribution functions and quark and gluon fragmentation functions were parametrized according to ref. 10. The parameter Λ determining the Q^2 dependence of $\alpha_s(Q^2)$ was set to 0.5 GeV.

For each of the four different ranges of z in fig. 3 we determine $\langle p_\perp^2\rangle^{NP}$ in eq. (1) by a fit. The remaining contribution, $\langle p_\perp^2\rangle^{QCD}$, describes well the W^2 dependence as can be seen from the curves in fig. 3.

Using the derived values given above for $\langle p_\perp^2\rangle^{NP}$ we compare the predicted Q^2 dependence with the data (fig. 4). Again there is good agreement.

The z dependence of $\langle p_\perp^2\rangle$ in principle contains information on the intrinsic transverse momenta $k_\perp$. The measured z dependence is explicitly displayed in fig. 5. A strong rise with z^2 ('seagull') is visible. We compare the data with $\langle p_\perp^2\rangle^{frag}$ (curve I) as obtained from the non perturbative quark to hadron cascade model[11], in which the $p_\perp^2$ distribution of the primary mesons in the cascade is assumed to be $\propto \exp(-p_\perp^2/2\sigma^2)$ with $\sigma = 0.35$ GeV and which does not contain an intrinsic $k_\perp$. The sum (curve III) of $\langle p_\perp^2\rangle^{frag}$ (curve I) and the contribution from the perturbative QCD model (curve II) does not yield the large $\langle p_\perp^2\rangle$ observed at high z.

If instead we parametrize[12]

$$\langle p_\perp^2\rangle = \langle p_\perp^2\rangle^{frag} + \langle k_\perp^2\rangle\, z^2 + \langle p_\perp^2\rangle^{QCD} \quad (2)$$

where now $\langle p_\perp^2\rangle^{NP}$ is written as the sum of $\langle p_\perp^2\rangle^{frag}$, the contribution from the fragmentation process plus $z^2\cdot\langle k_\perp^2\rangle$, that from intrinsic $k_\perp$, the data at $z^2 \gtrsim 0.2$ are well represented by $\langle k_\perp^2\rangle = (0.63\pm0.10)$ GeV^2 where the error is statistical only, as obtained from a fit according to eq. (2) to the measured $\langle p_\perp^2\rangle$ shown in fig. 5. The three data points at small z^2 are not included in this fit as they depend strongly on the unknown details of the fragmentation process, for instance the fraction of directly produced pions and kaons as compared to hadron resonances.

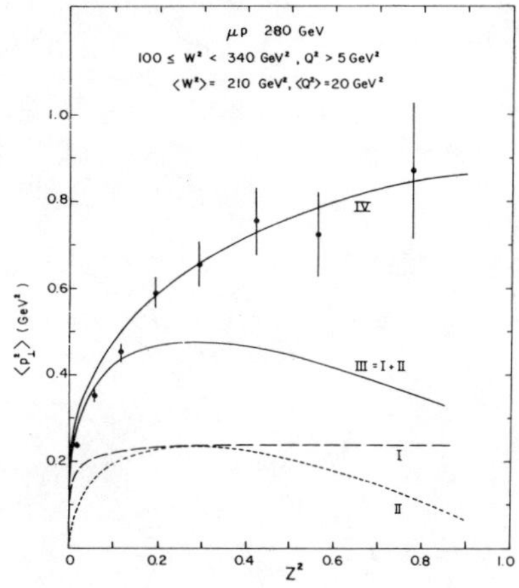

Fig. 5 - z^2-dependence of $<p_\perp^2>$ of charged hadrons. For the curves see text.

In conclusion the energy distributions of the hadrons behave very much as predicted by the Quark Parton Model. The behaviour of $<p_\perp^2>$ is consistent with perturbative QCD and an intrinsic $<k_\perp^2>$ of 0.6 GeV^2.

REFERENCES

1. EMC, J.J. Aubert et al., CERN-EP/80-134, to be submitted to Nucl. Instr. and Methods.
2. W.A. Loomis et al., Phys. Rev. D19, 2543 (1979).
3. C.S. Tao, Ph.D. Thesis, Harvard Univ., Cambridge, Mass., 1979.
4. C.H. Berger et al., Phys. Lett. 70B, 471 (1977).
5. T.P.McPharlin et al., Phys. Lett. 90B, 479 (1979).
6. W. Chi-min, TH. 2769-CERN.
7. H. Deden et al., CERN-EP/80-66 (1980).
8. G. Altarelli and G. Martinelli, Phys. Lett. 76B, 89 (1978);
   A. Méndez, Nucl. Phys. B145, 199 (1978);
   A. Méndez, A. Raychaudhuri and V.J. Stenger, Nucl. Phys. B148, 499 (1979);
   P. Mazzanti, R. Odorico and V. Roberto, Phys. Lett. 81B, 219 (1979).
9. P. Kroll (Wuppertal University), private communication.
10. R. Baier, J. Engels and B. Petersson, Z. Physik C, 2, 265 (1979).
11. T. Sjöstrand and B. Söderberg, Lund preprint LU-TP 78-18.
12. M. Gronau, W.S. Lam, T.F. Walsh and Y. Zarmi, Nucl. Phys. B104, 307 (1976);
    M. Gronau, Y. Zarmi, Phys.Rev. D18, 2341 (1978);
    R. Odorico, Phys. Lett. 89B, 89 (1979).

εN, μN, νN INTERACTIONS, II

U. AMALDI, CERN, ORGANIZER

BEBC AND GARGAMELLE DATA ON HADRONIC FINAL STATE IN ν INTERACTIONS

C. Matteuzzi
CERN, European Organization For Nuclear Research, Geneva, Switzerland

ABSTRACT

This paper summarizes the results submitted to this conference from three BEBC experiments and one Gargamelle experiment (papers # 536, 796, 466A) about the characteristic of final hadrons produced by neutrino and antineutrino.

INTRODUCTION

Three main subjects have been studied in the bubble chambers: (1) The mean charged multiplicity $\langle n_{ch} \rangle$ distributions, both in reactions $\bar{\nu}D_2$ [1] (Amsterdam-Bologne-Padova-Pisa-Saclay-Torinio-BEBC Collaboration) and νH_2 [2] (Aachen-Bonn-CERN-Munich-Oxford-BEBC Collaboration) using wideband $\overset{(-)}{\nu}$ beams from the CERN SPS. (2) The average transverse momentum $\langle p_T \rangle$ as a function of W^2 and Q^2 for $\overset{(-)}{\nu}$-nucleus interactions with narrow band $\overset{(-)}{\nu}$ beams. The range of Q^2 and W^2 extends to 100 GeV^2 and 200 GeV^2 respectively (Aachen-Bonn-CERN-Athens-IC London-Oxford-Saclay-BEBC Collaboration) [3]. (3) The effect of higher twist terms on the u quark fragmentation function [4] (CERN-Milano-Orsay-Gargamelle Collaboration) with a wideband ν beam.

The two bubble chambers are equipped with an External Muon Identifier. The lower cut on muon energy is put at $E_\mu > 3$ GeV for BEBC and at $E_\mu > 2$ GeV for Gargamelle. The variables used are the "standard" ones, they are defined in refs 1, 2 and 3. For lack of space I assume here that they are known.

MULTIPLICITY DISTRIBUTIONS

The average charged multiplicity $\langle n_{ch} \rangle$ is plotted in fig. 1

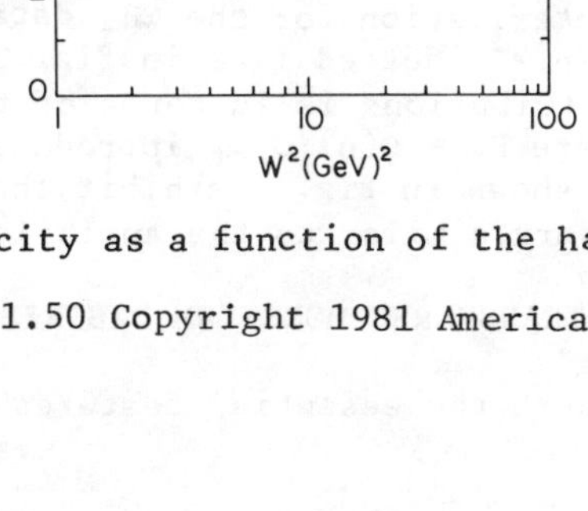

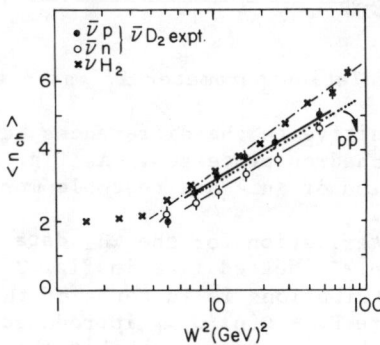

Fig. 1. Charged multiplicity as a function of the hadronic invariant mass W^2

as a function of W^2 for both experiments νD_2 and νH_2. The data is consistent with linear dependence on $\ln W^2$. A linear least-squares fit to the parameterization ($A + B \ln W^2$) obtains for the slope B the following results:

	B	W^2 range (GeV^2)	
$\bar{\nu}p$	1.29 ± 0.08	5-50	ref. 1
$\bar{\nu}n$	1.06 ± 0.02	5-50	
νp	1.31 ± 0.11	2-60	
νn	1.29 ± 0.07	2-60	
νp	1.33 ± 0.02		ref. 2

The correlation parameter $f_2^{--} = \langle n_-(n-1)\rangle - \langle n_-\rangle^2$ is also studied and in fig. 2 it plotted versus the mean negative multiplicity $\langle n_-\rangle$. This

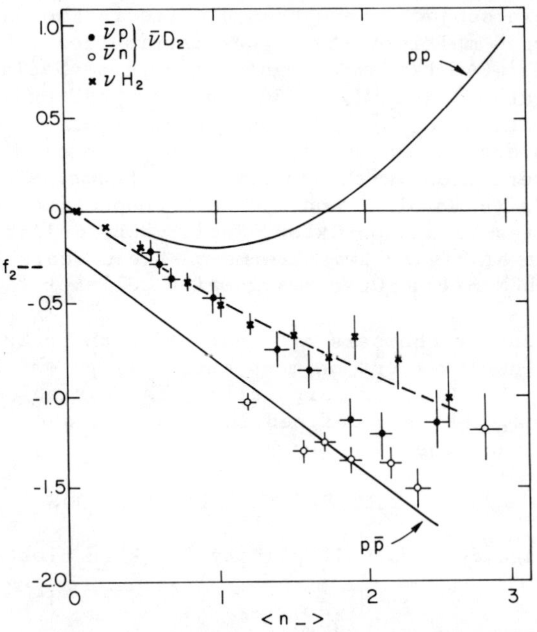

Fig. 2. Correlation parameter f_2^{--} as a function of $\langle n_-\rangle$

variable is very sensitive to the differences between annihilation processes and hadron-hadron processes. As fig. 2 shows, the ν data decreases as a function of $\langle n_-\rangle$ and resemble more closely those from $p\bar{p}$ annihilation.

The best parameterization for the νH_2 data is found to be $f_2^{--} = A + B\langle n_-\rangle + C\langle n_-\rangle^2$ (dotted line in fig. 2). Another way to study the multiplicity distributions is in terms of the normalized variables $n/\langle n\rangle$ and $\langle n\rangle P_n$ where $P_{n_s} = \sigma(n)/\sigma_{tot}$ introduced by Koba, Nielsen and Olesen [5]. The data shown in fig. 3 exhibit the predicted scaling and their behaviour is very similar to the annihilation data.

TRANSVERSE MOMENTUM DISTRIBUTIONS

I will resume here the essential features of the results presented in paper # 536.

The average value $\langle p_T^2 \rangle$ is shown in fig. 4 as a function of

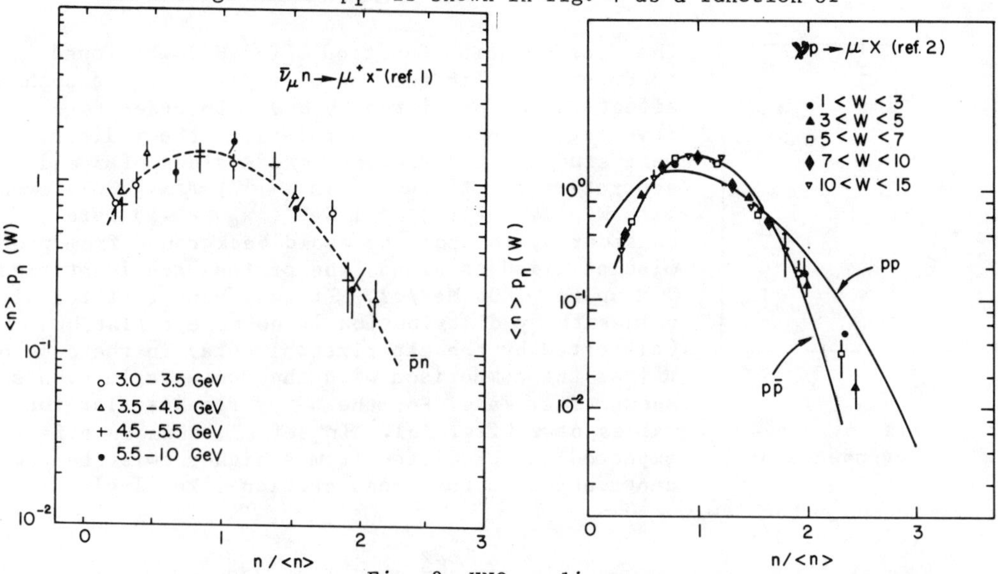

Fig. 3. KNO scaling

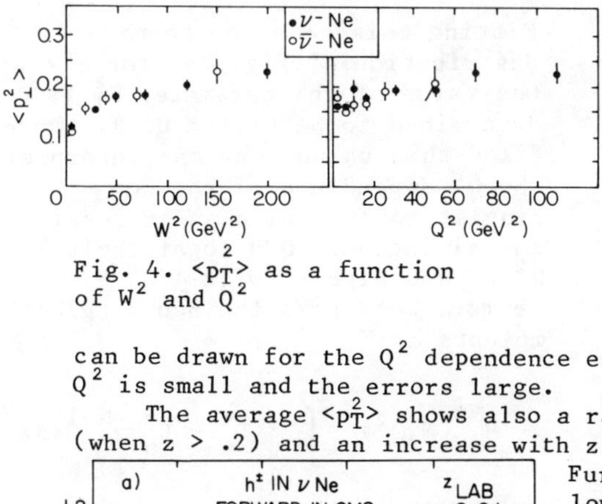

Fig. 4. $\langle p_T^2 \rangle$ as a function of W^2 and Q^2

W^2 and Q^2. Only those hadrons produced forward in the hadronic c.m.s. are plotted. At high W, the forward hadrons correspond to the production of the current fragmentation. Both in ν and $\bar{\nu}$ data, $\langle p_T^2 \rangle$ rises with W^2. An increase of $\langle p_T^2 \rangle$ as a function of W^2 is expected by the emission of hard gluons by the fragmenting quark in QCD [6]. A more uncertain conclusion can be drawn for the Q^2 dependence especially for $\bar{\nu}$, where the range in Q^2 is small and the errors large.

The average $\langle p_T^2 \rangle$ shows also a rapid rise with W^2 at fixed z (when z > .2) and an increase with z at fixed W^2 (fig. 5). Furthermore, it seems to depend on x in low Q^2 region Q^2 = 1-10 (fig. 6): if one takes only the region W > 4 however, no strong conclusion is allowed about this x dependence, because of the small statistics and the small x range. In the high Q^2 region as well as in selected high or low W regions, no x dependence of $\langle p_T^2 \rangle$ is visible.

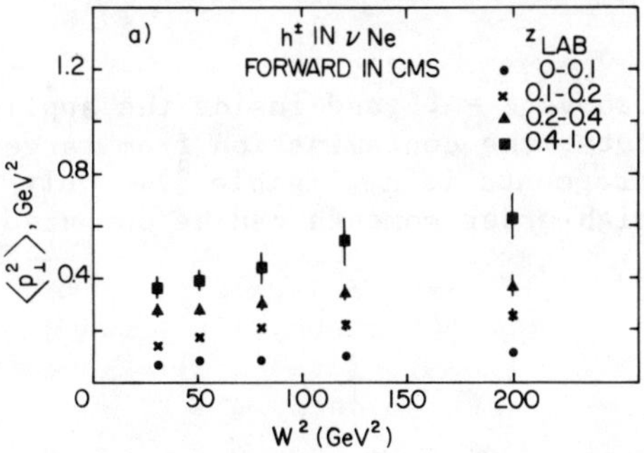

Fig. 5. Dependence of $\langle p_T^2 \rangle$ versus W^2 for different intervals of z.

y, z CORRELATIONS

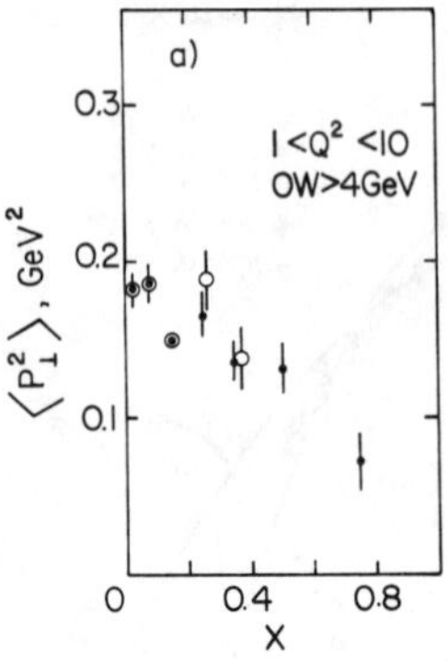

Fig. 6.
Dependence of $\langle p_T^2 \rangle$ versus x for low values of Q^2

The fragmentation function $D_u^+(z, Q^2)$ was found to depend on y in the low Q^2 region [4], and such an effect is not predicted by QCD. In order to investigate this y, z correlation, the y distributions were studied at different fixed values of z and separately for π^+ and π^- in $\nu N \to \mu^- \pi^\pm X$. Only events with $W^2 > 4$ GeV^2, $Q^2 > 1$ GeV^2, $x_{Bj} > .15$ were considered, in order to avoid background from protons misclassified as pions (the protons are identified in GGM up to $\sim$ 800 MeV/c). It was found that for high z values the y distribution is no more a flat one (distorted by the experimental cuts) in the case of π^+, as the comparison with the Monte-Carlo events shows (fig. 7a). For the π^-, y remains flat for all values of z (fig. 7b). In ref. 7, such an effect is expected and predicted from a higher twist term which contributes to the cross section like (1-y):

$$\frac{d\sigma}{dy} \alpha \ (1-z)^2 + \frac{K_T^2}{Q^2} (1-y)$$

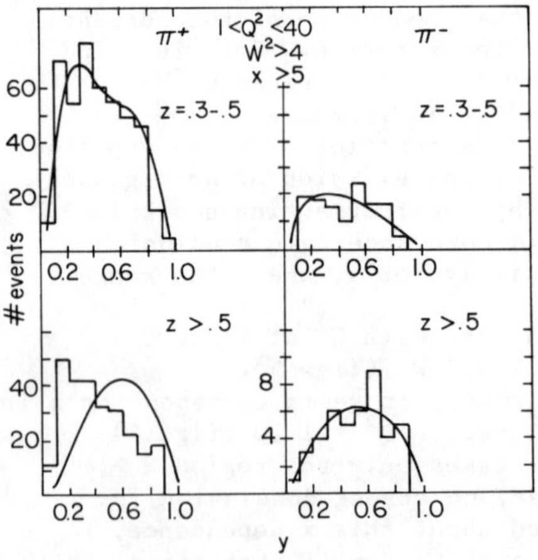

Fig. 7. y distribution for events in which (a) $z_{\pi^+} > 0.5$ and (b) $z_{\pi^-} > 0.5$

Fitting this equation to the distribution of fig. 7a, for $z > .5$ the value of the parameter K_T^2 is determined to be (1.7 ± 0.2). Once fixed this value, one can interpret the Q^2 dependence of the non singlet moments with an expression containing both QCD logarithmic $\ln Q^2/\Lambda^2$ and higher twist $1/Q^2$ terms. Lets take the non singlet moments as:

$$M^{NS}(N,Q^2) = \int_{.5}^{1} (D_u^+ - D_u^-) z^{N-1} \, dz$$

(above z = .5 and inside the applied cuts, the contamination from target fragments is negligible [8]). Only high order moments can be compared

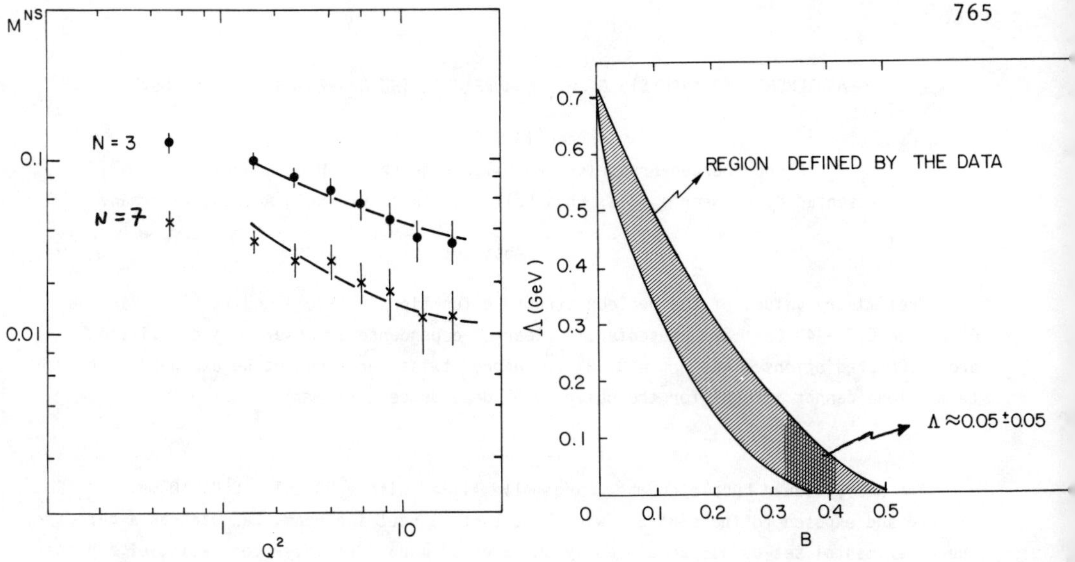

Fig. 8. Fit to the moments N = 3-7 with the expression

$$M^{NS} = \frac{K_N}{(\ell n \frac{Q^2}{\Lambda^2})} d_N \left[1 + \frac{0.37(N+1)(N+2)}{2\,Q^2} \right]$$

with the expected behaviour: fig. 8a shows the results of the fit for N = 3, 7. The fit is good ($\chi^2 = 1.2/_{d.o.f.}$) and it does not request a logarithmic scaling violation term, that is Λ^2 is found compatible with 0. If a N linear dependence is taken the fit is equally good and $\Lambda = (.050 \pm .050)$GeV (fig. 8b). This means that in the low Q^2 region effects of higher twist terms can be important and they have to be considered in analyzing the data.

REFERENCES

1. Paper # High energy antineutrino interactions in deuterium.
2. Paper # 796 Multiplicity distributions in neutrino-hydrogen interactions.
3. Paper # 536, Transverse momentum of hadrons produced in ν interactions on an isosolar target in BEBC.
4. Paper # 466 and CERN preprint 80-124.
5. Z. Koba, H.B. Nielsen, P. Olesen, Nucl. Phys. B40, 317 (1972).
6. G. Altarelli, G. Martinelli, Phys. Lett. 76B 89 (1978).
7. E.L. Berger, P. Prodsky, Phys. Rev. Lett. 42, 940, (1979); E.L. Berger, Phys. Lett. 89B, 241, (1980).
8. H. Lubatti, Proceedings of the Bergen Conference, 1979.

MEASUREMENT AND ANALYSIS OF $F_2^{\overset{(-)}{\nu}}$ and $xF_3^{\overset{(-)}{\nu}}$ IN THE Q^2-REGION 0.5 - 40 GeV^2

Gargamelle SPS-collaboration
Aachen-Bergen-Brussels-CERN-Milano-Strasbourg-U.C. London
Presented by H. Weerts, III. Physikalisches Institut, RWTH Aachen, W-Germany

Abstract

Preliminary values of the nucleon structure functions $F_2(x,Q^2)$ and $xF_3(x,Q^2)$ in the Q^2-region 0.5 - 40 GeV^2 are presented. A clear Q^2-dependence is observed and a fit to 2nd order QCD predictions gives $\Lambda_{\overline{MS}} = 0.28^{+.09}_{-.10}$. Higher twist terms cannot be excluded, but such terms alone cannot account for the observed Q^2-dependence.

The heavy liquid bubble chamber Gargamelle filled with a 90 mol % C_3H_8/10 mol % CF_3Br mixture and exposed to the CERN-SPS wide band (anti-) neutrino beam, was used as a detector.[1] The experimental set-up was completed by an external muon identifier consisting of 2 planes of MWPC's, allowing detection of muons down to an energy of 2 GeV.

Events were selected if they had a muon, identified by the EMI with momentum $P_\mu^- > 2$ GeV and $P_\mu^+ > 5$ GeV. Before a cut on the neutrino energy (E_ν) could be applied, the measured hadron energy had to be corrected for missing neutrals. This was done by using a method proposed by Heilmann[2], which is based on transverse momentum balance. After applying the cut in the neutrino energy: $15 < E_\nu < 150$ GeV, the sample consists of 2960 neutrino and 3790 antineutrino events.

The (anti-) neutrino flux was obtained by fitting a Monte Carlo generated spectrum to the observed muon spectrum in the shielding. The total cross-sections determined with this flux show a linear rise with energy and the slopes (σ/E) are:

$$\nu: 0.56 \pm .05 \qquad \bar{\nu}: .29 \pm .02 \qquad \left|10^{-38}\ cm^2/GeV\ Nucl\right|$$

Due to the preliminary nature of the absolute normalisation of the neutrino flux, the subsequent analysis, has been performed with a neutrino flux normalised to a total cross-section slope of $\sigma/E = 0.65\ 10^{-38}\ cm^2/GeV$ Nucl.

By adding or subtracting the double differential cross-sections $\frac{d^2\sigma}{dxdQ^2}$ for ν and $\bar{\nu}$ and assuming $A = 2xF_1/F_2 = 0.9$ [3] the structure functions F_2 and xF_3 can be determined. The results are shown in Fig. 1 and 2. For both structure functions a clear Q^2-dependence is observed.

The experimental results were compared to second order QCD predictions for the non-singlet part of the structure functions, given by Bialas and Buras[4] (BB) and Gonzalez-Arroyo, Lopez, Yndurain (GALY)[5]. To account for so called higher twist[6] effects (due to multi-quark scattering, transverse momentum spread of the quark, etc), that are expected to play a role in the explored Q^2-region, the structure functions were parameterised as:

$$F(x,Q^2,\Lambda,a) = F^{QCD}(x,Q^2,\Lambda)\ \left(1 + \frac{a}{Q^2(1-x)}\right) \qquad (1)$$

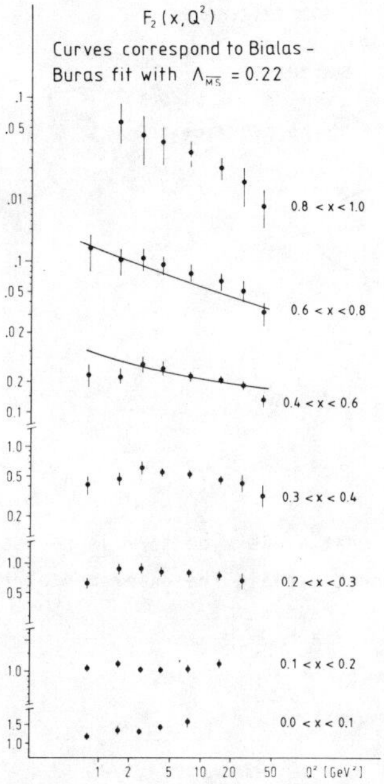

Fig. 1

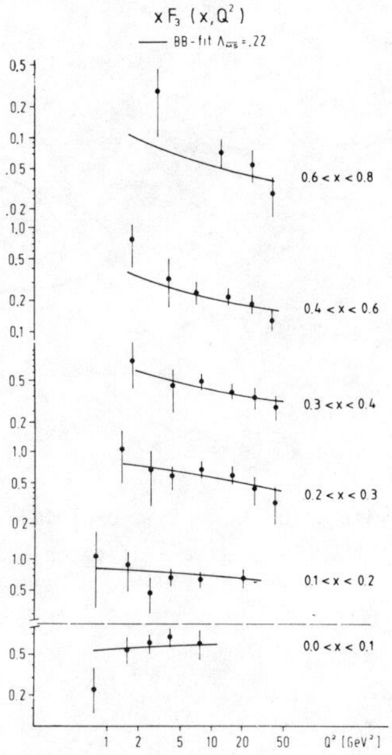

Fig. 2

where F^{QCD} comes from the BB-parametrization. The parameters to be determined are the QCD scale parameter Λ, and the magnitude of the higher twist contributions a. In the analysis, the range for xF_3 used was $0.02 < x < 0.8$ and for F_2: $0.4 < x < 0.8$. To test whether QCD alone can describe the data, various fits were done with a = 0. The first one uses the region $Q^2 > 2$ GeV^2 and the results for Λ are:

	F_2	χ^2/ND	xF_3	χ^2/ND
$\Lambda_{\overline{MS}}$(BB)	$.20 \pm .04$	0.38	$.26 \pm .06$	0.25
Λ_{MS}(GALY)	$.12 {+.13 \atop -.11}$	0.06	$.28 \pm .28$	0.13

The large error assigned to Λ_{MS} from the GALY fit is due to the milder Λ dependence of this formulation, compared to the BB-predictions. The best value for Λ was obtained by selecting a Q^2,x-region ($Q^2 > 5$ GeV^2, $.02 < x < .4$, xF_3 only) where higher twist contributions are expected to be minimal. With the BB-formalism: $\Lambda_{\overline{MS}} = 0.28 {+.09 \atop -.10}$.

To determine the magnitude of the higher twist contributions, $\Lambda_{\overline{MS}}$ and a from (1) have been determined simultaneously from a fit to xF_3 for $.02 < x < .8$. The 90% C.L.-contours are shown in Fig. 3 for 2 different Q^2-regions. The most significant result of this fit is the

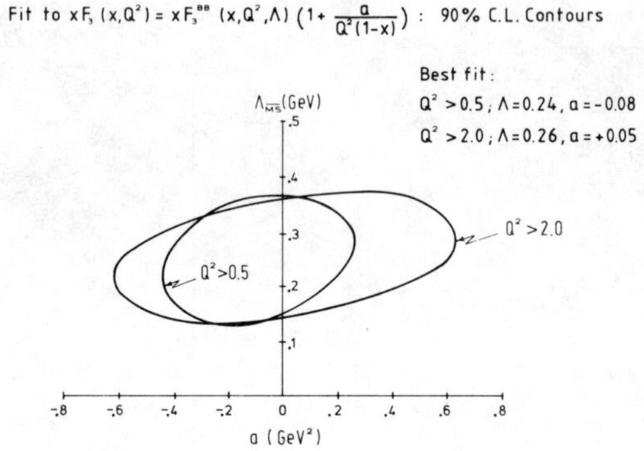

Fig. 3

confidence with which $\Lambda_{\overline{MS}} = 0$ is excluded. This implies that a QCD type term <u>is</u> necessary and that a higher twist term - <u>if</u> described by (1) - cannot explain the experimental results.

References:

[1] More details can be found in contributed paper nr. 705 to this Conference.

[2] H.G. Heilmann, internal report of WA21-BEBC collaboration, WA21-int-1.

[3] A = 0.9 is based on $R = (F_2 - 2xF_1)/2xF_1 = 0.1$ reported by the CDHS-group at ν'80, Erice, Italy

[4] A. Bialas, A.J. Buras, Phys. Rev. D21, 1825 (1980)

[5] A. Gonzâlez-Arroyo, C. Lôpez, F.J. Yndurain, Nucl. Phys. B153, 161 (1979)

[6] Higher twist means only twist-4 terms. Twist 6 and higher are neglected.

THE STRUCTURE AND THE AMOUNT OF THE $q\bar{q}$ SEA AND BROADENING OF CHARM JETS

CERN-Dortmund-Heidelberg-Saclay Collaboration*

presented by

J. KNOBLOCH

CERN, Geneva, Switzerland

INTRODUCTION

In this paper new results on the structure and the amount of the sea are presented. QCD effects on charmed quarks visualized in neutrino induced dimuon events are discussed.

The advent of high intensity wide band ν and $\bar{\nu}$ beams together with a high mass combined fucntion neutrino detector[1] made it possible to study with high statistics rare processes such as the $\nu(\bar{\nu})$ production of charmed quarks and their subsequent semileptonic decay with a muon in the final state.

It becomes also possible to study $\bar{\nu}$ scattering at high y ($y=E_{had}/E_{tot}>0.5$) which is largely dominated by scattering off the $\bar{q}$ sea.

In this analysis the following event sample has been used :

$\bar{\nu}$ charged current	100.000 events
$\nu\ \mu^-\mu^+$ events	10.000 events
$\bar{\nu}\ \mu^+\mu^-$ events	1.000 events

ANTIQUARK SEA $\bar{q}$ + $\bar{s}$

The $\bar{\nu}$ cross-section is given by a $(1-y)^2$ term for scattering off valence quarks and quarks from the sea, and a constant contribution for scattering off the $\bar{q}$-sea :

$$\frac{d\sigma^{\bar{\nu}}}{dy} \propto (\bar{q}+\bar{s}) + (1-y)^2 . (q-s) \qquad (1)$$

thus at high y one measures essentially the momentum distribution of sea quarks. Figure 1 shows the structure function of the sea got from $\bar{\nu}$ charged current for y>0.5. A small correction for the remaining contribution of valence quarks has been applied.

One observes at small x a strong rise with Q^2 which is more pronounced than that seen in scattering off valence quarks (xF_3)[2,3]. Like in the case of valence quarks the Q^2-dependence changes sign at $x\simeq 0.3$. $\int q(x)dx$ shows a rise from 0.55 to 0.7 between E_H=20GeV and E_H=140GeV (Fig.2a). There is some indication that the sea shrinks with E_H (Fig.2b).

STRANGE SEA

Opposite sign dimuon events produced in ν and $\bar{\nu}$ interactions on isoscaler targets are well described by the GIM-mechanism[4] as the production of a charmed quark and its subsequent semileptonic decay[5]. The differential cross-section is

$$\frac{\partial^2\sigma(\overset{(-)}{\nu} \to \overset{(-)}{c})}{\partial x\ \partial y} = 2\cos^2\Theta_c \overset{(-)}{s}(x) + \sin^2\Theta_c \left[\overset{(-)}{q}(x) - \overset{(-)}{s}(x)\right] \qquad (2)$$

* H.Abramowicz, I.Becker, F.Dydak, F.Eisele, T.Flottmann, C.Geweniger, J.G.H.de Groot, C.Guyot, J.T.He, H.Klasen, K.Kleinknecht, J.Knobloch, J.Królikowski, H.Lierl, J.May, J-P.Merlo, P.Palazzi, A.Para, B.Peyaud, B.Pszola, J.Rander, F.Ranjard, B.Renk, J.Rothberg, T.Z.Ruan, D.Schlatter, J.P.Schuller, J.Steinberger, H.Taureg, K.Tittel, R.Turlay, W.von Rüden, H.Wahl, H.J.Willutzki, J.Wotschack and W.M.Wu.

Thus $\bar{\nu}$ induced $\mu^+\mu^-$ events come in their large majority from the strange sea, while in the ν case about equal fraction stem from valence quarks and strange sea quarks.

The momentum distribution of strange sea quarks and of all sea quarks looks very much alike (Fig.3). A fit to the x-distribution of ν induced dimuons to determine the relative contribution α of sea and valence quarks of the form $F(x) = \alpha.\mathrm{sea}(x) + (1-\alpha).\mathrm{val}(x)$ (3)
was performed (Fig.4). val(x) was taken from a Buras-Gaemers[6] fit to charged current structure functions[7]. sea(x) is the y-distribution of $\bar{\nu}$ dimuon events. The fit was performed in four bins of $\nu = E_H + P_2$ having $\langle Q^2 \rangle$ = 8, 13, 21 and 40 GeV^2, respectively. The relative amount of strange quarks

$$\frac{2s}{(u+d)_v} = \frac{\alpha}{1-\alpha} \cdot \tan^2\Theta_c \text{ with } \tan^2\Theta_c = 0.057 \qquad (4)$$

is shown in Fig.5. The superimposed solid curve is the corresponding quantity $(\bar{q}+\bar{s})/\int xF_3$ for the overall sea. The apparent stronger increase of the strange sea can be explained by kinematic effects due to the mass difference of charmed and strange quarks-'slow rescaling' (dashed curve). The curves are normalized to the 30 GeV point.

GLUON RADIATION OF CHARM QUARKS

In the framework of QCD[8] gluon radiation off the struck quarks should broaden the jets that are observed in the final state with increasing W^2 :

$$\langle p_T^2 \rangle = Q^2.g(x,y).\alpha_s(Q^2) = c.\alpha_s(Q^2).W^2 \qquad (5)$$

Altarelli and Martinelli[8] calculate for electroproduction c=1/32. In the ν-case R.Moore[9] suggests c=1/28 for a charmed quark mass m_c=1.25 GeV. In ν-induced $\mu^+\mu^-$-events the μ^+ is a fragmentation product of a charmed quark. Thus it is at the same time an indication of the quark flavour and a measure of its tranverse momentum. Gluon radiation increasing with W^2 adds to the intrinsic transverse momentum k_T of the quark. With Z^* being the fraction of the quark-momentum seen in the μ^+ one expects p_T to develop like[10]

$$\langle p_{T_{out}}^2 \rangle = \tfrac{1}{2}\langle p_T^2 \rangle = \tfrac{1}{2}(Z^{*2}\left[\langle k_T^2 \rangle + c'W^2\right] + \text{const.}) \qquad (6)$$

Figure 6 shows $\langle p_{T_{out}}^2 \rangle$ in Z-bins as a function of W^2. A fit of (6) to the data yields

$$\langle k_T^2 \rangle = 0.6 \pm 0.2 (\pm 30\% \text{ syst}) \ (\mathrm{GeV}^2/c^2)$$

and $c' = c.\alpha_s(Q^2) = (1.3\pm0.3).10^{-2}$ ($\pm$20% syst). With $c = 3.6 . 10^{-2}$, $\alpha_s(Q^2 \simeq 20 (\mathrm{GeV}/c)^2 = 0.36\pm \sim 30\%$ well in line with other measurements presented at this conference[3,11].

REFERENCES

1. M. Holder et al., Nucl.Instr.and Methods 148, 235 (1978).
2. J.G.H. de Groot et al., Z.Phys. C1, 143 (1979).
3. J.G.H. de Groot, talk given at this conference.
4. S.L. Glashow, J. Ilioupoulos and L. Maiani, Phys.Rev. D2, 1285 (1970).
5. M. Holder et al., Phys.Lett. 69B, 377 (1977).
6. A.J. Buras and K.J.F. Gaemers, Nucl.Phys. B132, 249 (1978).
7. J.G.H. de Groot et al., Phys.Lett. 82B, 456 (1979).

8. G. Altarelli and G. Martinelli, Phys.Lett. 76B, 89 (1978).
9. R. Moore (Dortmund), private communication.
10. J. Rander, Contribution to the XVth Rencontre de Moriond, to be published (1980).
11. For e^+e^- results see e.g. B. Wiik's talk, this Conference.

FIGURES

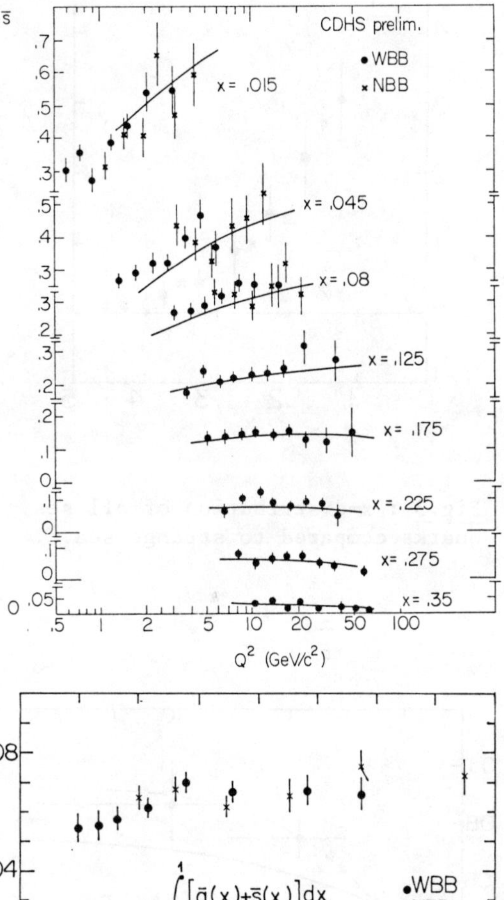

Fig.1 : Structure function of the sea. Full lines are from a fit to F_2 and xF_3 (ref.3).

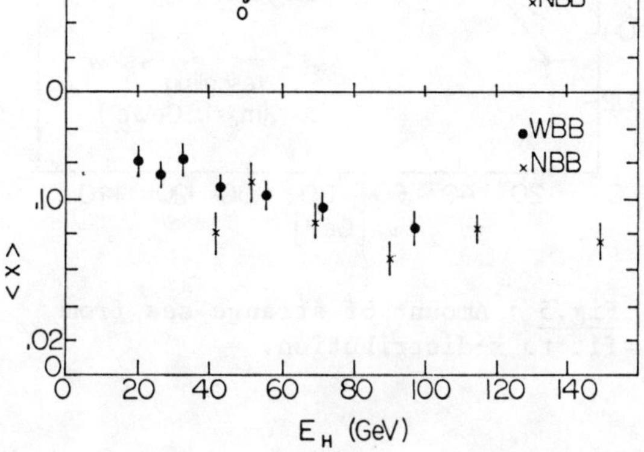

Fig.2a : Fractional momentum carried by sea quarks.

Fig.2b : Average x for sea quarks.

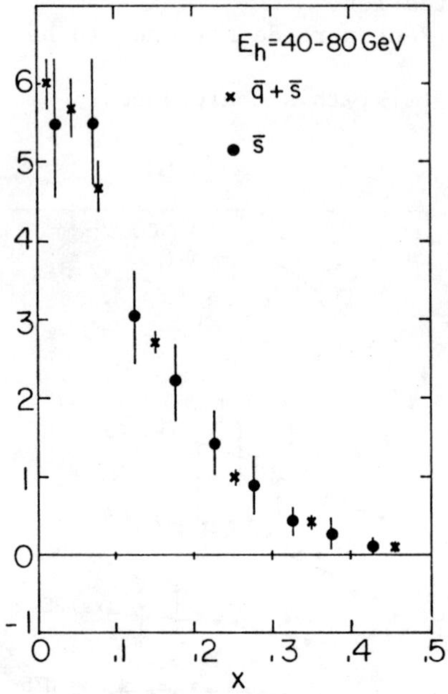

Fig.3 : x-distribution of all sea quarks compared to strange sea.

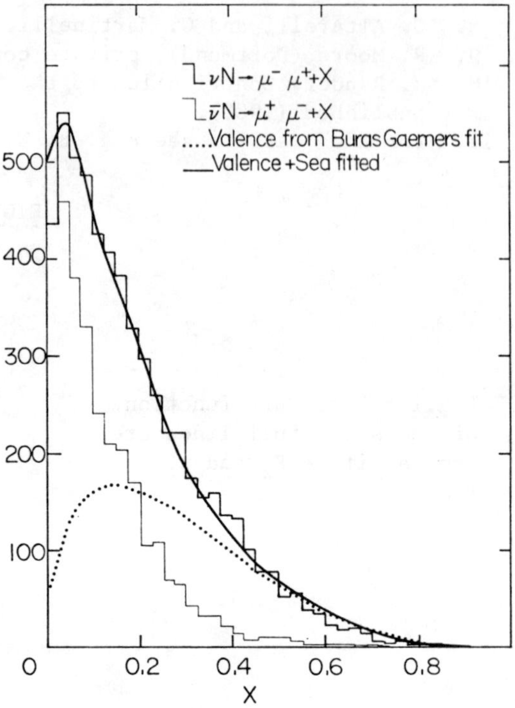

Fig.4 : x-distribution for ν and $\bar{\nu}$ $\mu^+\mu^-$ events.

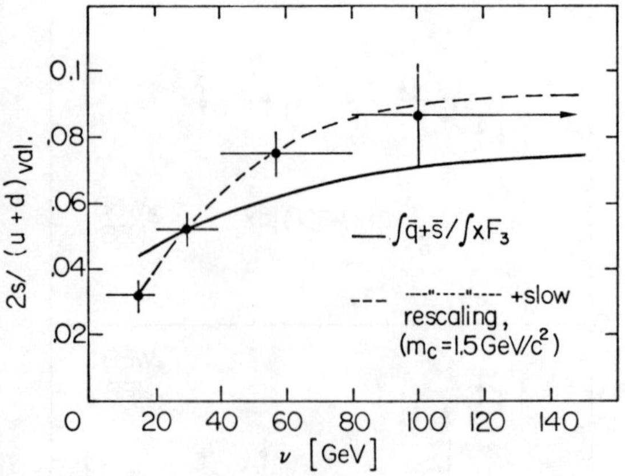

Fig.5 : Amount of strange sea from fit to x-distribution.

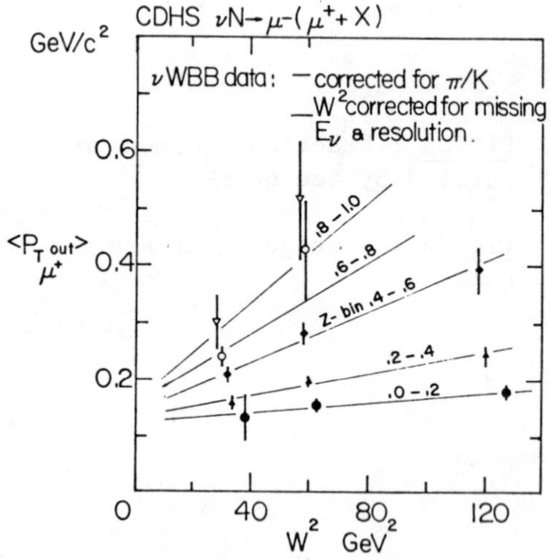

Fig.6 : Average p_T of μ^+ in ν induced $\mu^-\mu^+$ events.

DEEP INELASTIC MUON-NUCLEON SCATTERING AT HIGH Q^2

Bologna-CERN-Dubna-Munich-Saclay Collaboration

presented by M. Klein,

Dubna

The NA4 muon spectrometer (fig. 1) is made of ten identical iron toroids with interspersed liquid scintillation counters and multiwire proportional chambers to detect and track the scattered muon. The magnetized iron torus ensures good acceptance at high Q^2 confining tracks up to $Q^2/Q^2_{max} \sim 0.6$. A luminosity of 5.10^{27} cm^{-2} per incoming muon is achieved using a 40 m long carbon target. The trigger requires a set of four consecutive counter planes in coincidence with a beam particle anywhere in the apparatus. The typical trigger rate was 10^{-5} triggers/gated μ at beam intensities of about 10^7μ/pulse. A system of hodoscopes measures the individual primary energy with a resolution of 0.5%. By deflecting the μ beam into the spectrometer its momentum resolution was determined to be about 7%. By the same method, the absolute momentum calibration of the spectrometer has been checked to ⩽ 1% with respect to the primary energy.

A small fraction of the available data has been analyzed to determine the structure function $F_2(x,Q^2)$ at energies of 120,200 and 240 GeV. The 200 GeV data were taken with a beam polarity (and helicity) change from μ^+ to μ^-. No systematic differences between μ^+ and μ^- cross sections were observed within the statistical accuracy of ± 2%.

The F_2 determination relies on acceptance and resolution corrections based on the Monte-Carlo simulation of the experiment which includes the beam phase space, multiple scattering, energy loss, δ rays and hadronic showers. The data have been corrected for radiative effects using the parametization of Akhundov et al.[1)] Fig. 2 presents the F_2 values with statistical error bars using $R = \sigma_l/\sigma_t = 0$ without correction for Fermi motion. The data sets at different energies are in good agreement within an estimated relative normalization error of 5%. The Q^2 dependence of the structure functions is weak at all primary energies. Simple parametizations of the type $\sum_{i=1}^{3} ai(1-x)^{i+1}(1+c \ln Q^2/3 \ln 1/4x)$ give the Q^2 dependent term with an averaged c of 0.12 ± 0.03. The data shown in fig. 2 already have the statistical power to determine the pattern of scaling violations at high Q^2. The detailed contributions of systematic errors have not yet been finally assessed. Thus their quoted values reflect the current status of the analysis rather than the intrinsic limitations of the experiment.

1) A.A. Akhundov, D.Yu. Bardin and N.M. Shumeiko, Jad. Fiz. 26 (1977) 1251.

EXPERIMENTAL SET-UP (TOP-VIEW)

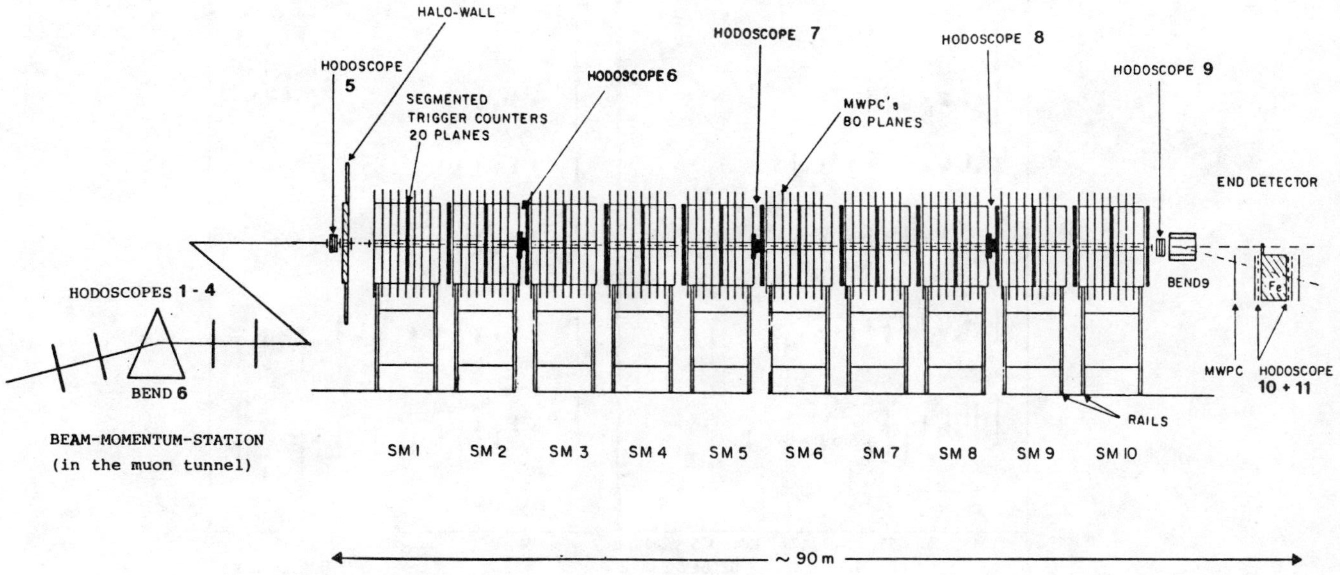

FIG.1

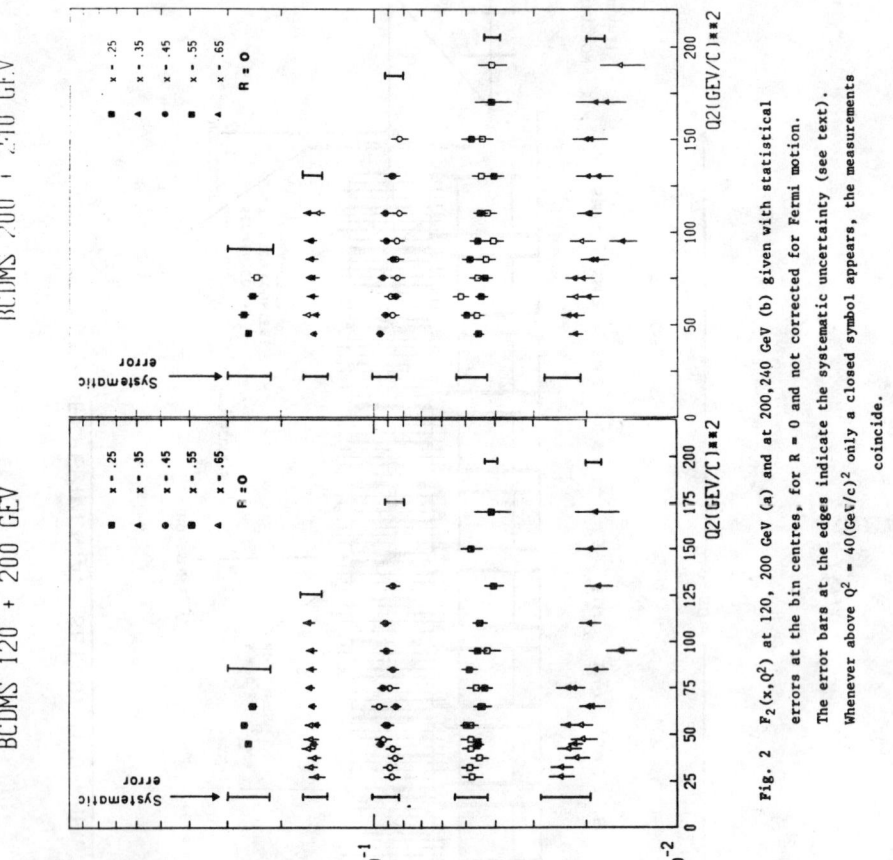

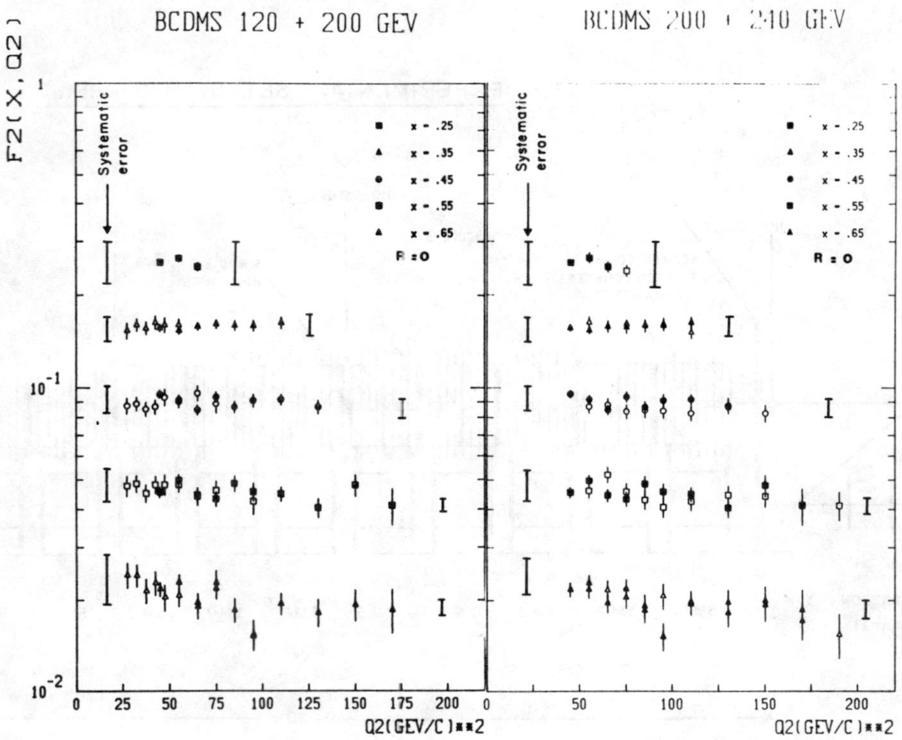

Fig. 2 $F_2(x,Q^2)$ at 120, 200 GeV (a) and at 200,240 GeV (b) given with statistical errors at the bin centres, for R = 0 and not corrected for Fermi motion. The error bars at the edges indicate the systematic uncertainty (see text). Whenever above $Q^2 = 40(GeV/c)^2$ only a closed symbol appears, the measurements coincide.

STRUCTURE FUNCTION MEASUREMENTS IN MUON-IRON AND MUON-PROTON SCATTERING, AND A QCD ANALYSIS.

The European Muon Collaboration
presented by P.R. Norton,
Rutherford Laboratory.

ABSTRACT

The structure function F_2 has been measured in the range $3.0 < Q^2 < 150\ GeV^2$ and $0.015 < x < 0.65$ for both hydrogen and iron targets. QCD fits to the scaling violation indicate that the scale-breaking parameter Λ is in the region of 100 MeV.

MEASUREMENT OF F_2

As part of an extensive study of high energy muon interactions we present data for the structure function F_2 measured on hydrogen at beam energies of 280 and 120 GeV and on iron at beam energies of 280, 250 and 120 GeV.

The apparatus is fully described elsewhere[1]. The muon beam was incident on a target of either 6m. of liquid hydrogen or 3.75m. of iron interleaved with scintillator. Scattered muons were measured in a system of drift and proportional chambers positioned on either side of a large aperture air-gap magnet. Muons were identified by their penetration through an iron hadron absorber. The trigger, formed from five scintillator counter hodoscopes in the spectrometer, demanded a muon originating from the target which scattered through at least 0.5^o (1° for iron). For the iron target a minimum energy deposition of 25 GeV (5 GeV for 120 GeV running) was also required.

Scattered muon tracks were geometrically fitted with the beam track to a common vertex and their momenta and angles obtained. The acceptance of the apparatus was determined by a Monte Carlo simulation which included the real beam phase space, multiple scattering, resolution and hodoscope and chamber efficiencies. Cuts were applied to both the data and Monte Carlo events:

- to remove events at low ν where the resolution on momentum measurement caused distortion in the measurement of x, and, for the iron, to cut beyond the minimum energy required by the trigger.
- to remove low momentum (<30-40 GeV) muons. In this region the radiative corrections are large and there is some contamination from pion decay.
- to define a fiducial target length.

F_2 was calculated by assuming a trial function in the Monte Carlo and iterating quickly to a stable solution. For the calculation of F_2, a fixed value of R $(=\sigma_\ell/\sigma_t)$ of 0.2 was assumed throughout. Radiative corrections were applied using the now standard formulae of Tsai[2]. For the iron, the effects of the Fermi motion were allowed for using a non-degenerate Fermi gas model with an approximately Gaussian momentum distribution[3].

The results for hydrogen are shown in fig. 1, and for iron in

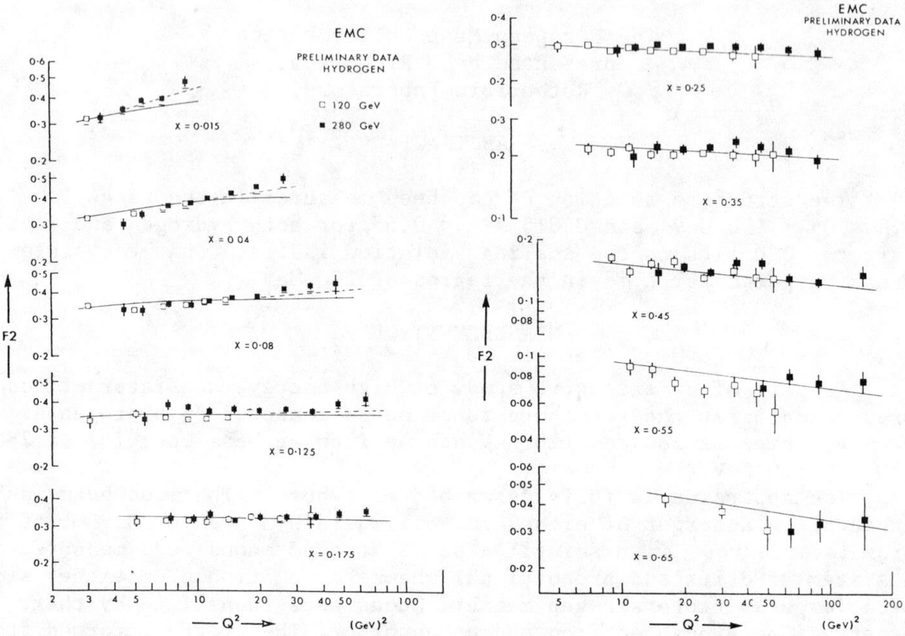

Fig. 1. The structure function F_2 measured on hydrogen.

fig. 2. F_2 is plotted as a function of Q^2 in different bins of x. For hydrogen, the two beam energies are plotted separately, but for the iron the three energies have been combined together. Scaling violation is seen clearly on both targets. F_2 rises with Q^2 at small x and falls with Q^2 at high x. This is more clearly seen in fig. 3, where the slope $b = d(\ln F_2)/d(\ln Q^2)$ is plotted as a function of x. The scaling violations on hydrogen and iron are very similar.

The effect of R for hydrogen is shown in fig. 1. The horizontal bars indicate the effect of assuming $R = 0$ instead of 0.2.

The errors shown in the plots are purely statistical. We estimate that the systematic errors are less than 10%, but can vary from point to point in some regions of Q^2 and x.

QCD ANALYSIS

The data have been compared with QCD models based on the Altarelli-Parisi equations[4]. The method of Gonzàles-Arroyo et al.[5] has been used for $x > 0.25$, where valence quarks are expected to be dominant and one is not sensitive to exact details of the glue and sea distributions or the threshold of charm production. The result is that for both targets we find $\Lambda \lesssim 100$ MeV. We have also used the method of Abbott and Barnett[6] over the whole x range. Again, for both targets, the best fit gives $\Lambda \lesssim 100$ MeV. These fits are shown by the

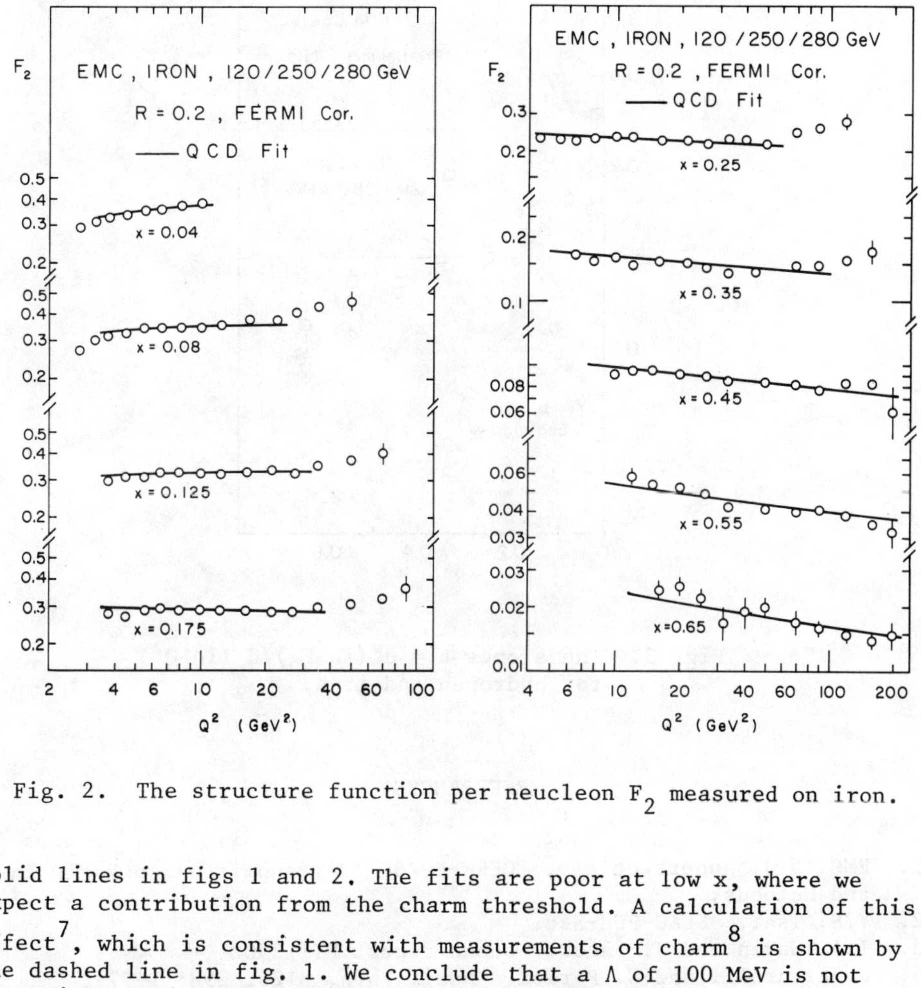

Fig. 2. The structure function per neucleon F_2 measured on iron.

solid lines in figs 1 and 2. The fits are poor at low x, where we expect a contribution from the charm threshold. A calculation of this effect[7], which is consistent with measurements of charm[8] is shown by the dashed line in fig. 1. We conclude that a Λ of 100 MeV is not inconsistent with the behaviour of F_2 at small x.

The errors on the determination of Λ are almost entirely systematic. From our knowledge of these systematic errors we ascribe an uncertainty of 100 MeV to Λ.

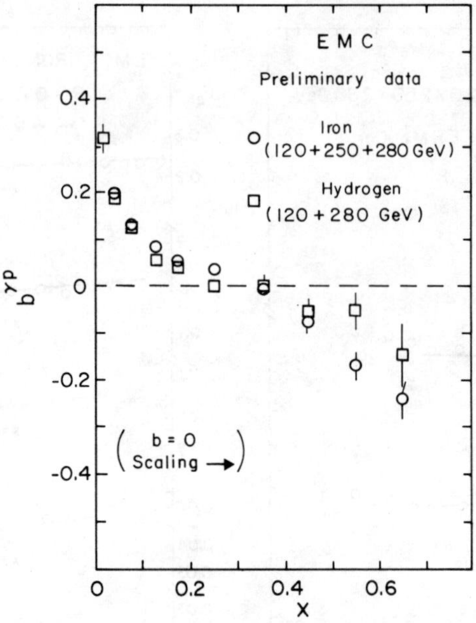

Fig. 3. The slopes $b = d(\ln F_2)/d(\ln Q^2)$ for hydrogen and iron.

REFERENCES

1. EMC, J.J. Aubert et al., CERN-EP/80-134, submitted to Nucl. Instr. and Methods.
2. Y.S. Tsai, SLAC-PUB-848.
3. I.A. Savin and J. Žáček, Dubna report EI-12502.
4. G. Altarelli and G. Parisi, Nucl. Phys. B126, 298 (1977).
5. A. Gonzàles-Arroyo et al., Nucl. Phys. B153, 161 (1979); Nucl. Phys. B159, 512 (1979).
6. L.F. Abbott and R.M. Barnett, SLAC-PUB-2325 (1979).
7. J.P. Leveille and T. Weiler, Nucl. Phys. B147, 147 (1979); M. Glück and E. Reya, Phys. Lett. 83B, 98 (1979).
8. EMC, J.J. Aubert et al., CERN-EP/80-62, to appear in Phys. Lett. A.R. Clark et al., LBL-10747, LBL-10879.

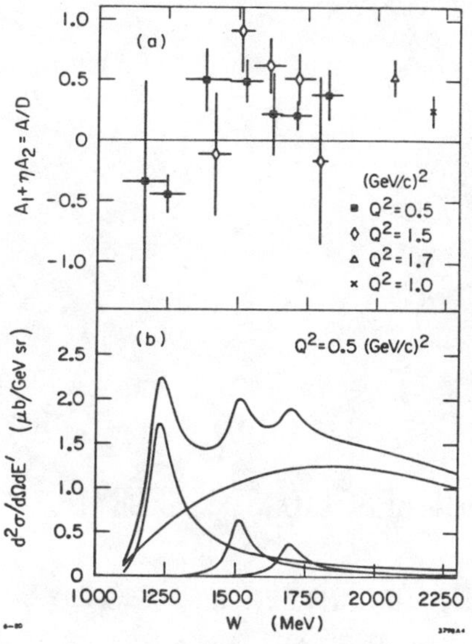

Fig. 3. (a) Asymmetry vs. missing mass W in the resonance region. (b) Differential cross section vs. W.

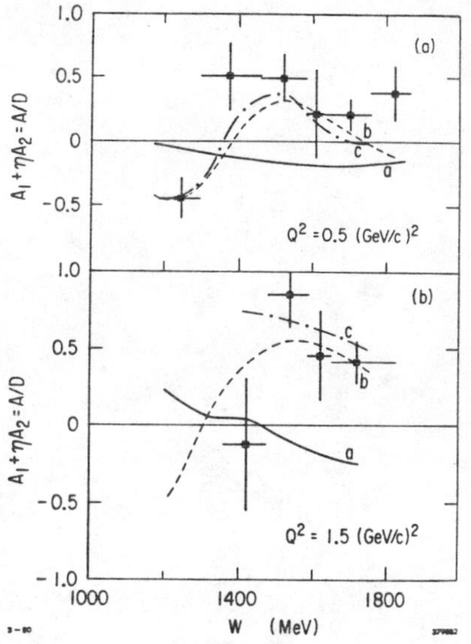

Fig. 4. (a) Asymmetry vs. W at Q^2 = 0.5 GeV^2 and comparison with multipole calculations of Ref. 6: "a" Born terms alone, "b" Born terms plus Δ(1232), "c" Born terms plus all resonances. (b) Same for Q^2 = 1.5 GeV^2.

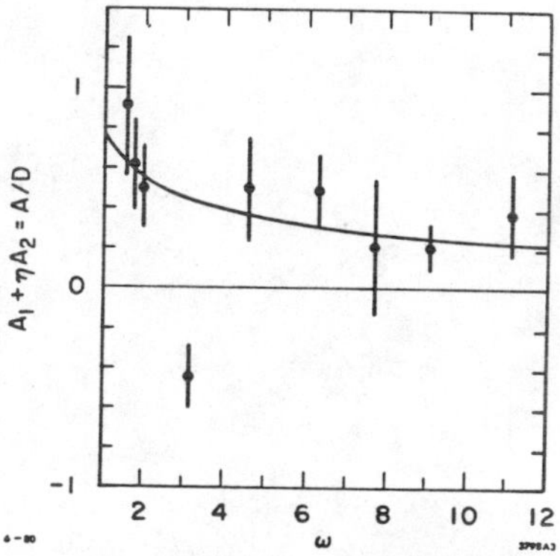

Fig. 5. Asymmetry vs. scaling variable ω. The curve 0.78 $\omega^{-1/2}$ is a fit to deep inelastic data (W > 2 GeV). The data points are the resonance region results (W < 2 GeV).

REFERENCES

1. M. J. Alguard et al., Phys. Rev. Lett. 37, 1261 (1976).
2. M. J. Alguard et al., Phys. Rev. Lett. 41, 70 (1978).
3. M. J. Alguard et al., Nucl. Inst. and Meth. 163, 29 (1979).
4. G. Baum et al., contributed paper No. 508 (June 1980).
5. G. Baum et al., SLAC-PUB-2557 (July 1980).
6. R.C.E. Devenish and V. Gerhardt, private communication.
7. E. D. Bloom and F. J. Gilman, Phys. Rev. D 4, 2901 (1971); V. Rittenberg and H. R. Rubinstein, Phys. Lett. B35, 50 (1971).

*Research supported in part by the U.S. DOE, the German BMFT, the MOE of Japan, and the Swiss NSF.

INTERPRETATIONS OF eN, μN, νN REACTIONS, THEORY

T. PETRONZIO, CERN, ORGANIZER

QUANTUM CHROMODYNAMICS AND DEEP-INELASTIC SCATTERING

Andrzej J. Buras
Fermi National Accelerator Laboratory, Batavia, Illinois 60510

ABSTRACT

Moments of deep-inelastic structure functions, parton distributions and parton fragmentation functions are discussed in the context of Quantum Chromodynamics with particular emphasis put on higher order corrections. A brief discussion of higher twist contributions is also given.

INTRODUCTION

It is an experimental fact* that the deep-inelastic structure functions depend on $x = (Q^2)/2\nu$ and Q^2; i.e., Bjorken scaling is violated. Quantum chromodynamics (QCD) predicts scaling violations in deep-inelastic scattering** with the pattern consistent with the experimental findings. In comparing QCD predictions with the experimental data one can either work with the moments of structure functions or with the structure functions themselves. Quite generally QCD predictions for the moments of the simplest (non-singlet, NS) structure functions can be written as follows

$$M_n^{NS}(Q^2) = \int_0^1 dx\, x^{n-2} F^{NS}(x,Q^2)$$

$$= \sum_{\substack{t=2\\ \text{even}}} \frac{A_n^{(t)}}{[Q^2]^{t-2}} \left[\alpha(Q^2)\right]^{d_n^{(t)}} \left[1 + R_n^{(t)} \frac{\alpha(Q^2)}{4\pi} + \ldots\right] \qquad (1)$$

Here the sum runs over various twist ($\underline{t}$) contributions: Leading twist ($t=2$), twist four ($t=4$) and so on. Furthermore $\alpha(Q^2)$ is the effective strong interaction coupling constant and $d_n^{(t)}$, $R_n^{(t)}$ and $A_n^{(t)}$ are numbers to be discussed below.

The expressions like (1) are rather formal and it is often convenient to cast them in a form of parton model formulas in which case the basic elements are the effective Q^2 dependent parton distributions and elementary parton cross-sections. In the case of semi-inclusive deep-inelastic scattering also the concept of the effective Q^2 dependent fragmentation functions is introduced.

* See the contributions to the Session "eN, μN and νN Interactions" and references therein.

** For recent reviews see refs. 1 and 2.

Here we shall discuss three topics:

i) Moments of the structure functions in the leading twist approximation ($t=2$ in Eq. 1) and with the next to leading order corrections taken into account ($R_n^{(2)}$ in Eq. 1).

ii) Parton distributions and parton fragmentation functions beyond the leading order in $\alpha(Q^2)$.

iii) Higher Twist ($t>2$) Contributions.

Our discussion includes the latest developments as well as some results obtained by various authors since the Tokyo Conference.

BASIC FORMULAE

In QCD and in the leading twist approximation the moments of any structure function are given as follows

$$M_n(Q^2) \equiv \int_0^1 dx\, x^{n-2} F(x,Q^2) = \sum_{i=NS,S,G} A_n^i(\mu^2) C_n^i\left(\frac{Q^2}{\mu^2}, g^2\right) \quad . \quad (2)$$

Here $A_n^i(\mu^2)$ are the hadronic matrix elements of non-singlet (NS), singlet (S) and gluon (G) operators and C_n^i are the corresponding coefficient functions in the Wilson operator product expansion. Furthermore g is the renormalized quark-gluon coupling constant and μ^2 is the subtraction scale at which the theory is renormalized. The important property of Eq. (2) is the factorization of non-perturbative pieces $A_n^i(\mu^2)$ from perturbatively calculable coefficient functions $C_n^i(Q^2/\mu^2, g^2)$.

Specializing Eq. (2) to non-singlet structure functions, and using renormalization group equations for $C_n^{NS}(Q^2/\mu^2, q^2)$ one obtains

$$M_n^{NS}(Q^2) = A_n^{NS}(\mu^2) \exp\left[-\int_{\bar{g}(\mu^2)}^{\bar{g}(Q^2)} dg' \frac{\gamma_n^{NS}(g')}{\beta(g')}\right] \cdot C_n^{NS}(1, \bar{g}^2(Q^2) \quad (3)$$

$$= A_n^{NS}(\mu^2) \left[\frac{\bar{g}^2(Q^2)}{\bar{g}^2(\mu^2)}\right]^{d_n^{NS}} \left[1 + \frac{\bar{g}^2(Q^2) - \bar{g}^2(\mu^2)}{16\pi^2} Z_n^{NS}\right] \cdot \left[1 + \frac{\bar{g}^2(Q^2)}{16\pi^2} B_n^{NS}\right] \quad (4)$$

where terms of order $\bar{g}^4$ have been neglected.

Furthermore

$$d_n^{NS} = \frac{\gamma_n^0}{2\beta_0} \quad ; \quad Z_n^{NS} = \frac{\gamma_n^{(1)}}{2\beta_0} - \frac{\gamma_n^0}{2\beta_0^2} \beta_1 \quad , \quad (5)$$

In obtaining Eqs. (4) and (5) the following expansions for the anomalous dimensions (γ_n^{NS}), β functions and the coefficient function $C_n^{NS}(1, \bar{g}^2)$ have been used:

$$\gamma_n^{NS}(\bar{g}) = \gamma_n^{0} \frac{\bar{g}^2}{16\pi^2} + \gamma_n^{(1)} \frac{\bar{g}^4}{(16\pi^2)^2} \quad , \tag{6}$$

$$\beta(\bar{g}) = -\beta_0 \frac{\bar{g}^3}{16\pi^2} - \beta_1 \frac{\bar{g}^5}{(16\pi^2)^2} \quad , \tag{7}$$

and

$$C_n^{NS}(1,\bar{g}^2) = 1 + \frac{\bar{g}^2}{16\pi^2} B_n^{NS} \quad . \tag{8}$$

Finally the Q^2 evolution of $\bar{g}^2(Q^2)$ is given as follows

$$\frac{\bar{g}^2(Q^2)}{16\pi^2} \equiv \frac{\alpha(Q^2)}{4\pi} = \frac{\left[1-(\beta_1/\beta_0^2)\ln\ln(Q^2/\Lambda^2)/\ln(Q^2/\Lambda^2)\right]}{\beta_0 \ln(Q^2/\Lambda^2)} \tag{9}$$

with Λ being the famous QCD scale parameter. The parameters γ_n^0, $\gamma_n^{(1)}$, β_0, β_1 and B_n^{NS} have been calculated by at least two groups: in refs. 3 and 4, 5 and 6, 7 and 8, 9 and 10, and 11 and 12, respectively.

To proceed further one can use either <u>formal</u> approach or <u>intuitive</u> approach.

In the formal approach one proceeds as follows. Since the left-hand side of Eq. (4) does not depend on μ^2 the r.h.s. of this equation can be put in the following form

$$M_n^{NS}(Q^2) = \tilde{A}_n^{NS}\left[\alpha(Q^2)\right]^{d_n^{NS}} \left[1 + \frac{\alpha(Q^2)}{4\pi} R_n^{NS}\right] \quad , \tag{10}$$

with

$$R_n^{NS} = Z_n^{NS} + B_n^{NS} \quad . \tag{11}$$

and $\tilde{A}_n^{NS}$ being independent of μ^2. Note that Eq. (10) represents just the twist two (t=2) contribution to Eq. (1).

In the <u>intuitive</u> approach setting $\mu^2=Q^2$ in Eq. (4) one obtains

$$M_n^{NS}(Q^2) = q_n^{NS}(Q^2) \cdot \sigma_n^{NS}(\bar{g}^2) \quad . \tag{12}$$

Here

$$q_n^{NS}(Q^2) \equiv A_n^{NS}(Q^2) = A_n^{NS}(\mu^2)\left[\frac{\bar{g}^2(Q^2)}{\bar{g}^2(\mu^2)}\right]^{d_n^{NS}} \left[1 + \frac{\bar{g}^2(Q^2)-\bar{g}^2(\mu^2)}{16\pi^2} Z_n^{NS}\right] \quad (13)$$

can be interpreted as the moments of an effective, Q^2 dependent non-singlet parton distribution $q^{NS}(x,Q^2)$ (e.g., valence quark distribution), and

$$\sigma_n^{NS}(\bar{g}^2) = 1 + \frac{\bar{g}^2(Q^2)}{16\pi^2} B_n^{NS} \quad (14)$$

may be regarded as the elementary parton cross-section. We shall now discuss these two approaches in more detail.

FORMAL APPROACH

We just list the most important properties of Eqs. 10 and 11.

1) $\gamma_n^{(1)}$ and B_n^{NS} depend on the renormalization scheme[5] used to calculate these quantities. This renormaliation prescription dependence of $\gamma_n^{(1)}$ and B_n^{NS} cancels in Eq. 11) if these quantities are calculated in the same scheme; i.e., the combination

$$\frac{\gamma_n^{(1)}}{2\beta_0} + B_n^{NS} \quad (15)$$

is renormalization prescription independent.

2) The parameters R_n^{NS} depend on the definition of $\alpha(Q^2)$.[13,11] If $\alpha(Q^2)$ is redefined to $\alpha'(Q^2)$ with

$$\alpha(Q^2) = \alpha'(Q^2) + r\left[\alpha'(Q^2)\right]^2 \qquad r - \text{const.} \quad (16)$$

then the expansion parameters in R_n^{NS} in Eq. (10) are changed to

$$\left[R_n^{NS}\right]' = R_n^{NS} + 4\pi r d_n^{NS} \quad . \quad (17)$$

Of course the final answer for $M_n^{NS}(Q^2)$ is independent of the definition of $\alpha(Q^2)$ since each change of the expansion parameters R_n^{NS} is compensated by the corresponding change of the values of $\alpha(Q^2)$ or equivalently values of Λ extracted from experiment. This is illustrated by the following example.

3) For the $\overline{MS}$[11] and Momentum Subtraction (MOM)[14] schemes, which have been discussed widely in the literature, Eqs. (16) and (17) read as follows

$$\alpha_{MOM} = \alpha_{\overline{MS}}\left[1 + 1.55\beta_0 \frac{\alpha_{\overline{MS}}}{4\pi}\right] , \qquad (16')$$

and

$$\left[R_n^{NS}\right]_{\overline{MS}} = \left[R_n^{NS}\right]_{MOM} + \beta_0\left[1.55\right]d_n^{NS} . \qquad (17')$$

Numerically we have:

i) $2<\left[R_n^{NS}\right]_{\overline{MS}}<16$ and $-4\leq\left[R_n^{NS}\right]_{MOM}\leq 2$ for $2\leq n\leq 8$ with both $\left[R_n^{NS}\right]_{\overline{MS}}$ and $\left[R_n^{NS}\right]_{MOM}$ increasing monotonically with n.

ii) If $\Lambda_{\overline{MS}}$=0.30 GeV then the MOM scheme with Λ_{MOM}=0.55 GeV leads to essentially indistinguishable results for $M_n^{NS}(Q^2)$ for $Q^2\geq 10$ GeV^2 The corresponding values of $\alpha(Q^2)$ are α_{MOM}=0.32 and $\alpha_{\overline{MS}}$=0.24 at Q^2=10 GeV^2.

iii) The quantity $1 + (\alpha(Q^2)/4\pi)R_n^{NS}$ in Eq. (10) varies for $2\leq n\leq 8$ and Q^2=10GeV^2 from 1.05 to 1.31 for $\overline{MS}$ scheme and from 0.92 to 1.08 for MOM scheme. Since in the leading order the quantity in question is equal to 1 we observe that MOM scheme seems to lead to a better expanssion in α that $\overline{MS}$ scheme. An opposite conclusion would be reached in the case of $1/\ln Q^2$ expansion.

4) One may think for a while that there is no point in doing next to leading and higher order calculations since at the end one can anyhow change the size of various terms in the expansion by redefining α. The point is however that by doing consistent higher-order calculations in various processes, such as deep-inelastic scattering, $e^+e^-\to$hadrons, photon-photon scattering etc. one can meaningfully compare QCD effects in these processes using one universal effective coupling constant $\alpha(Q^2)$ extracted, e.g., from deep-inelastic data. By studying higher order corrections to various processes one can find a universal definition of α for which the QCD perturbative expansions are behaving well. Such studies can be found in refs. 15,16,21. One finds that schemes with $\alpha_{\overline{MS}}\leq\alpha_i\leq\alpha_{MOM}$ lead to acceptable expansions. Another method for finding the optimal scheme for α has also been recently suggested.[17]

5) One can study properties of R_n which are independent of the definition of α.[18,19] One is so called Λ_n scheme.[13,11,18] Here one rewrites Eq. (10) as follows

$$M_n^{NS}(Q^2) = A_n^{NS}\left[\frac{1}{\ln(Q^2/\Lambda_n^2)}\right]^{d_n^{NS}}\left[\frac{1-(\beta_1/\beta_0^2)\ln\,\ln(Q^2/\Lambda^2)/\ln(Q^2/\Lambda^2)}{\beta_0\ln(Q^2/\Lambda^2)}\right]^{d_n^{NS}} \qquad (18)$$

with

$$\Lambda_n = \Lambda \exp\left[\frac{R_n^{NS}}{2\beta_0 d_n^{NS}}\right] \quad . \qquad (19)$$

The n dependence of Λ_n is independent of the definition of α_S (see eq. 17 and 19). Λ_n increases roughly by factor 2 and 3 for F_2^{NS} and F_3 structure functions respectively if n is varied from n=2 to n=8. In the leading order Λ is independent of n. The n dependence of Λ_n as given by Eq. (19) is in a very good agreement[20] with experimental data indicating the importance of next-to-leading-order corrections. Other quantities which are independent of the definition of α can be found in refs. 18 and 19.

6) As already stated in connection with the n dependence of Λ_n, the next-to-leading-order corrections to the Q^2 evolution are different for F_2^{NS} and F_3 structure functions. In the leading order F_2^{NS} and F_3 have the same Q^2 evolution. (R_n^{NS} depends on the structure function considered—see point 15 below.)

7) There are several new effects related to next-to-leading-order corrections. Most of them are small. They are summarized in ref. 21.

This completes the listing of the main properties of Eq. (10). One should also mention that,

8) The next-to-leading-order corrections to the singlet structure functions are known.[11,12]

9) Some of the next-to-leading-order corrections to deep-inelastic scattering on polarized targets have been calculated in ref. 22.

Finally:

10) It has been suggested in Ref. 23 to use the moments

$$B_{M,N}(Q^2) = \frac{(M+N+1)!}{M!N!}\int_0^1 x^n(1-x)^M F(x,Q^2)dx \quad , \qquad (20)$$

rather than the moments of Eq. (1), which for $N \gtrsim 4$ are mostly sensitive to $x>0.5$. With increasing M the moments of Eq. (20) become sensitive to small values of x and are particularly well suited for the study of gluon and sea distributions which are concentrated at small values of x.

INTUITIVE APPROACH

11) The novel feature of parton distributions beyond the leading order is that they can be defined in various ways. Two definitions have been discussed in the literature. They are as follows.

<u>Definition A.</u>[24] Moments of parton distributions are defined by the matrix elements of local operators normalized at Q^2. This is

the definition of Eq. (13) which constitutes the Q^2 evolution equation for so defined parton distributions.

Definition B.[25] The full higher order correction to F_2^{NS} is absorbed into the definition of parton distributions. Eqs. (12-14) are replaced by

$$M_n^{NS}(Q^2) = \left[q_n^{NS}(Q^2)\right]' \cdot \left[\sigma_n^{NS}(\bar{g}^2)\right]' \quad , \tag{12'}$$

with

$$\left[q_n^{NS}(Q^2)\right]' = A_n^{NS}(\mu^2)\left[\frac{\bar{g}^2(Q^2)}{\bar{g}^2(\mu^2)}\right]^{d_n^{NS}}\left[1 + \frac{\bar{g}^2(Q^2)-\bar{g}^2(\mu^2)}{16\pi^2} R_n^{NS}\right], \tag{13'}$$

R_n^{NS} given by Eq. (11), and

$$\left[\sigma_n^{NS}(\bar{g}^2)\right]' = \begin{cases} 1 & \text{for } F_2^{NS} \\ 1 + \dfrac{\bar{g}^2(Q^2)}{16\pi^2}\left[B_n^{NS}\right]_3 - \left[B_n^{NS}\right]_2 & \text{for } F_3 \end{cases} \quad . \tag{14'}$$

It should be remarked that since Z_n^{NS} and B_n^{NS} are separately renormalization prescription dependent so are the parton distributions of def. A. On the other hand the parton distributions defined by (13') are renormalization prescription independent.

12) Also fragmentation functions can be defined in various ways.[26] If we consider the process $e^+e^- \to h$ + anything then the moments (in the z variable) of relevant cross-sections take for the non-singlet contributions the form of Eqs. (12-14) and (12'-14') with the following replacements

$$q_n^{NS}(Q^2) \to D_n^{NS}(Q^2) \quad , \quad A_n^{NS}(\mu^2) \to V_n^{NS}(\mu^2)$$

$$Z_n^{NS} \to \left[Z_n^{NS}\right]_T \quad , \quad B_n^{NS} \to \left[B_n^{NS}\right]_T \quad . \tag{21}$$

Here $D_n^{NS}(Q^2)$ and $V_n^{NS}(\mu^2)$ are the moments of a non-singlet fragmentation function and the time-like cut vertices[27] respectively. Furthermore the index T stands for "time-like". In order to keep uniform notation we shall in the following use $[Z_n^{NS}]_S$, $[B_n^{NS}]_S$ for Z_n^{NS} and B_n^{NS} of Eqs. (13) and (14) with the index "S" standing for "space-like".

13) In the $\overline{MS}$ scheme one finds,[21] for $2\leq n\leq 8$

$$\left[Z_n\right]_{S,T} = \begin{cases} 1.5 - 2.5 & \text{for S} \\ 0.5 - 1.5 & \text{for T} \end{cases} \tag{22}$$

The inequality $\left[Z_n\right]_S \neq \left[Z_n\right]_T$ expresses the violation[6,28] of the Gribov-Lipatov relation beyond the leading order ($\left[\gamma_n^{(1)}\right]_S \neq \left[\gamma_n^{(1)}\right]_T$) as opposed to $\left[\gamma_n^{(0)}\right]_S = \left[\gamma_n^{0}\right]_T$. For large n $\left[Z_n\right]_S \to \left[Z_n\right]_T \sim \ln n$.

Furthermore both $B_{n\,T}$ and $B_{n\,S}$ increase with n as $(\ln n)^2$ with[6,26,29]

$$\left[B_n\right]_T \underset{\text{Large } n}{\approx} \left[B_n\right]_S + \frac{8}{3}\pi^2 \tag{23}$$

We are now in a position to compare both definitions of parton distributions and fragmentation functions.

14) i) Evolution Equations for parton distributions and parton fragmenation functions are essentially the <u>same</u> in the case of the def. A and are different in the case of def. B due to the substantial difference in the values of $\left[B_n\right]_S$ and $\left[B_n\right]_T$.

ii) Furthermore the evolution equations in question are in the case of def. A essentially the same as leading order equations ($\left[Z_n\right]_T$, $\left[Z_n\right]_S$ are small) except for the modified evolution of the effective coupling constant (see Eq. 9). The evolution equations in the case of def. B differ substantially at large n (Large x) from the corresponding leading order equations due to the large values of $\left[B_n\right]_S$ and $\left[B_n\right]_T$ at large n and due to the non-trivial behavior $(\ln n)^2$ of these parameters.

iii) Whereas the input distributions or structure functions ($A_n^{NS}(u^2=Q_0^2)$) at some $Q^2=Q_0^2$ in the case of the def. B will be for F_2^{NS} the same as in the leading order (i.e., the data does not change) the input distribution in the def. A will differ considerably at low Q^2 and large x from those used in the leading order phenomenology. The reason is that B_n's differ considerably from 1 for low Q^2 and large n.

Of course the final results for the structure functions should be independent of any particular definition since the differences in the parton distributions and parton fragmentation functions are compensated by the corresponding differences in the parton cross-sections. A detailed study of the effects discussed here has been done in ref. 30. It turns out that in the range $5\leq Q^2\leq 200$ GeV^2 and $0.02\leq x\leq 0.8$ one can find simple parametrizations for both definitions of parton distributions which represent to a high accuracy Eqs. (13) and (13'). These parametrizations are of the form of leading order parametrizations of ref. 31, i.e.,

$$x\, q^{NS}(x,Q^2) \sim x^{\eta_1(\bar{s})} (1-x)^{\eta_2(\bar{s})} \tag{24}$$

$$\eta_i(\bar{s}) = \eta_i^{(0)} + \eta_i' \bar{s} \;;\; \bar{s} = -\ln\left[\frac{\alpha(Q^2)}{\alpha(Q_0^2)}\right] . \tag{25}$$

In accordance with points ii) and iii) one has

$$\left[\eta_i^0\right]_{LO} = \left[\eta_i^0\right]_B \neq \left[\eta^0_i\right]_A \tag{26}$$

$$\left[\eta_i'\right]_{LO} \approx \left[\eta_i'\right]_A \neq \left[\eta_i'\right]_B$$

where the indices LO, A, B stand for leading order, definition A and definition B, respectively. For instance $[\eta_2^0]_B=2.71$ and $[\eta_2^0]_A=3.40$ whereas $[\eta_2']_A=0.76$ and $[\eta_2']_B=1.5$.

15) On the level of structure functions themselves two main properties of the next-to-leading-order corrections are worthwhile mentioning

i) If $\Lambda_{\overline{MS}}$ is chosen so that $\Lambda_{\overline{MS}}=\Lambda_{LO}$, where Λ_{LO} is the scale in the leading order expression($B_n^{NS}=0, Z_n^{NS}=0$) then a stronger increase (decrease) of structure functions at small (large) values of x is predicted by next-to-leading-order corrections relative to LO predications. If $\Lambda_{\overline{MS}}$ is decreased so that scaling violations for $x>0.4$ are similar to those predicted by leading order formulae still some additional increase due to next to leading order corrections is seen at small x.

ii) This increase at small x is more pronounced for F_3 then F_2^{NS}.[30]

There is some indication that this additional increase in F_3 at small x has been seen in the data.[32] Detailed comparison should however also include charm production effects[33] in F_3 which are of order $O\left((m_c^2-m_s^2)/Q^2\right)$ with m_c and m_s being charm and strange quark mass, respectively.

It should also be remarked that in refs. 20, 34 fits of structure functions to the data have been made with the general conclusion that the next-to-leading-order corrections improve the agreement of QCD with the data.

Final message to our experimental colleagues: In ref. 20,30,34 and 35 simple inversion methods of moments of structure functions or parton distributions have been developed. Therefore, the analysis of structure functions beyond the leading order should be now as easy as in the leading order.

HIGHER TWISTS

At low values of Q^2 one has to worry in addition to logarithmic scaling violations about power-like scaling violations. In QCD they are represented by higher twist contributions; the terms in Eq. (1) with t>2. Let us summarize what is known at present about these contributions

16) There are many operators of a given twist>2 contributing to Eq. (1) and consequently there are many unknown non-perturbative parameters $A_n^{(t)}$ (t>2) which have to be extracted from the data. This makes the phenomenology of higher twist contributions very complicated.[36] The situation might be considerably simplified in certain regions of phase-space (e.g., $x \to 1$) and for particular cross-sections in which case one can identify and calculate the dominant higher-twist contributions.[37]

17) The anomalous dimensions of some of the twist four (t=4) operators have been calculated in ref. 38. The two novel features as compared with $d_n^{(2)}$ are as follows. $d_n^{(4)}$ can be negative as opposed to $d_n^{(2)} \geq 0$. Furthermore, whereas $d_n^{(2)} \sim \ln n$ for large n. The $d_n^{(4)}$ may increase linearly with n. These two features indicate that the structure of logarithmic corrections to higher twist contributions might be much more complicated than in the case of the leading twist. It would be interesting to study numerically these effects.

18) Phenomenologically one can study the effects of higher order twist contributions in deep-inelastic scattering by using "QCD motivated" parametrizations of the terms t>2 in Eq. (1). Such an analysis has been done one year ago by Abbott and Barnett[39] who found that the deep-inelastic data can be fit by higher twist contributions alone. Their combined analysis of twist 2 and higher twist contributions indicated that the value of the parameter Λ is strongly dependent on the size of higher twist contributions. If the latter increase the Λ decreases. Recent analyses of Duke and Roberts[40] and Pennington and Ross[41] who combine all the existing data show however that the best fits to the data can be obtained if the higher twist contributions are small. Similar conclusion has been reached in ref. 42. Even if higher twist contributions may appear to be of little importance in the analysis of deep-inelastic structure functions for $Q^2 > 5$ GeV^2 and $x < 0.8$, they may be and they probably are important for $x \to 1$. This appears to be the case as discussed in ref. 37.

SUMMARY

A. Leading twist QCD with next-to-leading-order corrections taken into account is in a good agreement with experimental data for $x \leq 0.8$. However, more phenomenology of next-to-leading-order corrections (in particular for fragmentation functions) is needed. Recall that QCD predicts[29] non-trivial x and z non-factorization in semi-inclusive deep-inelastic scattering. Further tests of these predictions are of interest.

B. For large n or large x the next-to-leading-order corrections are large and still higher order corrections are probably non-negligible. Ways of including these higher order corrections have been suggested.[43]

C. For $x \to 1$, or $z \to 1$ higher twist effects are probably important.[37] These effects can also be important in longitudinal structure functions.[44]

D. Finally there is the outstanding question of calculating the x dependence of structure functions at fixed value of Q^2. This has been addressed in the context of specific models in refs. 42 and 45.

I would like to thank Dennis Duke for discussions.

REFERENCES

1. A.Peterman, Phys. Rep. 53C, 157 (1979).
2. A.J. Buras, Rev. Modern Phys. 52, 199 (1980).
3. H. Georgi and H.D. Politzer, Phys. Rev. D9, 416 (1974).
4. D.J. Gross and F. Wilczek, Phys. Rev. D9, 980 (1974).
5. E.G. Floratos, D.A. Ross, and C.T. Sachrajda, Nucl. Phys. B129, 66 (1977) and Erratum, Nucl. Phys. B139, 545.
6. G. Curci, W. Furmanski, and R.Petronzio, TH.2815-CERN (1980).
7. H.D. Politzer, Phys. Rev. Lett. 30, 1346 (1973).
8. D.J. Gross, and F. Wilczek, Phys. Rev. Lett. 30, 1323 (1973).
9. W. Caswell, Phys. Rev. Lett. 33, 244 (1974).
10. D.R.T. Jones, Nucl. Phys. B75, 531 (1974).
11. W.A. Bardeen, A.J. Buras, D.W.Duke and T. Muta, Phys. Rev. D18, 3998 (1978).
12. E.G. Floratos, D.A. Ross, and C.T. Sachrajda, Nucl. Phys. B152, 493, (1979).
13. M. Bace, Phys. Lett. 78B, 132 (1978); S. Wolfram, Caltech preprint 68-690 (1978) (unpublished).
14. R. Barbieri, L. Caneschi, G. Curci and E. d'Emilio, Phys. Lett. 81B, 207 (1979); W. Celmaster and R.J. Gonsalves, Phys. Rev. Lett. 42, 1435 (1979).
15. K. Harada and T. Muta, Phys. Rev. D22, No.3 (1980). L.F. Abbott, Phys. Rev. Lett. 44, 1569 (1980).
16. W. Celmaster and D. Sivers, ANL-HEP-PR-80-30.
17. P.M. Stevenson, University of Wisconsin preprints, 153, 155 (1980).
18. A. Para and C.T. Sachrajda, Phys. Letters 86B, 331 (1979).
19. M.R. Pennington and G.G. Ross, Phys. Letters 86B, 371 (1979).
20. D.W. Duke and R.G. Roberts, Nucl. Phys. B166, 243 (1980). H.L. Anderson et al. Fermilab-Pub-79/30-EXP.
21. A.J. Buras, talk presented at the Symposium on "Topical Questions in QCD", Copenhagen, June, 1980, Fermilab preprint, October 1980.
22. I. Kodaira, S. Matsuda, K. Sasaki and T. Uematsu, Nucl. Phys. B159, 99 (1979); I. Kodaira, S. Matsuda, T. Muta, and T. Uematsu, Phys. Rev. D20, 627 (1979).

23. R.G. Roberts, J. Wosiek, and K. Zalewski, in preparation.
24. L. Baulieu and C. Kounnas, Nucl. Phys. B141, 423 (1978). I. Kodaira and T. Uematsu, Nucl. Phys. B141, 497 (1978).
25. G. Altarelli, R.K. Ellis and G.Martinelli, Nucl. Phys. B143, 521 (1978).
26. G. Altarelli, R.K. Ellis, G. Martinelli and Pi, Nucl. Phys. B160, 30 (1979).
27. A. Mueller, Phys. Rev. D18, 3705 (1978).
28. J. Kalinowski, K. Konishi and T. Taylor, CERN preprint TH-2902; E.G. Floratos, R. Lacaze and C. Kounnas, Saclay preprints 80/77, 80/83.
29. N. Sakai, Phys. Lett. B85, 67 (1979); J. Ambjørn and N. Sakai, Nordita-80/14 (1980).
30. A.Białas and A.J. Buras, Phys. Rev. D21, 1825 (1980).
31. A.J. Buras and K.J.F. Gaemers, Nucl. Phys. B132, 249 (1978).
32. See the talk by H. Weerts and I.G.H. De Groot in these proceedings.
33. T. Gottschalk, ANL-HEP-PR=80-35 (1980).
34. A. Gonzalez-Arroyo, C. Lopez and F.J. Yndurain, Nucl. Phys. B153, 161 (1978).
35. R.P. Feynman, R. Field and D.A. Ross, unpublished; see R. Field, Cal-Tech-68-739 (1979).
36. H.D. Politzer, Cal-Tech 68-765.
37. E.L. Berger and S.J. Brodsky, Phys.Rev. Lett. 42, 940 (1979); G.R. Farrar and D.R. Jackson, Phys. Rev. Lett. 35, 1416 (1975). E.L. Berger, Phys. Lett. 89B, 241 (1980).
38. J. Gottlieb, Nucl. Phys. 139, 125 (1978). M. Okawa, University of Tokyo preprint, UT-337 (1980).
39. L.F. Abbott and R.N. Barnett, Ann. Phys. 125, 276 (1980).
40. D.W. Duke and R.G. Roberts, Rutherford Lab Preprint RL-80-016 (1980) to appear in Phys. Letters B.
41. M.R. Pennington and G.G. Ross, Oxford preprint (1980).
42. R.L. Jaffe and G.G. Ross, Phys. Letters 93B, 313 (1980).
43. S. Brodsky, SLAC preprint; D. Amati, A. Bassetto, M. Ciafaloni, G. Marchesini, and G. Veneziano, TH.2831-CERN (1980; L. Caneschi, University of Pisa preprint.
44. L.F. Abbott, E.L. Berger, R. Blankenbecler, and G. Kane, Phys. Letters 88B, 157 (1979).
45. A. De Rujula and F. Martin, MIT preprint CTP No.851 (1980).; see the talks by R.L. Jaffe and F. Martin in these proceedings.

ON THE SHAPE OF HADRON STRUCTURE FUNCTIONS*

François Martin
Stanford Linear Accelerator Center
Stanford University, Stanford, California 94305
and
CERN, 1211 Geneva 23, Switzerland

ABSTRACT

The hypothesis that, in the leading twist approximation and to all orders of perturbative QCD, there exists a momentum scale Q_0^2 at which hadrons are pure valence quark (or antiquark) bound states gives good results for nucleon, pion and kaon structure functions.

Perturbative QCD tells us how hadron structure functions, e.g., $F_2^{eN}(x,Q^2)$, evolve with Q^2 for high enough Q^2. But what about their shapes as functions of x? Our aim here is to investigate the possibility of predicting those shapes. To do so we consider the experimental structure functions in a Q^2 region where higher than twist two effects ($1/Q^2$ terms) are negligible ($Q^2 \gtrsim 1$-10 GeV^2) and extrapolate them at low Q^2 using perturbative QCD. We assume that this procedure leads to a momentum scale Q_0^2 at which the extrapolated hadron pictures correspond to pure valence quark or antiquark bound states with no glue and no sea[1-4] (qqq state for the nucleon and $q\bar{q}$ states for pion and kaon). Of course those perturbative QCD extrapolations of hadron structure functions have nothing to do with what is actually measured at the scale Q_0^2 because of the presence of very large higher twist effects at this scale. Our model consists in using perturbative QCD in the leading twist approximation with the following boundary conditions $xG(x,Q_0^2) = xq_s(x,Q_0^2) = 0$ (G and q_s stand respectively for gluon and light quark sea distributions). The valence boundary distributions $xq_v(x,Q_0^2)$ are the bound state distributions computed in the leading twist approximation ($1/Q_0^2$ terms neglected) as given, e.g., in the meson case, by the diagram of Fig. 1. Note that we never have to precise the exact value of Q_0^2.

We are able to predict the kaon structure functions in terms of the pion ones by using data on pion structure functions[5] together with the following formula[3]

$$M_{u_v^{K^+}}(n,Q^2) = M_{u_v^{\pi^+}}(n,Q^2) \times \frac{M_{u_v^{K^+}}(n,Q_0^2)}{M_{u_v^{\pi^+}}(n,Q_0^2)} \qquad (1)$$

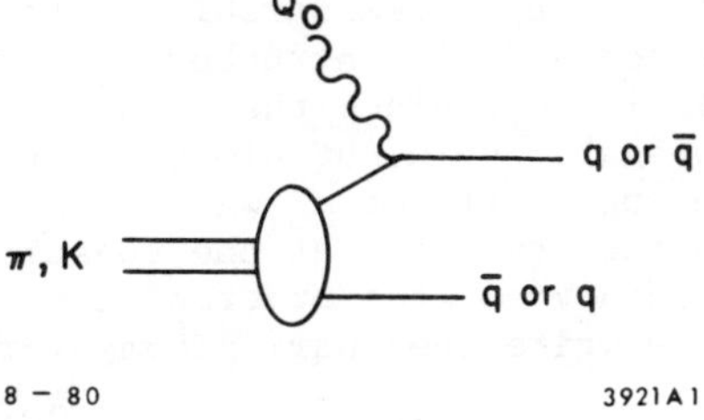

Fig. 1. Meson as a $q\bar{q}$ bound state.

* Work supported by the Department of Energy, contract DE-AC03-76SF00515.

valid to all orders of perturbative QCD in the leading twist approximation $\left(M_q(n,Q^2) = \int_0^1 x^{n-1}\, q(x,Q^2)\, dx\right)$. In principle the distributions $q_v(x,Q_0^2)$ should be given by the QCD bound state solution, but so far this problem has not been solved. We are therefore forced to make approximations.

(1) With a non relativistic approximation the distributions $q_v(x,Q_0^2)$ obtained are peaked at $x_0 = \mu/(m+\mu)$ with a width of order $1/RM$. μ, m and M are respectively the masses of the struck quark (or antiquark), the recoiling partons and the hadronic target, R^2 is the mean square radius of the hadron. For nucleon, pion and kaon x_0 equals respectively 0.33, 0.5 and 0.38 (0.38 corresponds to u_v in K^+ when using $\mu = m_u = 336$ MeV and $m = m_s = 540$ MeV) and $1/RM = 0.26$, 2.4 and 0.80. The resulting distributions $q_v(x,Q_0^2)$ (see Fig. 6 in Ref. 4) are quite different from those experimentally measured at $Q^2 \sim 20$ GeV^2. This means that a large amount of perturbative QCD corrections are needed to obtain $q_v(x,Q^2 = 20$ GeV$^2)$. This is a feasible possibility which is in fact realized when using the leading log approximation for the perturbative QCD corrections.[2,4] The shapes of the distributions $q_v(x,Q_0^2)$ are quite different from the experimental ones but the two following main features are not modified by the perturbative QCD corrections and are in fact experimentally observed: (a) the nucleon distribution is much more peaked and concentrated at small x than the pion one; and (b) the kaon distribution drops faster than the pion one at large x.

(2) Relativistic bound states: If we use a field theory with usual quark propagators, i.e., having poles at the quark constituent masses ($m_u \sim 336$ MeV), and massless gluon exchanges, we run into problems. The first problem is obviously that it is incompatible with confinement since it leads to free constituent quark states with the corresponding masses. The second problem has to do with structure functions. We get a nucleon distribution $u_v^N(x,Q_0^2)$ which is too narrowly peaked around $x_0 = 1/3$ (it does not reproduce $R_N = 0.8$ fm) and drops faster than the experimental one for $0.33 < x < 1$. We also obtain a pion distribution[6] $xu_v^{\pi^+}(x,Q_0^2) \sim Ax^3(1-x)^2$ which is steeper than the experimental one $(\sim 0.5\sqrt{x}\,(1-x))$ before being corrected by perturbative QCD. Therefore it is impossible for perturbative QCD, i.e., mainly gluon bremsstrahlung, to reproduce both the experimental nucleon and pion structure functions. One way out of these difficulties is to replace the first order perturbative pole of the quark propagator by a cut along the real axis.[3,7] A possibility is that the cut starts at m_0 which can be related to the current quark mass and that the pole at the constituent quark mass m is smeared out with a width γ^2 of the order of the strong interaction scale Λ^2. Explicitly if we write the quark propagator as

$$P(k) = \Pi(k^2)\not{k} + M(k^2) \tag{2}$$

we can choose the absorptive part of $\Pi(k^2)$ as

$$\text{Abs}\,\Pi(k^2) \sim \frac{\pi^{-1}\gamma^2\,\theta(k^2 - m_0^2)}{(k^2 - m^2)^2 + \gamma^2} \qquad (3)$$

with $\gamma^2 = \mathcal{O}(\Lambda^2)$. For the u and d quarks m_0 and m will be respectively of the order of 10 and 336 MeV whereas for the s quark they will be of the order of 150 and 540 MeV. Moreover we can choose the large k^2 behavior of our propagator to be given by perturbative QCD. Let us note that, since it has no more real pole, this propagator is now in agreement with confinement. Using this propagator, the nucleon, pion and kaon distributions $q_V(x,Q_0^2)$ obtained do not have the problem mentioned above. Note that their shapes are closer to the experimentally measured ones (corresponding to $Q^2 \sim 20$ GeV^2) than the non relativistic ones are. If, for u and d quarks we assume $m_0 = 0$, which is a good approximation, we find that $u_V^{\pi^+}(x,Q_0^2)$ is roughly proportional to (1-x) for $0.45 < x < 1$. On the other hand using $m_s = 150$ MeV for the s quark we find that $u_V^{K^+}(x,Q_0^2)$ is roughly a decreasing linear function of x for $0.4 < x < 0.85$ and behaves like $(1-x)^2$ for $0.85 < x < 1$. Therefore we do not have any more the $(1-x)^2$ behavior for the pion structure function.[3] For the nucleon we find that $u_V^N(x,Q_0^2)$ behaves like $(1-x)^3$ when x goes to 1.

Using a formula similar to (1) for nucleon and pion together with nucleon data[8] we can predict the pion structure functions. Agreement with pion data[5] is good both when using non relativistic calculations[4] or our relativistic model[3] for the input at Q_0^2. We can also predict the kaon structure functions using formula (1) together with pion data. Figure 2 shows our results for the ratio $[\bar{u}_V^{K^-}/\bar{u}_V^{\pi^-}](x,Q^2 = 20\ \text{GeV}^2)$ together with the data points of the CERN-NA3 experiment.[9] The solid line corresponds to a non relativistic approximation[4] and the dashed one to a relativistic calculation which uses the following parameters $(m_0,m) = (0,336\ \text{MeV})$ for the u and d quarks and = (150 MeV, 540 MeV) for the s quark. The dashed line is very sensitive to those parameters. So most probably an adjustment of the parameters can give a better agreement with data. Note also that this curve has been obtained assuming no spin correlation between the struck quark and the recoiling one.

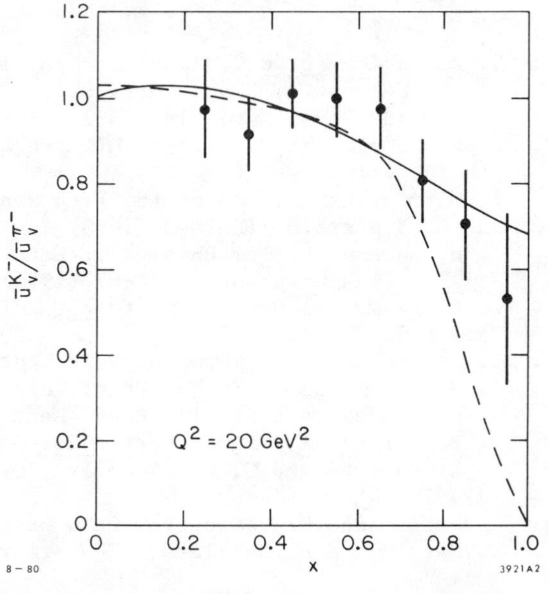

Fig. 2. The ratio of K^-/π^- structure functions versus x for $Q^2 \sim 20$ GeV^2. Data points are from Ref. 9. Curves are explained in the text.

It is possible to make predictions on a hadron structure function without referring to any other one if we use the leading log approximation of perturbative QCD.[1,2,4] In doing so we get the right partition of nucleon momentum carried between the valence, glue and sea.[1] Using the non relativistic input at Q_0^2, we obtain, in first approximation, good results for the valence distributions inside the nucleon, pion and kaon.[2,4] However the gluon and sea distributions obtained seem too steep near $x = 0$.[2,10] This may indicate the need for higher orders of perturbative QCD. Note that if we suppress the effect of the 3-gluon coupling for $Q^2 \lesssim 1\ \text{GeV}^2$ which corresponds to a freezing of the QCD running coupling constant $\alpha_s(Q^2)$ in this region[11] we obtain gluon and sea distributions which are in good agreement with the experimentally measured ones.[12]

We conclude that the idea of a hadron in which "real" gluons and sea quark-antiquark pairs all come from radiation processes (perturbative QCD) seems to work. This assumption used together with non relativistic or relativistic valence quark bound states gives good results when comparing nucleon, pion and kaon structure functions. In the relativistic case hadron structure functions can give information on the quark propagators in the non perturbative region. In particular the kaon structure functions can tell us about the s quark propagator.

This work was done in collaboration with A. De Rújula and P. Sorba. I acknowledge useful discussions with P. Binétruy, E. de Rafaël, R. Stora and the members of the CERN-NA3 experiment, especially D. Décamp, J. Lefrançois and P. Miné.

REFERENCES

1. V. A. Novikov et al., Ann. of Phys. 105, 276 (1977); G. Parisi and R. Petronzio, Phys. Lett. 62B, 331 (1976).
2. F. Martin, Phys. Rev. D19, 1382 (1979).
3. A. De Rújula and F. Martin, MIT preprint CTP#851 (1980), to be published in Phys. Rev. D.
4. F. Martin, talk given at the XVth Rencontre de Moriond (1980), and CERN preprint TH-2845 (1980).
5. C. B. Newman et al., Phys. Rev. Lett. 42, 951 (1979); J. Badier et al., CERN preprint EP-79-67 (1979).
6. Z. F. Ezawa, Il Nuovo Cimento 23, 271 (1974); T. N. Pham, Phys. Rev. D19, 707 (1979).
7. R. Oehme and W. Zimmermann, Phys. Lett. 79B, 93 (1978); B. M. McCoy and T. T. Wu, Phys. Lett. 72B, 219 (1977).
8. J. G. H. de Groot et al., Phys. Lett. 82B, 456 (1979).
9. J. Badier et al., Phys. Lett. 93B, 354 (1980).
10. I. Hinchliffe and C. H. Llewellyn Smith, Nucl. Phys. B128, 93 (1977).
11. G. Parisi and R. Petronzio, CERN preprint TH-2804 (1980).
12. F. Martin and B. Gaillard, CERN preprint TH-2803 (1980).

NORMALIZING THE RENORMALIZATION GROUP IN DEEP INELASTIC LEPTOPRODUCTION

R. L. Jaffe
Massachusetts Institute of Technology, Cambridge, Ma. 02139

The work I will describe was carried out in collaboration with G.G. Ross of Oxford University[1]. The now famous QCD renormalization group analysis of the moments of structure functions predicts their evolution with Q^2 but not their normalization:

$$\lim_{Q^2\to\infty} M(N,Q^2) = M(N,Q_0^2)\left(\frac{\alpha(Q_0^2)}{\alpha(Q^2)}\right)^{-\frac{\gamma_0^N}{2\beta_0}} + \text{higher order corrections} \tag{1}$$

for flavor non-singlets, with $\beta_0 = 11 - \frac{2}{3}N_f$ and $M(N,Q_0^2)$ unknown. It also relates $M(N,Q_0^2)$ to the matrix elements of local twist-two operators renormalized at Q_0^2:

$$P^{\mu_1}\ldots P^{\mu_n} M(N,Q_0^2) \equiv \frac{i^{n-1}}{n!} \langle P|\bar{q}(o)\gamma^{\mu_1}D^{\mu_2}\ldots D^{\mu_n} q(o) + \text{permutations}|P\rangle + O(m^2/Q_0^2) \tag{2}$$

(flavor indices suppressed). On the other hand the quark model has for years allowed us to estimate the matrix elements of local operators such as g_A/g_V, μ_N, σ_π, $\langle r^2\rangle$ with reasonable accuracy. Ross and I have attempted to use the quark-bag model to calculate the nucleon matrix elements of the local twist-two operators which determine $M(N,Q_0^2)$.

Our motivation was two fold: first we wish to connect the quark partons of deep inelastic scattering with the "constituent" quarks of the static quark model and thereby constrain both. This is an old hope which has surfaced periodically since 1970[2] but has never been fully realized. Second, if the analysis of twist-two succeeds, one would like to go on to calculate the matrix elements of the higher twist operators which govern the $O(m^2/Q^2)$ corrections to structure functions. These non-perturbative corrections must be understood before a detailed comparison of QCD with deep inelastic data is possible[3].

There are problems trying to marry perturbative QCD to the quark model. In a quark model a matrix element is a number, while in QCD matrix elements are Q^2-dependent. Put another way: a quark model treats the nucleon as (approximately) three quarks in some confining "bag", while according to QCD perturbation theory, if the nucleon were three quarks at $Q^2=\mu_0^2$ it would be more complicated at $Q^2>\mu_0^2$ because of gluon radiation. In principle one could incorporate QCD radiative corrections into the bag model description of the nucleon.

This has not been done primarily because confining boundary conditions make calculations much more difficult. Instead Ross and I proposed the following recipé: One is to interpret bag calculations of local matrix elements as pertaining to some low (but unknown) scale μ_0^2 (~1 GeV). For $Q^2>\mu_0^2$ the matrix elements evolve according to perturbative QCD. The motivation for this ansatz is the success of perturbative calculations of the baryon spectrum[4] indicating that once confinement is put in by hand, perturbation theory in $\alpha_s(\mu_0^2)$ is not a bad approximation.

Our strategy is first to calculate the bag matrix element of the spin-N, twist-two non-singlet operator $A_k^{NS}(N,\mu_0^2)$ (where k=2 for ep-en, k=3 for $\nu p+\nu n$ vector-axial interference). We then evolve this using QCD perturbation theory to a value of Q^2 large enough so the data can reasonably be expected to be dominated by twist-two. There we expect

$$M_k^{NS}(N,Q^2) = A_k^{NS}(N,Q^2) + O(m^2/Q^2) \tag{3}$$

Perhaps the simplest way to visualize this procedure is via the well known plot of the log of the Nth moment versus the log of the Mth. According to Eq.(1) to leading order in QCD and to leading twist, the plot should be a straight line with slope γ_0^N/γ_0^M. This fit is then extrapolated to low Q^2 using QCD perturbation theory (to as high an order as possible) to provide the "experimental" value of the twist-two operator matrix elements at low Q^2. The bag calculation gives a value for each moment, i.e. a single point on the log-log plot which should lie on the extrapolation of the data. Figure 1 shows such a plot for some of the low moments of $F_3^{\nu p+\nu n}$. The extra-

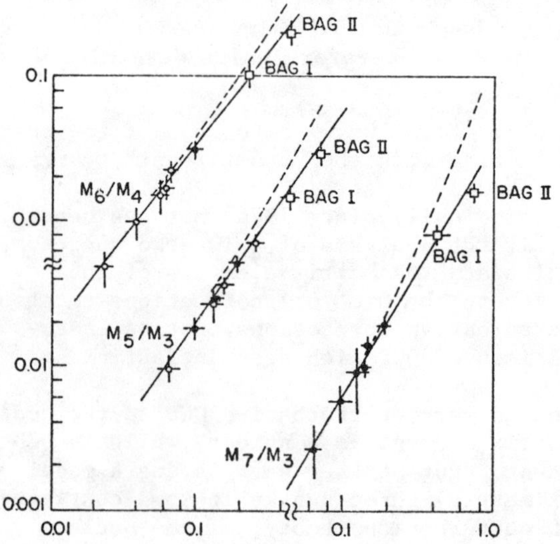

Figure 1. Non-singlet moments of F_3 plotted versus one another. BAG II and I correspond to calculations done with and without magnetic corrections to the proton mass respectively.

polation of lowest order QCD is shown by the solid line, higher order effects are shown by the dashed line. The bag points are shown with 20% error bars. The agreement is pretty good. To make a more detailed comparison consider

$$D_k(N) \equiv \ell n \left| A_k^{NS}(N,\mu_0^2)/M_k^{NS}(N,Q^2) \right|^{2\beta_0/\gamma_0^N} \tag{4}$$

To leading order in QCD $D_k(N)$ should be independent of N. In orders it develops a calculable N dependence. Combining the bag model predictions for $A_3^{NS}(N,\mu_0^2)$ with the data for $M_3^{NS}(N,Q^2)$ we obtain Fig.2.

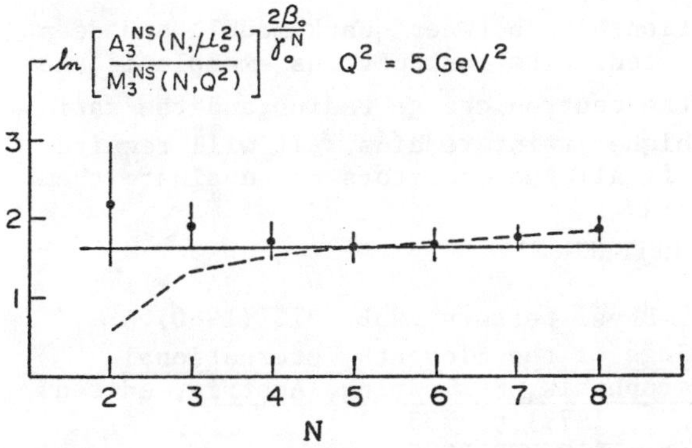

Figure 2. Comparison of bag predictions $A_3^{NS}(N,\mu_0^2)$ with $M_3^{NS}(N,Q^2)$.

Higher order corrections are only important for the 2nd and 3rd moments, making the comparison of the model with the data unreliable for these moments. Otherwise the agreement is excellent.

Although we have only studied twist-two nevertheless our work has implications for higher twist. Suppose $\Lambda_{QCD} = 0$ and all Q^2-dependence were from higher twist. Then the best estimates of the (Q^2 independent) twist-two operator matrix elements are to be found at the extreme lower left of the log-log plots where Q^2 is largest. However the bag prediction for the twist-two operator matrix element lies at the upper right. The two disagree by one to two orders of magnitude. We conclude that either the twist-two operator matrix element is Q^2 dependent or the bag model predictions for twist-two are very wrong.

I have yet to say how the bag predictions for $A_k^{NS}(N,\mu_0^2)$ are obtained. In fact a naive calculation of $\langle P|\bar{q}\gamma^{\mu_1}\partial^{\mu_2}\ldots\partial^{\mu_n}q|P\rangle$ [We replace D^μ by ∂^μ since we are ignoring gluon corrections by bag wavefunctions.] diverges. The problem is that the usual (cavity) approximation made in the bag model violates translation invariance. Therefore the cavity structure function, $f_{cav}(\zeta)$, fails to vanish for $\zeta>1$. Worse still because of the bag's sharp boundaries $f_{cav}(\zeta)$ falls only like a power of ζ as $\zeta\to\infty$. To obtain predictions for $A_k^{NS}(N,\mu_0^2)$ we must take moments of $f_{cav}(\zeta)$. These diverge for N>1. The solution to this problem is to restore translation invariance (at least along the null plane where it is needed for this calculation) to the model. This can be done in two dimensional space-time with a remarkably simple result: the structure function of a translation invariant cavity approximation to the bag is related to the naive cavity model by a change of variable:[5,6]

$$F(x) = \frac{1}{1-x}\,\theta(1-x)\, f_{cav}(-\log(1-x)) \tag{5}$$

The factor $(1-x)^{-1}$ preserves the area under $F(x)$ as required by Adler-esque sum rules. The replacement $x \to -\log(1-x)$ is a direct consequence of the Lorentz contraction of the recoiling bag[6]. There is some evidence that Eq.(5) is more general than the one dimensional bag model[7]. Ross and I have optimistically carried it over to three dimensions where it renders the moment integrals finite.

Where do we go from here? There are many interesting applications at the twist-two level: the spin dependent structure functions g_1 and g_2 can be predicted as can the structure functions of the π and K. Suspected relationships between quark models and deep inelastic phenomena can be tested. The most obvious example is the supposed connection between the neutron charge radius and the ratio F_2^{en}/F_2^{eP} as $x \to 1$. Finally, higher twist remains. It will require considerable energy to classify all the operators and evaluate them in the bag model.

REFERENCES

1. R.L. Jaffe and G.G. Ross, Phys. Letters, 93B, 313 (1980).
2. M. Gell-Mann, in Proceedings of the Eleventh International Universitätswochen für Kernphysik, Schladming, Austria, edited by P. Urban (Springer, N.Y., 1972) p. 733.
3. L. Abbott and R.M. Barnett, SLAC-PUB-2325.
   L. Abbot, W.B. Atwood and R.M. Barnett, SLAC-PUB-2400.
4. A. DeRujula, H. Georgi and S.L. Glashow, Phys. Rev. D12, 147 (1975).
   T.A. DeGrand, R.L. Jaffe, K. Johnson and J. Kiskis, Phys. Rev. D12, 2060 (1975).
5. A.C. Davis and E.J. Squires, Phys. Rev. D19, 388 (1978).
6. R.L. Jaffe, MIT Preprint MIT-CTP-870 to be published in Ann. Phys. (N.Y.).
7. J. Goldstone and R. Jackiw, private communication, R.L. Jaffe, unpublished.

STRUCTURE FUNCTIONS AND HIGH TWIST CONTRIBUTIONS IN PERTURBATIVE QUANTUM CHROMODYNAMICS*

Stanley J. Brodsky
Stanford Linear Accelerator Center
Stanford University, Stanford, California 94305
and
G. Peter Lepage†
Laboratory of Nuclear Studies
Cornell University, Ithaca, New York 14853

ABSTRACT

Perturbative QCD predictions are presented for the x near 1 behavior of hadronic structure functions. The available energy xW^2 is shown to control structure function evolution. In the case of meson structure functions, the $x \sim 1$ behavior is dominated by a high twist contribution to the longitudinal structure function $F_L \sim Cx^2/Q^2$ which can be rigorously computed and normalized.

One of the most important areas of study of perturbative quantum chromodynamics is the behavior of the hadronic wavefunctions at short distances or at far off-shell kinematics. This behavior can be tested not only in exclusive reactions such as form factors at large momentum transfer but also in deep inelastic scattering reactions at the edge of phase space. In this talk we will review the QCD predictions for the behavior of the hadronic structure functions $F_i(x,Q)$ in the end-point $x_{Bj} \sim 1$ region.[1] The endpoint region is particularly interesting because one must understand in detail (a) the contributions of exclusive channels, (b) the effect of high twist terms (power-law scale-breaking contributions) which can become dominant at large x, and (c) the essential role of the available energy W in controlling the logarithmic evolution of the structure functions. Note that as $x \sim 1$, essentially all of the hadron's momentum must be carried by one quark (or gluon), and thus each propagator which transfers this momentum becomes far-off shell: $k^2 \sim -(\vec{k}_\perp^2 + \mathcal{M}^2)/(1-x) \to -\infty$ [see Fig. 1]. Accordingly, if the spectator mass $\mathcal{M}$ is finite the leading power-law behavior in (1-x) is determined by the minimum number of gluon exchanges required to stop the hadronic spectators, and only the valence Fock states, $|qqq\rangle$ for baryons, and $|q\bar{q}\rangle$ for mesons, contribute to the leading power behavior. If one simply computes the connected tree graphs, as in Fig. 2, one finds that the results depend in an essential way on the quark's helicity:

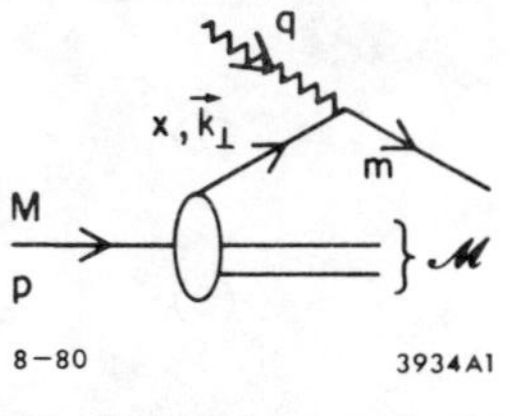

Fig. 1. Kinematics for inelastic structure functions.

$$G_{q/p}(x) \sim \begin{cases} (1-x)^3 & \text{parallel quark/nucleon helicity} \\ (1-x)^5 & \text{anti-parallel quark/nucleon helicity} \end{cases}$$

* Work supported by Department of Energy, contract DE-AC03-76SF00515.
† Work supported by the National Science Foundation.

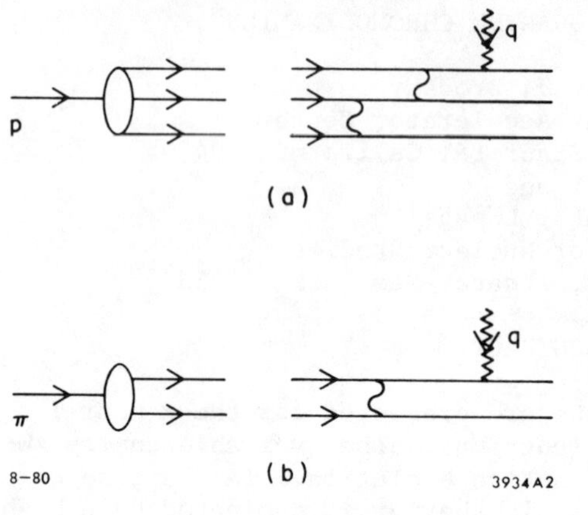

and $G_{q/\pi} \sim (1-x)^2$. If the nucleon wavefunctions satisfy the standard SU(6) spin-flavor symmetry, then the above results imply[2]

$$\frac{G_{u/p}}{G_{d/p}} \underset{x\to 1}{=} 5 \quad .$$

Fig. 2. Perturbative QCD tree diagrams for computing the $x \sim 1$ power behavior of baryon and meson structure functions.

Let us now consider how these results for the power-law behavior emerge within the complete perturbative structure of QCD. Including corrections from gluon radiation, vertex and self-energy corrections, and continued iteration of the gluon-exchange kernel, one finds for the nucleon's quark distribution[1]

$$G_{q/p}(x,Q) \underset{x\to 1}{=} (1-x)^3 \alpha_s^4(k_x^2) \left| \sum_{j=0}^{\infty} b_j \left(\log \frac{k_x^2}{\Lambda^2} \right)^{-\gamma_j^N} \right|^2 P_q(x,Q)$$
$$\times \left[1 + \mathscr{O}\left(\alpha_s(k_x^2), 1/Q\right) \right] \quad . \tag{1}$$

The powers of α_s and $(1-x)$ reflect the behavior of the hard scattering amplitude at the off-shell value $k_x^2 = (\langle k_\perp^2 \rangle + \mathscr{M}^2)/(1-x)$ where $\langle k_\perp^2 \rangle$ is set by the spectator transverse momentum integrations. The anomalous dimensions γ_j^N are the anomalous dimensions of the nucleon's valence Fock state wavefunction at short distances. Their contribution to $G_{q/p}(x,Q)$ are due to the evolution of the wavefunction integrated up to the transverse momentum scale $\vec{\ell}_\perp^2 < k_x^2$ as in the corresponding exclusive channel analyses.[1] The last factor $P_q(x,Q)$ represents the target-independent evolution of the structure function due to gluon emission from the struck quark: $(C_F = 4/3)$

$$P_q(x,Q) \sim (1-x)^{4C_F\xi(Q)} \tag{2}$$

where

$$\xi(Q) = \int_{Q_0^2}^{Q^2} \frac{dj_\perp^2}{j_\perp^2} \alpha_s(j_\perp^2) \sim \log\left(\frac{\log Q^2/\Lambda^2}{\log Q_0^2/\Lambda^2} \right) \quad . \tag{3}$$

The lower limit Q_0^2 of the gluon's transverse momentum integration is set by the mean value of the spectator quark's transverse momenta and masses. This hadronic scale sets the starting point for structure functions evolution.[3] Equation (1) then gives the light-cone momentum distribution for parallel-helicity quarks with x near 1 at the transverse momentum scale Q.

It should be emphasized that the actual momentum scale probed by various deep inelastic inclusive reactions depends in detail on the process under consideration; the actual upper limit of the transverse momentum integration is set by kinematics. For example, if we consider the contribution of Fig. 3 to the deep inelastic structure functions, the propagator (or energy denominator) associated with the top loop reduces to the usual Bjorken structure $2q\cdot p - Q^2/x + i\varepsilon$ only if $k_\perp^2 \ll (1-y)Q^2 \le (1-x)Q^2$ where $k_\perp$ is the quark's transverse momentum and $y \ge x$ is the light-cone variable indicated in the figure. The remaining structure factorizes into a form which defines $G_{q/p}(x/y, k_\perp)$. Thus the actual relation between the structure function and the momentum distribution for $x \sim 1$ is[1,4]

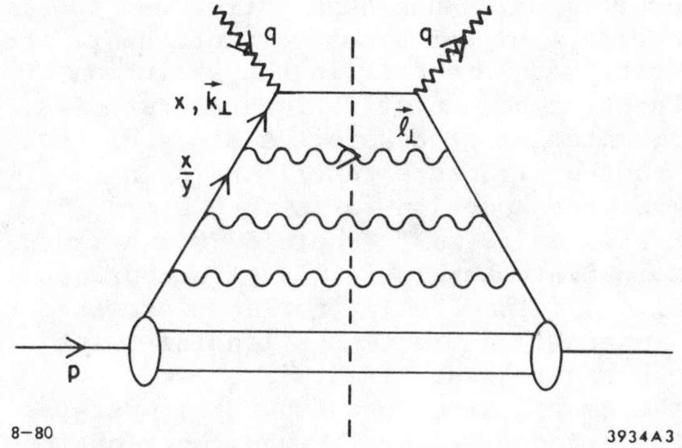

Fig. 3. Perturbative QCD diagrams for structure function evolution.

$$F_2(x,Q) = \sum_i e_i^2 \, x_{B_j}\left[G_{q_i/p}(x_{B_j},Q) + \delta G_{q_i/p}(x_{B_j},Q)\right] \tag{4}$$

where

$$\delta G_{q/p}(x,Q) = -2C_F \int_0^1 dy \, \frac{1+y^2}{1-y} \int_{(1-y)Q^2}^{Q^2} \frac{dk_\perp^2}{k_\perp^2} \frac{\alpha_s(k_\perp^2/y)}{4\pi} \times \left\{ G_{q/p}(x/y,k_\perp) \frac{\theta(y>x)}{y} - G_{q/p}(x,k_\perp)\right\} \tag{5}$$

corrects for the fact that the top loop is integrated to $k_\perp^2 < (1-y)Q^2$ ($\le x_{B_j}W^2$) not Q^2. [The argument of α_s is also crucial here.] Other inclusive reactions have to be individually examined: in the case of the Drell-Yan process $q_a\bar{q}_b \to \mu^+\mu^-$, the structure functions evolve to $(1-y_a)Q^2$ and $(1-y_b)Q^2$ not $Q^2 = (p_+ + p_-)^2$.

The actual evolution of structure functions in deep inelastic lepton scattering is thus controlled by the available energy $x_{B_j}W^2$, and is more moderate at $x_{B_j} \sim 1$ than would be expected from lowest order expectations. Analytic forms for the (1-x) behavior are readily computed.[1] The most important features are the following: (1) The $\delta G/G$ correction to leading order in α_s reproduces the critical $2C_F(\alpha_s(Q^2)/4\pi)\log^2 n$ terms in the structure function moments as calculated using the operator product expansion and renormalization group. In our analysis a series of terms of all orders in $(\alpha_s \log^2(1/1-x))^p$ or $(\alpha_s \log^2 n)^p$ arises simply from the fact that the natural evolution parameter for the structure functions $F_i(x,Q)$ and moments $\mathcal{M}_n(x,Q)$ is controlled by $(1-y)Q^2 < (1-x)Q^2$ and not Q^2; the basic momentum distributions $G(x,Q)$ do not contain the anomalous double-log terms and have a straightforward perturbative evolution. (2) The extended evolution equations based on Eqs. (4) and (5) have a number of phenomenological advantages. After taking into account the appropriate evolution limits, each deep inelastic process can be related to the basic distributions $G_q(x,Q)$, avoiding large kinematic corrections. The scale parameter Λ_n which has been introduced to eliminate the strong n-dependence of the higher order corrections to the moments is unnecessary. The fact that xW^2 controls the evolution suggest its use in structure function parameterizations and studies of moment factorization in fragmentation processes. A study of the application of this method to photon structure functions is in progress. (3) The exclusive-inclusive connection fails in QCD.[1,5] At fixed but large W^2, $F_{2N}(x,Q)$ falls as $(1-x)^{3+\delta}$ where $\delta > 0$, whereas, modulo logarithmic factors, exclusive channels in QCD give contributions $\sim(1-x)^3$ from the Q^{-4} scaling of the leading nucleon form factors. Thus exclusive channels will eventually dominate the leading twist contributions to inclusive cross sections at fixed W^2, $Q^2 \to \infty$.

A complete treatment of the hadron structure functions must take into account higher twist contributions. Although such contributions are suppressed by powers of $1/Q^2$, they can have fewer powers of $(1-x)$ and, accordingly, may be phenomenologically important in the large x domain.[6] In the case of nucleons, the $\ell + qq \to \ell' + qq$ subprocess (in which the lepton recoils against two quarks) leads to a structure function contribution $\sim(1-x)/Q^4$ since only one quark spectator is required. A large longitudinal structure function is also expected.[6,7,8] Although complete calculations of such terms have not been done, the presence of such terms can reduce the amount of logarithm scale-violation required from the leading twist contributions in phenomenological fits.[6,9]

The analysis of meson structure functions at $x \sim 1$ is similar to that of the baryon, with two striking differences: (1) The controlling power behavior of the leading twist contribution is $(1-x)^2$ from perturbative QCD.[2,10] The extra factor of $(1-x)$ -- compared to what would have been expected from spectator counting -- can be attributed to the mismatch between the quark spin and that of the meson. (2) The longitudinal meson structure function has an anomalous nonscaling component[7,11] which is finite at $x \to 1$: $F_L(x,Q) \sim Cx^2/Q^2$. This high twist term, which comes from the lepton scattering off an instantaneous fermion-line in light-cone perturbation theory, can be rigorously computed and <u>normalized</u> in perturbative QCD. The crucial fact is that the wavefunction evolution and spectator transverse momentum integrations in Fig. 4 can be written directly in terms of a corresponding calculation of the meson form factor. The result for the pion structure function to leading order in $\alpha_s(k_x^2)$ and $\alpha_s(Q^2)$ is[11,12]

$$F_L^\pi(x,Q) = \frac{2x^2}{Q^2} C_F \int_{m^2/(1-x)}^{Q^2} dk^2 \; \alpha_s(k^2)\, F_\pi(k^2) \qquad (6)$$

which numerically is $F_L \sim x^2/Q^2$ (GeV^2 units).

The dominance of the longitudinal structure functions in the fixed W limit for mesons is an essential prediction of perturbative

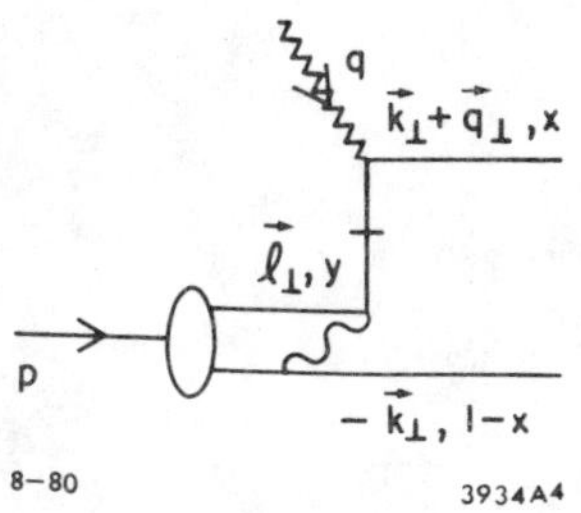

Fig. 4. Perturbative contribution to the meson longitudinal structure function $F_L \sim C/Q^2$.

QCD. Perhaps the most dramatic consequence is in the Drell-Yan process $\pi p \to \ell^+\ell^- X$; one predicts[7] that for fixed pair mass Q, the angular distribution of the ℓ^+ (in the pair rest frame) will change from the conventional $(1+\cos^2\theta_+)$ distribution to $\sin^2(\theta_+)$ for pairs produced at large x_L. A recent analysis of the Chicago-Illinois-Princeton experiment[14] at FNAL appears to confirm the QCD high twist prediction with about the expected normalization. Striking evidence for the effect has also been seen in a Gargamelle analysis[15] of the quark framentation functions in $\nu p \to \pi^+\mu^- X$. The results yield a quark fragmentation distribution into positive charged hadrons which is consistent with the predicted form: $dN^+/dzdy \sim B(1-z)^2 + (C/Q^2)(1-y)$ where the (1-y) behavior corresponds to a longitudinal structure function. It is also crucial to check that the $e^+e^- \to MX$ cross section becomes purely longitudinal $(\sin^2\theta)$ at large z at moderate Q^2. The implications of this high twist contribution for meson production at large p_T will be discussed elsewhere.[12]

REFERENCES

1. S. J. Brodsky and G. P. Lepage, SLAC-PUB-2447, Proc. of SLAC 1979 Summer Inst. on Particle Physics, and references therein.
2. G. R. Farrar and D. R. Jackson, Phys. Lett. 35, 1416 (1975). Recent BEBC data appear to be consistent with this result. See F. Sciulli, this meeting.
3. S. J. Brodsky, G. P. Lepage and T. Huang, SLAC-PUB-2540 (1980).
4. See also G. Curci and M. Greco, Phys. Lett. 92B, 175 (1980); D. Amati et al., CERN-TH-2831 (1980); and M. Ciafaloni, this meeting.
5. See also G. Parisi, Phys. Lett. 84B, 225 (1979).
6. R. Blankenbecler and I. Schmidt, Phys. Rev. D16, 1318 (1977); R. Blankenbecler et al., Phys. Rev. D12, 3469 (1975).
7. E. L. Berger and S. J. Brodsky, Phys. Rev. Lett. 42, 940 (1979); E. L. Berger, Phys. Lett. 89B, 241 (1980).
8. L. Abbott et al., Phys. Lett. 88B, 157 (1979).
9. L. Abbott et al., Phys. Rev. D22, 582 (1980).
10. Z. F. Ezawa, Nuovo Cimento 23A, 271 (1974). This result assumes the mass of the spectator is finite. If $\mathcal{M} \to 0$, perturbative and even-non-perturbative calculations allow $F_{2\pi}(x) \sim (1-x)$. See A. DeRujula and F. Martin, MIT-CTP-851 (1980); F. Martin, SLAC-PUB-2581, this conference; and S. J. Brodsky et al., Ref. 3.
11. Tests of this result will be discussed in E. L. Berger, S. J. Brodsky and G. P. Lepage (in preparation).
12. See also A. Duncan and A. Mueller, Phys. Lett. 90B, 159 (1980).
13. K. J. Anderson et al., Phys. Rev. Lett. 43, 1219 (1979).
14. CERN-MILAN-ORSAY Collaboration, CERN-EP/80-124 (1980); C. Matteuzzi et al., contribution to this meeting.